Engineering Design and Graphics with Autodesk® Inventor™ 2008

James D. Bethune

Boston University

Upper Saddle River, New Jersey
Columbus, Ohio

Library of Congress Control Number: 2007 792 1187

Editor in Chief: Vernon Anthony
Acquisitions Editor: Jill Jones-Renger
Editorial Assistant: Yvette Schlarman
Development Editor: Lisa S. Garboski, bookworks publishing services
Production Editor: Louise N. Sette
Production Supervision: Lisa S. Garboski, bookworks publishing services
Design Coordinator: Diane Ernsberger
Cover Designer: Jason Moore
Cover Image: Photo Disc
Art Coordinator: Karen Bretz
Production Manager: Deidra M. Schwartz
Director of Marketing: David Gesell
Marketing Manager: Jimmy Stephens
Marketing Coordinator: Alicia Dysert

This book was set by Aptara, Inc. It was printed and bound by Bind-Rite Graphics. The cover was printed by Coral Graphic Services, Inc.

Pearson Education Ltd.
Pearson Education Singapore Pte. Ltd.
Pearson Education Canada. Ltd.
Pearson Education—Japan
Pearson Education Australia Pty. Limited
Pearson Education North Asia Ltd.
Pearson Educación de Mexico, S. A. de C.V.
Pearson Education Malaysia Pte. Ltd.

10 9 8 7 6 5 4 3 2 1
ISBN-13: 978-0-13-159225-4
ISBN-10: 0-13-159225-4

THE NEW AUTODESK DESIGN INSTITUTE PRESS SERIES

Pearson/Prentice Hall has formed an alliance with Autodesk® to develop textbooks and other course materials that address the skills, methodology, and learning pedagogy for the industries that are supported by the Autodesk® Design Institute (ADI) software products. The Autodesk Design Institute is a comprehensive software program that assists educators in teaching technological design.

Features of the Autodesk Design Institute Press Series

JOB SKILLS—Coverage of computer-aided drafting job skills, compiled through research of industry associations, job websites, college course descriptions, and the Occupational Information Network database has been integrated throughout the ADI Press books.

PROFESSIONAL and **INDUSTRY ASSOCIATIONS INVOLVEMENT**—These books are written in consultation with and reviewed by professional associations to ensure they meet the needs of industry employers.

AUTODESK LEARNING LICENSES AVAILABLE—Many students ask how they can get a copy of the Inventor software for their home computer. Through a recent agreement with Autodesk®, Prentice Hall now offers the option of purchasing textbooks with either a 180-day or a 1-year student software license agreement for Inventor. This provides adequate time for a student to complete all the activities in the book. The software is functionally identical to the professional license, but is intended for student personal use only. It is not for professional use.

For more information about this book and the Autodesk Student Portfolio, contact your local Pearson Prentice Hall sales representative, or contact our National Marketing Manager, Jimmy Stephens, at 1-800-228-7854 x3725 or at Jimmy_Stephens@prenhall.com. For the name and number of your sales rep. please contact Prentice Hall Faculty Services at 1-800-526-0485.

FEATURES OF *ENGINEERING DESIGN AND GRAPHICS WITH AUTODESK® INVENTOR™ 2008*

This text presents a modern approach to using Inventor. That is, it addresses advances in technology and software evolution and introduces commands and procedures that reflect a modern, efficient use of Inventor 2008. Features include:

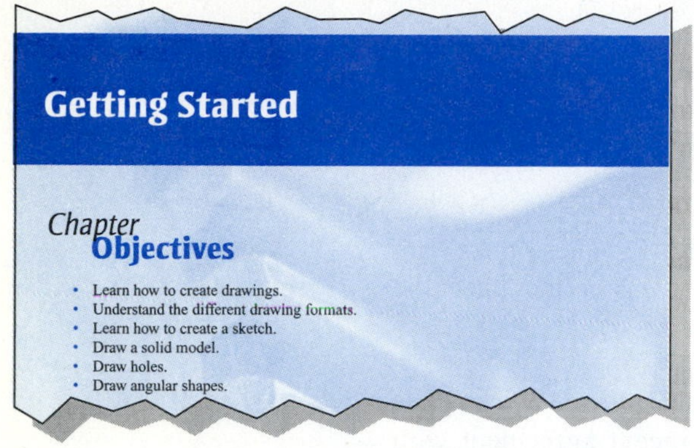

A "Getting Started" chapter at the beginning of the book allows users to get up to speed in no time to create and even plot Inventor drawings. Quick Start topics and concepts are linked to corresponding chapters later in the book, providing a motivational preview and allowing the user to delve into detailed topics of instruction as they choose, at their own pace.

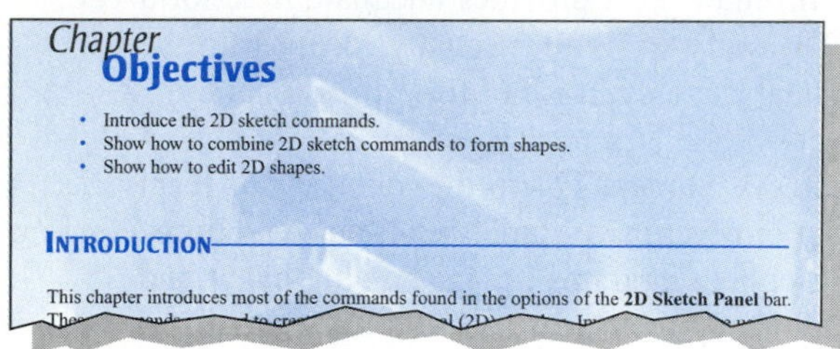

Chapter Objectives, a bulleted list of learning objectives for each chapter, provide users with a road map of important concepts and practices that will be introduced in the chapter.

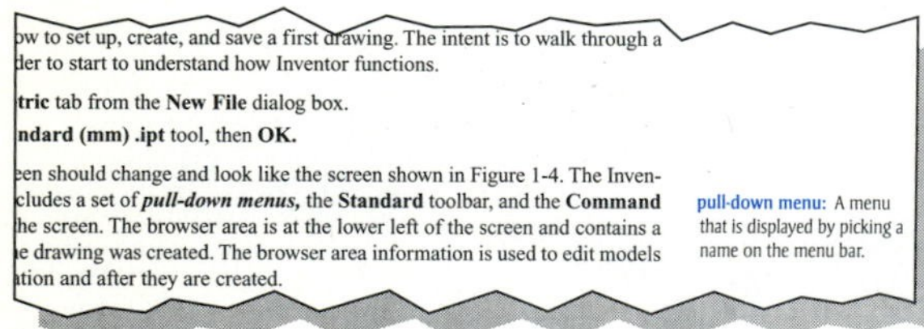

Key Terms are bold and italic within the running text and defined in the margin to help students understand and use the language of the computer-aided drafting world.

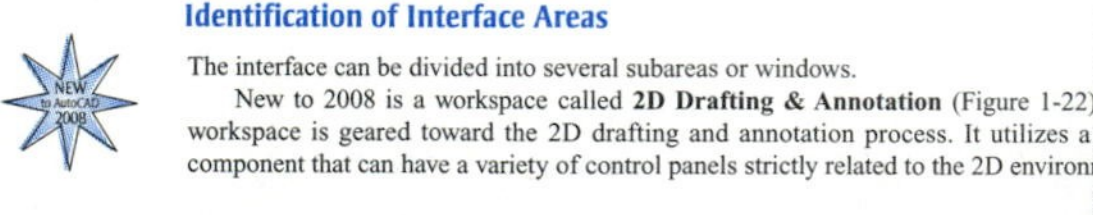

A New to Inventor 2008 icon flags features that are new to the 2008 version of the Inventor software, creating a quick "study guide" for instructors who need to familiarize themselves with the newest features of the software to prepare for teaching the course. Additional details about these new features can be found in the Online Instructor's Manual.

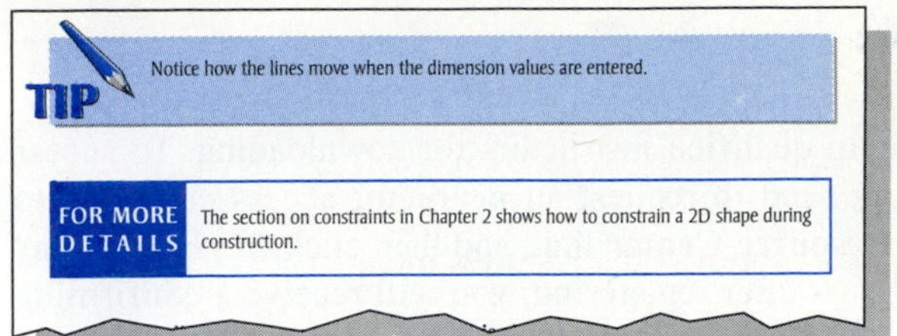

TIP

Notice how the lines move when the dimension values are entered.

FOR MORE DETAILS

The section on constraints in Chapter 2 shows how to constrain a 2D shape during construction.

TIP, NOTE, and FOR MORE DETAILS boxes highlight additional helpful information for the student.

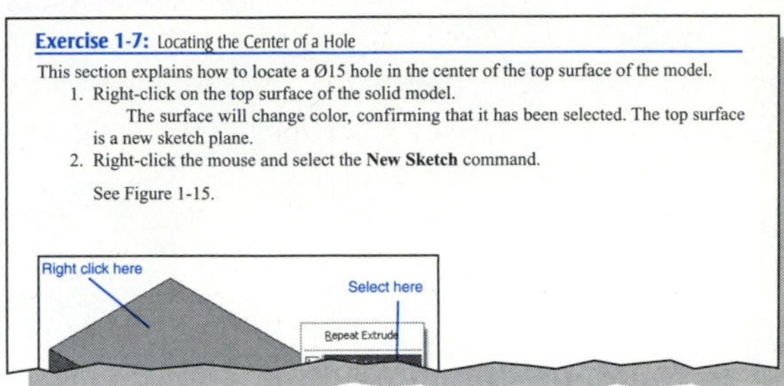

Exercise 1-7: Locating the Center of a Hole

This section explains how to locate a Ø15 hole in the center of the top surface of the model.

1. Right-click on the top surface of the solid model.

   The surface will change color, confirming that it has been selected. The top surface is a new sketch plane.
2. Right-click the mouse and select the **New Sketch** command.

   See Figure 1-15.

Exercises throughout the chapters provide step-by-step walk-through activities for the student, allowing immediate practice and reinforcement of newly learned skills.

SUMMARY

This chapter introduced the Inventor drawing screen and its components, including the **Standard** toolbar, pull-down menus, the panel bar, and the browser area. It also explained how to access and use the features of d strating how to set up, create, and save create a solid model.

CHAPTER PROJECT

Project 1-1:

Sketch the shapes shown in Figures P1-1 through P1-8, then create solid models using the specified thickness values.

End-of-Chapter material, easily located by shading on page edges, provides:

- Summaries
- Chapter Projects

to help students check their own understanding of important chapter concepts.

Instructor's Resources

An Online Instructor's Manual is available to qualified instructors for downloading. To access supplementary materials online, instructors need to request an instructor access code. Go to **www.prenhall.com,** click the **Instructor Resource Center** link, and then click **Register Today** for an instructor access code. Within 48 hours after registering, you will receive a confirming e-mail including an instructor access code. Once you have received your code, go to the site and log on for full instructions on downloading the materials you wish to use.

Preface

This book introduces Autodesk® Inventor™ 2008 and shows how to use Autodesk Inventor to create and document drawings and designs. The book is an extensive revision of the Release 10 version. The book now puts heavy emphasis on engineering drawings and on drawing components used in engineering drawings such as springs, bearings, cams, and gears. It shows how to create drawings using many different formats such as .ipt, .iam, ipn, and .idw for both English and metric units. It explains how to create drawings using **Design Accelerator** and how to extract parts from the **Content Center.**

All topics are presented using a step-by-step format so that the reader can work directly from the text to the screen. There are many easy-to-understand labeled illustrations. The book contains many sample problems that demonstrate the subject being discussed. Each chapter contains a variety of projects that serve to reinforce the material just presented and allow the reader to practice the techniques described.

Chapters 1 and 2 present 2D sketching commands and the **Extrude** command. These chapters serve as an introduction to the program.

Chapter 3 demonstrates the commands needed to create 3D models, including **Shell, Rib, Split, Loft, Sweep,** and **Coil.** Work points, work axes, and work planes are explained and demonstrated.

Chapter 4 shows how to create orthographic views from 3D models. The creation of isometric views, sectional views, and auxiliary views is also covered.

Chapter 5 shows how to create assembly drawings using both the bottom-up and the top-down process. The chapter includes presentation drawings and exploded isometric drawings with title blocks, parts lists, revision blocks, and tolerances blocks. There is an extensive step-by-step example that shows how to create an animated assembly, that is, a drawing that moves on the screen.

Chapter 6 covers threads and fasteners. Drawing conventions and callouts are defined for both inch and metric threads. The chapter shows how to calculate thread lengths and how to choose the appropriate fastener from Inventor's **Content Center.** The **Content Center** also includes an extensive listing of nuts, setscrews, washers, and rivets.

Chapter 7 shows how to apply dimensions to drawings. Both ANSI and ISO standards are demonstrated. Different styles of dimensioning, including ordinate, baseline, and Inventor's **Hole Table,** are presented. Applying dimension to a drawing is considered an important skill, so many examples and sample problems are included.

Chapter 8 is an extensive discussion of tolerancing, including geometric tolerances. The chapter first shows how to use Inventor to apply tolerances to a drawing. The chapter then shows how to calculate tolerances in various design situations. Positional tolerances for both linear and geometric applications are included. The chapter introduces the **Limits** and **Fits** options of the **Design Accelerator tool.** The information contained in this option eliminates the need for an appendix that includes fit tables.

Chapter 9 is new and shows how to draw springs using the **Standard.ipt** format and the **Coil** command. It also shows how to draw springs using the **Design Accelerator.** Compression, extension, torsion, and Belleville springs are included.

Chapter 10 is new and shows how to draw shafts using **Design Accelerator.** Chamfers, retaining rings, retaining ring grooves, keys and keyways, splines, pins, O-rings, and O-ring grooves are covered. The chapter contains many exercise problems.

Chapter 11 is new and shows how to match bearings to specific shafts using the **Content Center.** Plain, ball, and thrust bearings are presented. An explanation of tolerances between a shaft and bearing bore and between the bearing's outside diameter and the assembly housing is given. Both ANSI and ISO standards are presented.

Chapter 12 is a revision of Chapter 10 in the Release 10 book. The emphasis has been shifted to drawing gears and mounting them into assembly drawings. Spur, bevel, and worm gears are introduced. The chapter shows how to create gear hubs with setscrews, and keyways with keys, and how to draw assembly drawings that include gears. There are two new extensive assembly exercise problems.

Chapter 13 shows how to draw basic sheet metal parts including features such as tabs, reliefs, flanges, cuts, holes, and hole patterns.

Chapter 14 shows how to create and draw weldments. Only fillet and groove welds are covered.

Chapter 15 shows how to design and draw cams. Displacement diagrams and different types of followers are discussed.

Autodesk Learning License

Through a recent agreement with the Autodesk Inventor publisher, Autodesk®, Prentice Hall now offers the option of purchasing *Engineering Design and Graphics with Autodesk® Inventor™ 2008* with either a 180-day or a 1-year student software license agreement. This provides adequate time for a student to complete all the activities in this book. The software is functionally identical to the professional license, but it is intended **for student personal use only.** It is not for professional use. For more information about this book and the Autodesk Learning License, contact your local Pearson Prentice Hall sales representative, or contact our National Marketing Manager, Jimmy Stephens, at 1-800-228-7854, x3725 or at JimmyStephens@prenhall.com. For the name and number of your sales rep, please contact Prentice Hall Faculty Services at 1-800-526-0485.

Acknowledgments

I would like to thank the following individuals who reviewed this title: Jon Bridges, Lane Community College; Chuck Bales, Moraine Valley Community College; Joel Robert Brodeur, Montana State University–Northern; Mike Coler, Texas State Technical College–West Texas; and George Gibson, Athens Technical College.

Thanks to the editor, Jill Jones-Renger. Also thanks to Lisa Garboski, project manager.

Thanks to my family: David, Maria, Randy, Lisa, Hannah, Wil, Madison, Jack, and Luke, and now Sam and Ben.

A special thanks to Cheryl.

James D. Bethune
Boston University

Style Conventions in *Engineering Design and Graphics with Autodesk® Inventor™ 2008*

Text Element	Example
Key terms—Bold and italic on first mention (first letter lowercase) in the body of the text. Brief glossary definition in margin following first mention.	Views are created by placing **viewport** objects in the paper space layout.
Inventor commands—Bold and uppercase.	Start the **LINE** command.
Toolbar names, menu items and dialog box names—Bold and follow capitalization convention in Inventor toolbar or pull-down menu (generally first letter capitalized).	The **Layer Manager** dialog box The **File** pull-down menu
Toolbar buttons and dialog box controls/buttons/input items—Bold and follow capitalization convention of the name of the item or the name shown in the Inventor tooltip.	Choose the **Line** tool from the **Draw** toolbar. Choose the **Symbols and Arrows** tab in the **Modify Dimension Style** dialog box. Choose the **New Layer** button in the **Layer Properties Manager** dialog box. In the **Line and Arrows** tab, set the **Arrow size:** to **.125.**

Contents

Chapter 6 Threads and Fasteners

Chapter 7 Dimensioning

Chapter 8 Tolerancing

Chapter 9 Springs

Chapter 10 Shafts

Chapter 11 Bearings

Chapter 12 Gears

Chapter 13 Sheet Metal Drawings

Chapter 14 Weldment Drawings

Chapter 15 Cams

Appendix 653

Index 663

Getting Started

1

Chapter Objectives

- Learn how to create drawings.
- Understand the different drawing formats.
- Learn how to create a sketch.
- Draw a solid model.
- Draw holes.
- Draw angular shapes.

INTRODUCTION

This chapter presents a step-by-step introduction to Inventor 2008. When the program is first accessed, the **Open** ***dialog box*** will appear. See Figure 1-1. The **Open** dialog box appears, so that if you want to return to a project you are working on, you can access it directly. Click the X box in the upper right corner to close the **Open** dialog box so that you can start a new drawing.

dialog box: A temporary window that appears for entering user input.

Figure 1-1

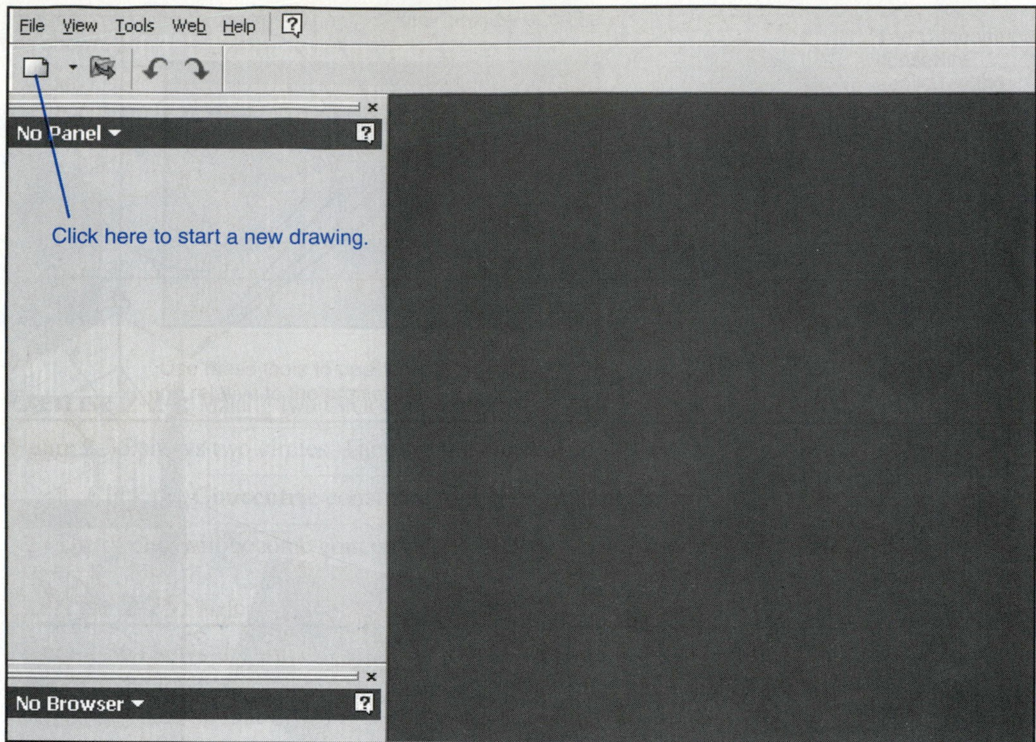

Figure 1-2

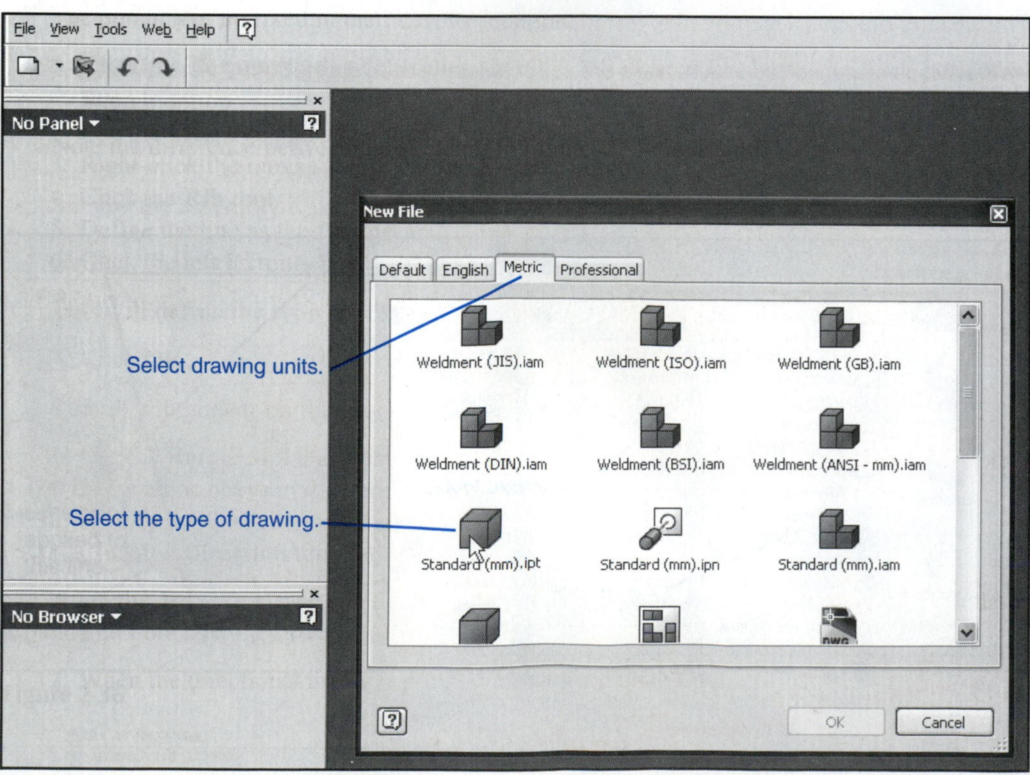

Figure 1-3

There are seven options for creating drawings using four different types of files. The files are categorized using four different extensions. The extensions are defined as follows.

.ipt: part files for either 3D model drawings or sheet-metal drawings. These files are for individual parts.

.iam: assembly drawings and weldments. Assembly drawings are formed by combining .ipt files.

.ipn: presentation files. These files are used to create exploded assembly drawings.

.idw: drawing layout files. These files are used to create orthographic views from already created assembly and presentation files.

Figure 1-2 shows a drawing screen after the **Open** dialog box has been closed. Click the **New** tool to start a new drawing. Figure 1-3 shows the **New File** dialog box.

There are several different drawing standards listed. This book uses the ANSI (American National Standards Institute) and ISO (International Standards Organization) standards. These drawing standards will be covered in detail in the chapters on dimensions and tolerances. The **Standard (mm).ipt** format will be used to create the first drawing.

Creating a First Sketch

This section shows how to set up, create, and save a first drawing. The intent is to walk through a simple drawing in order to start to understand how Inventor functions.

1. Select the **Metric** tab from the **New File** dialog box.
2. Select the **Standard (mm) .ipt** tool, then **OK.**

The drawing screen should change and look like the screen shown in Figure 1-4. The Inventor drawing screen includes a set of ***pull-down menus,*** the **Standard** toolbar, and the **Command** toolbar at the top of the screen. The browser area is at the lower left of the screen and contains a running list of how the drawing was created. The browser area information is used to edit models both during their creation and after they are created.

pull-down menu: A menu that is displayed by picking a name on the menu bar.

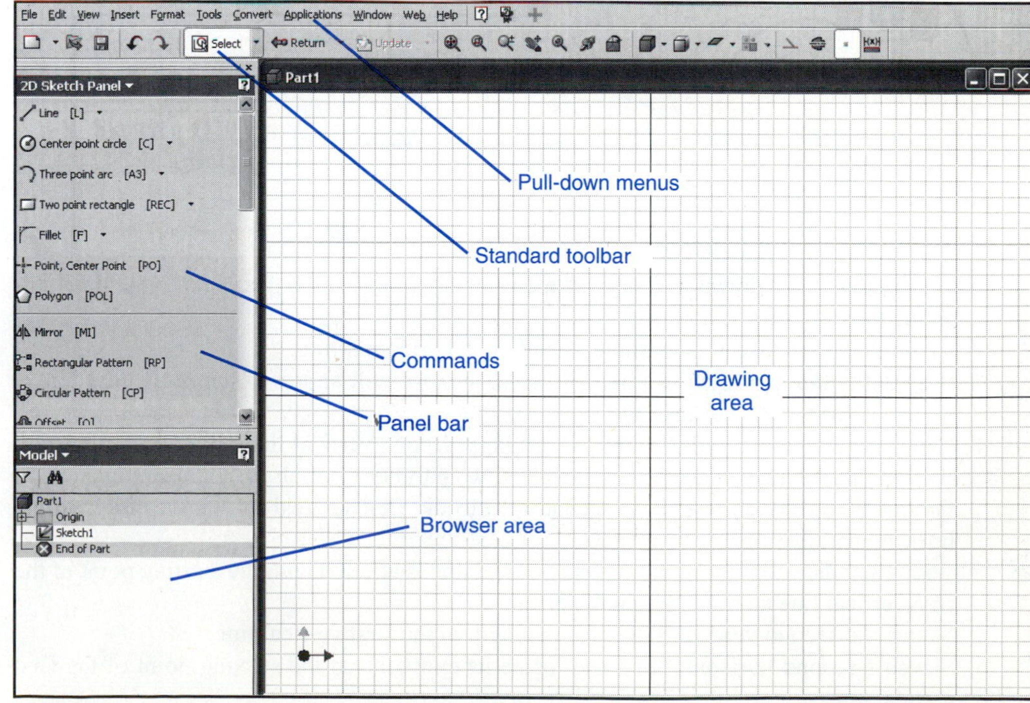

Figure 1-4

The panel bar contains command tools used to create drawings. The tool listing will change according to the operating mode selected.

Exercise 1-1: Sketching a 30 × 40 Rectangle

1. Select the **Line** tool on the **2D Sketch Panel** bar.

   Inventor does not use command line prompts, and there is no coordinate value input or axis reference. All work is done on the drawing screen. Each model generates its own set of reference values.

   Lines are first sketched, that is, drawn without dimensions, and then modified to the required size. See Figure 1-5.

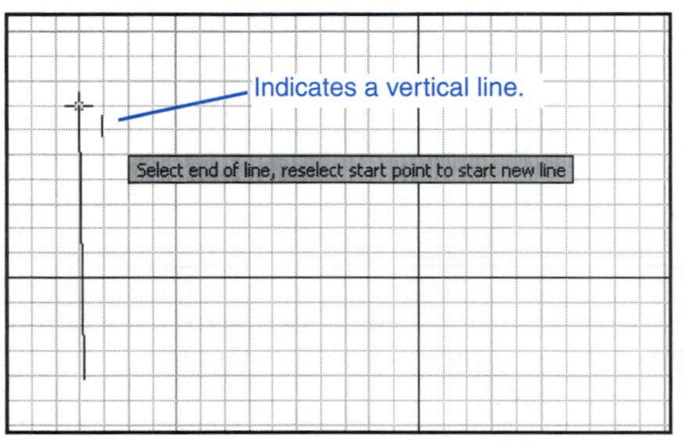

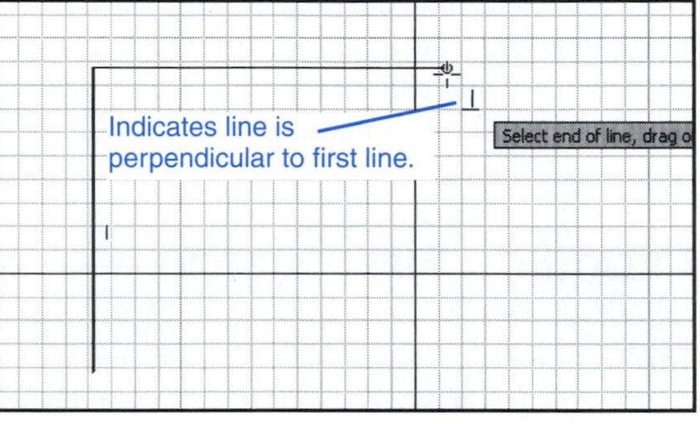

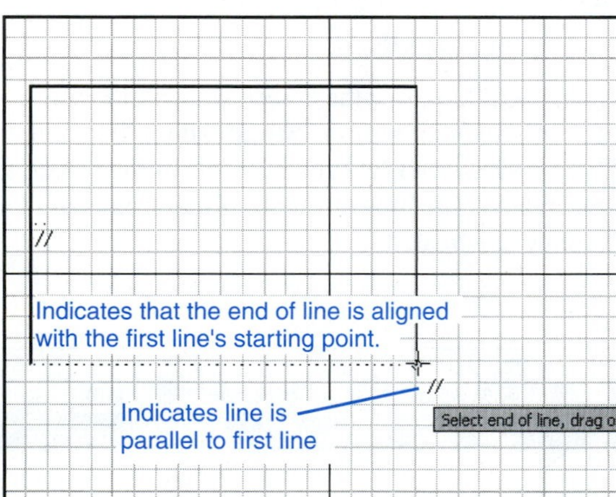

Figure 1-5

2. Sketch a vertical line anywhere on the screen.

   As the line is drawn, if it is vertical, a small symbol will appear next to the line indicating that the line is vertical.
3. Left-click the mouse and continue, sketching a horizontal line.

   As the line is sketched a perpendicular symbol will appear if the horizontal line is perpendicular to the vertical line.
4. Left-click the mouse and continue, sketching a second vertical line.

   As the second vertical line is sketched two parallel symbols will appear, one next to the line being sketched and the second next to the first vertical line, indicating that the lines are parallel.

   When the endpoint of the second vertical line is aligned with the starting point of the first vertical line a broken line will appear.
5. Sketch the second vertical line equal in length to the first vertical line.
6. Sketch a second horizontal line and locate its endpoint on the starting point of the first vertical line.

When the two points are aligned the cursor dot will change its color, and a small arc-like symbol will appear. See Figure 1-6.

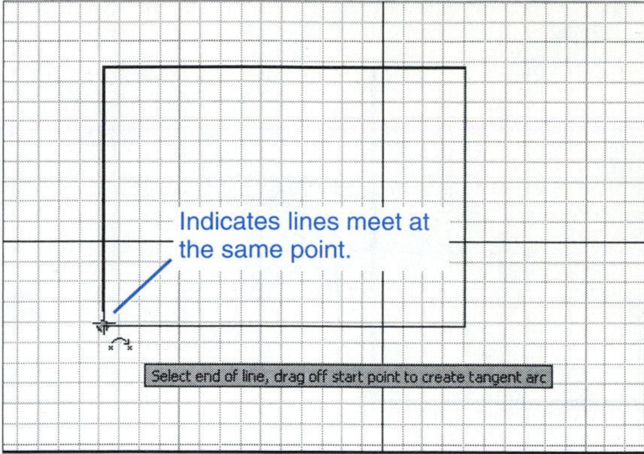

Figure 1-6

7. Right-click the mouse and select the **Done** option.

Exercise 1-2: Deleting Lines

Lines and other objects may be deleted from a sketch.

1. Select the line to be deleted.
2. Right-click the mouse.
   A dialog box will appear on the screen. See Figure 1-7.

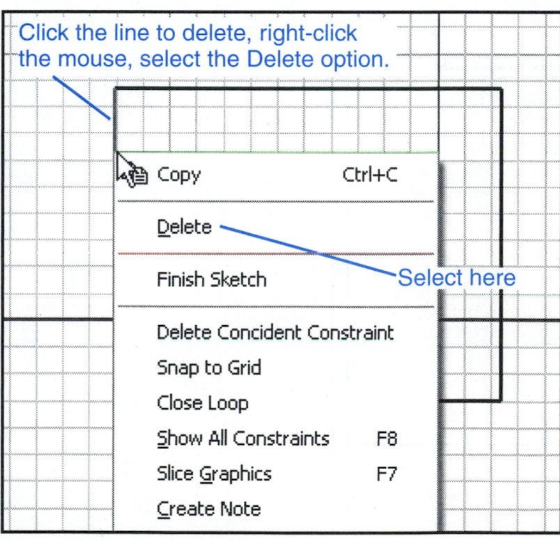

Figure 1-7

3. Select the **Delete** option.
   The line will disappear.

Exercise 1-3: Undoing a Command

The **UNDO** command will undo the last command entered.

1. Click on the **Undo** tool located at the top of the screen on the **Standard** toolbar.
   The line will reappear.

Exercise 1-4: Sizing the Rectangle

1. Select the **General Dimension** tool from the **2D Sketch Panel** bar.
2. Select the left vertical line, then move the created dimension to the left of the object and click the left mouse button.

A small dialog box will appear containing the distance on the sketched line. See Figure 1-8.

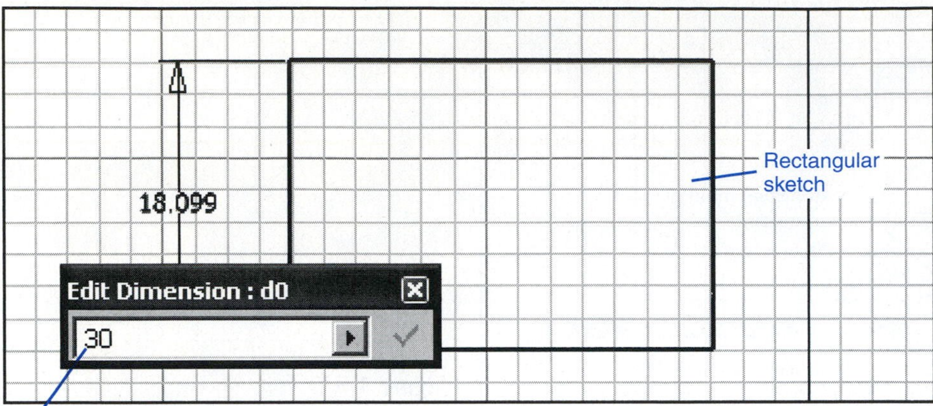

Enter the desired 30 value. The vertical sides of the rectangle will change to 30 mm.

Figure 1-8

3. Press the <**Del**> key to remove the value, and type in **30,** the required length.
4. Click on the check mark on the dialog box.
   The line will change length.
5. Repeat the procedure for one of the horizontal lines, changing the sketched value to **40.**
6. Right-click the mouse button and select the **Done** option.
   Figure 1-9 shows the resulting 30 × 40 rectangle.

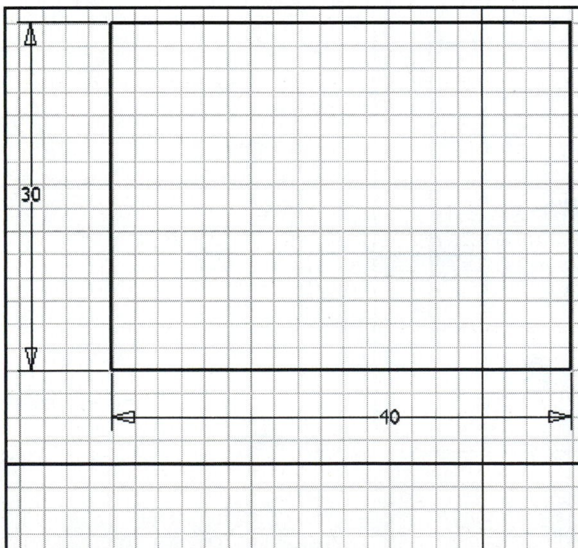

Figure 1-9

Creating a Solid Model

The 30 × 40 rectangle sketched in the previous section will now be used to create a 3D solid model.

Exercise 1-5: Changing to an Isometric View

1. Right-click the mouse and select the **Isometric View** option.
   See Figure 1-10.

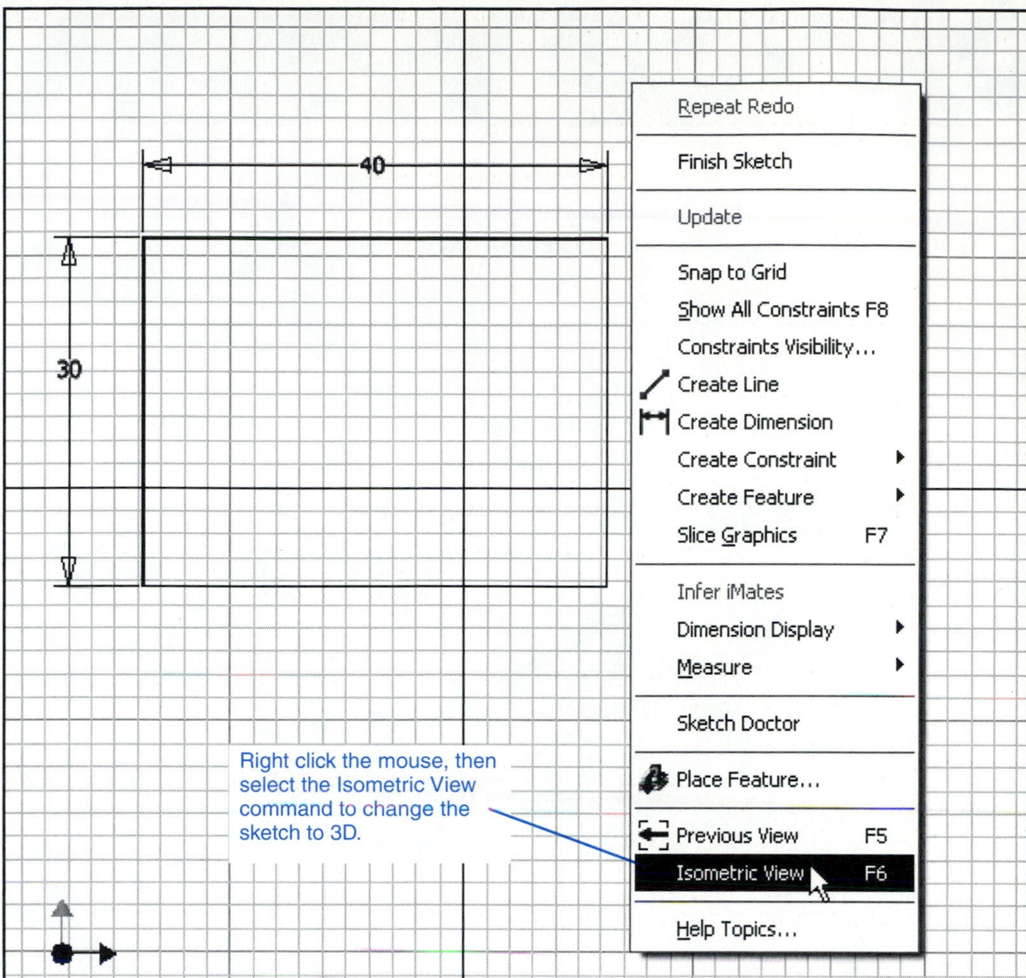

Figure 1-10

The screen will rotate into an isometric view orientation. Use the center mouse button to zoom the sketch to an acceptable size on the screen. See Figure 1-11.

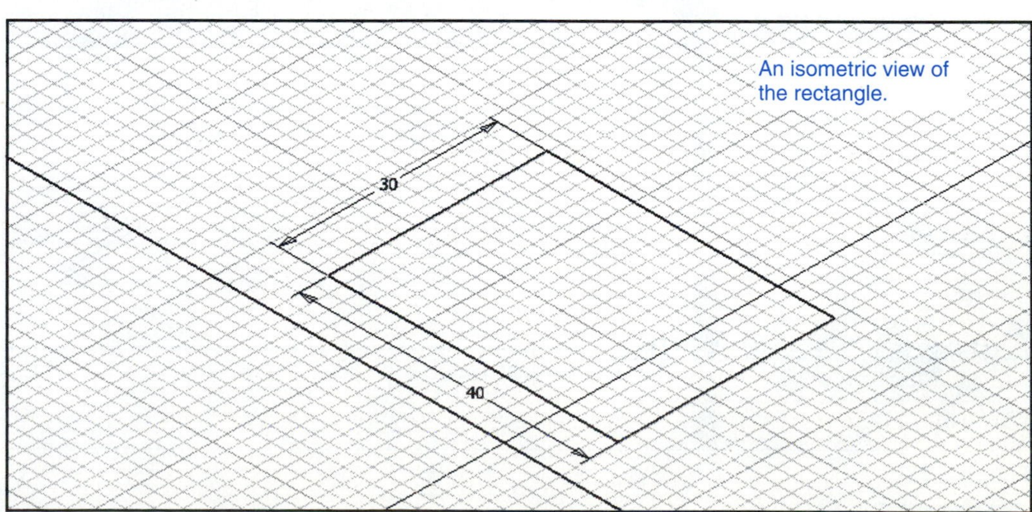

Figure 1-11

TIP Rotate the mouse wheel to zoom the drawing in and out. Hold the mouse wheel down to move the drawing around the screen.

Exercise 1-6: Creating a Solid Model

1. Right-click the mouse and select the **Finish Sketch** option. See Figure 1-12.

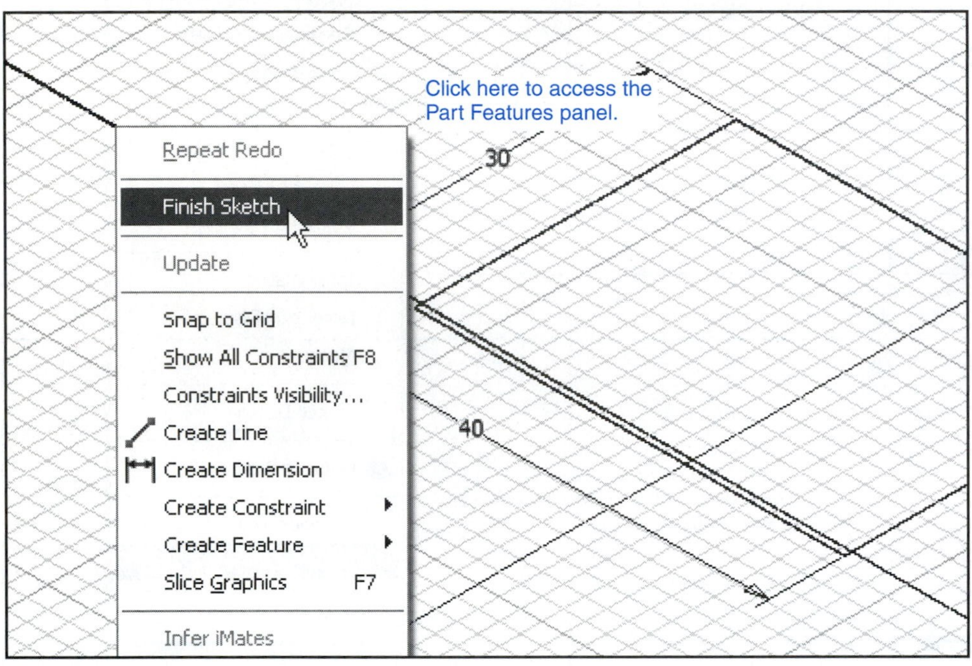

Figure 1-12

The panel bar will change to a listing of **Part Features** tools. See Figure 1-13.

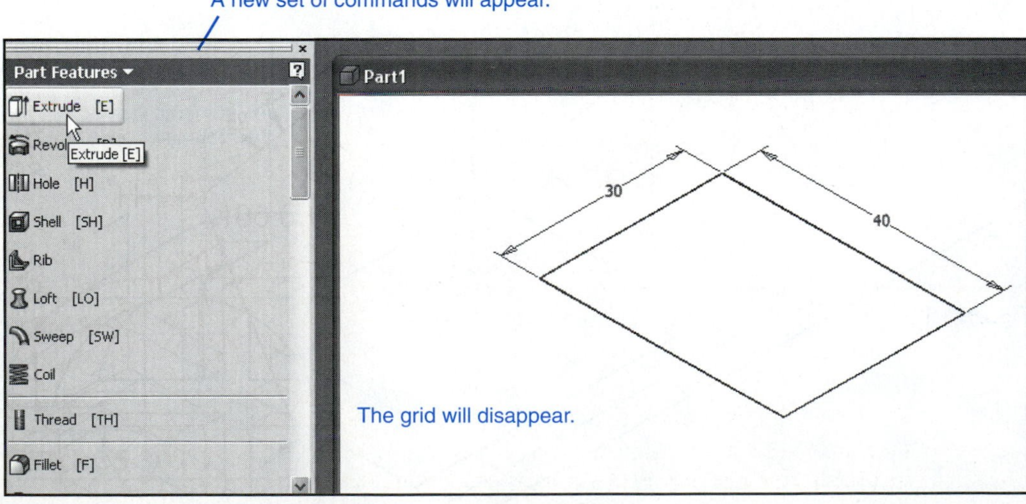

Figure 1-13

2. Select the **Extrude** tool from the panel bar.
   The **Extrude** dialog box will appear. See Figure 1-14.

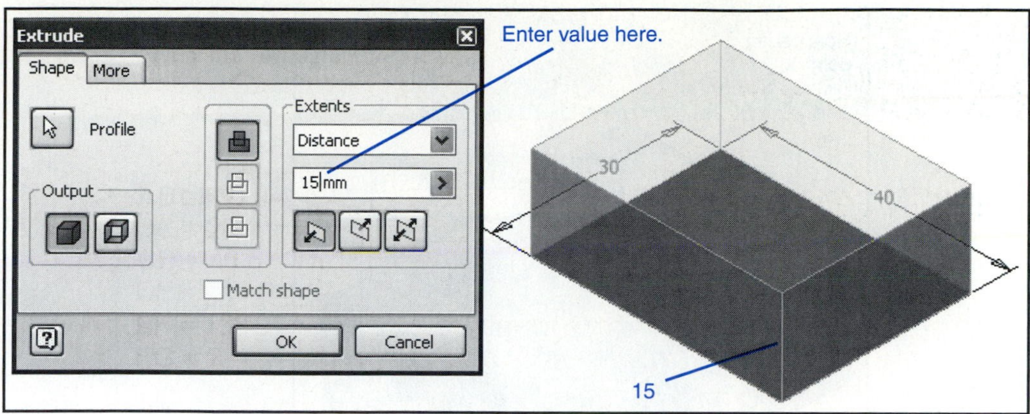

Figure 1-14

3. Change the **Extents** value to **15,** then select **OK.**
   Figure 1-14 shows the results.

Exercise 1-7: Locating the Center of a Hole

This section explains how to locate a Ø15 hole in the center of the top surface of the model.

1. Right-click on the top surface of the solid model.
   The surface will change color, confirming that it has been selected. The top surface is a new sketch plane.
2. Right-click the mouse and select the **New Sketch** command.

   See Figure 1-15.

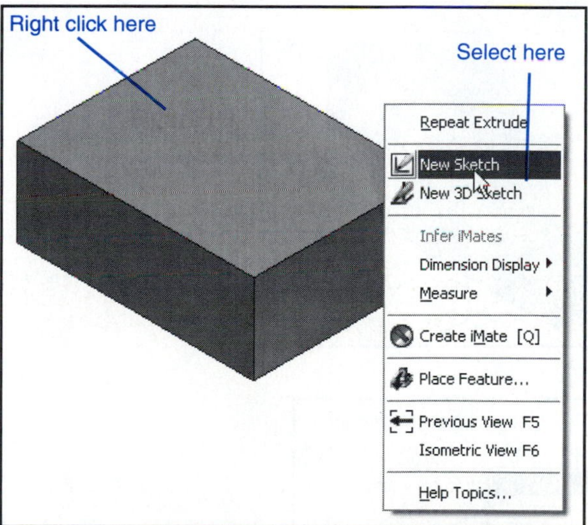

Figure 1-15

The panel bar will change to **2D Sketch** tools, and the screen grid will be aligned with the top surface of the model.

3. Select the **Point, Center Point** tool from the **2D Sketch Panel** bar.
4. Locate the point near the center of the top surface, then right-click the mouse and select the **Done** option.
5. Select the **General Dimension** tool from the panel bar.
6. Select the left edge line, then the hole's center point, and change the dimension value to **20**. See Figure 1-16.

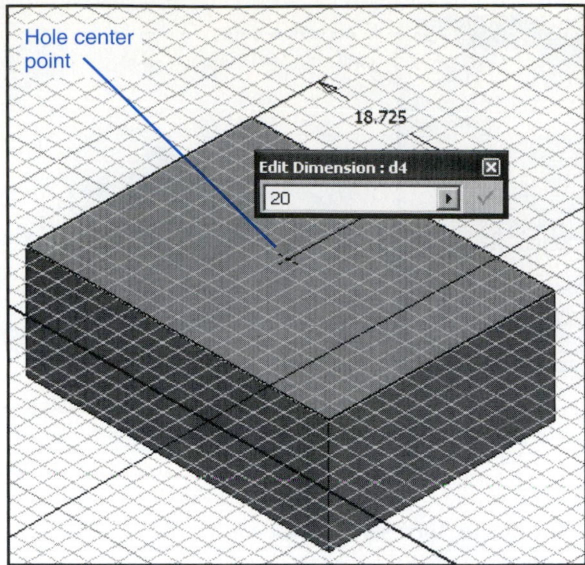

Figure 1-16

7. Repeat the procedure for the required vertical distance of **15**.
8. Right-click the mouse and select the **Done** option.
   See Figures 1-17 and 1-18.

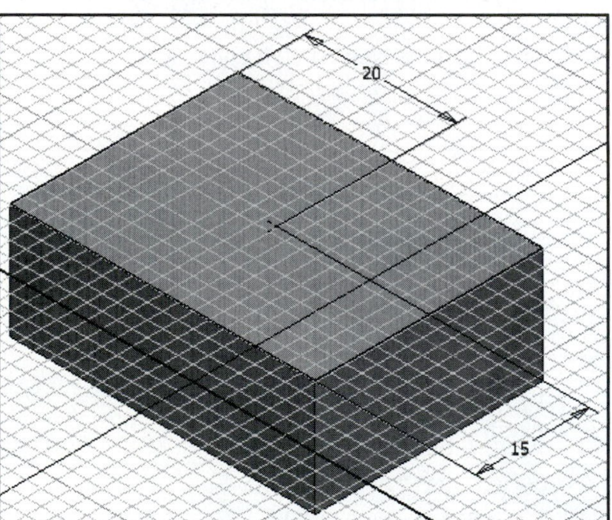

Figure 1-17

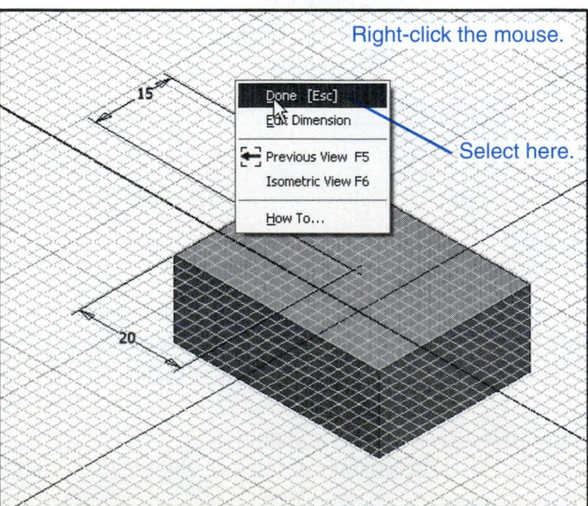

Figure 1-18

Exercise 1-8: Drawing a Ø15 Hole

1. Right-click the mouse and select the **Finish Sketch** option. See Figure 1-19.

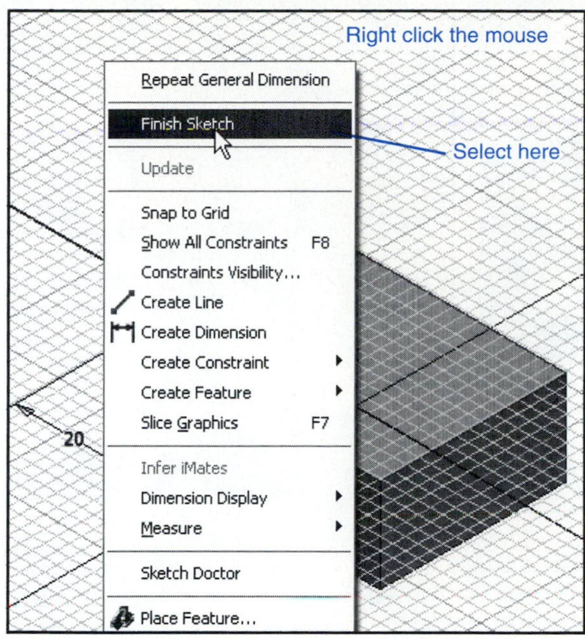

Figure 1-19

The panel bar will change to **Part Features** tools.

2. Click the **Hole** tool.
   The **Hole** dialog box will appear. See Figure 1-20.

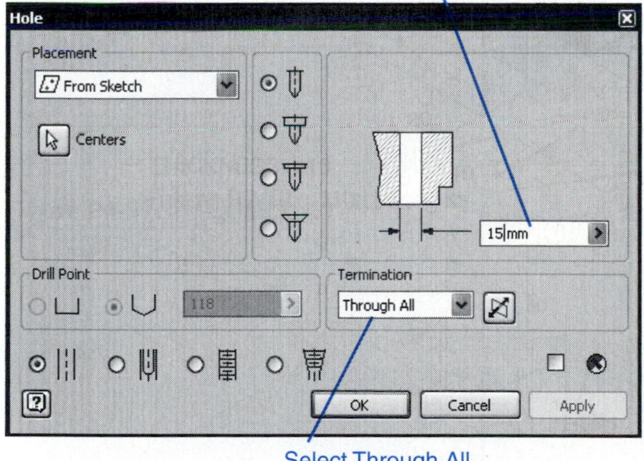

Figure 1-20

3. Locate the cursor between the hole's diameter value and the mm symbol shown in the preview box on the **Hole** dialog box, backspace out the value, and type in **15**.
4. Set the **Termination** for **Through All,** then click the **OK** box.

Figure 1-21 shows the resulting model.

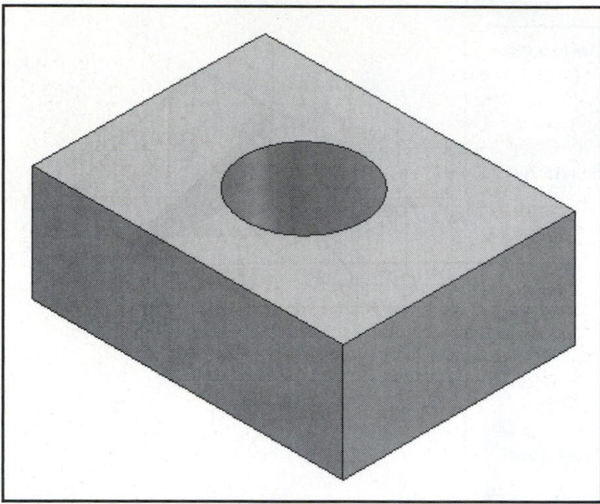

Figure 1-21

Exercise 1-9: Saving the Model

1. Click on the **File** pull-down menu, then select the **Save Copy As. . .** option.

   See Figure 1-22.

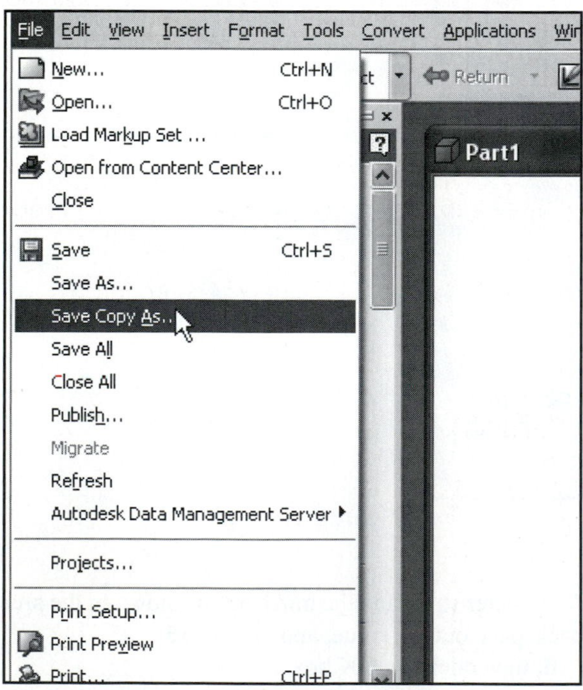

Figure 1-22

The **Save Copy As** dialog box will appear. See Figure 1-23.

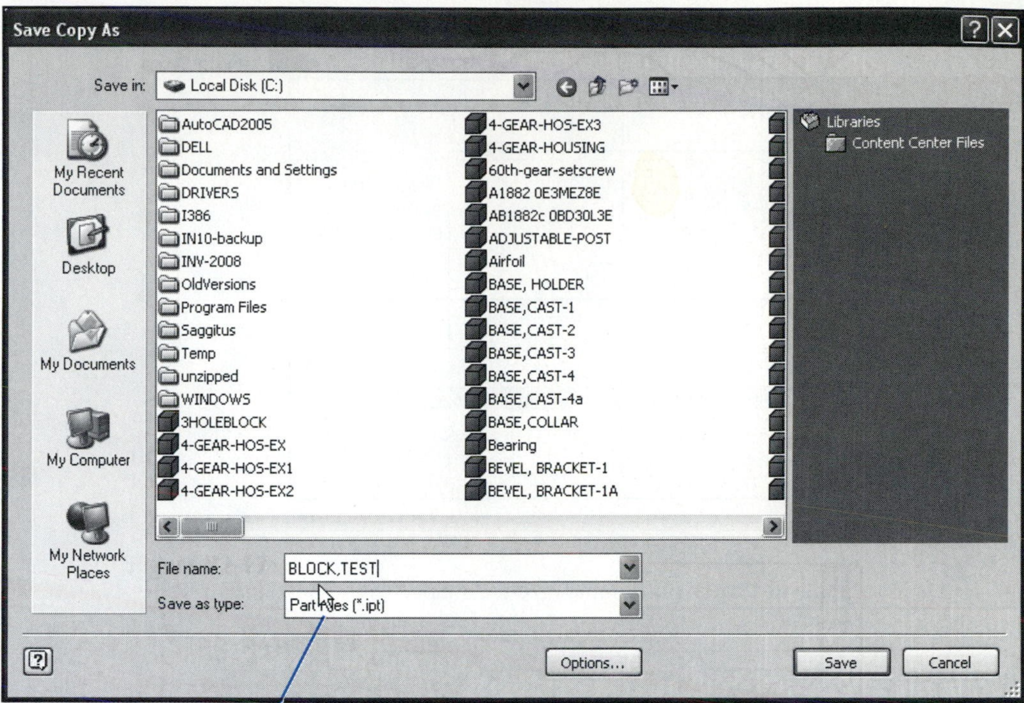

Figure 1-23

2. Select a directory and file name and save the model.
   In this example the model was saved in a directory called **Inventor** using the file name **BLOCK, TEST**.

Sample Problem SP1-1

This section shows how to draw the problem presented in Figure P1-2 in the Chapter Project. The problem's dimensions are in inches.

1. Start a new drawing, click the **English** tab, then select the **Standard (in).ipt** format.
   See Figure 1-24.

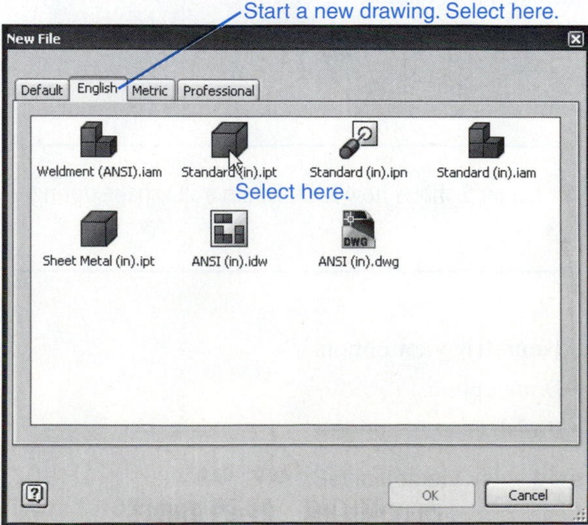

Figure 1-24

2. Sketch the approximate required shape.

   See Figure 1-25. Use the parallel and perpendicular screen symbols to make the shape as accurate as possible.

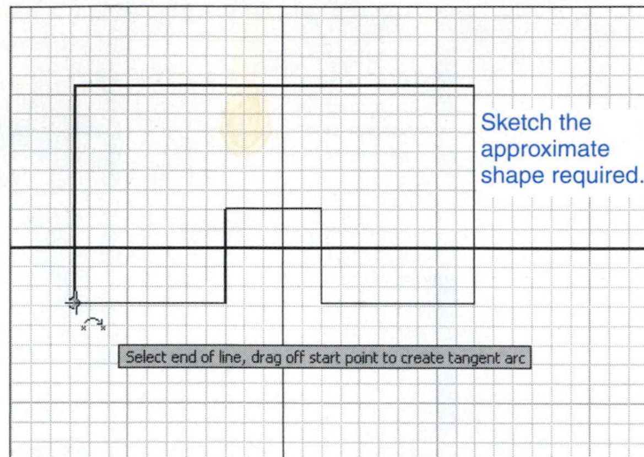

Figure 1-25

3. Use the **General Dimension** tool to size the shape.

   See Figure 1-26.

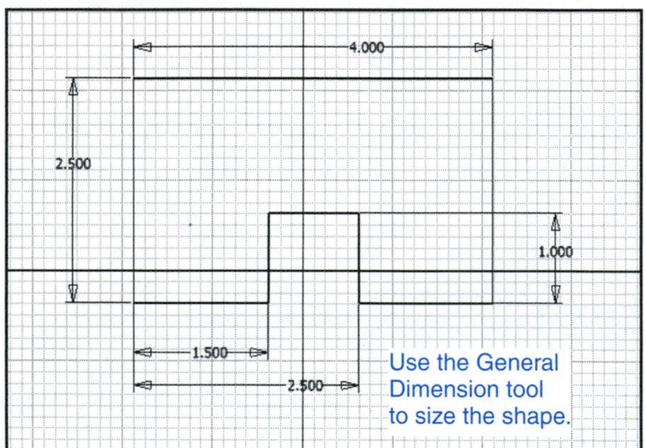

Figure 1-26

TIP Notice how the lines move when the dimension values are entered.

FOR MORE DETAILS The section on constraints in Chapter 2 shows how to constrain a 2D shape during construction.

4. Right-click the mouse and select the **Isometric view** option.
5. Right-click the mouse and select the **Done** option.
6. Right-click the mouse and select the **Finish Sketch** option.
7. Click the **Extrude** option and enter a thickness value of **0.500.**

   See Figure 1-27.

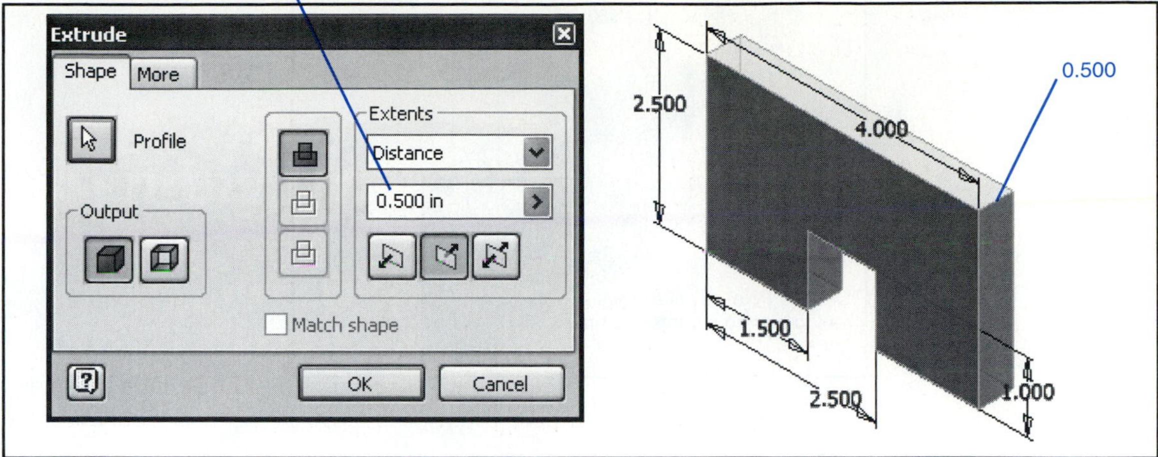

Figure 1-27

8. Click **OK.**

   Figure 1-28 shows the resulting shape.

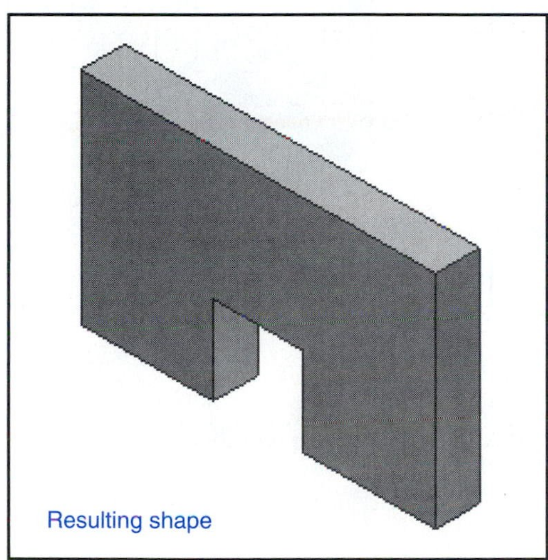

Figure 1-28

Figure 1-29

9. Right-click the mouse and select the **New Sketch** option.
10. Create a new sketch plane on the front surface of the part. See Figure 1-29.
11. Use the **Point, Center Point** tool to add hole center points to the part.
12. Use the **General Dimension** tool to locate the center points according to the given dimensions.
13. Right-click the mouse and select the **Done** option.
14. Right-click the mouse and select the **Finish Sketch** option.

    See Figure 1-30.

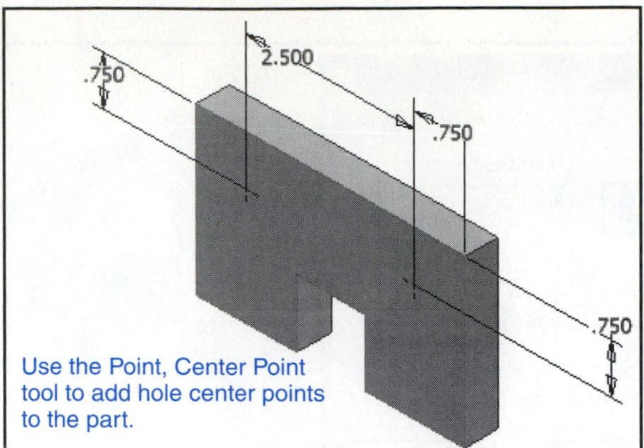

Figure 1-30

15. Select the **Hole** tool in the **Part Features** panel.
    The **Hole** dialog box will appear. See Figure 1-31.

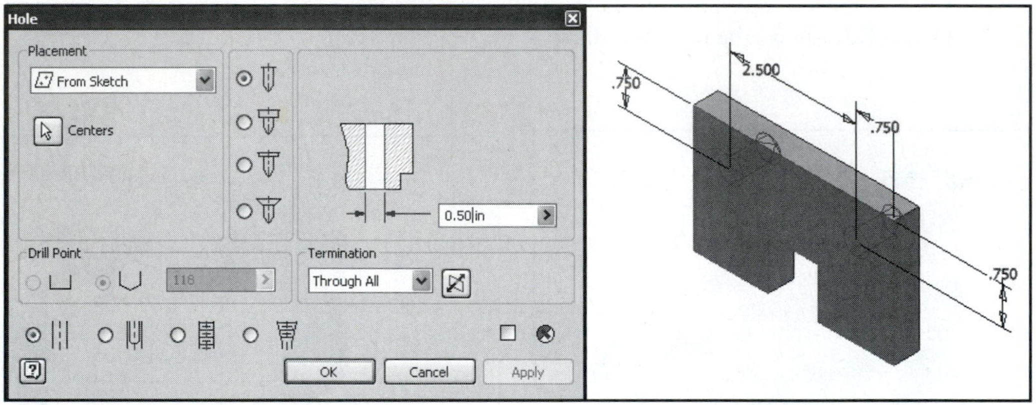

Figure 1-31

16. Set the hole's diameter to **0.500** and the termination to **Through All,** then click **OK**.
    Figure 1-32 shows the finished drawing.

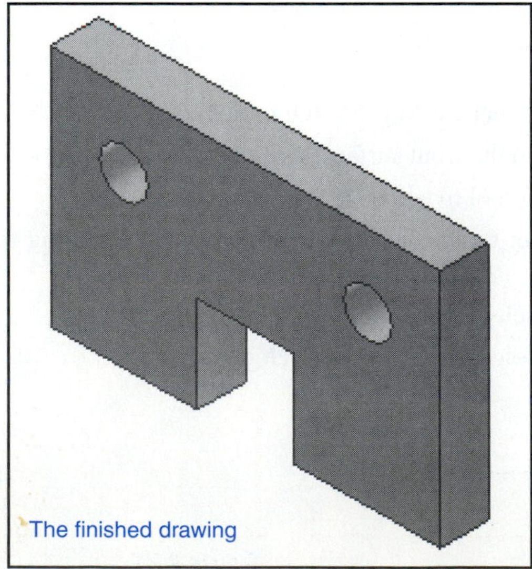

Figure 1-32

Exercise 1-10: Saving the Drawing

1. Click the **File** heading at the top left corner of the screen.
2. Select the **Save Copy As. . .** option.
3. Enter the file name; click **Save.**

   See Figure 1-33

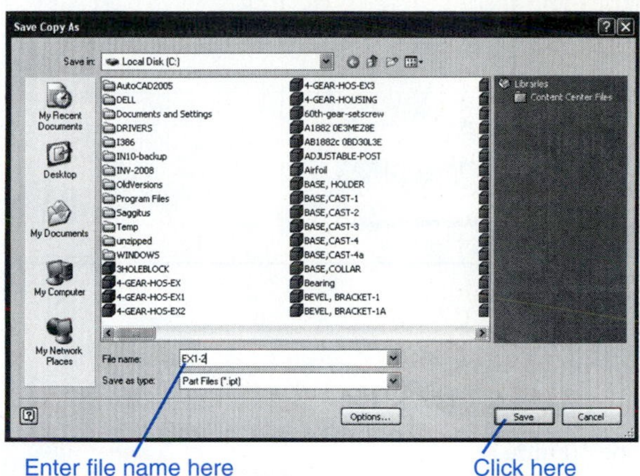

Figure 1-33

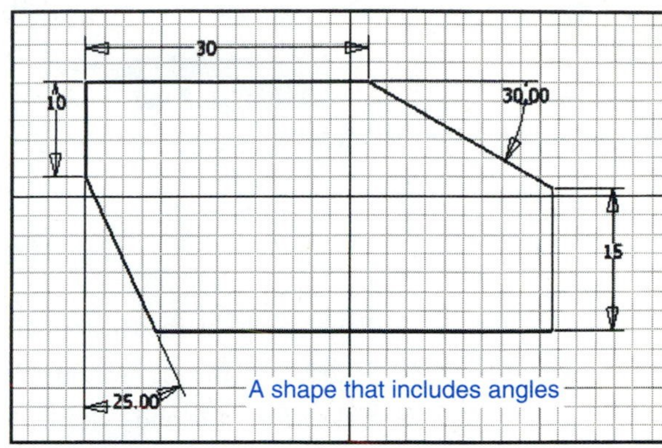

Figure 1-34

Drawing Angular Shapes

Figure 1-34 shows a shape that includes angles. The **General Dimension** tool is used to define angular values.

1. Start a new drawing using the **Metric Standard (mm).ipt** format.
2. Draw an approximate sketch of the shape.

   See Figure 1-35.
3. Use the **General Dimension** tool to start to size the shape.

   See Figure 1-36.
4. Use the **General Dimension** tool to size the angle. Click the horizontal line and the angular line, then move the cursor away from the lines.

   An angular dimension will appear. See Figure 1-37.
5. Click the angular dimension and enter the required value.
6. Size the second angular side of the shape.

   See Figure 1-38.

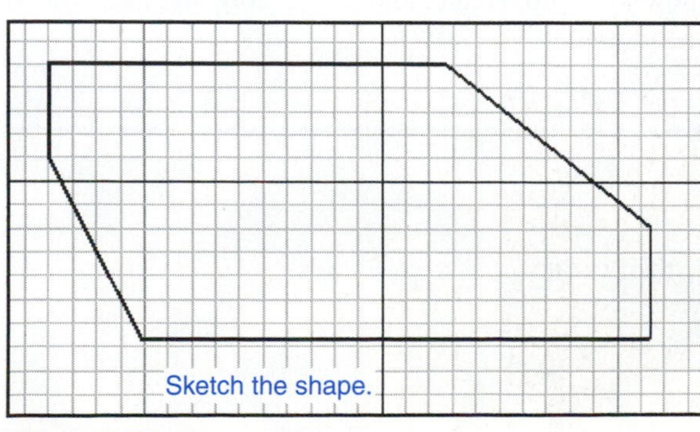

Figure 1-35

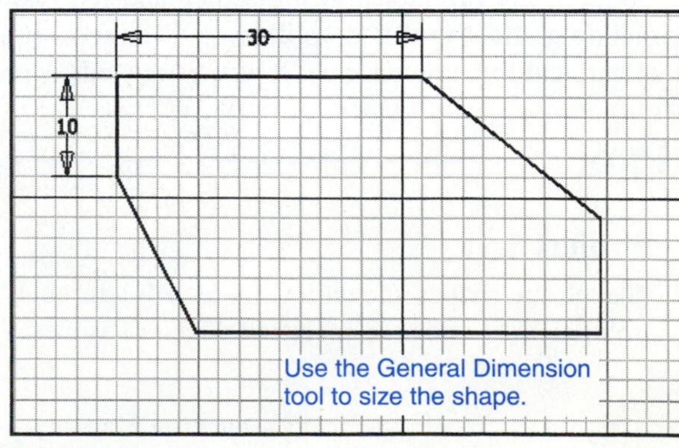

Figure 1-36

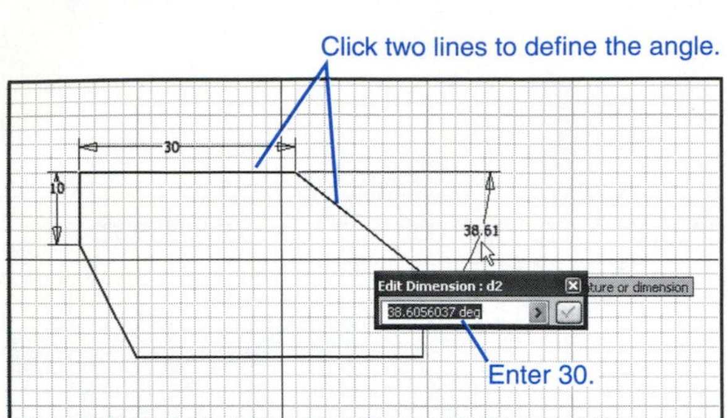

Click two lines to define the angle.
30
10
38.61
Edit Dimension : d2
38.6056037 deg
Enter 30.
Use the General Dimension tool to define the angle.
Figure 1-37

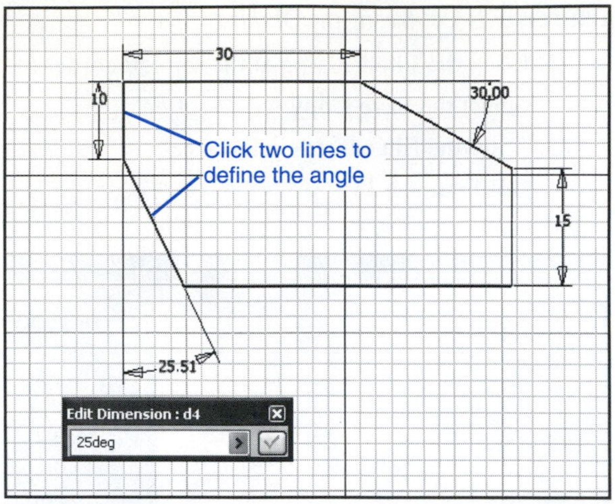

30
30.00
10
Click two lines to
define the angle
15
25.51
Edit Dimension : d4
25deg
Figure 1-38

7. Right-click the mouse and select the **Done** option.
8. Right-click the mouse and select the **Finish Sketch** option.
9. Select **Extrude** and extrude the shape 10 mm.
   See Figure 1-39.

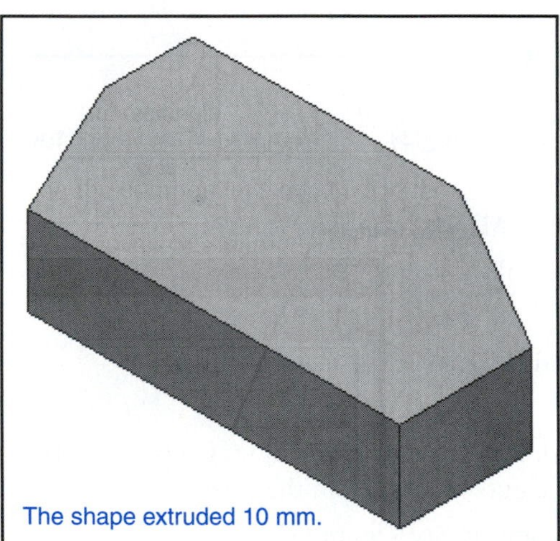

The shape extruded 10 mm.
Figure 1-39

Summary

This chapter introduced the Inventor drawing screen and its components, including the **Standard** toolbar, pull-down menus, the panel bar, and the browser area. It also explained how to access and use the features of dialog boxes in demonstrating how to set up, create, and save a drawing, and how to create a solid model.

Chapter Project

Project 1-1:

Sketch the shapes shown in Figures P1-1 through P1-8, then create solid models using the specified thickness values.

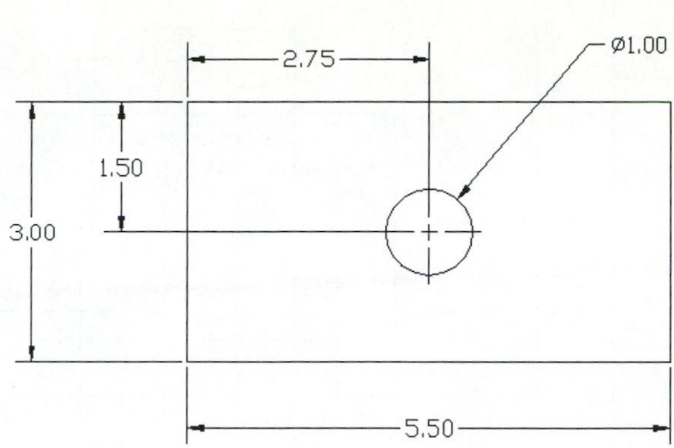

THICKNESS = 1.00
Figure P1-1 INCHES

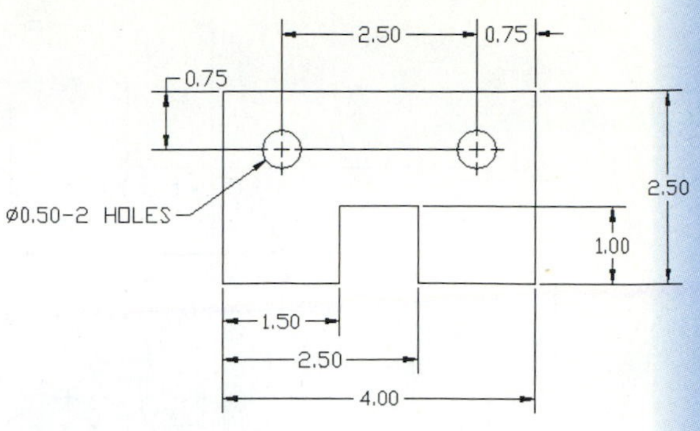

THICKNESS = 0.500
Figure P1-2 INCHES

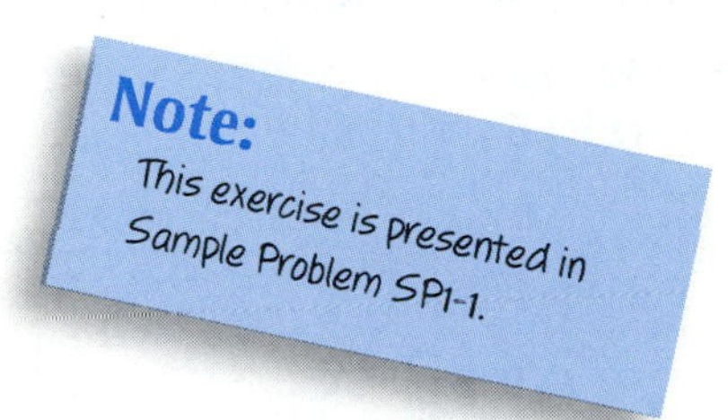

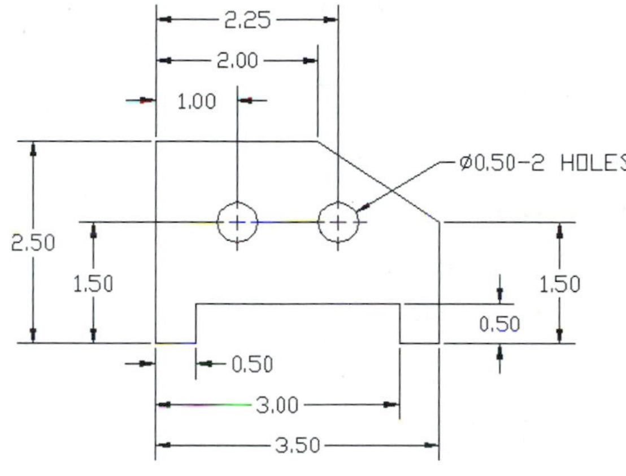

THICKNESS = 0.750
Figure P1-3 INCHES

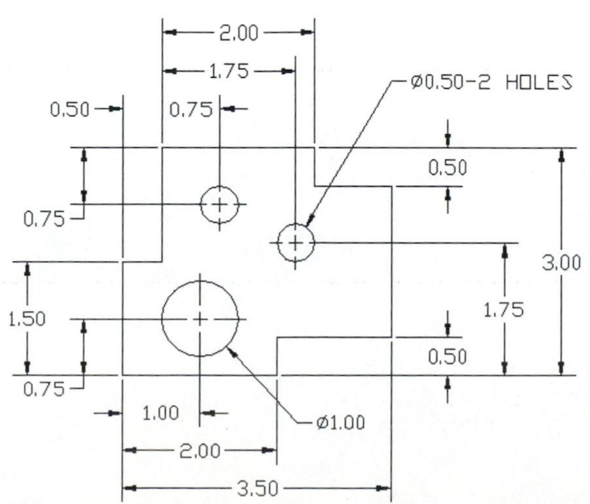

THICKNESS = 1.125
Figure P1-4 INCHES

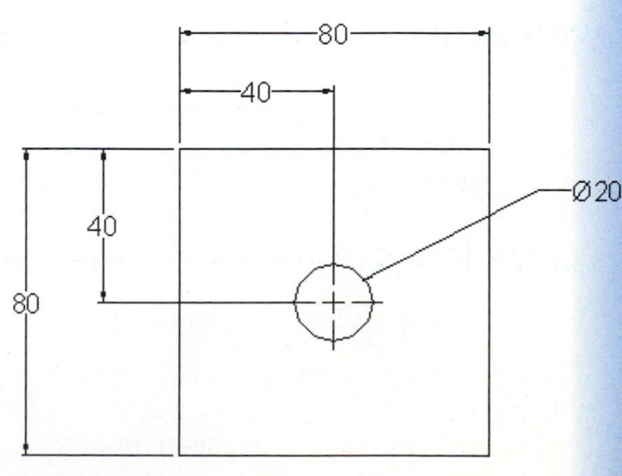

THICKNESS = 10
Figure P1-5 MILLIMETERS

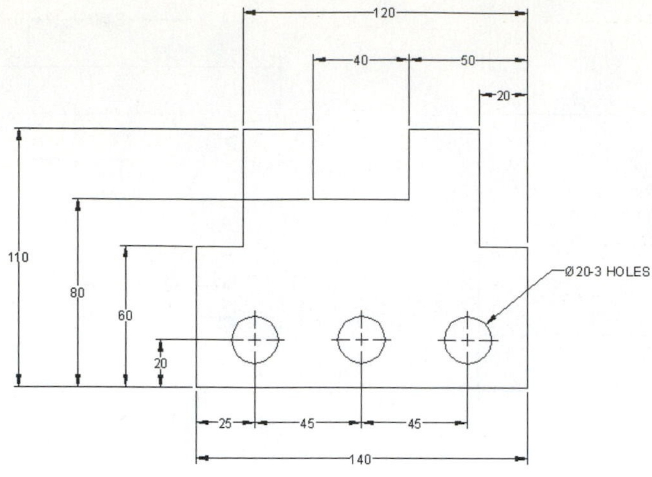

THICKNESS = 15
Figure P1-6 MILLIMETERS

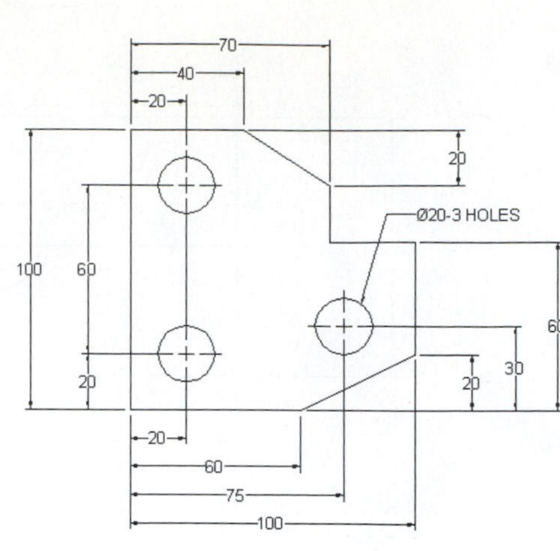

THICKNESS = 5
Figure P1-7 MILLIMETERS

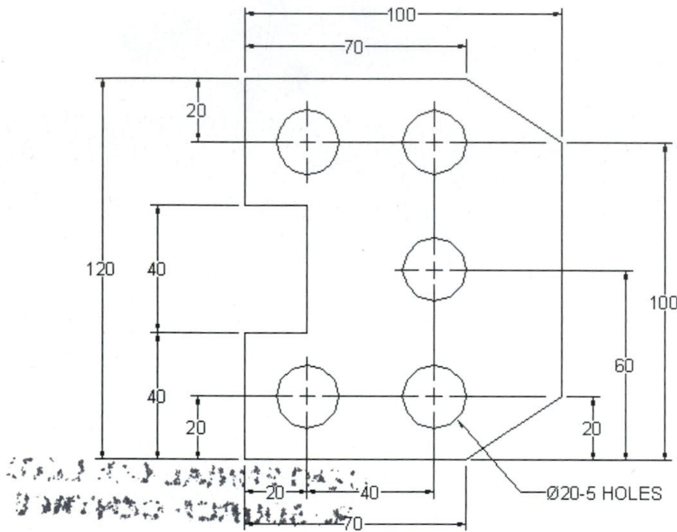

THICKNESS = 12
Figure P1-8 MILLIMETERS

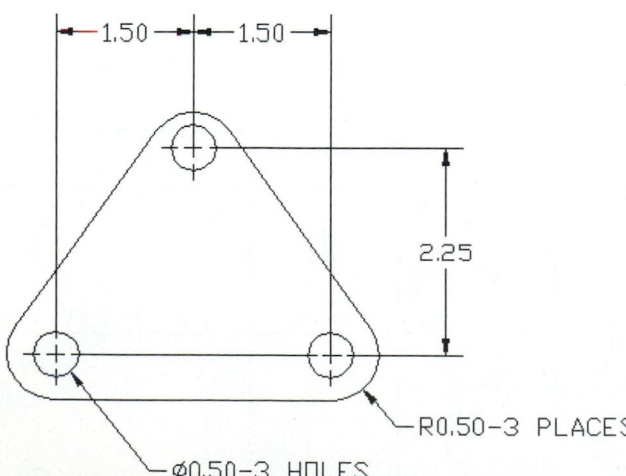

THICKNESS = .250
Figure P1-9 INCHES

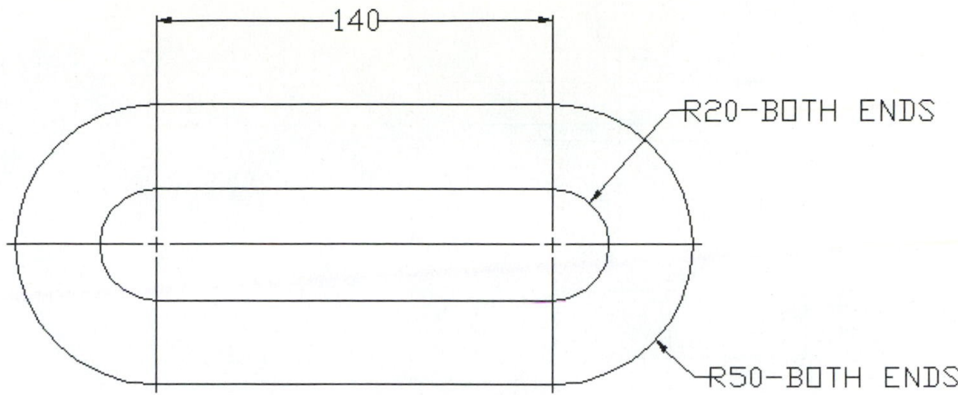

THICKNESS = 6.5

Figure P1-10 MILLIMETERS

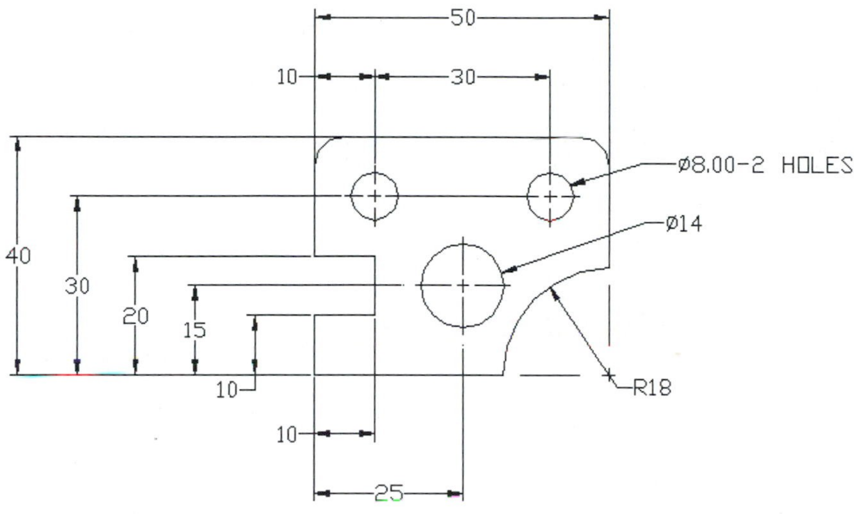

THICKNESS = 16

Figure P1-11 MILLIMETERS

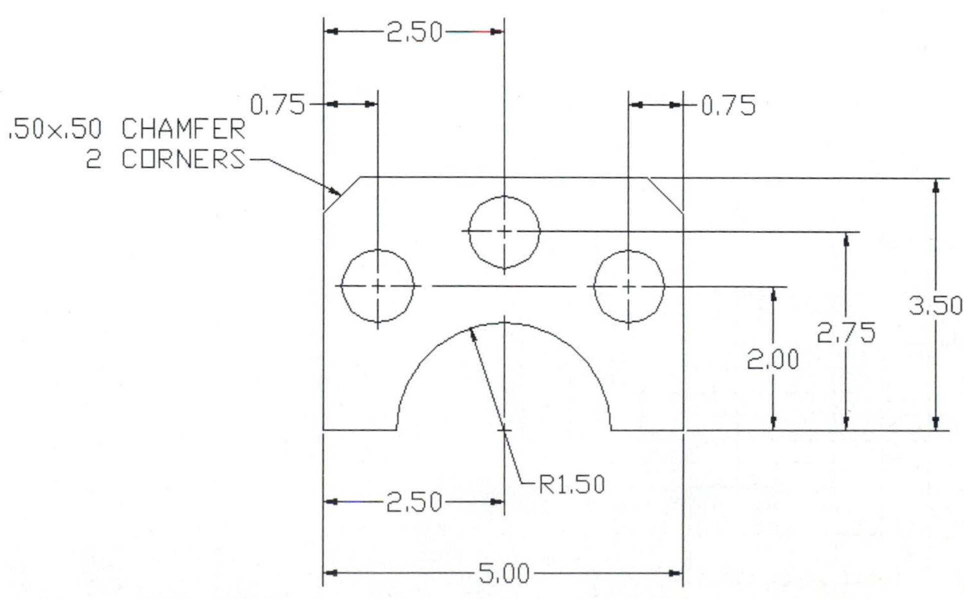

THICKNESS = .50

Figure P1-12 INCHES

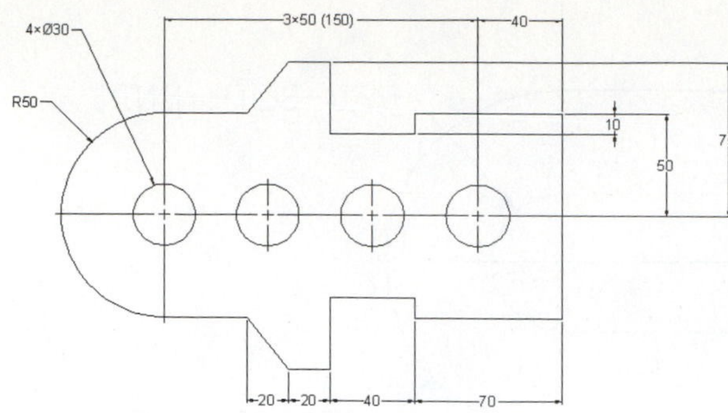

THICKNESS = 12

Figure P1-13 MILLIMETERS

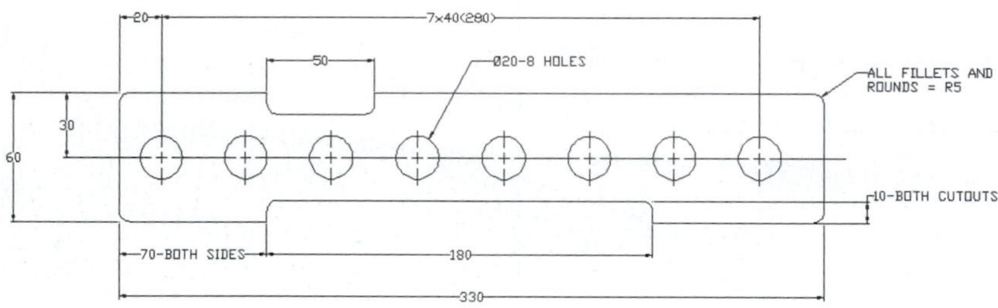

THICKNESS = 2

Figure P1-14 MILLIMETERS

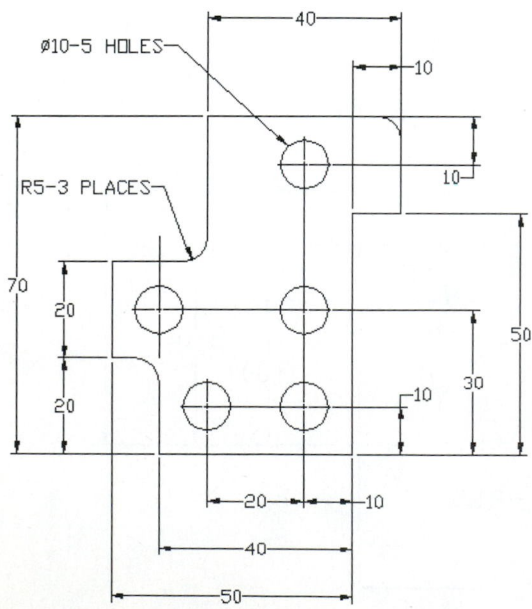

THICKNESS = 8.25

Figure P1-15 MILLIMETERS

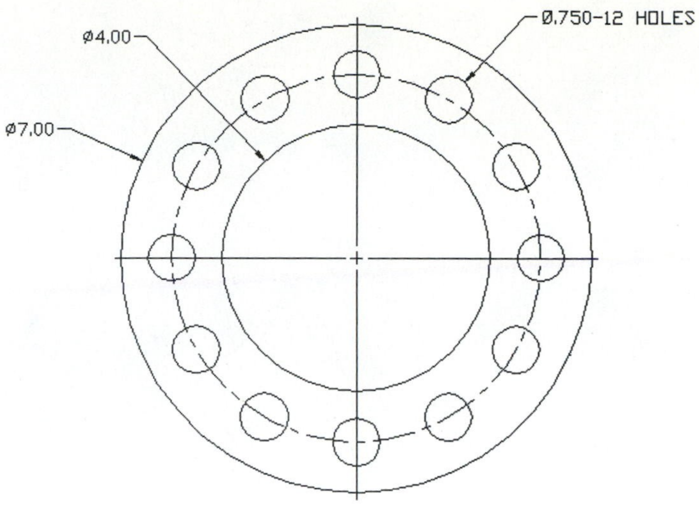

THICKNESS = .1875

Figure P1-16 INCHES

Two-Dimensional Sketching 2

Chapter Objectives

- Introduce the 2D sketch commands.
- Show how to combine 2D sketch commands to form shapes.
- Show how to edit 2D shapes.

INTRODUCTION

This chapter introduces most of the commands found in the options of the **2D Sketch Panel** bar. These commands are used to create two-dimensional (2D) sketches. Inventor models are usually based on an initial 2D sketch that is first extruded then manipulated using additional planes to develop the final model shape.

Figure 2-1 shows the sketch options available on the **2D Sketch Panel** bar.

LINE

1. Click the **Line** tool on the **2D Sketch Panel** bar.
2. Select any point on the screen.
3. Move the cursor around the screen.

As the cursor is moved a bar symbol will appear when the line is either horizontal or vertical. See Figure 2-2.

4. Click the line when an appropriate endpoint has been located.
5. Right-click the mouse and select the **Done** option.

Exercise 2-1: Sizing Lines

Figure 2-2 shows a vertical line that was sketched using the **LINE** command.

1. Click the **General Dimension** tool on the **2D Sketch Panel** bar.
2. Click the vertical line, move the cursor away from the line, and then double-click the left mouse button.

A dimension indicating the length of the line will be created.

3. Change the value to **25** and click the check mark on the dialog box.

The line's length will change to the indicated distance.

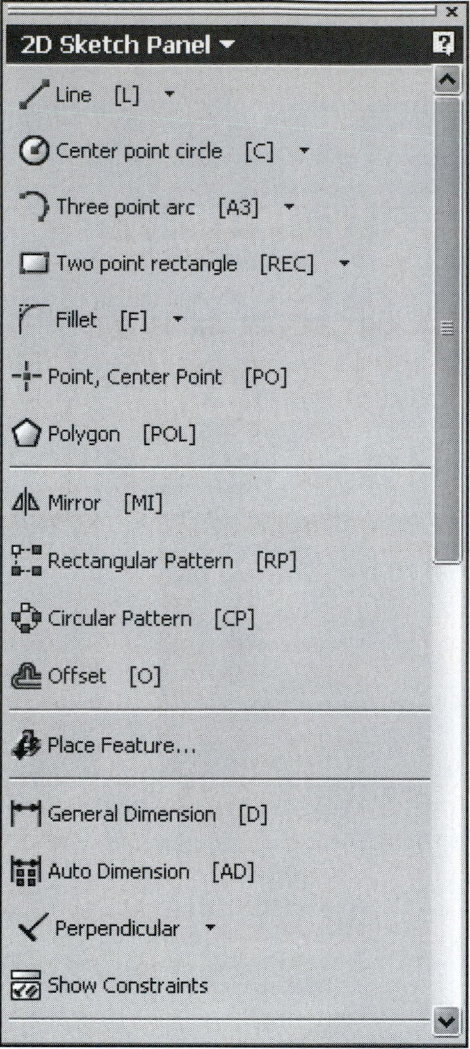

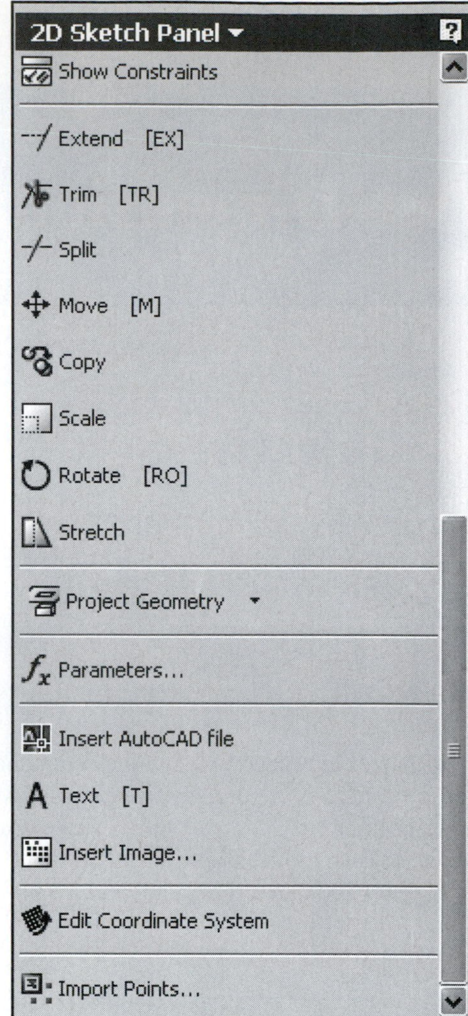

Figure 2-1

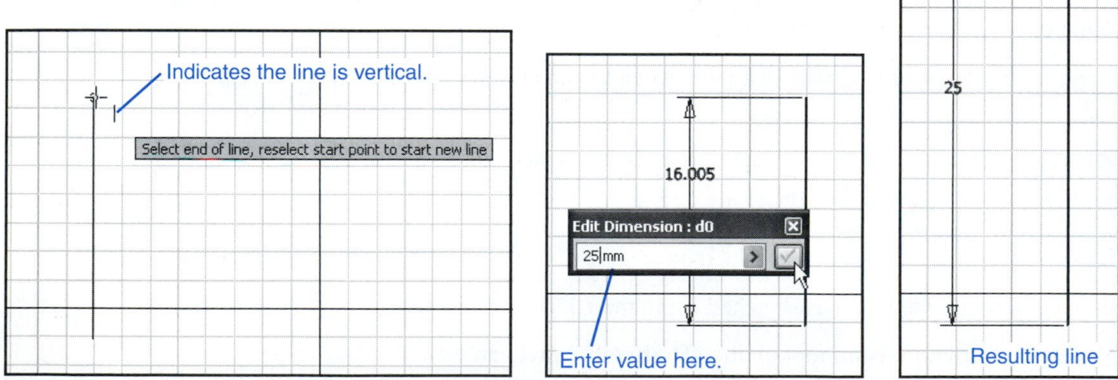

Figure 2-2

Exercise 2-2: Drawing Continuous Lines

Given the line created in Figure 2-2, draw a horizontal and a vertical line starting at the top endpoint of the existing line.

1. Click the **Line** tool and move the cursor to the top end of the 25-mm vertical line.

As the cursor is moved across the screen a yellow dot will appear with the cursor. When the cursor is aligned with the line's endpoint the dot will turn green. See Figure 2-3.

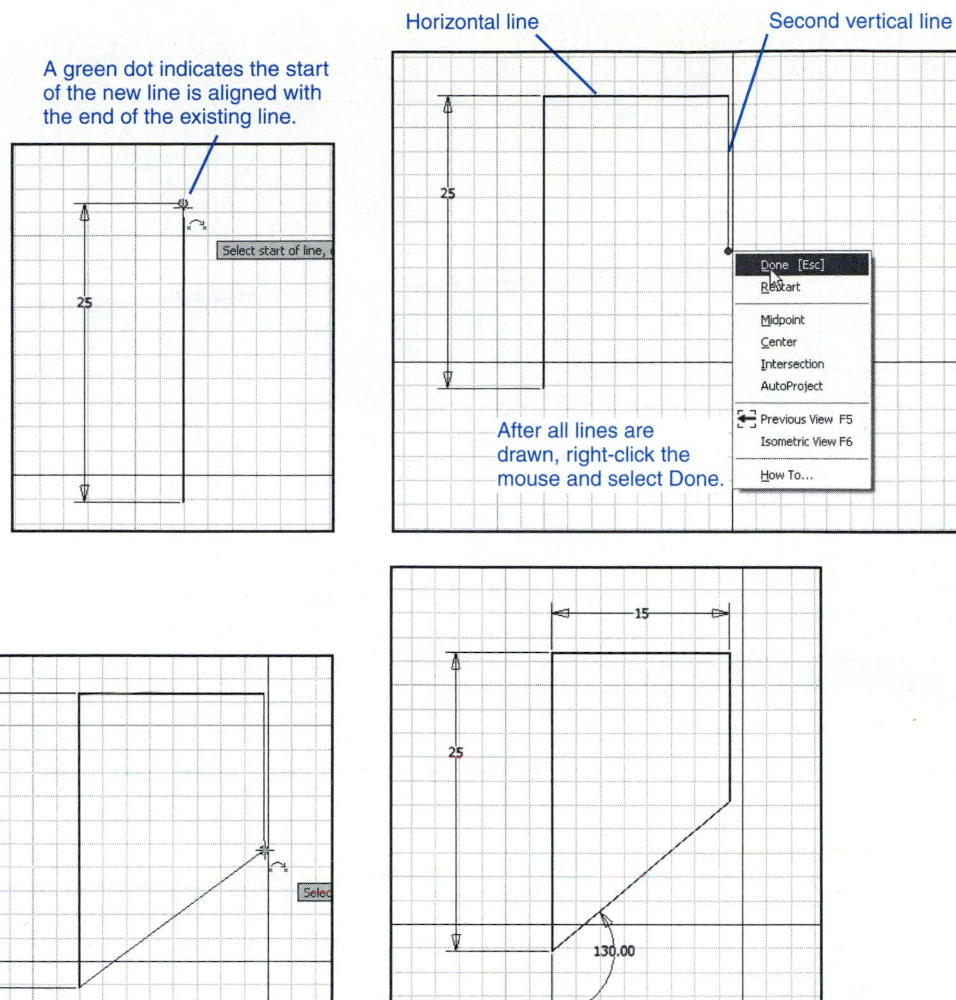

Figure 2-3

2. When the cursor is aligned with the top end of the 25-mm line, click the mouse and drag the cursor to the right, creating a horizontal line.

Note that the perpendicular symbol will appear when the line is horizontal and perpendicular to the vertical 25-mm line.

3. When the horizontal line is long enough left-click the mouse and drag the cursor downward, creating a second vertical line.
4. Right-click the mouse and select the **Done** option.
5. Click the **Line** tool again and draw a line between the endpoints of the two vertical lines.

See Figure 2-3.

6. Use the **General Dimension** tool and create an angular dimension for the slanted line.

Exercise 2-3: Drawing Lines at an Angle

1. Sketch two lines at an angle to each other.

See Figure 2-4.

2. Click the **General Dimension** tool.
3. Click each line.

An angular dimension will appear between the lines. Move the cursor around to verify that other angular values are available.

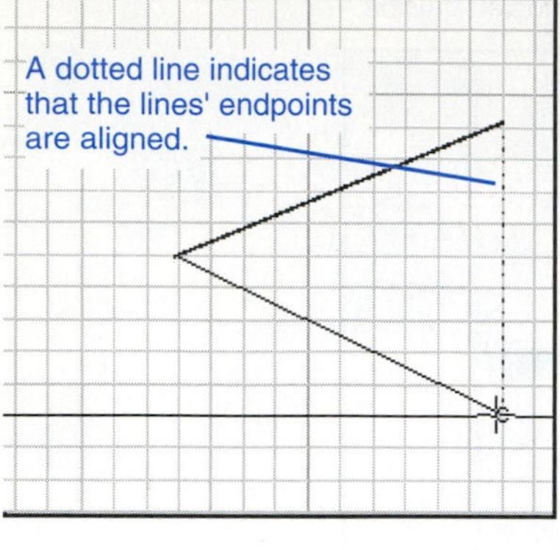

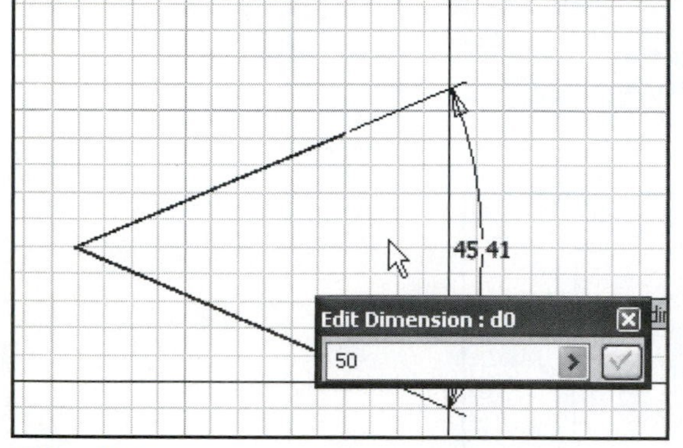

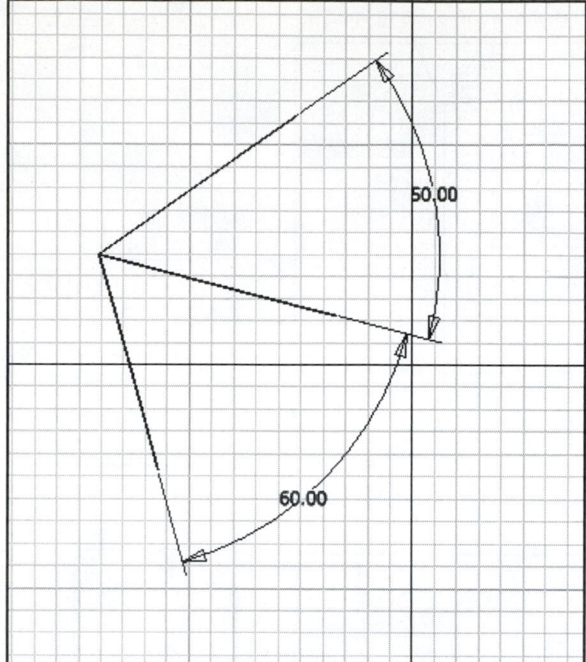

Figure 2-4

4. Locate the angular dimension and press the left mouse button twice.

A dialog box will appear defining the existing angle between the lines.

5. Change the value to **50** and click the check mark on the dialog box.

The angle between the lines will change to the entered value.

6. Sketch another line 60° to the lower angled line.

SPLINE

flyout: A command whose icon is under a related command.

open spline: A curved line whose ends do not meet.

closed spline: An enclosed curved line on which the start and endpoints are the same point.

The **Spline** tool is a ***flyout*** from the **Line** tool. An ***open spline*** is a curved line. A ***closed spline*** is an enclosed curved line on which the start and endpoints are the same point.

Exercise 2-4: Showing an Open Spline

1. Click on the **Spline** tool on the **2D Sketch Panel** bar.
2. Select four random points, then press the right mouse button.

A dialog box will appear.

3. Select the **Create** option.

See Figure 2-5.

Exercise 2-5: Drawing a Closed Spline

1. Click the **Spline** tool on the **2D Sketch Panel.**
2. Select three random points, then a fourth point aligned with the first point.
3. Click the fourth point

See Figure 2-6.

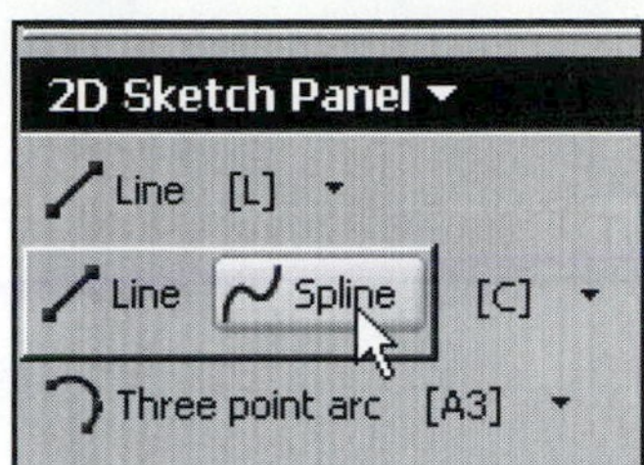

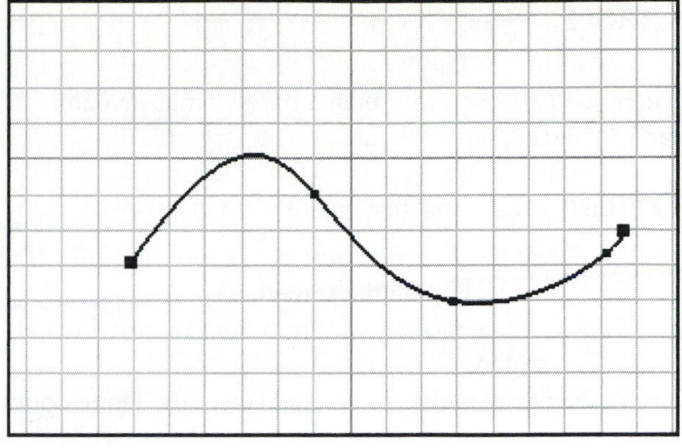

Figure 2-5

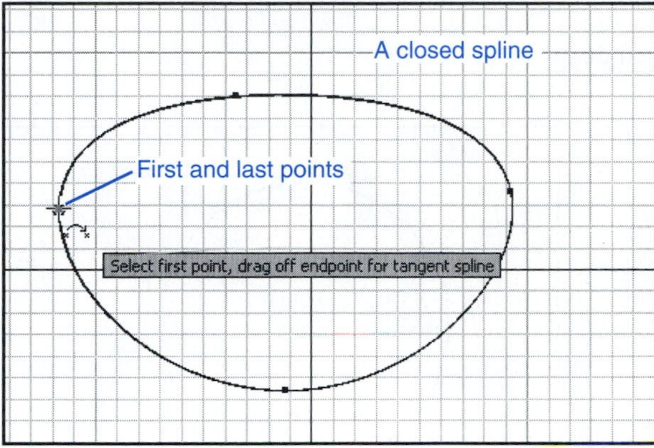

Figure 2-6

Exercise 2-6: Editing a Spline

A spline may be edited with the **General Dimension** tool.

1. Click the **General Dimension** tool and select any two points of the spline.

   See Figure 2-7.

2. Change the dimension value to **15** and click the check mark.

The distance will change. The vertical distance between points may also be edited.

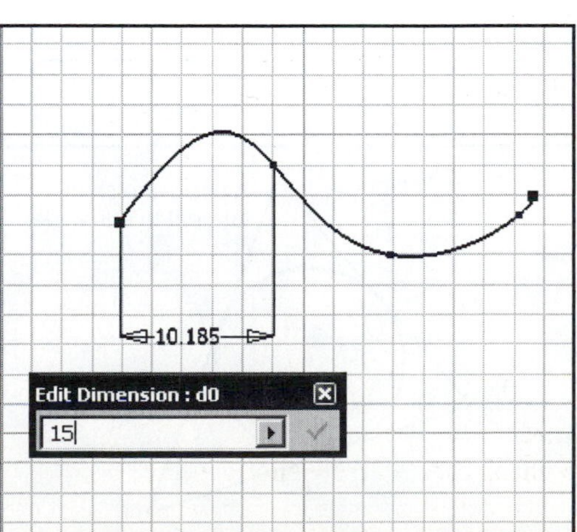

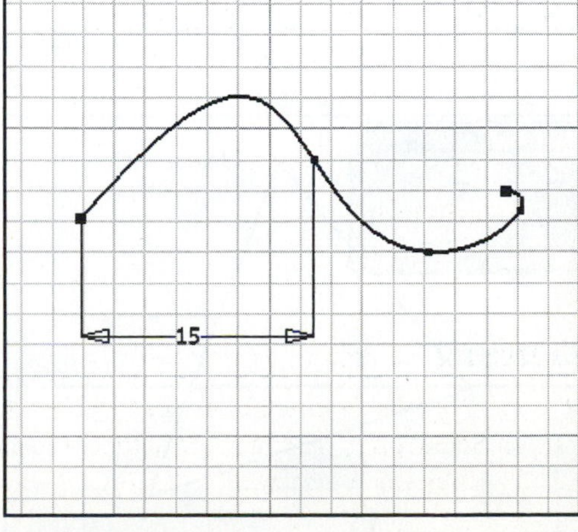

Figure 2-7

CIRCLE

There are two ways to sketch a circle using Inventor: select a center point, then define a diameter; and define three tangent points.

Exercise 2-7: Using the Center Point Option

1. Select the **Center point circle** tool from the **2D Sketch Panel** bar.
2. Select a point on the screen.
3. Move the cursor away from the center point and left-click when an approximate diameter is created.
4. Right-click the mouse and select the **Done** option.

   See Figure 2-8.

5. Select the **General Dimension** tool and use it to enter the desired diameter.

   The circle will change to the defined diameter value.

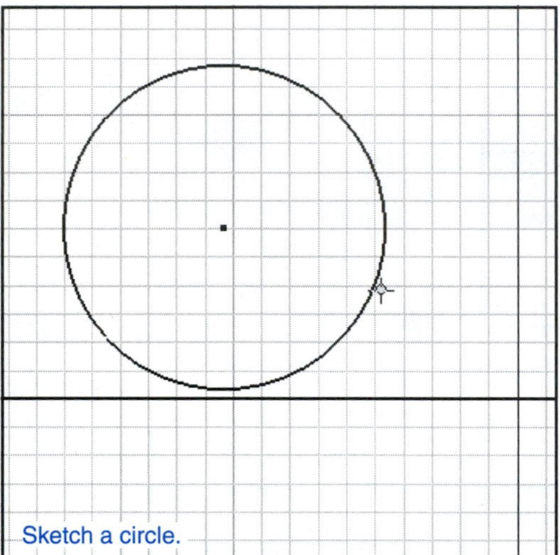

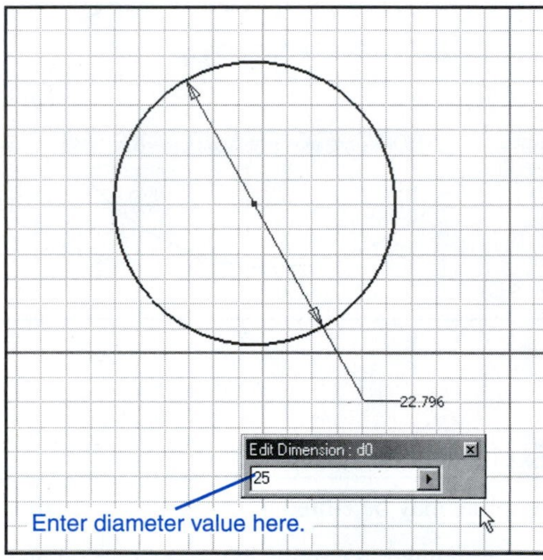

Figure 2-8

Exercise 2-8: Using the Tangent Circle Option

The **Tangent circle** option requires that some entities already exist on the screen. This option is a flyout from the **Center point circle** tool. In this example a triangle was drawn using the **Line** tool. See Figure 2-9.

1. Select the **Tangent circle** option from the **2D Sketch Panel** bar.
2. Select each of the three lines of the triangle.

A circle will appear tangent to the three lines.

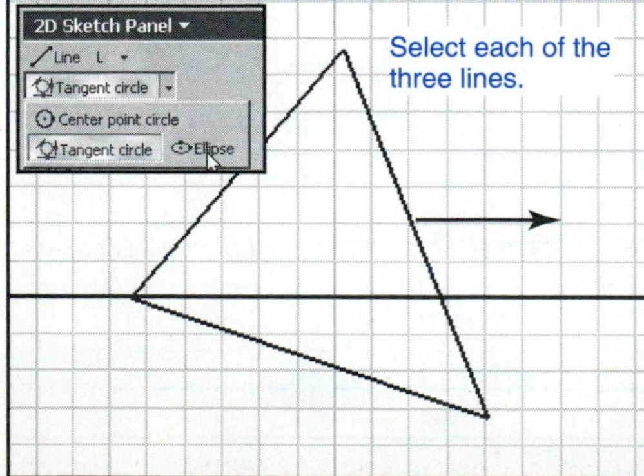

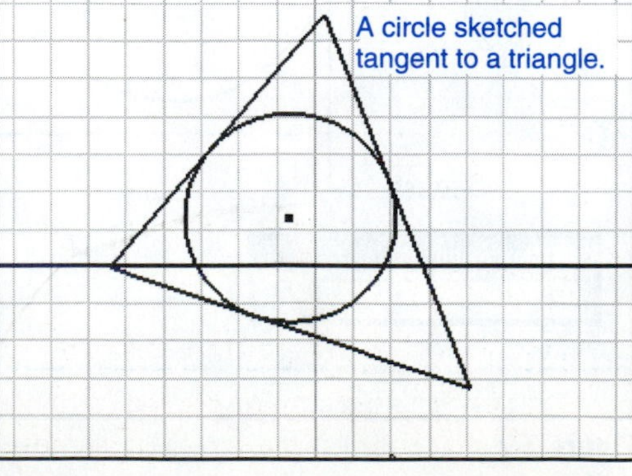

Figure 2-9

Ellipse

The **Ellipse** tool is a flyout from the **Center point circle** tool.

1. Click the **Ellipse** tool on the **2D Sketch Panel** bar.
2. Select a point on the screen.

This point will become the center point of the ellipse.

3. Move the cursor away from the point and select a point.

A *centerline* (a line with a pattern of long and short dashes) will extend from the selected point through the first point and equidistant to the other side of the point. See Figure 2-10.

4. Move the cursor above the line to define the elliptical shape.
5. Select a point, then right-click the mouse and select the **Done** option.

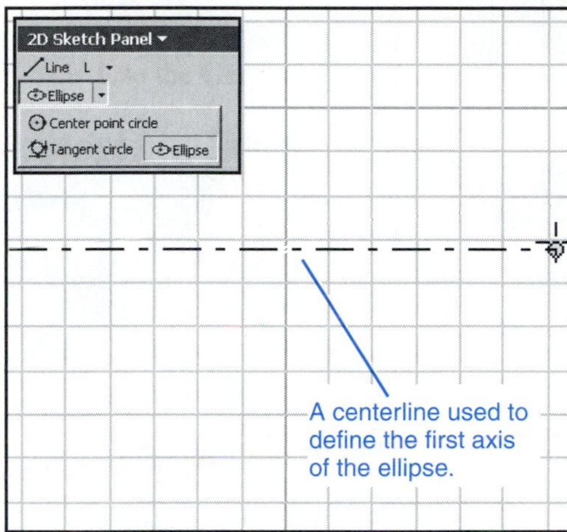

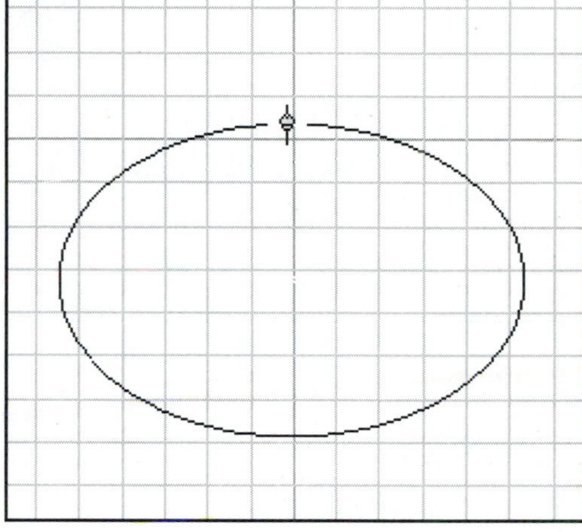

Figure 2-10

Exercise 2-9: Sizing the Ellipse

1. Click the **General Dimension** tool.
2. Select the left edge of the ellipse, then move the cursor to a location below the existing ellipse.
3. Locate the dimension and then click the dimension.
4. Enter the desired distance value.

See Figure 2-11.

5. Click the check mark on the dialog box.

The ellipse will change shape.

6. Click the **General Dimension** tool again and define the vertical elliptical value.
7. Right-click the mouse and select the **Done** option.

Arc

There are three ways to draw arcs using Inventor: select three points, define a tangent, and select a center point.

Exercise 2-10: Sketching a Three-Point Arc

1. Click the **Three point arc** tool.
2. Select a point on the screen, then move the cursor and select a second point.

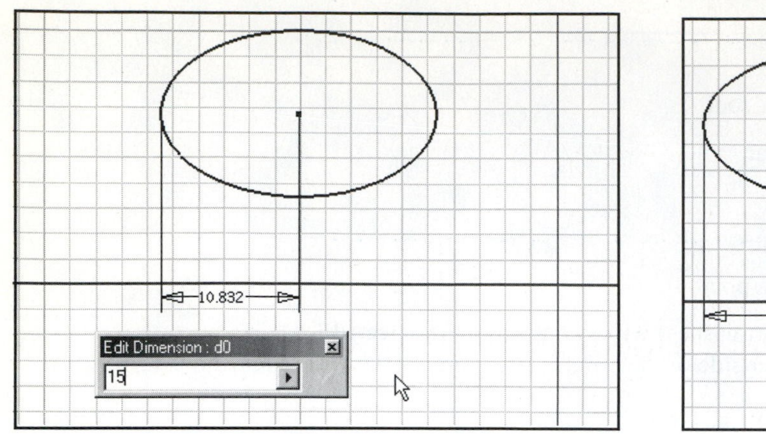

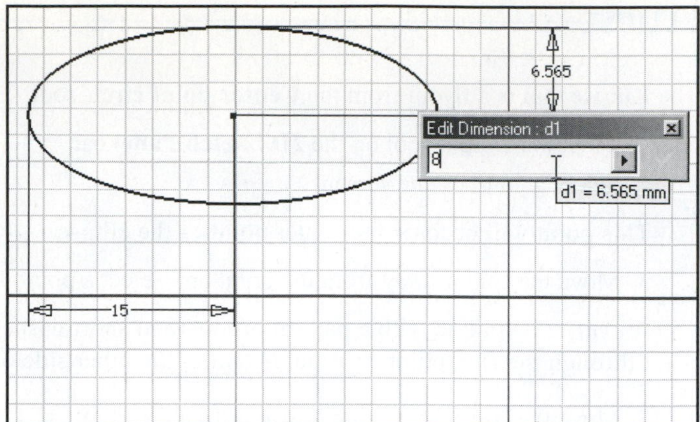

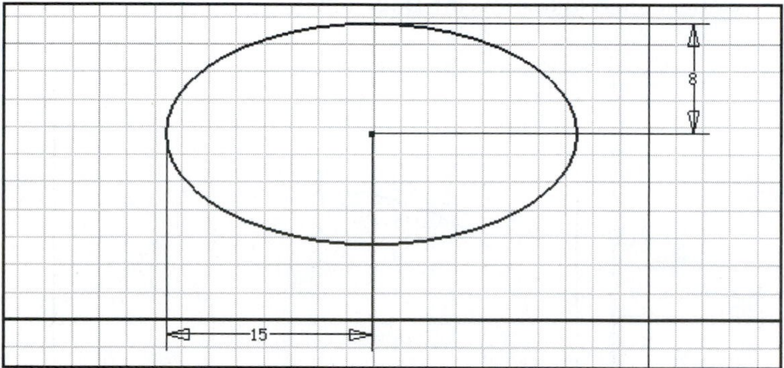

Figure 2-11

See Figure 2-12.

3. Select a third point.
4. Click the right mouse button and select the **Done** option.

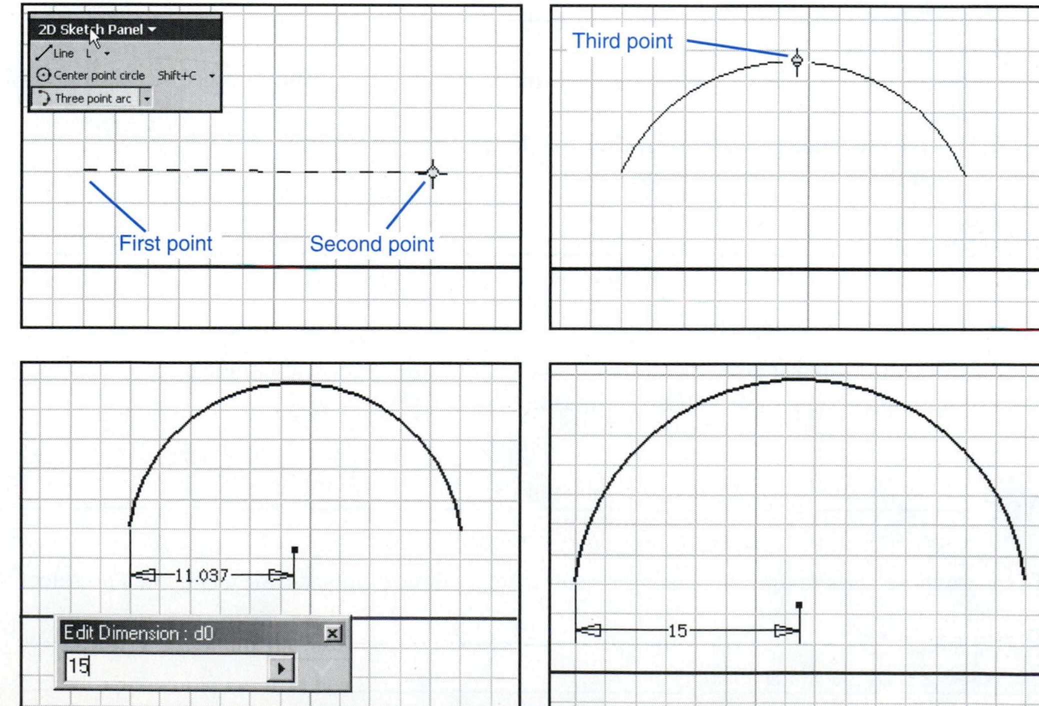

Figure 2-12

Exercise 2-11: Editing an Arc

1. Click the **General Dimension** tool.

Select one of the endpoints of the arc, then the center point.

2. Enter the appropriate dimensional value.

Exercise 2-12: Sketching a Tangent Arc

The **Tangent arc** command requires existing entities. The **Tangent arc** tool is a flyout from the **Center point circle** tool. Figure 2-13 shows two parallel lines.

1. Click the **Tangent arc** tool.
2. Select the endpoint of one of the lines, then select the endpoint of the other line.
3. Right-click the mouse and select the **Done** option.

The arc may be edited using the **General Dimension** tool.

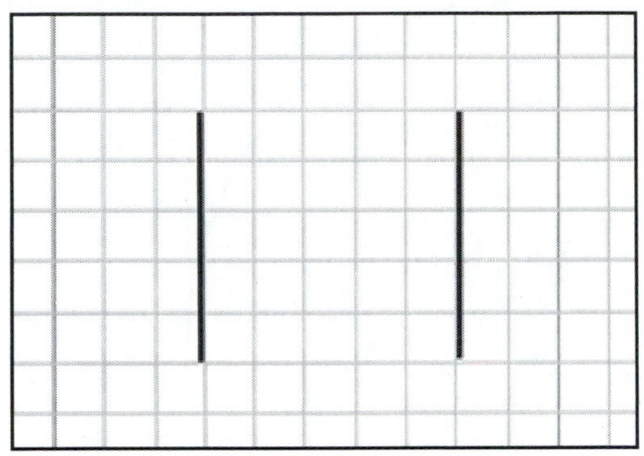

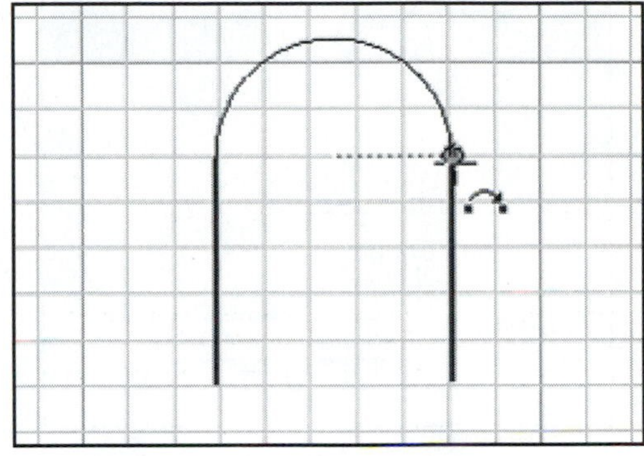

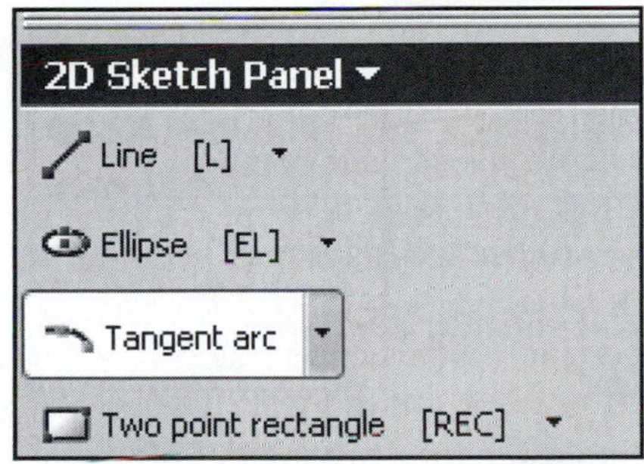

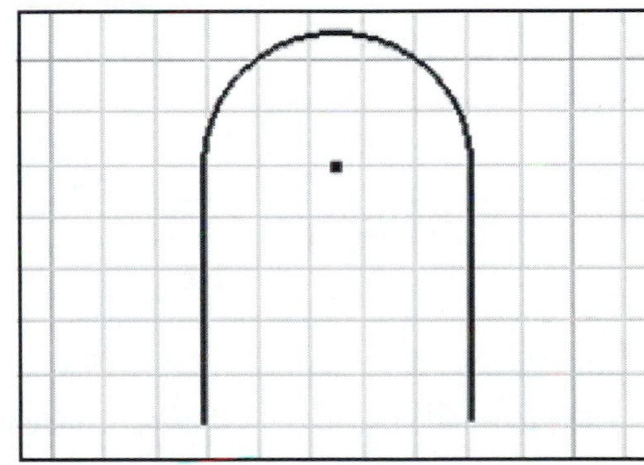

Figure 2-13

Exercise 2-13: Sketching a Center Point Arc

The **Center point arc** tool is a flyout from the **Center point circle** tool.

1. Click the **Center point arc** tool.
2. Select a point on the screen, then select a second point.

See Figure 2-14. The distance between the two points will define the radius of the arc.

3. Select a third point.
4. Right-click the mouse and select the **Done** option.

The arc may be edited using the **General Dimension** tool.

RECTANGLE

There are two ways to sketch a rectangle using Inventor: by selecting two points and by selecting three points.

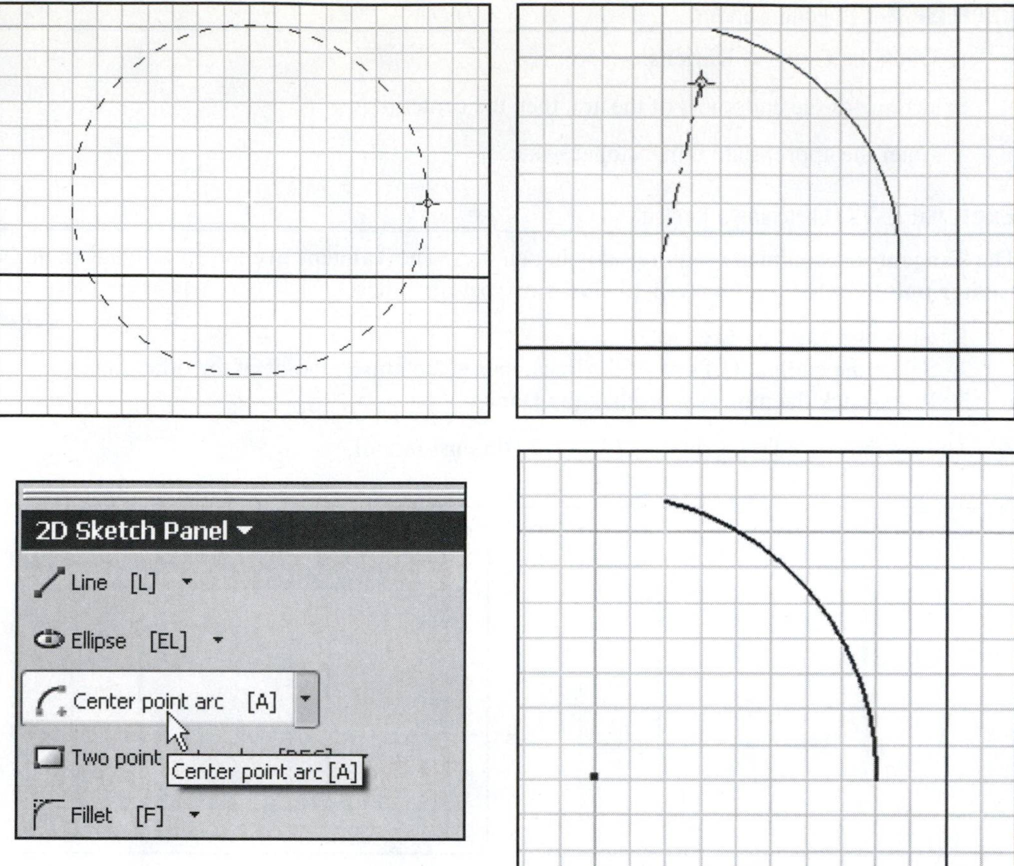

Figure 2-14

Exercise 2-14: Sketching a Two-Point Rectangle

1. Click the **Two point rectangle** tool.
2. Select a point on the screen.
3. Select a second point on the screen.
4. Right-click the mouse and select the **Done** option.

The rectangle may be edited using the **General Dimension** tool. See Figure 2-15.

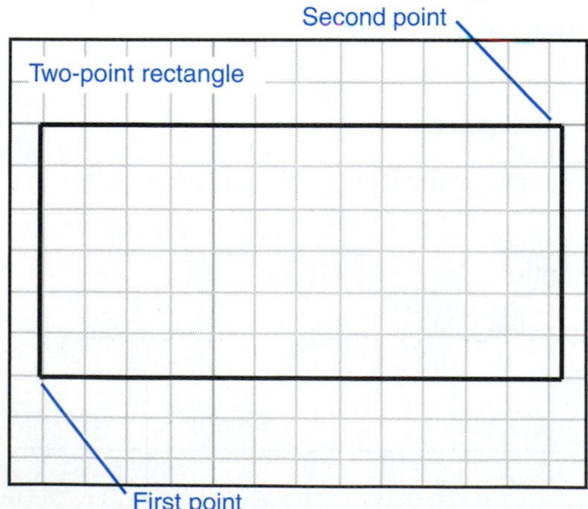

Figure 2-15

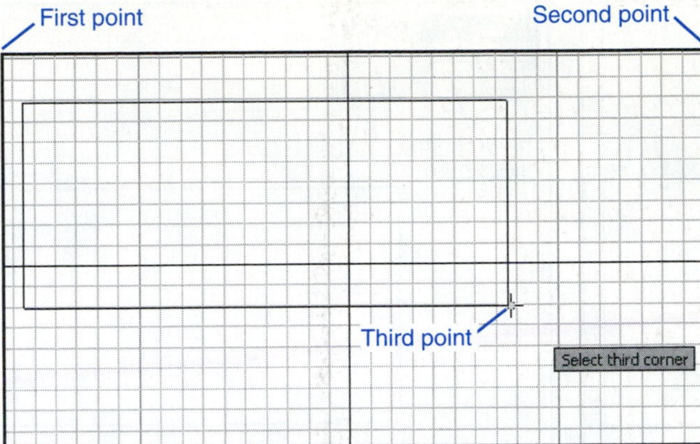

Figure 2-16

Exercise 2-15: Sketching a Three-Point Rectangle

The **Three point rectangle** tool is a flyout from the **Two point rectangle** tool.

1. Click the **Three point rectangle** tool.
2. Select a point on the screen.
3. Select a second point on the screen.

The distance between the two points will define one side of the rectangle. See Figure 2-16.

4. Select a third point.
5. Right-click the mouse and select the **Done** option.

The rectangle may be edited using the **General Dimension** tool.

Fillets can also be applied in 3D.

Fillet

A ***fillet*** is a rounded edge or corner that is added to an existing entity.

fillet: A rounded edge or corner on an entity.

1. Click the **Fillet** tool.

The **2D Fillet** dialog box will appear. See Figure 2-17.

2. Enter the radius value for the fillet.
3. Select two lines on the rectangle.

The fillet will be added to the sketch.

4. Create as many fillets as needed, then right-click the mouse and select the **Done** option.

Chamfer

A ***chamfer*** is an angled edge or corner that is added to an existing entity. The **Chamfer** tool is a flyout from the **Fillet** tool.

chamfer: An angled edge or corner on an entity.

Chamfers may be defined in one of three ways: with two equal distances, with two distances not equal, or with a distance and an angle.

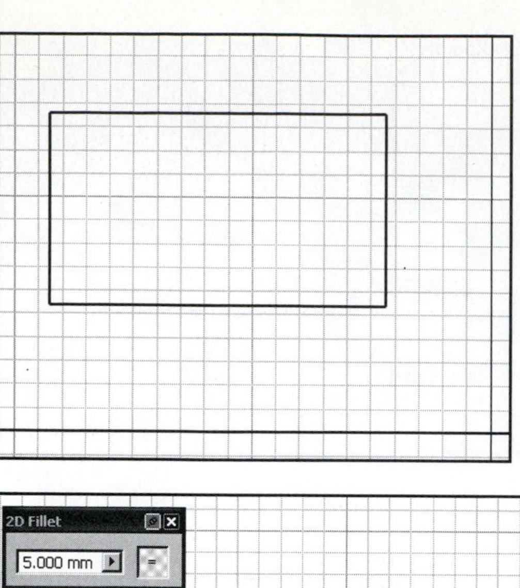

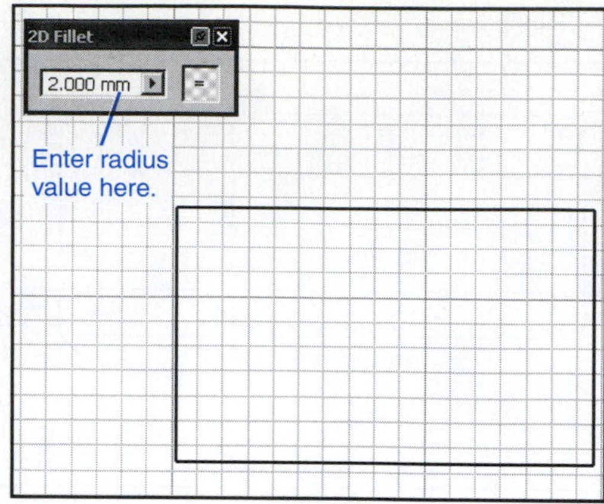

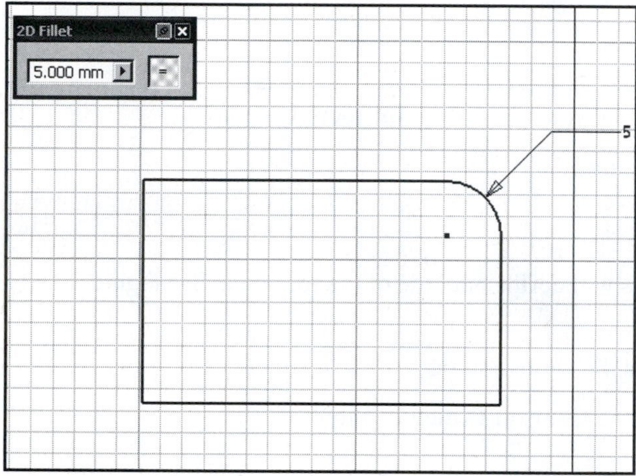

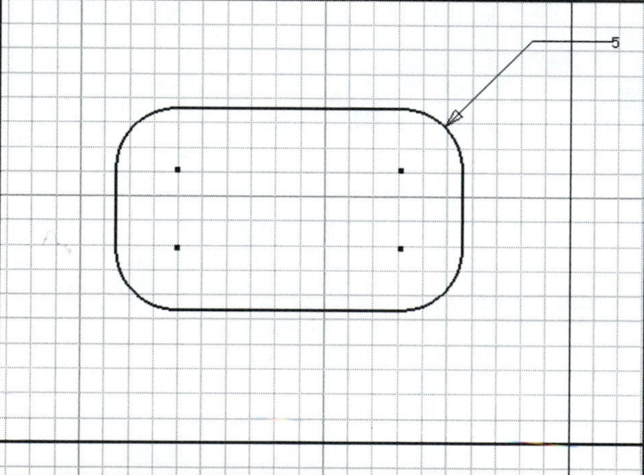

Figure 2-17

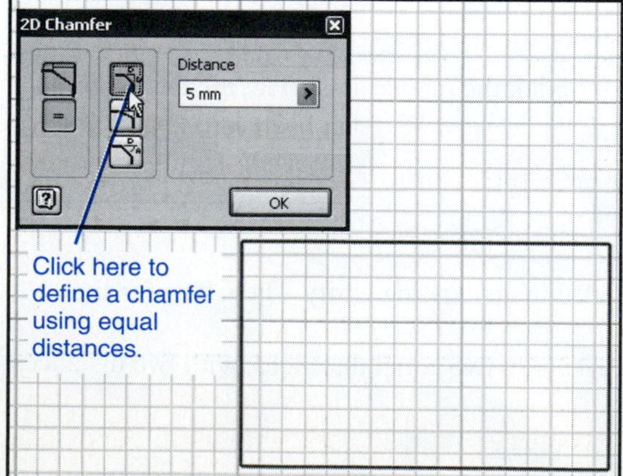

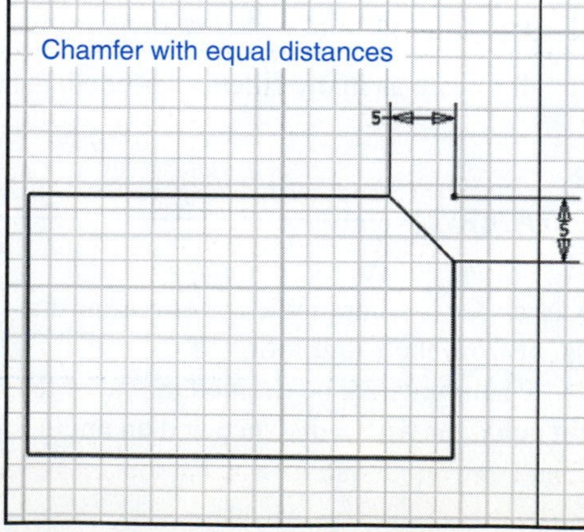

Figure 2-18

TIP Chamfers can also be applied in 3D.

Exercise 2-16: Defining a Chamfer with Two Equal Distances

1. Click the **Chamfer** tool.

The **2D Chamfer** dialog box will appear. See Figure 2-18.

2. Select the **Equal distances** box.
3. Enter the chamfer distance.
4. Select two lines on the rectangle, then click the **Done** box.

Exercise 2-17: Defining a Chamfer with Unequal Distances

1. Click the **Chamfer tool.**
2. Click the **Unequal distances** box, then click the **Done** box.

The **2D Chamfer** dialog box will change to allow the entry of two distances. See Figure 2-19.

3. Select two lines on the rectangle, then click the **Done** box.

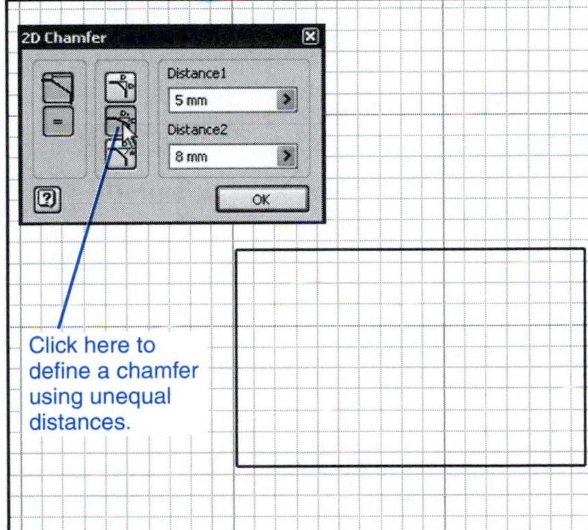

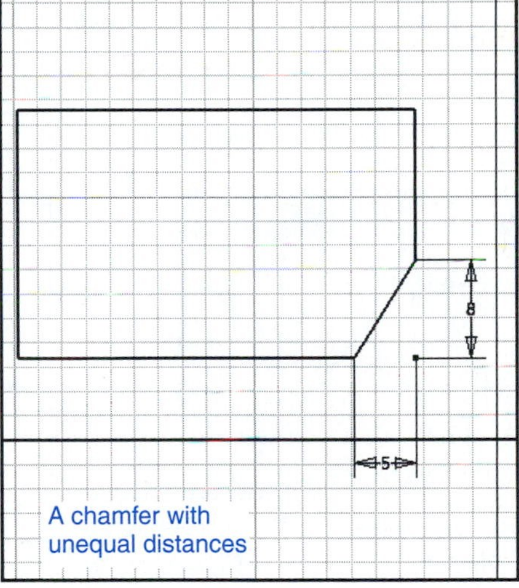

Figure 2-19

Exercise 2-18: Defining a Chamfer Using a Distance and an Angle

1. Click the **Chamfer** tool.
2. Click the **Distance and angle** button.

The dialog box will change to allow entry of a distance and an angle value. See Figure 2-20.

3. Select two lines on the rectangle, then click the **Done** box.

POLYGON

The **POLYGON** command can be used to sketch either an inscribed or a circumscribed polygon.

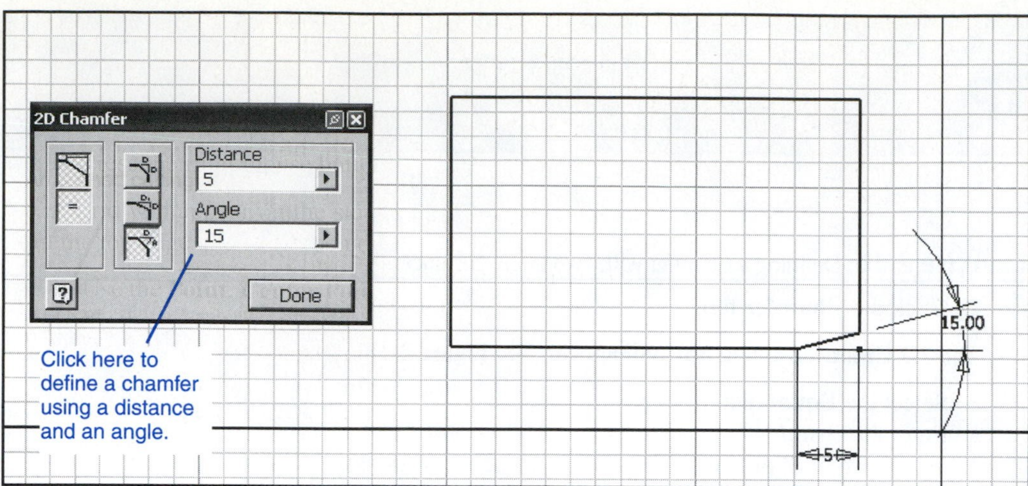

Figure 2-20

Exercise 2-19: Creating an Inscribed Polygon

1. Select the **Polygon** tool.

The **Polygon** dialog box will appear. See Figure 2-21.

2. Select the **Inscribe** box and specify the number of sides for the polygon.
3. Select a point on the screen, then move the cursor away from the point, sketching the polygon.
4. Right-click the mouse and select the **Done** option.
5. Click the **General Dimension** tool and enter the appropriate dimension for the polygon.

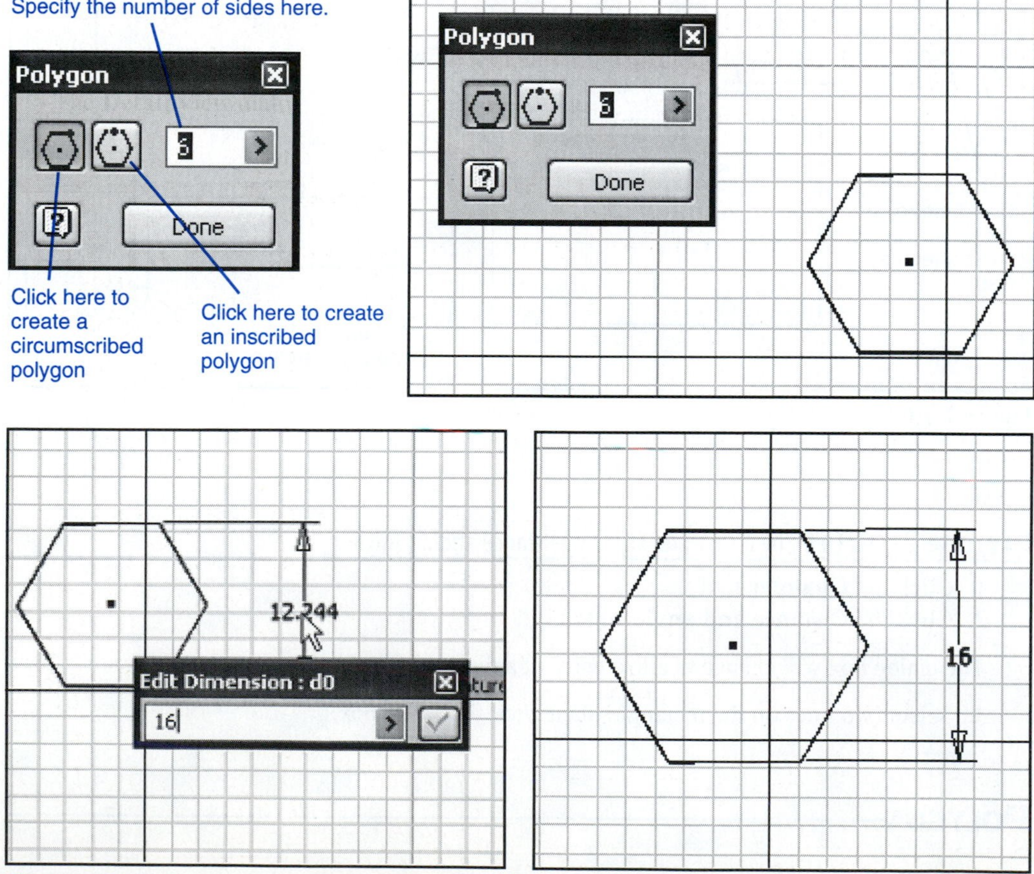

Figure 2-21

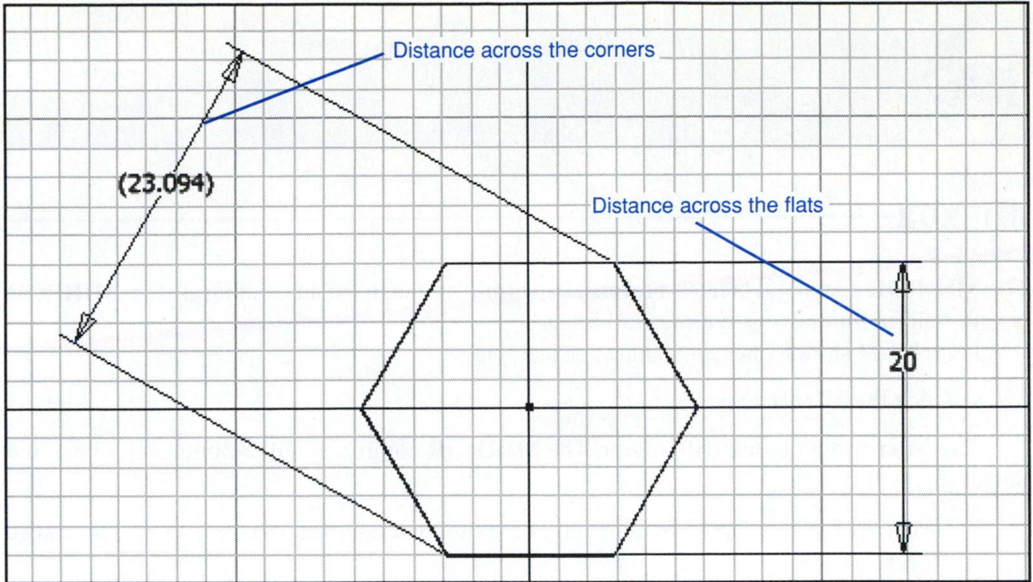

Figure 2-22

Polygons can be measured in one of two ways: across the flats and across the corners. See Figure 2-22.

Exercise 2-20: Creating a Circumscribed Polygon

1. Select the **Polygon** tool.

The **Polygon** dialog box will appear. See Figure 2-23.

2. Select the **Circumscribe** box and specify the number of sides for the polygon.
3. Select a point on the screen, then move the cursor away from the point, sketching the polygon.
4. Right-click the mouse and select the **Done** option.
5. Click the **General Dimension** tool and enter the appropriate dimension for the polygon.

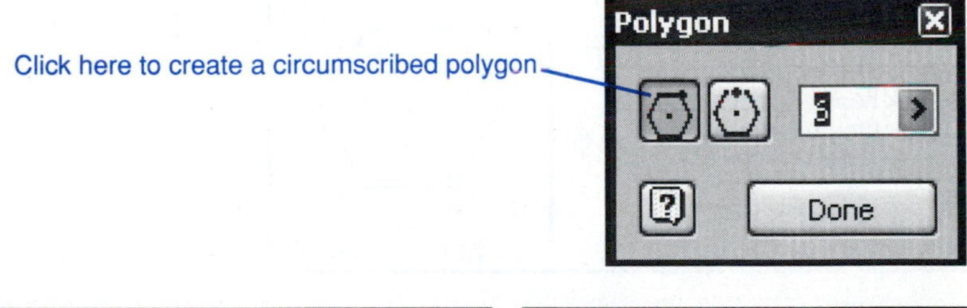

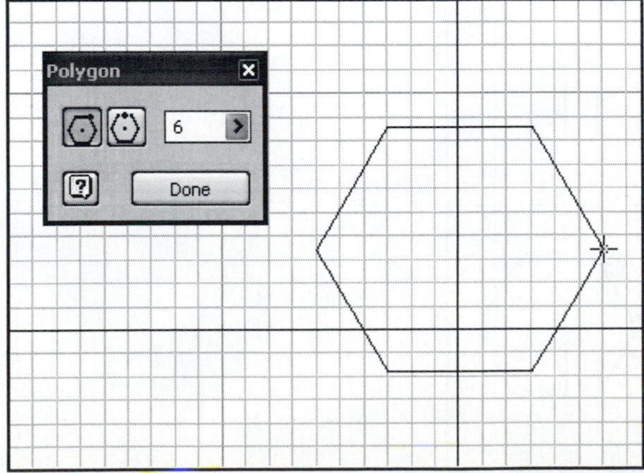

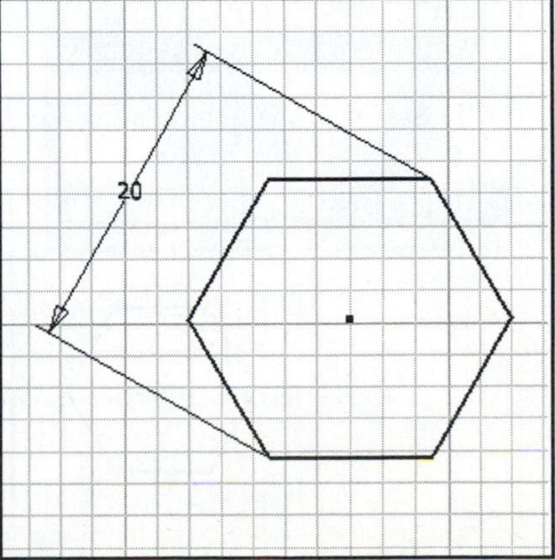

Figure 2-23

Mirror can also be applied in 3D drawings and assembly drawings.

Mirror

The **MIRROR** command creates a reverse copy (mirror image) of an existing sketch. **MIRROR** is very helpful for drawing symmetrical images.

Figure 2-24 shows a hexagon and a vertical line.

1. Click the **Mirror** tool.

The **Mirror** dialog box will appear. The **MIRROR** command will automatically be in the **Select** mode.

2. Select the hexagon either by selecting the six individual lines or by windowing the entire object.

The hexagon will change color, indicating that it has been selected. The **MIRROR** command will automatically switch to **Mirror line.**

3. Select the vertical line as the mirror line.

The line will change color, indicating that it has been selected.

4. Click **Apply** in the **Mirror** dialog box.

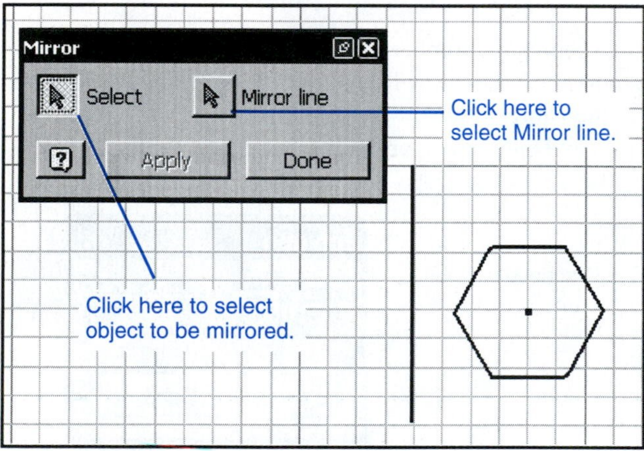

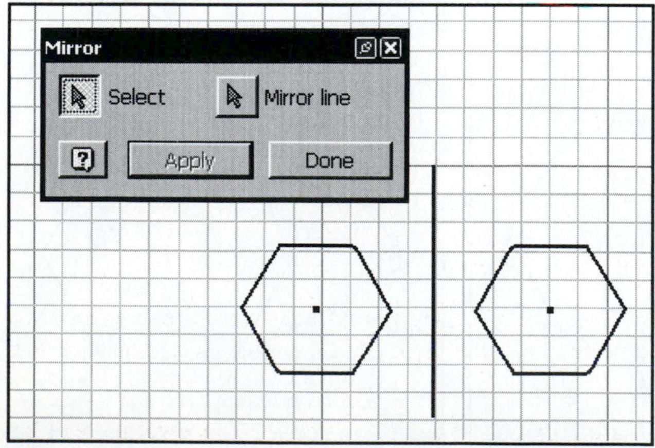

Figure 2-24

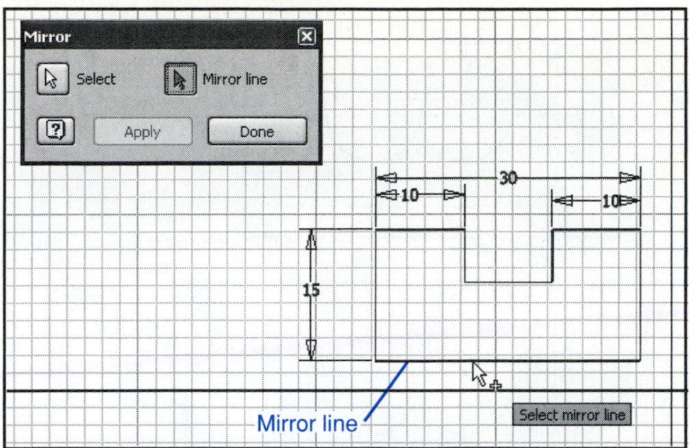

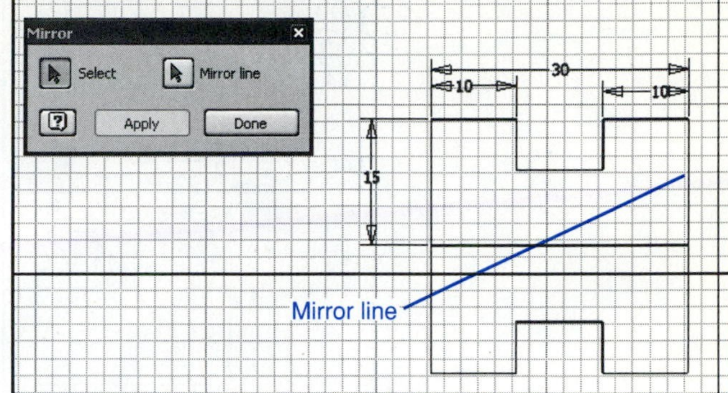

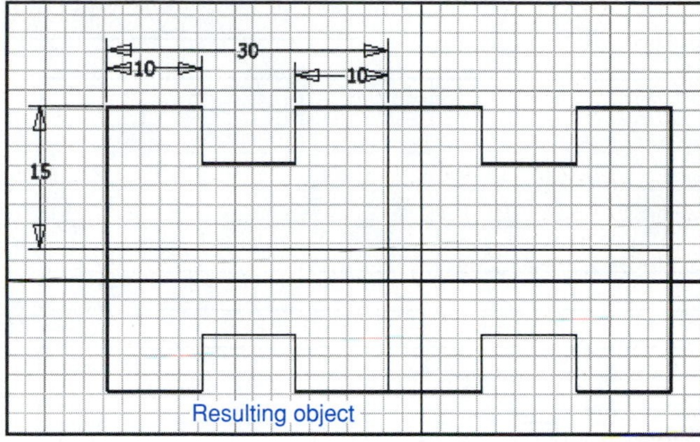

Figure 2-25

Figure 2-25 shows an object that is to be mirrored. The mirror line is one of the lines of the object. Any line in the object could be used as a mirror line. The object was then mirrored again using the right vertical line as the mirror line.

Rectangular patterns can also be created in 3D.

Rectangular Pattern

The **RECTANGULAR PATTERN** command is used to create rectangular arrays of rows and columns. Figure 2-26 shows a rectangular shape. The shape will be used to create a 3 × 4 pattern.

1. Click the **Rectangular Pattern** tool on the **2D Sketch Panel.**

The **Rectangular Pattern** dialog box will appear.

2. Click the **Geometry** box.
3. Window the rectangular shape.

The lines will change colors when they are selected.

4. Click the **Direction 1** arrow box, then select the lower horizontal line of the shape. This will define **Direction 1.** If the directional arrow points in the wrong direction, click the **Flip** box next to the **Arrow** box.

The default values for **Direction 1** are 2 items located 10 mm apart. These values will be applied to the shape. As the required values are added, the pattern will adjust.

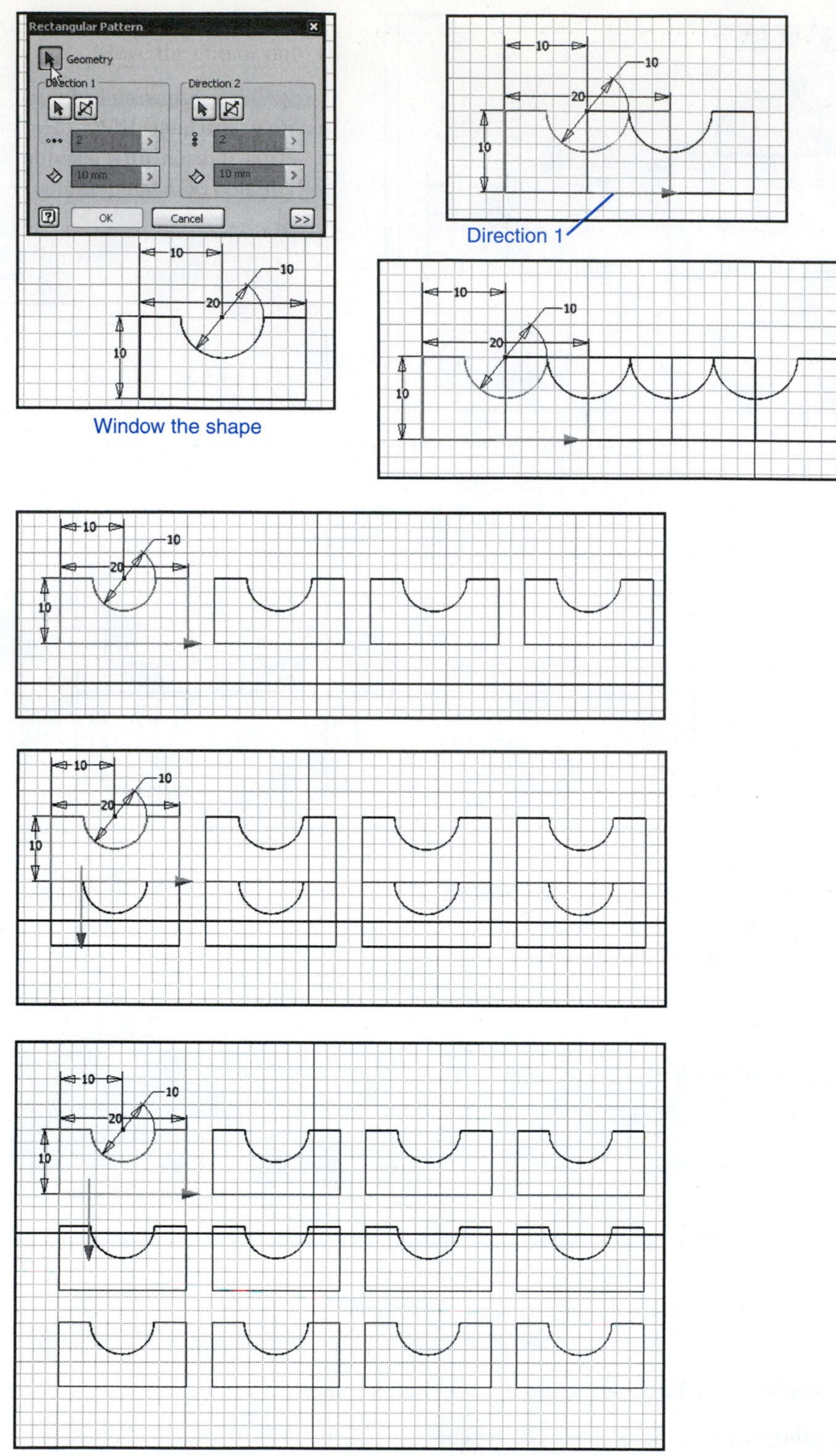

Figure 2-26

5. Enter a value of **4** items, then a distance apart of **25 mm.**
6. Click the arrow box for **Direction 2** and click the left vertical line of the shape.

Point the **Direction 2** arrow downward. If needed, use the **Flip** box to point the arrow downward.

7. Enter a number of items value of **3** and a distance apart value of **15 mm.**

Figure 2-26 shows the resulting pattern.

Circular Pattern

The **Circular Pattern** tool is used to array an object around a common center point. Figure 2-27 Shows a round object that contains a circle. The final circular pattern is to contain eight circles.

1. **Click Circular Pattern** tool on the **2D Sketch Panel**

The **Circular Pattern** dialog box will appear. See Figure 2-27.

2. Click the **Geometry** box and then click the **Ø10** circle.
3. Click the arrow box to the left of the word **Axis,** then click the center point of the round object.

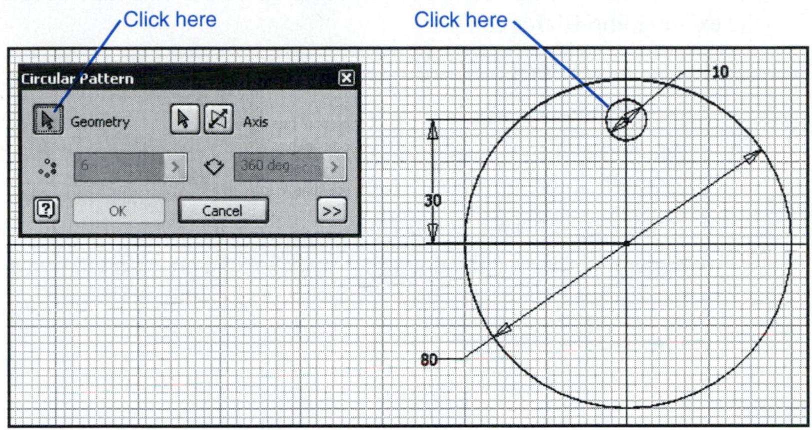

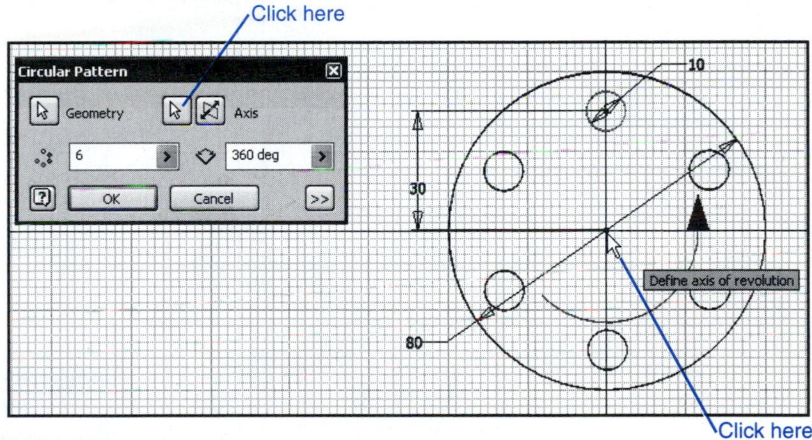

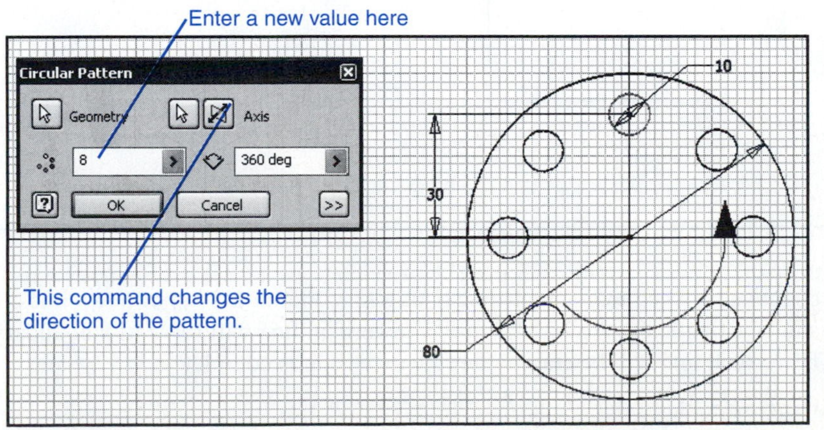

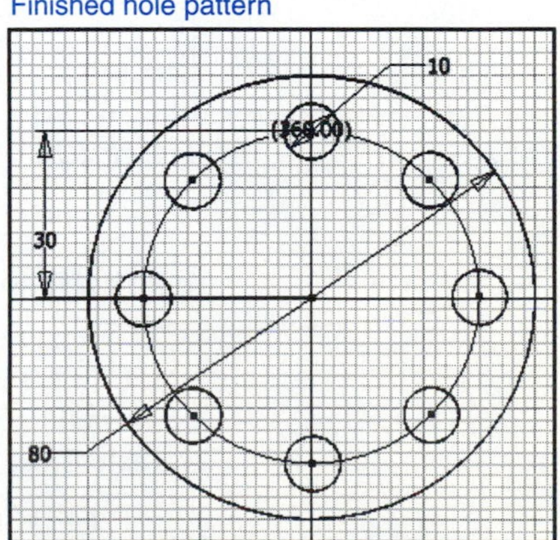

Figure 2-27

The default value for a circular pattern is 6 at 360°. Six circles will appear on the screen. The **Flip** box next to the word **Axis** is used to change the direction from counterclockwise to clockwise.

4. Change the number of items from **6** to **8** and click **OK.**

Eight circles will appear.

OFFSET

The **OFFSET** command is used to create lines parallel to existing lines at a specified distance. Circles and curves may also be offset. Figure 2-28 shows a line. Create a second line parallel to the existing line 10 mm away.

1. Click the **Offset** tool.
2. Select the line.
3. Move the cursor away from the line and select a new location for the second line, then right-click the mouse and select the **Done** option.
4. Click the **General Dimension** tool, then select the two lines.
5. Enter a dimensional value of **10** and click the check mark on the dialog box.

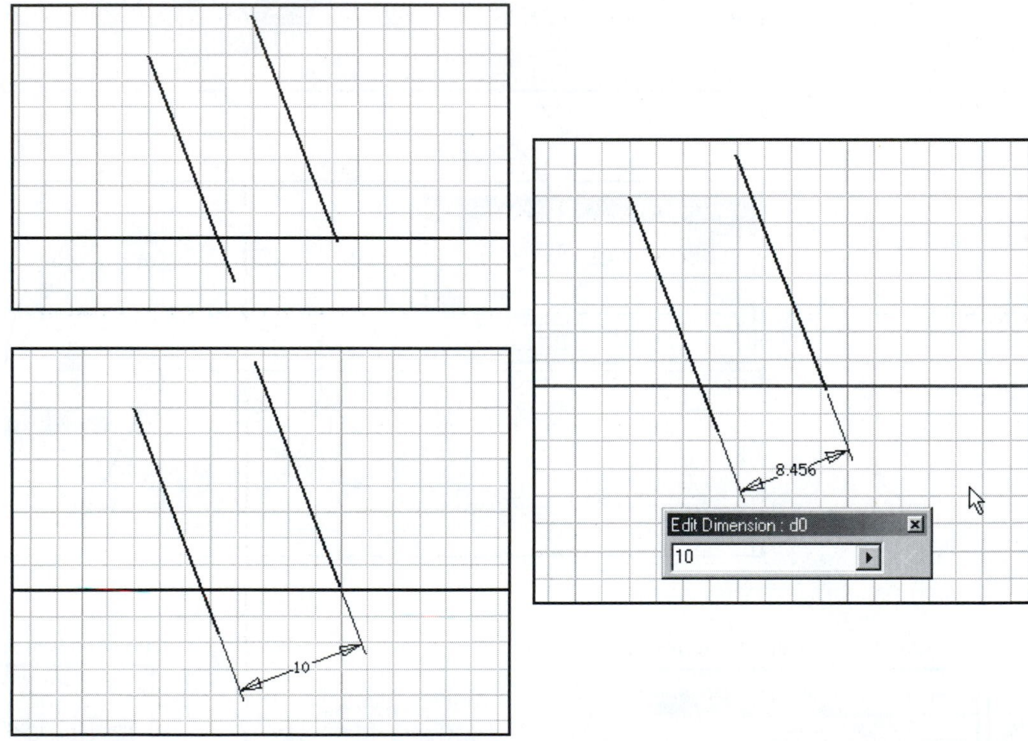

Figure 2-28

EXTEND

The **EXTEND** command is used to lengthen an existing line. Figure 2-29 shows two lines. The horizontal line is to be extended to the vertical line. The vertical line will serve as a boundary to the extension. The vertical line will be erased after the extension is complete.

1. Click the **Extend** tool.
2. Select the horizontal line.

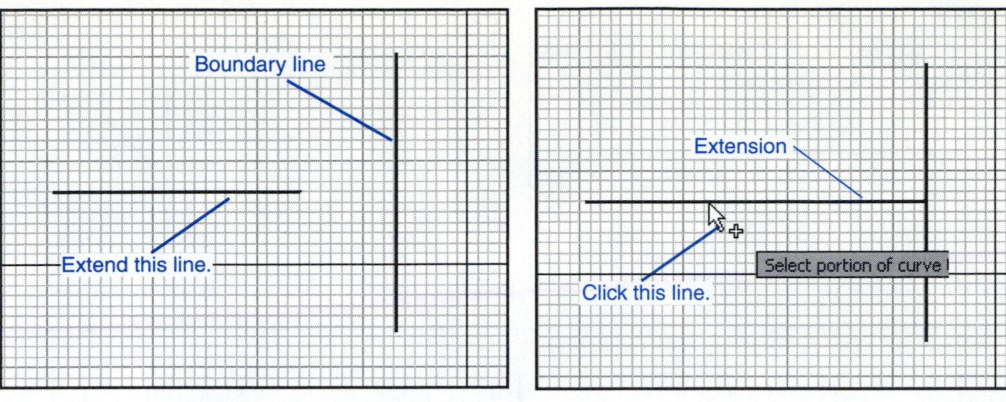

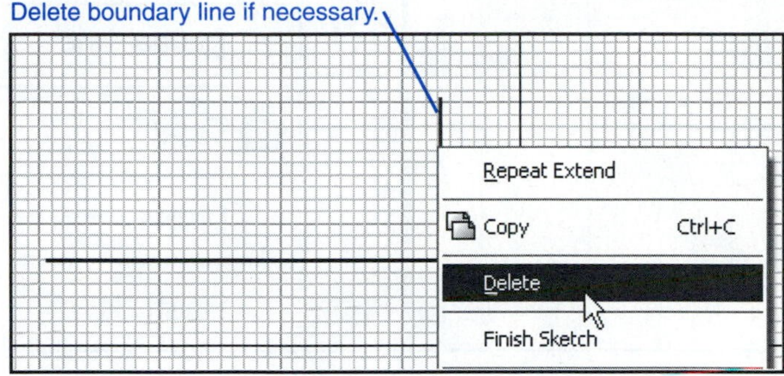

Figure 2-29

3. Right-click the mouse and select the **Done** option.
4. Click the vertical line, then right-click the mouse and select the **Delete** option.
5. Delete the boundary line if necessary.

TRIM

The **TRIM** command is used to shorten an existing line. Figure 2-30 shows two lines. The horizontal line will be shortened using the vertical line as a cutting line.

1. Click the **Trim** tool.
2. Move the cursor onto the horizontal line to the right of the vertical line.

The right portion of the horizontal line will change its color and line pattern.

3. Click the line.
4. Delete the vertical line if necessary.

TIP There are **Move** tools in both 3D and assembly drawings.

MOVE

The **MOVE** command is used to reposition an existing sketch or to copy an existing sketch. Figure 2-31 shows two rectangles. The lower rectangle is to be moved so that it is aligned with the upper rectangle.

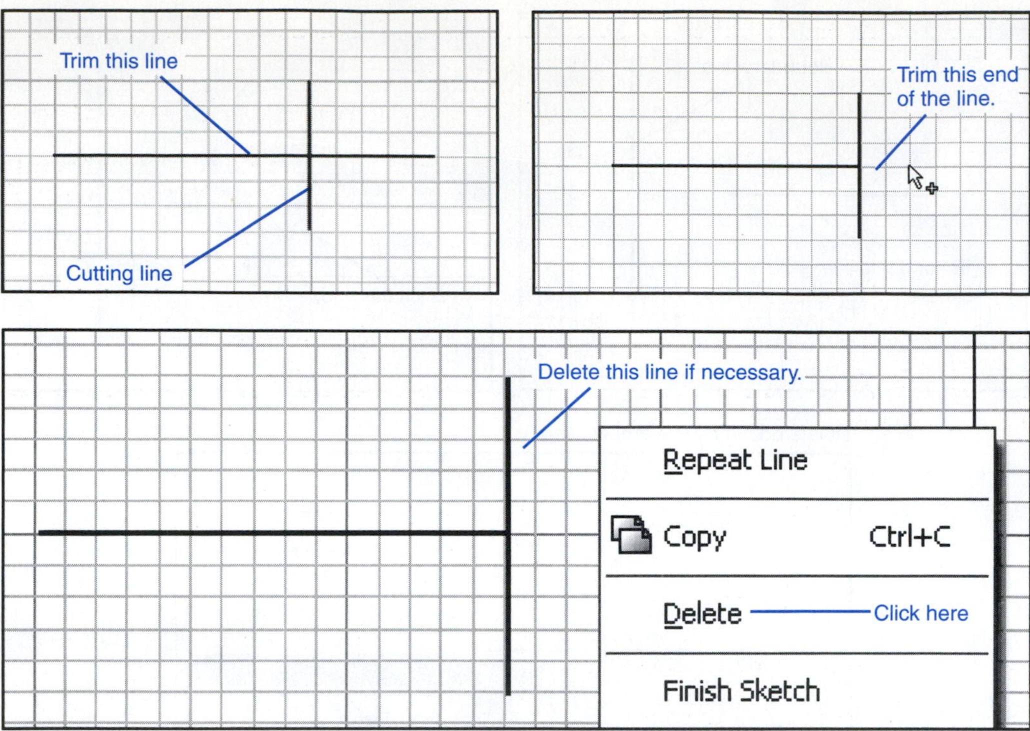

Figure 2-30

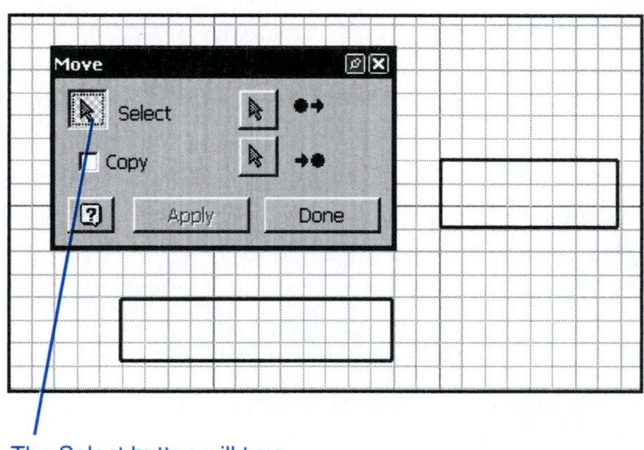

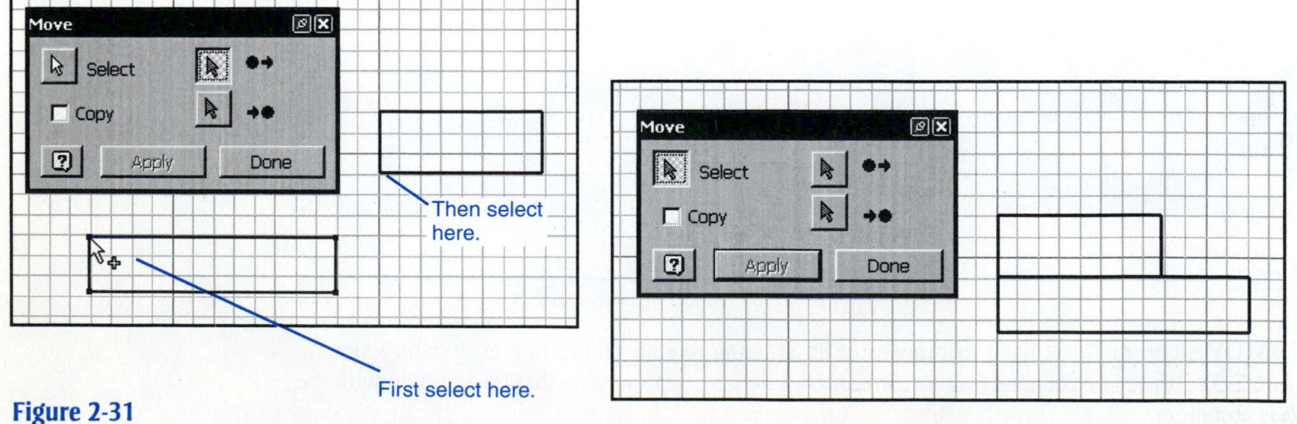

Figure 2-31

Exercise 2-21: Moving a Sketch

1. Click the **Move** tool on the **2D sketch panel.**

The **Move** dialog box will appear. The **MOVE** command is automatically in the **Select** mode.

2. Window the lower rectangle, press the right mouse button, and select the **Continue** option.
3. Select the top arrow, then select the upper left corner of the lower rectangle.

The **MOVE** command will automatically shift to the second arrow box on the **Move** dialog box.

4. Select the lower left corner of the upper rectangle.
5. Click the **Apply** box, then click the **Done** box.

Exercise 2-22: Copying a Sketch

Figure 2-32 shows a rectangle. This section shows how to make a copy of the rectangle.

1. Click the **Move** tool on the **2D Sketch Panel.**
2. Click the **Copy** box on the **Move** dialog box.

A check mark will appear.

3. Window the rectangle.
4. Click the arrow box on the **Move** dialog box, then click the lower left corner of the rectangle.
5. Drag the base point away from the rectangle.

A copy will appear and will move with the cursor.

6. Locate the copy and click the mouse.

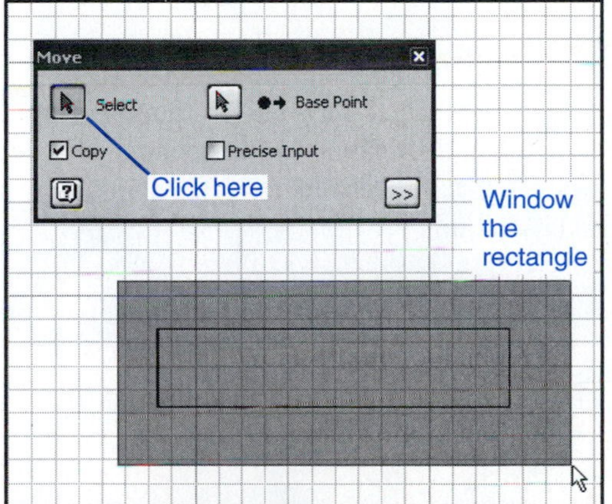

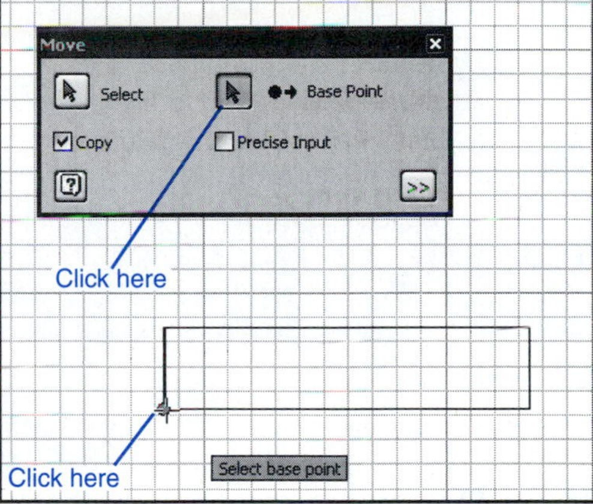

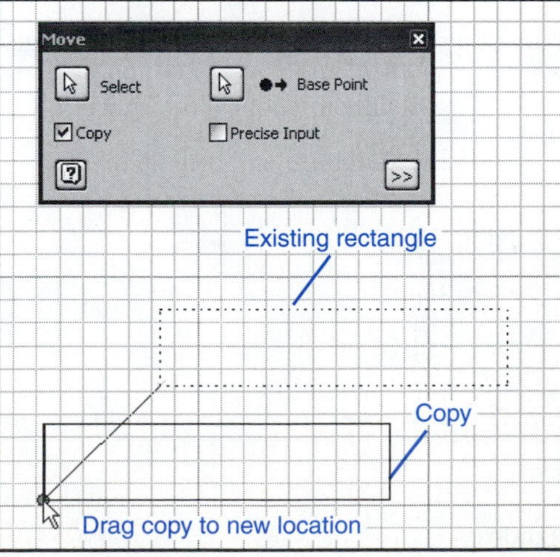

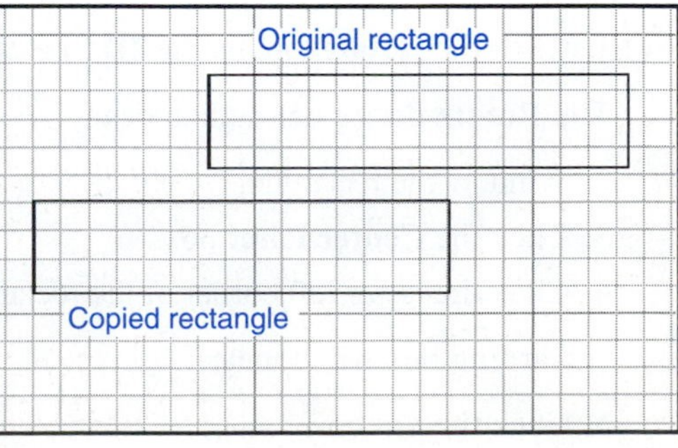

Figure 2-32

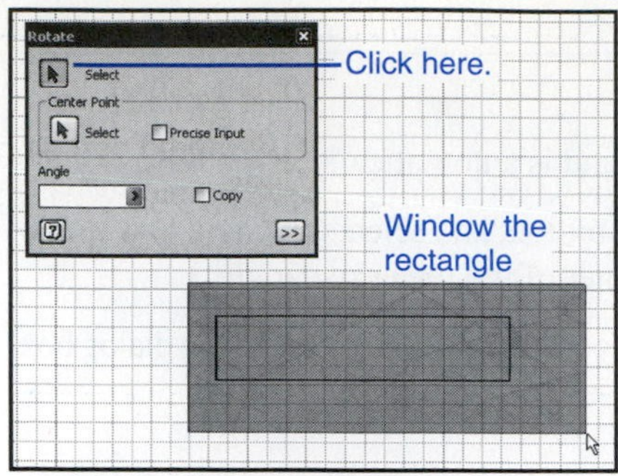

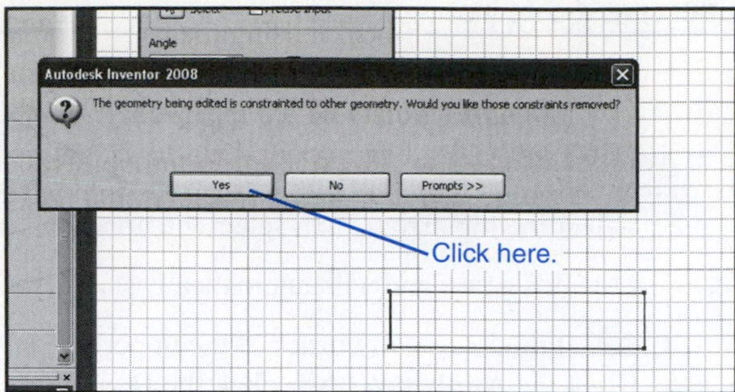

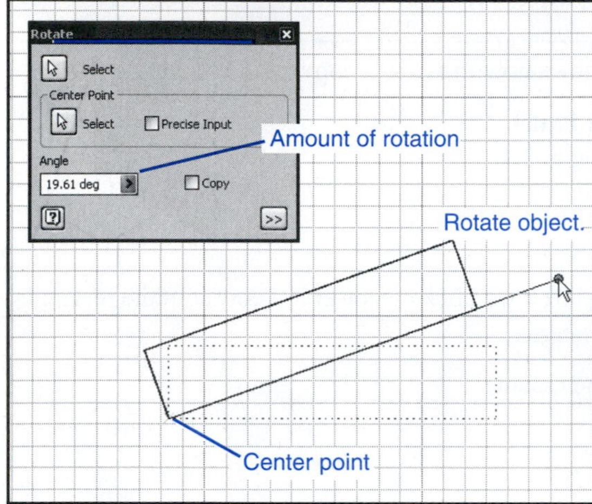

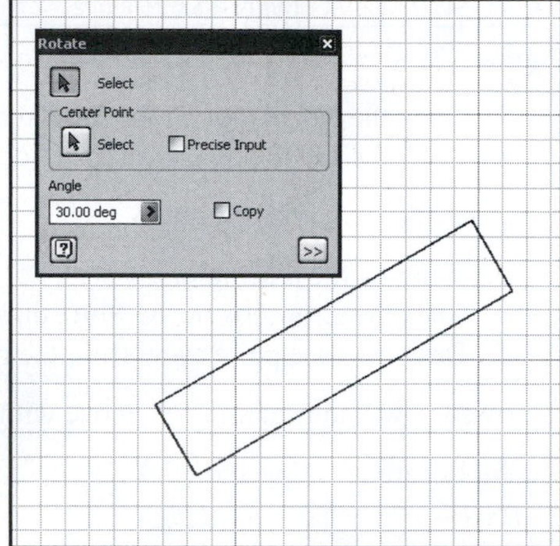

Figure 2-33

TIP There are **Copy** tools in both 3D and assembly drawings.

Rotate

The **ROTATE** command is used to rotate a sketch about a point. Figure 2-33 shows a rectangle. It is to be rotated 30° about its lower left corner.

1. Click the **Rotate** tool on the **2D Sketch Panel.**

The **Rotate** dialog box will appear.

2. Window the rectangle.
3. Click the **Center Point** box.
4. Click the lower left corner of the rectangle.

A warning box will appear.

5. Click the **Yes** box.
6. Rotate the rectangle by moving the cursor.
7. When the desired angle is reached click the mouse.

Constraints

The **Contraints** tool contains 12 tools that can be used to change and define the shape of a sketch. The constraint tools are accessed by using the flyouts located above the **Show Constraints** tool. The 12 constraints are **Perpendicular, Parallel, Tangent, Smooth, Coincident, Concentric, Colinear, Horizontal, Vertical, Equal, Fix,** and **Symmetric.**

Exercise 2-23: Using Horizontal, Vertical, Perpendicular, and Parallel Constraints

Figure 2-34 shows a sketch that is to be changed into a rectangle.

1. Click the **Vertical** constraint, then click the right line of the sketch.

   The line will become vertical.

2. Click the **Perpendicular** constraint, then click the top line of the sketch, then the vertical line.

   The top line will be made perpendicular to the vertical line.

3. Click the **Parallel** constraint, then click the top horizontal line, then the lower line.

   The lower line will become horizontal or parallel to the top line.

4. Click the **Perpendicular** constraint, click the top horizontal line, then the left slanted line.

Constraints

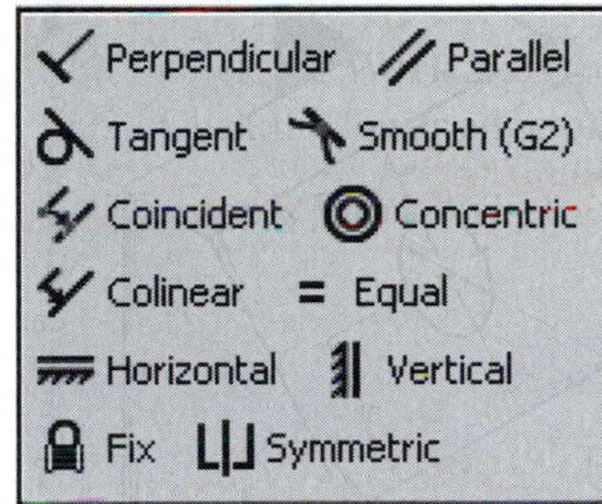

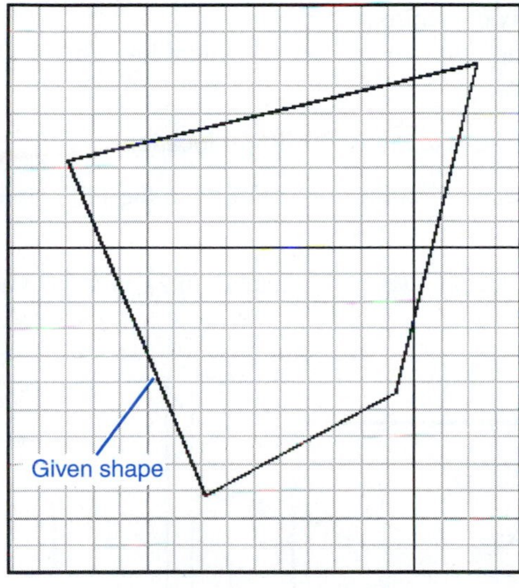

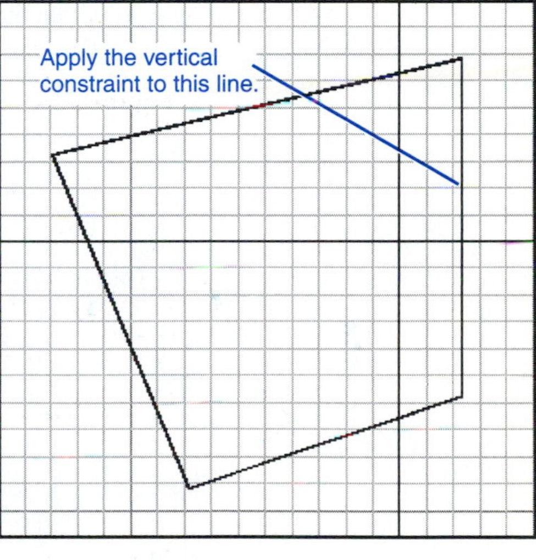

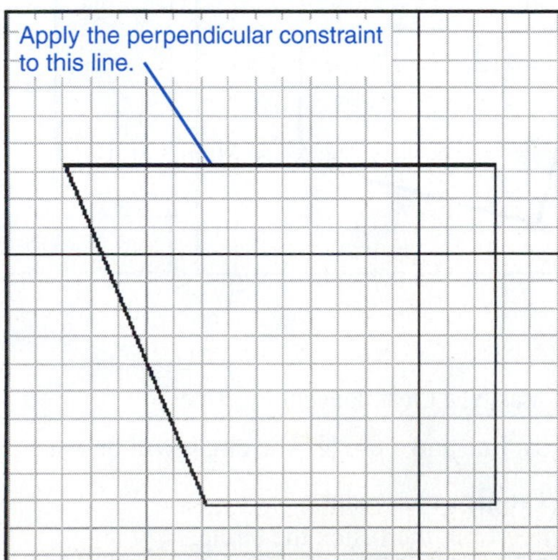

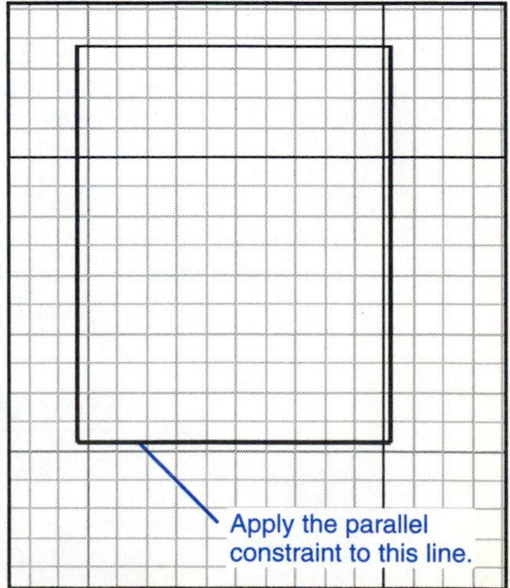

Figure 2-34

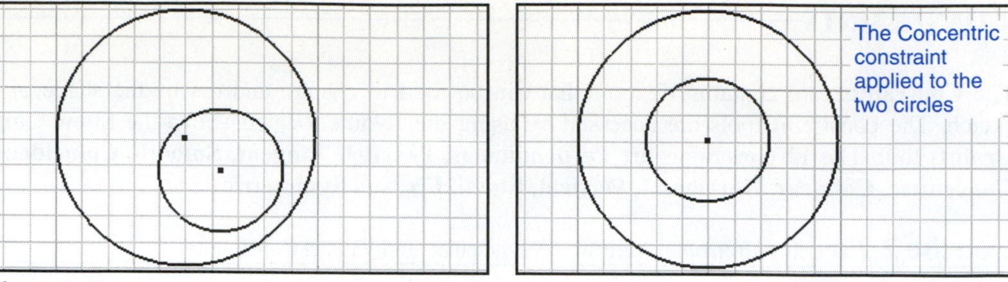

Figure 2-35

Exercise 2-24: Making Two Circles Concentric

Figure 2-35 shows two circles. They are not concentric.

1. Click the **Concentric** constraint tool, then each of the circles.

The circles will become concentric.

Exercise 2-25: Fixing a Point

Figure 2-36 shows the same sketch presented in Figure 2-34. When the parallel constraint was applied to the lower line in Figure 2-34 the top horizontal line became shortened. This horizontal line can be fixed so that it will remain the same length.

1. Click the **Fix** constraint.
2. Click the endpoints of the top horizontal line, then right-click the mouse and select the **Done** option.

The points will be fixed in their current location.

3. Click the **Perpendicular** constraint, then the left slanted line, then the fixed horizontal line.

Note the difference between the sketches in Figures 2-34 and 2-36.

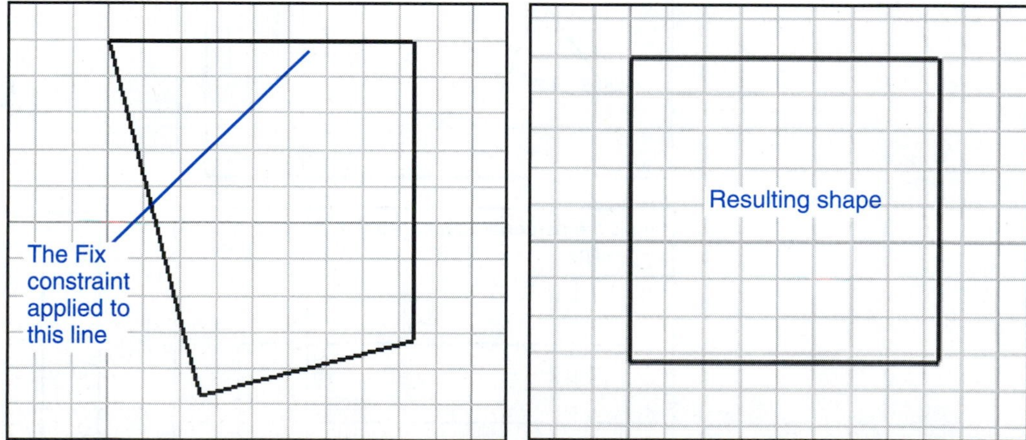

Figure 2-36

Exercise 2-26: Making a Circle Tangent to a Line

Figure 2-37 shows a line and a circle. The circle will be made tangent to the line.

1. Click the **Tangent** constraint.
2. Click the line first, then click the circle.

The circle will move to a point tangent to the line.

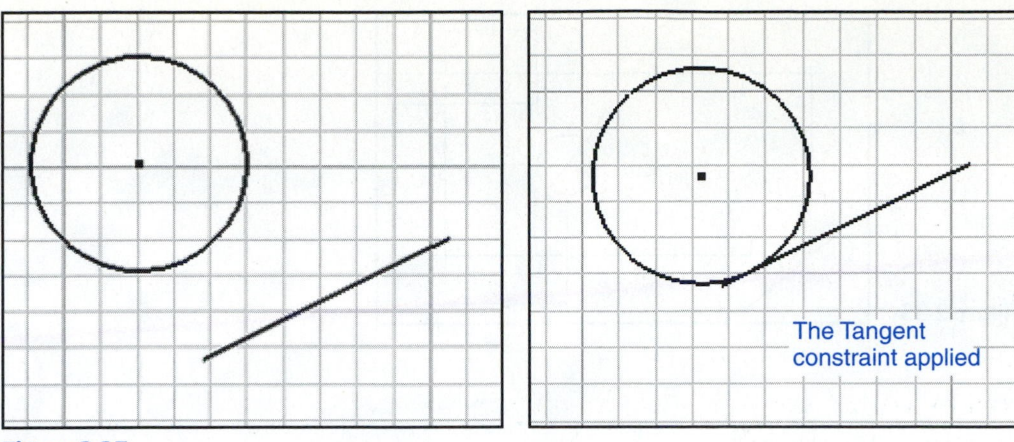

Figure 2-37

SHOW CONSTRAINTS

The **SHOW CONSTRAINTS** command will show the constraints that have been applied to a sketch. This command is helpful when unknown constraints interfere with the sketching process.

Figure 2-38 shows two circles that were constrained to be concentric and the rectangular shape created for Figure 2-34.

1. Click the **Show Constraints** tool.

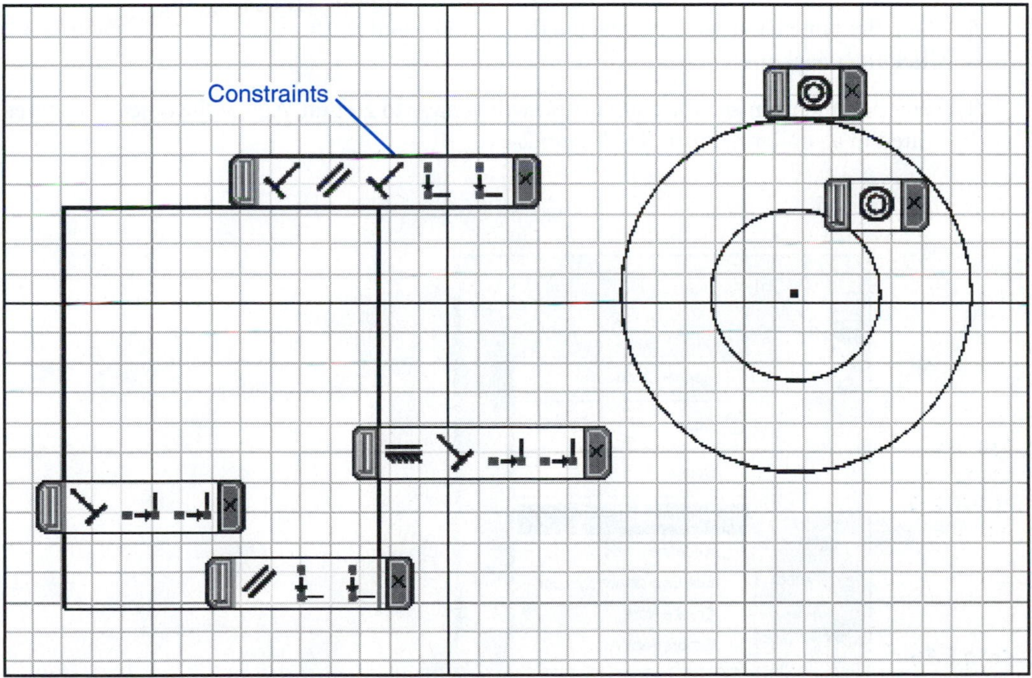

Figure 2-38

EDITING A SKETCH

Figure 2-39 shows a finished sketch. Any of the features may be changed by editing the dimensions.

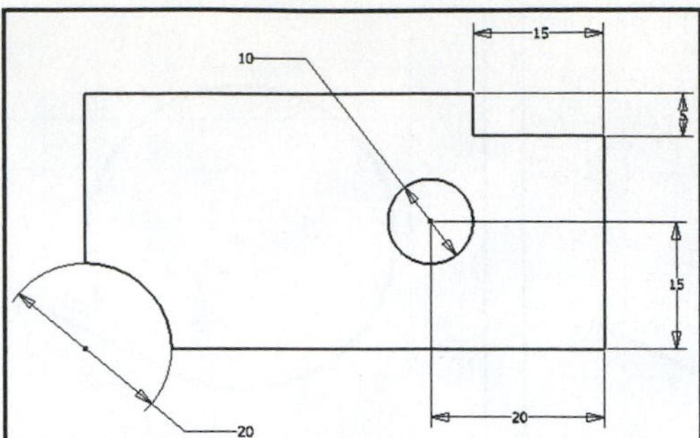

Figure 2-39

Exercise 2-27: Editing a Feature

Suppose the 5 × 15 cutout in the top right corner of the object is to be changed to a 5 × 20 cutout.

1. Right-click the word **Sketch** in the browser box.

   See Figure 2-40.

2. Select the **Edit Sketch** option.

   The drawing will return to the **2D Sketch Panel.**

3. Right-click the **15** dimension, then select the **Delete** option.

   The dimension will disappear.

4. Select the **General Dimension** tool and dimension the cutout.

   A value of **15** will appear. See Figure 2-41.

5. Click the dimension, then change the dimension value to **20.**
6. Click the check mark.

The cutout dimension and the cutout itself will change to 20 mm. Any of the object's features may be edited in a similar manner. See Figure 2-42.

Figure 2-40

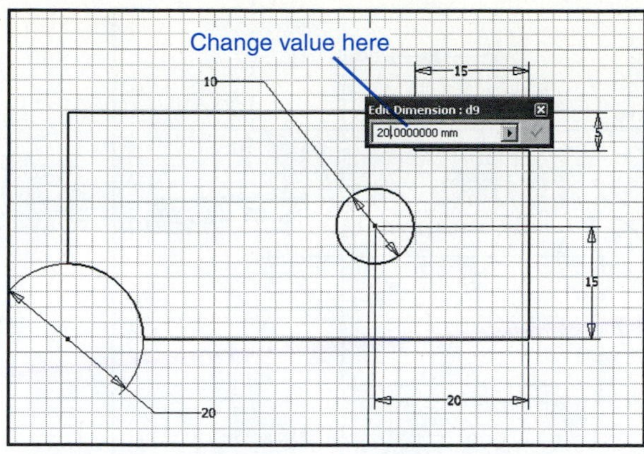

Figure 2-41

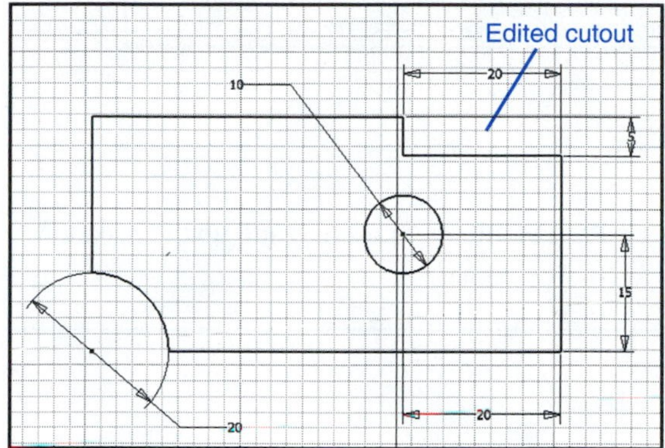

Figure 2-42

Exercise 2-28: Moving the Hole

Say the Ø10 hole is to be moved 10 to the left.

1. Delete the **20** dimension and replace it with **30.**

Note that the right vertical line moves 10 to the right. This means that the 30 distance from the edge was achieved by moving the vertical line and not the hole.

2. Apply a **Fix** constraint to the right vertical line.

If the left vertical line moves when the new 30 dimension is entered, apply a **Fix** to the left vertical line. See Figure 2-43.

3. Enter the **30** dimension.

The hole should now move 10 from the left from its original position. See Figure 2-44.

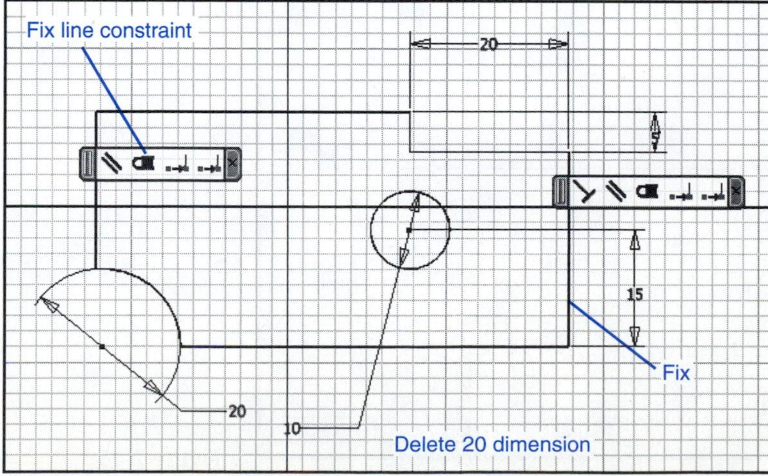

Figure 2-43

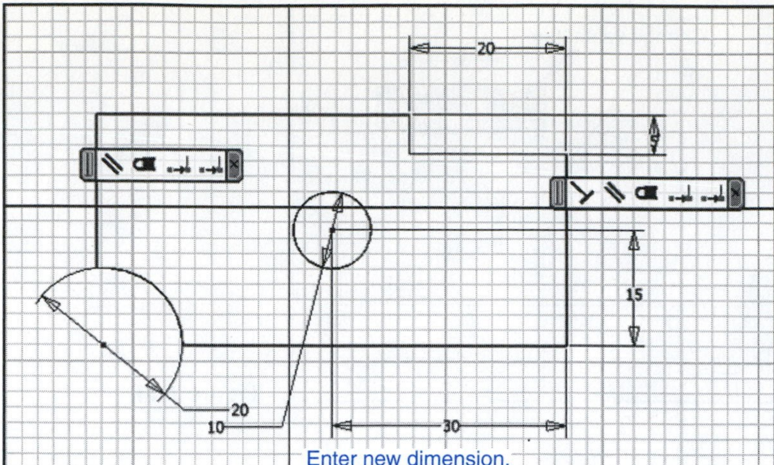

Figure 2-44

TEXT

The **TEXT** command, located on the **2D Sketch Panel** toolbox, is used to add text to a sketch.

Exercise 2-29: Creating Text

1. Select the **Text** tool.
2. Click the drawing screen and drag out a box.

The **Format Text** dialog box will appear. See Figure 2-45.

3. Specify the font style and text height.

In this example a Tahoma font at a height of 3.5 mm was specified.

4. Type in the text.
5. Click **OK.**

The text will appear on the screen. See Figure 2-46.

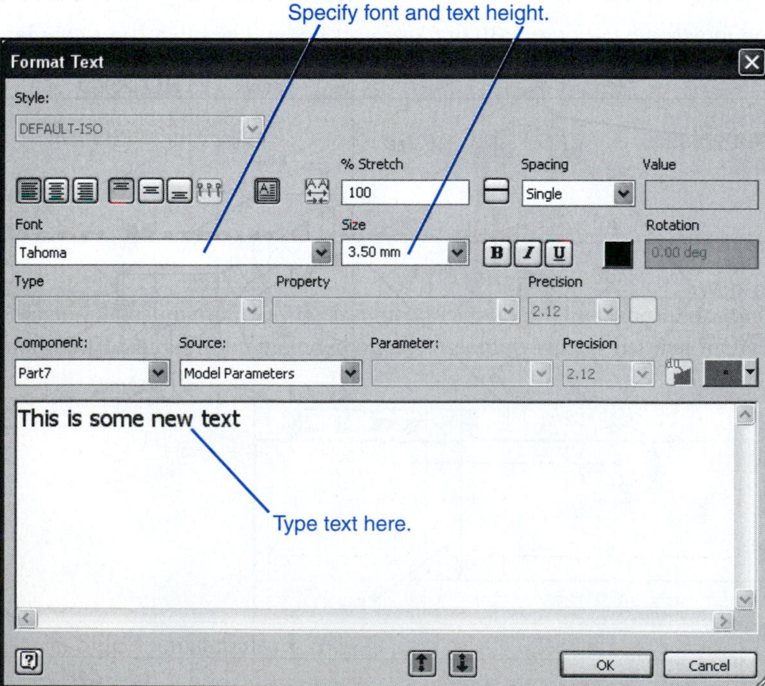

Figure 2-45

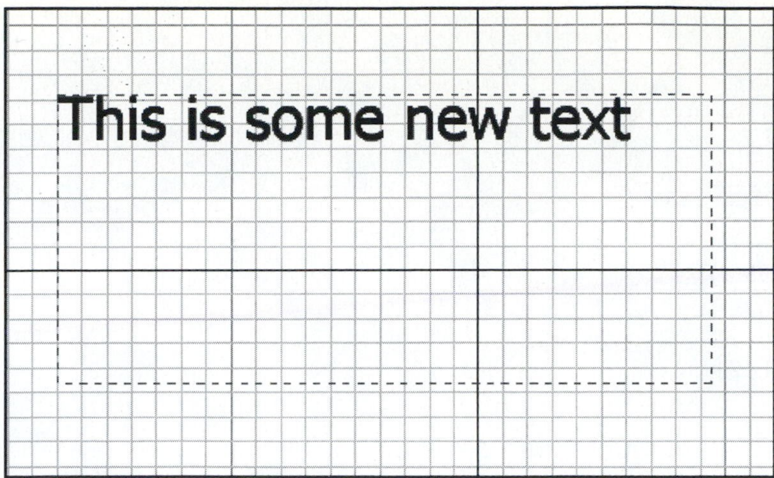

Figure 2-46

Summary

This chapter introduced most of the commands found in the **2D Sketch Panel** bar and demonstrated their use in creating 2D sketches. The **LINE** command was used to draw lines and draw them at angles, and lines were offset, extended, and trimmed. Open and closed splines were introduced. Circles, ellipses, arcs, rectangles, and polygons were drawn using several different options. Fillets and three different types of chamfers were added to existing entities.

The **MOVE** and **ROTATE** commands and the **Constraints** tool were used to manipulate and change the shape of existing sketches, and other editing tools were introduced. Finally, text was added to a sketch.

Chapter Project

Project 2-1:

Redraw the following objects using the given dimensions. Create solid models of the objects using the specified thicknesses.

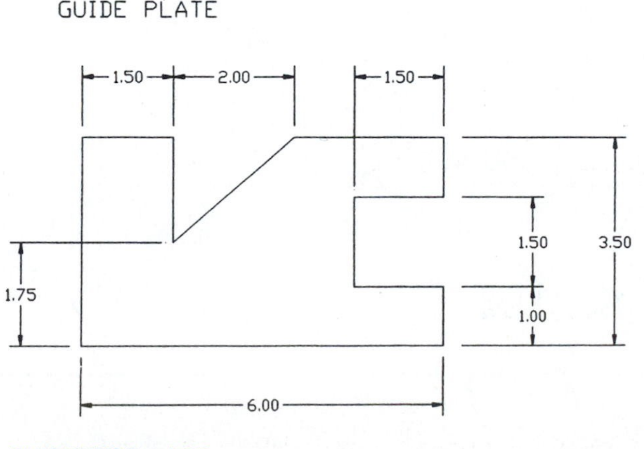

THICKNESS = 1.00
Figure P2-1 INCHES

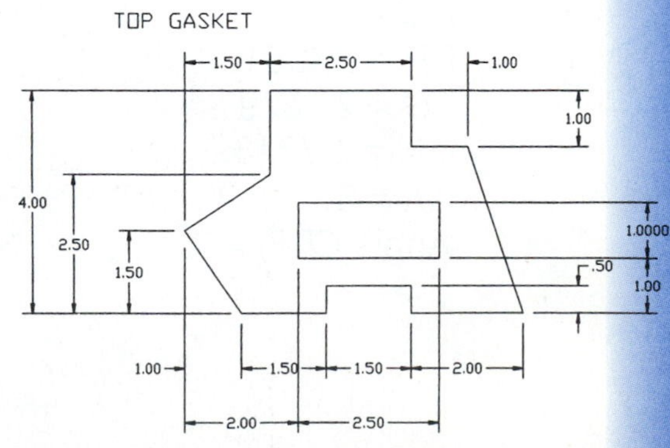

THICKNESS = .625
Figure P2-2 INCHES

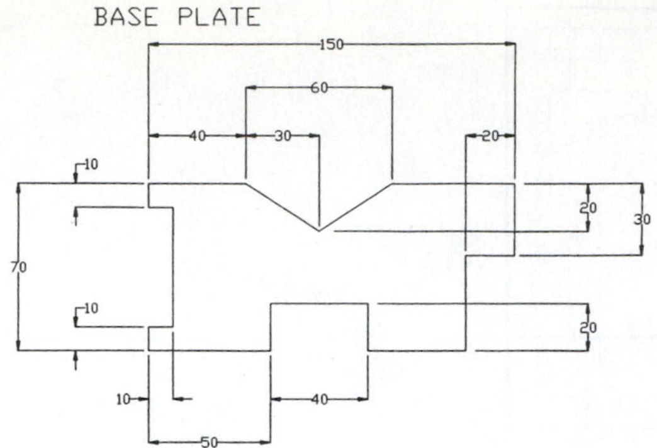

THICKNESS = 12

Figure P2-3 MILLIMETERS

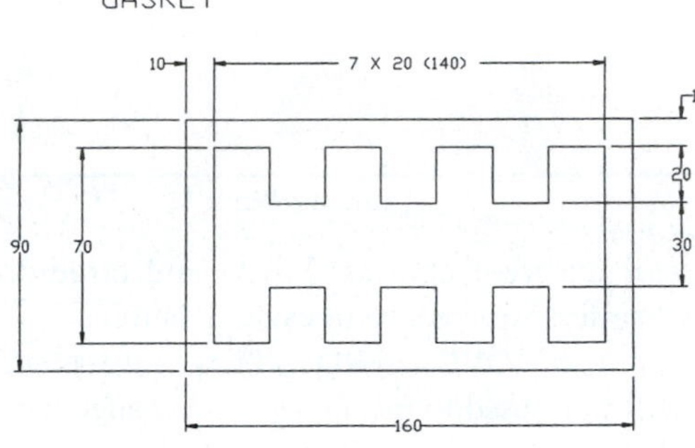

THICKNESS = 2

Figure P2-4 MILLIMETERS

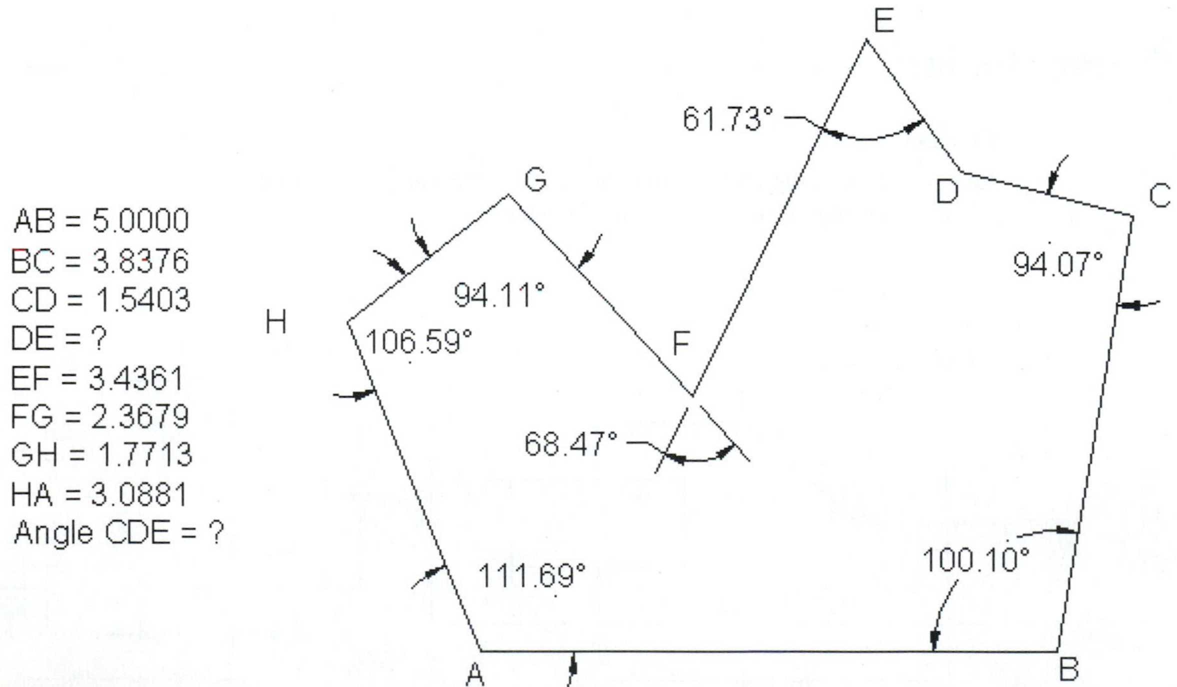

THICKNESS = 1.25

Figure P2-5 INCHES

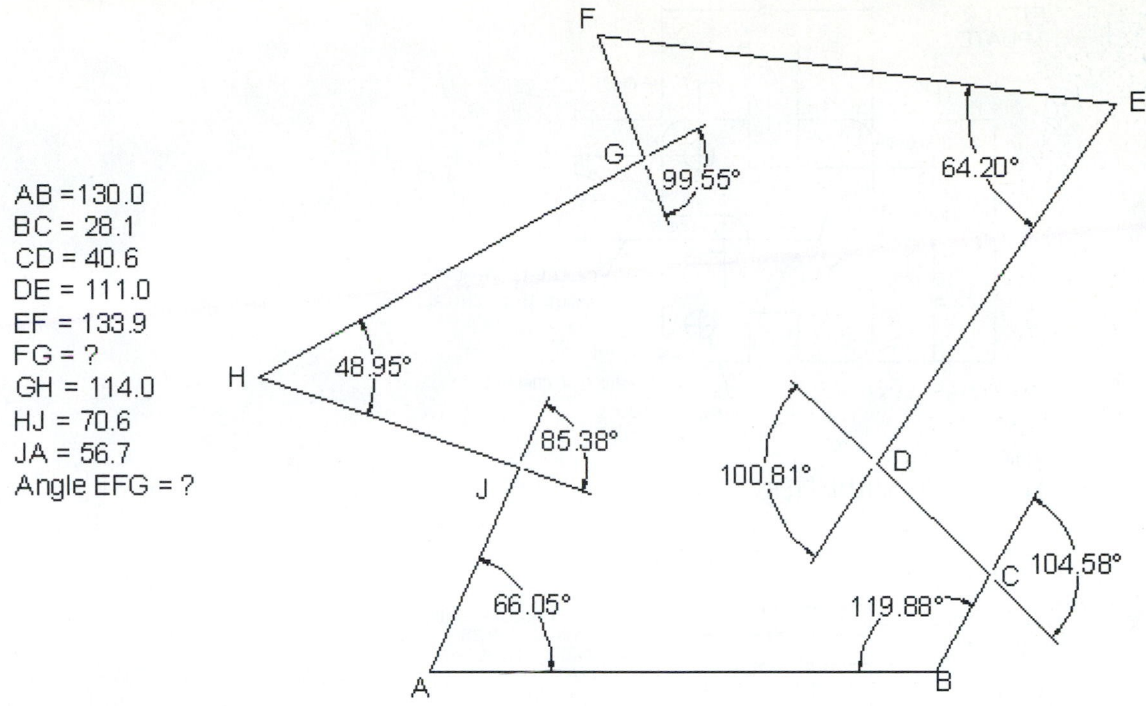

THICKNESS = 12

Figure P2-6 MILLIMETERS

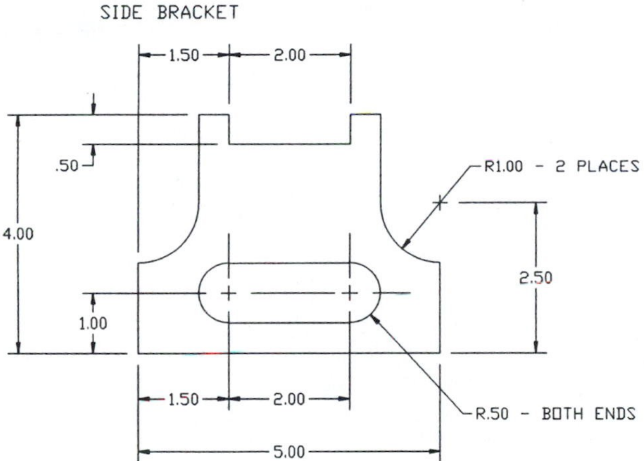

THICKNESS = 1.25

Figure P2-7 INCHES

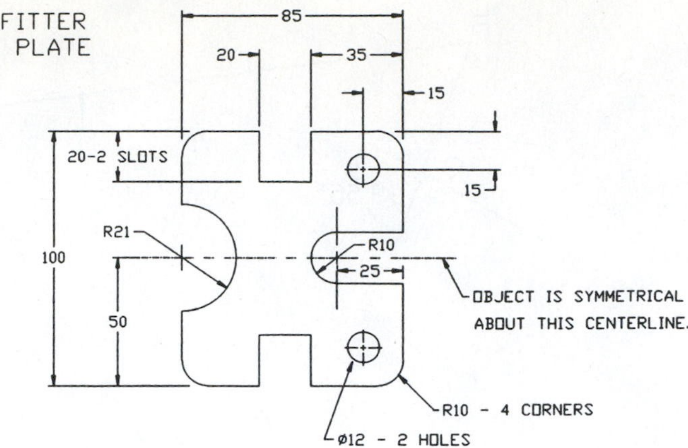

THICKNESS = 8

Figure P2-8 MILLIMETERS

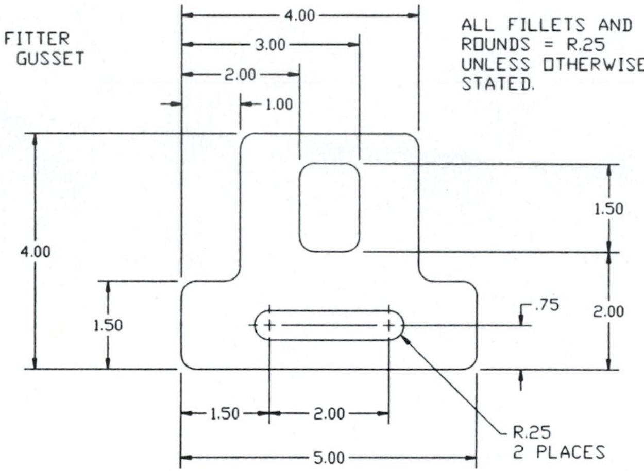

THICKNESS = .375

Figure P2-9 INCHES

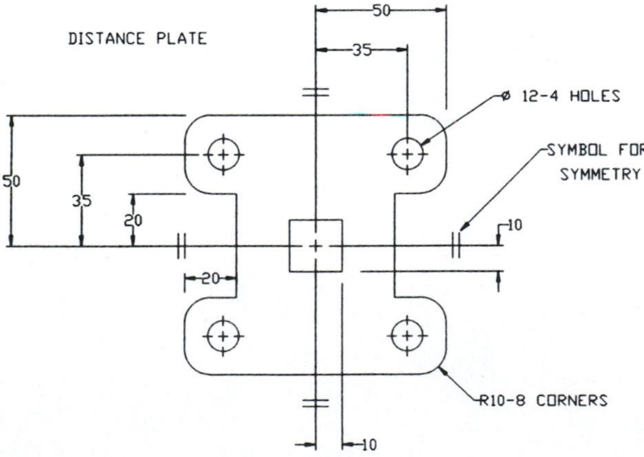

THICKNESS = 16

Figure P2-10 MILLIMETERS

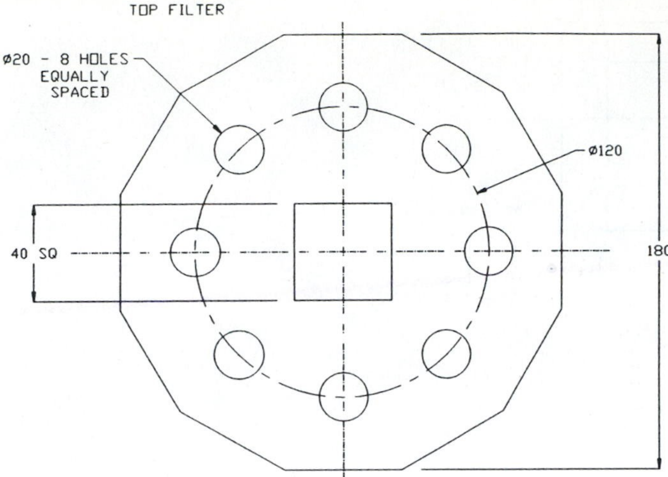

THICKNESS = 6

Figure P2-11 MILLIMETERS

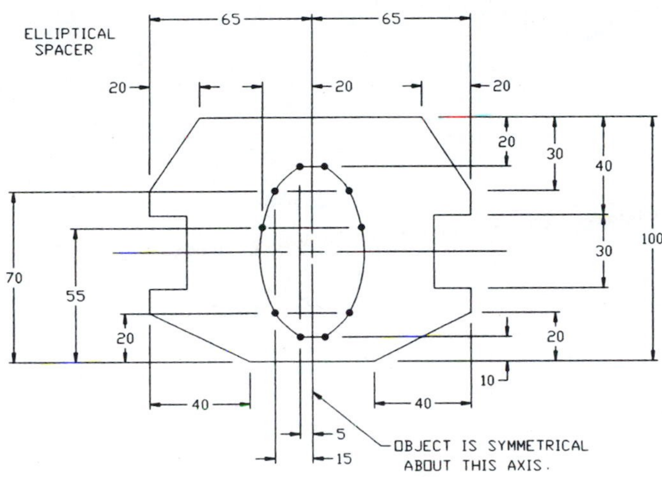

THICKNESS = 10

Figure P2-12 MILLIMETERS

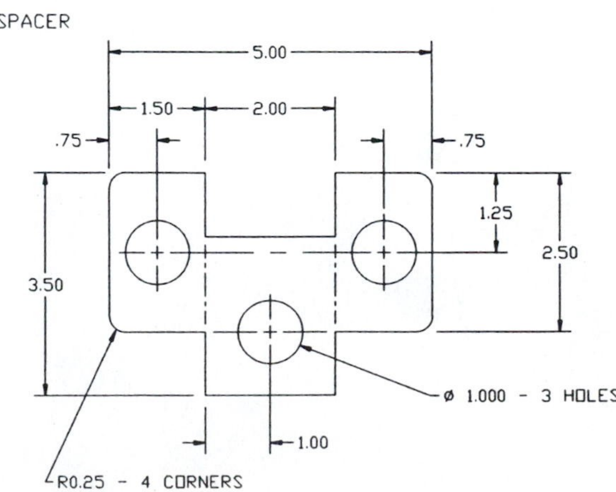

THICKNESS = .75

Figure P2-13 INCHES

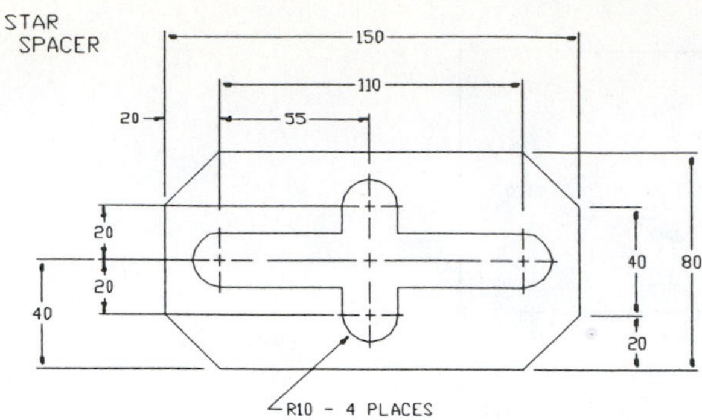

THICKNESS = 7.5

Figure P2-14 MILLIMETERS

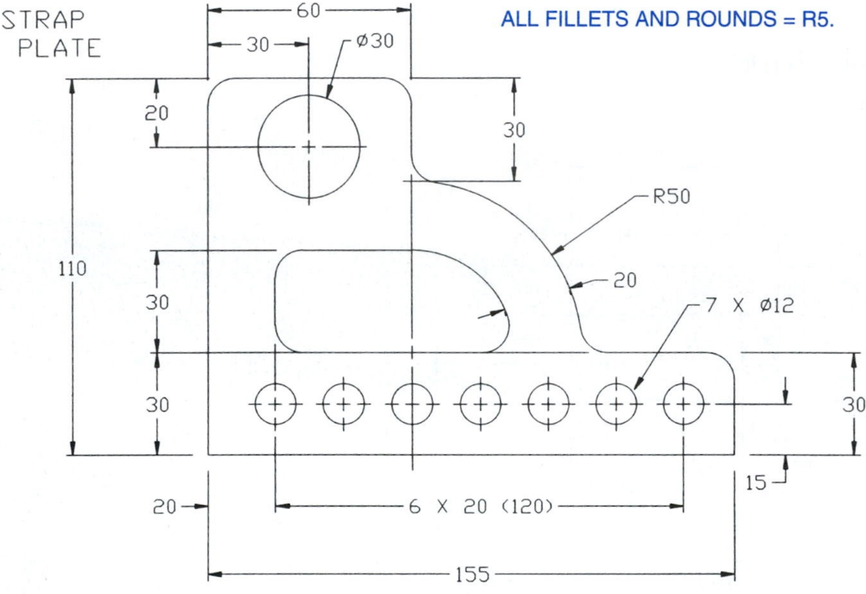

ALL FILLETS AND ROUNDS = R5.

THICKNESS = 5

Figure P2-15 MILLIMETERS

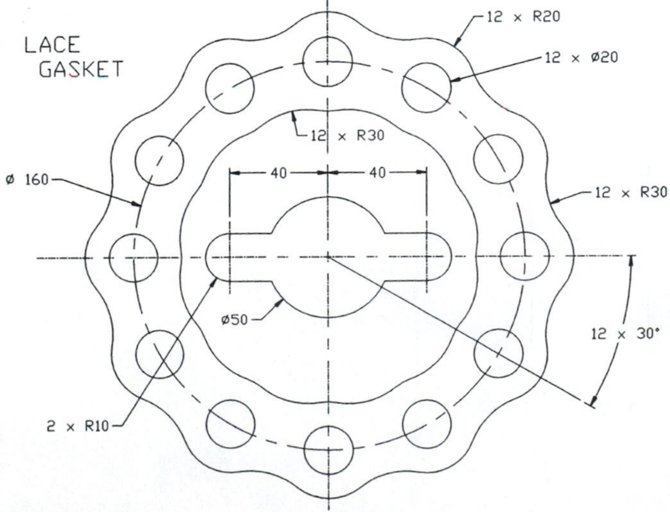

THICKNESS = 6; central area is 4 thick

Figure P2-16 MILLIMETERS

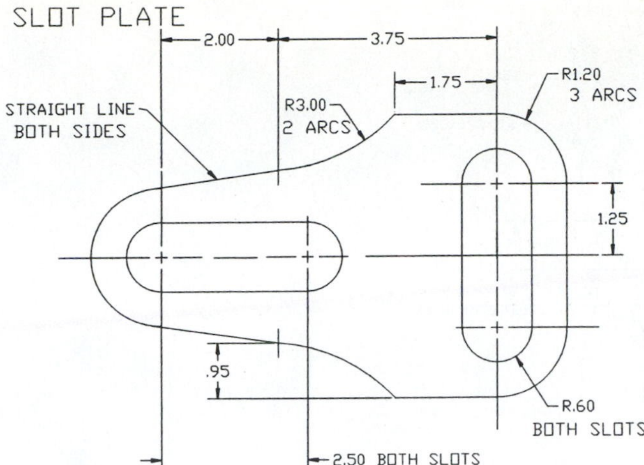

NOTE: Object is symmetrical
about its horizontal centerline.

THICKNESS = .875

Figure P2-17 INCHES

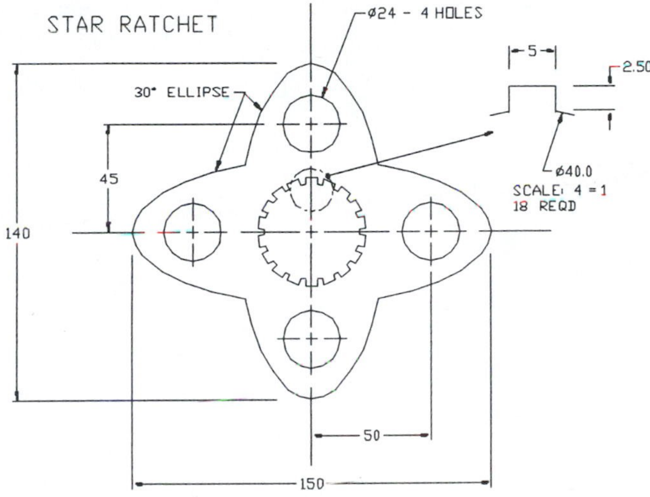

THICKNESS = 6

Figure P2-18 MILLIMETERS

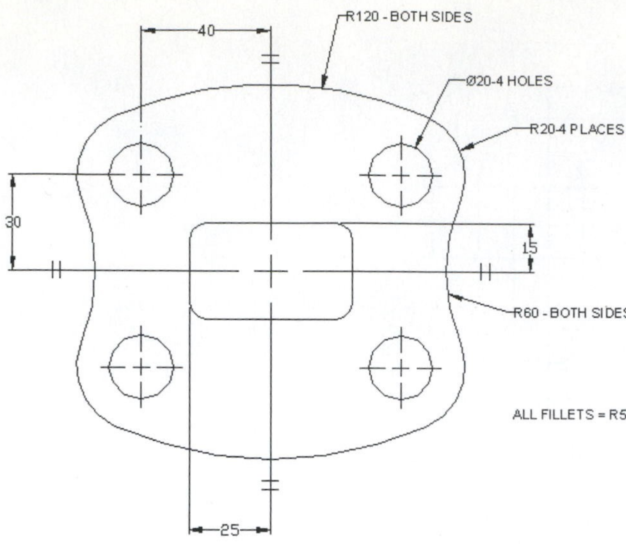

THICKNESS = 7.25

Figure P2-19 MILLIMETERS

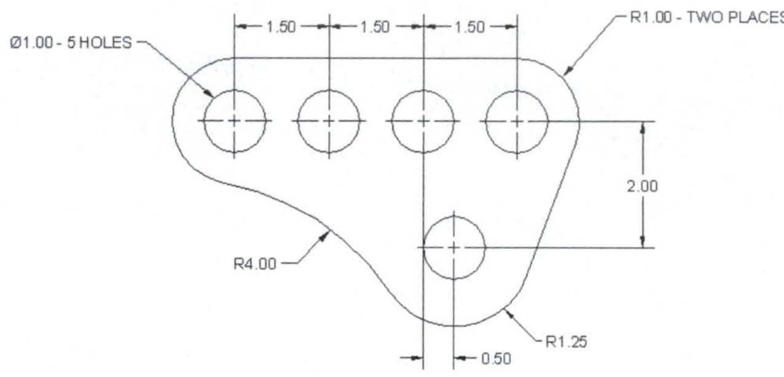

THICKNESS = .25

Figure P2-20 INCHES

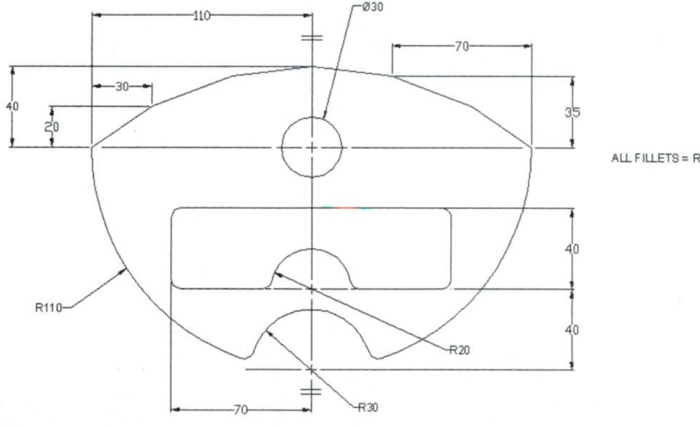

THICKNESS = 5

Figure P2-21 MILLIMETERS

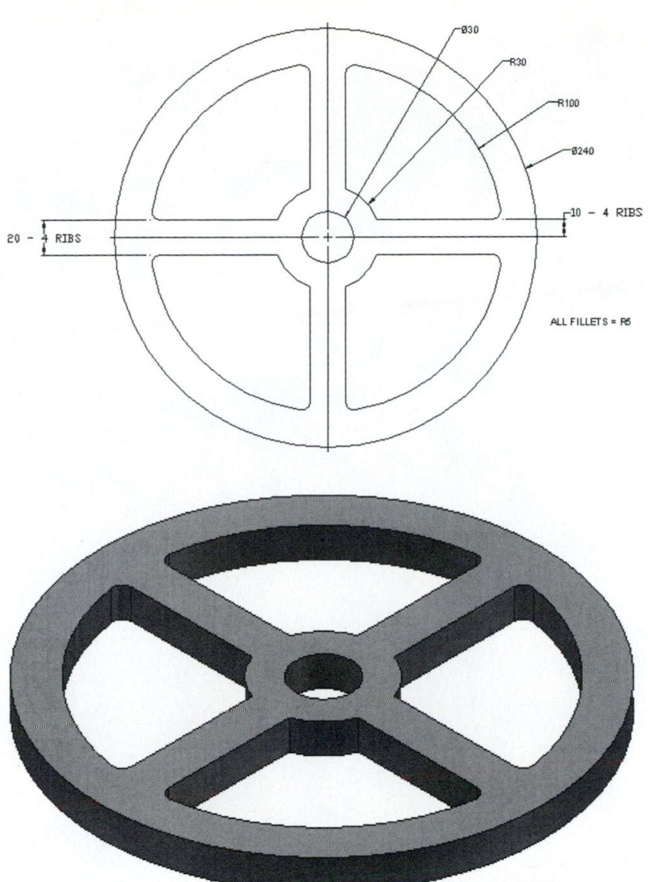

Thickness = 16

Figure P2-22 MILLIMETERS

Three-Dimensional Models 3

Chapter Objectives

- Show how to draw 3D models.
- Show how to use the **Part Features** tools.
- Show how to edit 3D models.
- Show how to create and use work planes.

INTRODUCTION

This chapter introduces and demonstrates how to create 3D models using the commands in the **Part Features** panel bar. These commands are used to convert 2D sketches into 3D solid models and to modify existing models.

The first part of the chapter demonstrates some feature-modifying commands. The second part of the chapter introduces sketch and work planes and shows how they are used to alter and refine 3D models.

Figure 3-1 shows the **Part Features** panel. The tools are used to create 3D models.

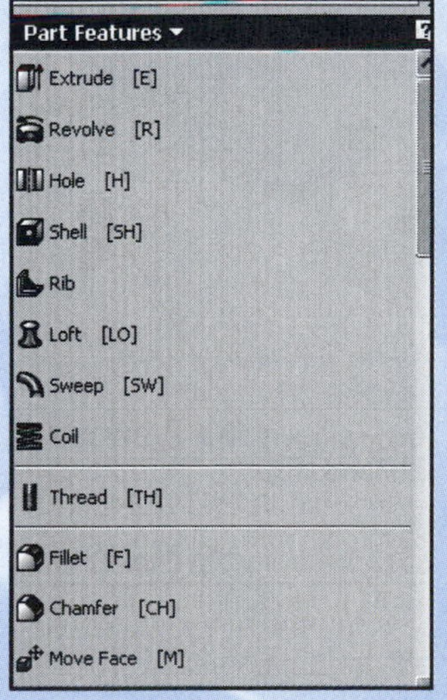

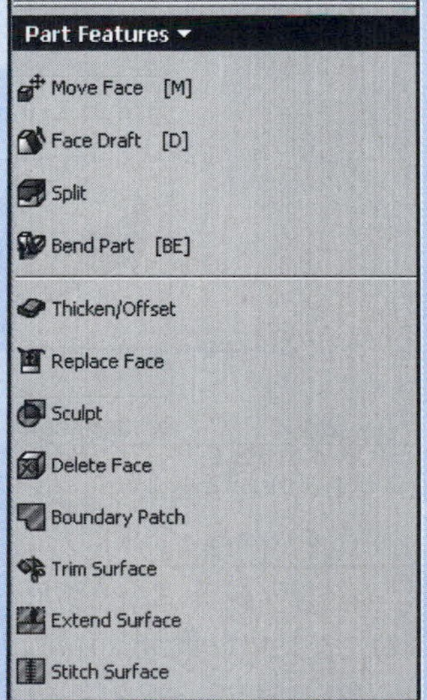

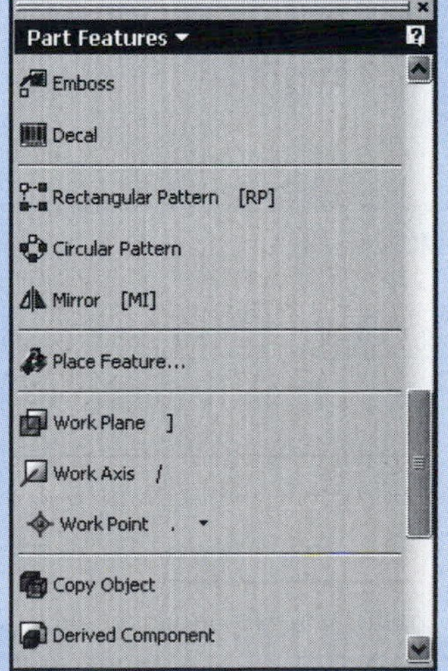

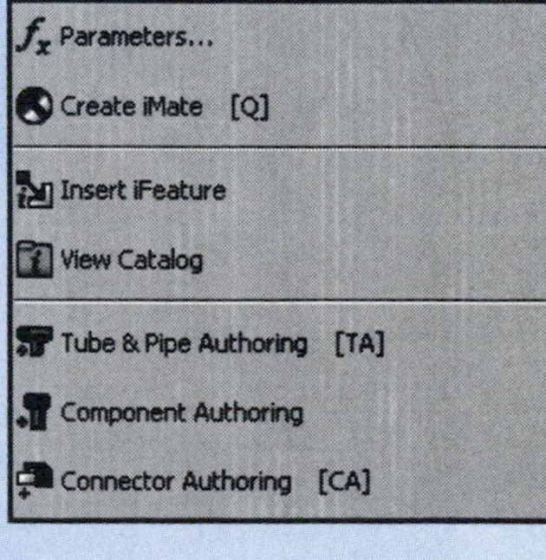

Figure 3-1

EXTRUDE

The **EXTRUDE** command is used to convert 2D sketches into solid models.

Figure 3-2 shows a 12 mm × 30 mm rectangle created using the **Standard.ipt** format and the **2D Sketch Panel Two point rectangle** tool. After the sketch is completed, right-click the mouse and select the **Done** option. Also, create an isometric view of the sketch.

1. Right-click the mouse and select the **Finish Sketch** option, then click the **Extrude** tool on the **Part Features** panel bar.

The **Extrude** dialog box will appear. See Figure 3-2.

2. Change the **Extents Distance** to **20,** then click **OK.**

Figure 3-2 shows the extruded rectangle.

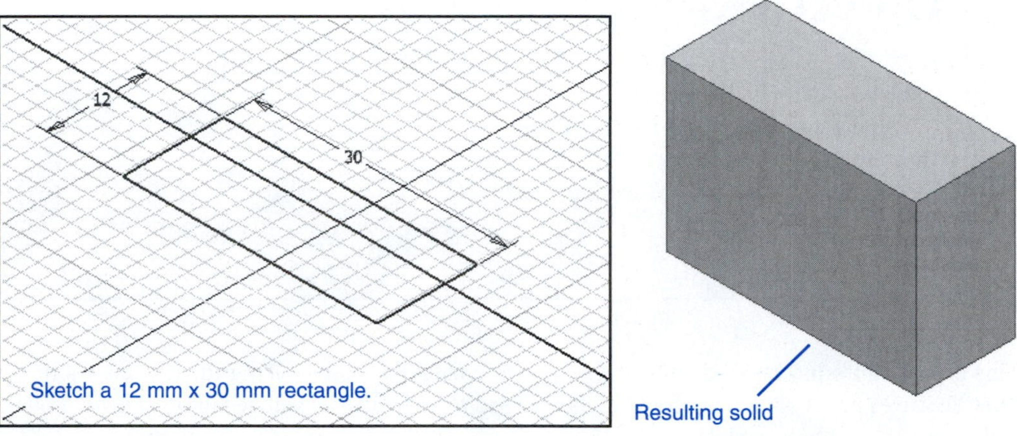

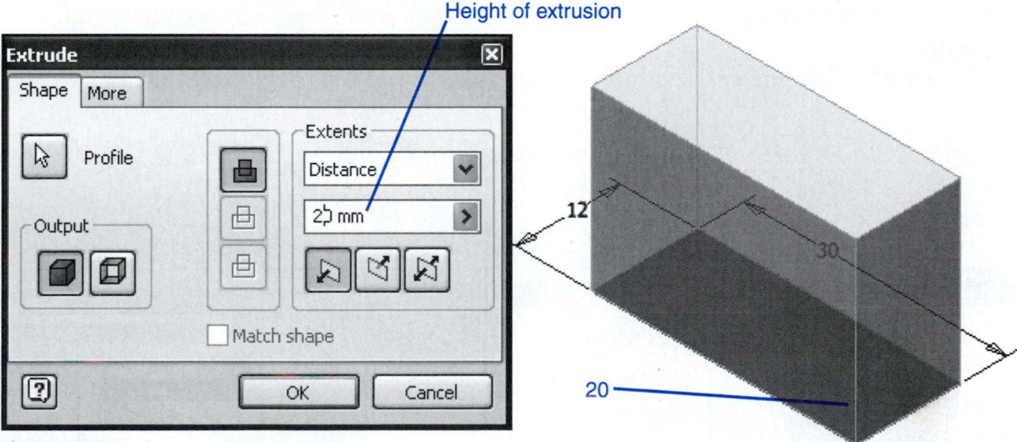

Figure 3-2

Exercise 3-1: Creating a Taper

1. Click the **Extrude** tool.
2. Click the **More** tab on the **Extrude** dialog box, then set the **Taper** angle for **15.**
3. Click **OK.**

Figure 3-3 shows the resulting tapered shape.

Exercise 3-2: Controlling the Direction of the Taper

1. Click the **Extrude** tool.
2. Set the **Taper** for **15** and click the middle button in the **Extents** area as shown.
3. Click **OK.**

Figure 3-4 shows the resulting tapered shape.

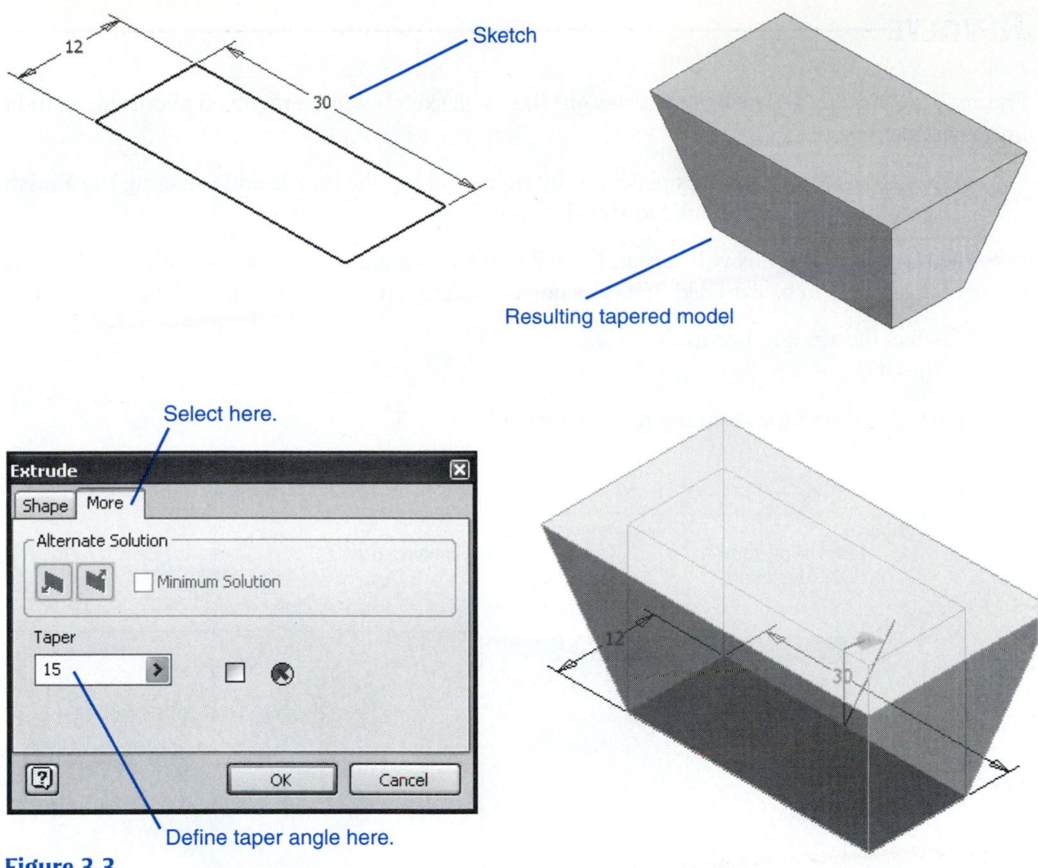

Figure 3-3

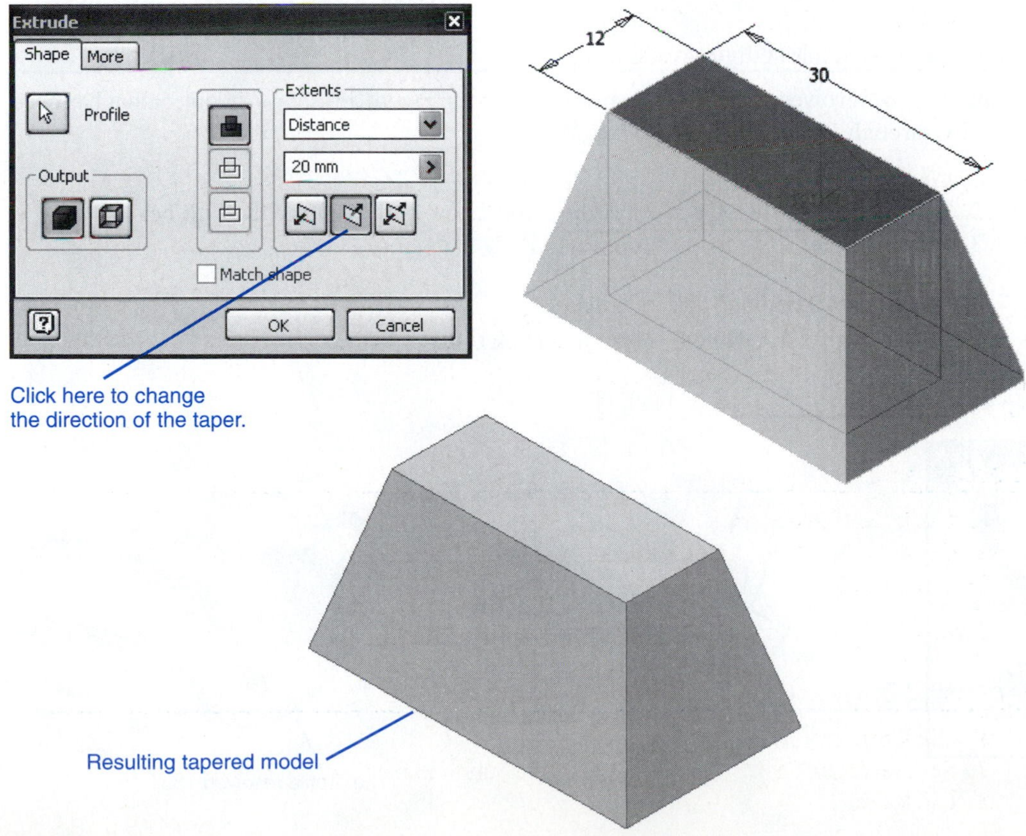

Figure 3-4

REVOLVE

Figure 3-5 shows a 2D profile and a straight line. The sketch will be revolved about the straight line to create a model.

1. Access the **Part Features** panel bar by right-clicking the mouse and selecting the **Finish Sketch** option. Then click the **Revolve** tool.

The **Revolve** dialog box will appear. The **REVOLVE** command will automatically select the 2D shape as the profile to be extruded. If it does not, click the **Profile** box and window the 2D profile.

2. Select the straight line as the axis.
3. Click **OK.**

Figure 3-5 shows the resulting revolved model.

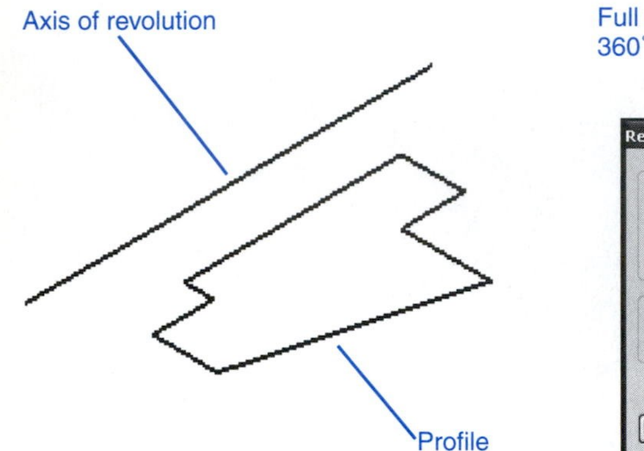

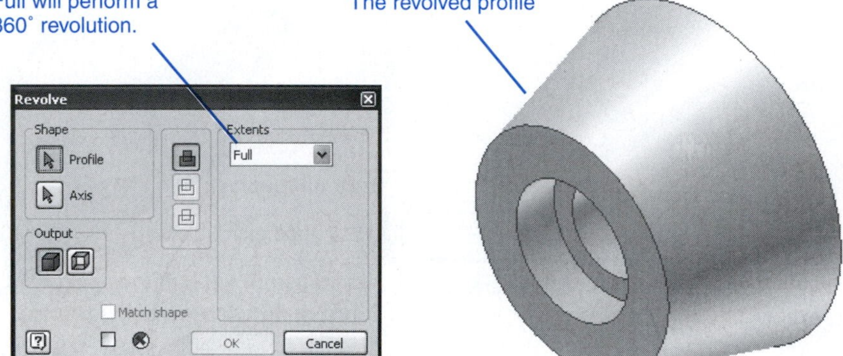

Figure 3-5

Exercise 3-3: Revolving Through Less than 360°

Shapes may be revolved through any number of degrees. Figure 3-6 shows an object that has been revolved through 180°.

1. Access the **Revolve** tool.
2. On the **Revolve** dialog box, click the scroll arrow to the right of the **Full** box.
3. Select **Angle.**

The **Angle** dialog box will appear.

4. Type in the required angle value, then click **OK.**

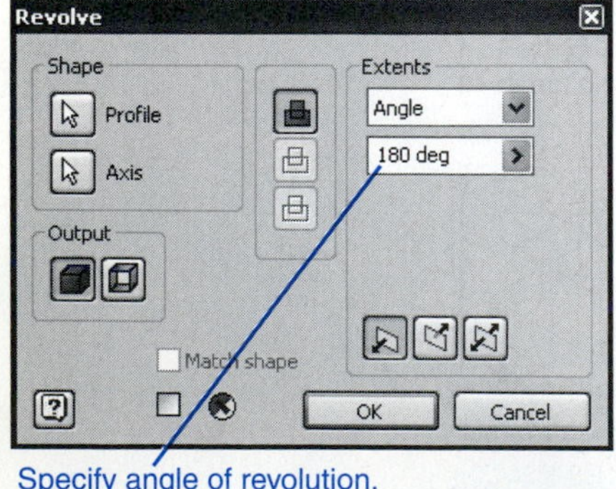

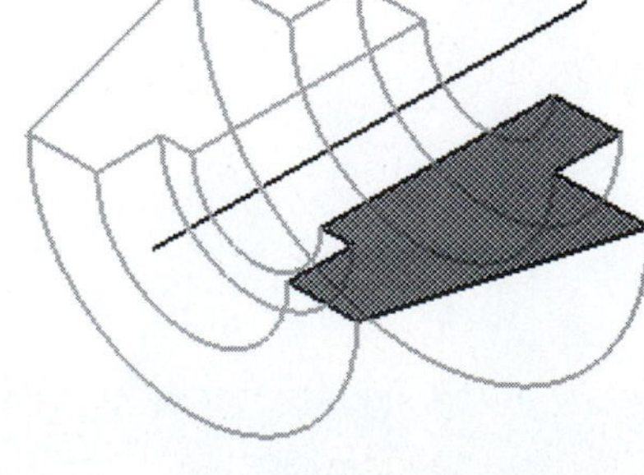

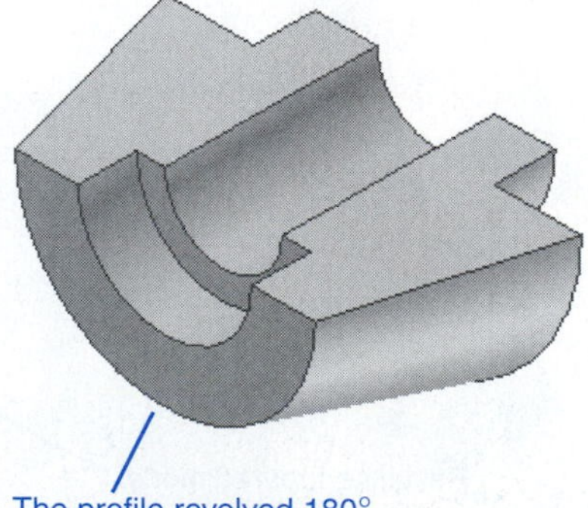

Figure 3-6

Exercise 3-4: Creating a Sphere

See Figure 3-7.

1. Draw a line and a circle. Use the **General Dimension** tool to size the circle.
2. Trim half the circle.
3. Right-click the mouse and select the **Done** option.
4. Right-click the mouse and select the **Finish Sketch** option.
5. Select the **Revolve** tool from the **Part Features** panel.
6. Select the semicircle as the profile and the line as the axis.

A preview will appear.

7. Click **OK.**

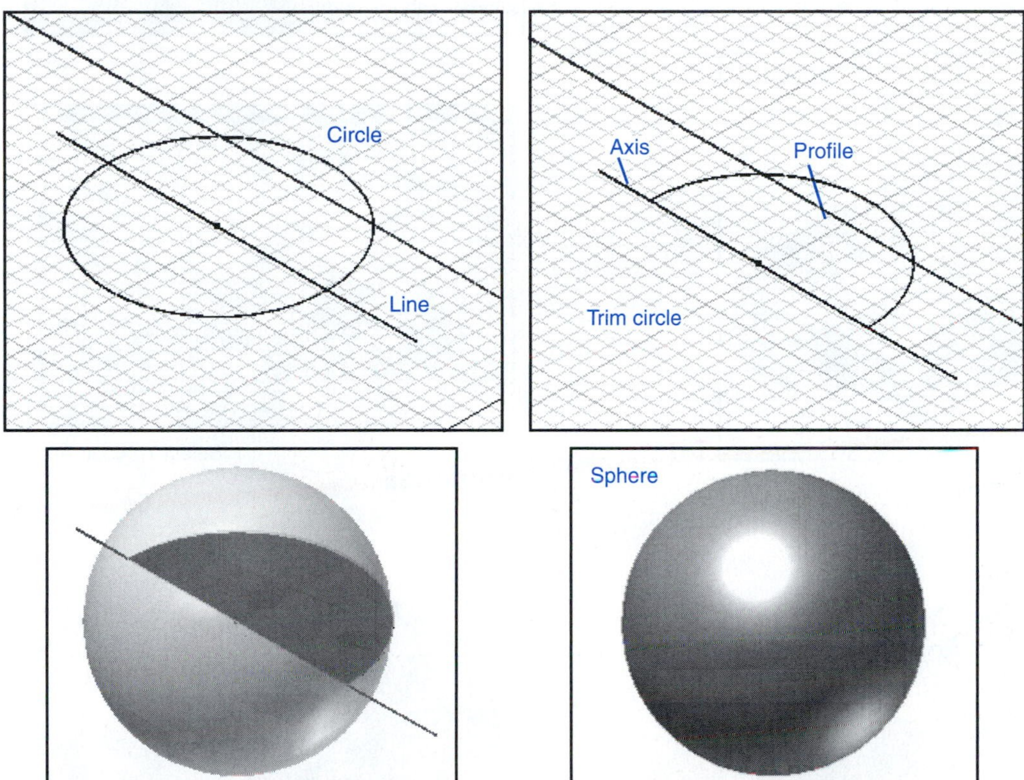

Figure 3-7

Holes

Holes may be added to a 3D solid model by first defining a sketch plane for the hole and then locating the hole's center point. Figure 3-8 shows a 20 × 40 rectangular model that has been extruded 16. Two small holes will be located in the top surface, and a large hole will be located in the front surface.

1. Click on the top surface of the model, right-click the mouse, and select the **New Sketch** option.

A grid will appear aligned with the top surface. The top surface is now the current sketch plane.

2. Click the **Point, Center Point** tool.

TIP: The different types of holes are discussed in Chapter 7.

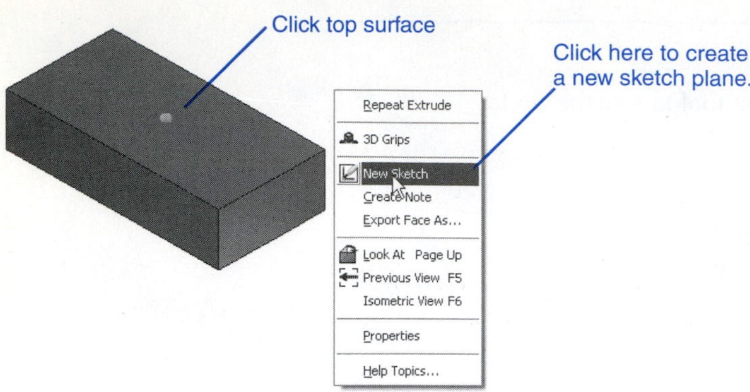

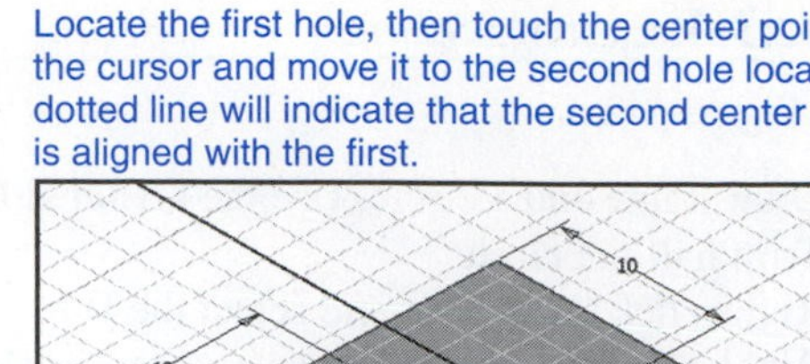

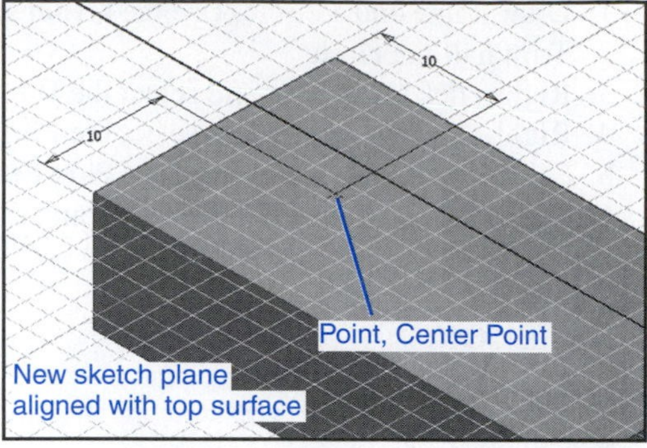

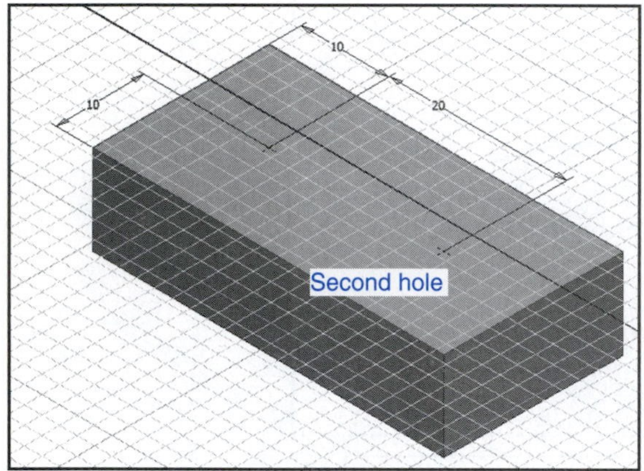

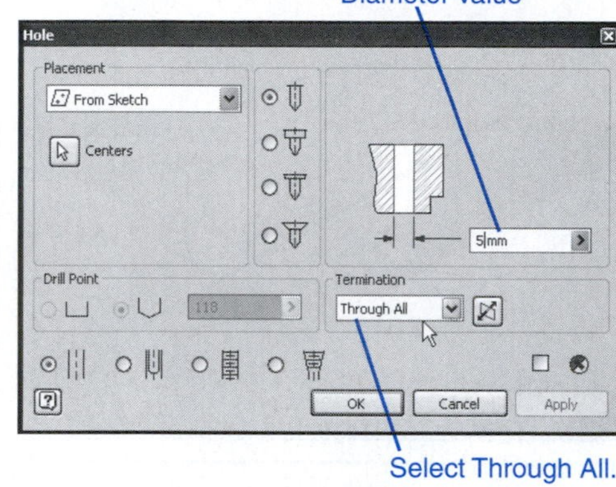

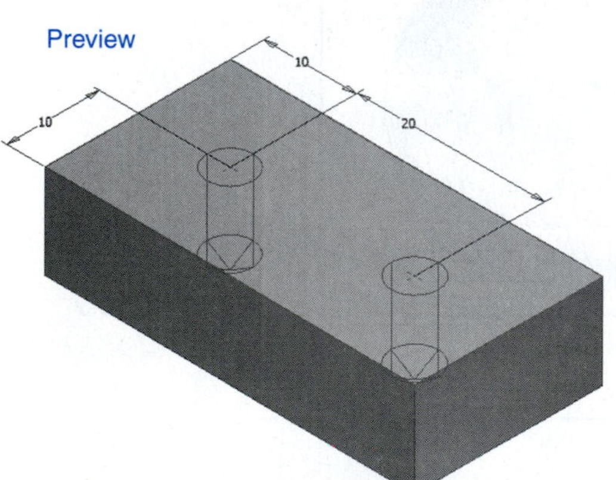

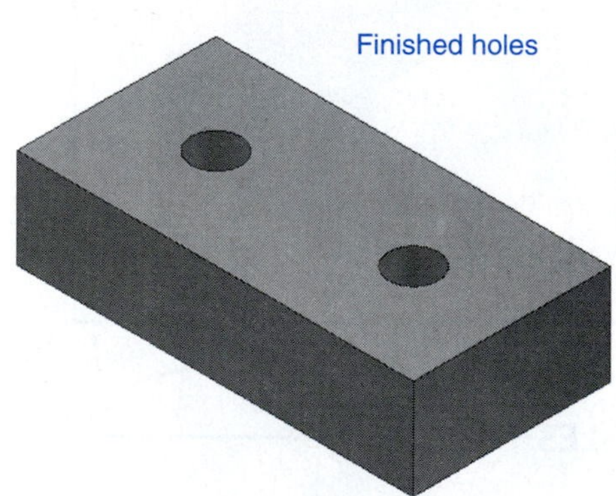

Figure 3-8

3. Locate one of the hole center points on the top surface, then right-click the mouse and select the **Done** option.
4. Click the **General Dimension** tool and locate the first hole.
5. Use the **General Dimension** tool to locate the second hole's center point, then right-click the mouse and select the **Finish Sketch** option.

Note:

The second hole's center point may be located from the first hole location. Click the **Point, Center Point** tool, then touch the cursor to the first hole's center point. Move the cursor to the approximate second hole's location. A dotted line will appear from the first center point location when the second location is aligned to it. Only one dimension is now needed to locate the second hole's center point.

The **Part Features** panel bar will reappear.

6. Select the **Hole** tool from the **Part Features** panel.

The **Hole** dialog box will appear. See Figure 3-8.

7. Set the **Termination** for **Through All** and set the holes' diameter for **5 mm,** then click **OK.**

The holes will appear.

Exercise 3-5: Locating a Hole in the Front Plane

1. Click the front surface, right-click the mouse, and select the **New Sketch** option.
2. Click the **Sketch** heading on the **Command** toolbar.

The **2D Sketch Panel** bar will appear. See Figure 3-9.

3. Use the **Point, Center Point** and **General Dimension** tools and locate a center point in the center of the surface.
4. Right-click the mouse and select the **Finish Sketch** option.
5. Click the **Hole** tool and add a **Ø8** hole through the model.
6. Click **OK.**

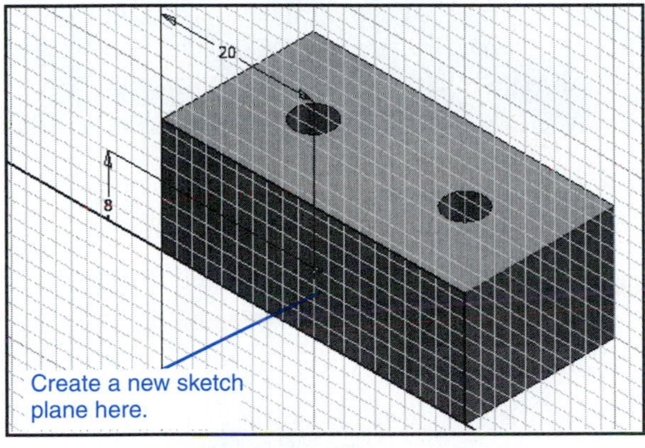

Use the Point, Center Point tool to define the hole's center point.

Resulting model

Hole diameter

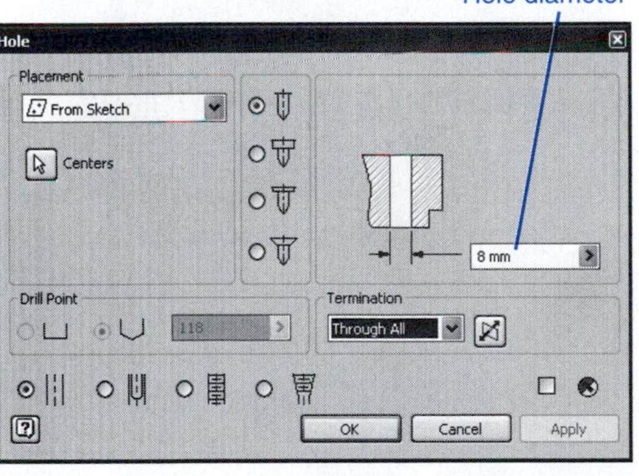

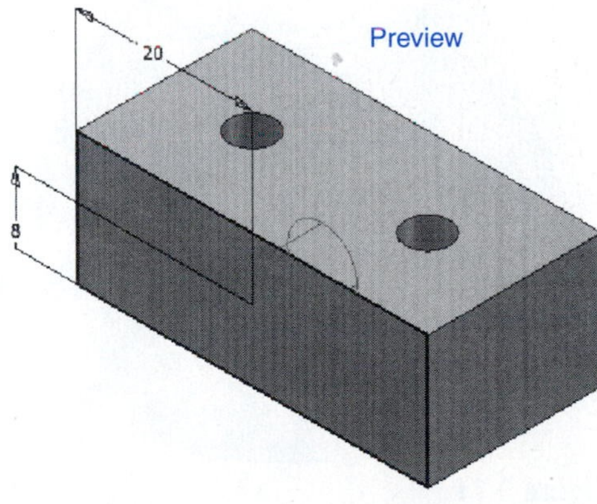

Figure 3-9

SHELL

The **SHELL** command is used to create thin-walled objects from existing models. Figure 3-10 shows a 12 × 30 × 20 model.

1. Click the **Shell** tool on the **Part Features** panel bar.

The **Shell** dialog box will appear. There are three different ways to define a shell, which are accessed by the three boxes on the right side of the **Shell** dialog box. The options are as follows:

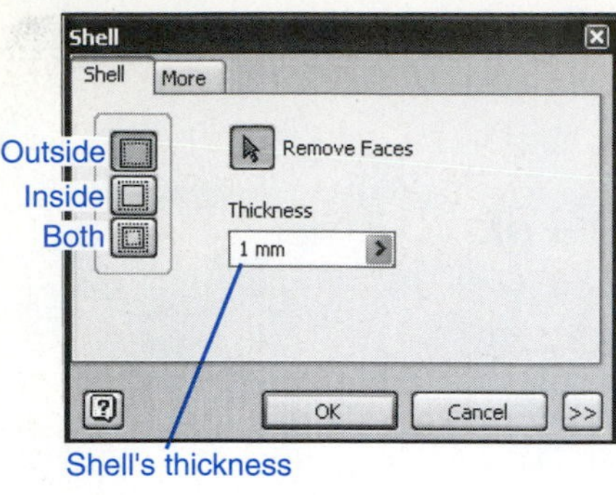

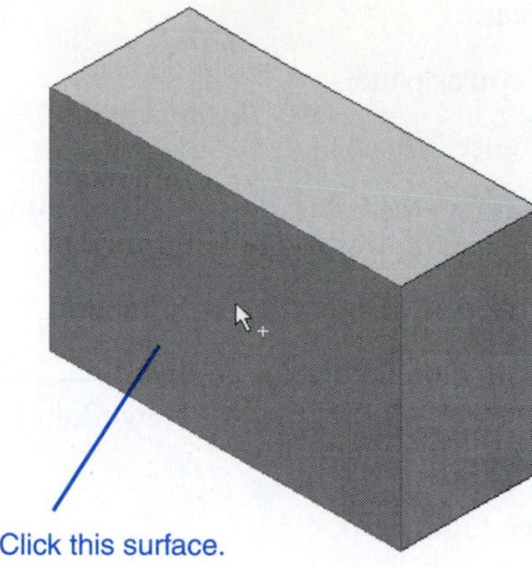

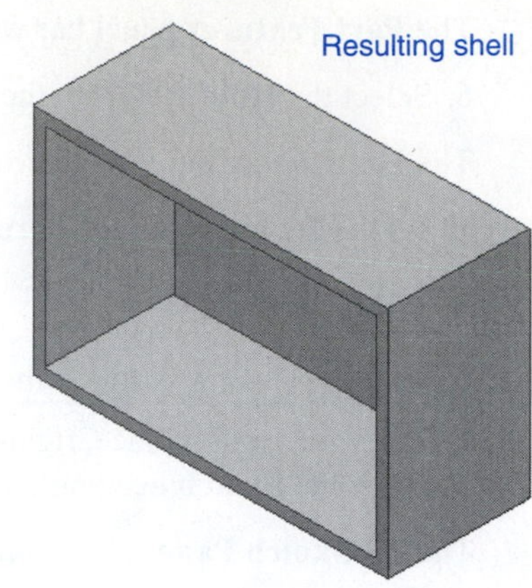

Figure 3-10

Inside: The external wall of the existing model will become the external wall of the shell.
Outside: The external wall of the existing model will become the internal wall of the shell.
Both Sides: The existing outside wall will become the center of the shell; half the thickness will be added to the outside and half to the inside.

2. Click on the front surface of the model, then click **OK.**

Shells may be created from any shaped model. Figure 3-11 shows a cone that has been used to create a hollow thin-walled cone.

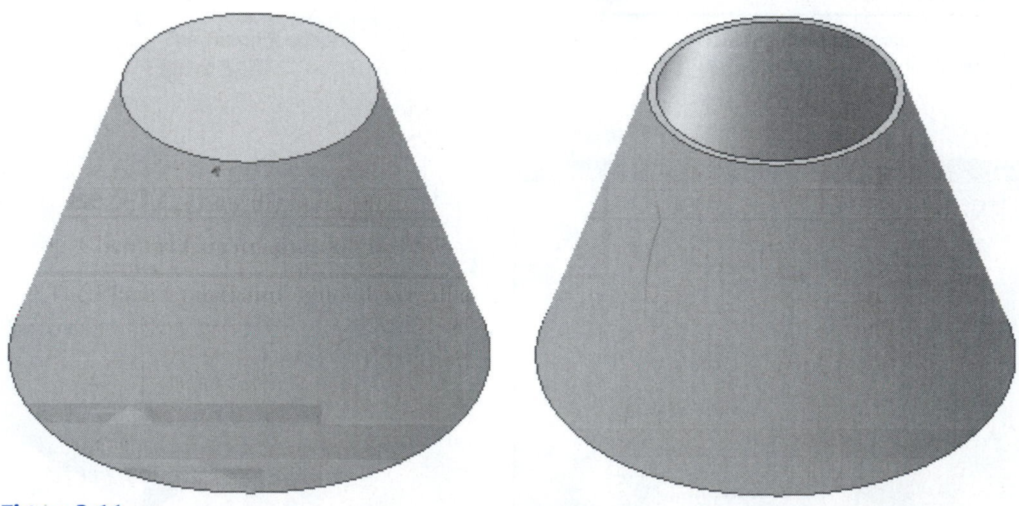

Figure 3-11

Exercise 3-6: Removing More Than One Surface

1. Click the **Shell** tool.
2. Click the surfaces to be removed.
3. Select the **Inside** option.
4. Click **OK.**

See Figure 3-12.

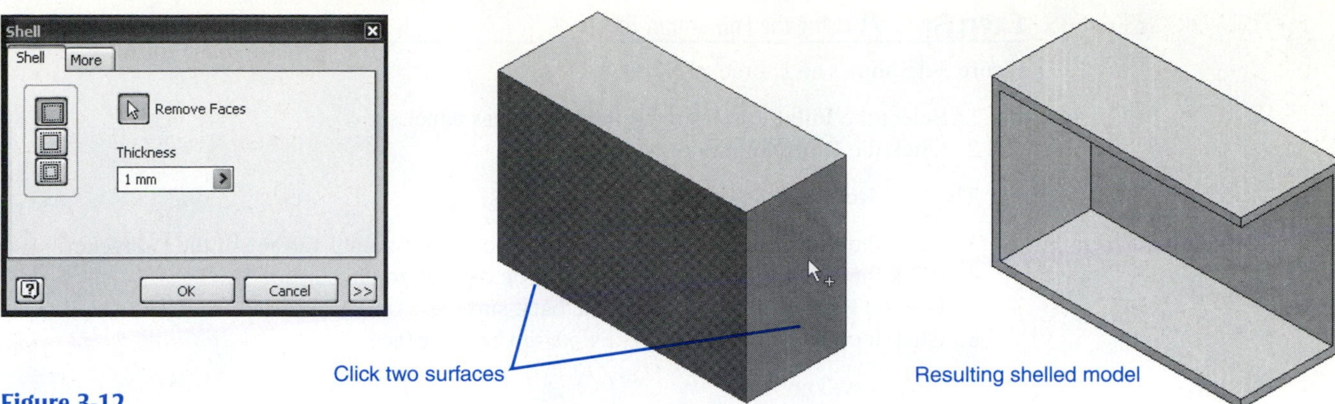

Figure 3-12

FILLET

The **FILLET** command is used to create rounded edges. Figure 3-13 shows a finished 12 × 30 × 20 3D model.

1. Click the **Fillet** tool on the **Part Features** panel bar.

The **Fillet** dialog box will appear.

2. Change the **Radius** value to 3.
3. Click the **Edges** option and select the edges to be filleted.
4. Click **OK.**

A fillet may be added to an internal edge such as shown in Figure 3-14. External fillets are called ***rounds.***

round: An external fillet.

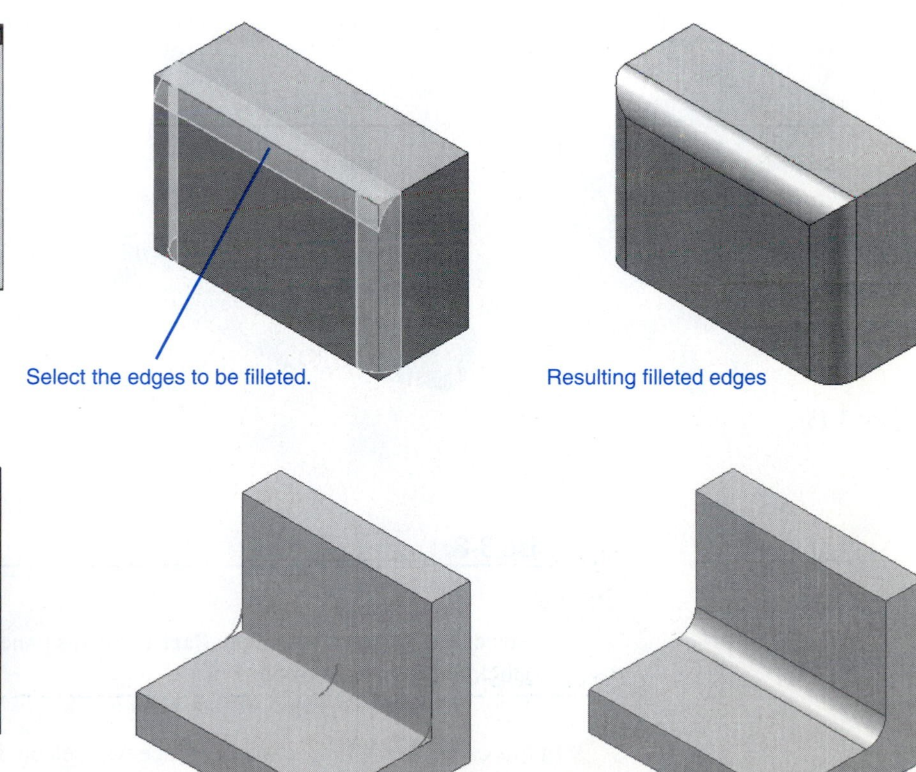

Define the radius of the fillet.

Figure 3-13

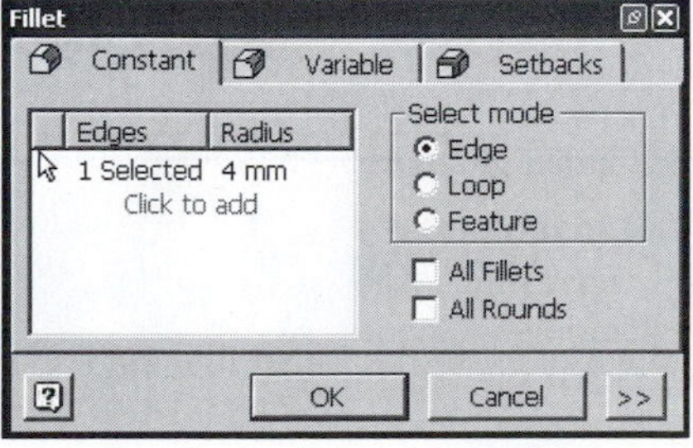

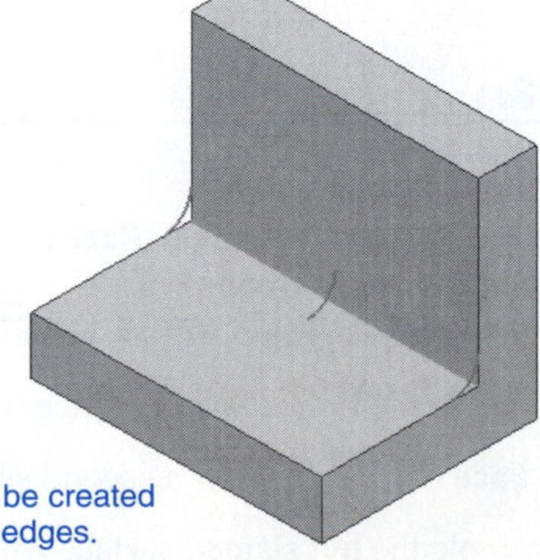

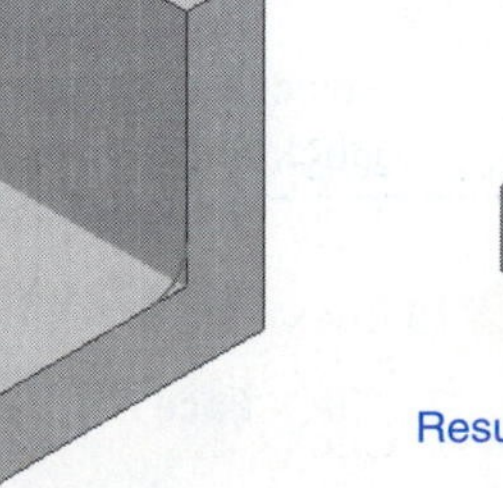

Figure 3-14

Exercise 3-7: Using the Full Round Tool

Figure 3-15 shows an L-bracket.

1. Select the **Fillet** tool from the **Part Features** panel.
2. Click the **Round** option.

The **Full Round** dialog box will appear.

3. Click the **Side Face Set 1** box. Select the lower horizontal surface of the L-bracket.
4. Click the **Center Face Set** box. Select the top horizontal surface.
5. Use the **Rotate** tool and expose the back surface.
6. Click the **Side Face Set 2** box. Select the back surface.

A preview will appear.

7. **Click Apply** and return the view to the Isometric view.

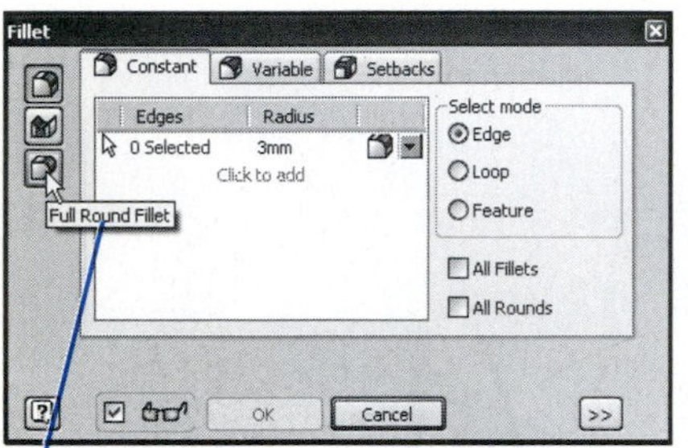

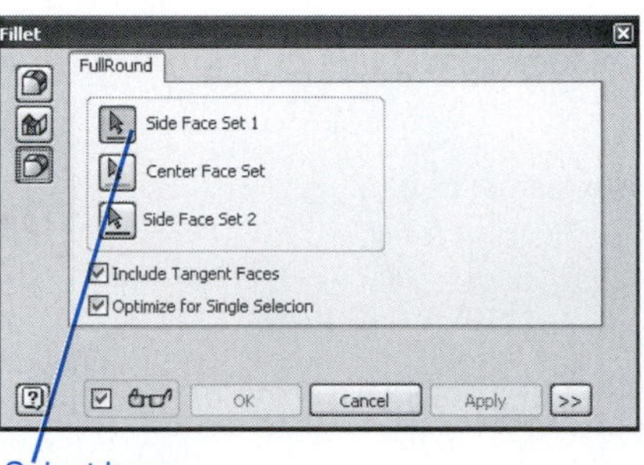

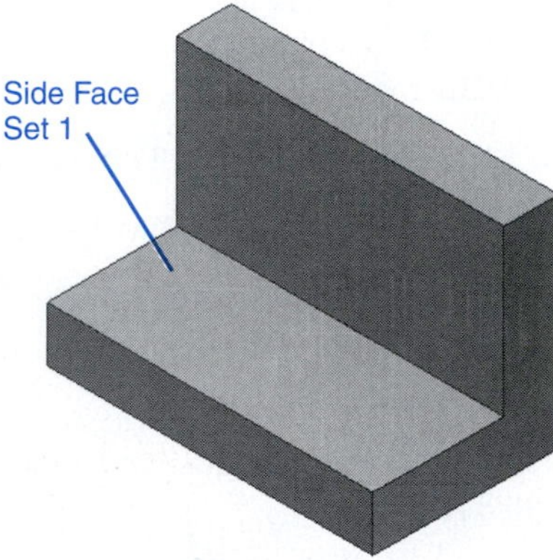

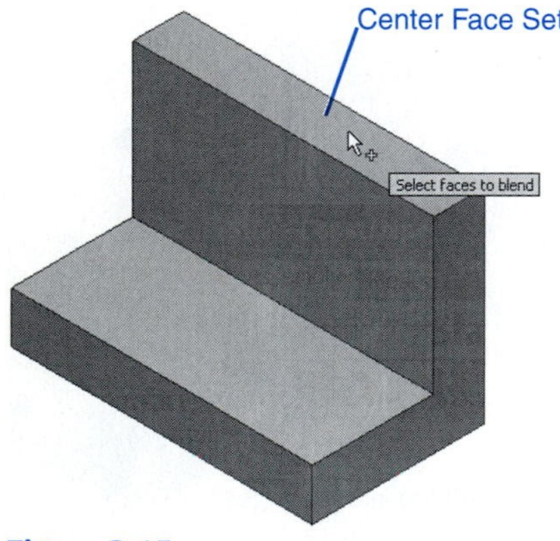

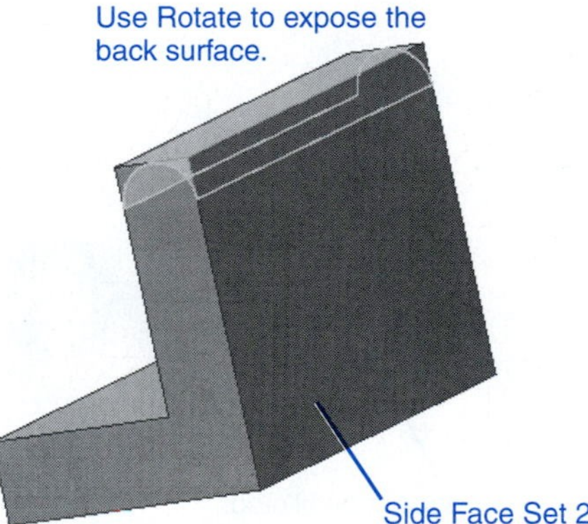

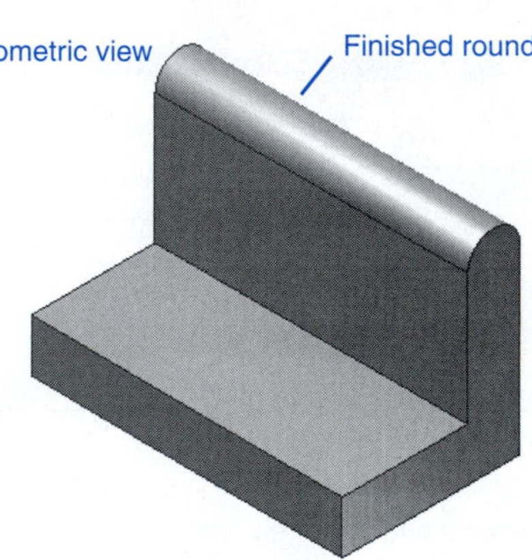

Figure 3-15

Exercise 3-8: Creating a Face Fillet

See Figure 3-16.

1. Select the **Fillet** tool from the **Part Features** panel.
2. Click the **Face** fillet option.
3. Click the **Face Set 1** option and select a face.

In this example the right slanted surface was selected.

4. Click **Face Set 2** and select a second surface.

In this example the left slanted surface was selected.

5. Click **Apply.**

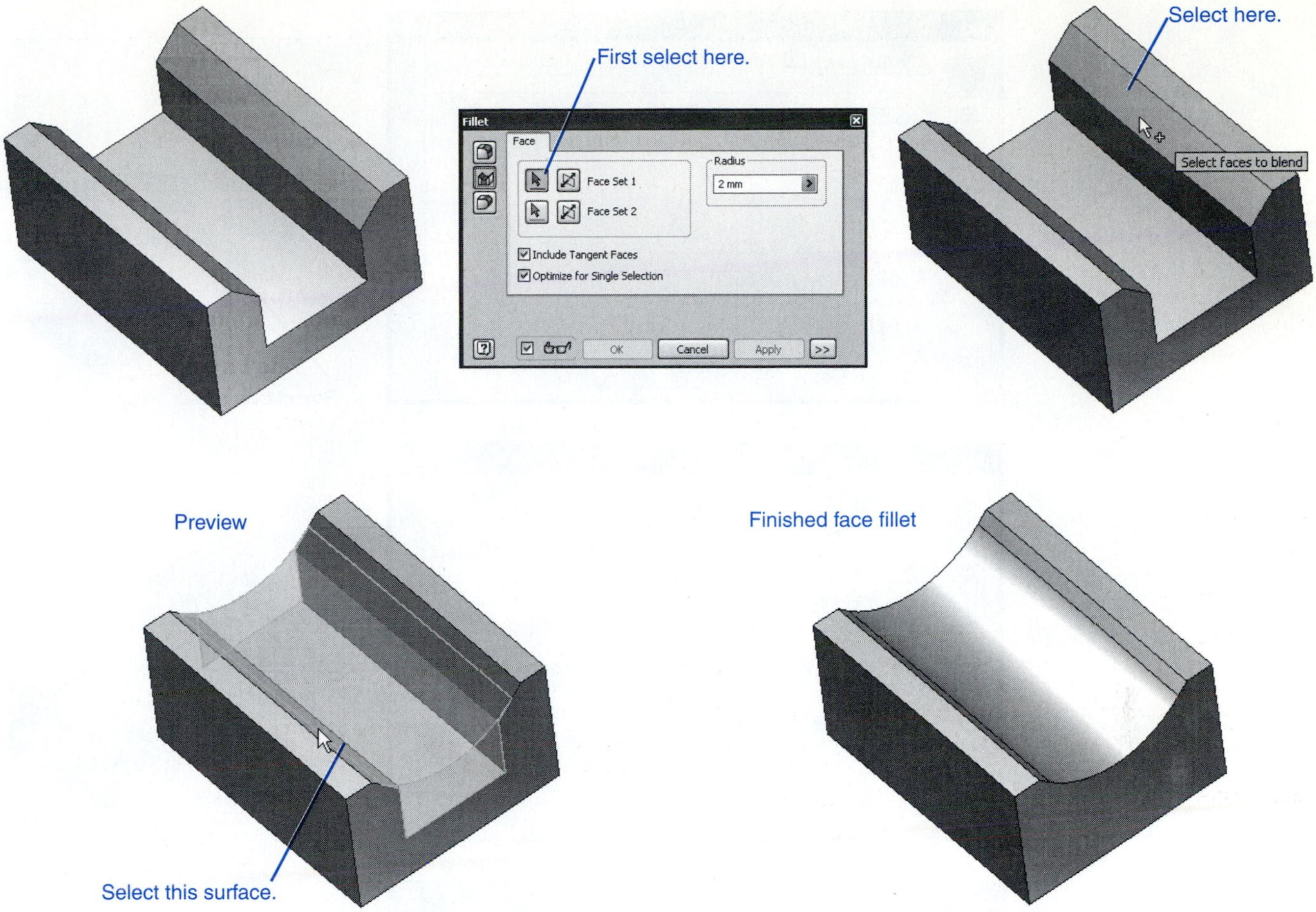

Figure 3-16

The face fillet will appear.

Exercise 3-9: Creating a Variable Fillet

See Figure 3-17.

1. Select the **Fillet** tool from the **Part Features** panel bar.
2. Click the **Variable** tab.
3. Select the edge for the fillet.
4. Define the **Start** radius.

   In this example a value of **1 mm** was selected.

5. Click the word **End** and define a value.

   In this example a value of **3 mm** was selected.

6. Click **OK.**

Fillets and chamfers can be created in 2D sketches.

CHAMFER

The **CHAMFER** command is used to create beveled edges. See Figure 3-18. Chamfers are defined by specifying linear setback distances or by specifying a setback distance and an angle. Most chamfers have equal setback distances or an angle of 45°.

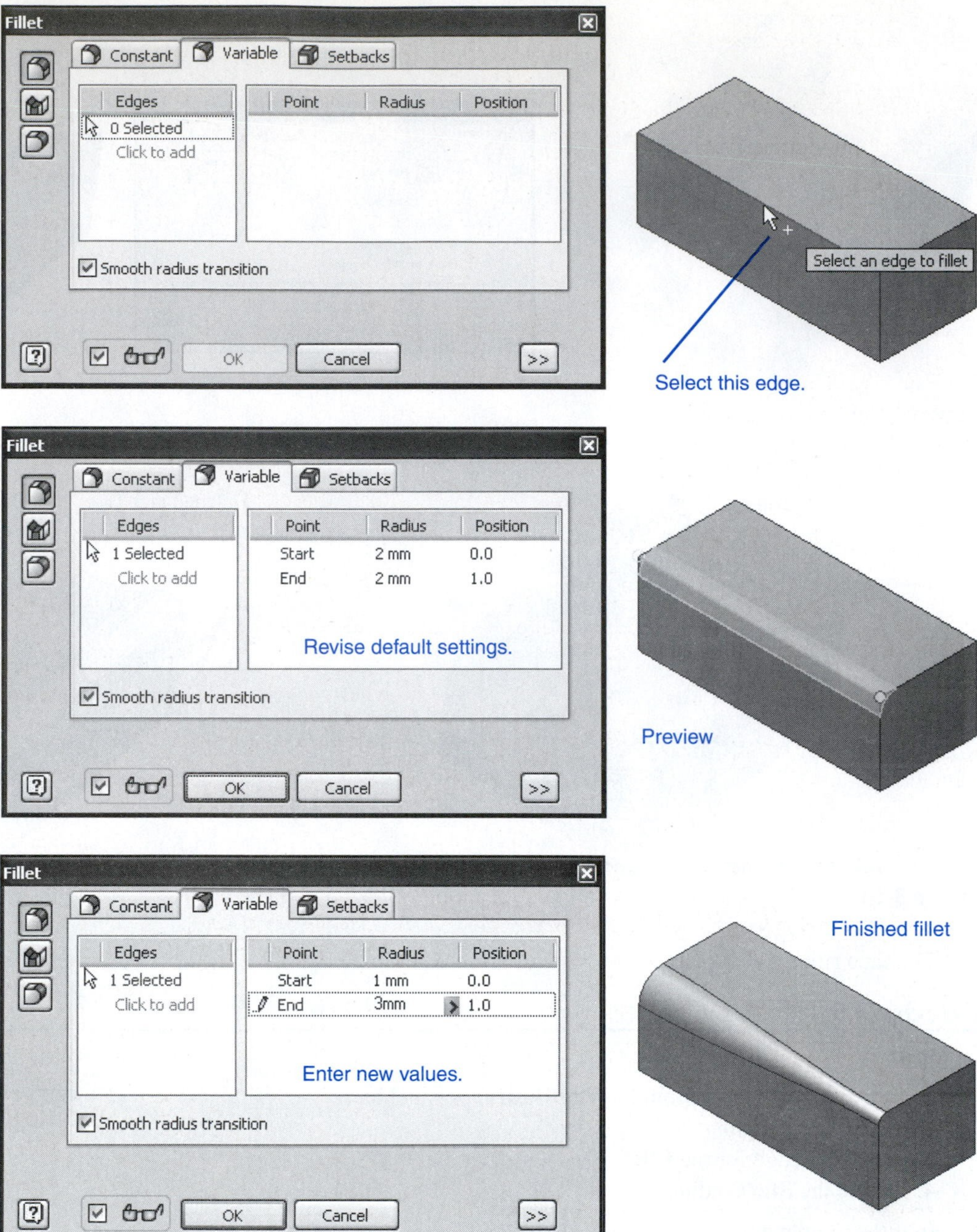

Figure 3-17

1. Click the **Chamfer** tool.

The **Chamfer** dialog box will appear. The first option box on the left side of the **Chamfer** dialog box is used to create chamfers with equal distances.

2. Set the distance for **1.**
3. Select the edges to be chamfered.
4. Click **OK.**

The chamfers drawn in Figure 3-18 were defined using two equal distances creating a 45° chamfer. Chamfers may also be defined using a distance and an angle. Figure 3-19 shows a 2 × 60° chamfer. Chamfers may also be defined using two unequal distances. Figure 3-20 shows a 1 × 4 chamfer.

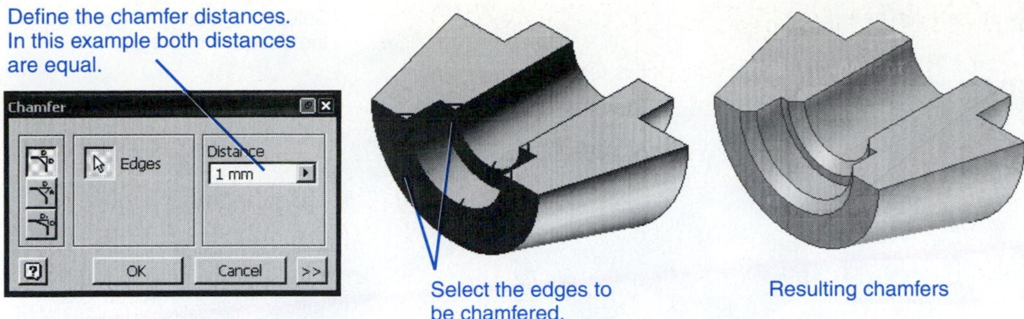

Figure 3-18

A chamfer defined by a distance and an angle

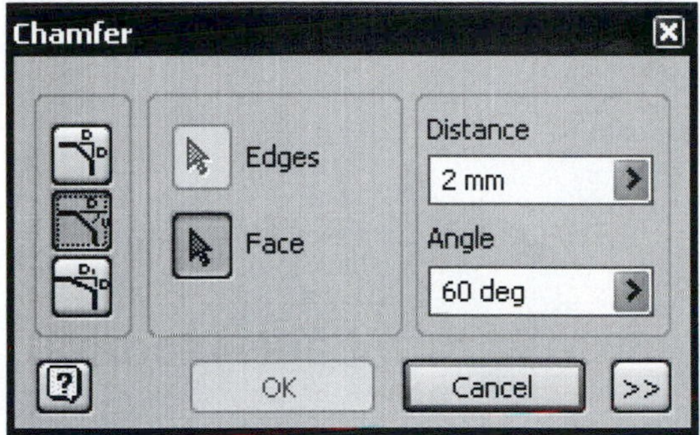

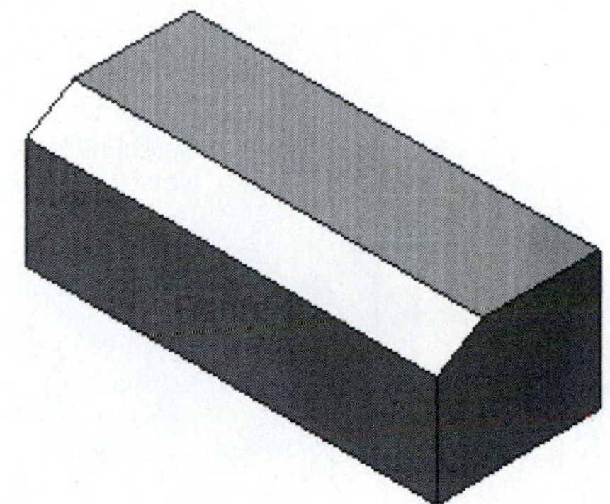

Figure 3-19

A chamfer defined by two unequal distances

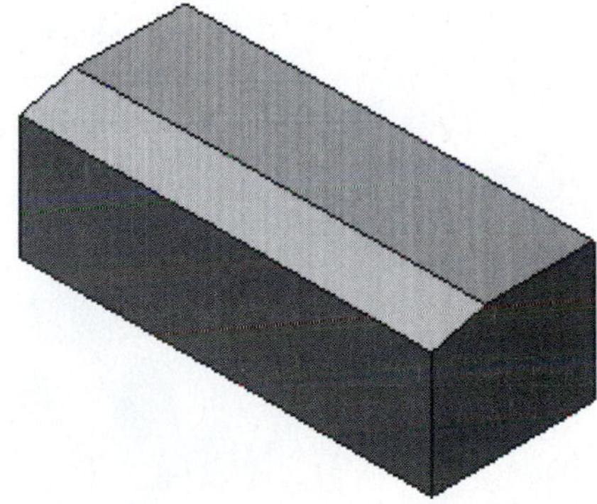

Figure 3-20

FACE DRAFT

The **Face Draft** tool is used to create angled surfaces. See Figure 3-21.

1. Click the **Face Draft** tool.

The **Face Draft** dialog box will appear.

2. Click the right front surface of the model.
3. Click the front edge of the model.
4. Change the **Draft Angle** value to **15.**
5. Click **OK.**
6. Undo the face draft and repeat the procedure this time clicking the back edge of the top surface.

Several surfaces can be drafted at the same time. See Figure 3-22.

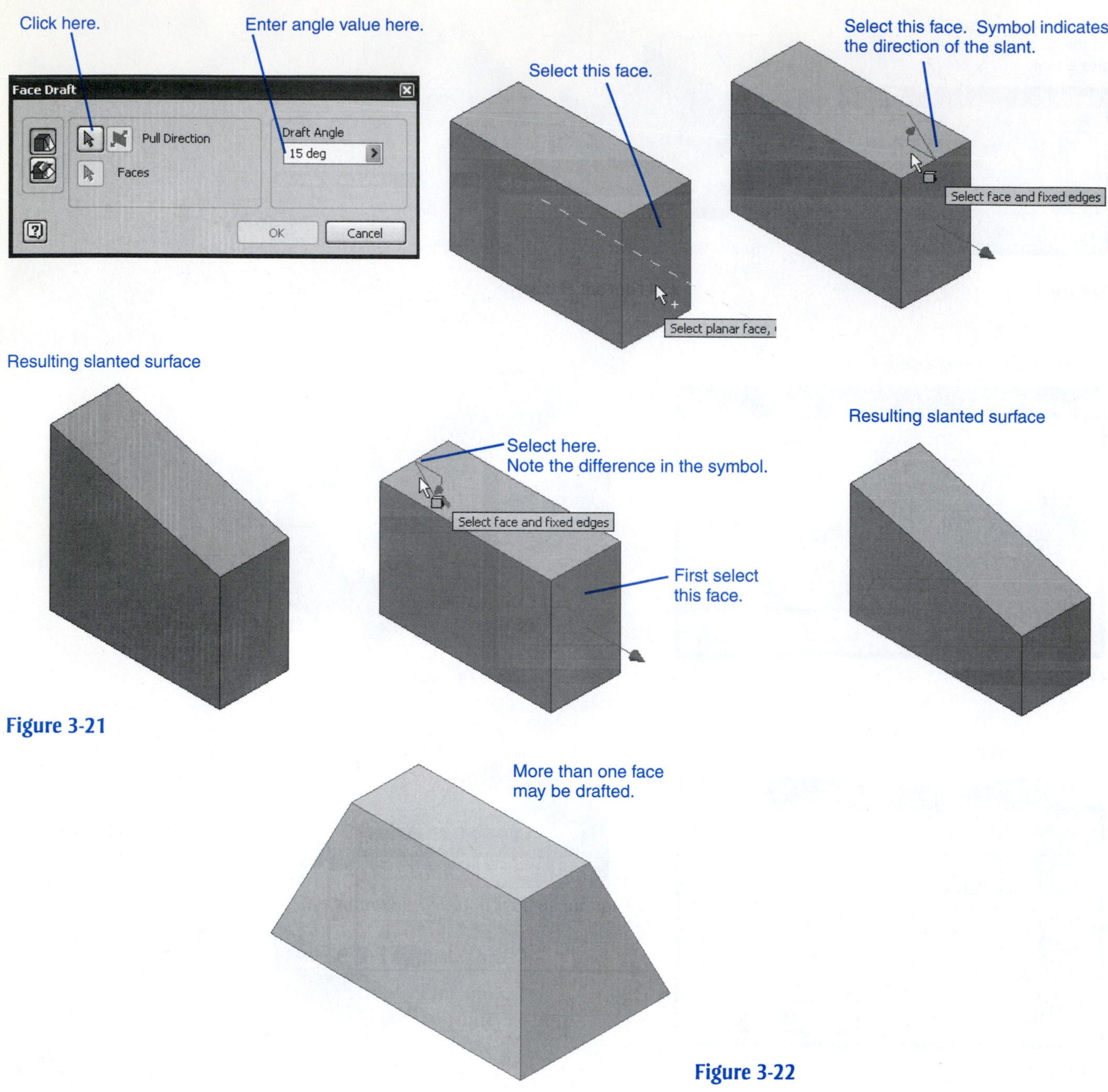

Figure 3-21

Figure 3-22

SPLIT

The **SPLIT** command is used to trim away a portion of a model. See Figure 3-23. A sketch line is used to define the location and angle of the split.

Exercise 3-10: Defining the Split Line

1. Click the left front surface of the 12 × 30 × 20 model.

The surface will change color.

2. Right-click the mouse and select the **New Sketch** option.

A grid will appear on the screen oriented to the selected face.

3. Click the **Line** tool and sketch a line across the front left surface.

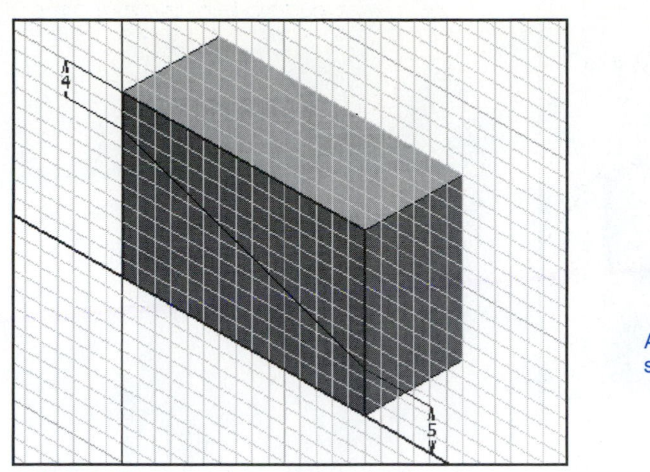

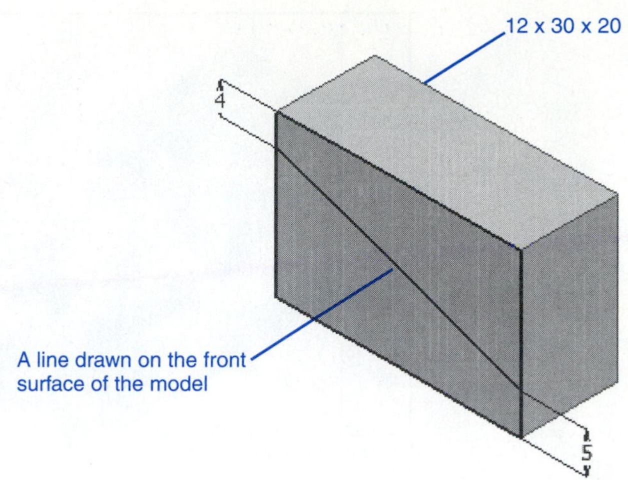

Figure 3-23

4. Right-click the mouse and select the **Done** option.
5. Use the **General Dimension** tool to locate the line as needed.

Exercise 3-11: Splitting the Model

1. Right-click the mouse and select the **Finish Sketch** option.

The **Split** dialog box will appear. See Figure 3-24.

2. Click the **Split Tool** box.
3. Click the sketch line.
4. Select the **Split Part** box under **Method** in the **Split** dialog box.
5. Use the **Remove** option to define which side of the model is to be removed.
6. Click **OK.**

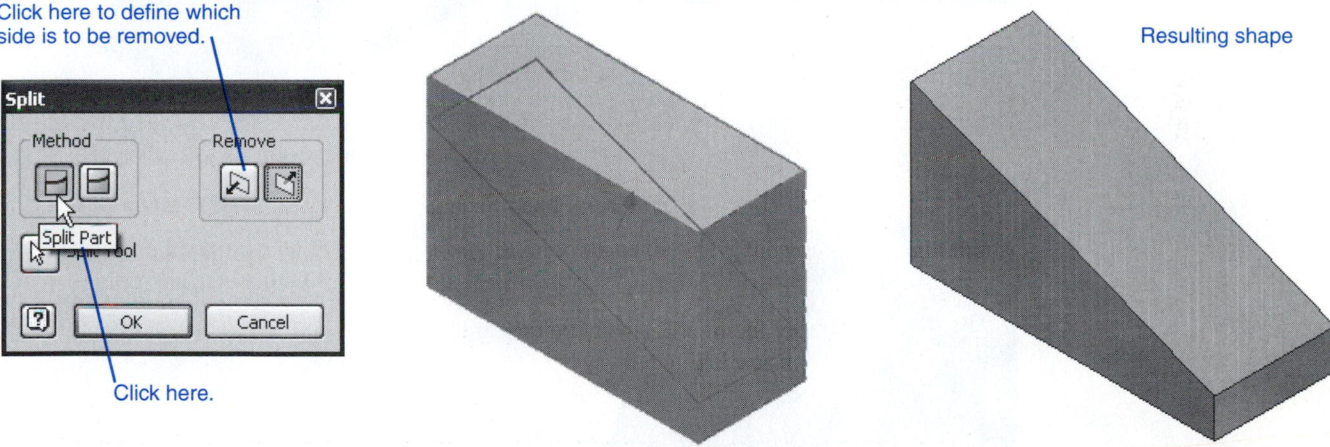

Figure 3-24

Figure 3-25 shows a split that was created using a sketched circle. The arrow that appears on the top surface indicates the direction of removal.

MIRROR

The **MIRROR** command is used to create mirror images of an existing model. See Figure 3-26.

1. Click the **Mirror** tool.

The **Mirror** dialog box will appear.

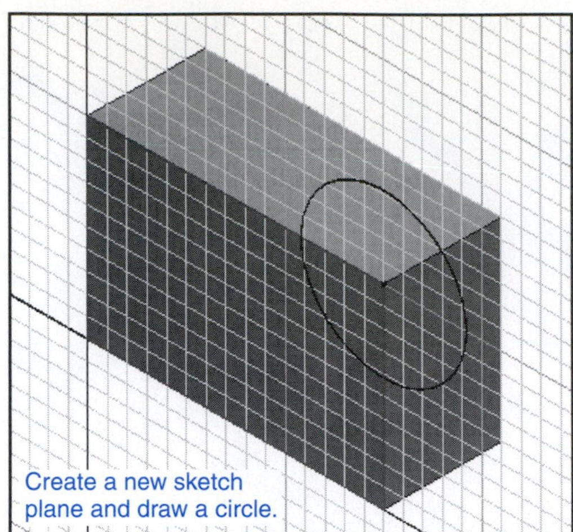
Create a new sketch
plane and draw a circle.

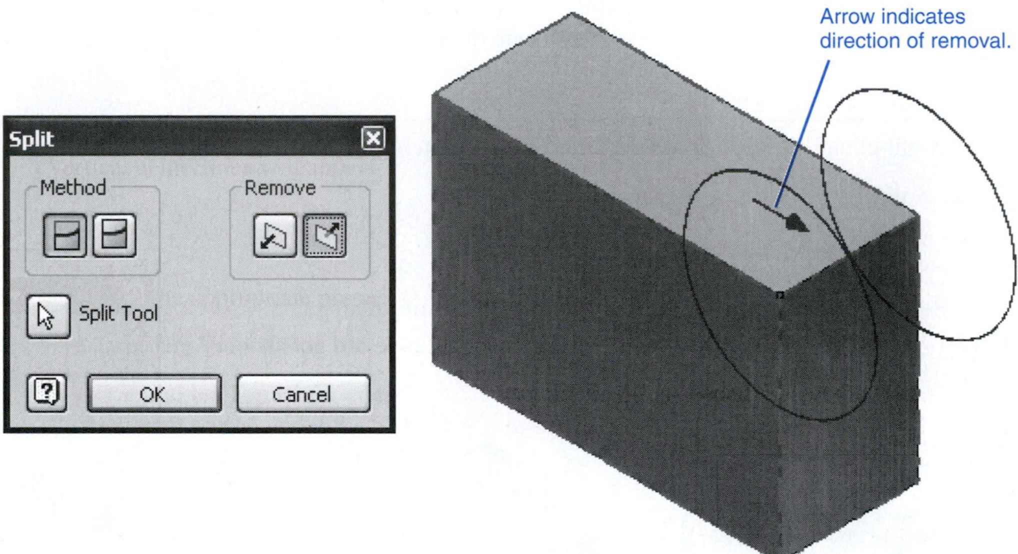
Arrow indicates
direction of removal.
Split
Method
Remove
Split Tool
OK
Cancel

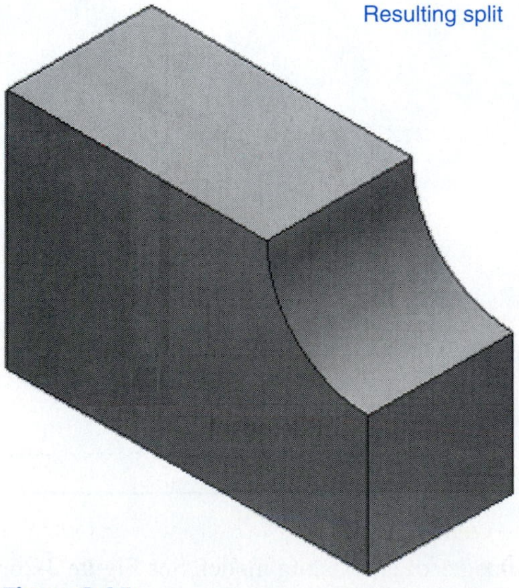
Resulting split

Figure 3-25

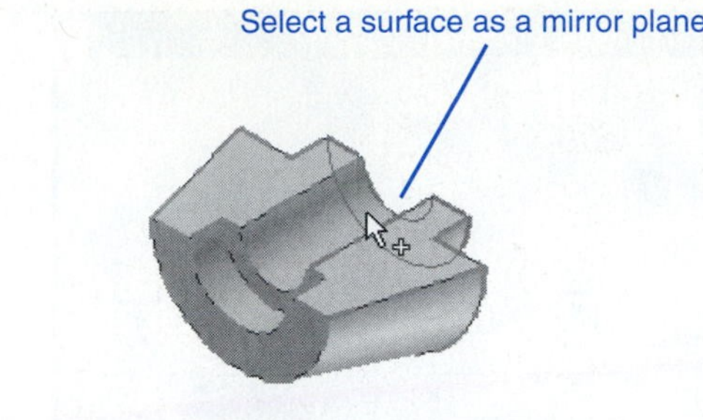

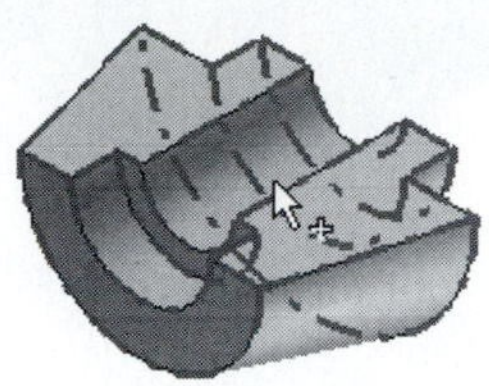

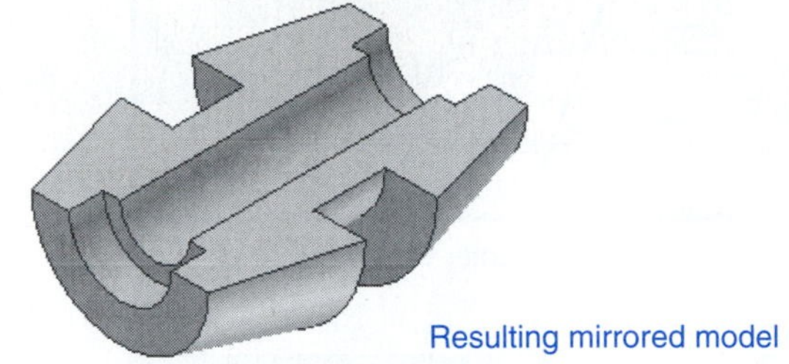

Figure 3-26

2. Click the model.
3. Click the **Mirror Plane** box.
4. Select one of the model's surfaces as a mirror plane.
5. Click **OK.**

RECTANGULAR PATTERN

The **RECTANGULAR PATTERN** command is used to create a rectangular array of an existing model feature. Figure 3-27 shows a 30 × 40 × 5 plate with a Ø5 hole located 5 mm from each edge.

1. Click the **Rectangular Pattern** tool located on the **Part Features** panel bar.

The **Rectangular Pattern** dialog box will appear. The **Features** box will automatically be active.

2. Click the hole.

The hole is the feature.

3. Click the **Direction 1** box, then click the back left edge of the model to define direction 1.

Use the **Direction 1** box to reverse the direction if necessary.

4. Set the count value for **3** and the **Spacing** value for **10.**
5. Click the **Direction 2** box, then click the front left edge of the model to define direction 2.
6. Set the count value for **4** and the **Spacing** for **10.**
7. Click **OK.**

CIRCULAR PATTERN

The **CIRCULAR PATTERN** command is used to create a polar array of an existing model feature. Figure 3-28 shows a Ø40 cylinder 5 mm high with two Ø5 holes. One hole is located in the center of the model; the second is located 15 mm from the center.

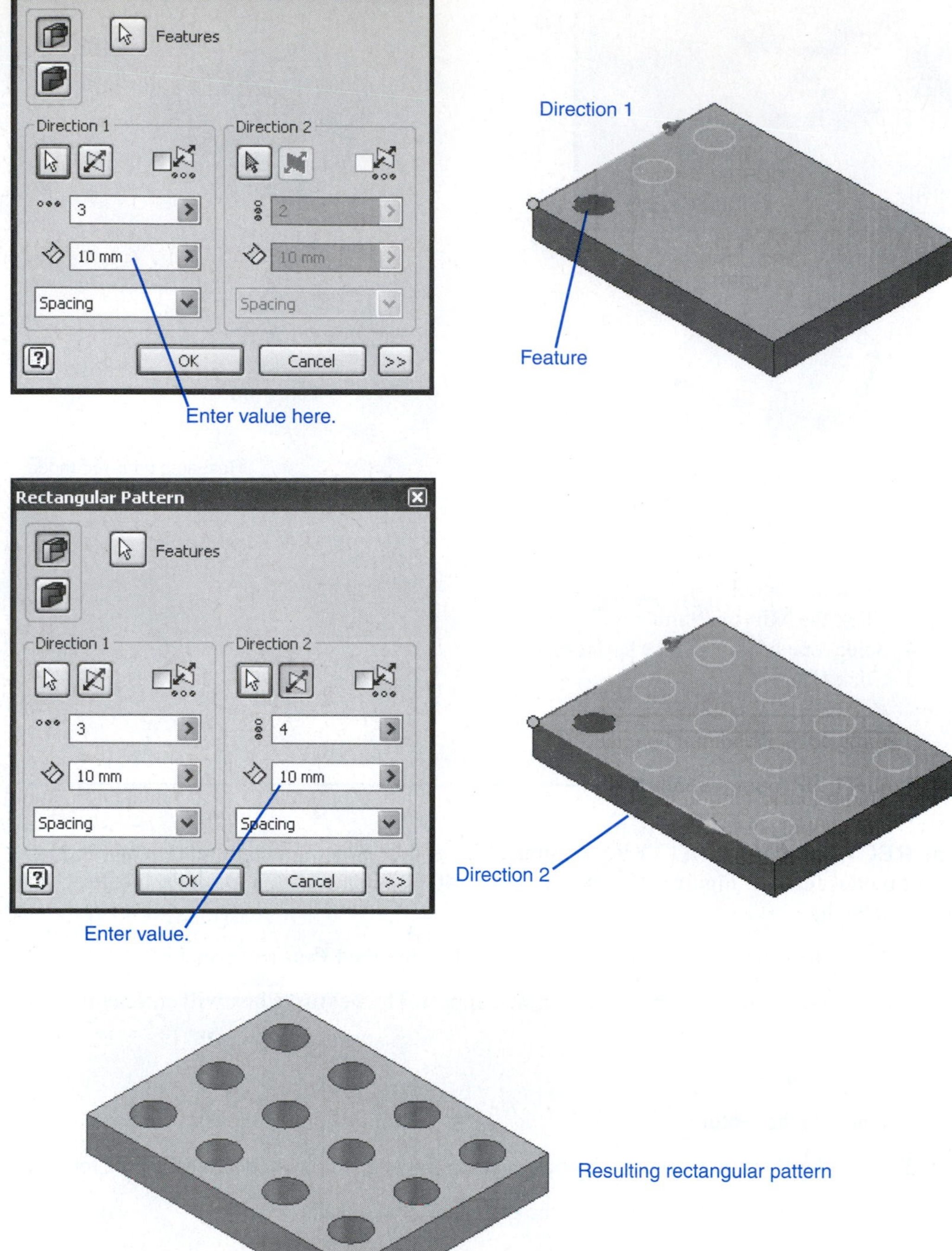

Figure 3-27

1. Click the **Circular Pattern** tool located on the **Part Features** panel bar.

The **Circular Pattern** dialog box will appear. The **Features** box will automatically be active.

2. Click the hole to be used to create the circular pattern.
3. Click the **Rotation Axis** button, then click the center hole.
4. Set the count value for **8** and the angle value for **360.**
5. Click **OK.**

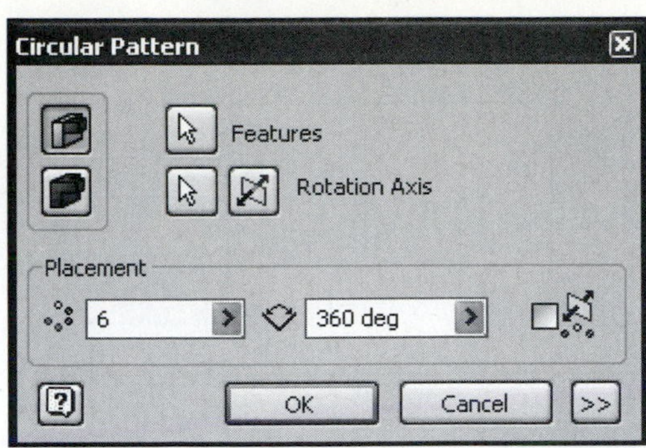

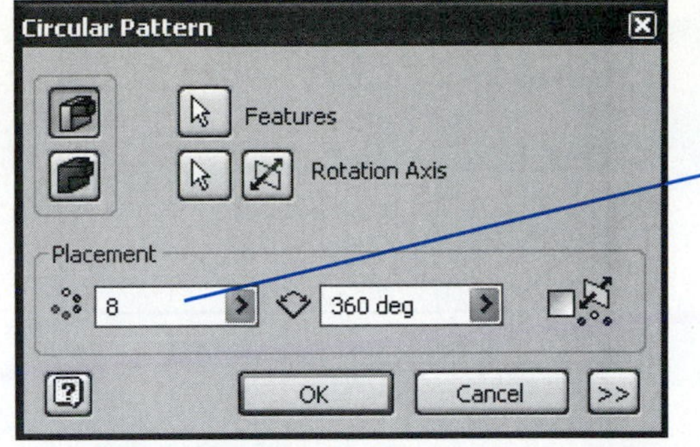

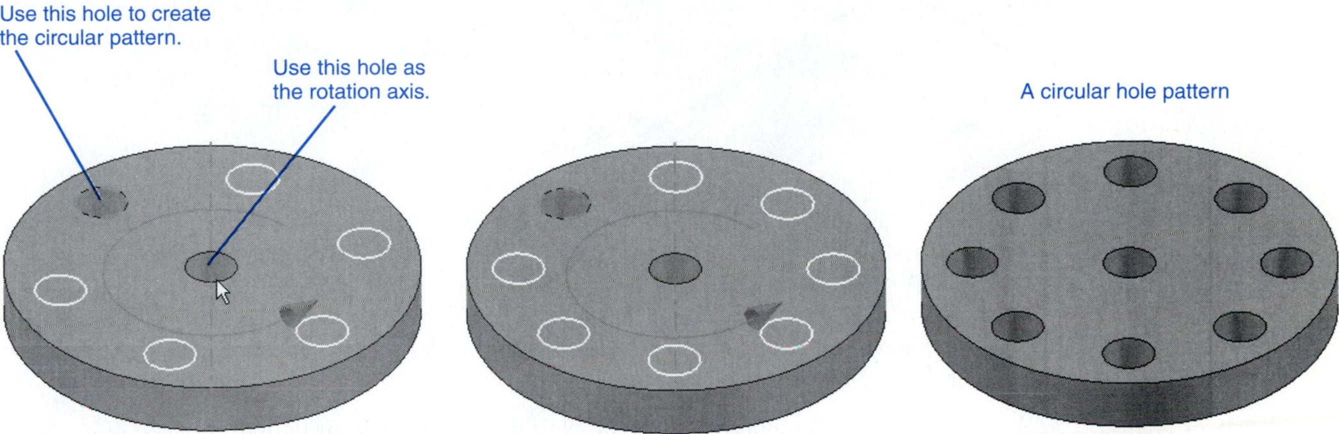

Figure 3-28

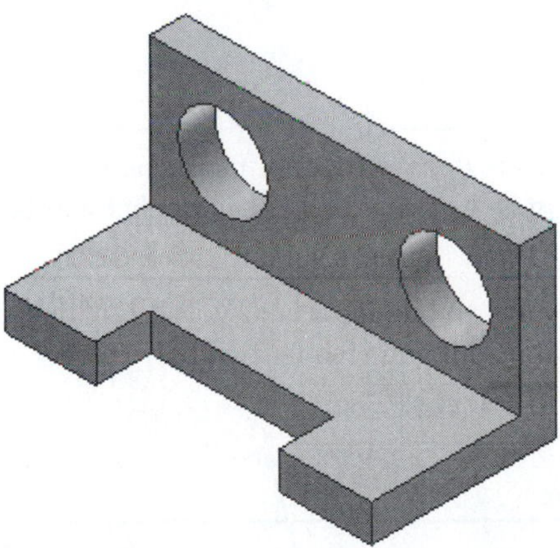

Figure 3-29

SKETCH PLANES

Sketches are created on ***sketch planes.*** Any surface on a model may become a sketch plane. As models become more complex they require the use of additional sketch planes.

sketch plane: A 2D plane drawn on any surface or work plane on a model used for sketching.

Figure 3-29 shows a model that was created using several different sketch planes. The model is a composite of basic geometric shapes added to one another.

Exercise 3-12: Creating the Base

1. Start a new drawing using the metric **Standard (mm).ipt** settings.
2. Click the **Two point rectangle** tool and sketch a **10 × 20** rectangle.

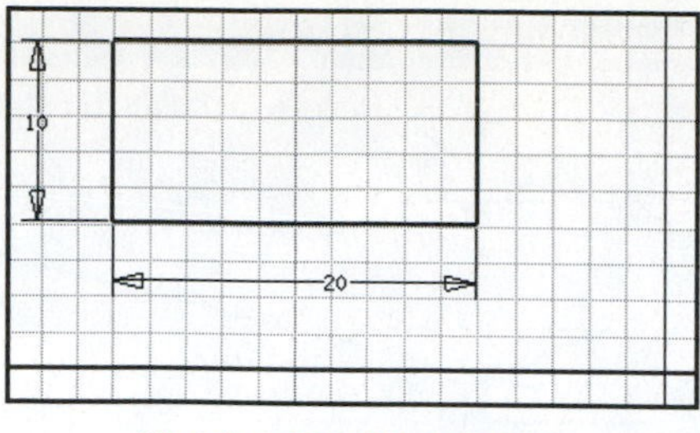

Sketch a 10 x 20 rectangle.

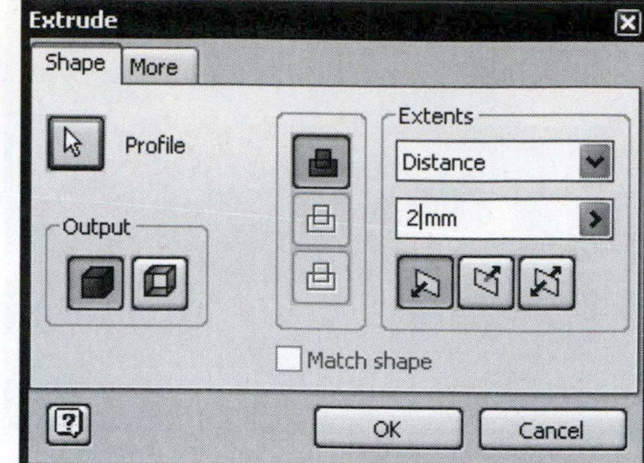

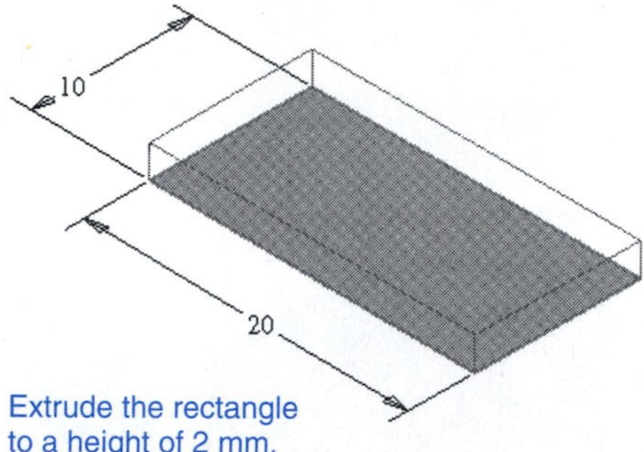

Extrude the rectangle to a height of 2 mm.

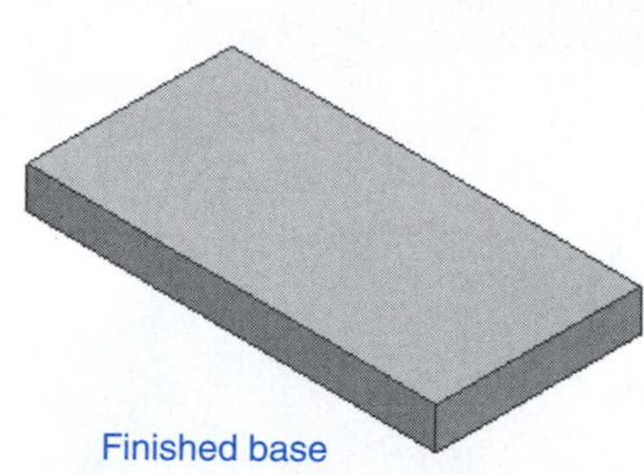

Finished base

Figure 3-30

See Figure 3-30. The object will automatically be drawn on a sketch plane aligned with the program's XY plane.

3. Click the right mouse button and select the **Isometric View** option.

Use the mouse wheel to zoom the rectangle as necessary.

4. Right-click the mouse and select the **Finish Sketch** option. Then, click the **Extrude** tool.

The **Extrude** dialog box will appear.

5. Set the extrusion height for **2 mm** and click **OK.**

Exercise 3-13: Creating the Vertical Portion

The rectangular vertical back portion of the model will be created by first defining a new sketch plane on the top surface of the base, then sketching and extruding a rectangle that will be joined to the existing base. See Figure 3-31.

1. Click the top surface of the base.

The surface will change color, indicating that it has been selected.

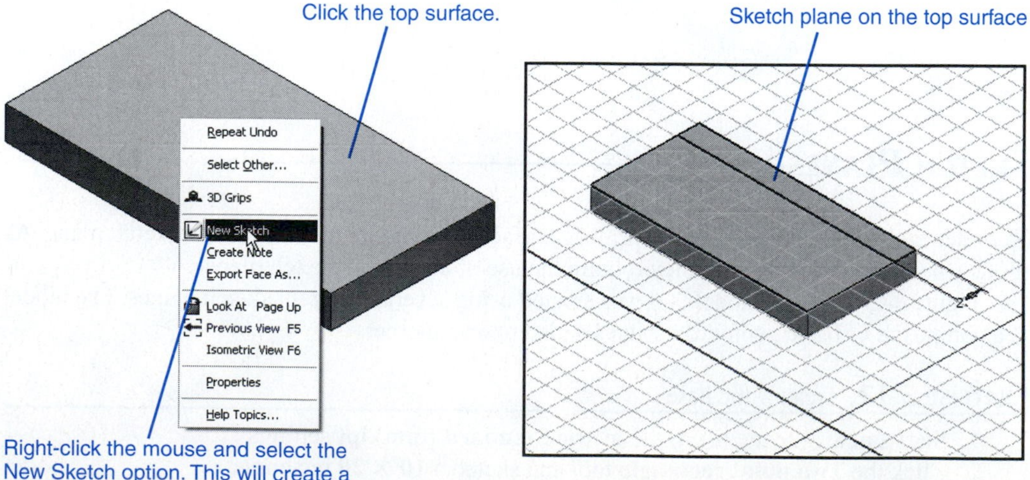

Figure 3-31

2. Right-click the mouse and select the **New Sketch** option.

A new grid pattern will appear aligned with the top surface of the base. This is a new sketch plane.

3. Click the **Two point rectangle** tool and sketch a **2 × 20** rectangle on the top surface so that it is aligned with the back edge of the base.

Note that the cursor changes from yellow to green when it is aligned with the plane's corner point.

4. Right-click the mouse and select the **Done** option and then the **Finish Sketch option.** Click the **Extrude** tool.
5. Select the 2 × 20 rectangle and set the extrusion height for **8,** then click **OK.**

Note that the surfaces are unioned together to form one object. See Figure 3-32.

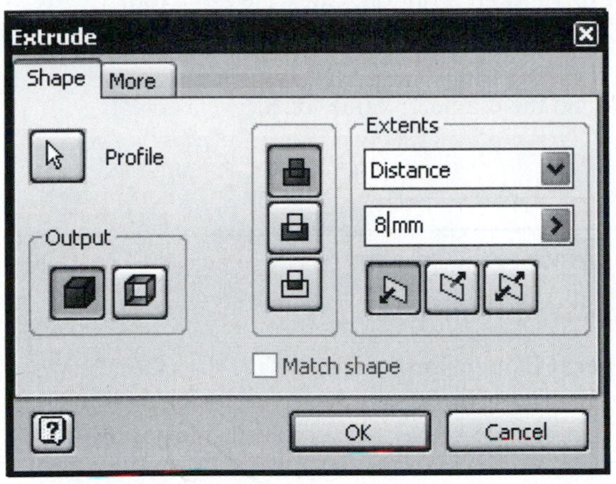

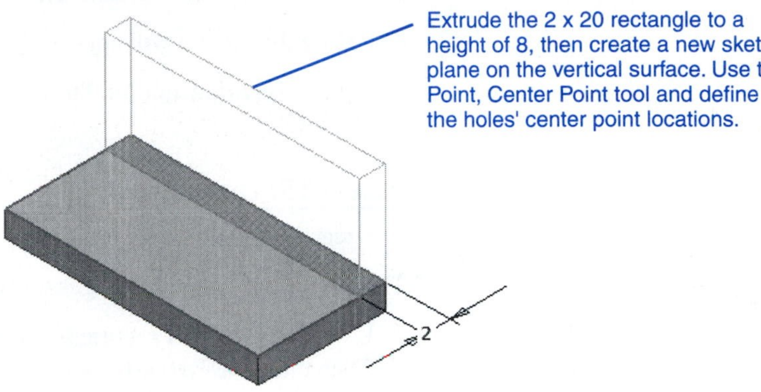

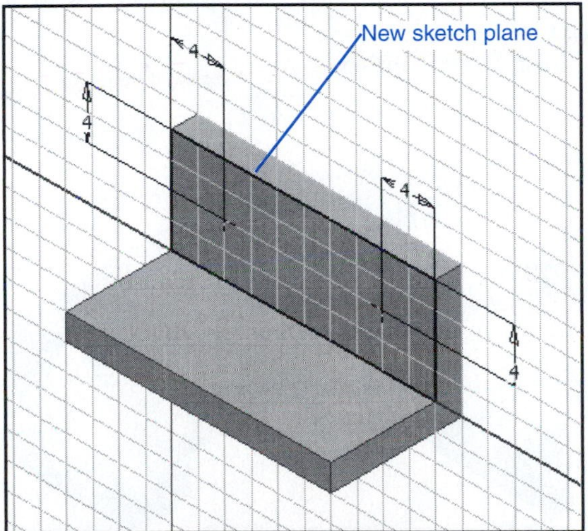

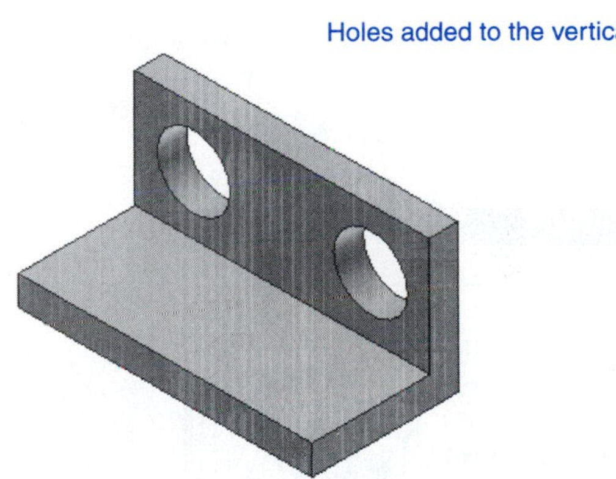

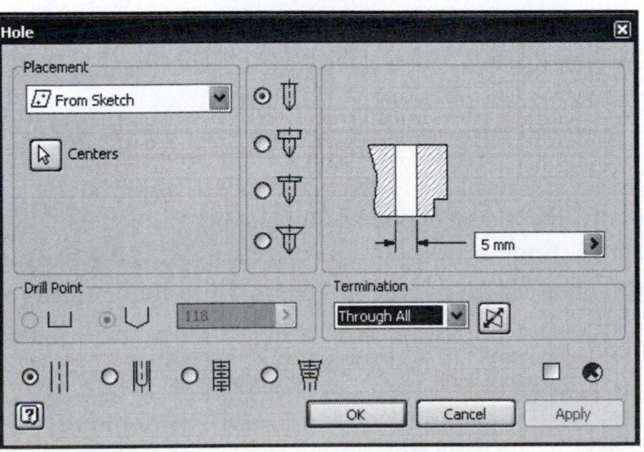

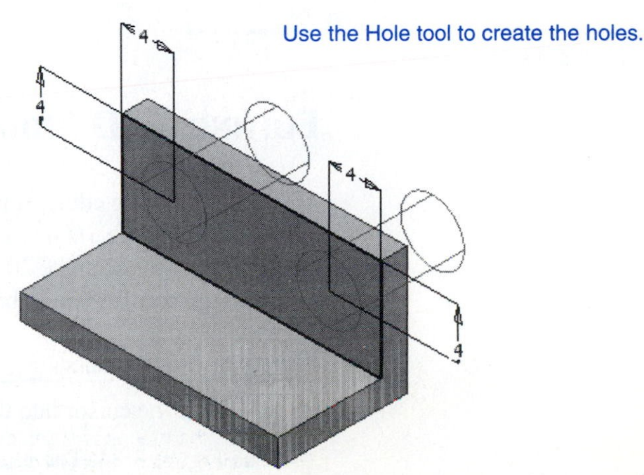

Figure 3-32

Exercise 3-14: Adding Holes to the Vertical Surface

1. Click the front edge of the vertical surface.

The surface will change color, indicating that it has been selected.

2. Press the right mouse button and select the **New Sketch** option.

A grid will appear on the surface. This is a new sketch plane. The holes are located 4 mm from the top edge and from each of the side edges. See Figure 3-32.

3. Use the **Point, Center Point** and the **General Dimension** tools to locate the center points for the two holes.
4. Right-click the mouse and select the **Done** option.
5. Right-click the mouse and select the **Finish Sketch** option. Click the **Hole** tool.

The **Hole** dialog box will appear.

6. Set the **Termination** for **Through All** and the diameter value for **5.**
7. Click **OK.**

Exercise 3-15: Creating the Cutout

1. Create a new sketch plane on the top surface of the base.

The cutout is 3 deep with edges 5 from each end of the model.

2. Use the **Two point rectangle** and **General Dimension** tools to define the cutout's size.
3. Then click the **Extrude** tool.

The **Extrude** dialog box will appear. See Figure 3-33.

4. Select the cutout rectangle and set the extrusion distance for **2,** the direction arrow for a direction into the model, and select the **Cut** option.
5. Right-click the mouse and select the **Finish Sketch option.**

Use the Cut option to remove the extruded rectangle.

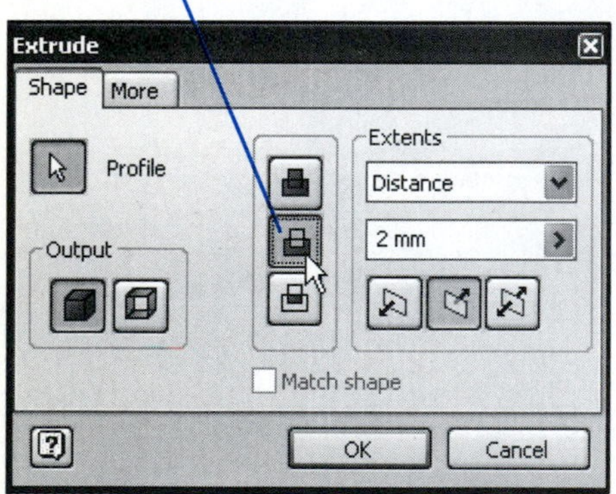

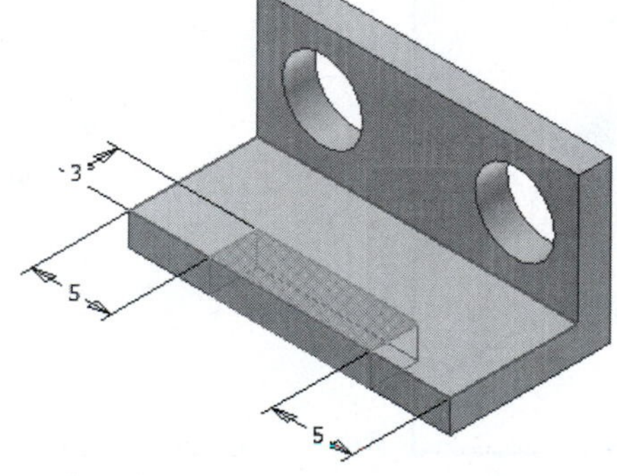

Figure 3-33

Editing a 3D Model

3D models may be edited; that is, dimensions and features may be changed at any time. For example, suppose the 3D model drawn in the last section and shown in Figures 3-29 and 3-33 requires some changes. The 20-mm length is to be changed to 25, the holes are to be changed from Ø5 to Ø3, and fillets are to be added on the front corners.

Exercise 3-16: Changing the Model's Length

1. Move the cursor into the browser box and click the **plus sign** (+) to the left of **Extrusion 1.**

See Figure 3-34. The plus sign will change to a minus sign (−), and **Sketch 1** will appear.

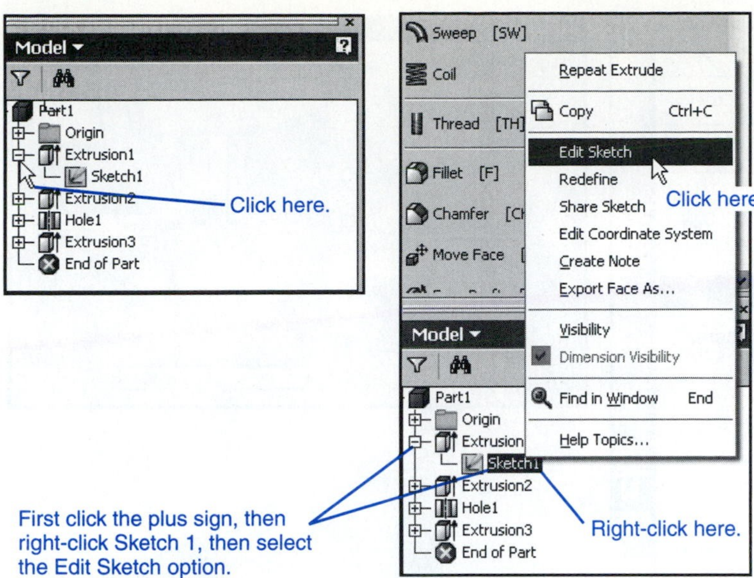

Figure 3-34

2. Right-click **Sketch 1,** then select the **Edit Sketch** option.

See Figure 3-35.

3. Right-click the **20 mm** dimension and select the **Delete** option.
4. Use the **General Dimension** tool and redimension the length, changing it to **25 mm.**
5. Right-click the mouse and select the **Done** option, then right-click again and select the **Finish Sketch** option.

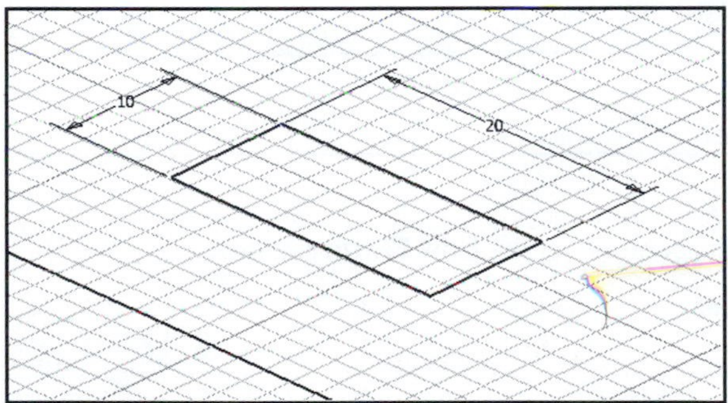

The Edit Sketch option will change the screen. The original 10 x 20 2D rectangle will appear.

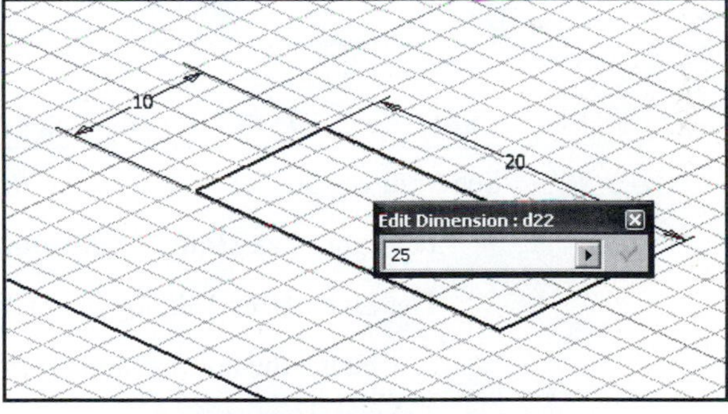

Delete the 2D dimension, then use the General Dimension tool to change the dimension value to 25.

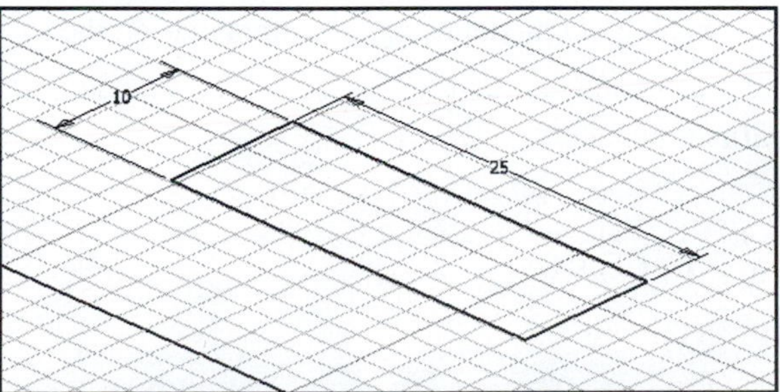

The revised 2D sketch. Right-click the mouse and select the Done option. Then right-click and select the Finish Sketch option.

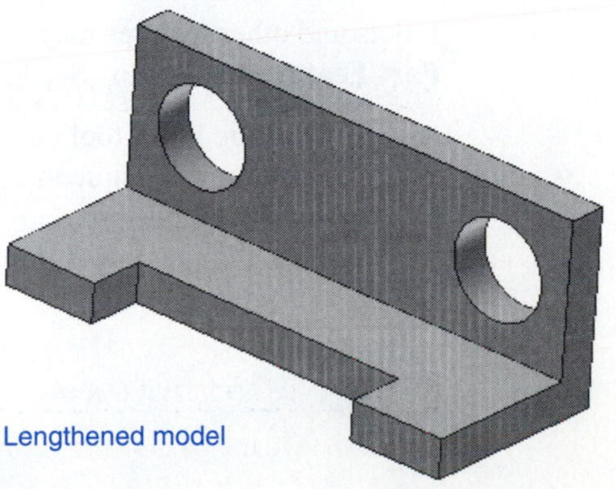

Lengthened model

Figure 3-35

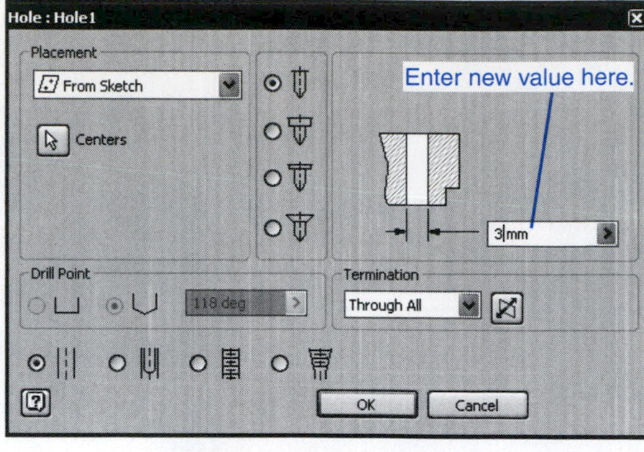

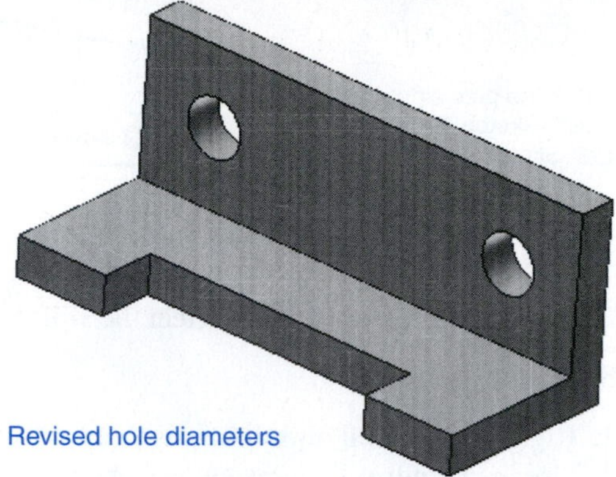

Figure 3-36

TIP Holes are *features*. The rectangle face used to create the object is a *sketch*.

Exercise 3-17: Changing the Hole's Diameters

See Figure 3-36.

1. Right-click **Hole 1** in the browser box and select the **Edit Feature** option.

   The **Hole: Hole 1** dialog box will appear.

2. Change the hole's diameter to **3 mm.**
3. Click **OK.**

Exercise 3-18: Adding a Fillet

Fillets and other features may be added to an existing 3D model using the commands on the **Part Features** panel box. See Figure 3-37.

1. Click the **Fillet** tool on the **Part Features** panel.
2. Let the radius value to **2 mm.**
3. Click the **Edge** box.
4. Click the four edges shown in Figure 3-37. Click OK.

TIP The default planes are listed in the browser box under the Origin heading.

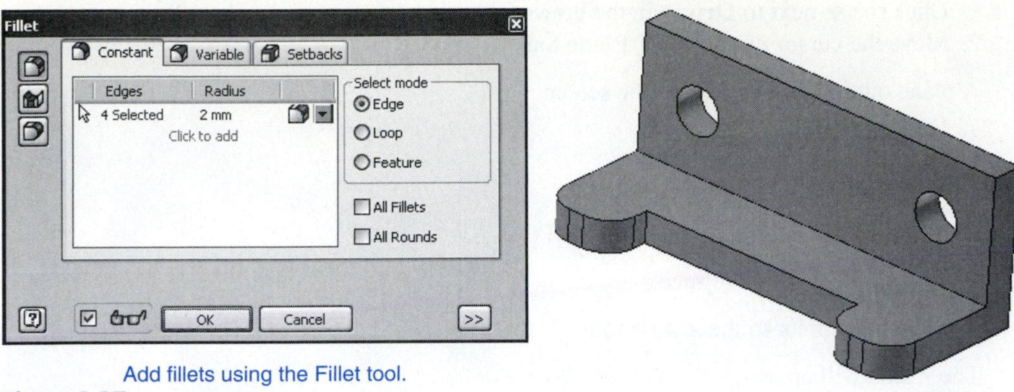

Add fillets using the Fillet tool.
Figure 3-37

DEFAULT PLANES AND AXES

Inventor includes three ***default planes*** and three ***default axes***. The three default planes are YZ, XZ, and XY, and the three axes are X, Y, and Z. The default planes and axes tools are accessed through the browser. See Figure 3-38.

default axes: In Inventor, the X, Y, and Z axes.

default planes: In Inventor, the YZ, XZ, and XY planes.

1. Click the + to the left of the **Origin** heading.

The default plane and axis headings will cascade down.

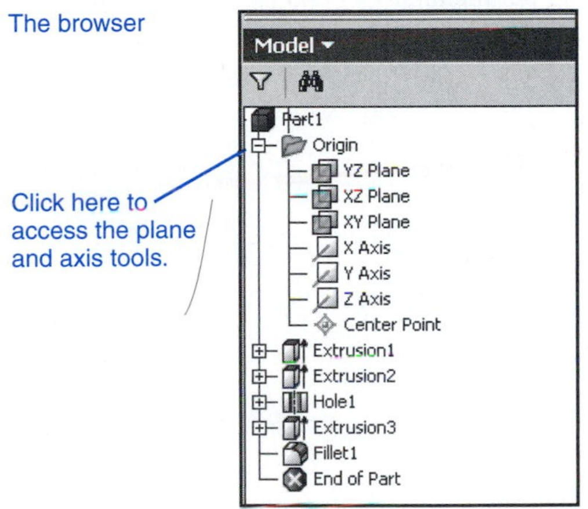

Figure 3-38

Exercise 3-19: Displaying the Default Planes and Axes

Figure 3-39 shows a Ø30 × 16 cylinder that was drawn with its center point on the 0,0,0 origin. The base of the cylinder is on the XY plane. Inventor sketches are automatically created on the default XY axis.

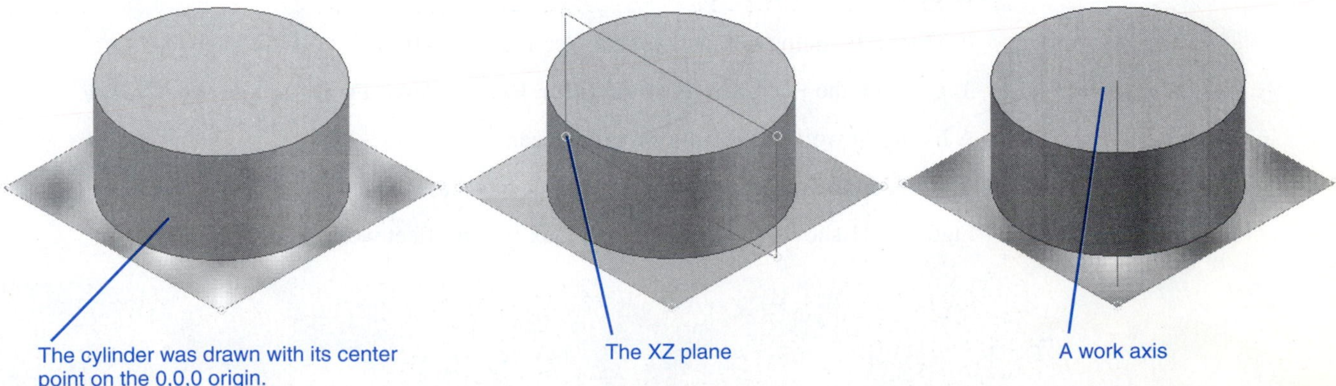

Figure 3-39

1. Click the + next to **Origin** in the browser box.
2. Move the cursor onto the **XY Plane** tool.

A plane outline will appear on the screen.

3. Click the **XY Plane** tool.

The plane will be filled with color.

4. Move the cursor to the **XZ Plane** tool.

The XZ plane will be outlined.

5. Move the cursor to the **Z Axis** tool.

The Z axis will appear.

6. Move the cursor through all the tools and note the planes and axes that appear.

WORK PLANES

work planes: A plane used for sketching that is created independent of the model.

Work planes are planes used for sketching, but unlike sketch planes, work planes are not created using the surfaces of models. Work planes are created independent of the model. Work planes may be created outside or within the body of a model. Work planes are used when no sketch plane is available.

Work planes may be defined using the following parameters:

Angle to a plane
Edge and face normal
Point and face normal
Point and face parallel
Sketch geometry
Tangent and face parallel
3-point
2-edge or 2-axis
Through point perpendicular
Tangent to face through

Work Plane Help

If you are not sure how to create a work plane, Inventor includes an animated help feature.

1. Click the **Work Plane** tool, then move the cursor to the browser box and right-click the **YZ plane** tool.

A dialog box will appear. See Figure 3-40.

2. Click the **How to . . .** option.

A **Show Me** dialog box will appear. See Figure 3-40.

3. Click on the + on **Autodesk Inventor, Parts – Work Features,** and **Work plane.**

A listing of work plane options will appear.

4. Click on the specific type of work plane to access the animated help.

Figure 3-41 shows the animated help box for an offset work plane.

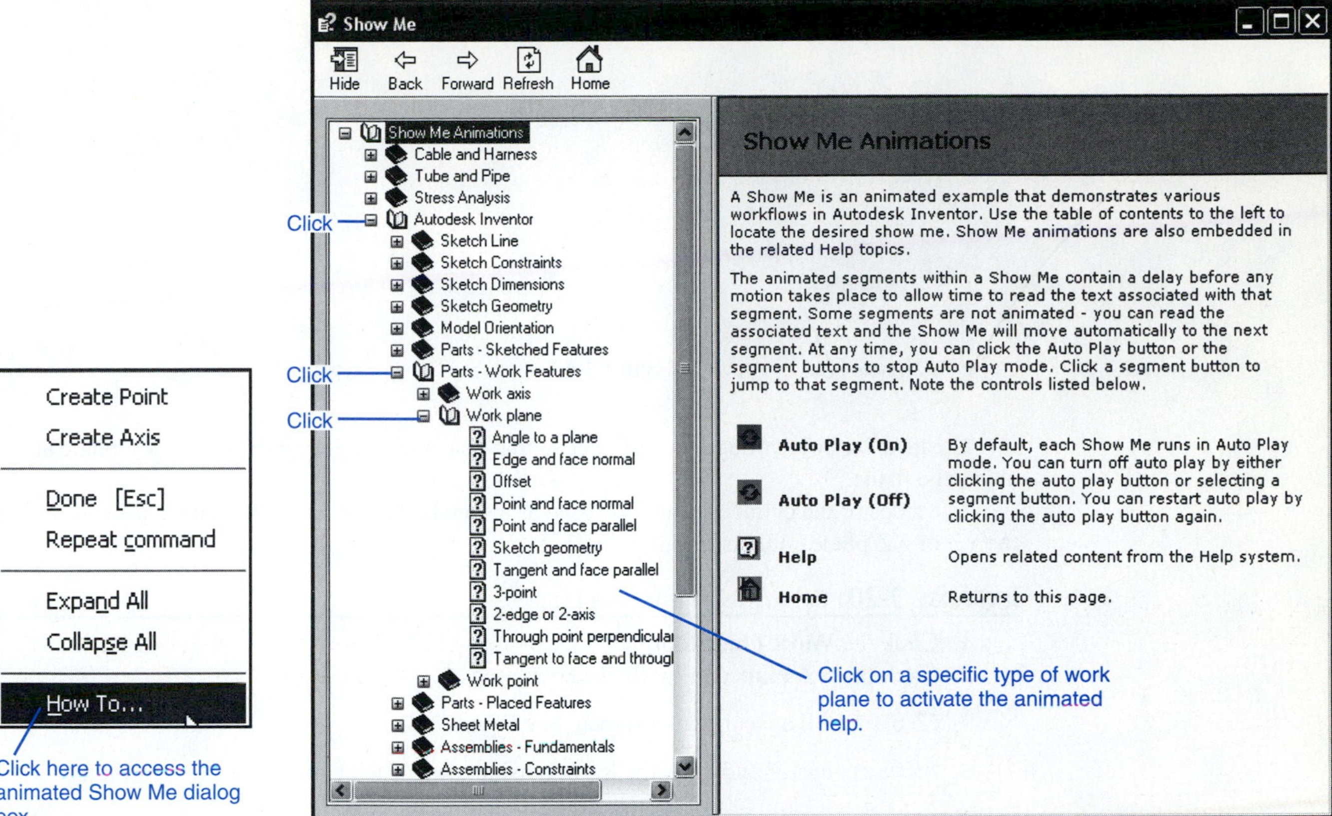

Figure 3-40

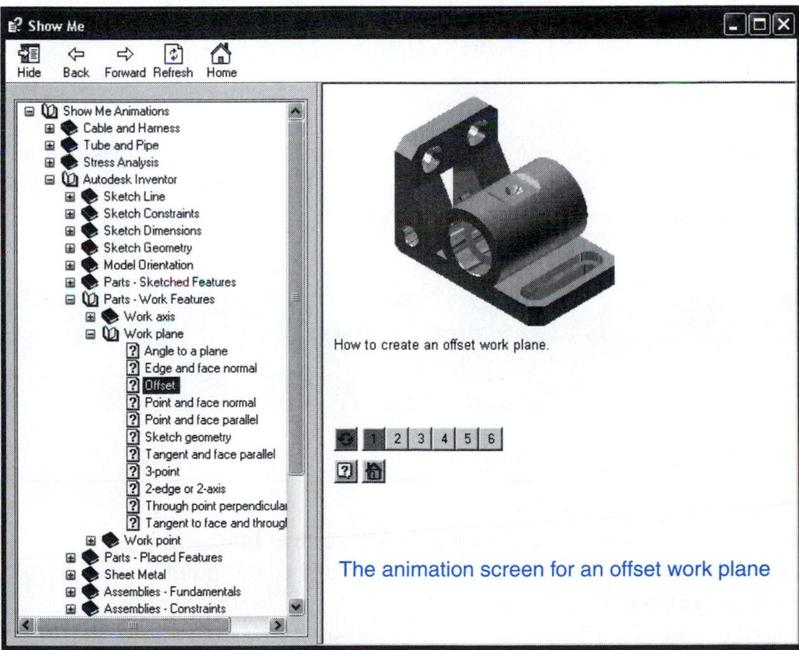

Figure 3-41

Sample Problem SP3-1

Figure 3-42 shows a Ø20 × 10 cylinder that was sketched but not aligned with the system's origin. The sketch was created on the default XY plane.

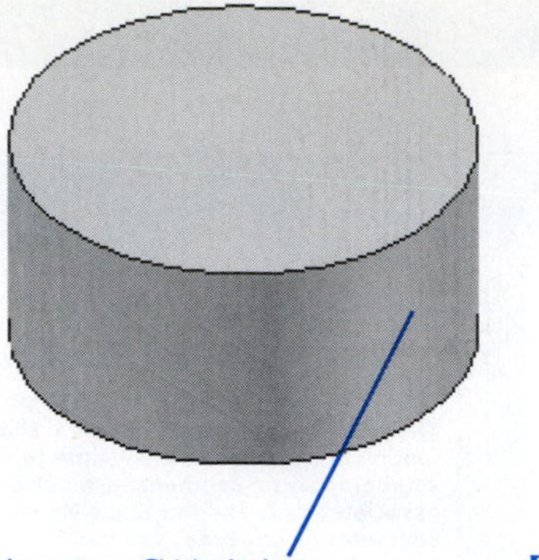

Figure 3-42

Create a Ø4 hole through the cylinder so that its centerline is parallel to the XY plane and 5 above the plane.

The sides of the cylinder cannot be used as a sketch plane, so a work plane is needed. Either the YZ or XZ plane could be used.

Exercise 3-20: Creating a Tangent Work Plane

1. Click the **Work Plane** tool.
2. Click the **YZ Plane** tool in the browser area.

A YZ plane will appear on the screen. See Figure 3-43.

3. Move the cursor and click the lower outside edge of the cylinder.

A work plane will be created tangent to the cylinder.

4. Right-click one of the corners of the work plane (yellow circles will appear) and select the **New Sketch** option.
5. Move the cursor to one of the work plane's corner points.

A small circle will appear.

6. Click the circle.

A grid will appear. The grid will include intersecting horizontal and vertical lines that are darker than the other grid lines. The horizontal line is aligned with the XY plane, and the vertical line is parallel to the YZ plane tangent to the edge of the cylinder.

TIP

Use its **look at** tool to create a 2D view of the sketch plane.

Exercise 3-21: Creating the Hole Through the Cylinder

1. Click the **Circle** tool.
2. Sketch a hole with its center point located on the darker vertical line.
3. Use the **General Dimension** tool to create a **Ø4** circle with a center point located **5** from the top surface of the cylinder.
4. Click the arrow to the right of the 2D sketch heading and select the **Part Features** option. Click the **Extrude** tool.

The **Extrude** dialog box will appear.

5. Set the extrusion distance for **22** in a direction that passes through the cylinder, and select the **Cut** option.
6. Click **OK.**

The **Point, Center point** and **Hole** tools could be use to create the hole.

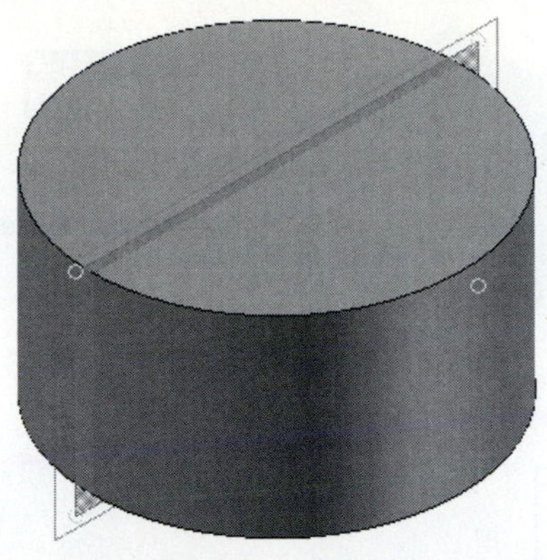

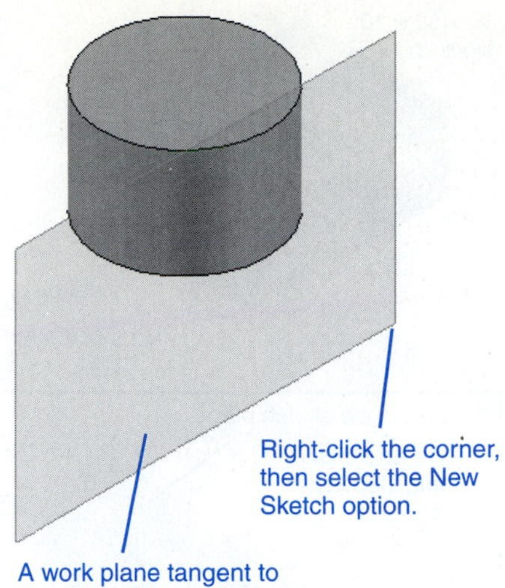

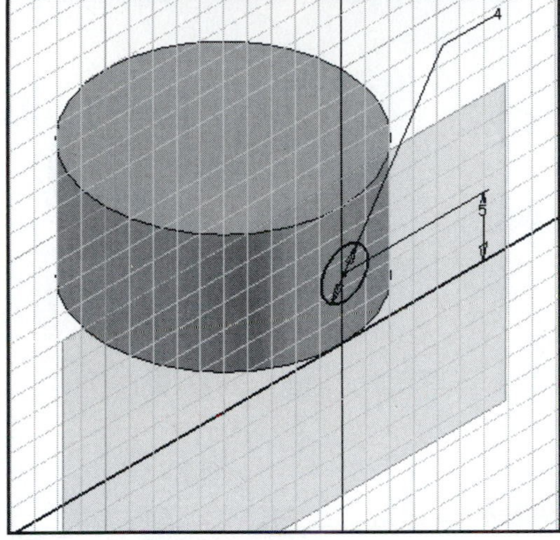

Create a sketch plane on the tangent work plane and sketch a hole.

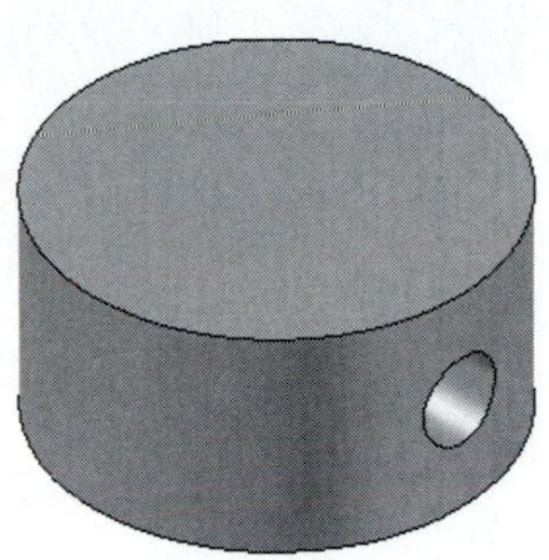

The finished model

Figure 3-43

Hiding Work Planes

Work planes may be hidden by right-clicking one of the corners of the work plane and selecting the **Visibility** option.

Restoring a Work Plane

To restore a hidden work plane, right-click the work plane's reference in the browser box and select the **Visibility** option.

ANGLED WORK PLANES

Work planes may be created at an angle to a model. For example, suppose a hole must be drilled through the 30 × 50 × 10 block shown in Figure 3-44 at a 45° angle. Only extrusions perpendicular to a plane can be created, so a 45° plane is needed.

Exercise 3-22: Creating an Angled Work Plane

1. Select the **Work Axis** tool and create a work axis on the edge of the block as shown.

See Figure 3-44. (To create a work axis, click the **Work Axis** tool, then click the edge location for the axis.)

2. Select the **Work Plane** tool, then click the work axis, then the front right vertical surface of the block.
3. Enter an angle value, then click the check mark.

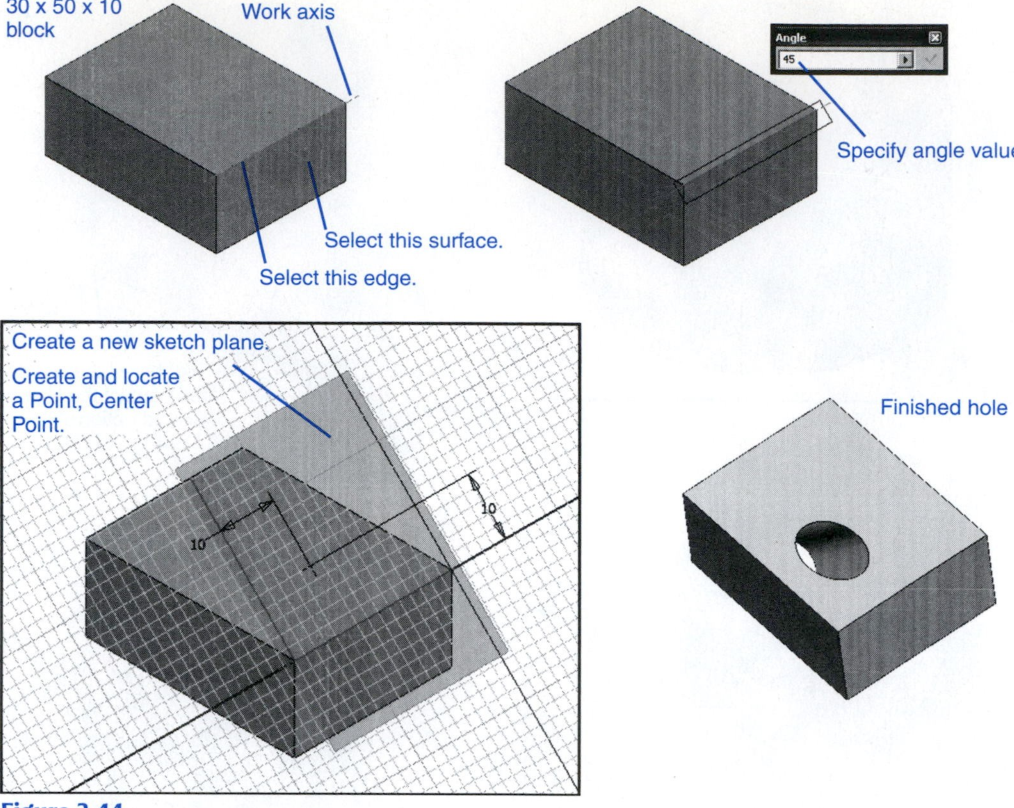

Figure 3-44

In this example, a value of **45°** was entered.

4. Right-click one of the work plane's corner points and select the **New Sketch Plane** option.
5. Use the **Point, Center Point** and **General Dimension** tools to create and locate a hole's center point.
6. Right-click the mouse and select the **Done** option, then right-click again and select the **Finish Sketch** option.
7. Select the **Hole** tool and specify the hole's diameter and length.

In this example a diameter of **10 mm** and a depth of **Through All** were used.

8. Click **OK.**
9. Hide the work plane.

Offset Work Planes

Figure 3-45 shows an object in which a small cylinder passes through a larger cylinder. An offset work plane was used to create the object.

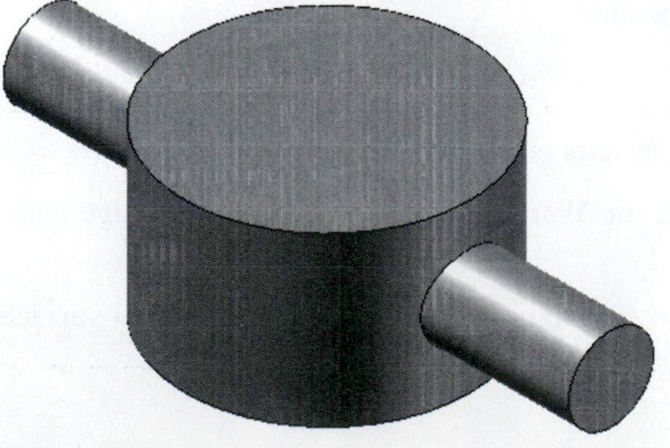

Figure 3-45

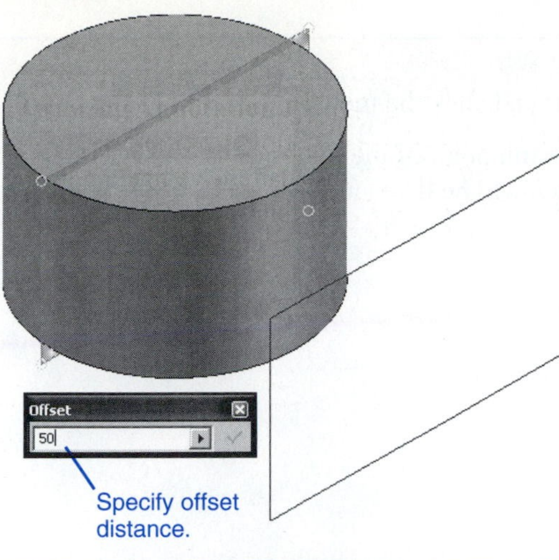

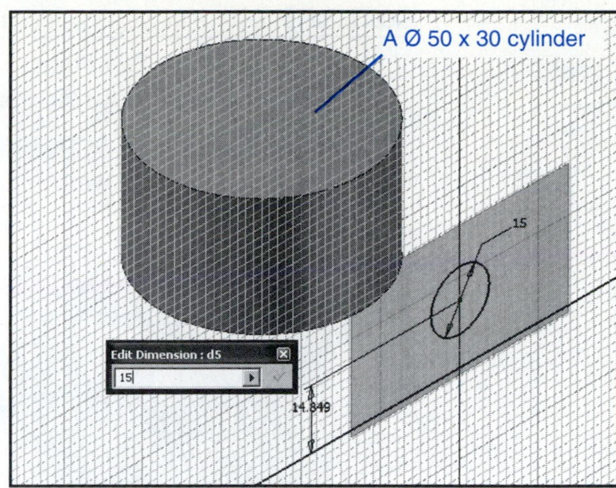

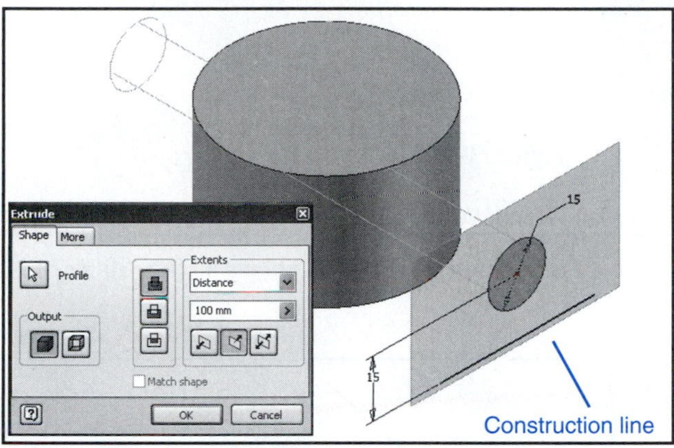

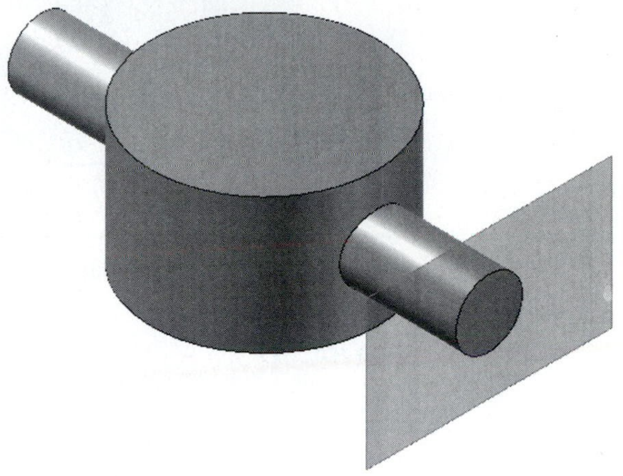

Figure 3-46

Exercise 3-23: Creating an Offset Work Plane

See Figure 3-46.

1. Draw a **Ø50 × 30** cylinder.
2. Select the **Work Plane** tool and **YZ plane** and create a work plane through the center of the cylinder.
3. Click on one of the work plane's corner points, and holding the left mouse button down, drag the work plane away from the cylinder.
4. Specify the offset distance.

In this example a value of **50 mm** was used.

5. Right-click one of the work plane's corner points and select the **New Sketch Plane** option.
6. Draw a line along the thick centerline on the grid.

The thick grid line is aligned to the bottom surface of the large cylinder.

7. Use the **Point**, **Center Point** tool and the **General Dimension** tool to create a **Ø15** hole as shown.
8. Right-click the mouse and select the **Done** option, then right-click again and select the **Finish Sketch** option.
9. Select the **Extrude** tool, then select the circle as the profile and extrude it **100 mm** through the large cylinder.
10. Click **OK.**

Work Points

Work points are defined points on a model. They are used to help locate work planes and work axes.

work point: A defined point on a model used to help locate work planes and work axes.

Exercise 3-24: Defining a Work Point

1. Click the **Work Point** tool.
2. Select the location for the work points and click the mouse.

In the example shown in Figure 3-47 the midpoint of the left edge was selected along with the lower front corner. The work points created will be listed in the browser box.

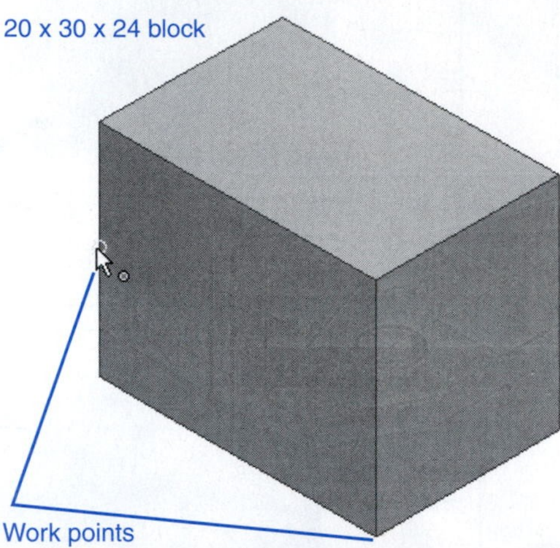

Figure 3-47

Exercise 3-25: Creating an Oblique Work Plane Using Work Points

An oblique work plane may be created using work points. Figure 3-48 shows a 20 × 30 × 24 rectangular block.

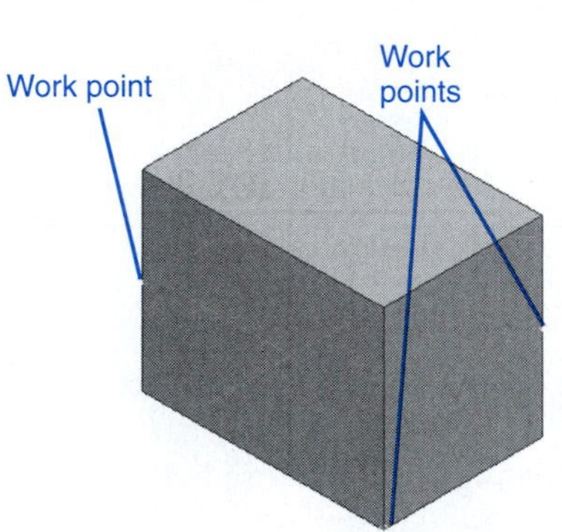

A work plane created using three work points.

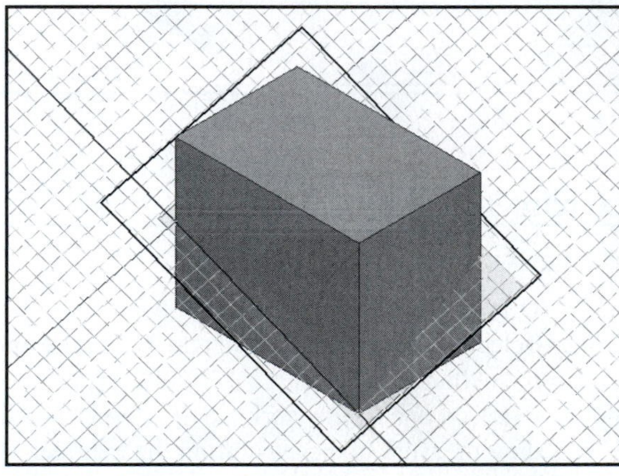

A sketch plane created on the work plane, and a rectangle sketched on the sketch plane

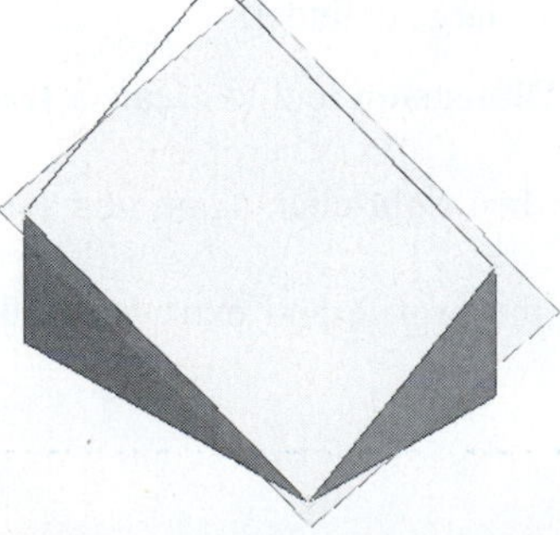

Extrude the rectangle, then use the cut option to trim the model.

An oblique surface

Figure 3-48

1. Create three work points on the prism, two on the midpoints of the vertical edges and one at the lower corner as shown in Figure 3-48.
2. Click on the **Work Plane** tool, then click the three work points.
3. Left-click one of the work plane's corner points, right-click the mouse, and select the **New Sketch Plane** option.
4. Click the **Two point rectangle** tool and draw a very large rectangle on the new sketch plane.

The rectangle may be any size that exceeds the size of the prism.

5. Access the **Extrude** tool and cut out the top portion of the prism.

WORK AXES

A ***work axis*** is a defined line. Work axes are used to help define work planes and to help define the geometric relationship between assembled models.

work axis: A defined line on a model.

Exercise 3-26: Creating a Work Axis

1. Click the **Work Axis** tool.
2. Click the edge line that is to be defined as a work axis.

Figure 3-49 shows a model with two defined work axes.

Exercise 3-27: Drawing a Work Axis at the Center of a Cylinder

Figure 3-50 shows a cylinder.

1. Click the **Work Axis** tool.
2. Click the lower edge of the cylinder.

The words **Work Axis** 1 will appear in the browser.

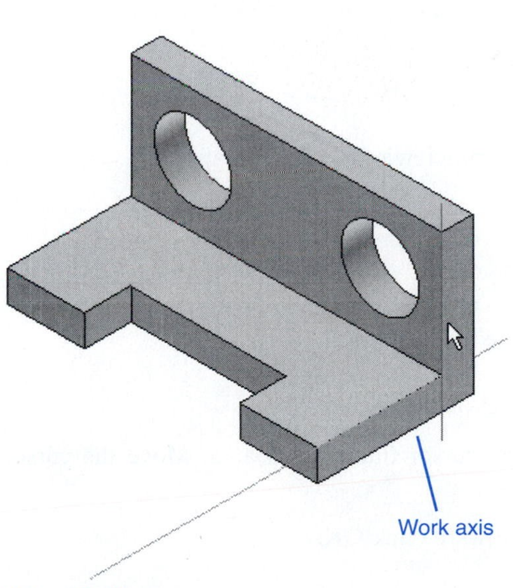

Figure 3-49

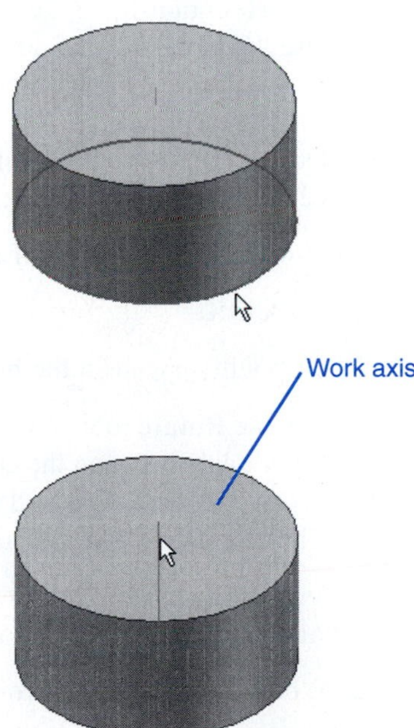

Figure 3-50

RIBS (WEBS)

A ***rib*** is used to add strength to a model. Ribs or webs are typically used with cast or molded parts. Figure 3-51 shows an L-bracket. Ribs 5 mm thick are to be added to each end of the bracket.

rib: An element added to a model to give it strength.

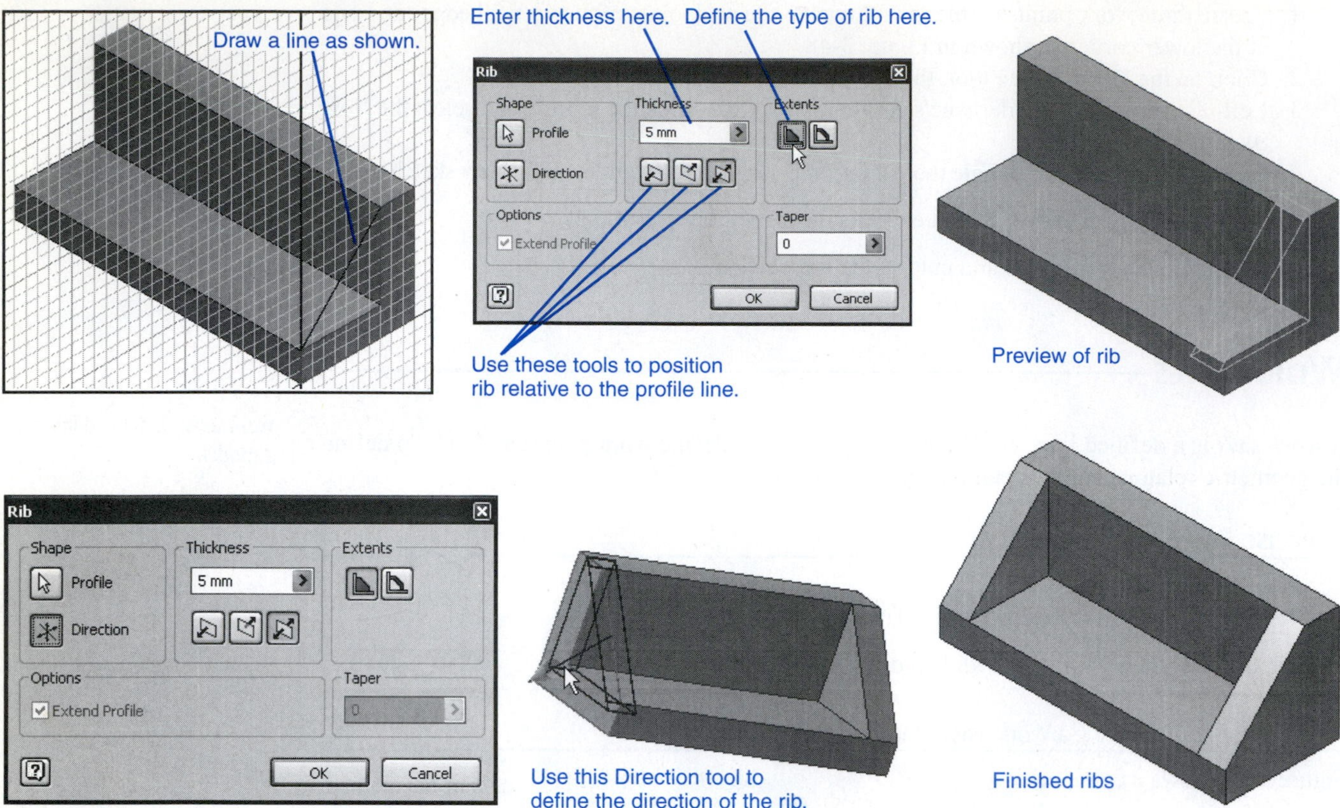

Figure 3-51

1. Click the right end surface of the bracket, click the right mouse button, and select the **New Sketch** option.
2. Use the **Line** tool and add a line as shown if Figure 3-51.
3. Right-click the mouse and select the **Finish Sketch** caption.
4. Click the **Rib** tool.
5. Define the line as the **Profile.**
6. Click the left **Extents** box.

This will define the type of rib to be drawn. A preview of the rib will appear.

7. Click **OK.**

The rib will appear on the bracket.

8. Use the **Rotate** tool and rotate the bracket so that the other end of the bracket is visible.
9. Draw a line between the corners as was done for the first rib.
10. Specify the thickness and the **Extents.**
11. Click the **Direction** tool and define the direction for the rib.

After the **Direction** box is clicked, move the cursor into the rib area. Move the cursor around and note how the directional arrow changes.

12. When the directional arrow is pointing downward click **OK.**

LOFT

The **LOFT** command is used to create a solid between two or more sketches. Figure 3-52 shows a loft surface created between a circle and a rectangle. Both the circle and the rectangle are first drawn on the same XY plane. This allows the **General Dimension** tool to be used to ensure the alignment between the two sketches. The rectangle is then projected onto another work plane. The **Loft** tool is then used to create a surface between the two planes.

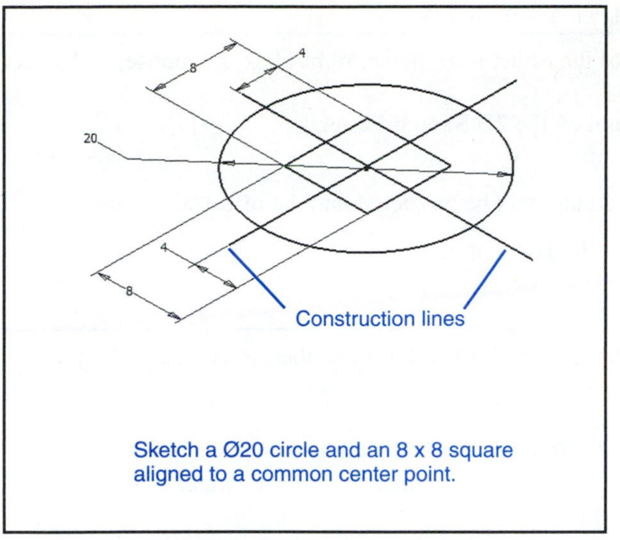

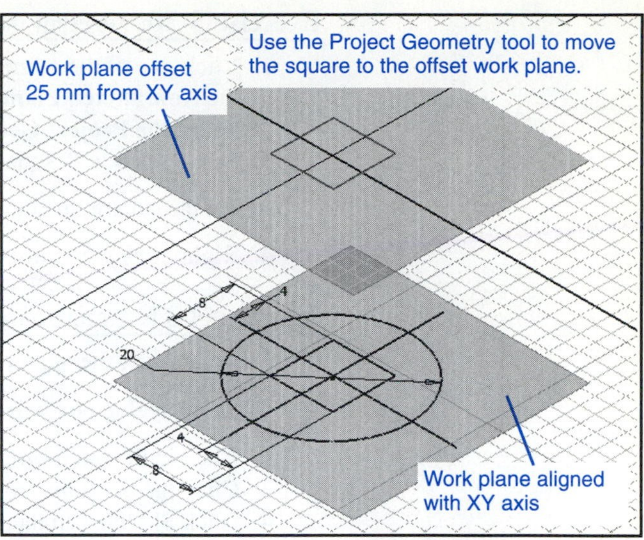

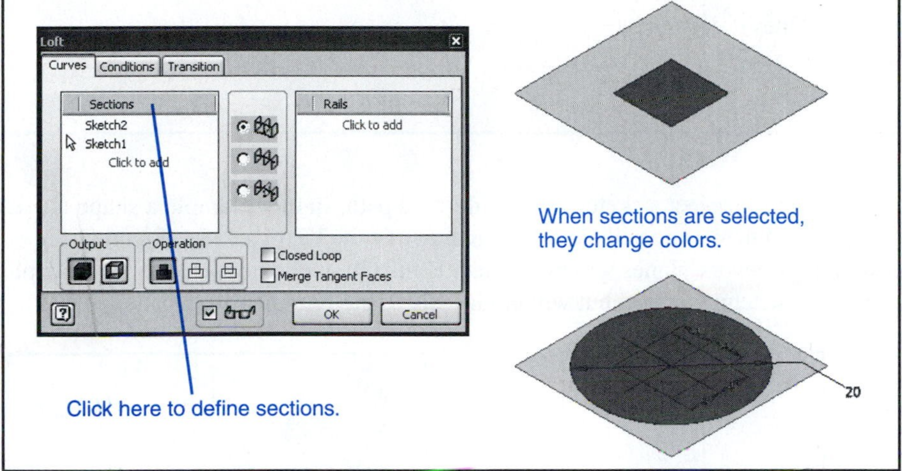

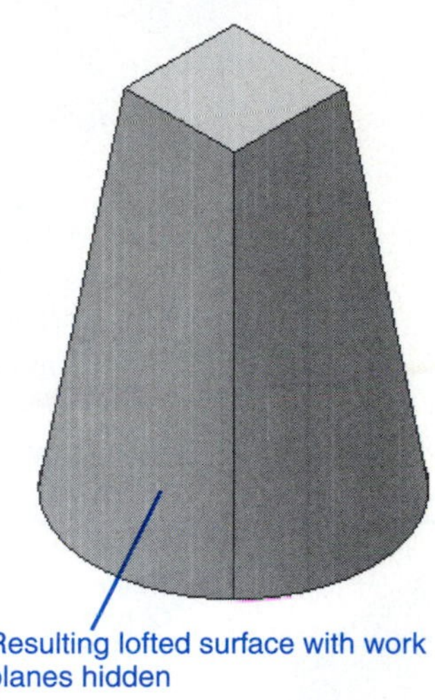

Figure 3-52

Exercise 3-28: Sketching the Circle and the Rectangle

1. Sketch a **Ø20** circle and an **8 × 8** square aligned to a common center point.
2. Create a work plane aligned with the XY plane.

Use the **Fix** constraint and construction lines if necessary.

Exercise 3-29: Creating an Offset Work Plane

1. Click the **Work Plane** tool, then click the **XY Plane** tool in the browser area.

A new plane will appear.

2. Click one of the corner points of the new plane and move the cursor upward.

An **Offset** dialog box will appear.

3. Set the offset distance for **25** and click the check mark on the dialog box.

Check the browser area to verify that two work planes have been created.

Exercise 3-30: Projecting the Rectangle

1. Click one of the corner points of the offset work plane, right-click the mouse, and select the **New Sketch Plane** option.
2. Select the **Project Geometry** tool on the **2D Sketch Panel.**
3. Select the **8 × 8** square.

Select the square line by line. The square will be projected into the offset work plane.

4. Right-click the mouse and select the **Done** option.

Exercise 3-31: Creating a Loft

1. Right-click the mouse and select the **Finish Sketch** option, then select the **Loft** option.

The **Loft** dialog box will appear.

2. Click in the **Sections** area of the **Loft** dialog box, then click the square.
3. Click the **Sections** area of the **Loft** dialog box again and click the circle.
4. Click **Sketch 1** in the **Loft** dialog box.
5. Click **OK.**

Hide the work planes if desired.

Sweep

The **Sweep** tool is used to project a sketch along a defined path. In this example a shape is created in the XZ plane and then projected along a path drawn in the YZ plane. See Figure 3-53.

For this example two work planes will be created. Unlike in previous examples the example will not start on a 2D sketching screen but will go directly to the **Part Features** panel.

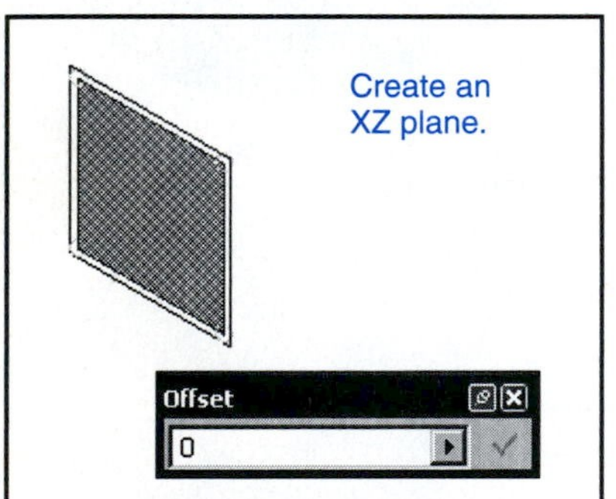

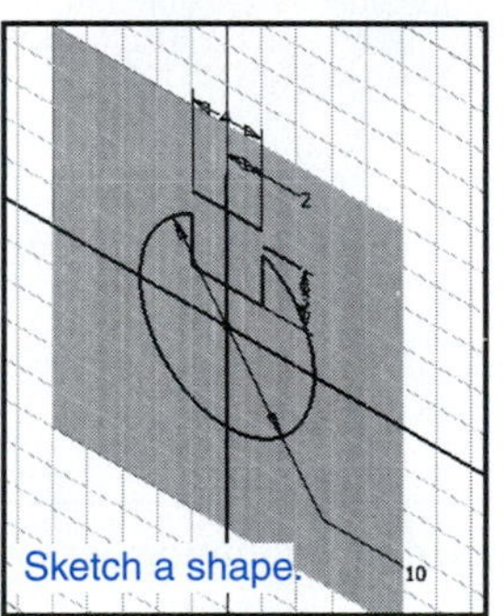

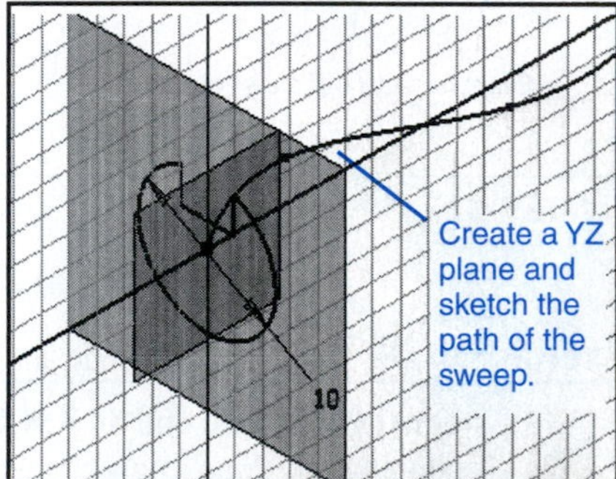

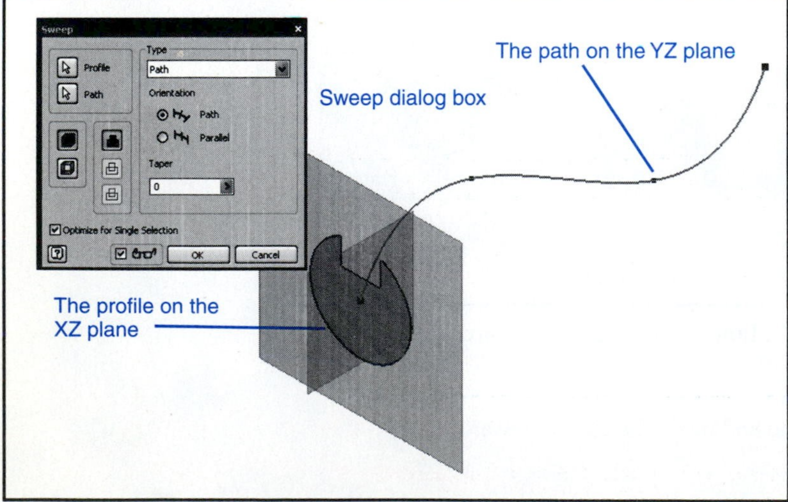

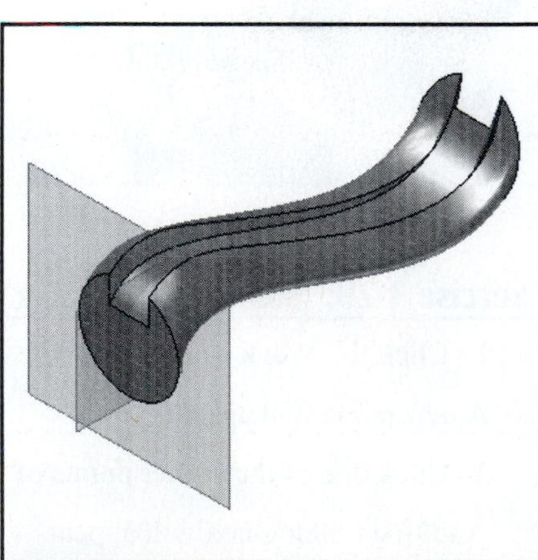

Figure 3-53

Exercise 3-32: Creating the Sketch

1. Start a new drawing using the **Standard (mm).ipt** format.

The drawing will go to the **2D Sketch** tools.

2. Right-click the mouse and select the **Finish sketch** option.

The screen should be clear.

3. Set the view for **Isometric,** then click the **Work Plane** tool, then the **XZ Plane** tool in the browser area.

A plane will appear.

4. Click one of the plane's corners and offset the work plane **0.000,** then click the check mark on the **Offset** dialog box.
5. Right-click one of the plane's corner points and select the **New Sketch** option.
6. Sketch a **Ø10** circle with a **4 × 3** keyway as shown.
7. Right-click the mouse and select the **Finish Sketch** option.

Exercise 3-33: Creating the Path

1. Click the **Work Plane** tool, then the **YZ Plane** tool in the browser.
2. Click one the plane's corner points and create a second work plane offset **0.000 mm.**
3. Right-click one of the YZ work plane's corner points and select the **New Sketch** option.
4. Click the **Spline** tool and sketch a spline starting at the hole's center point. Click the right mouse button and select the **Create** option, then right-click the mouse again and select the **Done** option.

In this example a random spline was used.

Exercise 3-34: Creating the Sweep

1. Right-click the mouse and select the **Finish Sketch** option, then click the **Sweep** tool.

The **Sweep** dialog box will appear. The circular sketch will automatically be selected as the profile.

2. Click the spline to define it as the path.
3. Click **OK.**

COIL

A coil is similar to a sweep, but the path is a helix. A sketch is drawn, then projected along a helical-shaped path.

Exercise 3-35: Creating the Sketch

1. Sketch the shape shown in Figure 3-54 on the XY plane.
2. Sketch a line below the shape as shown.

This line will serve as the axis of rotation.

3. Right-click the mouse, and click the **Finish Sketch** option.

Exercise 3-36: Creating the Coil

1. Click the **Coil** tool on the **Part Features** panel bar.

The **Coil** dialog box will appear. The sketched profile will be selected automatically.

2. Select the sketch line as the axis.

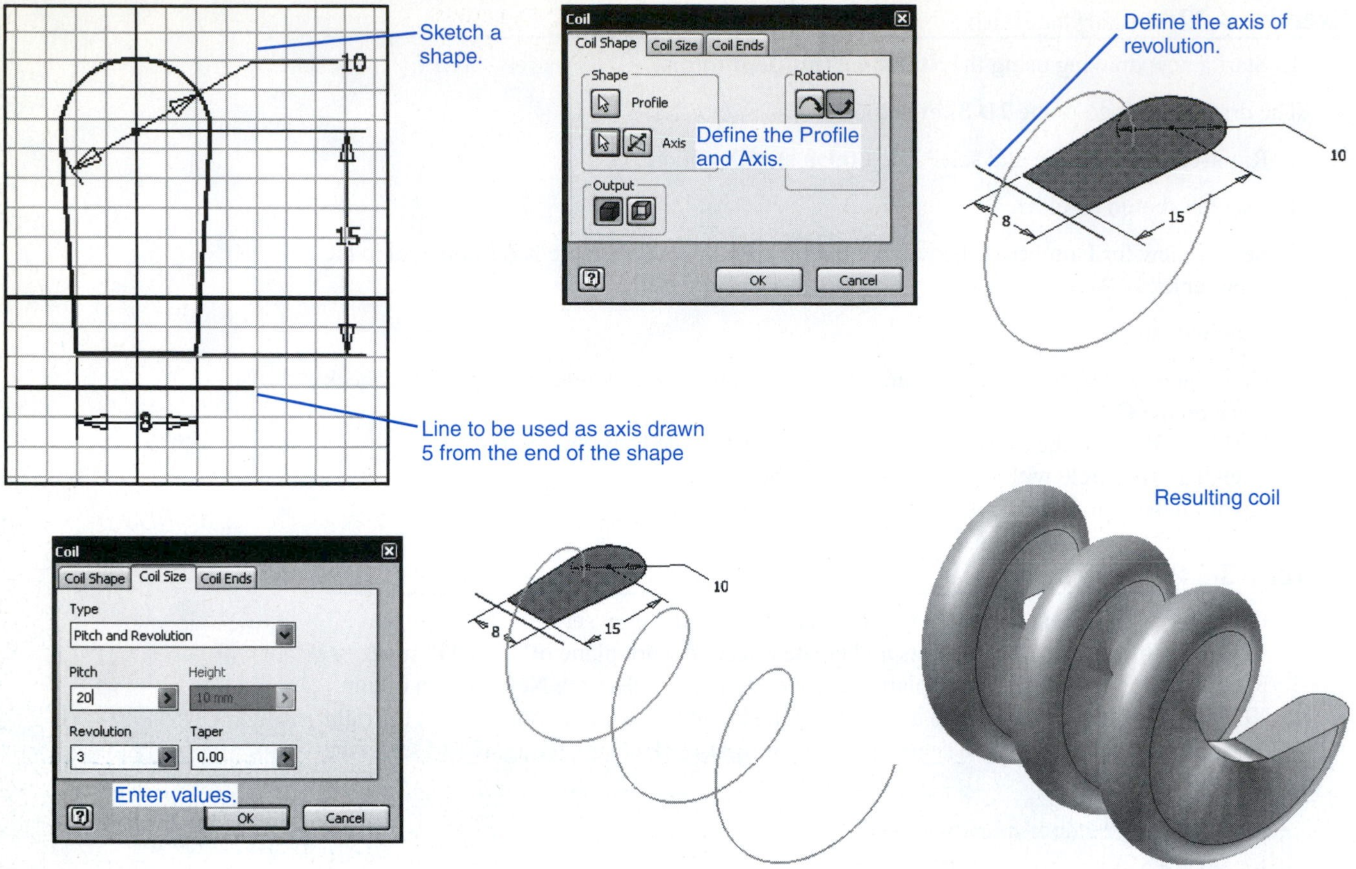

Sketch a shape.

Define the Profile and Axis.

Define the axis of revolution.

Line to be used as axis drawn 5 from the end of the shape

Enter values.

Resulting coil

Figure 3-54

3. Click the **Coil Size** tab.

The dialog box will change.

4. Set the **Type** for **Pitch and Revolution,** the **Pitch** for **20,** and the **Revolution** for **3.**
5. Click **OK.**

See Figure 3-54.

MODEL MATERIAL

A material designation may be assigned to a model. The material designation becomes part of the model's file and will be included on any assembly's parts list that includes the model.

Exercise 3-37: Defining a Model's Material

1. Right-click on the model's name in the browser box and select the **Properties** option.

See Figure 3-55. The **Properties** dialog box will appear.

2. Select the **Physical** tab and then the scroll arrow on the right side of the **Material** box.

See Figure 3-56.

3. Select a material.

In this example, mild steel was selected.

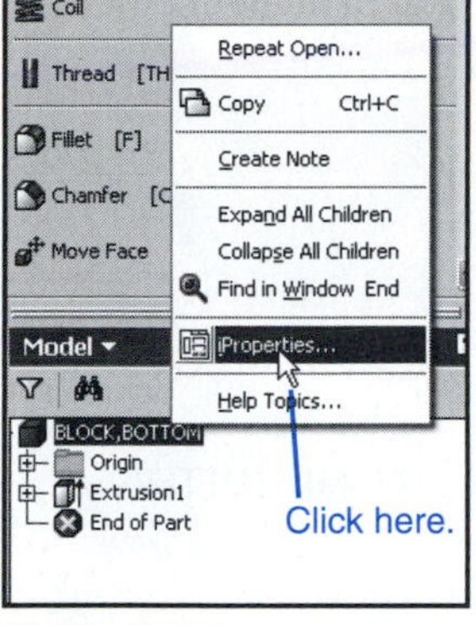

Figure 3-55

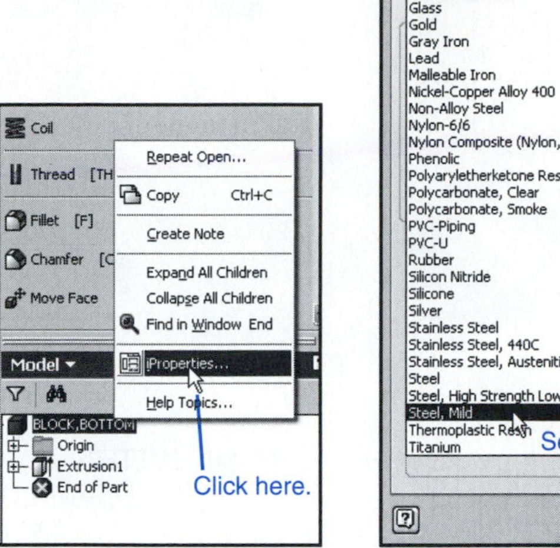

Figure 3-56

SUMMARY

The first part of the chapter demonstrated how to convert 2D sketches into 3D models and then modify features using some of the commands in the **Part Features** panel bar. Exercises included extruding, revolving, lofting, and mirroring models as well as trimming away portions, and creating shells. Fillets, chamfers, and holes in both rectangular and circular arrangements were also added to models.

The second part of the chapter introduced sketch and work planes and work axes and demonstrated how to use them to refine 3D models.

CHAPTER PROJECT

Project 3-1:

Redraw the following objects as solid models based on the given dimensions. Make all models from mild steel.

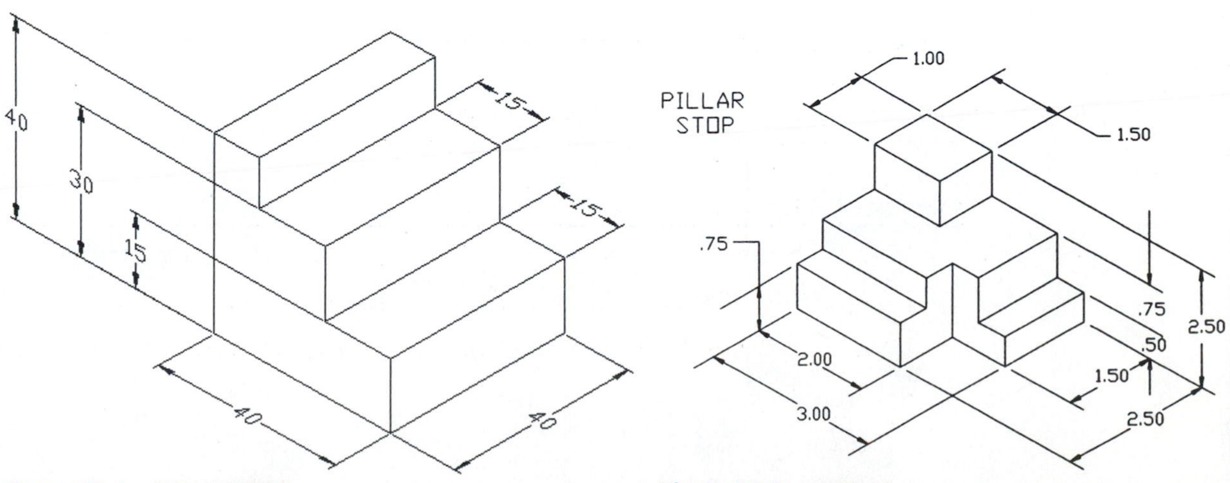

Figure P3-1 MILLIMETERS

Figure P3-2 INCHES

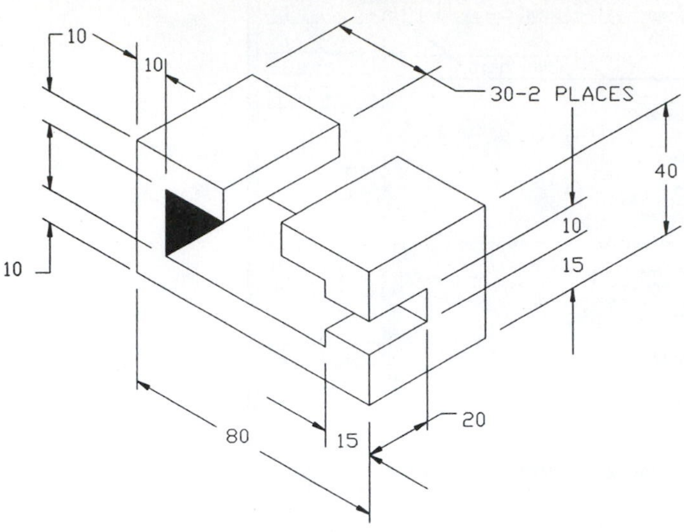

Figure P3-3 MILLIMETERS

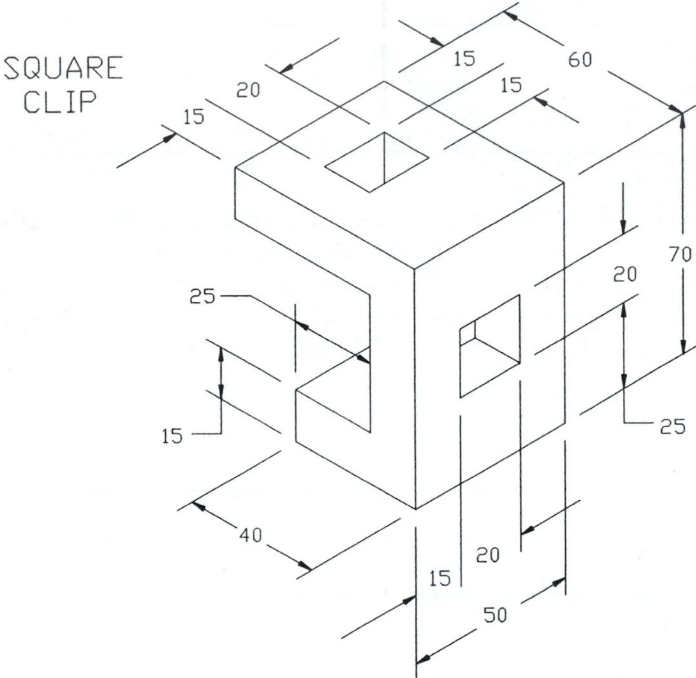

Figure P3-4 MILLIMETERS

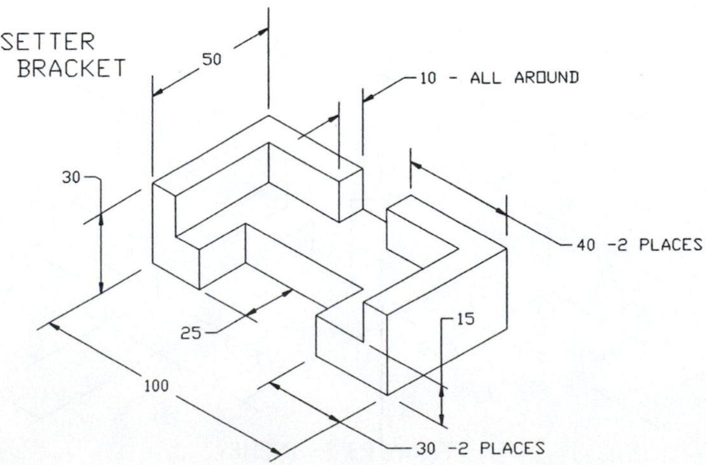

Figure P3-5 MILLIMETERS

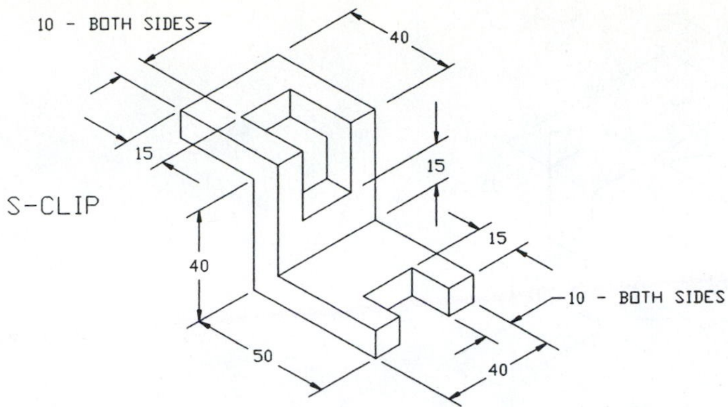

Figure P3-6 MILLIMETERS

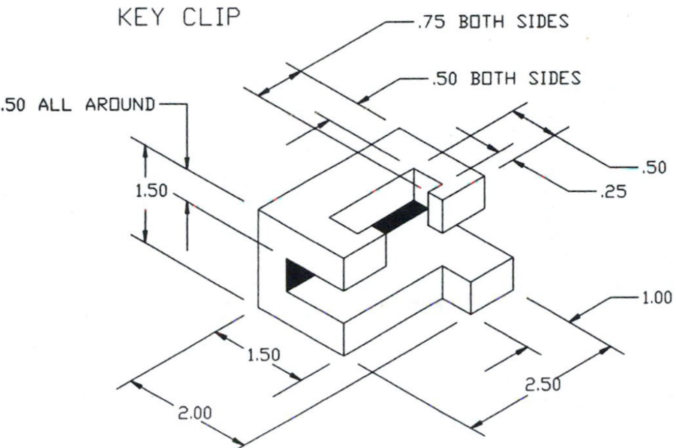

Figure P3-7 INCHES

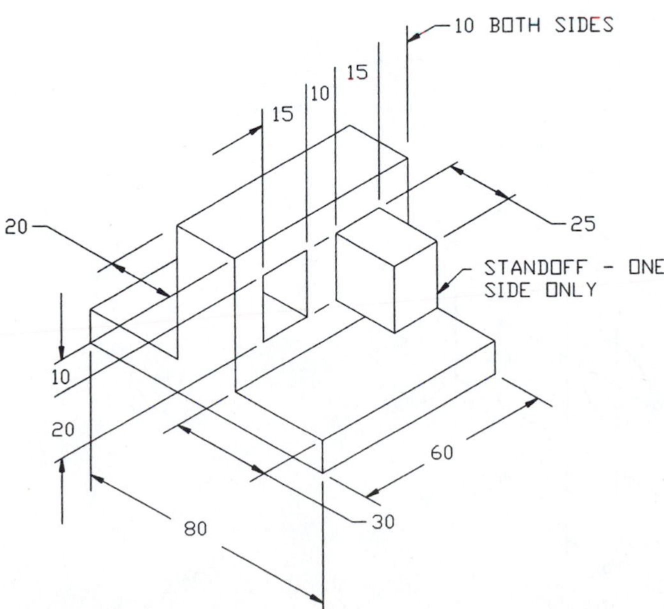

Figure P3-8 MILLIMETERS

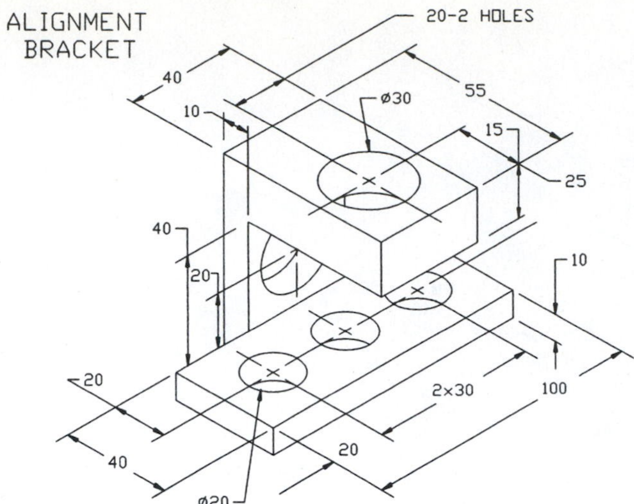

Figure P3-9 MILLIMETERS

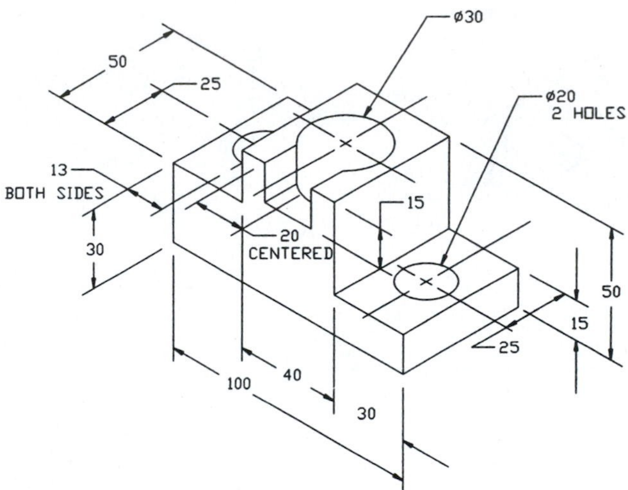

Figure P3-10 MILLIMETERS

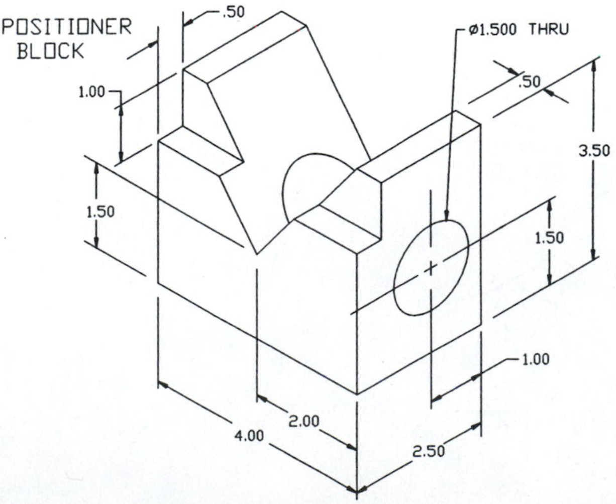

Figure P3-11 INCHES

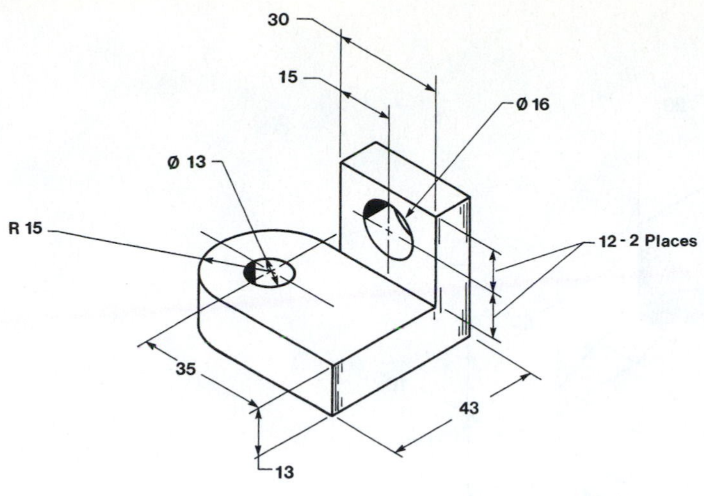

Figure P3-12 MILLIMETERS

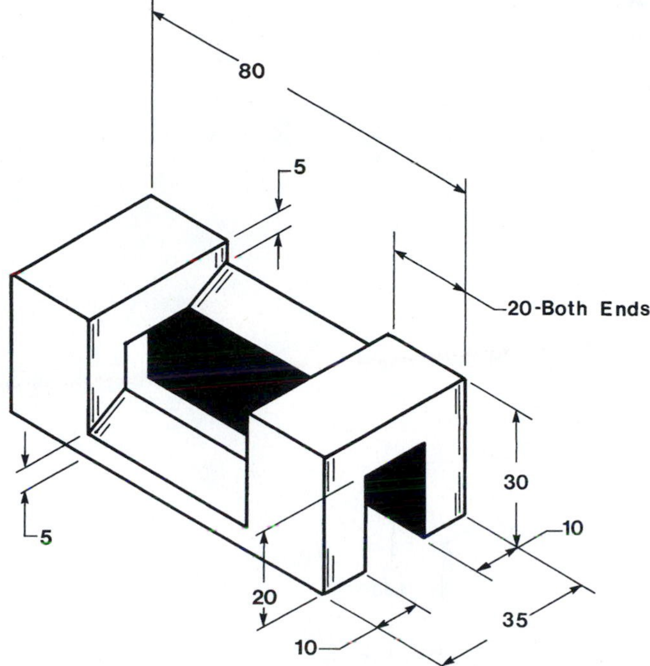

Figure P3-13 MILLIMETERS

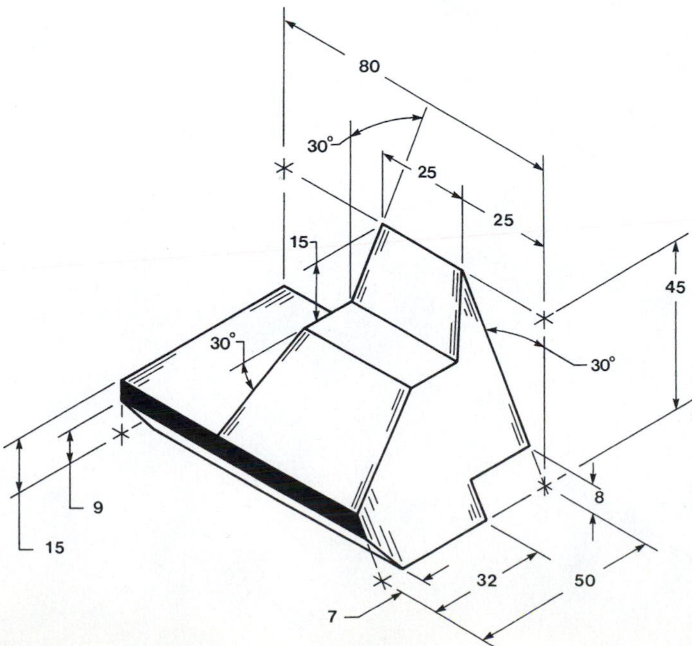

Figure P3-14 MILLIMETERS

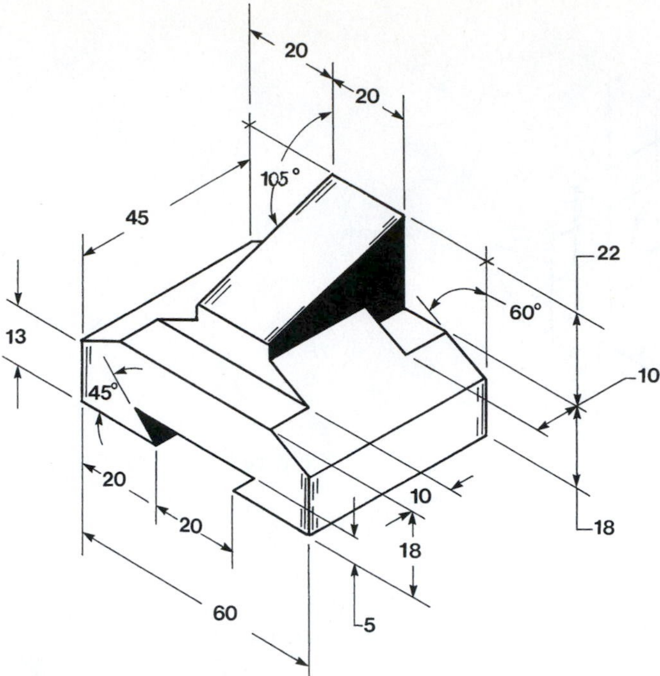

Figure P3-15 MILLIMETERS

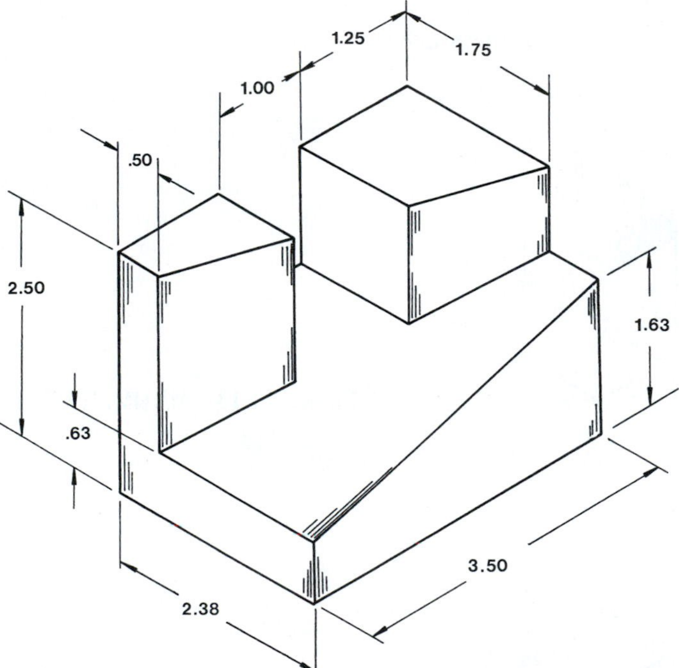

Figure P3-16 INCHES

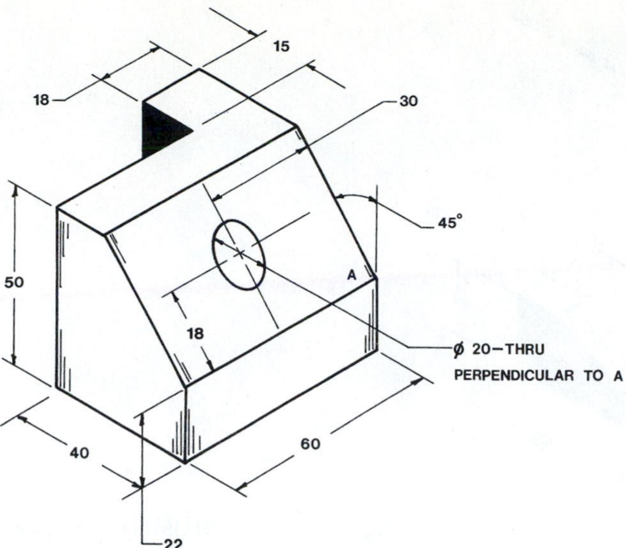

Figure P3-17 MILLIMETERS

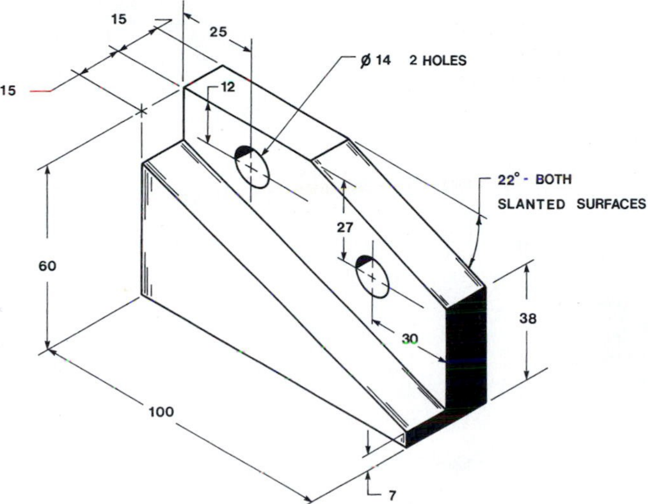

Figure P3-18 MILLIMETERS

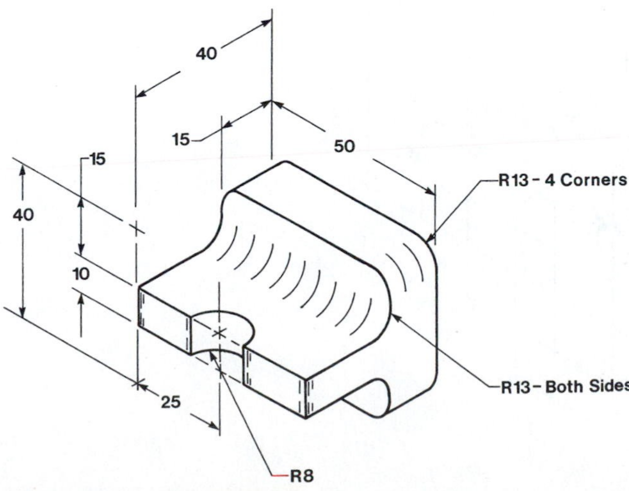

Figure P3-19 MILLIMETERS

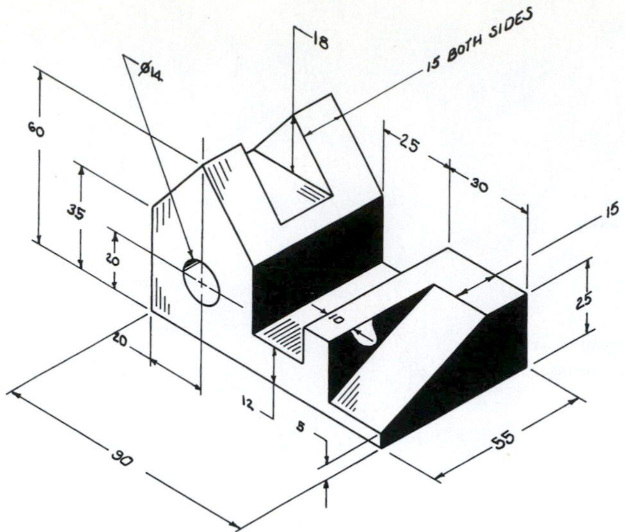

Figure P3-20 MILLIMETERS

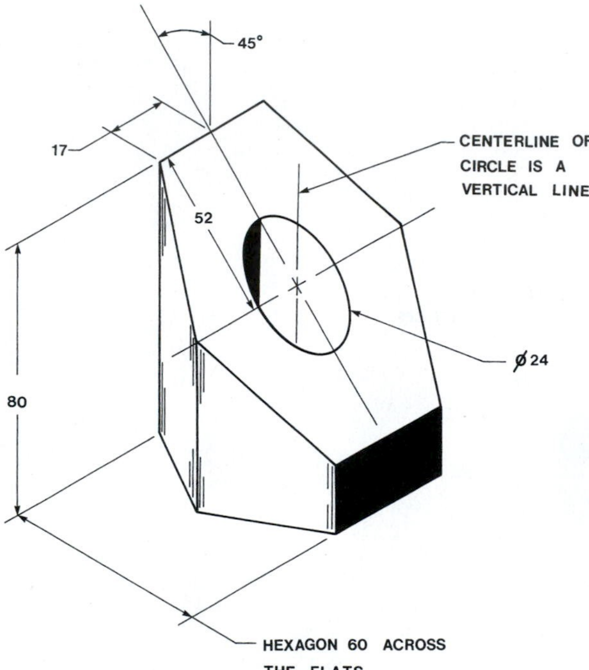

Figure P3-21 MILLIMETERS

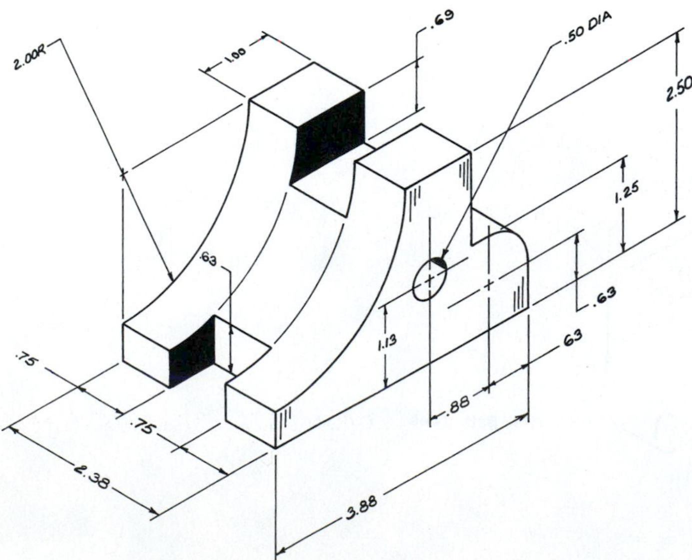

Figure P3-22 INCHES

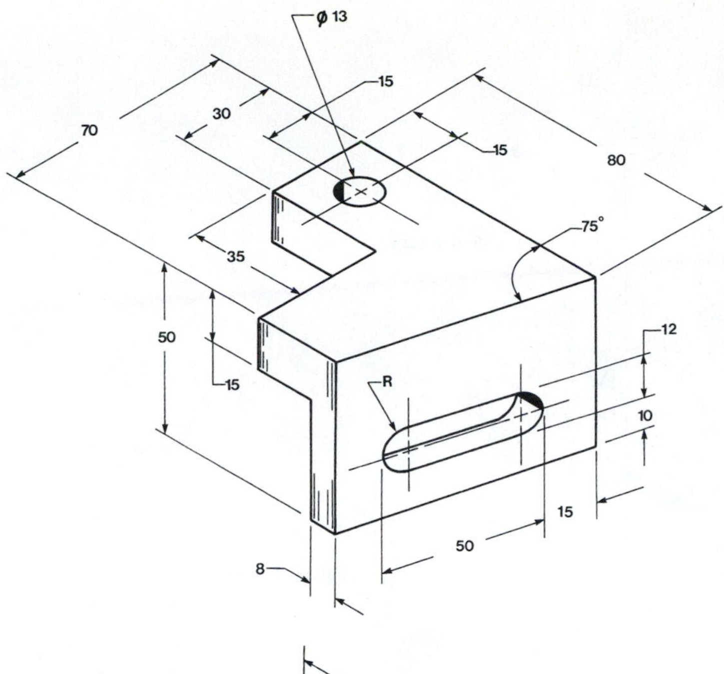

Figure P3-23 MILLIMETERS

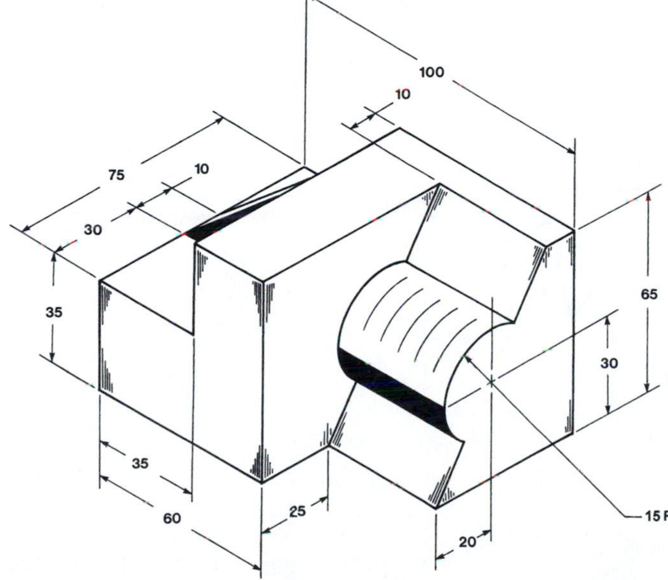

Figure P3-24 MILLIMETERS

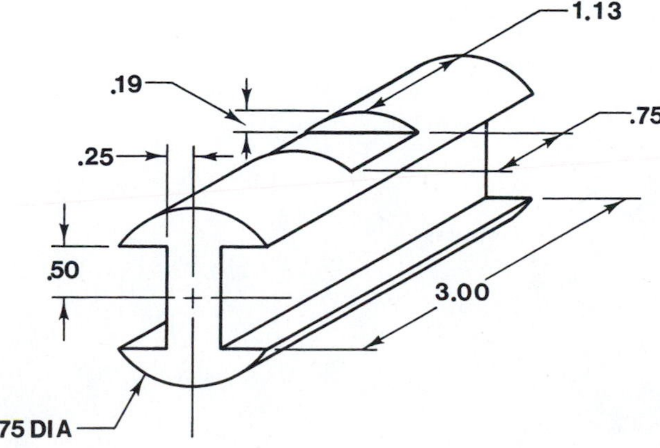

Figure P3-25 INCHES (SCALE: 4 = 1)

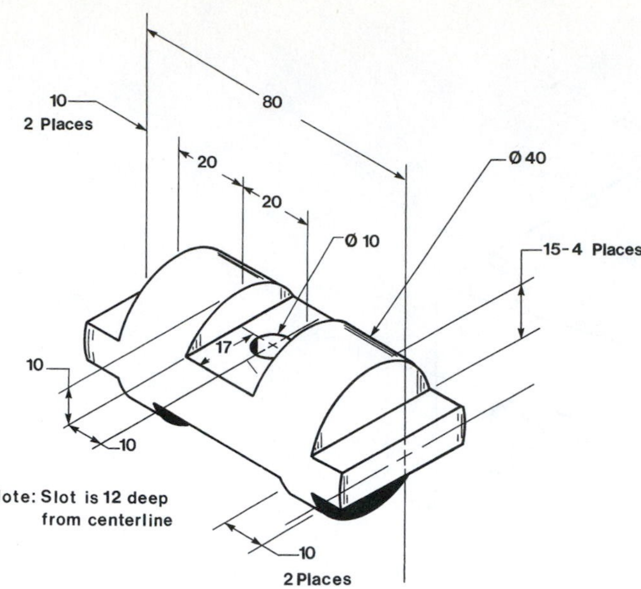

Figure P3-26 MILLIMETERS

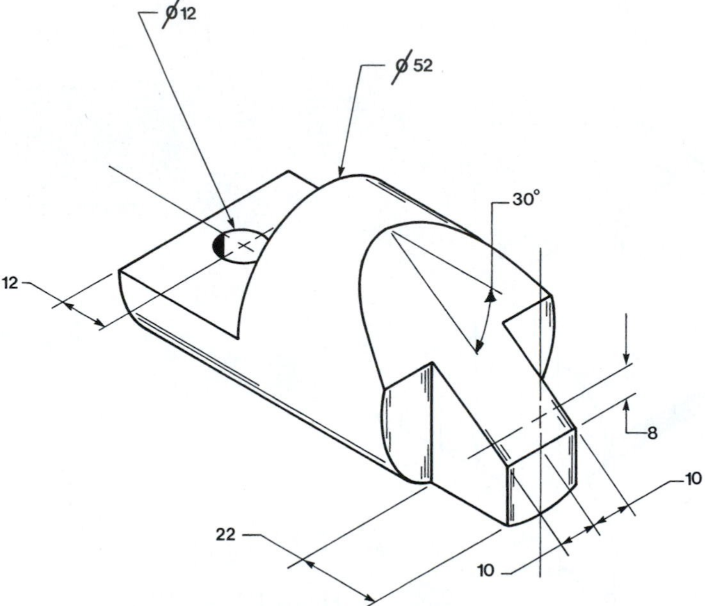

Figure P3-27 MILLIMETERS

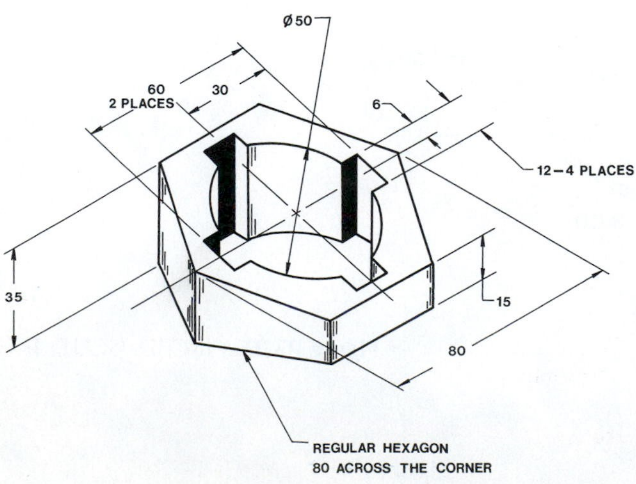

Figure P3-28 MILLIMETERS

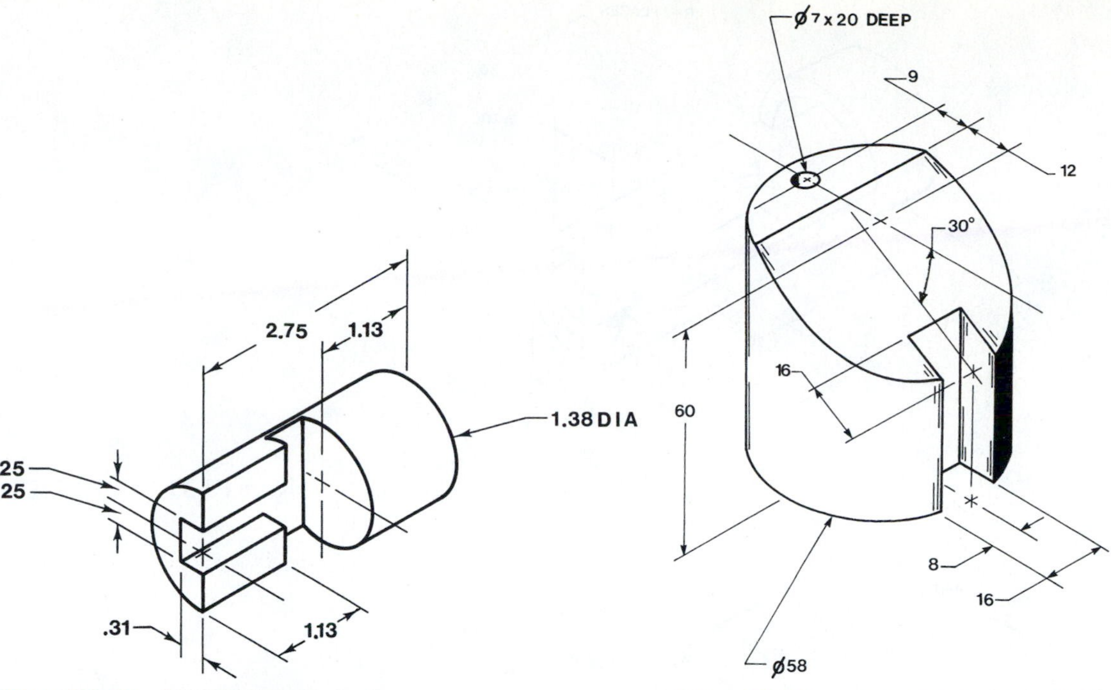

Figure P3-29 INCHES (SCALE: 4=1)

Figure P3-30 MILLIMETERS (Scale: 2=1)

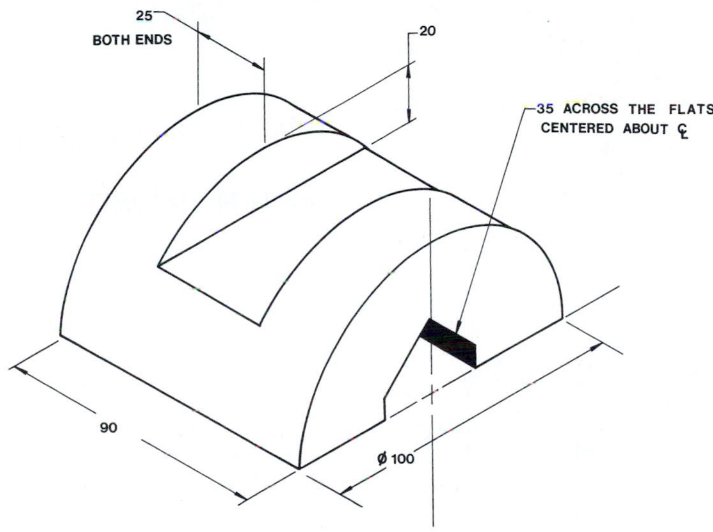

Figure P3-31 MILLIMETERS

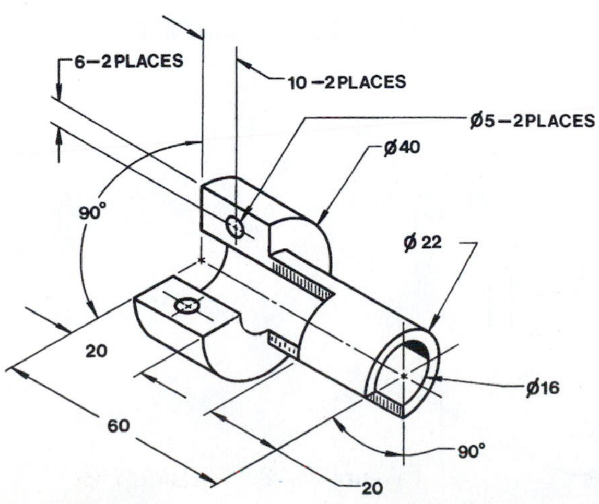

Figure P3-32 MILLIMETERS

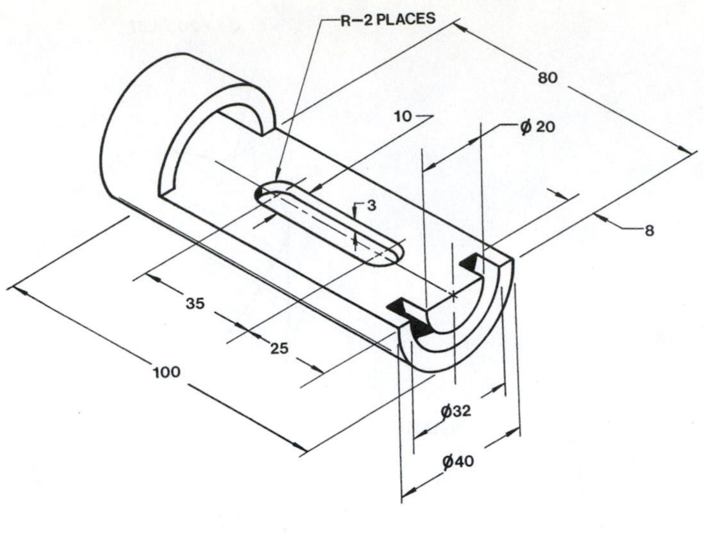

Figure P3-33 MILLIMETERS

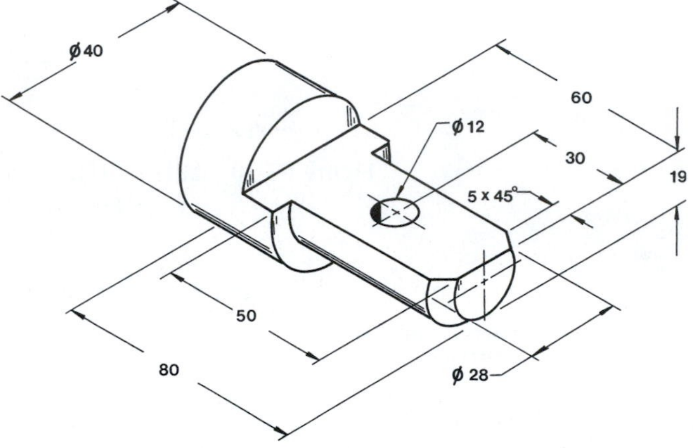

Figure P3-34 MILLIMETERS

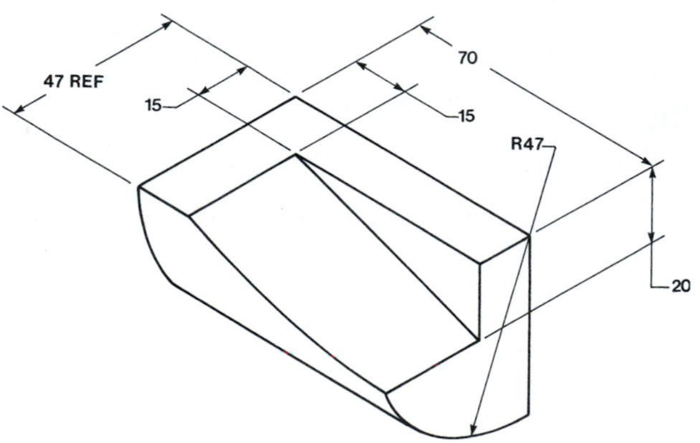

Figure P3-35 MILLIMETERS

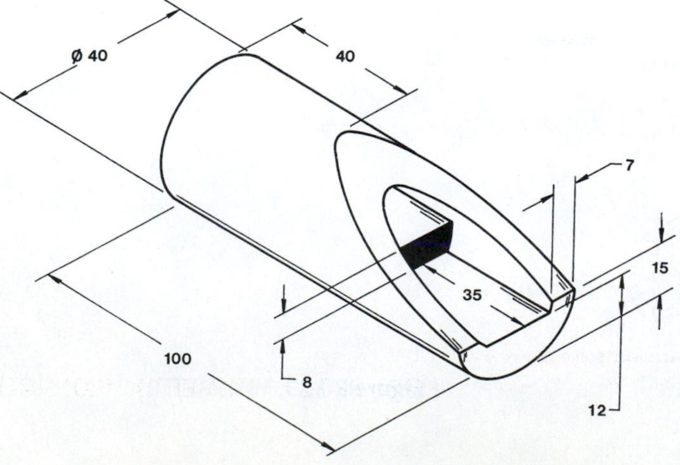

Figure P3-36 MILLIMETERS

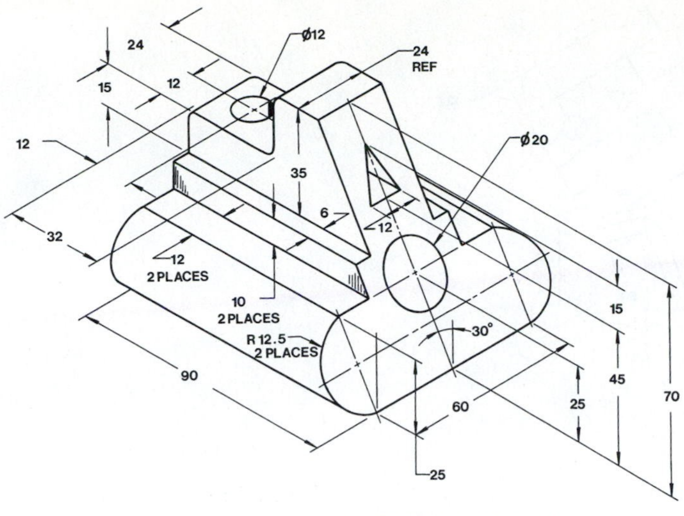

Figure P3-37 MILLIMETERS

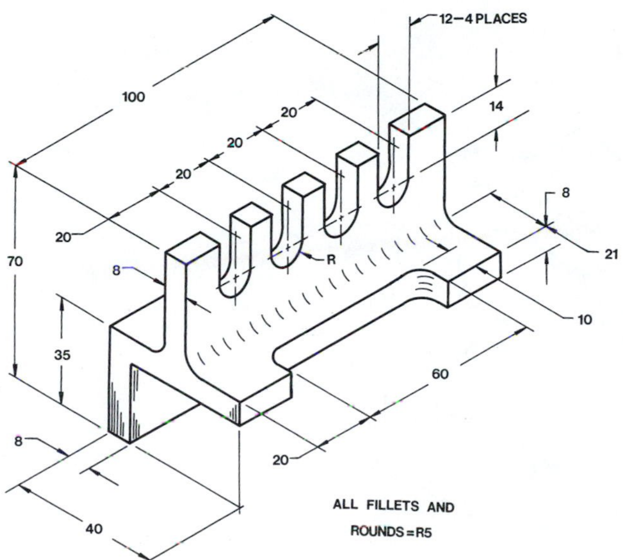

Figure P3-38 MILLIMETERS

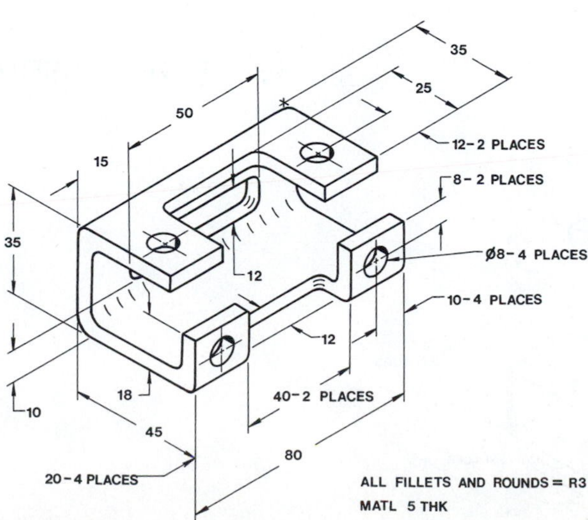

Figure P3-39 MILLIMETERS (CONSIDER A SHELL)

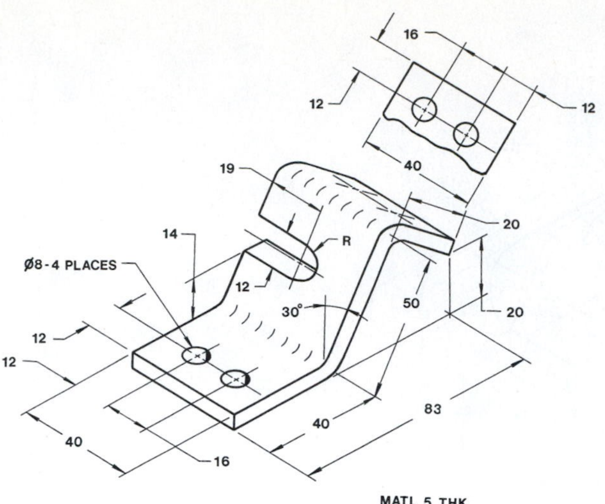

Figure P3-40 MILLIMETERS

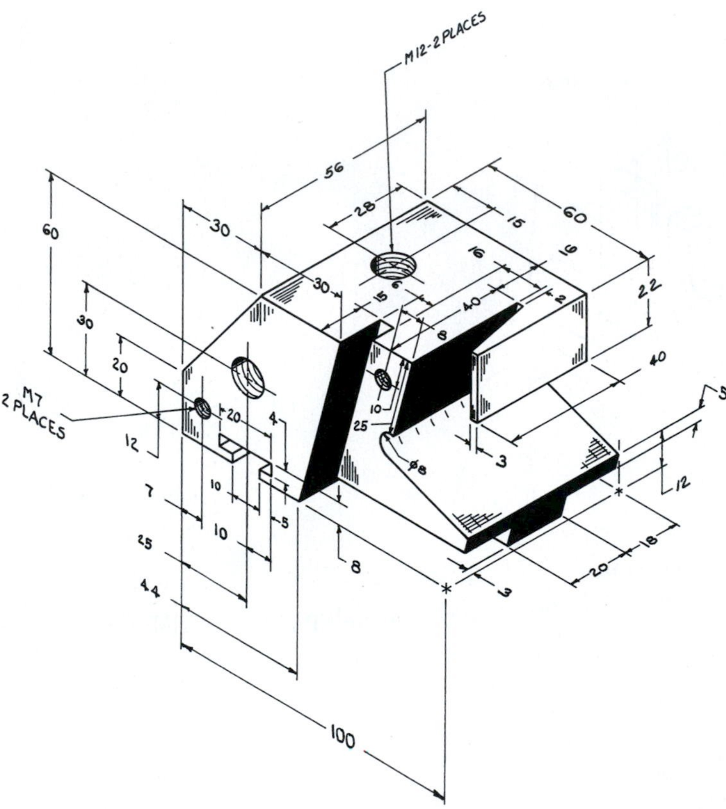

Figure P3-41 MILLIMETERS

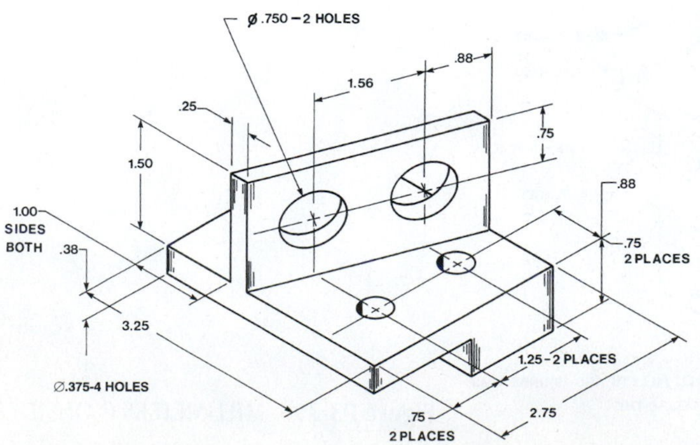

Figure P3-42 INCHES

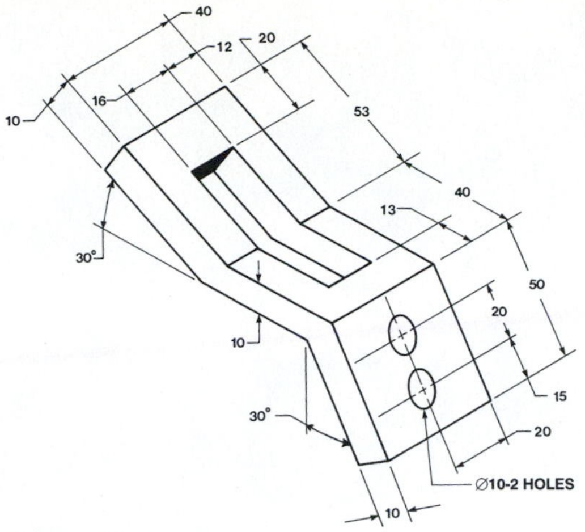

Figure P3-43 MILLIMETERS

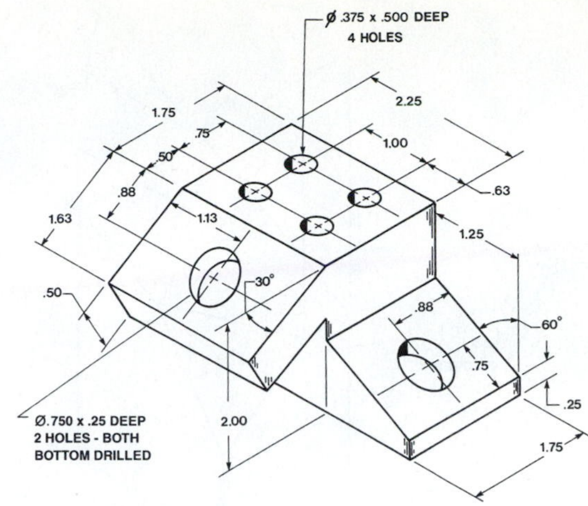

Figure P3-44 INCHES

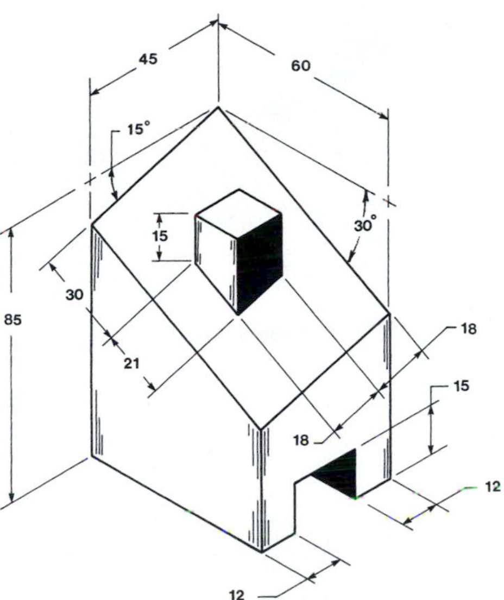

Figure P3-45 MILLIMETERS

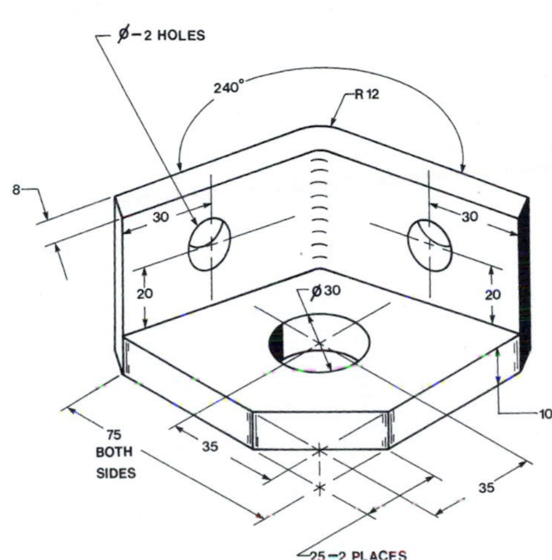

Figure P3-46 MILLIMETERS

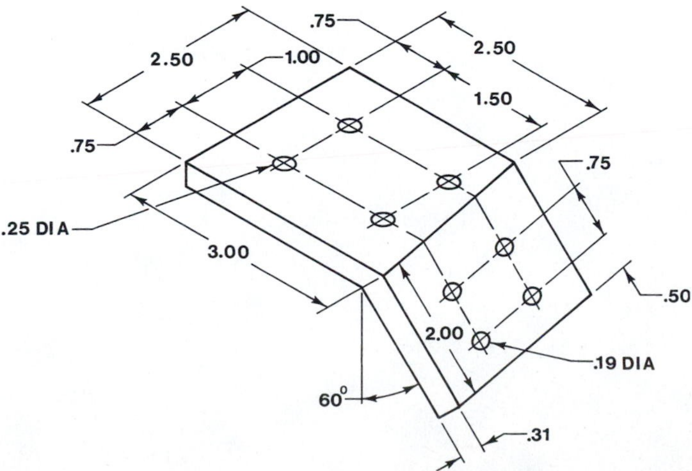

Figure P3-47 INCHES

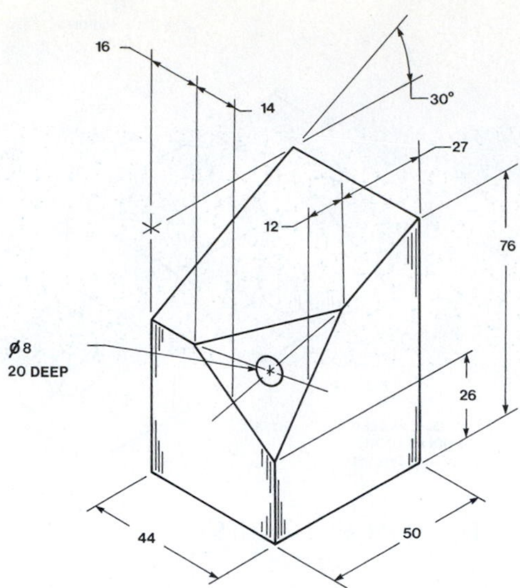

Figure P3-48 MILLIMETERS

Orthographic Views

4

Chapter Objectives

- Introduce orthographic views.
- Introduce ANSI standards and conventions.
- Show how to draw sectional and auxiliary views.

INTRODUCTION

Orthographic views may be created directly from 3D Inventor models. ***Orthographic views*** are two-dimensional views used to define a three-dimensional model. Unless the model is of uniform thickness, more than one orthographic view is necessary to define the model's shape. Standard practice calls for three orthographic views: a front, top, and right side view, although more or fewer views may be used as needed.

orthographic view: A two-dimensional view used to define a three-dimensional model.

Modern machines can work directly from the information generated when a solid 3D model is created, so the need for orthographic views—blueprints—is not as critical as it once was; however, there are still many drawings in existence that are used for production and reference. The ability to create and read orthographic views remains an important engineering skill.

This chapter presents orthographic views using third-angle projection in accordance with ANSI standards. ISO first-angle projections are also presented.

FUNDAMENTALS OF ORTHOGRAPHIC VIEWS

Figure 4-1 shows an object with its front, top, and right-side orthographic views projected from the object. The views are two-dimensional, so they show no depth. Note that in the projected right plane there are three rectangles. There is no way to determine which of the three is closest and which is farthest away if only the right-side view is considered. All views must be studied to analyze the shape of the object.

Figure 4-2 shows three orthographic views of a book. After the views are projected they are positioned as shown. The positioning of views relative to one another is critical. The views must be aligned and positioned as shown.

Normal Surfaces

Normal surfaces are surfaces that are at 90° to each other. Figures 4-3, 4-4, and 4-5 show objects that include only normal surfaces and their orthographic views.

normal surfaces: Surfaces that are 90° to each other.

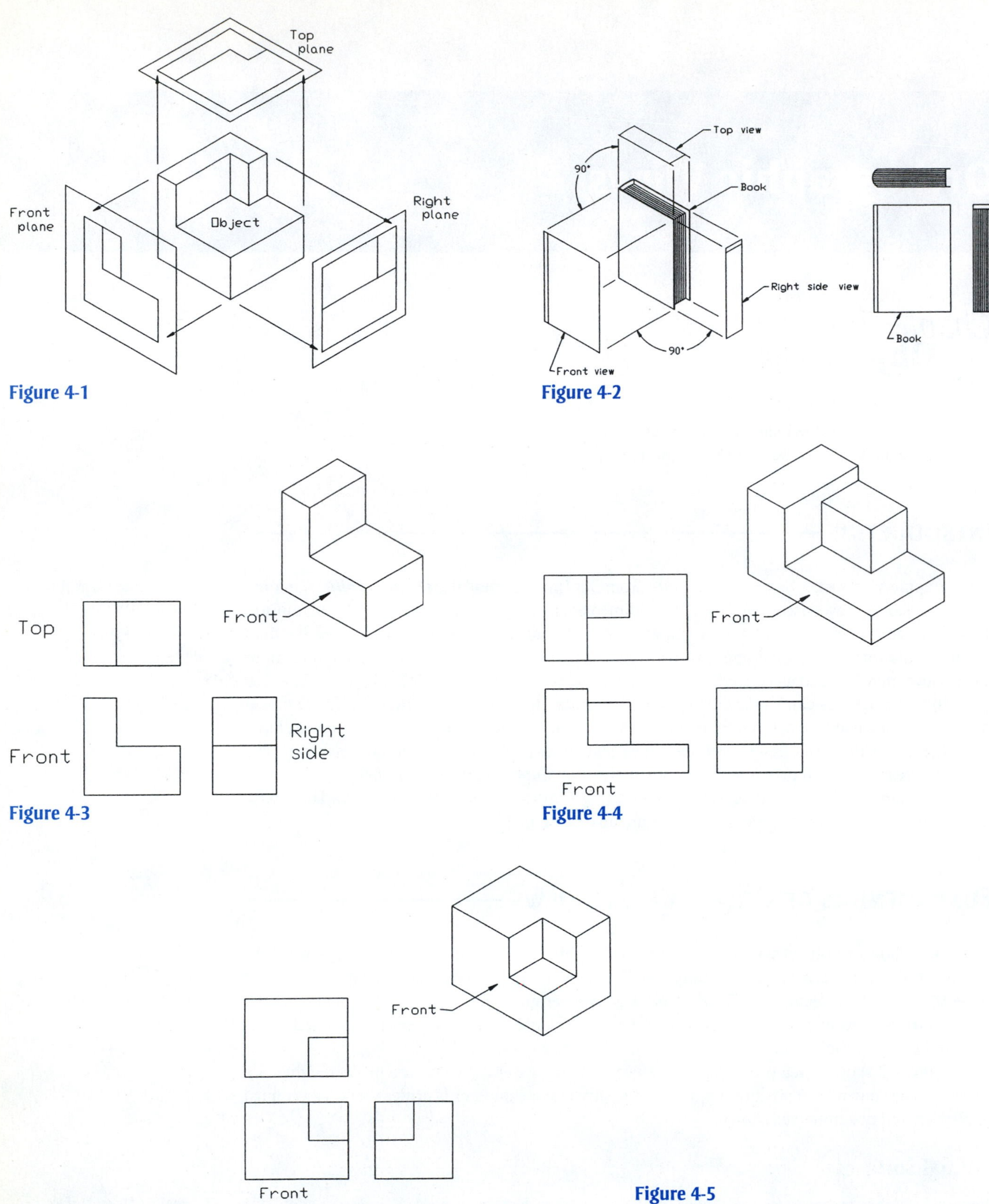

Figure 4-1

Figure 4-2

Figure 4-3

Figure 4-4

Figure 4-5

Hidden Lines

Hidden lines are used to show surfaces that are not directly visible. All surfaces must be shown in all views. If an edge or surface is blocked from view by another feature, it is drawn using a hidden line. Figures 4-6 and 4-7 show objects that require hidden lines in their orthographic views.

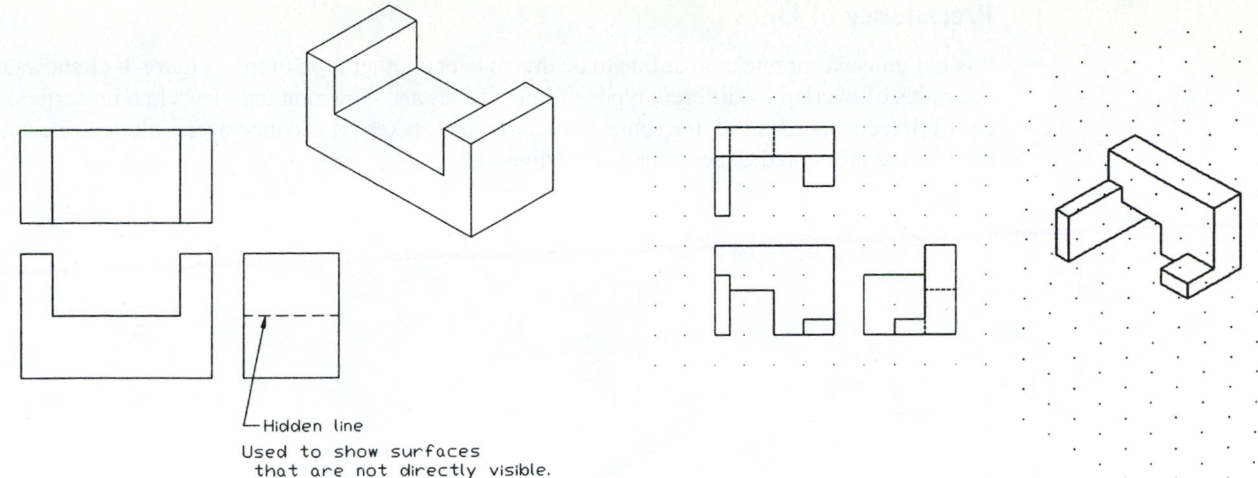

Figure 4-6

Figure 4-7

Figure 4-8 shows an object that contains an edge line, A-B. In the top view, line A-B is partially hidden and partially visible. The hidden portion of the line is drawn using a hidden-line pattern, and the visible portion of the line is drawn using a solid line.

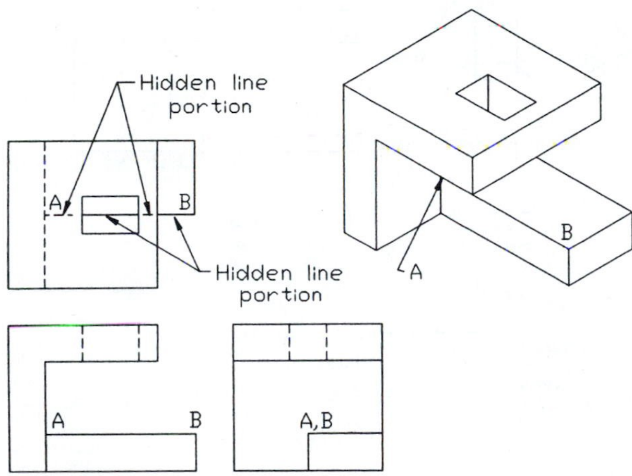

Figure 4-8

Figures 4-9 and 4-10 show objects that require hidden lines in their orthographic views.

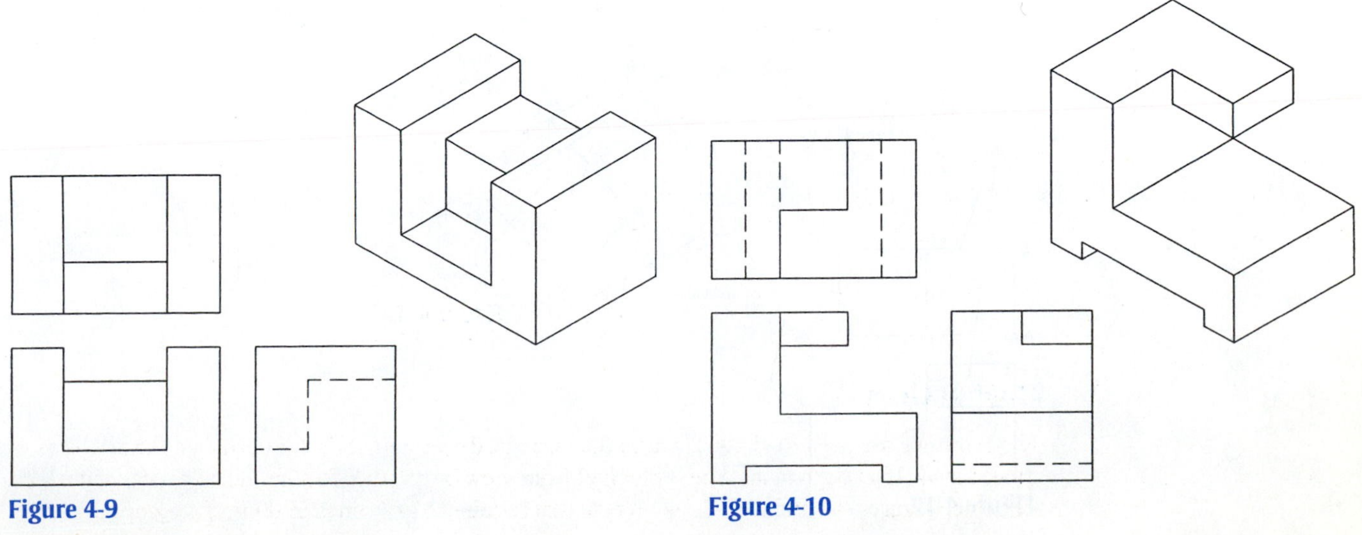

Figure 4-9

Figure 4-10

Precedence of Lines

It is not unusual for one type of line to be drawn over another type of line. Figure 4-11 shows two examples of overlap by different types of lines. Lines are shown on the views in a prescribed order of precedence. A solid line (object or continuous) takes precedence over a hidden line, and a hidden line takes precedence over a centerline.

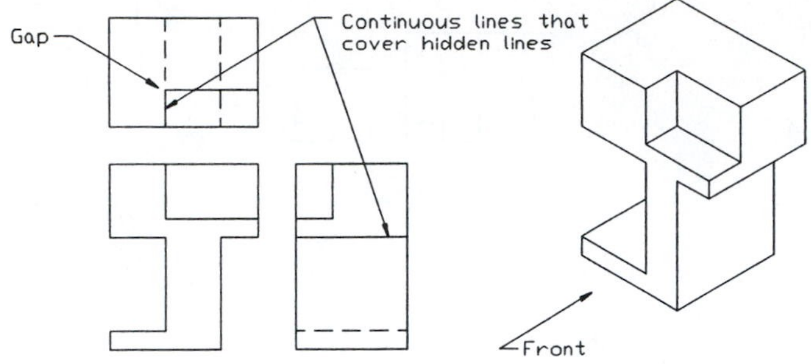

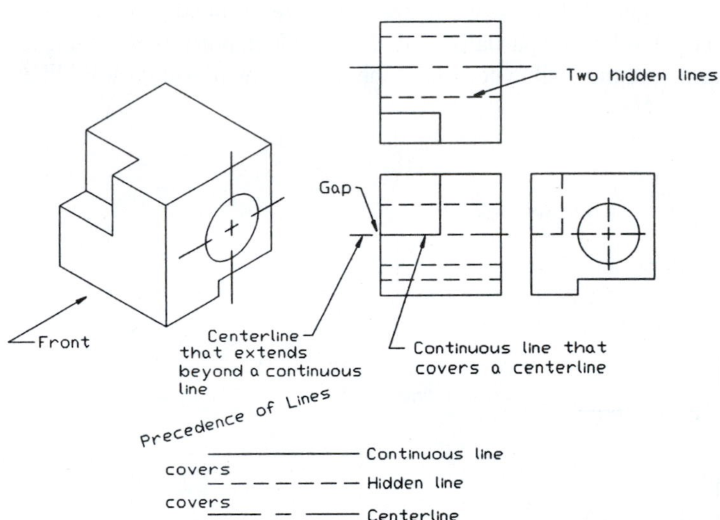

Figure 4-11

Slanted Surfaces

slanted surfaces: Surfaces that are at an angle to each other.

Slanted surfaces are surfaces drawn at an angle to each other. Figure 4-12 shows an object that contains two slanted surfaces. Surface ABCD appears as a rectangle in both the top and front

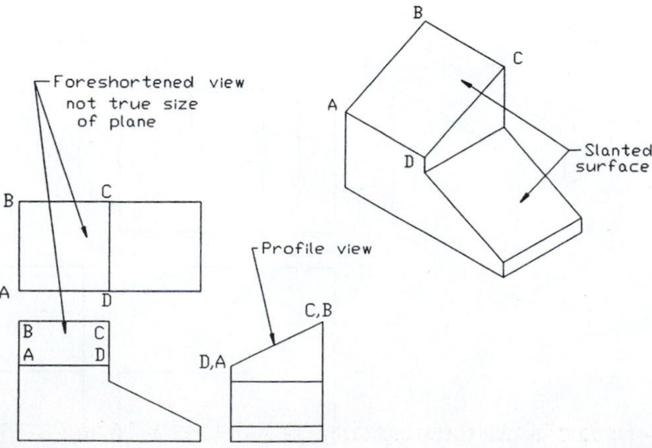

Figure 4-12

views. Neither rectangle represents the true shape of the surface. Each is smaller that the actual surface. Also, none of the views show enough of the object to enable the viewer to accurately define the shape of the object. The views must be used together for a correct understanding of the object's shape.

Figures 4-13 and 4-14 show objects that include slanted surfaces. Projection lines have been included to emphasize the importance of correct view location. Information is projected between the front and top views using vertical lines and between the front and side views using horizontal lines.

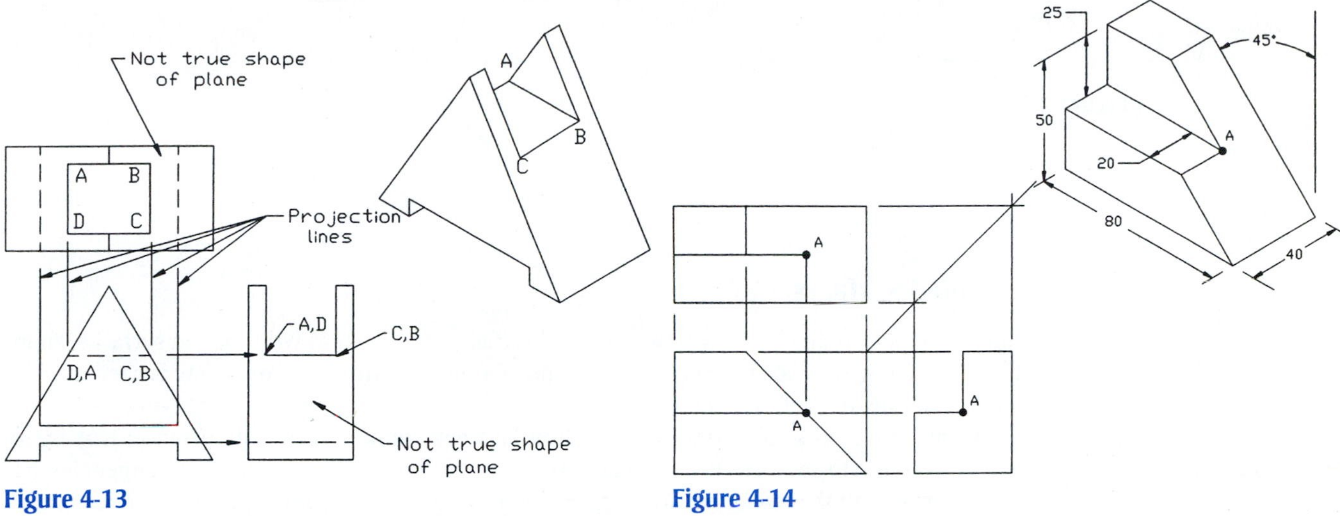

Figure 4-13

Figure 4-14

Compound Lines

A ***compound line*** is formed when two slanted surfaces intersect. Figure 4-15 shows an object that includes a compound line.

compound line: A line that is neither perpendicular nor parallel to the X, Y, or Z axis.

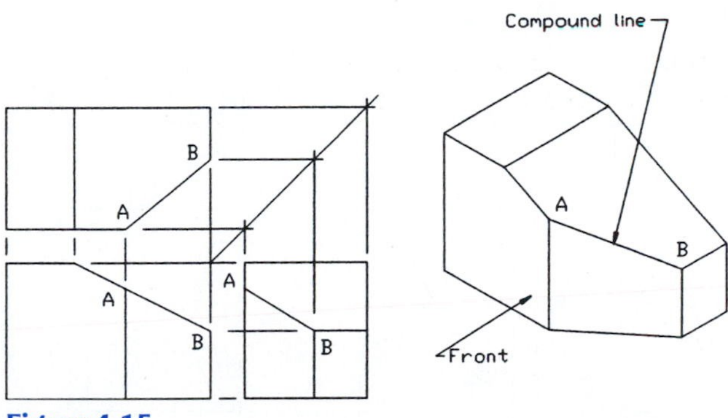

Figure 4-15

Oblique Surfaces

An ***oblique surface*** is a surface that is slanted in two different directions. Figures 4-16 and 4-17 show objects that include oblique surfaces.

oblique surface: A surface that is slanted in two different directions.

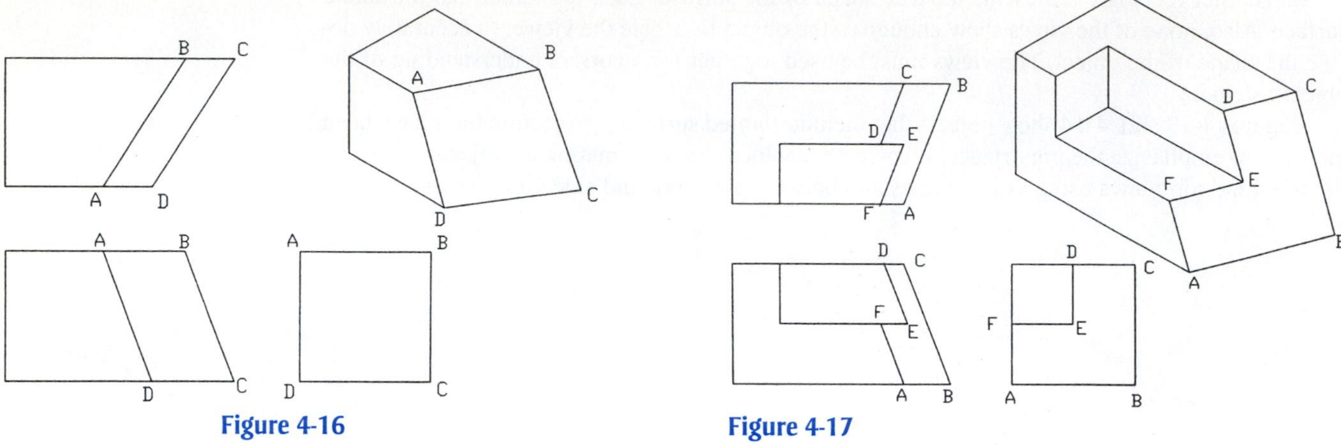

Figure 4-16

Figure 4-17

Rounded Surfaces

Figure 4-18 shows an object with two rounded surfaces. Note that as with slanted surfaces, an individual view is insufficient to define the shape of a surface. More than one view is needed to accurately define the surface's shape.

Convention calls for a smooth transition between rounded and flat surfaces; that is, no lines are drawn to indicate the tangency. Inventor includes a line to indicate tangencies between surfaces in the isometric drawings created using the multiview options but does not include them in the orthographic views. Tangency lines are also not included when models are rendered.

Figure 4-19 shows the drawing conventions for including lines for rounded surfaces. If a surface includes no vertical portions or no tangency, no line is included.

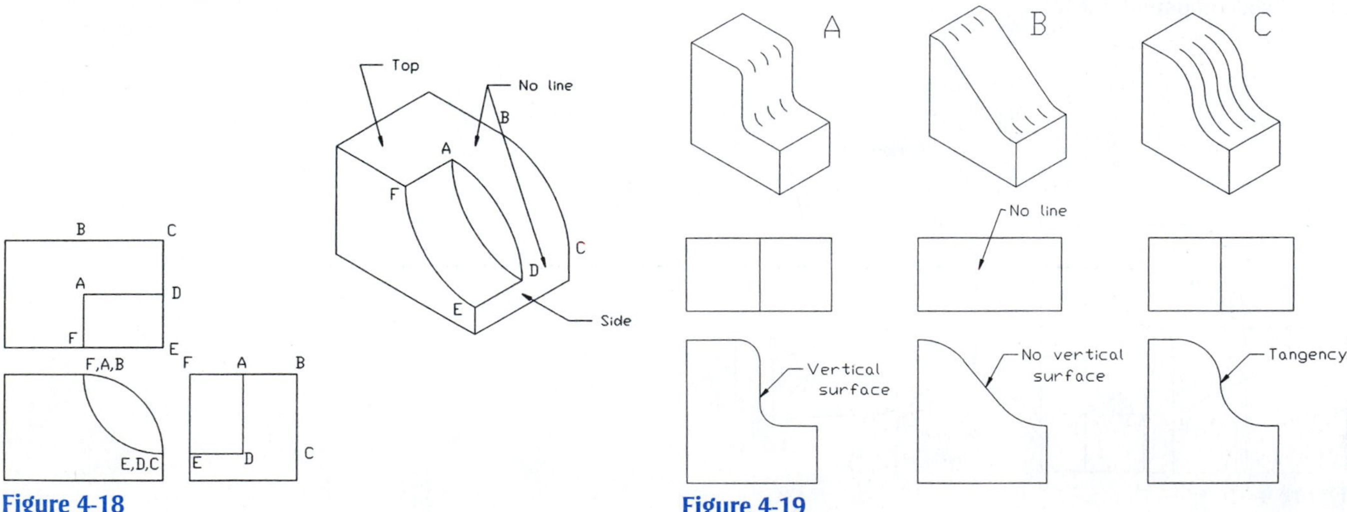

Figure 4-18

Figure 4-19

Figure 4-20 shows an object that includes two tangencies. Each is represented by a line. Note in Figure 4-20 that Inventor will add tangent lines to the 3D model. These lines will not appear in the orthographic views.

Figure 4-21 shows two objects with similar configurations; however, the boxlike portion of the lower object blends into the rounded portion exactly on its widest point, so no line is required.

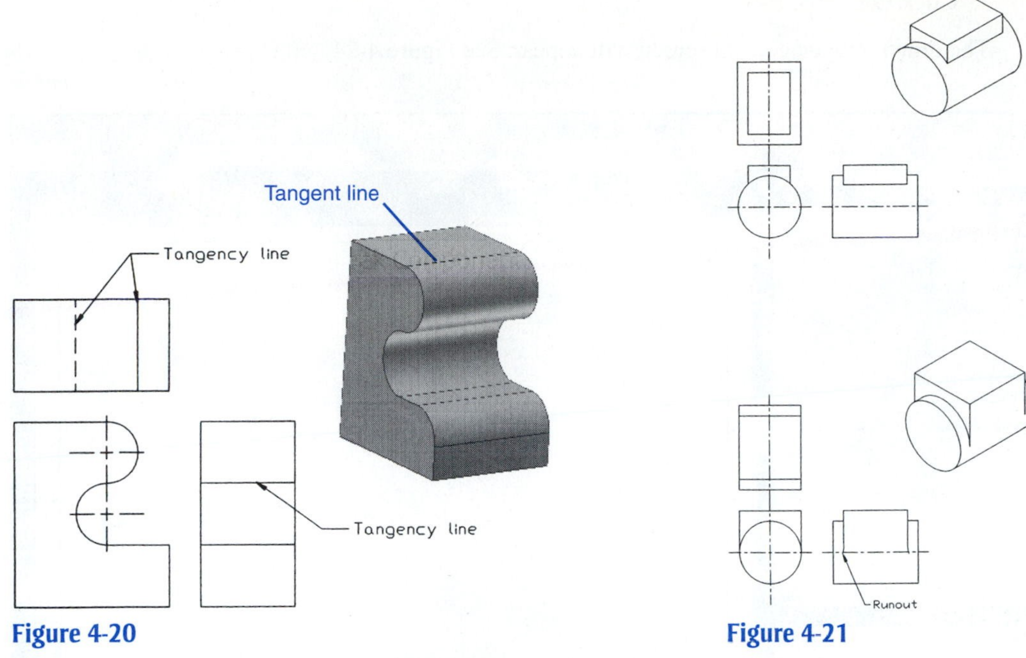

Figure 4-20

Figure 4-21

Orthographic Views With Inventor

Inventor will create orthographic views directly from models. Figure 4-22 shows a completed three-dimensional model. It was created using an existing file, **BLOCK, 3-HOLES.** It will be used throughout this chapter to demonstrate orthographic presentation views.

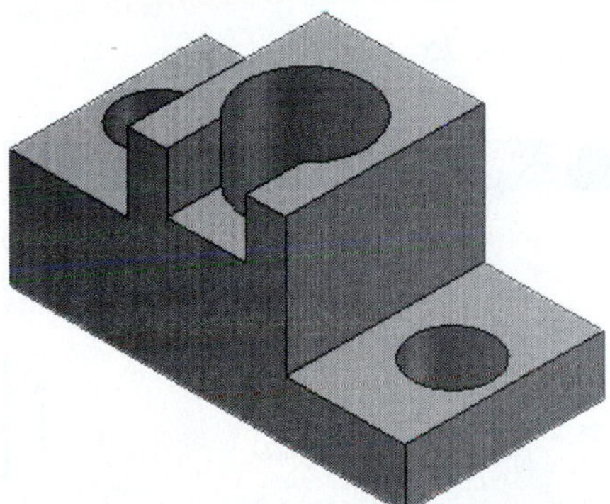

Figure 4-22

Exercise 4-1: Creating an Orthographic View

1. Start a new drawing, click the **Metric** tab, and select the **ANSI (mm).idw** option.

See Figure 4-23.

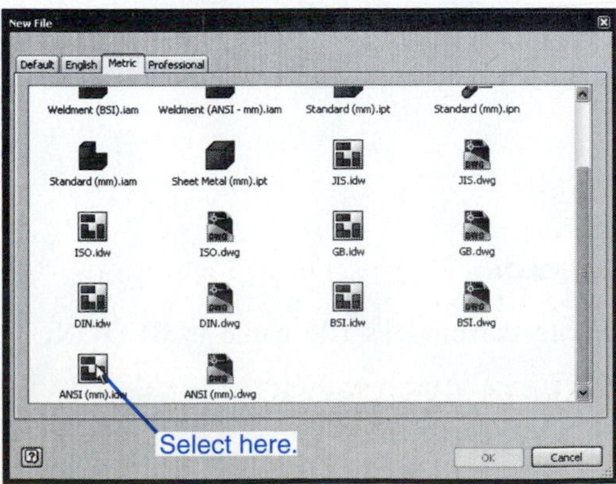

Figure 4-23

2. Click **OK.**

The drawing management screen will appear. See Figure 4-24.

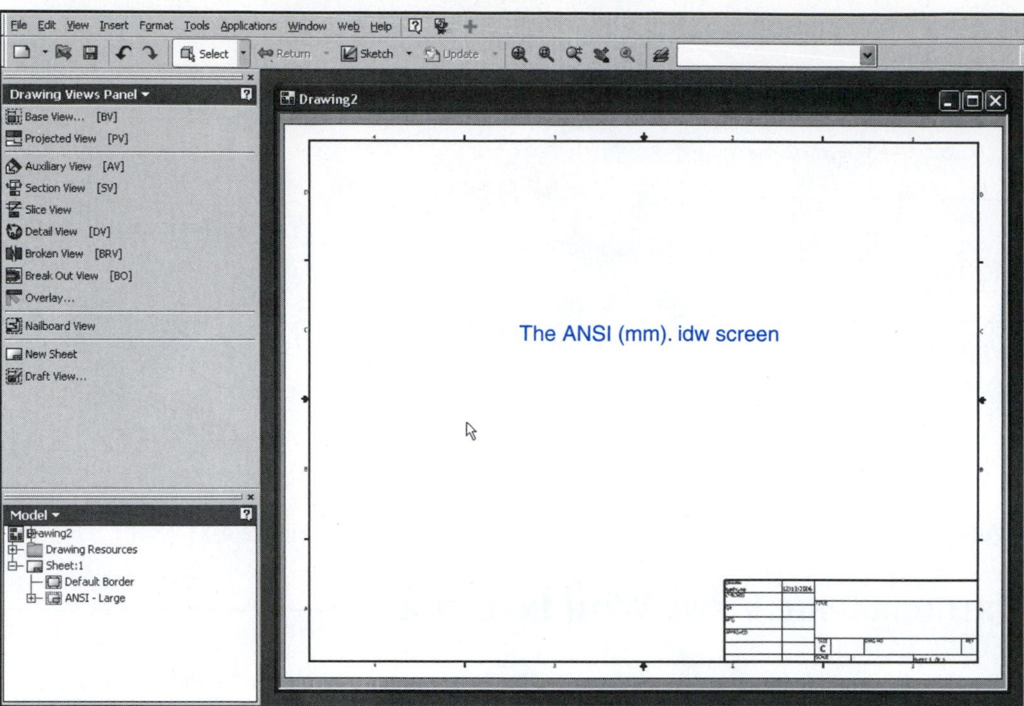

Figure 4-24

3. Click the **Base View…** tool in the **Drawing Views Panel** bar.

The **Drawing View** dialog box will appear. See Figure 4-25.

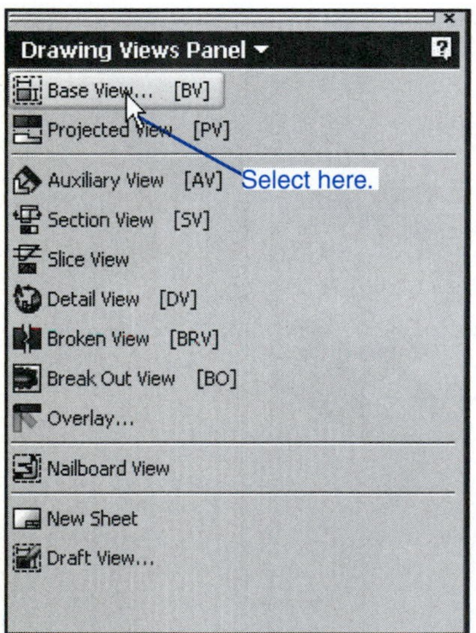

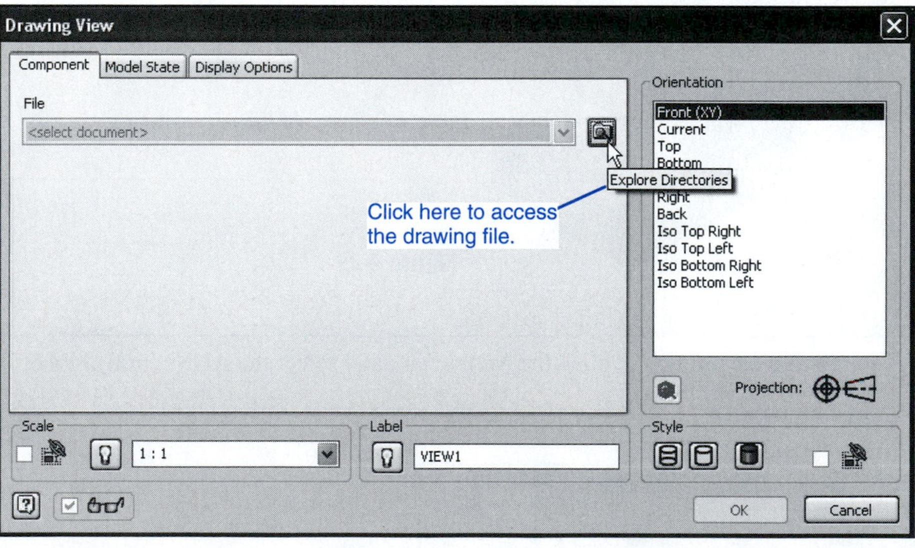

Figure 4-25

4. Click the **Explore Directories** button.

The **Open** dialog box will appear. See Figure 4-26.

5. Select the desired model. In this example the model's file name is **BLOCK, 3-HOLES**.

The **Drawing View** dialog box will appear. See Figure 4-27.

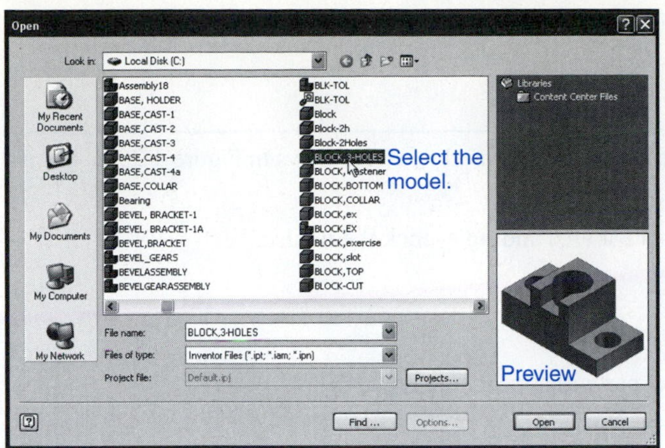

Figure 4-26

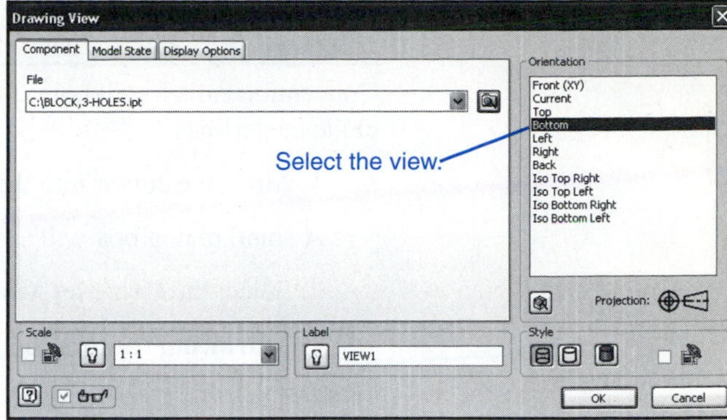

Figure 4-27

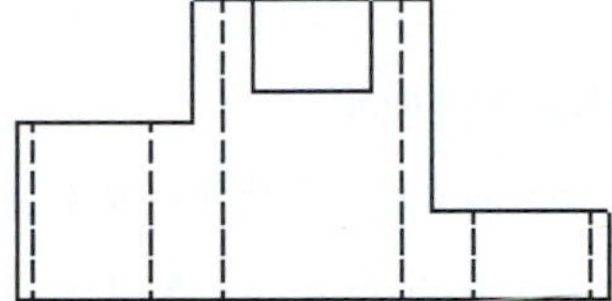
Resulting orthographic view Figure 4-28

6. Select the **Bottom** option, locate the view on the drawing screen, and click the location.

Figure 4-28 shows the resulting orthographic view.

The screen will include a border and a title block. The lettering in the title block may appear illegible. This is normal. The text will be legible when printed. The section on title blocks will explain how to work with title blocks.

Exercise 4-2: Creating Other Orthographic Views

1. Click the **Projected View** tool on the **Drawing Views Panel** bar.
2. Click the view already on the drawing screen.
3. Move the cursor upward from the view.

A second view will appear.

4. Select a location, click the left mouse button to place the view, then click the right mouse button and select the **Create** option.

Figure 4-29 shows the resulting two orthographic views. The initial view was created using the **Bottom** option. This is a relative term based on the way the model was drawn. The initial view

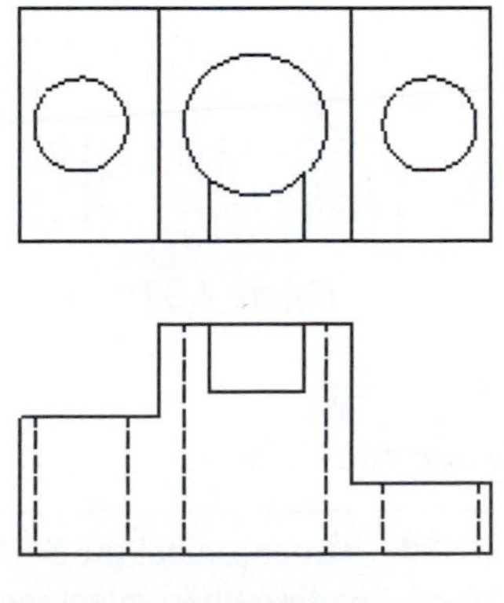
Orthographic views projected from the created view Figure 4-29

can be defined as the front view, and the second view created from that front view is also, by definition, the top view.

Exercise 4-3: Adding Centerlines

Convention calls for all holes to be defined using centerlines. The views in Figure 4-29 do not include centerlines.

1. Move the cursor into the panel bar area and right-click the mouse.

A small dialog box will appear.

2. Select the **Drawing Annotation** option.

The **Drawing Annotation Panel** bar will appear. See Figure 4-30.

3. Click the **Center Mark** tool.
4. Move the cursor into the drawing screen and click the edges of the holes in the top view.

See Figure 4-31.

5. Click the **Centerline Bisector** tool.

The **Centerline Bisector** tool is a flyout from the **Center Mark** tool.

6. Click each side of the hole projections in the front view.

Vertical centerlines will appear. See Figure 4-32.

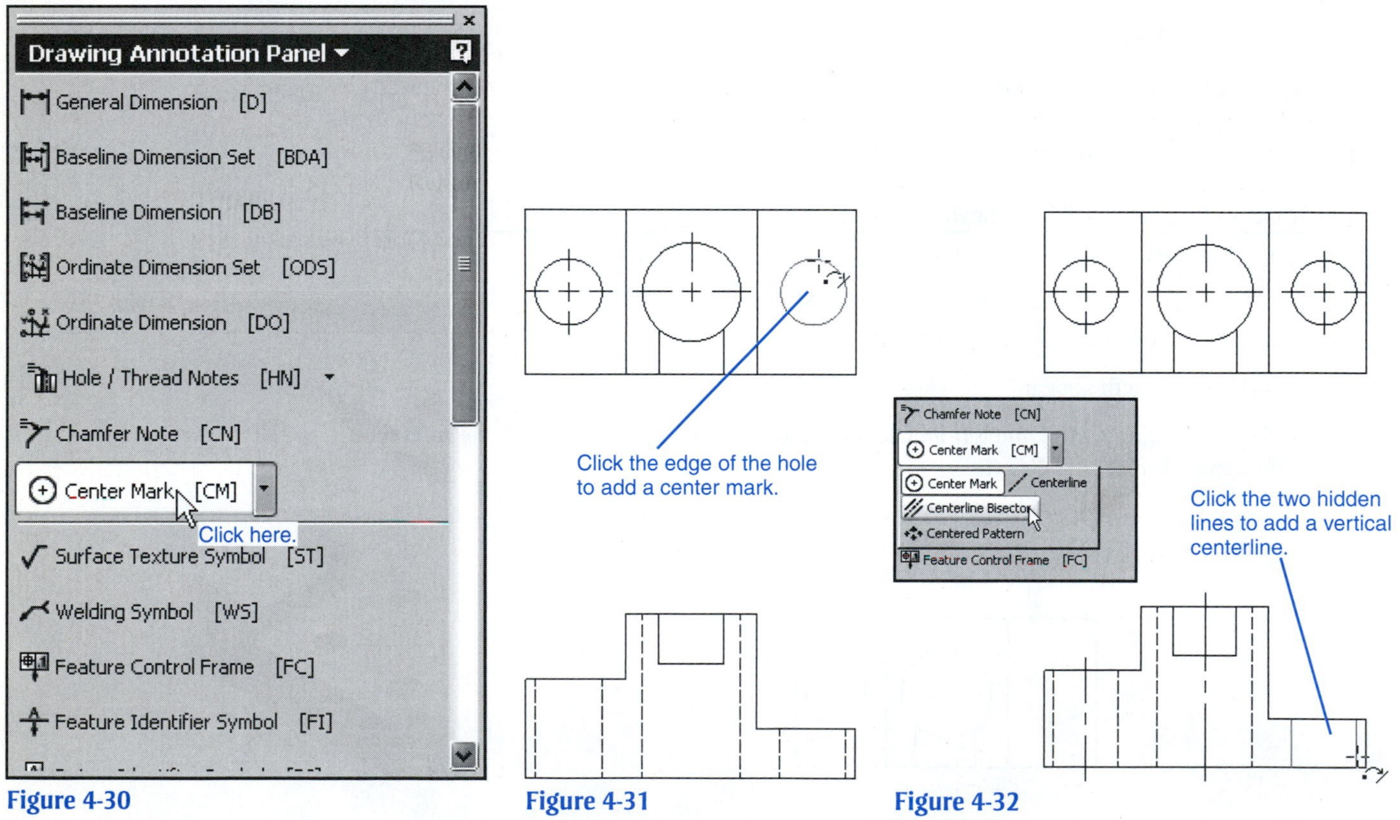

Figure 4-30

Figure 4-31

Figure 4-32

Exercise 4-4: Editing the Size of a Centerline

1. Select the **Format** heading at the top of the screen, select the **Styles Editor** option, click the + sign to the left of the **Center Mark** heading, then select the **Center Mark (ANSI)** option.

2. Change the center mark values as needed.

See Figure 4-33.

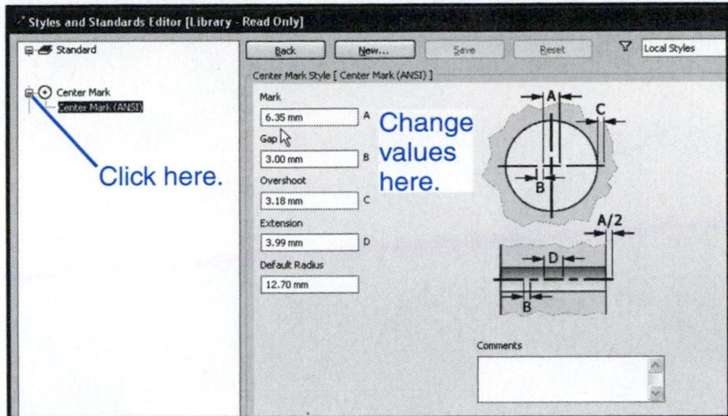

Figure 4-33

Exercise 4-5: Changing the Background Color of the Drawing Screen

1. Click the **Format** heading at the top of the screen.

Select the **Active Standards** option.

The **Document Settings** dialog box will appear. See Figure 4-34.

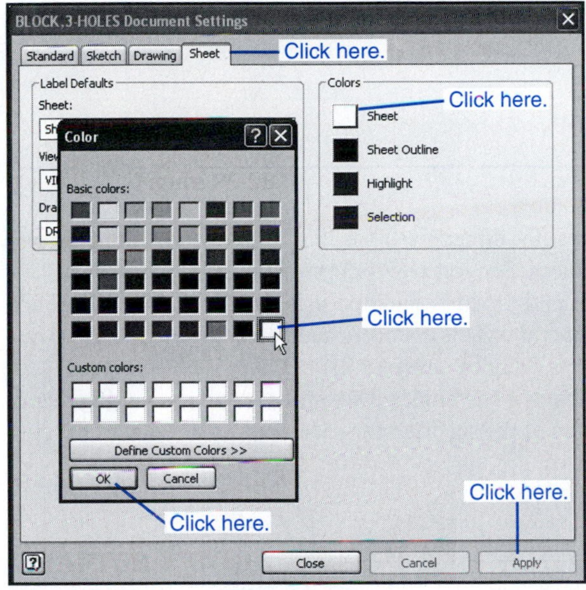

Figure 4-34

3. Click the **Sheet** tab, then the heading **Sheet** in the **Colors** box.

If the centerlines are too large or too small, edit their size.

The **Color** dialog box will appear.

4. Click the desired color, then **OK.**

The **Document Settings** dialog box will appear.

5. Click the **Apply** box, then **OK.**

The sheet's background color will be changed.

Isometric Views

An isometric view may be created from any view on the screen. The resulting orientation will vary according to the view selected. In this example the front view is selected.

1. Click the **Projected View** tool.
2. Click the **Front** view.
3. Move the cursor to the right of the front view and select a location for the isometric view.
4. Move the cursor slightly and click the right mouse button.
5. Select the **Create** option.

Figure 4-35 shows the resulting isometric view.

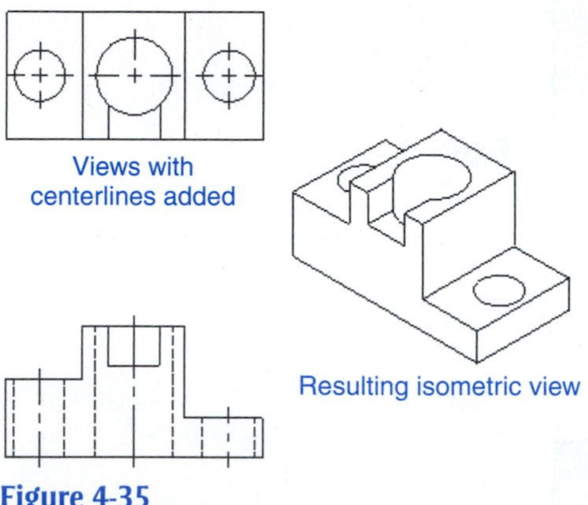

Figure 4-35

Section Views

section view: A view used to expose an internal surface of a model.

cutting plane: A plane used to define the location of a section view.

Some objects have internal surfaces that are not directly visible in normal orthographic views. ***Section views*** are used to expose these surfaces. Section views do not include hidden lines.

Any material cut when a section view is defined is hatched using section lines. There are many different styles of hatching, but the general style is evenly spaced 45° lines. This style is defined as ANSI 31 and will be applied automatically by Inventor.

Figure 4-36 shows a three-dimensional view of an object. The object is cut by a cutting plane. ***Cutting planes*** are used to define the location of the section view. Material to one side of the cutting plane is removed, exposing the section view.

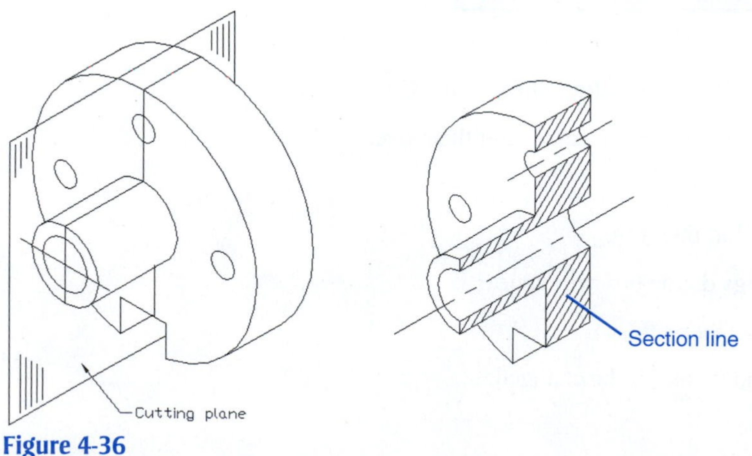

Figure 4-36

Figure 4-37 shows the same object presented using two dimensions. The cutting plane is represented by a cutting plane line. The cutting plane line is defined as A-A, and the section view is defined as view A-A.

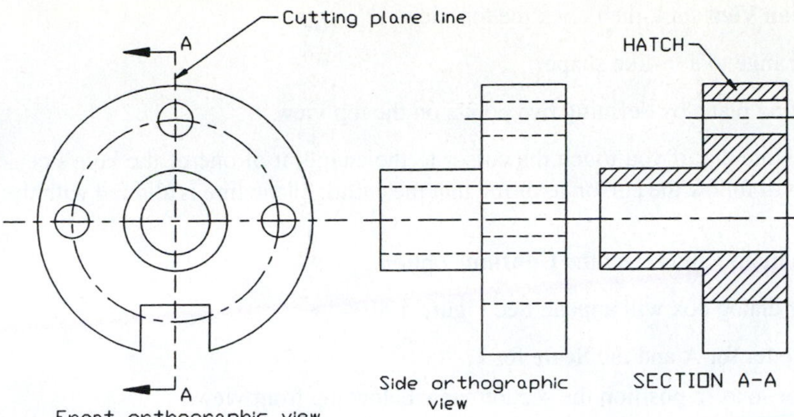

Figure 4-37

All surfaces directly visible must be shown in a section view. In Figure 4-38 the back portion of the object is not affected by the section view and is directly visible from the cutting plane. The section view must include these surfaces. Note how the rectangular section blocks out part of the large hole. No hidden lines are used to show the hidden portion of the large hole.

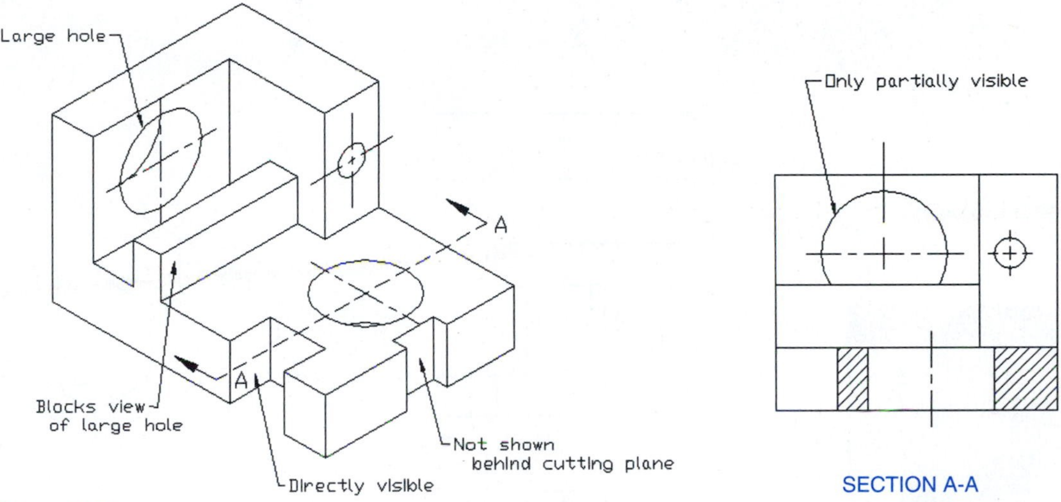

Figure 4-38

Exercise 4-6: Drawing a Section View Using Inventor

Figure 4-39 shows the front and top views of the object defined in Figure P3-10. A section view will be created by first defining the cutting plane line in the top view, then projecting the section view below the front view.

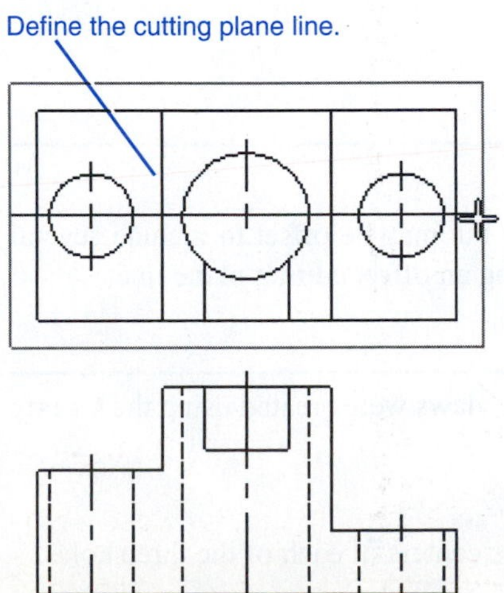

Figure 4-39

1. Click the **Section View** tool, then click the top view.

The cursor will change to a +-like shape.

2. Define the cutting plane by defining two points on the top view.

See Figure 4-39. Note that if you touch the cursor to the endpoint of one of the hole's centerlines, a dotted line will follow the cursor, assuring that the cutting plane line is aligned with the hole's centerlines.

3. Right-click the mouse and select the **Continue** option.

The **Section View** dialog box will appear. See Figure 4-40.

4. Set the **Label** letter for **A** and the **Scale** for **1.**
5. Move the cursor so as to position the section view below the front view.
6. Click the section view location.
7. Add the appropriate centerlines using the **Centerline Bisector** tool.

Figure 4-41 shows the resulting section view. Notice that the section view is defined as A-A, and the scale is specified. The arrows of the cutting plane line are directed away from the section view. The section view is located behind the arrows.

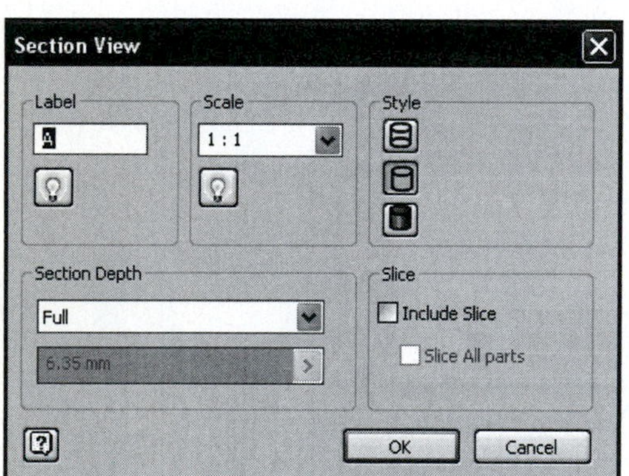

Figure 4-40

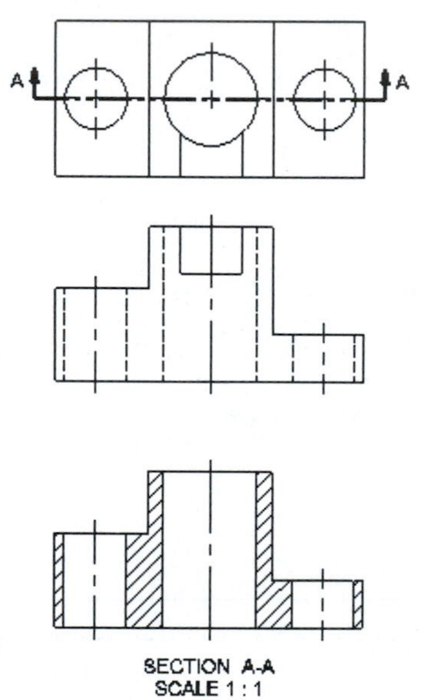

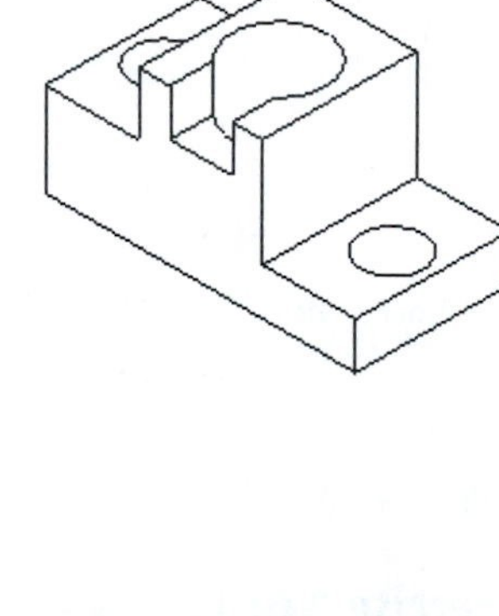

Figure 4-41

Offset Section Views

Cutting plane lines need not pass directly across an object but may be offset to include several features. Figure 4-42 shows an object that has been cut using an offset cutting plane line.

Exercise 4-7: Creating an Offset Cutting Plane

Figure 4-43 shows the front and top views of an object. The views were created using the **Create View, Project View,** and **Centerline** tools.

1. Click the **Section View** tool, then click the top view.
2. Draw a cutting plane across the top view through the centers of each of the three holes.
3. Locate the section view below the front view.

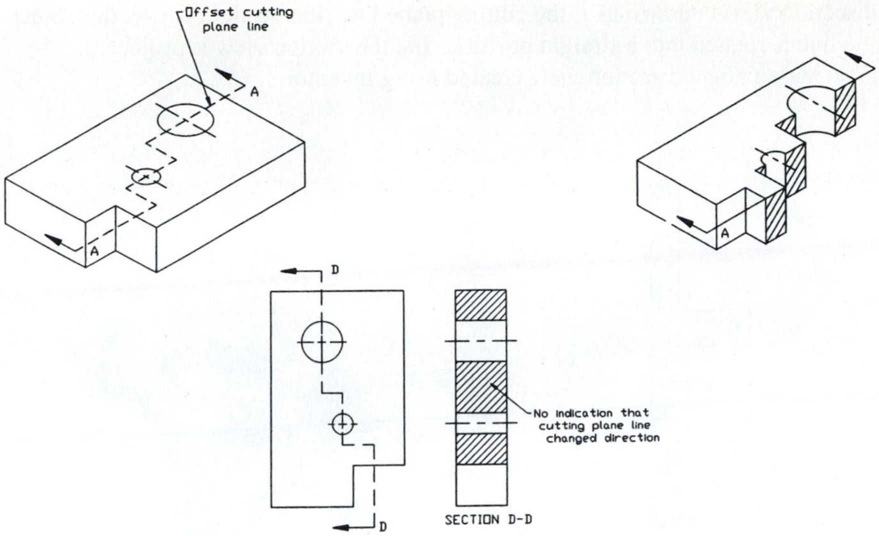

Offset cutting plane line

No indication that cutting plane line changed direction

SECTION D-D

Figure 4-42

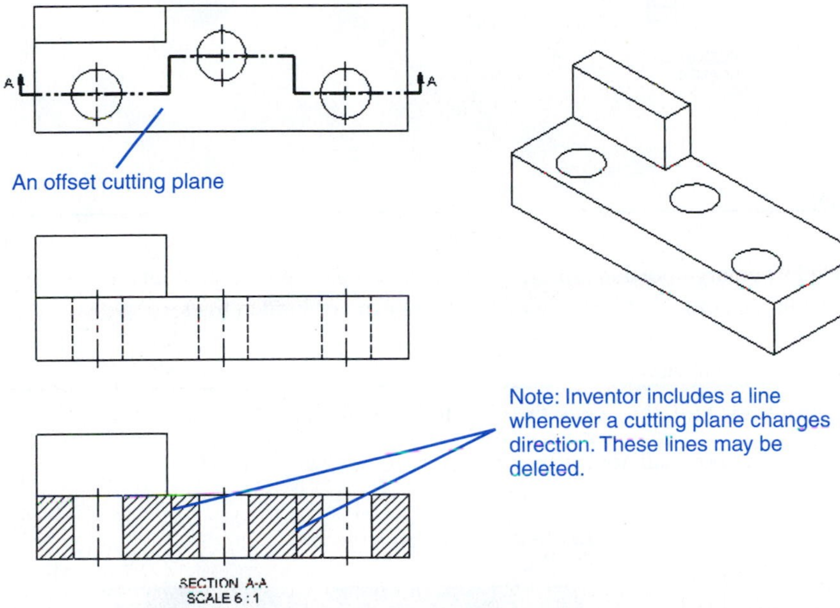

An offset cutting plane

Note: Inventor includes a line whenever a cutting plane changes direction. These lines may be deleted.

SECTION A-A
SCALE 6 : 1

Figure 4-43

Aligned Section Views

Figure 4-44 shows an example of an aligned section view. Aligned section views are most often used on circular objects and use an angled cutting plane line to include more features in the section view, like an offset cutting plane line.

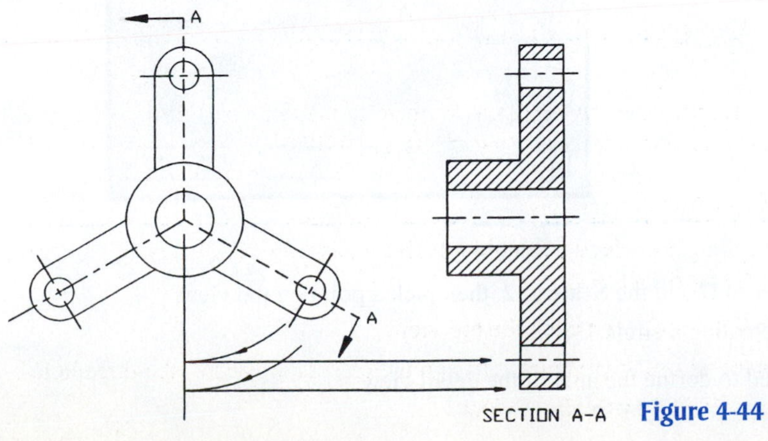

Figure 4-44

An aligned section view is drawn as if the cutting plane line runs straight across the object. The cutting plane line is rotated into a straight position, and the section view is projected.

Figure 4-45 shows an aligned section view created using Inventor.

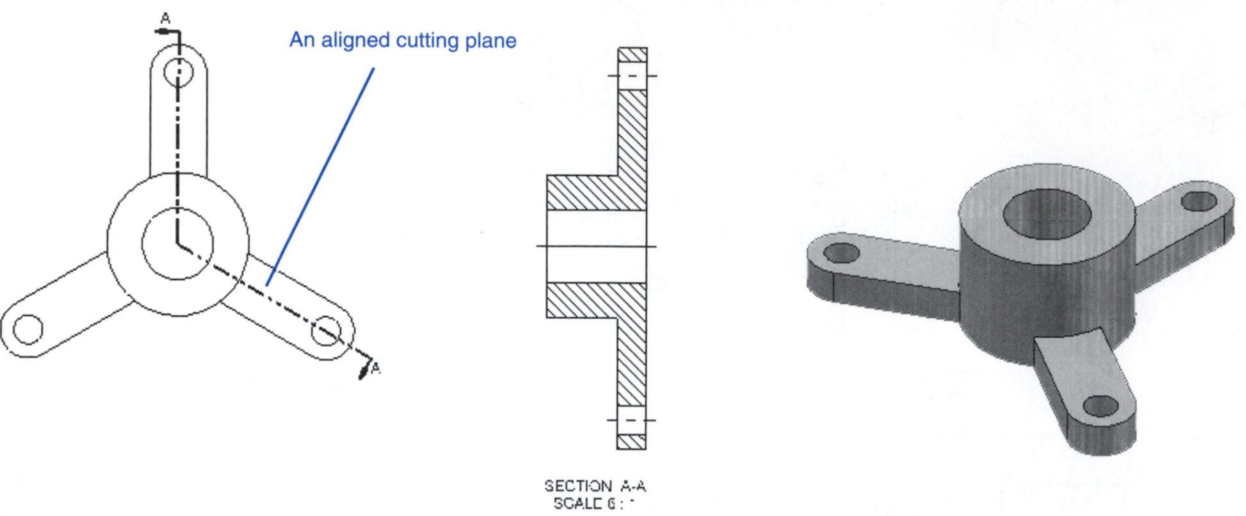

Figure 4-45

Detail Views

detail view: An enlarged view of a portion of a model.

Detail views are used to enlarge portions of an existing drawing. The enlargements are usually made of areas that could be confusing because of many crossing or hidden lines.

Exercise 4-8: Creating a Detail View

1. Click the **Detail View** tool, then click the view to be enlarged.

The **Detail View** dialog box will appear. See Figure 4-46.

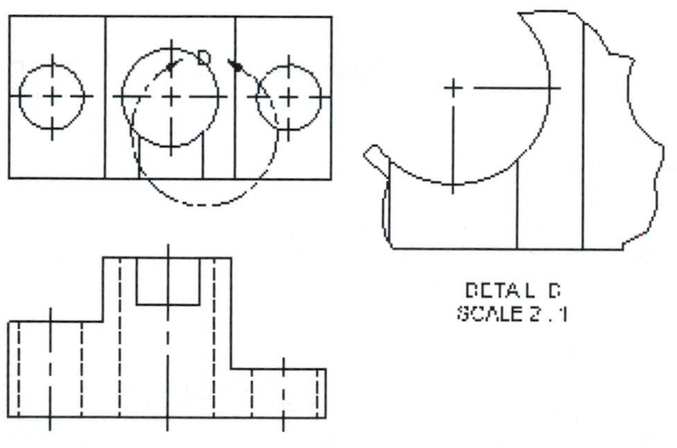

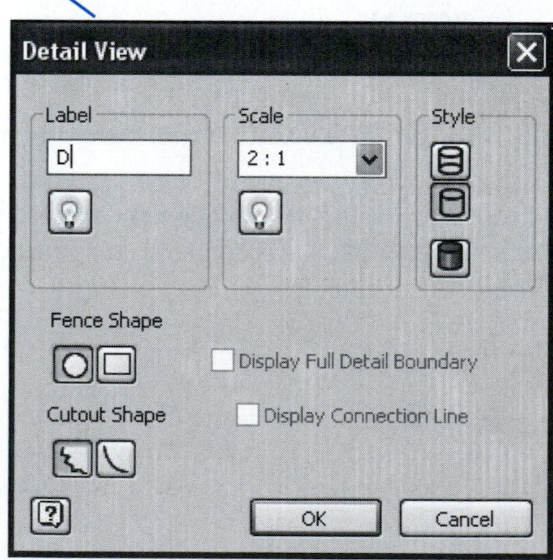

Figure 4-46

2. Set the **Label** letter to **D** and the **Scale** to **2**, then pick a point on the view.
3. Move the cursor, creating a circle.

The circle will be used to define the area of the detail view.

4. When the circle is of appropriate diameter click the left mouse button and move the cursor away from the view.
5. Locate the detail view and click the location.

Note that the shape of the detail view may be changed.

Broken Views

It is often convenient to break long continuous shapes so that they take up less drawing space. Figure 4-47 shows a long L-bracket that has a continuous shape; that is, its shape is constant throughout its length. Figure 4-48 shows an orthographic view of the same L-bracket.

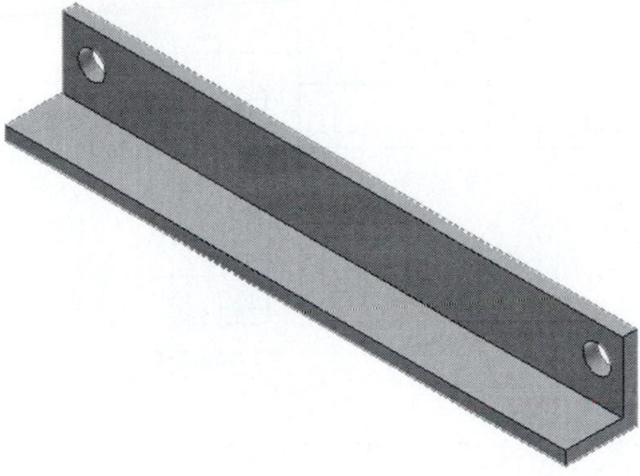

Figure 4-47

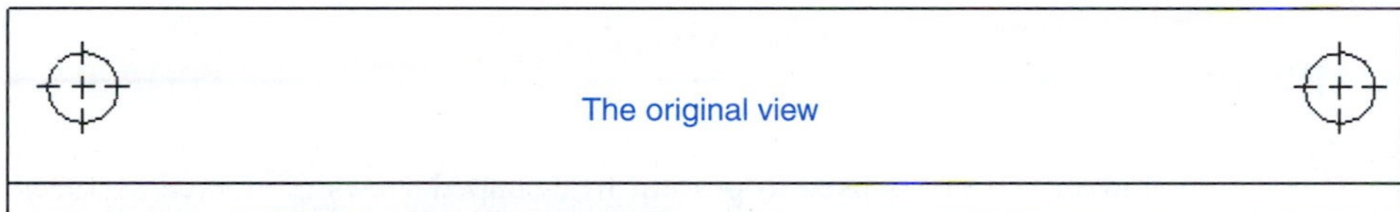

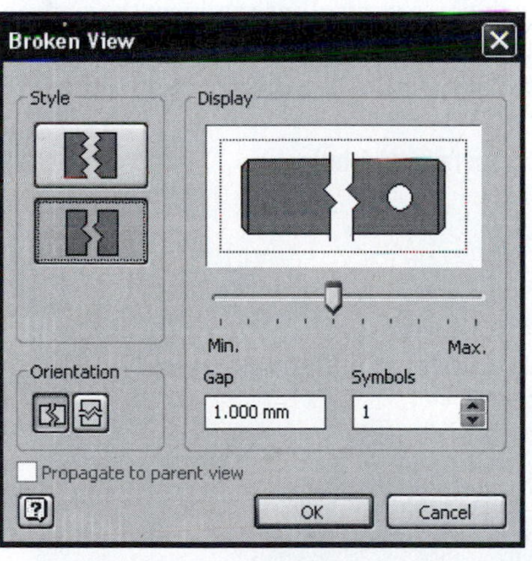

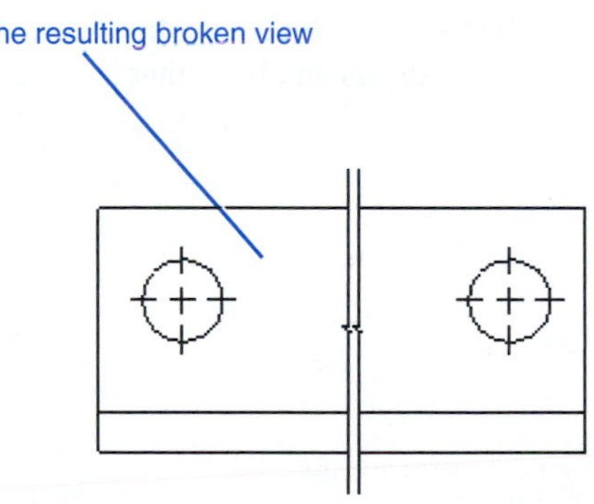

Figure 4-48

Exercise 4-9: Create a Broken View

1. Click the **Broken View** tool, then the orthographic view.

The **Broken View** dialog box will appear.

2. Select the orientation of the break and the gap distance between the two portions of the L-bracket.
3. Click a point near the left end of the L-bracket, then move the cursor to the right and click a second point near the right end of the L-bracket.

Figure 4-48 shows the resulting broken view.

Multiple Section Views

It is acceptable to take more than one section view of the same object in order to present a more complete picture of the object. Figures 4-49 and 4-50 show objects that use more than one section view.

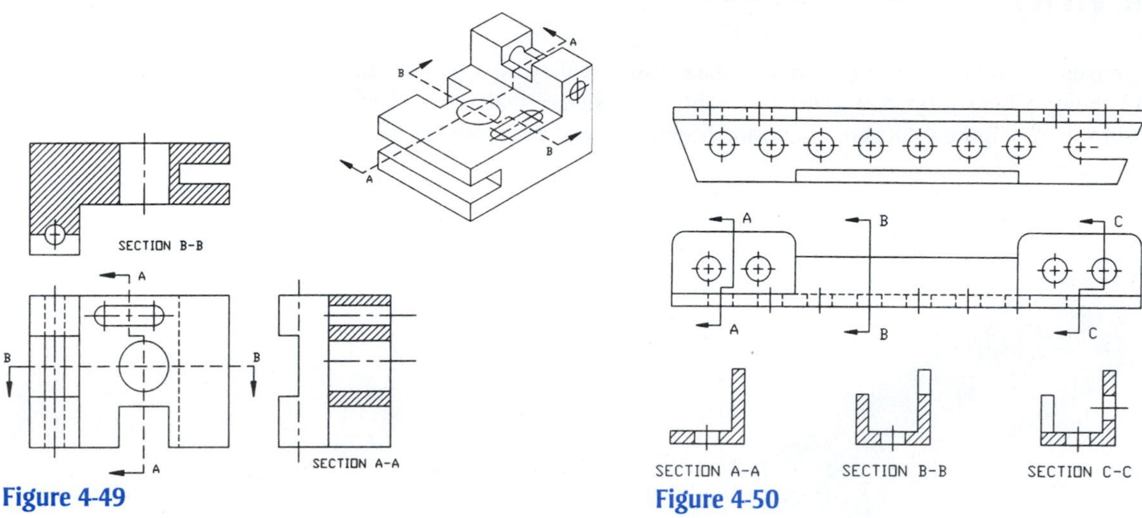

Figure 4-49

Figure 4-50

AUXILIARY VIEWS

auxiliary view: An orthographic view used to present the true shape of a slanted surface.

Auxiliary views are orthographic views used to present true-shaped views of slanted surfaces. Figure 4-51 shows an object with a slanted surface that includes a hole drilled perpendicular to the slanted surface. Note how the right-side view shows the hole as an ellipse and that the surface A-B-C-D is foreshortened; that is, it is not shown at its true size. Surface A-B-C-D does appear at its true shape and size in the auxiliary view. The auxiliary view was projected at 90° from the slanted surface so as to generate a true-shaped view.

Figure 4-52 shows an object that includes a slanted surface and hole.

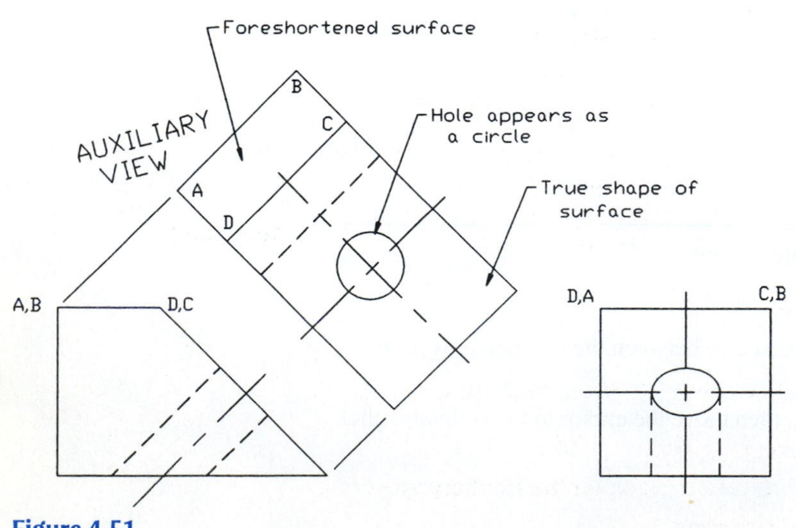

Figure 4-51

An object with a slanted surface created using Inventor.

Figure 4-52

Exercise 4-10: Drawing an Auxiliary View

1. Click the **Create View** and **Project View** tools and create front and right-side views as shown in Figure 4-53.

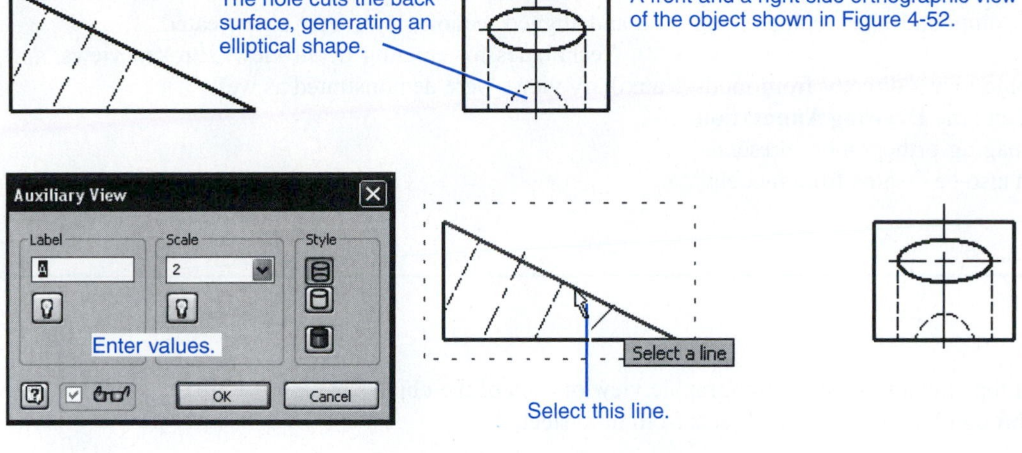

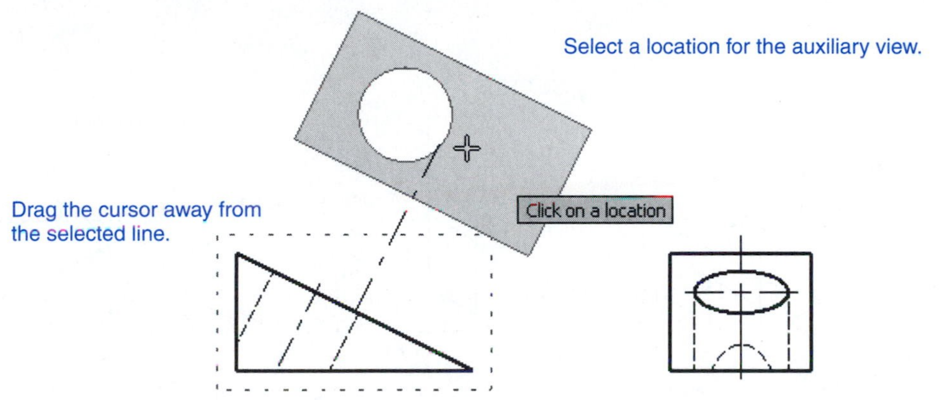

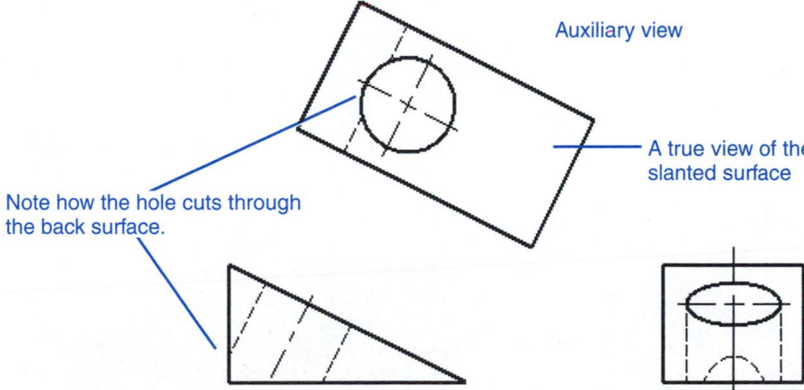

Figure 4-53

Click the **Auxiliary View** tool, then the front view.

2. The **Auxiliary View** dialog box will appear.
3. Enter the appropriate settings, then click the slanted edge line in the front view.

In this example, a scale of 2:1 was used.

4. Move the cursor away from the front view and select a location for the auxiliary view.
5. Click the left mouse button and create the auxiliary view.

SUMMARY

This chapter introduced orthographic drawings using third-angle projection in accordance with ANSI standards. Conventions were demonstrated for objects with normal surfaces, hidden lines, slanted surfaces, compound lines, oblique surfaces, and rounded surfaces.

Inventor creates orthographic views directly from models. The **Drawing Views Panel** and the **Drawing Annotation Panel** were introduced for managing orthographic presentation views. Isometric views can also be created from models.

Section views are used to expose internal surfaces that are not directly visible in normal orthographic views. Cutting planes were used to define the location of section views. Offset and aligned section views were also created.

Techniques for creating detail views, broken views, and auxiliary views were demonstrated as well.

CHAPTER PROJECTS

Project 4-1:

Draw a front, a top, and a right-side orthographic view of each of the objects in Figures P4-1 through P4-24. Make all objects from mild steel.

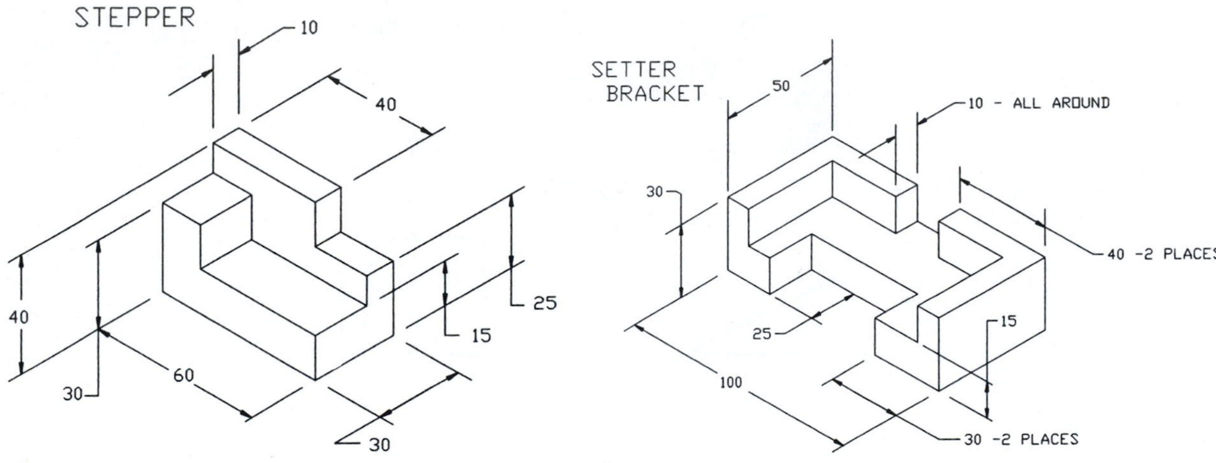

Figure P4-1 MILLIMETERS

Figure P4-2 MILLIMETERS

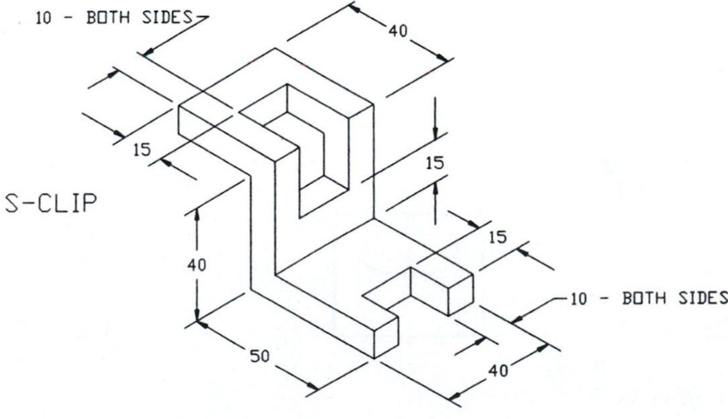

Figure P4-3 MILLIMETERS

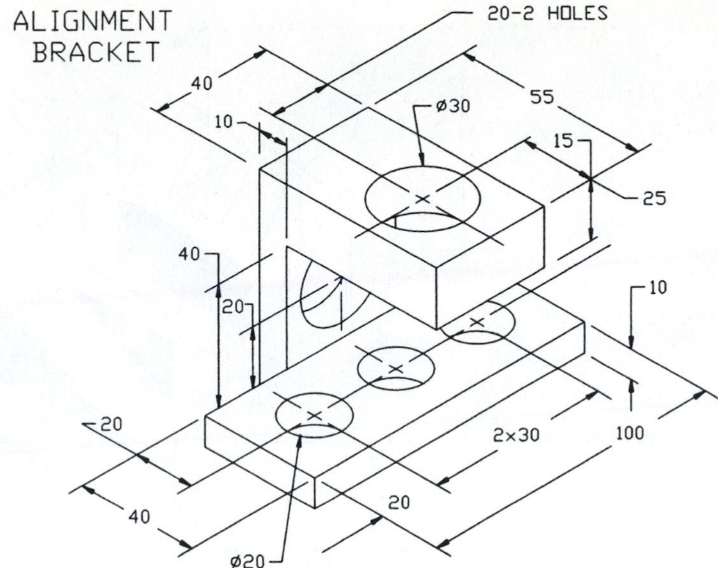

Figure P4-4 MILLIMETERS

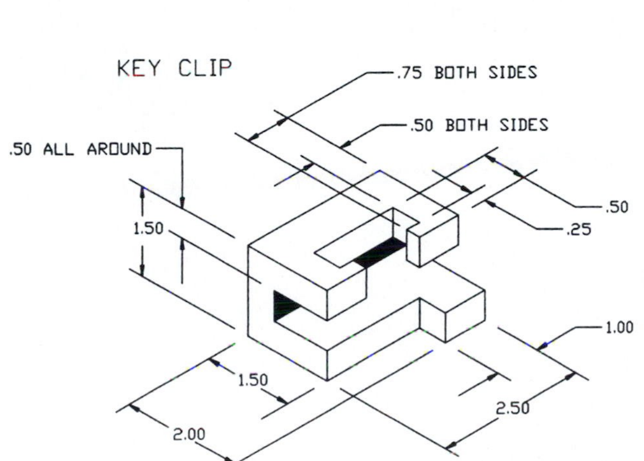

Figure P4-5 INCHES

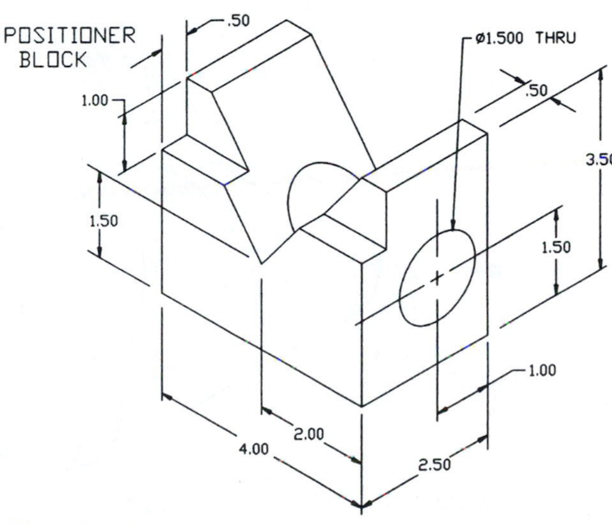

Figure P4-6 INCHES

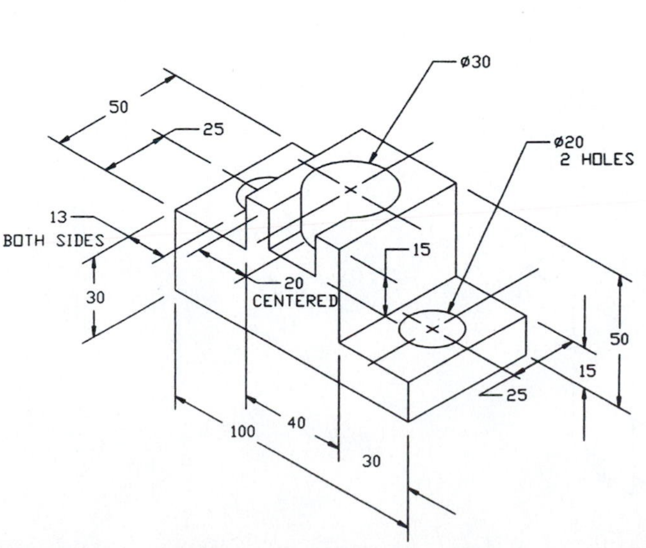

Figure P4-7 MILLIMETERS

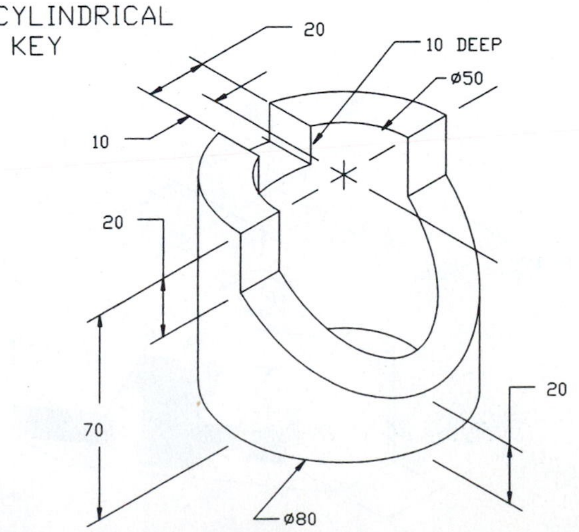

Figure P4-8 MILLIMETERS

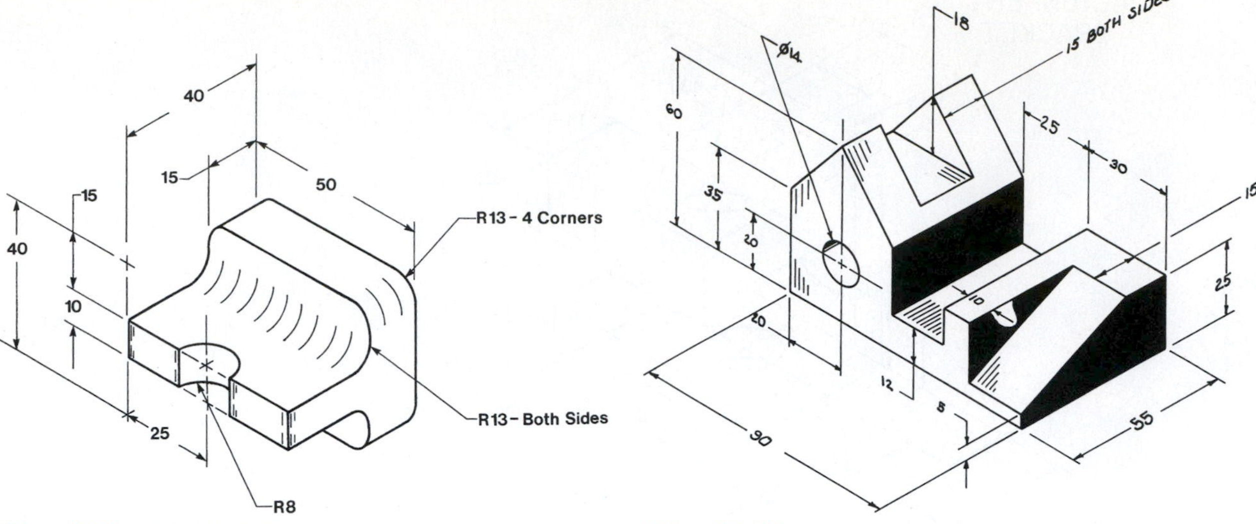

Figure P4-9 MILLIMETERS

Figure P4-10 MILLIMETERS

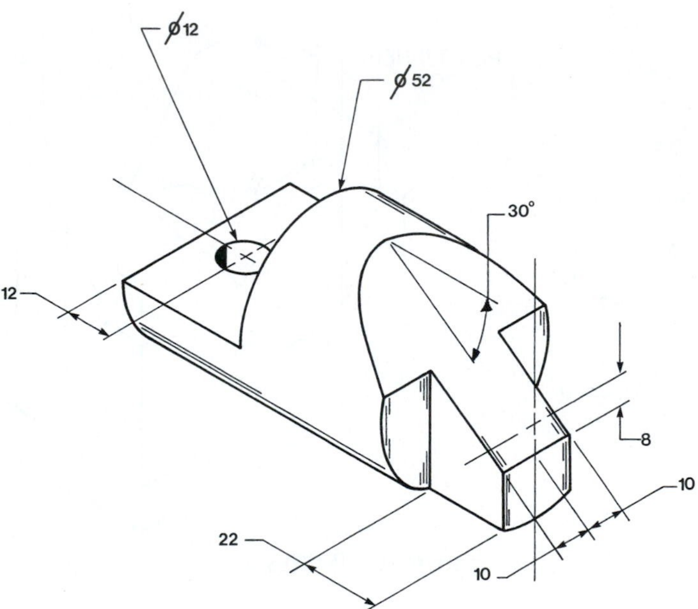

Figure P4-11 MILLIMETERS

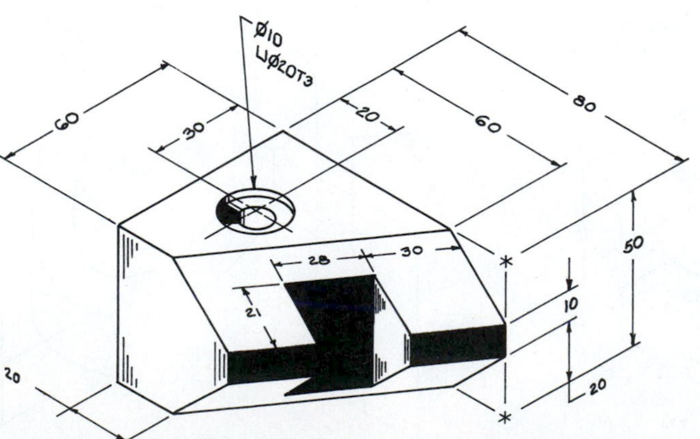

Figure P4-12 MILLIMETERS

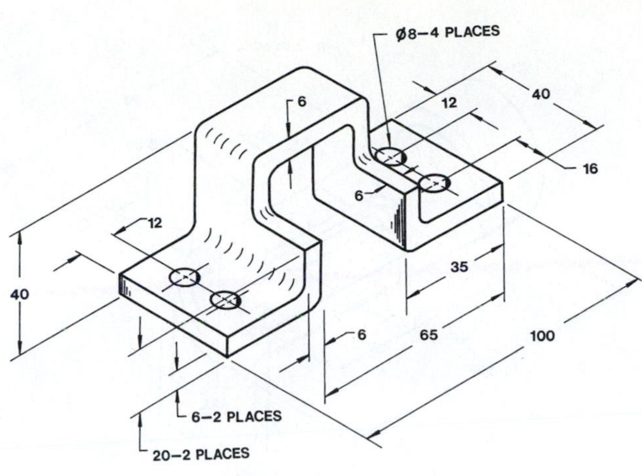

Figure P4-13 MILLIMETERS

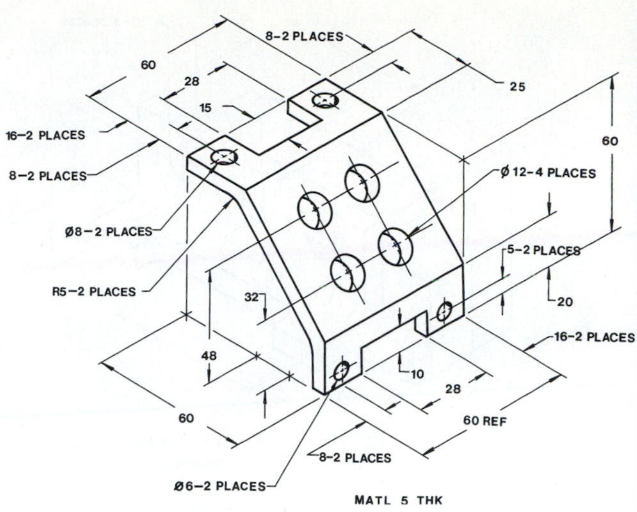

Figure P4-14 MILLIMETERS

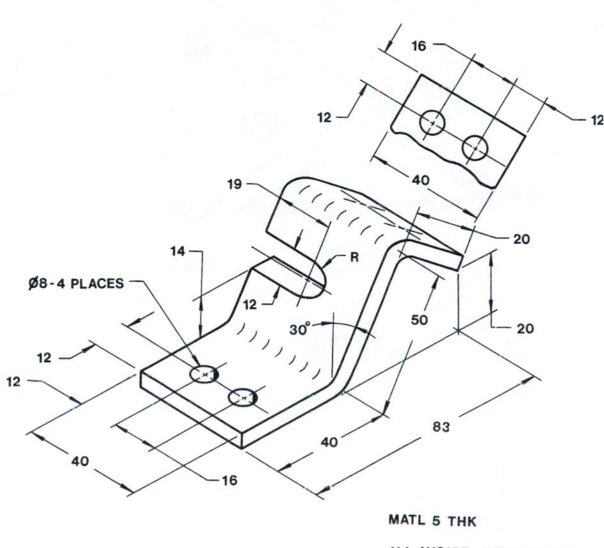

Figure P4-15 MILLIMETERS

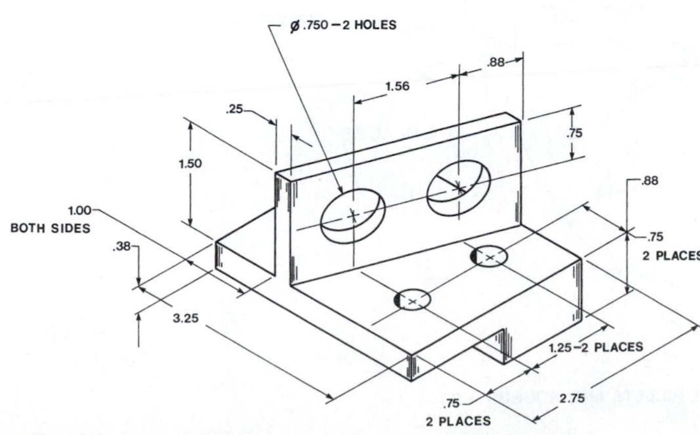

Figure P4-16 INCHES

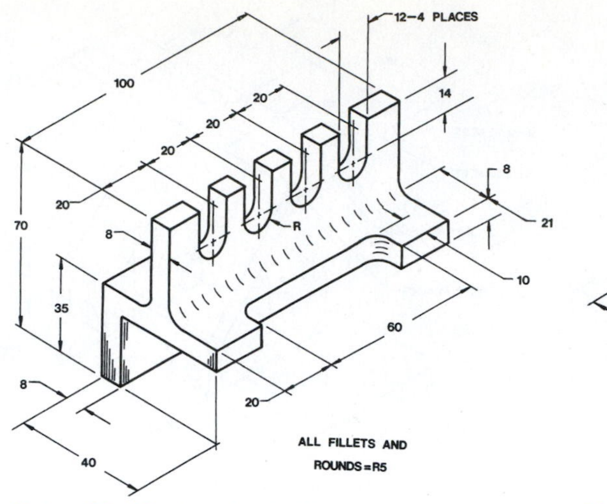

Figure P4-17 MILLIMETERS

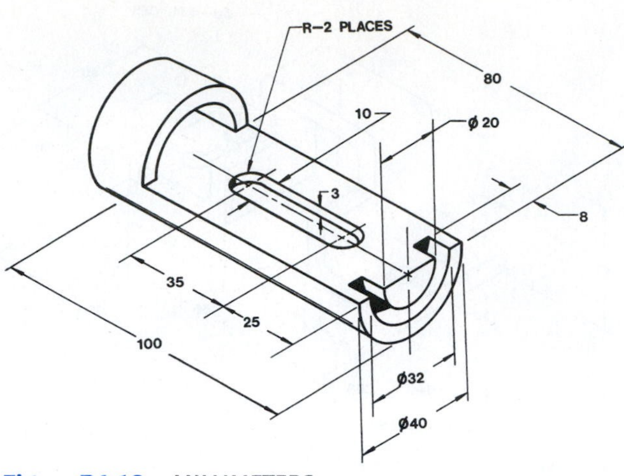

Figure P4-18 MILLIMETERS

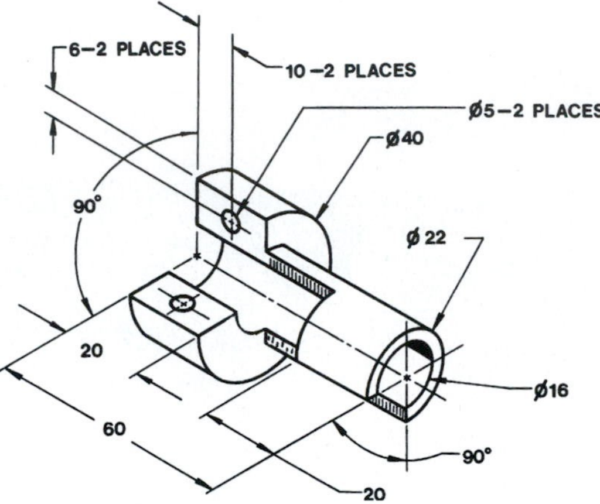

Figure P4-19 MILLIMETERS

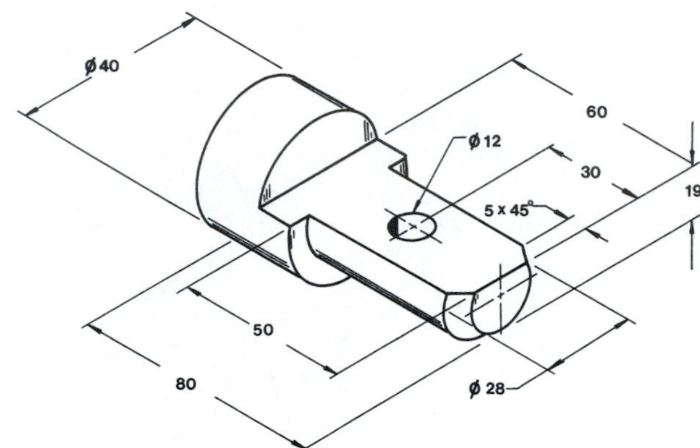

Figure P4-20 MILLIMETERS

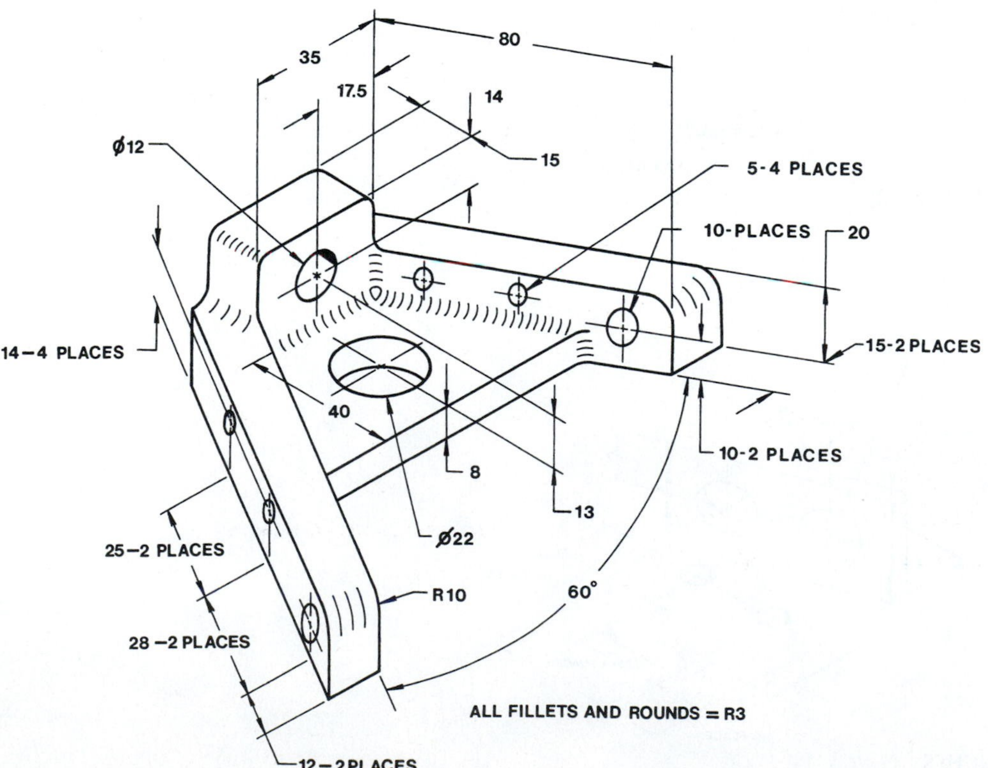

Figure P4-21 MILLIMETERS

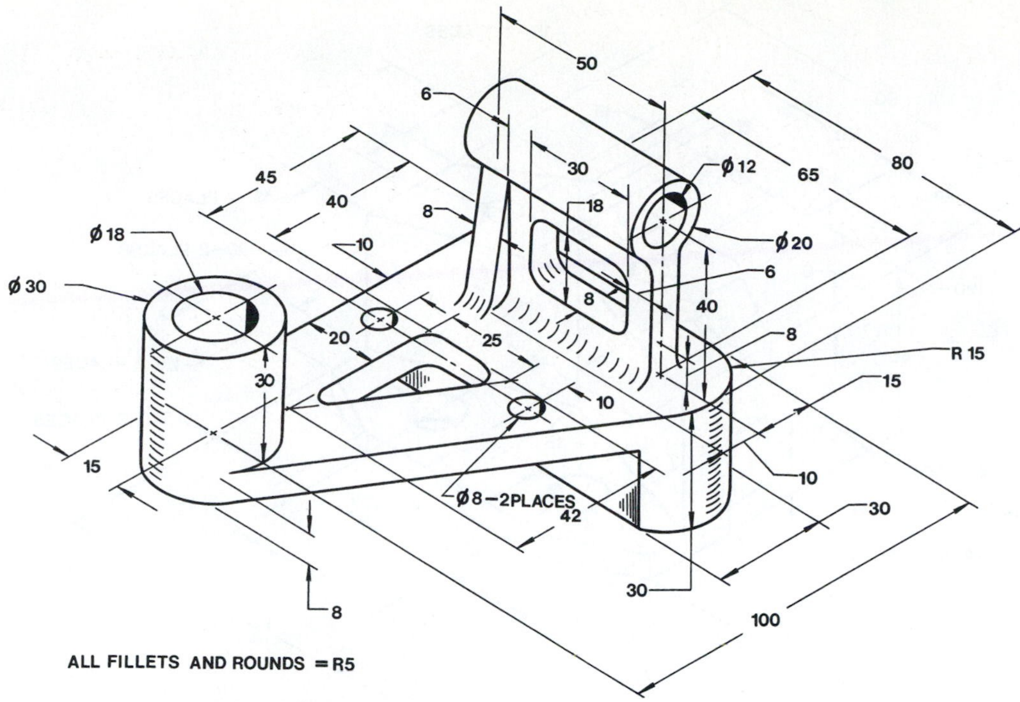

Figure P4-22 MILLIMETERS

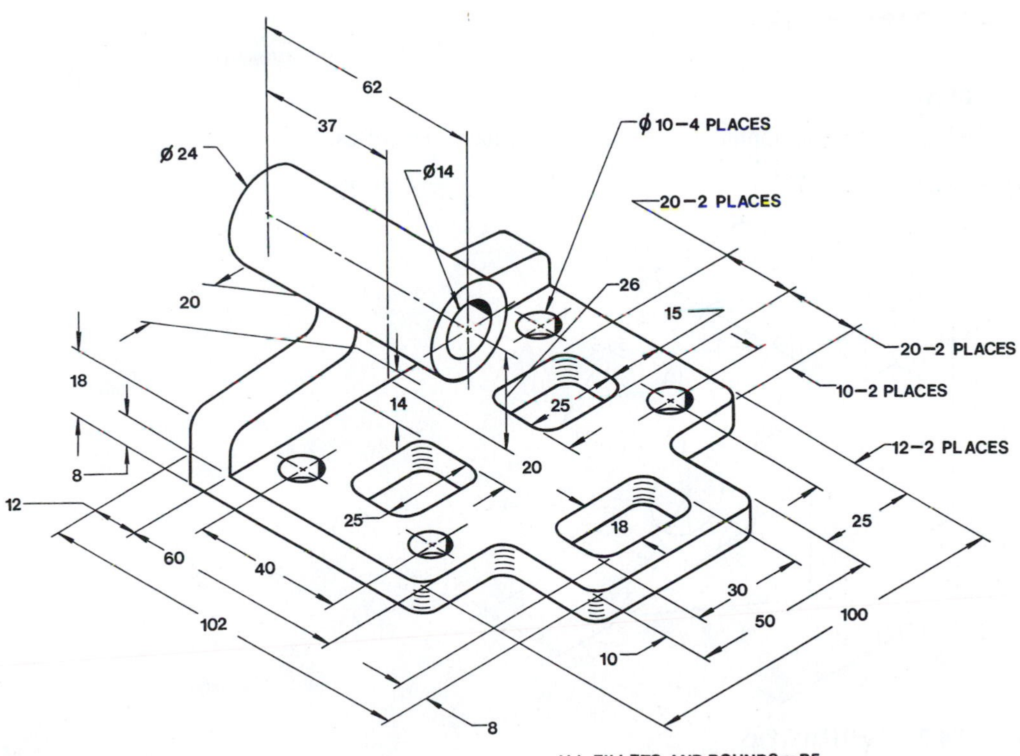

Figure P4-23 MILLIMETERS

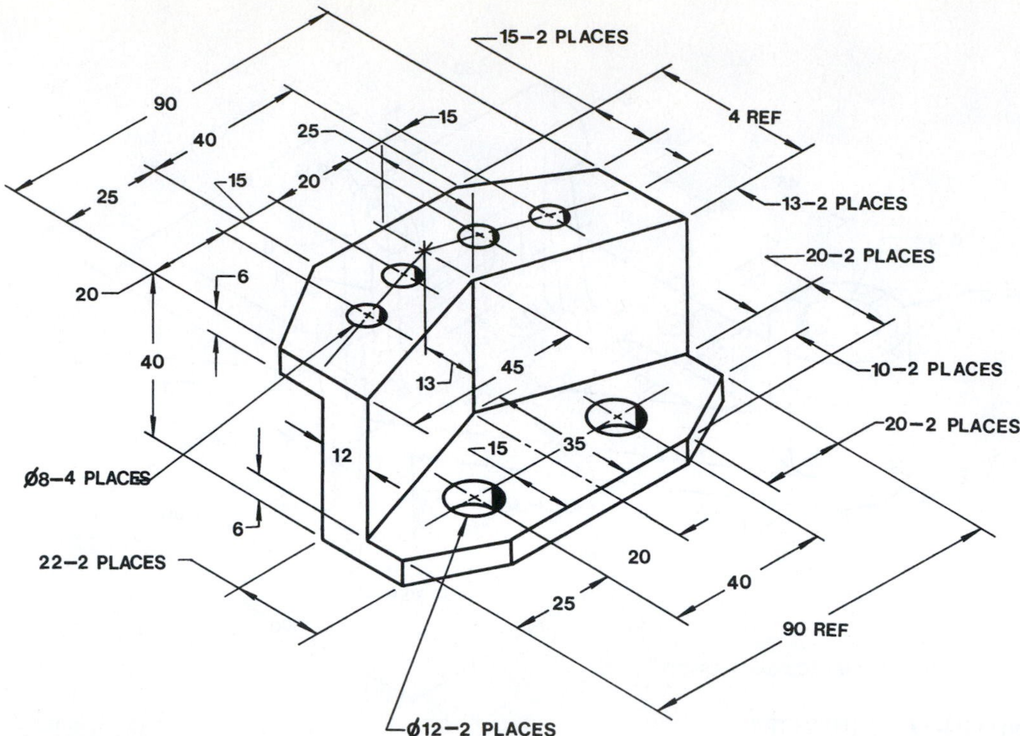

Figure P4-24 MILLIMETERS

Project 4-2:

Draw at least two orthographic views and one auxiliary view of each of the objects in Figures P4-25 through P4-36.

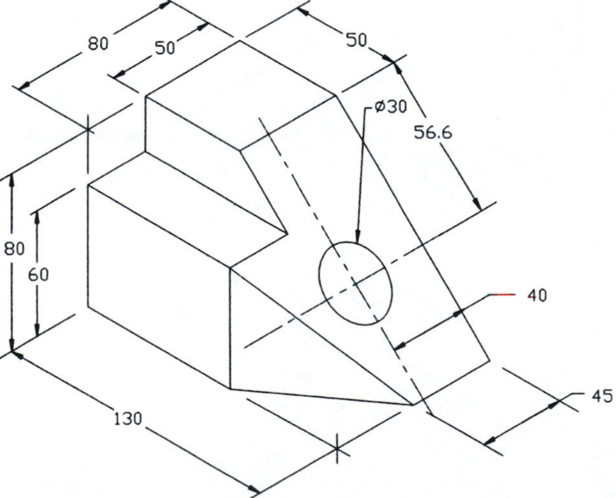

Figure P4-25 MILLIMETERS

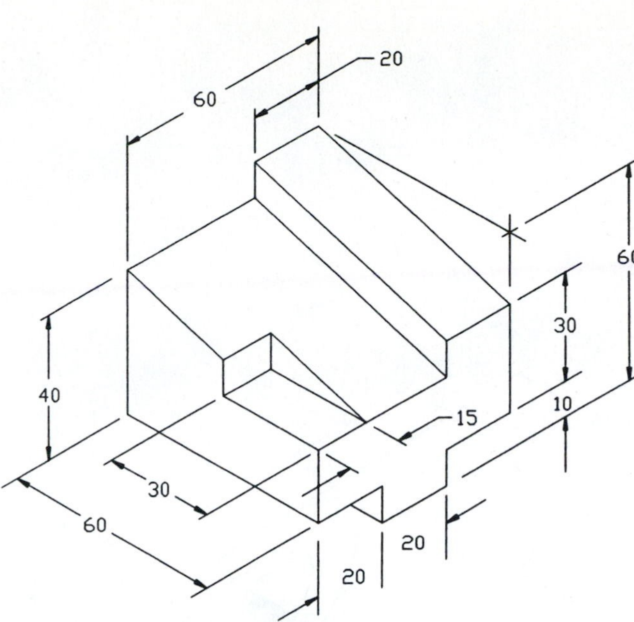

Figure P4-26 MILLIMETERS

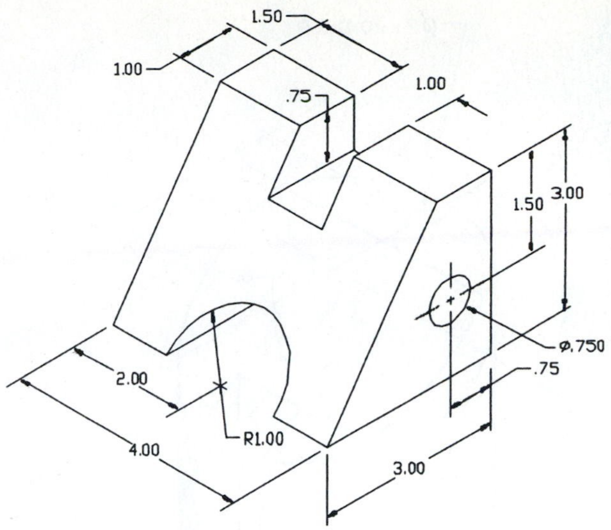

Figure P4-27 INCHES

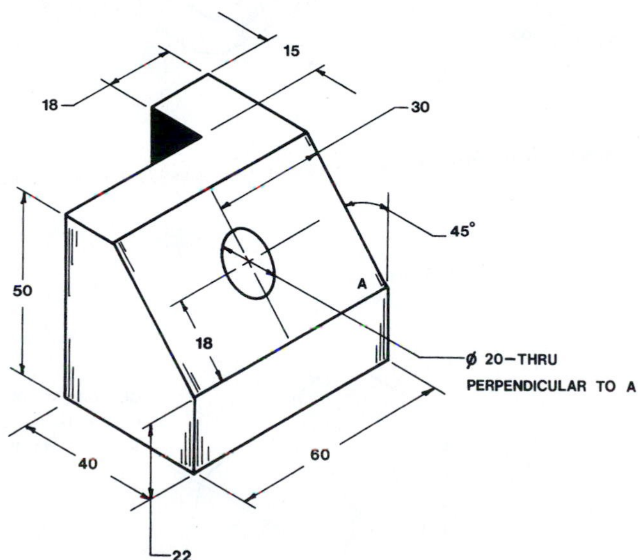

Figure P4-28 MILLIMETERS

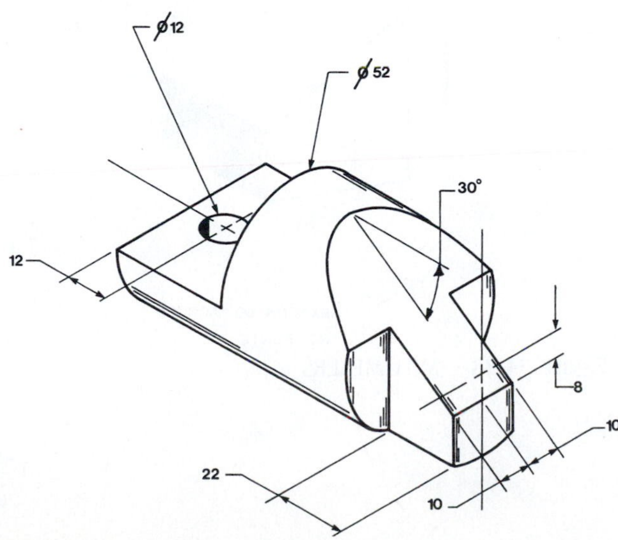

Figure P4-29 MILLIMETERS

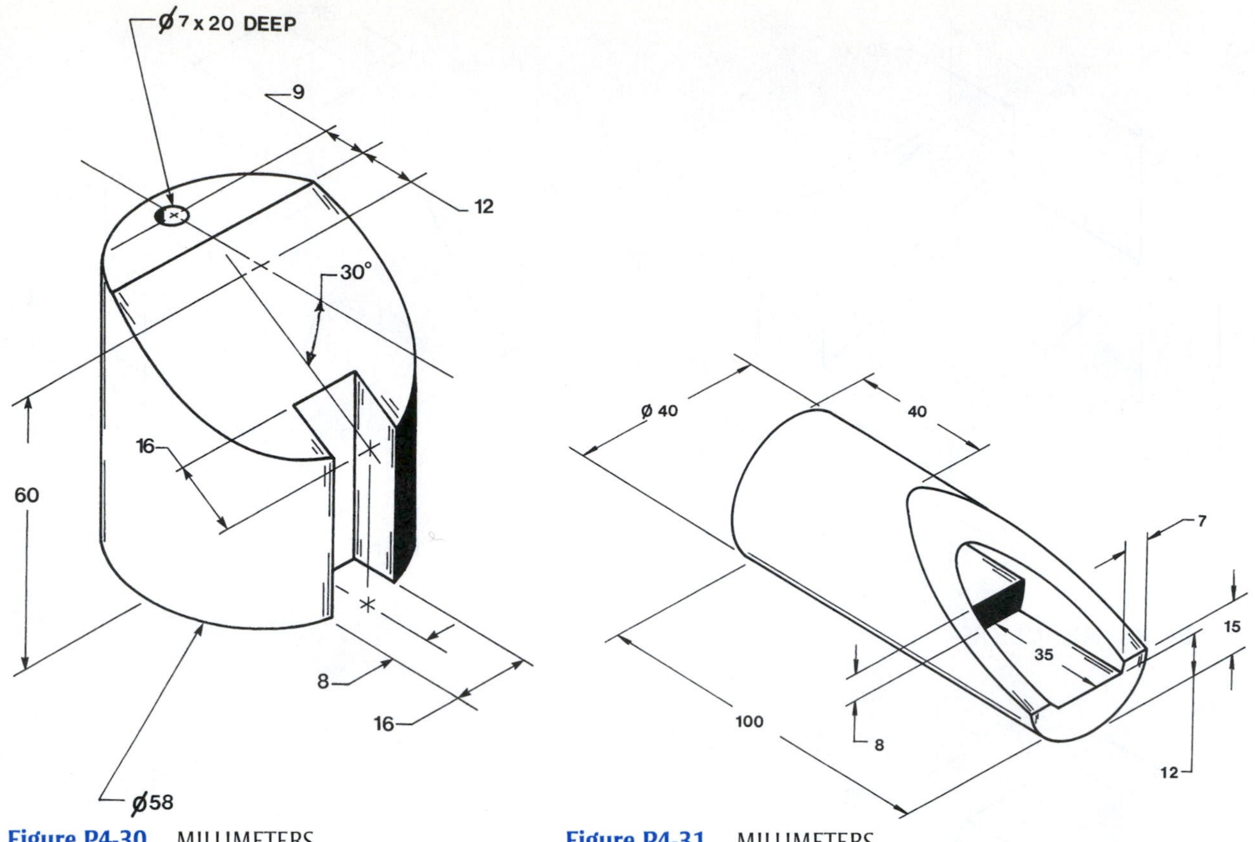

Figure P4-30 MILLIMETERS

Figure P4-31 MILLIMETERS

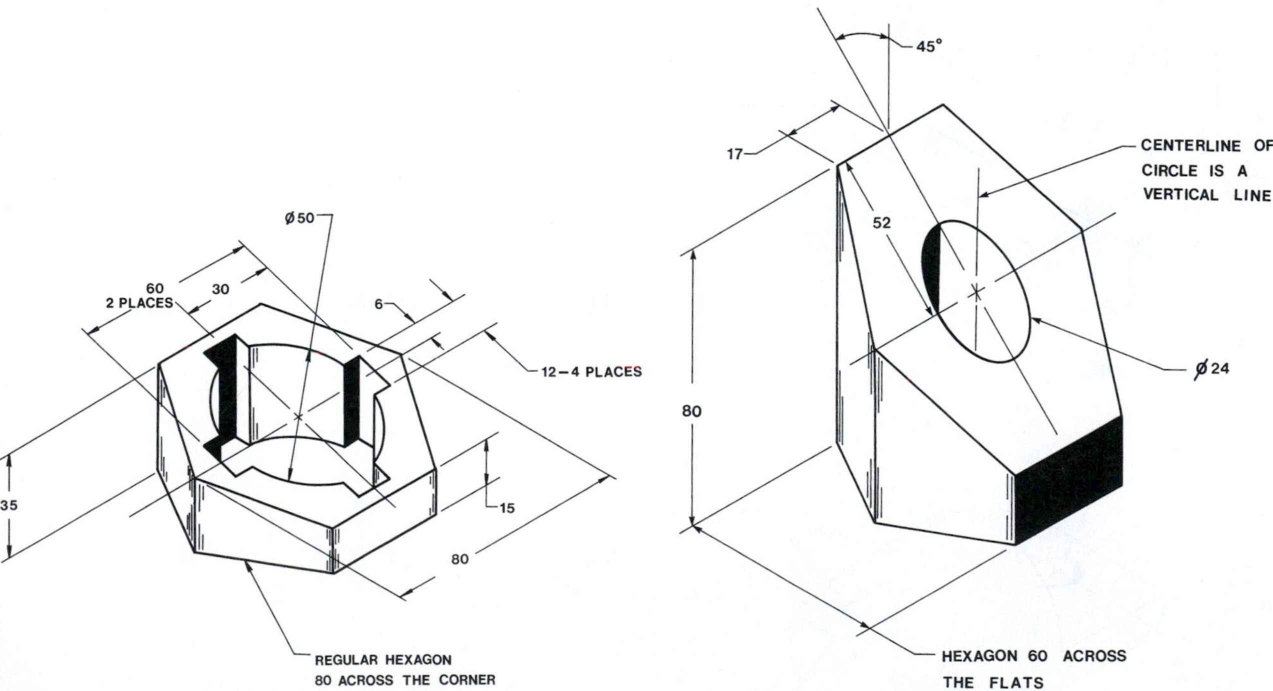

Figure P4-32 MILLIMETERS

Figure P4-33 MILLIMETERS

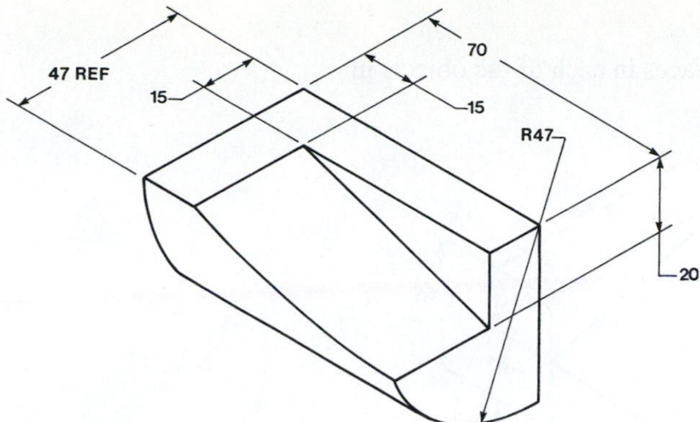

Figure P4-34 MILLIMETERS

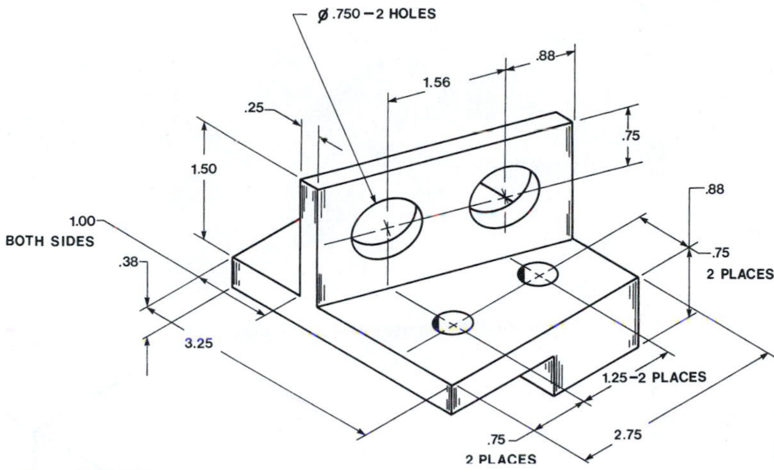

Figure P4-35 INCHES

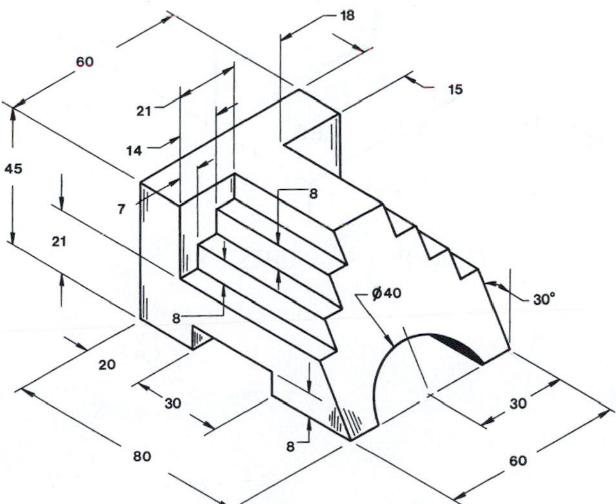

Figure P4-36 MILLIMETERS

Project 4-3:

Define the true shape of the oblique surfaces in each of the objects in Figures P4-37 through P4-40.

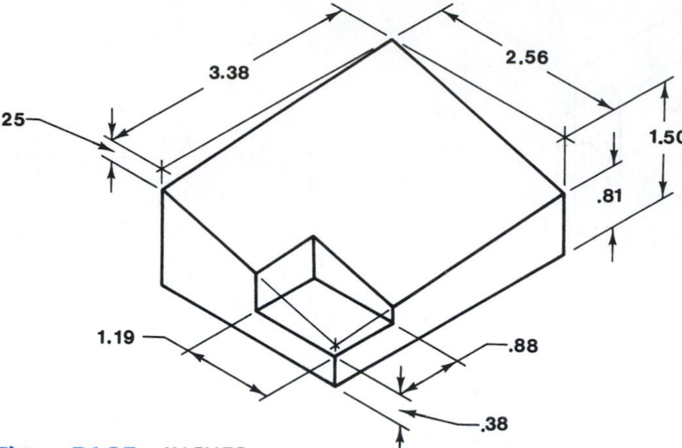

Figure P4-37 INCHES

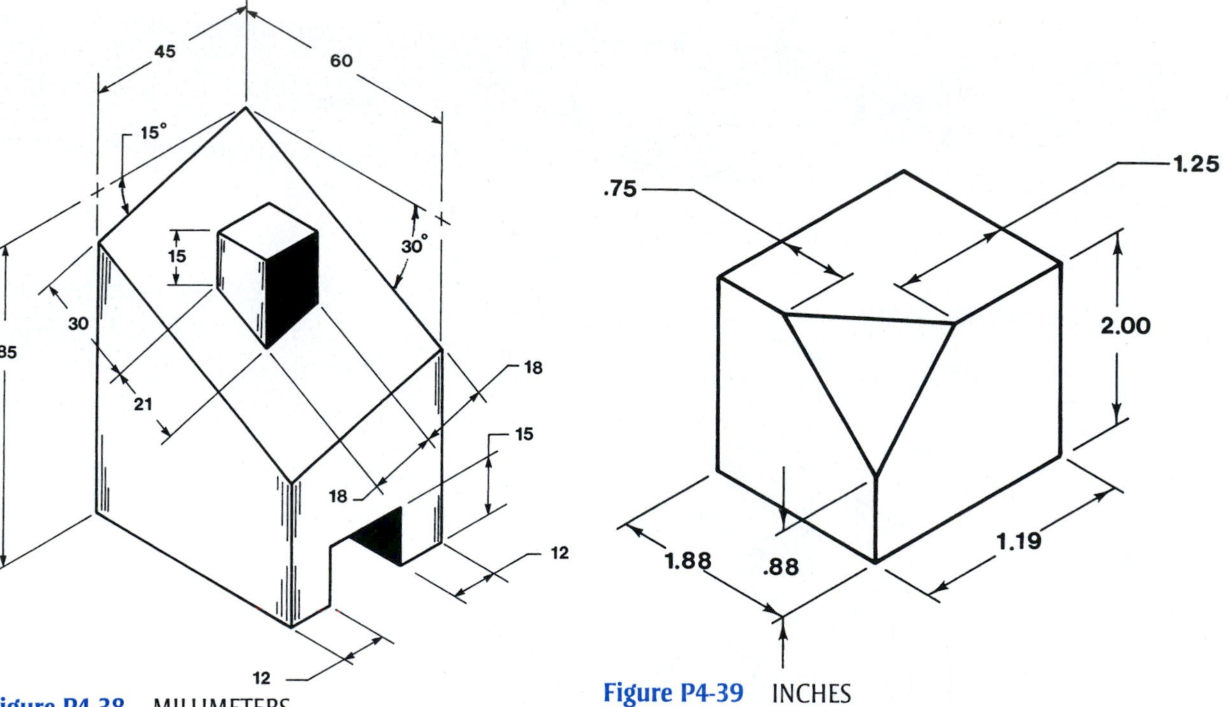

Figure P4-38 MILLIMETERS

Figure P4-39 INCHES

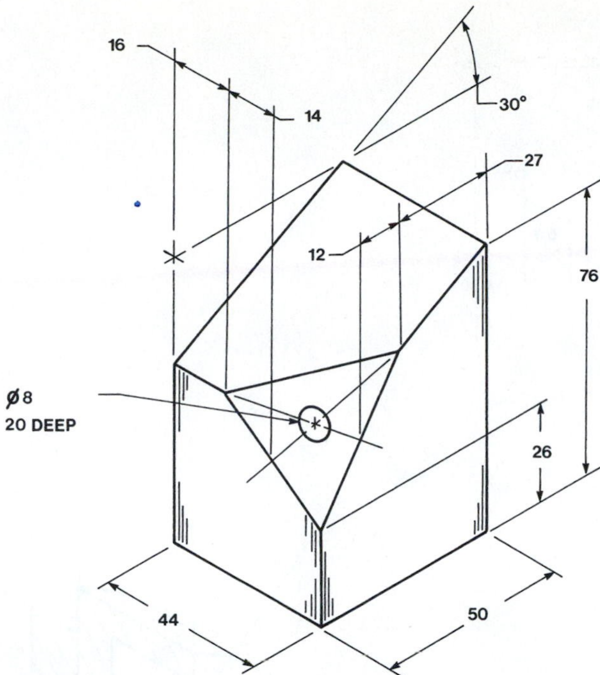

Figure P4-40 MILLIMETERS

Project 4-4:

Draw each of the objects shown in Figures P4-41 through P4-44 as models, then draw a front view and an appropriate sectional view of each.

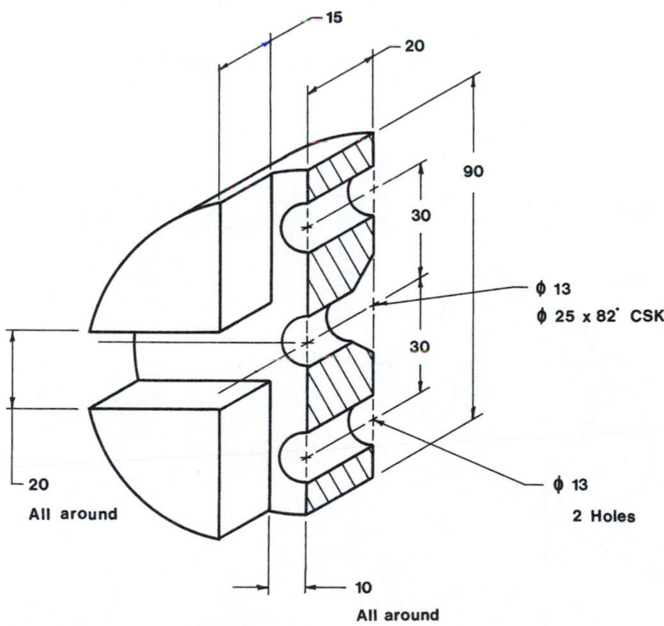

Figure P4-41 MILLIMETERS

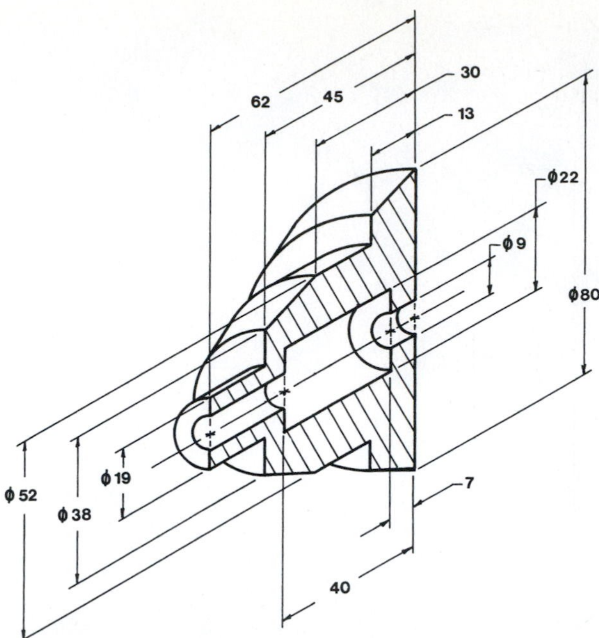

Figure P4-42 MILLIMETERS

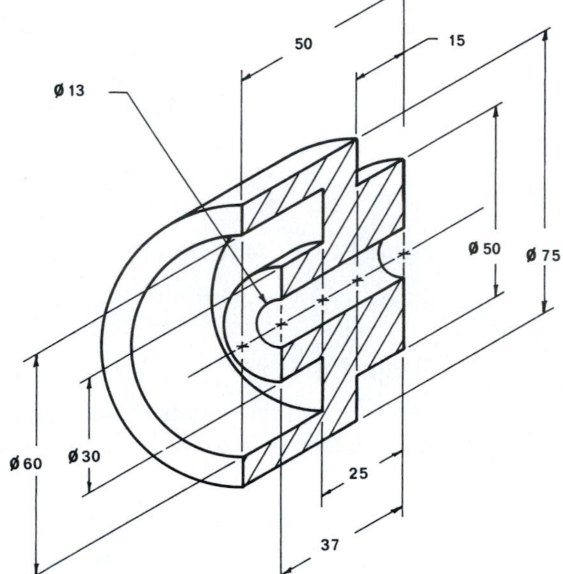

Figure P4-43 MILLIMETERS

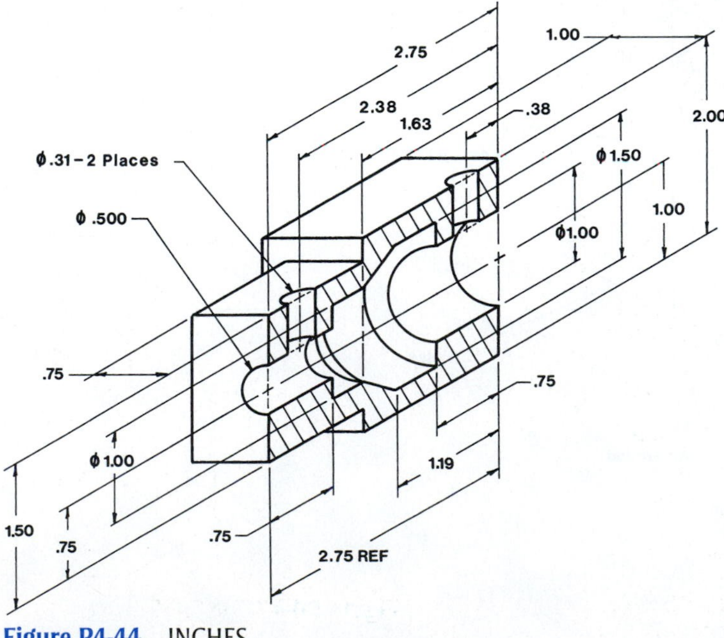

Figure P4-44 INCHES

Project 4-5:

Draw at least one orthographic view and the indicated sectional view for each object in Figures P4-45 through P4-50.

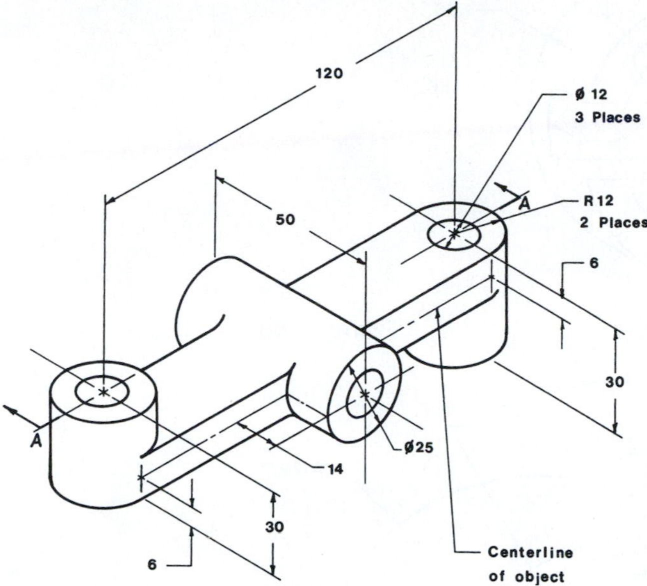

Figure P4-45 MILLIMETERS

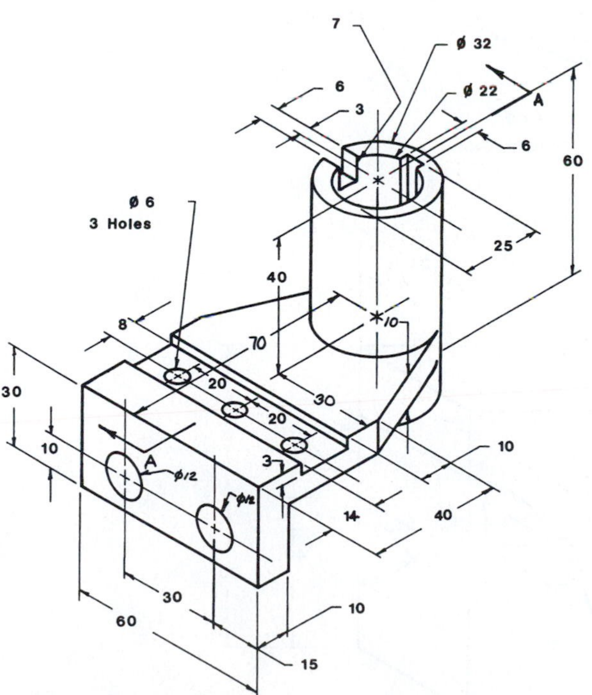

Figure P4-46 MILLIMETERS

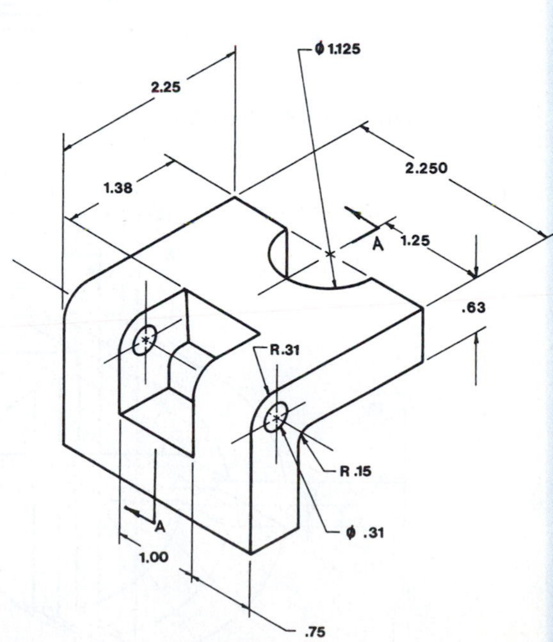

Figure P4-47 INCHES

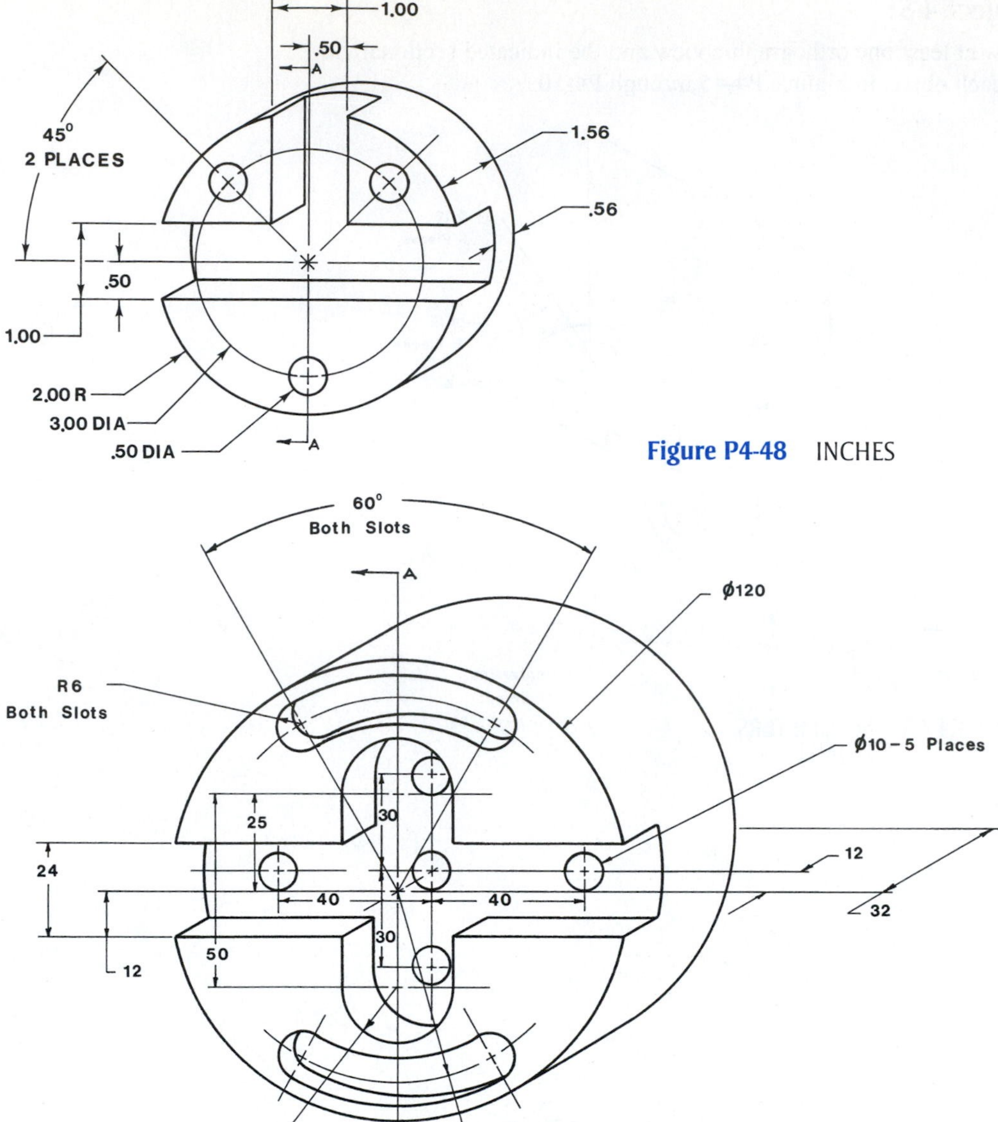

Figure P4-48 INCHES

Figure P4-49 MILLIMETERS

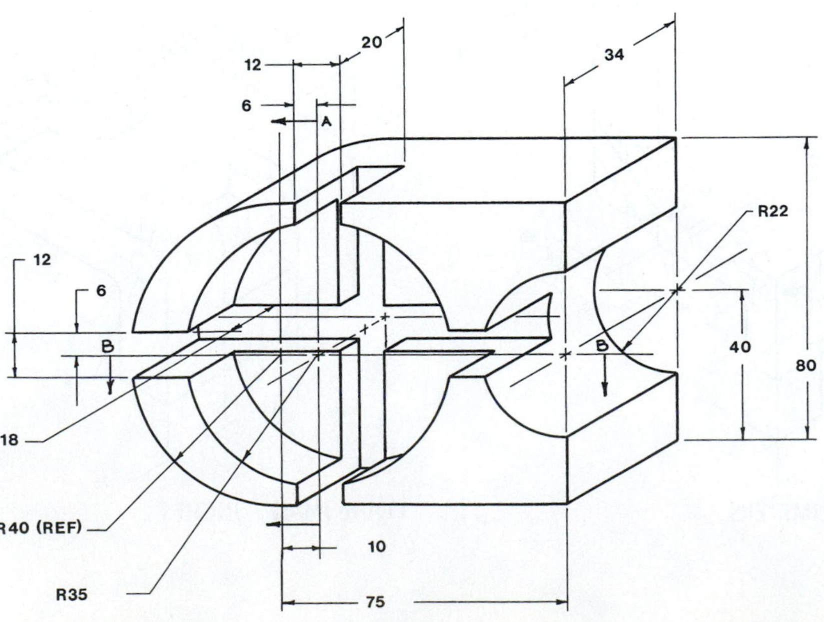

Figure P4-50 MILLIMETERS

Project 4-6:

Given the orthographic views in Figures P4-51 and P4-52, draw a model of each, then draw the given orthographic views and the appropriate sectional views.

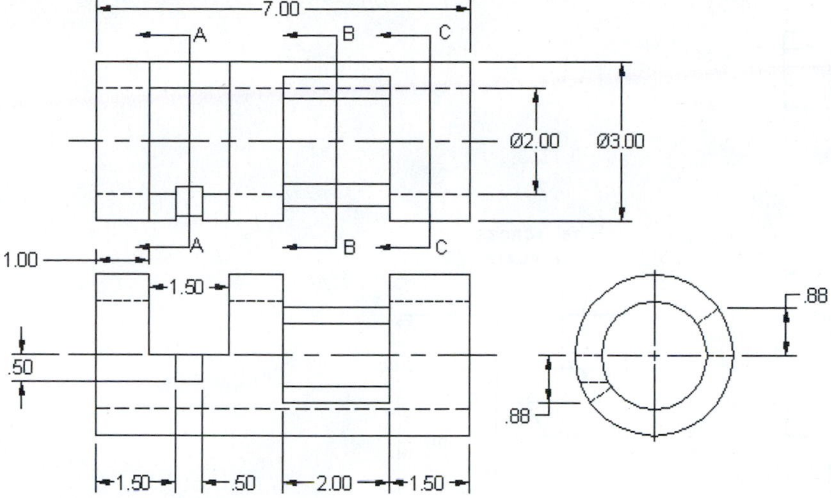

Figure P4-51 INCHES

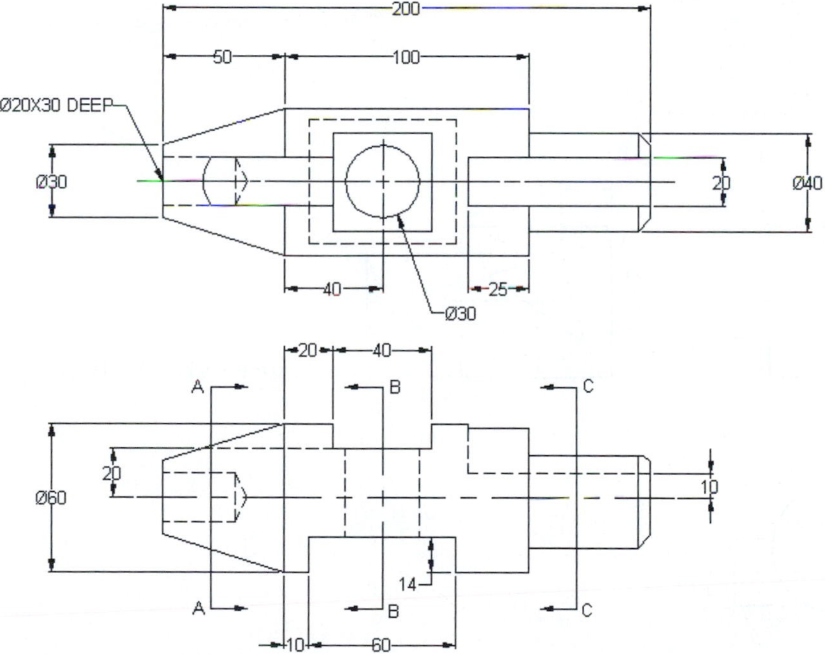

Figure P4-52 MILLIMETERS

Project 4-7:

Draw a 3D model and a set of multiviews for each object shown in Figures P4-53 through P4-60.

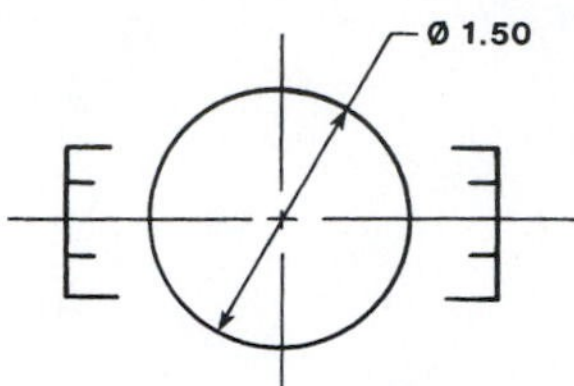

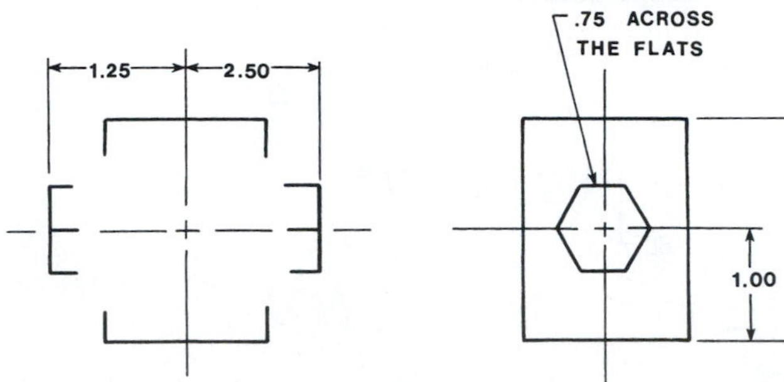

Figure P4-53 INCHES

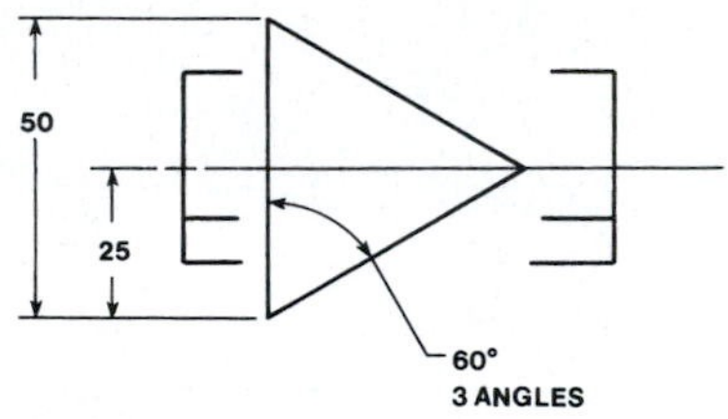

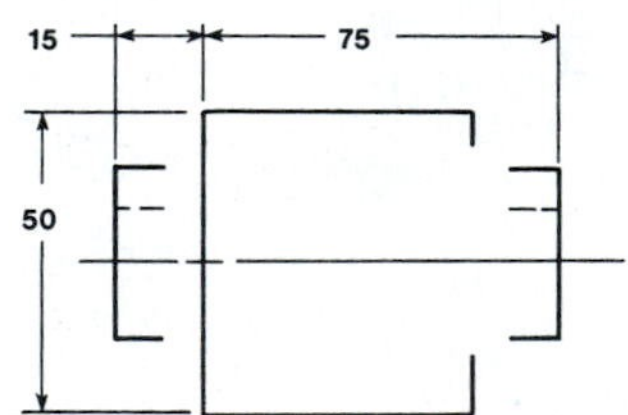

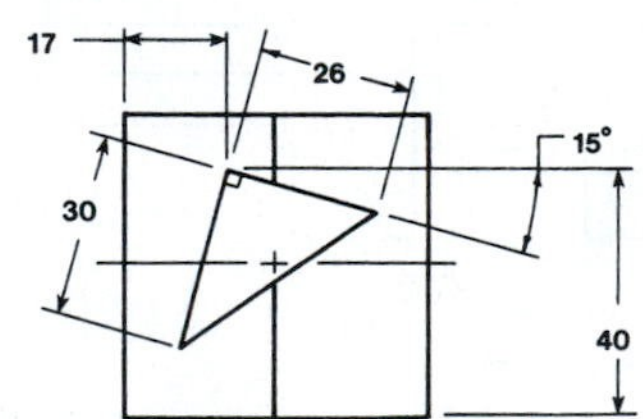

Figure P4-54 MILLIMETERS

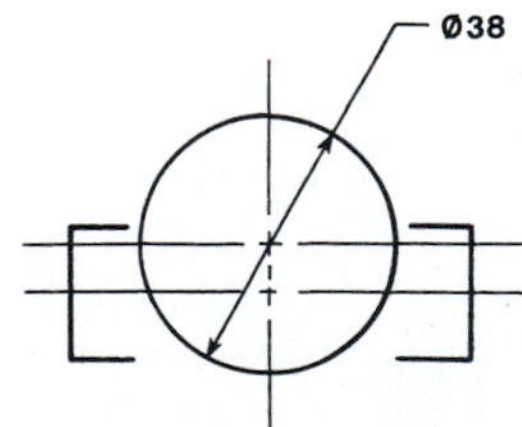

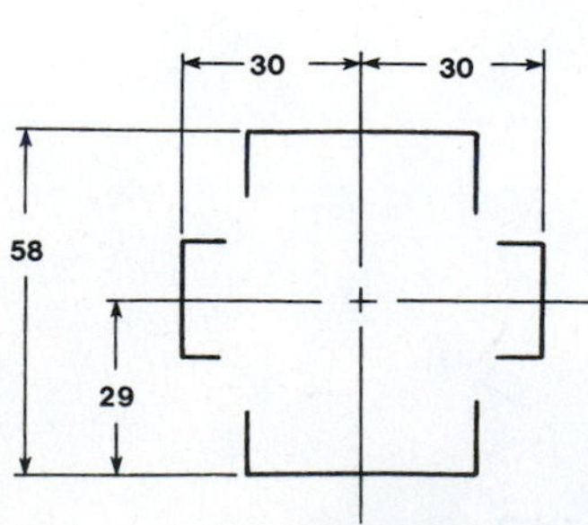

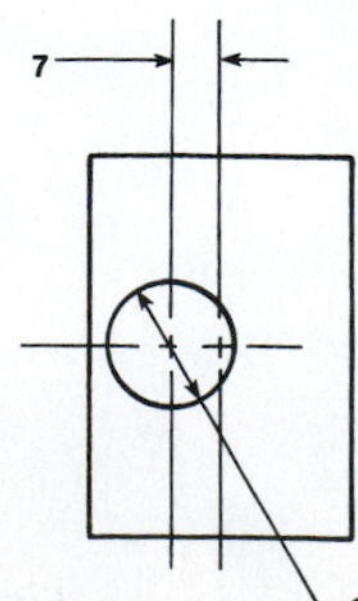

Figure P4-55 MILLIMETERS

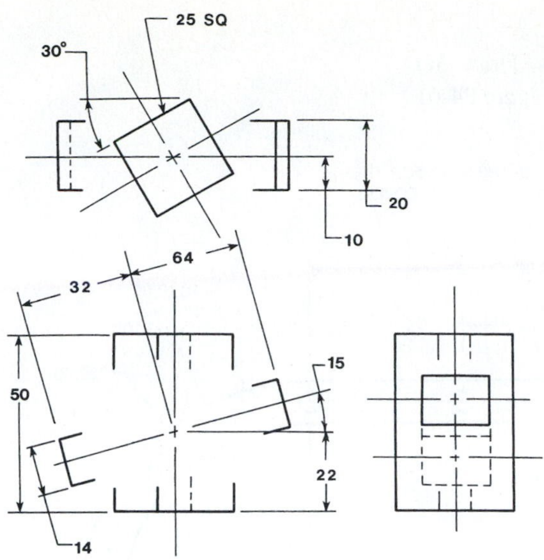

Figure P4-56 MILLIMETERS

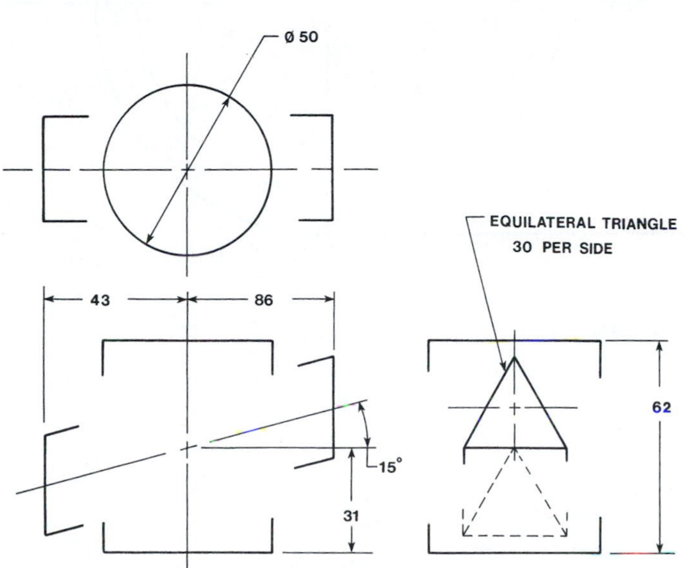

Figure P4-57 MILLIMETERS

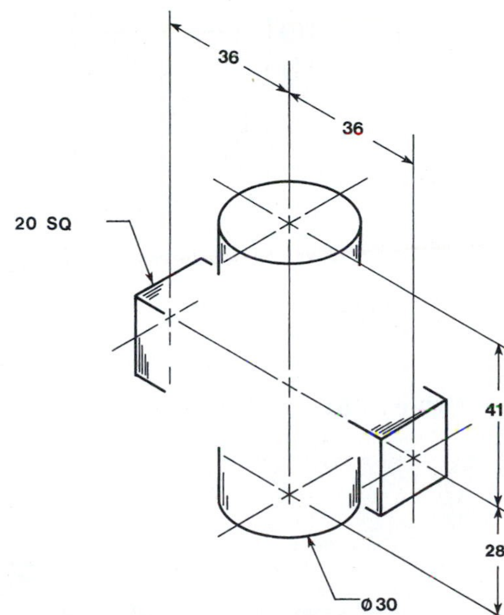

Figure P4-58 MILLIMETERS

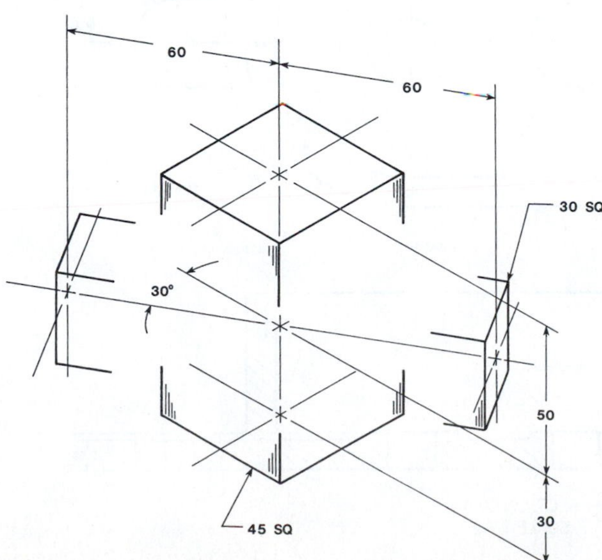

Figure P4-59 MILLIMETERS

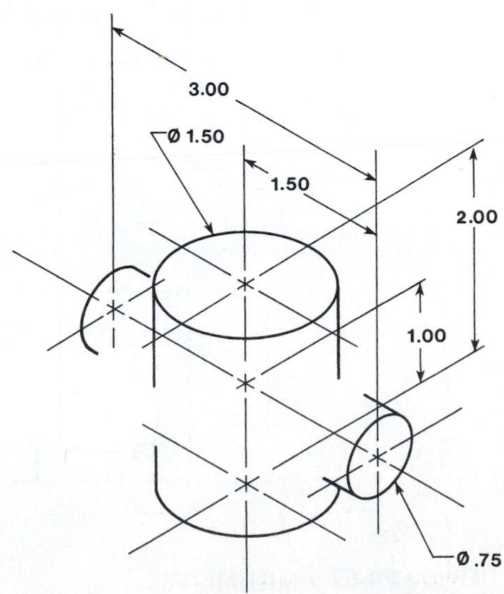

Figure P4-60 INCHES

Project 4-7:

Figures P4-61 through P4-66 are orthographic views. Draw 3D models from the given views. The hole pattern defined in Figure P4-61 also applies to Figure P4-62.

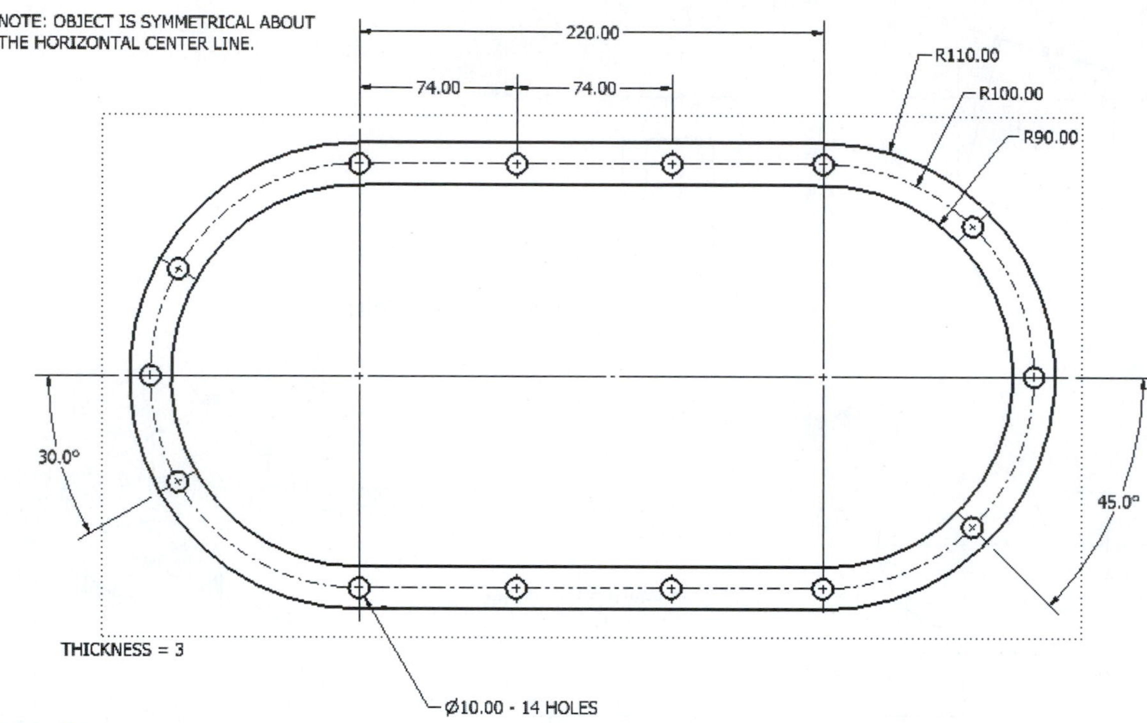

Figure P4-61 MILLIMETERS

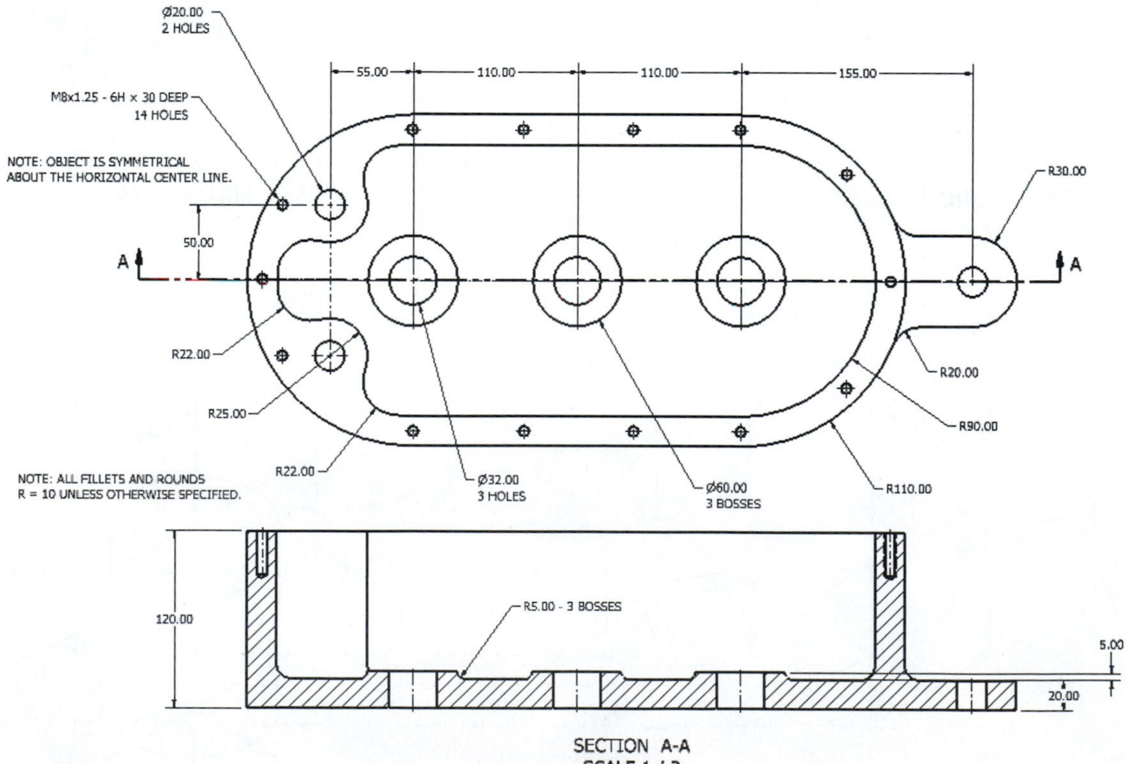

Figure P4-62 MILLIMETERS

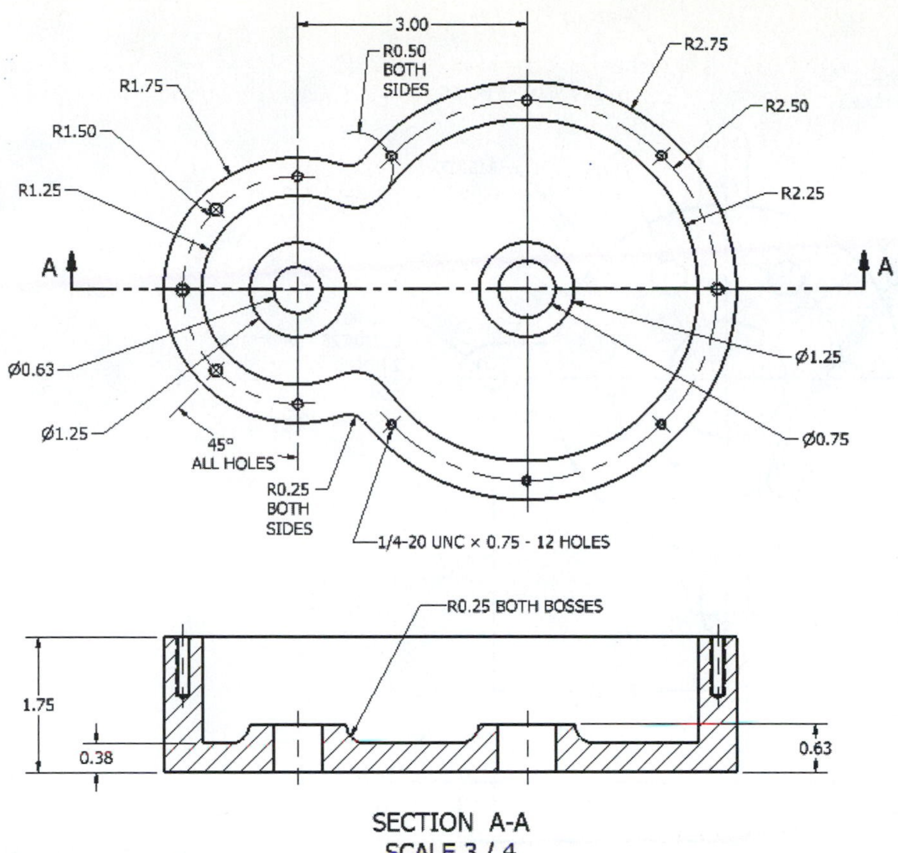

Figure P4-63 INCHES

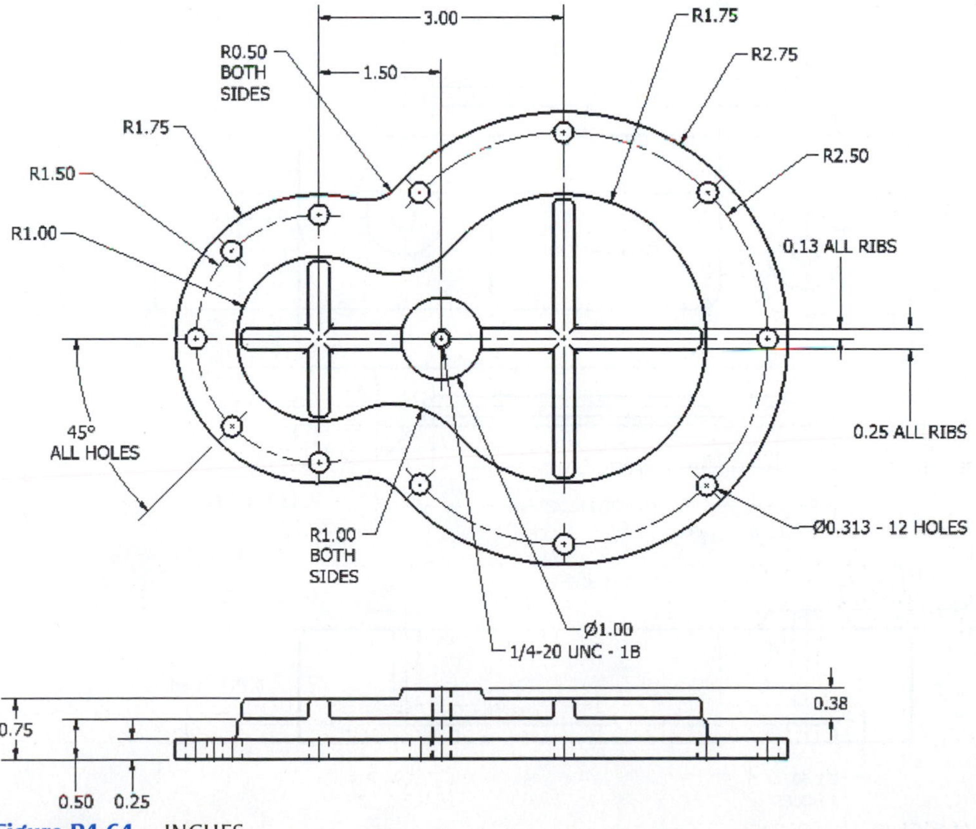

Figure P4-64 INCHES

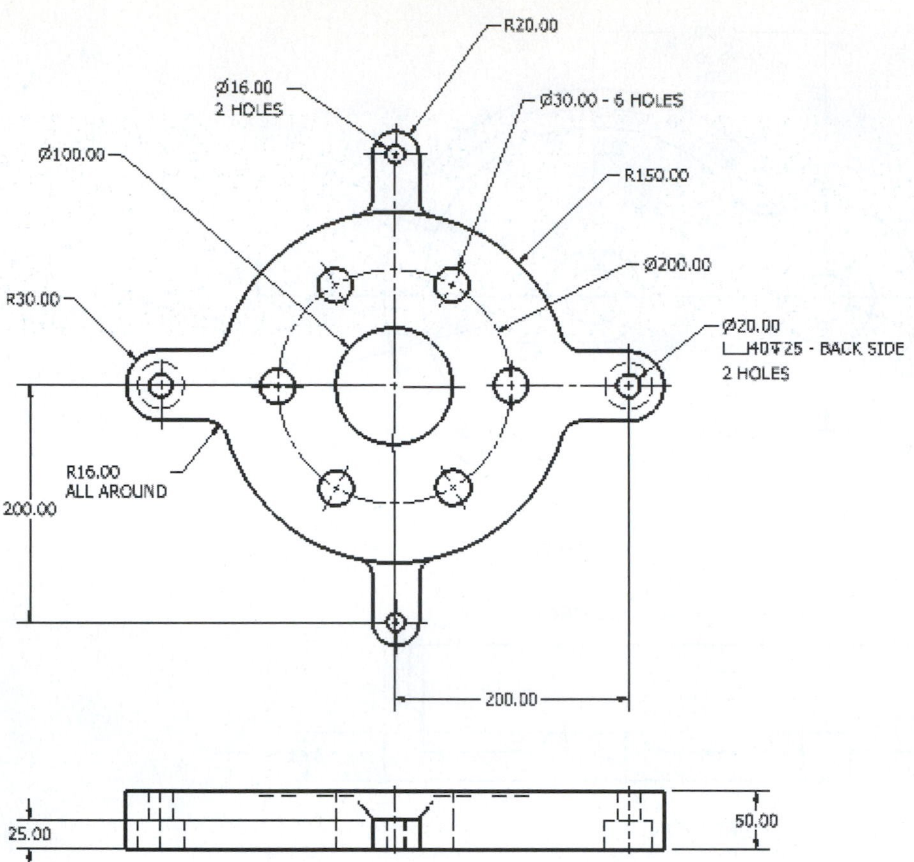

Figure P4-65 MILLIMETERS

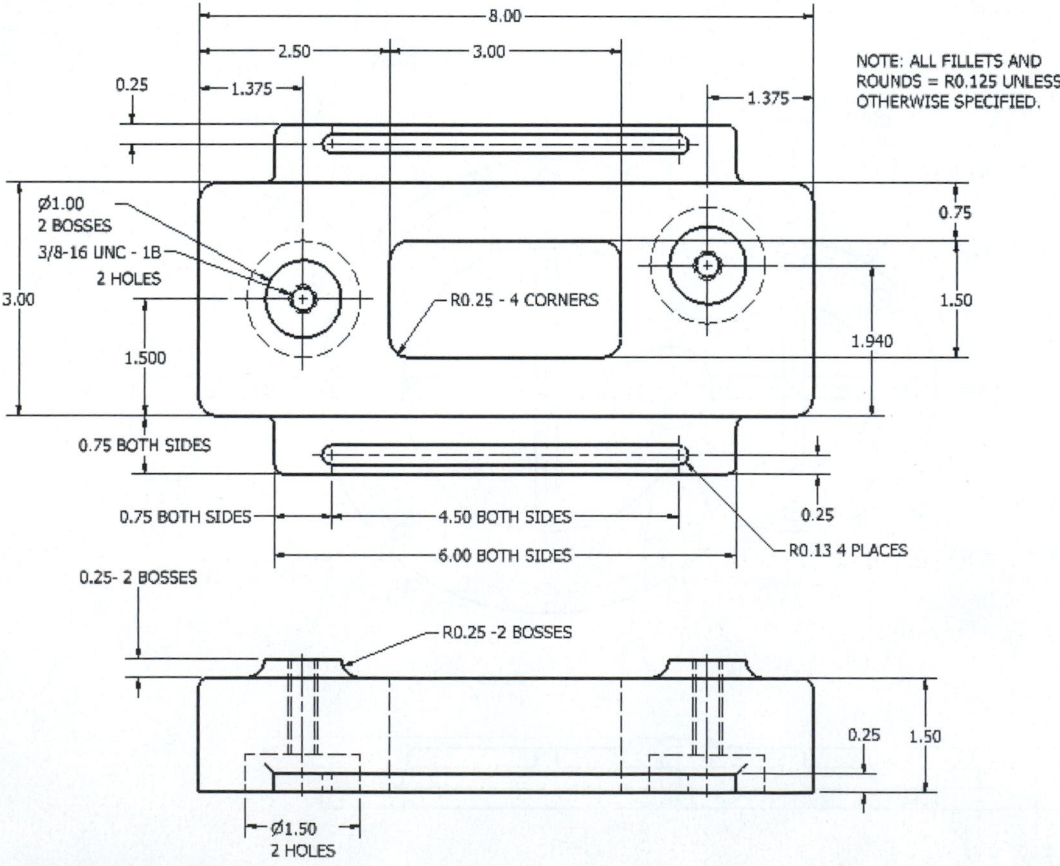

Figure P4-66 INCHES

Assembly Drawings 5

Chapter Objectives

- Show how to create assembly drawings.
- Show how to create a family of drawings.
- Show how to animate assembly drawings.
- Show how to edit assembly drawings.

Introduction

This chapter explains how to create assembly drawings. It uses a group of relatively simple parts to demonstrate the techniques required. The idea is to learn how to create assembly drawings and then gradually apply the knowledge to more difficult assemblies. For example, the next chapter introduces threads and fasteners and includes several exercise problems that require the use of fasteners when creating assembly drawings. Assembly drawings will be included throughout the remainder of the book.

This chapter also shows how to create bills of materials, isometric assembly drawings, title blocks, and other blocks associated with assembly drawings. The chapter also shows how to animate assembly drawings.

Bottom-Up and Top-Down Assemblies

There are three ways to create assembly drawings: bottom up, top down, or a combination of the two. A ***bottom-up*** approach uses drawings that already exist. Model drawings are pulled from files and compiled to create an assembly. The ***top-down*** approach creates model drawings from the assembly drawing. It is also possible to pull drawings from a file and then create more drawings as needed to complete the assembly.

bottom-up approach: Creating an assembly drawing by compiling files from existing drawings.

top-down approach: Creating an assembly drawing by creating model drawings on the assembly drawing.

Starting an Assembly Drawing

Assembly drawings are created using the .iam format. In this example the bottom-up approach will be used. A model called SQBLOCK already exists. The SQBLOCK figure was created from a 30 mm × 30 mm × 30 mm cube with a 15 mm × 15 mm × 30 mm cutout.

Exercise 5-1: Starting an Assembly Drawing:

1. Click on the **New** tool (the **New File** dialog box appears), select the **Metric** tab, then **Standard (mm).iam.**

See Figure 5-1. The **Assembly Panel** bar will appear. See Figure 5-2.

2. Click the **Place Component** tool.

The **Open** dialog box will appear. See Figure 5-3.

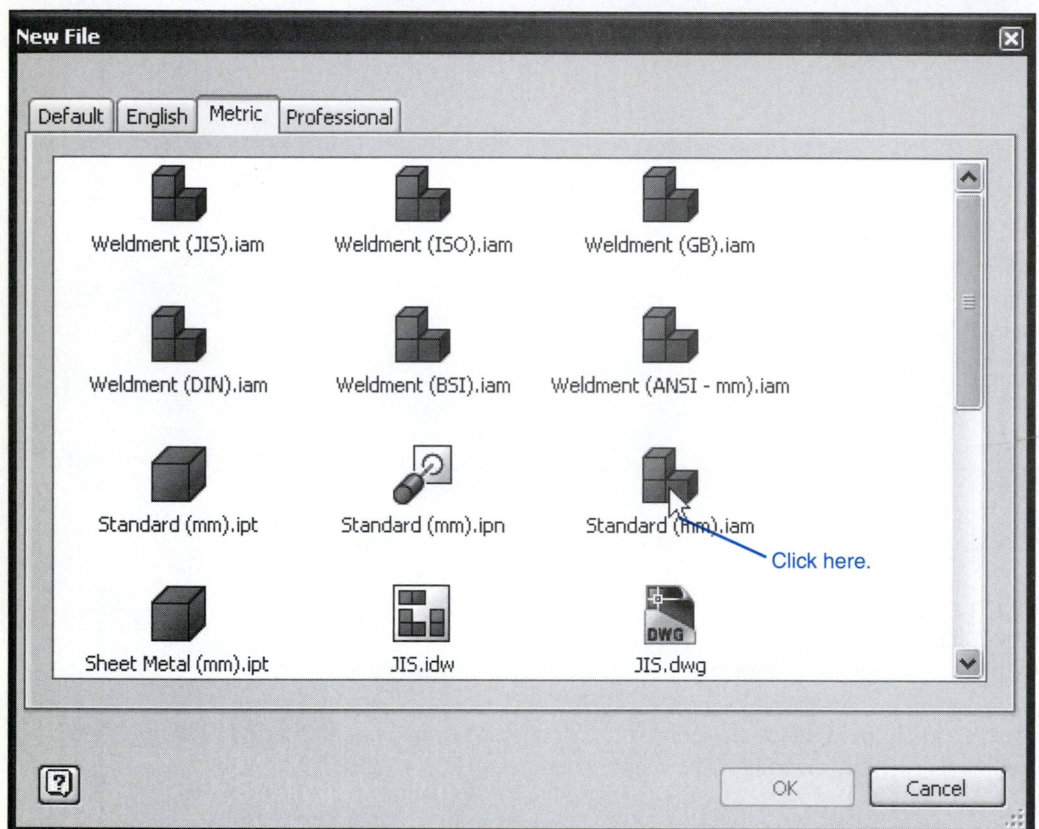

Figure 5-1

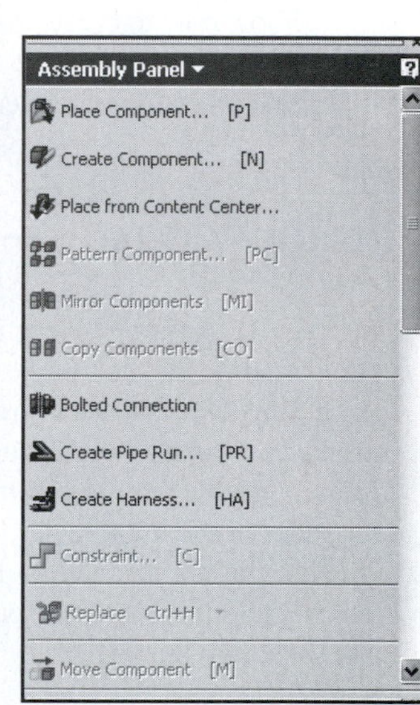

Figure 5-2

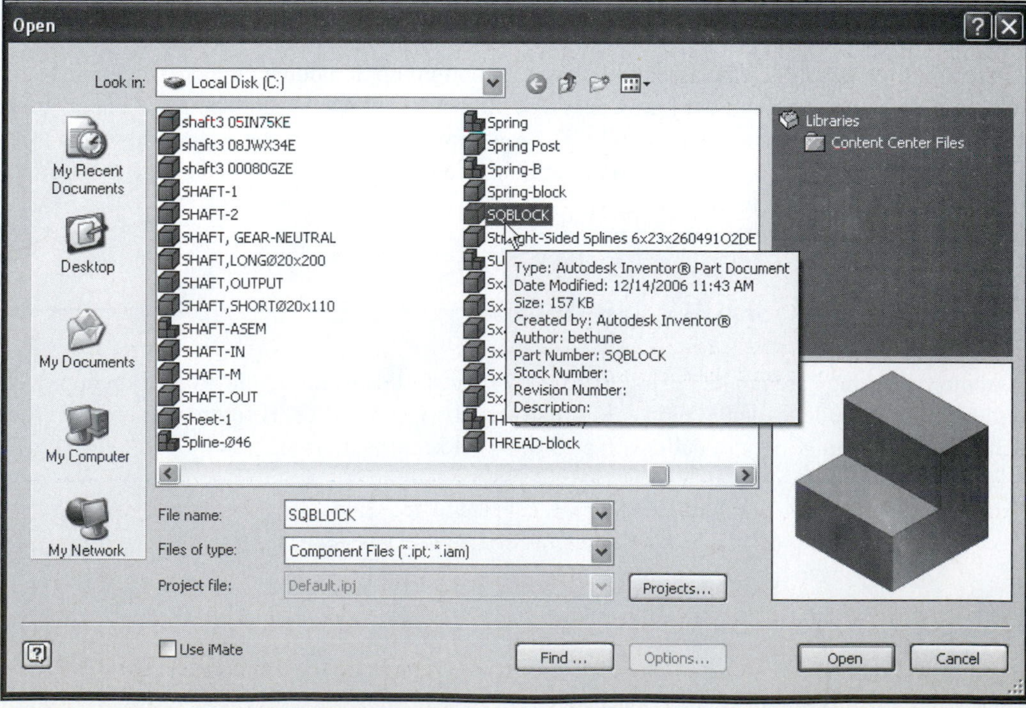

Figure 5-3

3. Click the desired file name, then **OK.**

In this example the **SQBLOCK** file was selected. The selected model (component) will appear on the screen.

4. Zoom the component to an appropriate size, then left-click the mouse to locate the component.

A second copy of the component will automatically appear.

5. Move the second component away from the first. Left-click the mouse to locate the second component, then right-click the mouse and select the **Done** option.

See Figure 5-4.

Note:
The mouse wheel is used to zoom the drawing. Moving the mouse while holding the wheel down will move the drawing.

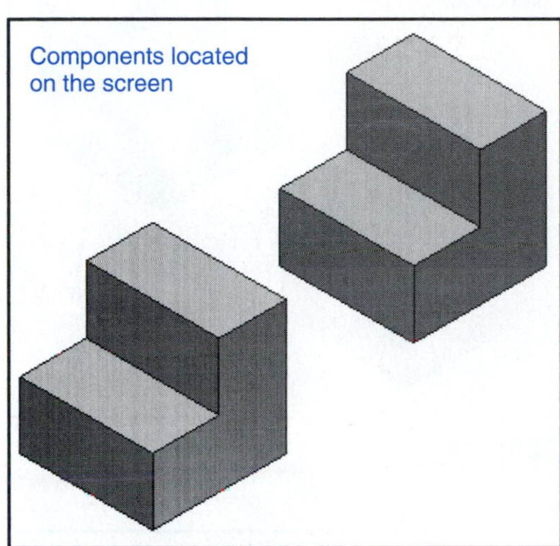

Figure 5-4

Degrees of Freedom

Components are either free to move or they are grounded. ***Grounded components*** will not move when assembly tools are applied. The first component will automatically be grounded. Grounded components are identified by a pushpin icon in the browser box. See Figure 5-5.

grounded component: A component of a drawing that will not move when assembly tools are applied.

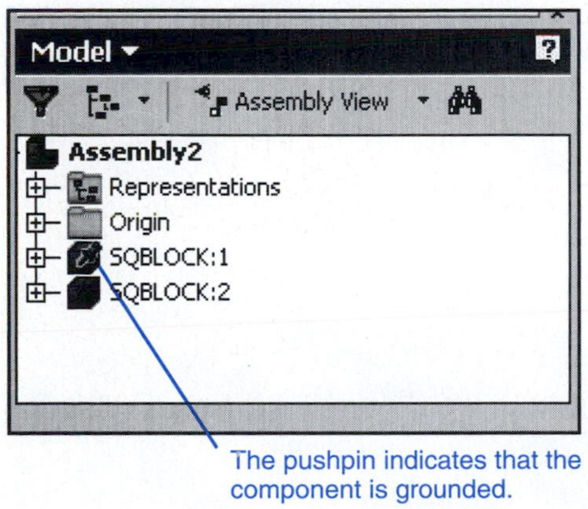

Figure 5-5

Exercise 5-2: Displaying the Degrees of Freedom

Components that are not grounded will have degrees of freedom. The available degrees of freedom for a component may be seen by using the **Degrees of Freedom** option.

1. Click the **View** heading at the top of the screen.
2. Click the **Degrees of Freedom** option.

See Figure 5-6. The available degrees of freedom will appear on the components. See Figure 5-7. Note that in Figure 5-7 the first component does not have any degrees of freedom; it is grounded.

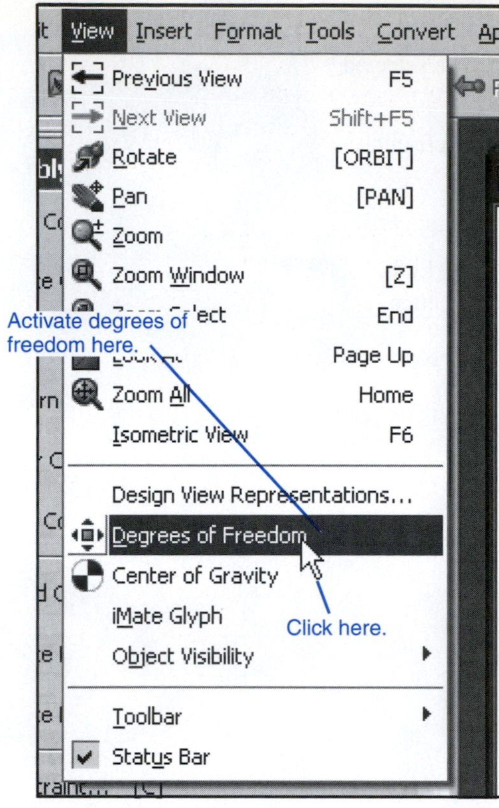

Figure 5-6

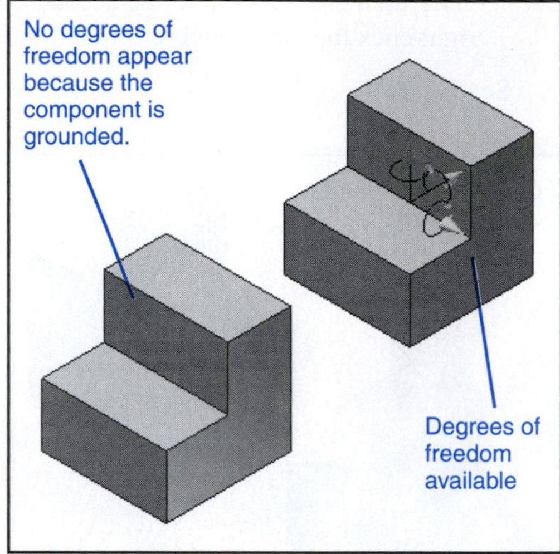

Figure 5-7

Exercise 5-3: Ungrounding a Component

1. Right-click on the component's heading in the browser box.

A dialog box will appear. See Figure 5-8.

2. Click the **Grounded** option.

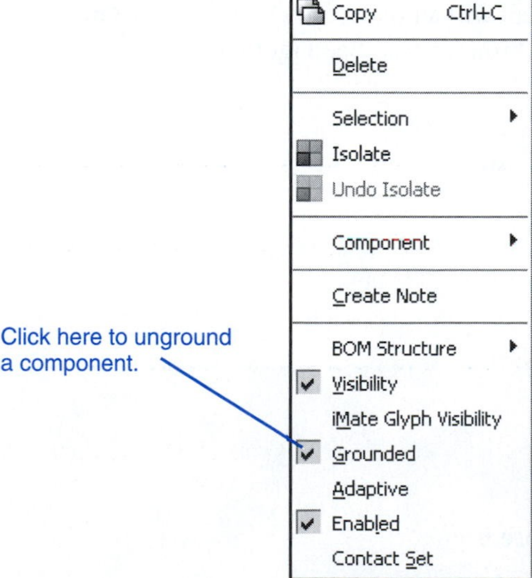

Figure 5-8

Moving Components and Rotating Components

The **Move Component** and **Rotate Component** tools found on the **Assembly Panel** bar are used, as their names imply, to move and rotate individual components. Use the **Rotate** tool to rotate the entire assembly.

Exercise 5-4: Move a Component

1. Click the **Move Component** tool, then click the component to be moved.
2. Hold the left mouse button down and move the component about the screen.
3. When the desired location is reached, release the left button.
4. Right-click the mouse and select the **Done** option.

Exercise 5-5: Rotating a Component

1. Click the **Rotate Component** tool, then click the component to rotate.

A circle will appear around the component. See Figure 5-9.

2. Click and hold the left mouse button outside the circle and move the cursor.

The component will rotate. It is suggested that various points outside the circle be tried to see how the component can be rotated.

3. When the desired orientation is achieved, press the right mouse button and select the **Done** option.

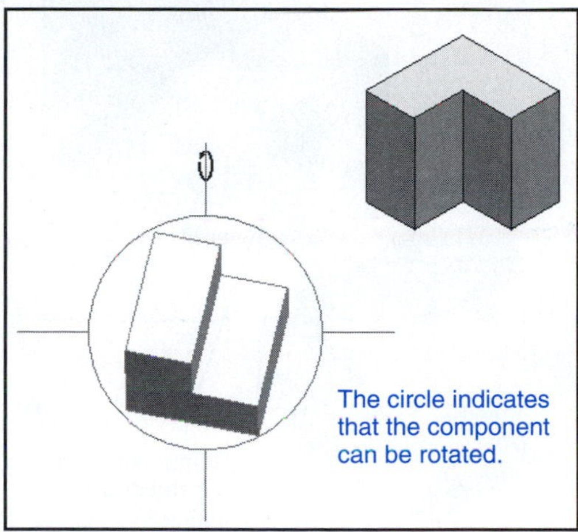

Figure 5-9

CONSTRAINT

The **Constraint** tool is used to locate components relative to one another. Components may be constrained using the **Mate, Flush, Angle, Tangent,** or **Insert** options.

Exercise 5-6: Using the Mate Option

1. Click the **Constraint** tool.

The **Place Constraint** dialog box will appear. See Figure 5-10. The **Mate** option will automatically be selected.

2. Click the front face of the left **SQBLOCK** as shown.
3. Click the front face of the right **SQBLOCK** as shown.
4. Click the **Apply** box on the **Place Constraint** dialog box or right-click the mouse and select the **Apply** option.

See Figure 5-11. The blocks will be joined at the selected surfaces. The blocks may not be perfectly aligned when assembled. This situation may be corrected using the **Flush** option.

The **Mate** option may also be used to align centerlines of holes, shafts, and fasteners and the edges of models. See Figure 5-12.

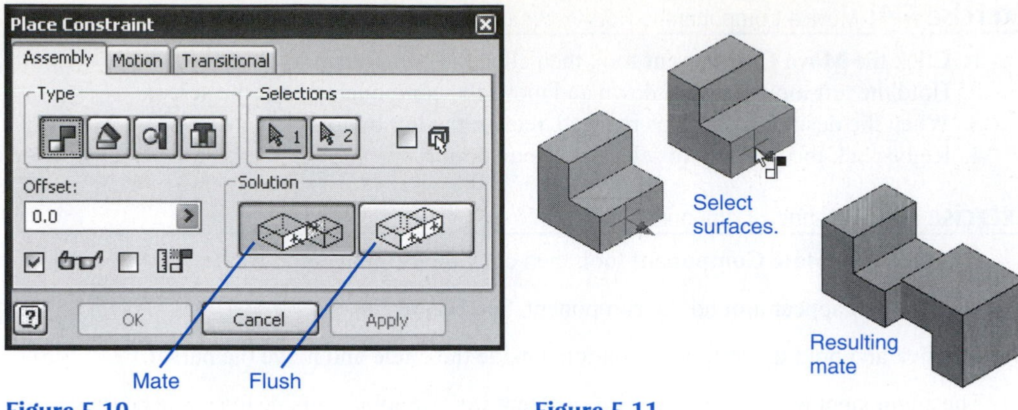

Figure 5-10

Figure 5-11

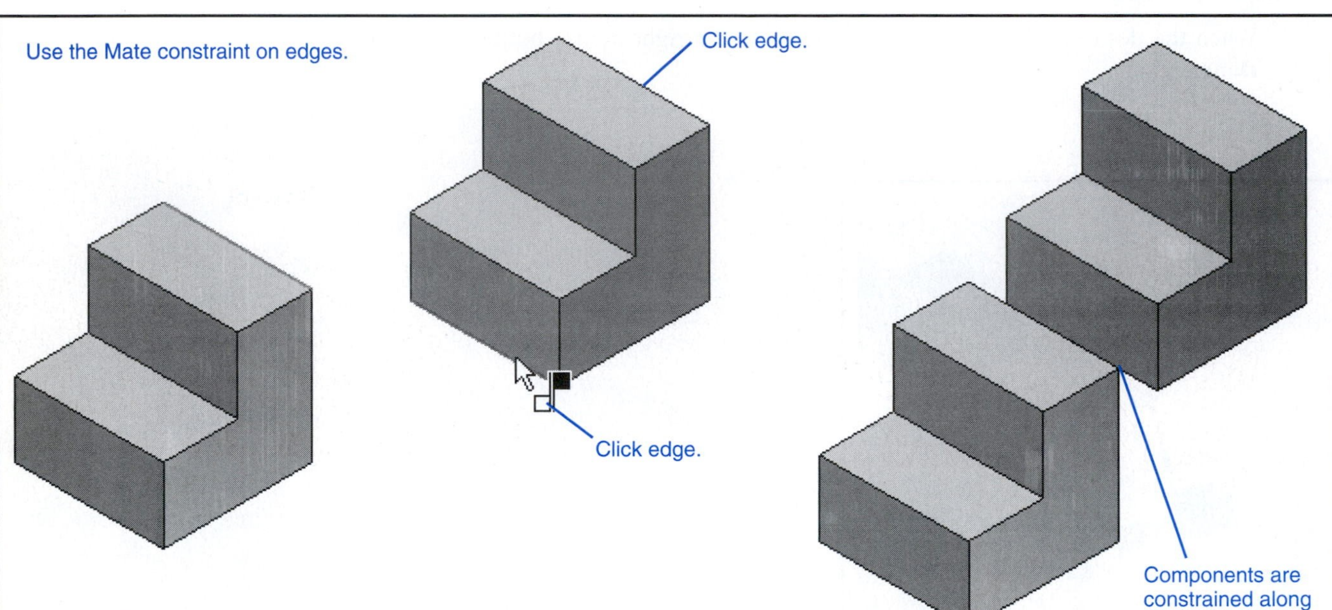

Figure 5-12

Exercise 5-7: Using the Flush Option

1. Move and rotate the components approximately into the position shown in Figure 5-13.
2. Click on the **Flush** button on the **Place Constraint** dialog box.
3. Click the top surface of each block as shown.
4. Click the **Apply** button.
5. Make other surfaces flush as needed to align the two blocks.

Exercise 5-8: Using the Offset Option

Figure 5-14 shows two SQBLOCKS.

1. Click the **Constraint** tool and select the **Mate** option.
2. Use the **Mate** tool and mate the components' edges as shown.
3. Enter a value of **10** into the **Offset** box.

The two mated edges will move apart 10 mm.

TIP

Offset values may be negative. Negative values create an offset in the direction opposite that of positive values.

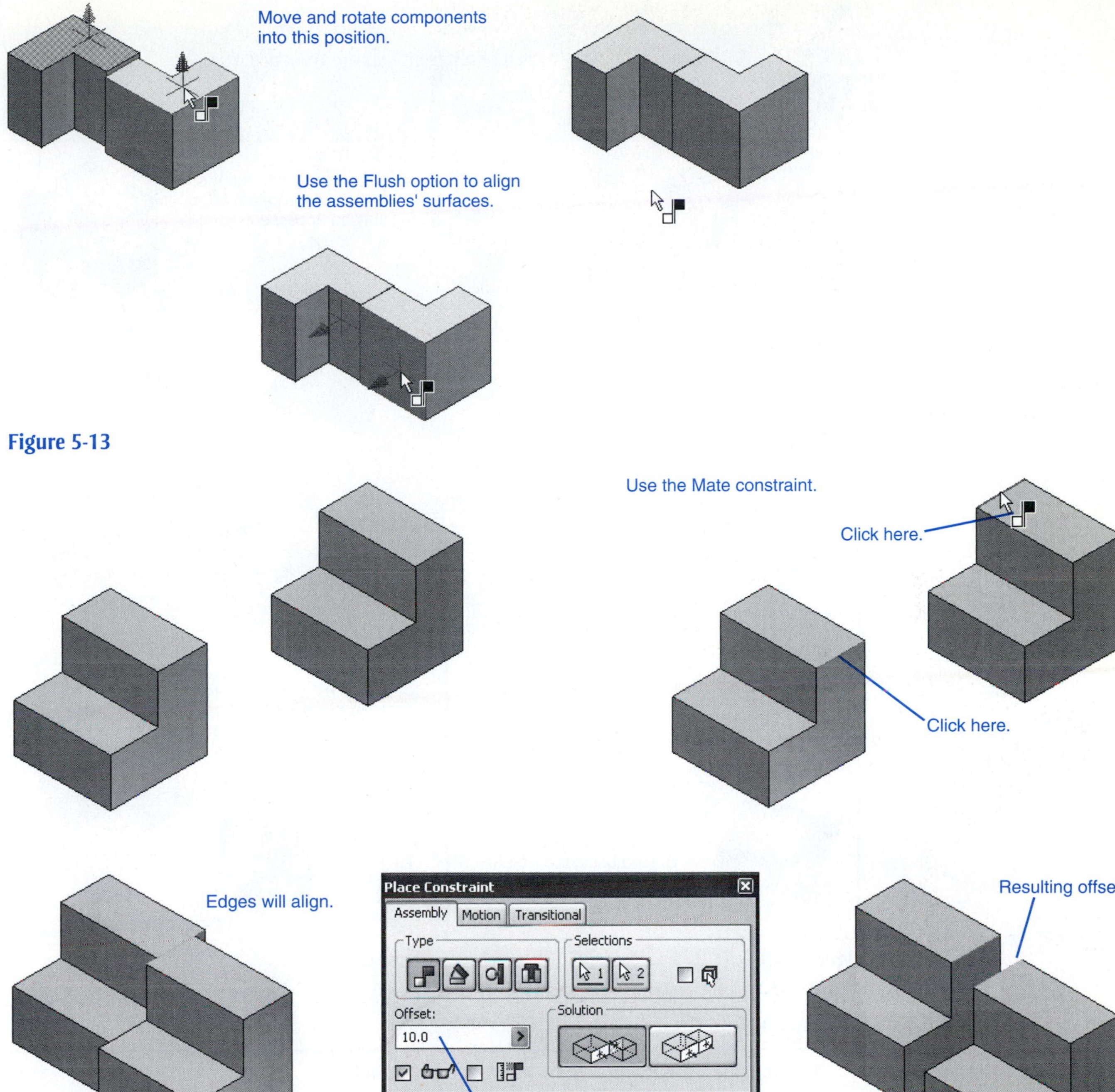

Figure 5-14

Exercise 5-9: Positioning Objects

Sometimes components are not oriented so they can be joined as desired. In these cases first rotate or move one of the components as needed, then use the **Constraint** commands. See Figure 5-15. The left component has been rotated using the **Rotate Component** tool.

1. Click the **Constraint** tool in the **Assembly Panel** bar.
2. Click the edge lines of the two components as shown.
3. Click the **Apply** box.

See Figure 5-16.

4. Use the **Flush** option to align the components.
5. Click the **Apply** button.

TIP Components cannot be moved or rotated if they are grounded.

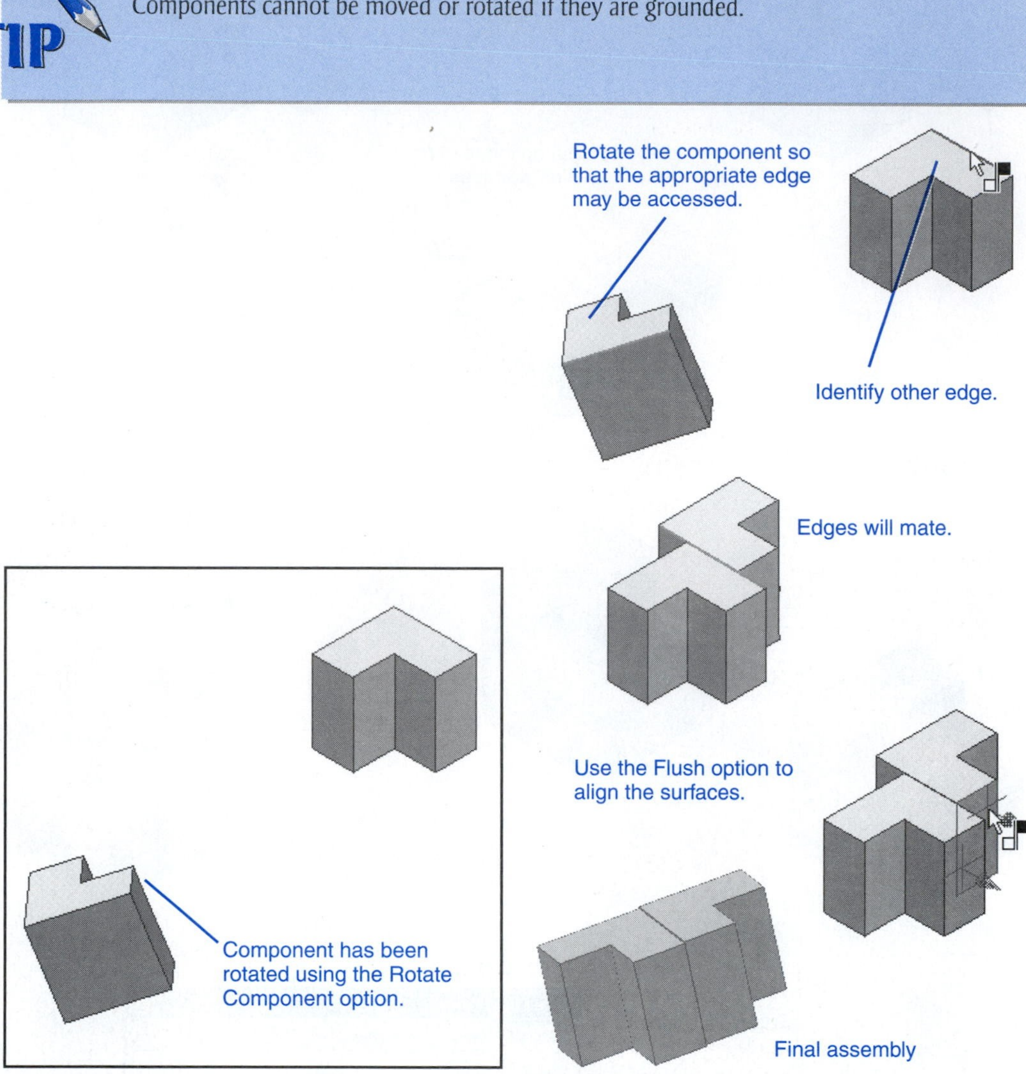

Figure 5-15

Figure 5-16

Exercise 5-10: Using the Angle Option

1. Click the **Constraint** tool on the **Assembly Panel.**

The **Place Constraint** dialog box will appear. See Figure 5-17.

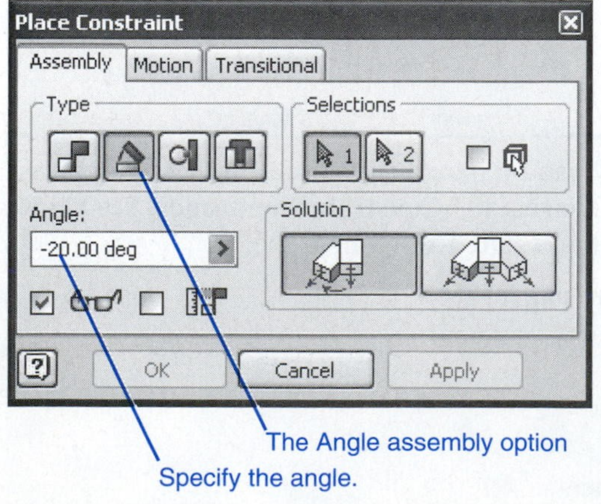

Figure 5-17

2. Use the **Mate** constraint and align the edges of the SQBLOCKs shown in Figure 5-18.

3. Access the **Angle** constraint, set the angle for −**20.00,** and click the two front surfaces of the SQBLOCKs as shown.

4. Click the **Apply** box.

5. Click the **Flush** constraint and align the two surfaces as shown.

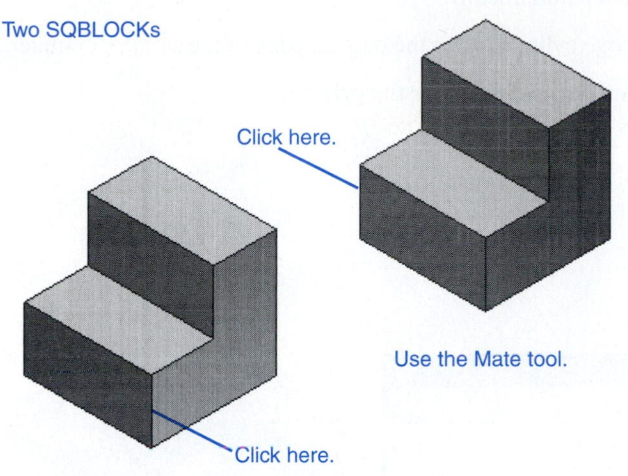

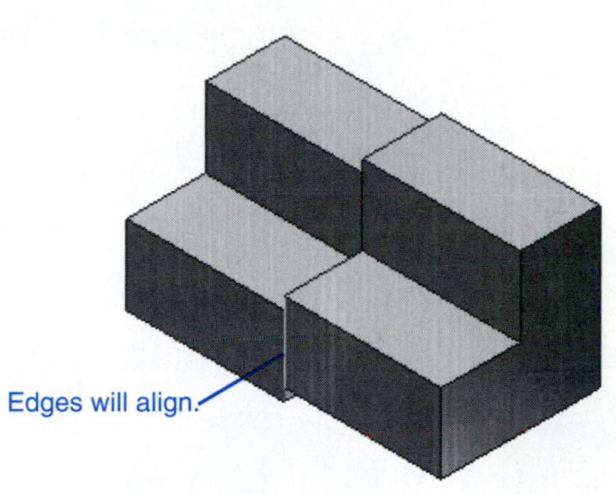

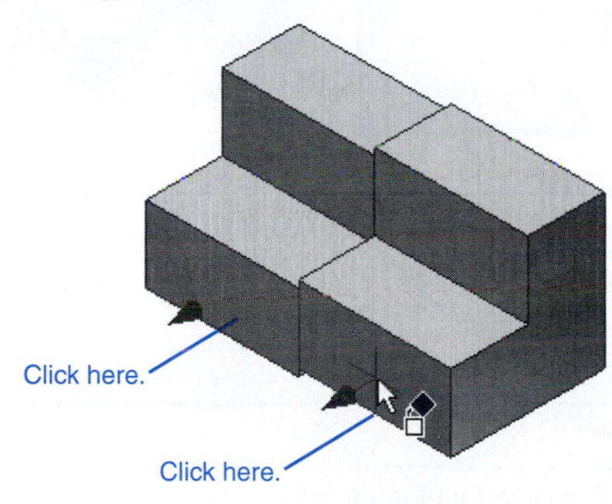

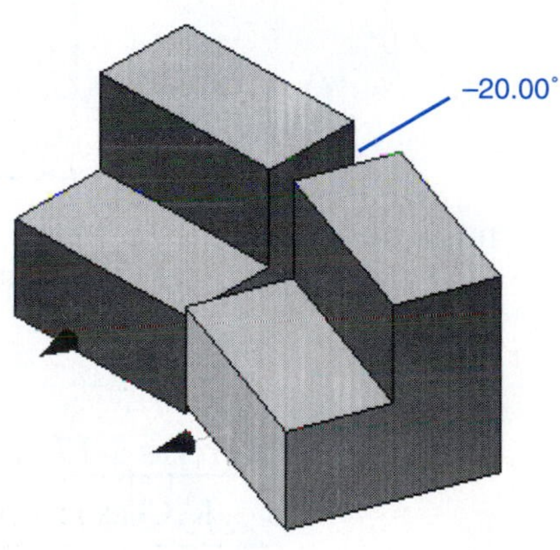

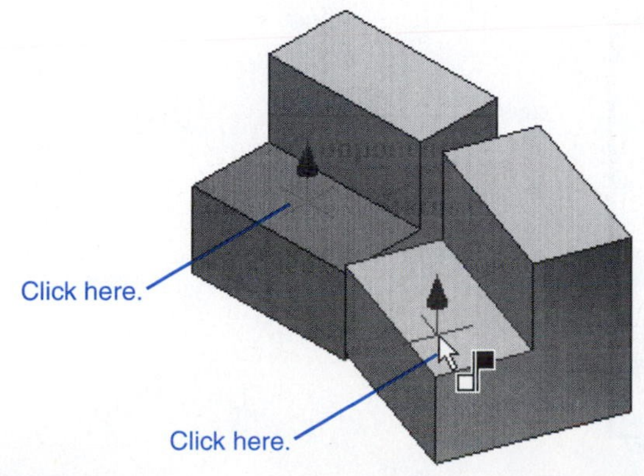

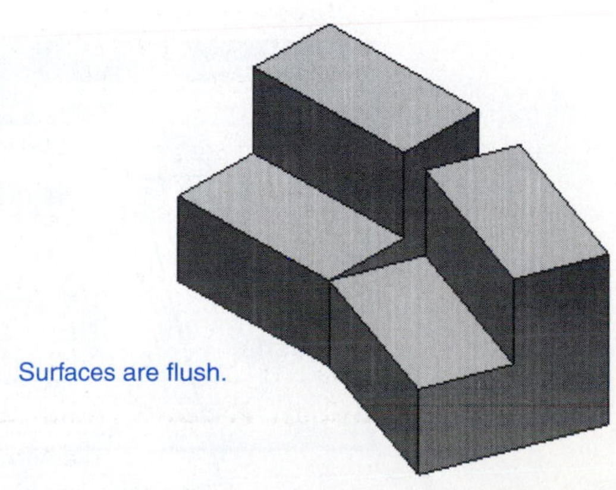

Figure 5-18

Exercise 5-11: Using the Tangent Option

Figure 5-19 shows two cylinders. The smaller cylinder has dimensions of Ø10 × 20, and the larger cylinder has dimensions of Ø20 × 20 with a Ø10 centered longitudinal hole.

1. Click the **Constraint** tool.

The **Place Constraint** dialog box will appear. See Figure 5-20.

2. Click the **Tangent** box under the **Type** heading.

The **Outside** option will be selected automatically.

3. Select the outside edge of the large cylinder, then the outside edge of the smaller cylinder.

Figure 5-21 shows the resulting tangent constraint for the cylinders.

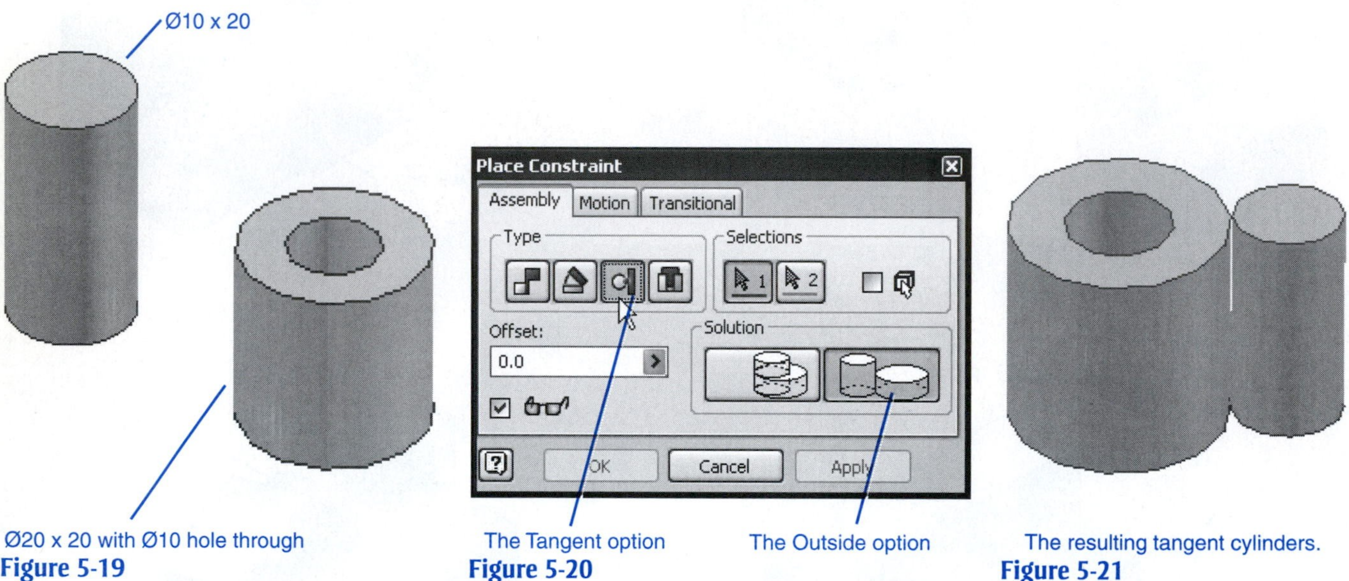

Figure 5-19 Figure 5-20 Figure 5-21

Exercise 5-12: Using the Insert Option

1. Click the **Constraint** tool.

The **Place Constraint** dialog box will appear. See Figure 5-22.

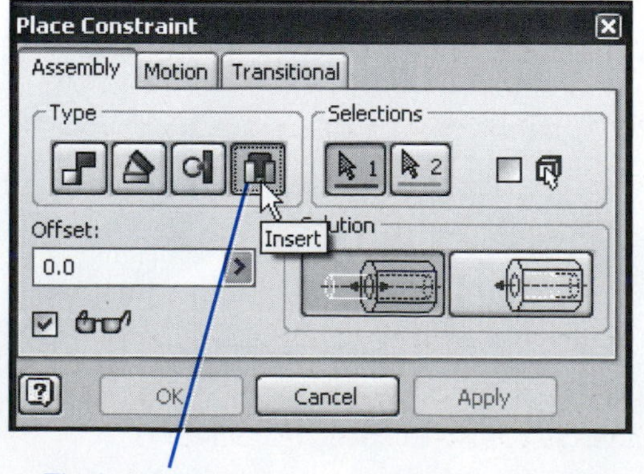

Figure 5-22

2. Click the **Insert** box under the **Type** heading, then click the **Aligned** box under the **Solution** heading.

Note that the aligned box is the right-hand box.

3. Click the top surface of each cylinder as shown.

See Figure 5-23.

4. Click the **Apply** button.

Figure 5-23 also shows the result of using the **Opposed** option under the **Solution** heading.

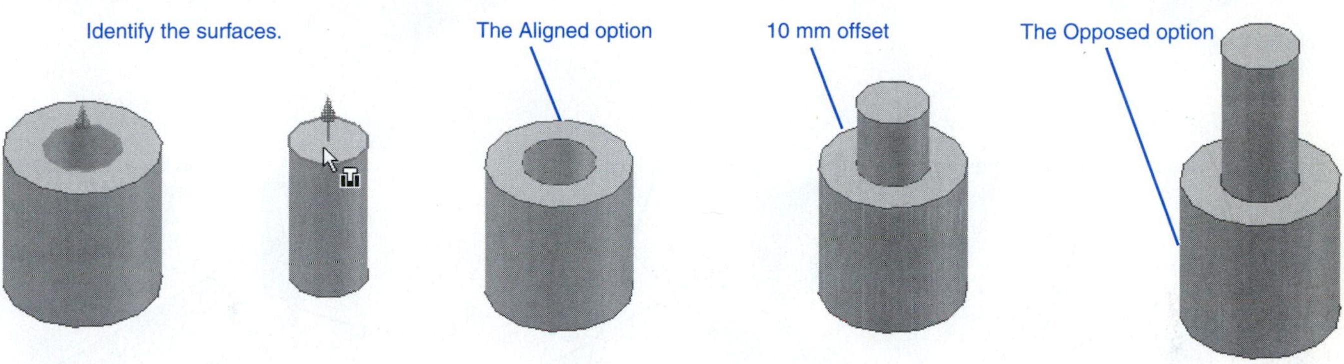

Figure 5-23

SAMPLE ASSEMBLY PROBLEM SP5-1

Figure 5-24 shows three models that will be used to create an assembly drawing. The dimensions for the models are given in Figure 5-24.

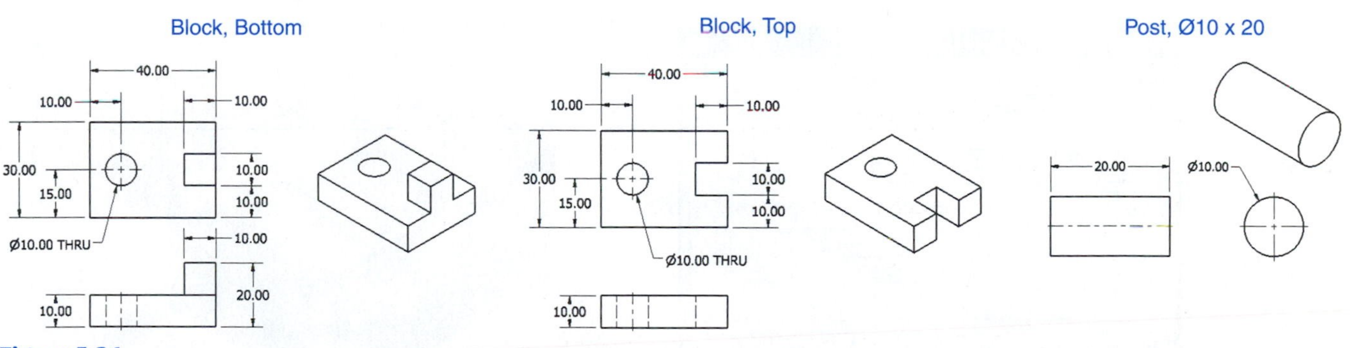

Figure 5-24

1. Create a new drawing using the **Standard (mm).iam** format.
2. Use the **Place Component** tool and place the Block, Bottom; Block, Top; and Post Ø10 × 20 on the drawing screen.

This will be a bottom-up assembly.

3. Click the **Constraint** tool and use the **Mate** option to align the two edges as shown.
4. Use the **Flush** constraint to align the front surfaces of the two blocks.
5. Use the **Insert** tool to locate the Post in the holes in the blocks. See Figure 5-25.
6. Save the assembly.

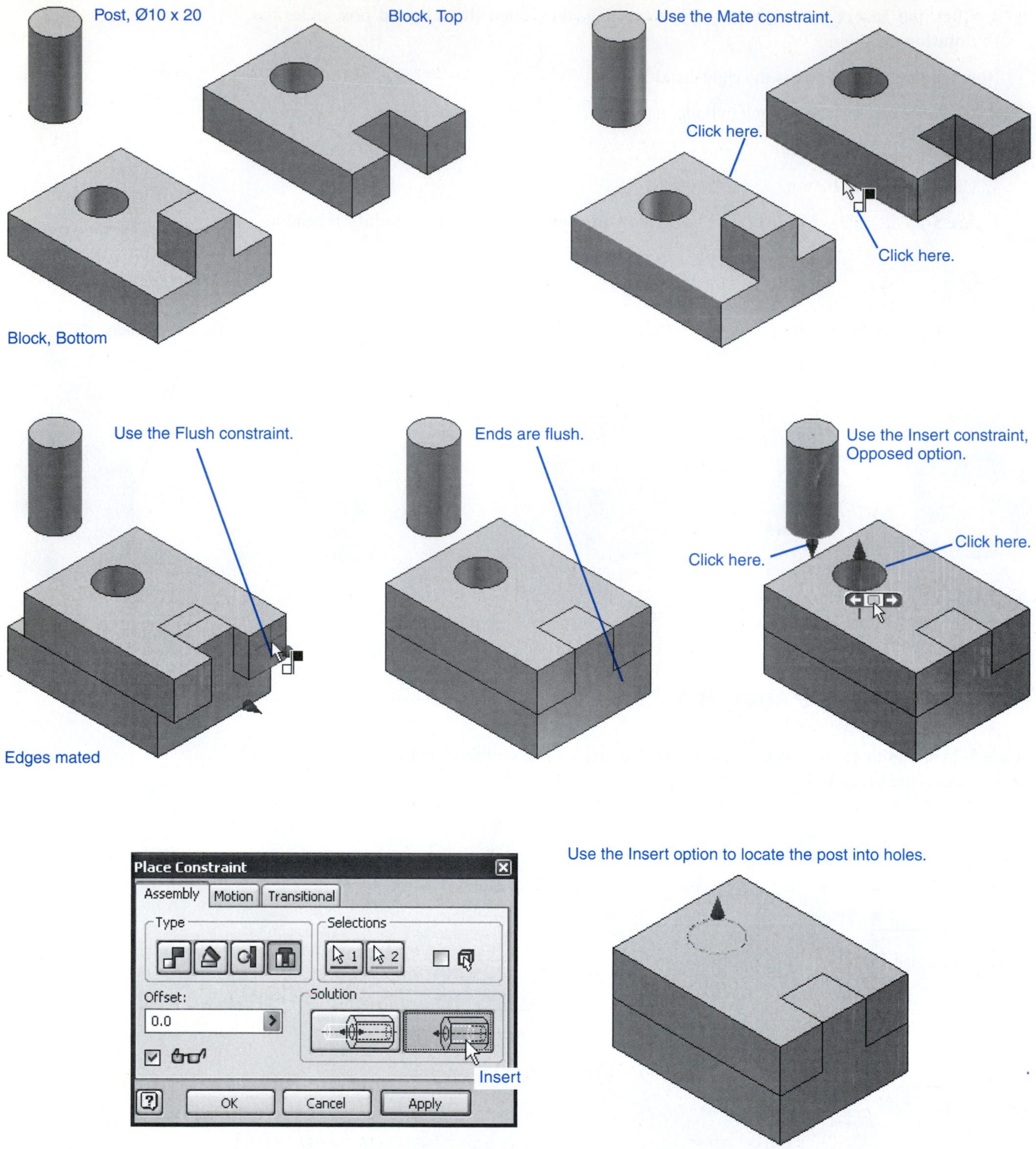

Post, Ø10 x 20

Block, Top

Block, Bottom

Use the Mate constraint.

Click here.

Click here.

Use the Flush constraint.

Edges mated

Ends are flush.

Use the Insert constraint, Opposed option.

Click here.

Click here.

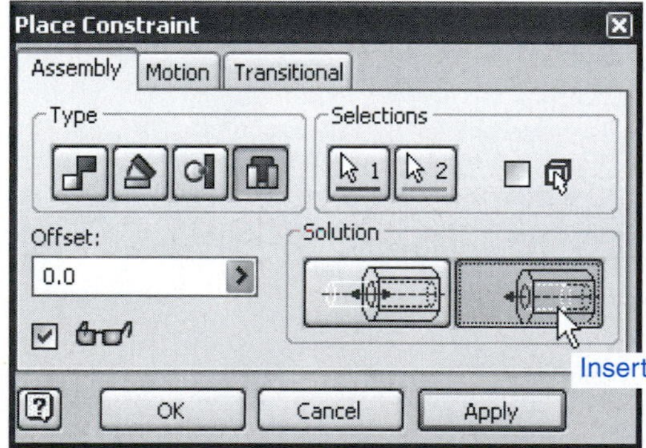

Use the Insert option to locate the post into holes.

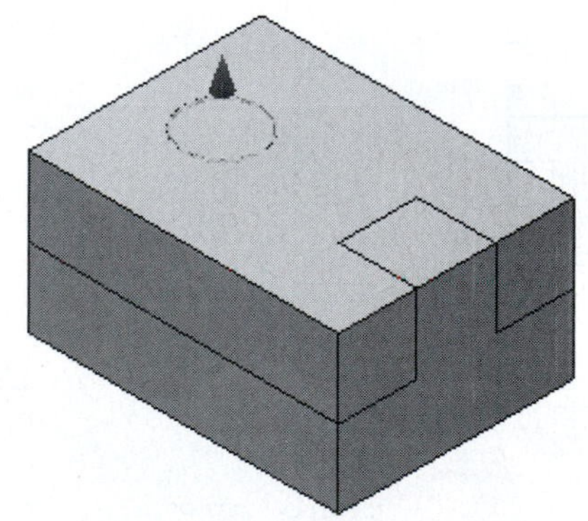

Figure 5-25

Exercise 5-13: Saving an Assembly

1. Create the assembly, then click the **Save Copy As…** heading under the **File** pull-down menu.

   The file will be saved as an .iam file.

PRESENTATION DRAWINGS

Presentation drawings are used to create exploded assembly drawings that can then be animated to show how the assembly is to be created from its components.

presentation drawing: An exploded assembly drawing that can be animated to show how the assembly is to be created.

Exercise 5-14: Creating a Presentation Drawing

1. Click on the **New** tool.

The **New File** dialog box will appear. See Figure 5-26.

2. Click the **Standard (mm).ipn tool,** then **OK.**

The **Presentation Panel** bar will appear. See Figure 5-27.

3. Click the **Create View** tool in the **Presentation Panel** bar.

The **Select Assembly** dialog box will appear. See Figure 5-28.

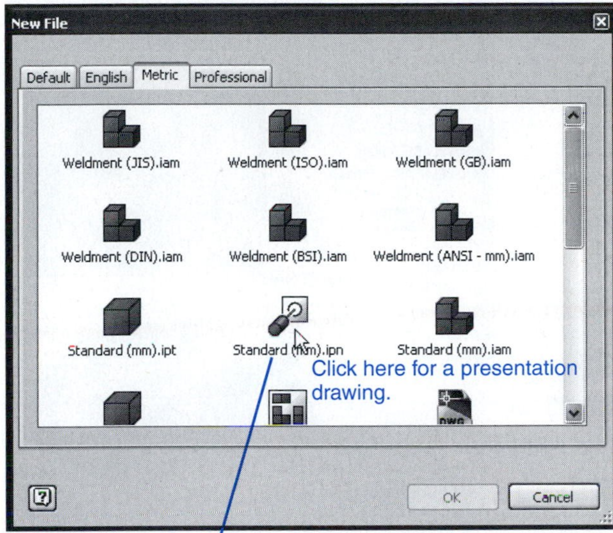

Figure 5-26

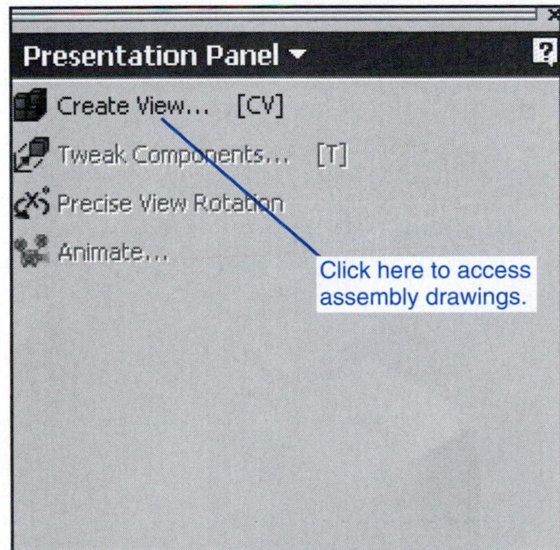

Figure 5-27

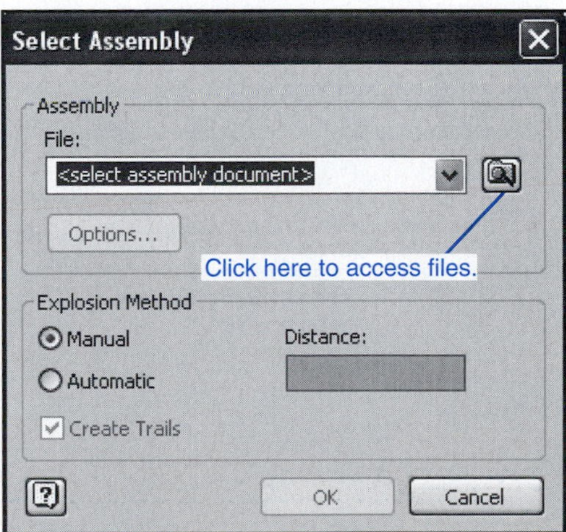

Figure 5-28

4. Click the **Explore Directories** box.

The **Open** dialog box will appear, listing all the existing assembly drawings. See Figure 5-29.

5. Select the appropriate assembly drawing, then click **Open.**

The **Select Assembly** dialog box will reappear listing the selected assembly under the **File** heading. See Figure 5-30.

6. Click **OK.**

The assembly will appear. See Figure 5-31.

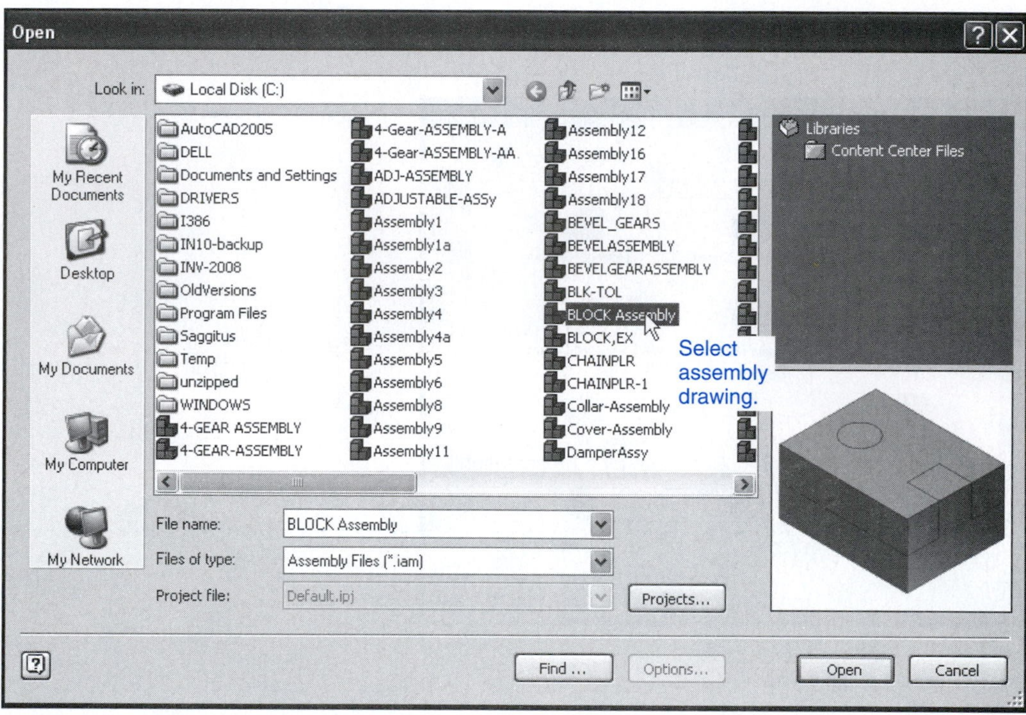

Figure 5-29

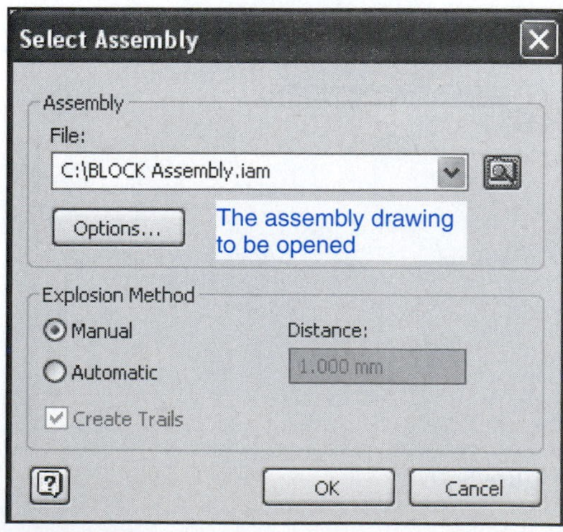

Figure 5-30

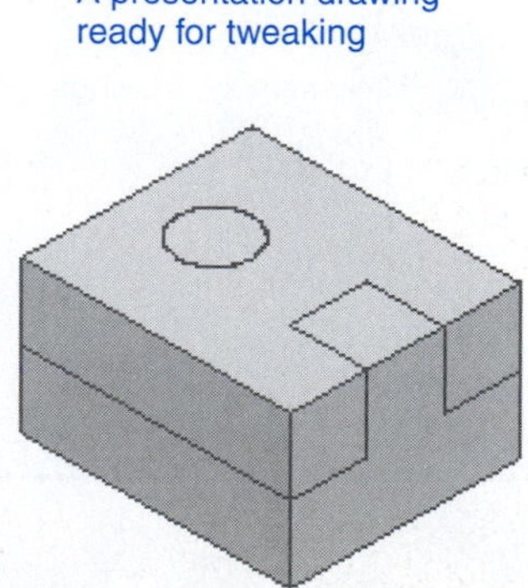

Figure 5-31

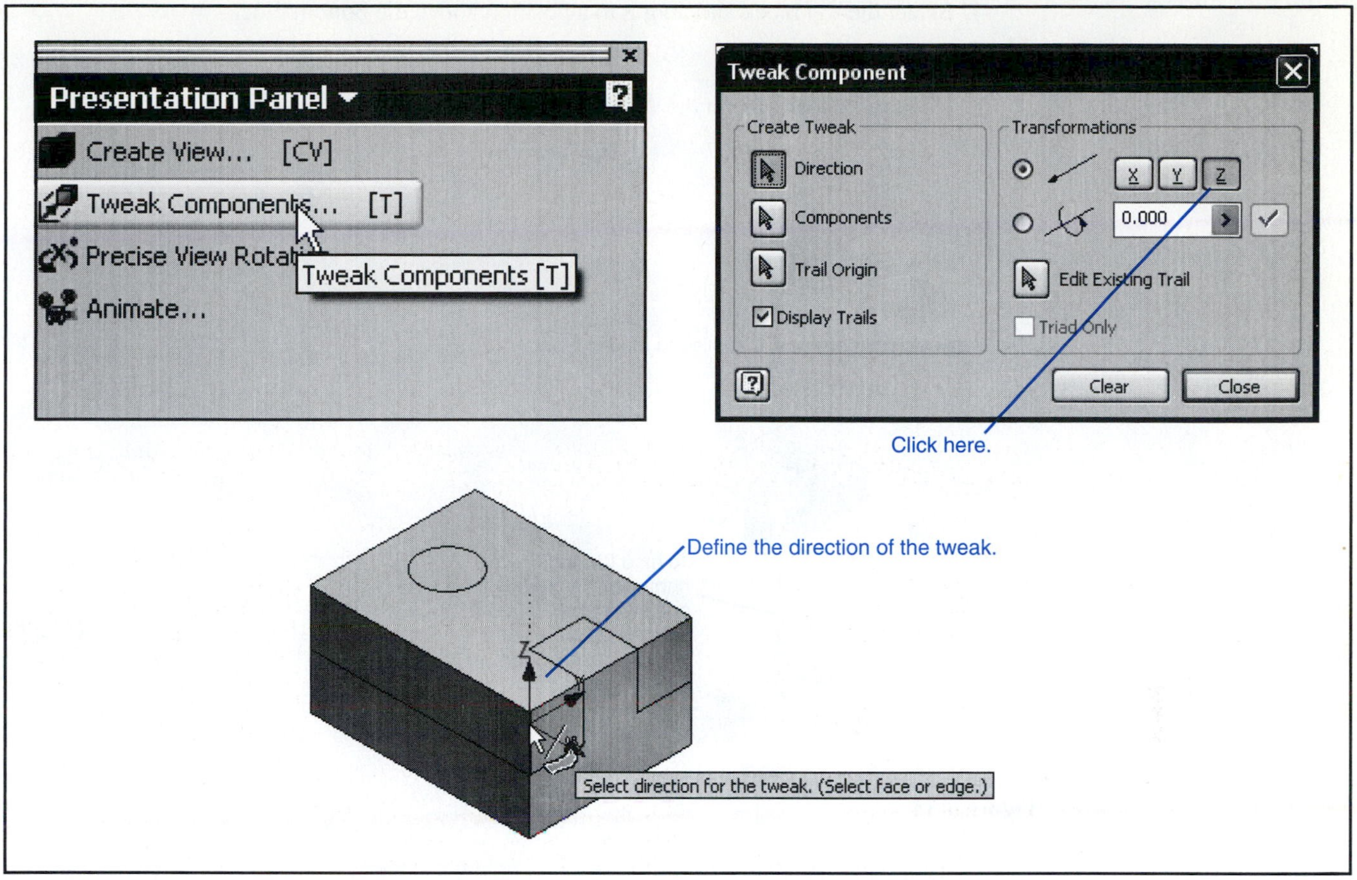

Figure 5-32

Exercise 5-15: Creating an Exploded Assembly Drawing

1. Click the **Tweak Components** tool in the **Presentation Panel** bar.

The **Tweak Component** dialog box will appear. See Figure 5-32. The **Direction** option will automatically be selected.

2. Select the direction of the tweak by selecting one of the assembly's vertical edge lines.

The **Tweak Component** dialog box will switch to the **Components** option. In this example the Z axis was selected.

3. Select the peg, then hold the left mouse button down and drag the peg to a position above the assembly.

See Figure 5-33.

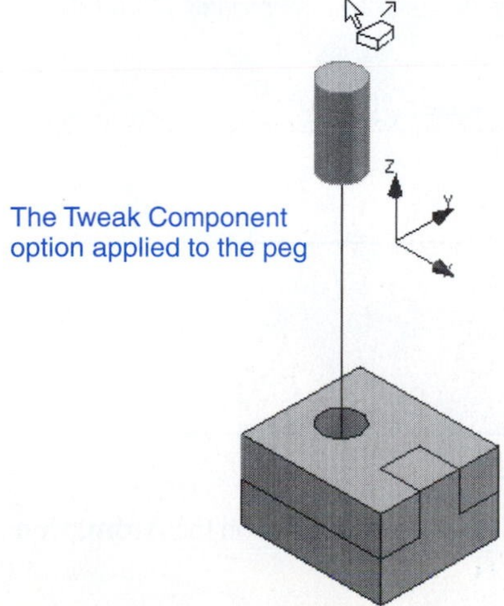

Figure 5-33

4. Select the top block and drag it to a position above the bottom block.

See Figure 5-34.

5. Click the **Clear** box on the **Tweak Component** dialog box, then click the **Close** box.

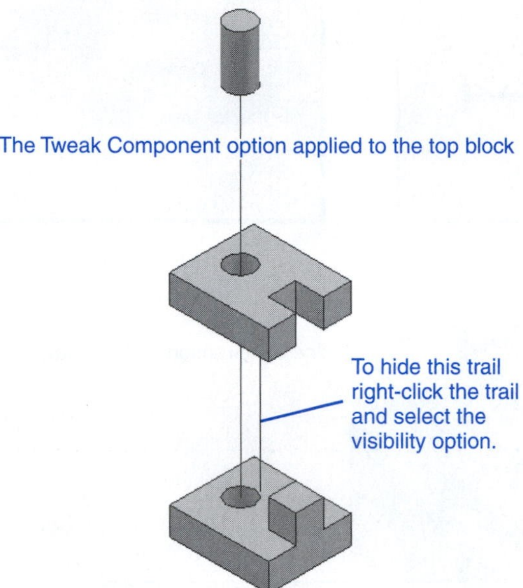

Figure 5-34

Exercise 5-16: Saving the Presentation Drawing

1. Click the **Save Copy As...** heading on the **File** pull-down menu.

The **Save Copy As** dialog box will appear.

2. Enter the file name and click the **Save** box.

The drawing will be saved as an .ipn drawing.

Exercise 5-17: Hiding a Trail

1. Right-click the trail.
2. Select the **Visibility** option.

Note that if you delete a trail, the tweaking will also be deleted, and the models will return to their original assembled positions..

ANIMATION

Presentation drawings can be animated using the **Animate** tool.

Exercise 5-18: Animating a Presentation Drawing

1. Click on the **Animate** tool on the **Presentation Panel** bar.

The **Animation** dialog box will appear. See Figure 5-35. The control buttons on the **Animation** dialog box are similar to those found on CD players.

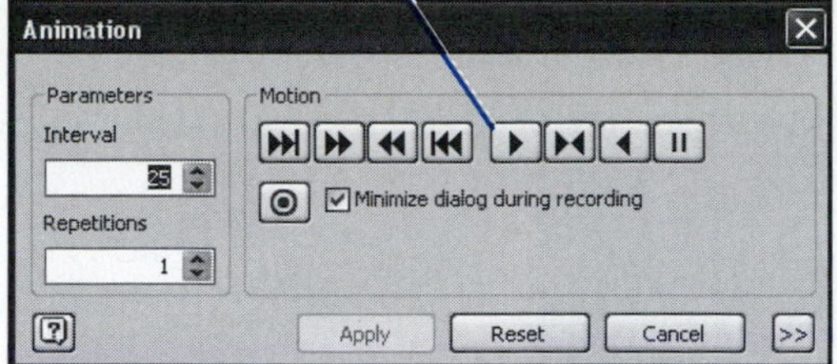

Figure 5-35

2. Click the **Play forward** button.

The assembly will be slowly reassembled in the reverse of the order used to tweak the components.

3. Click the **Reset** button to re-create the original presentation drawing.

Isometric Drawings

Isometric drawings can be created directly from presentation drawings. Assembly numbers (balloons) can be added to the isometric drawings and a parts list will automatically be created.

Exercise 5-19: Creating an Isometric Drawing

1. Click on the **New** tool, then the **Metric** tab.

The **New File** dialog box will appear.

2. Select the **ANSI (mm).idw** tool, then click **OK.**

The **Drawing Management** tools will appear in the panel bar. See Chapter 4 for a further explanation of the **Drawing Management** tools.

3. Click the **Base View** tool.

The **Drawing View** dialog box will appear. See Figure 5-36.

4. Click the **Explore Directories** button.

The **Open** dialog box will appear. See Figure 5-37.

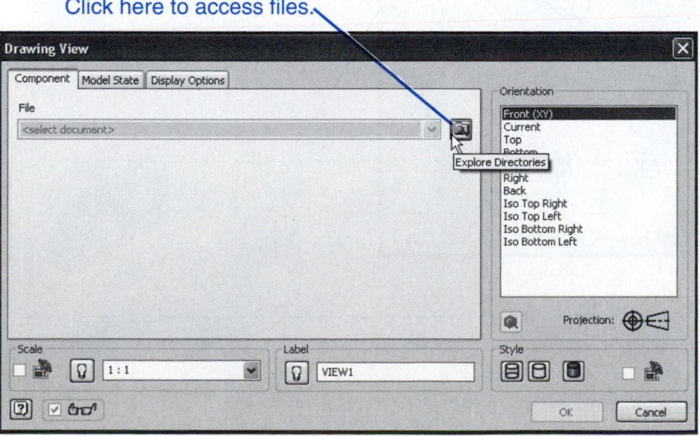

Figure 5-36

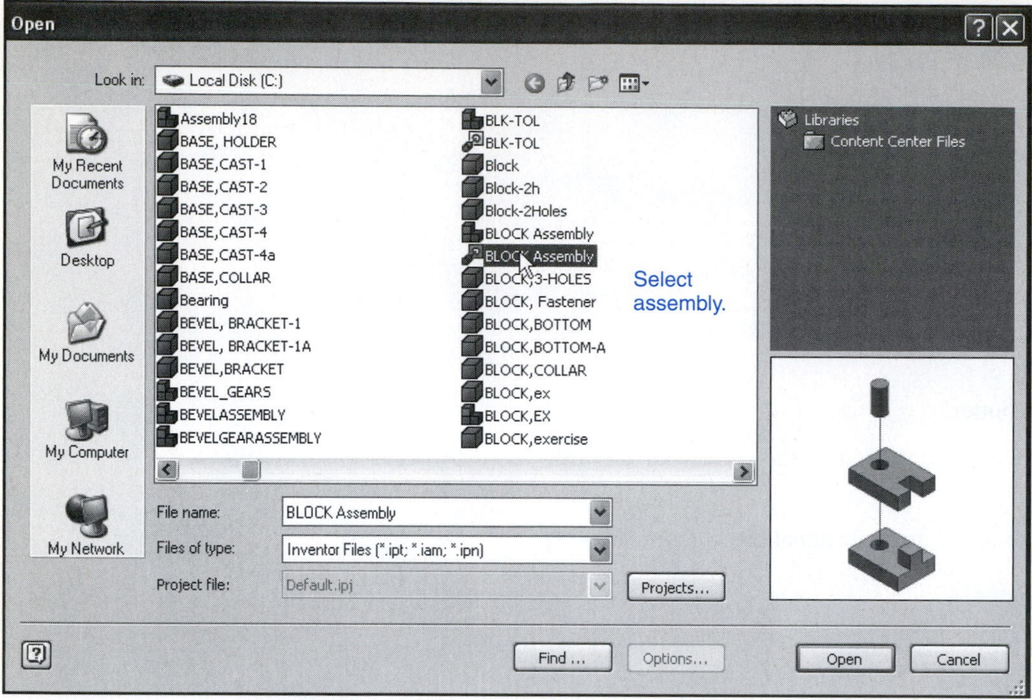

Figure 5-37

5. Select the appropriate presentation drawing (file type is .ipn), then click **Open.**

The **Drawing View** dialog box will reappear. See Figure 5-38.

6. Select the **Iso Top Right** orientation and set the **Scale** as needed. Select the **Hidden Lines Removed** option under the **Style** heading.

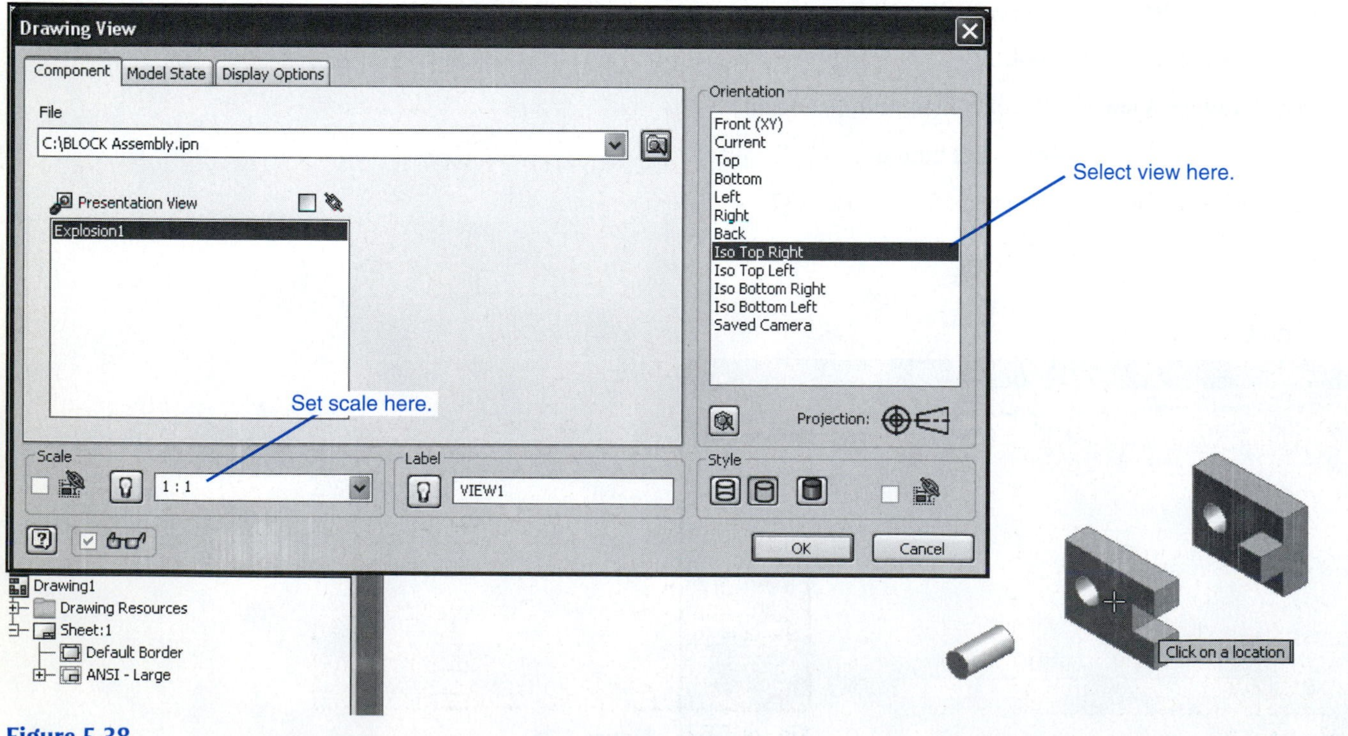

Figure 5-38

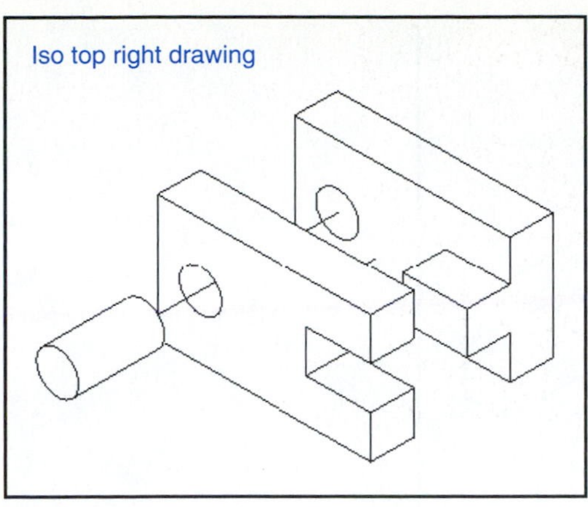

Figure 5-39

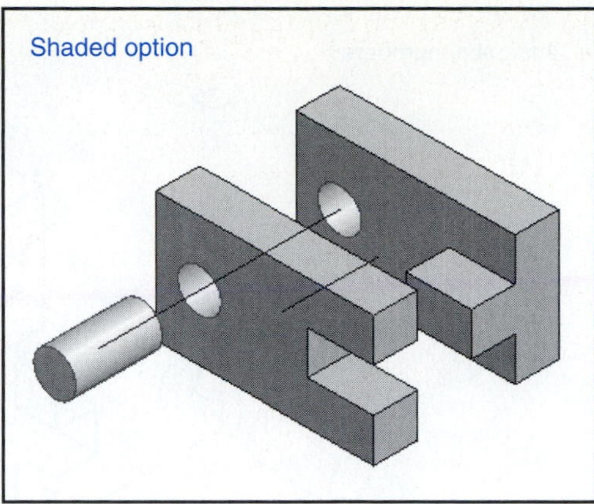

Figure 5-40

Figure 5-39 shows the resulting isometric view. Figure 5-40 shows the isometric drawing created using the **Shaded** option.

Assembly Numbers

Assembly numbers are added to an isometric drawing using the **Balloon** tool.

Exercise 5-20: Adding Balloons

1. Locate the cursor in the panel bar area and right-click the mouse.
2. Click the **Drawing Annotation** option.

The **Drawing Annotation Panel** bar will appear. See Figure 5-41.

3. Click the topmost edge line of the bottom block.

The **BOM Properties** dialog box will appear. See Figure 5-42.

4. Click **OK,** then drag the cursor away from the selected edge line.
5. Locate a position away from the component and click the left mouse button. Move the cursor in a horizontal direction and click the left mouse button again.
6. Right-click the mouse and select the **Continue** option.
7. Add balloons to the other components.
8. Move the cursor to the center of the screen and click the right mouse button, then select the **Done** option.

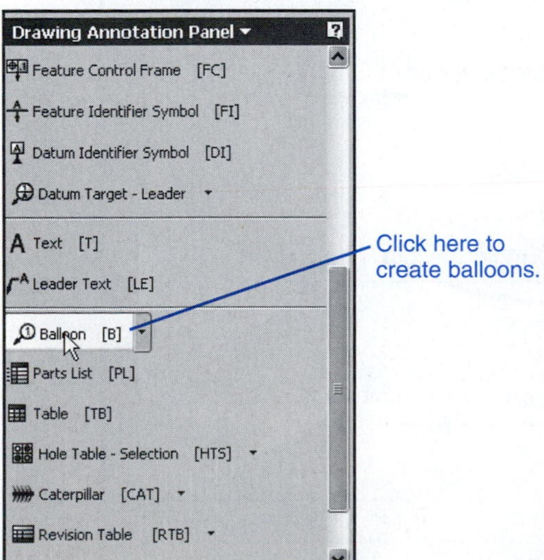

Figure 5-41

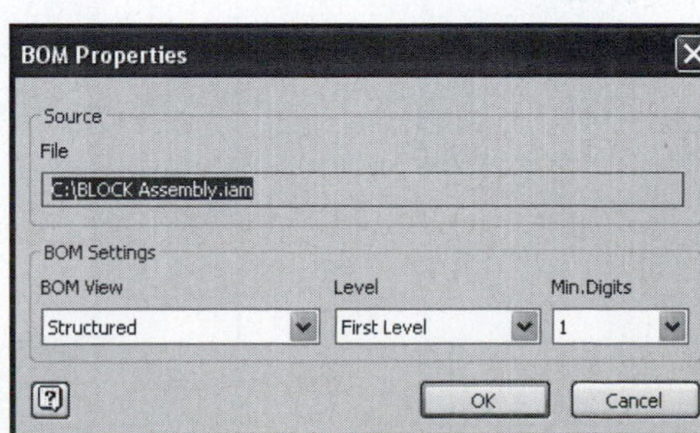

Figure 5-42

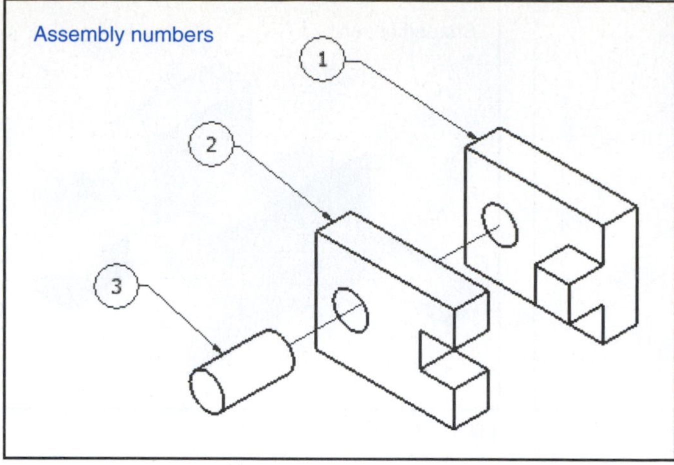

Figure 5-43

See Figure 5-43. The balloon numbers will be in the order the parts were added to the drawing.

TIP Making the balloon leaders lines the same angle will give the drawing a well-organized appearance.

Exercise 5-21: Editing Balloons

In general, the biggest parts have the lowest numbers. The assigned numbers can be edited.

1. Right-click the balloon to be edited and select the **Edit Balloon** option. See Figure 5-44. The **Edit Balloon** dialog box will appear. See Figure 5-45.
2. Make any needed changes in the **Edit Balloon** dialog box and click **OK.**

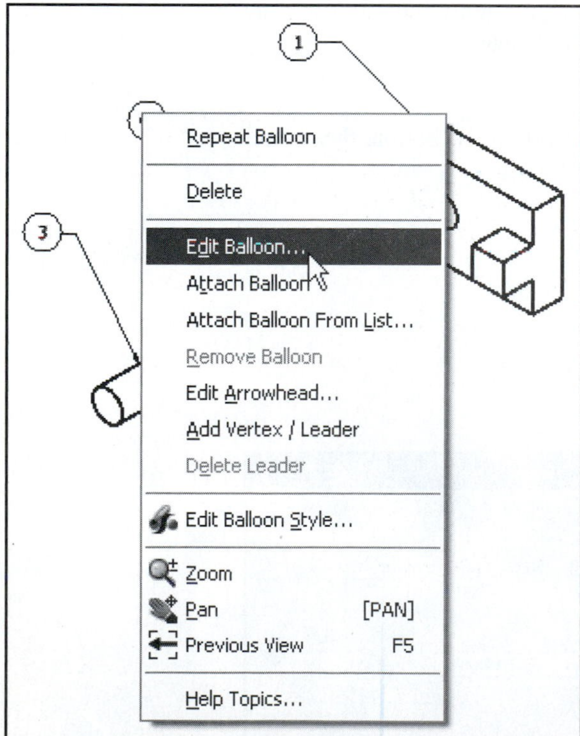

Figure 5-44

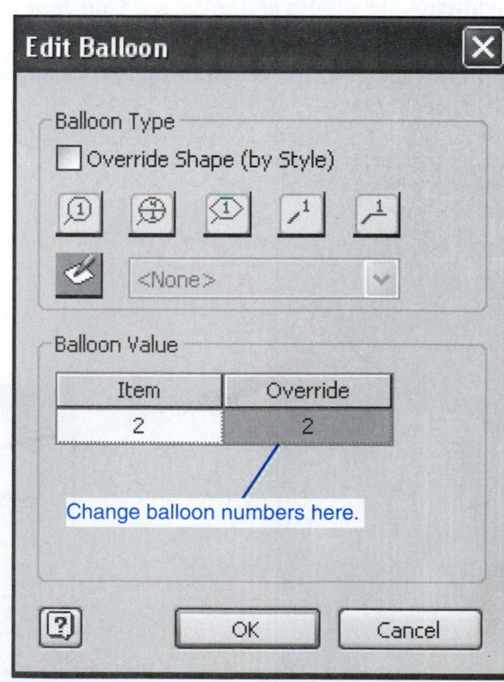

Figure 5-45

The terms *part list* and *BOM* are interchangeable.

Parts List

A parts list can be created from an isometric drawing after the balloons have been assigned using the **Parts List** tool on the **Drawing Annotation Panel** bar.

The **Drawing Annotation Panel** bar is accessed by moving the cursor into the panel bar area, then right-clicking the mouse and selecting the **Drawing Annotation** option.

Exercise 5-22: Creating a Parts List

1. Click the **Parts List** tool on the **Drawing Annotation Panel** bar.

The **Parts List** dialog box will appear. See Figure 5-46.

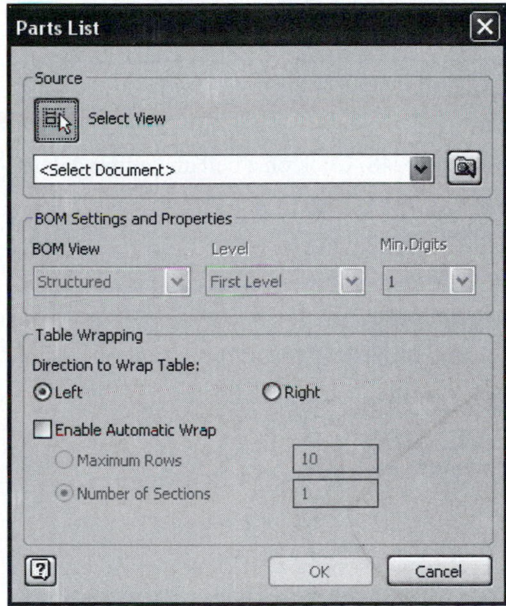

Figure 5-46

2. Move the cursor into the area around the isometric drawing.

A broken red line will appear when the cursor is in the area.

3. Click the left mouse button, then click the **OK** button on the **Parts List** dialog box. Move the cursor away from the isometric drawing area.

An outline of the parts list will appear and move with the cursor.

4. Select a location for the parts list and left-click the mouse.

Figure 5-47 shows the resulting drawing. The parts list was generated using information from the original model drawings and the presentation drawings.

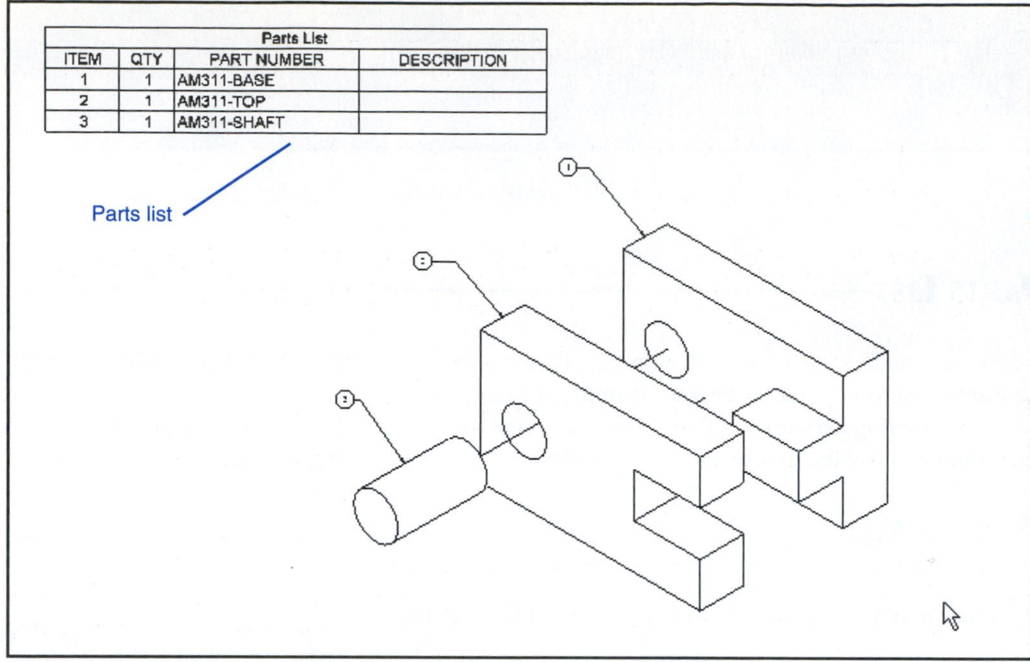

Parts List			
ITEM	QTY	PART NUMBER	DESCRIPTION
1	1	AM311-BASE	
2	1	AM311-TOP	
3	1	AM311-SHAFT	

Figure 5-47

Exercise 5-23: Editing a Parts List

1. Move the cursor onto the parts list and right-click the mouse.
2. Click the **Edit Parts List** option.

The **Edit Parts List** dialog box will appear. See Figure 5-48. Click on a cell and either delete or add text. Figure 5-49 shows an edited parts list.

The original parts

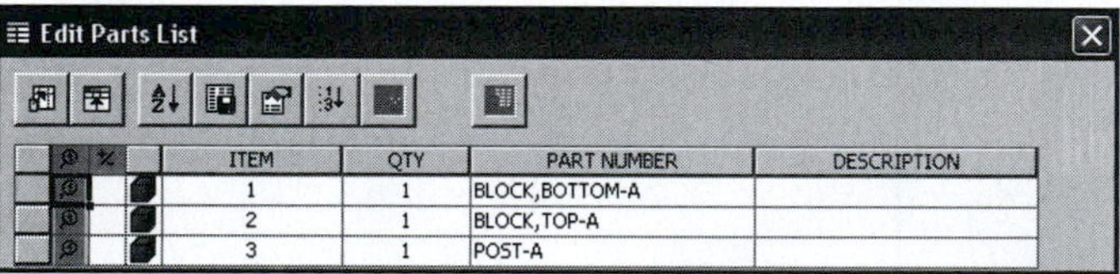

ITEM	QTY	PART NUMBER	DESCRIPTION
1	1	BLOCK,BOTTOM-A	
2	1	BLOCK,TOP-A	
3	1	POST-A	

Figure 5-48

The Column Chooser

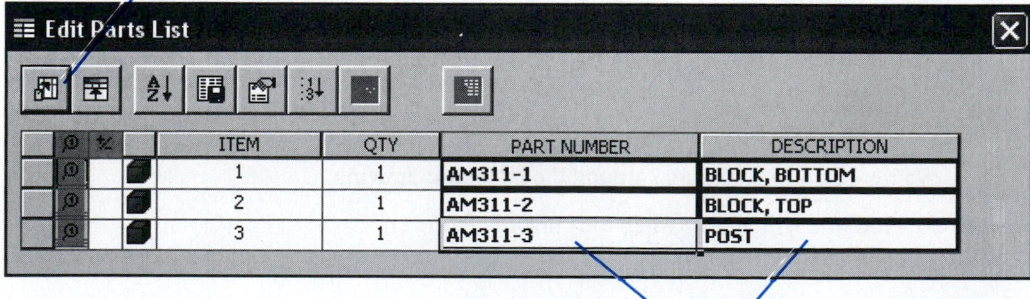

ITEM	QTY	PART NUMBER	DESCRIPTION
1	1	AM311-1	BLOCK, BOTTOM
2	1	AM311-2	BLOCK, TOP
3	1	AM311-3	POST

The PART NUMBER and DESCRIPTION columns have been edited

Figure 5-49

Naming Parts

Each company or organization has its own system for naming parts. In the example in this book the noun, modifier format was used.

Exercise 5-24: Adding a New Column

Say two additional columns were required for the parts list shown in Figure 5-49: Material and Notes.

1. Click the **Column Chooser** button at the top of the **Edit Parts List** dialog box.

The **Parts List Column Chooser** dialog box will appear. See Figure 5-50.

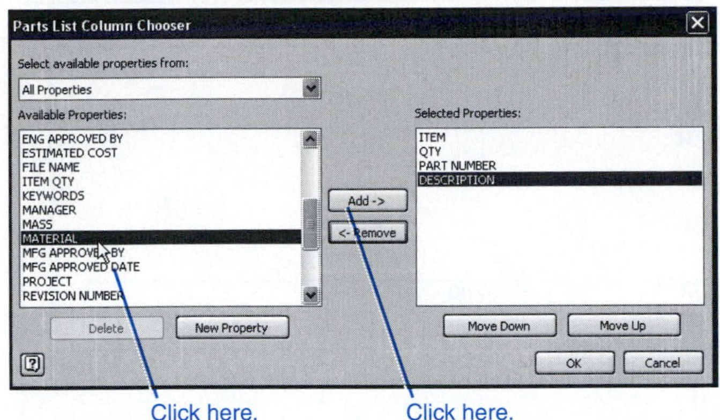

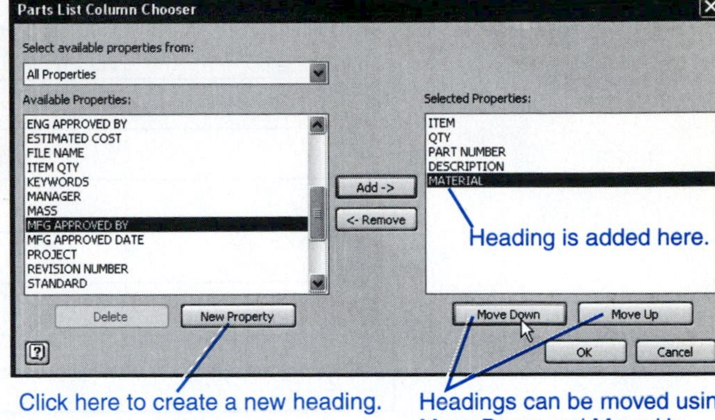

Figure 5-50

2. Scroll down the **Available Properties** listing to see if the new colum headings are listed.
3. **Material** is listed, so click on the listing, then click the **Add** box in the middle of the screen.

The heading **Material** will appear in the **Selected Properties** area. Use the **Move Down** and **Move Up** boxes to sequence the columns heading.

The heading **Notes** is not listed, so it must be defined.

4. Click on the **New Property** box.

The **Define New Property** dialog box will appear. See Figure 5-51.

5. Type in the name of the new column, then click **OK.**

In this example a **NOTES** column was added.

Note that only uppercase letters are used to define column headings.

6. Click **OK** on the **Parts List Column Chooser** dialog box.

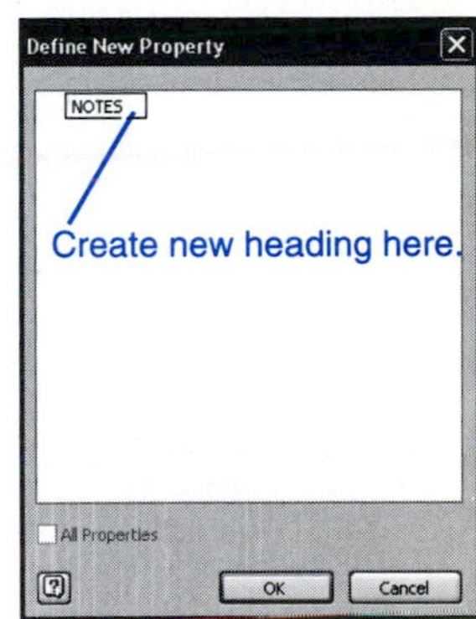

Figure 5-51

Figure 5-52 shows the revised column in the parts list. If no material is defined, the word **Default** will appear. The material for a model will be assigned to the model drawing and brought forward into the parts list. The **Material** column can be edited like the other columns.

Figure 5-53 shows the edited parts list on the drawing.

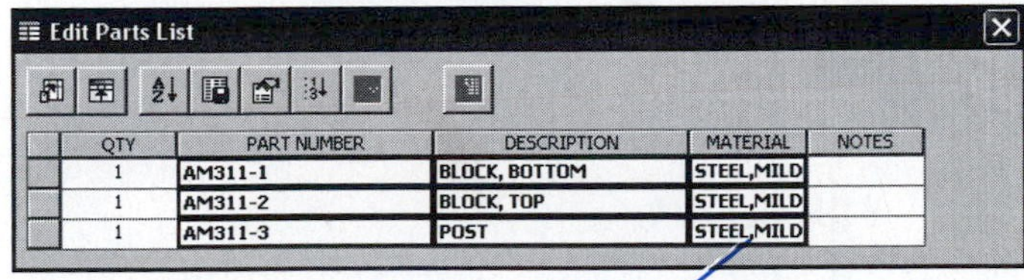

Figure 5-52

Parts List					
ITEM	QTY	PART NUMBER	DESCRIPTION	MATERIAL	NOTE
1	1	AM311-1	Block,Bottom	Steel, Mild	
2	1	AM311-2	Block, Top	Steel, Mild	
3	1	AM311-3	Ø20PEG	Steel, Mild	

The material specifications were entered as the objects were created. See Chapter 3. If no material is defined, the word "Default" will appear.

Figure 5-53

Title Block

All drawings include a title block, usually located in the lower right corner of the drawing sheet, as Figure 5-54 shows. Text may be added to a title block under existing headings, or new headings may be added.

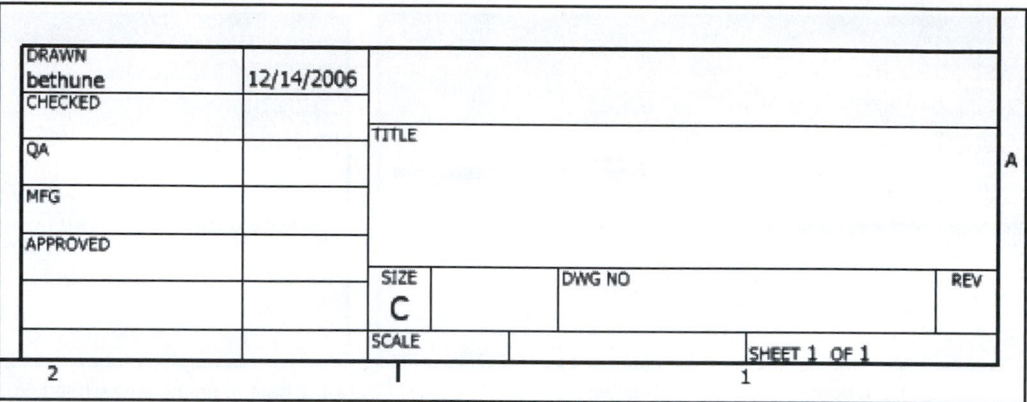

Figure 5-54

Exercise 5-25: Adding Text to a Title Block

1. Right-click the drawing name in the **Model** browser box, then click the **Properties** option.

See Figure 5-55. The drawing's **Properties** dialog box will appear. Text can be typed into the **Properties** dialog box and will appear on the title block. Figure 5-56 shows the **Summary** input.

2. Click the **Summary** tab on the **Properties** dialog box and enter the appropriate information.

In this example, the Title, Author, and Company were added.

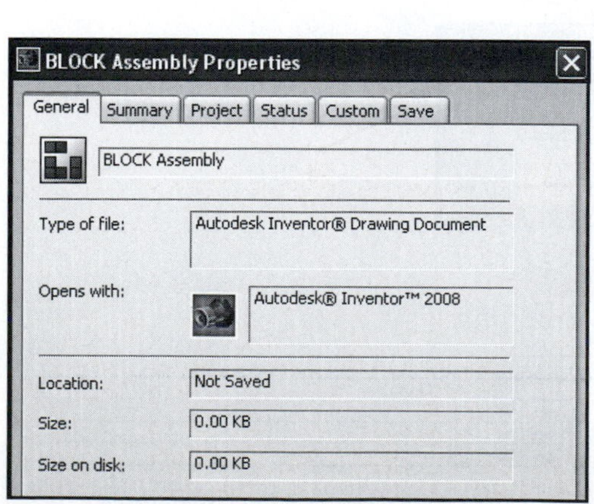

Figure 5-55

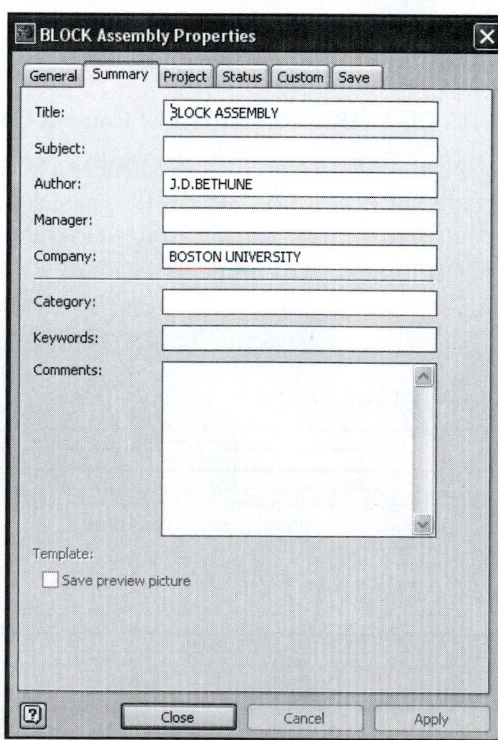

Figure 5-56

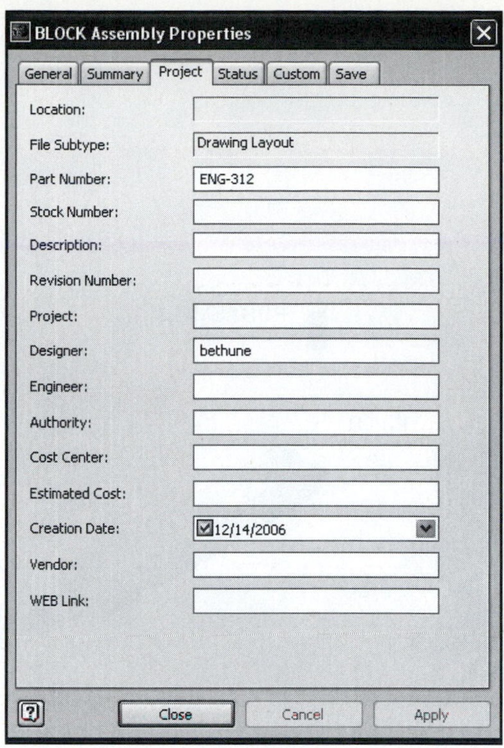

Figure 5-57

3. Click the **Project** tab and add the **Part Number.** See Figure 5-57. In this example the number **ENG-312** was added. This is an assembly drawing number. Each individual part has its own number.
4. Click **Apply** and **Close.**

See Figure 5-57. Figure 5-58 shows the completed title block.

The title block included with Inventor is only one possible format. Each company and organization will have its own specifications.

Revised title block

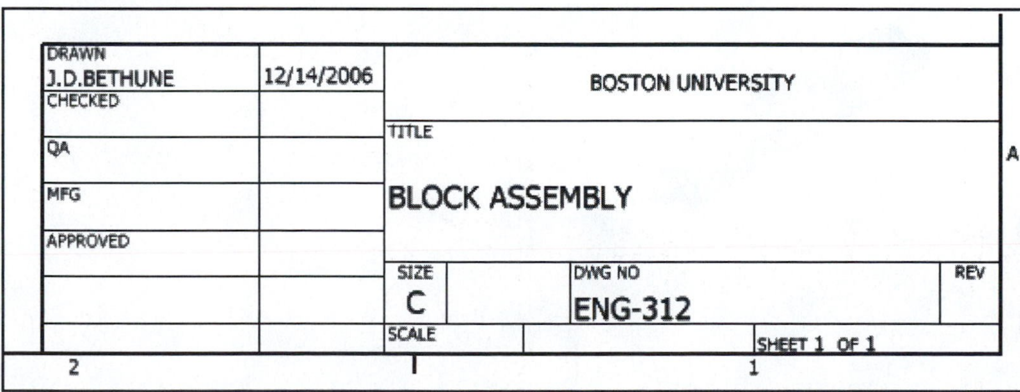

Figure 5-58

Subassemblies

Figure 5-59 shows a slightly more complicated assembly than the BLOCK assembly used in the previous sections. It is called a **PIVOT** assembly. Figure 5-60 shows the components needed to create the assembly. This will be a bottom-up assembly.

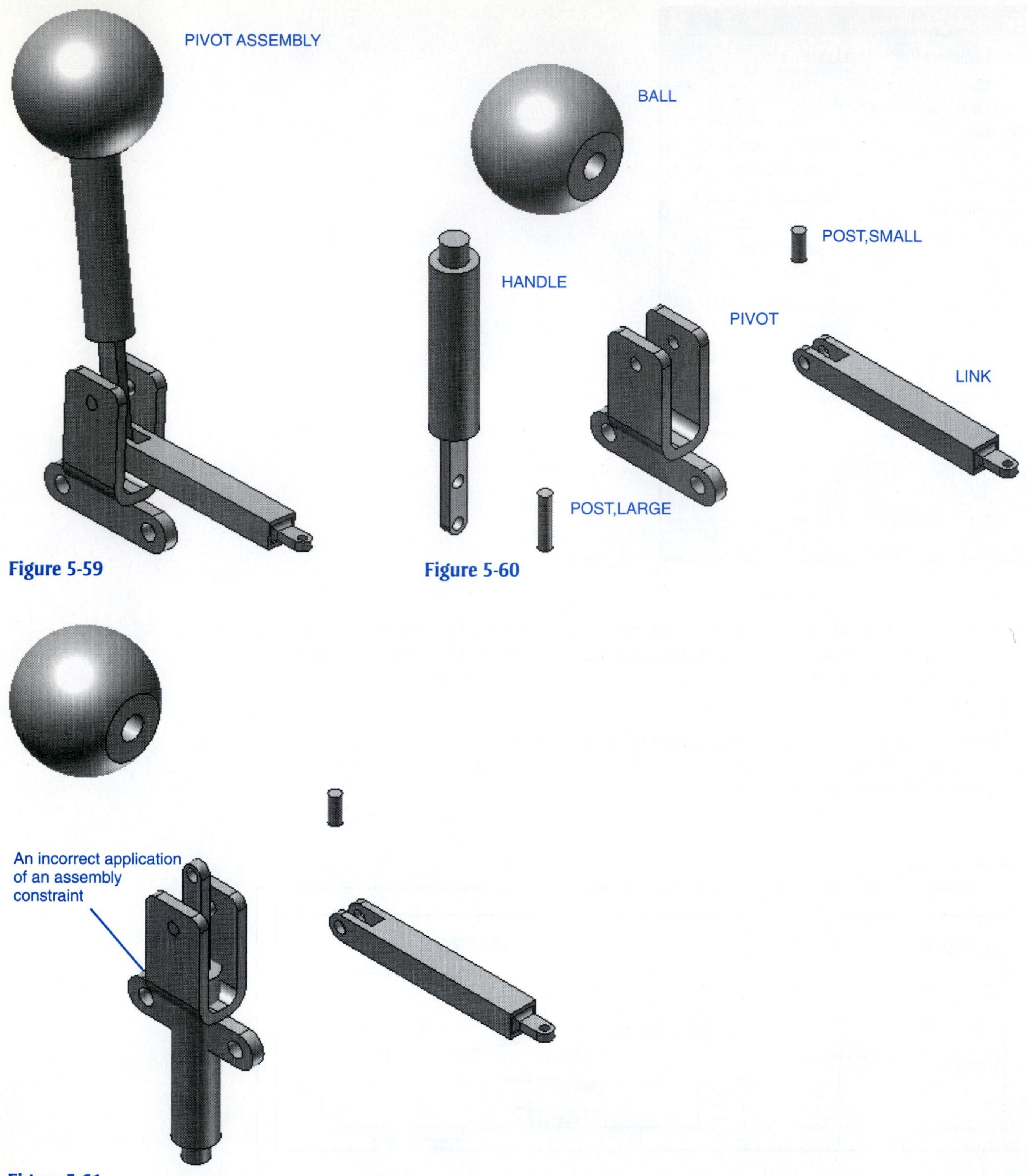

Figure 5-59

Figure 5-60

Figure 5-61

It is sometimes difficult to control the assembly constraints. The parts seem to move randomly about the screen when constraints are added. Figure 5-61 shows an example of an incorrect application of an assembly constraint. The assembly sequence presented here is one of many different ones that could be used.

The PIVOT is grounded because it was the first component entered on the screen.

1. Right-click **HANDLE** in the browser box and ground the handle.

More than one component many be grounded at one time. See Figure 5-62.

2. Use the **Insert** constraint and insert the POST,LARGE into the top hole of the HANDLE. Use the **Offset** option to center the post.
3. Use **Insert** and position the POST,SMALL into the HANDLE.

See Figure 5-63.

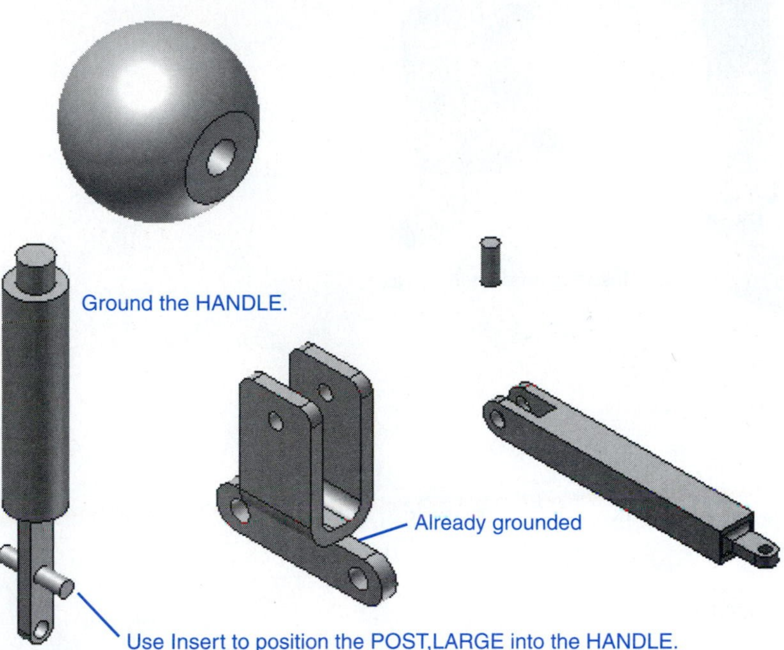

Figure 5-62

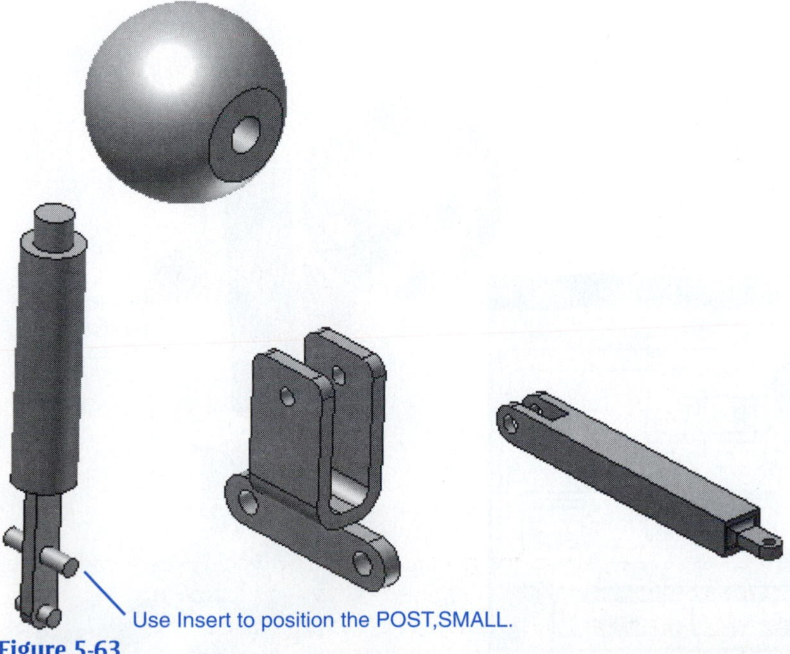

Figure 5-63

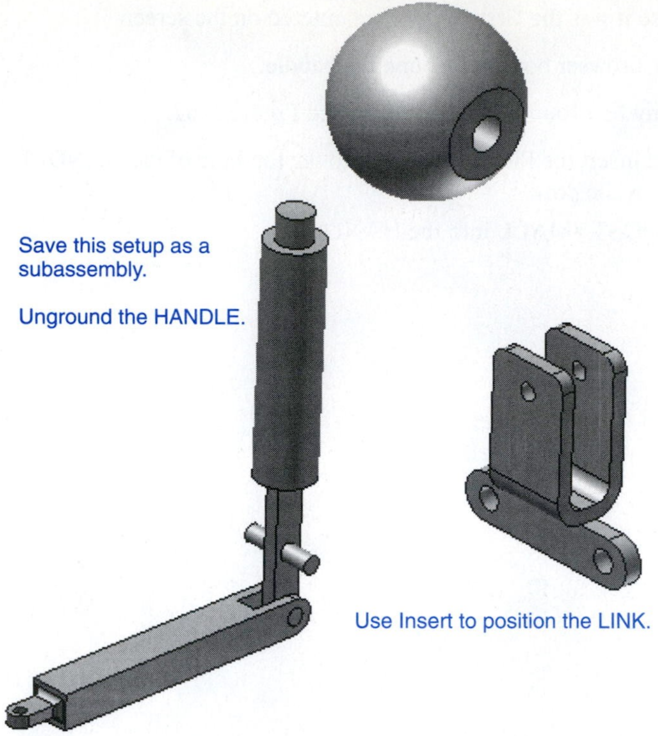

Figure 5-64

4. Use **Insert** and position the LINK onto the POST,SMALL.

See Figure 5-64.

5. Unground the HANDLE.
6. Save the assembly drawing as **PIVOT ASSEMBLY.**
7. Use **Insert** to position the subassembly into the PIVOT.

See Figure 5-65.

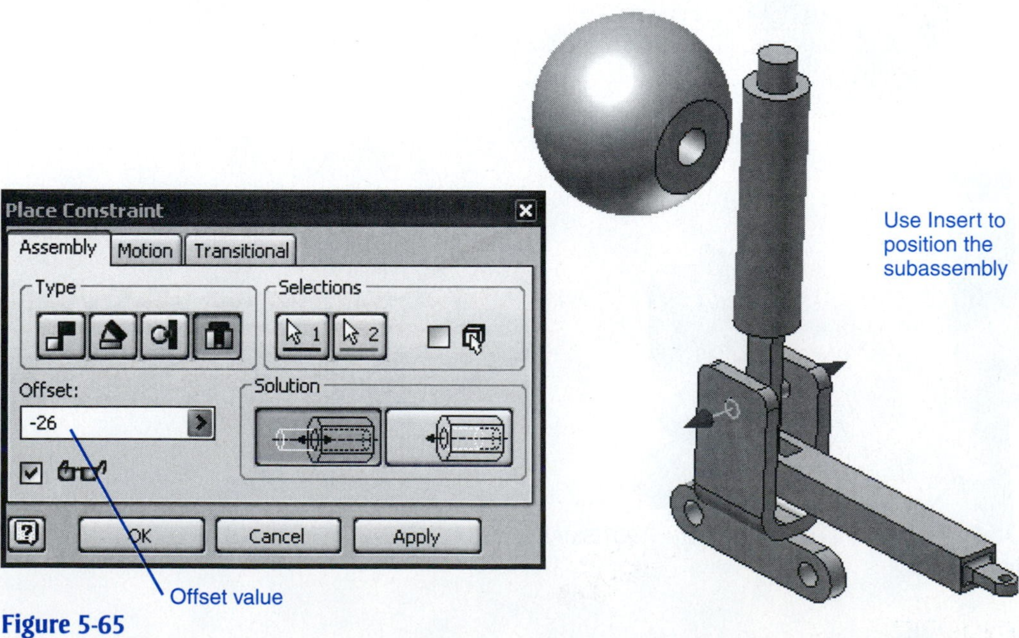

Figure 5-65

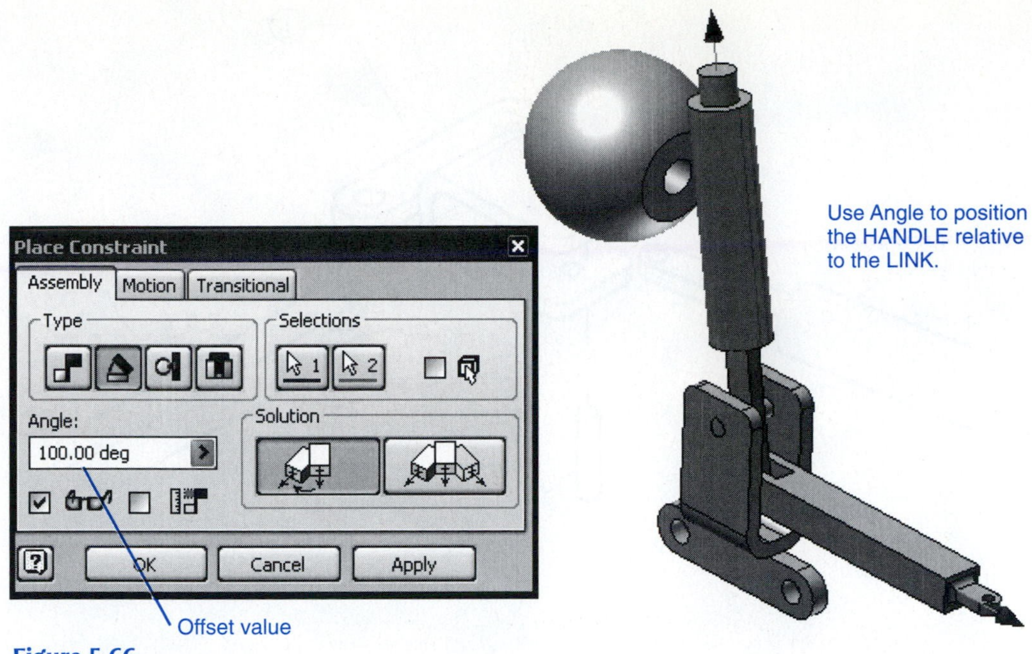

Figure 5-66

8. Use the **Angle** constraint to position the HANDLE relative to the LINK.

In this example, an angle of 100° was used. See Figure 5-66.

9. Use **Insert** to position the BALL on top of the HANDLE.

See Figure 5-59.

Figure 5-67 shows a presentation drawing of the PIVOT ASSEMBLY, and Figure 5-68 shows an exploded isometric drawing of the assembly and a parts list.

Figure 5-67

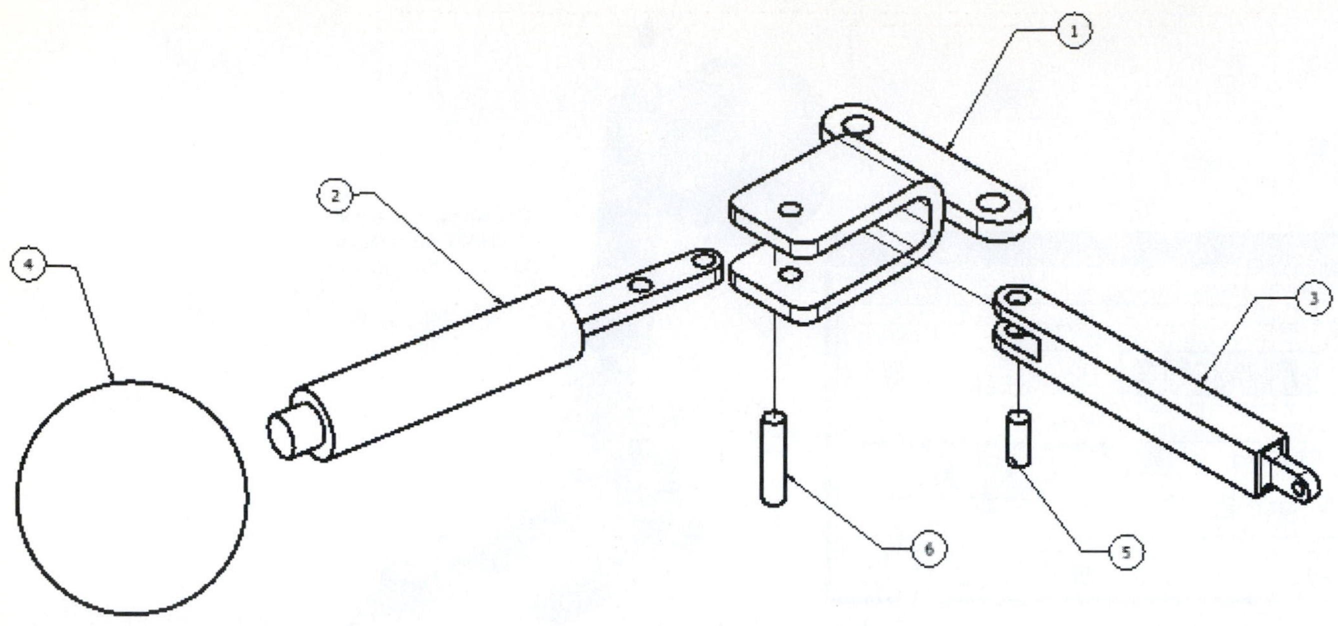

Parts List				
ITEM	PART NUMBER	DESCRIPTION	MATERIAL	QTY
1	ENG-A43	BOX,PIVOT	SAE1020	1
2	ENG-A44	POST,HANDLE	SAE1020	1
3	ENG-A45	LINK	SAE1020	1
4	AM300-1	HANDLE	STEEL	1
5	EK-132	POST-Ø6x14	STEEL	1
6	EK-131	POST-Ø6x26	STEEL	1

Figure 5-68

DRAWING SHEETS

Some assemblies are so large they require larger paper sheet sizes. Drawings are prepared on predefined standard-size sheets of paper. Each standard size has been assigned a letter value. Figure 5-69 shows the letter values and the sheet size assigned to each. All these sizes and more are available within Inventor.

Standard Drawing Sheet Sizes
Inches

A = 8.5 × 11
B = 11 × 17
C = 17 × 22
D = 22 × 34
E = 34 × 44

Standard Drawing Sheet Sizes
Millimetres

A4 = 210 × 297
A3 = 297 × 420
A2 = 420 × 594
A1 = 594 × 841
A0 = 841 × 1189

Figure 5-69

Figure 5-70 shows a drawing done on a C-size drawing sheet. Note the letter C in the title block.

Exercise 5-26: Changing a Sheet Size

1. Locate the cursor on the **Sheet:1** heading in the browser box and right-click the mouse.
2. Select the **Edit Sheet** option.

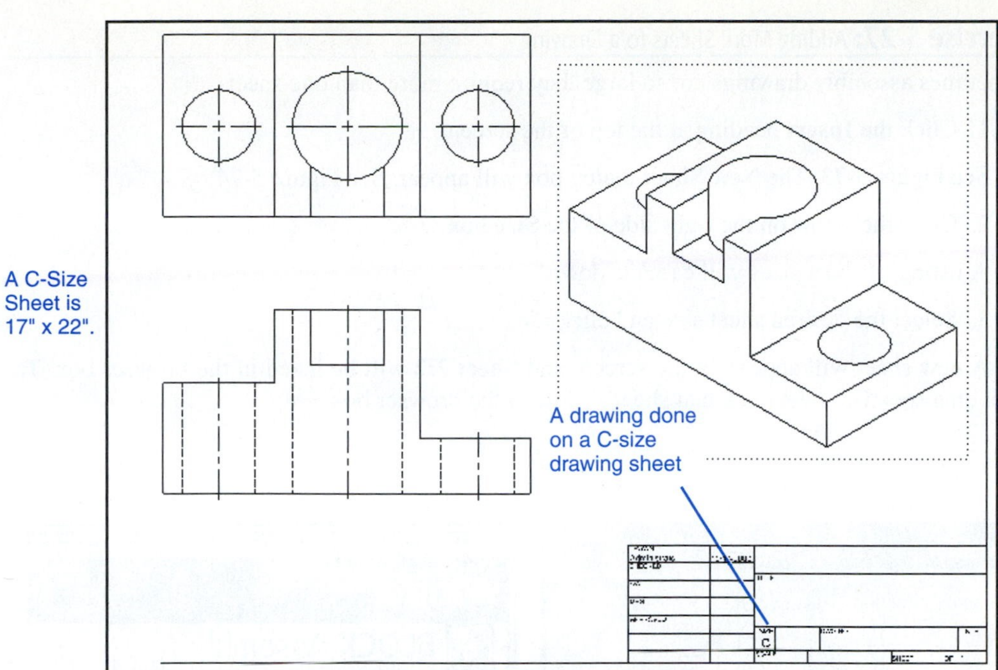

Figure 5-70

The **Edit Sheet** dialog box will appear. See Figure 5-71.

3. Select the **D** option and click **OK.**

See Figure 5-72. Note that the letter C has been replaced with the letter D.

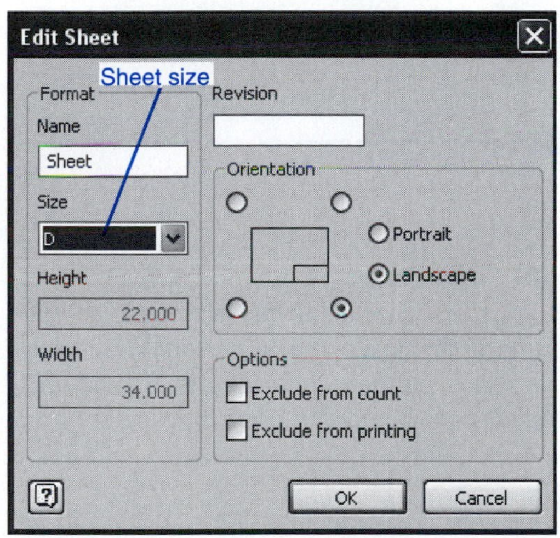

Figure 5-71

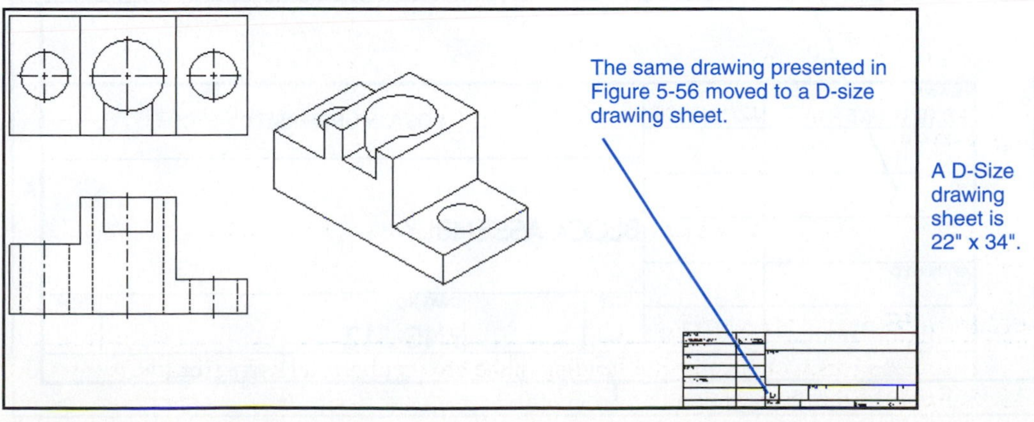

Figure 5-72

Exercise 5-27: Adding More Sheets to a Drawing

Sometimes assembly drawings are so large they require more than one sheet.

1. Click the **Insert** heading at the top of the screen.

See Figure 5-73. The **New Sheet** dialog box will appear. See Figure 5-74.

2. Click the arrow on the right side of the **Size** box.

A listing of sheet sizes will cascade down.

3. Select the desired sheet size and click **OK.**

A new sheet will appear on the screen, and **Sheet 2:2** will be listed in the browser box. To work on a specific sheet click that sheet's name in the browser box.

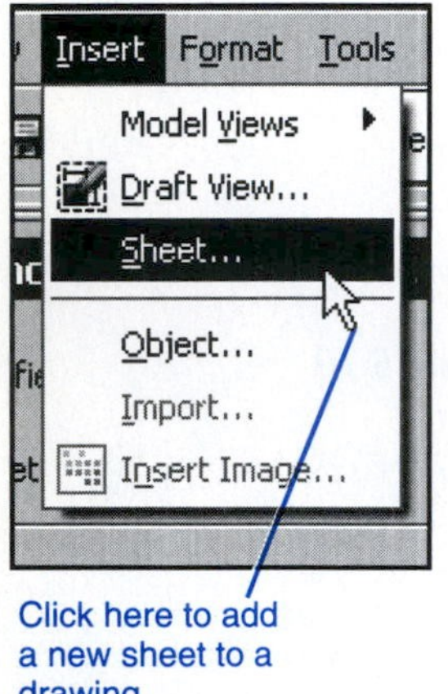

Figure 5-73

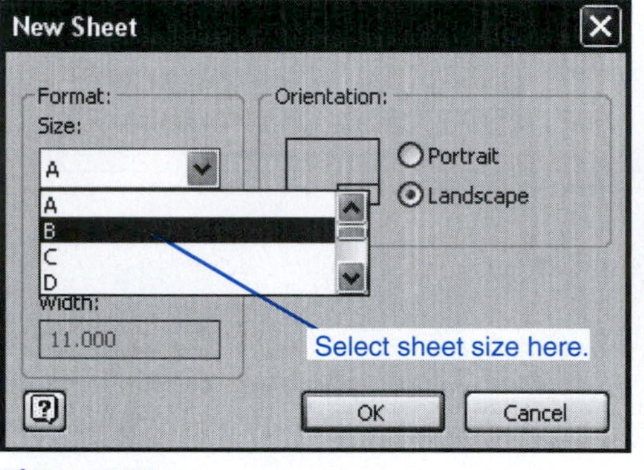

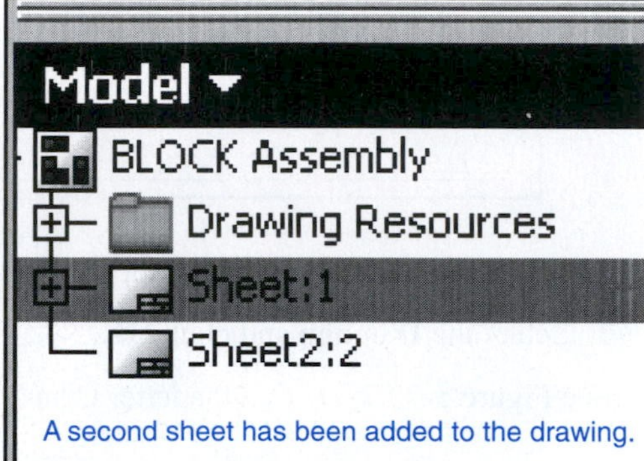

Figure 5-74

Other Types of Drawing Blocks

Release Blocks

release block: The area in a title block where required approval signatures are entered.

Figure 5-75 shows an enlarged view of the title block. The area on the left side of the block is called a ***release block.*** After a drawing is completed it is first checked. If the drawing is acceptable, the checker will initial the drawing and forward it to the next approval person. Which person(s) and which department approve new drawings varies, but until a drawing is "signed off," that is, all required signatures have been entered, it is not considered a finished drawing.

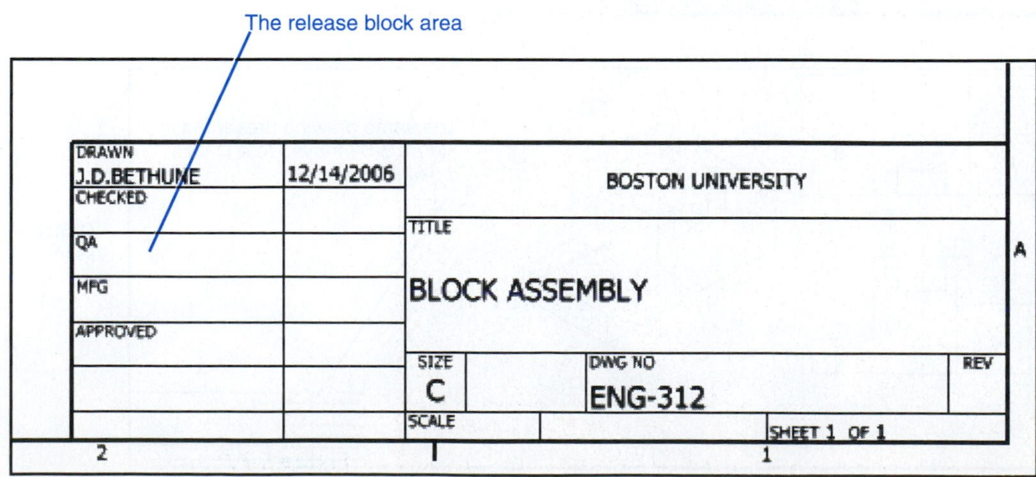

Figure 5-75

Revision Blocks

Figure 5-76 shows a sample ***revision block.*** It was created using the **Revision** Table tool located on the **Drawing Annotation Panel bar.** Drawings used in industry are constantly being changed. Products are improved or corrected, and drawings must reflect and document these changes.

revision block: The area in a drawing where changes are listed by number with a brief description of the change.

Drawing changes are listed in the revision block by number. Revision blocks are usually located in the upper right corner of the drawing.

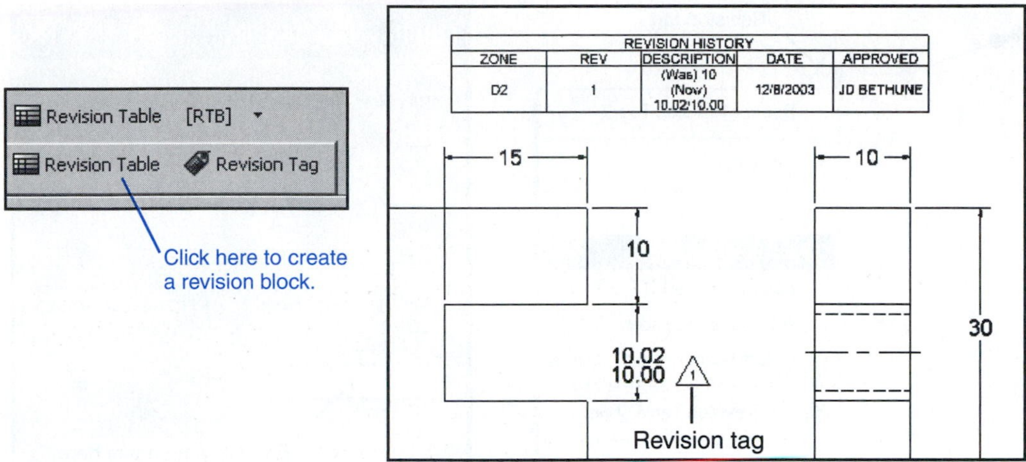

Figure 5-76

Exercise 5-28: Creating a Revision Block

1. Click the **Revision Table** tool on the **Drawing Annotation Panel** bar.

A **Revision Table** dialog box will appear on the screen. See Figure 5-77. Revisions are usually numbered starting with 1. The default **Start Value** shown in the **Revision Table** dialog box is 1. The numbers shown in the **Revision tag** should correspond to numbers listed under the **REV** heading in the revision block.

2. Click **OK.**

The revision block will appear on the screen. Revision block are usually located in the upper right corner of the drawing.

Each drawing revision is listed by number in the revision block. A brief description of the change is also included. It is important that the description be as accurate and complete as possible. The zone on the drawing where the revision is located is also specified.

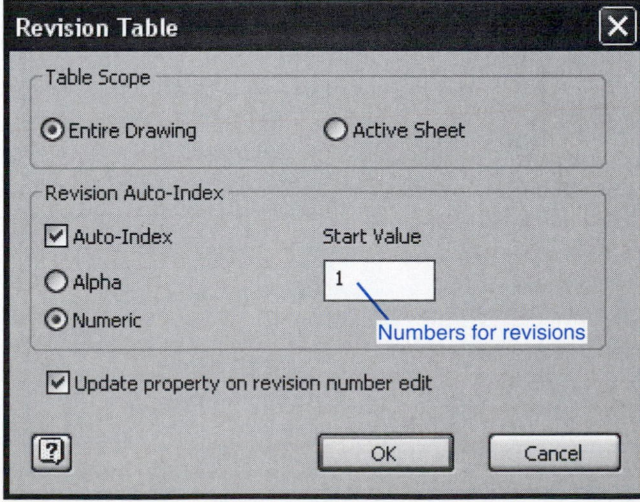

Figure 5-77

The revision number is added to the field of the drawing in the area where the change was made. The revision letter is located within a "flag" to distinguish it from dimensions and drawing notes. The flag is created using the **Revision Tag** tool located on the **Drawing Annotation Panel** bar. See Figure 5-76. The **Revision Tag** tool is a flyout from the **Revision Table** tool.

To change the number within a revision tag, right-click the tag and select the **Edit Tag** option. A text dialog box will appear, and the tag number may be changed. See Figure 5-78.

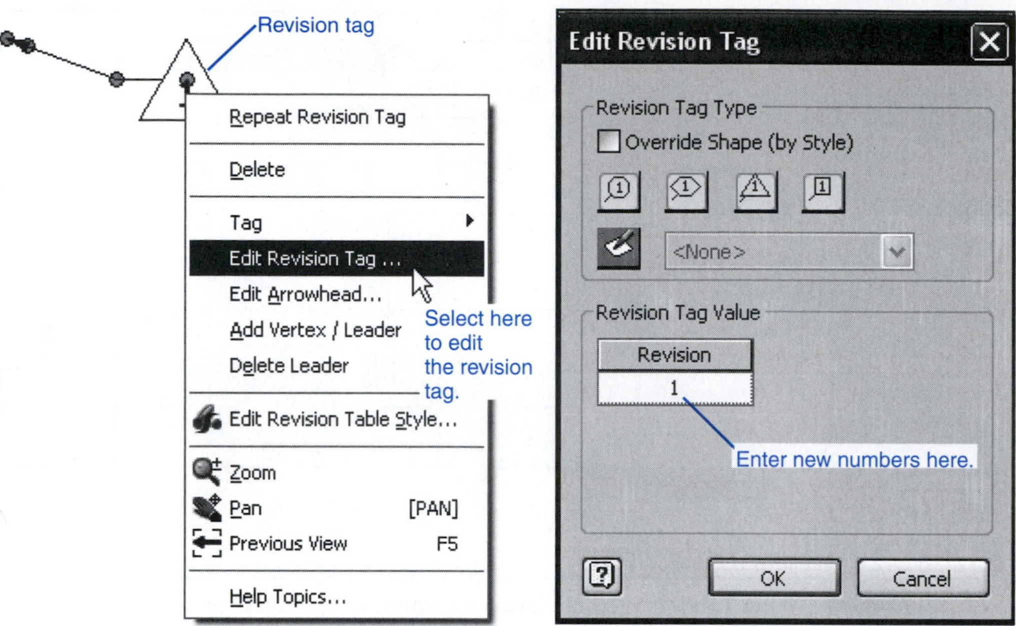

Figure 5-78

Exercise 5-29: Editing the Revision Block

1. Move the cursor onto the revision block.

Filled green circles will appear around the revision block. See Figure 5-79

2. Right-click the mouse and select the **Edit** option.

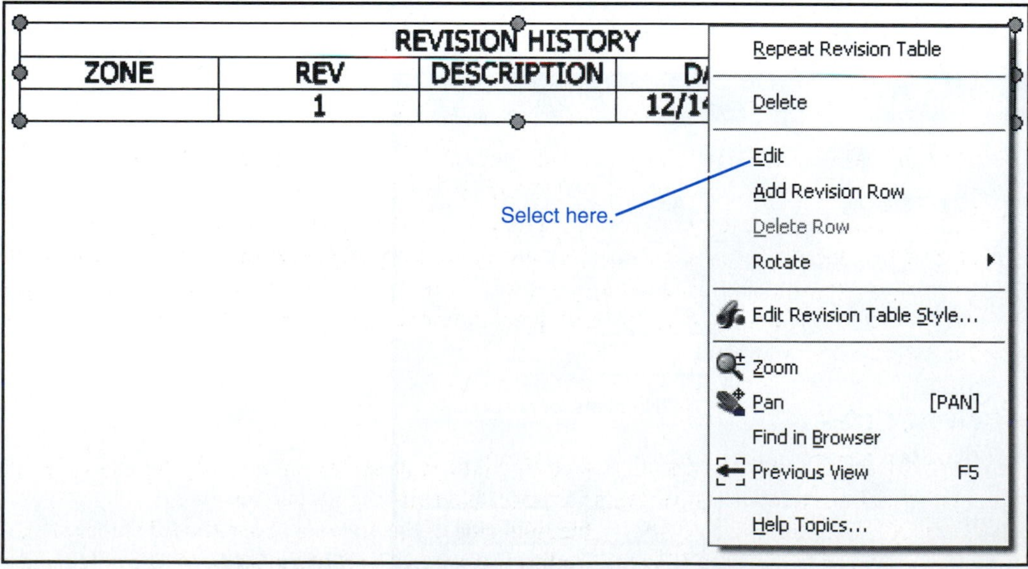

Figure 5-79

The **Edit Revision Table** dialog box will appear. The block's headings may be edited or rearranged as needed. See Figure 5-80.

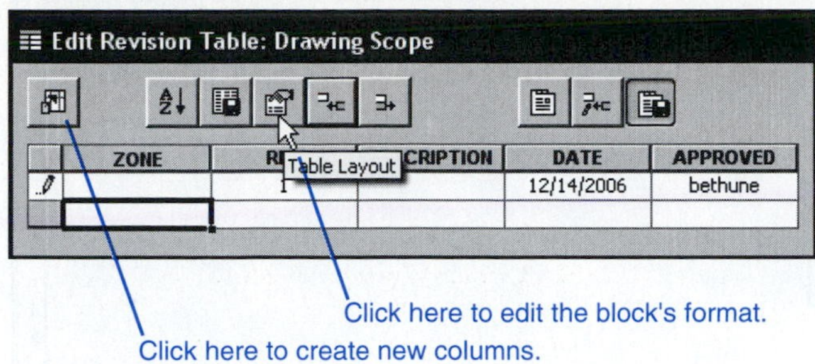

Figure 5-80

ECOs

Most companies have systems in place that allow engineers and designers to make quick changes to drawings. These change orders are called *engineering change orders* (ECOs), *engineering orders* (EOs), or *change orders* (COs), depending on the company's preference. Change orders are documented on special drawing sheets that are usually stapled to a print of the drawing. Figure 5-81 shows a sample change order attached to a drawing.

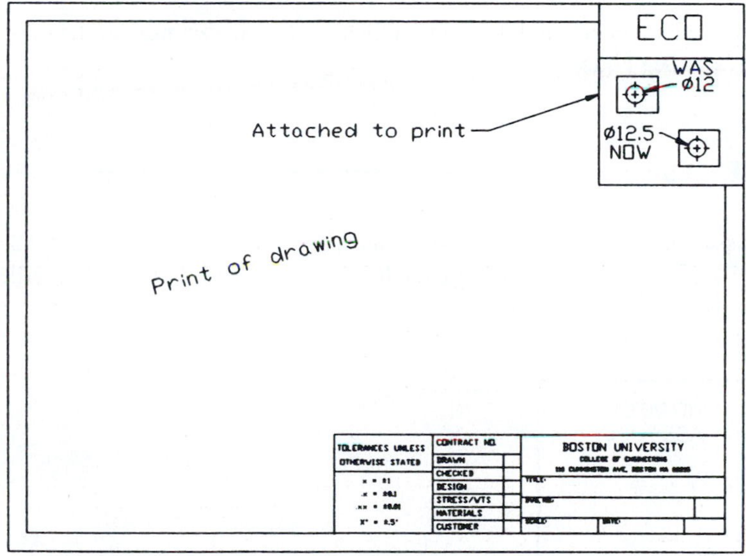

Figure 5-81

After a number of change orders have accumulated, they are incorporated into the drawing. This process is called a ***drawing revision,*** which is different from a revision to the drawing. Drawing revisions are usually identified by a letter located somewhere in the title block. The revision letters may be included as part of the drawing number or in a separate box in the title block. Whenever you are working on a drawing make sure you have the latest revision and all appropriate change orders. Companies have recording and referencing systems for listing all drawing revisions and drawing changes.

drawing revision: A version of a drawing into which change orders have been incorporated.

Drawing Notes

Drawing notes are used to provide manufacturing information that is not visual, for example, finishing instructions, torque requirements for bolts, and shipping instructions.

Drawing notes are usually listed on the right side of the drawing above the title block. Drawing notes are listed by number. If a note applies to a specific part of the drawing, the note number

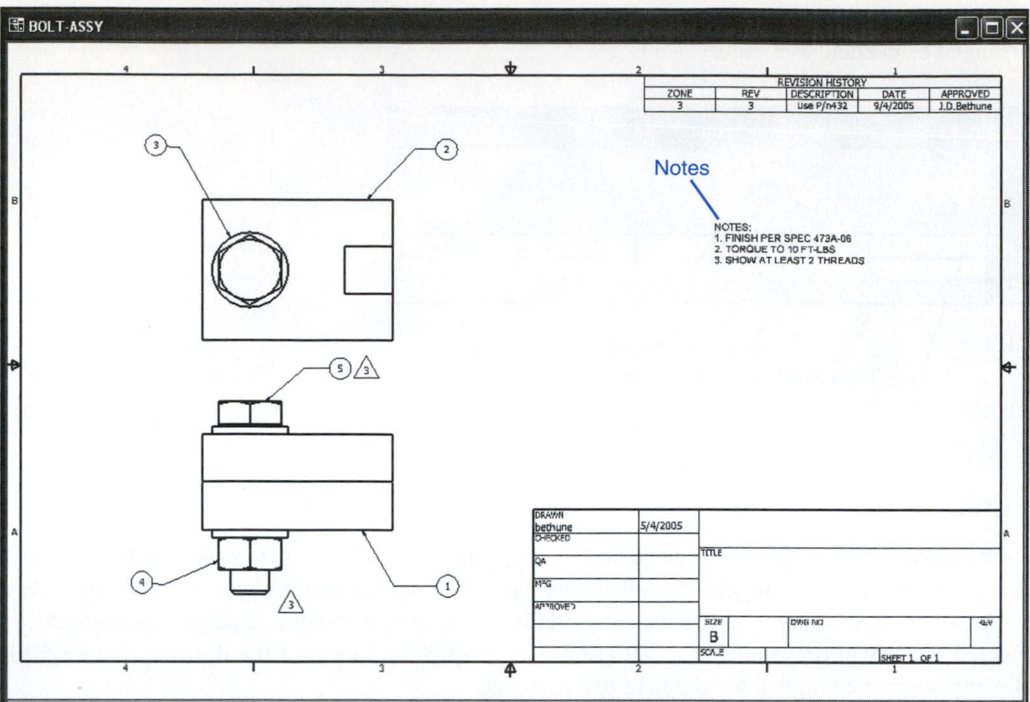

Figure 5-82

in enclosed in a triangle. The note numbers enclosed in triangles are also drawn next to the corresponding areas of the drawing. See Figure 5-82.

Top-Down Assemblies

A *top-down assembly* is an assembly that creates new parts as the assembly is created. Figure 5-83 shows a ROTATOR ASSEMBLY that was created using the top-down method. This section will explain how the assembly was created.

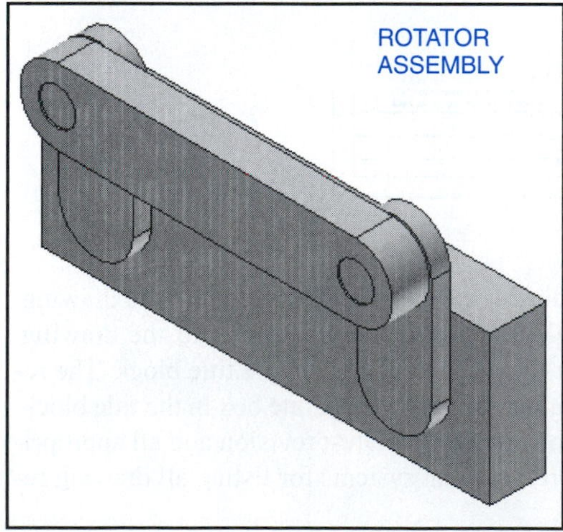

Figure 5-83

Exercise 5-30: Starting an Assembly

1. Click on the **New** tool.
2. Click the **Metric** tab, then select **Standard (mm).iam.**

The **Assembly Panel** will appear. See Figure 5-84.

3. Left-click on the heading **Assembly** in the browser box.

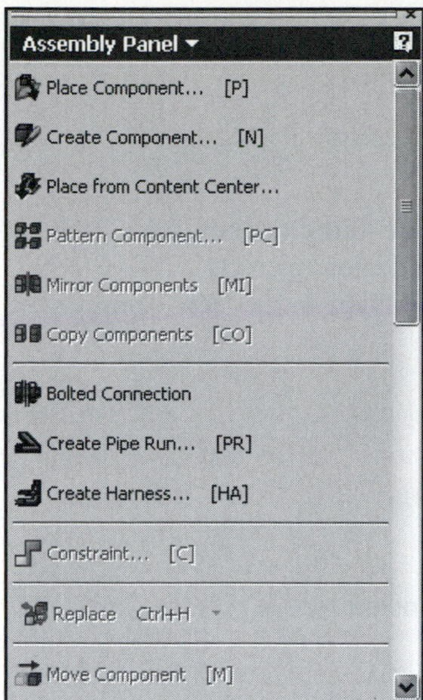

Figure 5-84

4. Click on the **File** heading at the top left of the screen and select the **Save All** tool from the cascading menu.

The **Save As** dialog box will appear.

5. Name the assembly, then click **Save.**

In this example the assembly was named **ROTATOR ASSEMBLY.**

Exercise 5-31: Changing the Sketch Plane

The parts created for this assembly were created on the XZ plane. This gives the assembly a more realistic orientation.

1. Click the **Tools** heading at the top of the screen and select **Application Options.**
2. Click the **Part** tab.
3. Select the **Sketch on x-z plane** button, then **Apply** and **OK.**

See Figure 5-85.

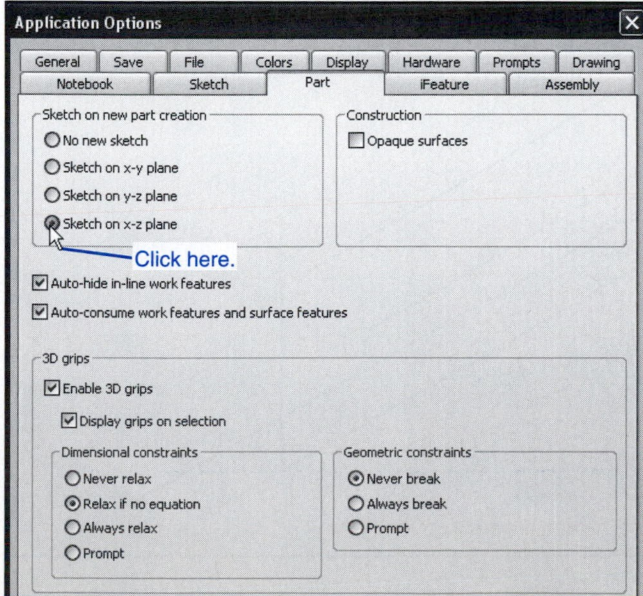

Figure 5-85

Exercise 5-32: Creating a Part

1. Click the **Create Component** tool on the **Assembly Panel** bar.

The **Create In-Place Component** dialog box will appear. See Figure 5-86.

2. Change the file name to **PLATE.** Select a file location.
3. Click the **Browse Templates** box located to the right of the **Template** box.
4. Click the **Metric** tab, then select the **Standard (mm).ipt** option, then **OK.**
5. The **Create In-Place Component** dialog box will appear. Type in the **New Component Name, PLATE,** then click **OK.**

A cursor will appear with a 3D box next to it.

6. Right-click the mouse and select the **Isometric View** option.
7. Click the left mouse button.

The **2D Sketch Panel** will appear in the panel box area.

8. Use the **Two point rectangle, Line,** and **General Dimension** tools to create the PLATE shown in Figure 5-87.

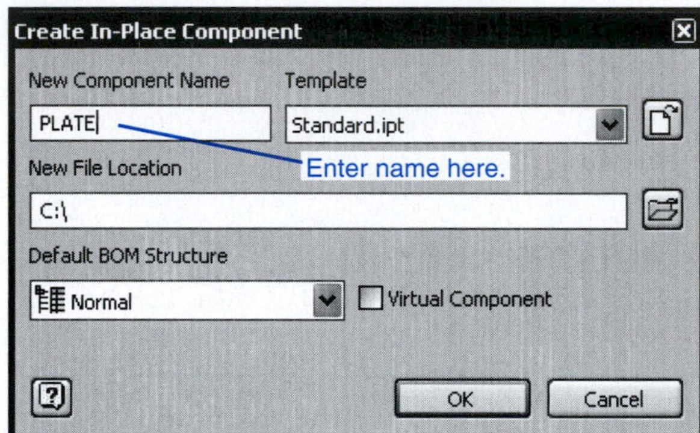

Figure 5-86

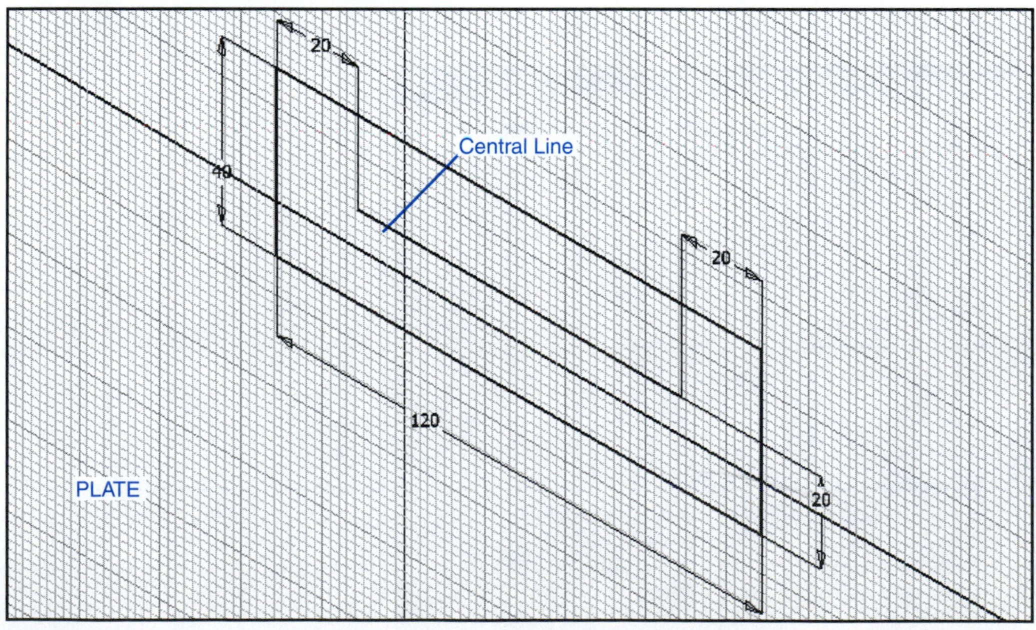

Figure 5-87

Exercise 5-33: Adding Work Points and a Work Axis

1. Right-click the mouse and select **Done,** then the **Finish Sketch** option.

The **Part Features** panel will reappear.

Do *not* select the **Finish Edit** option.

2. Click the **Work Point** option and locate work points at both ends of the central line.
3. Click the **Work Axis** tool and add a work axis between the two work points.

The work points and work axis will be listed in the browser area.

4. Click on **Origin** under **PLATE** in the browser area.
5. Select the **Work Axis** tool, then click the **XZ Plane** in the browser area and add a work axis through both work points.
6. Right-click the mouse and select the same **Finish Edit** options.

See Figure 5-88.

7. Save the drawing.

The work axis perpendicular to the XZ plane must appear and both sides of the PLATE.

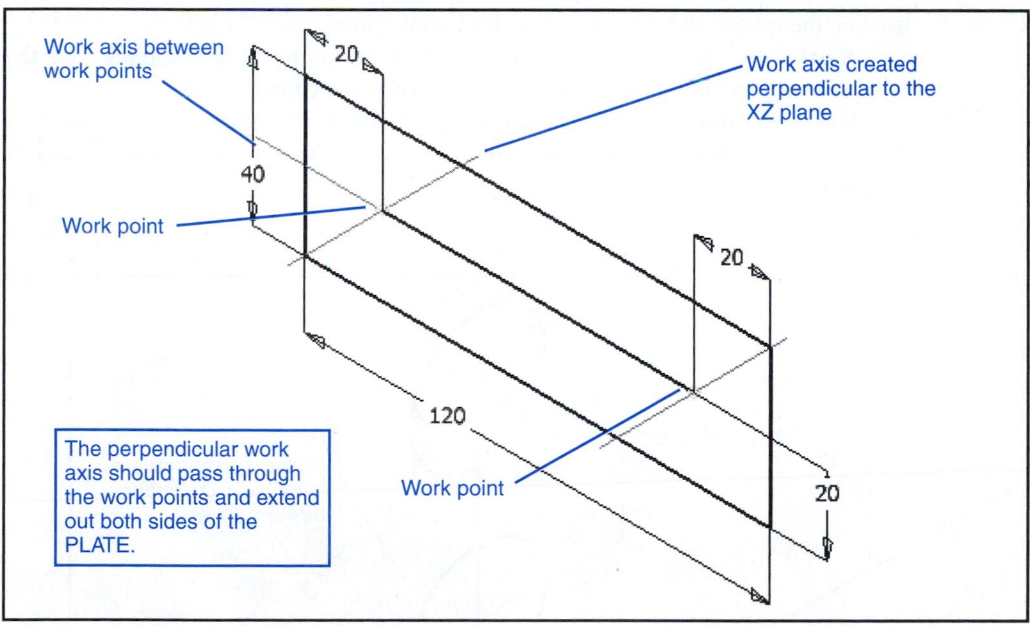

Figure 5-88

Exercise 5-34: Creating LINK-L

1. Click on the **Look At** tool on the **Standard** toolbar, then click one of the lines on the PLATE.

The drawing will return to a two-dimensional view.

2. Click on the **Create Component tool** and create file **LINK-L** using the **standard (mm).ipt** format. Click **OK.**

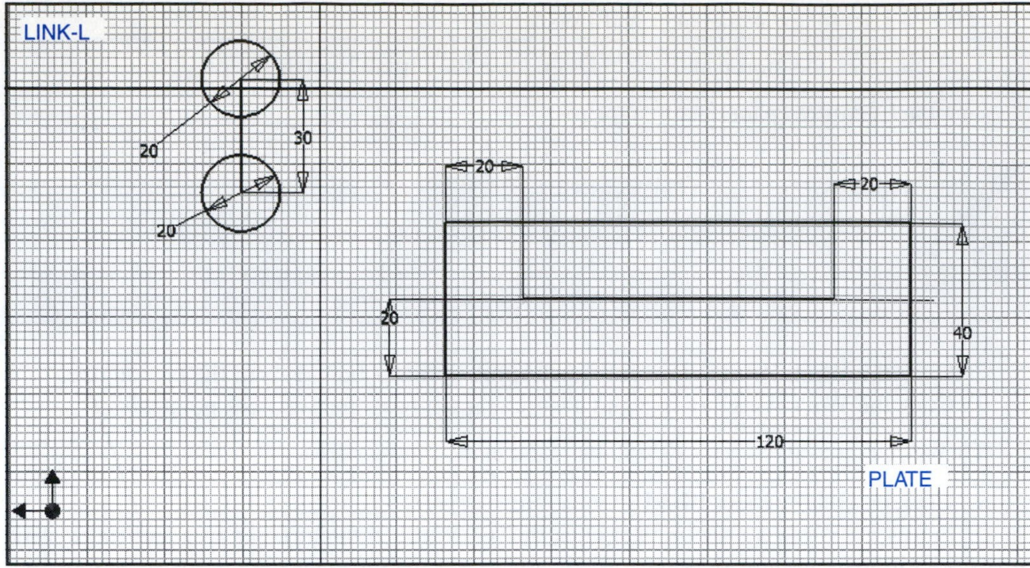

Figure 5-89

3. Click the drawing screen to access the **2D Sketch Panel.** The drawing grid will reappear.
4. Use the **Circle** and **Line** tools to create LINK-L as shown in Figure 5-89.
5. Change to the isometric view.

TIP

When the isometric view is created the drawing will reverse. The LINK-L will appear on the right. See Figure 5-90.

6. Right-click the mouse and select the **Finish Sketch** option.
7. Use the **Work Point** tool and create work points at the center of both circles. Use the **Work Axis** tool to create a work axis between the two hole centers.
8. Use the **Work Axis** tool and the **XZ Plane** option listed under **Origin** under **LINK-1** in the browser area to create two work axes through the two work points, perpendicular to the XZ plane.

See Figure 5-90.

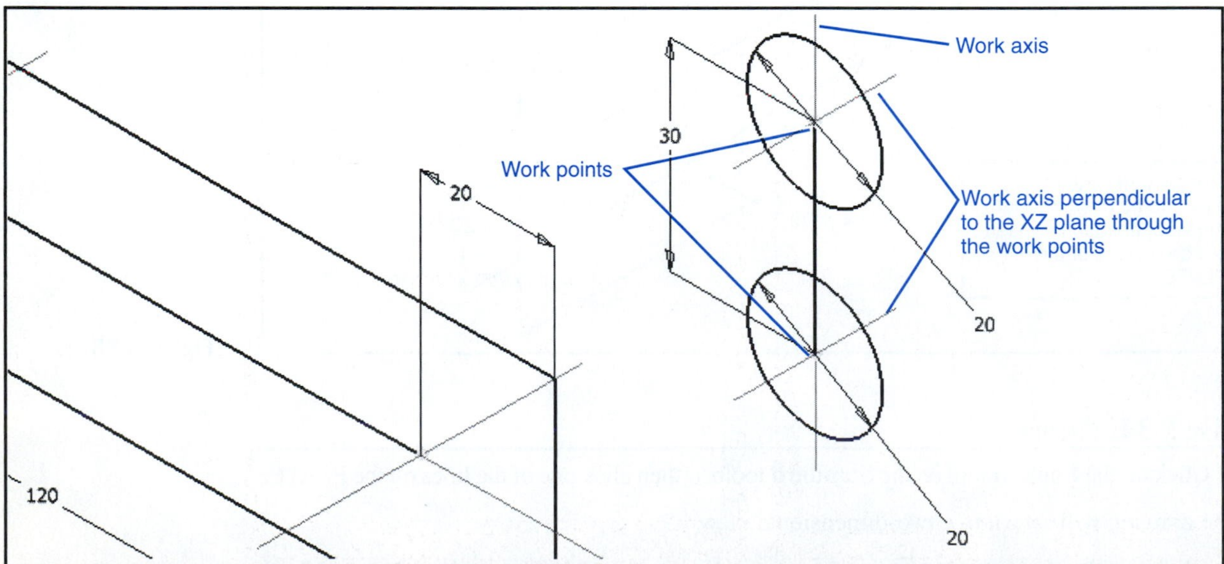

Figure 5-90

9. Right-click the mouse and select **Finish Edit.**
10. Click the **Look At** tool, then click one of the lines on the PLATE.

Exercise 5-35: Copying a Component

Figure 5-91 shows the drawing screen with the PLATE and LINK-L components. The LINK-L component will be copied and its name changed to LINK-R.

1. Click the **Copy Components** tool on the **Assembly Panel** bar.

The **Copy Components: Status** dialog box will appear. See Figure 5-92.

2. Click the **LINK-L** heading in the browser box.

The **LINK-L** heading will appear in the dialog box.

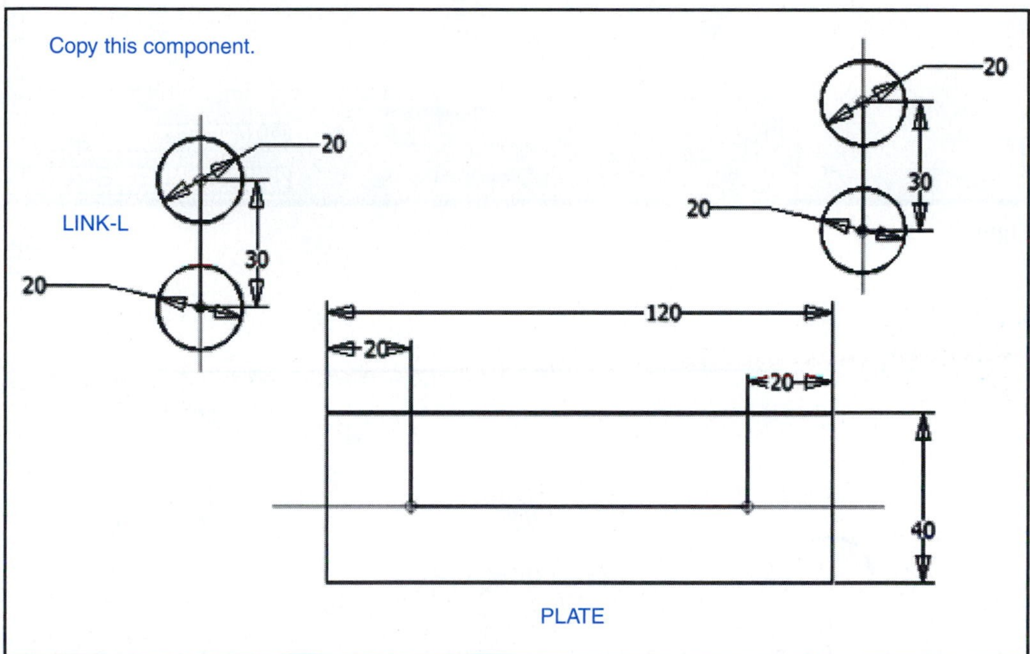

Figure 5-91

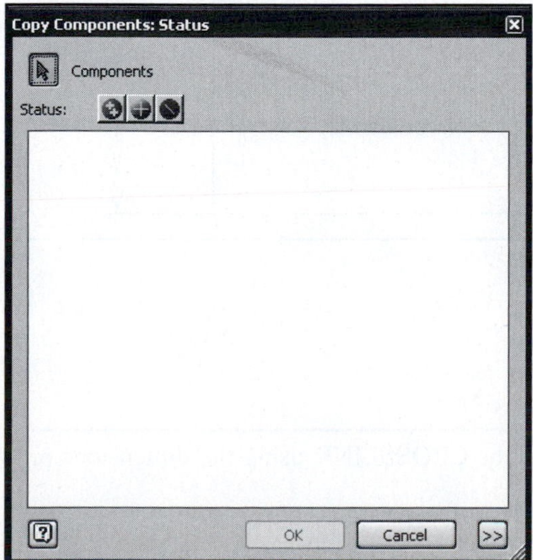

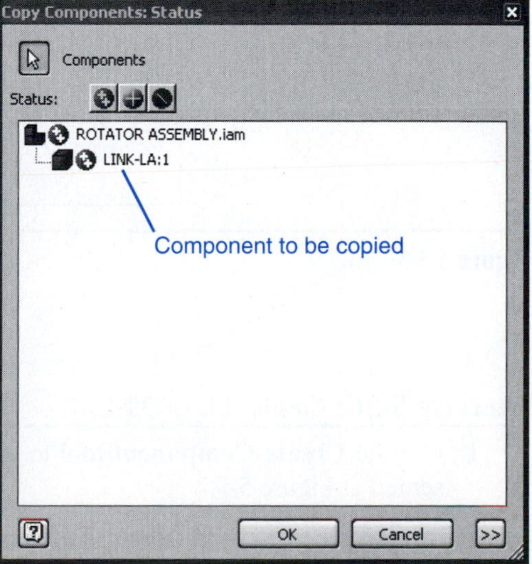

Figure 5-92

3. Click **OK.**

The **Copy Components: File Names** dialog box will appear. See Figure 5-93.

4. Change the name LINK-L to **LINK-R,** then click **OK.**
5. Position LINK-R to the right of the PLATE. See Figure 5-94.

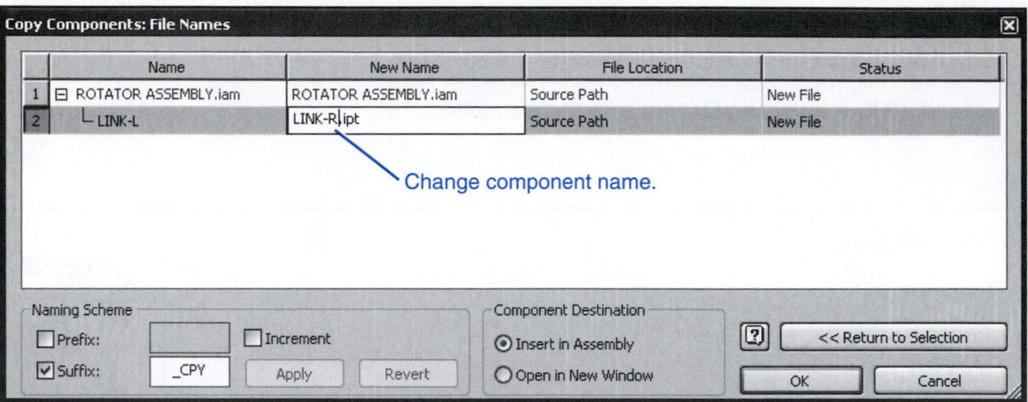

Figure 5-93

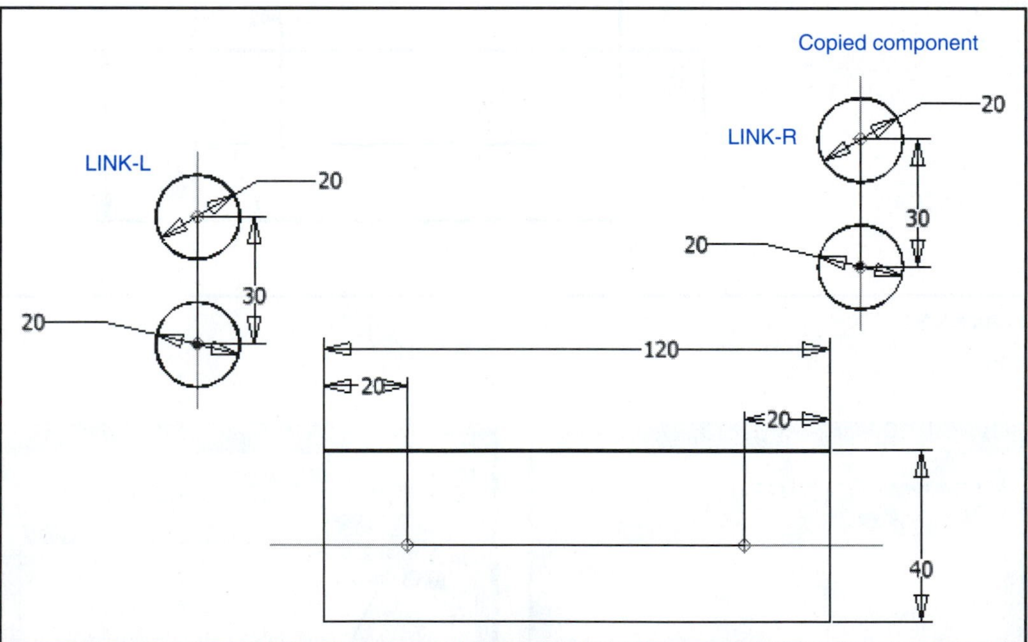

Figure 5-94

Exercise 5-36: Creating the CROSSLINK

1. Use the **Create Component** tool to create the CROSSLINK using the dimensions presented in Figure 5-95.

The drawing screen should look like Figure 5-96

2. Right-click the mouse and select the **Isometric View** option.

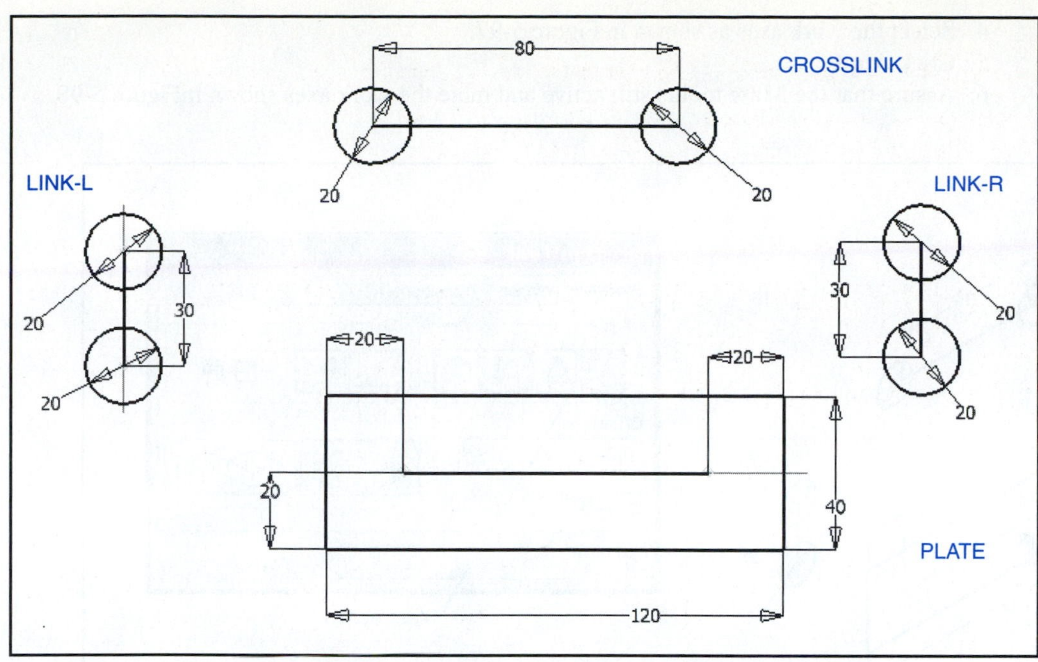

Figure 5-95

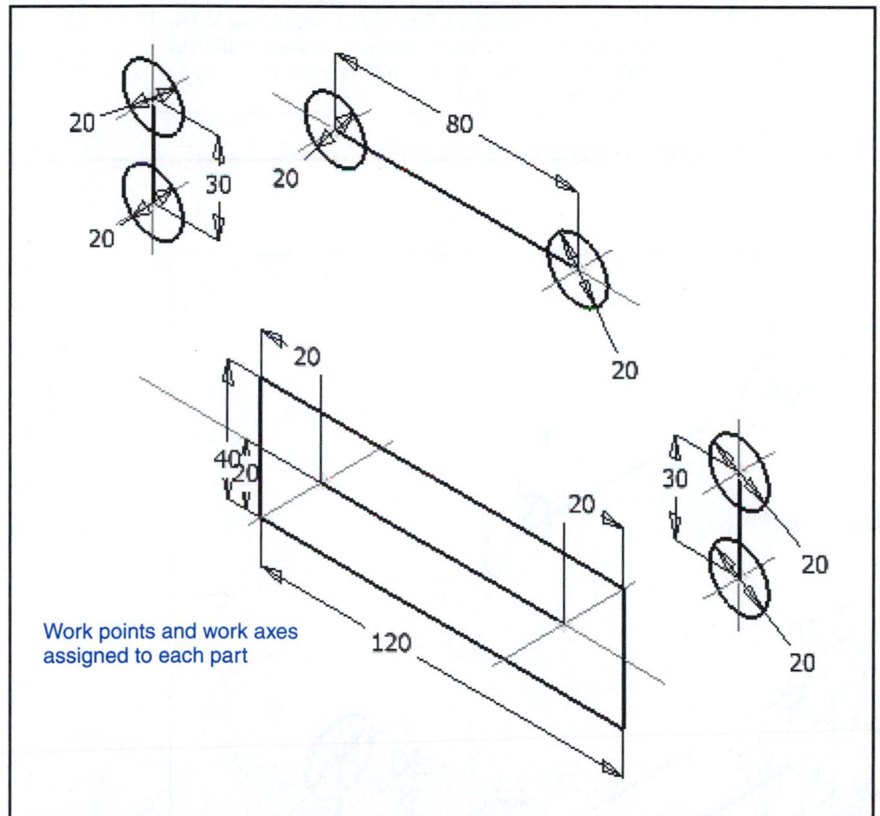

Figure 5-96

Exercise 5-37: Assembling the Parts

1. Double-click the word **ROTATOR ASSEMBLY** in the browser area to return to the **Assembly Panel.**
2. Click the **Constraint** tool.

The **Place Constraint** dialog box will appear.

3. Select the **Mate** option.

The **Mate** option may automatically be selected.

4. Select the work axes as shown in Figure 5-97.
5. Click **Apply.**
6. Assure that the **Mate** tool is still active and mate the work axes shown inFigure 5-98.

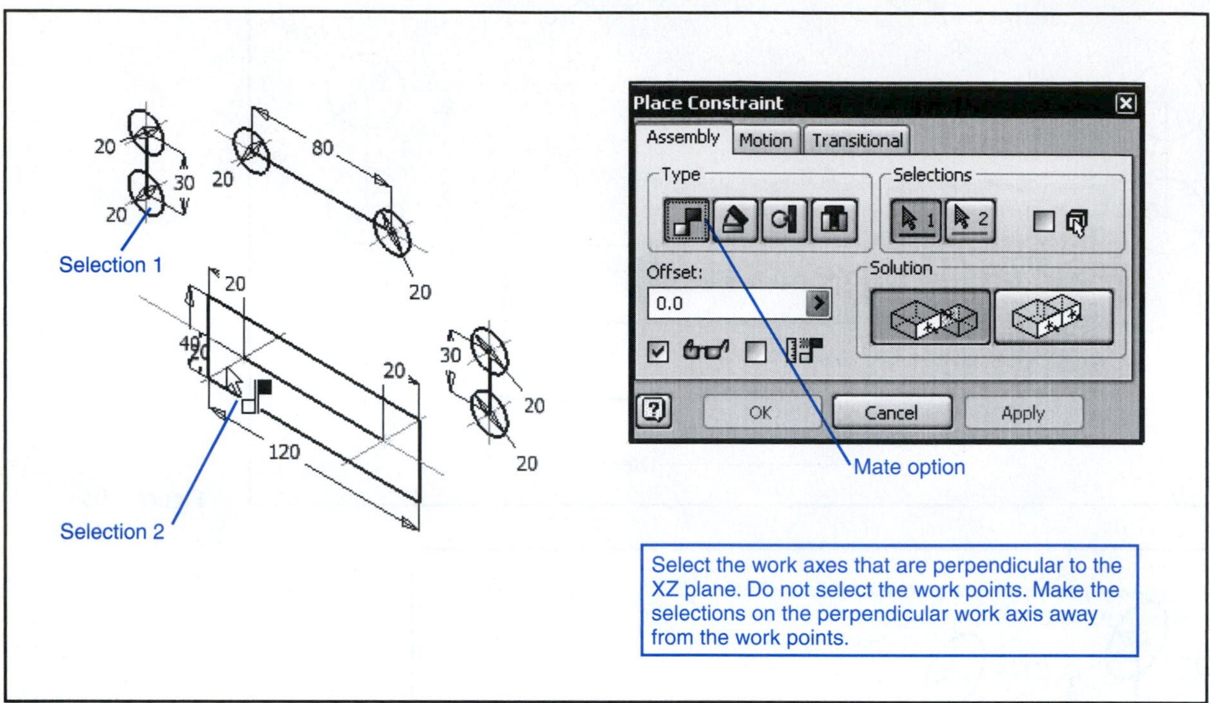

Figure 5-97

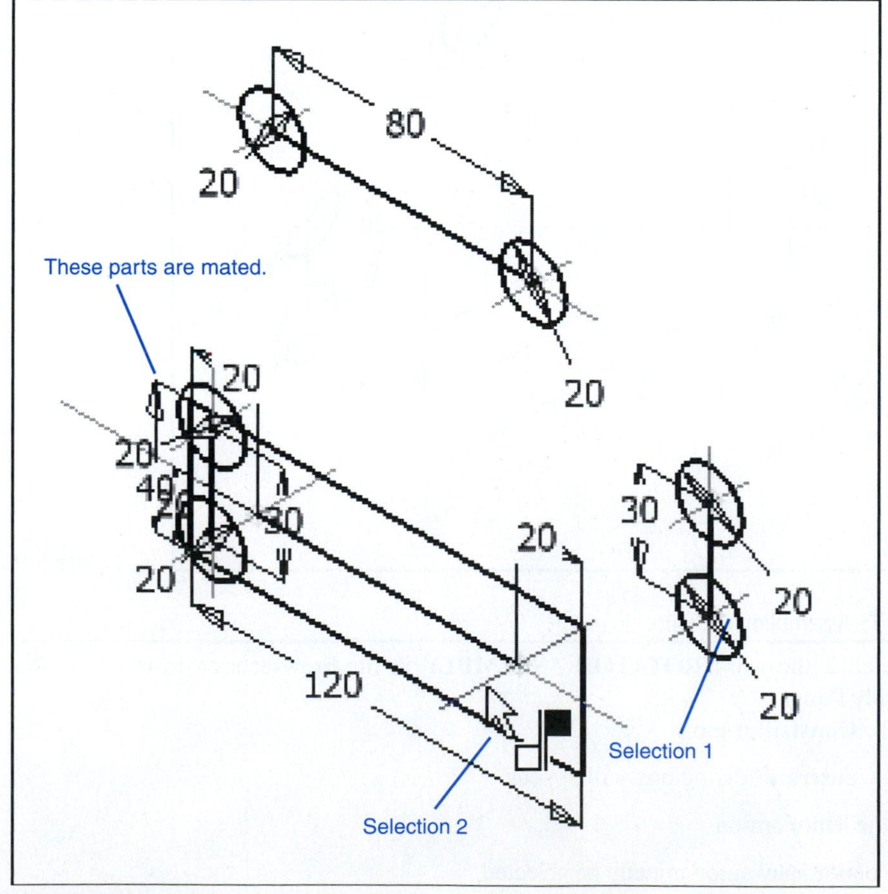

Figure 5-98

7. Assure that the **Mate** tool is still active and mate the work axes shown in Figures 5-99 and 5-100.
8. Use the **Look At** tool to create a two-dimensional view of the assembly.

Select the work axes that are perpendicular to the XZ plane. Do not select the work points. Make the selections on the perpendicular work axis away from the work points.

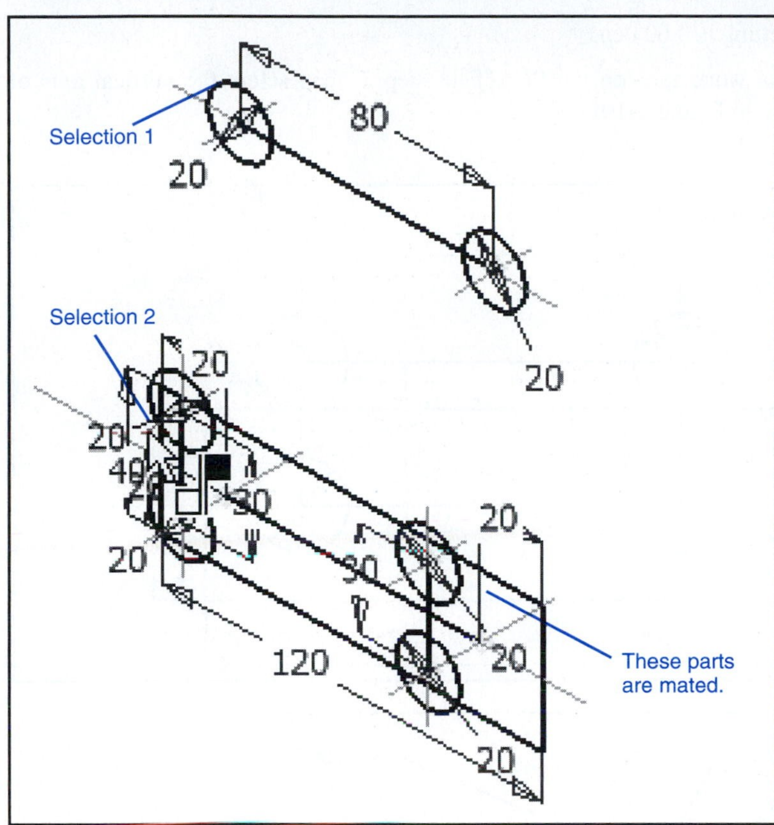

Figure 5-99

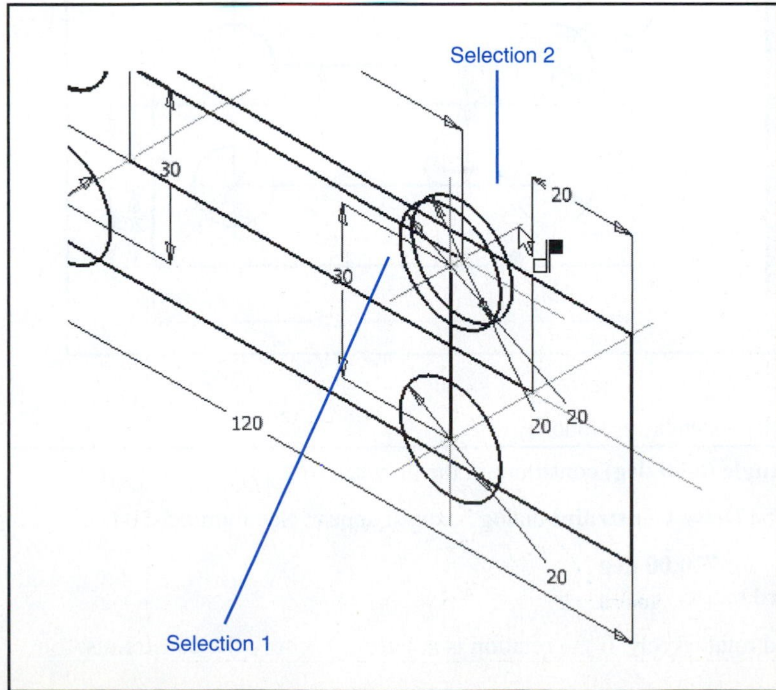

Figure 5-100

Figure 5-101 Shows the resulting assembly drawing.

Exercise 5-38: Animating the LINKs

1. Double-click the word **ROTATOR ASSEMBLY** in the browser area to access the **Assembly Panel.**
2. Access the PLATE and create a work axis on the edge of the PLATE as shown in Figure 5-101.
3. Click the **Constraint** tool.

The **Place Constraint** dialog box will appear.

4. Select the **Angle** option.

The default angle setting is 0.00 deg.

5. Select the vertical work axes on the PLATE in step 2, then select the vertical axis of LINK-R as shown in Figure 5-102.
6. Apply the constraints.

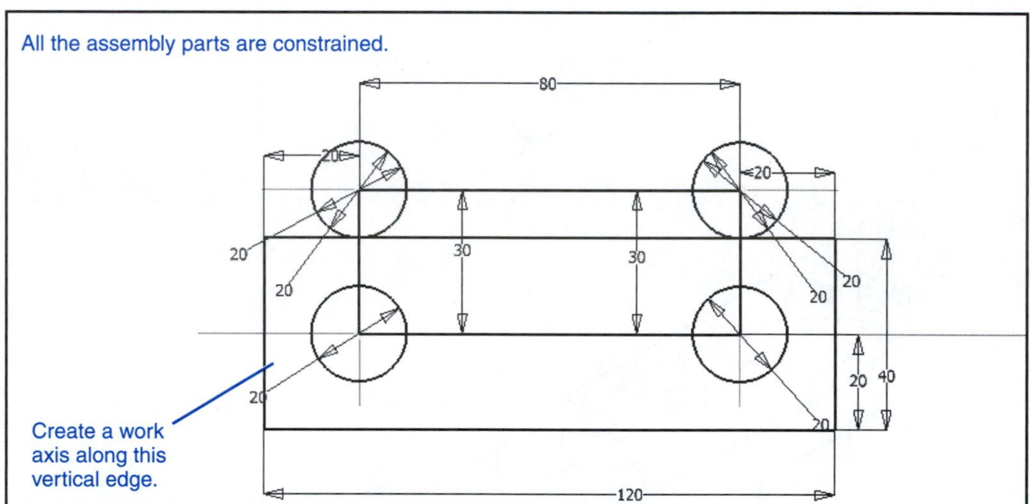

Figure 5-101

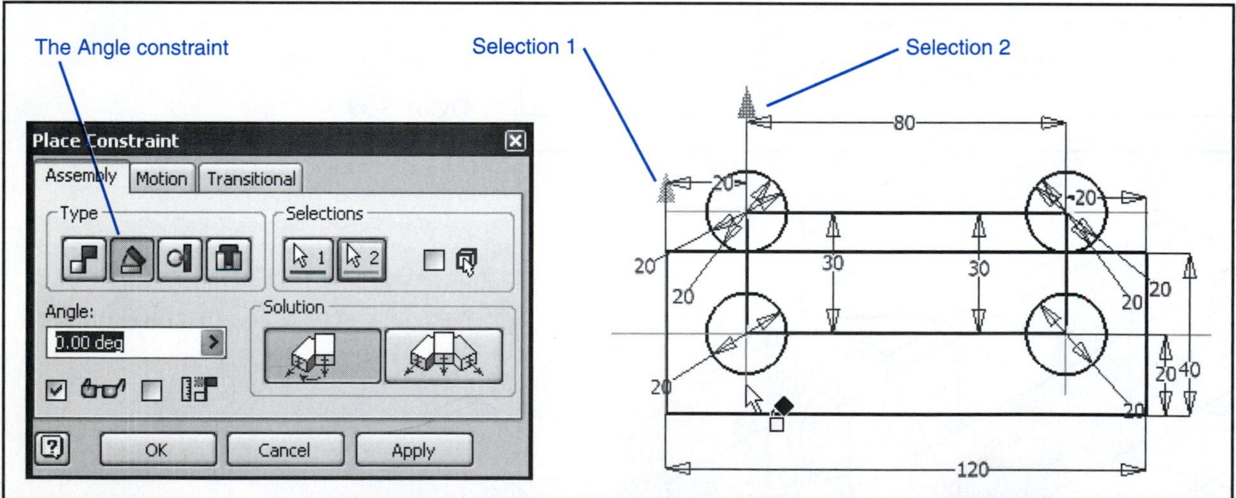

Figure 5-102

Exercise 5-39: Setting the Assembly in Motion

1. Right-click the **Angle (0.00 deg)** constraint in the browser area.

See Figure 5-103. The **Drive Constraint** dialog box will appear. See Figure 5-104.

2. Set the **End** angle for **720.00 deg**
3. Click the **Forward** button.

The assembly should rotate freely. If the rotation is not correct, check the constraints.

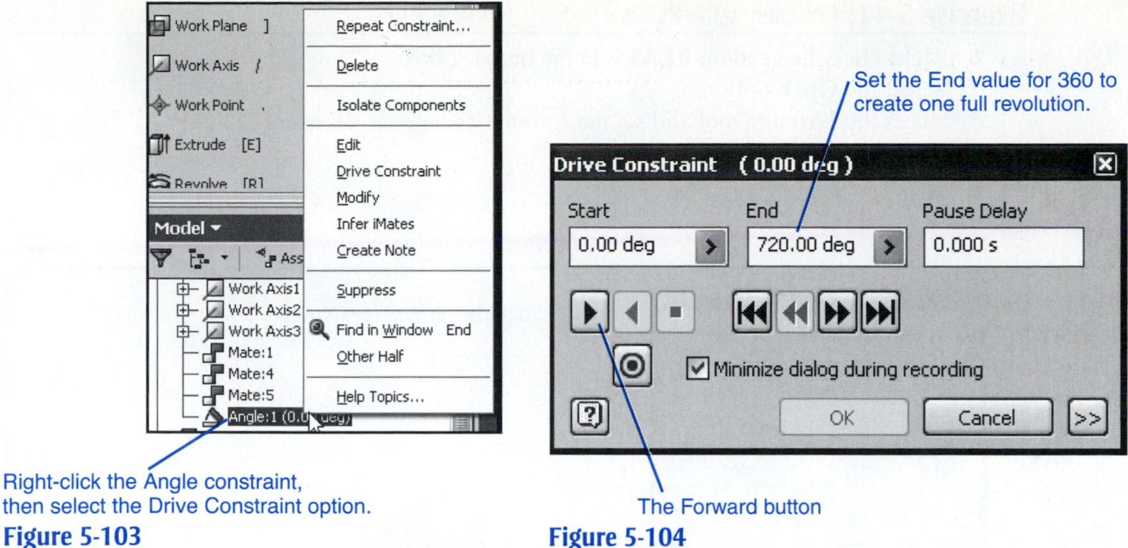

Figure 5-103

Figure 5-104

Exercise 5-40: Controlling the Speed of the Rotation

1. Click the << button on the **Drive Constraint** dialog box.

The box will expand. See Figure 5-105

2. Set the **Increment** value to **5.00 deg.**
3. Click the **Forward** button.

The higher the **Increment** value, the faster the assembly will rotate. The **Repetitions** setting is used to control the number of revolutions generated.

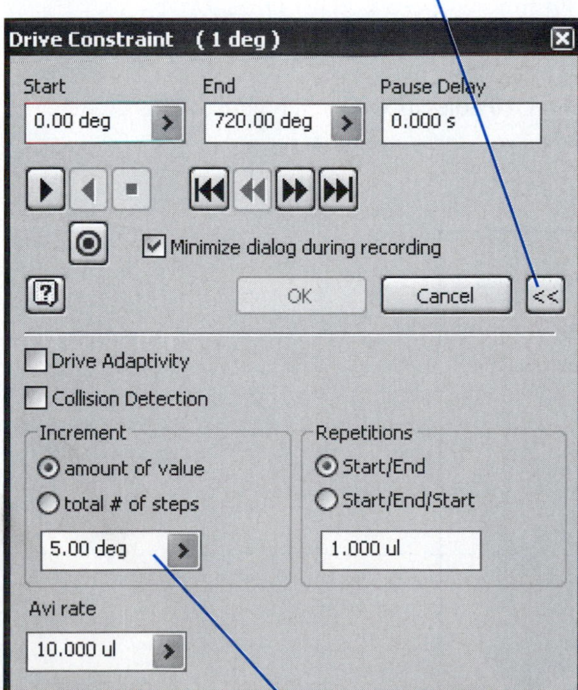

Figure 5-105

Exercise 5-41: Completing the PLATE

1. Right-click the heading **PLATE** in the browser box.
2. Select the **Open** option.
3. Select the **Extrude** tool and set the **Extents** distance to **15 mm.**

See Figure 5-106.

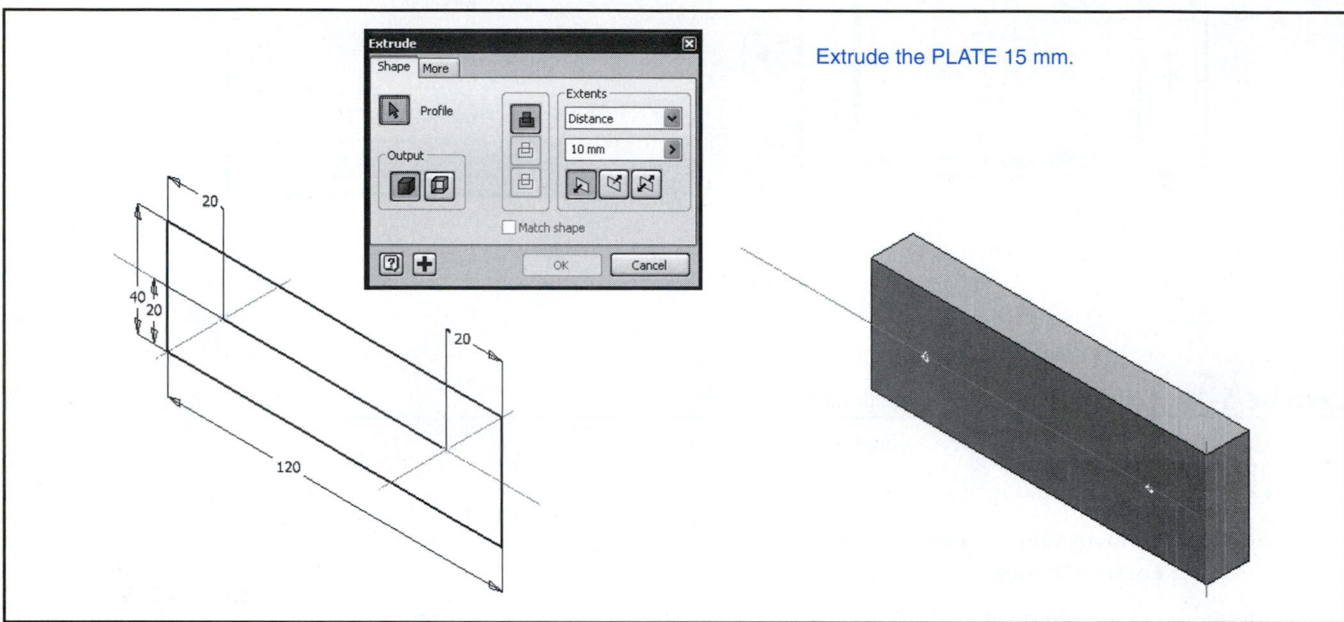

Figure 5-106

4. Click **OK.**
5. Right-click the mouse and select the **New Sketch** option**.**
6. Click the **Hole** tool.
7. Create two **Point, Center Point.** on the two work points, right-click the mouse, and select the **Done** option.
8. Click the **Return** tool, then create two **Ø10** holes as shown in Figure 5-107.
9. Save the PLATE changes

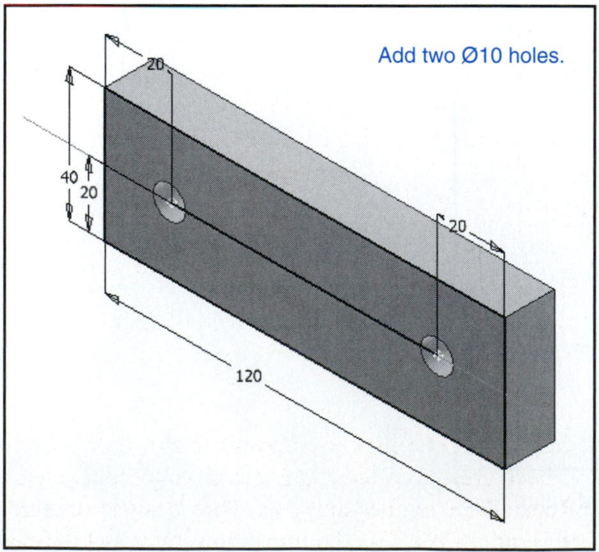

Figure 5-107

Exercise 5-42: Completing LINK-L and LINK-R

1. Right-click **LINK-L** in the browser area and select the **Open** option.
2. Right-click the word **Sketch** under the **LINK-L** heading and select the **Edit Sketch** option.
3. Draw two tangent lines and two **Ø10** circles as shown in Figure 5-108.
4. Click the **Return** tool and extrude LINK-L **5 mm.**
5. Extrude the top circle **10** forward, and extrude the lower circle **20** through and back as shown.
6. Edit LINK-R to the same dimensions and features.
7. Save the LINKs.

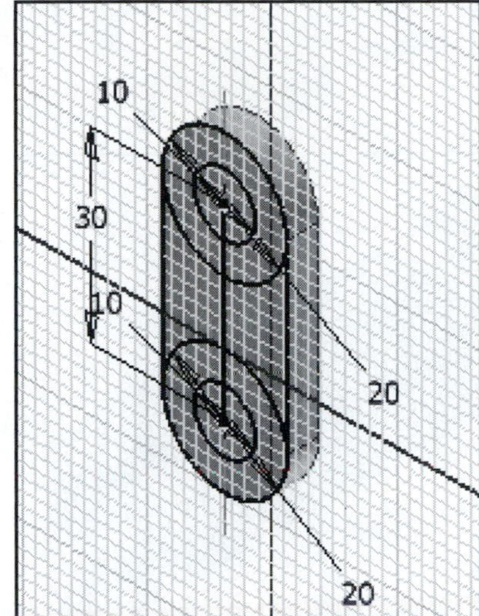

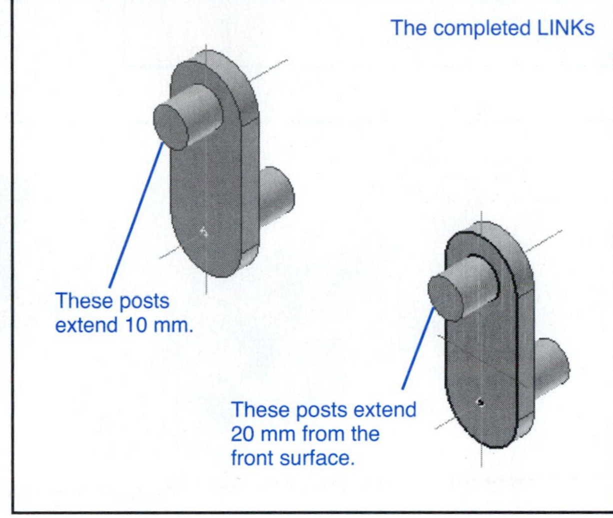

Figure 5-108

Exercise 5-43: Completing the CROSSLINK

1. Right-click **CROSSLINK** in the browser area and select the **Open** option.
2. Add two tangent lines and extrude the CROSSLINK **10 mm.**

See Figure 5-109.

3. Create a **New Sketch** and add two **Ø10** holes as shown.

Save the CROSSLINK.

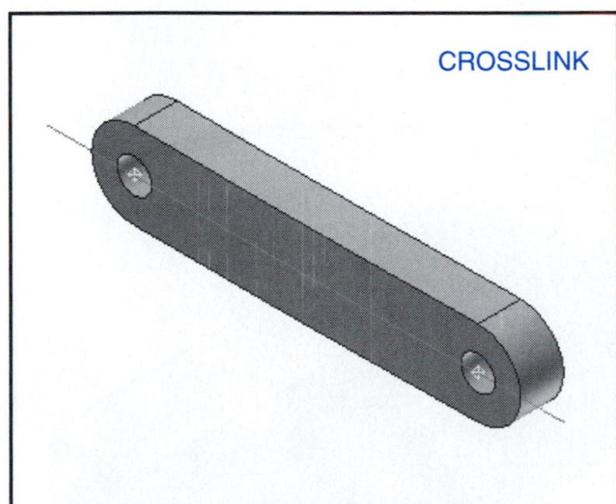

Figure 5-109

Aligning the Assembly

Now that the components have thickness, they may not be aligned correctly; that is, they interfere with each other. The alignment could have been created as the components were constrained in the last section by defining offset values. Rotate the assembly and see if there is any interference or if the components are too far apart. See Figure 5-110. Use the **Flush** constraint and define a **1-mm** clearance between the components. The finished assembly should look like Figure 5-111.

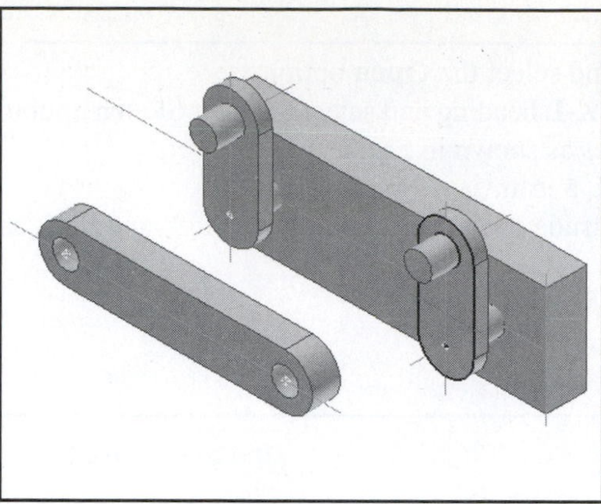

Figure 5-110

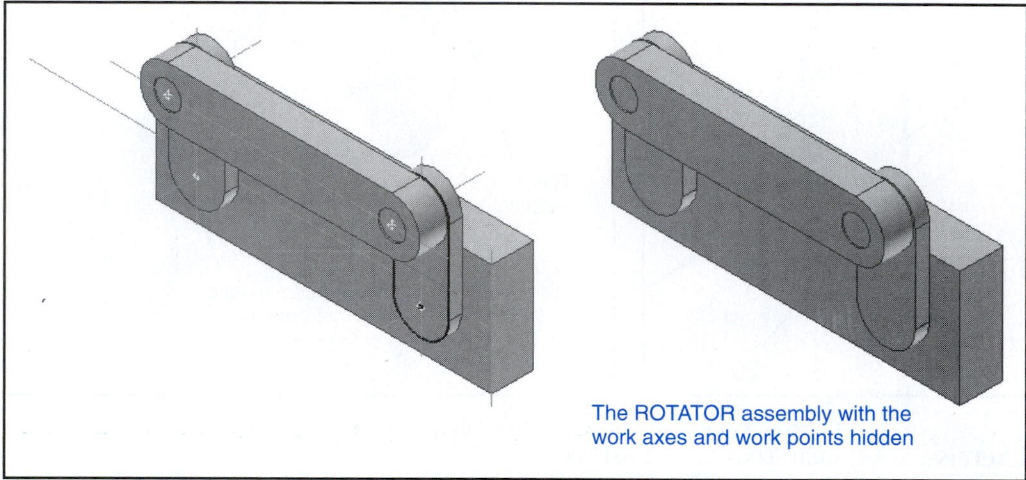

Figure 5-111

Presentations

Figure 5-112 shows a presentation drawing of the ROTATOR, and Figure 5-113 shows an exploded isometric drawing created using the .idw format.

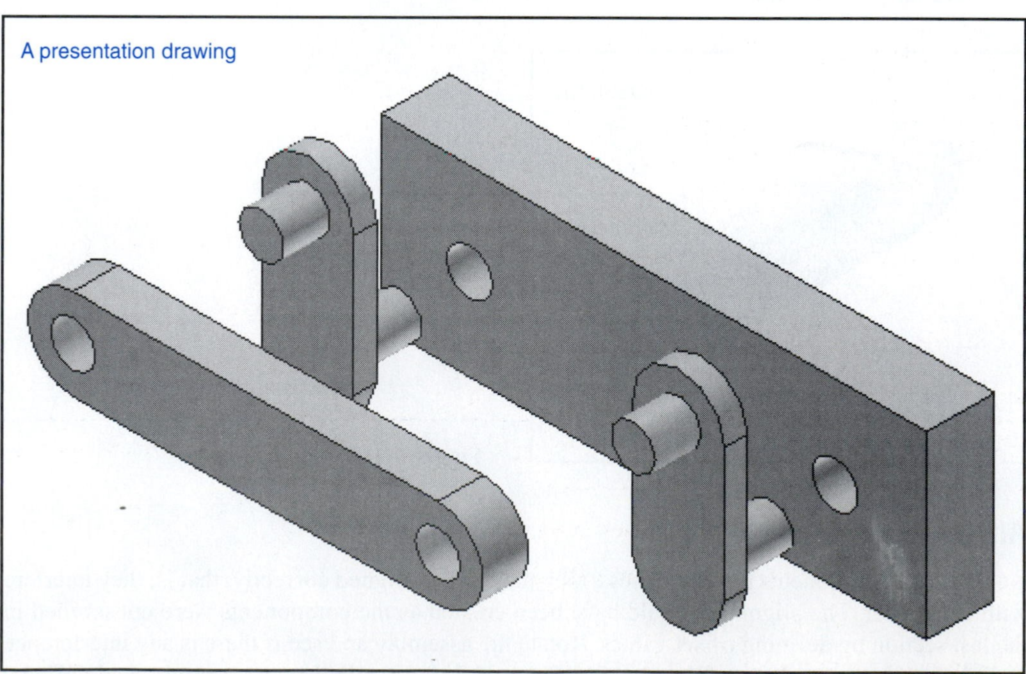

Figure 5-112

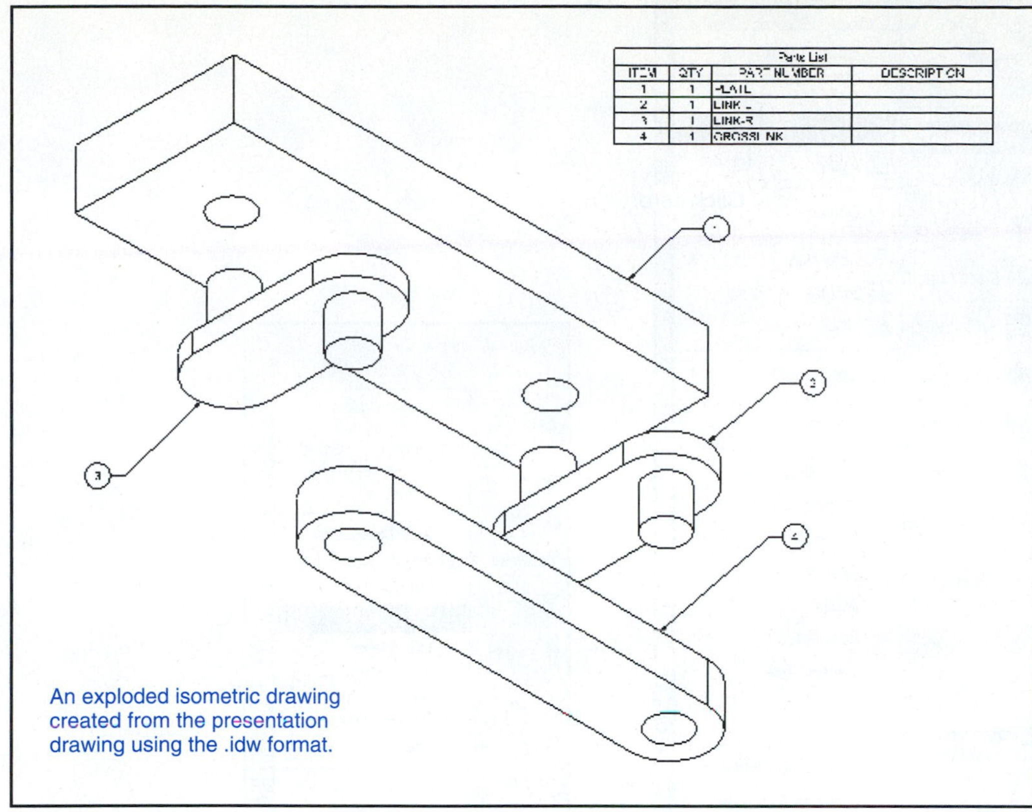

Parts List			
ITEM	QTY	PART NUMBER	DESCRIPTION
1	1	PLATE	
2	1	LINK-L	
3	1	LINK-R	
4	1	CROSSLINK	

An exploded isometric drawing created from the presentation drawing using the .idw format.

Figure 5-113

EDITING A PART WITHIN AN ASSEMBLY DRAWING

Figure 5-114 shows an assembly drawing. The assembly is called 114-ASSEMBLY and is made from three components: 114-PLATE, 114-BRACKET, and 114-POST.

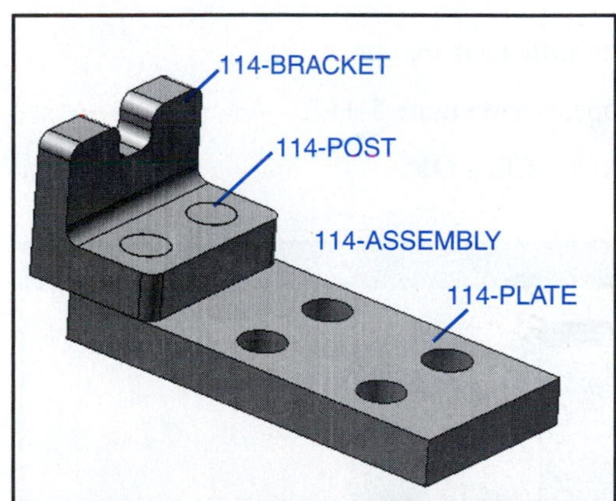

Figure 5-114

Exercise 5-44: Editing a Feature

It has been determined that the Ø10-mm holes in the bracket are too small. They are to be increased to Ø11 mm. The holes are features.

1. Right-click the **114-bracket** heading in the browser box.

See Figure 5-115. A list of options will appear. See Figure 5-116.

2. Click the **OPEN** option.

The browser box display will change.

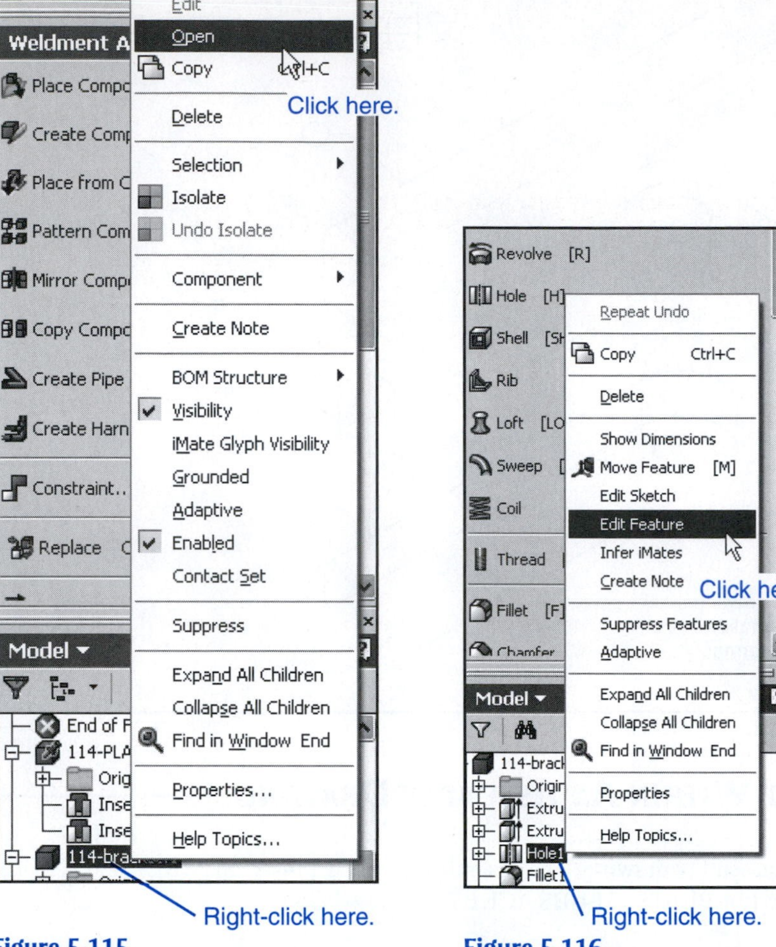

Figure 5-115

Figure 5-116

3. Right-click the **Hole1** heading and select the **Edit Feature** option.

See Figure 5-116. The **Hole** dialog box will appear. See Figure 5-117.

4. Change the hole's diameter value from **10** to **11.** Click **OK.**

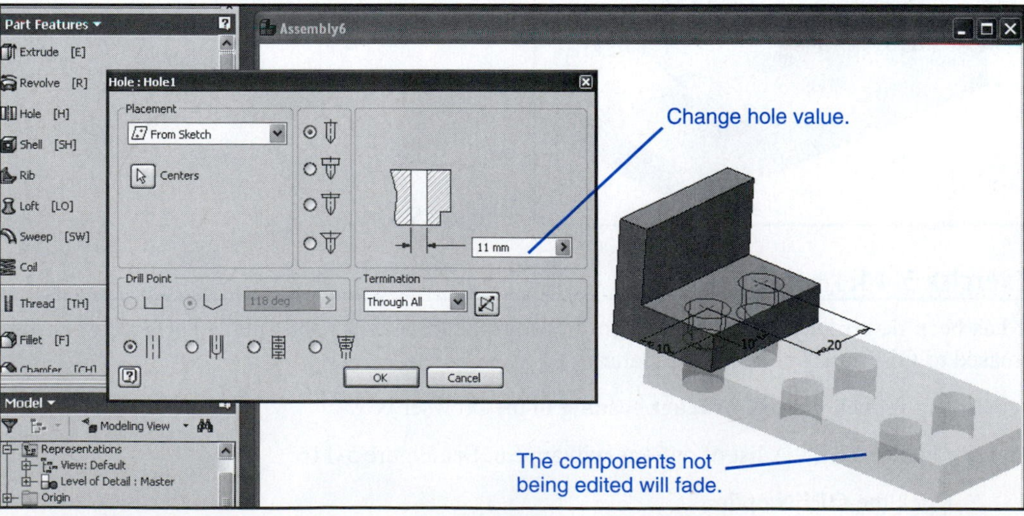

Figure 5-117

The hole's location dimensions could also be edited at this time.

The bracket will appear on the screen. Click the **Close** button (the **X** in the upper right corner of the bracket's screen). A warning box will appear. See Figure 5-118.

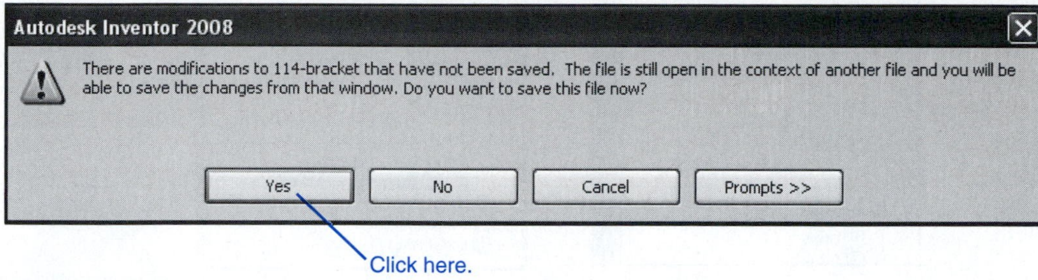

Figure 5-118

5. Click the **Yes** box.

The assembly will appear.

6. Click the bracket.

Figure 5-119 shows the edited bracket. Note that the holes appear larger than they did in Figure 5-114.

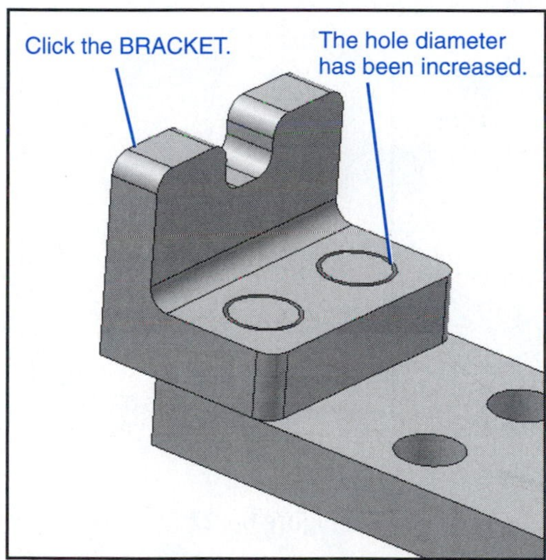

Figure 5-119

Exercise 5-45: Editing a Sketch

It has been determined that the length of the plate is to be increased from 90 mm to 100 mm. This change is a change to the initial sketch.

1. Right-click the **114-PLATE** heading in the browser box.

A list of options will appear. See Figure 5-120.

2. Select the **Open** option.

The browser box will change.

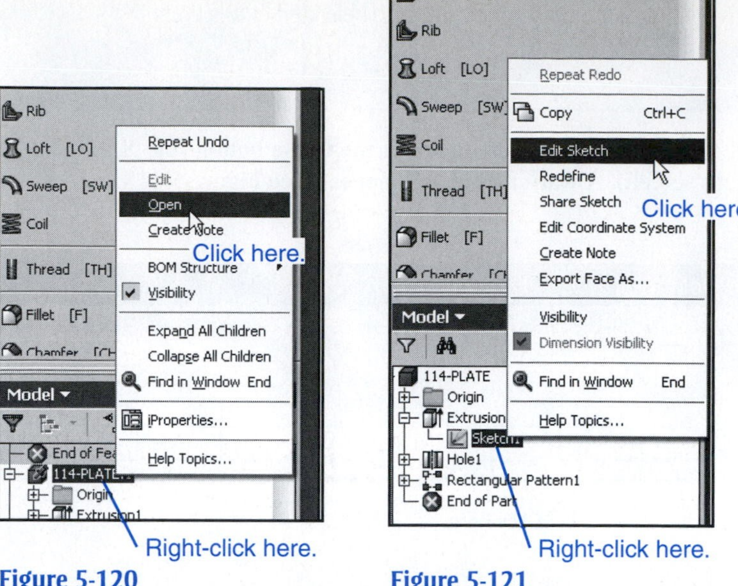

Figure 5-120 **Figure 5-121**

3. Click the **Extrusion** heading in the browser box, then right-click the **Sketch** heading.

See Figure 5-121. The original rectangular sketch will appear. See Figure 5-122.

4. Use the **Fix** constraint and fix the back line. See Figure 5-123.

This constraint will assure that the dimensional expansion occurs at the front of the plate and will not affect the assembly between the plate and the bracket.

5. Delete the **90** dimension and create a new **100** dimension.

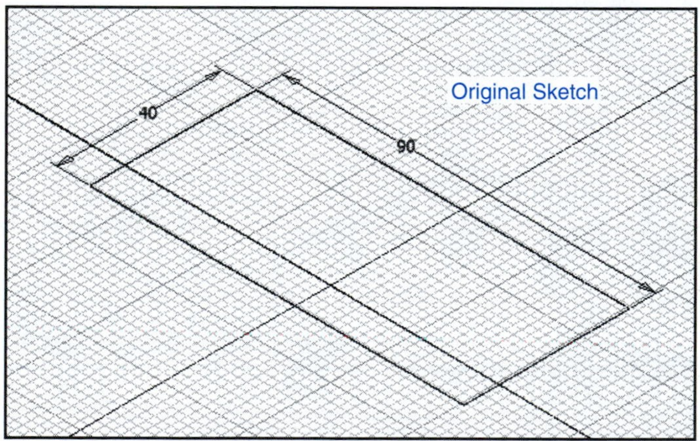

Figure 5-122

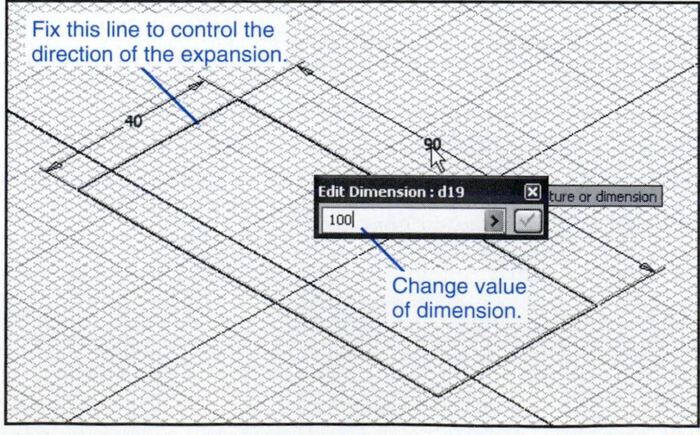

Figure 5-123

6. Right-click the mouse and select the **Finish Edit** option.

The plate will appear on the screen.

7. Click the **Close** box in the upper right corner of the plate's screen.

A warning box will appear. See Figure 5-124.

8. Click the **Yes** box.

The assembly will appear on the screen. See Figure 5-125.

9. Click the plate.

Note the increase in the plate's length.

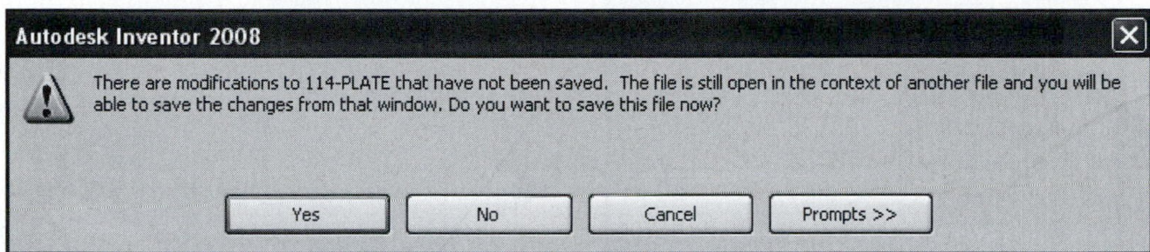

Figure 5-124

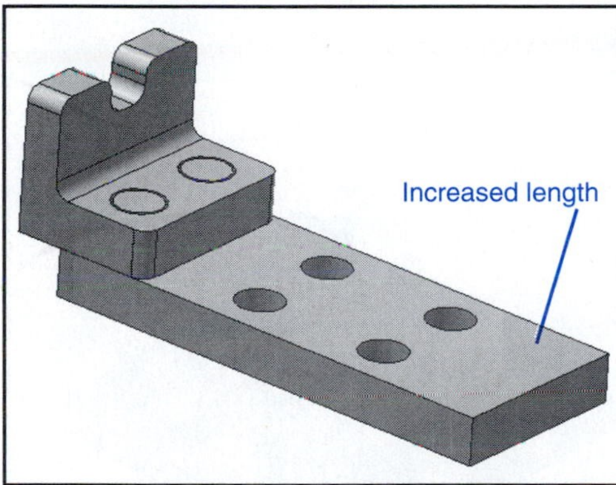

Figure 5-125

Patterning Components

Figure 5-126 shows an assembly where a post is inserted into a plate. Posts are to be inserted into all 16 holes in the plate.

Exercise 5-46: Creating a Pattern of Components

1. Select the **Pattern Component** tool.

The **Pattern Component** dialog box will appear. See Figure 5-126.

2. Select the **Post** as the component.
3. Select the **Rectangular box** tab.
4. Define the **Column** direction, the number of components, and the distance between components value.

In this example the holes in the plate are **30 mm** apart.

5. Define the **Row** values.
6. Click **OK.**

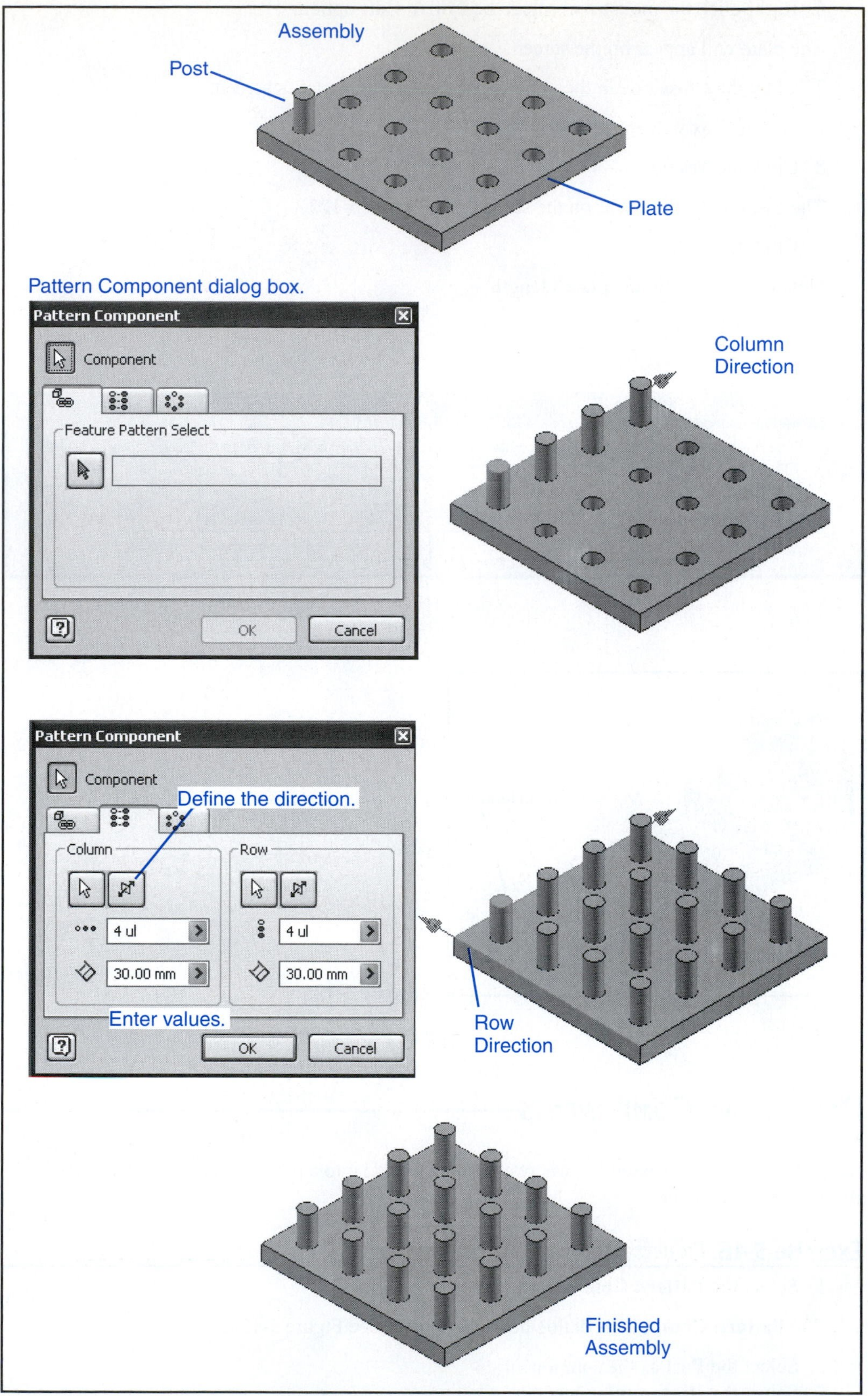

Figure 5-126

Mirroring Components

Components within an assembly may be mirrored and added to the assembly. See Figure 5-127.

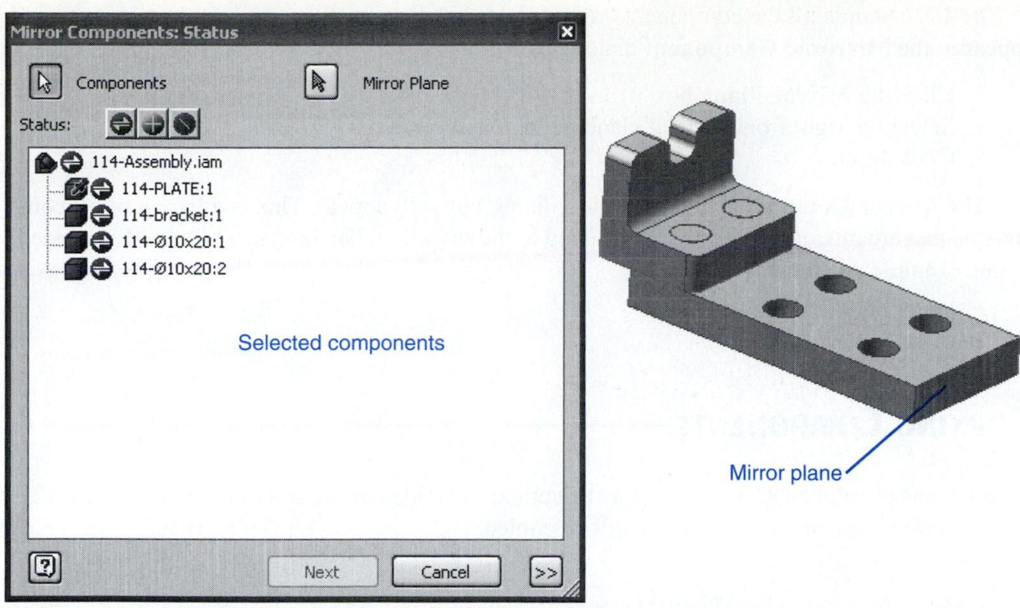

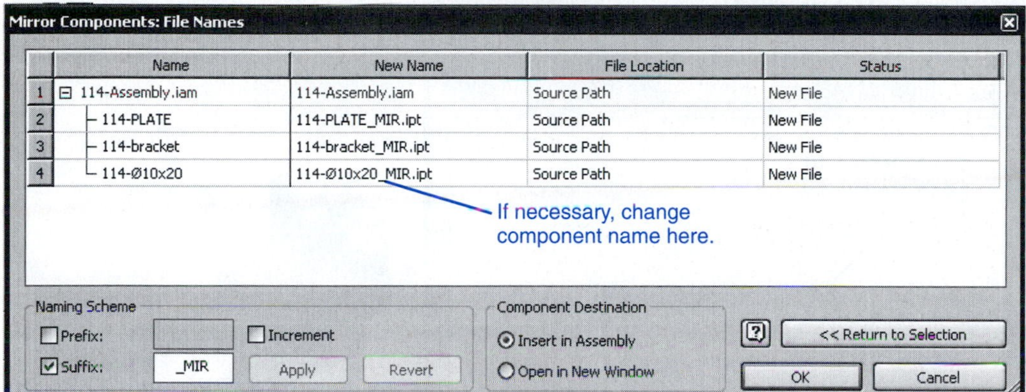

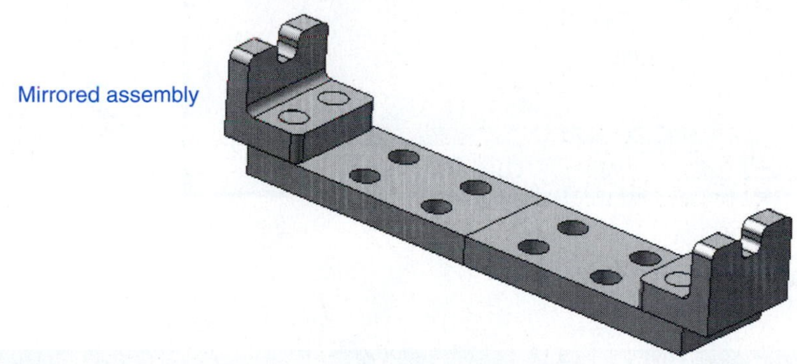

Figure 5-127

Exercise 5-47: Mirroring an Assembly

1. Click the **Mirror Component** tool.

The **Mirror Components: Status** dialog box will appear.

2. Click the components to be mirrored.

TIP The assembly shown in Figure 5-128 is the 114-Assembly used on page 209. Note how the dimension change from 90 mm to 100 mm affects the material under the copied bracket.

In this example all the components were selected. A listing of the selected components will appear in the **Mirrored Component** dialog box.

3. Click the **Mirror Plane** box.
4. Select the right front vertical plane as the mirror plane.
5. Click **Next.**

The **Mirror Components: File Names** dialog box will appear. This is a listing of all components that are mirrored and have been added to the assembly. The box may be edited as needed. In this example no changes were made.

6. Click **OK.**

Copying Components

Components already on an assembly may be copied and added to the assembly. See Figure 5-128. In this example the bracket and posts will be copied.

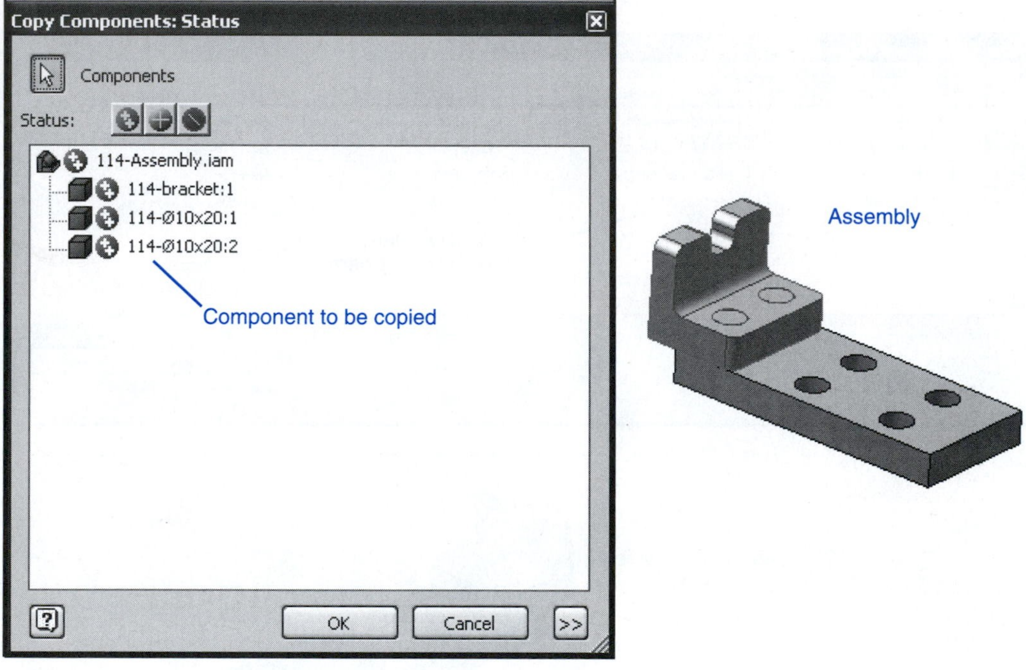

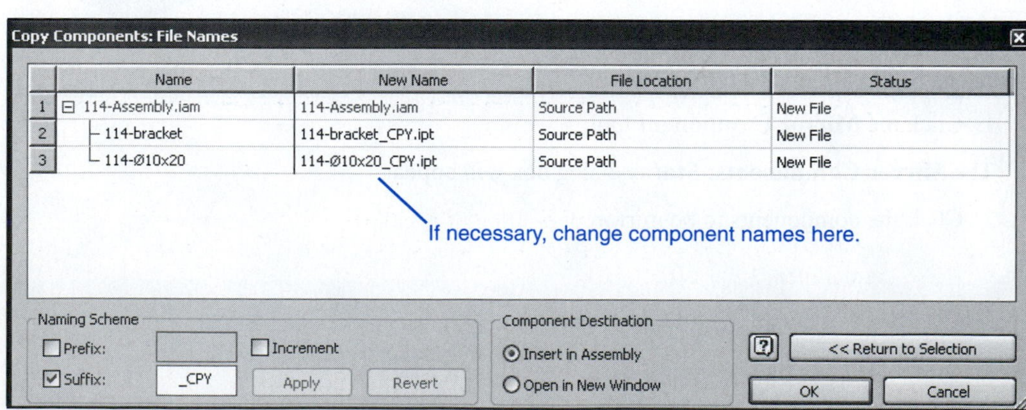

Figure 5-128

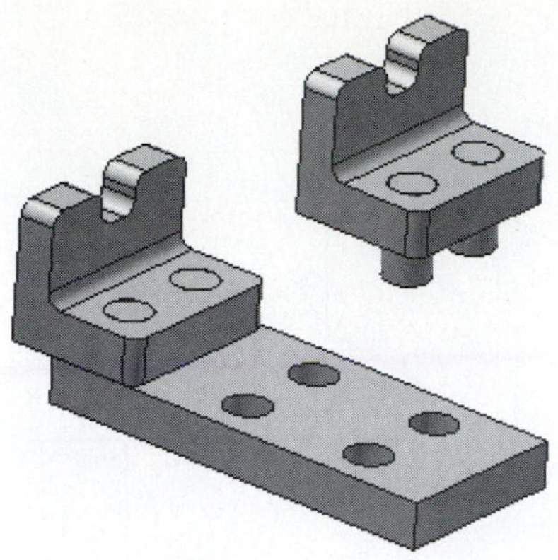

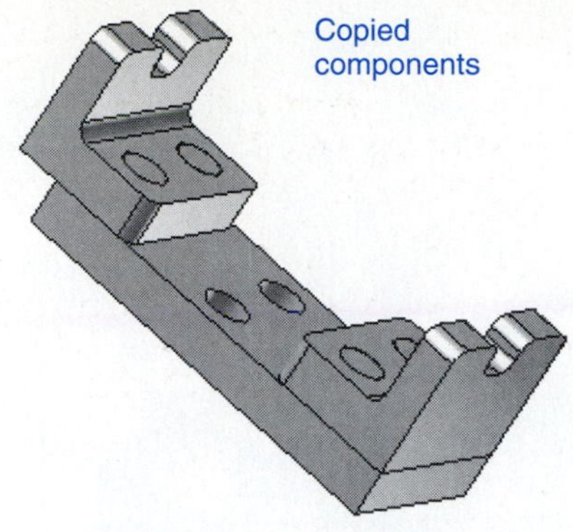

Figure 5-128 *(Continued)*

Exercise 5-48: Copying a Component

1. Click the **Copy Component** tool.

The **Copy Components** dialog box will appear.

2. Select the bracket and two posts.
3. Select **OK.**

The **Copy Components: File Names** dialog box will appear. This is a listing of all components that have been copied and added to the assembly.

4. Move the copied component away from the assembly.
5. Use the **Constraint** tool to position the copied component into the assembly.

Summary

This chapter explained how to create assembly drawings from individual parts. The tools in the **Assembly Panel** bar were introduced and used in a bottom-up approach to create the first assembly drawing from an existing model. The method for grounding components was demonstrated. The **Move** and **Rotate** tools as well as all the options of the **Constraint** tool were used to manipulate components of the drawing.

The tools in the **Presentation Panel** bar were used to explode an assembly drawing to show how the assembly is created from its components. The presentation drawing was then animated. Assembly numbers were added and edited to isometric views of the presentation drawing.

The other elements of a presentation drawing—the parts list or bill of materials, the title block, release blocks, and revision blocks were also demonstrated.

The top-down approach to creating an assembly drawing was also illustrated, and various techniques for editing assembly drawings were explained, including patterning, mirroring, and copying components.

Chapter Projects

Project 5-1:

A dimensioned block is shown in Figure P5-1. Redraw this block and save it as SQBLOCK. See page 159. Use the SQBLOCK to create assemblies as shown.

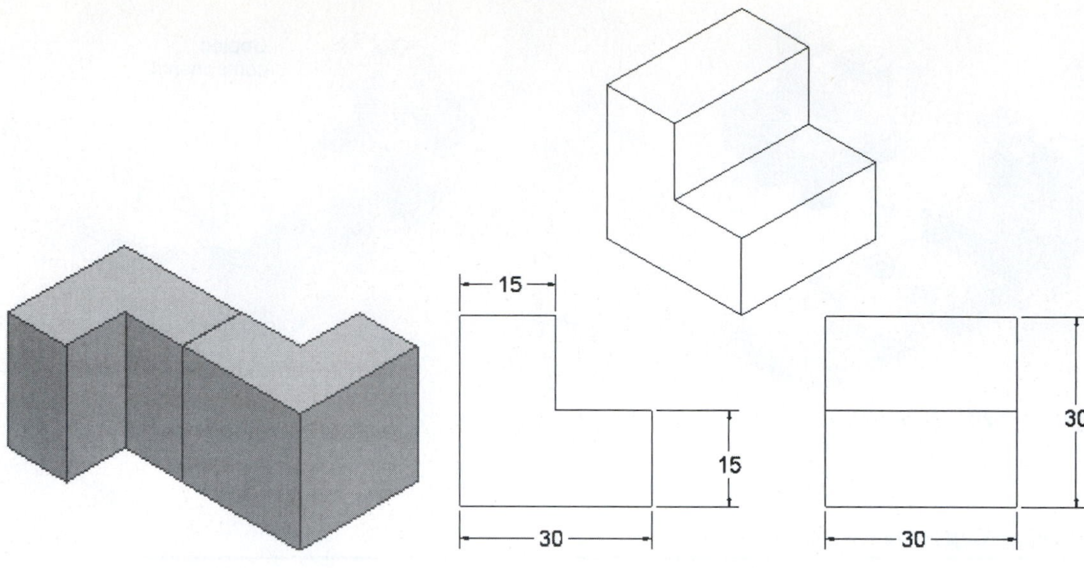

Figure P5-1A MILLIMETERS

Figure P5-1B

20° angle between the two SQBLOCKs

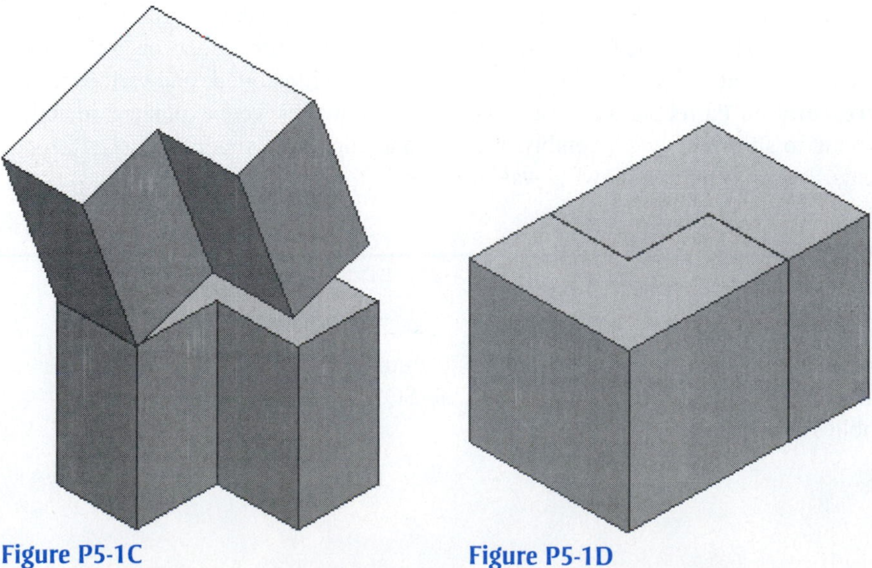

Figure P5-1C

Figure P5-1D

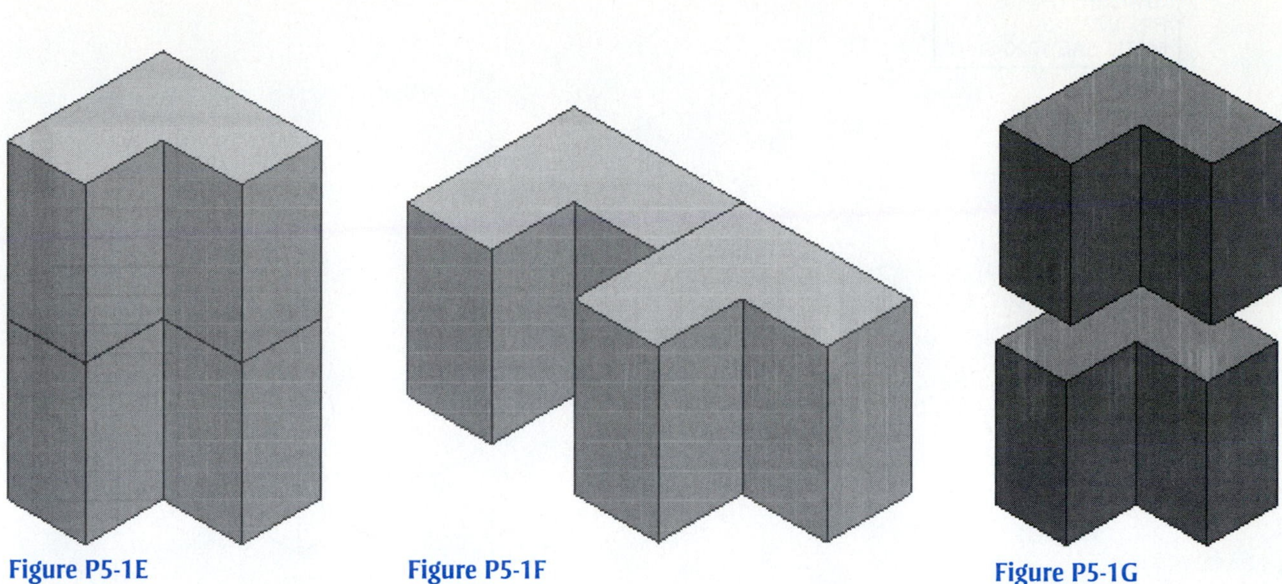

Figure P5-1E

Figure P5-1F

Figure P5-1G

Pages 219 through 222 show a group of parts. These parts are used to create the assemblies presented as problems in this section. Use the given descriptions, part numbers, and materials when creating BOMs for the assemblies.

Project 5-2:

Redraw the following models and save them as **Standard (mm).ipn** files. All dimensions are in millimeters.

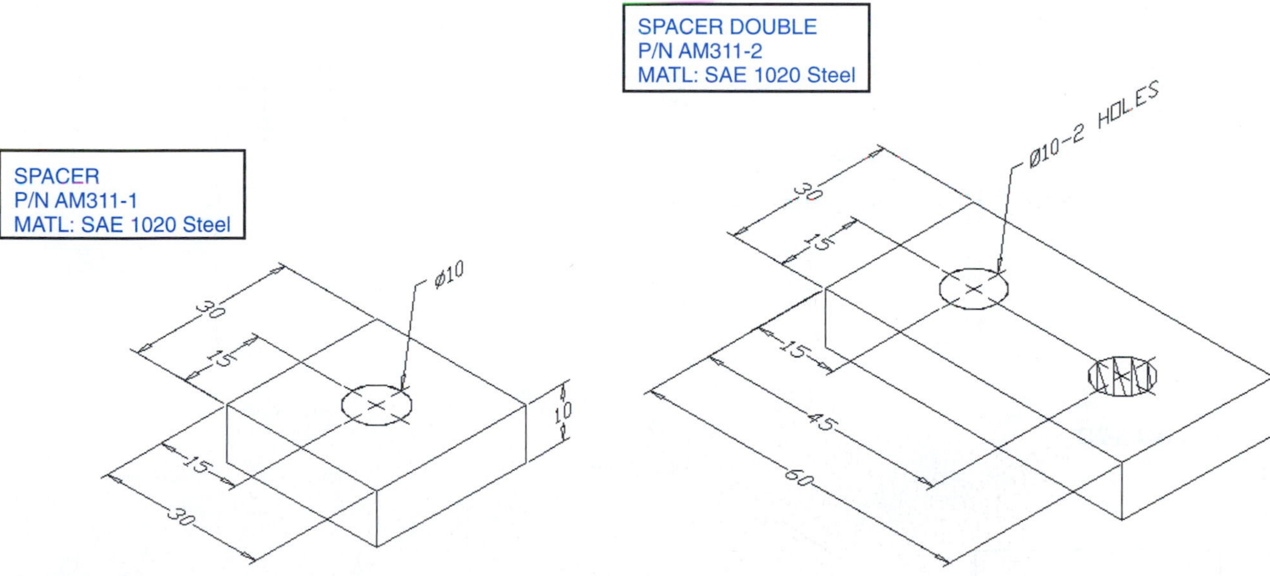

Figure P5-2A

Figure P5-2B

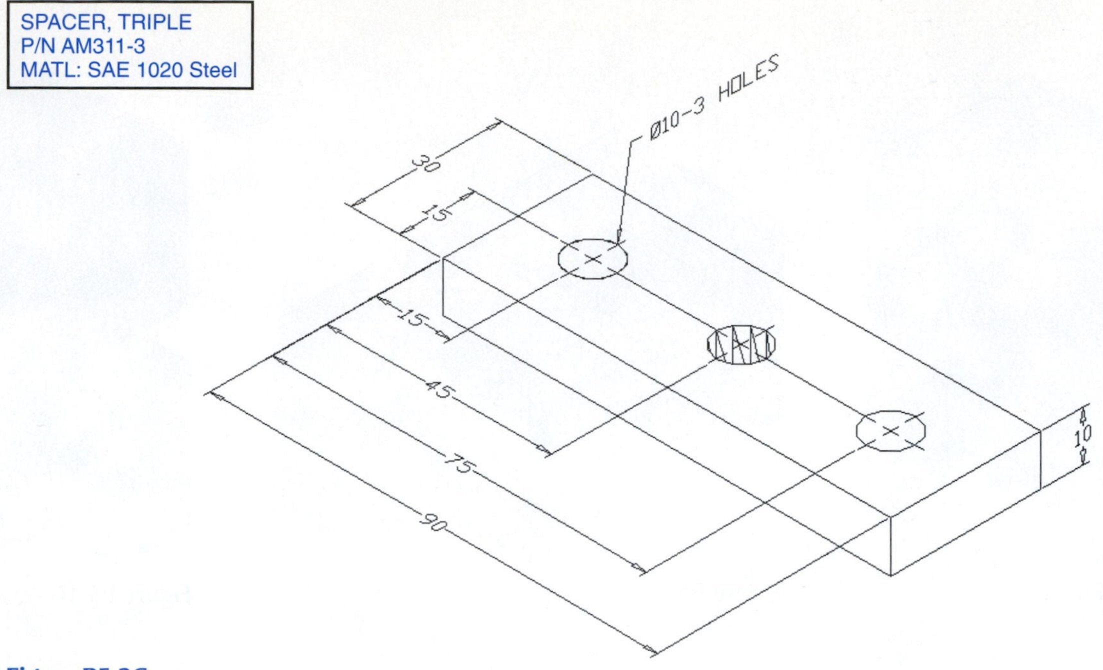

Figure P5-2C

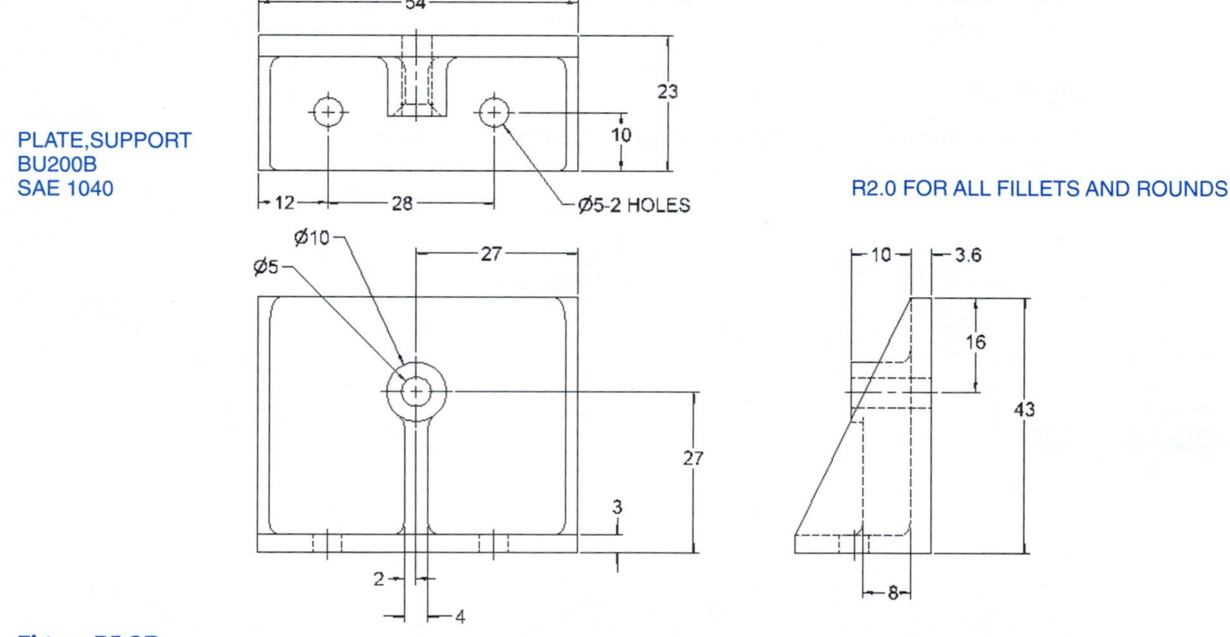

Figure P5-2D

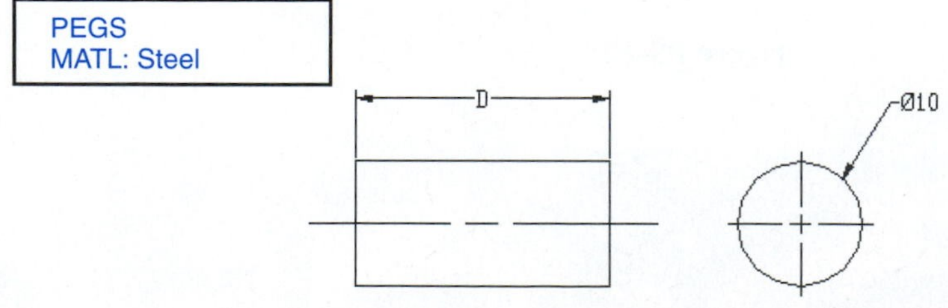

DESCRIPTION	PART NO.	D
PEG, SHORT	PG20-1	20
PEG	PG30-1	30
PEG, LONG	PG40-1	40

ALL DISTANCES IN MILLIMETERS

Figure P5-2E

L-BRACKET
P/N BK20-1
MATL: SAE 1040 Steel

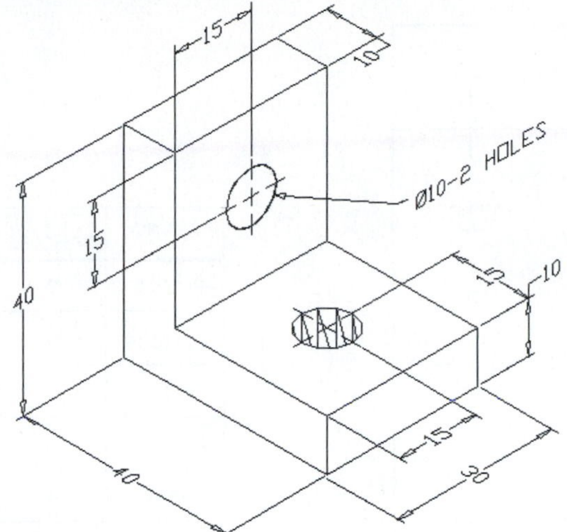

Figure P5-2F

Z-BRACKET
P/N BK20-2
MATL: SAE 1040 Steel

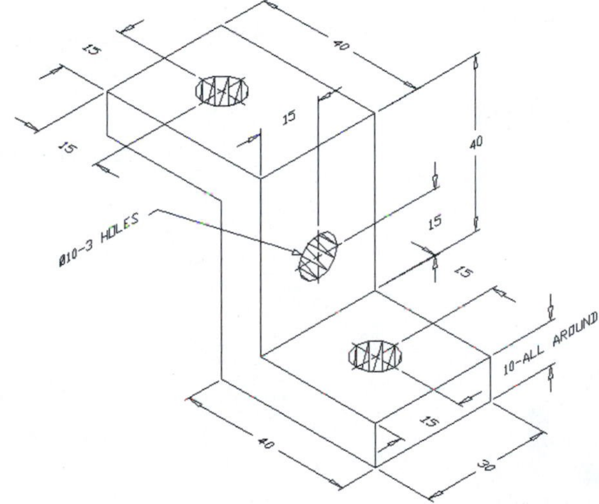

Figure P5-2G

C-BRACKET
P/N BK20-3
SAE 1040 Steel

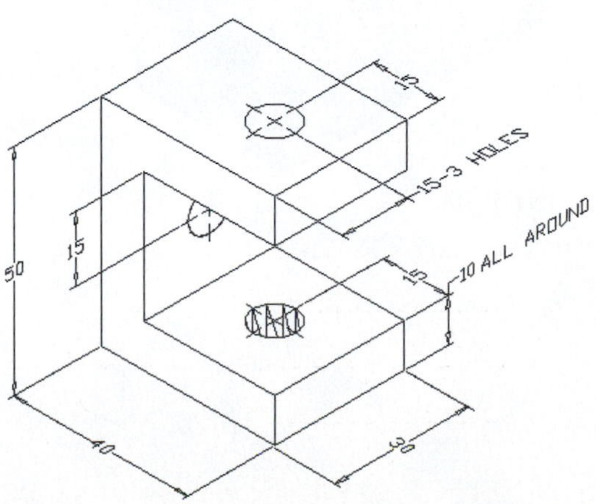

Figure P5-2H

PLATE, BASE
SAE 1020 Steel

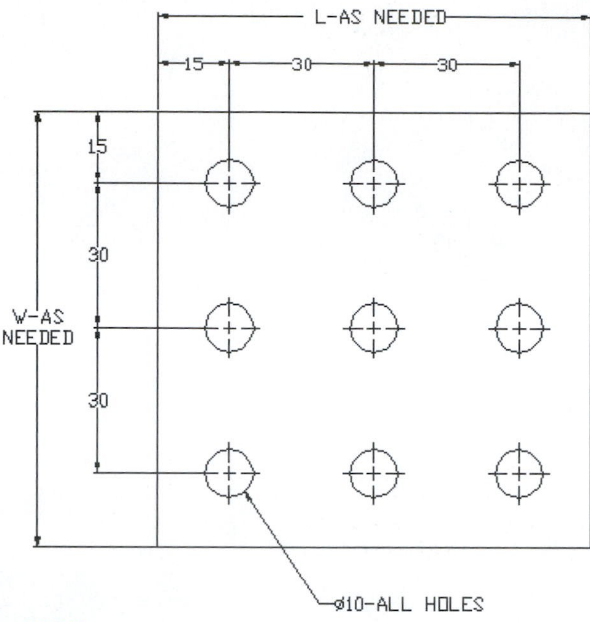

PART NO.	TOTAL NO OF HOLES	L	W	HOLE PATTERN
PL110-9	9	90	90	3x3
PL110-16	16	120	120	4x4
PL110-6	6	60	90	2x3
PL110-8	8	60	120	2x4
PL80-4	4	60	60	2x2

Figure P5-21

Project 5-3:

Draw an exploded isometric assembly drawing of Assembly 1. Create a BOM.

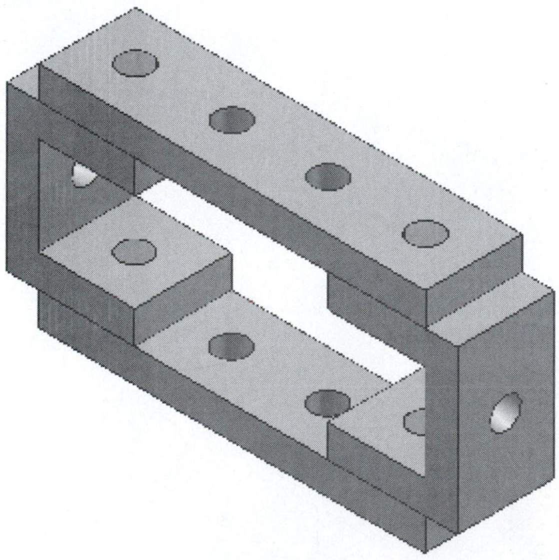

Figure P5-3A MILLIMETERS

ASSEMBLY 1

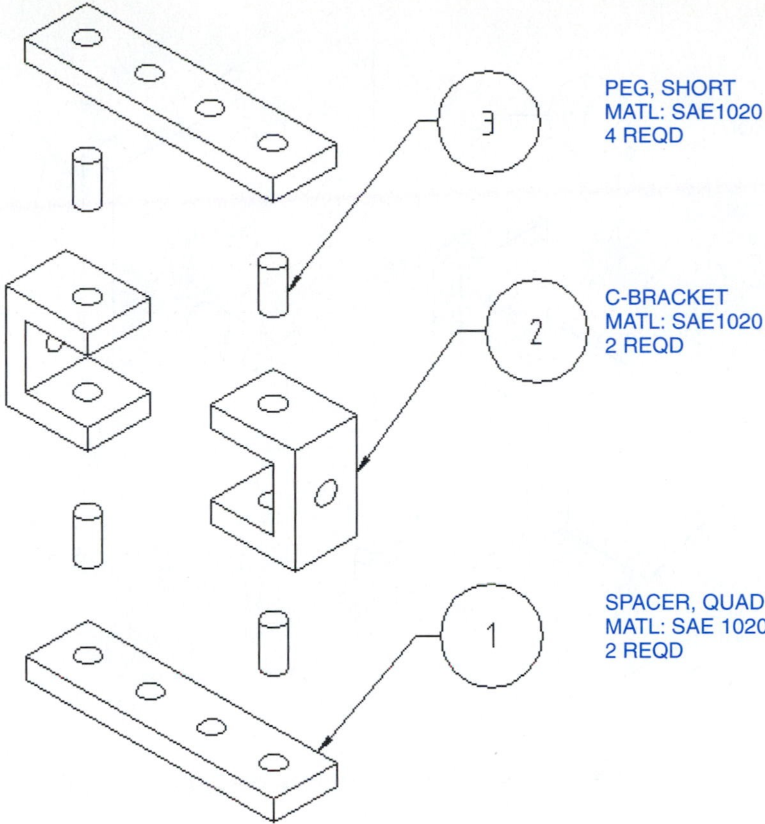

Figure P5-3B

Project 5-4:

Draw an exploded isometric assembly drawing of Assembly 2. Create a BOM.

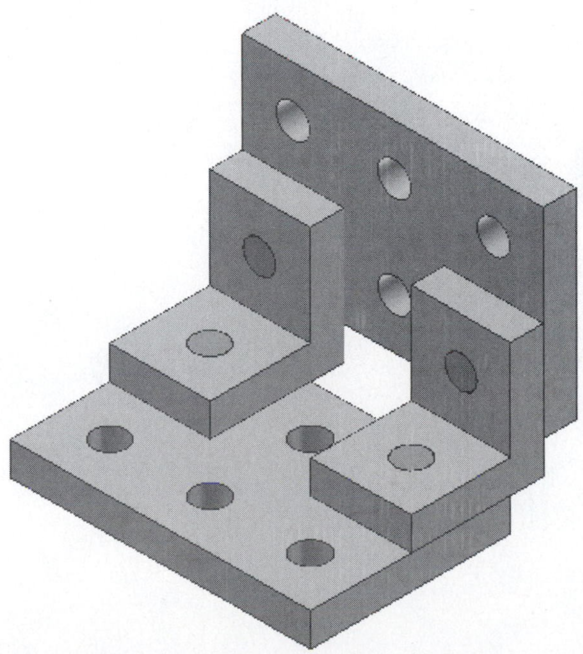

Figure P5-4A MILLIMETERS

ASSEMBLY 2

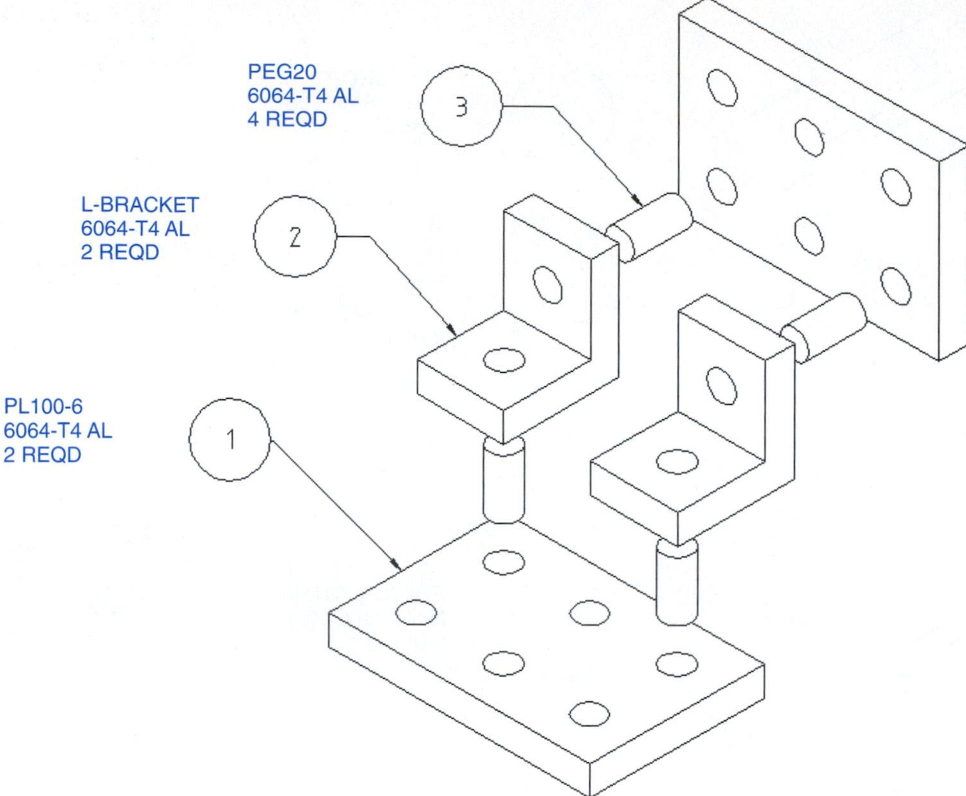

Figure P5-4B

Project 5-5:

Draw an exploded isometric assembly drawing of Assembly 3. Create a BOM.

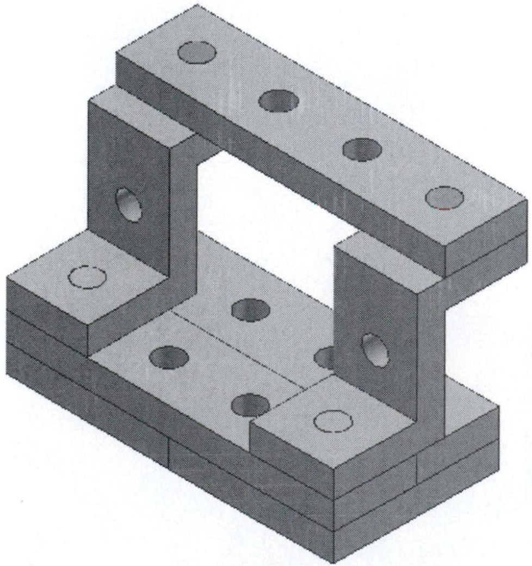

Figure P5-5A MILLIMETERS

ASSEMBLY 3

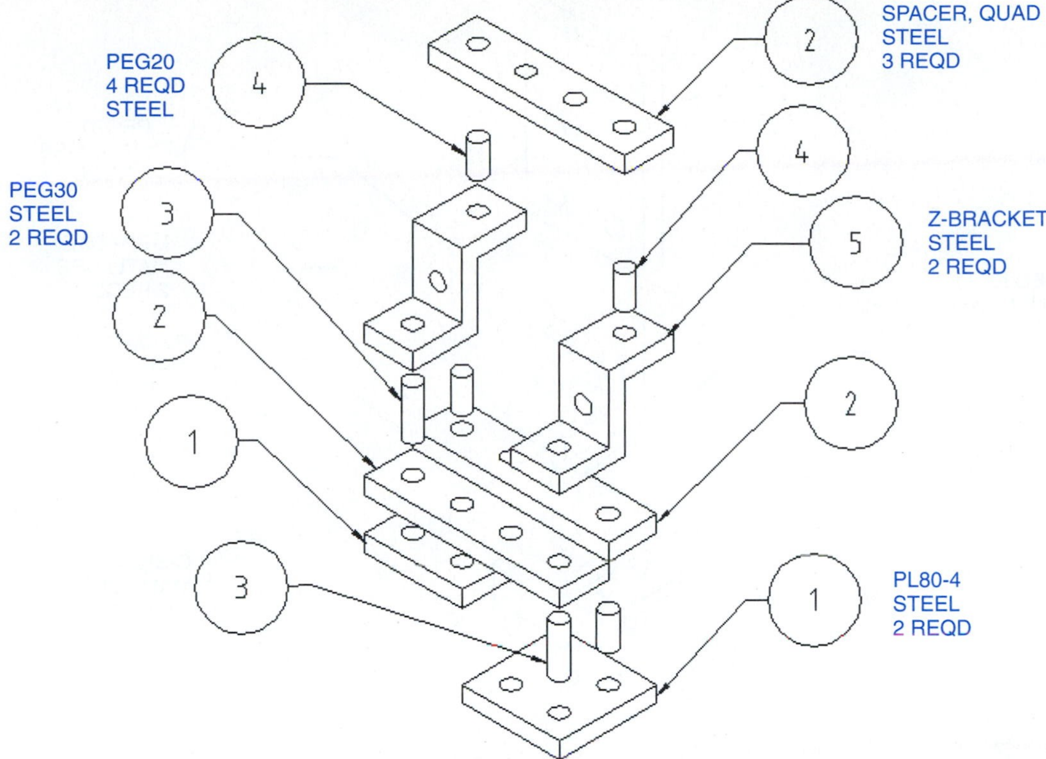

Figure P5-5B

Project 5-6:

Draw an exploded isometric assembly drawing of Assembly 4. Create a BOM.

Figure P5-6A MILLIMETERS

ASSEMBLY 4

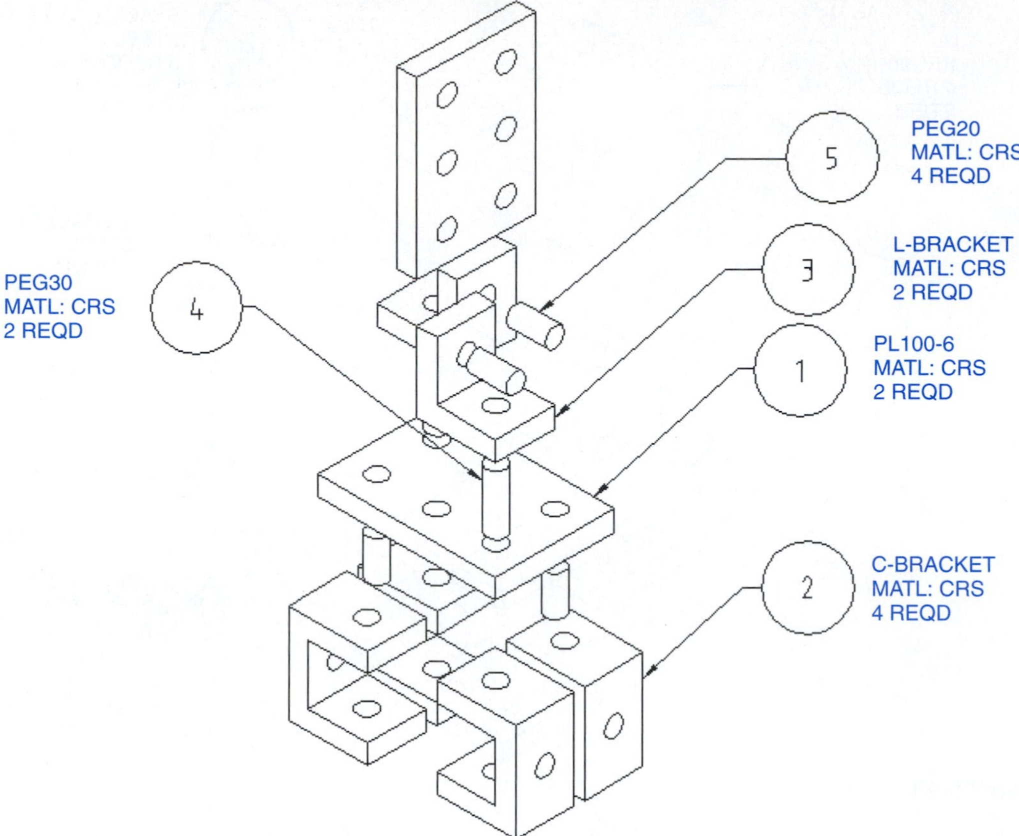

Figure P5-6B

Project 5-7:

Draw an exploded isometric assembly drawing of Assembly 5. Create a BOM.

Figure P5-7A MILLIMETERS

ASSEMBLY 5

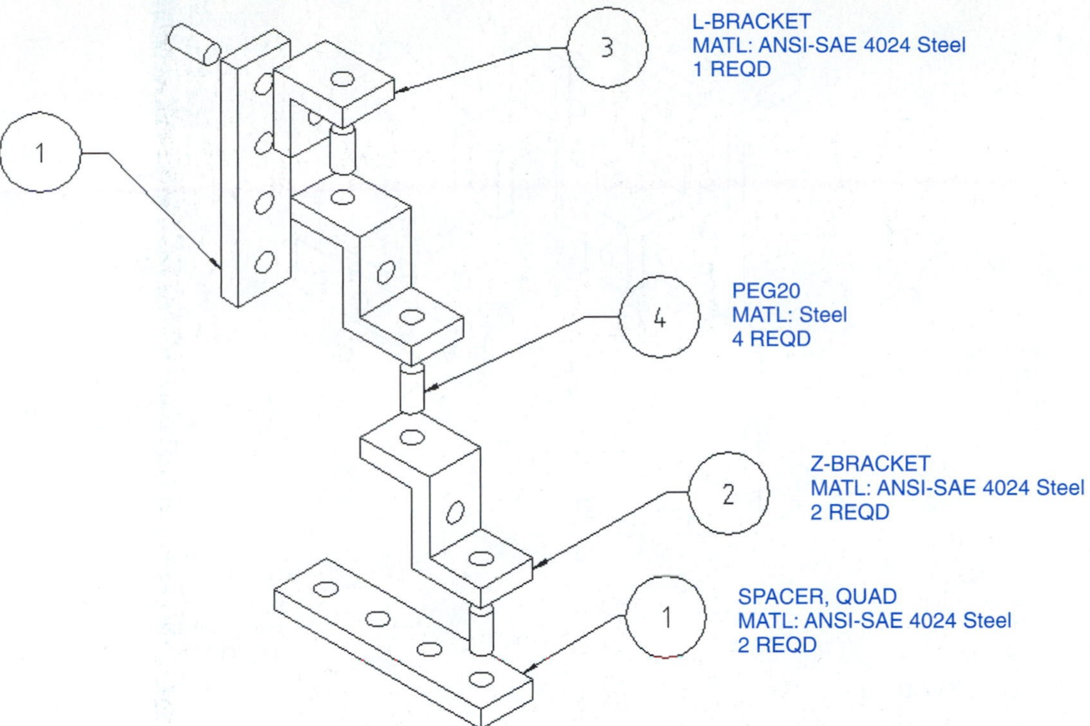

Figure P5-7B

Project 5-8:

Draw an exploded isometric assembly drawing of Assembly 6. Create a BOM.

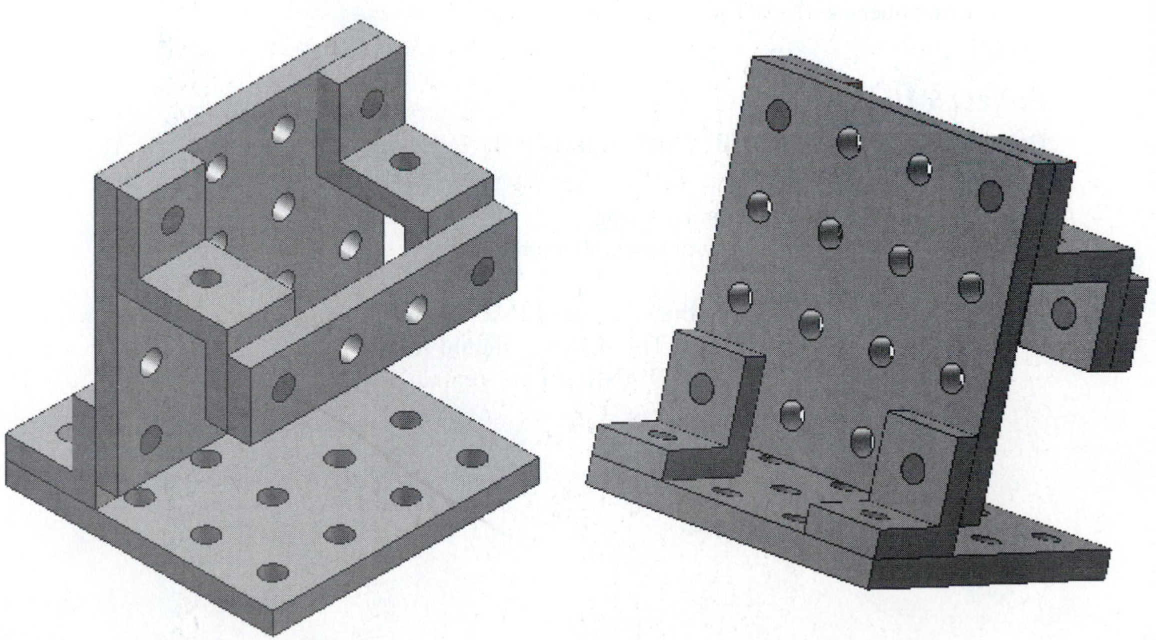

Figure P5-8A

ASSEMBLY 6

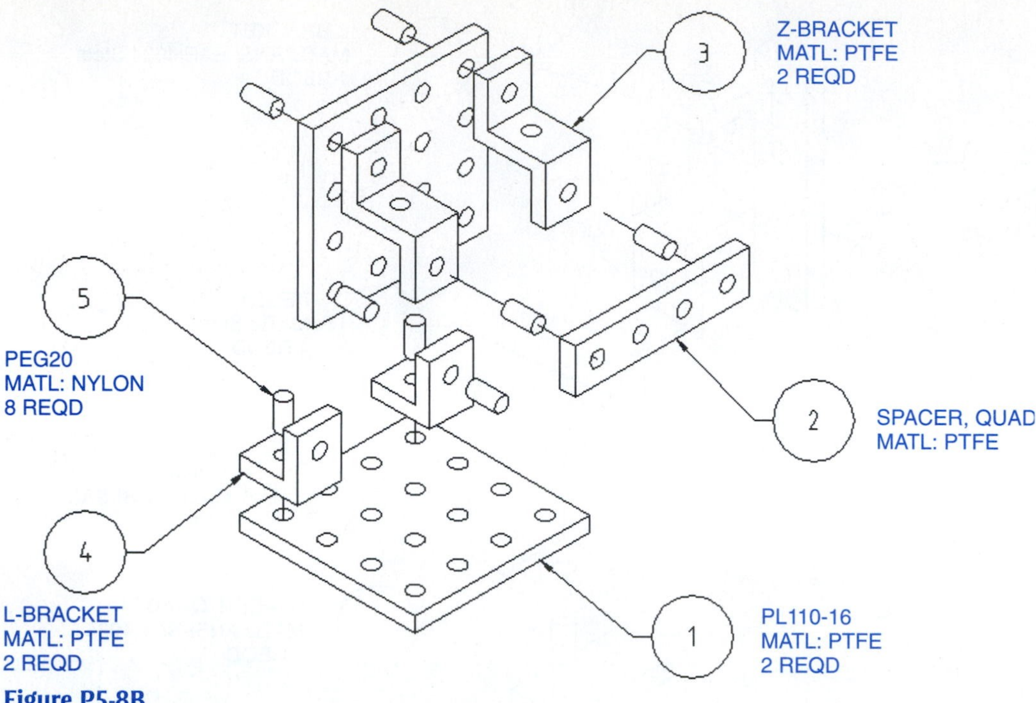

Figure P5-8B

Project 5-9:

Create an original assembly based on the parts shown on pages 219–222. Include a scene, an exploded isometric drawing with assembly numbers, and a BOM. Use at least 12 parts.

Project 5-10:

Draw the ROTATOR ASSEMBLY shown. Include the following:

A. An assembly drawing
B. An exploded presentation drawing
C. An isometric drawing with assembly numbers
D. A parts list
E. An animated assembly drawing; the LINKs should rotate relative to the PLATE. The LINKs should carry the CROSSLINK. The CROSSLINK should remain parallel during the rotation.

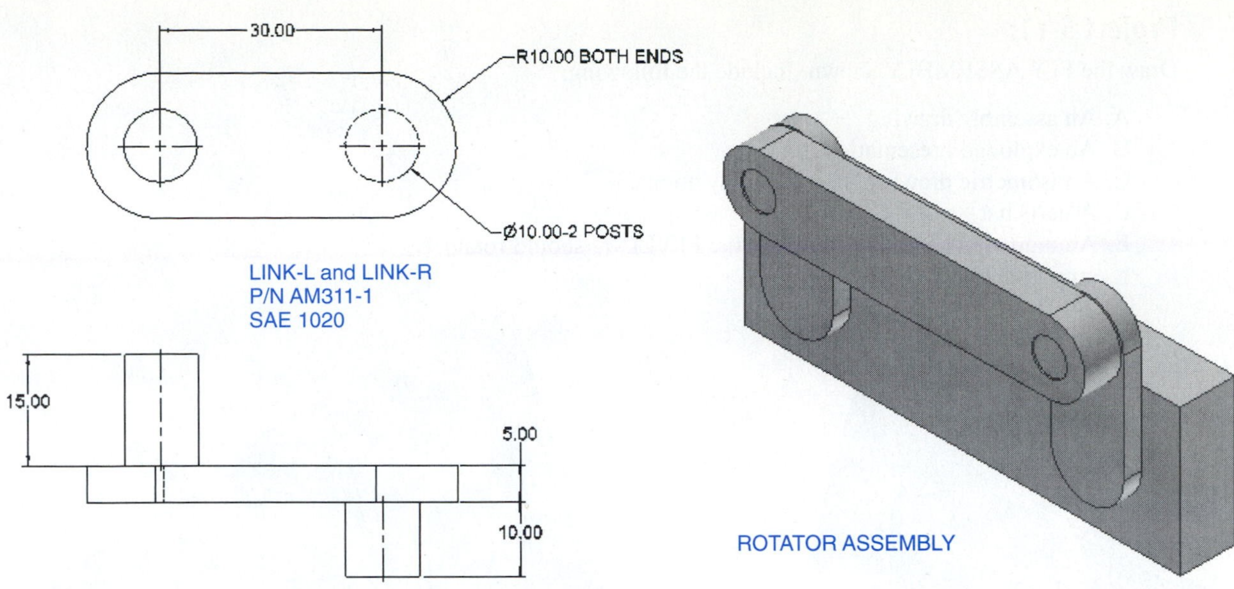

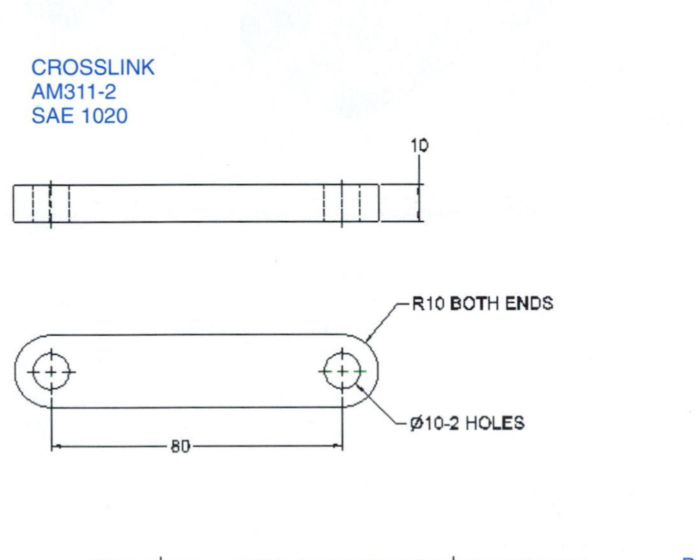

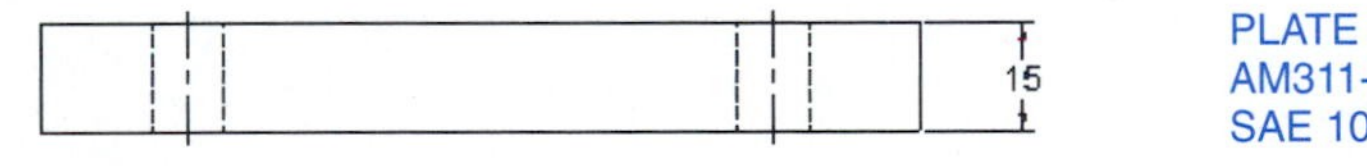

PLATE
AM311-1
SAE 1020

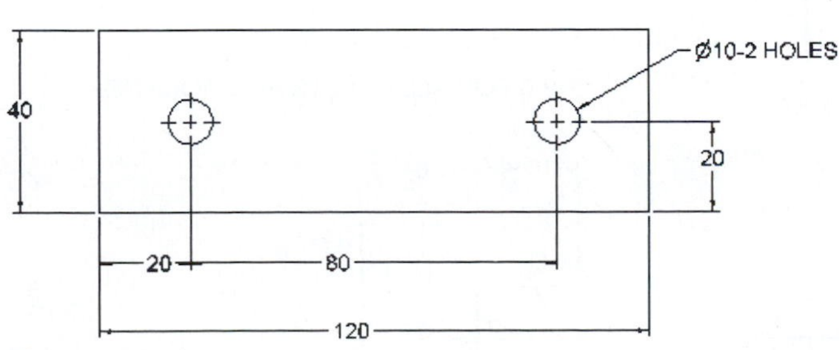

Figure P5-10

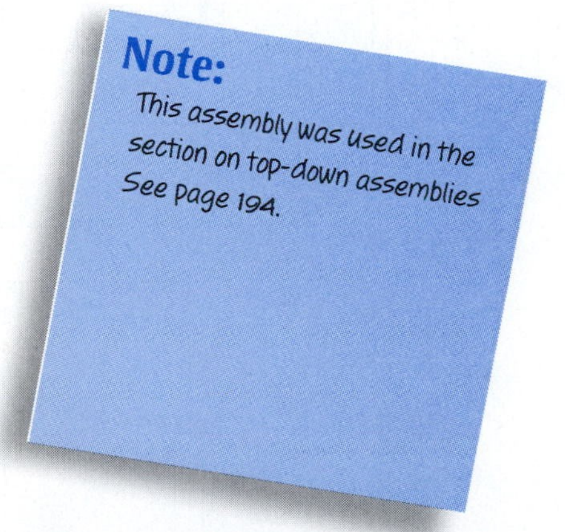

Project 5-11:

Draw the FLY ASSEMBLY shown. Include the following:

A. An assembly drawing
B. An exploded presentation drawing
C. An isometric drawing with assembly numbers
D. A parts list
E. An animated assembly drawing; the FLYLINK should rotate around the SUPPORT base.

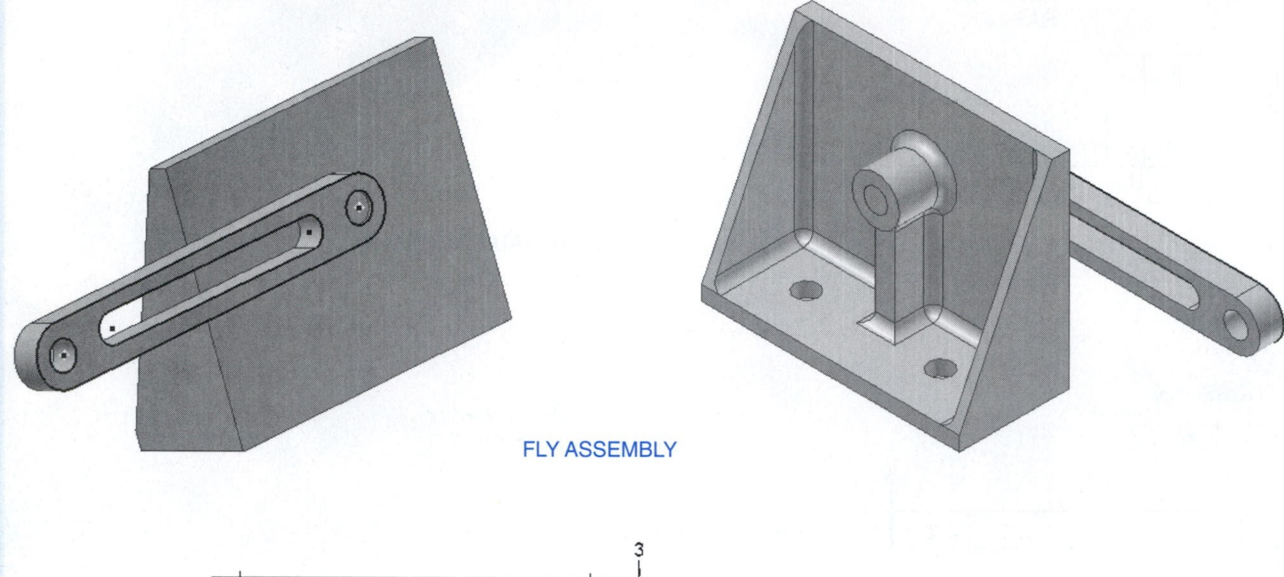

FLY ASSEMBLY

FLYLINK
BU200A
SAE 1040

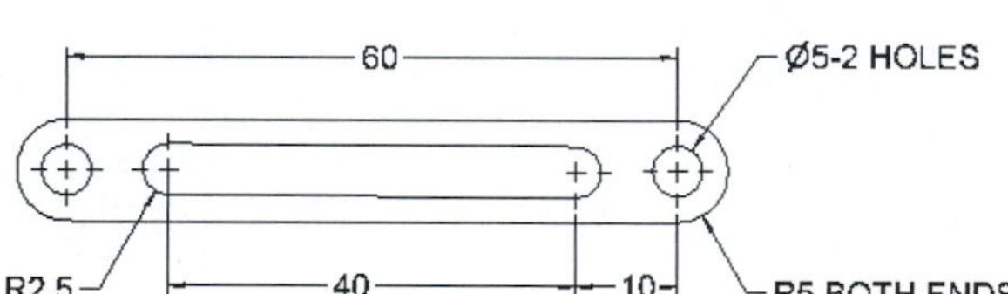

PEGØ5
BU-200C
SAE1040

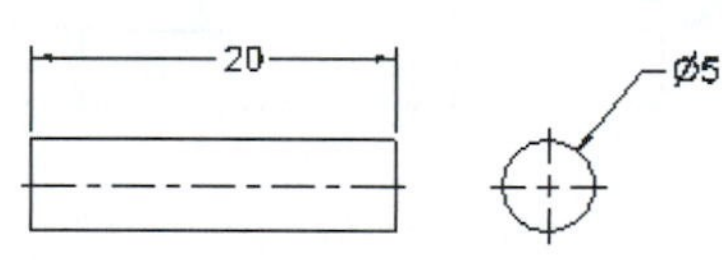

PLATE,SUPPORT
BU200B
SAE 1040

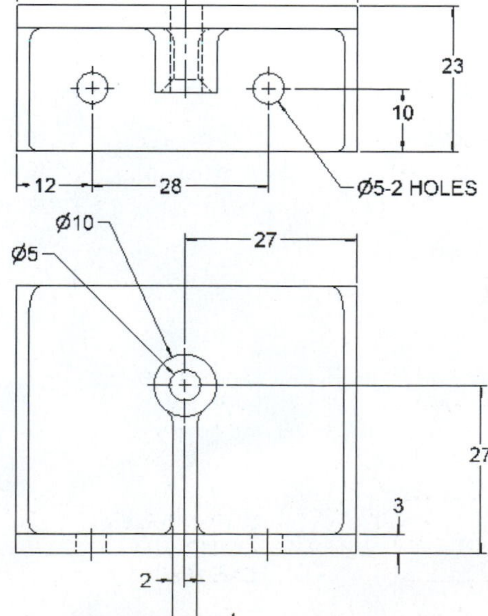

R2.0 FOR ALL FILLETS AND ROUNDS

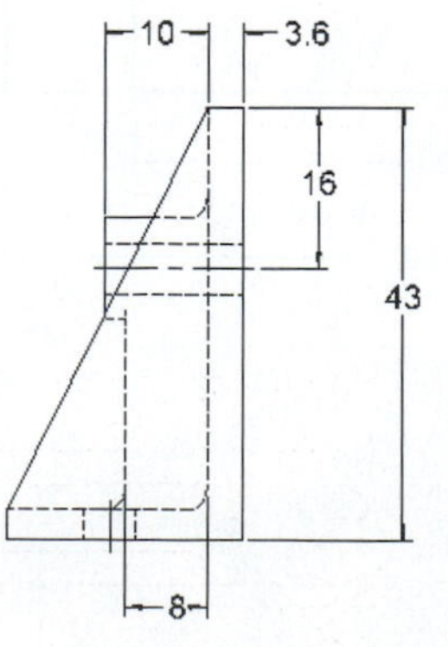

Figure P5-11

Project 5-12:

Draw the ROCKER ASSEMBLY shown. Include the following:

A. An assembly drawing
B. An exploded presentation drawing
C. An isometric drawing with assembly numbers
D. A parts list
E. An animated assembly drawing

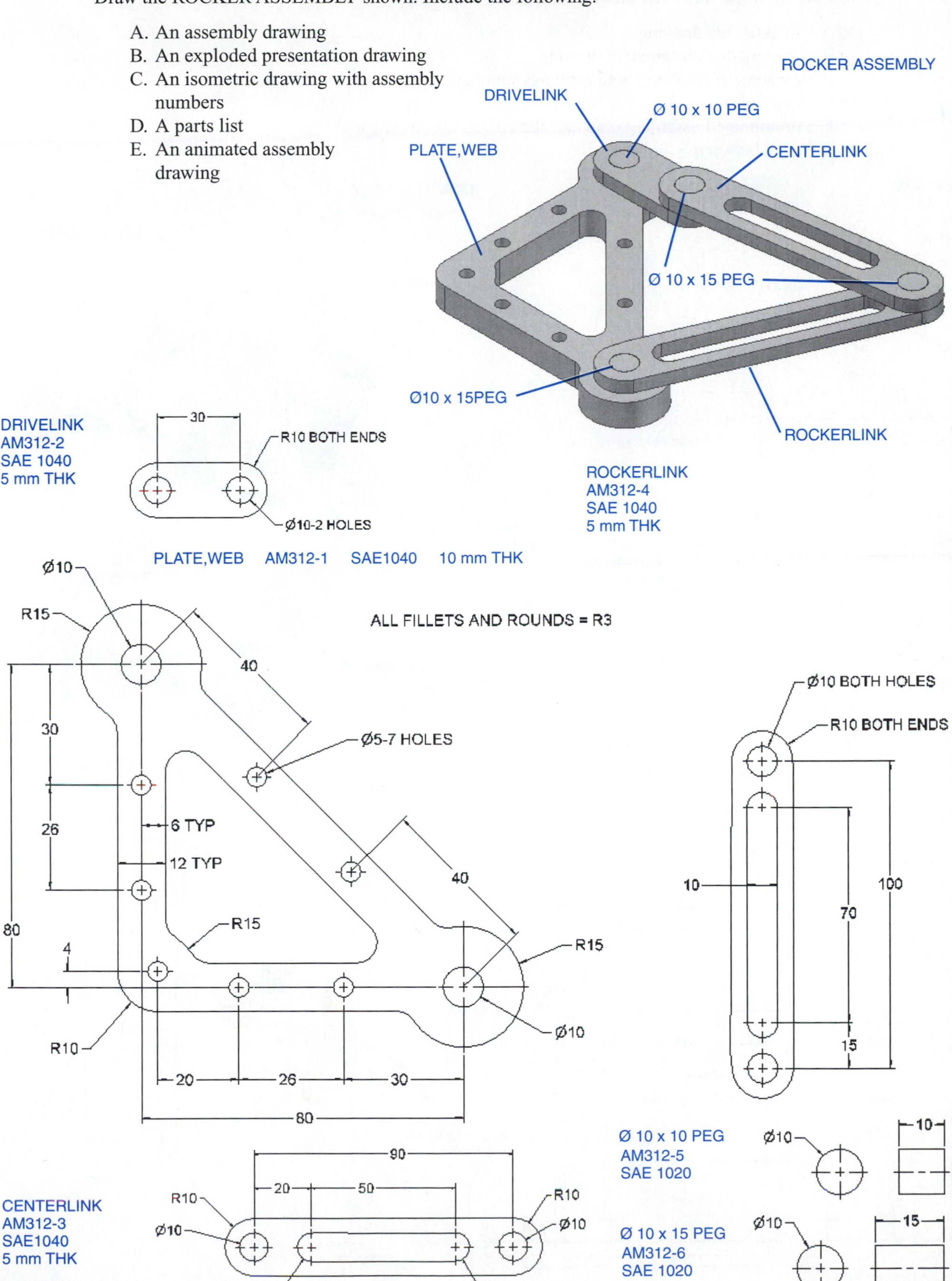

Figure P5-12

Project 5-13:

Draw the LINK ASSEMBLY shown. Include the following:

A. An assembly drawing
B. An exploded presentation drawing
C. An isometric drawing with assembly numbers
D. A parts list
E. An animated assembly drawing; the HOLDER ARM should rotate between −30° and +30°.

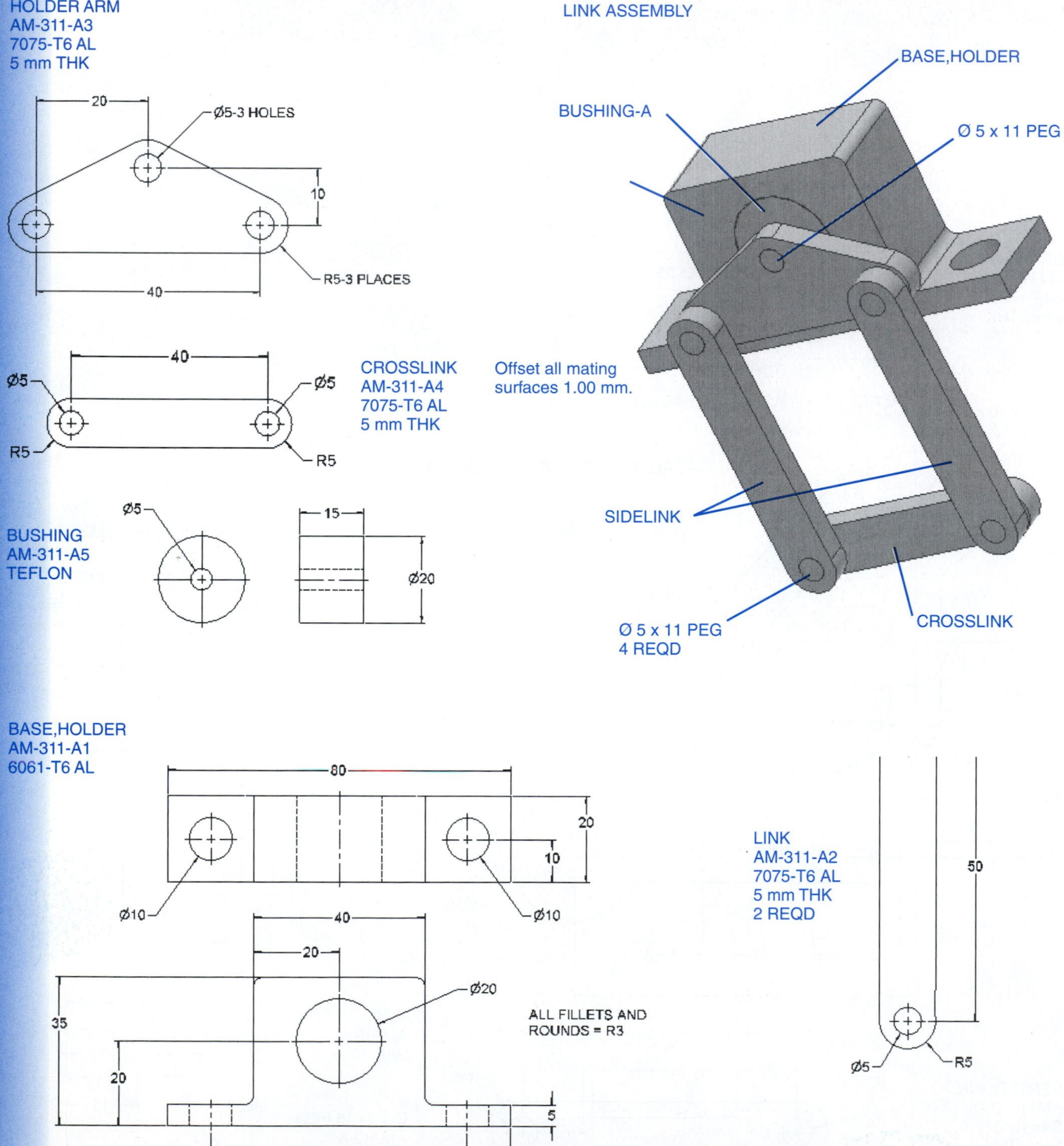

Figure P5-13

Project 5-14:

Draw the PIVOT ASSEMBLY shown using the dimensioned components given. Include the following:

A. A presentation drawing
B. A 3D exploded isometric drawing
C. A parts list

TIP This assembly was used in the section on subassemblies. See page 183.

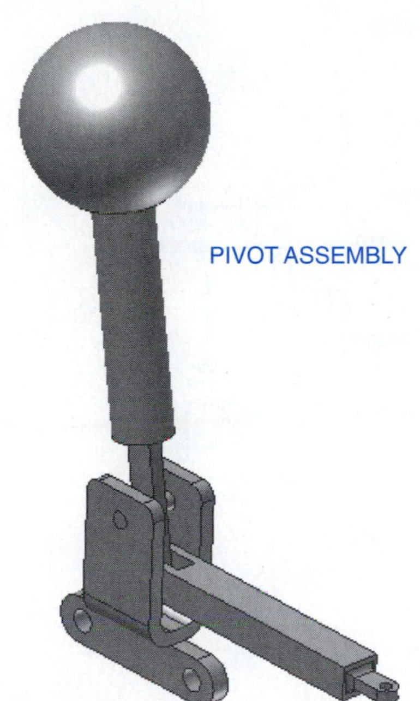

Figure P5-14A MILLIMETERS

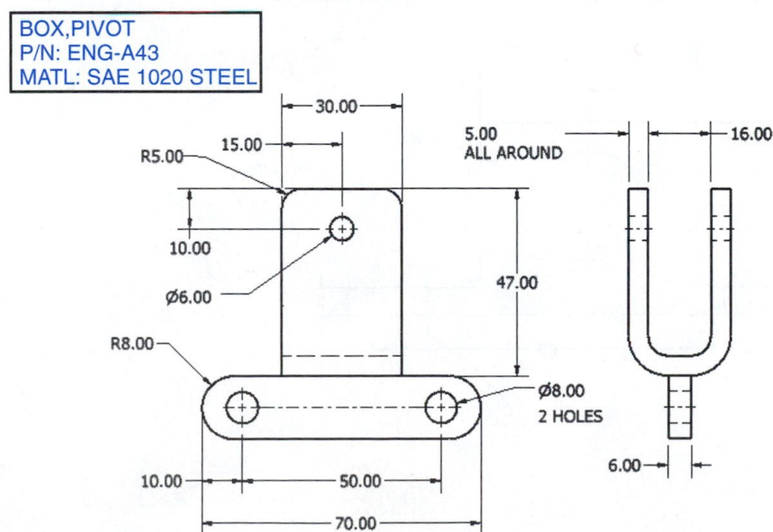

Figure P5-14B

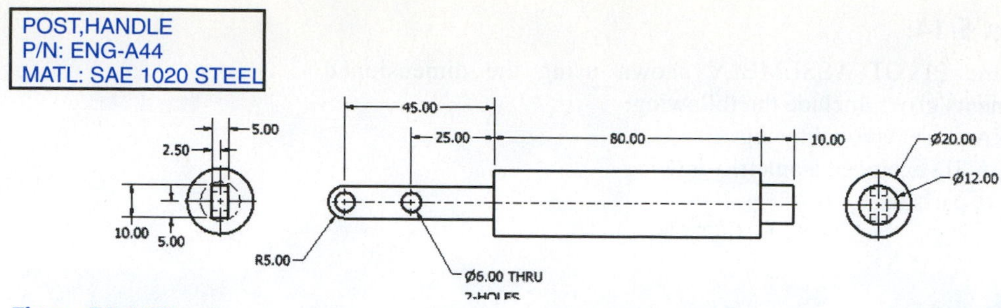

Figure P5-14C

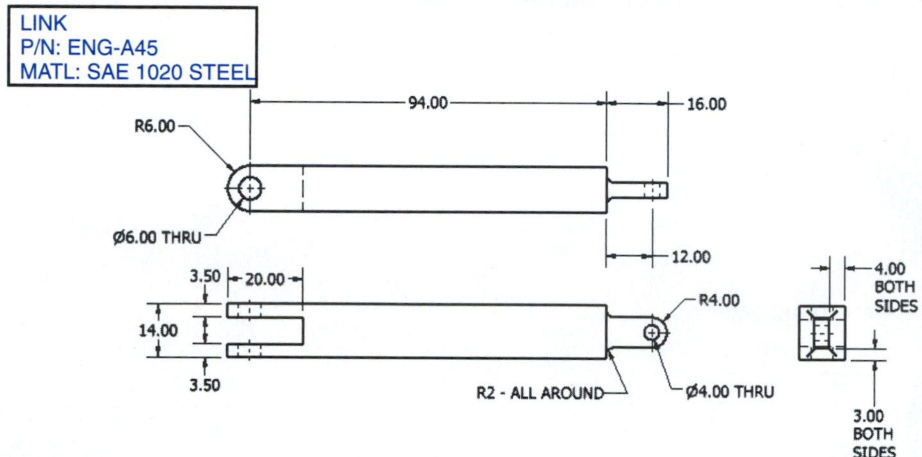

Figure P5-14D

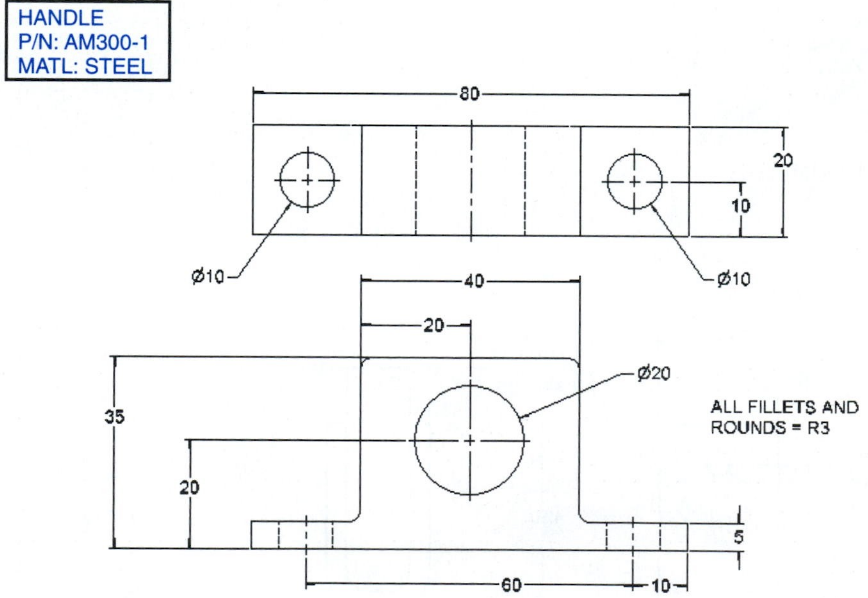

Figure P5-14E

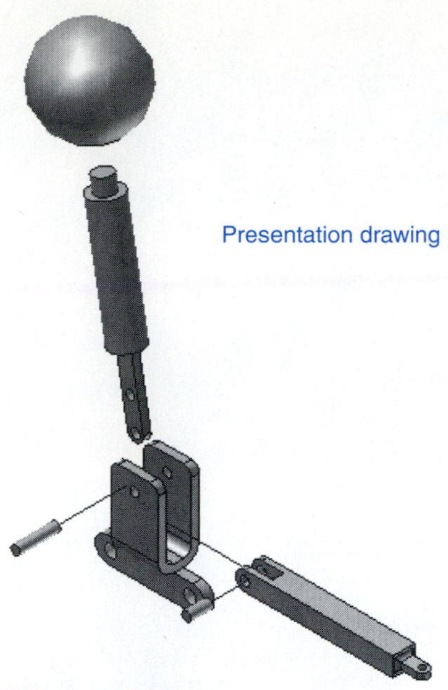

Figure P5-14F

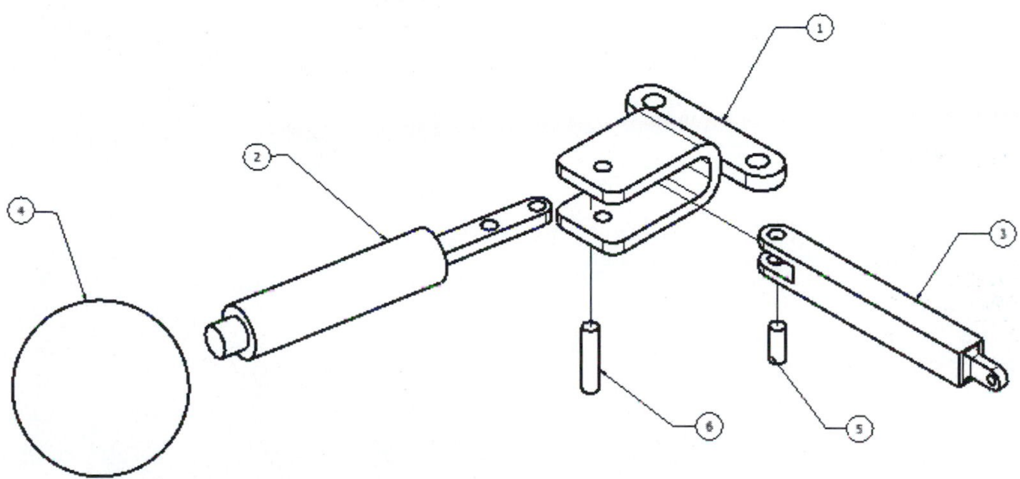

Figure P5-14G

Parts List				
ITEM	PART NUMBER	DESCRIPTION	MATERIAL	QTY
1	ENG-A43	BOX,PIVOT	SAE1020	1
2	ENG-A44	POST,HANDLE	SAE1020	1
3	ENG-A45	LINK	SAE1020	1
4	AM300-1	HANDLE	STEEL	1
5	EK-132	POST-Ø6x14	STEEL	1
6	EK-131	POST-Ø6x26	STEEL	1

Exercise P5-14H

Threads and Fasteners 6

Chapter Objectives

- Explain thread terminology and conventions.
- Show how to draw threads.
- Show how to size both internal and external threads.
- Show how to use standard-sized threads.
- Show how to use and size washers, nuts, and setscrews.

INTRODUCTION

This chapter explains how to draw threads and washers. It also explains how to select fasteners and how to design using fasteners, washers, and keys.

Threads are created in Inventor using either the **Hole** or the **Thread** tool located on the **Part Features** panel bar. See Figure 6-1. Predrawn fasteners may be accessed using the **Content Center** tool. The **Content Center** library is explained later in the chapter.

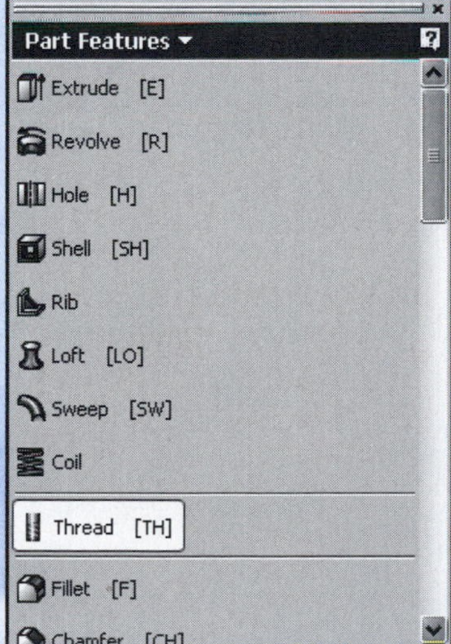

Figure 6-1

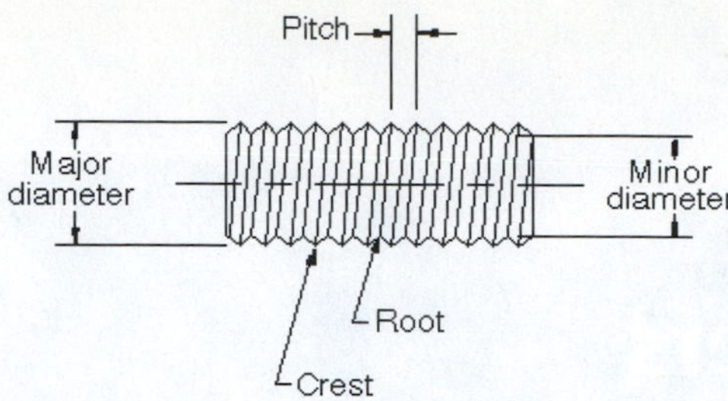

Figure 6-2

THREAD TERMINOLOGY

crest: The peak of a thread.

root: The valley of a thread.

major diameter: The distance across a thread from crest to crest.

minor diameter: The distance across a thread from root to root.

pitch: The linear distance along a thread from crest to crest.

Figure 6-2 shows a thread. The peak of a thread is called the ***crest,*** and the valley portion is called the ***root.*** The ***major diameter*** of a thread is the distance across the thread from crest to crest. The ***minor diameter*** is the distance across the thread from root to root.

The ***pitch*** of a thread is the linear distance along the thread from crest to crest. Thread pitch is usually referred to in terms of a unit of length such as 20 threads per inch or 1.6 mm per thread.

THREAD CALLOUTS—METRIC UNITS

Threads are specified on a drawing using drawing callouts. See Figure 6-3. The M at the beginning of a drawing callout specifies that the callout is for a metric thread. Holes that are not threaded use the Ø symbol.

The number following the M is the major diameter of the thread. An M10 thread has a major diameter of 10 mm. The pitch of a metric thread is assumed to be a coarse thread unless otherwise stated. The callout M10 × 30 assumes a coarse thread, or a thread length of 1.5 mm per thread. The number 30 is the thread length in millimeters. The "×" is read as "by," so the thread is called a "ten by thirty."

The callout M10 × 1.25 × 30 specifies a pitch of 1.25 mm per thread. This is not a standard coarse thread size, so the pitch must be specified.

Figure 6-4 shows a listing of standard metric thread sizes. Other sizes may be located by scrolling through the given **Nominal Sizes.** Inventor lists metric threads according to ANSI (American National Standards Institute) Metric M Profile standards.

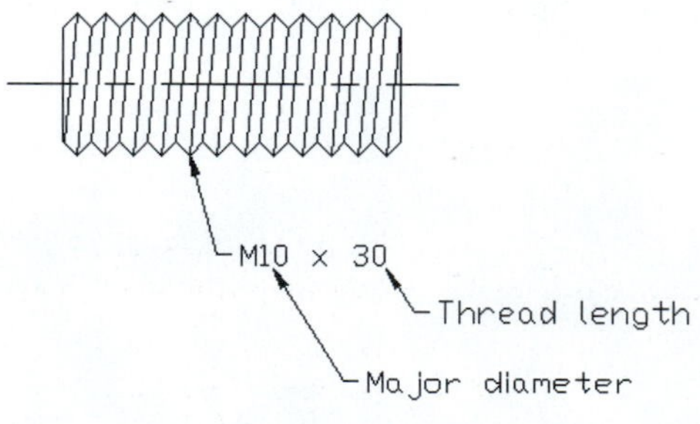

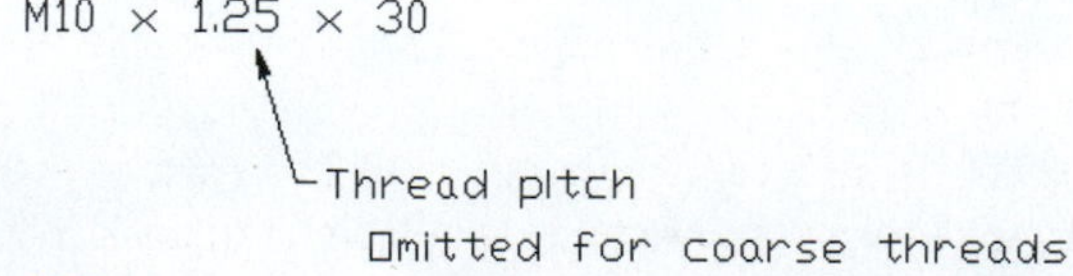

Figure 6-3

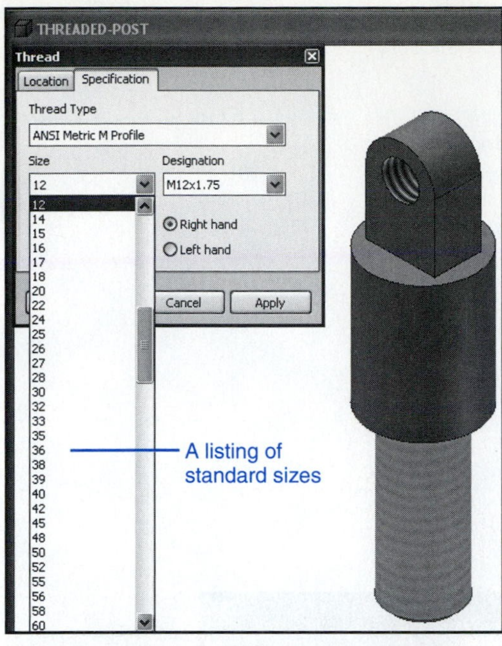

Figure 6-4

Whenever possible use preferred thread sizes for designing. Preferred thread sizes are readily available and are usually cheaper than nonstandard sizes. In addition, tooling such as wrenches is also readily available for preferred sizes.

Thread Callouts—ANSI Unified Screw Threads

ANSI Unified Screw Threads (English units) always include a thread form specification. Thread form specifications are designated by capital letters, as shown in Figure 6-6, and are defined as follows.

UNC—Unified National Coarse

UNF—Unified National Fine

UNEF—Unified National Extra Fine

UN—Unified National, or constant-pitch threads

An ANSI (English units) thread callout starts by defining the major diameter of the thread followed by the pitch specification. The callout .500-13 UNC means a thread whose major diameter is .500 in. with 13 threads per inch. The thread is manufactured to the UNC standards.

There are three possible classes of fit for a thread: 1, 2, and 3. The different class specifications specify a set of manufacturing tolerances. A class 1 thread is the loosest and a class 3 the most exact. A class 2 fit is the most common.

The letter A designates an external thread, B an internal thread. The symbol × means "by" as in 2 × 4, "two by four." The thread length (3.00) may be followed by the word LONG to prevent confusion about which value represents the length.

Drawing callouts for ANSI (English unit) threads are sometimes shortened, such as in Figure 6-5. The callout .500-13 UNC-2A × 3.00 LONG is shortened to .500-13 × 3.00. Only a coarse thread has 13 threads per inch, and it should be obvious whether a thread is internal or external, so these specifications may be dropped. Most threads are class 2, so it is tacitly accepted that all threads are class 2 unless otherwise specified. The shortened callout form is not universally accepted. When in doubt, use a complete thread callout.

A listing of standard ANSI (English unit) threads, as presented in Inventor, is shown in Figure 6-6. Some of the drill sizes listed use numbers and letters. The decimal equivalents to the numbers are listed in Figure 6-6.

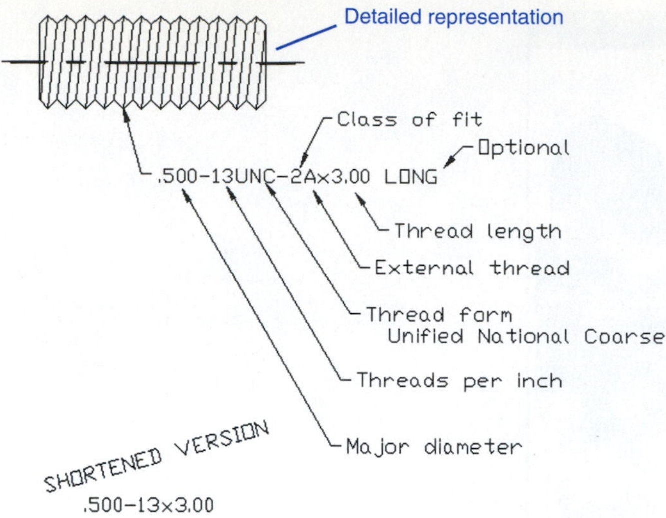

Figure 6-5

The decimal equivalents for threads specifed by numbers

#1	Ø.073
#2	Ø.086
#3	Ø.090
#4	Ø.112
#5	Ø.126
#6	Ø.138
#8	Ø.164
#10	Ø.190
#12	Ø.216

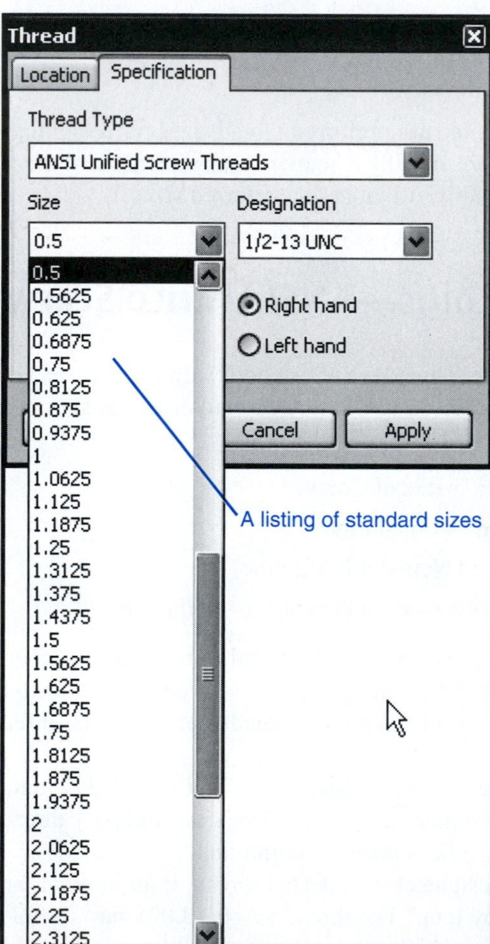

Figure 6-6

Thread Representations

There are three ways to graphically represent threads on a technical drawing: detailed, schematic, and simplified. Figure 6-5 shows an external detailed representation, and Figure 6-7 shows both the external and internal simplified and schematic representations.

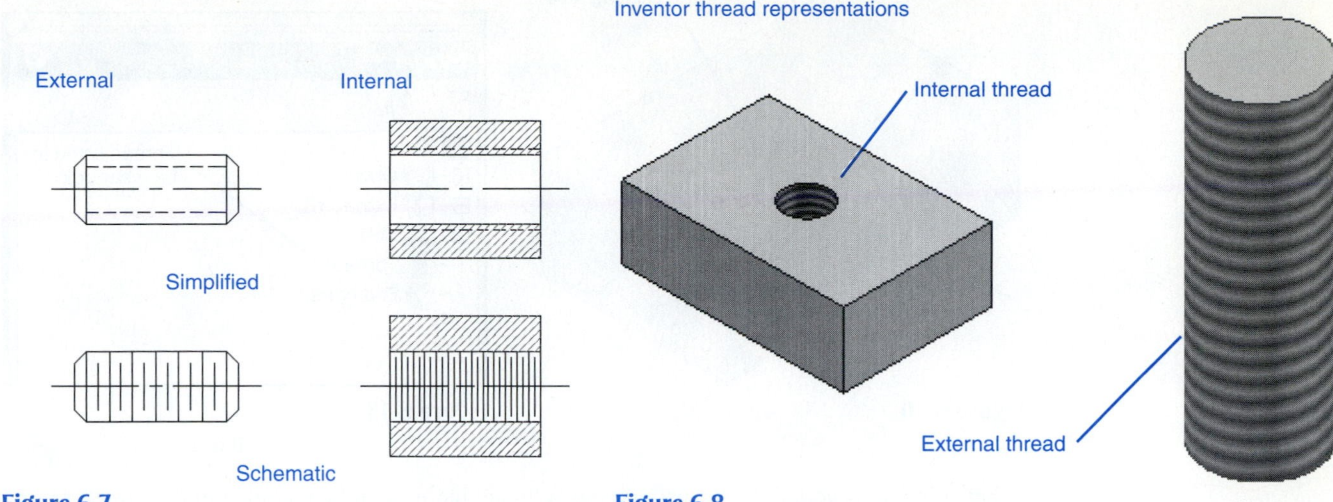

Figure 6-7

Figure 6-8

Figure 6-8 shows an internal and an external thread created using Inventor. The threads will automatically be sized to the existing hole. Threads may be created only around existing holes and cylinders.

INTERNAL THREADS

Figure 6-9 shows an object with a Ø6.0 hole drilled through its center.

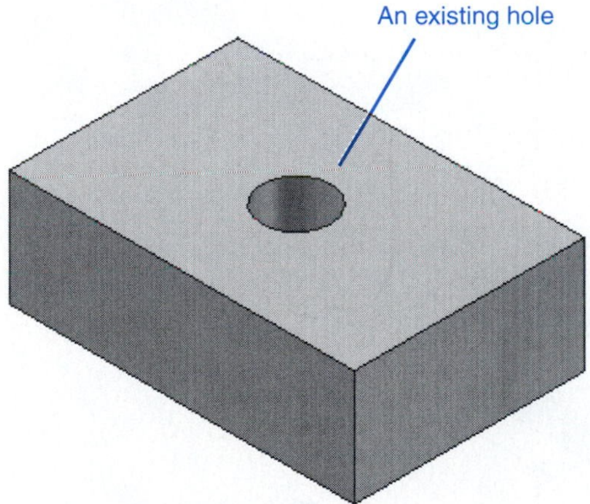

Figure 6-9

Exercise 6-1: Adding Threads to an Existing Hole

1. Click on the **Thread** tool on the **Part Features** panel bar.

The **Thread** dialog box will appear. See Figures 6-4 and 6-6.

2. Click on the existing hole.

The threads will automatically be created to match the hole's diameter.

3. Click **OK.**

Figure 6-10 shows the resulting threaded hole.

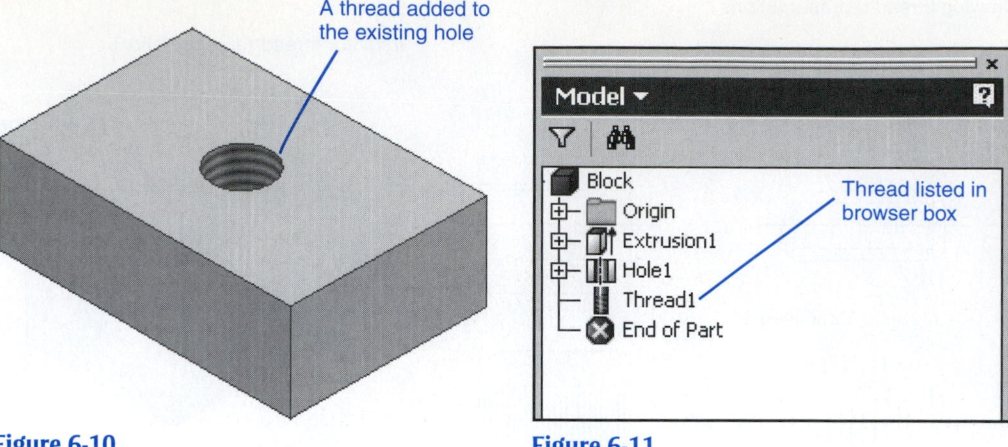

Figure 6-10

Figure 6-11

When a thread is added to an existing hole, a thread listing will be included in the browser box. See Figure 6-11. The listing will confirm that a thread has been added, but it will not define the size or type of thread.

Exercise 6-2: Determining the Thread Size

1. Right-click on the **Hole** listing in the browser box.

A dialog box will appear. See Figure 6-12.

2. Select the **Show Dimensions** option.

Figure 6-13 shows the resulting dimensions. The dimensions define the hole's diameter, and because Inventor will match the thread size to the existing hole diameter, the thread is an M6.

Right-click the mouse and select the Show Dimensions option.

Repeat Thread
Copy Ctrl+C
Delete
Show Dimensions
Move Feature [M]
Edit Sketch
Edit Feature
Infer iMates
Create Note
Suppress Features
Adaptive
Expand All Children
Collapse All Children
Find in Window End
Properties
Help Topics...

Click here.

Figure 6-12

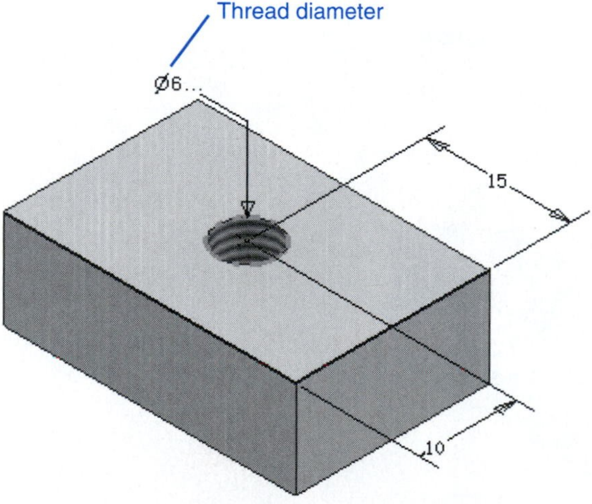

Figure 6-13

THREADED BLIND HOLES

The internally threaded holes presented in the last section passed completely through the material. This section shows how to draw holes that do not pass completely through the object but have a defined depth.

Figure 6-14 shows a tapped hole. It is drawn using the simplified representation. Note that there are three separate portions to the hole representation: the threaded portion, the unthreaded portion, and the 120° conical point. Only the threaded portion of the tapped hole is used to define the hole's depth. The unthreaded portion and the conical point are shown but are not included in the depth calculation.

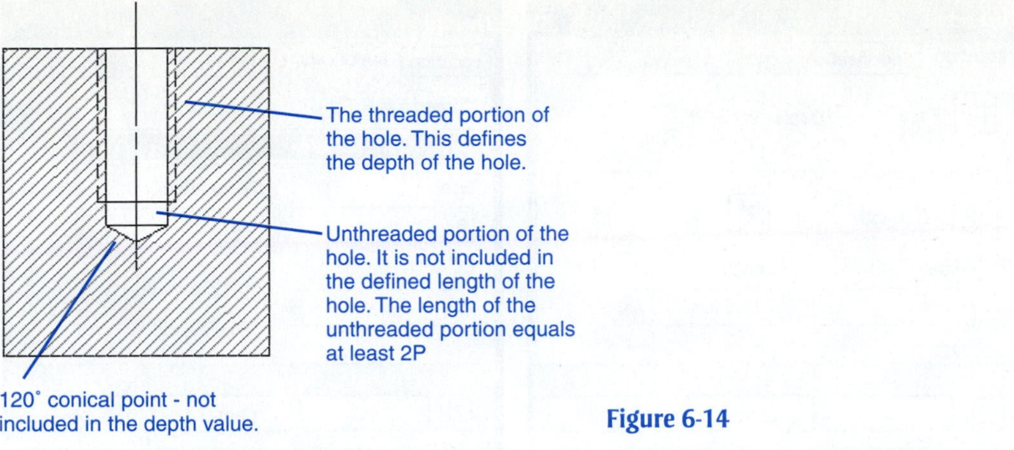

Figure 6-14

A tapped hole is manufactured by first drilling a pilot hole that is slightly smaller than the major diameter of the threads. The threads are then cut into the side of the pilot hole. The tapping bit has no cutting edges on its bottom surface, so if it strikes the bottom of the hole, the bit can be damaged. Convention calls for the unthreaded portion of the pilot hole to extend about the equivalent of two thread pitches (2P) beyond the end of the threaded portion. The conical portion is added to the bottom surface of the pilot hole.

Drawing a Threaded Blind Hole—Metric

Figure 6-15 shows an object with an existing Ø8 × 16 deep hole. Inventor will automatically apply an M8 thread to the Ø8 hole. From Figure 6-24 the pitch for an M8 thread is 1.25 mm per thread for a coarse thread.

A distance equal to two pitches (2P) is recommended between the bottom of the hole and the end of the threads. One pitch equals 1.25 mm, so 2P = 2.50 mm. The existing hole is 16 mm, so the thread depth is 13.5 mm.

1. Click on the **Thread** tool on the **Part Features** panel bar.

The **Thread** dialog box will appear. See Figure 6-16.

2. Set the **Length** for **13.5** and the **Specification** for an **M8 × 1.25.**
3. Click the directional button located in the middle of the dialog box to ensure that the thread starts at the top surface plane.

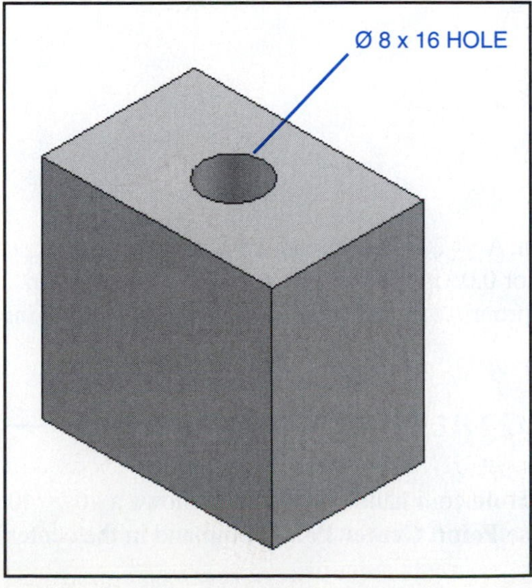

Figure 6-15

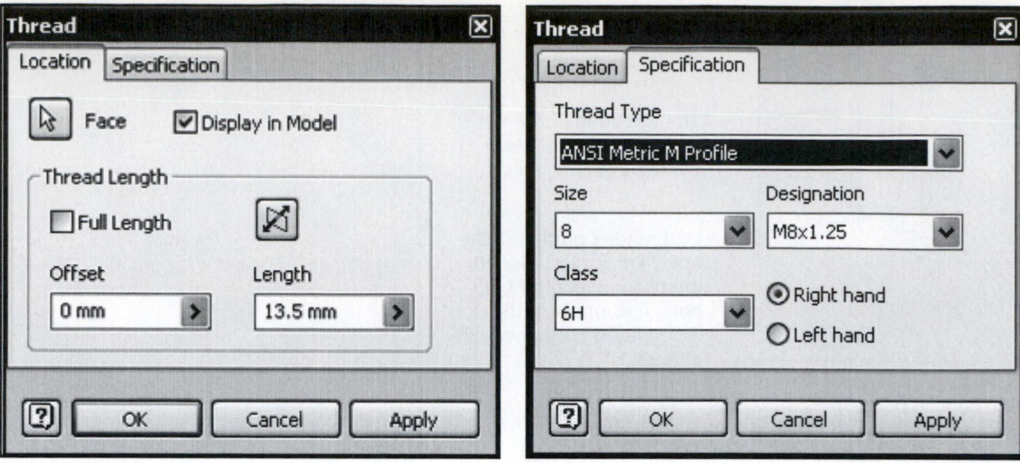

Figure 6-16

Figure 6-17 shows a sectional view of the threaded blind hole. Note how hidden lines are used to represent threads in both the top and the section views.

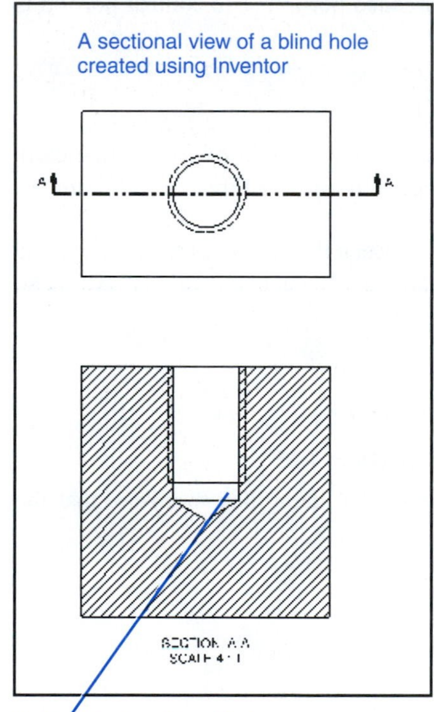

Figure 6-17

Drawing a Blind Hole—ANSI Threads

Pitch for ANSI threads is defined as threads per inch. A ¼-20 UNC thread has a pitch of 20 threads per inch. The length in inches of one thread is 1/20 or 0.05 in. Therefore, 2P = 2(0.05) = 0.10 in.

If a block includes a hole 1.50 deep, then the appropriate thread length is 1.50 − 0.10 = 1.40 in.

Creating Threaded Holes Using the Hole Command

Threaded holes may be created directly using the **Hole** command. Figure 6-18 shows a 40 × 40 × 30 mm block with a hole center defined using the **Point, Center Point** command in the center of the top surface.

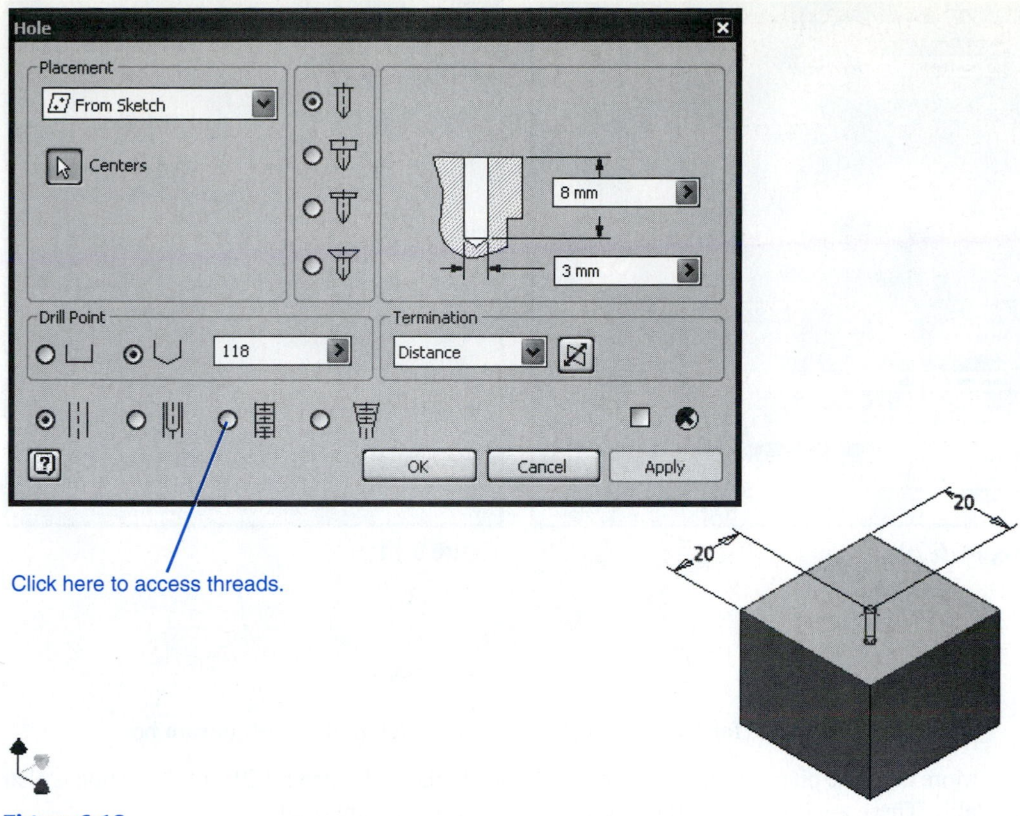

Figure 6-18

Exercise 6-3: Creating a Threaded Through Hole

1. Click on the **Hole** command on the **Part Features** panel.

The **Hole** dialog box will appear. See Figure 6-18.

2. Click the **Termination** box and select the **Through All** option.
3. Click the **Tapped Hole** button.
4. Click on **Thread Type.** Select the **ANSI Metric M Profile** option.

See Figure 6-19.

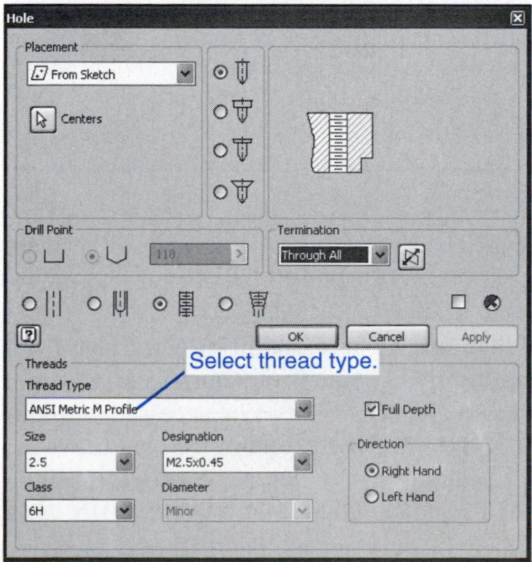

Figure 6-19

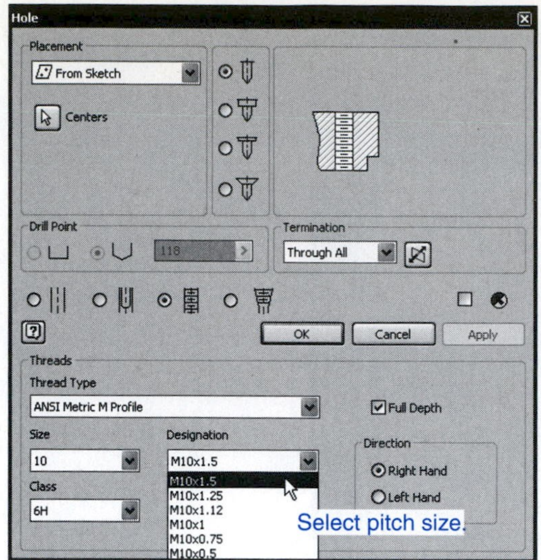

Figure 6-20

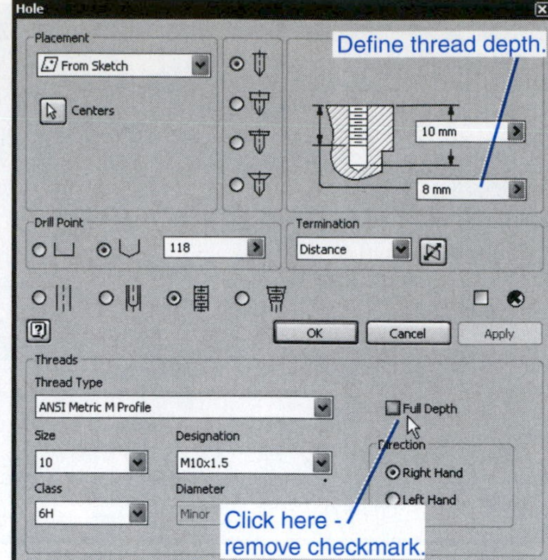

Figure 6-21

5. Set the thread **Size** for **10.** Select the **M10 × 1.5** pitch in the **Designation** box.

More than one pitch size is available. In addition to the 1.50 pitch a 1.25 or 0.75 option is also available. These are Fine and Extra Fine designations. See Figure 6-20.

6. Click **OK.**

Exercise 6-4: Creating a Blind Threaded Hole

Create a 40 × 40 × 30 mm block with hole center defined in the center of the top surface.

1. Click on the **Hole** command.

The **Hole** dialog box will appear. See Figure 6-21.

2. Set the **Termination** for **Distance,** and click the **Tapped Hole** button.
3. Set the **Size** for **10** and the **Designation** for **M10 × 1.5.**
4. Click the **Full Depth** box so that no check appears.
5. Set the thread depth for **8** and the hole depth for **10.**

The hole depth should, with rare exceptions, be greater than the thread depth.

6. Click **OK.**

STANDARD FASTENERS

Fasteners, such as screws and bolts, and their associated hardware, such as nuts and washers, are manufactured to standard specifications. Using standard-sized fasteners in designs saves production costs and helps assure interchangeability.

Inventor includes a library of standard parts that may be accessed using the **Place from Content Center** tool on the **Assembly Panel.** Clicking on the **Place from Content Center** tool accesses the **Content Center** dialog box. The **Content Center** may also be accessed by right-clicking the mouse and selecting the **Place from Content Center** option. See Figure 6-22. Figure 6-23 shows the **Place from Content Center** dialog box. Click the **Fastener** heading under **Category View.** Click **Fasteners, Bolts, Hex Head, Shoulder,** and select the **DIN 7968** shoulder bolt. A listing of available diameters and lengths will appear. See Figure 6-24. Click the **Table View** tab to access a table of sizes and dimensions that apply to the selected bolt. See Figure 6-25.

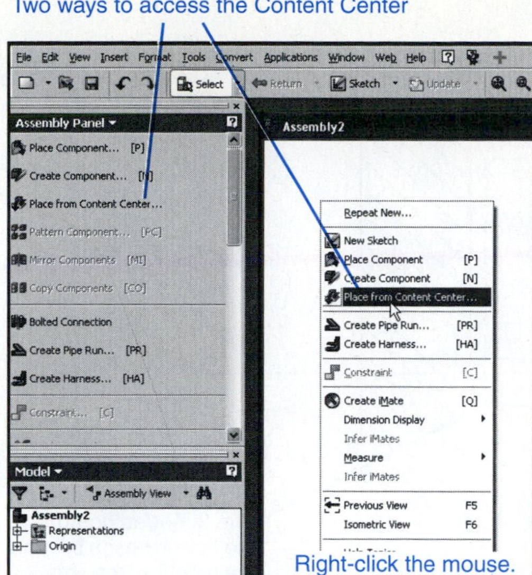

Figure 6-22

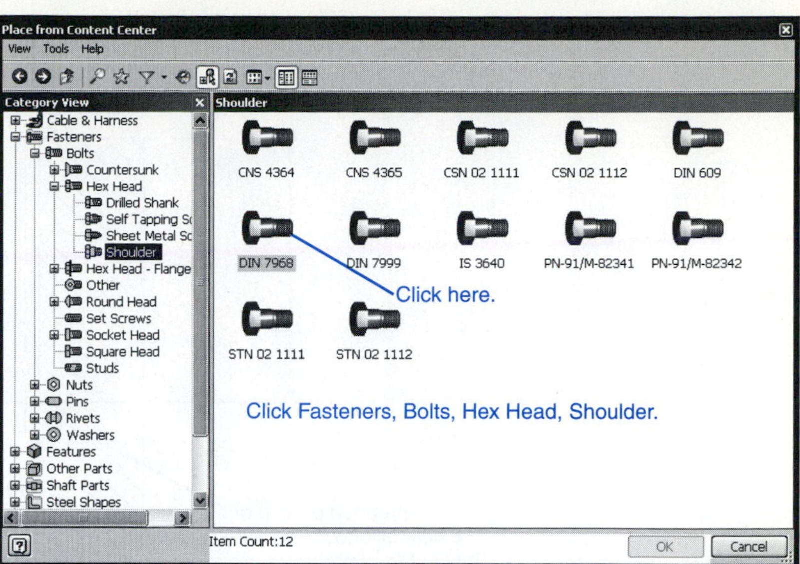

Figure 6-23

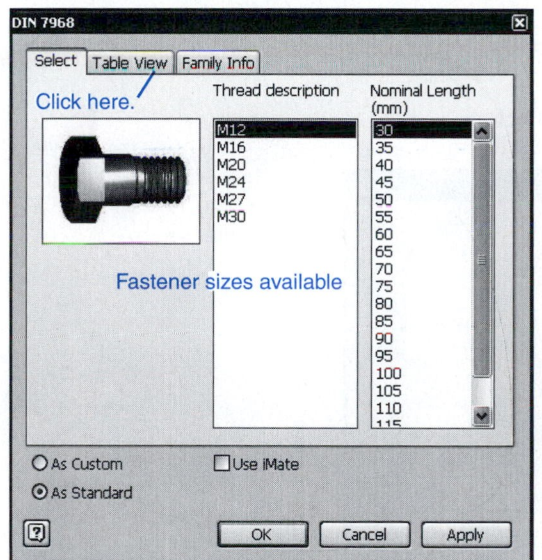

Figure 6-24

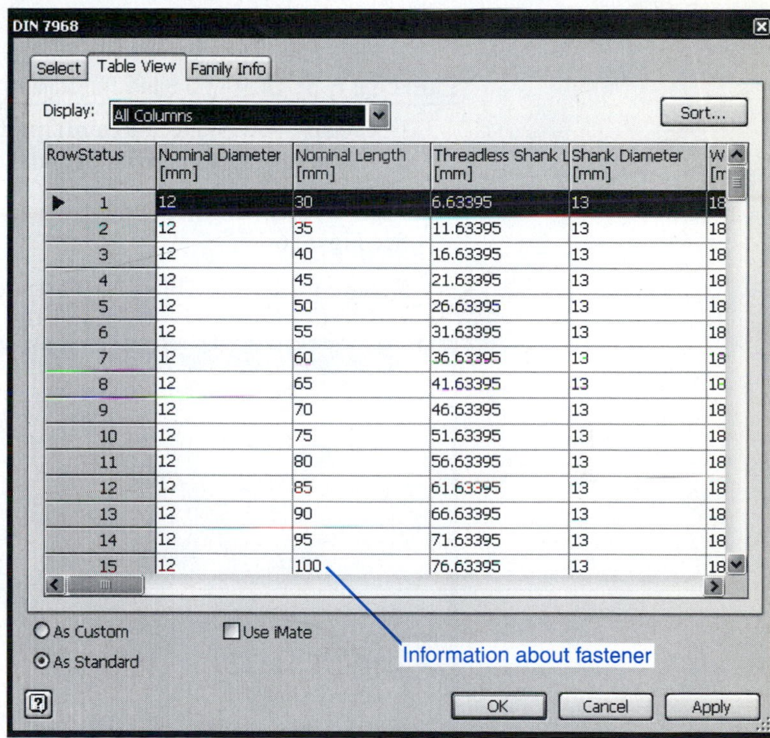

Figure 6-25

Sizing a Threaded Hole to Accept a Screw

Say we wish to determine the length of thread and the depth of hole needed to accept an M10 × 25 hex head screw and to create a drawing of the screw in the threaded hole. The length of the threaded hole must extended two pitches (2P) beyond the end of the screw, and the untapped portion of the hole must extend two pitches (2P) beyond the threaded portion of the hole. This requirement ensures that all the fastener threads will be in contact with hole's threads and that fasteners will not bottom out. In this example the thread pitch equals 1.50 mm. Therefore, 2P = 2(1.50) = 3.0 mm. This is the minimum distance and can be increased but never decreased. See Figure 6-26.

The two-pitch length requirement for the distance between the end of the screw and the end of the threaded portion of the hole determines that the thread depth should be 25.0 + 3.0 = 28.0 mm. The two-pitch length requirement between the end of the threaded portion of the hole and the bottom of the hole requires that the hole must have a depth of 28.0 + 3.0, or 31.0 mm.

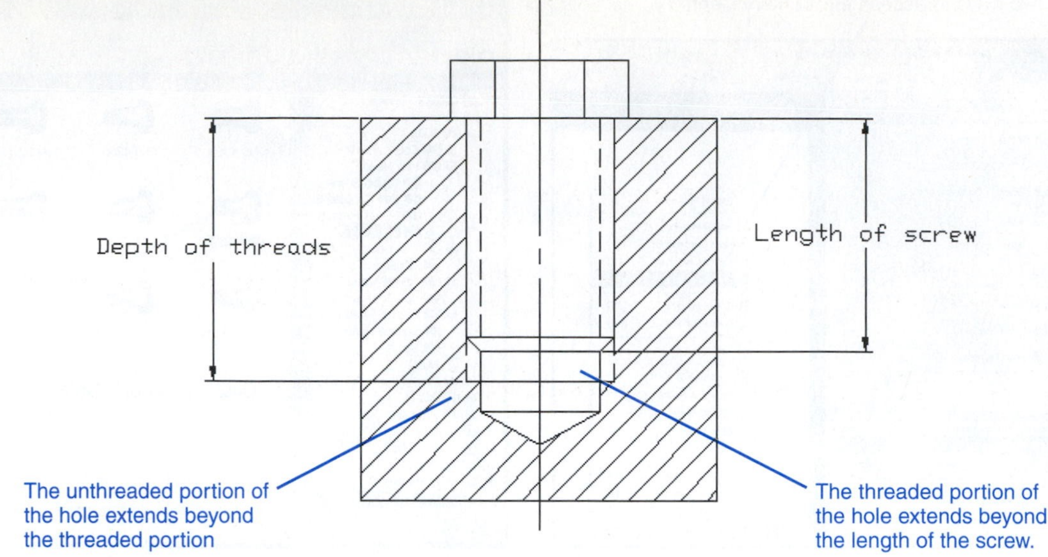

Figure 6-26

It is important that complete hole depths be specified, as they will serve to show any interference with other holes or surfaces.

Exercise 6-5: Drawing a Blind Threaded Hole

1. Create a new **Standard (mm).ipt** drawing and create a 40 × 40 × 60 block.
2. Use the **Point, Center Point** and the **Hole** commands to create an **M10 × 28** deep thread and a **31** deep hole.

See Figure 6-27.

3. Save the threaded block as **ThreadedBlock.**

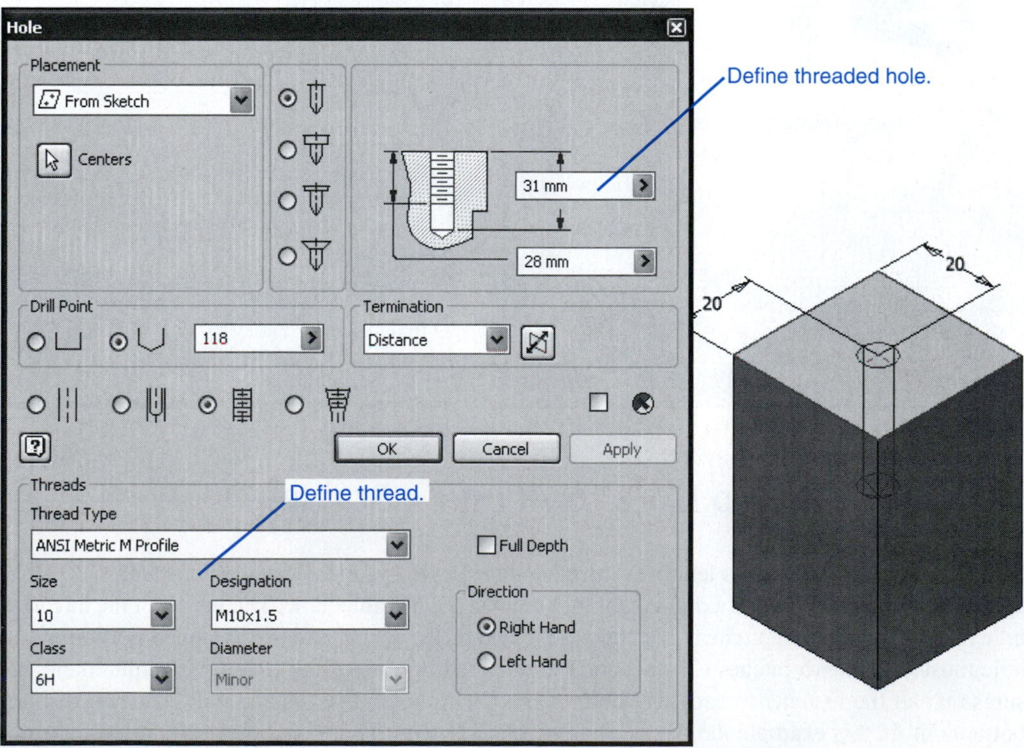

Figure 6-27

Exercise 6-6: Accessing an M10 × 25 Hex Head Bolt

1. Create a new **Standard (mm).iam** drawing called **M10ASSEMBLY.**
2. Use the **Place Component** command and locate one copy of the **ThreadedBlock** block on the drawing screen.

3. Click the **Place from Content Center** tool.

The **Place from Content Center** dialog box will appear. See Figure 6-28.

4. Click the plus sign to the left of the **Fasteners** option.

See Figure 6-29.

5. Click the **Bolts** option, then click the **Hex Head** option.
6. Scroll through the options and select the **AS 1110-Metric** option.

A picture of the AS 1110-Metric fastener will appear in the dialog box along with a listing of available thread sizes and lengths. See Figure 6-30.

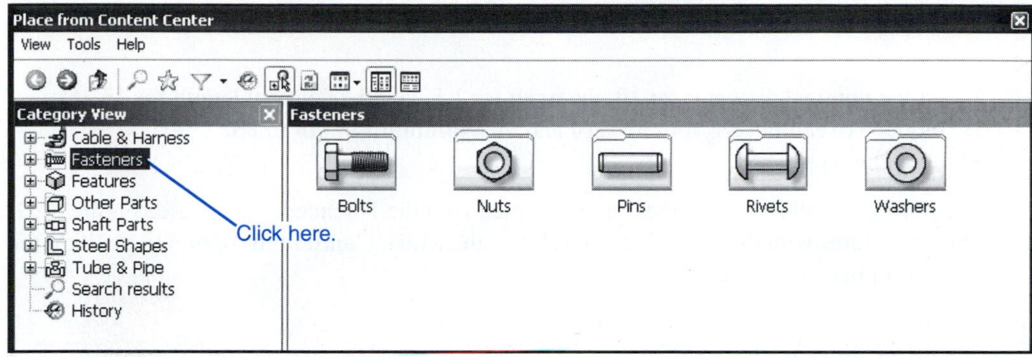

Figure 6-28

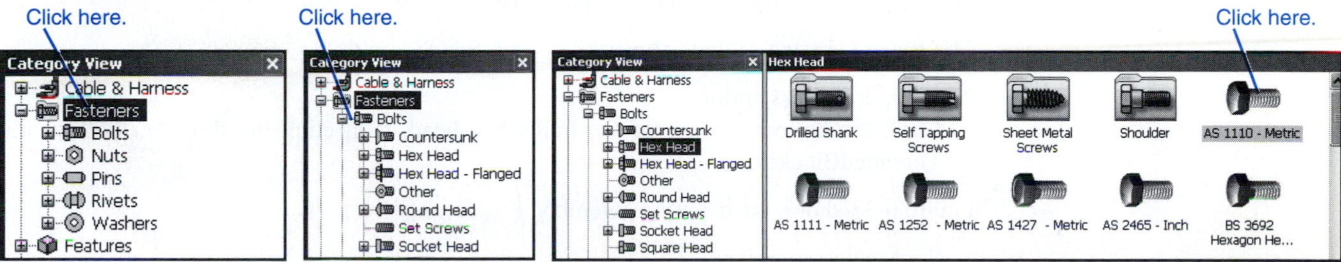

Figure 6-29

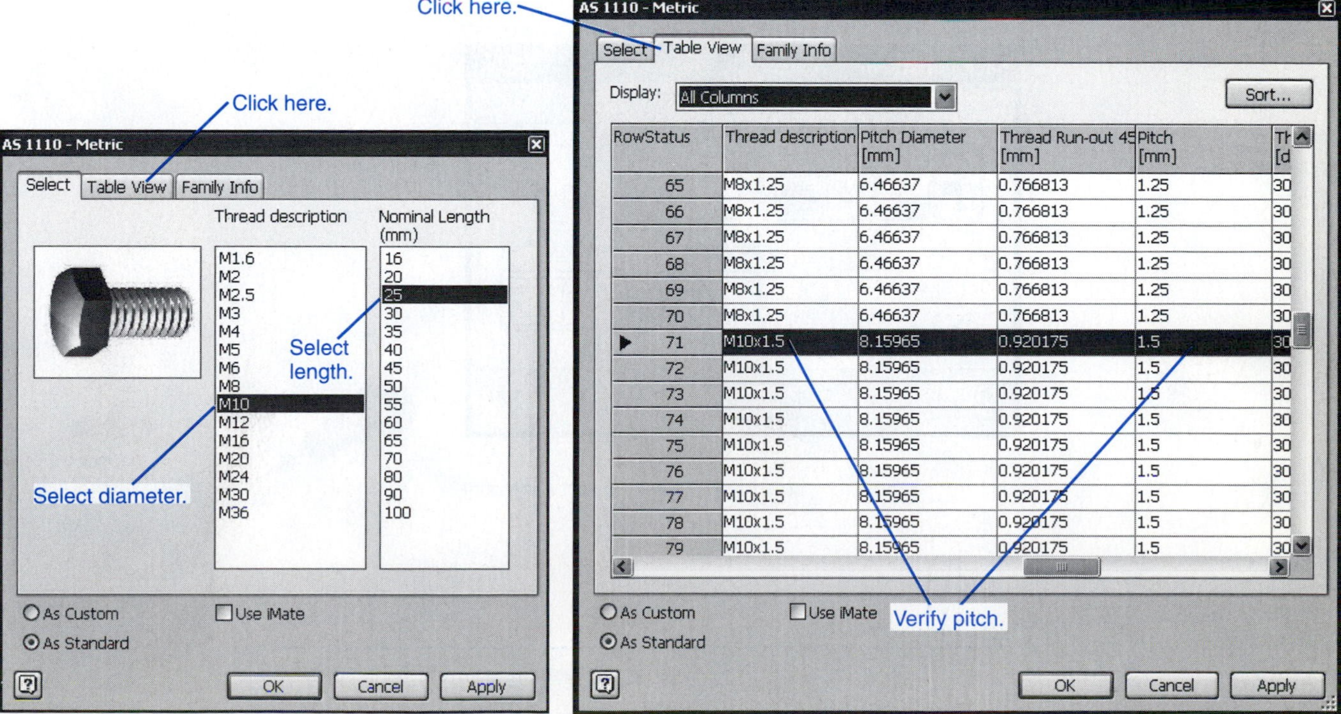

Figure 6-30

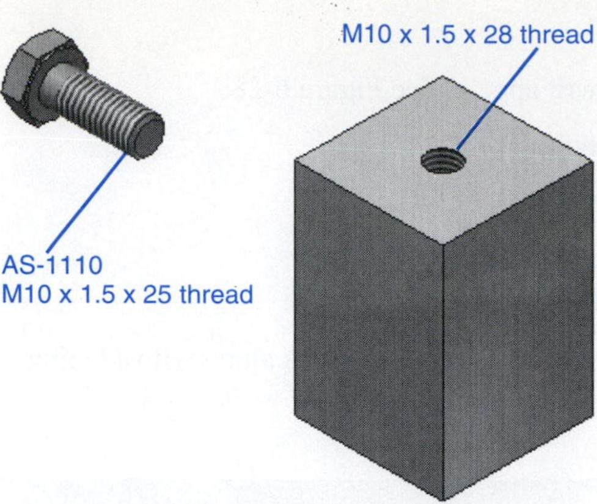

Figure 6-31

7. Set the nominal diameter for **10,** the pitch for **1.5,** and the nominal length for **25.**
8. Set the **Thread description** to **M10** and the **Nominal Length** to **25.**
9. Click the **Apply** box.

The M10 bolt will appear on the drawing screen with the ThreadedBlock:1. See Figure 6-31. If the bolt interferes with the ThreadedBlock:1, use the **Move Component** option to position the bolt away from the block. See Figure 6-31.

Exercise 6-7: Inserting the M10 Bolt

1. Click the **Constraint** tool on the panel bar.

The **Place Constraint** dialog box will appear. See Figure 6-32.

2. Click the **Insert** option
3. Click the bottom surface of the bolt's hex head, then click the threaded hole in the ThreadedBlock:1.

Figure 6-33 shows the resulting assembly.

4. Save the assembly.

Figure 6-34 shows a top view and a sectional view of the M10ASSEMBLY created using the **ISO.idw** option. Note the open area between the bottom of the M10 bolt and the bottom of the hole.

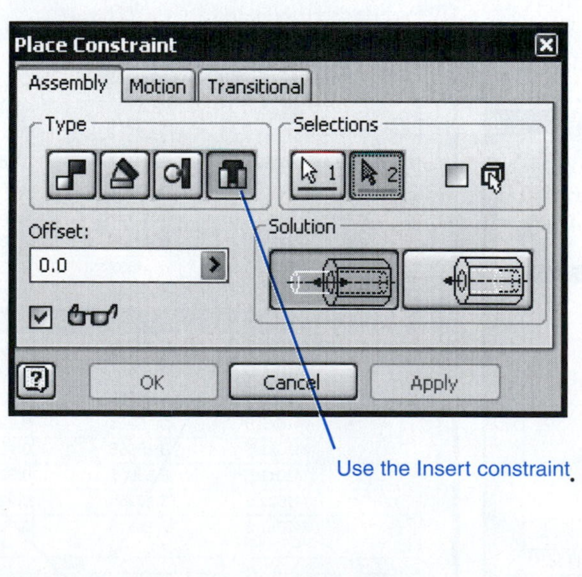

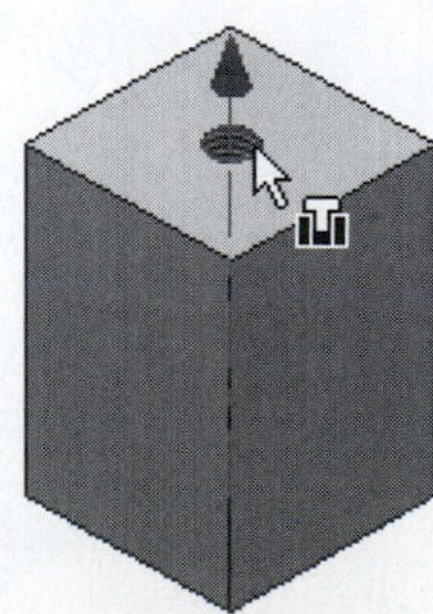

Figure 6-32

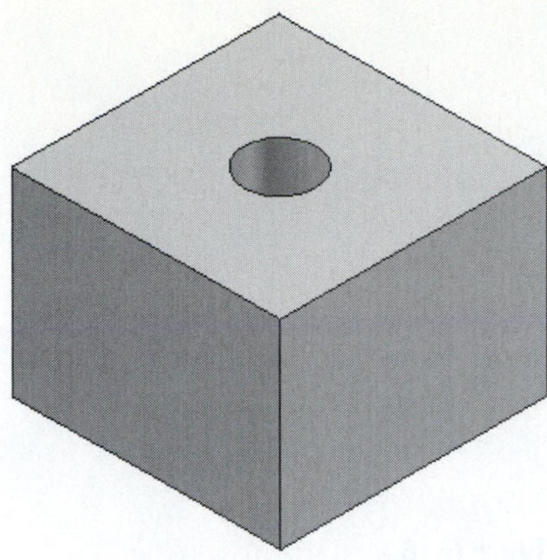

Figure 6-33

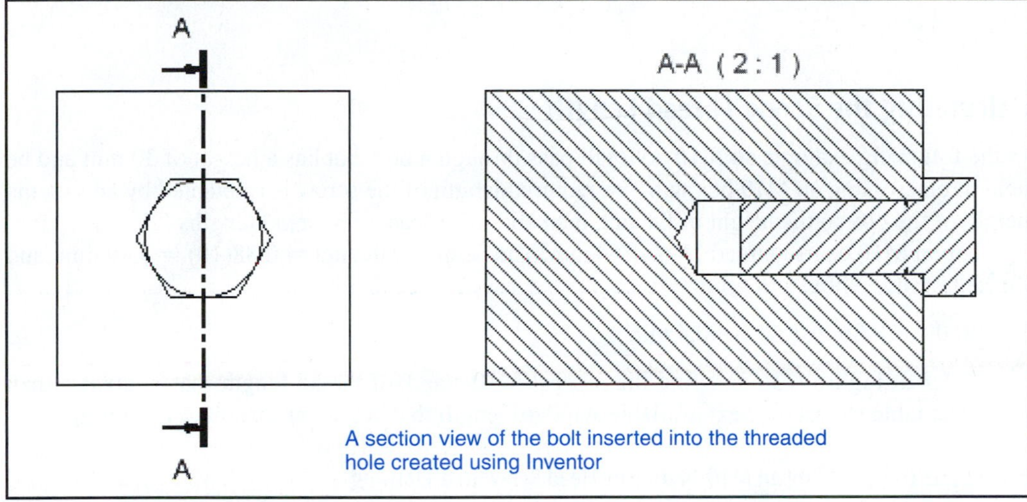

Figure 6-34

Screws and Nuts

Screws often pass through an object or group of objects and are secured using nuts. The threads of a nut must match the threads of the screw.

Nuts are manufactured at a variety of heights depending on their intended application. Nuts made for heavy loads are thicker than those intended for light loads. In general, nut thickness can be estimated as 0.88 of the major diameter of the thread. For example, if the nut has an M10 thread, the thickness will be about 0.88(10) = 8.8 mm.

It is good practice to specify a screw length that allows for at least two threads beyond the nut. This will ensure that all threads of the nut are in contact with the screw threads. Strength calculations for screw threads are based on 100% contact with the nut. See Figure 6-35.

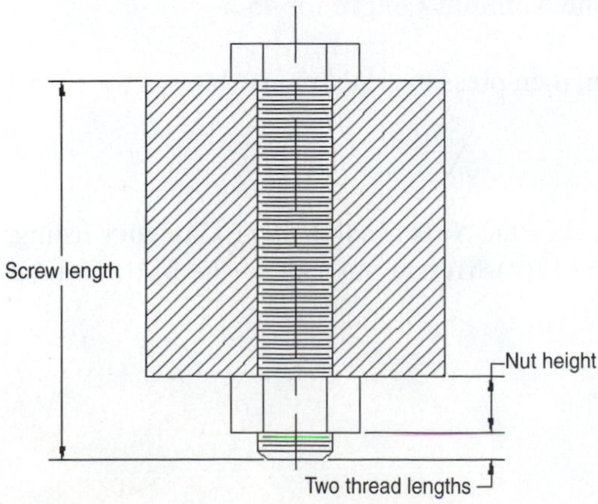

Figure 6-35

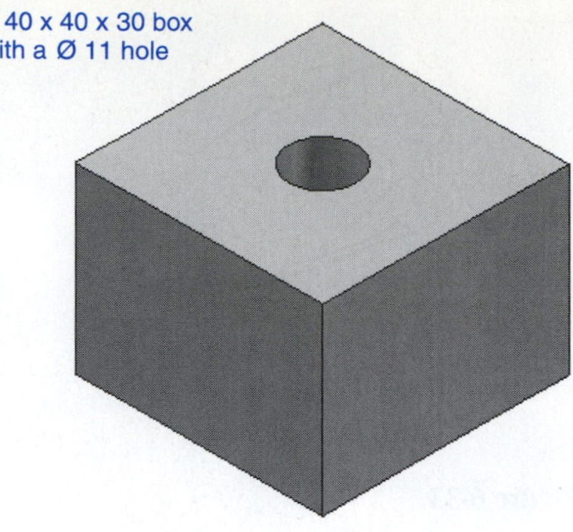

Figure 6-36

Calculating the Screw Thread Length

In the following example an M10 bolt will pass through a box that has a height of 30 mm and be held in place using an M10 hex nut. The required length of the screw is calculated by adding the height of the box to the height of the nut, then adding at least two thread lengths (2P).

For an M10 coarse thread 2P = 1.60 mm. The height of the nut = 0.88(10) = 8.80 mm, and the height of the box = 30 mm:

30.00 + 8.80 + 1.60 = 40.4 mm

Refer to Figure 6-30 and find the nearest M10 standard thread length that is greater than 40.4. The table shows the next available standard length that is greater than 40.4 is 45 mm.

Exercise 6-8: Adding an M10 × 45 Hex Head Screw to a Drawing

1. Draw a **40 × 40 × 30** box.
2. Locate a **Ø11** hole in the center of the top surface of the box.

The hole does not have threads. It is a clearance hole and so should be slightly larger than the M10 thread. See Figure 6-36.

3. Save the block as **Ø11BLOCK.**

Exercise 6-9: Adding a Bolt and Nut to the Drawing

1. Start a new assembly drawing called **M10NUT.** Use the **Standard (mm).iam** format.
2. Use the **Place Component** tool and place a copy of the Ø11BLOCK on the drawing screen.
3. Use the **Save All** command on the **File** pull-down menu to save and name the new assembly.
4. Click the **Place from Content Center** tool.
5. Click the **Bolt** option, the **Hex Head** option, and the **AS 1110-Metric** bolt.
6. Set the **Thread description** for **M10** and the **Nominal Length** for **45.**
7. Click the **Apply** box.
8. Locate the bolt, press the left mouse button, then press the right mouse button. Select the **Done** option.

See Figure 6-37.

9. Click the **Place from Content Center** tool, click the **Nuts** heading in the **Category** listing, click the **Hex** option, then select the **AS1112(4)-Metric** nut. Select the **M10 Thread description.**

See Figure 6-38.

10. Click the **Apply** box.

See Figure 6-39.

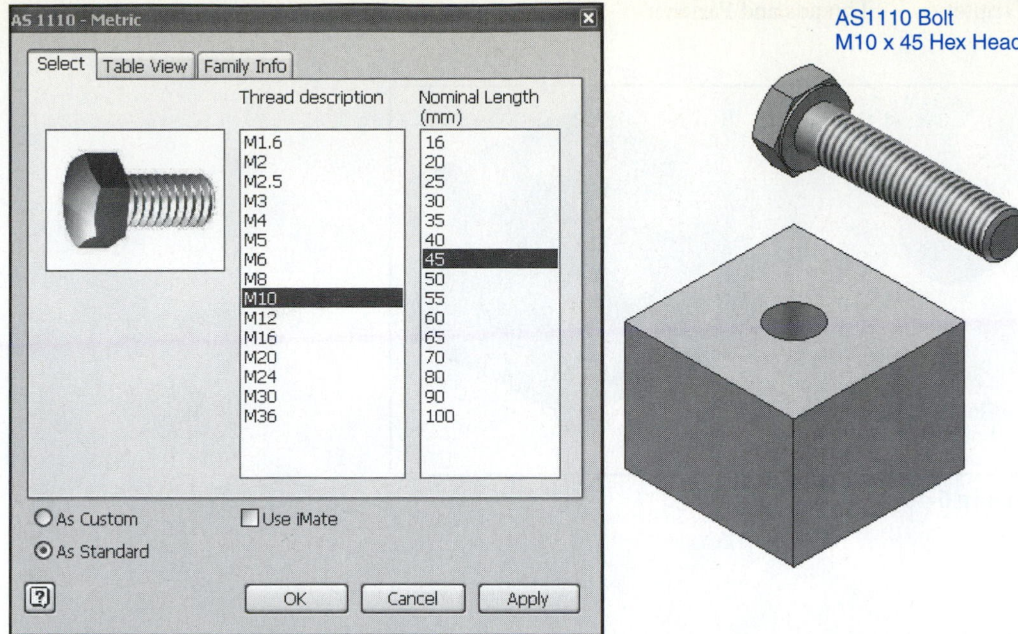

Figure 6-37

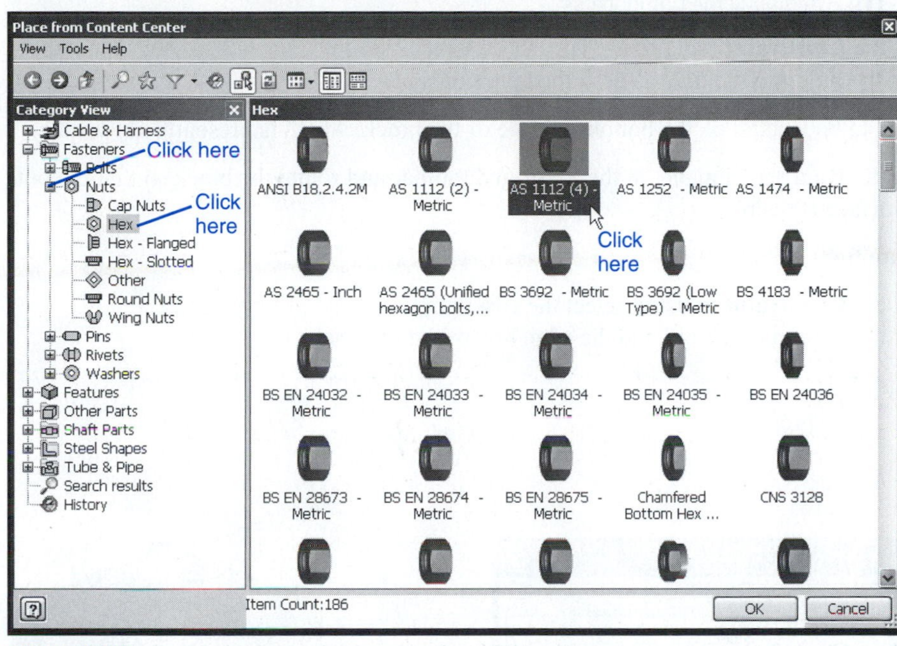

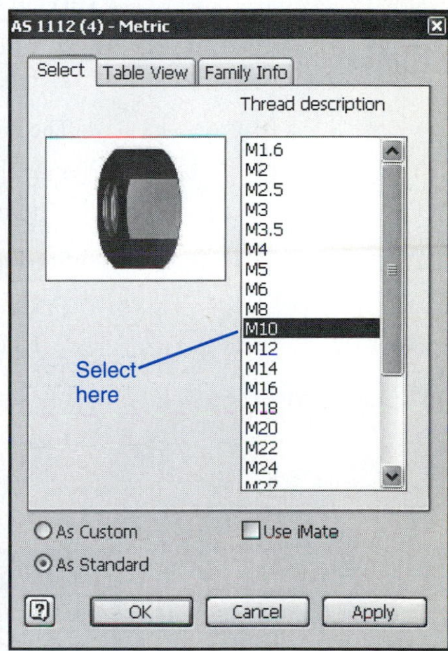

Figure 6-38

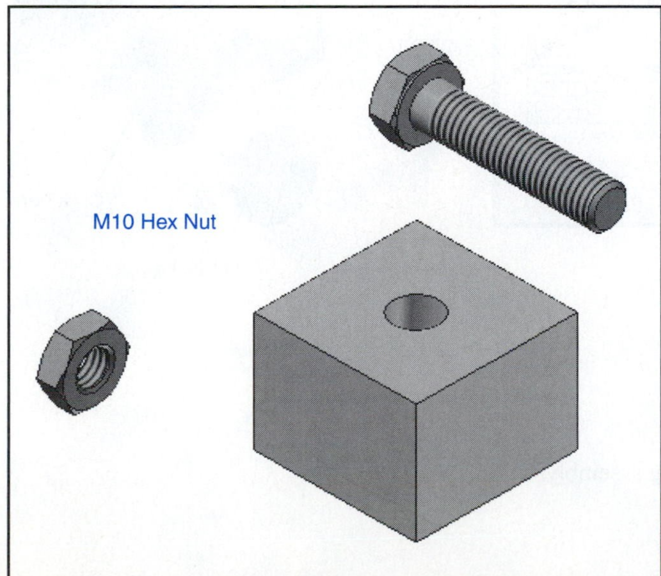

Figure 6-39

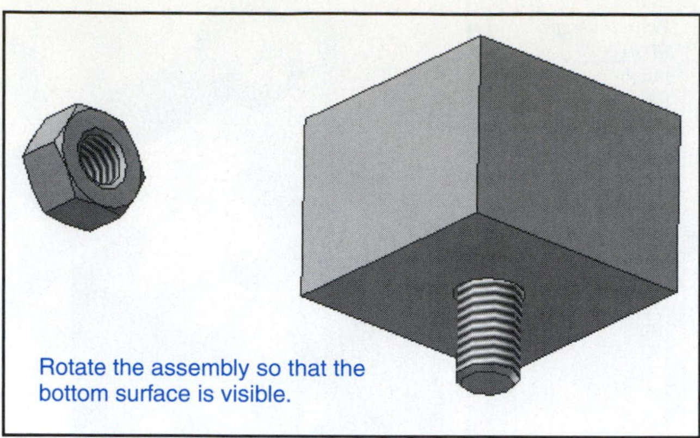

Figure 6-40

Exercise 6-10: Assembling the Components

1. Click the **Constraint** tool.
2. Insert the bolt into the hole as described previously.

The nut is to be located on the bottom surface of the block, which is presently not visible.

3. Click the **Rotate** command on the **Standard** toolbar and rotate the block so that the bottom surface is visible.

See Figure 6-40.

4. Click the **Constraint** tool and select the **Insert** option.
5. Click the bottom of the nut and the edge line of the Ø11 hole.

See Figure 6-41.

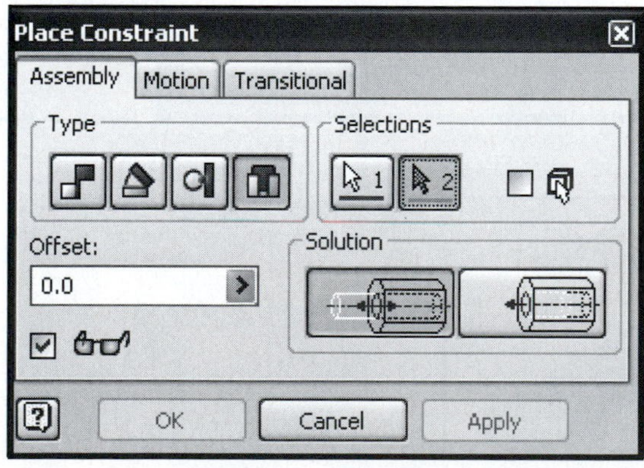

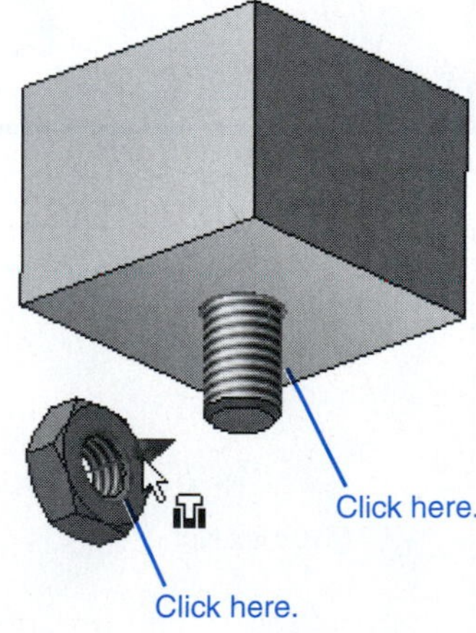

Figure 6-41

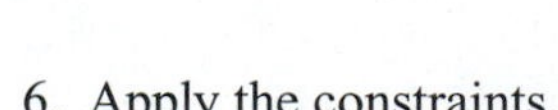

6. Apply the constraints.

Figure 6-42 shows the resulting assembly.

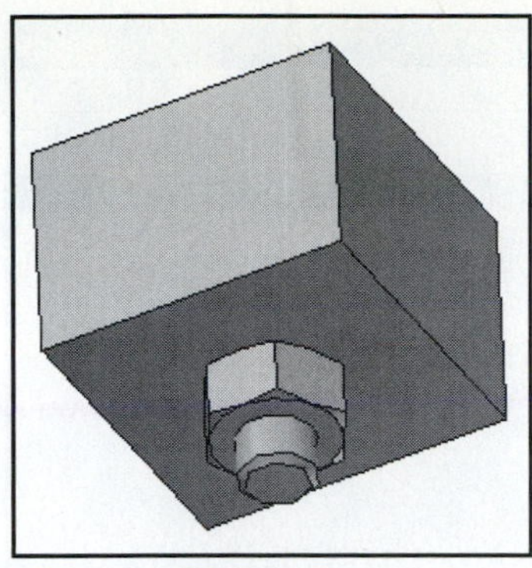

Resulting assembly

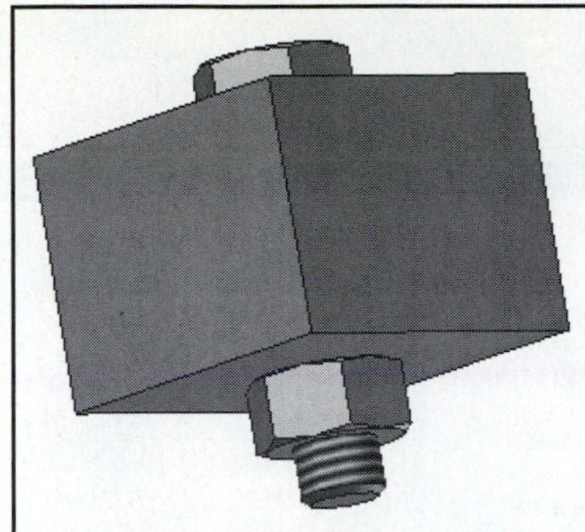

Figure 6-42

TYPES OF THREADED FASTENERS

There are many different types of screws generally classified by their head types. Figure 6-43 shows six of the most commonly used types.

The choice of head type depends on the screw's application and design function. A product design for home use would probably use screws that had slotted heads, as most homes possess a blade screwdriver. Hex head screws can be torqued to higher levels than slotted pan heads but

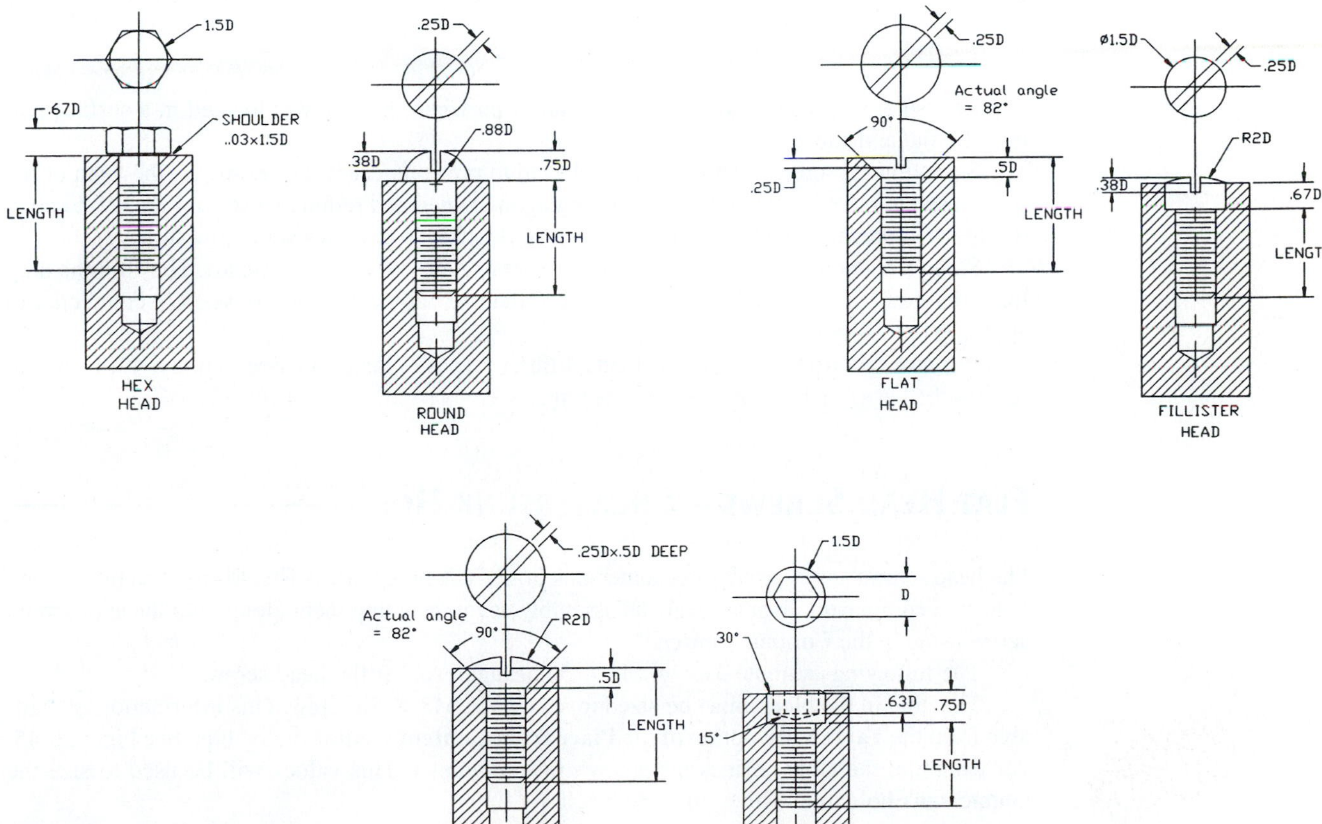

Note:
The dimensions listed are for reference only. See manufacturer's specifications for the actual sizes.

Figure 6-43

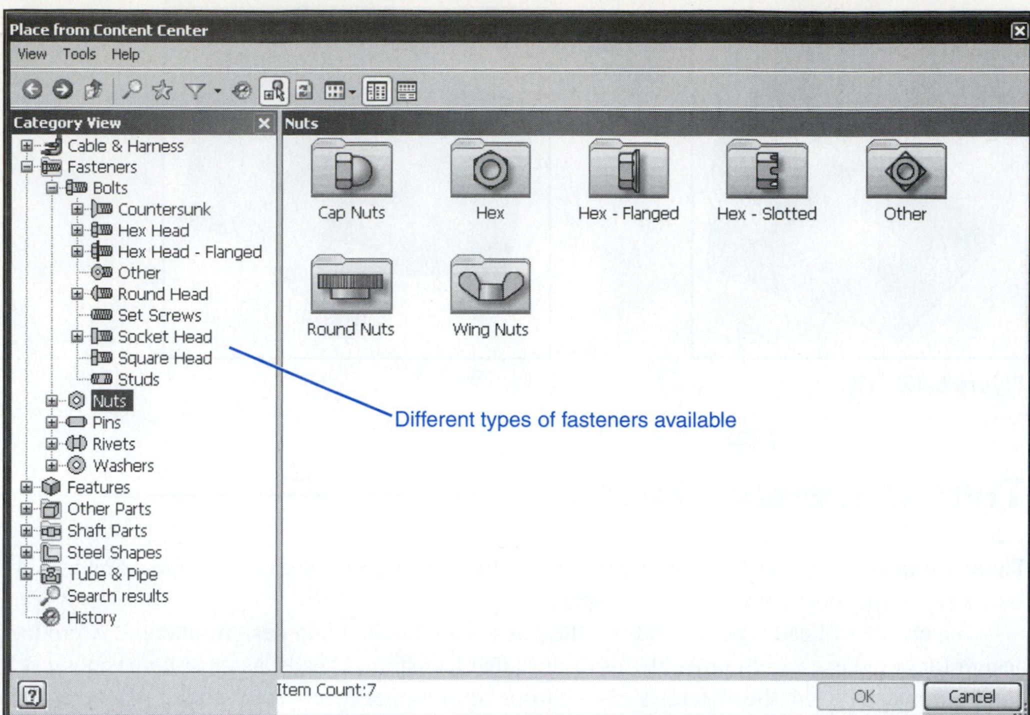

Figure 6-44

require socket wrenches. Flat head screws are used when the screw is located in a surface that must be flat and flush.

Sometimes a screw's head shape is selected to prevent access. For example, the head of the screw used to open most fire hydrants is pentagon-shaped and requires a special wrench to open it. This is to prevent unauthorized access that could affect a district's water pressure.

Screw connections for oxygen lines in hospitals have left-handed threads. They are the only lines that have left-handed threads, to ensure that no patient needing oxygen is connected to anything but oxygen.

Inventor's **Content Center** lists many different types of fasteners. See Figure 6-44. There are many subfiles to each of the fastener headings.

Flat Head Screws—Countersunk Holes

Flat head screws are inserted into countersunk holes. The procedure is first to create a countersunk hole on a component, then to create an assembly using the component along with the appropriate screw listed in the **Content Center.**

The following example uses an M8 × 50 hexagon socket flat head screw.

The hole in the block must be sized to accept the M8 × 50 screw. This information is available from the **Table View** option of the **Place from Content Center** dialog box. See Figure 6-45. For example, the table defines the screw's head diameter. This values will be used to size the countersunk hole in the 40 × 40 × 80 block.

Exercise 6-11: Creating a Countersunk Hole

1. Create a **40 × 40 × 80** block.
2. Locate a hole's center point in the center (20 × 20) of the top surface of the block using the **Point, Center Point** command.
3. Go to the **Part Features** panel bar and click the **Hole** tool.
4. Click the **Countersink** and **Tapped** boxes.

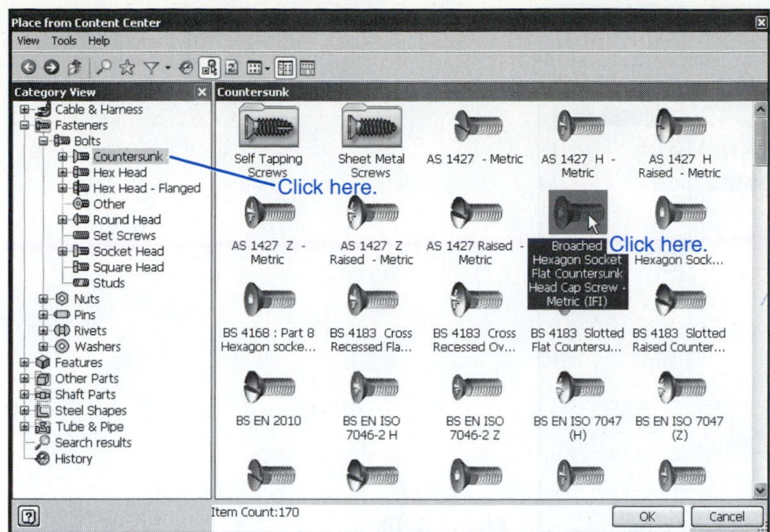

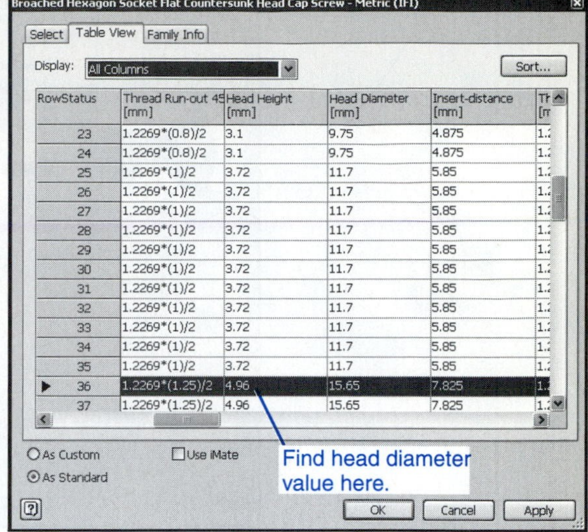

Figure 6-45

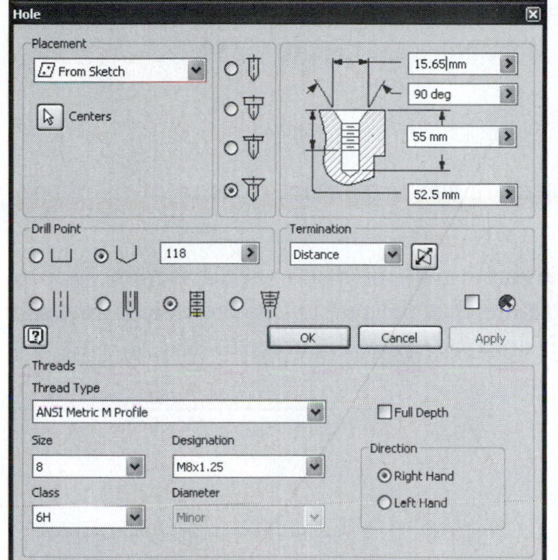

Figure 6-46

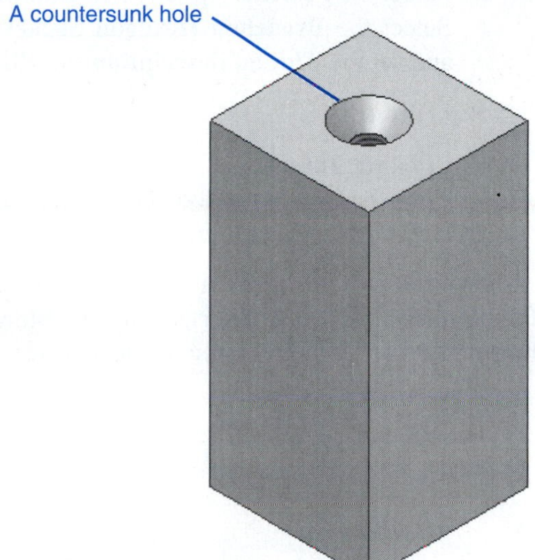

Figure 6-47

See Figure 6-46.

The pitch length of an M8 thread is 1.25, so two thread lengths (2P) equals 2.50.

The hole's threads must be at least 50.00 + 2.50 = 52.50, and the pilot hole must be at least 52.50 + 2.50 = 55.00 deep.

5. Click the **Full Depth** box (remove check mark) and set the **Thread Type** option for **ANSI Metric M Profile,** the thread depth for **52.5,** the hole depth for **55,** and the head diameter for **15.65.**

The value 15.65 came from the **Table View** portion of the **Place from Content Center** dialog box. Figure 6-47 shows the countersunk hole located in the 40 × 40 × 80 block.

6. Save the block.

Exercise 6-12: Creating and Inserting a Flat Head Screw

1. Create an assembly drawing using the **Standard (mm).iam** format.
2. Use the **Place Component** tool and locate one copy of the block on the drawing screen.
3. Use the **Save all** command to save and name the assembly.

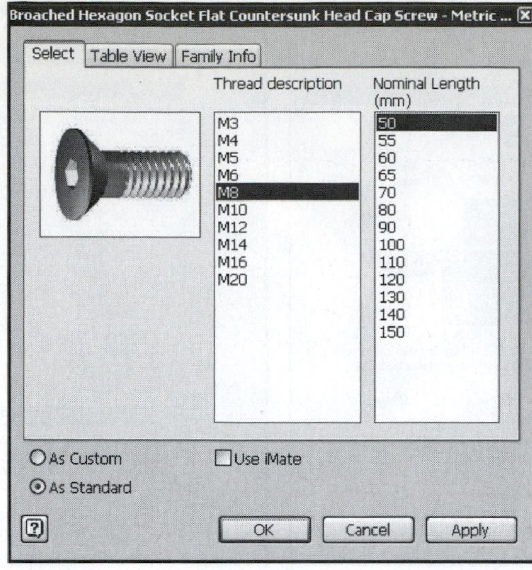

Figure 6-48

4. Click the **Place from Content Center** tool.
5. Select the **Fastener** option, then **Countersink.**
6. Select the **Broached Hexagon Socket Flat Countersunk Head Cap Screw—Metric** and set the **Thread description** for **M8** and the **Nominal Length** for **50.**

See Figure 6-48.

7. Click the **Insert** box.
8. Click the **Constraint** tool, then select the **Insert** option on the **Place Constraint** dialog box.
9. Insert the screw into the block.

Figure 6-49 shows the resulting assembly. Figure 6-50 shows a top and a section view of the countersunk screw inserted into the block. Note that the portion of the hole below the bottom of the M8 × 50 screw is clear; that is, it does not show the unused threads. This is a

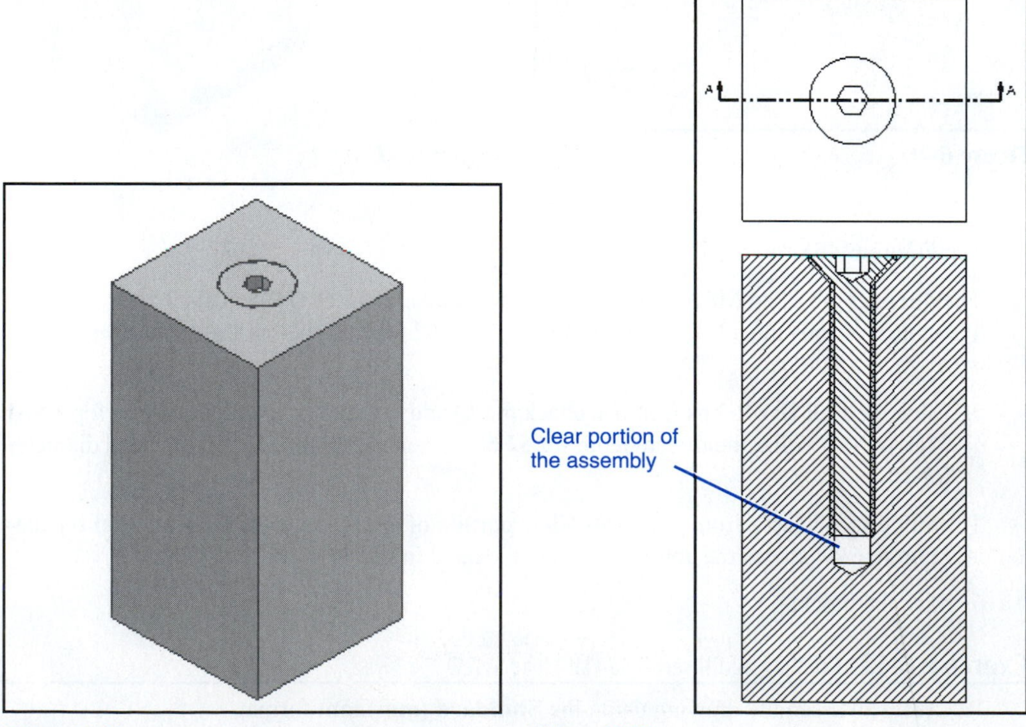

Figure 6-49

Figure 6-50

drawing convention that is intended to add clarity to the drawing. Inventor will automatically omit the unused threads.

COUNTERBORES

A counterbored hole is created by first drilling a hole, then drilling a second larger hole aligned with the first. Counterbored holes are often used to recess the heads of fasteners.

Say we wish to fit a 3/8-16 UNC × 1.50 LONG hex head screw into a block that includes a counterbored hole, and that after assembly the head of the screw is to be below the surface of the block.

Determining the Counterbore Depth

Figure 6-51 shows the **Table View** portion of the **Place from Content Center** dialog box. It lists the head height and distance across the flats for a ⅜-16 UNC (AS 2465) hex head screw. Other values can be obtained from manufacturers' catalogs, many of which are posted on the Web, or by using the approximations presented in Figure 6-43.

In this example the head height is .243 in. (See Figure 6-51.) The counterbore must have a depth greater than the head height. A distance of .313 (5/16) was selected.

Determining the Thread Length

The screw is 1.50 in. long and has a pitch of 16 threads per inch. Each thread is therefore 1/16 or .0625 in. It is recommended that there be at least two threads beyond the end of the screw. Two thread pitches would be 2(.0625) = .125 in.

The thread depth is 1.50 + .125 or 1.625 in. minimum; however, the thread is created below the counterbore, so for Inventor the value must include the depth of the counterbore. The thread depth is 1.625 + .313 = 1.938 in.

Determining the Depth of the Hole

The hole should extend at least two pitch lengths beyond the threaded portion of the hole, plus the depth of the counterbore, so 1.938 + .125 = 2.063 in.

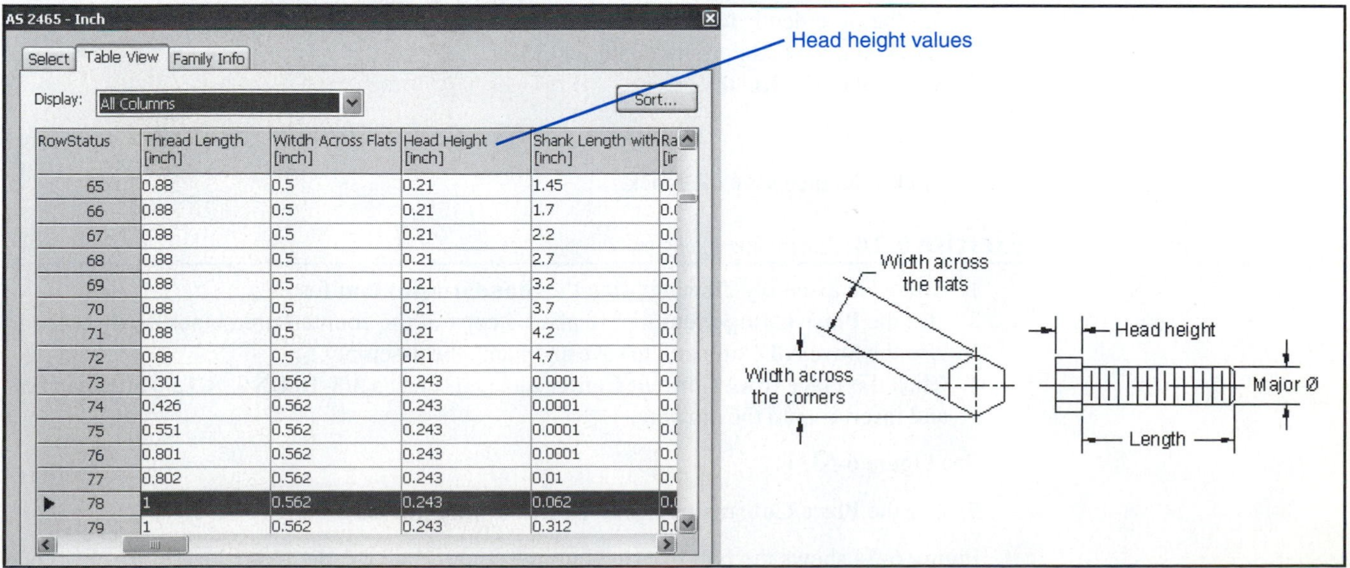

RowStatus	Thread Length [inch]	Witdh Across Flats [inch]	Head Height [inch]	Shank Length with [inch]	Ra [ir
65	0.88	0.5	0.21	1.45	0.(
66	0.88	0.5	0.21	1.7	0.(
67	0.88	0.5	0.21	2.2	0.(
68	0.88	0.5	0.21	2.7	0.(
69	0.88	0.5	0.21	3.2	0.(
70	0.88	0.5	0.21	3.7	0.(
71	0.88	0.5	0.21	4.2	0.(
72	0.88	0.5	0.21	4.7	0.(
73	0.301	0.562	0.243	0.0001	0.(
74	0.426	0.562	0.243	0.0001	0.(
75	0.551	0.562	0.243	0.0001	0.(
76	0.801	0.562	0.243	0.0001	0.(
77	0.802	0.562	0.243	0.01	0.(
▶ 78	1	0.562	0.243	0.062	0.(
79	1	0.562	0.243	0.312	0.(

Figure 6-51

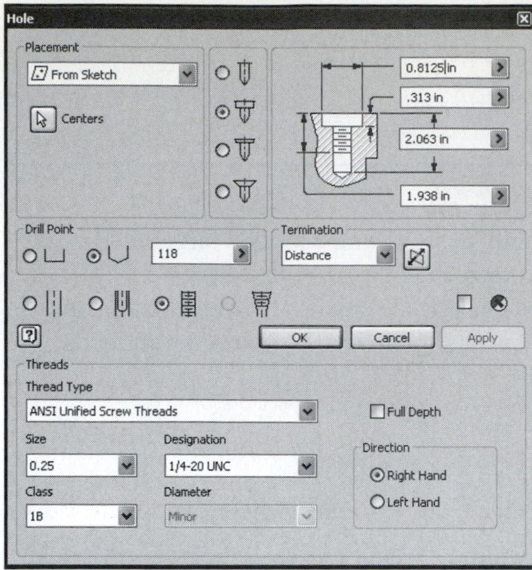

Figure 6-52

Determining the Counterbore's Diameter

The distance across the corners of the screw is listed in the **Table View** portion of the **Place from Content Center** dialog box as .650 in. The counterbored hole must be at least this large plus an allowance for the tool (socket wrench) needed to assemble and disassemble the screw. In general, the diameter is increased 0.125 in. or ⅛ in. [3.2 mm] all around to allow for tooling needs.

If space is a concern, then designers will change the head type of a screw so that the tooling will not add to the required diameter. For example, a socket head type screw may be used if design requirements permit.

The minimum counterbore diameter is .650 + .125 = .775. For this example 0.8125 or 13⁄16 in. was selected.

Exercise 6-13: Drawing a Counterbored Hole

1. Draw a **3.00 × 3.00 × 5.00** block.
2. Locate a hole's center point in the center of the 3.00 × 3.00 surface.
3. Access the **Part Features** menu and click the **Hole** tool.

The **Hole** dialog box will appear.

4. Click the **Counterbore, Tapped,** and **Full Depth** (turned off) bolt.
5. Set the hole depth for **2.063,** the thread depth for **1.938,** the counterbore diameter for **.8125,** and the counterbore depth for **.313.**
6. Select **3/8-16 UNC** threads.

See Figure 6-52.

7. Click **OK,** then save the block.

Exercise 6-14: Assembling the Screw

1. Create an assembly drawing using the **Standard (in).iam** format.
2. Use the **Place Component** tool to place a copy of the counterbored block on the screen.
3. Use the **Save All** command to save and name the assembly.
4. Click the **Place from Content Center** tool and select a **3/8-16 UNC × 1.50** hex head bolt and insert it onto the drawing.

See Figure 6-53.

5. Use the **Place Constraint** tool to insert the bolt into the hole.

Figure 6-54 shows the bolt inserted into the counterbored hole. Note the tooling clearance around the hex head and the clearance between the top of the bolt and the top surface of the block.

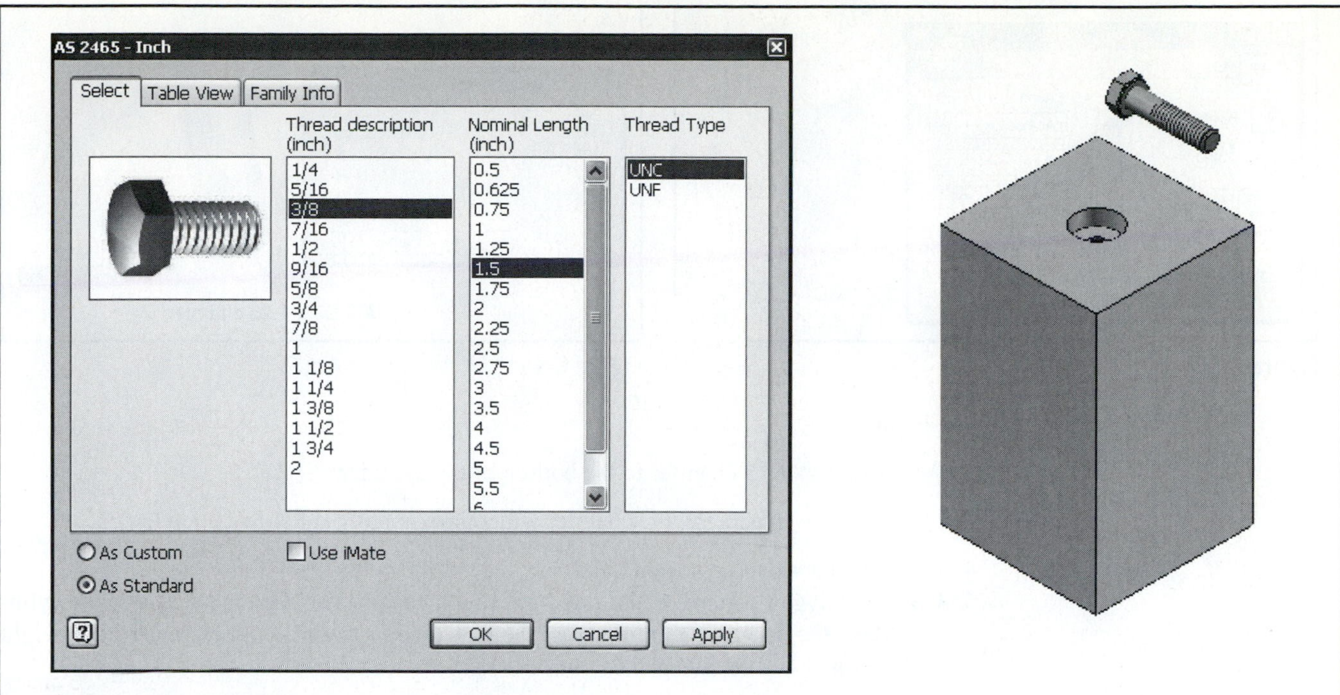

Figure 6-53

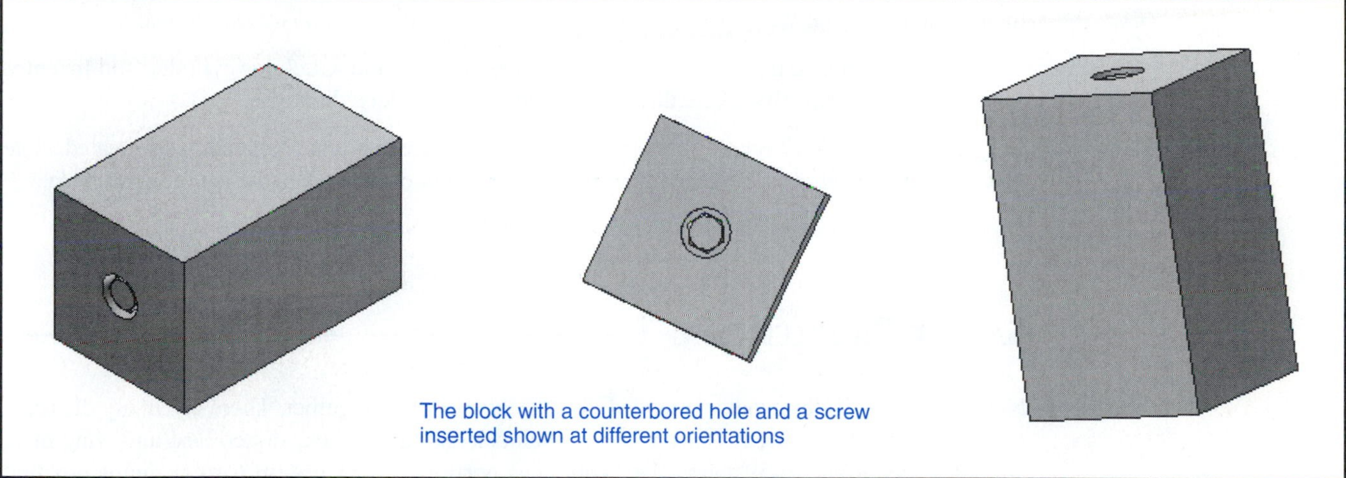

Figure 6-54

Drawing Fasteners Not Included in the Content Center

The **Content Center** contains a partial listing of fasteners. There are many other sizes and styles available. Inventor can be used to draw specific fasteners that can then be saved and used on assemblies.

Say we wish to draw an M8 × 25 hex head screw and that this size is not available in the **Content Center.**

Exercise 6-15: Drawing an M8 × 25 Hex Head Screw

1. Create a **Standard(mm).ipt** drawing.
2. Draw a **Ø8 × 25** cylinder. Draw the cylinder with its top surface on the XY plane so that it extends in the negative Z direction.

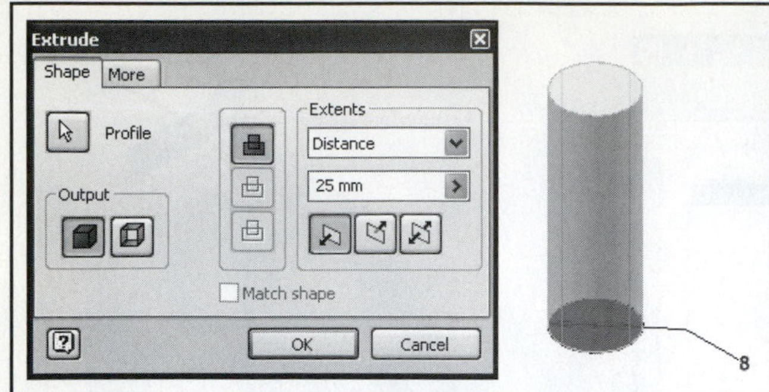

Figure 6-55

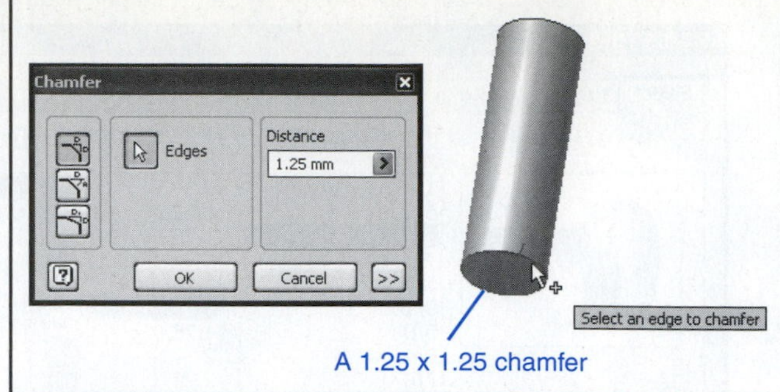

Figure 6-56

See Figure 6-55.

3. Add a **1.25 × 1.25** chamfer to the bottom of the cylinder.

See Figure 6-56. In general, the chamfer will approximately equal one pitch length.

4. Create a new sketch plane on the top surface of the cylinder.
5. Use the **Polygon** command and draw a hexagon centered on the top surface of the cylinder. Make the hexagon **12 mm** across the flats (see Chapter 2) and extrude it to a height of **5.4 mm** above the XY plane.

See Figure 6-57.

The head height and distance across the flats were determined using the general values defined in Figure 6-43. Specific values may be obtained from manufacturers, many of whom list their products on the Web, or from reference books such as *Machinery's Handbook.*

6. Add threads to the cylinder, using the **Thread** command. Click the cylinder, and Inventor will automatically create threads to match the cylinder's diameter.

See Figure 6-58. Check the thread specification to assure that correct threads were created. Note that the coarse pitch of 1.25 was automatically selected, but other pitch values are also available.

7. Save the drawing as **M8 × 1.25 × 25 Hex Head Screw.**

Sample Problem SP6-1

Nuts are used with externally threaded objects to hold parts together. There are many different styles of nuts. The **Content Center** library includes listings for hex, slotted hex, and wing nuts, among others. Figure 6-59 shows the **Table View** portion of the **Content Center** dialog box that includes nut heights.

The threads of a nut must be exactly the same as the external threads inserted into them. For example, if a screw with an M8 × 1.25 thread is selected, an M8 × 1.25 nut thread must be selected.

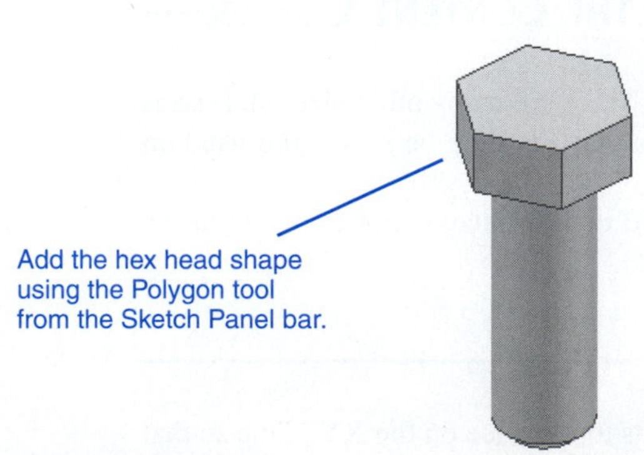

Figure 6-57

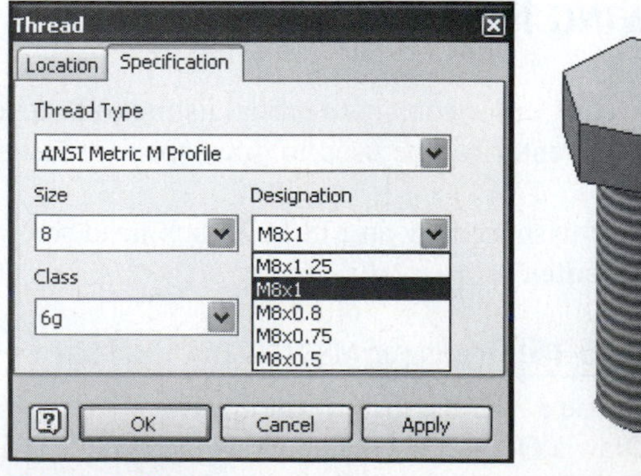

Figure 6-58

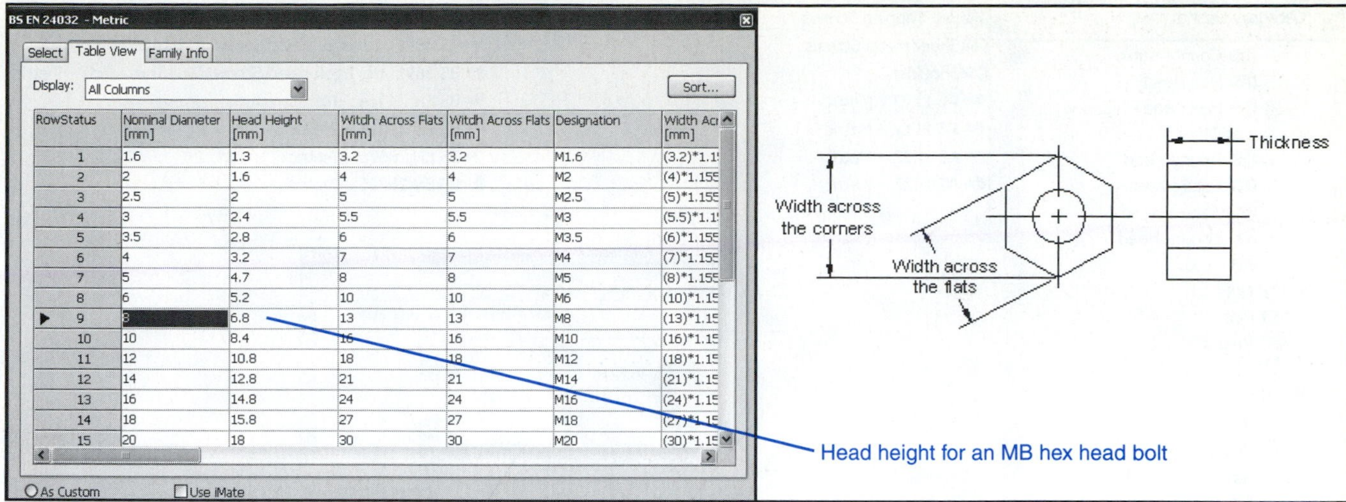

RowStatus	Nominal Diameter [mm]	Head Height [mm]	Witdh Across Flats [mm]	Witdh Across Flats [mm]	Designation	Width Acı [mm]
1	1.6	1.3	3.2	3.2	M1.6	(3.2)*1.1
2	2	1.6	4	4	M2	(4)*1.155
3	2.5	2	5	5	M2.5	(5)*1.155
4	3	2.4	5.5	5.5	M3	(5.5)*1.1
5	3.5	2.8	6	6	M3.5	(6)*1.155
6	4	3.2	7	7	M4	(7)*1.155
7	5	4.7	8	8	M5	(8)*1.155
8	6	5.2	10	10	M6	(10)*1.15
9	8	6.8	13	13	M8	(13)*1.15
10	10	8.4	16	16	M10	(16)*1.15
11	12	10.8	18	18	M12	(18)*1.15
12	14	12.8	21	21	M14	(21)*1.15
13	16	14.8	24	24	M16	(24)*1.15
14	18	15.8	27	27	M18	(27)*1.15
15	20	18	30	30	M20	(30)*1.15

Figure 6-59

The head height of the nut must be considered when determining the length of a bolt. It is good practice to have at least two threads extend beyond the nut to help assure that the nut is fully secured. Strength calculations are based on all nut threads' being engaged, so having threads extend beyond a nut is critical.

Figure 6-60 shows two blocks, each 25 mm thick with a center hole of Ø9.00 mm. The holes are clearance holes and do not include threads. The blocks are to be held together using an M8 hex head screw and a compatible nut.

Determining the Minimum Thread Length Required

Each block is 25 mm thick, for a total of 50 mm. The nut height, from Figure 6-59, for an M8 hex nut is 6.80, so the minimum thread length that will pass through both parts and the nut is 56.80 mm. Two threads must extend beyond the nut to assure that is fully secured. From Figure 6-59 the length of an M8 thread is given as 1.25 mm, so two threads equal 2.50 mm. Therefore, the minimum thread length must be 50.00 + 6.80 + 2.50 = 59.3 mm.

Bolts are manufactured in standard lengths, some of which are listed in the **Content Center** library. If the required thread length was not available from the library, manufacturers' catalogs would have to be searched and a new screw drawing created.

Exercise 6-16: Selecting a Screw

1. Click the **Place from Content Center** tool.
2. Select the **Fasteners** option, then **Bolts,** then **Hex Head.** Select **AS 1110-Metric** screw.

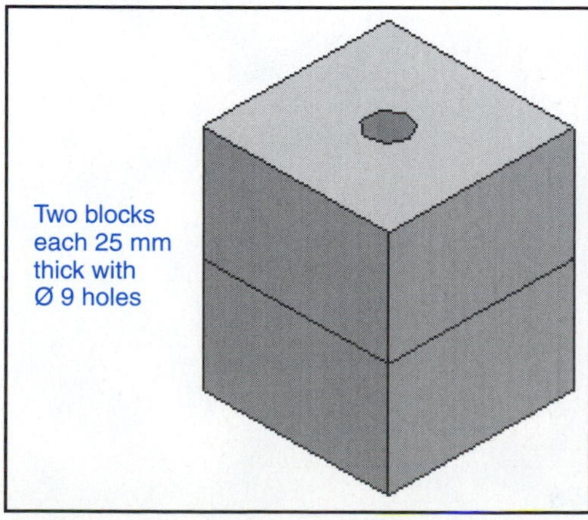

Figure 6-60

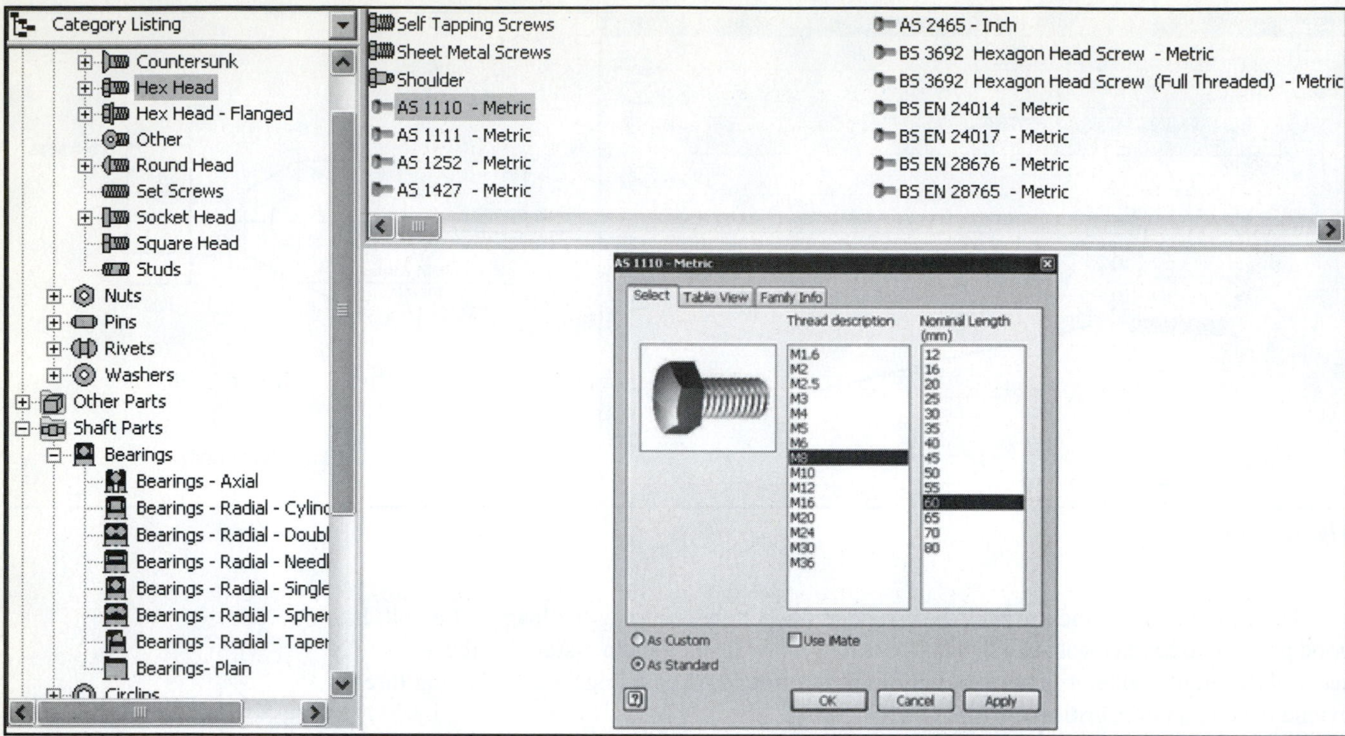

Figure 6-61

See Figure 6-61. The standard thread length that is closest to, but still greater than, 59.3 is 60 mm. The 60 mm length is selected and applied to the drawing.

3. Define the values for an **M8 × 60** hex head screw and insert it into the drawing.

See Figure 6-62.

4. Insert the screw into the two assembled parts.

See Figure 6-63. Note that the screw extends beyond the bottom of the two assembled parts.

Exercise 6-17: Selecting a Nut

1. Click the **Place from Content Center** and select the **Fasteners** option, then **Nuts,** then **Hex,** then the **Hex Nut Metric** listing.

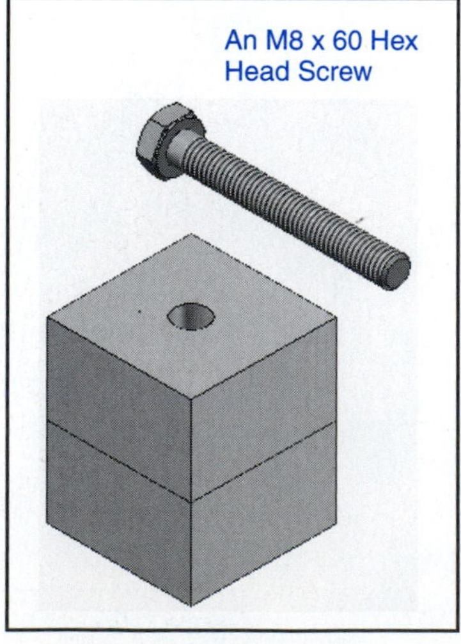

Figure 6-62

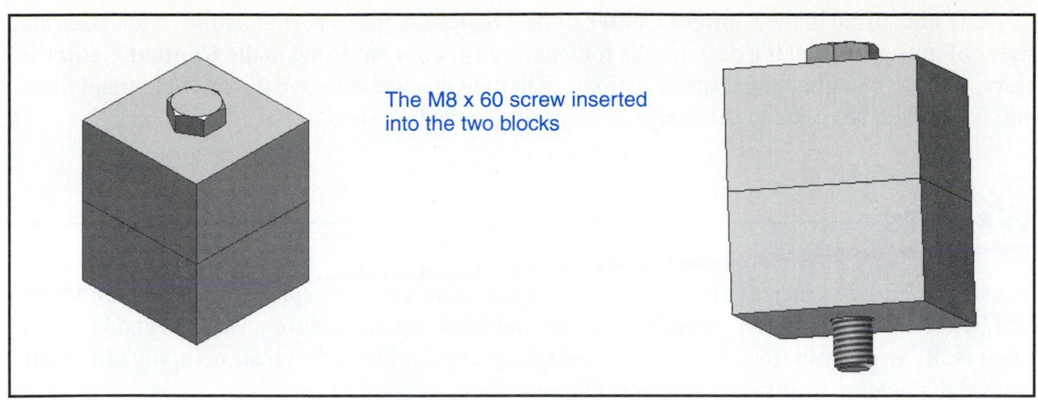

Figure 6-63

See Figure 6-64.

2. Insert a copy of the nut into the drawing area.

Figure 6-65 shows the nut added to the drawing screen.

3. Use the **Place Constraint** tool and insert the nut onto the screw so that it is flush with the bottom surface of the blocks.

Figure 6-66 shows the nut inserted onto the screw.

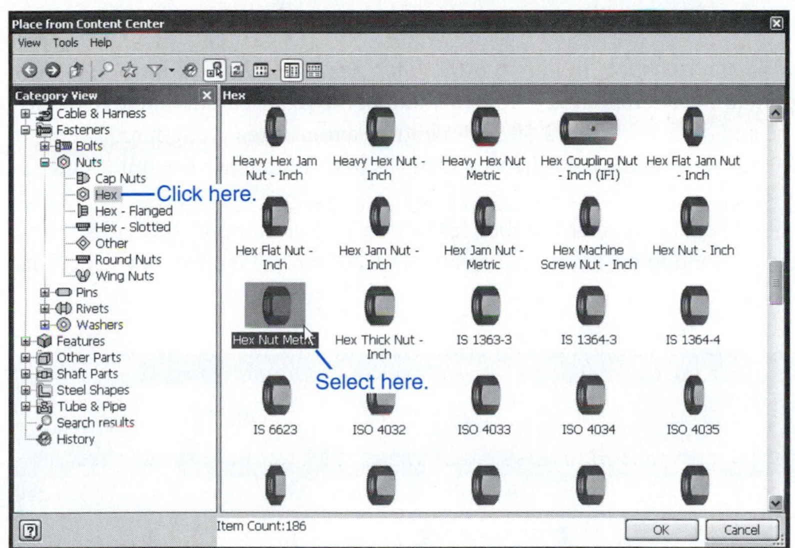

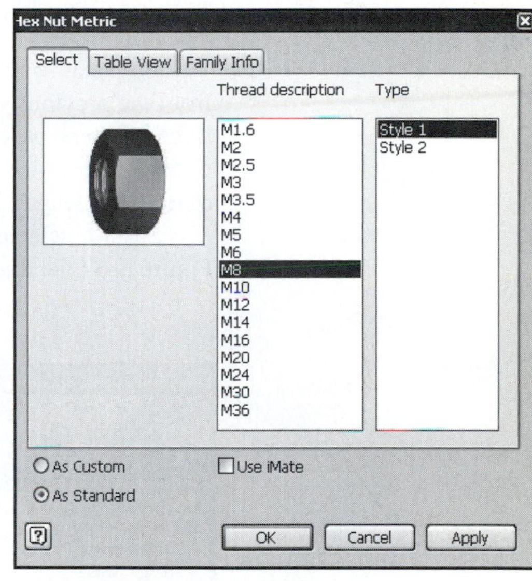

Figure 6-64

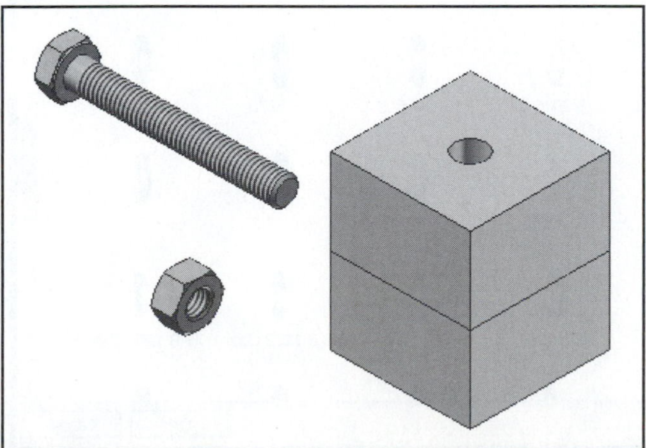

Figure 6-65

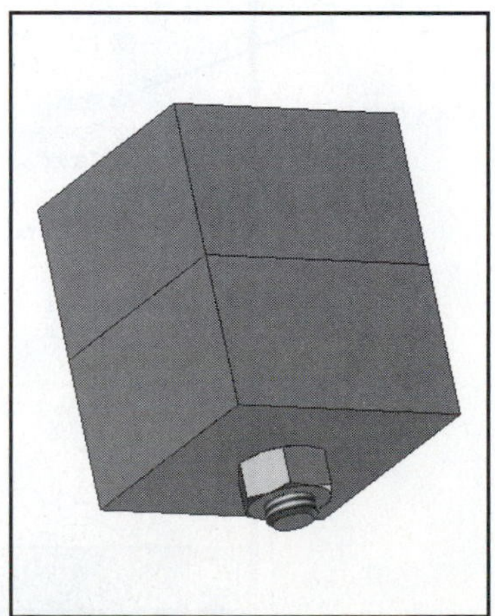

Figure 6-66

The nuts listed in the **Content Center** library represent only a partial listing of the sizes and styles of nuts available. If a design calls for a nut size or type not listed in the **Content Center** library, refer to manufacturers' specifications, then draw the nut and save it as an individual drawing. It can then be added to the design drawings as needed.

WASHERS

washer: A flat thin ring used to increase the bearing area under a fastener or as a spacer.

Washers are used to increase the bearing area under fasteners or as spacers. Washers are identified by their inside diameter, outside diameter, and thickness. In addition, washers can be designated N, R, or W for narrow, regular, and wide, respectively. These designations apply only to the outside diameters; the inside diameter is the same.

Inventor lists washers by their nominal diameter. The nominal diameters differ from the actual inside diameter by a predetermined clearance allowance. For example, a washer with a nominal diameter of 8 has an actual inside diameter of 8.40, or 0.40 mm greater than the 8 nominal diameter. This means that washer sizes can easily be matched to thread sizes using nominal sizes. A washer with a nominal diameter designation of 8 will fit over a thread designated M8.

There are different types of washers including, among others, plain and tapered. The **Place from Content Center** dialog box includes a listing of both plain and tapered washers. Figure 6-67 shows the **Table View** portion of the **Place from Content Center** dialog box for a plain washer ISO 7089.

Inserting Washers onto a Fastener

We again start with the two blocks shown in Figure 6-60. Each has a height of 25 mm. We know from the previous section that the nut height is 6.80 mm and that the requirement that at least two threads extend beyond the end of the nut adds 2.50 mm, yielding a total thread length requirement of 59.3. We now have to add the thickness of the two washers and recalculate the minimum required bolt length.

Say we selected a plain regular washer number ISO 7089 with a nominal size of 8. From Figure 6-67 the thickness is found to be 1.6 mm or a total of 3.20 mm for the two washers. This

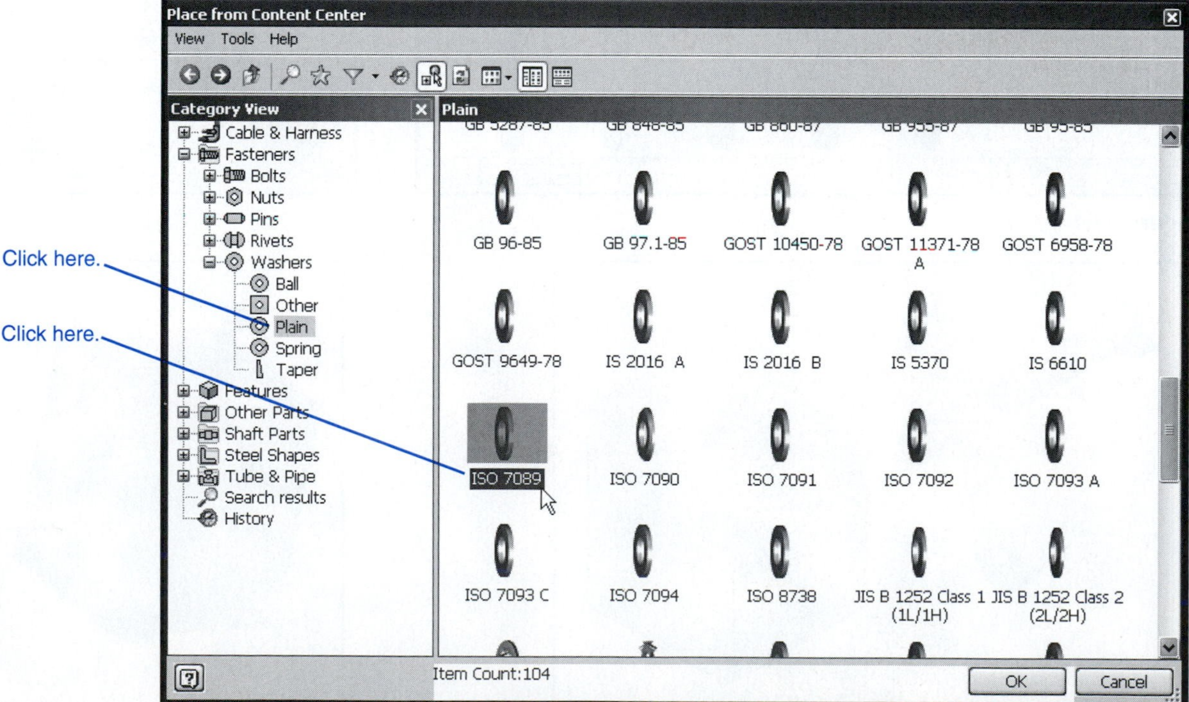

Figure 6-67

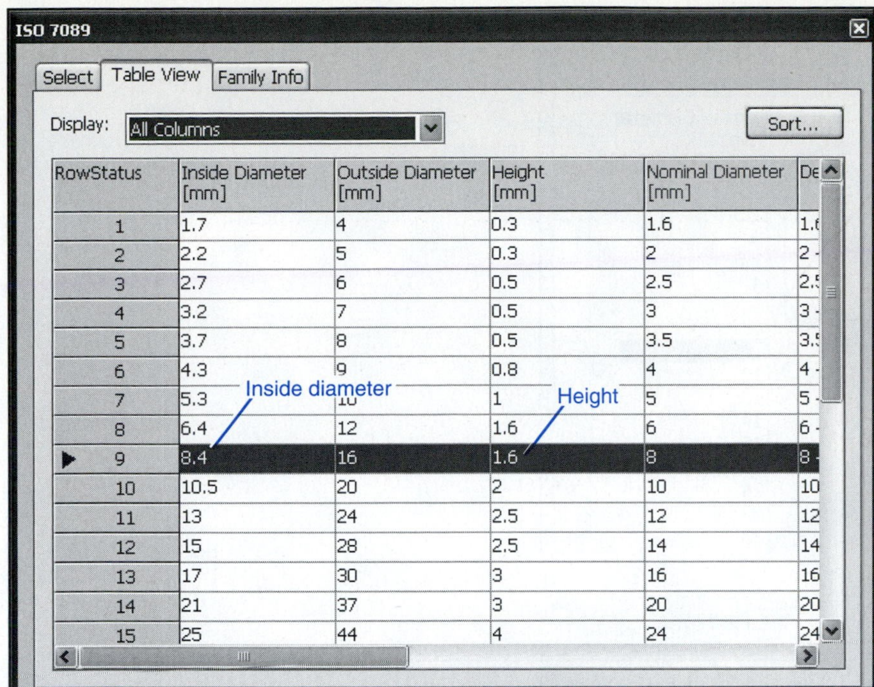

RowStatus	Inside Diameter [mm]	Outside Diameter [mm]	Height [mm]	Nominal Diameter [mm]	De
1	1.7	4	0.3	1.6	1.
2	2.2	5	0.3	2	2
3	2.7	6	0.5	2.5	2.
4	3.2	7	0.5	3	3
5	3.7	8	0.5	3.5	3.
6	4.3	9	0.8	4	4
7	5.3	10	1	5	5
8	6.4	12	1.6	6	6
9	8.4	16	1.6	8	8
10	10.5	20	2	10	10
11	13	24	2.5	12	12
12	15	28	2.5	14	14
13	17	30	3	16	16
14	21	37	3	20	20
15	25	44	4	24	24

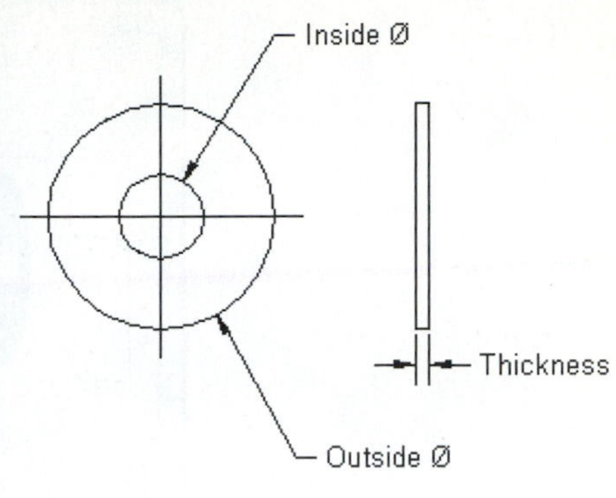

Figure 6-67 *(Continued)*

extends the minimum bolt length requirement to 59.3 + 3.20 = 62.5. The nearest standard thread bolt length listed in the **Place from Content Center** dialog box that is greater than the 62.5 requirement is 65.

Exercise 6-18: Adding Washers to an Assembly

1. Use the **Place Component** tool and locate two copies of the block on the drawing, then align the blocks.
2. Use the **Save All** command to save and name the assembly.
3. Click the **Place from Content Center** tool and select an **M8 × 65 Hex Head** screw and an **M8 Hex Nut** and insert them into the drawing.

See Figure 6-68.

4. Access the **Place from Content Center** dialog box and select the **Washers** option, then **Plain,** then an **ISO 7089** washer.

See Figure 6-69. Figure 6-67 shows the **Table View** values for the washer.

5. Insert two copies into the drawing.
See Figure 6-70.

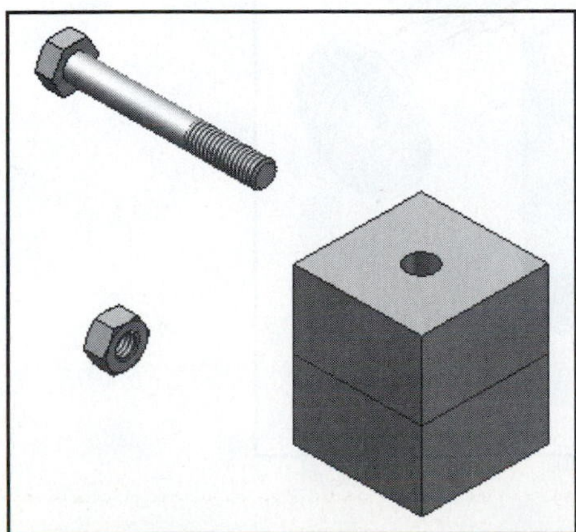

Figure 6-68

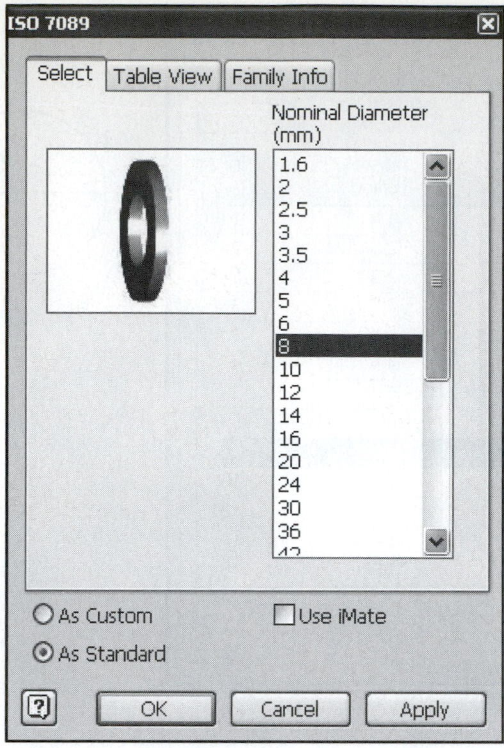

Figure 6-69

6. Use the **Place Constraint Insert** option tool and align the washers with the holes in the blocks.

See Figure 6-71.

7. Use the **Place Constraint** tool and insert the M8 × 65 screw and the nut.

Figure 6-72 shows the resulting assembly.

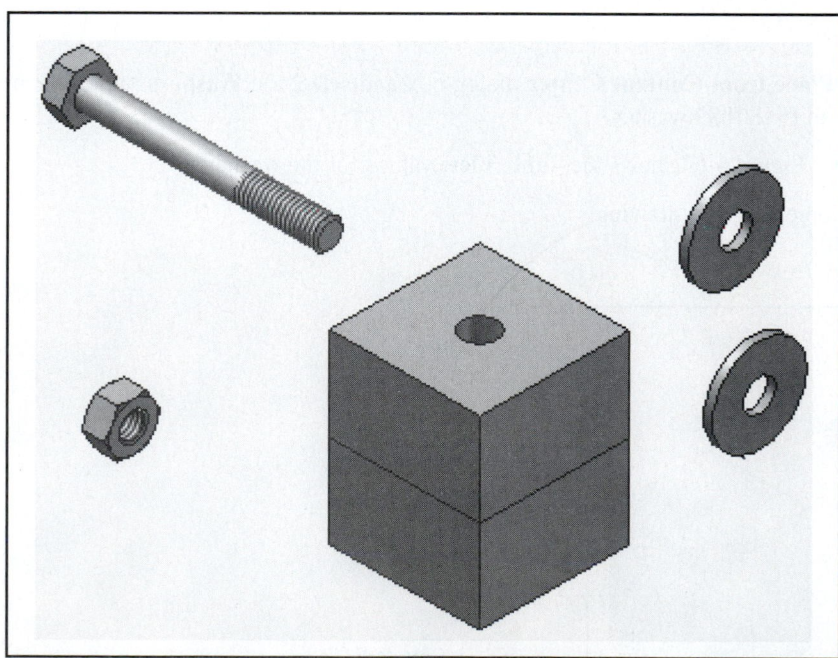

Figure 6-70

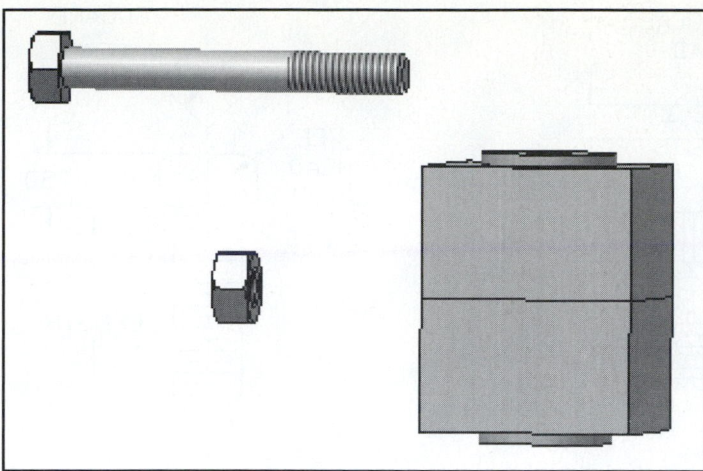

Figure 6-71

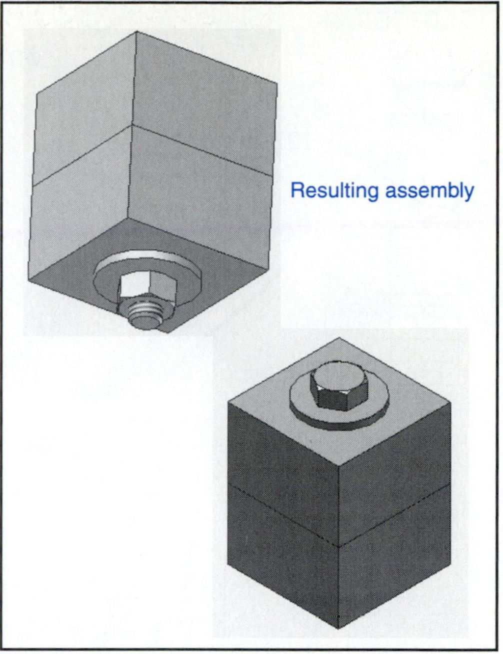

Figure 6-72

The washers listed in the **Content Center** dialog box library represent only a partial listing of the washers available. If a design calls for a washer size or type not listed in the library, refer to manufacturers' specifications, then draw the washer and save it as an individual drawing. It can then be added to the design drawings as needed.

SETSCREWS

Setscrews are fasteners used to hold parts like gears and pulleys to rotating shafts or other moving objects to prevent slippage between the two objects. See Figure 6-73.

Most setscrews have recessed heads to help prevent interference with other parts.

Many different head styles and point styles are available. See Figure 6-74. The dimensions shown in Figure 6-74 are general sizes for use in this book. For actual sizes, see the manufacturer's specifications.

Figure 6-75 shows a collar with two 10(.19)-32UNF threaded holes.

setscrew: A fastener used to hold parts to rotating shafts or other moving objects to prevent slippage between the two objects.

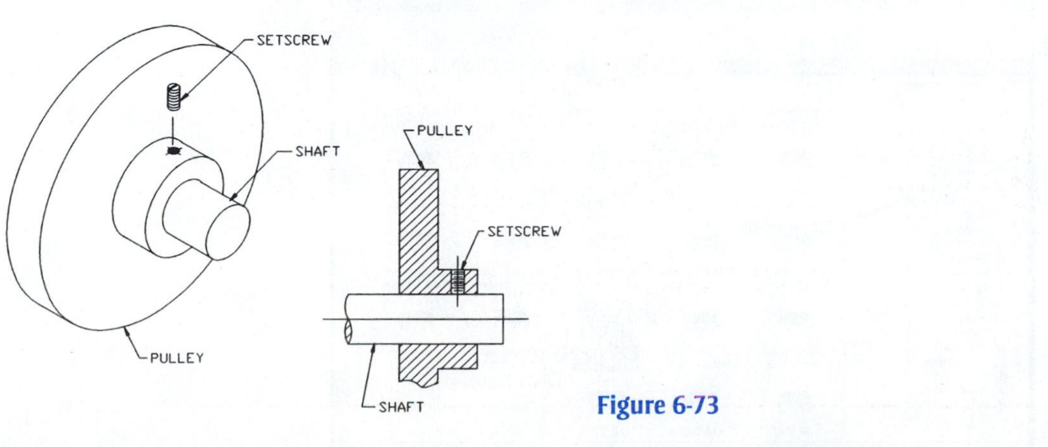

Figure 6-73

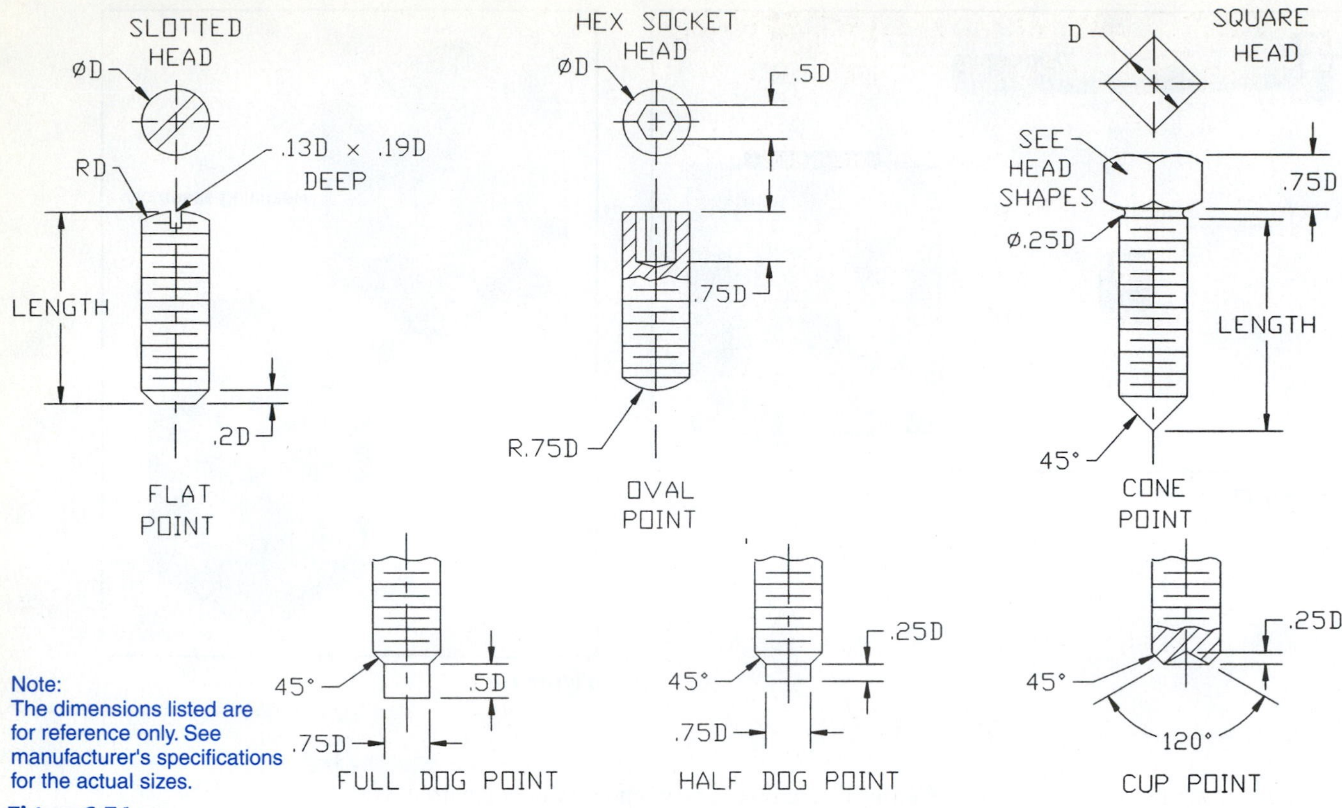

Note:
The dimensions listed are for reference only. See manufacturer's specifications for the actual sizes.

Figure 6-74

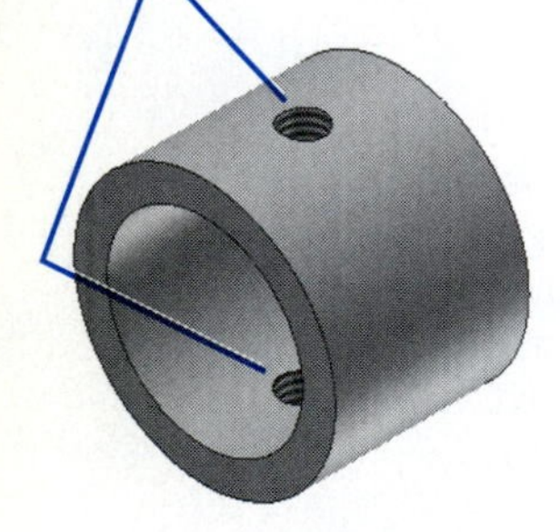

Collar I.D. = .750
D.O. = 1.000
LENGTH = .750

Figure 6-75

Exercise 6-19: Adding a Setscrew

1. Create an assembly drawing using the **Standard (mm).iam** format.
2. Use the **Place Component** tool and place one copy of the collar on the drawing.
3. Use the **Save All** tool to save and name the assembly.
4. Access the **Place from Content Center** dialog box and select the **Fasteners** option, then **Bolts,** then **Set Screws.**
5. Select the **Hexagon Socket Unbrako Dog Point - Inch** option and set the (nominal diameter) **Thread description** for **#10,** the **Nominal Length** for **.32,** and the **Thread Type** for **UNF.**

See Figure 6-76.

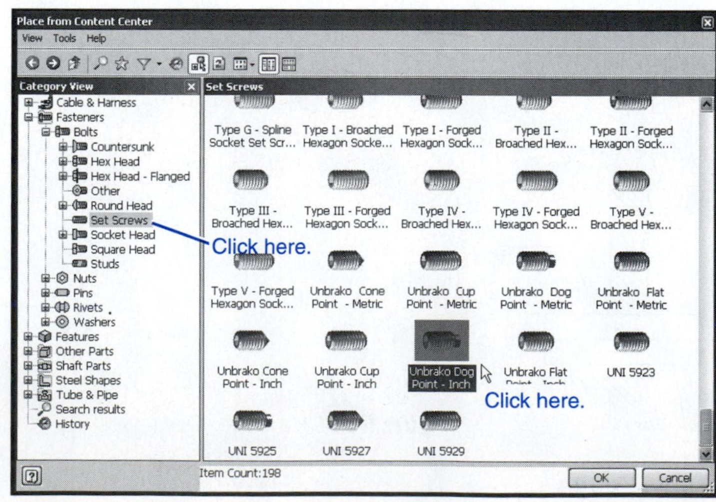

Figure 6-76 *(Continued)*

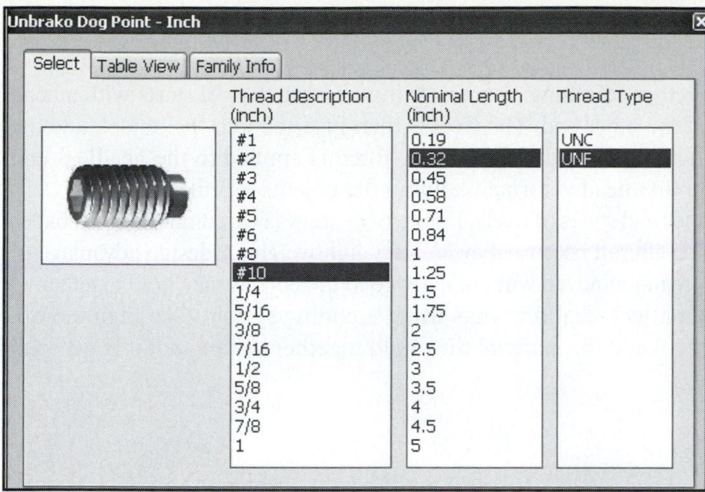

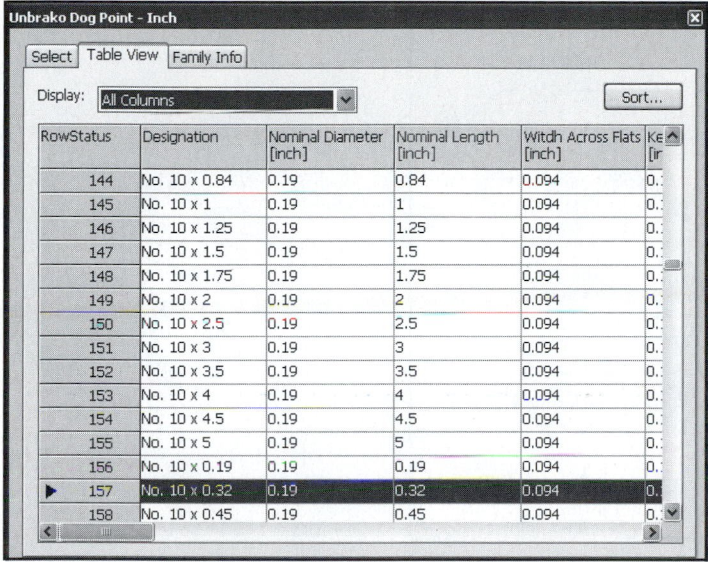

RowStatus	Designation	Nominal Diameter [inch]	Nominal Length [inch]	Witdh Across Flats [inch]	Ke [in
144	No. 10 x 0.84	0.19	0.84	0.094	0.:
145	No. 10 x 1	0.19	1	0.094	0.:
146	No. 10 x 1.25	0.19	1.25	0.094	0.:
147	No. 10 x 1.5	0.19	1.5	0.094	0.:
148	No. 10 x 1.75	0.19	1.75	0.094	0.:
149	No. 10 x 2	0.19	2	0.094	0.:
150	No. 10 x 2.5	0.19	2.5	0.094	0.:
151	No. 10 x 3	0.19	3	0.094	0.:
152	No. 10 x 3.5	0.19	3.5	0.094	0.:
153	No. 10 x 4	0.19	4	0.094	0.:
154	No. 10 x 4.5	0.19	4.5	0.094	0.:
155	No. 10 x 5	0.19	5	0.094	0.:
156	No. 10 x 0.19	0.19	0.19	0.094	0.:
▶ 157	No. 10 x 0.32	0.19	0.32	0.094	0.:
158	No. 10 x 0.45	0.19	0.45	0.094	0.:

Figure 6-76

6. Insert a copy of the setscrew on the drawing.

See Figure 6-77.

7. Use the **Place Constraint** tool and insert the setscrew into one of the threaded holes.

Figure 6-78 shows the resulting assembly.

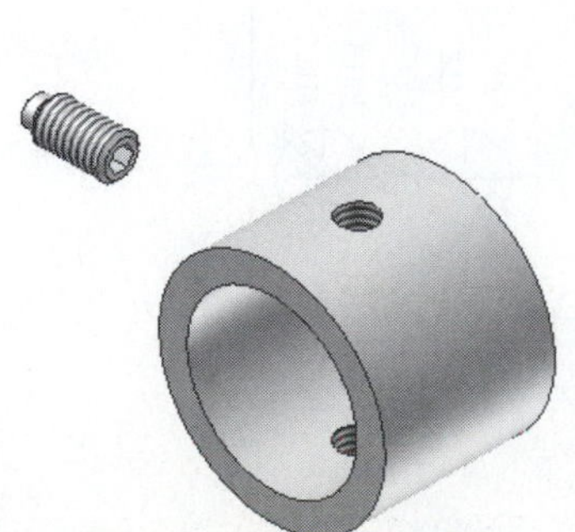

Figure 6-77

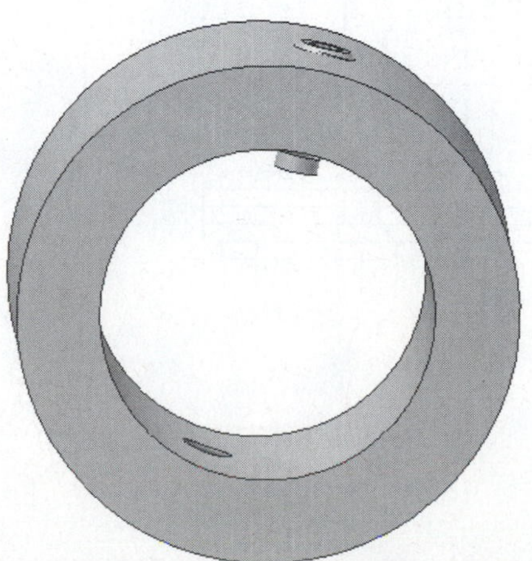

Figure 6-78

Rivets

rivet: A metal fastener with a head and a straight shaft for holding together overlapping and adjoining objects.

Rivets are fasteners that hold together adjoining or overlapping objects. A rivet starts with a head at one end and a straight shaft at the other end. The rivet is then inserted into the object, and the headless end is "bucked" or otherwise forced into place. A force is applied to the headless end that changes its shape so that another head is formed holding the objects together.

There are many different shapes and styles of rivets. Figure 6-79 shows five common head shapes for rivets. Hollow rivets are used on aircraft because they are very lightweight. A design advantage of rivets is that they can be drilled out and removed without damage to the objects they hold together.

Rivet types are represented on technical drawings using a coding system. See Figure 6-80. Since rivets are sometimes so small and the material they hold together so thin that it is difficult

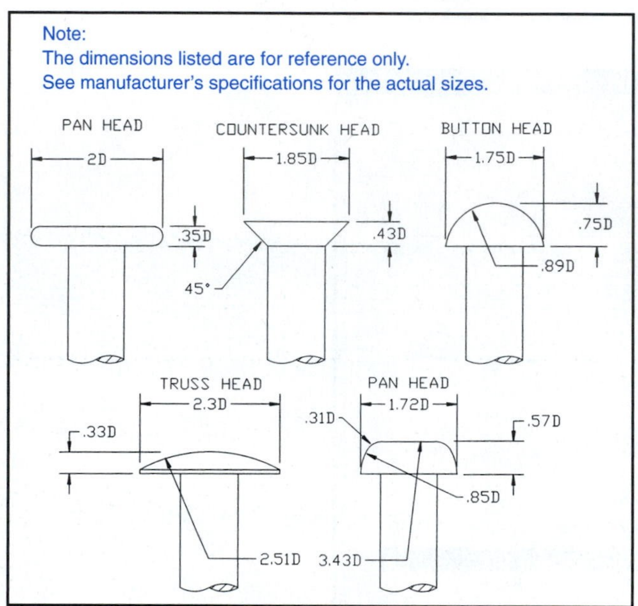

Figure 6-79

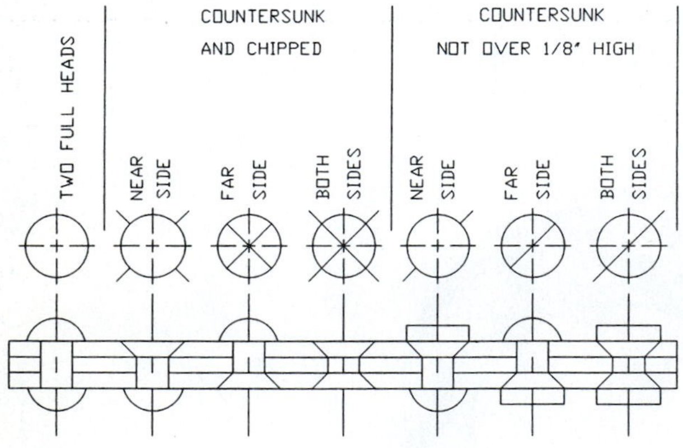

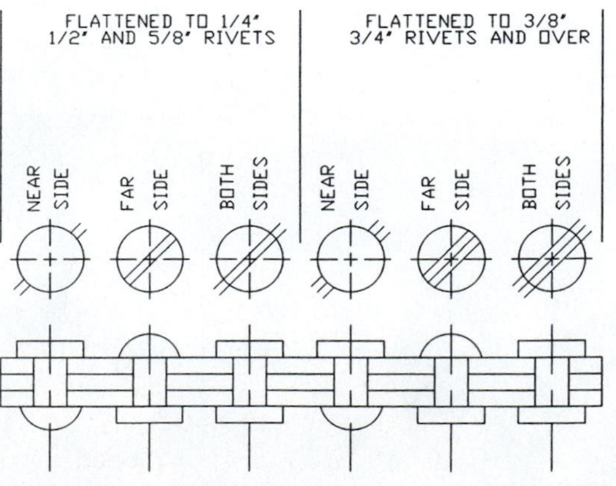

Figure 6-80

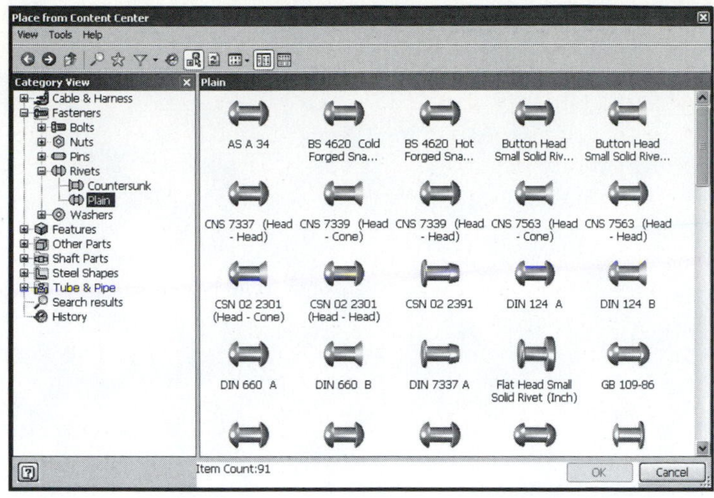

Figure 6-81

to clearly draw the rivets, some companies draw only the rivet's centerline in the side view and identify the rivets using a drawing callout.

Figure 6-81 shows the **Place from Content Center** dialog box for a BS 4620 Cold Forged Snap Head Rivet - (Metric).

SUMMARY

This chapter explained how to draw threads and washers as well as how to select fasteners and how to design using fasteners, washers, and keys.

Thread terminology was explained and illustrated, and the different thread form specifications and ways of graphically representing threads were described, including threaded through holes, internal threads, and blind threaded holes. ANSI standards and conventions were followed.

The **Place from Content Center** tool was used to specify different types of fasteners in drawings, including bolts and screws coupled with nuts. Countersunk screws and counterbored holes were described and illustrated.

CHAPTER PROJECTS

Project 6-1: Millimeters

Figure P6-1 shows three blocks. Assume that the blocks are each 30 × 30 × 10 and that the hole is ∅9. Assemble the three blocks so that their holes are aligned and they are held together by a hex head bolt secured by an appropriate hex nut. Locate a washer between the bolt head and the top block and between the nut and the bottom block. Create all drawings using either an A4 or A3 drawing sheet, as needed. Include a title block on all drawing sheets.

A. Define the bolt.
B. Define the nut.
C. Define the washers.
D. Draw an assembly drawing including all components.
E. Create a BOM for the assembly.
F. Create a presentation drawing of the assembly.
G. Create an isometric exploded drawing of the assembly.
H. Create an animation drawing of the assembly.

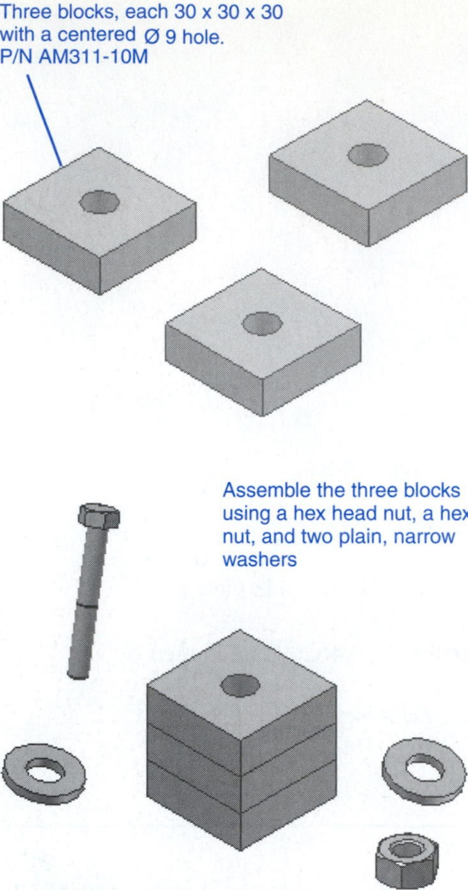

Figure P6-1

Project 6-2: Millimeters

Figure P6-2 shows three blocks, one 30 × 30 × 50 with a centered M8 threaded hole, and two 30 × 30 × 10 blocks with centered ∅9 holes. Join the two 30 × 30 × 10 blocks to the 30 × 30 × 50 block using an M8 hex head bolt. Locate a regular plain washer under the bolt head.

A. Define the bolt.
B. Define the thread depth.

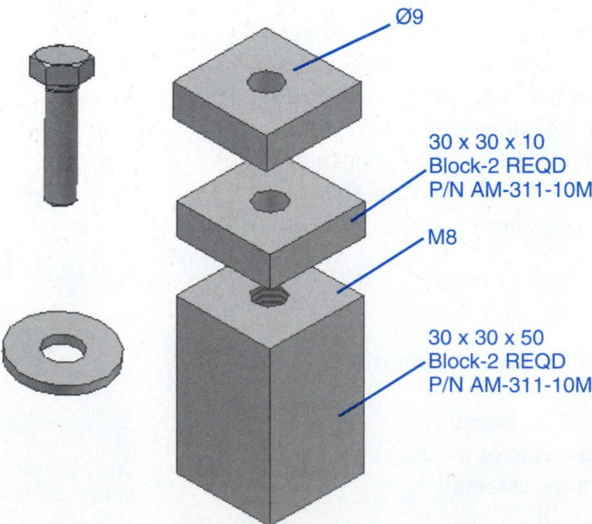

Figure P6-2

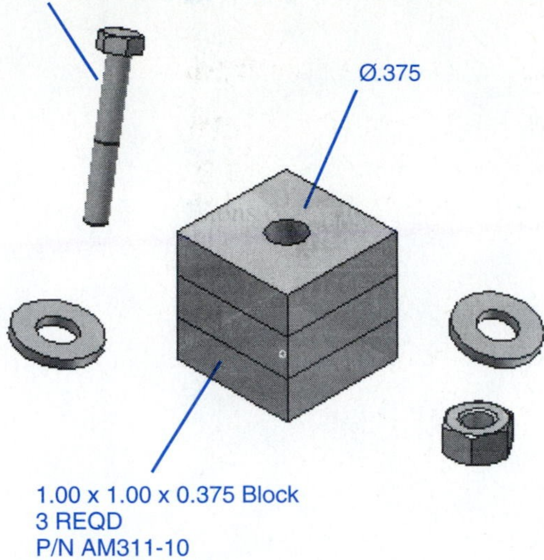

Figure P6-3

C. Define the hole depth.
D. Define the washer.
E. Draw an assembly drawing including all components.
F. Create a BOM for the assembly.
G. Create a presentation drawing of the assembly.
H. Create an isometric exploded drawing of the assembly.
I. Create an animation drawing of the assembly.

Project 6-3: Inches

Figure P6-3 shows three blocks. Assume that each block is 1.00 × 1.00 × 0.375 and that the hole is ∅.375. Assemble the three blocks so that their holes are aligned and that they are held together by a 5/16-18 UNC indented regular hex head bolt secured by an appropriate hex nut. Locate a washer between the bolt head and the top block and between the nut and the bottom block. Create all drawings using either an A4 or A3 drawing sheet, as needed. Include a title block on all drawing sheets.

A. Define the bolt.
B. Define the nut.
C. Define the washers.
D. Draw an assembly drawing including all components.
E. Create a BOM for the assembly.
F. Create a presentation drawing of the assembly.
G. Create an isometric exploded drawing of the assembly.
H. Create an animation drawing of the assembly.

Project 6-4: Inches

Figure P6-4 shows three blocks, one 1.00 × 1.00 × 2.00 with a centered threaded hole, and two 1.00 × 1.00 × .375 blocks with centered ∅.375 holes. Join the two 1.00 × 1.00 × .375 blocks to the 1.00 × 1.00 × 2.00 block using a 5/16-18 UNC hex head bolt. Locate a regular plain washer under the bolt head.

A. Define the bolt.
B. Define the thread depth.
C. Define the hole depth.
D. Define the washer.
E. Draw an assembly drawing including all components.
F. Create a BOM for the assembly.

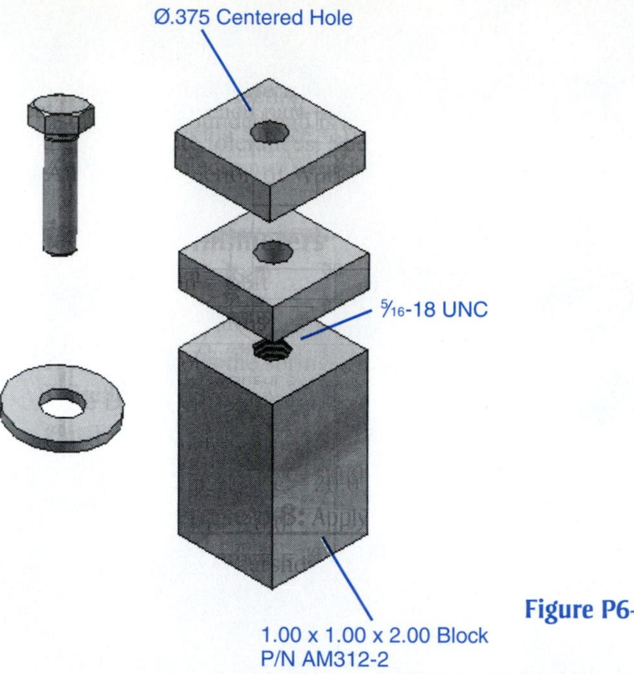

Figure P6-4

G. Create a presentation drawing of the assembly.
H. Create an isometric exploded drawing of the assembly.
I. Create an animation drawing of the assembly.

Project 6-5: Inches or Millimeters

Figure P6-5 shows a centering block. Create an assembly drawing of the block and insert three setscrews into the three threaded holes so that they extend at least .25 in. or 6 mm into the center hole.

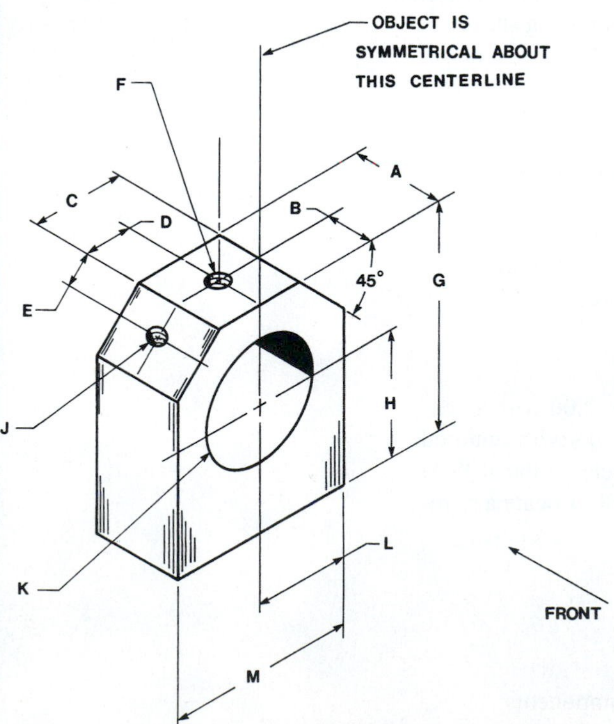

DIMENSION	INCHES	mm
A	1.00	26
B	.50	13
C	1.00	26
D	.50	13
E	.38	10
F	.190-32 UNF	M8X1
G	2.38	60
H	1.38	34
J	.164-36 UNF	M6
K	Ø1.25	Ø30
L	1.00	26
M	2.00	52

Figure P6-5

A. Use the inch dimensions.
B. Use the millimeter dimensions.
C. Define the setscrews.
D. Draw an assembly drawing including all components.
E. Create a BOM for the assembly.
F. Create a presentation drawing of the assembly.
G. Create an isometric exploded drawing of the assembly.
H. Create an animation drawing of the assembly.

Project 6-6: Millimeters

Figure P6-6 shows two parts; a head cylinder and a base cylinder. The head cylinder has outside dimensions of ∅40 × 20, and the base cylinder has outside dimensions of ∅40 × 50. The holes in both parts are located on a ∅24 bolt circle. Assemble the two parts using hex head bolts.

A. Define the bolt.
B. Define the holes in the head cylinder, the counterbore diameter and depth, and the clearance hole diameter.
C. Define the thread depth in the base cylinder.

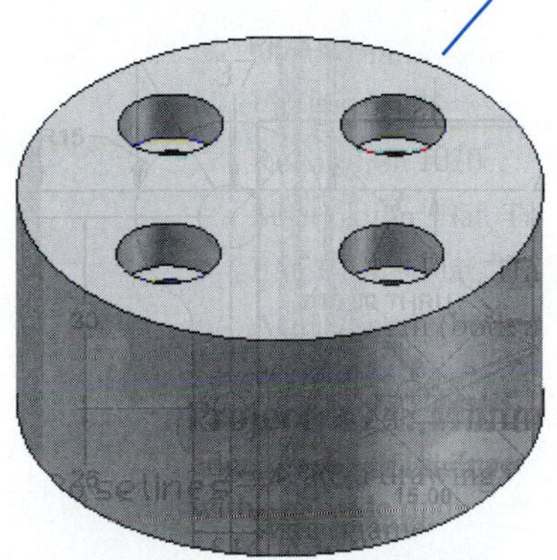

Figure P6-6

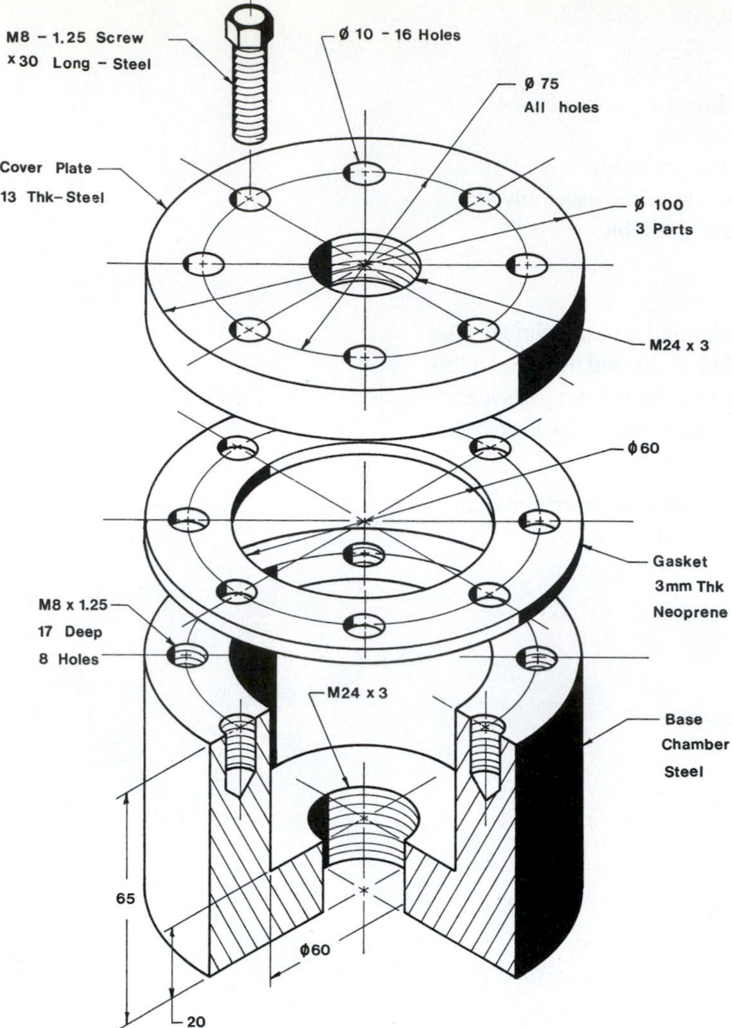

Figure P6-7

D. Define the hole depth in the base cylinder.
E. Draw an assembly drawing including all components.
F. Create a BOM for the assembly.
G. Create a presentation drawing of the assembly.
H. Create an isometric exploded drawing of the assembly.
I. Create an animation drawing of the assembly.

Project 6-7: Millimeters

Figure P6-7 shows a pressure cylinder assembly.

A. Draw an assembly drawing including all components.
B. Create a BOM for the assembly.
C. Create a presentation drawing of the assembly.
D. Create an isometric exploded drawing of the assembly.
E. Create an animation drawing of the assembly.

Project 6-8: Millimeters

Figure P6-7 shows a pressure cylinder assembly.

A. Revise the assembly so that it uses M10 × 35 hex head bolts.
B. Draw an assembly drawing including all components.
C. Create a BOM for the assembly.

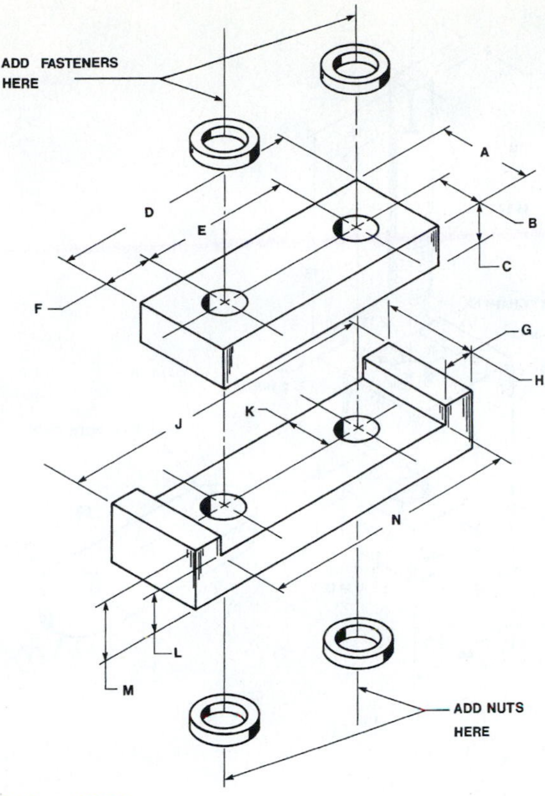

DIMENSION	INCHES	mm
A	1.25	32
B	.63	16
C	.50	13
D	3.25	82
E	2.00	50
F	.63	16
G	.38	10
H	1.25	32
J	4.13	106
K	.63	16
L	.50	13
M	.75	10
N	3.38	86

Figure P6-9

D. Create a presentation drawing of the assembly.
E. Create an isometric exploded drawing of the assembly.
F. Create an animation drawing of the assembly.

Project 6-9: Inches and Millimeters

Figure P6-9 shows a C-block assembly.

Use one of the following fasteners assigned by your instructor.

1. M12 hex head
2. M10 square head
3. 1/4-20 UNC hex head
4. 3/8-16 UNC square head
5. M10 socket head
6. M8 slotted head
7. 1/4-20 UNC slotted head
8. 3/8-16 UNC socket head

A. Define the bolts.
B. Define the nuts.
C. Define the washers.
D. Draw an assembly drawing including all components.
E. Create a BOM for the assembly.
F. Create a presentation drawing of the assembly.
G. Create an isometric exploded drawing of the assembly.
H. Create an animation drawing of the assembly.

Project 6-10: Millimeters

Figure P6-10 shows an exploded assembly drawing. There are no standard parts, so each part must be drawn individually.

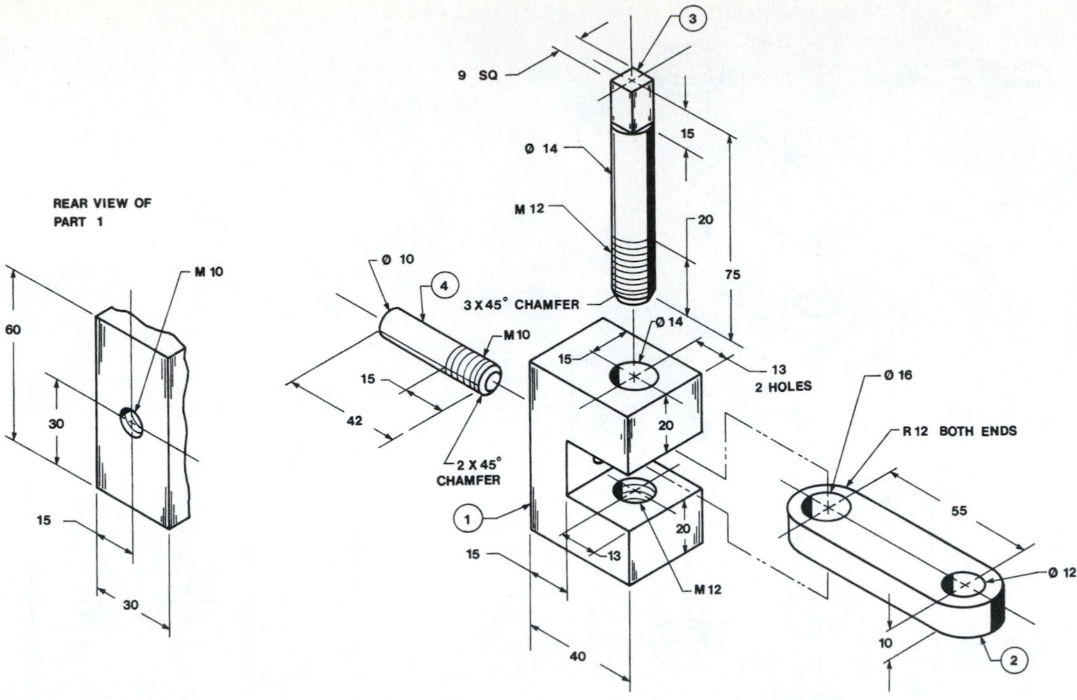

Figure P6-10

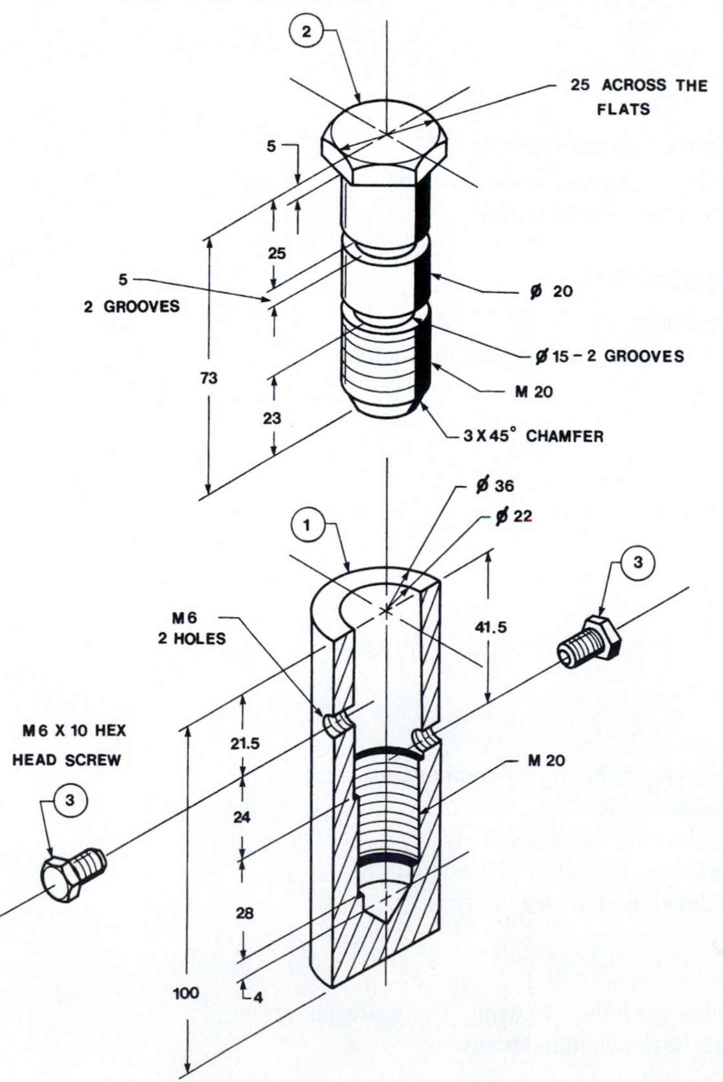

Figure P6-11

A. Draw an assembly drawing including all components.
B. Create a BOM for the assembly.
C. Create a presentation drawing of the assembly.
D. Create an isometric exploded drawing of the assembly.
E. Create an animation drawing of the assembly.

Project 6-11: Millimeters

Figure P6-11 shows an exploded assembly drawing.

A. Draw an assembly drawing including all components.
B. Create a BOM for the assembly.
C. Create a presentation drawing of the assembly.
D. Create an isometric exploded drawing of the assembly.
E. Create an animation drawing of the assembly.

Project 6-12: Inches or Millimeters

Figure P6-12 shows an exploded assembly drawing. No dimensions are given. If parts 3 and 5 have either M10 or ⅜-16 UNC threads, size parts 1 and 2. Based on these values estimate and create the remaining sizes and dimensions.

A. Draw an assembly drawing including all components.
B. Create a BOM for the assembly.

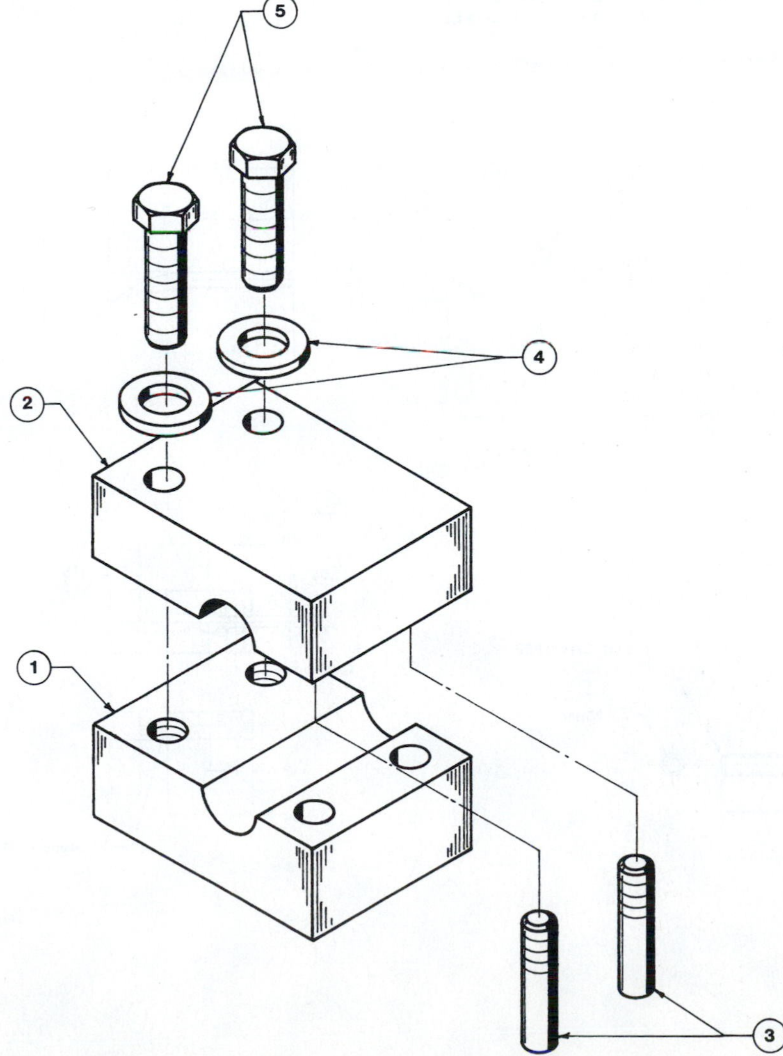

Figure P6-12

SIMPLIFIED SURFACE GAGE

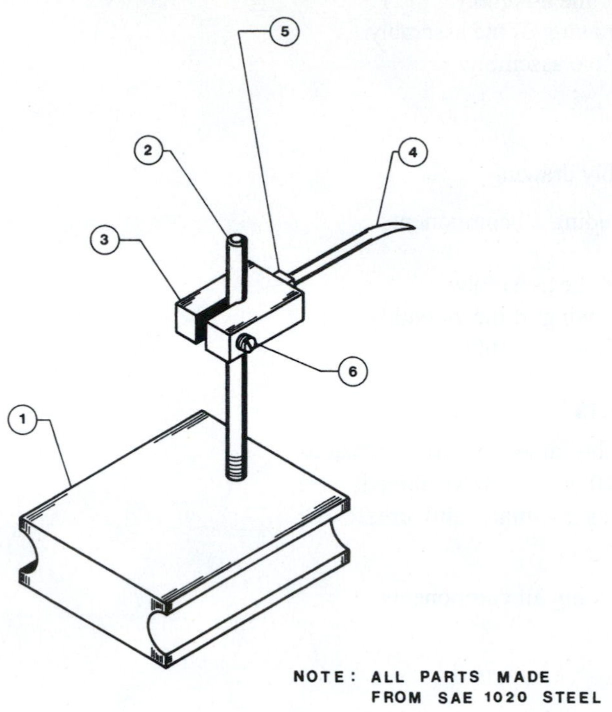

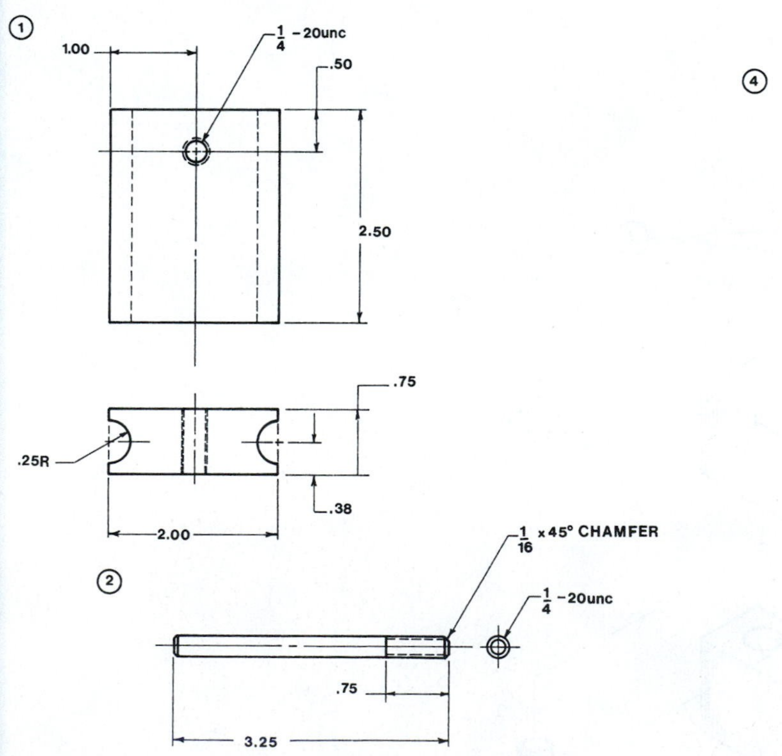

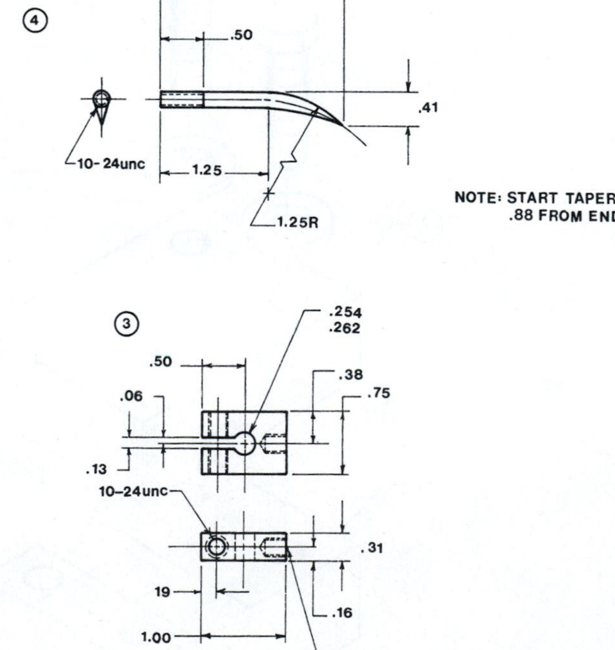

Figure P6-13

C. Create a presentation drawing of the assembly.
D. Create an isometric exploded drawing of the assembly.
E. Create an animation drawing of the assembly.

Project 6-13: Inches

Figure P6-13 shows an assembly drawing and detail drawings of a surface gauge.

A. Draw an assembly drawing including all components.
B. Create a BOM for the assembly.
C. Create a presentation drawing of the assembly.
D. Create an isometric exploded drawing of the assembly.
E. Create an animation drawing of the assembly.

Project 6-14: Millimeters

Figure P6-14 shows an assembly made from parts defined on pages 219 through 222. Assemble the parts using M10 threaded fasteners.

A. Define the bolt.
B. Define the nut.
C. Draw an assembly drawing including all components.
D. Create a BOM for the assembly.
E. Create a presentation drawing of the assembly.
F. Create an isometric exploded drawing of the assembly.
G. Create an animation drawing of the assembly.
H. Consider possible interference between the nuts and ends of the fasteners both during and after assembly. Recommend an assembly sequence.

Project 6-15: Millimeters

Figure P6-15 shows an assembly made from parts defined on pages 219 through 222. Assemble the parts using M10 threaded fasteners.

A. Define the bolt.
B. Define the nut.
C. Draw an assembly drawing including all components.
D. Create a BOM for the assembly.
E. Create a presentation drawing of the assembly.
F. Create an isometric exploded drawing of the assembly.

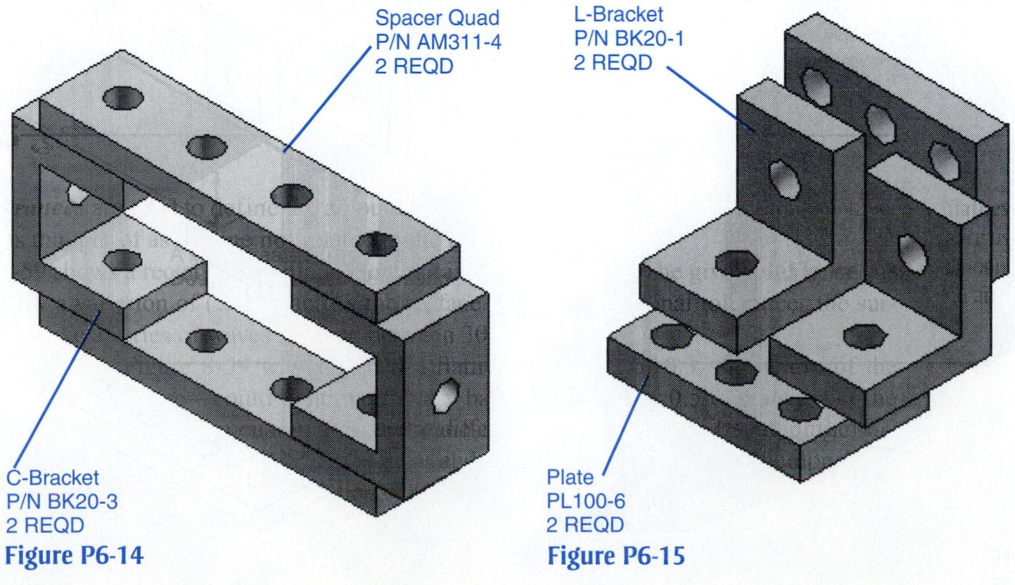

Figure P6-14

Figure P6-15

G. Create an animation drawing of the assembly.
H. Consider possible interference between the nuts and ends of the fasteners both during and after assembly. Recommend an assembly sequence.

Project 6-16: Millimeters

Figure P6-16 shows an assembly made from parts defined on pages 219 through 222. Assemble the parts using M10 threaded fasteners.

A. Define the bolt.
B. Define the nut.
C. Draw an assembly drawing including all components.
D. Create a BOM for the assembly.
E. Create a presentation drawing of the assembly.
F. Create an isometric exploded drawing of the assembly.
G. Create an animation drawing of the assembly.
H. Consider possible interference between the nuts and ends of the fasteners both during and after assembly. Recommend an assembly sequence.

Project 6-17: Access Controller

Design an access controller based on the information given in Figure P6-17. The controller works by moving an internal cylinder up and down within the base so the cylinder aligns with output holes A and B. Liquids will enter the internal cylinder from the top, then exit the base through holes A and B. Include as many holes in the internal cylinder as necessary to create the following liquid exit combinations.

1. A open, B closed
2. A open, B open
3. A closed, B open

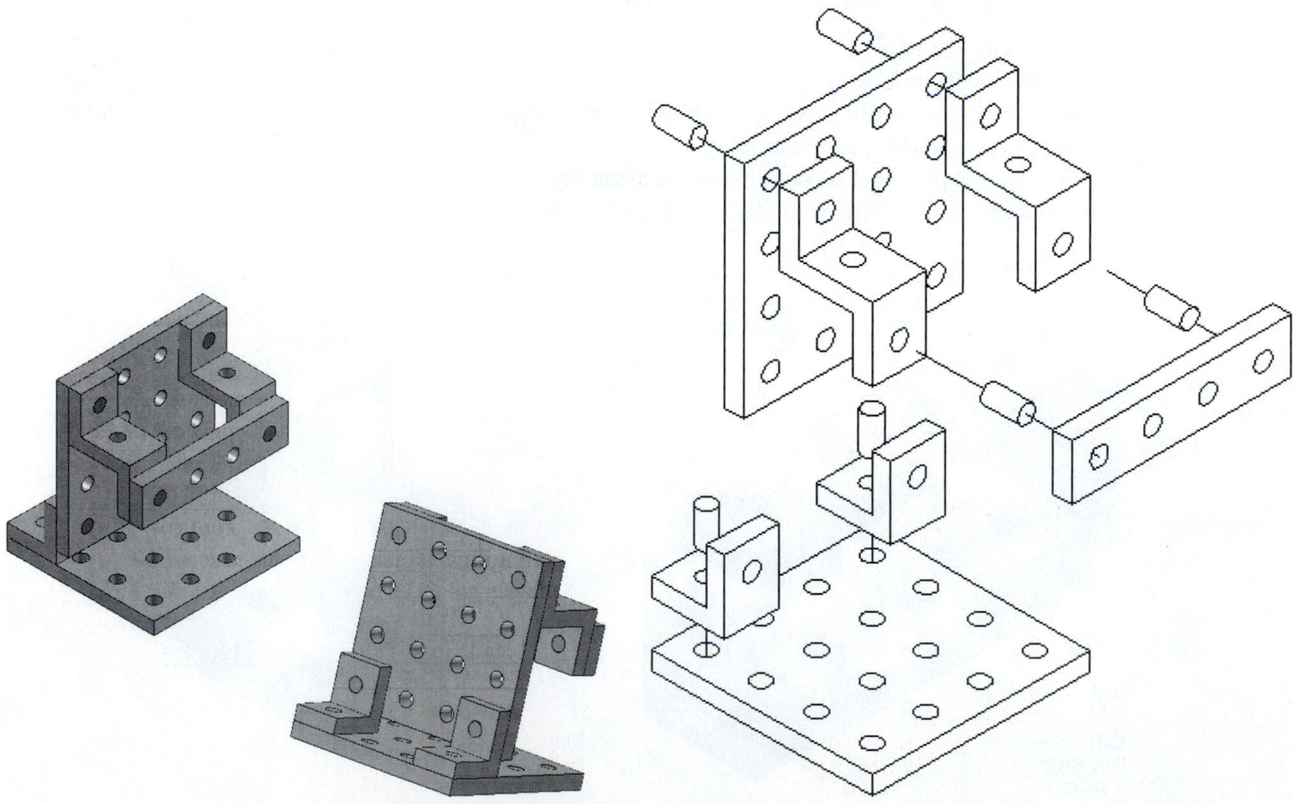

Figure P6-16

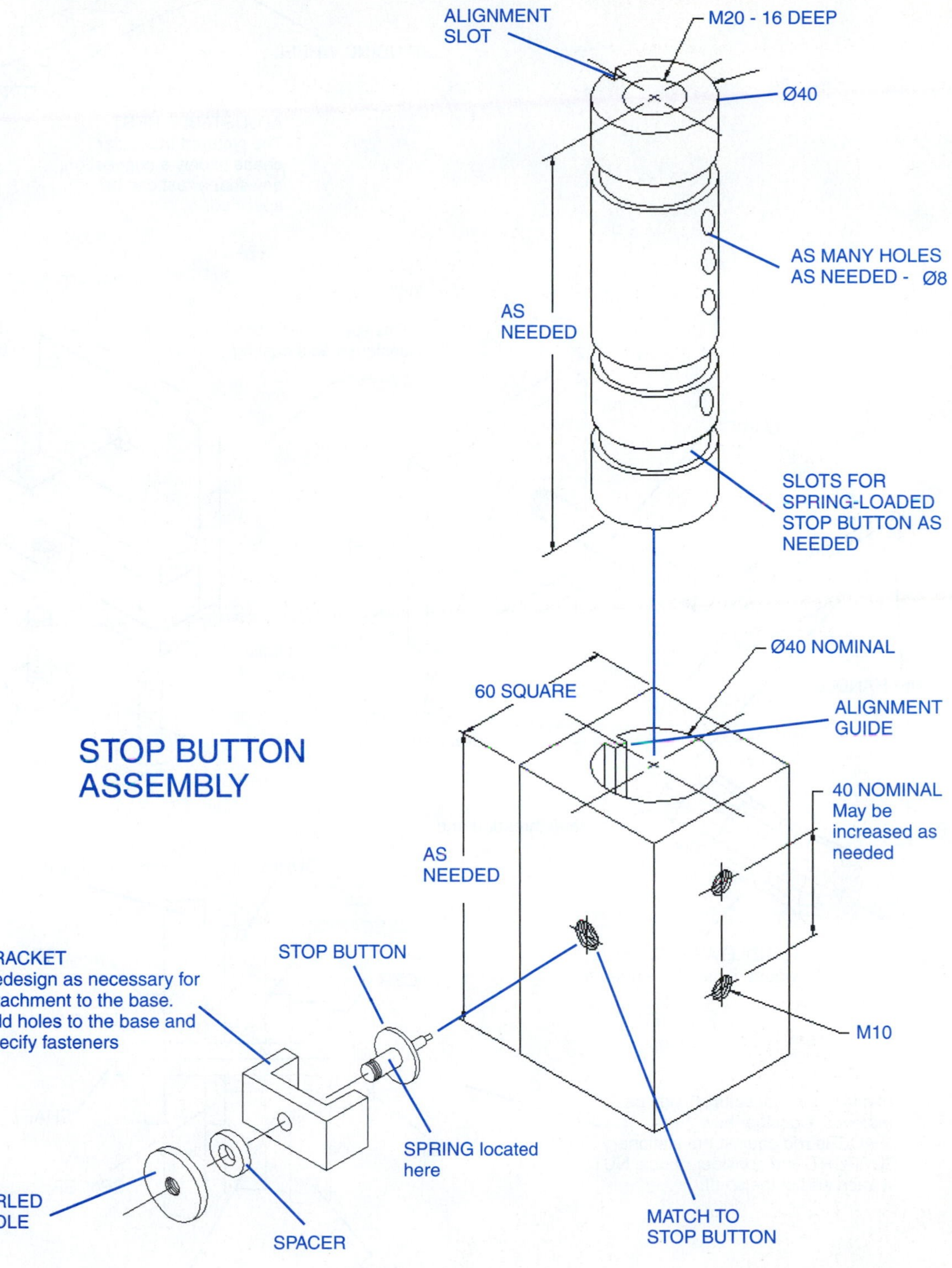
INTERNAL CYLINDER
ALIGNMENT SLOT
M20 - 16 DEEP
Ø40
AS MANY HOLES AS NEEDED - Ø8
AS NEEDED
SLOTS FOR SPRING-LOADED STOP BUTTON AS NEEDED
Ø40 NOMINAL
60 SQUARE
ALIGNMENT GUIDE
STOP BUTTON ASSEMBLY
40 NOMINAL May be increased as needed
AS NEEDED
BRACKET Redesign as necessary for attachment to the base. Add holes to the base and specify fasteners
STOP BUTTON
M10
SPRING located here
KNURLED HANDLE
SPACER
MATCH TO STOP BUTTON
BASE

Figure P6-17

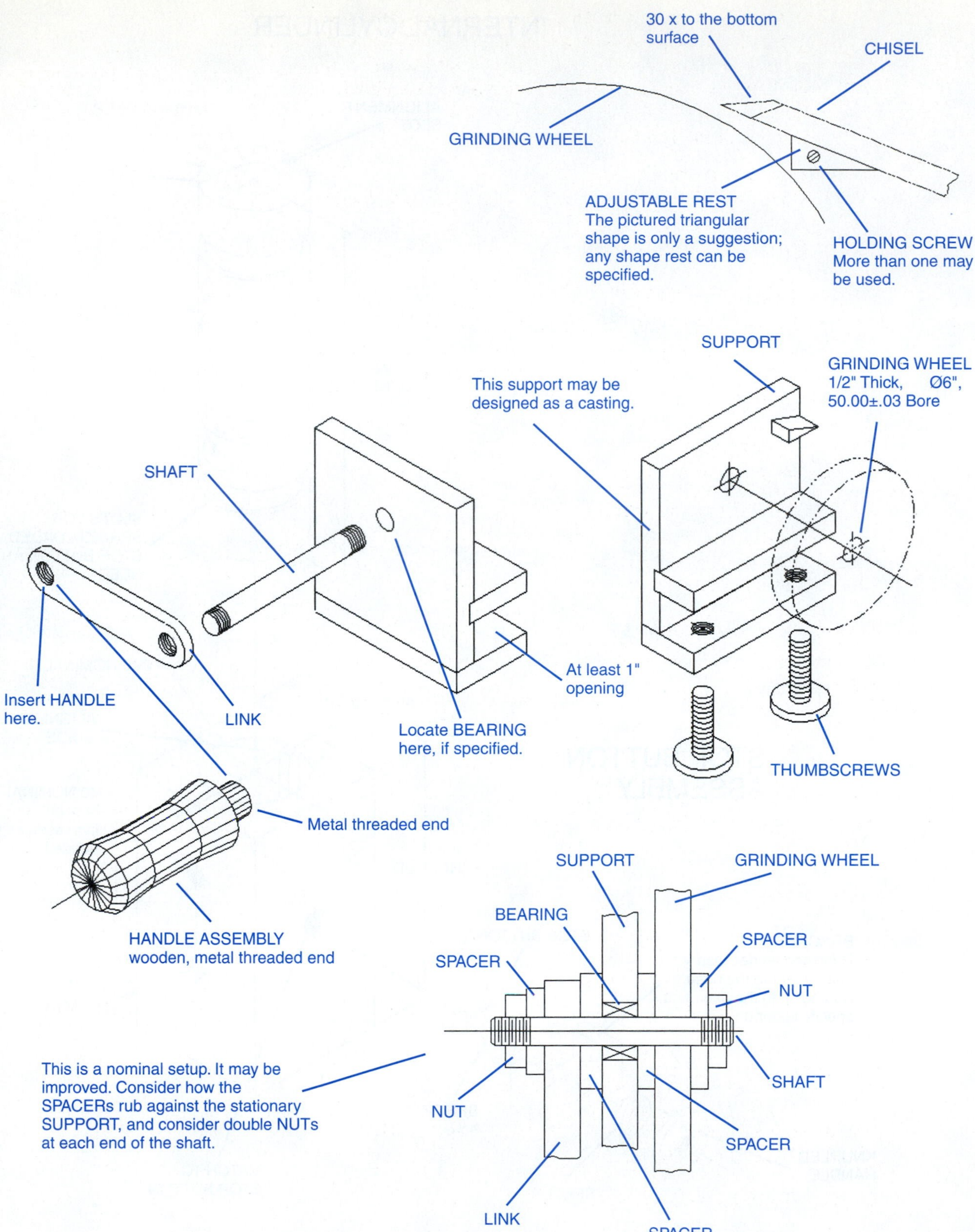
30 x to the bottom surface
CHISEL
GRINDING WHEEL
ADJUSTABLE REST
The pictured triangular shape is only a suggestion; any shape rest can be specified.
HOLDING SCREW
More than one may be used.
SUPPORT
GRINDING WHEEL
1/2" Thick, Ø6", 50.00±.03 Bore
This support may be designed as a casting.
SHAFT
At least 1" opening
Insert HANDLE here.
LINK
Locate BEARING here, if specified.
THUMBSCREWS
Metal threaded end
HANDLE ASSEMBLY
wooden, metal threaded end
SUPPORT
GRINDING WHEEL
BEARING
SPACER
SPACER
NUT
This is a nominal setup. It may be improved. Consider how the SPACERs rub against the stationary SUPPORT, and consider double NUTs at each end of the shaft.
SHAFT
NUT
SPACER
LINK
SPACER

Figure P6-18

The internal cylinder is to be held in place by an alignment key and a stop button. The stop button is to be spring-loaded so that it will always be held in place. The internal cylinder will be moved by pulling out the stop button, repositioning the cylinder, then reinserting the stop button.

Prepare the following drawings.

A. Draw an assembly drawing.
B. Draw detail drawings of each nonstandard part. Include positional tolerances for all holes.
C. Prepare a parts list.

Project 6-18: Grinding Wheel

Design a hand-operated grinding wheel as shown in Figure P6-18 specifically for sharpening a chisel. The chisel is to be located on an adjustable rest while it is being sharpened. The mechanism should be able to be clamped to a table during operation using two thumbscrews. A standard grinding wheel is Ø6.00″ and 1/2″ thick, and has an internal mounting hole with a 50.00±.03 bore.

Prepare the following drawings.

A. Draw an assembly drawing.
B. Draw detail drawings of each nonstandard part. Include positional tolerances for all holes.
C. Prepare a parts list.

Project 6-19: Millimeters

Given the assembly shown in Figure P6-19 on page 288, add the following fasteners.

1. Create an assembly drawing.
2. Create a parts list including assembly numbers.
3. Create a dimensioned drawing of the Support Block and specify a dimension for each hole including the thread size and the depth required.

Fasteners:

A.

1. M10 3 35 HEX HEAD BOLT
2. M10 3 35 HEX HEAD BOLT
3. M10 3 30 HEX HEAD BOLT
4. M10 3 25 HEX HEAD BOLT

B.

1. M10 3 1.5 3 35 HEX HEAD BOLT
2. M8 3 35 ROUND HEAD BOLT
3. M10 3 30 HEXAGON SOCKET HEAD CAP SCREW
4. M6 3 30 SQUARE BOLT

Project 6-20: Inches

1. Create an assembly drawing.
2. Create a parts list including assembly numbers.
3. Create a dimensioned drawing of the Base and specify a dimension for each hole including the thread size and the depth required.

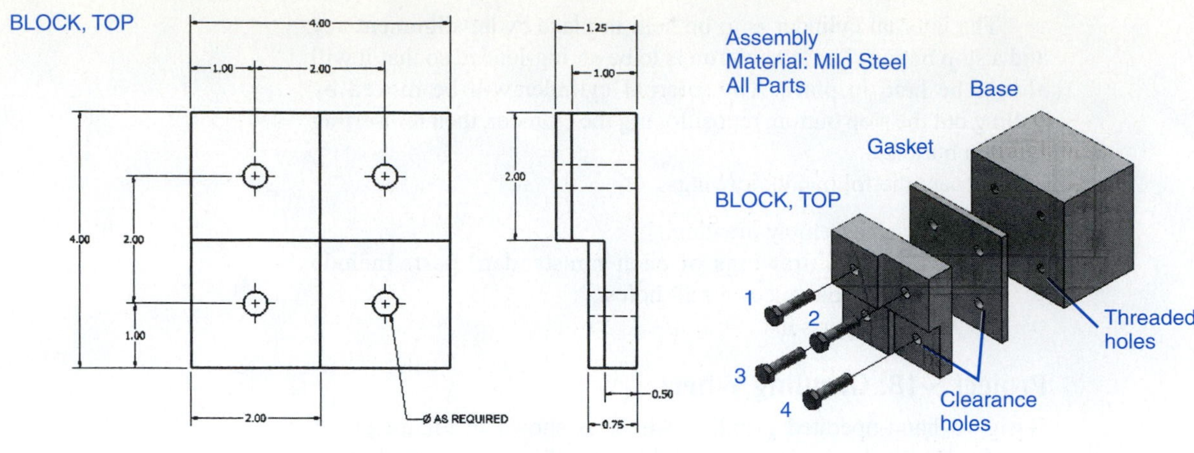

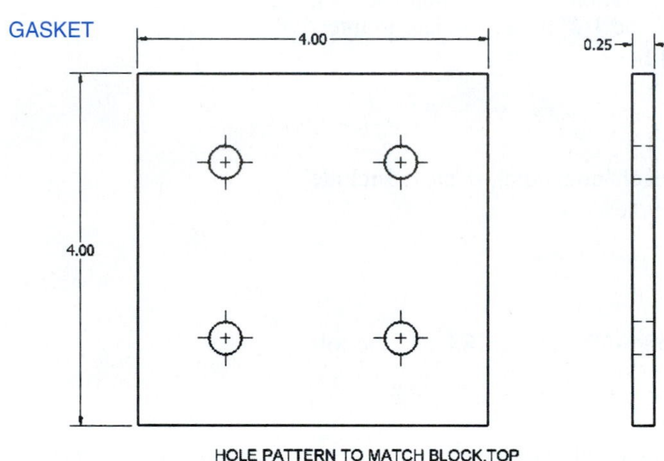

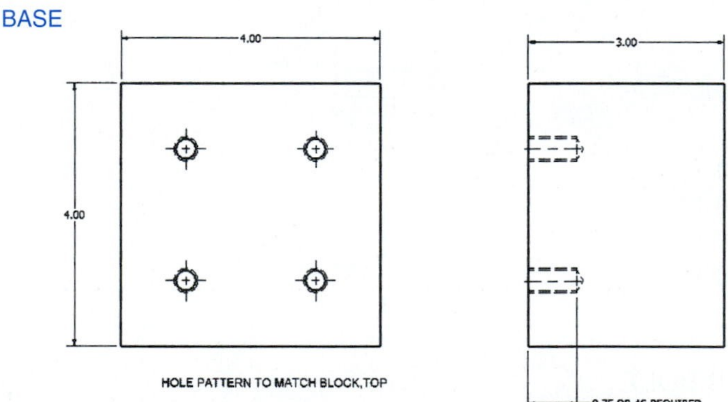

Figure P6-19

Fasteners:

A.

1. 3/8-16UNC 3 2.50 HEX HEAD SCREW
2. 1/4-20 UNC 3 2.00 HEX HEAD SCREW
3. 7/16-14 UNC 3 1.75 HEX HEAD SCREW
4. 5/16-18 UNC 3 2.25 HEX HEAD SCREW

B.

1. 1/4-28 UNF 3 2.00 HEX HEAD SCREW
2. #8(.164)-32 UNC 3 2.00 INDENTED LARGE HEX HEAD SCREW

3. 3/8-16 UNC 3 1.75 CROSS RECESSED PAN HEAD MACHINE SCREW
4. 5/16-18 UNC 3 1.75 HEXAGON SOCKET HEAD CAP SCREW

Project 6-21: Millimeters

Given the collar shown in Figure P6-21, add the following setscrews.

1. Create an assembly drawing.
2. Create a parts list.
3. Create a dimensioned drawing of the collar. Specify a thread specification for each hole as required by the designated set screw.

Holes:

A.

1. M4 3 6-AS1421 DOG POINT-METRIC
2. M3 × 3 BROACHED HEXAGON SOCKET SET SCREW - FLAT POINT - METRIC
3. M2.5 × 4 BS4168: PART 3 HEXAGON SOCKET SET SCREW - CONE POINT - METRIC
4. M4 × 5 FORGED HEXAGON SOCKET SET SCREW - HALF DOG POINT - METRIC

B.

1. M2 × 4 JIS B 1117 TRUNCATED CONE POINT - METRIC
2. M3 × 6 JIS B 1117 LONG DOG POINT - METRIC
3. M4 × 5 SS-ISO 4766 SLOTTED HEADLESS SET SCREW
4. M1.6 × 4 JIS B 1117 FLAT POINT HEXAGON SOCKET SET SCREW

Project 6-22: Inches

Given the collar shown in Figure P6-22, add the following setscrews.

1. Create an assembly drawing.
2. Create a parts list.
3. Create a dimensioned drawing of the collar. Specify a thread specification for each hole as required by the designated setscrew.

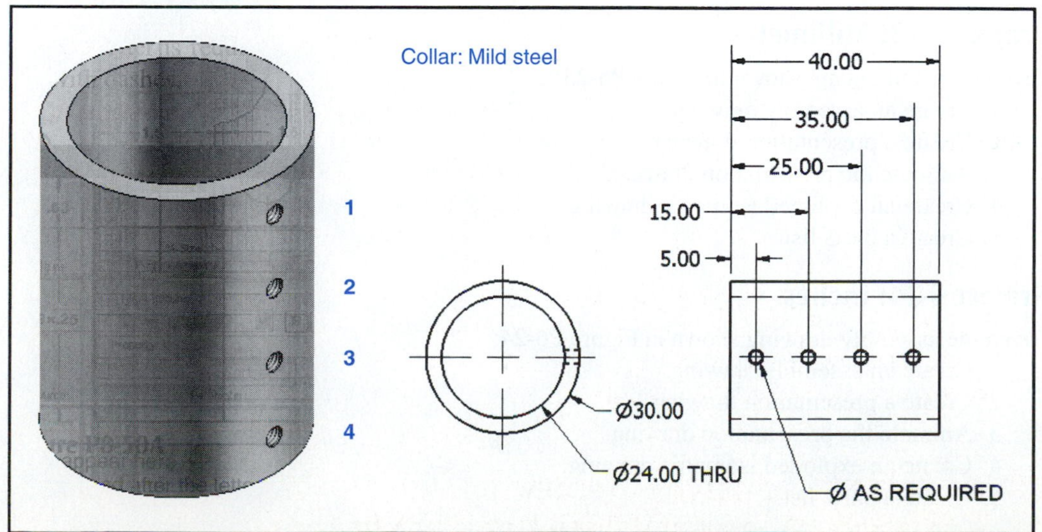

Figure P6-21

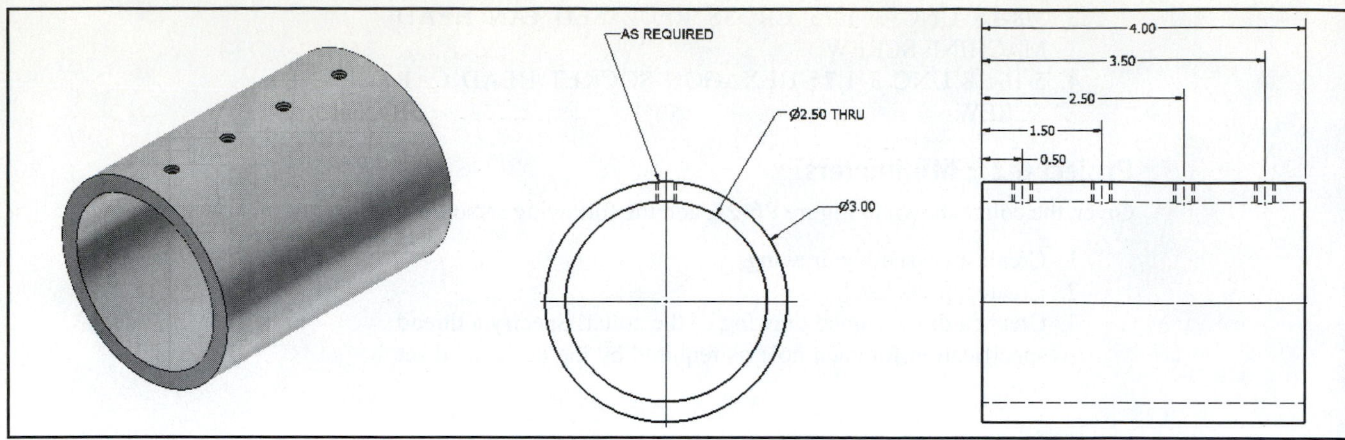

Figure P6-22

Holes:

A.

1. #10 (0.190) × .375 SQUARE HEAD SET SCREW - DOG POINT - INCH
2. #6 (0.138) × .125 SLOTTED HEADLESS SET SCREW - FLAT POINT - INCH
3. #8 (0.164) × .375 TYPE C - SPLINE SOCKET SET SCREW - CUP POINT - INCH
4. #5(0.126) × .45 HEXAGON SOCKET SET SCREW - UNBRAKO CONE POINT - INCH

B.

1. #6 (0.138) × .25 TYPE D - SPLINE SOCKET SET SCREW - CUP POINT - INCH
2. #8 (0.164) × .1875 SLOTTED HEADLESS SET SCREW - DOG POINT - INCH
3. #10 (0.190) × .58 HEXAGON SOCKET SET SCREW - FLAT POINT - INCH
4. #6(0.138) × .3125 SPLINE SOCKET SET SCREW - HALF DOG POINT - INCH

Project 6-23: Millimeters

Given the components shown in Figure P6-23:

1. Create an assembly drawing.
2. Create a presentation drawing.
3. Animate the presentation drawing.
4. Create an exploded isometric drawing.
5. Create a parts list.

Project 6-24: Inches

Given the assembly drawing shown in Figure P6-24:

1. Create an assembly drawing.
2. Create a presentation drawing.
3. Animate the presentation drawing.
4. Create an exploded isometric drawing.
5. Create a parts list.

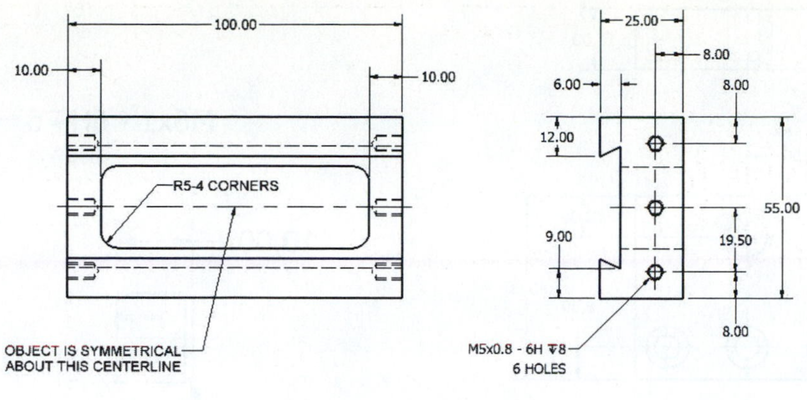
100.00
10.00
10.00
R5-4 CORNERS
OBJECT IS SYMMETRICAL ABOUT THIS CENTERLINE
25.00
8.00
6.00
8.00
12.00
55.00
9.00
19.50
8.00
M5x0.8 - 6H ⧩8
6 HOLES

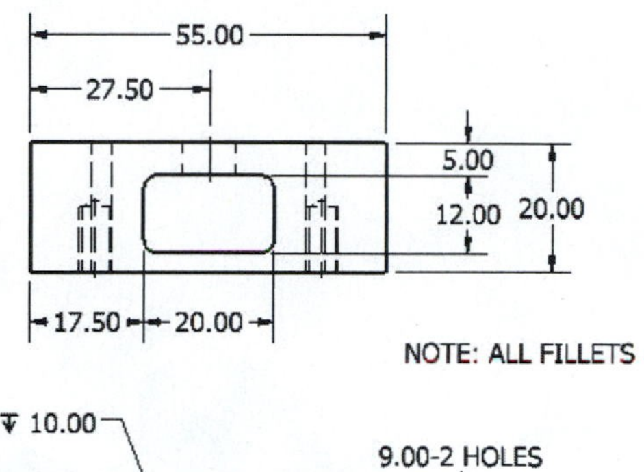
55.00
27.50
5.00
12.00
20.00
17.50
20.00
NOTE: ALL FILLETS = R2

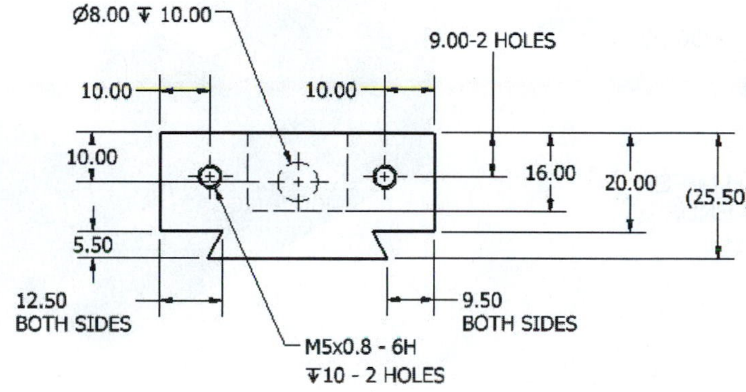
Ø8.00 ⧩ 10.00
9.00-2 HOLES
10.00
10.00
10.00
16.00
20.00
(25.50)
5.50
12.50 BOTH SIDES
9.50 BOTH SIDES
M5x0.8 - 6H ⧩10 - 2 HOLES

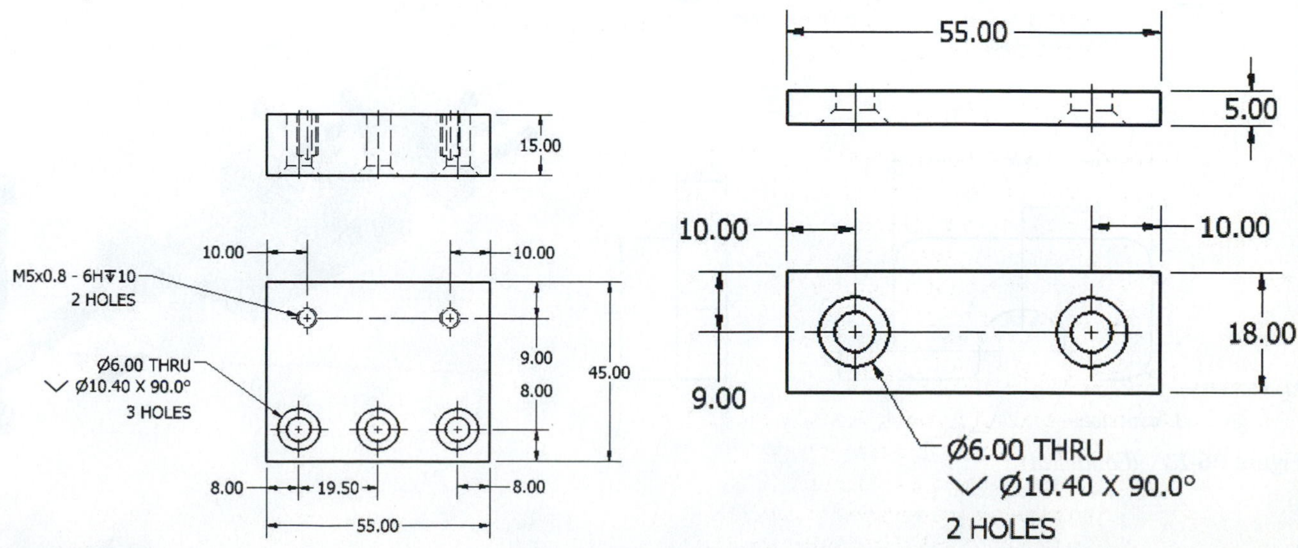
15.00
10.00
10.00
M5x0.8 - 6H⧩10 2 HOLES
Ø6.00 THRU ⌵ Ø10.40 X 90.0° 3 HOLES
9.00
8.00
45.00
8.00
19.50
8.00
55.00
55.00
5.00
10.00
10.00
18.00
9.00
Ø6.00 THRU ⌵ Ø10.40 X 90.0° 2 HOLES

Figure P6-23

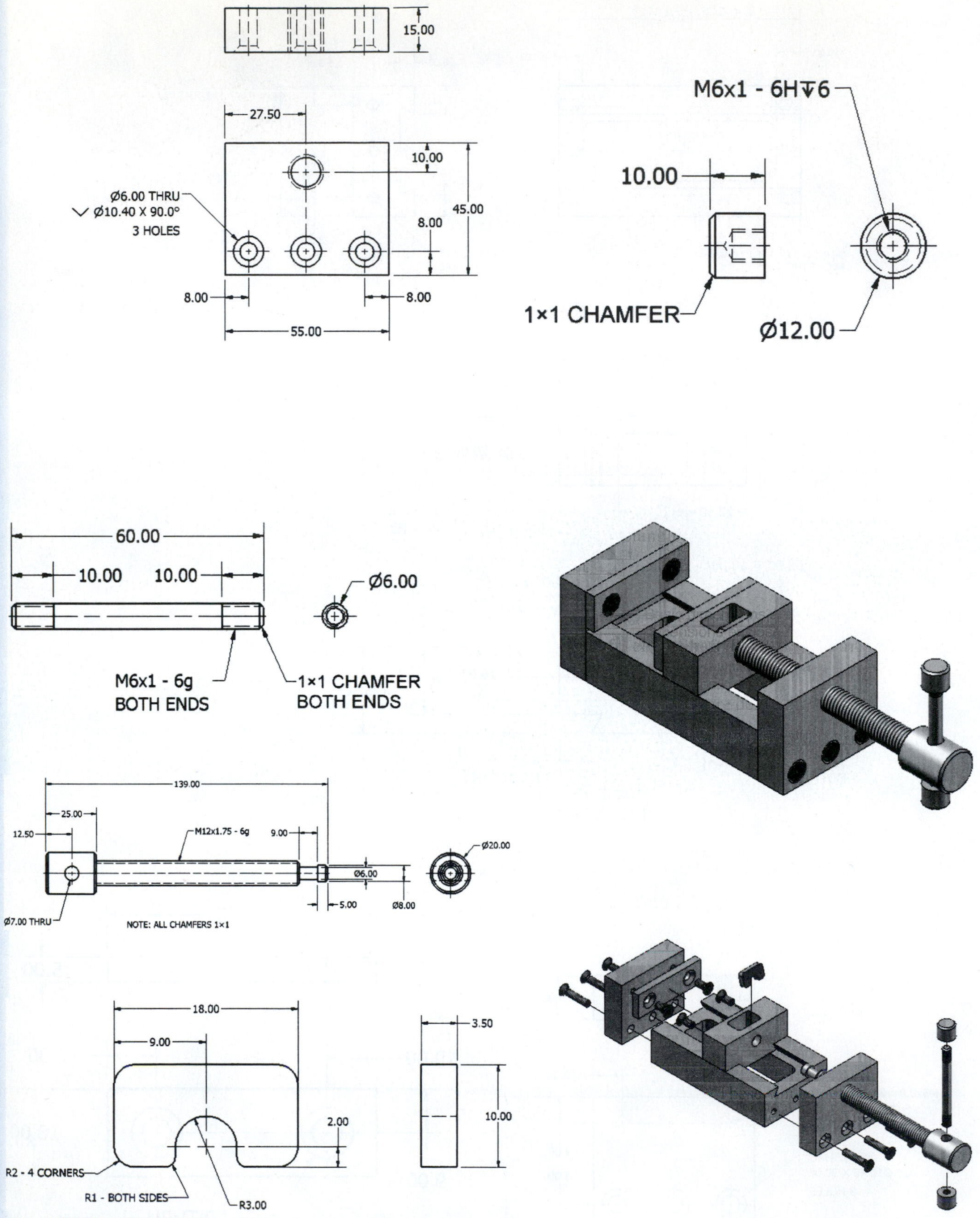

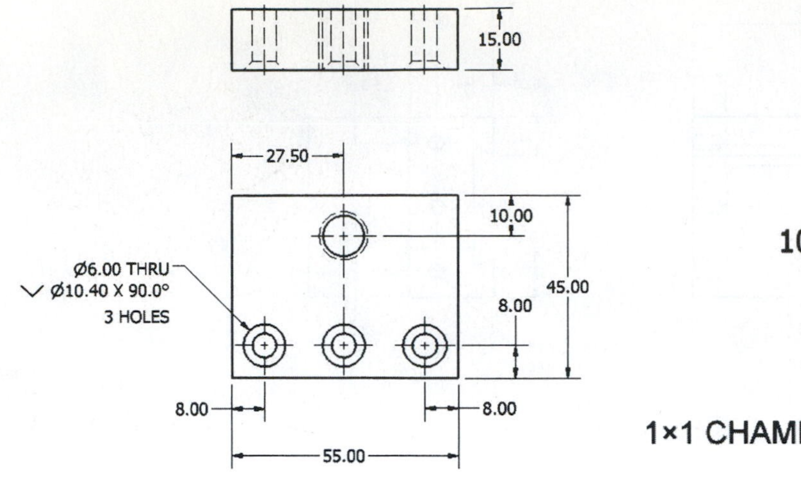

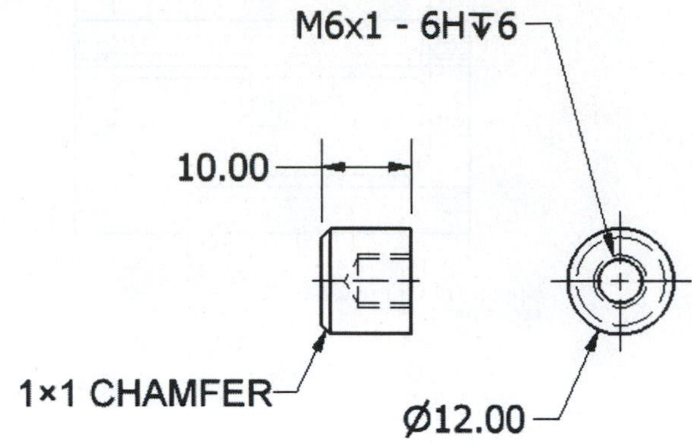

60.00
10.00 10.00
Ø6.00
M6x1 - 6g
BOTH ENDS
1×1 CHAMFER
BOTH ENDS

139.00
25.00
12.50
M12x1.75 - 6g
9.00
Ø20.00
Ø6.00
5.00
Ø8.00
Ø7.00 THRU
NOTE: ALL CHAMFERS 1×1

18.00
9.00
3.50
2.00
10.00
R2 - 4 CORNERS
R1 - BOTH SIDES
R3.00

Figure P6-23 *(Continued)*

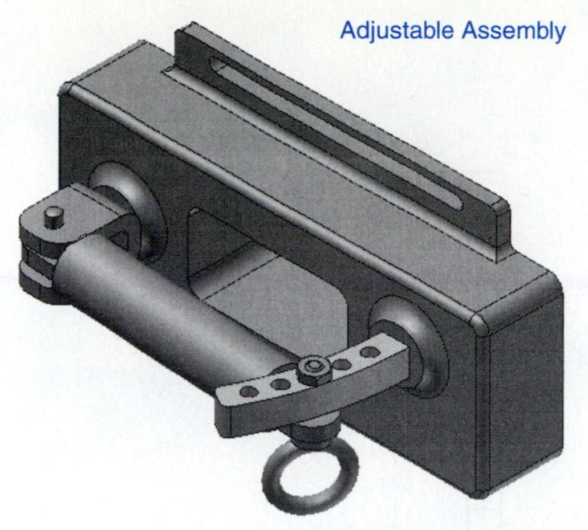

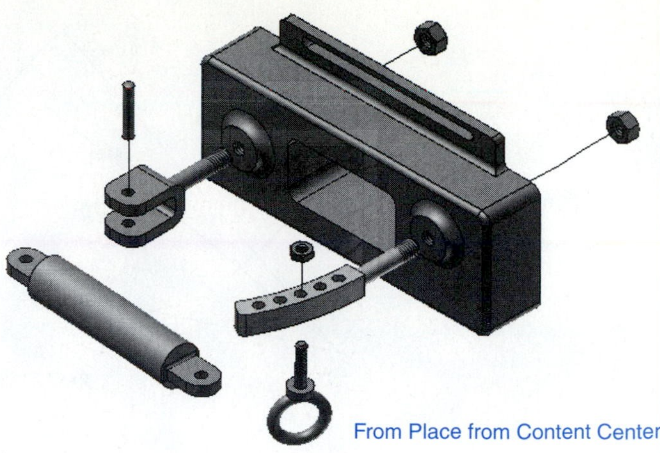

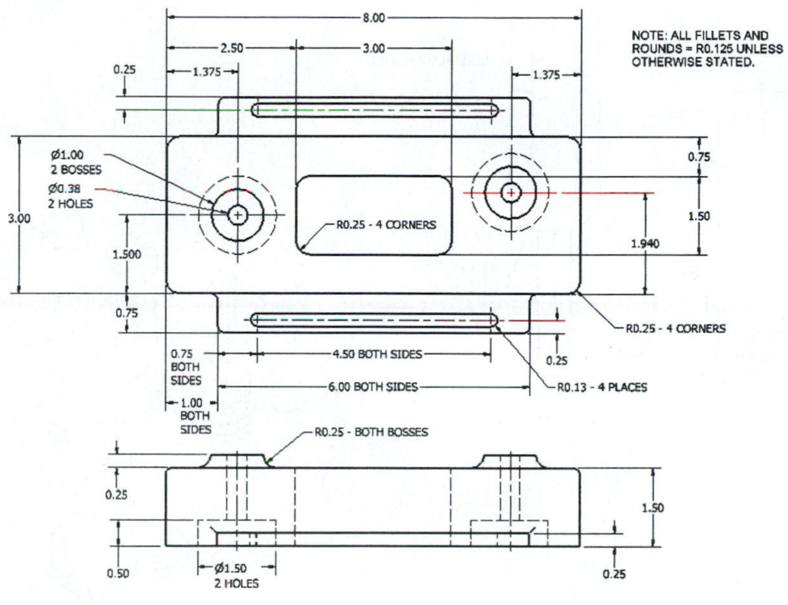

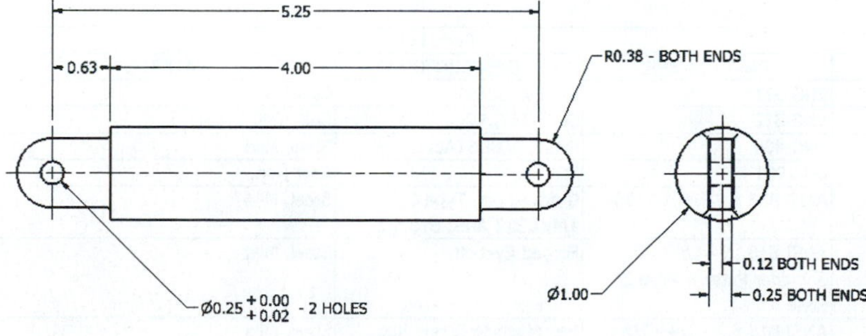

NOTE: ALL FILLETS = R 0.125

Figure P6-24

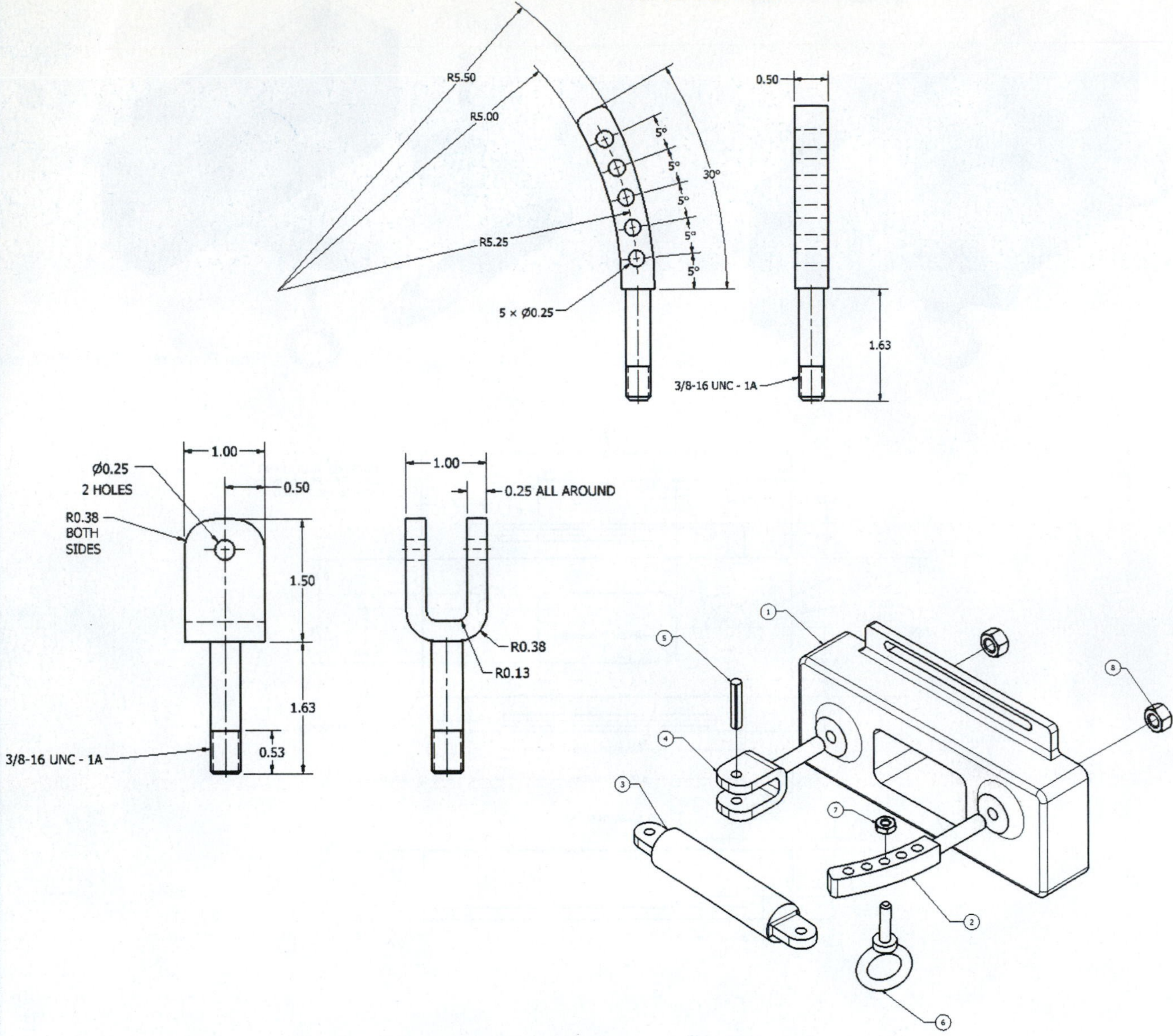

Parts List				
ITEM	PART NUMBER	DESCRIPTION	MATERIAL	QTY
1	ENG-311	BASE#4, CAST	Cast Iron	1
2	ENG-312	SUPPORT, ROUNDED	SAE 1040 STEEL	1
3	ENG-404	POST, ADJUSTABLE	Steel, Mild	1
4	BU-1964	YOKE	Cast Iron	1
5	ANSI B18.8.2 1/4x1.3120	Grooved pin, Type C - 1/4x1.312 ANSI B18.8.2	Steel, Mild	1
6	ANSI B18.15 - 1/4 - 20. Shoulder Pattern Type 2 - Style A	Forged Eyebolt	Steel, Mild	1
7	ANSI B18.6.3 - 1/4 - 20	Hex Machine Screw Nut	Steel, Mild	1
8	ANSI B18.2.2 - 3/8 - 16	Hex Nut	Steel, Mild	2

Figure 6-24 *(Continued)*

Dimensioning

7

Chapter Objectives

- Teach how to dimension objects.
- Present ANSI standards and conventions.
- Show how to dimension different shapes and features.
- Introduce 3D dimensioning.

Introduction

Inventor uses two types of dimensions: model and drawing. ***Model dimensions*** are created as the model is being constructed and may be edited to change the shape of a model. ***Drawing dimensions*** can be edited, but the changes will not change the shape of the model. If the shape of a model is changed, the drawing dimensions associated with the revised surfaces will change to reflect the new values.

model dimension: A dimension created as a model is being constructed; it may be edited to change the shape of a model.

drawing dimension: A dimension attached to a specified distance on a drawing that can be edited without changing the shape of the model.

Dimensions are usually applied to a drawing using either American National Standards Institute (ANSI) or International Organization for Standardization (ISO) standards. If English units are selected when a new drawing is started, the ANSI inch standards (ANSI (in).idw) will be invoked. If metric units are selected, the ISO standards (ISO.idw) may be invoked. This book uses ANSI standards for both inch and metric units.

Figure 7-1 shows a drawing that includes only the parametric dimensions. The **Metric** option was selected before the model was drawn. The model dimensions were created automatically as the model was created. The **General Dimension** tool was used to edit the sketch dimensions.

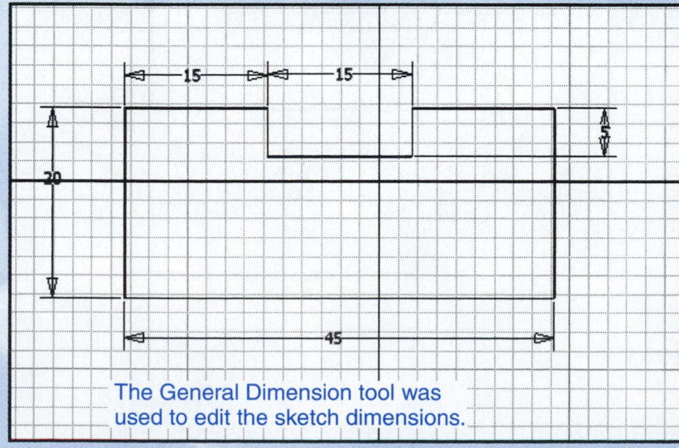

Figure 7-1

Terminology and Conventions—ANSI

Some Common Terms

Figure 7-2 shows both ANSI and ISO style dimensions. The terms apply to both styles.

Dimension Lines: In mechanical drawings, lines between extension lines that end with an arrowhead and include a numerical dimensional value located within the line.

Extension Lines: Lines that extend away from an object and allow dimensions to be located off the surface of an object.

Leader Lines: Lines drawn at an angle, not horizontal or vertical, that are used to dimension specific shapes such as holes. The start point of a leader line includes an arrowhead. Numerical values are drawn at the end opposite the arrowhead.

Linear Dimensions: Dimensions that define the straight-line distance between two points.

Angular Dimensions: Dimensions that define the angular value, measured in degrees, between two straight lines.

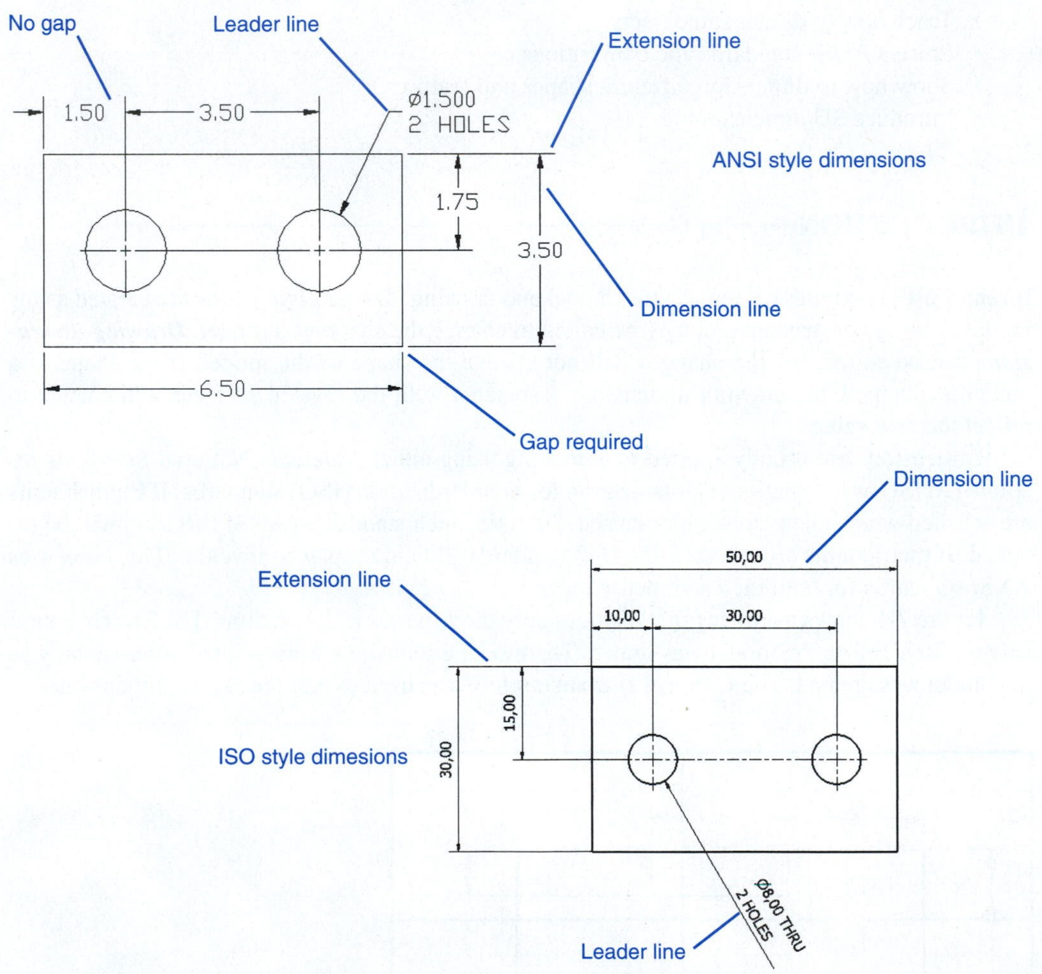

Figure 7-2

Some Dimensioning Conventions

See Figure 7-3.

1. Dimension lines should be drawn evenly spaced; that is, the distance between dimension lines should be uniform. A general rule of thumb is to locate dimension lines about ½ in. or 15 mm apart.

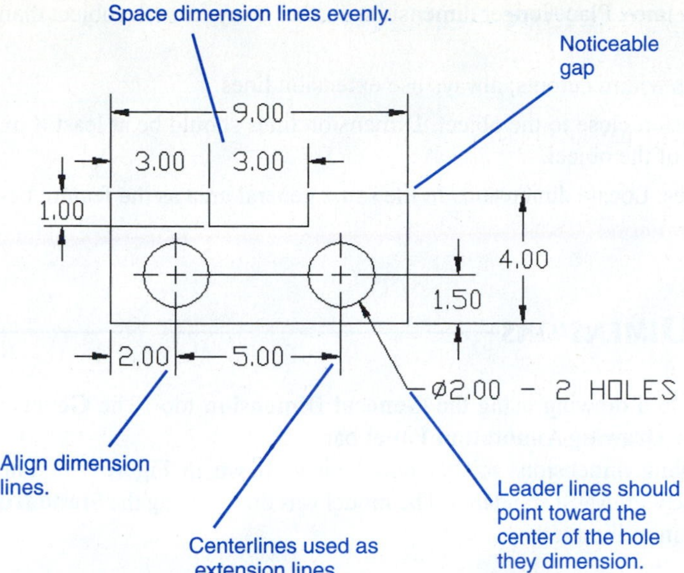

Figure 7-3

2. There should a noticeable gap between the edge of a part and the beginning of an extension line. This serves as a visual break between the object and the extension line. The visual difference between the line types can be enhanced by using different colors for the two types of lines.
3. Leader lines are used to define the size of holes and should be positioned so that the arrowhead points toward the center of the hole.
4. Centerlines may be used as extension lines. No gap is used when a centerline is extended beyond the edge lines of an object.
5. Align dimension lines whenever possible to give the drawing a neat, organized appearance.

Some Common Errors to Avoid

See Figure 7-4.

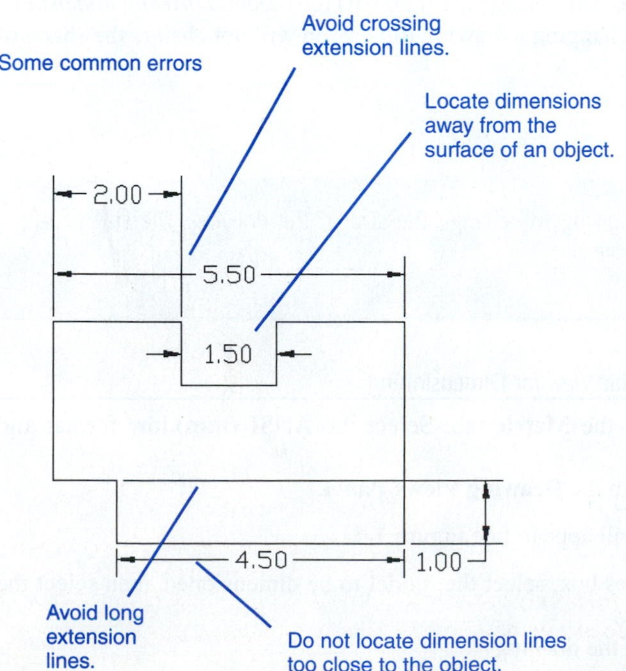

Figure 7-4

1. Avoid crossing extension lines. Place longer dimensions farther away from the object than shorter dimensions.
2. Do not locate dimensions within cutouts; always use extension lines.
3. Do not locate any dimension close to the object. Dimension lines should be at least ½ in. or 15 mm from the edge of the object.
4. Avoid long extension lines. Locate dimensions in the same general area as the feature being defined.

CREATING DRAWING DIMENSIONS

Drawing dimensions are added to a drawing using the **General Dimension** tool. The **General Dimension** tool is located on the **Drawing Annotation Panel** bar.

This section will add drawing dimensions to the model view shown in Figure 7-5. The dimensions will be in compliance with ANSI standards. The model was drawn using the **Standard (mm).ipt** format. All values are in millimeters.

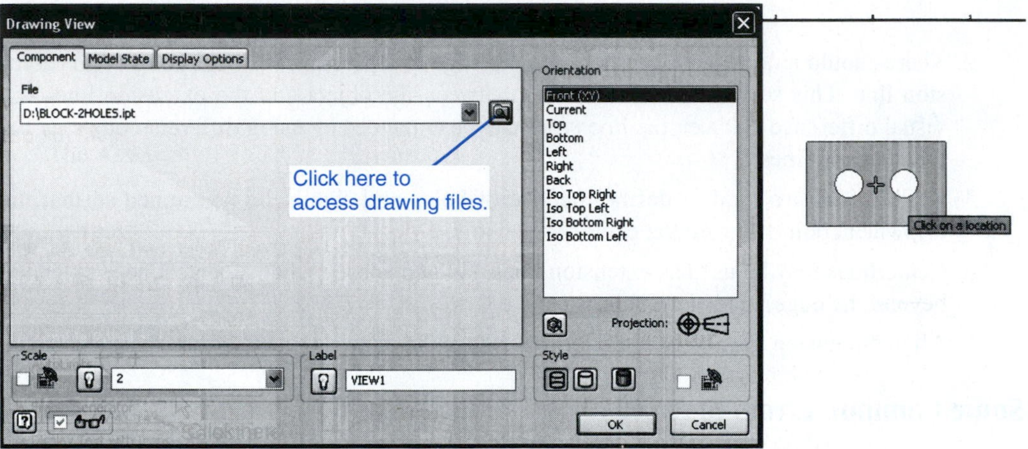

Figure 7-5

Drawing dimensions are different from model dimensions. *Model dimensions* are created as the model is created and can be used to edit (change the shape of) the model. *Drawing dimensions* are attached to a specified distance. Changing a drawing dimension will not change the shape of the model.

TIP Changing the scale of a drawing will change the size of the drawing. The size of the dimension text will not change.

Exercise 7-1: Creating an Orthographic View for Dimensioning

1. Click on the **New** tool, then the **Metric** tab. Select the **ANSI (mm).idw** format and click **OK.**
2. Select the **Base View** tool from the **Drawing Views Panel.**

The **Drawing View** dialog box will appear. See Figure 7-5.

3. Select the **Explore Directories** box, select the model to be dimensioned, then select the appropriate orthographic view.
4. If needed, change the scale of the drawing.
5. Click **OK.**

Exercise 7-2: Accessing the Drawing Annotation Panel Bar

1. Move the cursor into the panel bar area, and right-click the mouse.

A small dialog box will appear. See Figure 7-6.

2. Select the **Drawing Annotation Panel** option.

The panel bar will change to the **Drawing Annotation Panel** bar shown in Figure 7-7. The **General Dimension** tool is the first tool listed.

A model created using the Standard (mm) .ipt format. All dimensions are in millimeters.

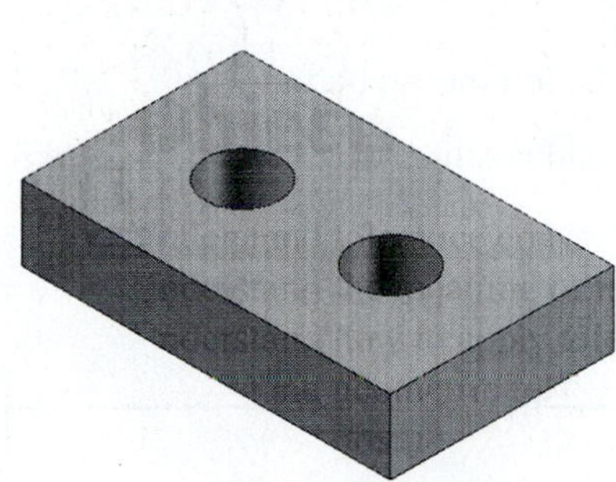

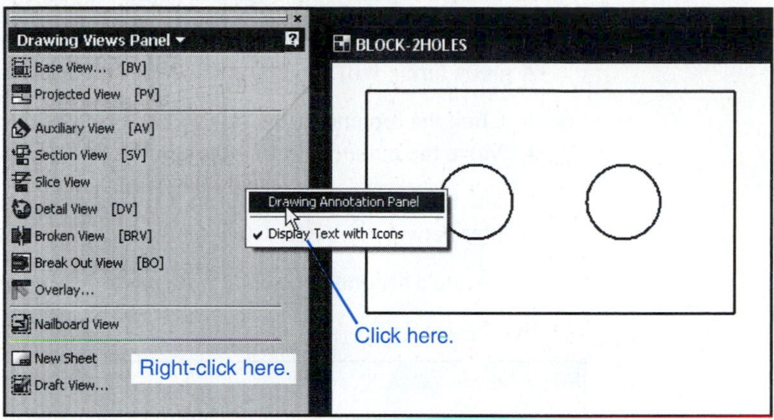

Figure 7-6

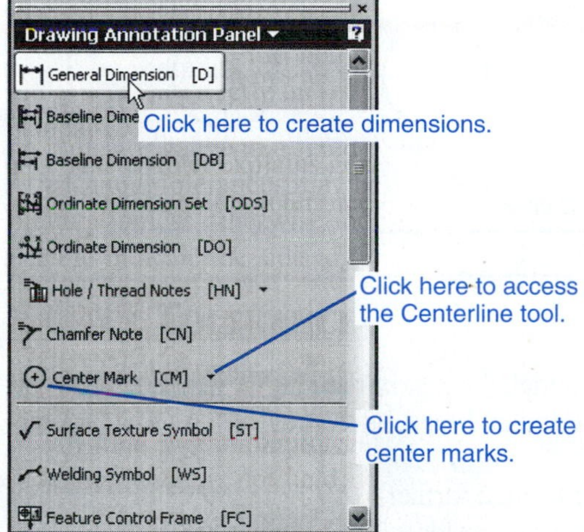

Figure 7-7

Adding Centerlines to Holes

Figure 7-8 shows a front orthographic view of the model to be dimensioned.

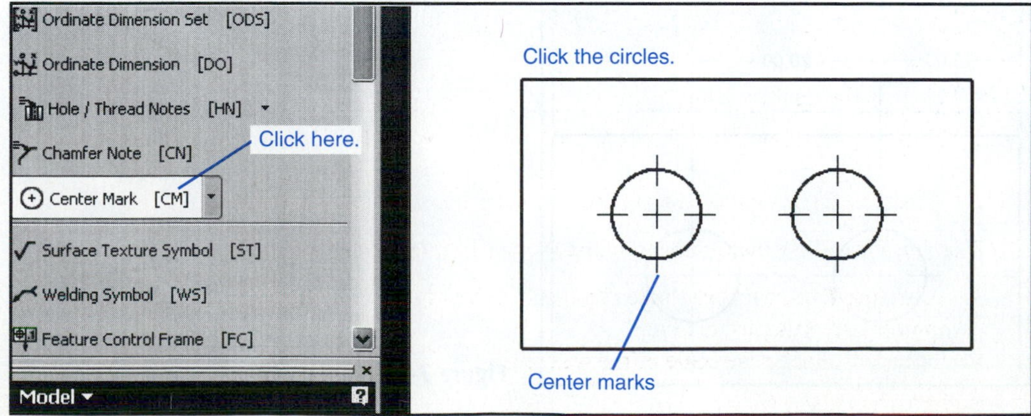

Figure 7-8

Centerlines were added to the model using the **Center Mark** tool located on the **Drawing Annotation Panel** bar.

1. Click the **Center Mark** tool.
2. Click the edge of the circles.
3. Use the **Centerline** option to draw a centerline between the hole. See Figure 7-9.

Exercise 7-3: Adding Horizontal Dimensions

1. Click the **General Dimension** tool.
2. Move the cursor into the drawing area and first click the upper left corner of the model.

A green circle will appear on the corner, indicating that it has been selected.

3. Click the top end of the left hole's vertical centerline.
4. Move the cursor, locating the dimension, then press the left mouse button.

Locate the dimension away from the edge of the model and position the text in the approximate center between the two extension lines.

5. Locate a second horizontal dimension between the two vertical hole centerlines.

See Figure 7-10.

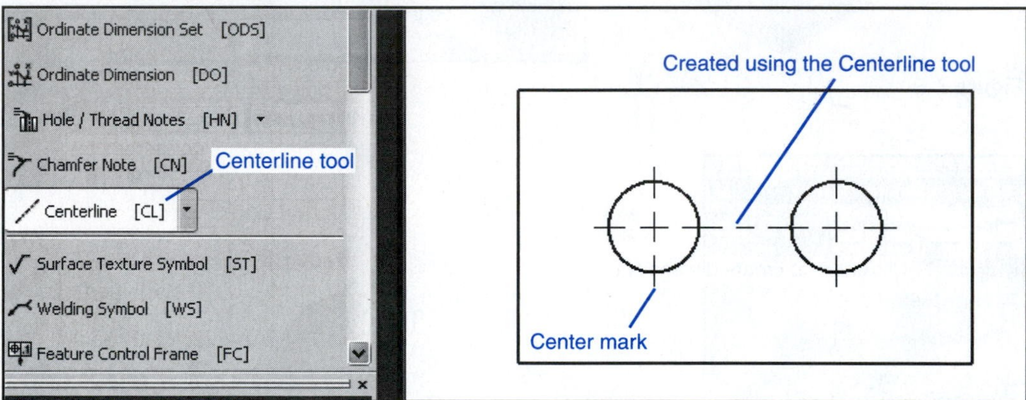

Figure 7-9

Overall Dimensions

overall dimensions: Dimensions that define the outside sizes of a model: the maximum length, width, and height.

Overall dimensions define the outside sizes of a model, the maximum length, width, and height. It is important that overall dimensions be easy to find and read, as they are often used to determine the stock sizes needed to produce the model.

Convention calls for overall dimensions to be located farther away from the model than any other dimensions. In Figure 7-10 the 50 overall dimension was located above the other two horizontal

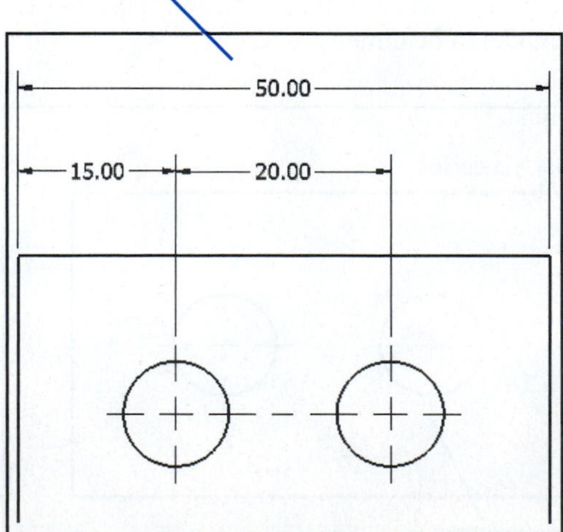

Figure 7-10

dimensions, that is, farther away from the model. The dimension could also have been located below the model.

Note that the spacing between the model's edge and the two horizontal dimensions is approximately equal to the distance between the overall dimension and the two horizontal dimensions.

Vertical Dimensions

ANSI standards call for the text of vertical dimensions to be written *unidirectionally.* This means that the text should be written horizontally and be read from left to right. Figure 7-11 shows two

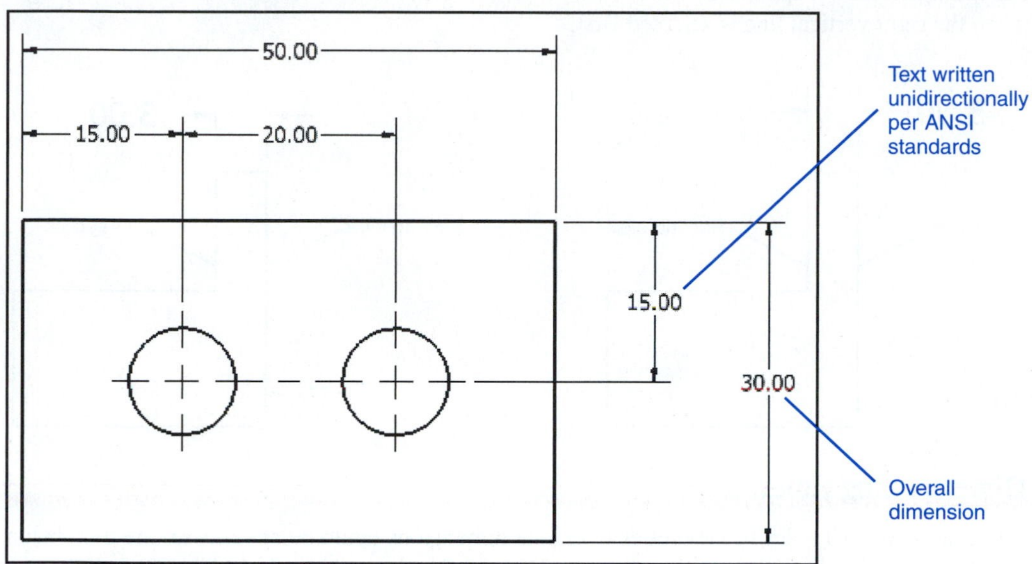

Figure 7-11

vertical dimensions added to the model. Both use unidirectional text. Note also that the overall height dimension is located the farthest away from the model's edge.

Exercise 7-4: Creating Unidirectional Text

1. Click the **Format** heading at the top of the screen, then select the **Styles Editor** option.

The **Styles and Standards Editor** dialog box will appear.

2. Click the **+ sign** to the left of the **Dimension** listing.
3. Select the **Default - mm (ANSI)** option.

The **Styles and Standards Editor** dialog box will change. See Figure 7-12.

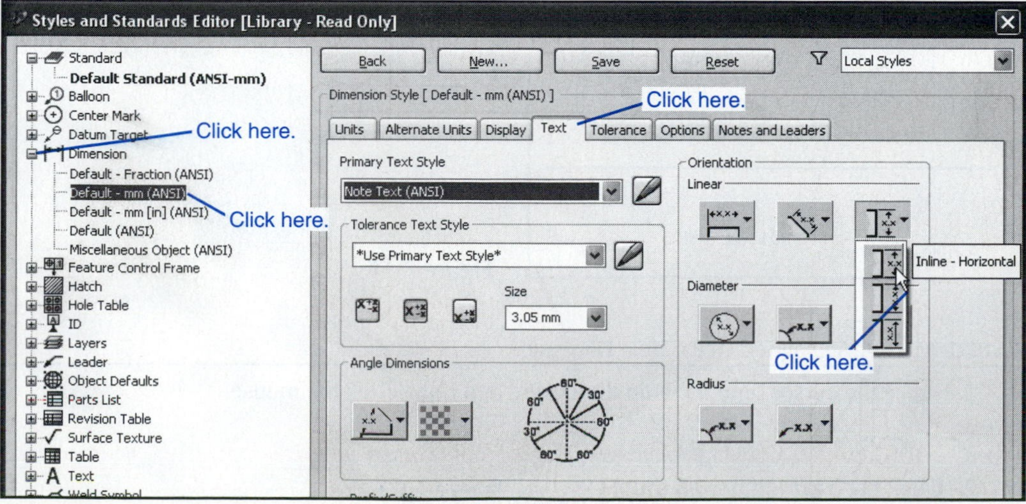

Figure 7-12

4. Select the **Text** tab.
5. Go to the **Orientation** box and select the **Inline - Horizontal** option as shown.
6. Click the **Save** box, then click **Done.**

All vertical dimensions will now be written using unidirectional text.

Positioning Dimension Text

You can control the location of dimension text by the sequence used to define the distance to be dimensioned. See Figure 7-13. The text will appear on the side of the first selected edge or point. For example, the text appears on the left when the left vertical line is selected first and on the right when the right vertical line is selected first.

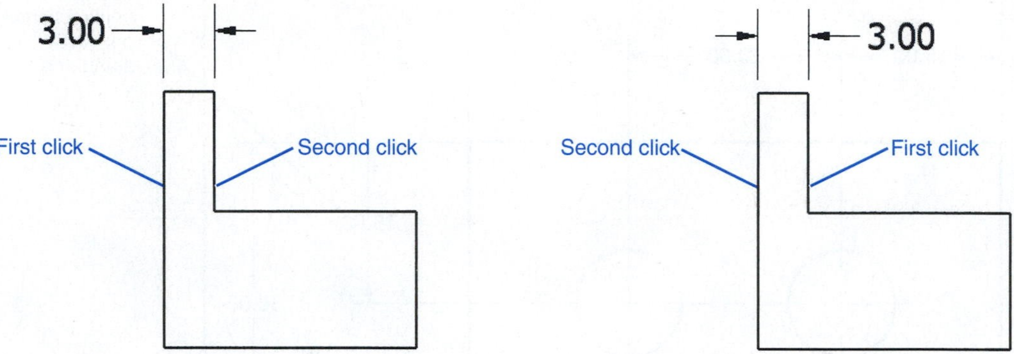

Figure 7-13

Dimensioning Holes

There are two ∅10 holes in the model. Two hole dimensions could be applied, or one dimension could be used with the additional note 10 - 2 HOLES. In general, it is desirable to use as few dimensions as possible to clearly and completely define the model's size. This helps prevent a cluttered and confusing drawing.

1. Click the **Hole/Thread Notes** tool, then the edge of one of the holes.
2. Hold the left mouse button down and drag the dimension away from the model.
3. Locate the hole dimension and left-click the mouse.
4. Right-click the mouse and select the **Done** option.

See Figure 7-14.

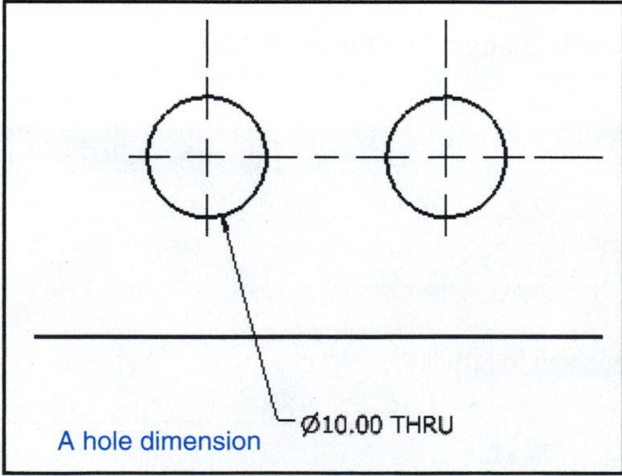

Figure 7-14

Exercise 7-5: Adding Text to the Hole Dimension

1. Move the cursor onto the hole dimension and right-click the mouse.

See Figure 7-15.

2. Click the **Text** option.

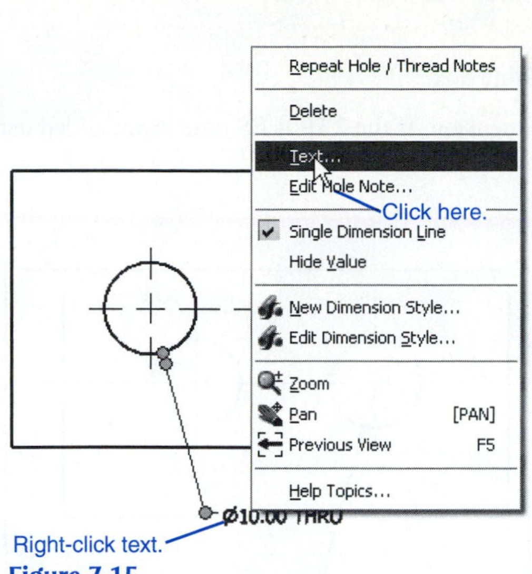

Figure 7-15

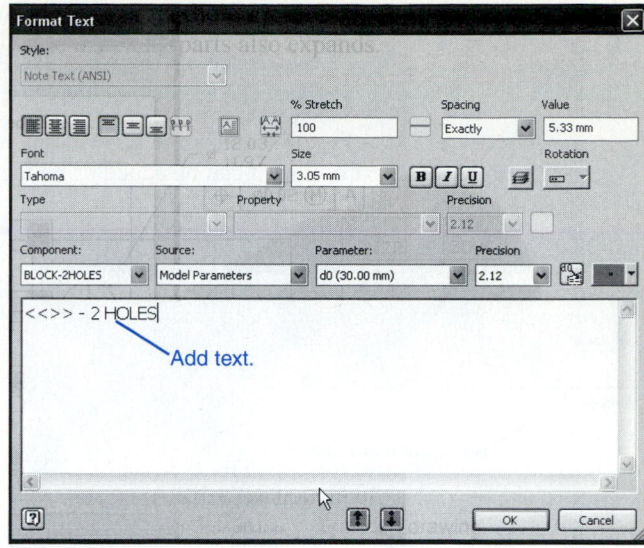

Figure 7-16

The **Format Text** dialog box will appear. See Figure 7-16. The <<>> symbol represents the existing text.

3. Locate the cursor to the right of the <<>> symbol and type **- 2 HOLES.**
4. Click the **OK** box.

Figure 7-17 shows the resulting dimension.

TIP Drawing convention calls for all drawing text to use uppercase letters.

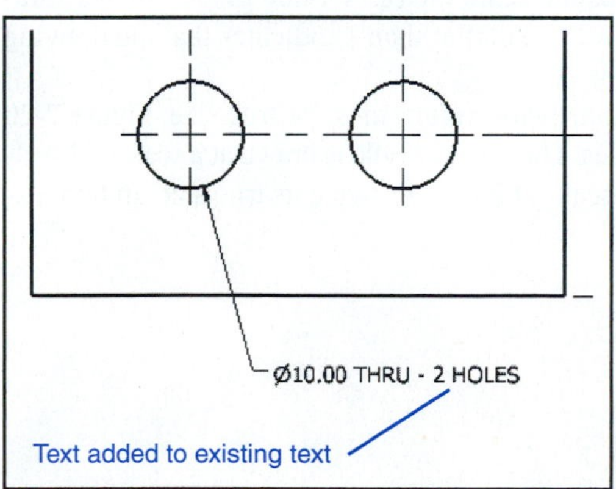

Figure 7-17

Exercise 7-6: Editing a Hole Dimension

The hole dimension shown in Figure 7-17 includes the word **THRU** after the dimension value. Including the word THRU after a hole dimension is an optional practice, and not all companies include it. The word may be removed using the **Edit Hole Note** option.

1. Move the cursor to the hole dimension and right-click the mouse.
2. Select the **Edit Hole Note** option.

The **Edit Hole Note** dialog box will appear. See Figure 7-18.

3. Backspace out the word **THRU.**
4. Click the **Save** box, then click the **Done** box.

Figure 7-19 shows the resulting hole dimension. If the 2 HOLES note is not added using the **Text** option, it can be added using the **Edit Hole Note** option.

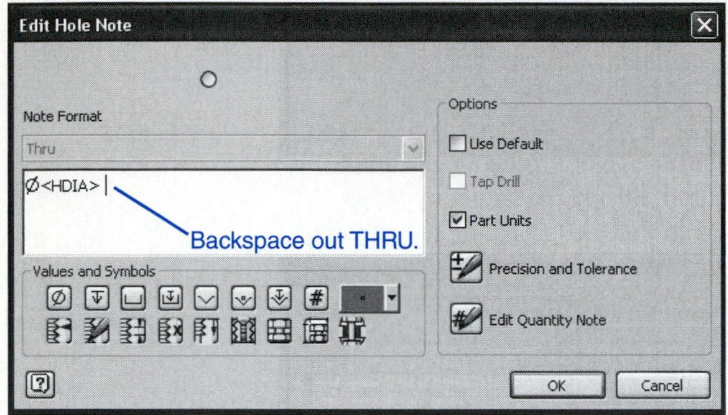

Figure 7-18

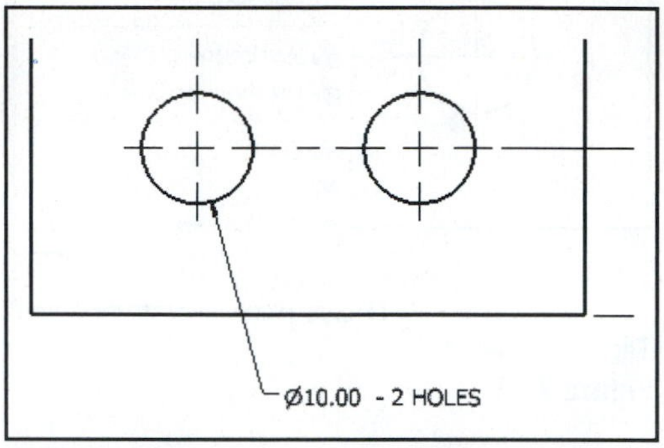

Ø10.00 - 2 HOLES

Figure 7-19

Drawing Scale

Drawings are often drawn "to scale" because the actual part is either too big to fit on a sheet of drawing paper or too small to be seen. For example, a microchip circuit must be drawn at several thousand times its actual size to be seen.

Drawing scales are written using the following formats:

SCALE: 1=1

SCALE: FULL

SCALE: 1000=1

SCALE: .25=1

In each example the value on the left indicates the scale factor. A value greater than 1 indicates that the drawing is larger than actual size. A value smaller than 1 indicates that the drawing is smaller than actual size.

Regardless of the drawing scale selected the dimension values must be true size. Figure 7-20 shows the same rectangle drawn at two different scales. The top rectangle is drawn at a scale of 1 = 1, or its true size. The bottom rectangle is drawn at a scale of 2 = 1, or twice its true size. In both examples the 3.00 dimension remains the same.

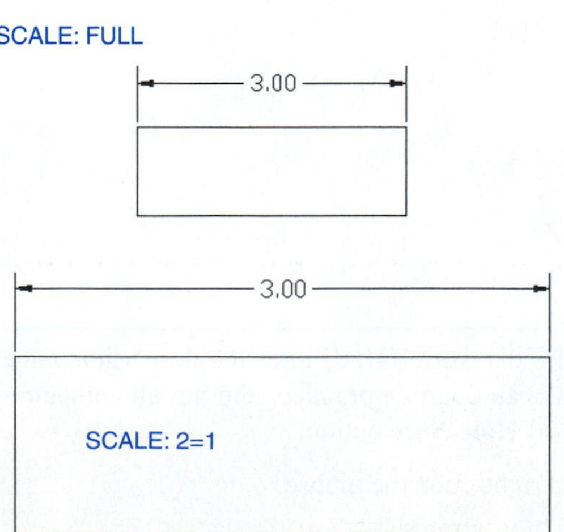

Figure 7-20

UNITS

It is important to understand that dimensional values are not the same as mathematical units. Dimensional values are manufacturing instructions and always include a tolerance, even if the tolerance value is not stated. Manufacturers use a predefined set of standard dimensions that are applied to any dimensional value that does not include a written tolerance. Standard tolerance values differ from organization to organization. Figure 7-21 shows a chart of standard tolerances.

In Figure 7-22 a distance is dimensioned twice: once as 5.50 and a second time as 5.5000. Mathematically these two values are equal, but they are not the same manufacturing instruction. The 5.50 value could, for example, have a standard tolerance of ± .01, whereas the 5.5000 value could have a standard tolerance of ± .0005. A tolerance of ± .0005 is more difficult and, therefore, more expensive to manufacture than a tolerance of ± .01.

Figure 7-23 shows examples of units expressed in millimeters and in decimal inches. A zero is not required to the left of the decimal point for decimal inch values less than one. Millimeter values do not require zeros to the right of the decimal point. Millimeter and decimal inch values never include symbols; the units will be defined in the title block of the drawing.

TOLERANCES UNLESS OTHERWISE STATED

X	± 1
.X	± .1
.xx	± .01
.XXX	± .005
X°	± 1°
.X°	± .1°

Figure 7-21

These dimensions are not the same. They have different tolerance requirements.

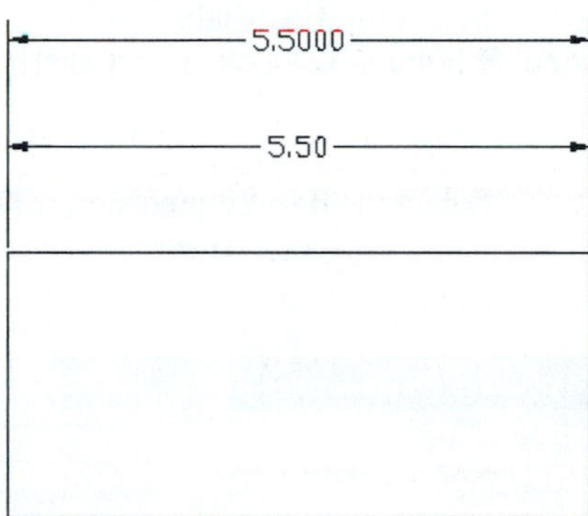

Figure 7-22

Millimeters

0.25 0.5 0.033
32 14.5 3

Zero required

Inches

No zero required

.25 .05 .033
32.00 14.50 3.000

Figure 7-23

Exercise 7-7: Preventing a 0 from Appearing to the Left of the Decimal Point

1. Click on the **Format** heading at the top of the screen, then select the **Styles Editor** option.

The **Styles and Standards Editor** dialog box will appear. See Figure 7-24.

2. Click the **+ sign** to the left of **Dimension,** and select **Default - mm (ANSI).**
3. Select the **Units** tab.
4. Remove the check mark from the **Leading Zero** box in the **Display** area.

This procedure creates a new dimension style that suppresses all leading zeros.

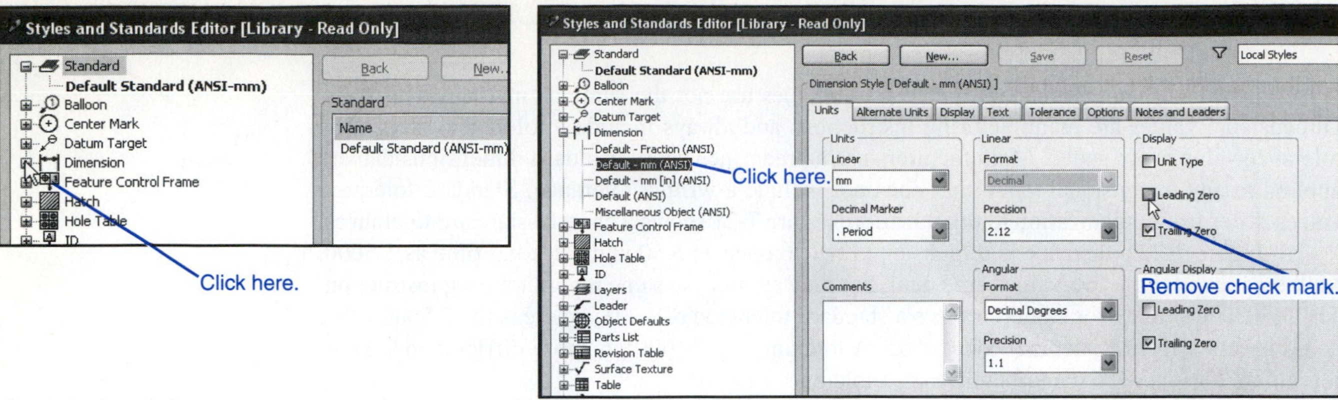

Figure 7-24

Exercise 7-8: Changing the Number of Decimal Places in a Dimension Value

1. Click on the **Format** heading at the top of the screen, then select the **Styles Editor** option.

The **Styles and Standards Editor** dialog box will appear. See Figure 7-25.

2. Click the **+ sign** to the left of **Dimension,** and select **Default - mm (ANSI).**
3. Select the **Units** tab.
4. Click the scroll arrow on the right side of the **Precision** box.

A listing of available precision settings will cascade down.

5. Select the desired precision value.
6. Click **Save** then **Done.**

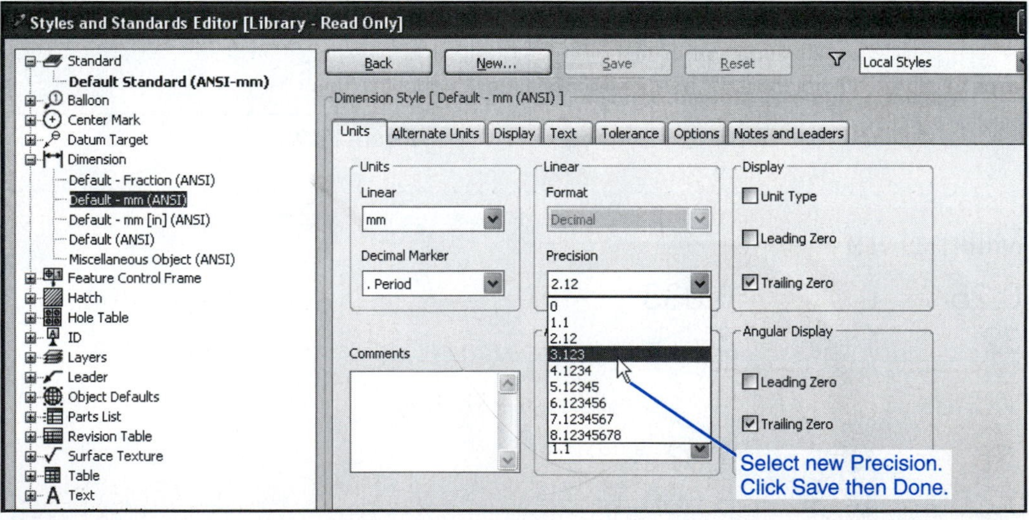

Figure 7-25

Aligned Dimensions

aligned dimension: A dimension that is parallel to a slanted edge or surface.

Aligned dimensions are dimensions that are parallel to a slanted edge or surface. They are not horizontal or vertical. The unit values for aligned dimensions should be horizontal or unidirectional.

Exercise 7-9: Creating an Aligned Dimension

Figure 7-26 shows an orthographic view created using the **ANSI (mm).idw** format. The **General Dimension** tool was used to add the aligned dimension.

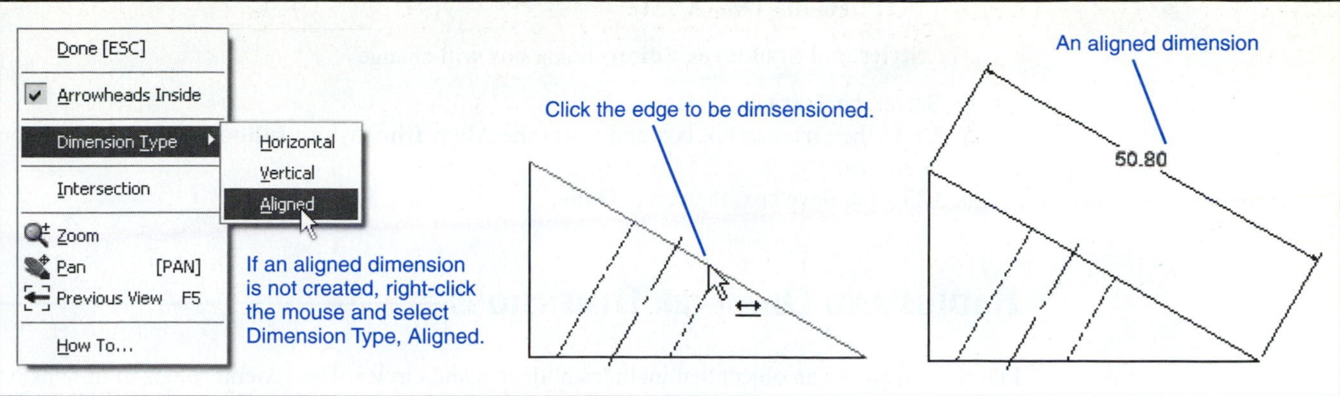

Figure 7-26

1. Access the **Drawing Annotation Panel** bar by moving the cursor into the panel bar area and right-clicking the mouse, then select the **Drawing Annotation** option.
2. Click the **General Dimension** tool, then move the cursor onto the slanted edge to be dimensioned.

The edge will change color, indicating that it has been selected.

3. Move the cursor away from the edge and left-click the mouse.
4. Move the mouse as necessary to locate the aligned dimension.

See Figure 7-26.

TIP If an aligned dimension is not created, right-click the mouse and select **Dimension Type, Aligned** before locating the dimension.

Exercise 7-10: Creating a Unidirectional Aligned Dimension

Aligned dimensions can be made unidirectional by using the **Orientation** option on the **Styles and Standards Editor** dialog box.

1. Click on the **Format** heading at the top of the screen, then select the **Styles Editor** tool.

The **Styles and Standards Editor** dialog box will appear. See Figure 7-27.

2. Click the **+ sign** to the left of the **Dimensions** listing.

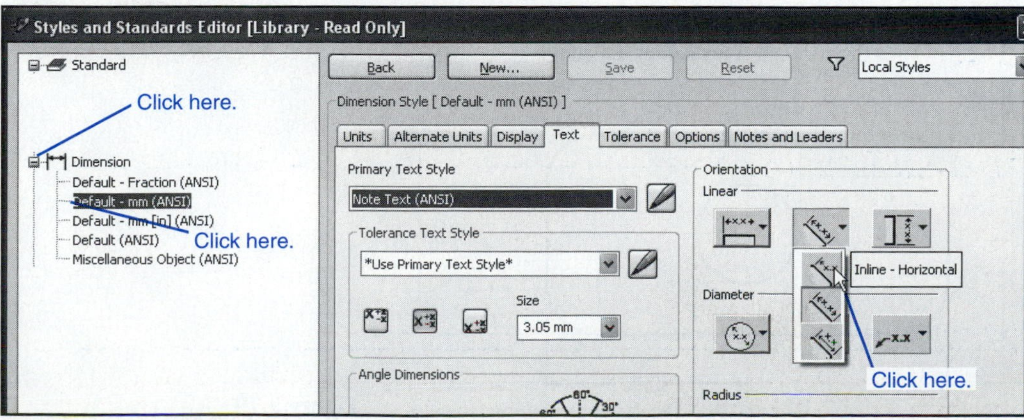

Figure 7-27

3. Select **Default - mm (ANSI).**

The **Styles and Standards Editor** dialog box will change.

4. Select the **Text** tab.
5. Go to the **Orientation** box and select the **Align Dimension: Inline - Horizontal** option as shown.
6. Click the **Save** box, then click **Done.**

RADIUS AND DIAMETER DIMENSIONS

Figure 7-28 shows an object that includes both arcs and circles. The general rule is to dimension arcs using a radius dimension, and circles using diameter dimensions. This convention is consistent with the tooling required to produce the feature shape. Any arc greater than 170° is considered a circle and is dimensioned using a diameter.

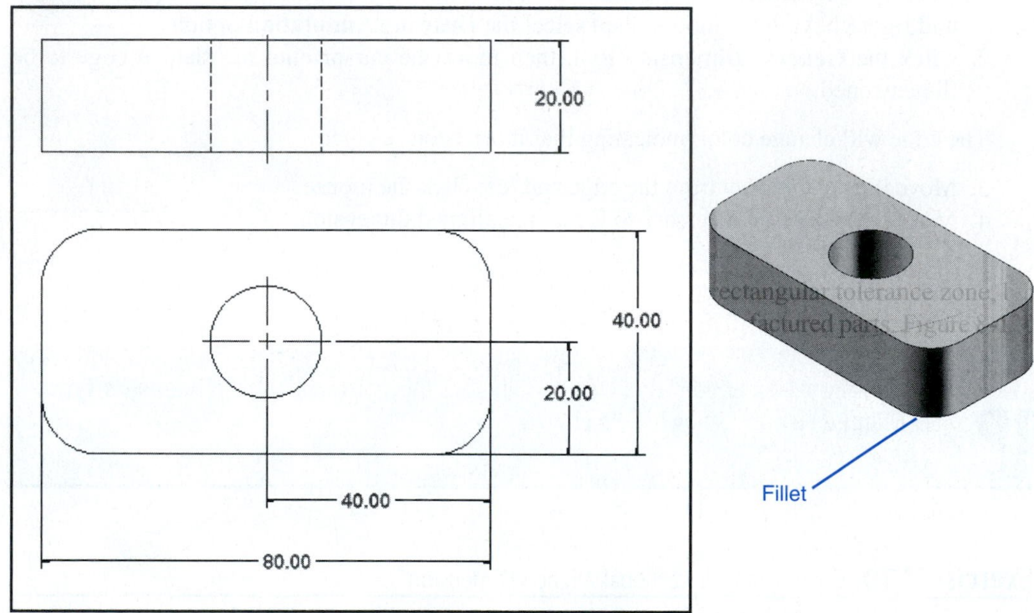

Figure 7-28

Exercise 7-11: Creating a Radius Dimension

1. Click the **Leader Text** tool on the **Drawing Annotation Panel** bar.
2. Click one of the filleted corners and drag the cursor away from the corner.
3. Create a short horizontal segment on the leader line.

See Figure 7-29.

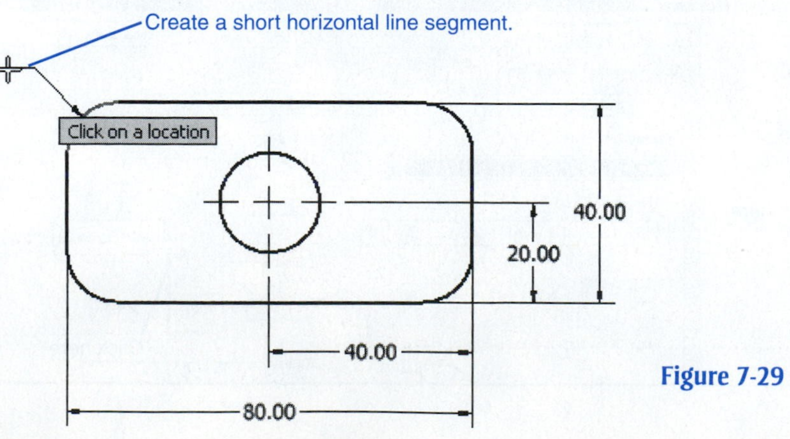

Figure 7-29

4. Right-click the mouse and select the **Continue** option.

The **Format Text** dialog box will appear. See Figure 7-30.

5. Enter **R10.00,** then click **OK.**

Figure 7-31 shows the resulting dimension. There are four equal arcs on the object, and they all must be dimensioned. It would be better to add the words 4 CORNERS to the radius dimension than to include four radius dimensions.

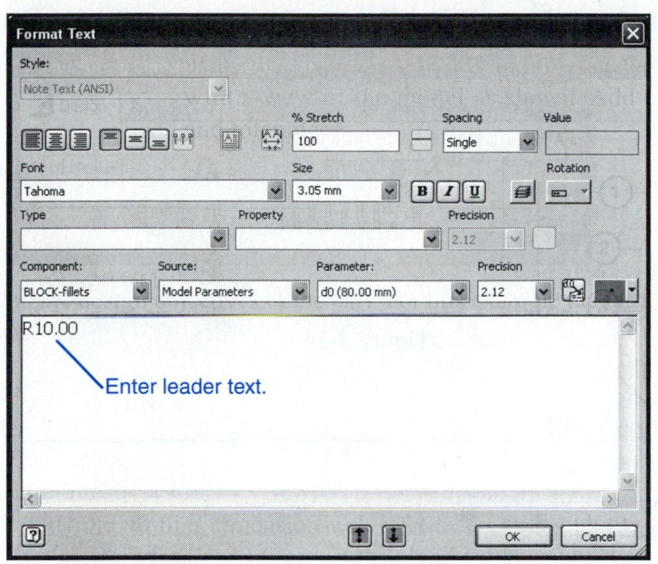

Figure 7-30

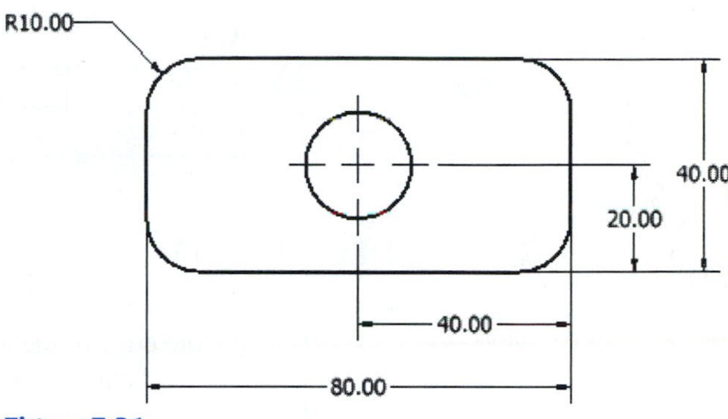

Figure 7-31

Exercise 7-12: Adding Text to an Existing Drawing Dimension

1. Move the cursor to the fillet dimension.

Filled colored circles will appear, indicating that the dimension has been selected.

2. Right-click the mouse and select the **Edit Leader Text** option.

The **Format Text** dialog box will appear with the existing text. See Figure 7-32.

3. Type **4 CORNERS** under the existing text and click the **OK** box.

Figure 7-33 shows the resulting dimension.

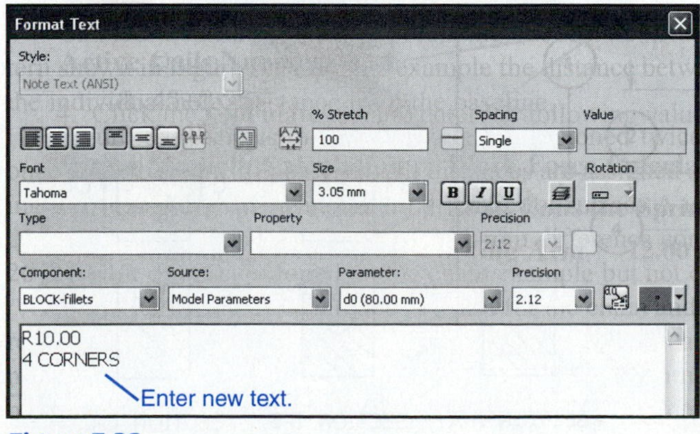

Figure 7-32

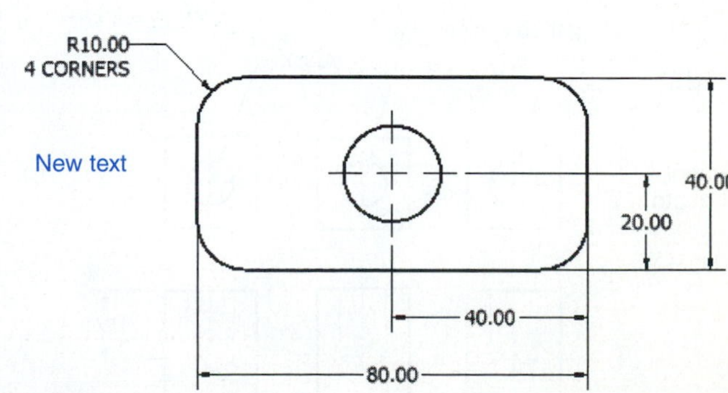

Figure 7-33

Dimensioning Holes

Holes are dimensioned by stating their diameter and depth, if any. Holes that go completely through an object are defined using only a diameter dimension. See Figure 7-34. The word **THRU** may be added if desired.

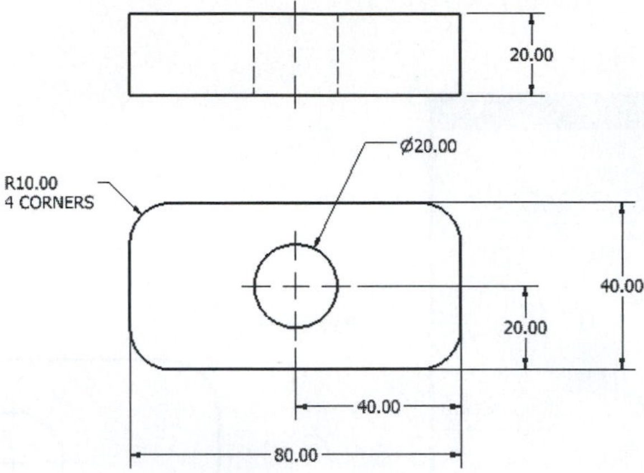

Figure 7-34

Exercise 7-13: Dimensioning a Through Hole

1. Click on the **Hole/Thread Note** tool.
2. Click the edge of the hole and drag the dimension.
3. Locate and enter the dimension.
4. Remove the word **THRU** from the hole dimension.

The **General Dimension** tool may also be used.

Dimensioning Individual Holes

Figure 7-35 shows three different methods that can be used to dimension a hole that does not go completely through an object. Depth values may be added using the **Power Dimensioning** dialog box, or the **Edit Text** or **Edit Format** options.

Figure 7-36 shows two methods of dimensioning holes in sectional views. The single line note version is the preferred method.

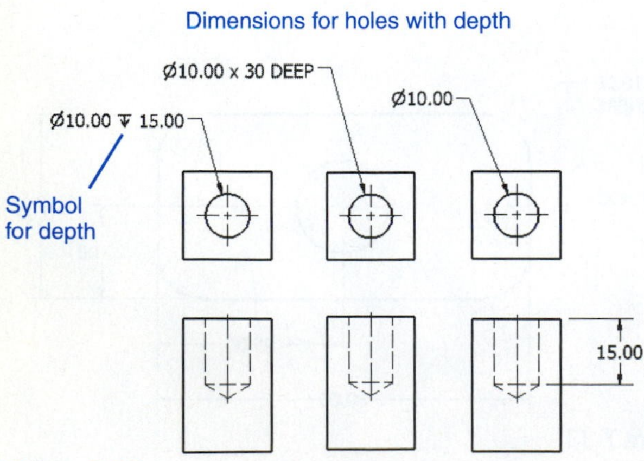

Figure 7-35

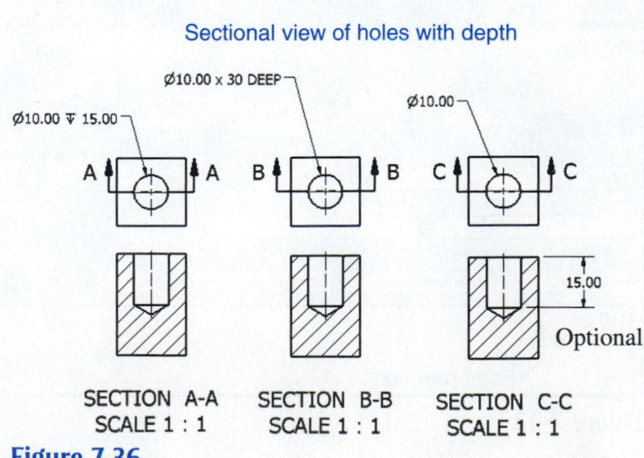

Figure 7-36

Dimensioning Hole Patterns

Figure 7-37 shows two different hole patterns dimensioned. The circular pattern includes the note ∅10 - 4 HOLES. This note serves to define all four holes within the object.

Figure 7-37 also shows a rectangular object that contains five holes of equal diameter, equally spaced from one another. The notation 5 × ∅10 specifies 5 holes of 10 diameter. The notation 4 × 20 (=80) means 4 equal spaces of 20. The notation (=80) is a reference dimension and is included for convenience. Reference dimensions are explained in Chapter 9.

Figure 7-38 shows two additional methods for dimensioning repeating hole patterns. Figure 7-39 shows a circular hole pattern that includes two different hole diameters. The hole

Dimensions for hole patterns

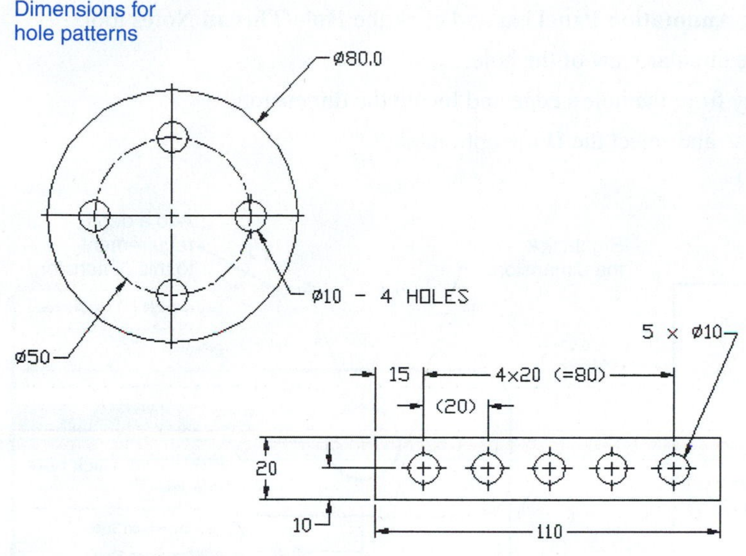

Figure 7-37

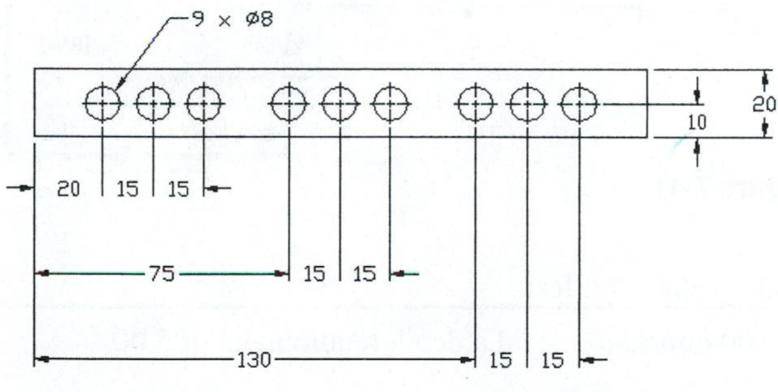

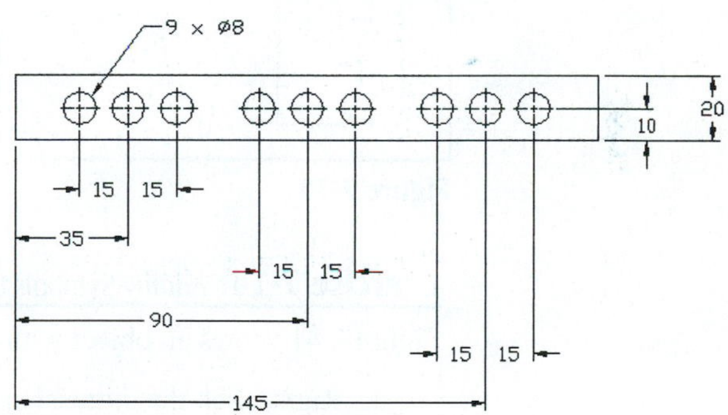

Figure 7-38

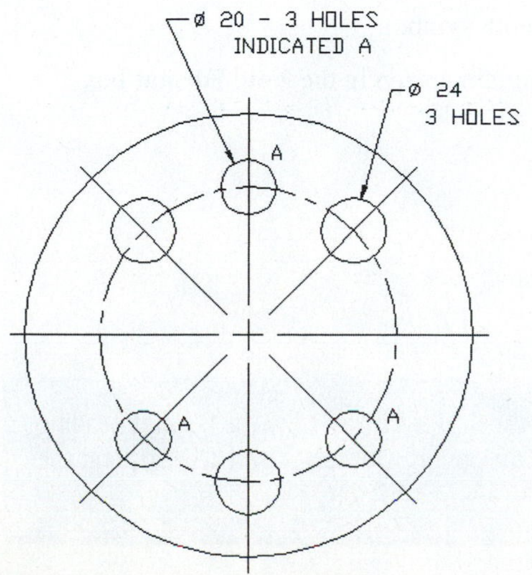

Figure 7-39

diameters are not noticeably different and could be confused. One group is defined by indicating letter (A); the other is dimensioned in a normal manner.

Using Symbols with Dimensions

In an attempt to eliminate language restrictions from drawings, ANSI standards permit the use of certain symbols for dimensions. Like musical notation that can be universally read by people who speak different languages, symbolic dimensions can be read by different people regardless of which language they speak. For example, the symbol ∅ replaces the notation DIA.

Figure 7-40 shows a ∅10 hole that has a depth of 15. Inventor will automatically apply symbolic dimensions. The shown dimension is created as follows.

1. Access the **Drawing Annotation Panel** bar and click the **Hole/Thread Notes** tool.
2. Click the edge of the circular view of the hole.
3. Drag the cursor away from the hole's edge and locate the dimension.
4. Right-click the mouse and select the **Done** option.

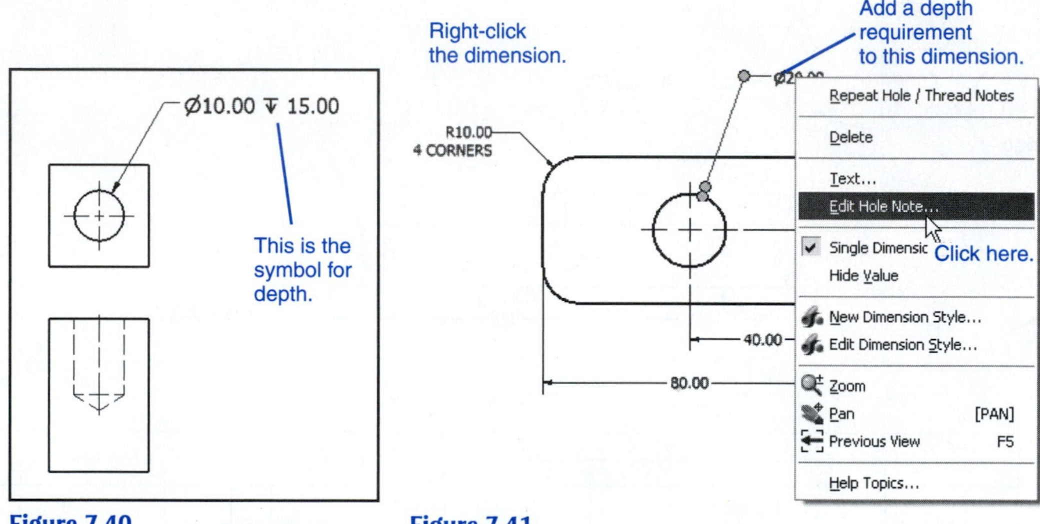

Figure 7-40 **Figure 7-41**

Exercise 7-14: Adding Symbols to Existing Dimension Text

Figure 7-41 shows an object with a ∅20.00 dimension. Add a depth requirement of 5.00.

1. Right-click the dimension and select the **Edit Hole Note** option.

The **Edit Hole Note** dialog box will appear. See Figure 7-42.

2. Click the **Insert Symbol** box and click the depth symbol.

The depth symbol will appear next to the existing dimension in the **Note Format** box.

3. Type **5.00,** then click **OK.**

Figure 7-43 shows the resulting dimension.

4. Select a symbol.
5. Click the **Save** box, then click **Done.**

Drawing symbols can also be accessed using the **Styles Editor.** Click the **Format** heading at the top of the screen, click the **Dimension** option, **Default - mm (ANSI),** and the **Insert Symbol** box. See Figure 7-44.

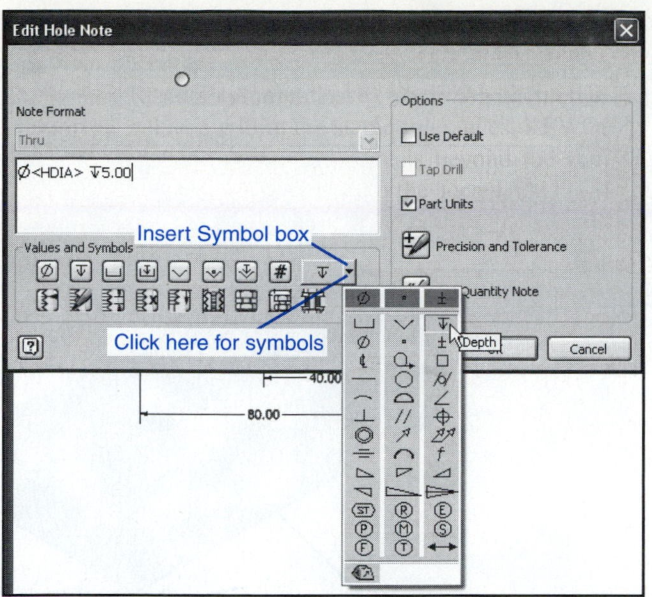

Figure 7-42

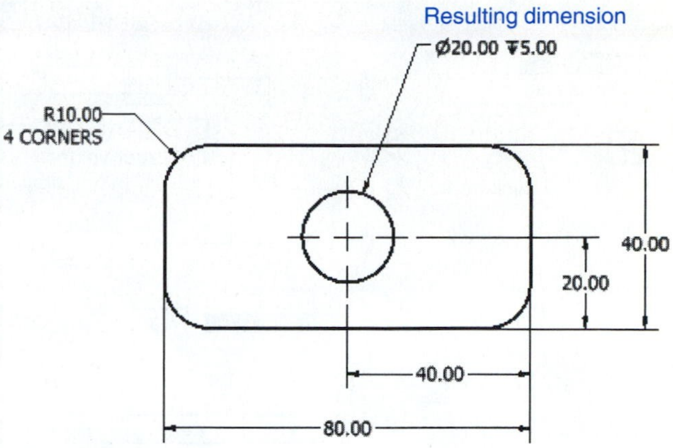

Figure 7-43

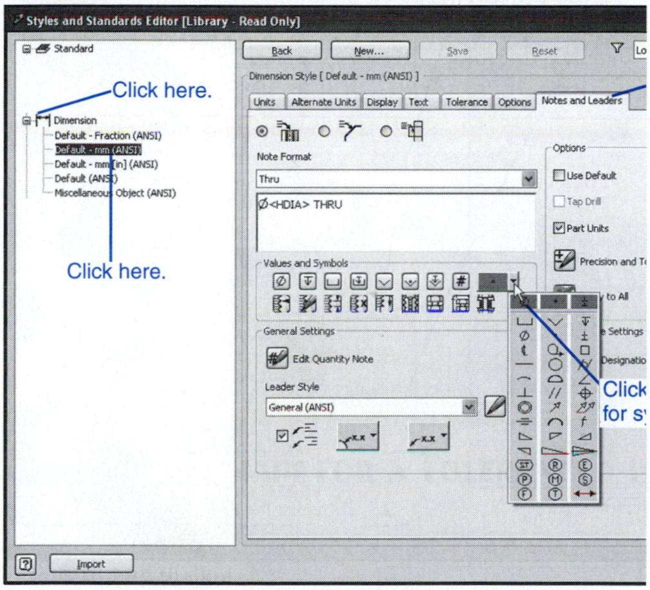

Figure 7-44

Dimensioning Counterbored, Countersunk Holes

Chapter 6 explained how to draw counterbored and countersunk holes. This section shows how to dimension countersunk and counterbored holes.

Figure 7-45 shows a 70 × 30 × 20 block that includes both a counterbored and a countersunk hole. The sizes for the holes were determined based on the fastener information given in Chapter 6. The **Hole** dialog boxes used to create the holes are also shown.

The information used to create the holes will automatically be used to create the hole's dimension. Prefixes and suffixes may be added to the dimensions, and the dimension value may be changed using the **Edit Hole Note** option. Hole dimension values may also be changed by editing the hole's feature value on the original model.

Exercise 7-15: Dimensioning a Counterbored Hole

The counterbored hole has a clear hole that goes completely through the block.

1. Access the **Drawing Annotation Panel** bar and click the **Hole/Thread Notes** tool.
2. Click the edge of the circular view of the hole.

3. Drag the cursor away from the hole's edge and locate the dimension.
4. Right-click the mouse and select the **Done** option.

Note how the dimensions match exactly the values used to create the hole. See Figures 7-45 and 7-46.

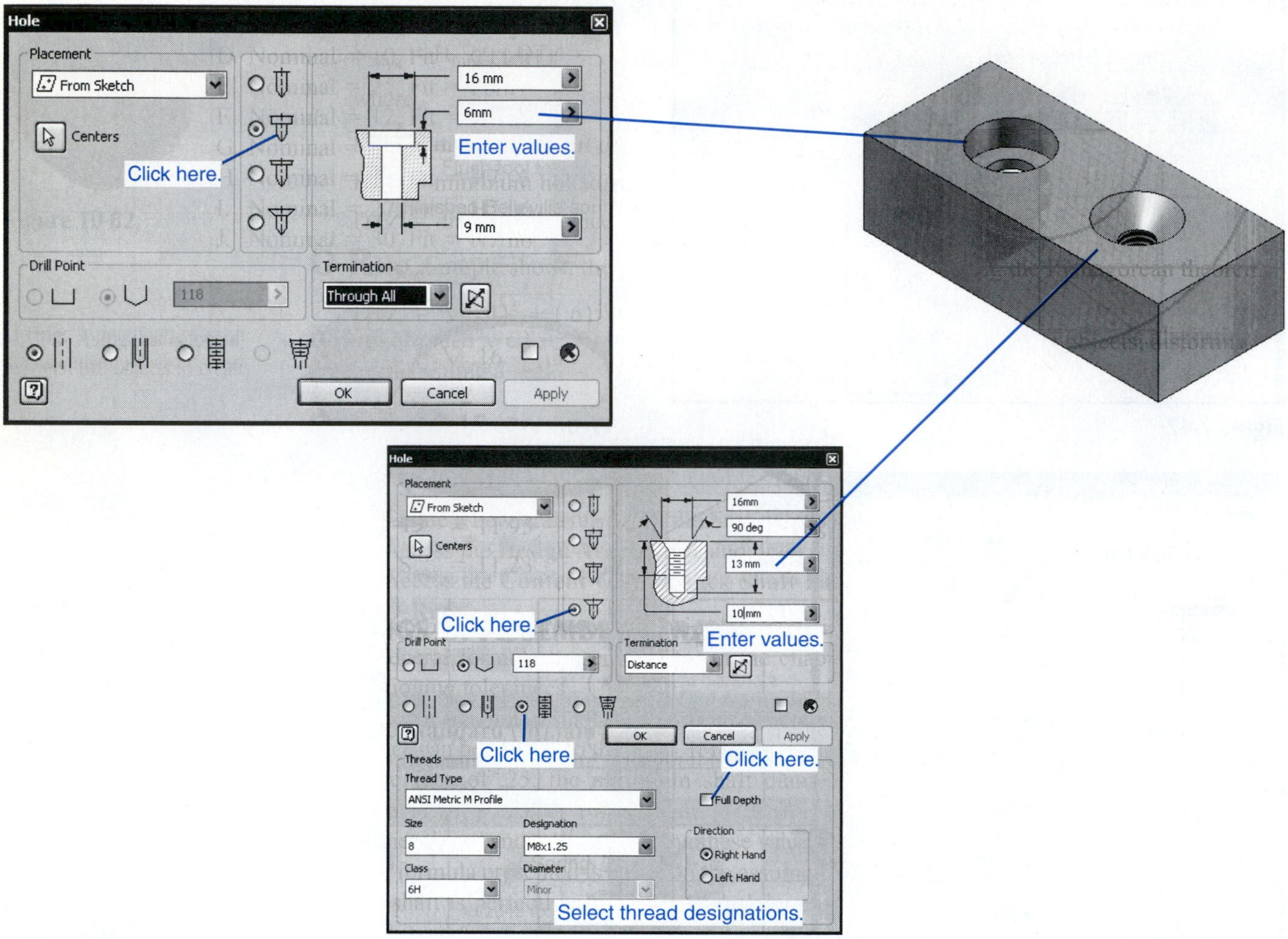

Figure 7-45

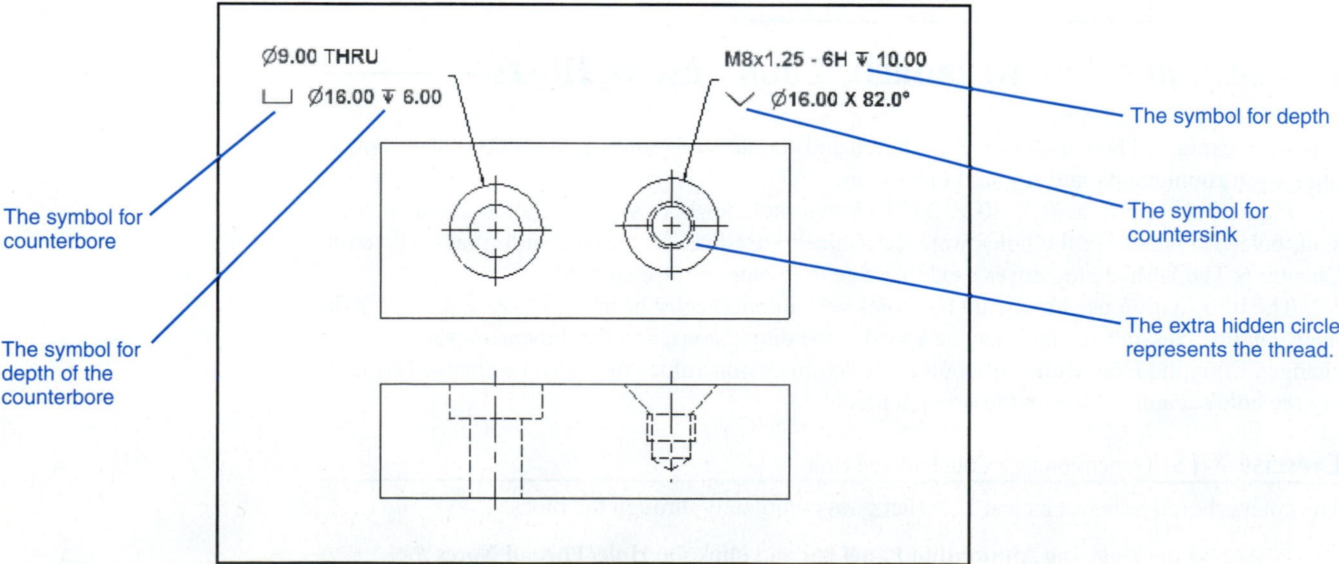

Figure 7-46

Exercise 7-16: Dimensioning a Countersunk Hole

The countersunk hole has M8 threads and a thread depth of 10 mm.

1. Create an orthographic view of the model using the **ANSI (mm). idw** format.
2. Add centerlines.
3. Access the **Drawing Annotation Panel** bar and click the **Hole/Thread Notes** tool.
4. Click the edge of the circular view of the hole.
5. Drag the cursor away from the hole's edge and locate the dimension.
6. Right-click the mouse and select the **Done** option.

Note how the dimensions match exactly the values used to create the hole.

ANGULAR DIMENSIONS

Figure 7-47 shows a model that includes a slanted surface. The dimension value is located beyond the model between two extension lines. Locating dimensions between extension lines is preferred to locating the value between an extension line and the edge of the model.

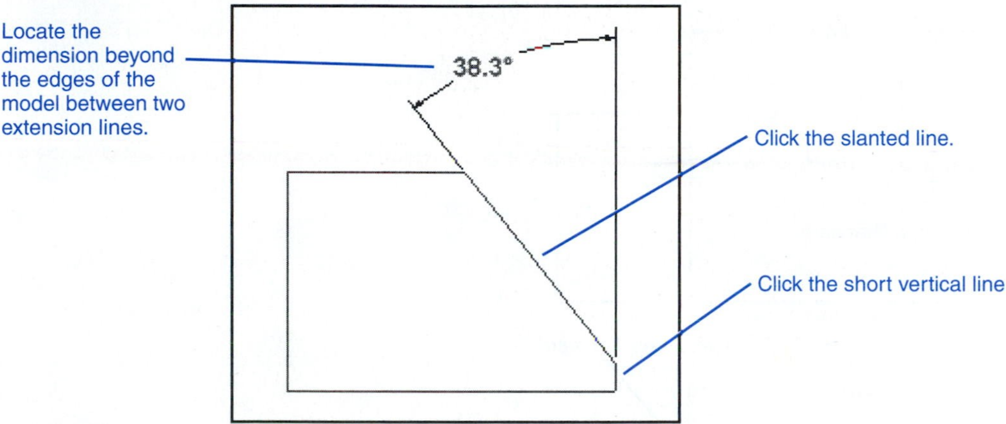

Figure 7-47

Exercise 7-17: Creating an Angular Dimension

1. Create an orthographic view of the model using the **ANSI (mm). idw** format.
2. Access the **Drawing Annotation Panel** bar and click the **General Dimension** tool.
3. Click the slanted line, then click the short vertical line on the right side of the model.
4. Drag the cursor away from the hole's edge and locate the dimension.
5. Right-click the mouse and select the **Done** option.

Avoiding Overdimensioning

Figure 7-48 shows a shape dimensioned using an angular dimension. The shape is completely defined. Any additional dimension would be an error. It is tempting, in an effort to make sure a shape is completely defined, to add more dimensions, such as a horizontal dimension for the short horizontal edge at the top of the shape. This dimension is not needed and is considered *double dimensioning.*

Figure 7-49 also shows the same front view dimensioned using only linear dimensions. The choice of whether to use angular or linear dimensions depends on the function of the model and which distances are more critical.

Figure 7-50 shows an object dimensioned two different ways. The dimensions used in the top example do not include a dimension for the width of the slot. This dimension is allowed to *float,* that is, allowed to accept any tolerance buildup. The dimensions used in the bottom example dimension the width of the slot but not the upper right edge. In this example the

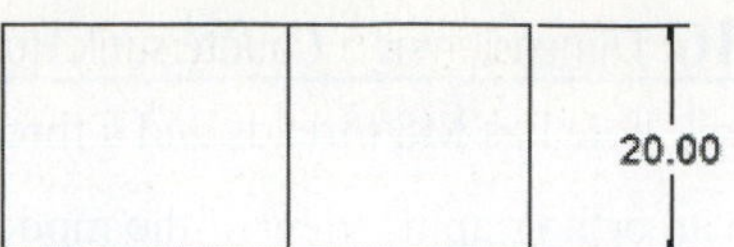

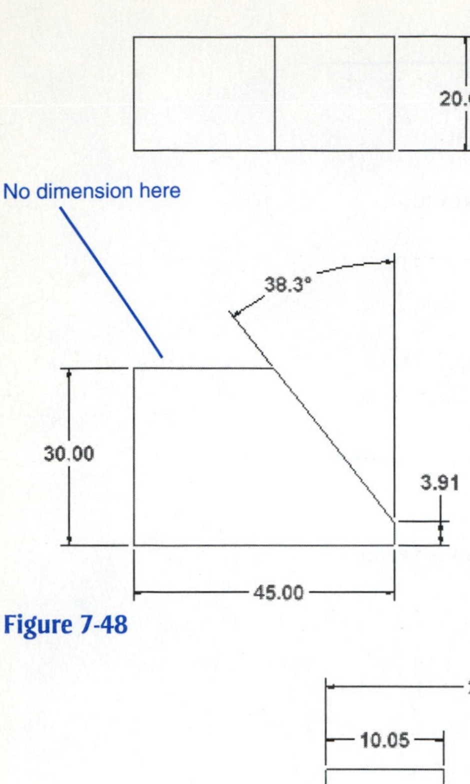

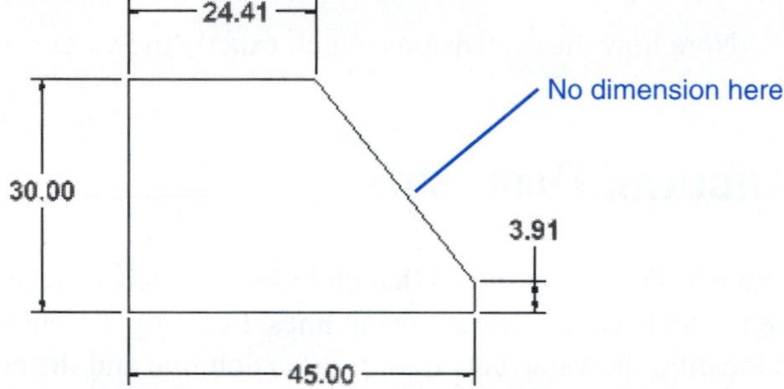

Figure 7-48

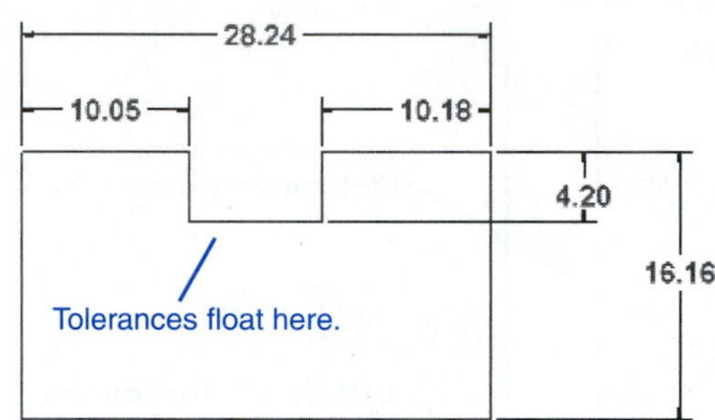

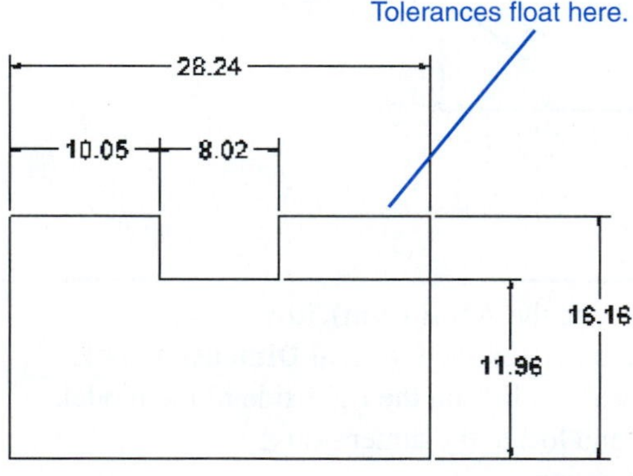

Figure 7-49

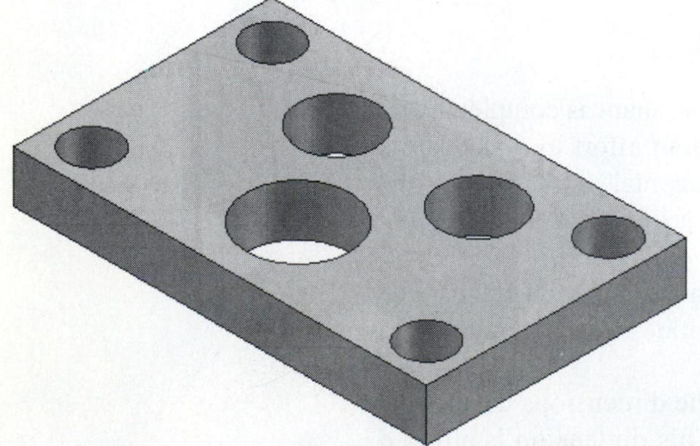

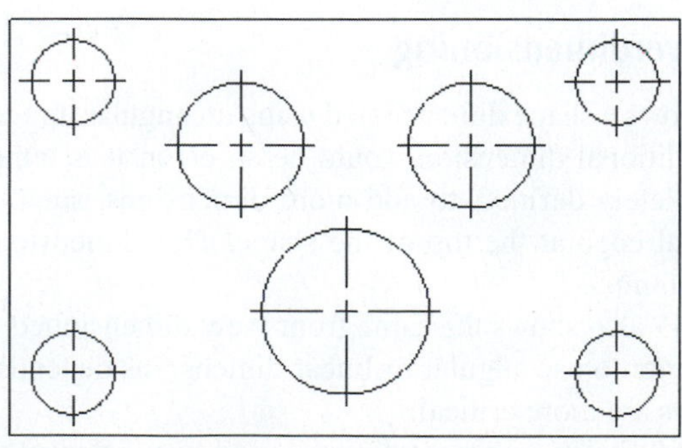

Figure 7-50

upper right edge is allowed to float or accept any tolerance buildup. The choice of which edge to float depends on the function of the part. If the slot were to interface with a tab on another part, then it would be imperative that it be dimensioned and toleranced to match the interfacing part.

ORDINATE DIMENSIONS

ordinate dimension: A dimension based on an X, Y coordinate system that consists simply of horizontal and vertical leader lines.

Ordinate dimensions are dimensions based on an X,Y coordinate system. Ordinate dimensions do not include extension lines, dimension lines, or arrowheads, but simply horizontal and vertical leader lines drawn directly from the features of the object. Ordinate dimensions are particularly useful when dimensioning an object that includes many small holes.

Figure 7-50 shows a model that is to be dimensioned using ordinate dimensions. Ordinate dimensions values are calculated from the X,Y origin, which, in this example, is the lower left corner of the front view of the model.

Exercise 7-18: Creating Ordinate Dimensions

1. Create an orthographic view of the model using the **ANSI(mm). idw** format.
2. Access the **Drawing Annotation Panel** bar, then click on the **Ordinate Dimension Set** tool.

See Figure 7-51.

3. Move the cursor into the drawing area and click the lower left corner of the model.
4. Drag the cursor away from the edge of the model and position the first ordinate dimension.

See Figure 7-52.

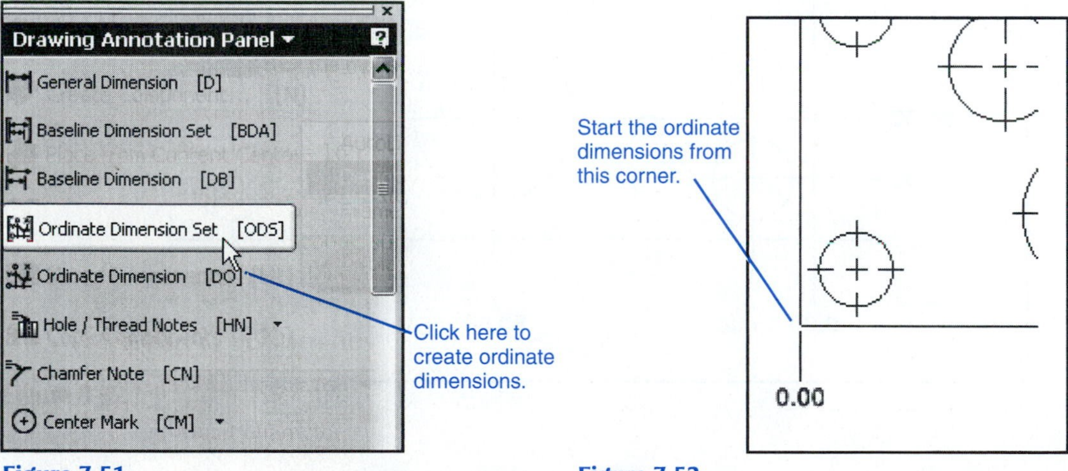

Figure 7-51

Figure 7-52

5. Click the lower end of the vertical centerline of the first hole and position the dimension so that it is in line with the first dimension.

See Figure 7-53.

Note how the extension line from the first hole's centerline curves so that the dimension value may be located in line with the first dimension. Inventor will automatically align ordinate dimensions.

6. Add all the vertical dimensions.

Again start with the lower left corner of the model, then click the appropriate horizontal centerlines. See Figure 7-54.

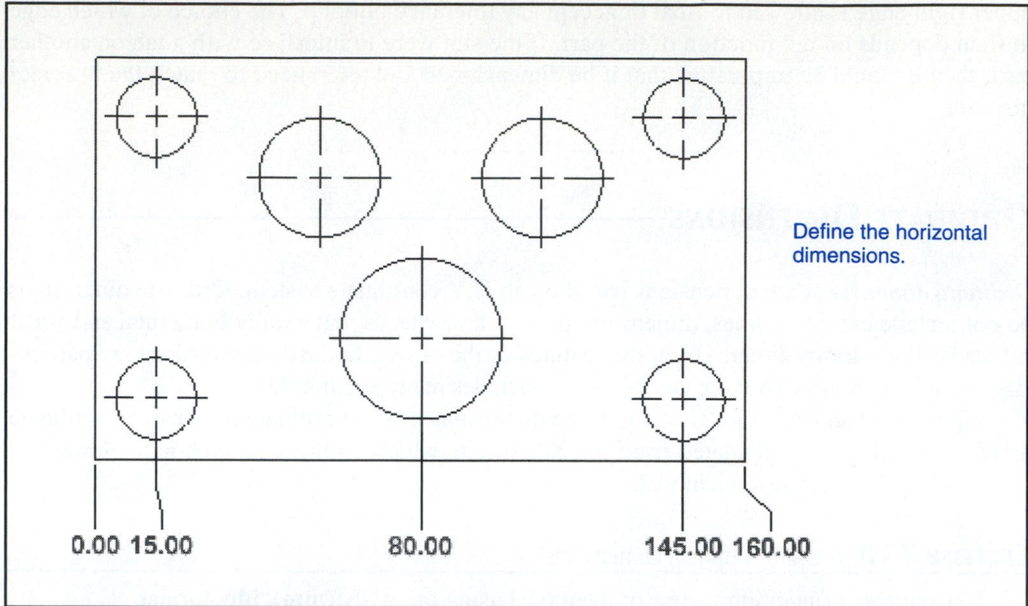

Figure 7-53

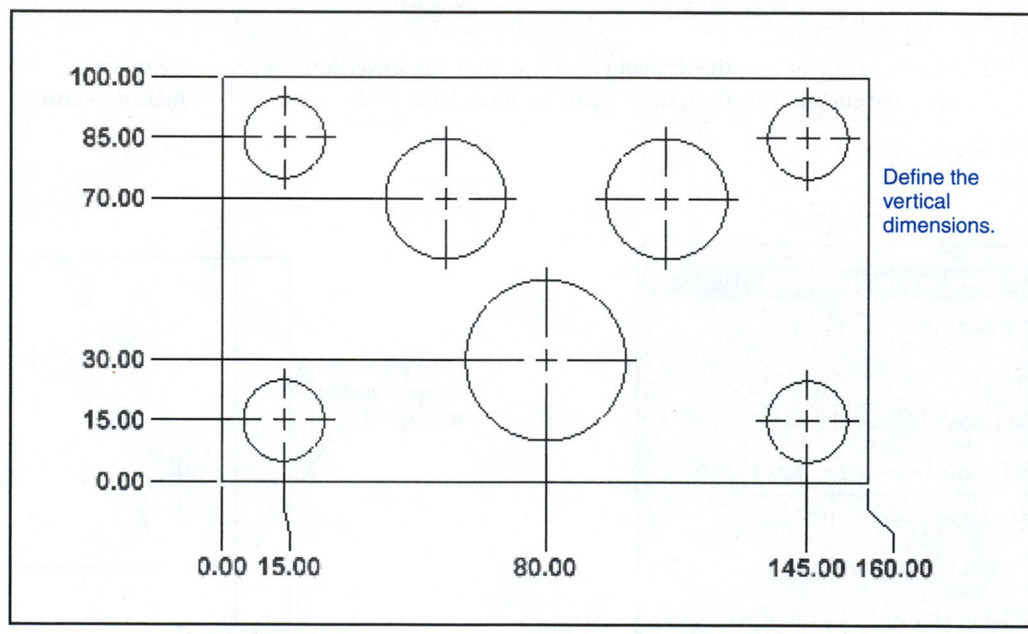

Figure 7-54

Exercise 7-19: Adding Hole Dimensions

1. Click the **Hole/Thread Notes** on the **Drawing Annotation Panel** bar.
2. Add the appropriate hole dimensions.

See Figure 7-55.

Baseline Dimensions

baseline dimensions: A series of dimensions that originate from a common baseline or datum line.

Baseline dimensions are a series of dimensions that originate from a common baseline or datum line. Baseline dimensions are very useful because they help eliminate the tolerance buildup that is associated with chain-type dimensions. The **Baseline Dimension** tool is a flyout of the **General Dimension** tool located on the **Drawing Annotation Panel** bar. See Figure 7-56.

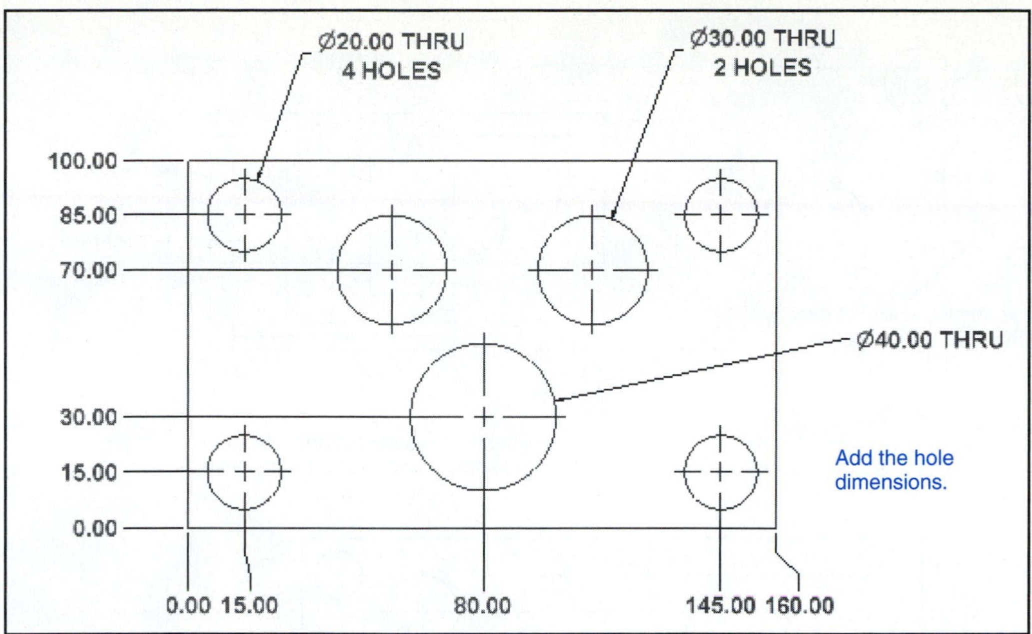

Figure 7-55

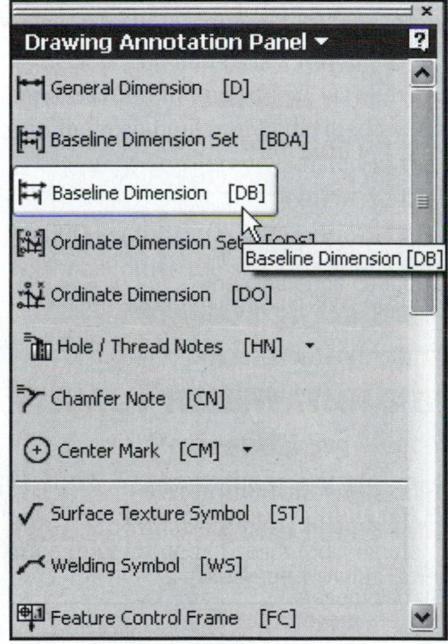

Figure 7-56

Exercise 7-20: Using the Baseline Dimension Tool

See Figure 7-57.

1. Click the **Baseline Dimension** tool on the **Drawing Annotation Panel** bar.
2. Click the left vertical edge of the model.
3. Click the lower end on each vertical centerline and the right edge vertical line.
4. Right-click the mouse and select the **Continue** option.
5. Move the cursor and click the top of the next hole's vertical centerline.
6. Click the centerlines of the remaining holes.
7. Press the right mouse button and select the **Continue** option, then create the vertical dimensions in a similar maner.
8. Add the hole dimensions using the **Hole/Thread Notes** tool.

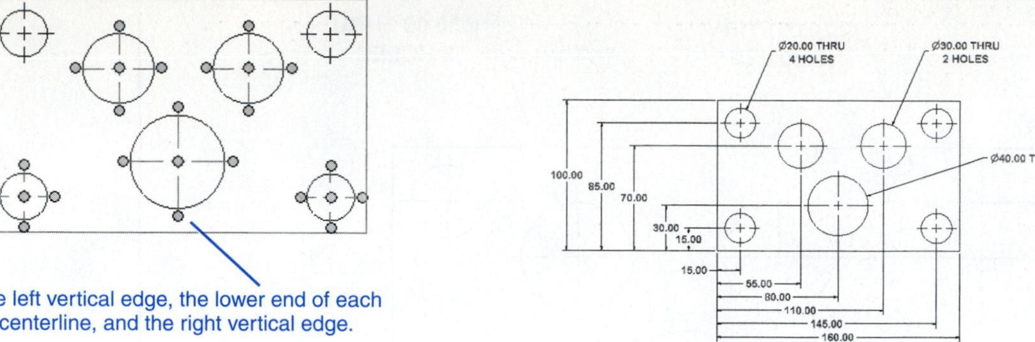

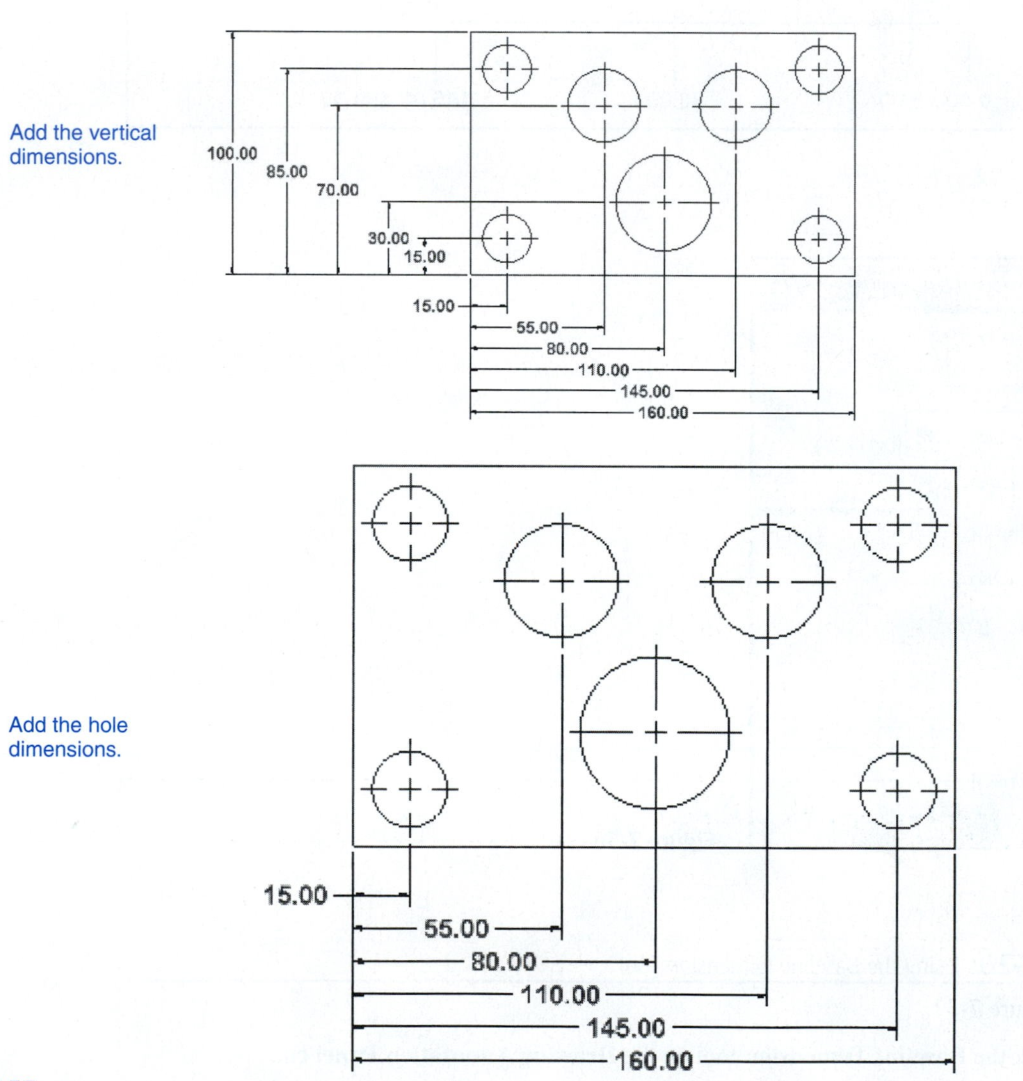

Figure 7-57

HOLE TABLES

Inventor will generate hole tables that list holes' diameters and locations. There are two options: list all the holes, or list selected holes.

Exercise 7-21: Listing All Holes in a Table

1. Click the **Hole Table - Selection** tool on the **Drawing Annotation Panel** bar.
2. Select the **View** option.

See Figure 7-58.

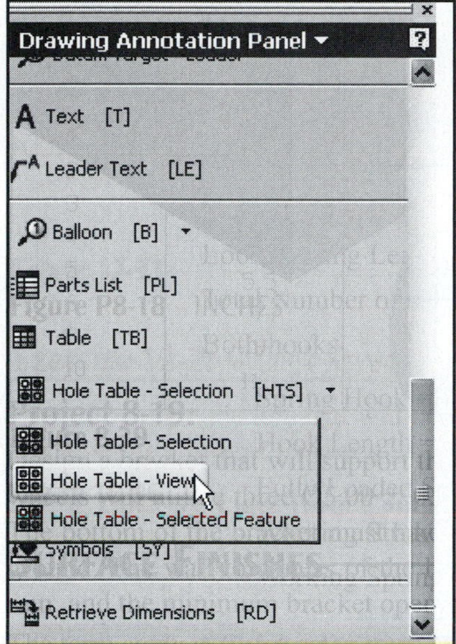

Figure 7-58

3. Move the mouse into the drawing area and left-click the mouse.

A +-shaped cursor will appear.

4. Click on the lower left corner of the model.

This will define the origin for the table's X and Y values.

5. Right-click the mouse and select the **Create** option.

A rectangle representing the table will appear on the drawing screen. See Figure 7-59.

6. Locate the table on the drawing, then right-click the mouse and select the **Done** option.

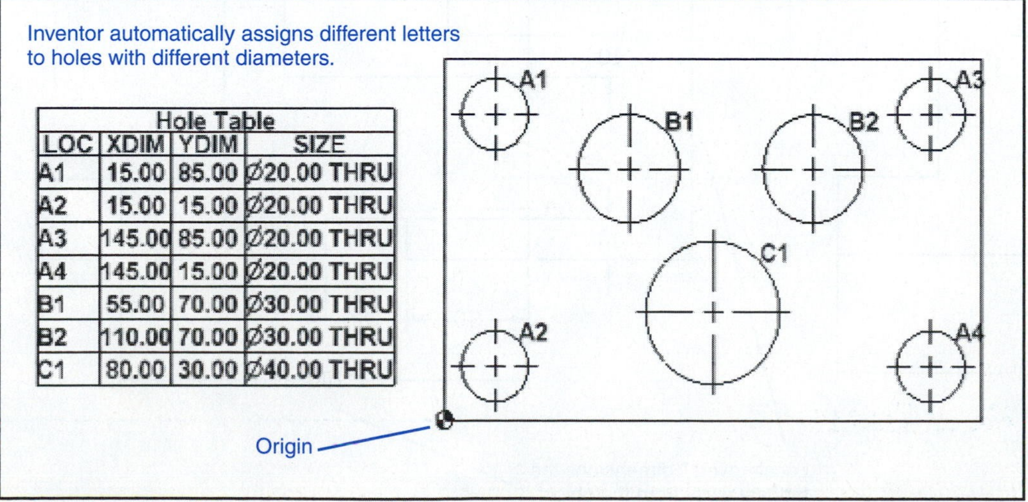

Hole Table			
LOC	XDIM	YDIM	SIZE
A1	15.00	85.00	Ø20.00 THRU
A2	15.00	15.00	Ø20.00 THRU
A3	145.00	85.00	Ø20.00 THRU
A4	145.00	15.00	Ø20.00 THRU
B1	55.00	70.00	Ø30.00 THRU
B2	110.00	70.00	Ø30.00 THRU
C1	80.00	30.00	Ø40.00 THRU

Figure 7-59

Locating Dimensions

There are eight general rules concerning the location of dimensions. See Figure 7-60.

1. Locate dimensions near the features they are defining.
2. Do not locate dimensions on the surface of the object.

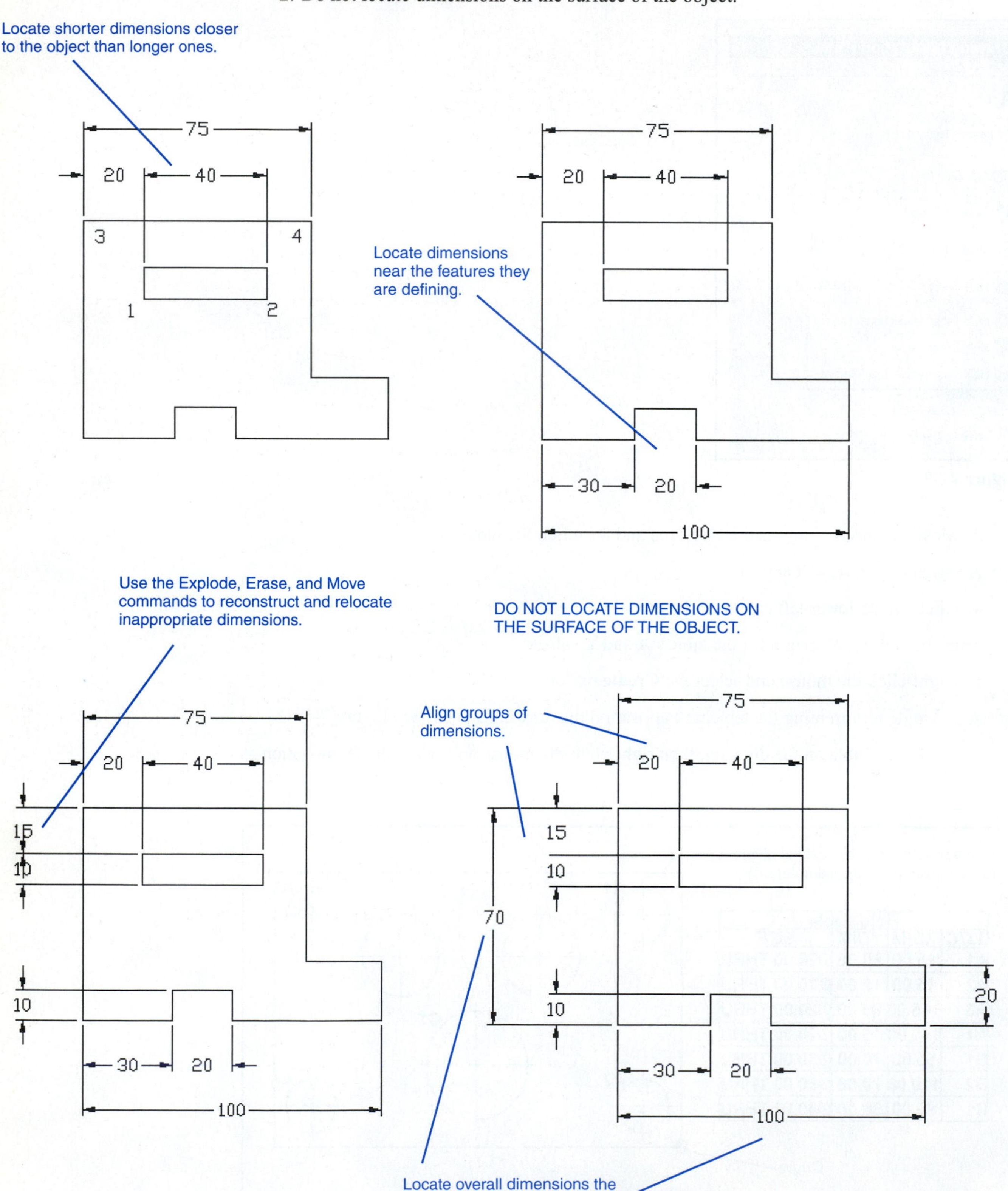

Figure 7-60

3. Align and group dimensions so that they are neat and easy to understand.
4. Avoid crossing extension lines.

Sometimes it is impossible not to cross extension lines because of the complex shape of the object, but whenever possible, avoid crossing extension lines.

5. Do not cross dimension lines.
6. Locate shorter dimensions closer to the object than longer ones.
7. Always locate overall dimensions the farthest away from the object.
8. Do not dimension the same distance twice. This is called double dimensioning and will be discussed in Chapter 8.

Fillets and Rounds

Fillets and rounds may be dimensioned individually or by a note. In many design situations all the fillets and rounds are the same size, so a note as shown in Figure 7-61 is used. Any fillets or rounds that have a different radius from that specified by the note are dimensioned individually.

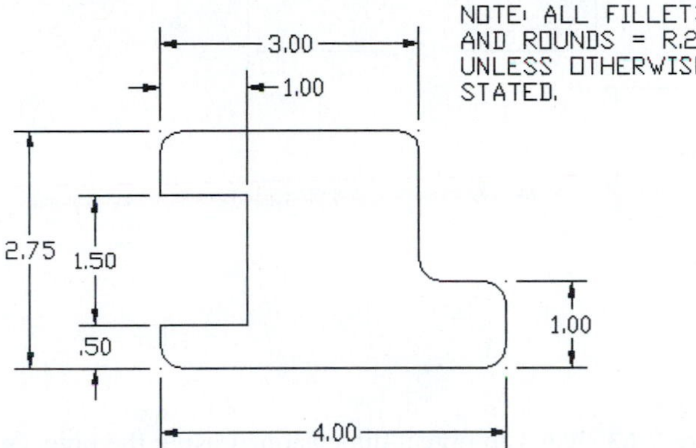

Figure 7-61

Rounded Shapes—Internal

Internal rounded shapes are called ***slots.*** Figure 7-62 shows three different methods for dimensioning slots. The end radii are indicated by the note R - 2 PLACES, but no numerical value is given. The width of the slot is dimensioned, and it is assumed that the radius of the rounded ends is exactly half of the stated width.

slot: An internal rounded shape.

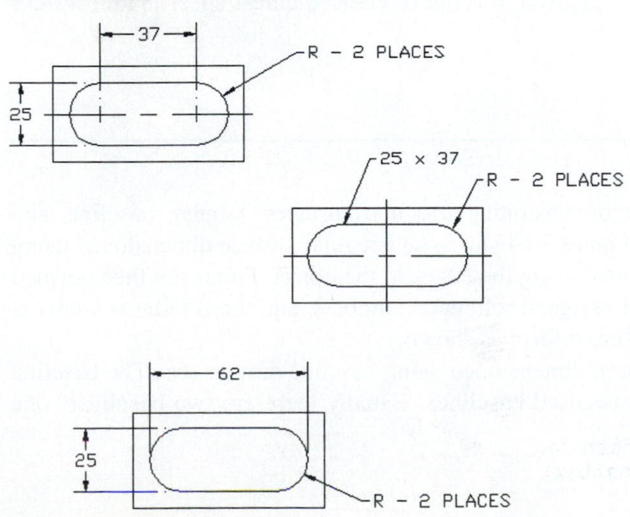

Figure 7-62

Rounded Shapes—External

Figure 7-63 shows two shapes with external rounded ends. As with internal rounded shapes, the end radii are indicated but no value is given. The width of the object is given, and the radius of the rounded end is assumed to be exactly half of the stated width.

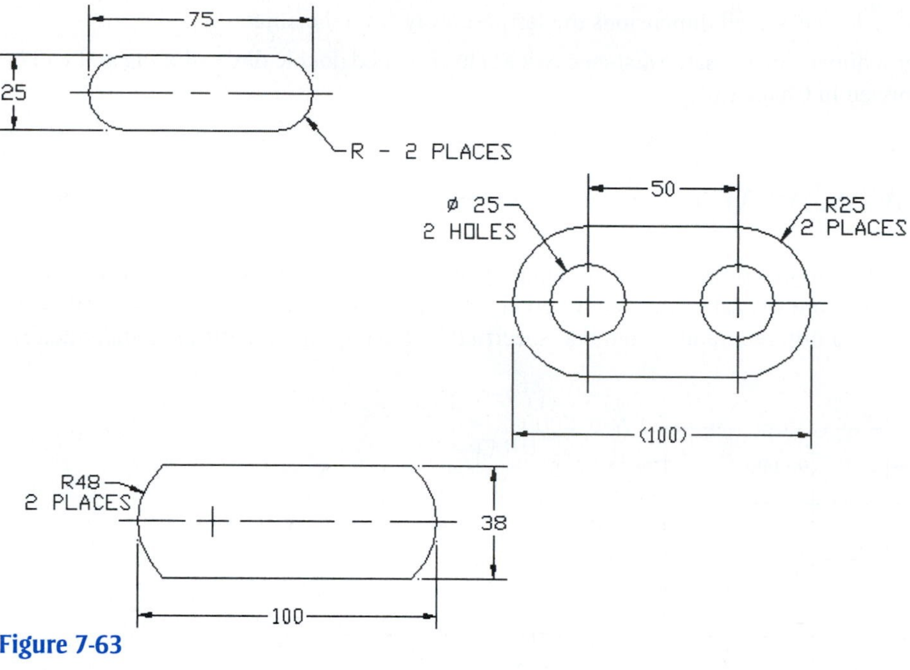

Figure 7-63

The second example shown in Figure 7-63 shows an object dimensioned using the object's centerline. This type of dimensioning is done when the distance between the holes is more important than the overall length of the object; that is, the tolerance for the distance between the holes is more exact than the tolerance for the overall length of the object.

The overall length of the object is given as a reference dimension (100). This means the object will be manufactured based on the other dimensions, and the 100 value will be used only for reference.

Objects with partially rounded edges should be dimensioned as shown in Figure 7-63. The radii of the end features are dimensioned. The centerpoint of the radii is implied to be on the object centerline. The overall dimension is given; it is not referenced unless specific radii values are included.

Irregular Surfaces

There are three different methods for dimensioning irregular surfaces: tabular, baseline, and baseline with oblique extension lines. Figure 7-64 shows an irregular surface dimensioned using the tabular method. An XY axis is defined using the edges of the object. Points are then defined relative to the XY axis. The points are assigned reference numbers, and the reference numbers and XY coordinate values are listed in chart form as shown.

Figure 7-65 shows an irregular curve dimensioned using baseline dimensions. The baseline method references all dimensions to specified baselines. Usually there are two baselines, one horizontal and one vertical.

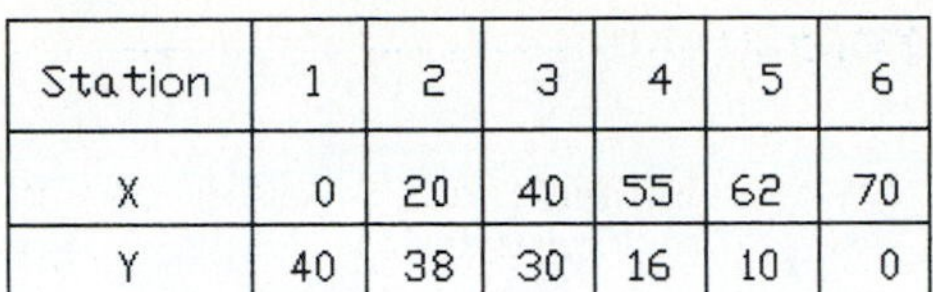

Station	1	2	3	4	5	6
X	0	20	40	55	62	70
Y	40	38	30	16	10	0

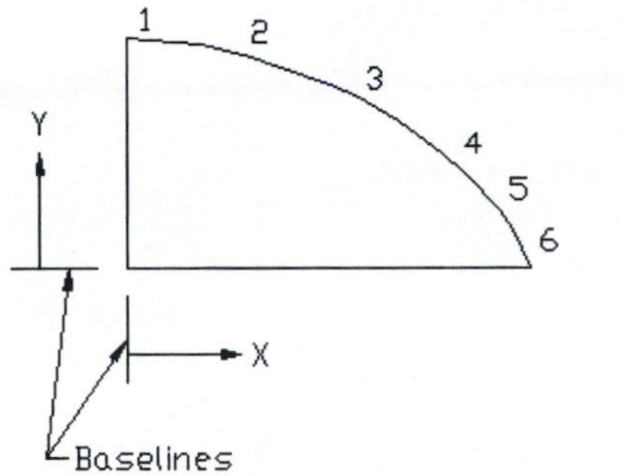

Figure 7-64

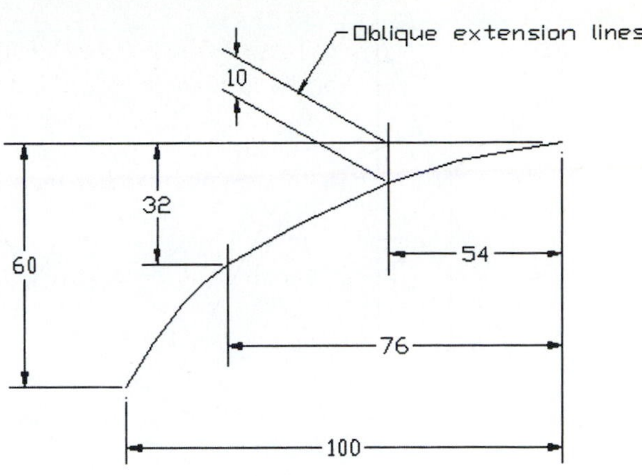

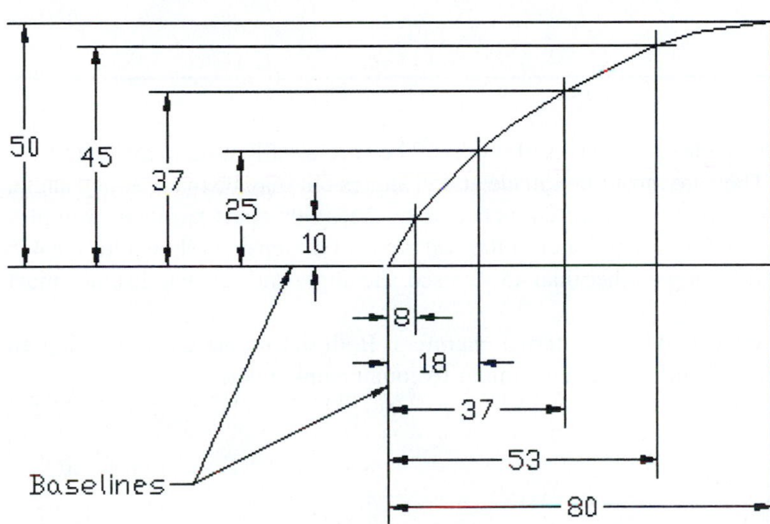

Figure 7-65

It is considered poor practice to use a centerline as a baseline. Centerlines are imaginary lines that do not exist on the object and would make it more difficult to manufacture and inspect the finished objects.

Baseline dimensioning is very common because it helps eliminate tolerance buildup and is easily adaptable to many manufacturing processes. Inventor has a special **Baseline Dimension** command for use in creating baseline dimensions.

POLAR DIMENSIONS

Polar dimensions are similar to polar coordinates. A location is defined by a radius (distance) and an angle. Figure 7-66 shows an object that includes polar dimensions. The holes are located on a circular centerline, and their positions from the vertical centerline are specified using angles.

polar dimension: A dimension defined by a radius and an angle.

Figure 7-67 shows an example of a hole pattern dimensioned using polar dimensions.

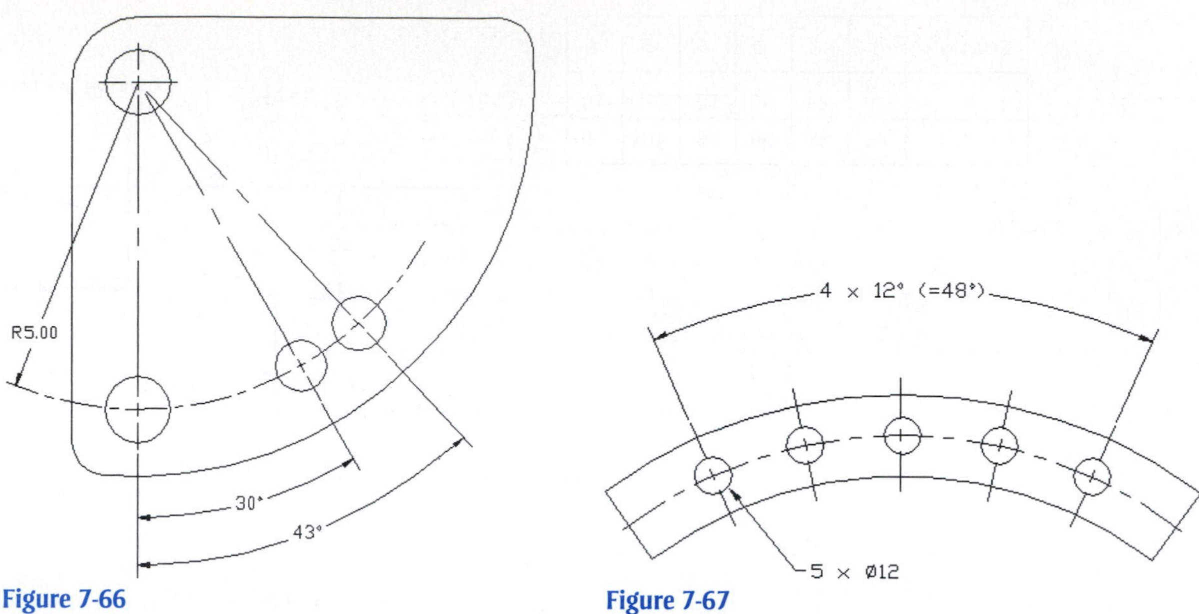

Figure 7-66

Figure 7-67

CHAMFERS

chamfer: An angular cut made on the edge of an object.

Chamfers are angular cuts made on the edges of objects. They are usually used to make it easier to fit two parts together. They are most often made at 45° angles but may be made at any angle. Figure 7-68 shows two objects with chamfers between surfaces 90° apart and two examples between surfaces that are not 90° apart. Either of the two types of dimensions shown for the 45° dimension may be used. If an angle other than 45° is used, the angle and setback distance must be specified.

Figure 7-69 shows two examples of internal chamfers. Both define the chamfer using an angle and diameter. Internal chamfers are very similar to countersunk holes.

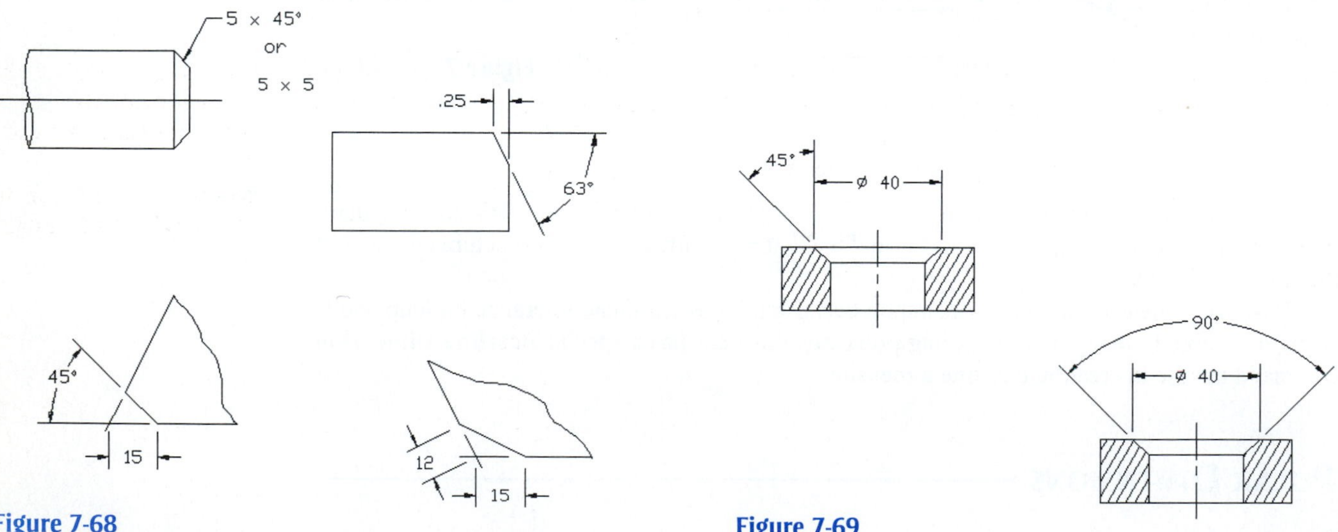

Figure 7-68

Figure 7-69

knurl: A series of small ridges on a metal surface used to make it easier to grip a shaft, or to roughen a surface before it is used in a press fit; may be diamond or straight.

KNURLING

There are two types of ***knurls:*** diamond and straight. Knurls are used to make it easier to grip a shaft, or to roughen a surface before it is used in a press fit.

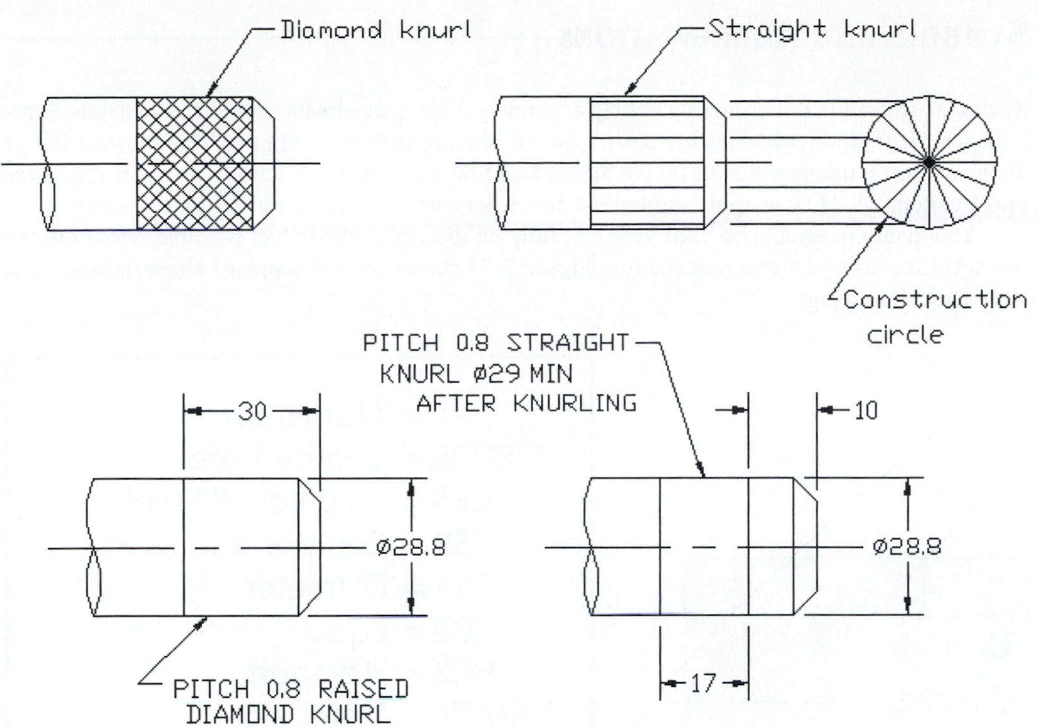

Figure 7-70

Knurls are defined by their pitch and diameter. See Figure 7-70. The pitch of a knurl is the ratio of the number of grooves on the circumference to the diameter. Standard knurling tools sized to a variety of pitch sizes are used to manufacture knurls for both English and metric units.

Diamond knurls may be represented by a double hatched pattern or by an open area with notes. The **Hatch** command is used to draw the double hatched lines. Straight knurls may be represented by straight lines in the pattern shown or by an open area with notes. The straight-line pattern is created by projecting lines from a construction circle. The construction points are evenly spaced on the circle.

KEYS AND KEYSEATS

Keys are small pieces of material used to transmit power. For example, Figure 7-71 shows how a key can be fitted between a shaft and a gear so that the rotary motion of the shaft can be transmitted to the gear.

There are many different styles of keys. See the keys listed on the **Content Center** under **Shaft Parts.** The key shown in Figure 7-71 has a rectangular cross section and is called a *square key.* Keys fit into grooves called ***keyseats*** or ***keyways.***

Keyways are dimensioned from the bottom of the shaft or hole as shown.

key: A small piece of material used to transmit power, such as rotary motion from a shaft to a gear.

keyseat: A groove into which a key fits; also called *keyway.*

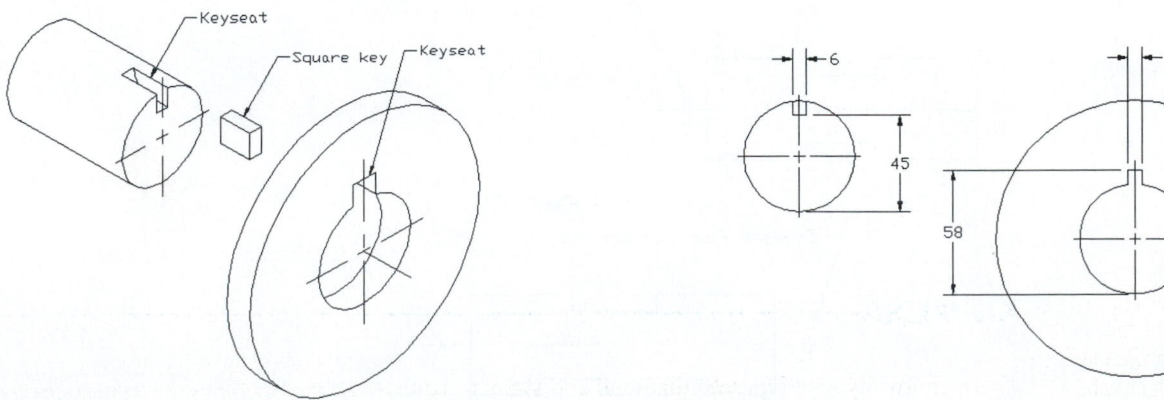

Figure 7-71

Symbols and Abbreviations

Symbols are used in dimensioning to help accurately display the meaning of the dimension. Symbols also help eliminate language barriers when reading drawings. Figure 7-72 shows a list of dimensioning symbols available on the **Styles and Standards Editor** dialog box under the **Notes and Leaders** tab. How to apply symbols to dimensions was explained earlier in the chapter.

Abbreviations should be used very carefully on drawings. Whenever possible, write out the full word including correct punctuation. Figure 7-73 shows several standard abbreviations used on technical drawings.

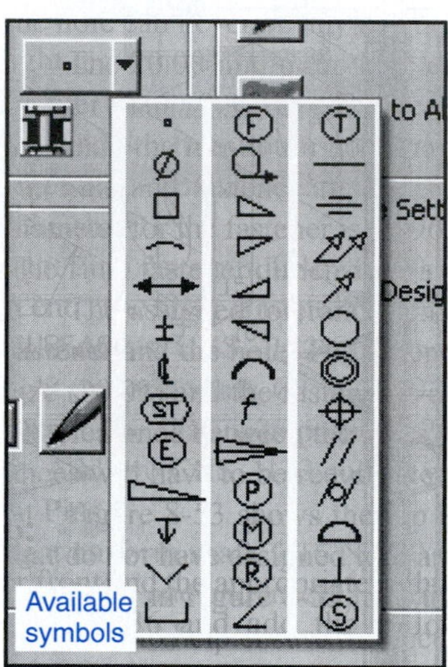

Figure 7-72

AL = Aluminum
C'BORE = Counterbore
CRS = Cold Rolled Steel
CSK = Countersink
DIA = Diameter
EQ = Equal
HEX = Hexagon
MAT'L = Material
R = Radius
SAE = Society of Automotive Engineers
SFACE = Spotface
ST = Steel
SQ = Square
REQD = Required

Figure 7-73

Symmetrical and Centerline Symbols

An object is symmetrical about an axis when one side is an exact mirror image of the other. Figure 7-74 shows a symmetrical object. The two short parallel lines symbol or the note OBJECT IS SYMMETRICAL ABOUT THIS AXIS (centerline) may be used to designate symmetry.

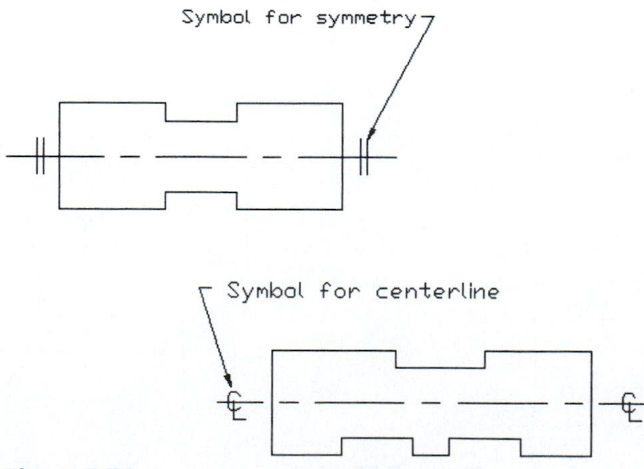

Figure 7-74

If an object is symmetrical, only half the object need be dimensioned. The other dimensions are implied by the symmetry note or symbol.

Centerlines are slightly different from the axis of symmetry. An object may or may not be symmetrical about its centerline. See Figure 7-74. Centerlines are used to define the center of both individual features and entire objects. Use the centerline symbol when a line is a centerline, but do not use it in place of the symmetry symbol.

Dimensioning to a Point

Curved surfaces can be dimensioned using theoretical points. See Figure 7-75. There should be a small gap between the surface of the object and the lines used to define the theoretical point. The point should be defined by the intersection of at least two lines.

There should also be a small gap between the extension lines and the theoretical point used to locate the point.

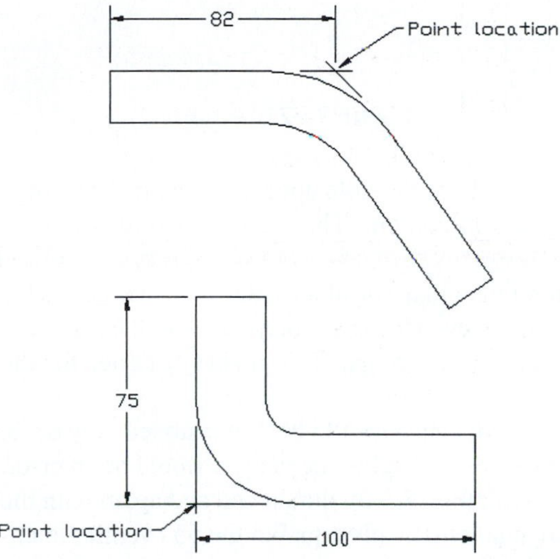

Figure 7-75

Dimensioning Sectional Views

Sectional views are dimensioned, as are orthographic views. See Figure 7-76. The sectional lines should be drawn at an angle that allows the viewer to clearly distinguish between the sectional lines and the extension lines.

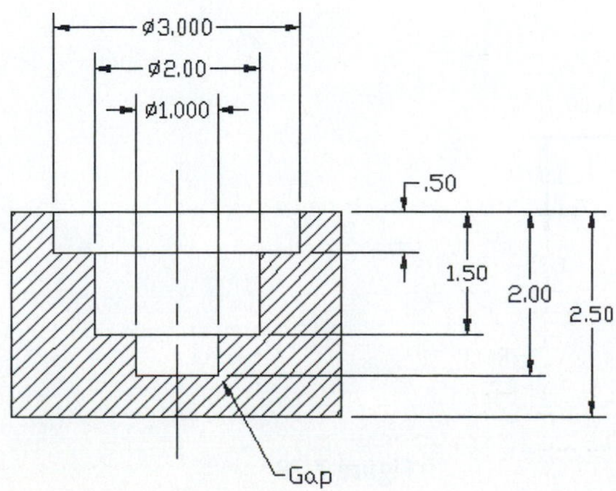

Figure 7-76

Dimensioning Orthographic Views

Dimensions should be added to orthographic views where the features appear in contour. Holes should be dimensioned in their circular views. Figure 7-77 shows three views of an object that has been dimensioned.

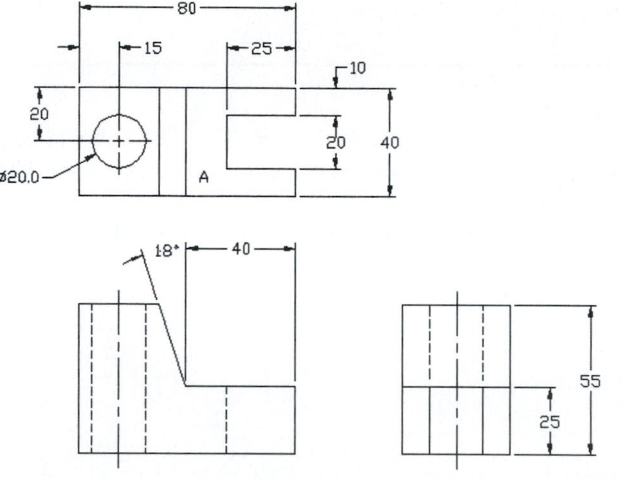

Figure 7-77

The hole dimensions are added to the top view, where the hole appears circular. The slot is also dimensioned in the top view because it appears in contour. The slanted surface is dimensioned in the front view.

The height of surface A is given in the side view rather than run along extension lines across the front view. The length of surface A is given in the front view. This is a contour view of the surface.

It is considered good practice to keep dimensions in groups. This makes it easier for the viewer to find dimensions.

Be careful not to double-dimension a distance. A distance should be dimensioned only once. If a 30 dimension were added above the 25 dimension on the right-side view, it would be an error. The distance would be double-dimensioned: once with the 25 + 30 dimension and again with the 55 overall dimension. The 25 + 30 dimensions are mathematically equal to the 55 overall dimension, but there is a distinct difference in how they affect the manufacturing tolerances. Double dimensions are explained more fully in Chapter 8.

Dimensions Using Centerlines

Figure 7-78 shows an object dimensioned from its centerline. This type of dimensioning is used when the distance between the holes relative to each other is critical.

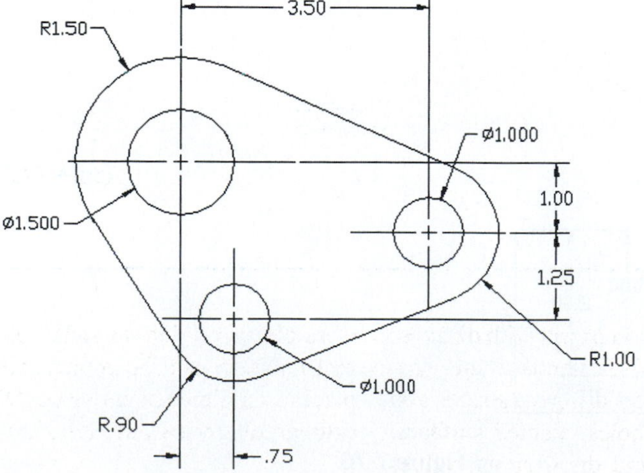

Figure 7-78

3D Dimensions

Inventor 2008 can be used to create dimensions on a three-dimensional object. Figure 7-79 shows an object drawn using the ANSI (mm) idw format. The drawing shows a front orthographic view of the object and an isometric view project from the front view. Figure 7-80 shows the isometric view dimensioned. The **General Dimension** tool was used to create the 3D dimensions. The procedure for creating 3D dimensions is the same as for 2D dimensioning.

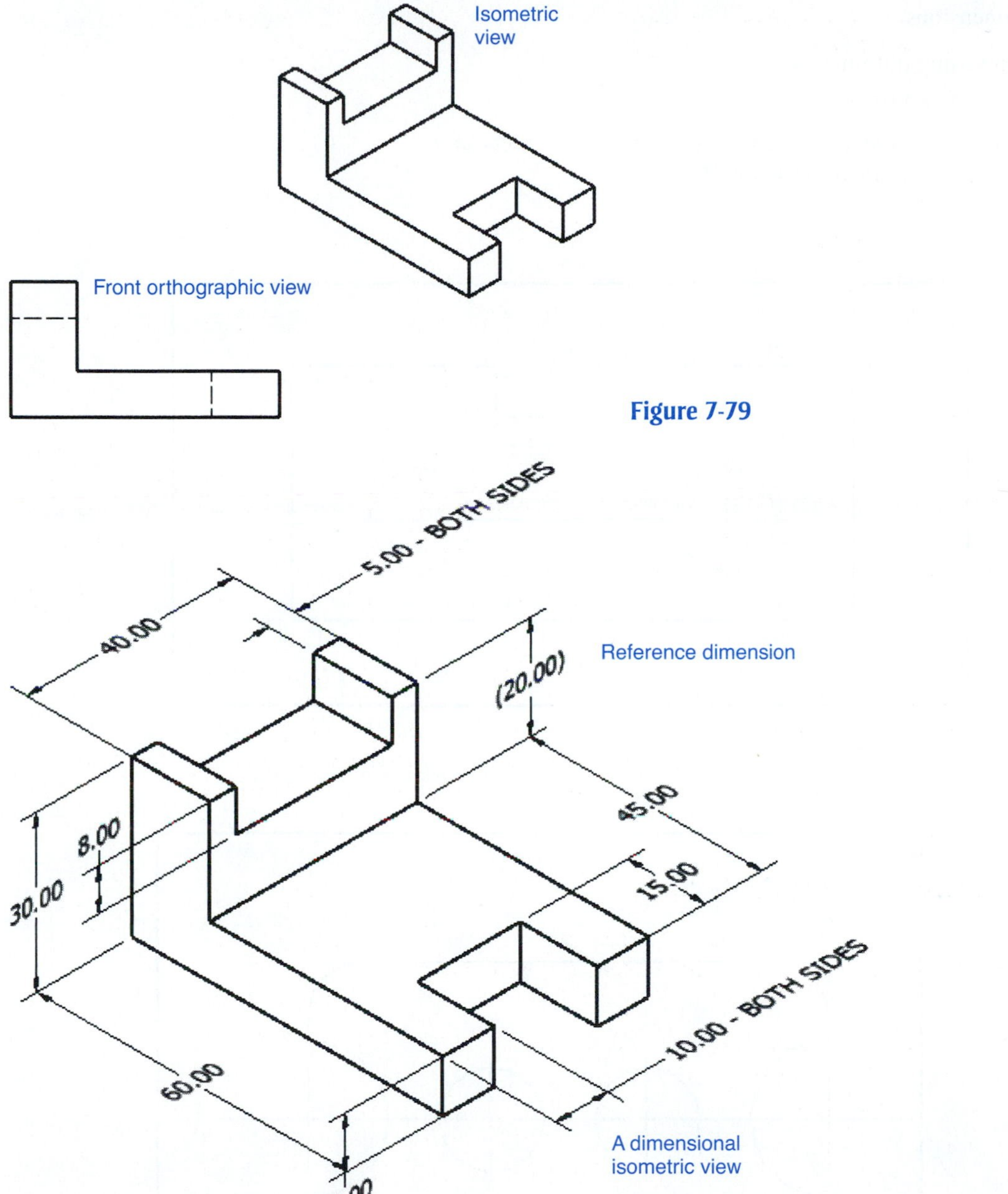

Figure 7-79

Figure 7-80

Summary

This chapter explained how to dimension object with different shapes and features. It presented the ANSI standards and conventions and showed how to dimension different shapes and features, including various types of holes, slanted surfaces, fillets and rounds, Drawing scales and dimensional values were discussed, including the use of leading zeros with decimal notation and the appropriate use of numbers of decimal places in a dimensional value. Dimensioning of sectional and orthographic views was also illustrated.

Chapter Projects

Project 7-1:

Measure and redraw the shapes in Figures P7-1 through P7-18. The dotted grid background has either .50-in. or 10-mm spacing. All holes are through holes. Specify the units and scale of the drawing. Create a model by using the **Extrude** tool. Create a set of multiviews (front, top, side, and isometric views) using the .idw format and add the appropriate dimensions.

A. Measure using millimeters.
B. Measure using inches.

All dimensions are within either .25 in. or 5 mm. All fillets and rounds are R.50 in., R.25 in. or R10 mm, R5 mm.

THICKNESS:
40 mm
1.50 in.

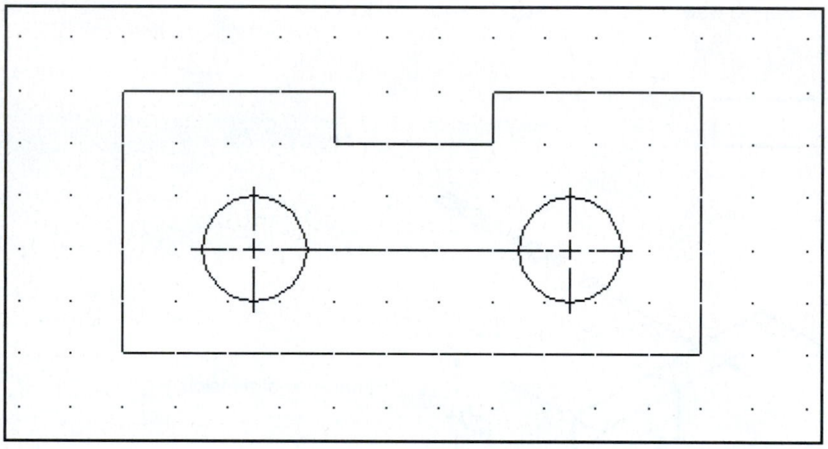

Figure P7-1

THICKNESS:
20 mm
.75 in.

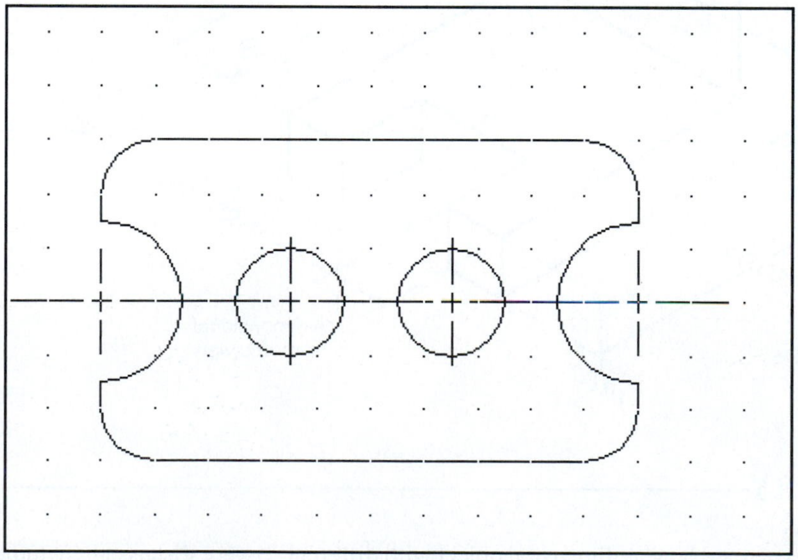

Figure P7-2

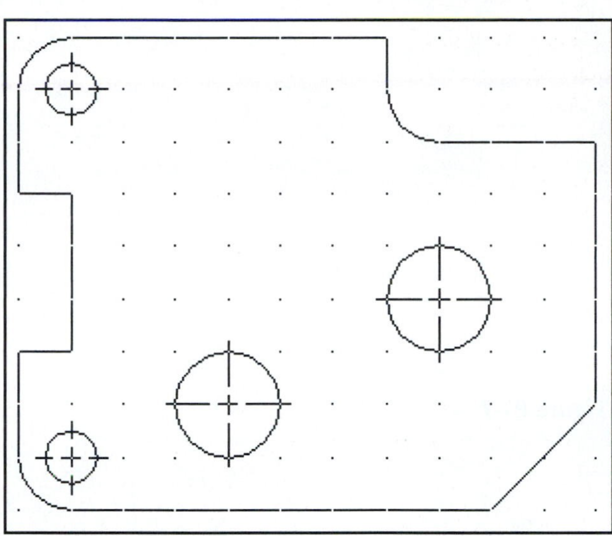

THICKNESS:
35 mm
1.25 in.

Figure P7-3

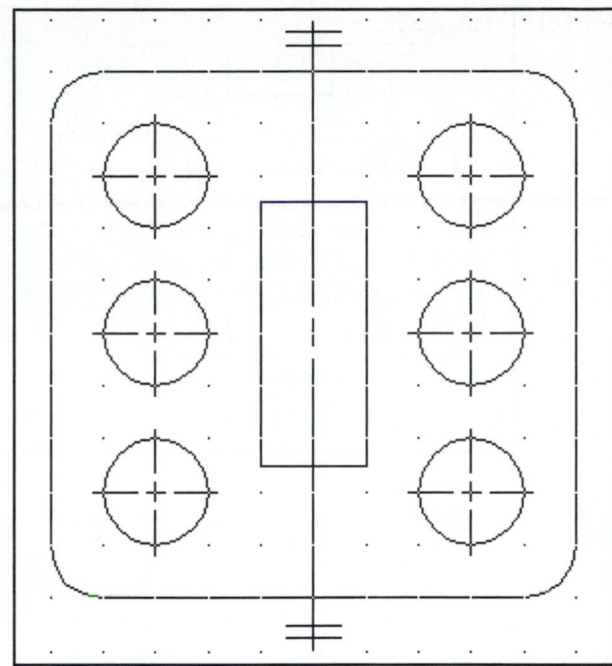

THICKNESS:
15 mm
.50 in.

Figure P7-4

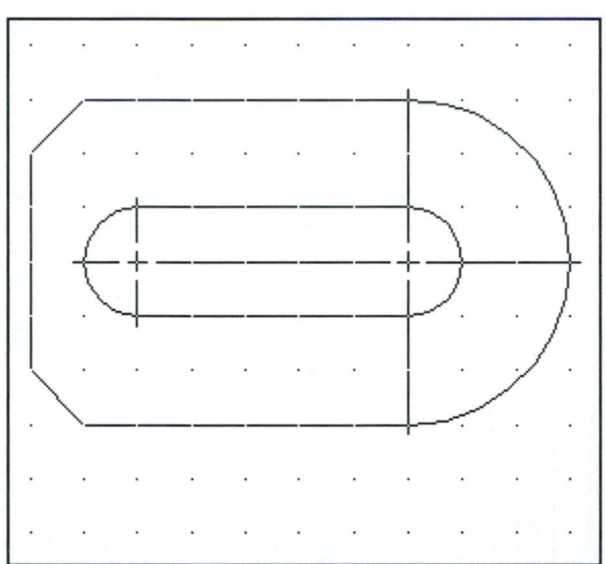

THICKNESS:
10 mm
.50 in.

Figure P7-5

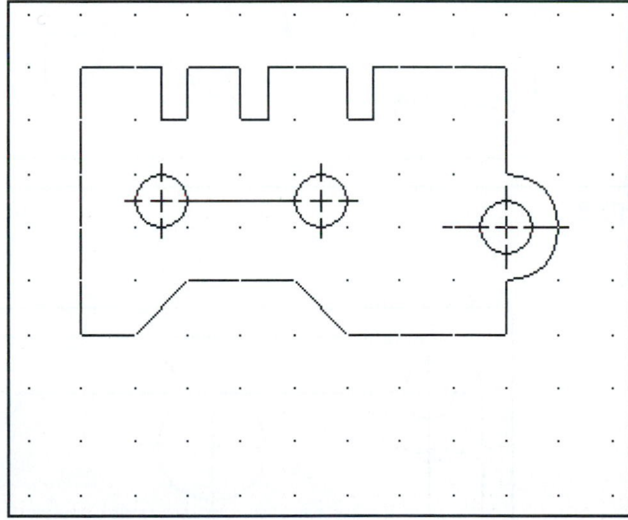

THICKNESS:
5 mm
.25 in.

Figure P7-6

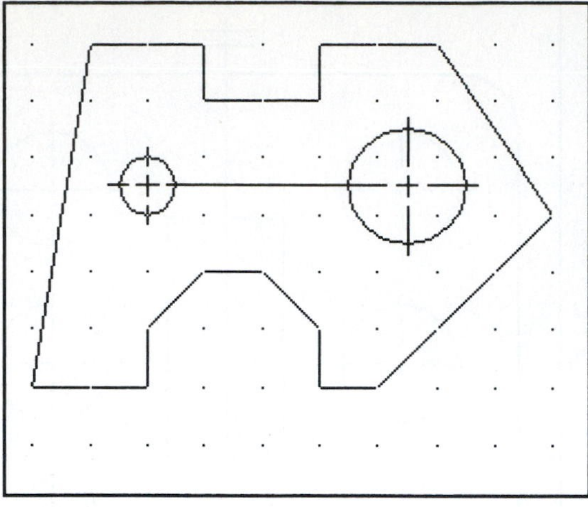

THICKNESS:
10 mm
.25 in.

Figure P7-7

THICKNESS:
8 mm
.25 in.

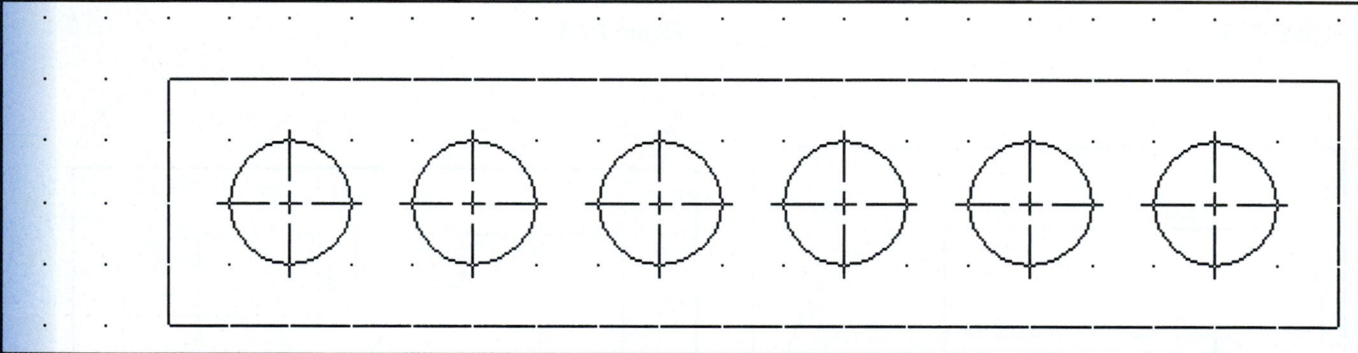

Figure P7-8

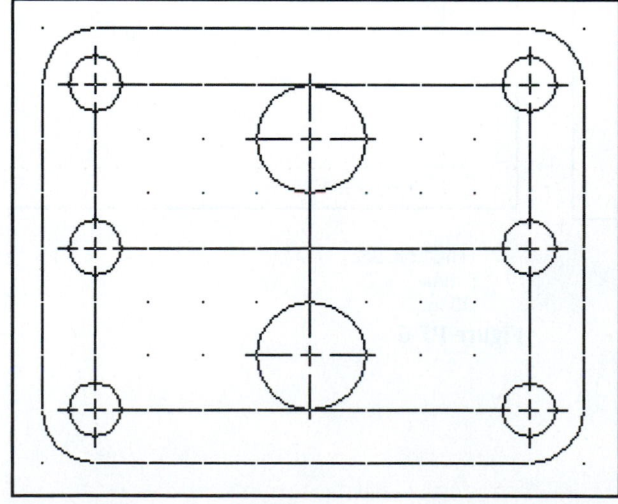

THICKNESS:
20 mm
.75 in.

Figure P7-9

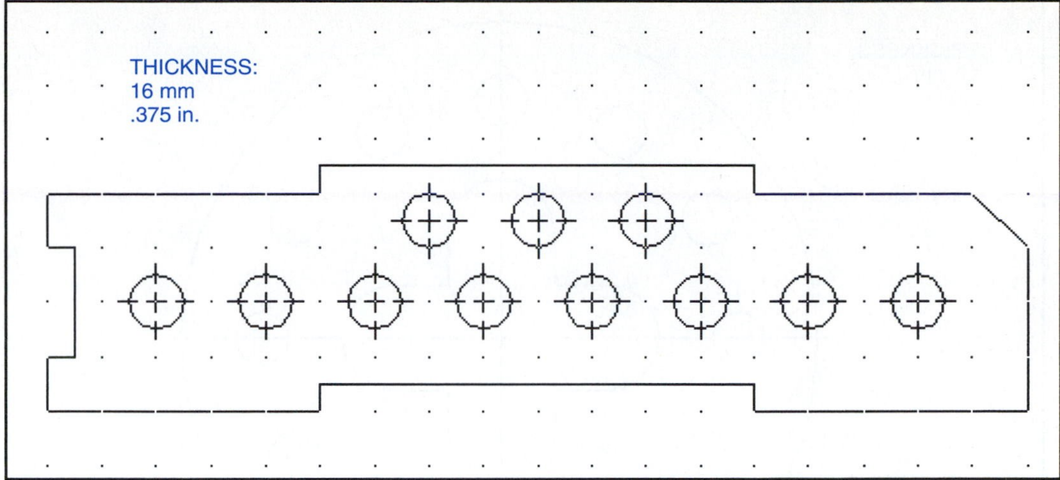

Figure P7-10

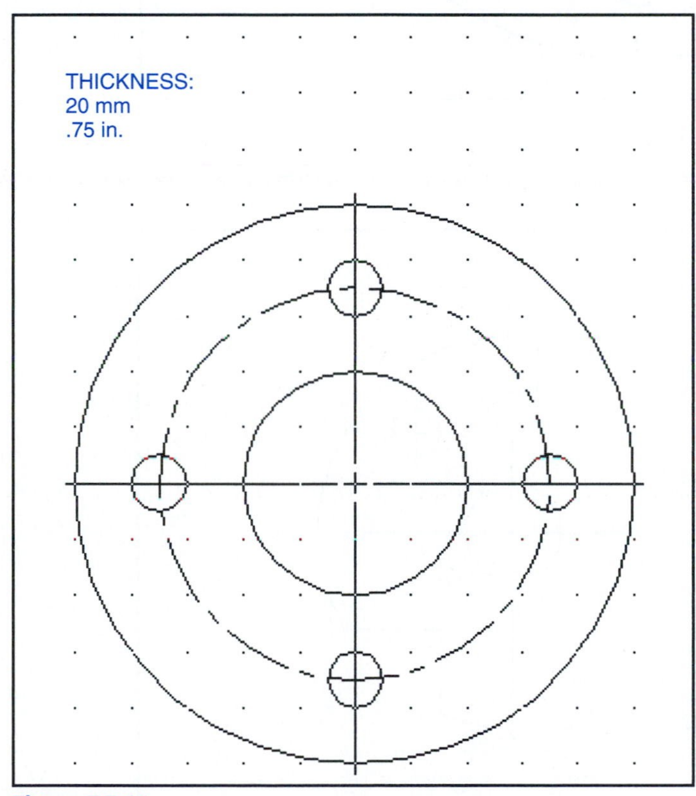

Figure P7-11

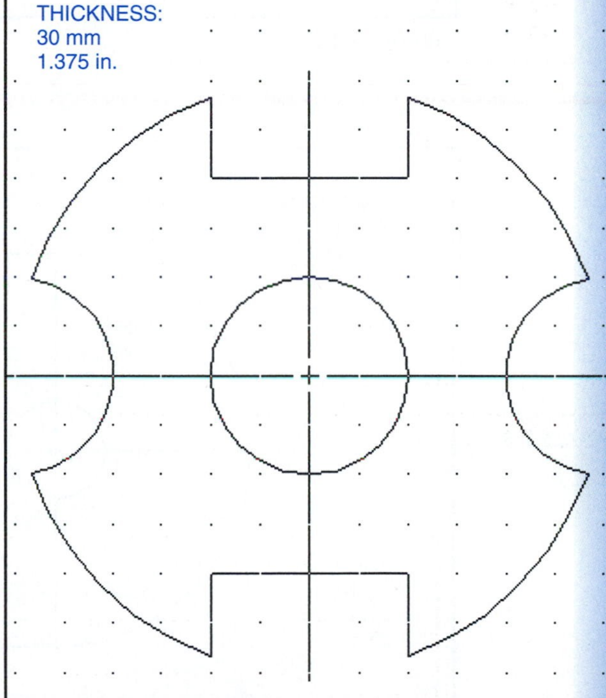

Figure P7-12

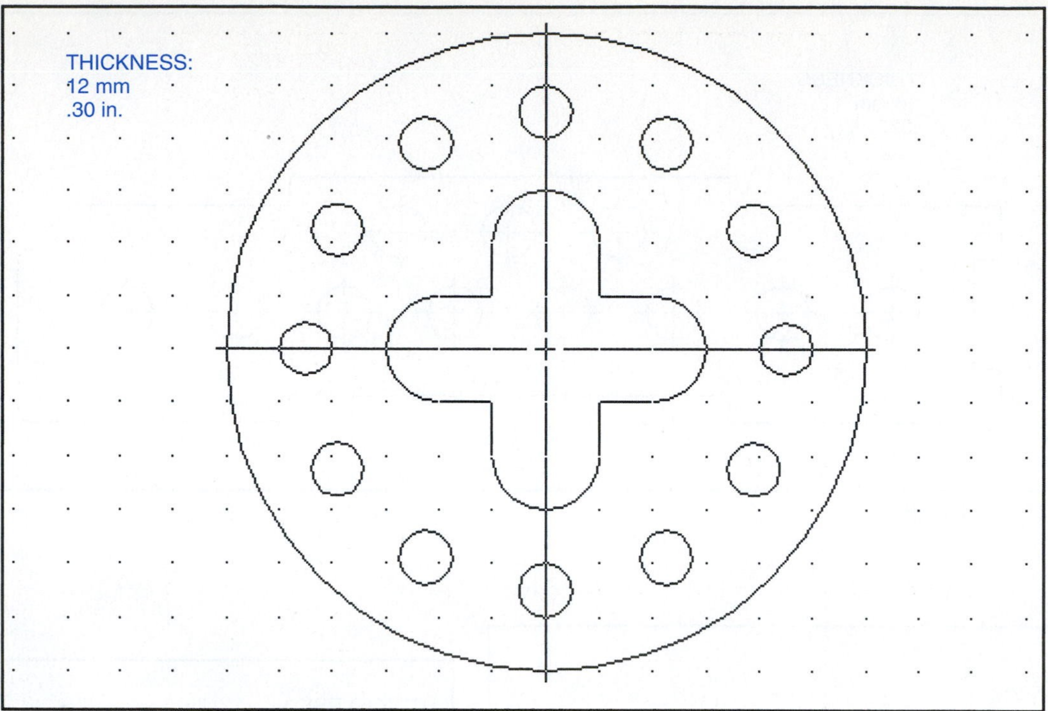

Figure P7-13

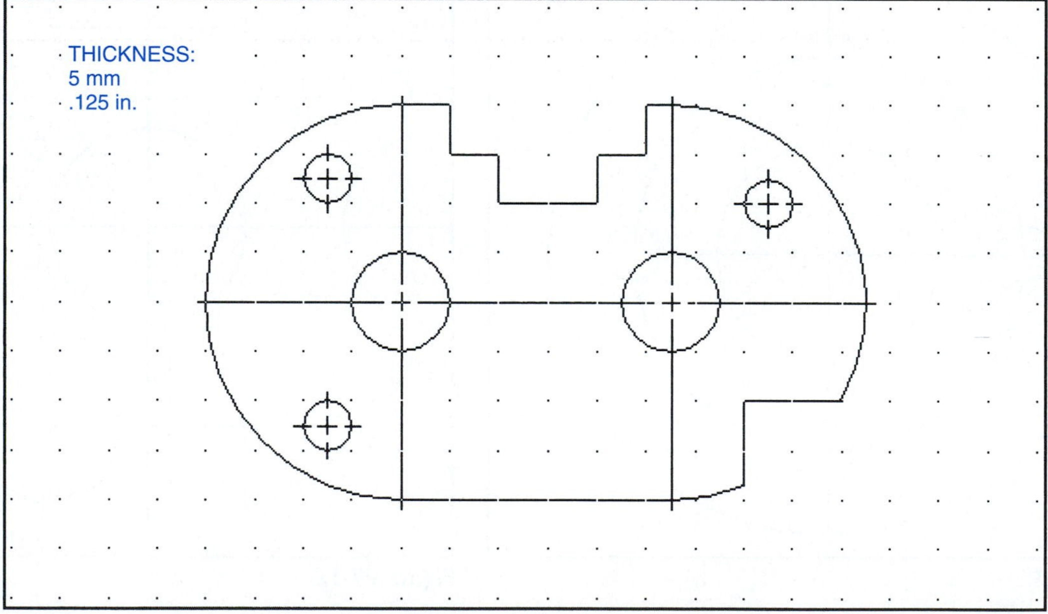

Figure P7-14

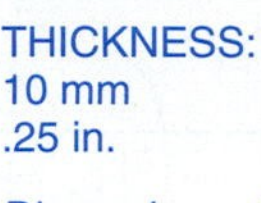

THICKNESS:
10 mm
.25 in.

Dimension using
baseline dimensions.

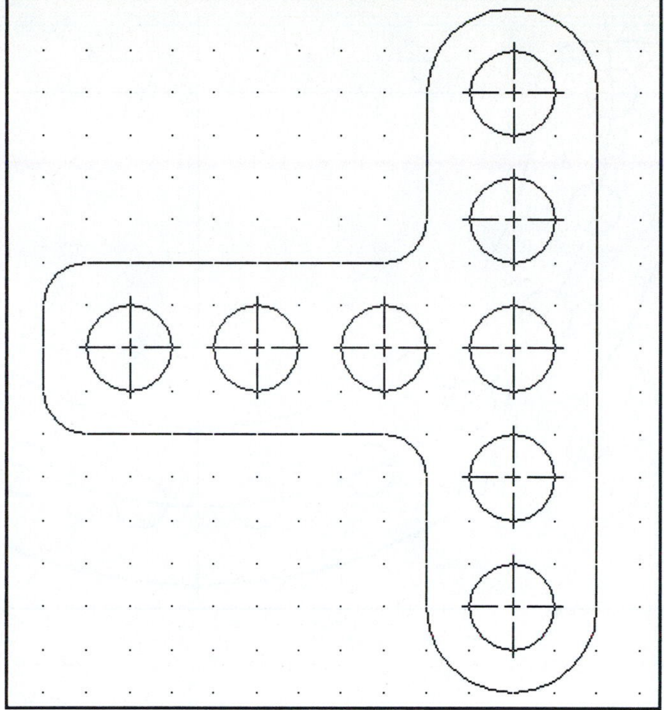

Figure P7-15

THICKNESS:
15 mm
.50 in.

Dimension using
A. Baseline dimensions.
B. Ordinate dimensions.
C. Chain dimensions.
D. Hole table.

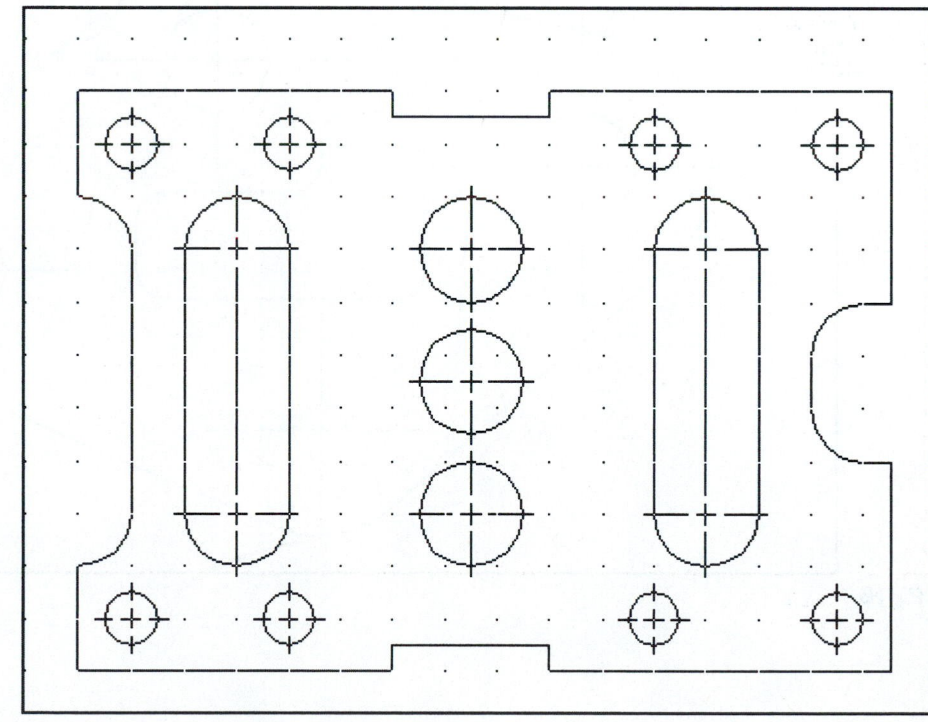

Figure P7-16

THICKNESS:
5 mm
.19 in.

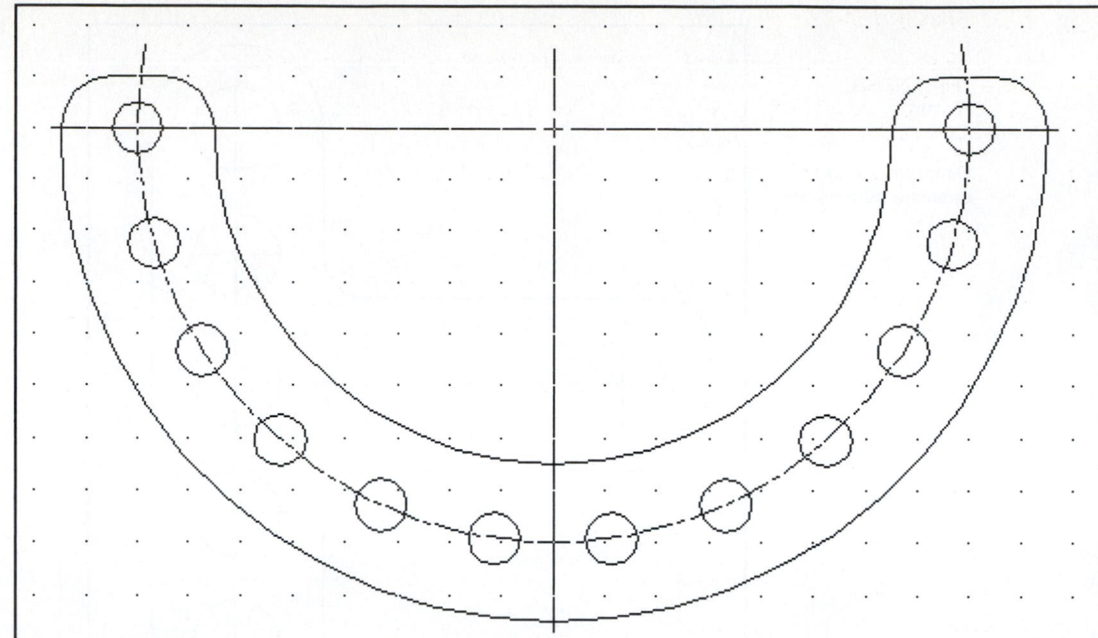

Figure P7-17

THICKNESS:
15 mm
.625 in.

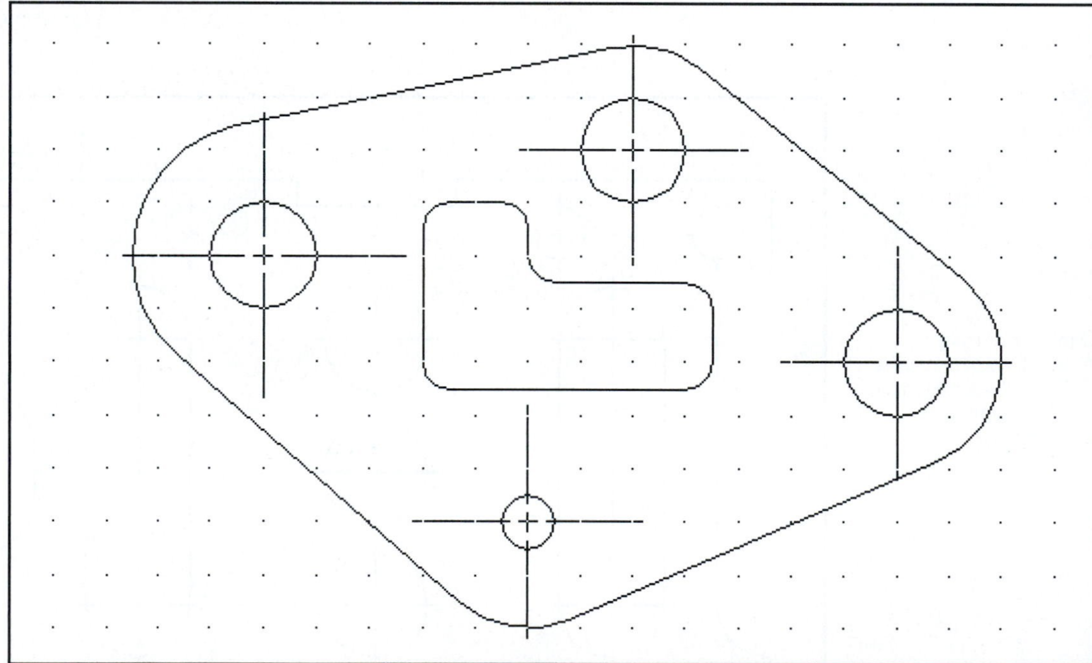

Figure P7-18

Project 7-2:

Draw models of the objects shown in Figures P7-19 through P7-40.

1. Create orthographic views of the objects. Dimension the orthographic view.
2. Create 3D models of the objects. Dimension the 3D models.

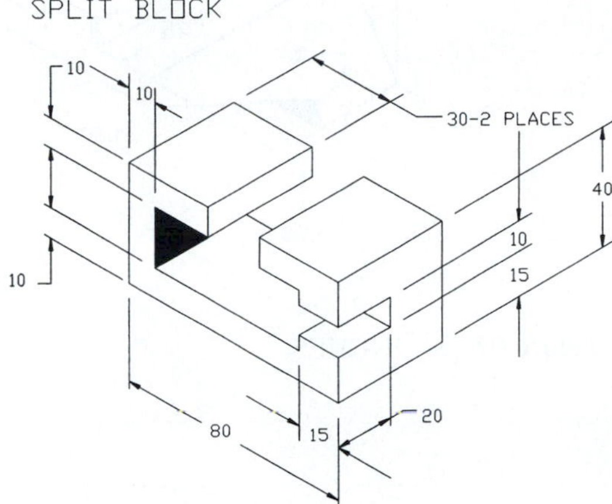

Figure P7-19 MILLIMETERS

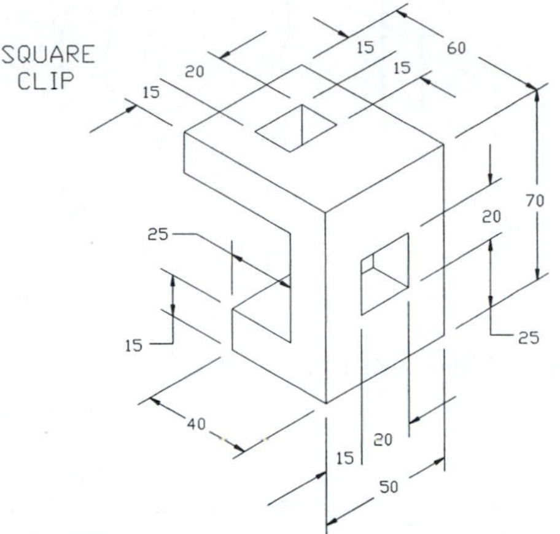

Figure P7-20 MILLIMETERS

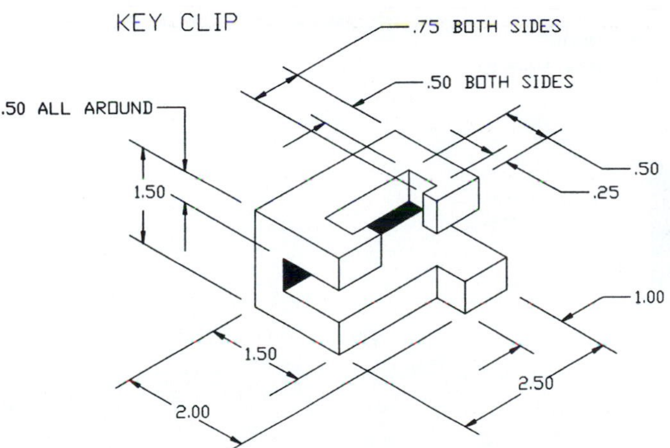

Figure P7-21 INCHES

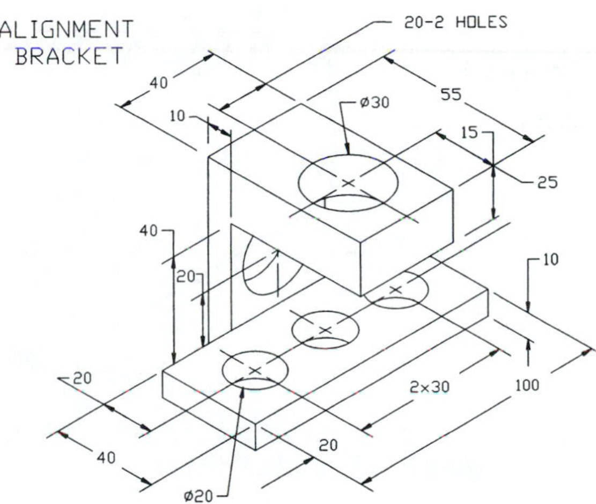

Figure P7-22 MILLIMETERS

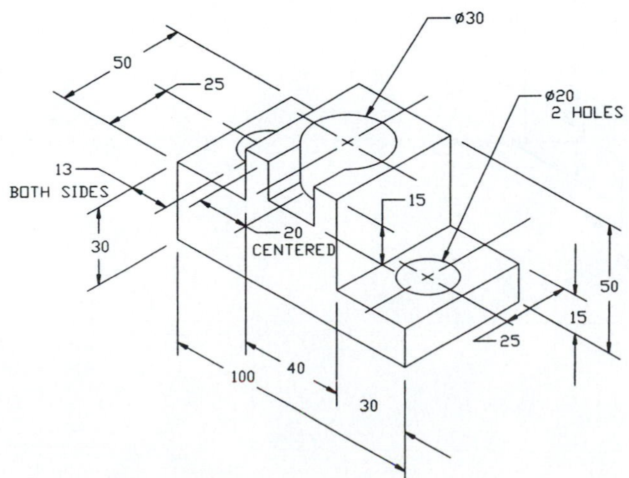

Figure P7-23 MILLIMETERS

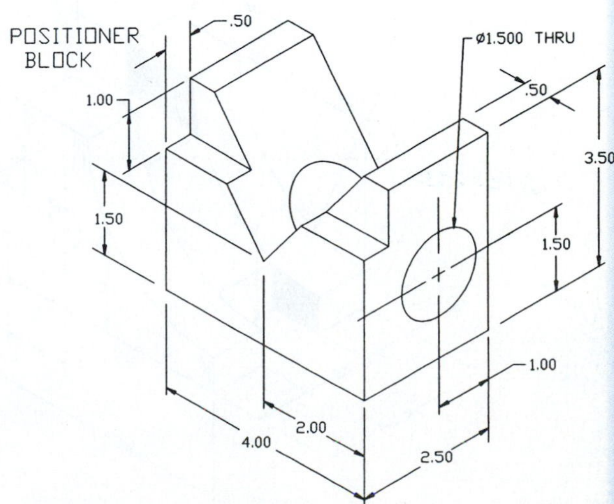

Figure P7-24 INCHES

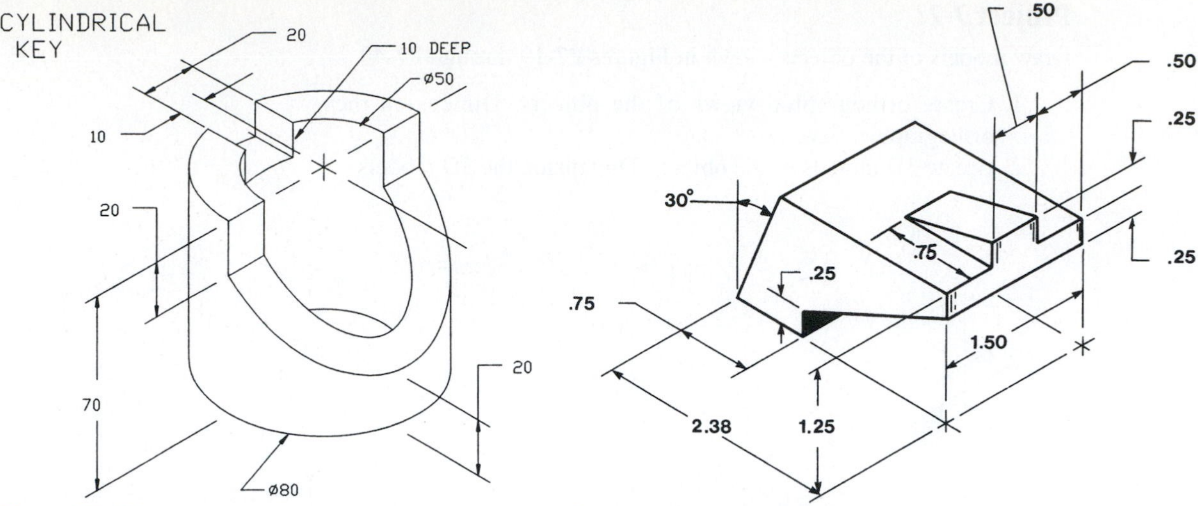

Figure P7-25 MILLIMETERS

Figure P7-26 INCHES

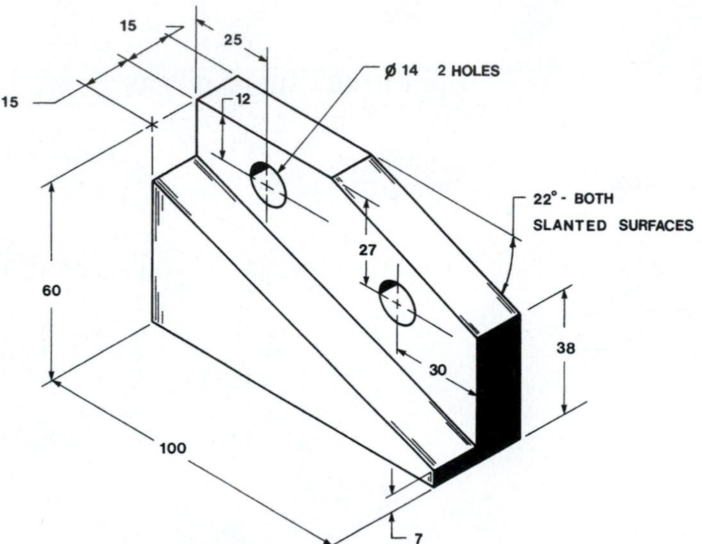

Figure P7-27 MILLIMETERS

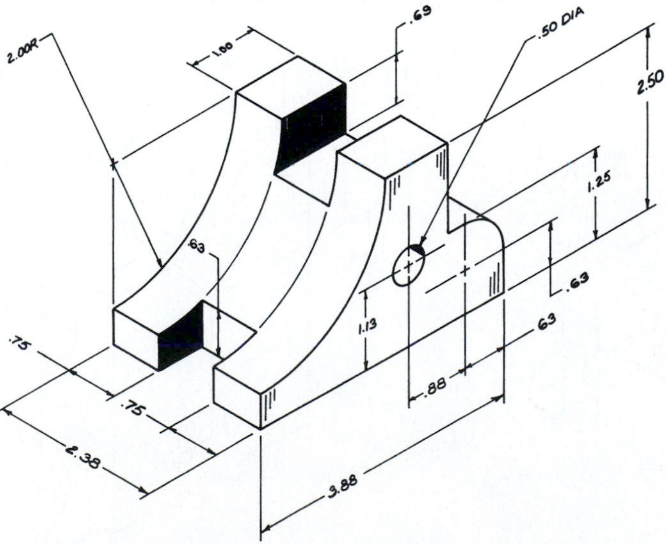

Figure P7-28 INCHES

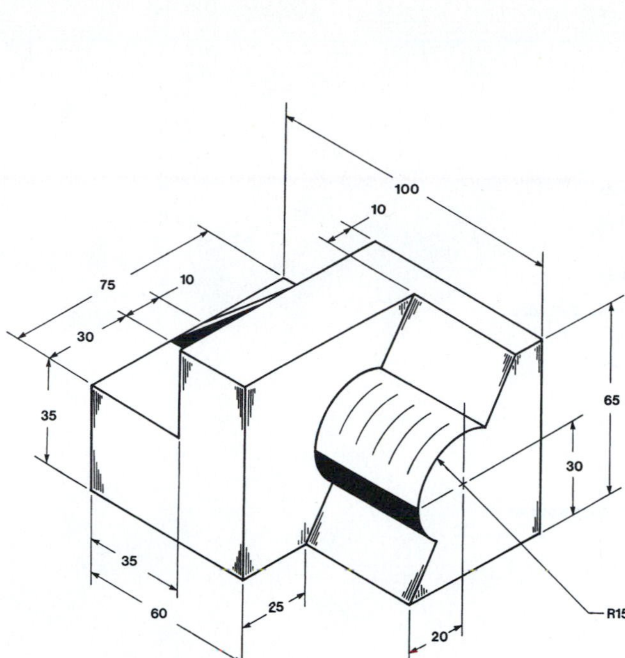

Figure P7-29 MILLIMETERS

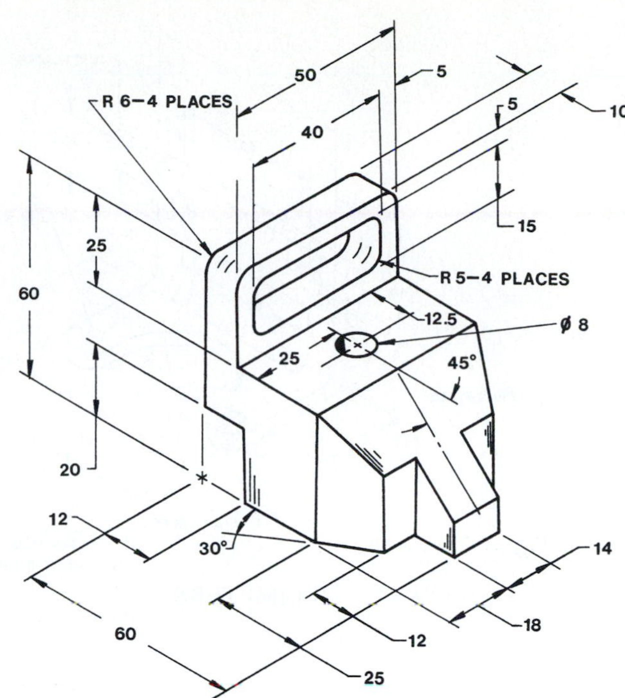

Figure P7-30 MILLIMETERS

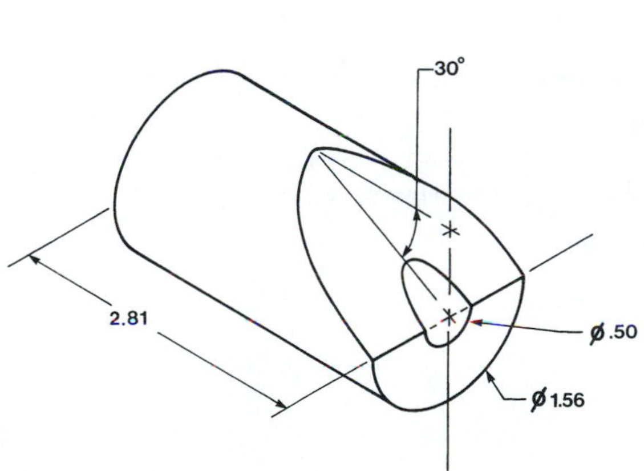

Figure P7-31 INCHES

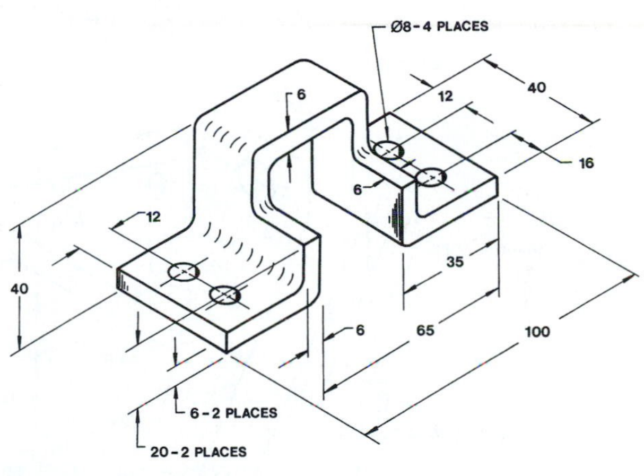

Figure P7-32 MILLIMETERS

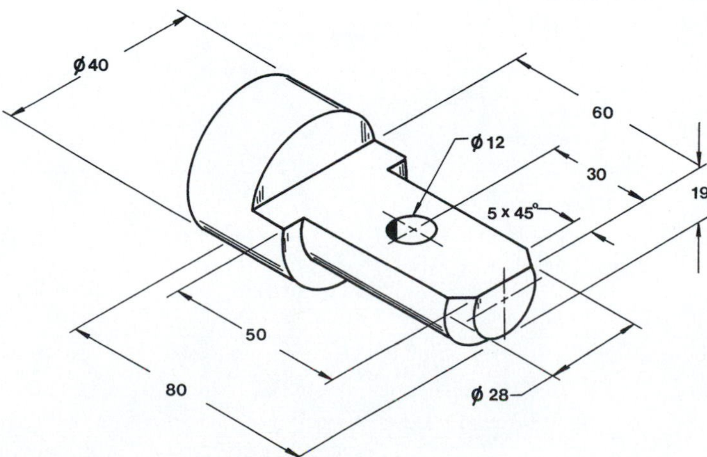

Figure P7-33 MILLIMETERS

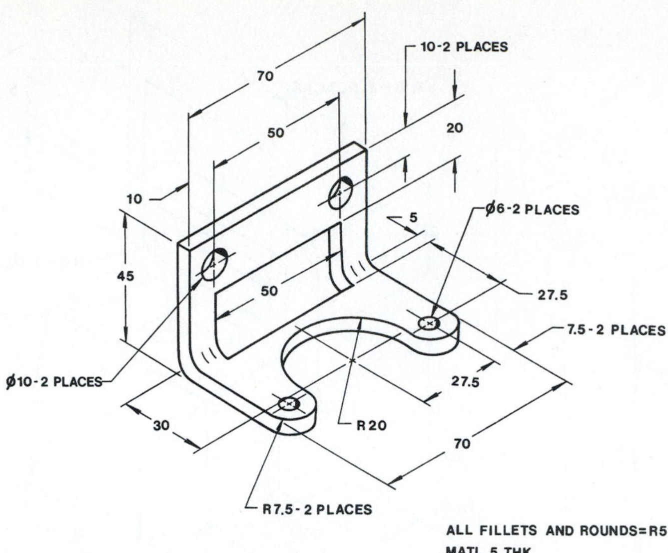

Figure P7-34 MILLIMETERS

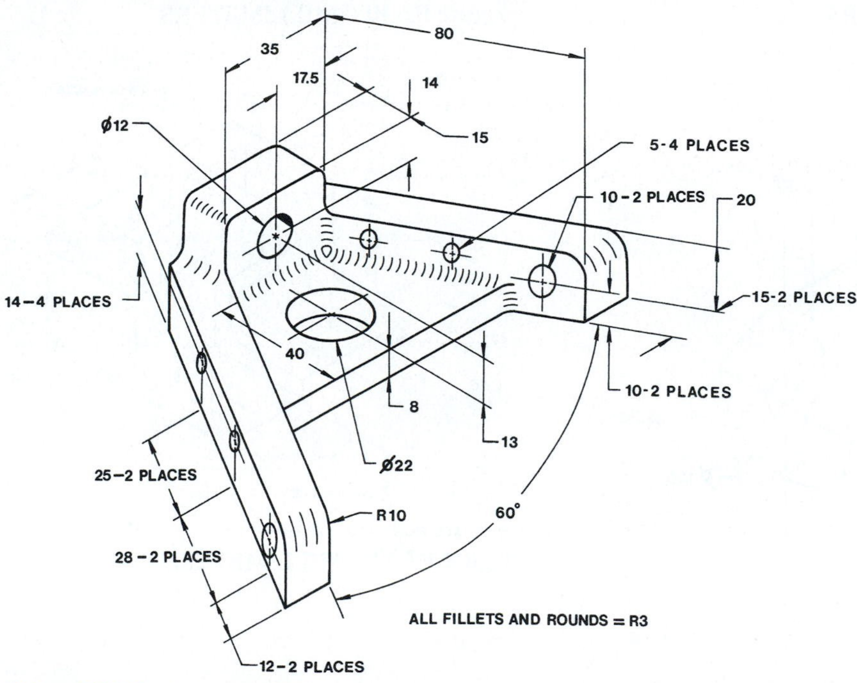

Figure P7-35 MILLIMETERS

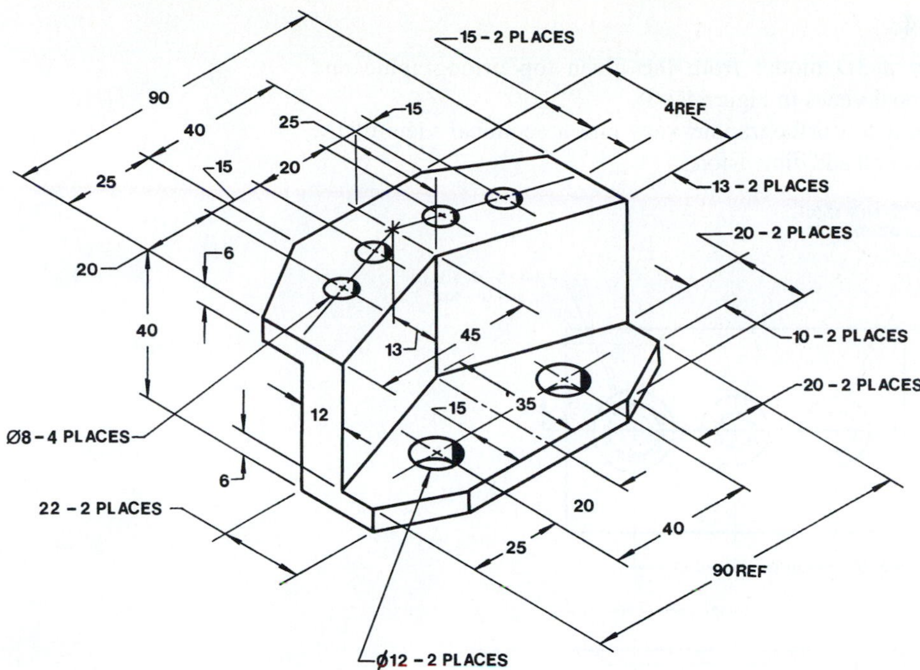

Figure P7-36 MILLIMETERS

Project 7-3:

1. Draw a 3D model from the given top orthographic and sectional views in Figure P7-37.
2. Draw a top orthographic view and a sectional view of the object and add dimensions.

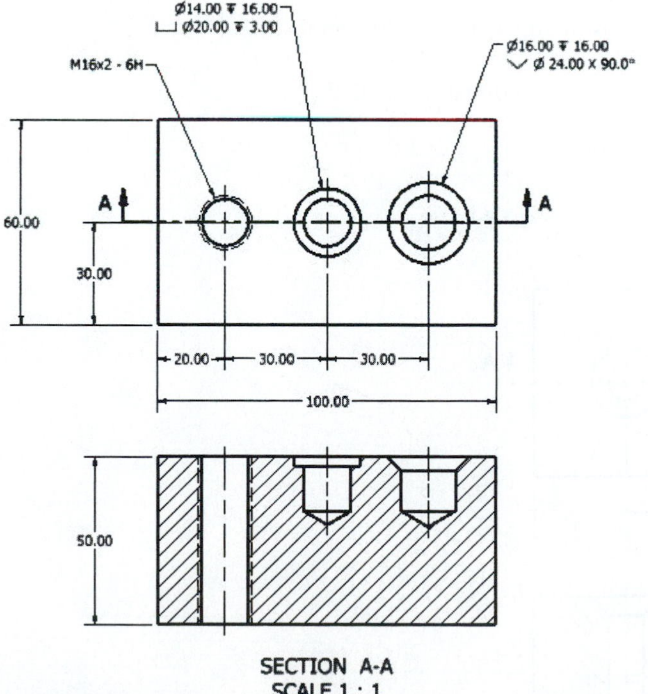

Figure P7-37 MILLIMETERS

Project 7-4:

1. Draw a 3D model from the given top orthographic and sectional views in Figure P7-38.
2. Draw a top orthographic view and a sectional view of the object and add dimensions.

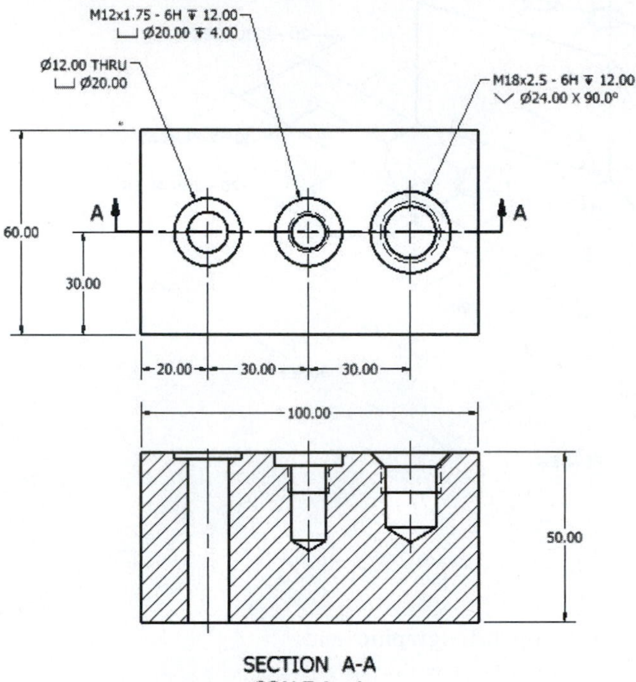

Figure P7-38 MILLIMETERS

Project 7-5:

1. Draw a 3D model from the given top orthographic and sectional views in Figure P7-39.
2. Draw a top orthographic view and a sectional view of the object and add dimensions.

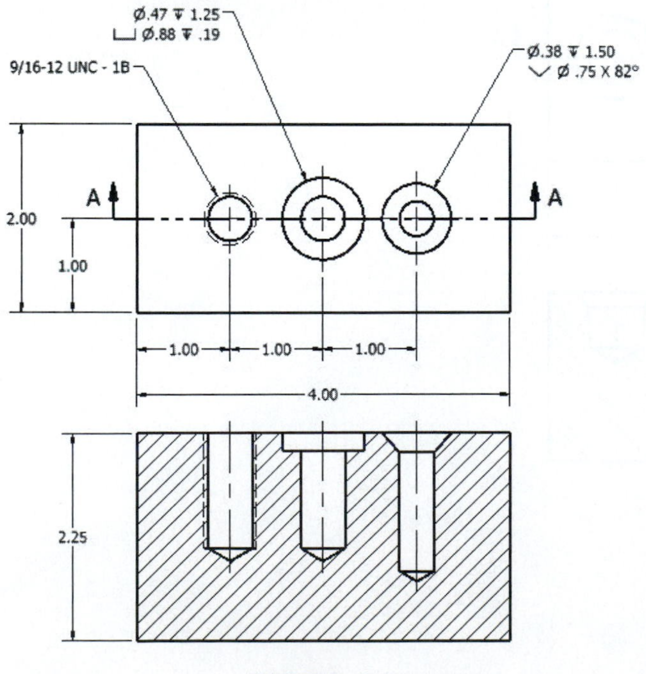

Figure P7-39 INCHES

Project 7-6:

1. Draw a 3D model from the given top orthographic and sectional views in Figure P7-40.
2. Draw a top orthographic view and a sectional view of the object and add dimensions.

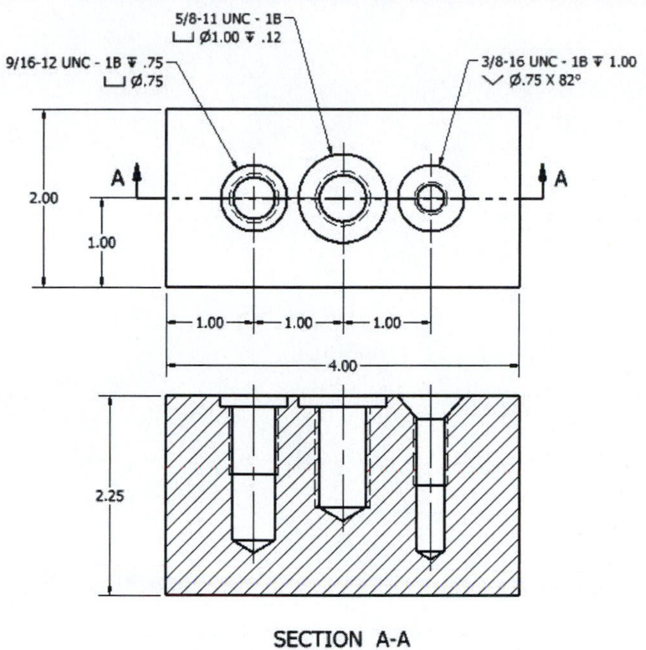

Figure P7-40 INCHES

Tolerancing

8

Chapter Objectives

- Understand tolerance conventions.
- Understand the meaning of tolerances.
- Understand how to apply tolerances.
- Understand geometric tolerances.
- Understand positional tolerances.

Introduction

Tolerances define the manufacturing limits for dimensions. All dimensions have tolerances either written directly on the drawing as part of the dimension or implied by a predefined set of standard tolerances that apply to any dimension that does not have a stated tolerance.

tolerance: The manufacturing limits for dimensions.

This chapter explains general tolerance conventions and how they are applied using Inventor. It includes a sample tolerance study and an explanation of standard fits and surface finishes.

Direct Tolerance Methods

There are two methods used to include tolerances as part of a dimension: *plus and minus,* and *limits.* Plus and minus tolerances can be expressed in either bilateral (deviation) or unilateral (symmetric) form.

A ***bilateral tolerance*** has both a plus and a minus value, whereas a ***unilateral tolerance*** has either the plus or the minus value equal to 0. Figure 8-1 shows a horizontal dimension of 60 mm that includes a bilateral tolerance of plus or minus 1 and another dimension of 60.00 mm that

bilateral tolerance: A tolerance with both a plus and a minus value.

unilateral tolerance: A tolerance with either the plus or the minus value equal to zero.

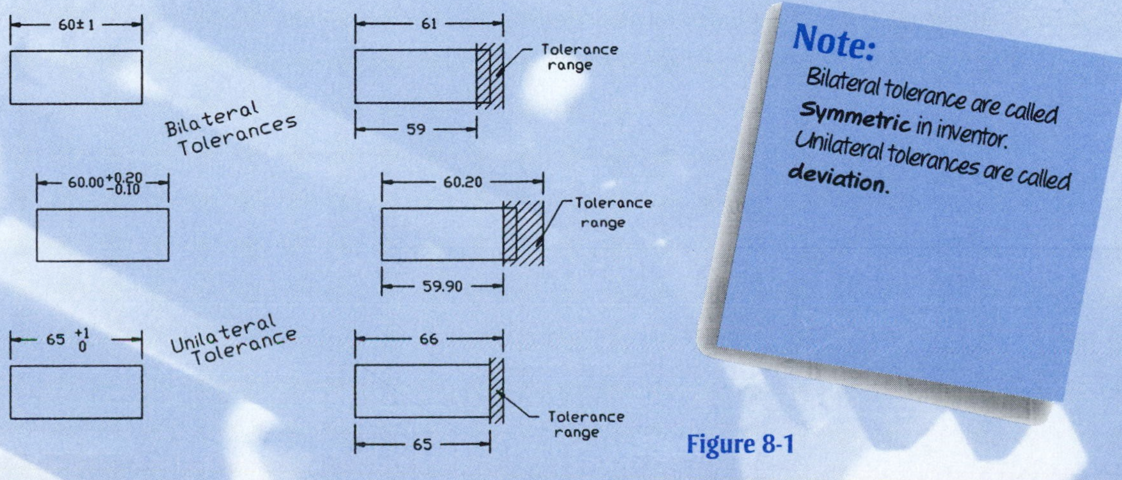

Figure 8-1

includes a bilateral tolerance of plus 0.20 or minus 0.10. Figure 8-1 also shows a dimension of 65 mm that includes a unilateral tolerance of plus 1 or minus 0.

Plus or minus tolerances define a range for manufacturing. If inspection shows that all dimensioned distances on an object fall within their specified tolerance range, the object is considered acceptable; that is, it has been manufactured correctly.

The dimension and tolerance of 60 0.1 means that the part must be manufactured within a range no greater than 60.1 nor less than 59.9. The dimension and tolerance 65 + 1, −0 defines the tolerance range as 65.0 to 66.0.

Figure 8-2 shows some bilateral and unilateral tolerances applied using decimal inch values. Inch dimensions and tolerances are written using a slightly different format than millimeter dimensions and tolerances, but they also define manufacturing ranges for dimension values. The horizontal bilateral dimension and tolerance 2.50 .02 defines the longest acceptable distance as 2.52 in. and the shortest as 2.48. The unilateral dimension 2.50 + .02 −.00 defines the longest acceptable distance as 2.52 and the shortest as 2.50.

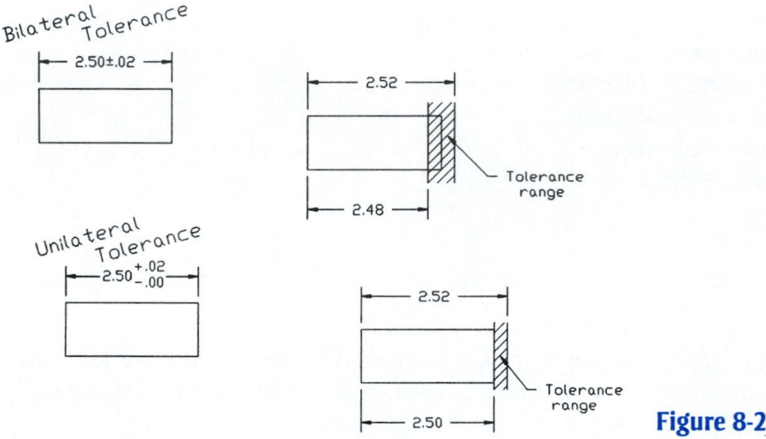

Figure 8-2

TOLERANCE EXPRESSIONS

Dimension and tolerance values are written differently for inch and millimeter values. See Figure 8-3. Unilateral dimensions for millimeter values specify a zero limit with a single 0. A zero limit for inch values must include the same number of decimal places given for the dimension value. In the example shown in Figure 8-3, the dimension value .500 has a unilateral tolerance with minus zero tolerance. The zero limit is written as .000, three decimal places for both the dimension and the tolerance.

Both values in a bilateral tolerance for inch values must contain the same number of decimal places; for millimeter values the tolerance values need not include the same number of decimal places as the dimension value. In Figure 8-3 the dimension value 32 is accompanied by tolerances of +0.25 and −0.10. This form is not acceptable for inch dimensions and tolerances. An equivalent inch dimension and tolerance would be written 32.00 + .25/−.10.

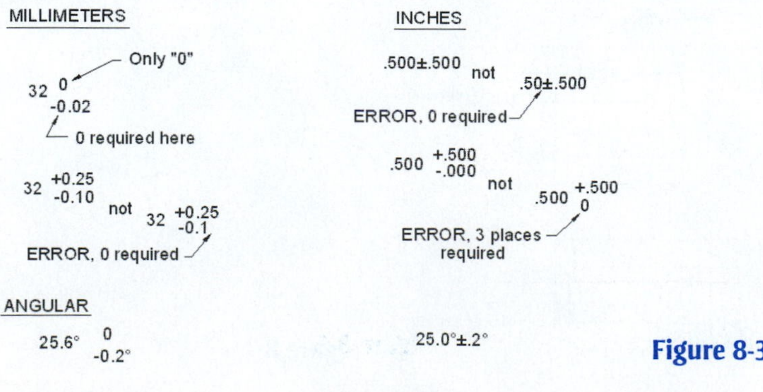

Figure 8-3

Degree values must include the same number of decimal places in both the dimension value and the tolerance values for bilateral tolerances. A single 0 may be used for unilateral tolerances.

Understanding Plus And Minus Tolerances

A millimeter dimension and tolerance of 12.0 +0.2/−0.1 means the longest acceptable distance is 12.2000 . . . 0, and the shortest is 11.9000 . . . 0. The total range is 0.3000 . . . 0.

After an object is manufactured, it is inspected to ensure that the object has been manufactured correctly. Each dimensioned distance is measured and, if it is within the specified tolerance, is accepted. If the measured distance is not within the specified tolerance, the part is rejected. Some rejected objects may be reworked to bring them into the specified tolerance range, whereas others are simply scrapped.

Figure 8-4 shows a dimension with a tolerance. Assume that five objects were manufactured using the same 12.0 +0.2/−0.1 dimension and tolerance. The objects were then inspected and the results were as listed. Inspected measurements are usually expressed to at least one more decimal place than that specified in the tolerance. Which objects are acceptable and which are not? Object 3 is too long and object 5 is too short because their measured distances are not within the specified tolerances.

Figure 8-5 shows a dimension and tolerance of 3.50 +.02 in. Object 3 is not acceptable because it is too short, and object 4 is too long.

GIVEN (mm)

$12^{+0.2}_{-0.1}$

MEANS

TOL MAX = 12.2
TOL MIN = 11.9
TOTAL TOL = 0.3

OBJECT	AS MEASURED	ACCEPTABLE?
1	12.160	OK
2	12.020	OK
3	12.203	TOO LONG
4	11.920	OK
5	11.895	TOO SHORT

Figure 8-4

GIVEN (inches)

3.50±.02

MEANS

TOL MAX = 3.52
TOL MIN = 3.48
TOTAL TOL = .04

OBJECT	AS MEASURED	ACCEPTABLE?
1	3.520	OK
2	3.486	OK
3	3.470	TOO SHORT
4	3.521	TOO LONG
5	3.515	OK

Figure 8-5

Creating Plus And Minus Tolerances

Plus and minus tolerances may be created using Inventor using the **Tolerance** option associated with existing dimensions, or by setting the plus and minus values using the **Dimension Styles** tool.

Create a drawing using the **Metric** tab and the **ANSI (mm).ipt** format. See Figure 8-6. Dimension the drawing.

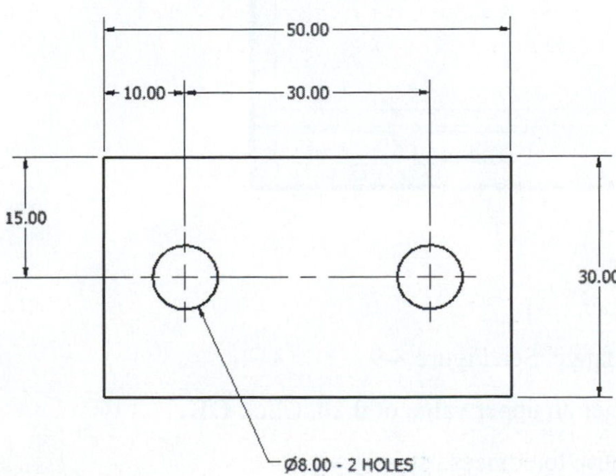

Figure 8-6

Exercise 8-1: Creating Plus and Minus Tolerances

1. Click the **30.00** dimension.

The dimension will change colors.

2. Right-click the mouse and select the **Edit** option.

See Figure 8-7. The **Edit Dimension** dialog box will appear. See Figure 8-8.

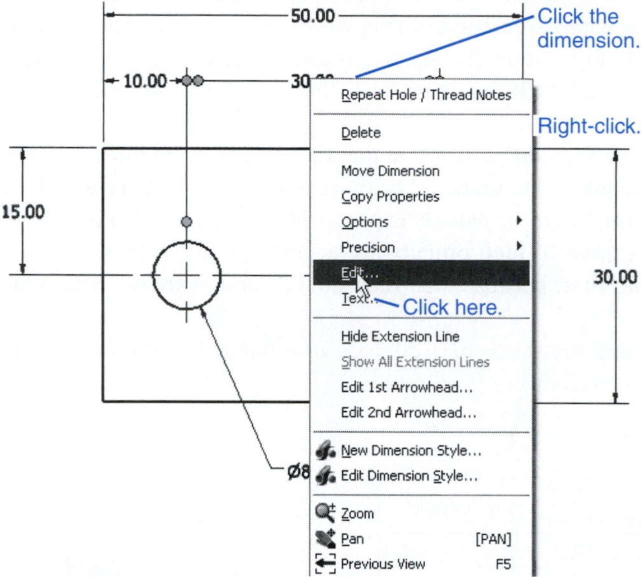

Figure 8-7

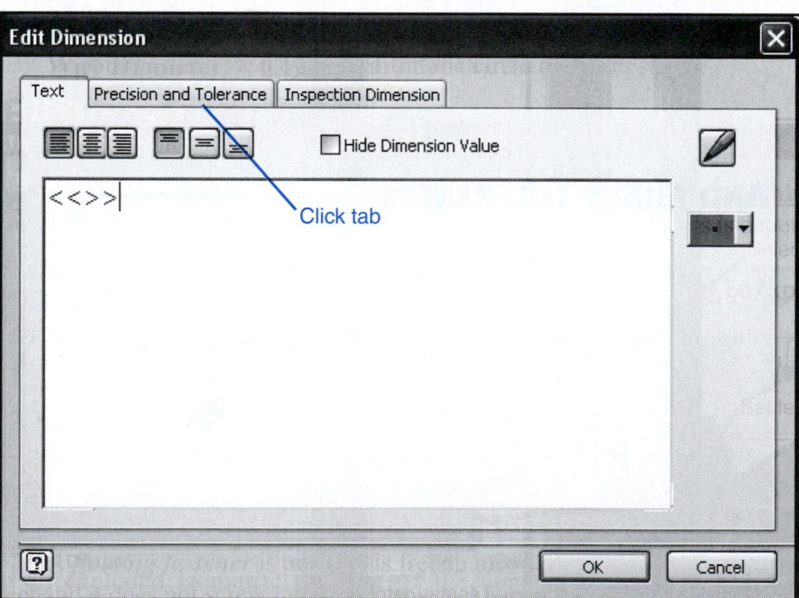

Figure 8-8

3. Click the **Precision and Tolerance** tab.

The **Edit Dimension** dialog box will change. See Figure 8-9.

4. Select the **Symmetric** option and enter an upper value of **0.20.** Click **OK.**

Figure 8-10 shows the resulting symmetric tolerances.

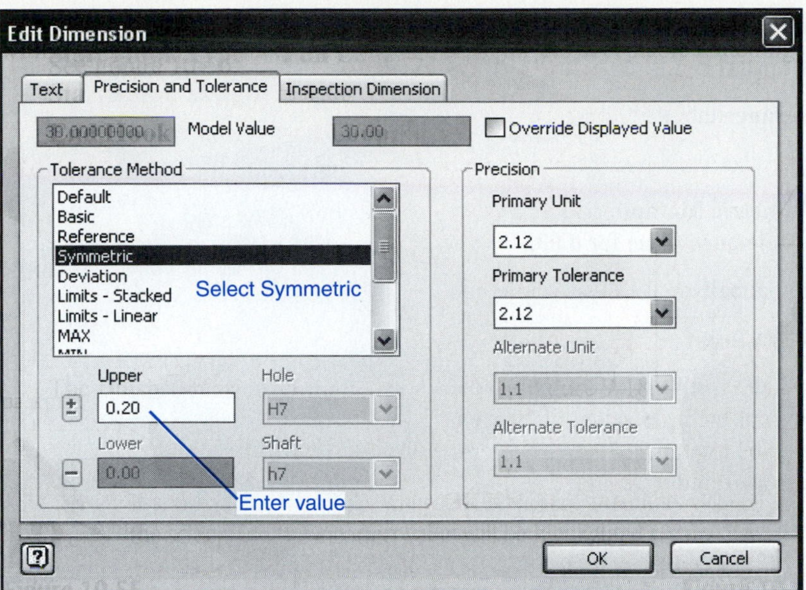

Figure 8-9

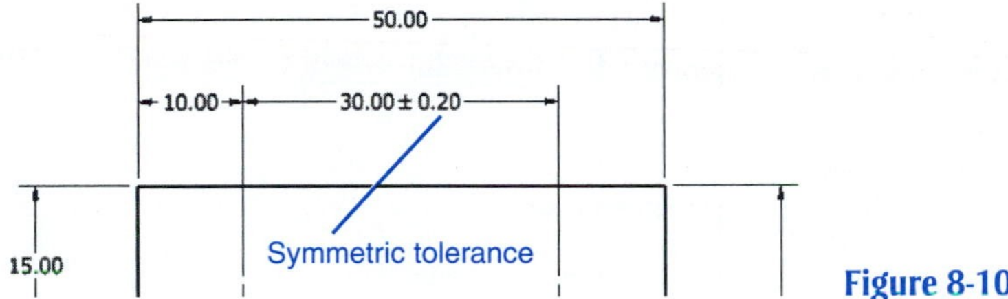

Figure 8-10

Exercise 8-2: Creating Plus and Minus Tolerances Using Styles Editor

1. Click the **Format** heading at the top of the screen and select the **Styles Editor** option.

The **Styles and Standards Editor** dialog box will appear. See Figure 8-11.

2. Click the **+ sign** to the left of the **Dimension** listing.

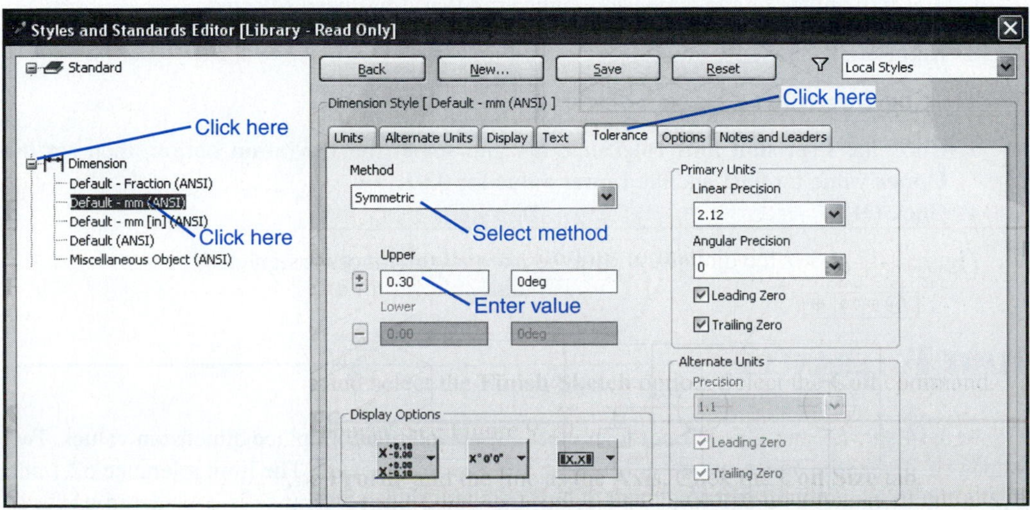

Figure 8-11

3. Select the appropriate standard.

The **Styles and Standards Editor** dialog box will change.

4. Select the **Tolerance** tab.

See Figure 8-11.

5. Select the **Symmetric Method.**
6. Set the **Upper** tolerance value for **0.03.**

Inventor will automatically make this a + toleranc

7. Click **Save,** then **Close.**

Figure 8-12 shows the resulting tolerances. Note that all dimensions, except the hole value, have been changed to include a ±0.03 tolerance. Standard tolerances will be discussed later in the chapter.

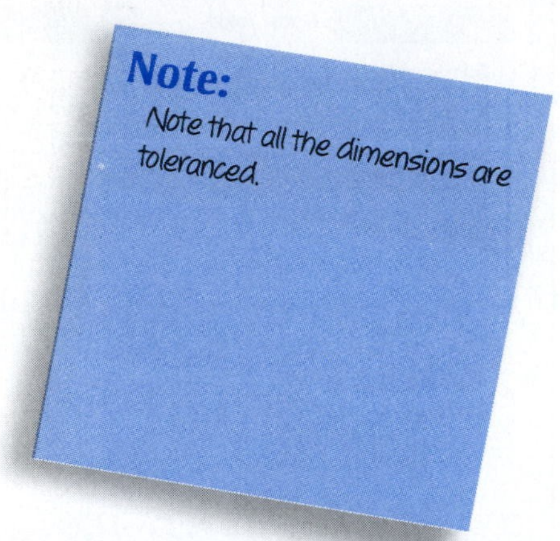

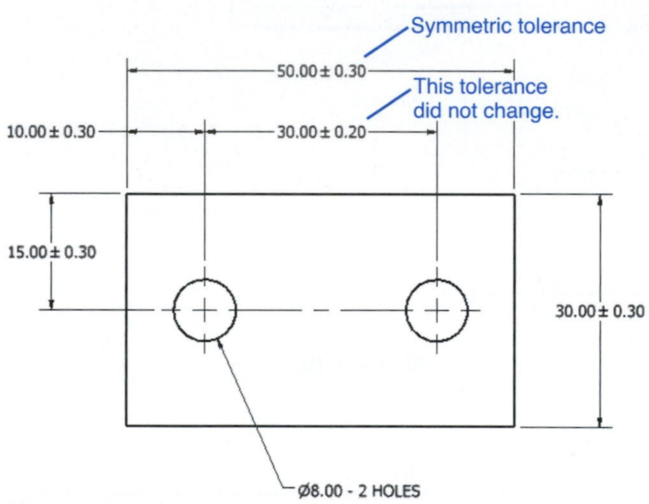

Figure 8-12

Exercise 8-3: Creating Unequal Plus and Minus Tolerances

1. Click on the dimension to be toleranced.

Circles will appear on the dimension, indicating that it has been selected.

2. Right-click the mouse and select the **Edit** option.

The **Edit Dimension** dialog box will appear. See Figure 8-13.

3. Click the **Precision and Tolerance** tab and select the **Deviation** option, then set the **Upper** value for **0.01** and the **Lower** value for **0.02.**
4. Click **OK.**

Figure 8-13 shows the dimension with the unequal tolerances assigned.

LIMIT TOLERANCES

Figure 8-14 shows examples of limit tolerances. Limit tolerances replace dimension values. Two values are given: the upper and lower limits for the dimension value. The limit tolerance 62.1 and 61.9 is mathematically equal to 62 ± 0.1, but the stated limit tolerance is considered easier to read and understand.

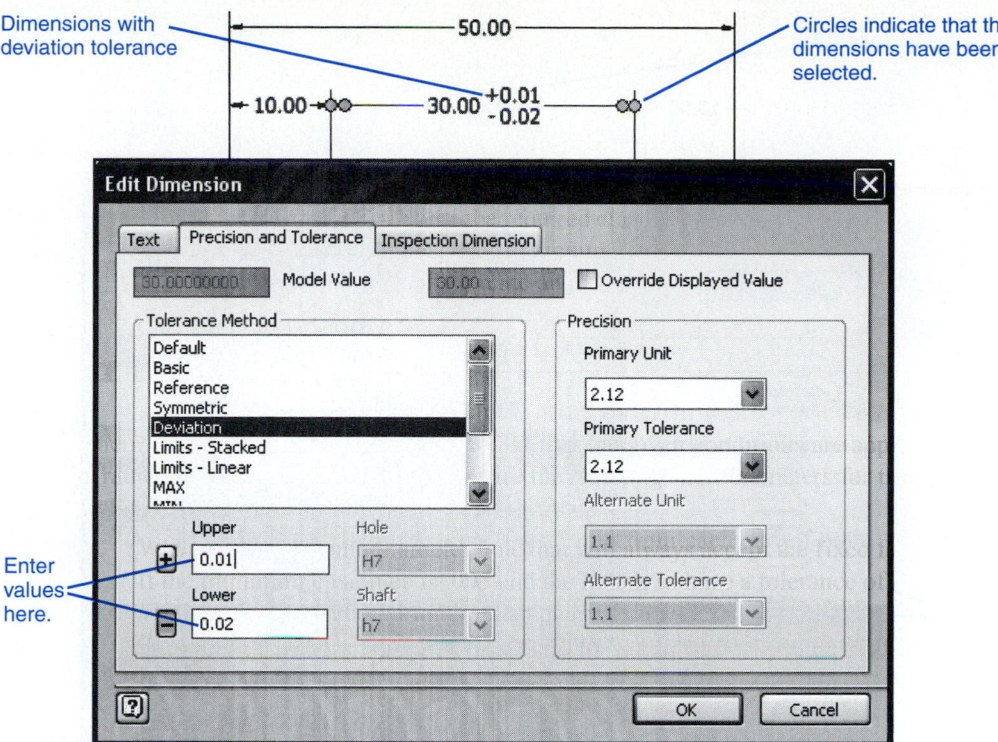

Figure 8-13

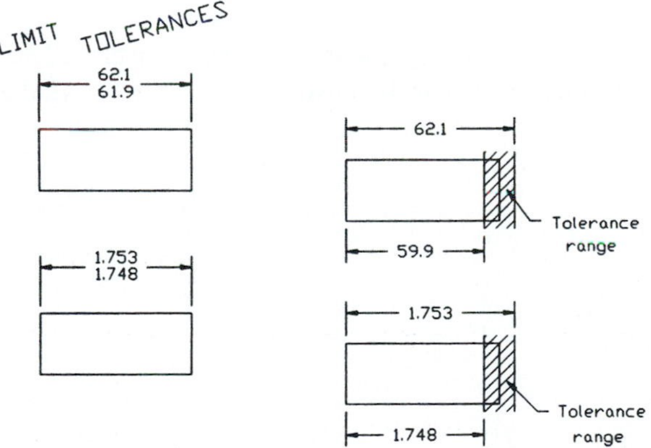

Figure 8-14

Limit tolerances define a range for manufacture. Final distances on an object must fall within the specified range to be acceptable.

Exercise 8-4: Creating Limit Tolerances

1. Click on the dimension to be toleranced, right-click the mouse, and select the **Edit** option.
2. Click the **Precision and Tolerance** tab.
3. Select the **Limits-Stacked** option, then set the **Upper** value for **30.03** and the **Lower** value for **29.96.**
4. Click **OK.**

Figure 8-15 shows the dimension with a limit tolerance assigned.

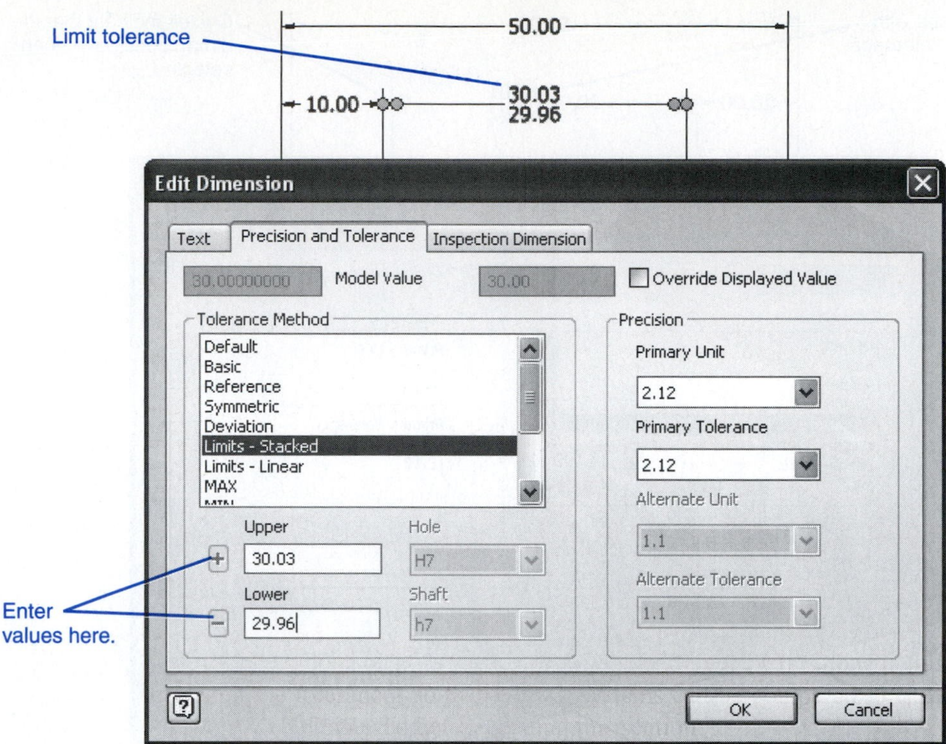

Figure 8-15

Angular Tolerances

Figure 8-16 shows an example of an angular dimension with a symmetric tolerance. The procedures explained for applying different types of tolerances to linear dimensions also apply to angular dimensions.

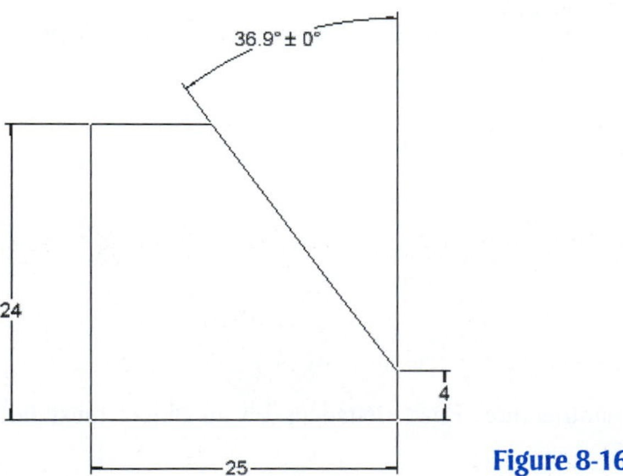

Figure 8-16

Exercise 8-5: Creating Angular Tolerances

Figure 8-17 shows a model with a slanted surface. The model has been dimensioned, but no tolerances have been assigned. This example will assign a stacked limits tolerance to the angular dimension.

1. Click the angular dimension, right-click, and select the **Edit** option.

   The **Edit Dimension** dialog box will appear.

2. Click the **Precision and Tolerance** tab.

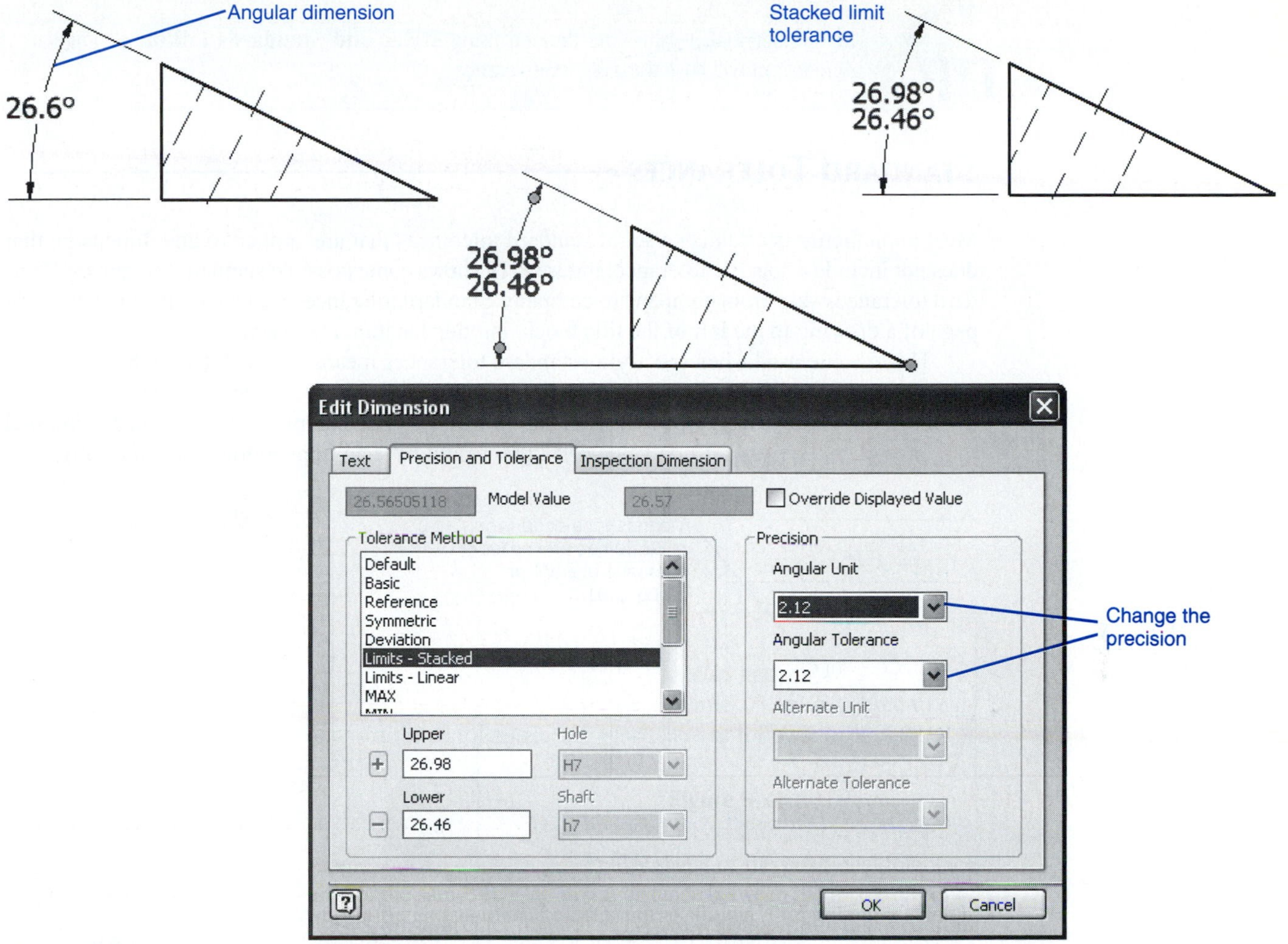

Figure 8-17

3. Change the precision of both the **Unit** and **Tolerance** to **2** places (0.00).
4. Select the **Limits - Stacked** option and set the upper and lower values for the tolerance.
5. Click **OK.**

Angular tolerances also can be assigned using the **Dimension Styles** tool. Figure 8-18 shows a **Styles and Standards Editor** dialog box.

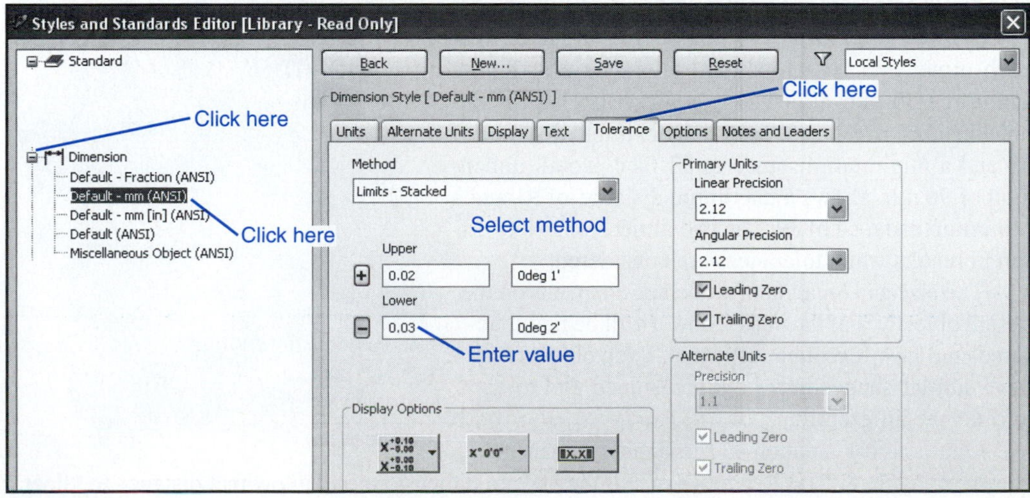

Figure 8-18

Remember that if values are defined using **Styles and Standards Editor,** all angular dimensions will have the assigned tolerance.

Standard Tolerances

Most manufacturers establish a set of standard tolerances that are applied to any dimension that does not include a specific tolerance. Figure 8-19 shows some possible standard tolerances. Standard tolerances vary from company to company. Standard tolerances are usually listed on the first page of a drawing to the left of the title block, but this location may vary.

The X value used when specifying standard tolerances means any X stated in that format. A dimension value of 52.00 would have an implied tolerance of ±.01 because the stated standard tolerance is .XX ± .01, so any dimension value with two decimal places has a standard implied tolerance of ±.01. A dimension value of 52.000 would have an implied tolerance of ±.001.

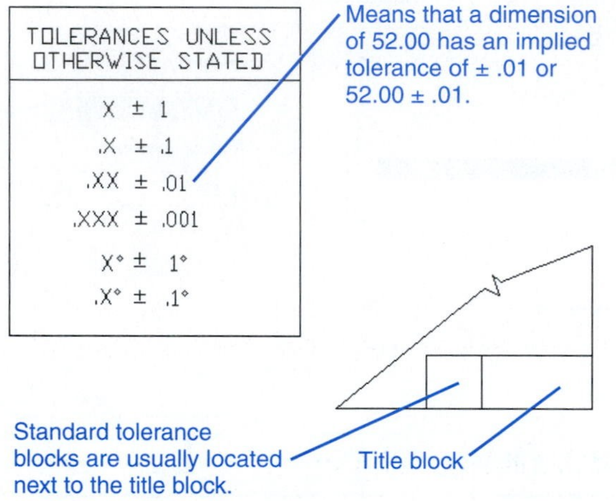

Figure 8-19

Double Dimensioning

double dimensioning: An error in which the same distance is dimensioned twice.

It is an error to dimension the same distance twice. This mistake is called ***double dimensioning.*** Double dimensioning is an error because it does not allow for tolerance buildup across a distance.

Figure 8-20 shows an object that has been dimensioned twice across its horizontal length, once using three 30-mm dimensions and a second time using the 90-mm overall dimension. The two dimensions are mathematically equal but are not equal when tolerances are considered. Assume that each dimension has a standard tolerance of ±1 mm. The three 30-mm dimensions could create an acceptable distance of 90 ± 3 mm, or a maximum distance of 93 and a minimum distance of 87. The overall dimension of 90 mm allows a maximum distance of 91 and a minimum distance of 89. The two dimensions yield different results when tolerances are considered.

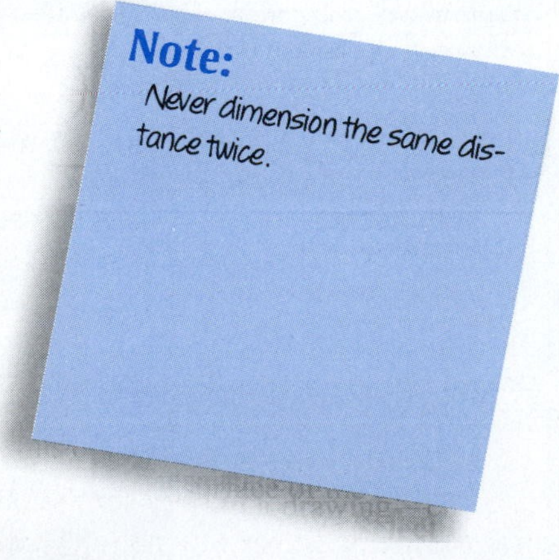

The size and location of a tolerance depends on the design objectives of the object, how it will be manufactured, and how it will be inspected. Even objects that have similar shapes may be dimensioned and toleranced very differently.

One possible solution to the double dimensioning shown in Figure 8-20 is to remove one of the 30-mm dimensions and allow that distance to "float," that is, absorb the cumulated tolerances. The choice of which 30-mm dimension to eliminate depends

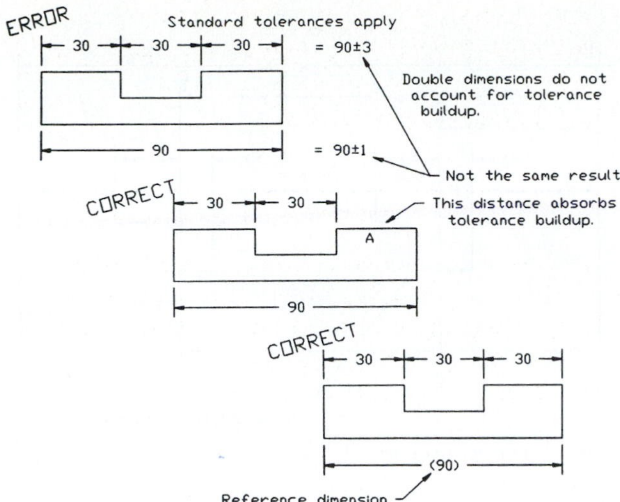

Figure 8-20

on the design objectives of the part. For this example the far-right dimension was eliminated to remove the double-dimensioning error.

Another possible solution to the double-dimensioning error is to retain the three 30-mm dimensions and to change the 90-mm overall dimension to a reference dimension. A reference dimension is used only for mathematical convenience. It is not used during the manufacturing or inspection process. A reference dimension is designated on a drawing using parentheses: (90).

If the 90-mm dimension was referenced, then only the three 30-mm dimensions would be used to manufacture and inspect the object. This would eliminate the double-dimensioning error.

Chain Dimensions and Baseline Dimensions

There are two systems for applying dimensions and tolerances to a drawing: chain and baseline. Figure 8-21 shows examples of both systems. ***Chain dimensions*** dimension each feature to the feature next to it. ***Baseline dimensions*** dimension all features from a single baseline or datum.

chain dimension: A dimension in which each feature is dimensioned to the feature next to it.

baseline dimension: A dimension in which all features are dimensioned from a single baseline or datum.

Chain and baseline dimensions may be used together. Figure 8-21 also shows two objects with repetitive features; one object includes two slots, and the other, three sets of three holes. In each example, the center of the repetitive feature is dimensioned to the left side of the object, which serves as a baseline. The sizes of the individual features are dimensioned using chain dimensions referenced to centerlines.

Baseline dimensions eliminate tolerance buildup and can be related directly to the reference axis of many machines. They tend to take up much more area on a drawing than do chain dimensions.

Chain dimensions are useful in relating one feature to another, such as the repetitive hole pattern shown in Figure 8-21. In this example the distance between the holes is more important than the individual hole's distance from the baseline.

Figure 8-22 shows the same object dimensioned twice, once using chain dimensions and once using baseline dimensions. All distances are assigned a tolerance range of 2 mm, stated using limit tolerances. The maximum distance for surface A is 28 mm using the chain system and 27 mm using the baseline system. The 1-mm difference comes from the elimination of the first 26–24 limit dimension found on the chain example but not on the baseline.

The total tolerance difference is 6 mm for the chain and 4 mm for the baseline. The baseline reduces the tolerance variations for the object simply because it applies the tolerances and dimensions differently. So why not always use baseline dimensions? For most applications, the baseline system is probably better, but if the distance between the individual features is more critical than the distance from the feature to the baseline, use the chain system.

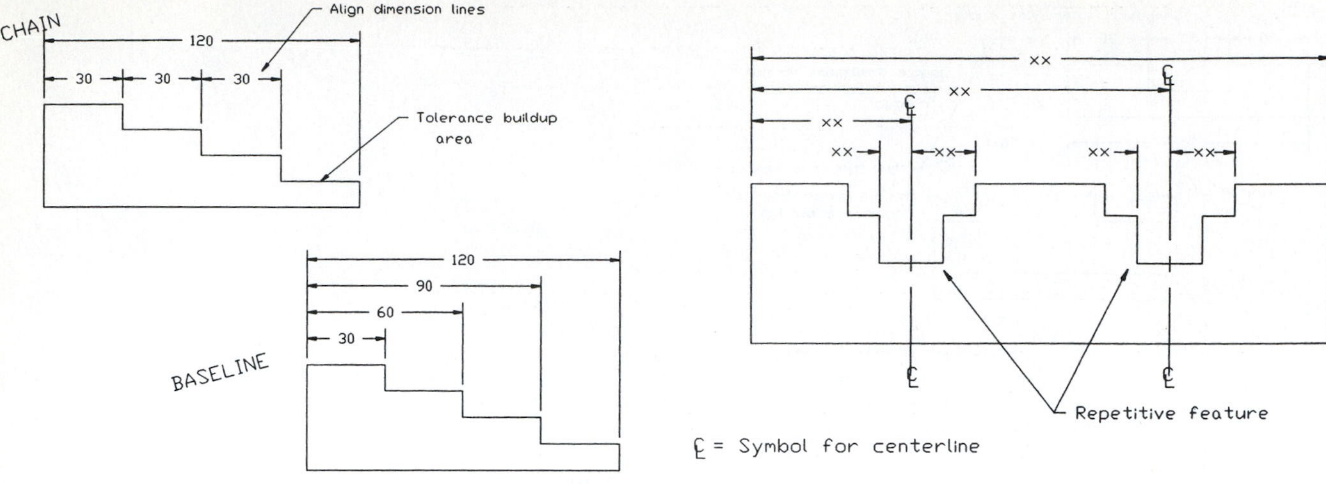

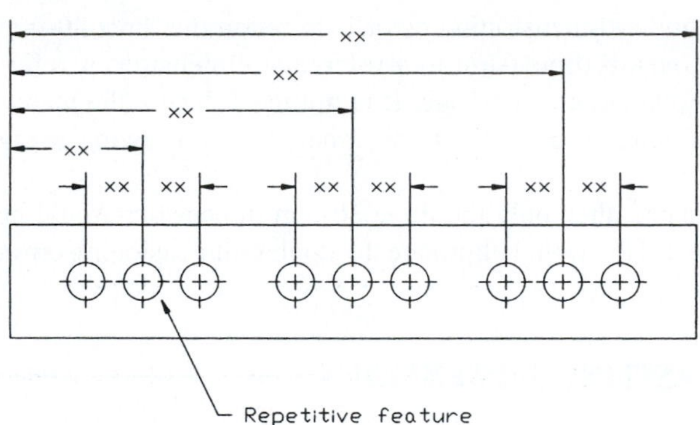

Figure 8-21

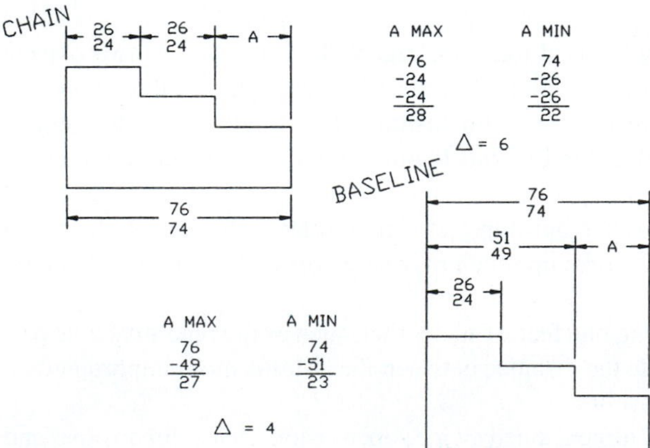

Figure 8-22

Baseline Dimensions Created Using Inventor

See Figure 8-23. See also Chapter 7.

Note in the example of baseline dimensioning shown in Figure 8-23 that each dimension is independent of the other. This means that if one of the dimensions is manufactured incorrectly, it will not affect the other dimensions.

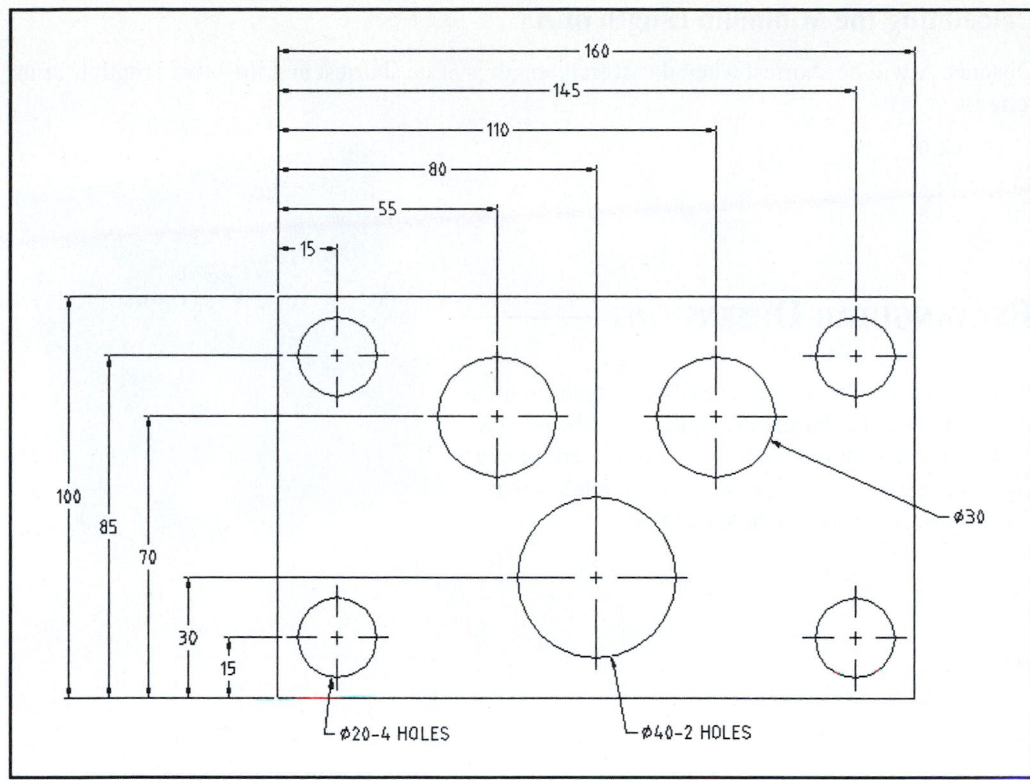

Figure 8-23

TOLERANCE STUDIES

The term ***tolerance study*** is used when analyzing the effects of a group of tolerances on one another and on an object. Figure 8-24 shows an object with two horizontal dimensions. The horizontal distance A is not dimensioned. Its length depends on the tolerances of the two horizontal dimensions.

tolerance study: An analysis of the effects of a group of tolerances on one another and on an object.

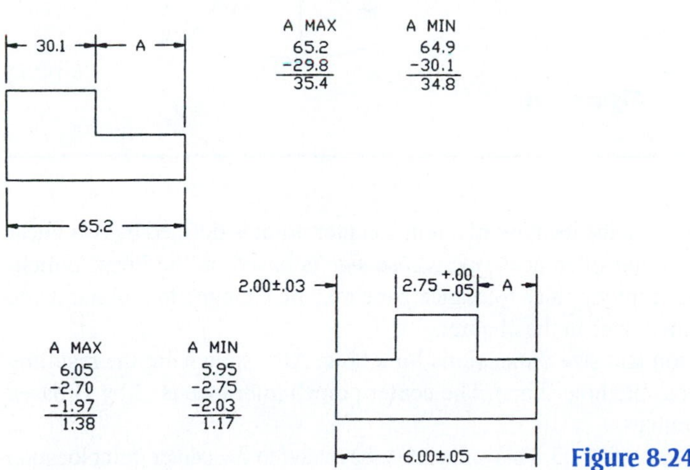

Figure 8-24

Calculating the Maximum Length of A

Distance A will be longest when the overall distance is at its longest and the other distance is at its shortest.

$$\begin{array}{r} 65.2 \\ -29.8 \\ \hline 35.4 \end{array}$$

Calculating the Minimum Length of A

Distance A will be shortest when the overall length is at its shortest and the other length is at its longest.

$$\begin{array}{r} 64.9 \\ -30.1 \\ \hline 34.8 \end{array}$$

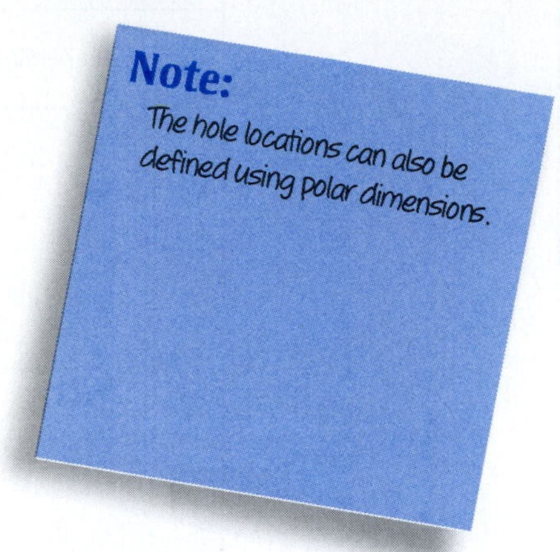

Rectangular Dimensions

Figure 8-25 shows an example of rectangular dimensions referenced to baselines. Figure 8-26 shows a circular object on which dimensions are referenced to a circle's centerlines. Dimensioning to a circle's centerline is critical to accurate hole location.

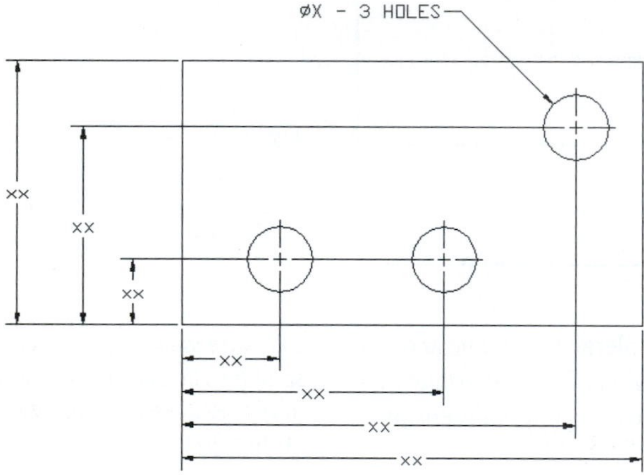

An example of rectangular coordinate dimensions.

Figure 8-25

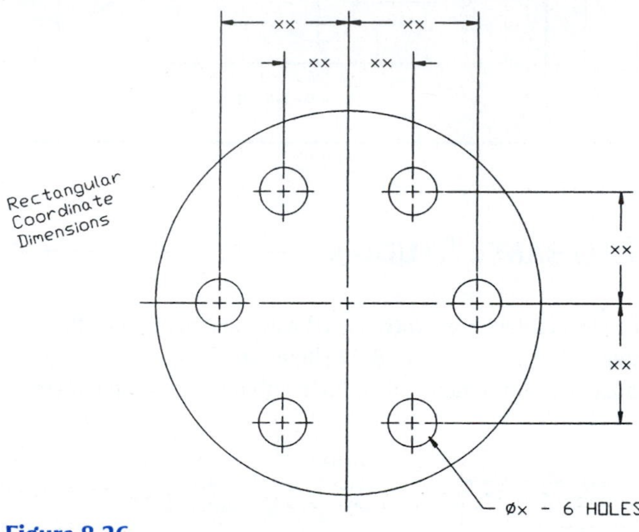

Figure 8-26

Hole Locations

When rectangular dimensions are used, the location of a hole's center point is defined by two linear dimensions. The result is a rectangular tolerance zone whose size is based on the linear dimension's tolerances. The shape of the center point's tolerance zone may be changed to circular using positioning tolerancing as described later in the chapter.

Figure 8-27 shows the location and size dimensions for a hole. Also shown are the resulting tolerance zone and the overall possible hole shape. The center point's tolerance is .2 by .3 based on the given linear locating tolerances.

The hole diameter has a tolerance of $\pm.05$. This value must be added to the center point location tolerances to define the maximum overall possible shape of the hole. The maximum possible hole shape is determined by drawing the maximum radius from the four corner points of the tolerance zone.

This means that the left edge of the hole could be as close to the vertical baseline as 12.75 or as far as 13.25. The 12.75 value was derived by subtracting the maximum hole diameter value 12.05 from the minimum linear distance 24.80 ($24.80 - 12.05 = 12.75$). The 13.25 value was derived by subtracting the minimum hole diameter 11.95 from the maximum linear distance 25.20 ($25.20 - 11.95 = 13.25$).

Figure 8-28 shows a hole's tolerance zone based on polar dimensions. The zone has a sector shape, and the possible hole shape is determined by locating the maximum radius at the four corner points of the tolerance zone.

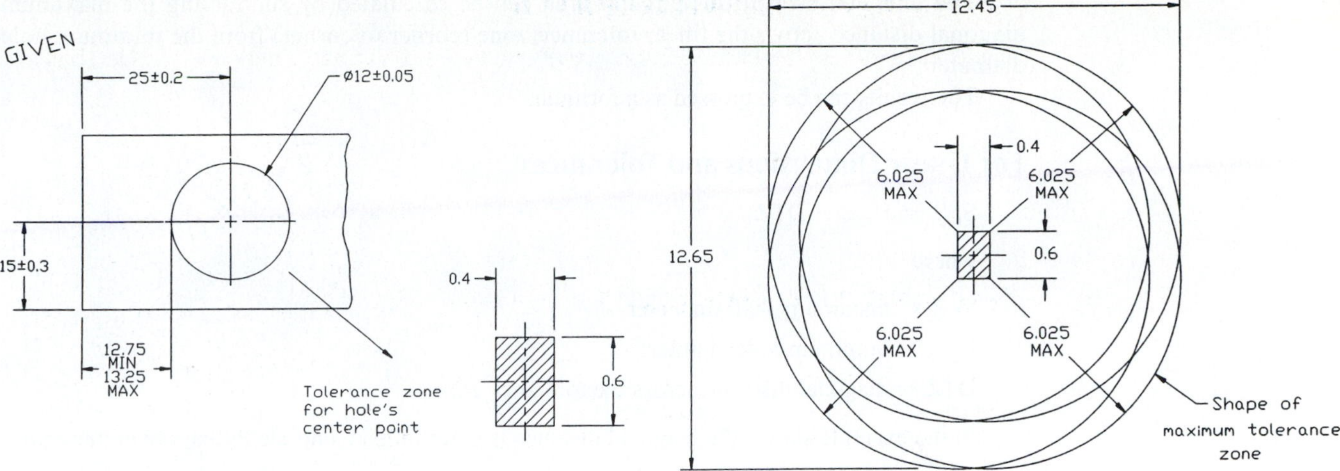

Figure 8-27

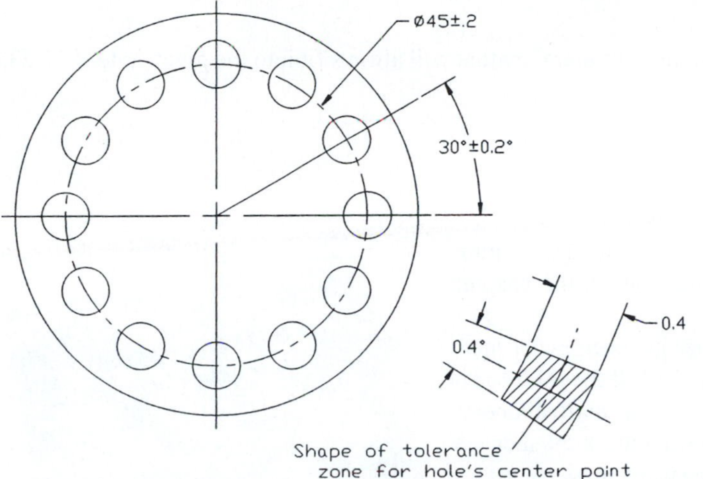

Figure 8-28

Choosing a Shaft for a Toleranced Hole

Given the hole location and size shown in Figure 8-27, what is the largest diameter shaft that will always fit into the hole?

Figure 8-29 shows the hole's center point tolerance zone based on the given linear locating tolerances. Four circles have been drawn centered at the four corners on the linear tolerance zone that represent the smallest possible hole diameter. The circles define an area that represents the maximum shaft size that will always fit into the hole, regardless of how the given dimensions are applied.

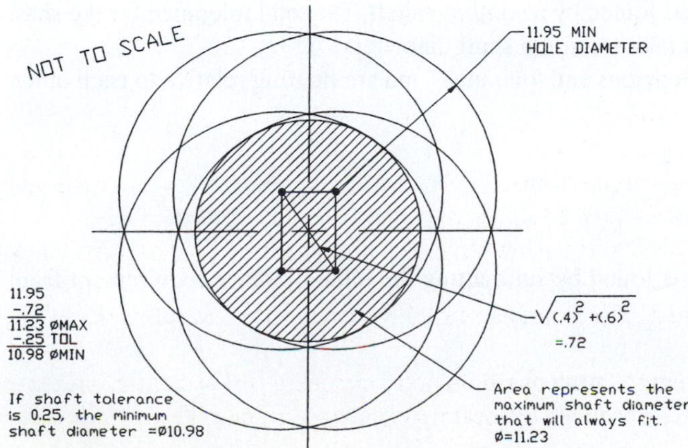

Figure 8-29

The diameter size of this circular area can be calculated by subtracting the maximum diagonal distance across the linear tolerance zone (corner to corner) from the minimum hole diameter.

The results can be expressed as a formula.

For Linear Dimensions and Tolerances

$$S_{max} = H_{min} - DTZ$$

where

S_{max} = maximum shaft diameter

H_{min} = minimum hole diamter

DTZ = diagonal distance across the tolerance zone

In the example shown the diagonal distance is determined using the Pythagorean theorem:

$$DTZ = \sqrt{(.4)^2 + (.6)^2}$$
$$= \sqrt{.16 + .36}$$
$$DTZ = .72$$

This means that the maximum shaft diameter that will always fit into the given hole is 11.23.

$$S_{max} = H_{min} - DTZ$$
$$= 11.95 - .72$$
$$S_{max} = 11.23$$

This procedure represents a restricted application of the general formula presented later in the chapter for positioning tolerances.

Once the maximum shaft size has been established, a tolerance can be applied to the shaft. If the shaft had a total tolerance of .25, the minimum shaft diameter would be 11.23 − .25, or 10.98. Figure 8-29 shows a shaft dimensioned and toleranced using these values.

The formula presented is based on the assumption that the shaft is perfectly placed on the hole's center point. This assumption is reasonable if two objects are joined by a fastener and both objects are free to move. When both objects are free to move about a common fastener, they are called ***floating objects.***

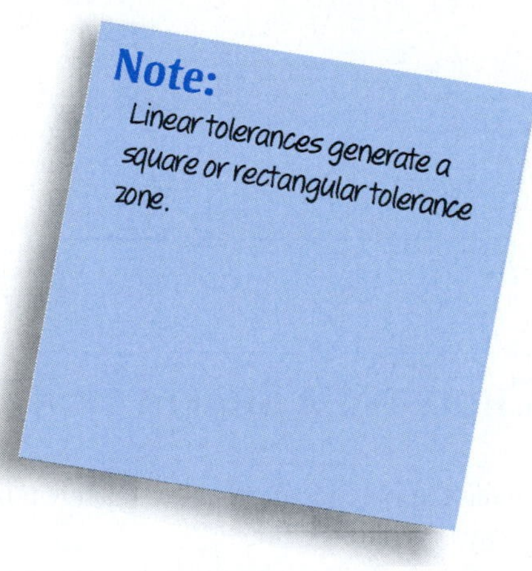

floating objects: Objects that are free to move about a common fastener.

Sample Problem SP8-1

Parts A and B in Figure 8-30 are to be joined by a common shaft. The total tolerance for the shaft is to be .05. What are the maximum and minimum shaft diameters?

Both objects have the same dimensions and tolerances and are floating relative to each other.

$$S_{max} = H_{min} - DTZ$$
$$= 15.93 - .85$$
$$S_{max} = 15.08$$

The shaft's minimum diameter is found by subtracting the total tolerance requirement from the calculated maximum diameter:

$$15.80 - .05 = 15.03$$

Therefore,

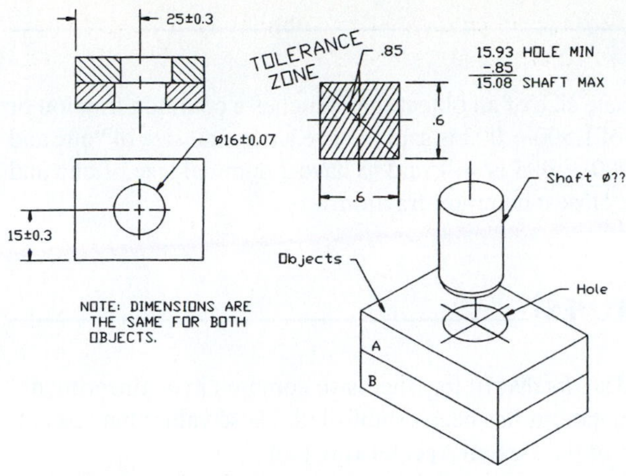

Figure 8-30

Shaft max = 15.08

Shaft min = 15.03

Sample Problem SP8-2

The procedure presented in Sample Problem SP8-1 can be worked in reverse to determine the maximum and minimum hole size based on a given shaft size.

Objects AA and BB as shown in Figure 8-31 are to be joined using a bolt whose maximum diameter is .248. What is the minimum hole size for objects that will always accept the bolt? What is the maximum hole size if the total hole tolerance is .005?

$S_{max} = H_{min} - DZT$

In this example H_{min} is the unknown factor, so the equation is rewritten as

$H_{min} = S_{max} + DZT$

$= .248 + .010$

$H_{min} = .258$

This is the minimum hole diameter, so the total tolerance requirement is added to this value:

$.258 + .005 = .263$

Therefore,

Hole max = .263

Hole min = .258

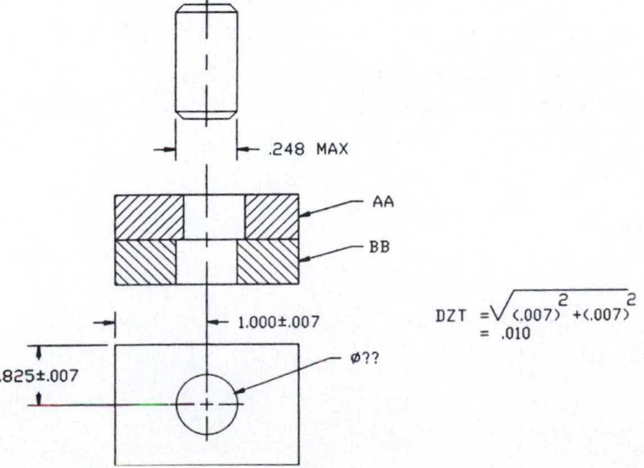

Figure 8-31

Nominal Sizes

nominal size: The approximate size of an object that matches a common fraction or whole number.

The term ***nominal*** refers to the approximate size of an object that matches a common fraction or whole number. A shaft with a dimension of 1.500 + .003 is said to have a nominal size of "one and a half inches." A dimension of 1.500 + .000/− .005 is still said to have a nominal size of one and a half inches. In both examples 1.5 is the closest common fraction.

Standard Fits (Metric Values)

Calculating tolerances between holes and shafts that fit together is so common in engineering design that a group of standard values and notations has been established. These values may be calculated using the **Limits and Fits** option of the **Design Accelerator** tool.

There are three possible types of fits between a shaft and a hole: clearance, transitional, and interference. There are several subclassifications within each of these categories.

clearance fit: A fit that always defines the maximum shaft diameter as smaller than the minimum hole diameter.

interference fit: A fit that always defines the minimum shaft diameter as larger than the maximum hole diameter.

A ***clearance fit*** always defines the maximum shaft diameter as smaller than the minimum hole diameter. The difference between the two diameters is the amount of clearance. It is possible for a clearance fit to be defined with zero clearance; that is, the maximum shaft diameter is equal to the minimum hole diameter.

An ***interference fit*** always defines the minimum shaft diameter as larger than the maximum hole diameter; that is, the shaft is always bigger than the hole. This definition means that an interference fit is the converse of a clearance fit. The difference between the diameter of the shaft and the hole is the amount of interference.

An interference fit is primarily used to assemble objects together. Interference fits eliminate the need for threads, welds, or other joining methods. Using an interference for joining two objects is generally limited to light load applications.

It is sometimes difficult to visualize how a shaft can be assembled into a hole with a diameter smaller than that of the shaft. It is sometimes done using a hydraulic press that slowly forces the two parts together. The joining process can be augmented by the use of lubricants or heat. The hole is heated, causing it to expand, the shaft is inserted, and the hole is allowed to cool and shrink around the shaft.

transition fit: A fit that may be either a clearance or an interference fit.

A ***transition fit*** may be either a clearance or an interference fit. It may have a clearance between the shaft and the hole or an interference.

The notations are based on Standard International Tolerance values. A specific description for each category of fit follows.

Clearance Fits

H11/c11 or C11/h11 = loose running fit

H8/d8 or D8/h8 = free running fit

H8/f7 or F8/h7 = close running fit

H7/g6 or G7/h6 = sliding fit

H7/h6 = locational clearance fit

Transitional Fits

H7/k6 or K7/h6 = locational transition fit

H7/n6 or N7/h6 = locational transition fit

Interference Fits

H7/p6 or P7/h6 = locational transition fit

H7/s6 or S7/h6 = medium drive fit

H7/u6 or U7/h6 = force fit

Using Inventor's Design Accelerator

Limits and fits can be derived using Inventor's **Design Accelerator Limits and Fits** calculator. The limit values are also presented in visual form so you can better visualize the type of fit being created.

Exercise 8-6: Accessing the Design Accelerator

1. Move the cursor into the **Assembly Panel** area and right-click the mouse or click the arrowhead to the right of the heading **Assembly Panel.**

A small dialog box will appear. See Figure 8-32.

2. Click the **Design Accelerator** option.

The **Design Accelerator** panel will appear.

See Figure 8-33.

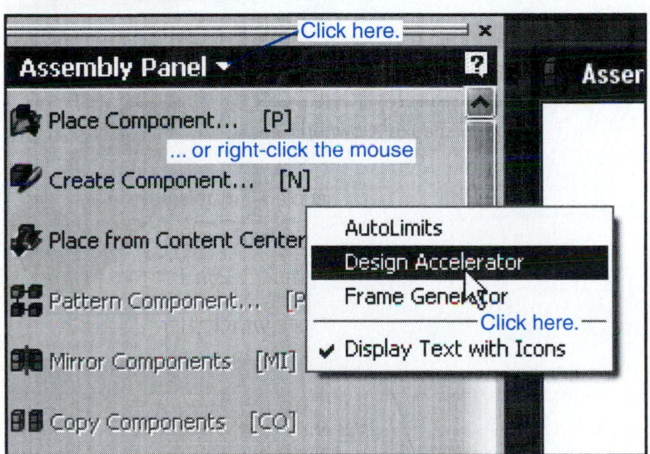

Figure 8-32

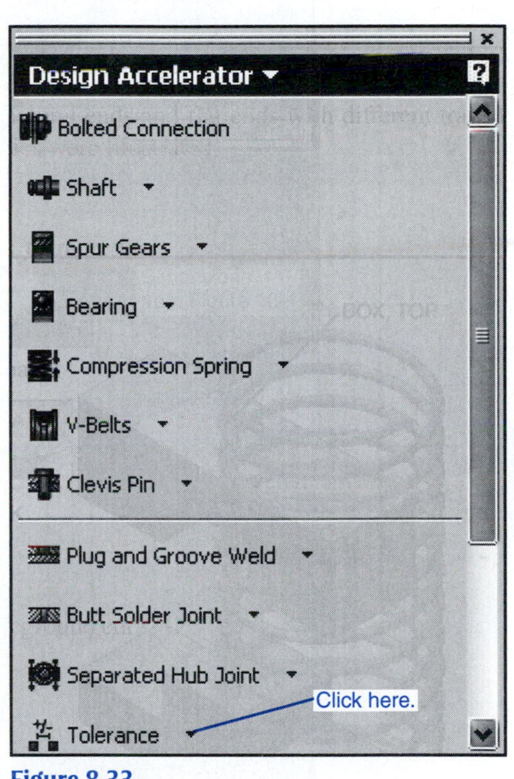

Figure 8-33

3. Scroll down the **Design Accelerator** panel and click on the flyout arrow next to the **Tolerance** heading.

See Figure 8-34.

4. Click the **Limits and Fits** heading.

The **Limits and Fits Mechanical Calculator** dialog box will appear. See Figure 8-35.

Hole and Shaft Basis

The **Limits and Fits Mechanical Calculator** shown in Figure 8-30 applies tolerances starting with the nominal hole sizes, called *hole-basis tolerances;* the other option applies tolerances starting with the shaft nominal sizes, called *shaft-basis tolerances.* The choice of which set of

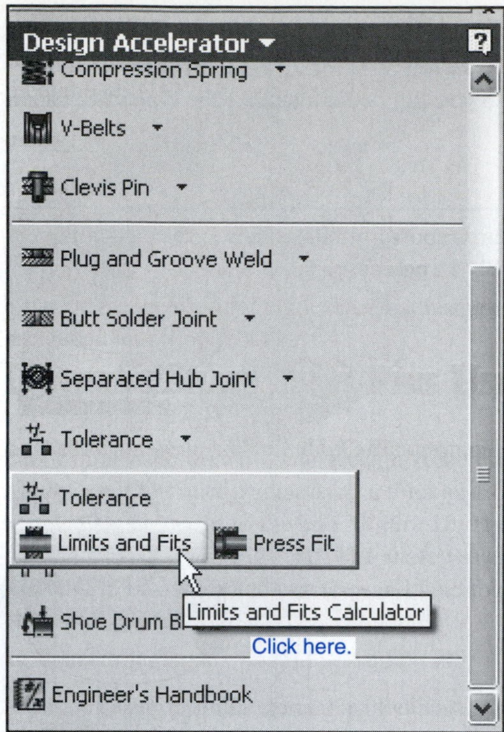

Figure 8-34

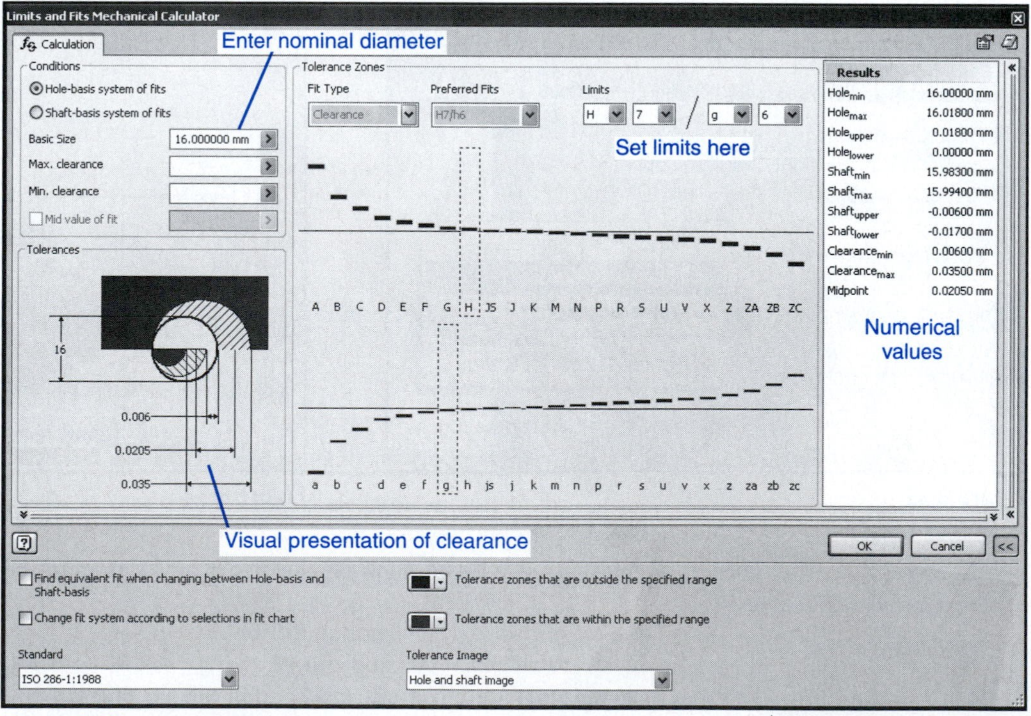

Figure 8-35

values to use depends on the design application. In general, hole-basis numbers are used more often because it is more difficult to vary hole diameters manufactured using specific drill sizes than shaft sizes manufactured using a lathe. Shaft sizes may be used when a specific fastener diameter is used to assemble several objects.

Figure 8-36 shows calculations done for a nominal diameter of 16 mm using the **Hole-basis** system. An H7/g6 fit was selected.

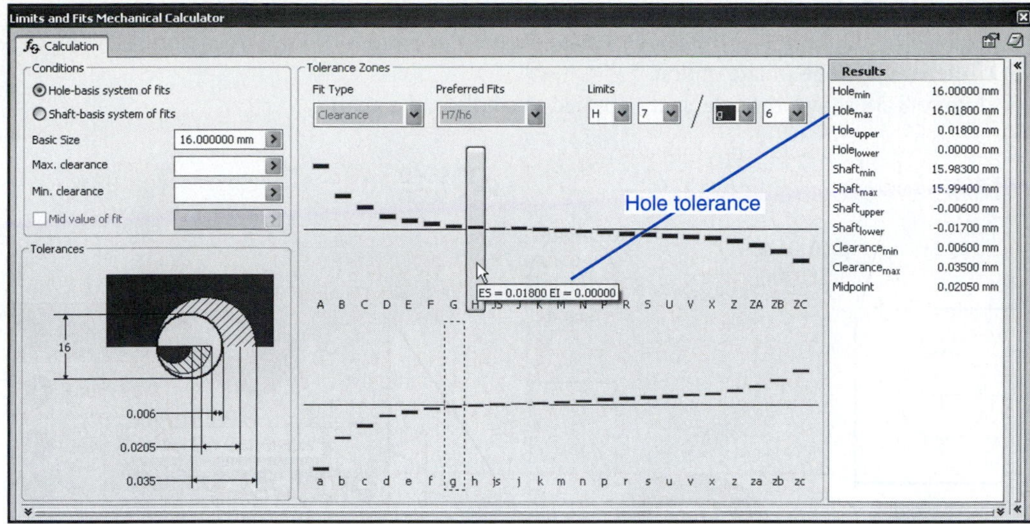

Figure 8-36

Exercise 8-7: Determining the Hole/Shaft Sizes

1. Access the **Limits and Fits Mechanical Calculator.**
2. Define the **Basic Size** and select the condition.

In this example **Ø16.00** and **Hole-basis** were entered.

The limit designation of H7/g6 is based on an international standard for hole and shaft tolerances. The hole's tolerance is always first and uses an uppercase letter designation. The shaft's tolerance is second and uses a lowercase letter designation. The numerical values for the H7/g6 tolerance can be found in one of two ways: by reading the values in the **Results** box or by moving the cursor into the rectangular area defined by a broken line. The hole's limits will appear. See Figure 8-36.

Note that tolerances listed in the **Results** box and in the **Tolerance Zone** area are the same. The $Hole_{min}$ is 16.0000 mm in the **Results** box and EI = 0.0000 in the **Tolerance Zone** area. The 16.0000 value or 0.0000 value comes from the Hole-basis designation. The nominal hole value of 16.00 becomes the starting point for all other tolerance designations. Had the **Shaft-basis** condition been selected for the Ø16.00 nominal size, the hole and shaft tolerances would be different.

Visual Presentations of the Hole and Shaft Tolerances

Figure 8-37 shows the visual presentation of the H7/g6 tolerances along with the **Results** box values. The clearance max value is **0.03500,** as stated in the **Results** box. Note how the 0.035 clearance max value is presented visually in the **Tolerances** box.

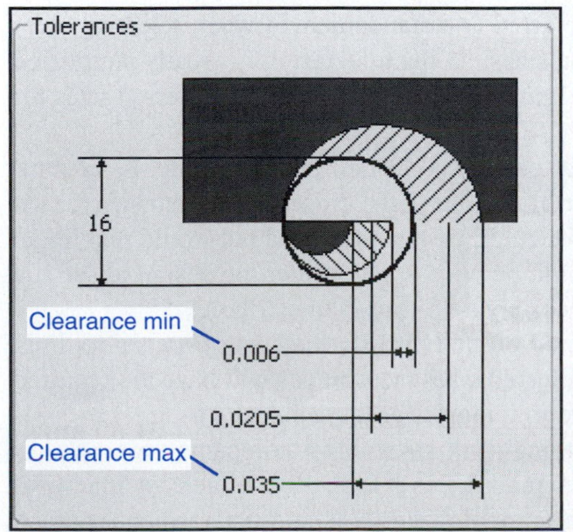

Results	
$Hole_{min}$	16.00000 mm
$Hole_{max}$	16.01800 mm
$Hole_{upper}$	0.01800 mm
$Hole_{lower}$	0.00000 mm
$Shaft_{min}$	15.98300 mm
$Shaft_{max}$	15.99400 mm
$Shaft_{upper}$	-0.00600 mm
$Shaft_{lower}$	-0.01700 mm
$Clearance_{min}$	0.00600 mm
$Clearance_{max}$	0.03500 mm
Midpoint	0.02050 mm

Same values

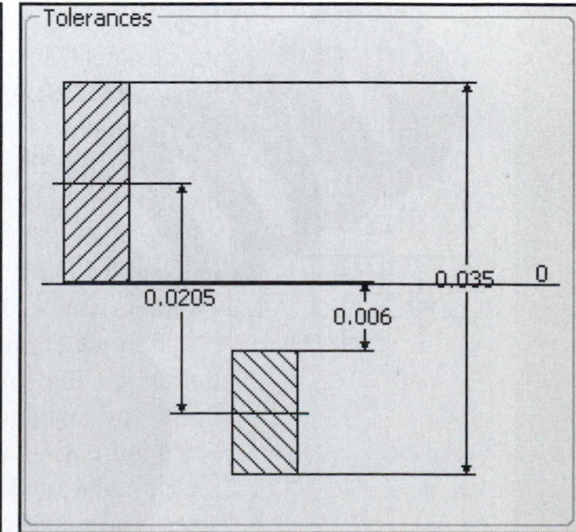

Figure 8-37

Figure 8-37 also shows a second visual presentation of the tolerances. This presentation can be obtained by clicking the **Tolerance Image** box at the bottom of the screen and selecting the **ISO Tolerance range** image option.

Figure 8-38 shows the tolerances applied to a hole and a shaft.

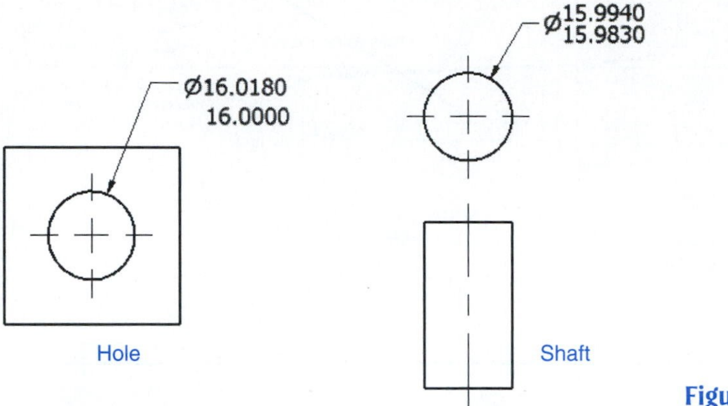

Figure 8-38

Standard Fits (Inch Values)

Inch values are accessed in the **Design Accelerator** tool by selecting the ANSI standards.

Fits defined using inch values are classified as follows:

RC = running and sliding fits

LC = clearance locational fits

LT = transitional locational fits

LN = interference fits

FN = force fits

Each of these general categories has several subclassifications within it defined by a number, for example, Class RC1, Class RC2, through Class RC8. The letter designations are based on International Tolerance Standards, as are metric designations.

Charts of tolerance values can be found in the appendix.

Preferred and Standard Sizes

It is important that designers always consider preferred and standard sizes when selecting sizes for designs. Most tooling is set up to match these sizes, so manufacturing is greatly simplified when preferred and standard sizes are specified. Figure 8-39 shows a listing of preferred sizes for metric values.

Consider the case of design calculations that call for a 42-mm-diameter hole. A 42-mm-diameter hole is not a preferred size. A diameter of 40 mm is the closest preferred size, and a 45-mm diameter is a second choice. A 42-mm hole could be manufactured but would require an unusual drill size that may not be available. It would be wise to reconsider the design to see if a 40-mm-diameter hole could be used, and if not, possibly a 45-mm-diameter hole.

A production run of a very large quantity could possibly justify the cost of special tooling, but for smaller runs it is probably better to use preferred sizes. Machinists will have the required drills, and maintenance people will have the appropriate tools for these sizes.

Figure 8-40 shows a listing of standard fractional drill sizes. Most companies now specify metric units or decimal inches; however, many standard items are still available in fractional sizes, and many older objects may still require fractional-sized tools and replacement parts. A more complete listing is available in the appendix.

PREFERRED SIZES			
First Choice	Second Choice	First Choice	Second Choice
1	1.1	12	14
1.2	1.4	16	18
1.6	1.8	20	22
2	2.2	25	28
2.5	2.8	30	35
3	3.5	40	45
4	4.5	50	55
5	5.5	60	70
6	7	80	90
8	9	100	110
10	11	120	140

Figure 8-39

Fraction	Decimal Equivalent	Fraction	Decimal Equivalent	Fraction	Decimal Equivalent
7/64	.1094	21/64	.3281	11/16	.6875
1/8	.1250	11/32	.3438	3/4	.7500
9/64	.1406	23/64	.3594	13/16	.8125
5/32	.1562	3/8	.3750	7/8	.8750
11/64	.1719	25/64	.3906	15/16	.9375
3/16	.1875	13/32	.4062	1	1.0000
13/64	.2031	27/64	.4219		
7/32	.2188	7/16	.4375	PARTIAL LIST of standard Twist Drill Sizes (Fractional Sizes)	
1/4	.2500	29/64	.4531		
17/64	.2656	15/32	.4688		
9/32	.2812	1/2	.5000		
19/64	.2969	9/16	.5625		
5/16	.3125	5/8	.6250		

Figure 8-40

Surface Finishes

The term ***surface finish*** refers to the accuracy (flatness) of a surface. Metric values are measured using micrometers (μm), and inch values are measured in microinches (μin.).

surface finish: The accuracy (flatness) of a surface.

The accuracy of a surface depends on the manufacturing process used to produce the surface. Figure 8-41 shows a listing of manufacturing processes and the quality of the surface finish they can be expected to produce.

Surface finishes have several design applications. ***Datum surfaces,*** or surfaces used for baseline dimensioning, should have fairly accurate surface finishes to help assure accurate measurements. Bearing surfaces should have good-quality surface finishes for better load distribution, and parts that operate at high speeds should have smooth finishes to help reduce friction. Figure 8-42 shows a screw head sitting on a very wavy surface. Note that the head of the screw is actually in contact with only two wave peaks, meaning all the bearing load is concentrated on the two peaks. This situation could cause stress cracks and greatly weaken the surface. A better-quality surface finish would increase the bearing contact area.

datum surface: The surface used for baseline dimensioning.

Figure 8-42 also shows two very rough surfaces moving in contact with each other. The result will be excess wear to both surfaces because the surfaces touch only on the peaks, and these peaks will tend to wear faster than flatter areas. Excess vibration can also result when interfacing surfaces are too rough.

Surface finishes are classified into three categories: surface texture, roughness, and lay. ***Surface texture*** is a general term that refers to the overall quality and accuracy of a surface.

Roughness is a measure of the average deviation of a surface's peaks and valleys. See Figure 8-43.

Lay refers to the direction of machine marks on a surface. See Figure 8-44. The lay of a surface is particularly important when two moving objects are in contact with each other, especially at high speeds.

surface texture: The overall quality and accuracy of a surface.

roughness: A measure of the average deviation of a surface's peaks and valleys.

lay: The direction of machine marks on a surface.

Surface Control Symbols

Surface finishes are indicated on a drawing using surface control symbols. See Figure 8-45. The general surface control symbol looks like a check mark. Roughness values may be included with the symbol to specify the required accuracy. Surface control symbols can also be used to specify the manufacturing process that may or may not be used to produce a surface.

Surface Roughness Average Obtained by Common Production Methods

Process	50 µm (2000 µin)	25 (1000)	12.5 (500)	6.3 (250)	3.2 (125)	1.6 (63)	0.8 (32)	0.4 (16)	0.2 (8)	0.1 (4)	0.05 (2)	0.025 (1)
Flame Cutting	▨	■	▨									
Snagging	▨	■	■	▨								
Sawing	▨	■	■	■	■	▨						
Planning, Shaping		▨	■	■	■	▨	▨					
Drilling			▨	■	■	▨						
Chemical Milling			▨	■	■	▨						
Elect Discharge Machine			▨	▨/■	■	▨						
Milling		▨	▨	■	■	■	▨	▨				
Broaching				▨	■	■	▨					
Reaming				▨	■	■	▨					
Electron Beam				■	■	■	▨	▨				
Laser				■	■	■	▨	▨				

Surface Roughness Average Obtained by Common Production Methods

Process	50 µm (2000 µin)	25 (1000)	12.5 (500)	6.3 (250)	3.2 (125)	1.6 (63)	0.8 (32)	0.4 (16)	0.2 (8)	0.1 (4)	0.05 (2)	0.025 (1)
Electrochemical			▨	▨	■	■	■	■	▨	▨		
Boring, Turning		▨	▨	■	■	■	■	■	▨	▨		
Barrel Finishing					▨	▨	■	■	▨	▨		
Electronic Grinding							▨/■	■	▨			
Roller Burnishing							▨	■	▨			
Grinding				▨	▨	■	■	■	■	▨	▨	
Honing					▨	▨	■	■	■	▨	▨	
Electropolishing						▨	■	■	■	▨	▨	▨
Polishing							▨	■	■	▨	▨	▨
Lapping							▨	■	■	■	▨	▨
Superfinish							▨	▨	■	■	▨	▨

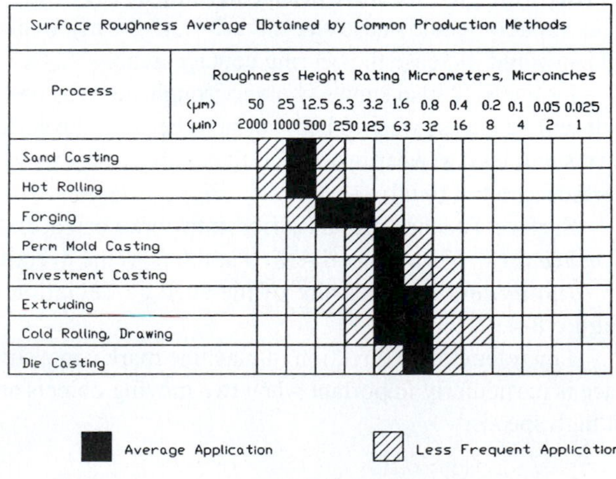

Figure 8-41

Figure 8-45 shows two applications of surface control symbols. In the first example, a 0.8-µm (32 µin.) surface finish is specified on the surface that serves as a datum for several horizontal dimensions. A 0.8-µm surface finish is generally considered the minimum acceptable finish for datums.

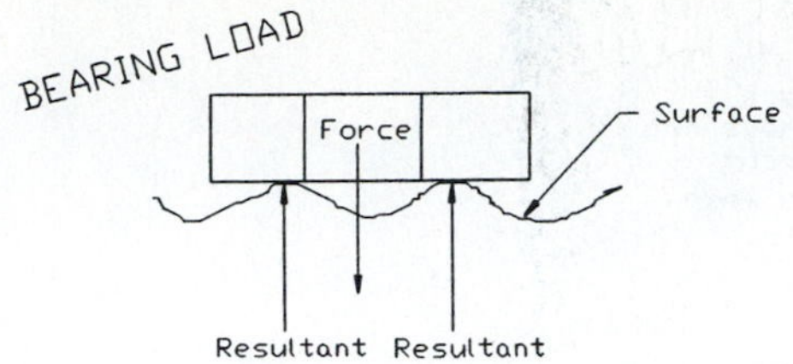

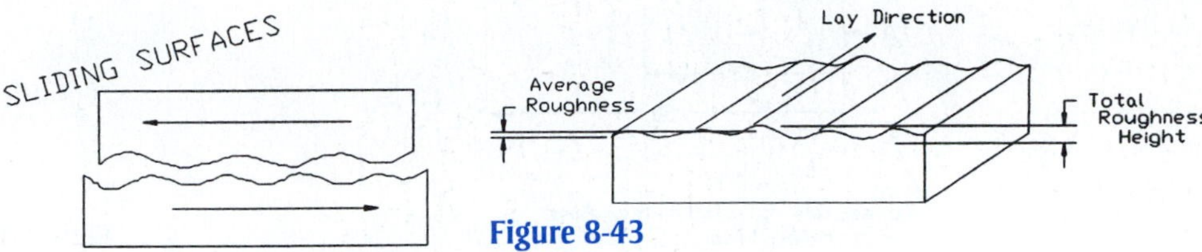

Figure 8-43

Figure 8-42

DIRECTION	SYMBOL	DIRECTION	SYMBOL
Parallel	$\sqrt{=}$	Concentric	$\sqrt{C}$
Perpendicular	$\sqrt{\perp}$	Radial	$\sqrt{R}$
Angular	$\sqrt{X}$		
No Direction	$\sqrt{M}$		

60°

Basic Surface Texture Symbol

32 Surface Finish Specified (32 microinches)

Material Removed by Machining

2.0 Material Removal Specified (2.0 millimeters)

No Material May Be Removed

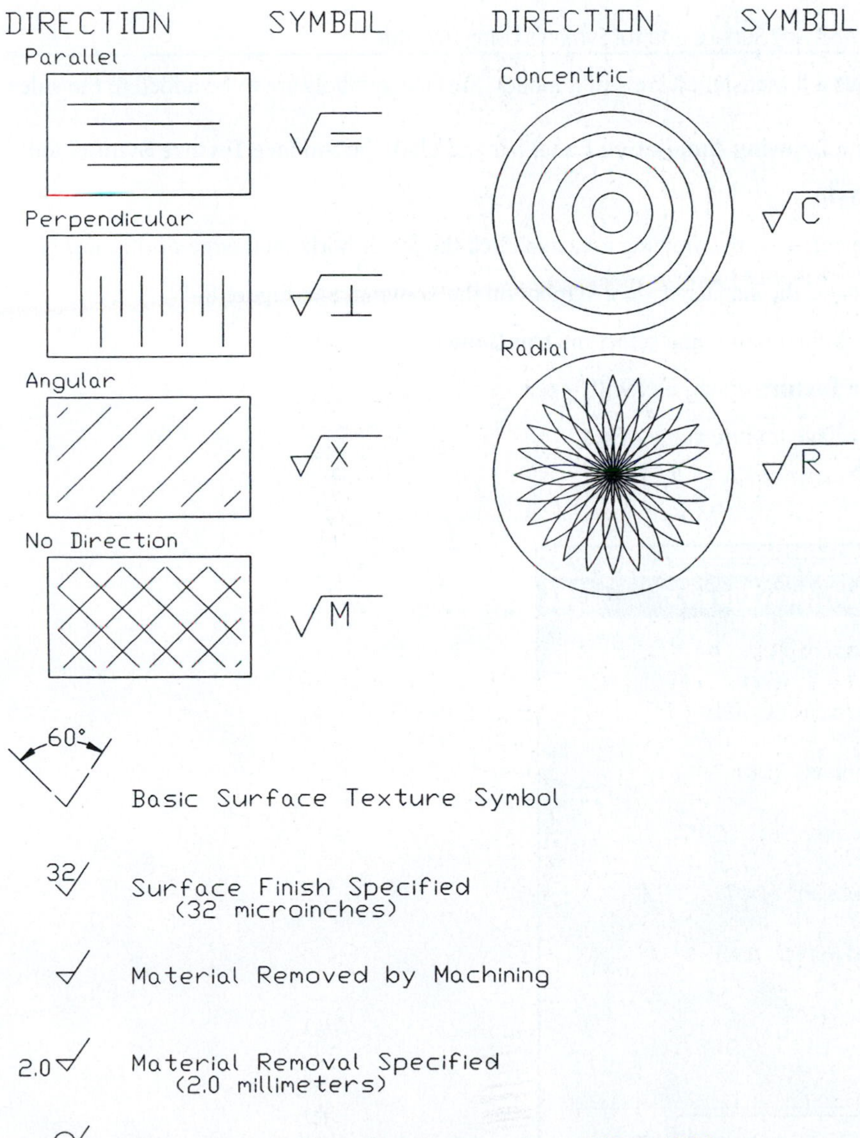

Figure 8-44

A second finish mark with a value of 0.4 μm is located on an extension line that refers to a surface that will be in contact with a moving object. The extra flatness will help prevent wear between the two surfaces.

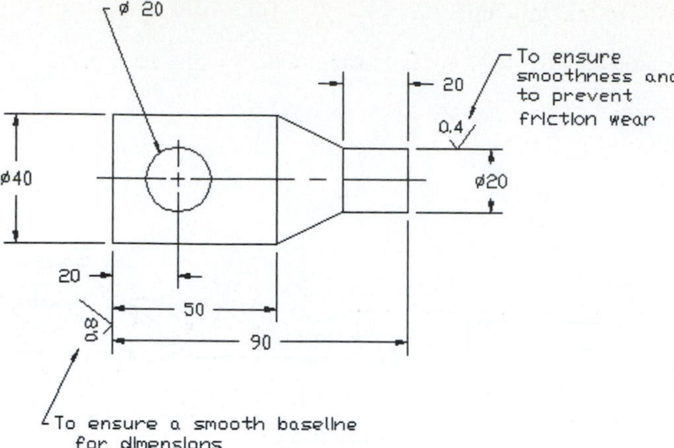

Figure 8-45

Exercise 8-8: Applying Surface Control Symbols Using Inventor

Figure 8-47 shows a dimensioned view of a model. Surface symbols are to be added to the sides of the slot.

1. Access the **Drawing Annotation Panel** bar and click the **Surface Texture Symbol** tool.

   See Figure 8-46.

2. Move the cursor to the drawing area and click the lower horizontal edge of the slot.

   This step locates the surface texture symbol on the drawing. See Figure 8-47.

3. Right-click the mouse, and select the **Continue** option.

   The **Surface Texture** dialog box will appear.

4. Enter a surface texture value of **0.8.**
5. Click **OK.**

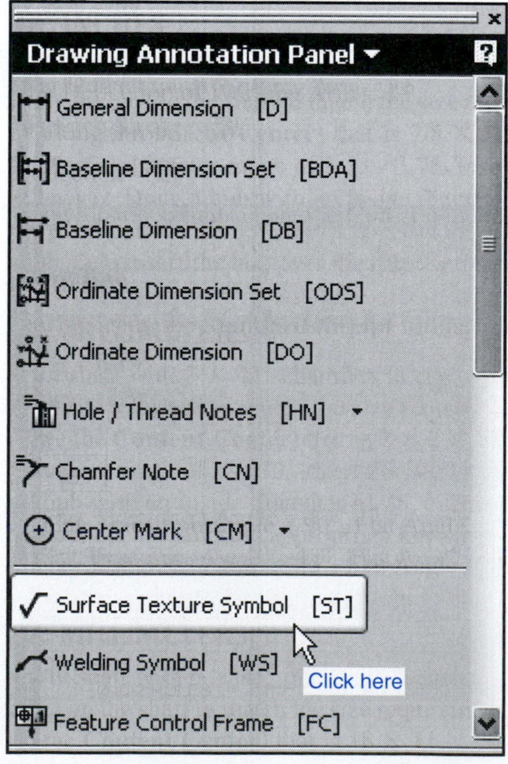

Figure 8-46

The surface symbol will be added to the drawing.

The cursor will remain in **Surface Texture** mode so that other symbols may be applied.

6. Add a second symbol to the upper edge of the slot.
7. Right-click the mouse and select the **Done** option.

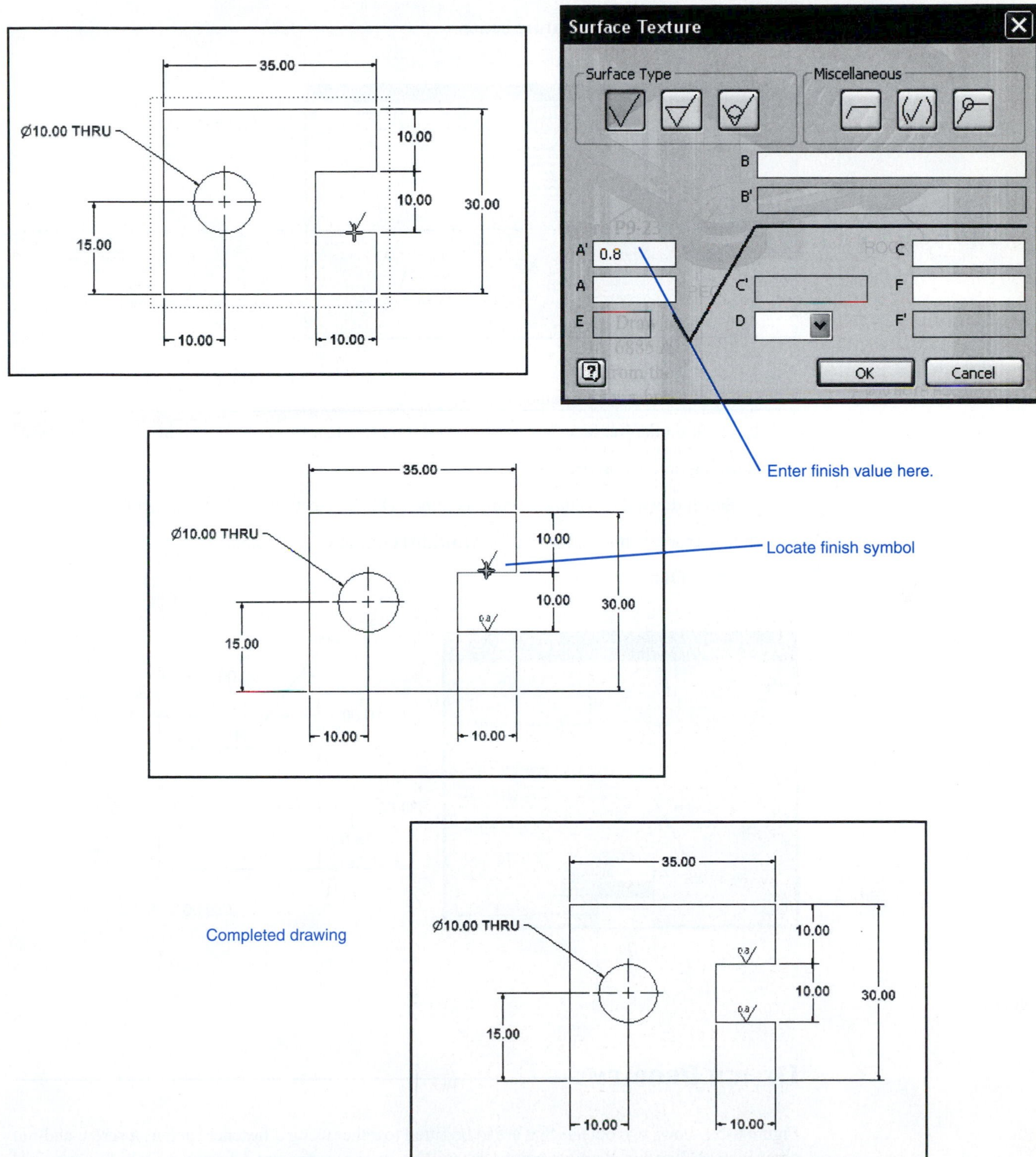

Figure 8-47

Available Lay Symbols

Inventor includes a group of lay symbols that can be added to the drawing using the **Surface Texture Symbol** tool. The definition of the symbols is located on the **Styles and Standards Editor** dialog box.

1. Click the **Format** heading at the top of the screen, then the **Styles and Standards Editor** option.

See Figure 8-48.

2. Select the **Surface Texture** option.

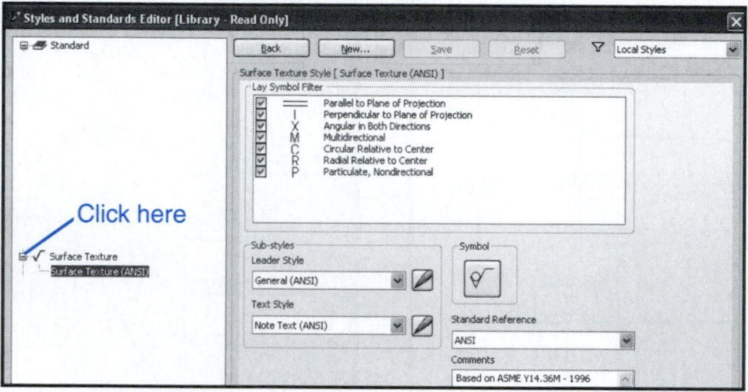

Figure 8-48

Exercise 8-9: Adding Lay Symbols

1. Click on the **Surface Texture Symbol** tool, locate a symbol, then right-click the mouse.

A dialog box will appear. See Figure 8-49.

2. Scroll down the available symbols in box **D** and select an appropriate symbol.

In this example the symbol **M** for **Multidirectional** was selected.

3. Click **OK.**

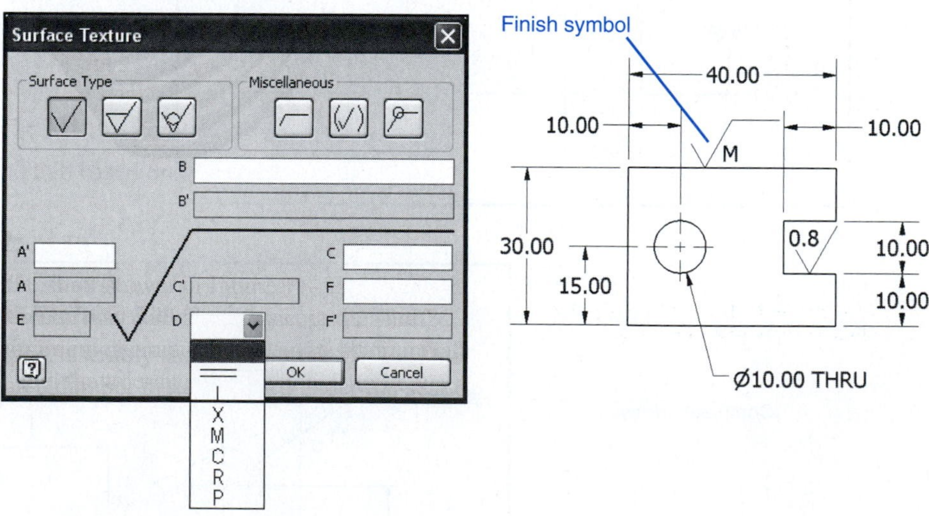

Figure 8-49

Design Problems

Figure 8-50 shows two objects that are to be fitted together using a fastener such as a screw-and-nut combination. For this example a cylinder will be used to represent a fastener. Only two nominal dimensions are given. The dimensions and tolerances were derived as follows.

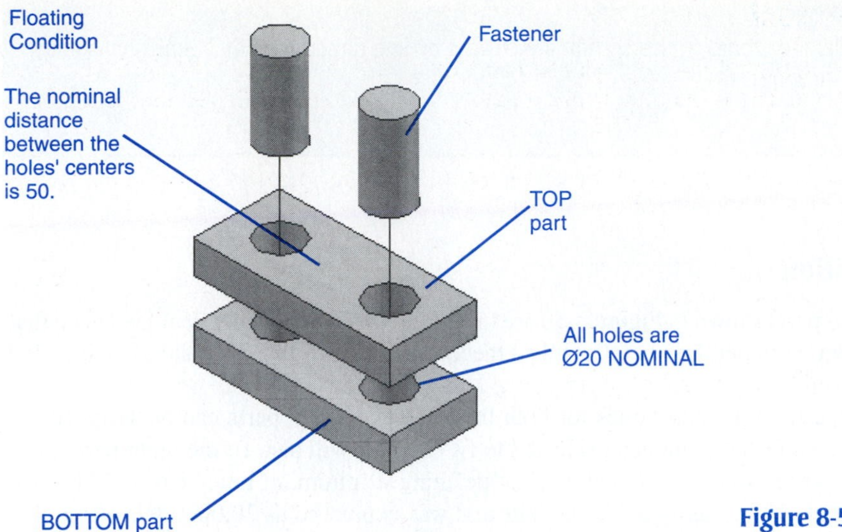

Figure 8-50

The distance between the centers of the holes is given as 50 nominal. The term *nominal* means that the stated value is only a starting point. The final dimensions will be close to the given value but do not have to equal it.

Assigning tolerances is an iteration process; that is, a tolerance is selected and other tolerance values are calculated from the selected initial values. If the results are not satisfactory, go back and modify the initial value and calculate the other values again. As your experience grows you will become better at selecting realistic initial values.

In the example shown in Figure 8-50, start by assigning a tolerance of ±.01 to both the top and bottom parts for both the horizontal and vertical dimensions used to locate the holes. This means that there is a possible center point variation of .02 for both parts. The parts must always fit together, so tolerances must be assigned based on the worst-case condition, or when the parts are made at the extreme ends of the assigned tolerances.

Figure 8-51 shows a greatly enlarged picture of the worst-case condition created by a tolerance of ±.01. The center points of the holes could be as much as .028 apart if the two center points were located at opposite corners of the tolerance zones. This means that the minimum hole diameter must always be at least .028 larger than the maximum stud diameter. In addition, there should be a clearance tolerance assigned so that the hole and stud are never exactly the same size. Figure 8-52 shows the resulting tolerances.

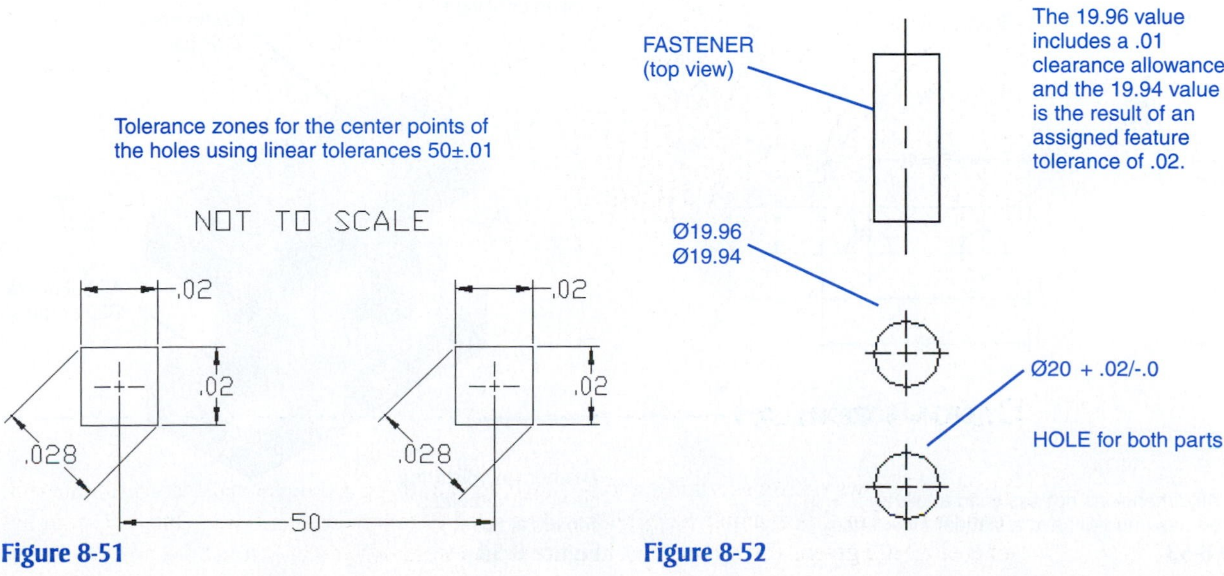

Figure 8-51

Figure 8-52

The tolerance zones in this section are created by line dimensions that generated square tolerance zones.

Floating Condition

floating condition: A situation in which the location of one fastener does not depend on the location of the other.

The top and bottom parts shown in Figure 8-50 are to be joined by two independent fasteners; that is, the location of one fastener does not depend on the location of the other. This situation is called a ***floating condition.***

This means that the tolerance zones for both the top and bottom parts can be assigned the same values and that a fastener diameter selected to fit one part will also fit the other part.

The final tolerances were developed by first defining a minimum hole size of 20.00. An arbitrary tolerance of .02 was assigned to the hole and was expressed as 20.00 + .02/−0. so that the hole can never be any smaller than 20.00.

The 20.00 minimum hole diameter dictates that the maximum fastener diameter can be no greater than 19.97, or .03 (the rounded-off diagonal distance across the tolerance zone—.028) less than the minimum hole diameter. A .01 clearance was assigned. The clearance ensures that the hole and fastener are never exactly the same diameter. The resulting maximum allowable diameter for the fastener is 19.96. Again, an arbitrary tolerance of .02 was assigned to the fastener. The final fastener dimensions are therefore 19.96 to 19.94.

The assigned tolerances ensure that there will always be at least .01 clearance between the fastener and the hole. The other extreme condition occurs when the hole is at its largest possible size (20.02) and the fastener is at its smallest (19.94). This means that there could be as much as .08 clearance between the parts. If this much clearance is not acceptable, then the assigned tolerances will have to be reevaluated.

Figure 8-53 shows the top and bottom parts dimensioned and toleranced. Any dimensions that do not have assigned tolerances are assumed to have standard tolerances.

Note, in Figure 8-53, that the top edge of each part has been assigned a surface finish. This was done to help ensure the accuracy of the 20 ± .01 dimension. If this edge surface was rough, it could affect the tolerance measurements.

This example will be done later in the chapter using geometric tolerances. Geometric tolerance zones are circular rather than rectangular.

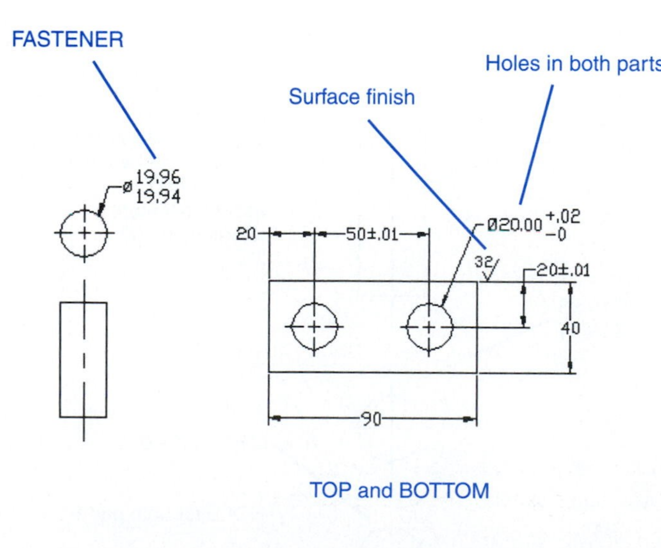

Figure 8-53

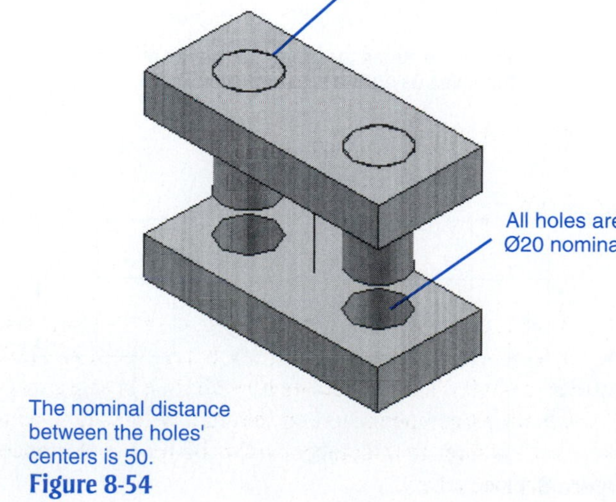

Figure 8-54

Fixed Condition

Figure 8-54 shows the same nominal conditions presented in Figure 8-45, but the fasteners are now fixed to the top part. This situation is called the ***fixed condition.*** In analyzing the tolerance zones for the fixed condition, two position tolerances must be considered: the positional tolerances for the holes in the bottom part, and the positional tolerances for the fixed fasteners in the top part. This relationship may be expressed in an equation as follows:

fixed condition: A situation in which the location of fasteners is fixed to a part.

$$S_{max} + DTSZ = H_{min} - DTZ$$

where:

S_{max} = maximum shaft (fastener) diameter

H_{min} = minimum hole diameter

DTSZ = diagonal distance across the shaft's center point tolerance zone

DTZ = diagonal distance across the hole's center point tolerance zone

If a dimension and tolerance of 50 ± .01 is assigned to both the center distance between the holes and the center distance between the fixed fasteners, the values for DTSZ and DTZ will be equal. The formula can then be simplified as follows.

$$S_{max} = H_{min} - 2(DTZ)$$

where DTZ equals the diagonal distance across the tolerance zone. If a hole tolerance of 20.00 + .02/−0 is also defined, the resulting maximum shaft size can be determined, assuming that the calculated distance of .028 is rounded off to .03. See Figure 8-55.

$$S_{max} = 20.00 - 2(0.03)$$
$$= 19.94$$

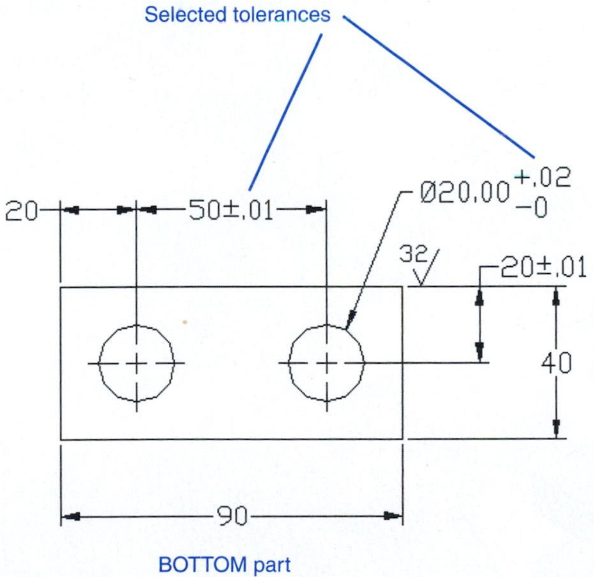

Figure 8-55

This means that 19.94 is the largest possible shaft diameter that will just fit. If a clearance tolerance of .01 is assumed to ensure that the shaft and hole are never exactly the same size, the maximum shaft diameter becomes 19.93. See Figure 8-56.

A feature tolerance of .02 on the shaft will result in a minimum shaft diameter of 19.91. Note that the .01 clearance tolerance and the .02 feature tolerance were arbitrarily chosen. Other values could have been used.

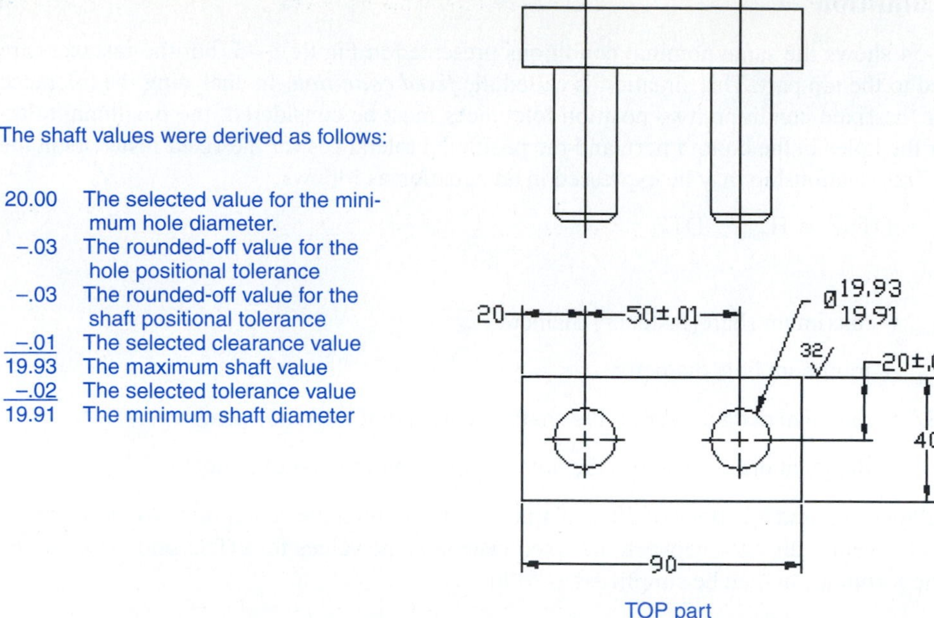

Figure 8-56

Designing a Hole Given a Fastener Size

The previous two examples started by selecting a minimum hole diameter and then calculating the resulting fastener size. Figure 8-57 shows a situation in which the fastener size is defined, and the problem is to determine the appropriate hole sizes. Figure 8-58 shows the dimensions and tolerances for both top and bottom parts.

Requirements:

Clearance, minimum = .003

Hole tolerances = .005

Positional tolerance = .002

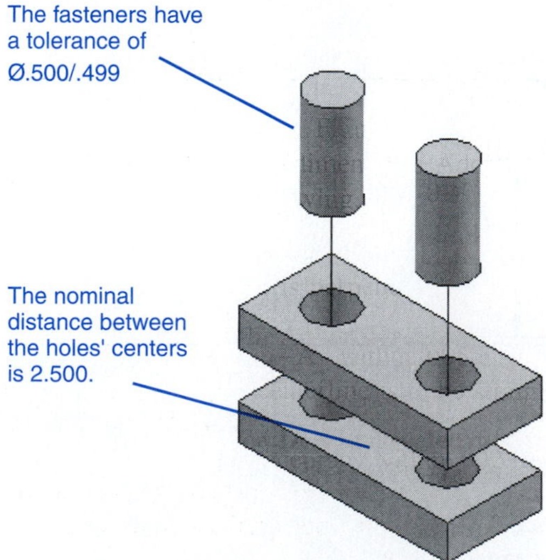

Figure 8-57

Geometric Tolerances

geometric tolerancing: A dimensioning and tolerancing system based on the geometric shape of an object.

Geometric tolerancing is a dimensioning and tolerancing system based on the geometric shape of an object. Surfaces may be defined in terms of their flatness or roundness, or in terms of how perpendicular or parallel they are to other surfaces.

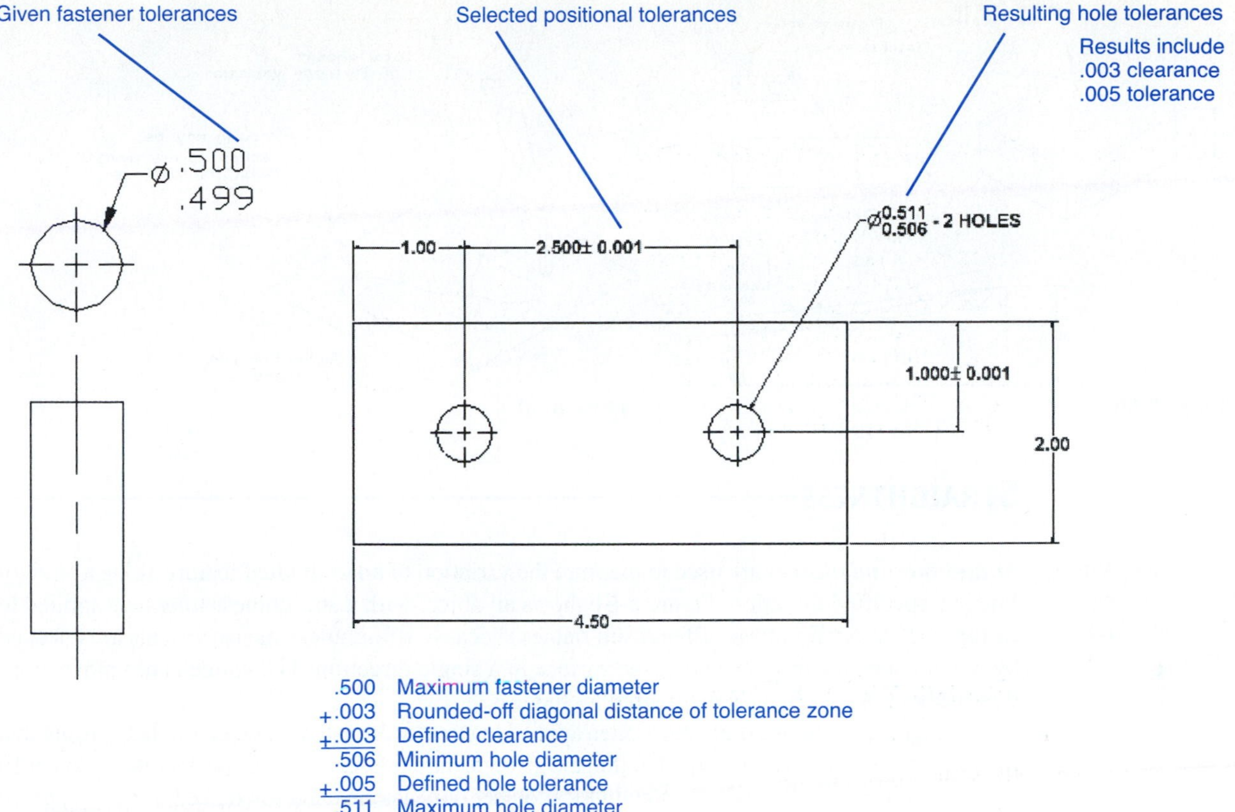

Figure 8-58

Geometric tolerances allow a more exact definition of the shape of an object than do conventional coordinate-type tolerances. Objects can be toleranced in a manner more closely related to their design function or so that their features and surfaces are more directly related to each other.

TOLERANCES OF FORM

Tolerances of form are used to define the shape of a surface relative to itself. There are four classifications: flatness, straightness, roundness, and cylindricity. Tolerances of form are not related to other surfaces but apply only to an individual surface.

tolerance of form: A tolerance used to define the shape of a surface relative to itself.

FLATNESS

Flatness tolerances are used to define the amount of variation permitted in an individual surface. The surface is thought of as a plane not related to the rest of the object.

flatness tolerance: A tolerance used to define the amount of variation permitted on an individual surface.

Figure 8-59 shows a rectangular object. How flat is the top surface? The given plus or minus tolerances allow a variation of (± 0.5) across the surface. Without additional tolerances the surface could look like a series of waves varying between 30.5 and 29.5.

If the example in Figure 8-59 was assigned a flatness tolerance of 0.3, the height of the object—the feature tolerance—could continue to vary based on the 30 ± 0.5 tolerance, but the surface itself could not vary by more than 0.3. In the most extreme condition, one end of the surface could be 30.5 above the bottom surface and the other end 29.5, but the surface would still be limited to within two parallel planes 0.3 apart as shown.

To better understand the meaning of flatness, consider how the surface would be inspected. The surface would be acceptable if a gauge could be moved all around the surface and never vary by more than 0.3. See Figure 8-60. Every point in the plane must be within the specified tolerance.

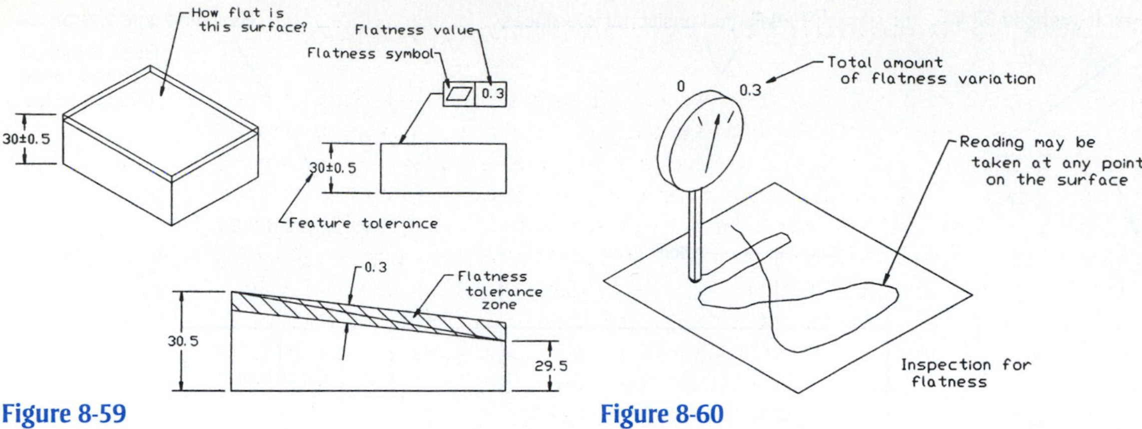

Figure 8-59

Figure 8-60

STRAIGHTNESS

straightness tolerance: A tolerance used to measure the variation of an individual feature along a straight line in a specified direction.

Straightness tolerances are used to measure the variation of an individual feature along a straight line in a specified direction. Figure 8-61 shows an object with a straightness tolerance applied to its top surface. Straightness differs from flatness because straightness measurements are checked by moving a gauge directly across the surface in a single direction. The gauge is not moved randomly about the surface, as is required by flatness.

Straightness tolerances are most often applied to circular or matching objects to help ensure that the parts are not barreled or warped within the given feature tolerance range and, therefore, do not fit together well. Figure 8-62 shows a cylindrical object dimensioned and toleranced using a standard feature tolerance. The surface of the cylinder may vary within the specified tolerance range as shown.

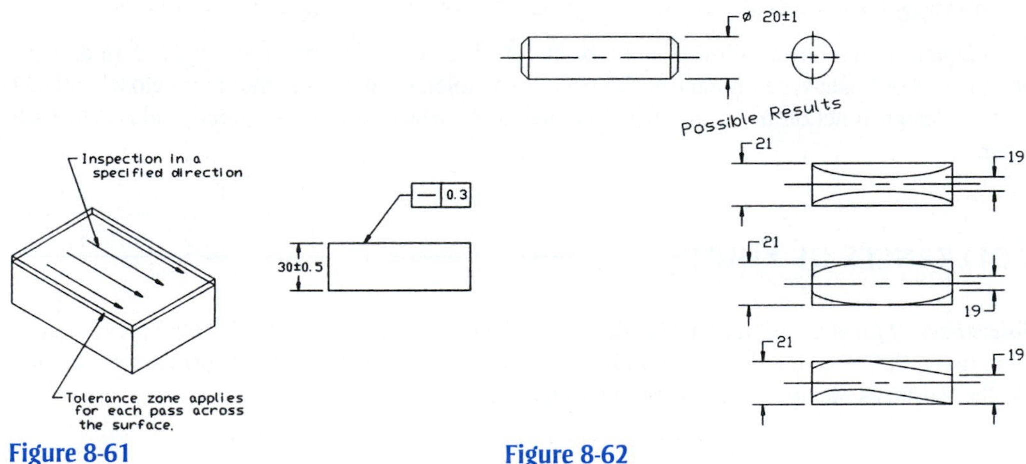

Figure 8-61

Figure 8-62

Figure 8-63 shows the same object shown in Figure 8-62 dimensioned and toleranced using the same feature tolerance but also including a 0.05 straightness tolerance. The straightness tolerance limits the surface variation to 0.05 as shown.

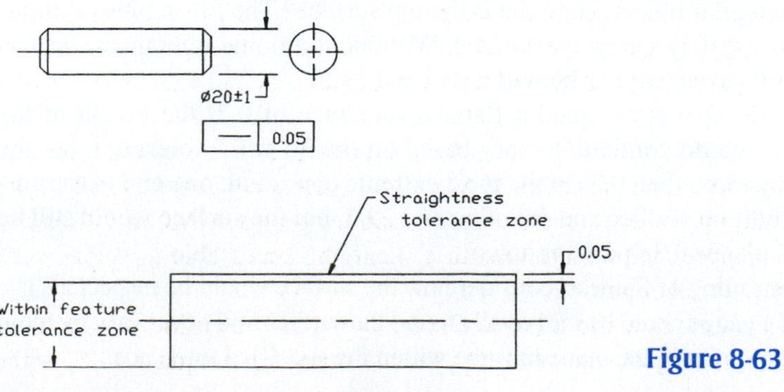

Figure 8-63

Straightness (RFS And MMC)

Figure 8-64 again shows the same cylinder shown in Figures 8-62 and 8-63. This time the straightness tolerance is applied about the cylinder's centerline. This type of tolerance permits the feature tolerance and geometric tolerance to be used together to define a *virtual condition.* A virtual condition is used to determine the maximum possible size variation of the cylinder or the smallest diameter hole that will always accept the cylinder. See Section 8-19.

The geometric tolerance specified in Figure 8-64 is applied to any circular segment along the cylinder, regardless of the cylinder's diameter. This means that the 0.05 tolerance is applied equally when the cylinder's diameter measures 19 or when it measures 21. This application is called RFS, *regardless of feature size.* RFS conditions are specified in a tolerance either by an S with a circle around it or implied tacitly when no other symbol is used. In Figure 8-59 no symbol is listed after the 0.05 value, so it is assumed to be applied RFS.

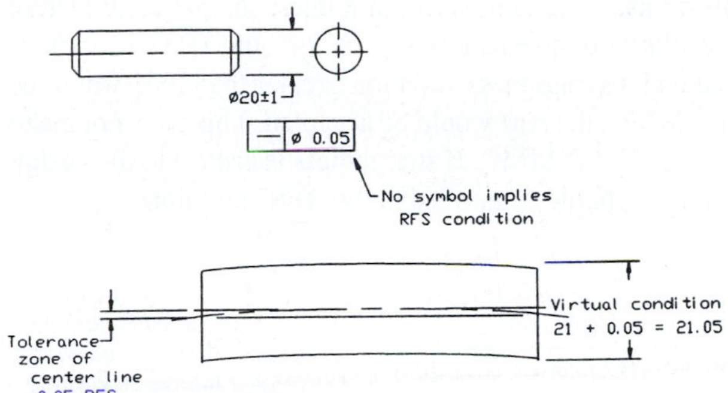

Figure 8-64

Figure 8-65 shows the cylinder dimensioned with an MMC condition applied to the straightness tolerance. MMC stands for *maximum material condition* and means that the specified straightness tolerance (0.05) is applied only at the MMC condition or when the cylinder is at its maximum diameter size (21).

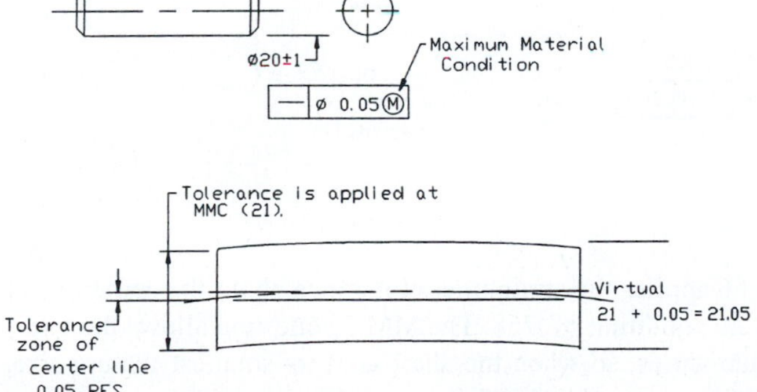

Measured Size	Allowable Tolerance Zone	Virtual Condition
21.0	0.05	21.05
20.9	0.15	21.15
20.8	0.25	21.25
.	.	.
20.0	1.05	22.05
.	.	.
19.0	2.05	23.05

Figure 8-65

A shaft is an external feature, so its largest possible size or MMC occurs when it is at its maximum diameter. A hole is an internal feature. A hole's MMC condition occurs when it is at its smallest diameter. The MMC condition for holes will be discussed later in the chapter along with positional tolerances.

Applying a straightness tolerance at MMC allows for a variation in the resulting tolerance zone. Because the 0.05 flatness tolerance is applied at MMC, the virtual condition is still 21.05, the same as with the RFS condition; however, the tolerance is applied only at MMC. As the cylinder's diameter varies within the specified feature tolerance range the acceptable tolerance zone may vary to maintain the same virtual condition.

The table in Figure 8-65 shows how the tolerance zone varies as the cylinder's diameter varies. When the cylinder is at its largest size or MMC, the tolerance zone equals 0.05, or the specified flatness variation. When the cylinder is at its smallest diameter, the tolerance zone equals 2.05, or the total feature size plus the total flatness size. In all variations the virtual size remains the same, so at any given cylinder diameter value, the size of the tolerance zone can be determined by subtracting the cylinder's diameter value from the virtual condition.

Note:
Geometric tolerances applied at MMC allow the tolerance zone to grow.

Figure 8-66 shows a comparison between different methods used to dimension and tolerance a .750 shaft. The first example uses only a feature tolerance. This tolerance sets an upper limit of .755 and a lower limit of .745. Any variations within that range are acceptable.

The second example in Figure 8-66 sets a straightness tolerance of .003 about the cylinder's centerline. No conditions are defined, so the tolerance is applied RFS. This limits the variations in straightness to .003 at all feature sizes. For example, when the shaft is at its smallest possible feature size of .745, the .003 still applies. This means that a shaft measuring .745 that had a straightness variation greater than .003 would be rejected. If the tolerance had been applied at MMC, the part would be accepted. This does not mean that straightness tolerances should always be applied at MMC. If straightness is critical to the design integrity or function of the part, then straightness should be applied in the RFS condition.

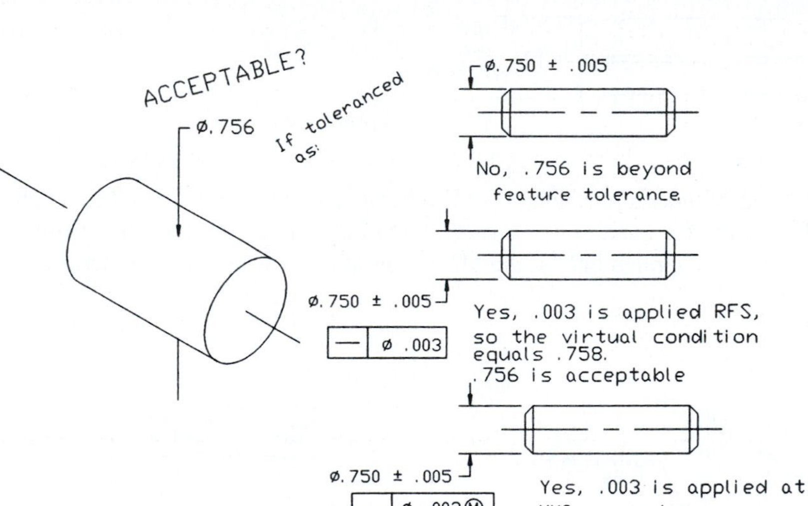

The RFS condition does not allow the tolerance zone to grow, as does the same tolerance applied at MMC.

Figure 8-66

The third example in Figure 8-66 applies the straightness tolerance about the centerline at MMC. This tolerance creates a virtual condition of .758. The MMC condition allows the tolerance to vary as the feature tolerance varies, so when the shaft is at its smallest feature size, .745, a straightness tolerance of .003 is acceptable (.005 feature tolerance + .003 straightness tolerance).

If the tolerance specification for the cylinder shown in Figure 8-66 was 0.000 applied at MMC, it would mean that the shaft would have to be perfectly straight at MMC or when the shaft was at its maximum value (.755); however, the straightness tolerance can vary as the feature size varies, as discussed for the other tolerance conditions. A 0.000 tolerance means that the MMC and the virtual conditions are equal.

Figure 8-67 shows a very long .750 diameter shaft. Its straightness tolerance includes a length qualifier that serves to limit the straightness variations over each inch of the shaft length and to prevent excess waviness over the full length. The tolerance .002/1.000 means that the total straightness may vary over the entire length of the shaft by .003 but that the variation is limited to .002 per 1.000 of shaft length.

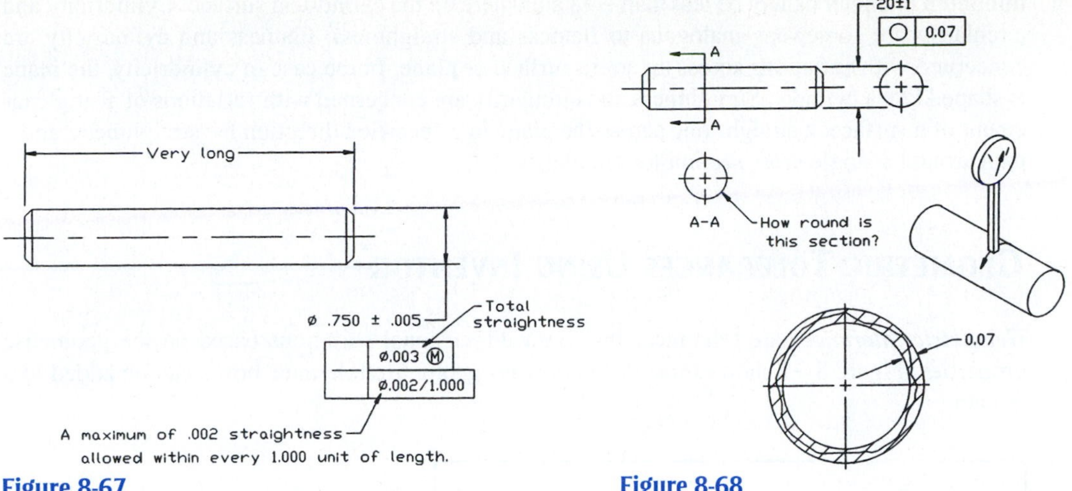

Figure 8-67

Figure 8-68

Circularity

circularity tolerance: A tolerance used to limit the amount of variation in the roundness of a surface of revolution.

A ***circularity tolerance*** is used to limit the amount of variation in the roundness of a surface of revolution. It is measured at individual cross sections along the length of the object. The measurements are limited to the individual cross sections and are not related to other cross sections. This means that in extreme conditions the shaft shown in Figure 8-68 could actually taper from a diameter of 21 to a diameter of 19 and never violate the circularity requirement. It also means that qualifications such as MMC cannot be applied.

Figure 8-68 shows a shaft that includes a feature tolerance and a circularity tolerance of 0.07. To understand circularity tolerances, consider an individual cross section or slice of the cylinder. The actual shape of the outside edge of the slice varies around the slice. The difference between the maximum diameter and the minimum diameter of the slice can never exceed the stated circularity tolerance.

Circularity tolerances can be applied to tapered sections and spheres, as shown in Figure 8-69. In both applications, circularity is measured around individual cross sections, as it was for the shaft shown in Figure 8-68.

Cylindricity

cylindricity tolerance: A tolerance used to define a tolerance zone both around individual circular cross sections of an object and also along its length.

Cylindricity tolerances are used to define a tolerance zone both around individual circular cross sections of an object and also along its length. The resulting tolerance zone looks like two concentric cylinders.

Figure 8-70 shows a shaft that includes a cylindricity tolerance that establishes a tolerance zone of .007. This means that if the maximum measured diameter is determined to be .755, the

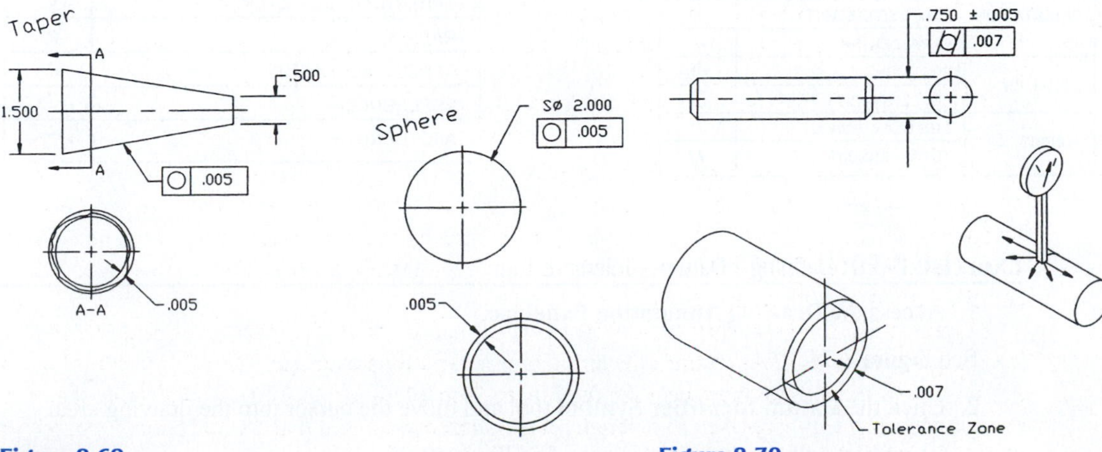

Figure 8-69

Figure 8-70

minimum diameter cannot be less than .748 anywhere on the cylindrical surface. Cylindricity and circularity are somewhat analogous to flatness and straightness. Flatness and cylindricity are concerned with variations across an entire surface or plane. In the case of cylindricity, the plane is shaped like a cylinder. Straightness and circularity are concerned with variations of a single element of a surface: a straight line across the plane in a specified direction for straightness, and a path around a single cross section for circularity.

GEOMETRIC TOLERANCES USING INVENTOR

geometric tolerance: A tolerance that limits dimensional variations based on the geometric properties of an object.

Geometric tolerances are tolerances that limit dimensional variations based on the geometric properties. Figure 8-71 shows three different ways geometric tolerance boxes can be added to a drawing.

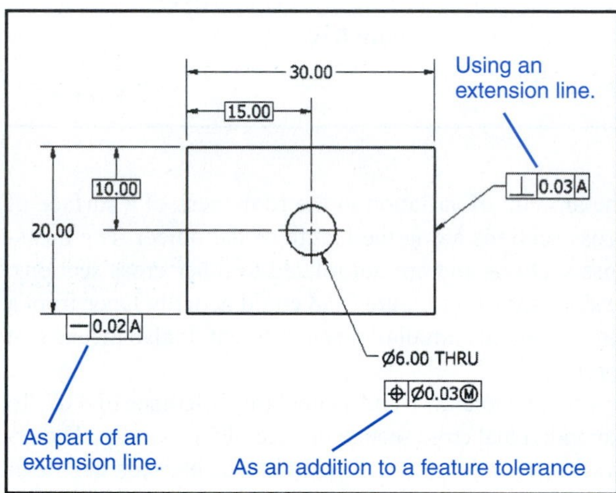

Figure 8-71

Figure 8-72 shows lists of geometric tolerance symbols. Figure 8-73 shows an object dimensioned using geometric tolerances. The geometric tolerances were created as follows.

	TYPE OF TOLERANCE	CHARACTERISTIC	SYMBOL
FOR INDIVIDUAL FEATURES	FORM	STRAIGHTNESS	—
		FLATNESS	▱
		CIRCULARITY	○
		CYLINDRICITY	⌭
INDIVIDUAL OR RELATED FEATURES	PROFILE	PROFILE OF A LINE	⌒
		PROFILE OF A SURFACE	⌓
RELATED FEATURES	ORIENTATION	ANGULARITY	∠
		PERPENDICULARITY	⊥
		PARALLELISM	//
	LOCATION	POSITION	⌖
		CONCENTRICITY	◎
	RUNOUT	CIRCULAR RUNOUT	↗
		TOTAL RUNOUT	⌰

TERM	SYMBOL
AT MAXIMUM MATERIAL CONDITION	Ⓜ
REGARDLESS OF FEATURE SIZE	Ⓢ
AT LEAST MATERIAL CONDITION	Ⓛ
PROJECTED TOLERANCE ZONE	Ⓟ
DIAMETER	Ø
SPHERICAL DIAMETER	SØ
RADIUS	R
SPHERICAL RADIUS	SR
REFERENCE	( )
ARC LENGTH	⌒

Figure 8-72

Exercise 8-10: Defining a Datum—Tolerance Tool

1. Access the **Drawing Annotation Panel** bar.

See Figure 8-74.

2. Click the **Datum Identifier Symbol** tool and move the cursor into the drawing area.

A datum box will appear on the cursor. See Figure 8-75.

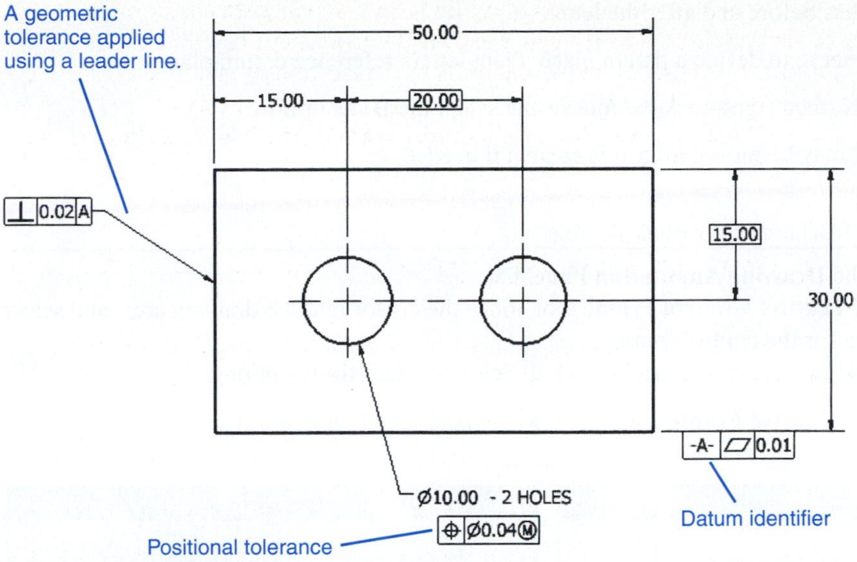

Figure 8-73

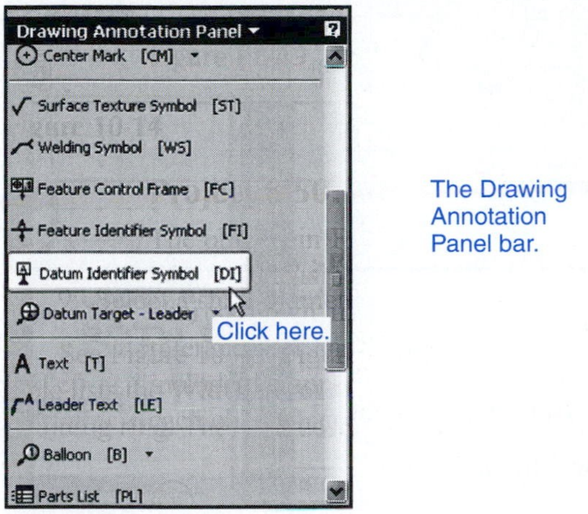

Figure 8-74

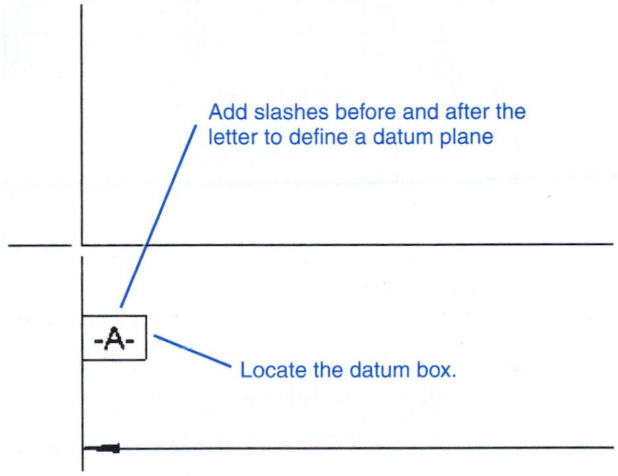

Figure 8-75

3. Position the datum identifier and press the left mouse button, then the right button.

A dialog box will appear.

4. Click the **Continue** option.

The **Format Text** dialog box will appear. See Figure 8-76. The letter **A** will automatically be selected. If another letter is required, backspace out the existing letter and type in a new one. Flank the letter with dashes.

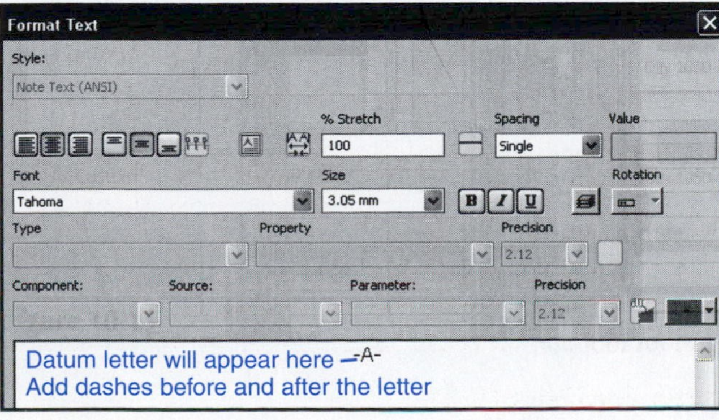

Figure 8-76

5. Add dashes before and after the letter.

The dashes serve to define a datum place. Plain letters reference datum planes.

6. Click **OK,** then right-click the mouse and select the **Done** option.

The symbol may be moved after it is created if needed.

Exercise 8-11: Defining a Perpendicular Tolerance

1. Access the **Drawing Annotation Panel** bar.
2. Click the **Feature Control Frame** tool, move the cursor into the drawing area, and select a location for the control frame.
3. Left-click the mouse, then right-click it. Select the **Continue** option.

The **Feature Control Frame** dialog box will appear. See Figure 8-77.

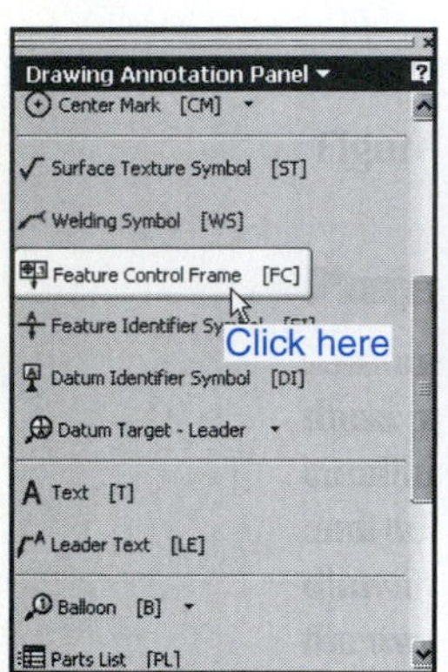

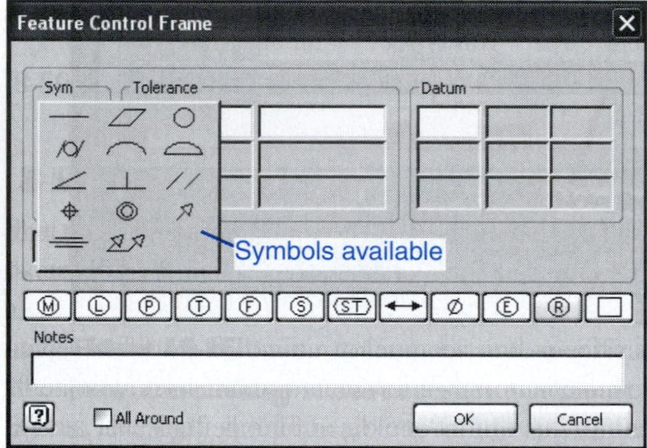

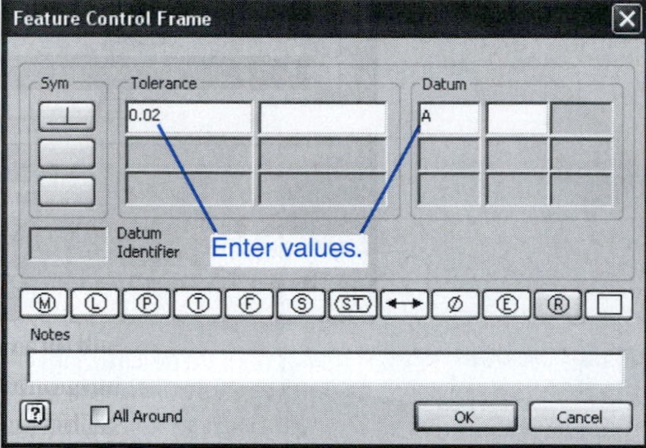

Figure 8-77

4. Click the **Sym** box and select the perpendicular symbol.
5. Set the **Tolerance 1** value for **.02** and the **Datum 1** value for **A.**

Figure 8-78 shows the resulting feature control frame.

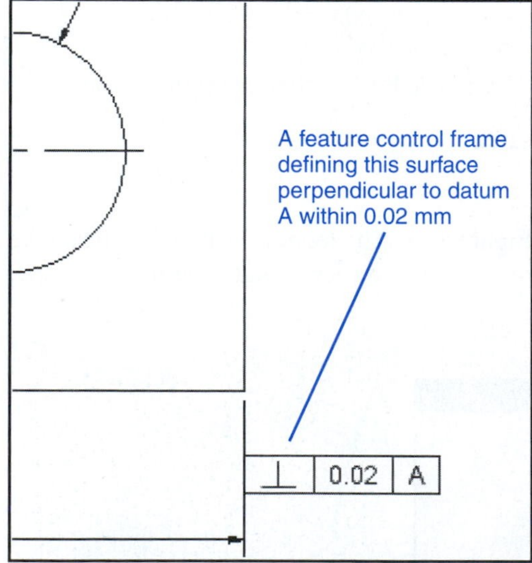

Figure 8-78

Exercise 8-12: Defining a Straightness Tolerance with a Leader Line

1. Access the **Drawing Annotation Panel** bar and select the **Feature Control Frame** tool.

See Figure 8-79.

2. Move the cursor into the drawing area.

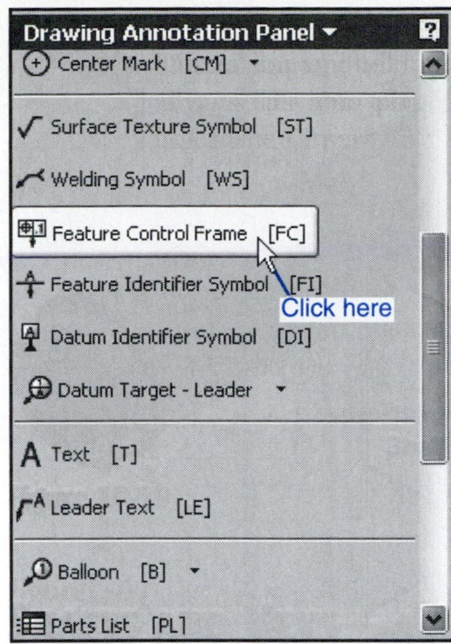

Figure 8-79

A feature control box will appear.

3. Left-click the edge line of the part, then move the frame away from the edge.
4. Select a location for the frame, left-click, then right-click the mouse and select the **Continue** option.
5. Edit the **Feature Control Frame** dialog box as shown in Figure 8-80.

Figure 8-80 shows the resulting feature control.

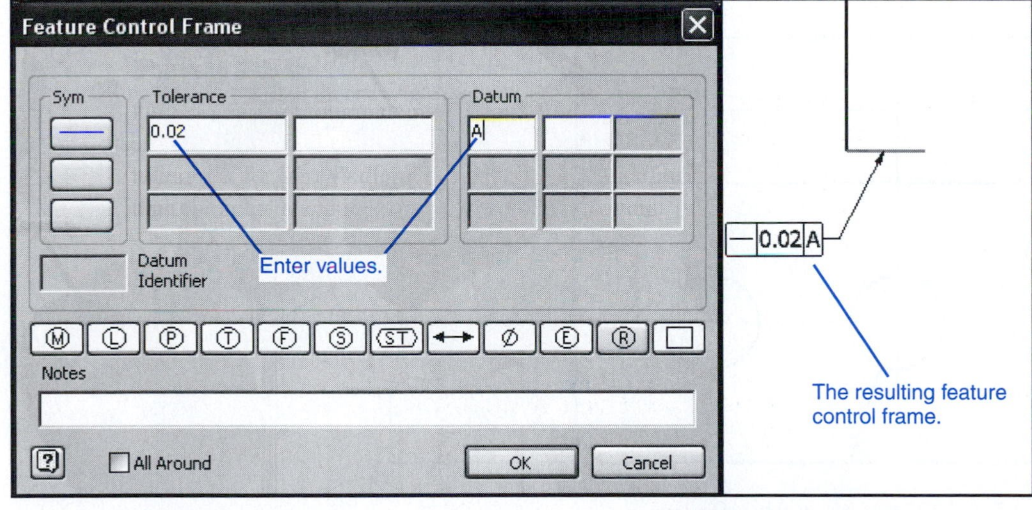

Figure 8-80

Positional Tolerance

A ***positional tolerance*** is used to locate and tolerance a hole in an object. Positional tolerances require basic locating dimensions for the hole's center point. Positional tolerances also require a feature tolerance to define the diameter tolerances of the hole, and a geometric tolerance to define the position tolerance for the hole's center point.

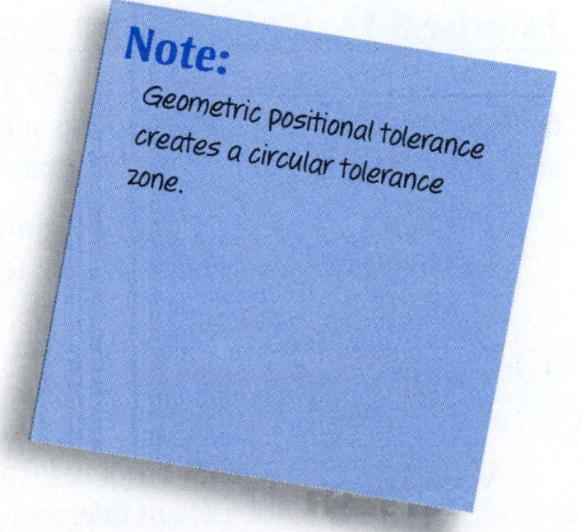

positional tolerance: A tolerance used to locate and tolerance a hole in an object.

Exercise 8-13: Creating a Basic Dimension

The 15 and 20 dimensions in Figure 8-73 used to locate the center position of the hole are basic dimensions. ***Basic dimensions*** are dimensions enclosed in rectangles.

basic dimension: A dimension enclosed in a rectangle.

1. Create dimensions using the **General Dimension** tool.
2. Right-click the existing dimension and select the **Edit** option.

The **Edit Dimension** dialog box will appear. See Figure 8-81.

3. Click the **Precision and Tolerance** tab.
4. Select the **Basic** option, then click **OK.**

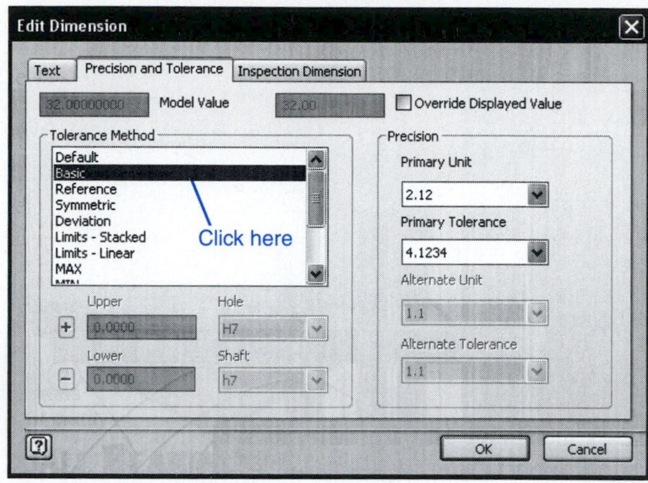

Figure 8-81

The selected dimension will be enclosed in a rectangle. This is a basic dimension. See Figure 8-82.

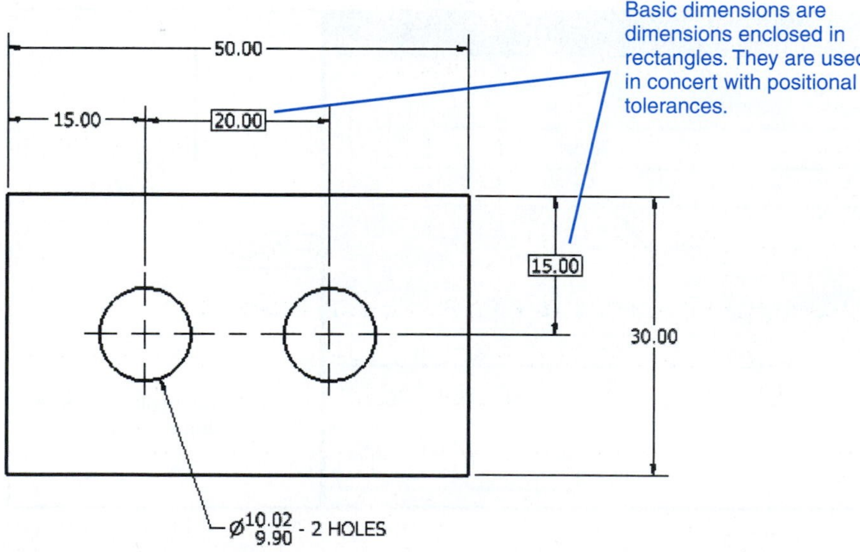

Figure 8-82

Exercise 8-14: Adding a Positional Tolerance to a Hole's Feature Tolerance

Figure 8-83 shows a feature tolerance for a hole. A feature tolerance defines the hole's size limits. In the example shown in Figure 8-83 the hole is defined as 10.2 to 9.9. These values define the tolerance of the hole. In addition, the location of the hole's center point must be defined and toleranced.

1. Click the **Feature Control Frame** tool.
2. Move the cursor and locate the feature control frame below the hole's feature control dimensions.
3. Click the left mouse button, then the right mouse button. Select the **Continue** option.

The **Feature Control Frame** dialog box will appear. See Figure 8-84.

4. Select the positional symbol, then move to the **Tolerance** box and locate the cursor to the left of the 0.000 default tolerance value.
5. Select the centerline symbol, **Ø.**

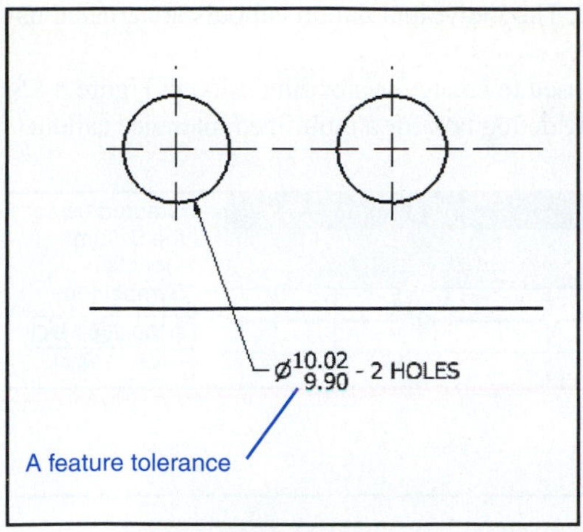

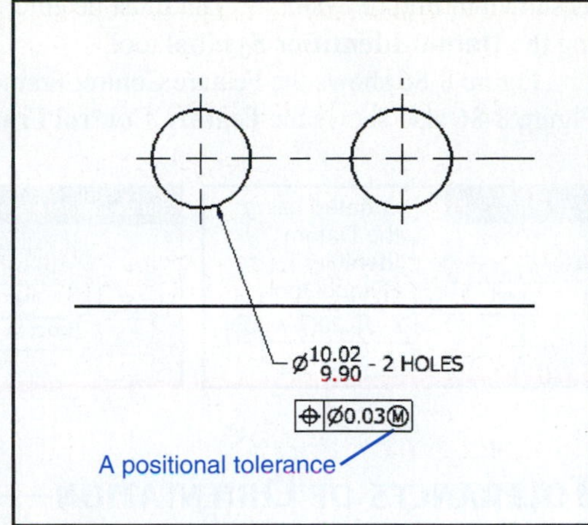

Figure 8-83

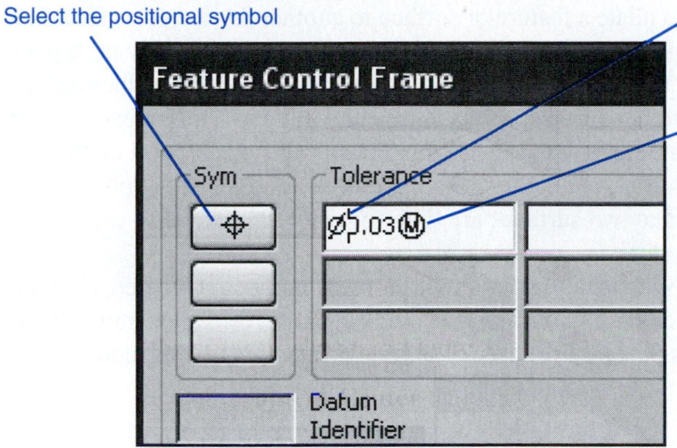

Figure 8-84

6. Create a tolerance value of **0.03,** then click the maximum material condition symbol, an M with a circle around it.
7. Click **OK.**
8. Right-click the mouse and select the **Done** option.

Creating More Complex Geometric Tolerance Drawing Callouts

Figure 8-85 shows a model that has several, more complex, geometric tolerance drawing callouts. These callouts are created using the procedure and the same **Feature Control Frame** dialog box

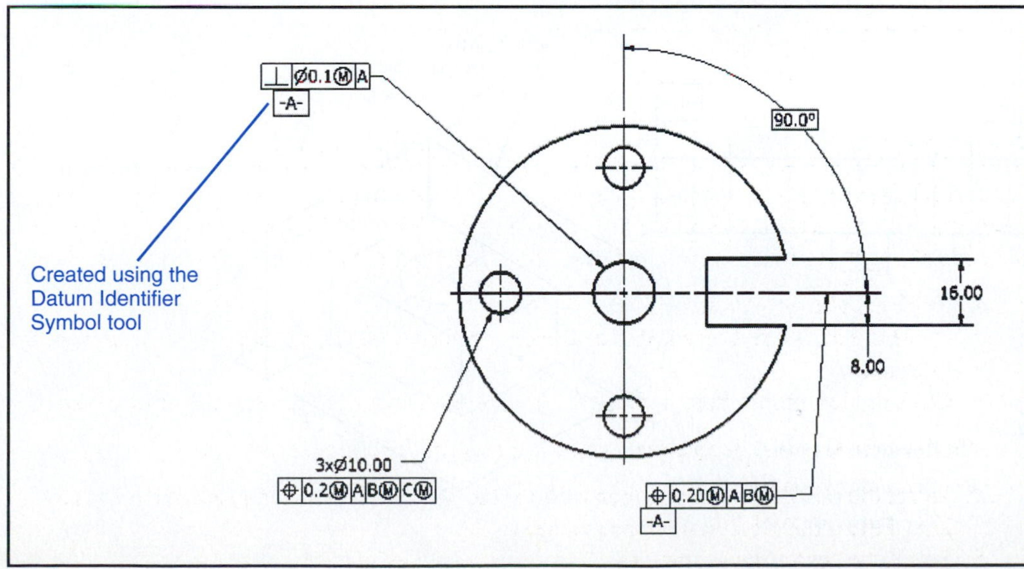

Figure 8-85

as shown in Figure 8-84, but data must be entered. The individual datum callouts are created using the **Datum Identifier Symbol** tool.

Figure 8-86 shows the Feature Control Frame used to create the slot dimension in Figure 8-85. Figure 8-86 also shows the **Feature Control Frame** dialog box for a multilined tolerance callout.

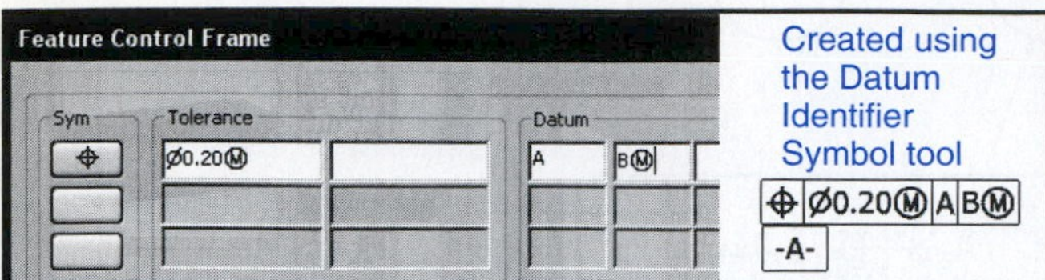

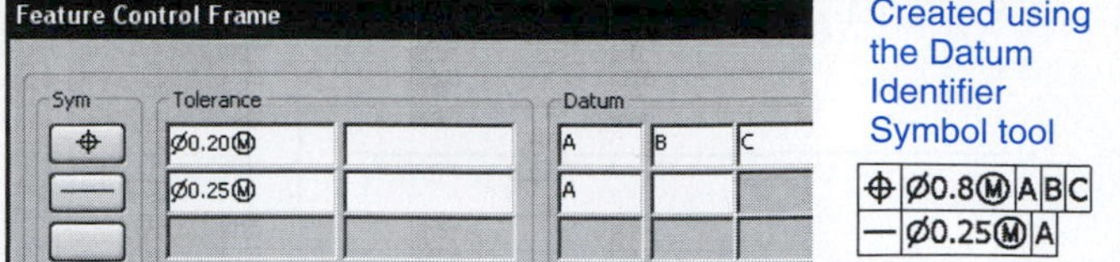

Figure 8-86

TOLERANCES OF ORIENTATION

tolerance of orientation: A tolerance used to relate a feature of surface to another feature or surface.

Tolerances of orientation are used to relate a feature or surface to another feature or surface. Tolerances of orientation include perpendicularity, parallelism, and angularity. They may be applied using RFS or MMC conditions, but they cannot be applied to individual features by themselves. To define a surface as parallel to another surface is very much like assigning a flatness value to the surface. The difference is that flatness applies only within the surface; every point on the surface is related to a defined set of limiting parallel planes. Parallelism defines every point in the surface relative to another surface. The two surfaces are therefore directly related to each other, and the condition of one affects the other.

Orientation tolerances are used with locational tolerances. A feature is first located, then it is oriented within the locational tolerances. This means that the orientation tolerance must always be less than the locational tolerances. The next four sections will further explain this requirement.

DATUMS

datum: A point, axis, or surface used as a starting reference point for dimensions and tolerances.

A ***datum*** is a point, axis, or surface used as a starting reference point for dimensions and tolerances. Figures 8-87 and 8-88 show a rectangular object with three datum planes labeled –A–, –B–, and –C–. The three datum planes are called the primary, secondary, and tertiary datums, respectively. The three datum planes are, by definition, exactly 90° to one another.

Figure 8-89 shows a cylindrical datum frame that includes three datum planes. The X and Y planes are perpendicular to each other, and the base A plane is perpendicular to the datum axis between the X and Y planes.

Datums are defined on a drawing by letters enclosed in rectangular boxes, as shown. The defining letters are written flanked by dashes: –A–, –B–, and –C–.

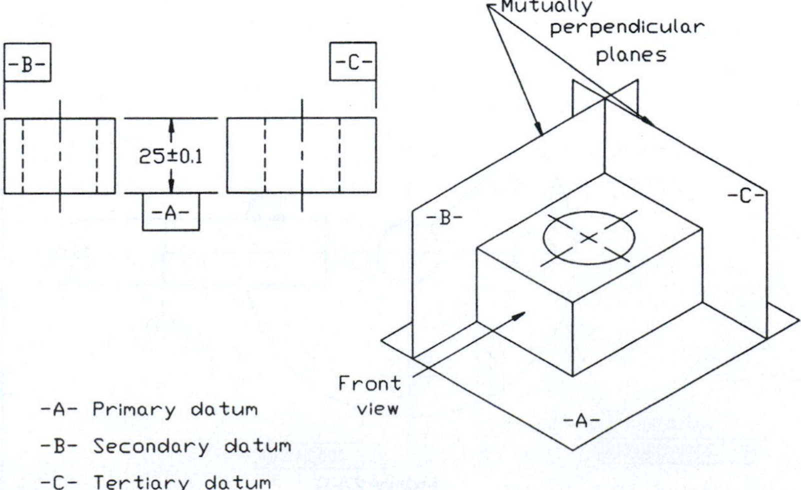

Figure 8-87

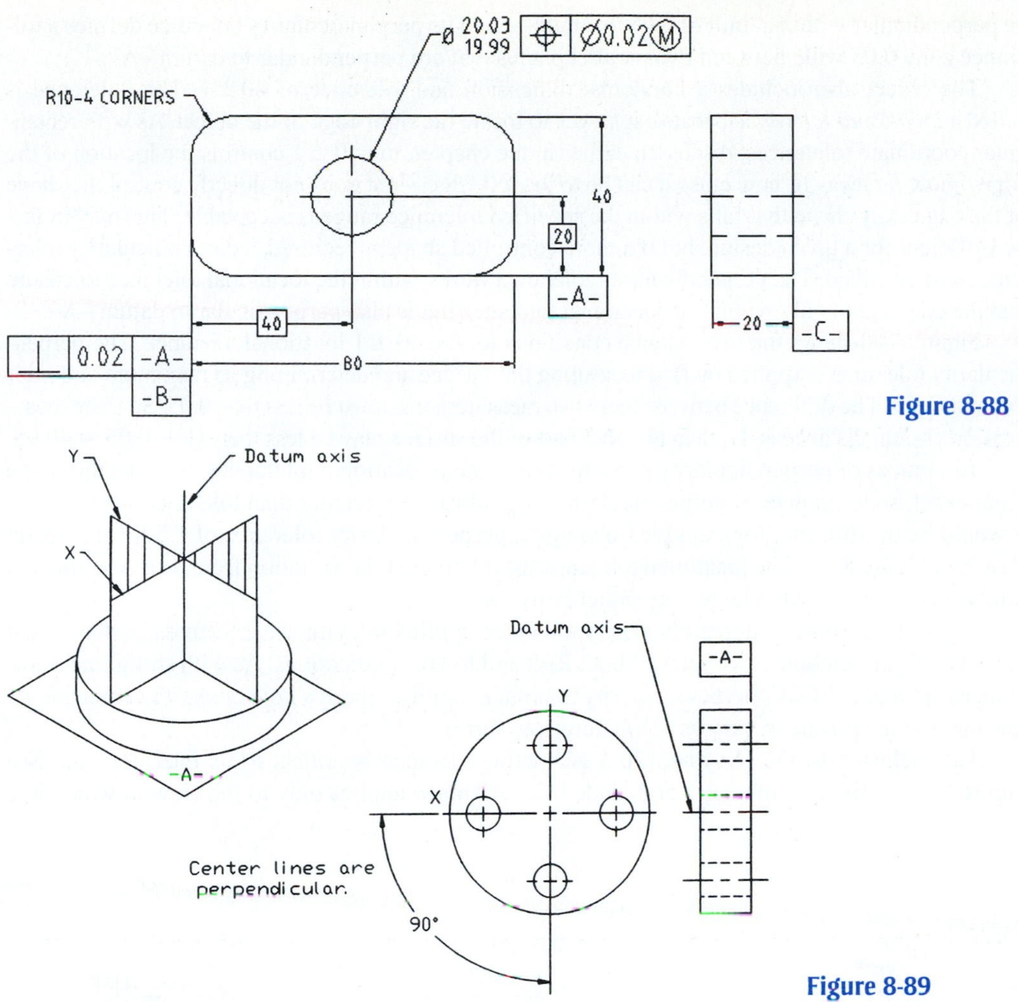

Figure 8-88

Figure 8-89

Datum planes are assumed to be perfectly flat. When assigning a datum status to a surface, be sure that the surface is reasonably flat. This means that datum surfaces should be toleranced using surface finishes, or created using machine techniques that produce flat surfaces.

PERPENDICULARITY

Perpendicularity tolerances are used to limit the amount of variation for a surface or feature within two planes perpendicular to a specified datum. Figure 8-90 shows a rectangular object. The bottom surface is assigned as datum –A–, and the right vertical edge is toleranced so that it must

perpendicularity tolerance: A tolerance used to limit the amount of variation for a surface or feature with two planes perpendicular to a specified datum.

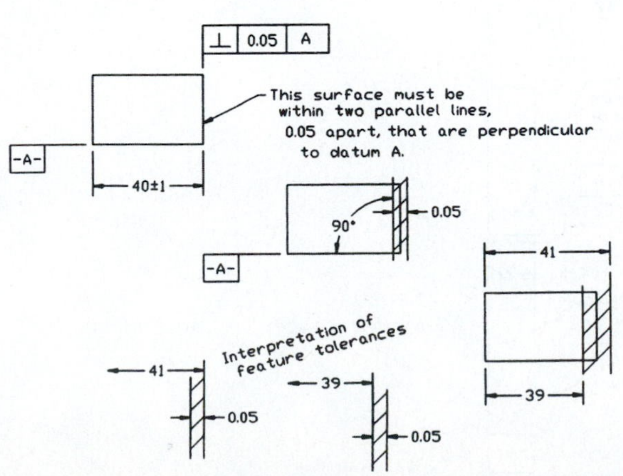

Figure 8-90

be perpendicular within a limit of 0.05 to datum –A–. The perpendicularity tolerance defines a tolerance zone 0.05 wide between two parallel planes that are perpendicular to datum –A–.

The object also includes a horizontal dimension and tolerance of 40 ± 1. This tolerance is called a *locational tolerance* because it serves to locate the right edge of the object. As with rectangular coordinate tolerances, discussed earlier in the chapter, the 40 ± 1 controls the location of the edge—how far away or how close it can be to the left edge—but does not directly control the shape of the edge. Any shape that falls within the specified tolerance range is acceptable. This may in fact be sufficient for a given design, but if a more controlled shape is required, a perpendicularity tolerance must be added. The perpendicularity tolerance works within the locational tolerance to ensure that the edge is not only within the locational tolerance but is also perpendicular to datum –A–.

Figure 8-90 shows the two extreme conditions for the 40 ± 1 locational tolerance. The perpendicularity tolerance is applied by first measuring the surface and determining its maximum and minimum lengths. The difference between these two measurements must be less than 0.05. So if the measured maximum distance is 41, then no other part of the surface may be less then 41 − 0.05 = 40.95.

Tolerances of perpendicularity serve to complement locational tolerances, to make the shape more exact, so tolerances of perpendicularity must always be smaller than tolerances of location. It would be of little use, for example, to assign a perpendicularity tolerance of 1.5 for the object shown in Figure 8-91. The locational tolerance would prevent the variation from ever reaching the limits specified by such a large perpendicularity tolerance.

Figure 8-92 shows a perpendicularity tolerance applied to cylindrical features: a shaft and a hole. The figure includes examples of both RFS and MMC applications. As with straightness tolerances applied at MMC, perpendicularity tolerances applied about a hole or shaft's centerline allow the tolerance zone to vary as the feature size varies.

The inclusion of the Ø symbol in a geometric tolerance is critical to its interpretation. See Figure 8-93. If the Ø symbol is not included, the tolerance applies only to the view in which it is

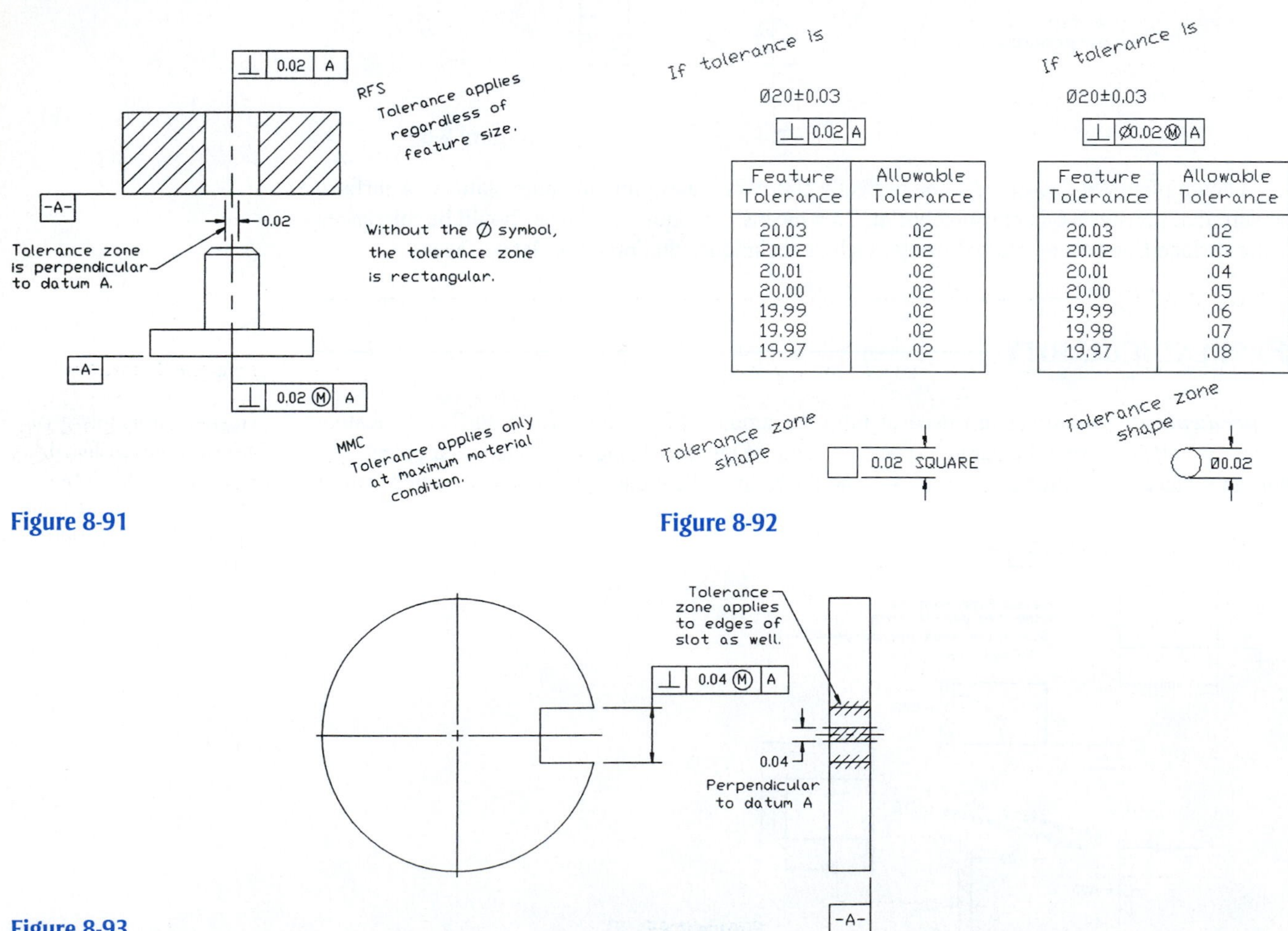

Figure 8-91

Figure 8-92

Figure 8-93

written. This means that the tolerance zone is shaped like a rectangular slice, not a cylinder, as would be the case if the ø symbol were included. In general it is better to always include the Ø symbol for cylindrical features because it generates a tolerance zone more like that used in positional tolerancing.

Figure 8-93 shows a perpendicularity tolerance applied to a slot, a noncylindrical feature. Again, the MMC specification is always for variations in the tolerance zone.

PARALLELISM

Parallelism tolerances are used to ensure that all points within a plane are within two parallel planes that are parallel to a referenced datum plane. Figure 8-94 shows a rectangular object that is toleranced so that its top surface is parallel to the bottom surface within 0.02. This means that every point on the top surface must be within a set of parallel planes 0.02 apart. These parallel tolerancing planes are located by determining the maximum and minimum distances from the datum surface. The difference between the maximum and minimum values may not exceed the stated 0.02 tolerance.

parallelism tolerance: A tolerance used to ensure that all points within a plane are within two parallel planes that are parallel to a referenced datum plane.

In the extreme condition of maximum feature size, the top surface is located 40.5 above the datum plane. The parallelism tolerance is then applied, meaning that no point on the surface may be closer than 40.3 to the datum. This is an RFS condition. The MMC condition may also be applied, thereby allowing the tolerance zone to vary as the feature size varies.

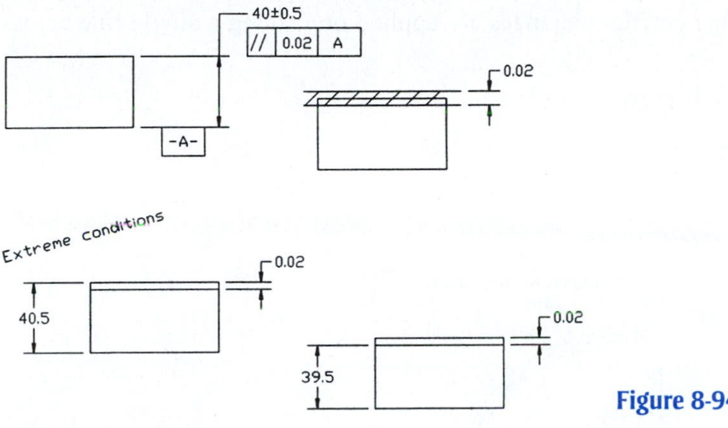

Figure 8-94

ANGULARITY

Angularity tolerances are used to limit the variance of surfaces and axes that are at an angle relative to a datum. Angularity tolerances are applied like perpendicularity and parallelism tolerances as a way to better control the shape of locational tolerances.

angularity tolerance: A tolerance used to limit the variance of surfaces and axes that are at an angle relative to a datum.

Figure 8-95 shows an angularity tolerance and several ways it is interpreted at extreme conditions.

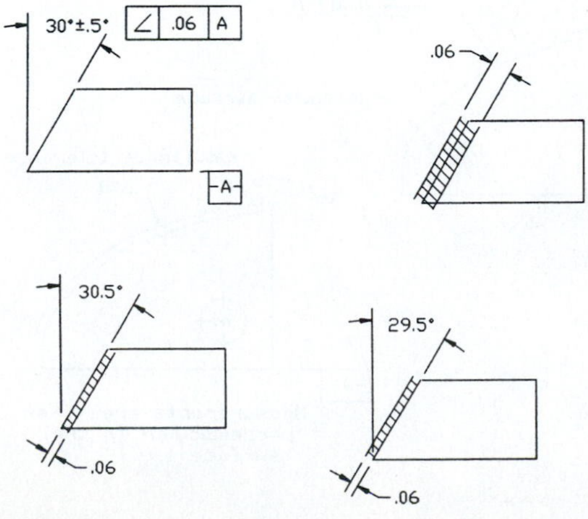

Figure 8-95

PROFILES

profile tolerance: A tolerance used to limit the variations of irregular surfaces.

Profile tolerances are used to limit the variations of irregular surfaces. They may be assigned as either bilateral or unilateral tolerances. There are two types of profile tolerances: surface and line. *Surface* profile tolerances limit the variation of an entire surface, whereas a *line* profile tolerance limits the variations along a single line across a surface.

Figure 8-96 shows an object that includes a surface profile tolerance referenced to an irregular surface. The tolerance is considered a bilateral tolerance because no other specification is given. This means that all points on the surface must be located between two parallel planes 0.08 apart that are centered about the irregular surface. The measurements are taken perpendicular to the surface.

Unilateral applications of surface profile tolerances must be indicated on the drawing using phantom lines. The phantom line indicates the side of the true profile line of the irregular surface on which the tolerance is to be applied. A phantom line above the irregular surface indicates that the tolerance is to be applied using the true profile line as 0 and then the specified tolerance range is to be added above that line. See Figures 8-97 and 8-98.

Profiles of line tolerances are applied to irregular surfaces, as shown in Figure 8-98. Profiles of line tolerances are particularly helpful when tolerancing an irregular surface that is constantly changing, such as the surface of an airplane wing.

Surface and line profile tolerances are somewhat analogous to flatness and straightness tolerances. Flatness and surface profile tolerances are applied across an entire surface, whereas straightness and line profile tolerances are applied only along a single line across the surface.

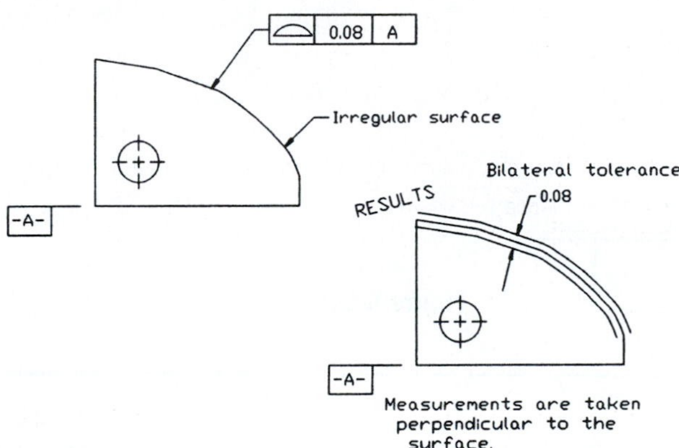

Figure 8-96

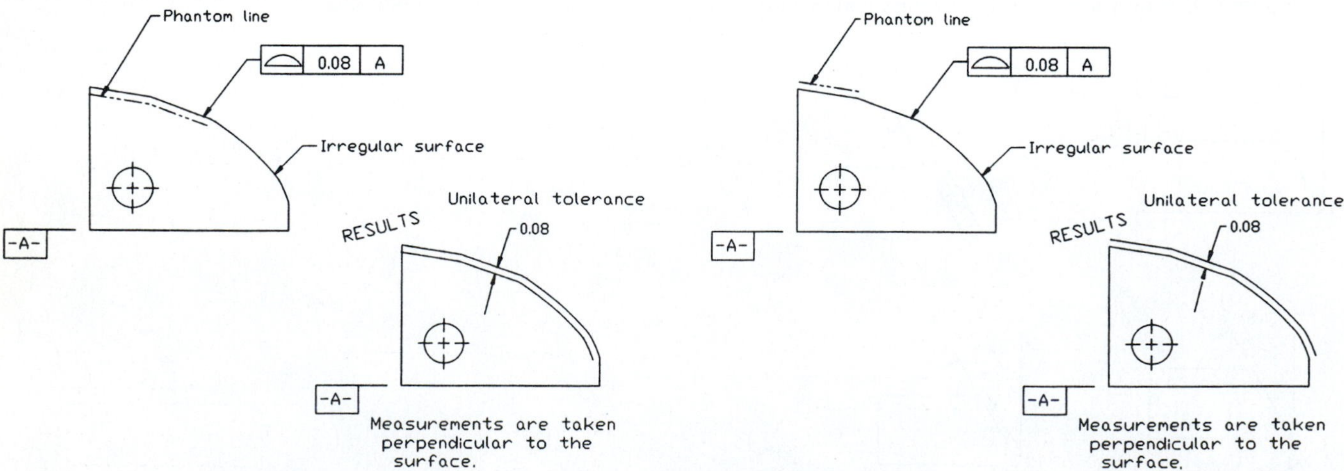

Figure 8-97

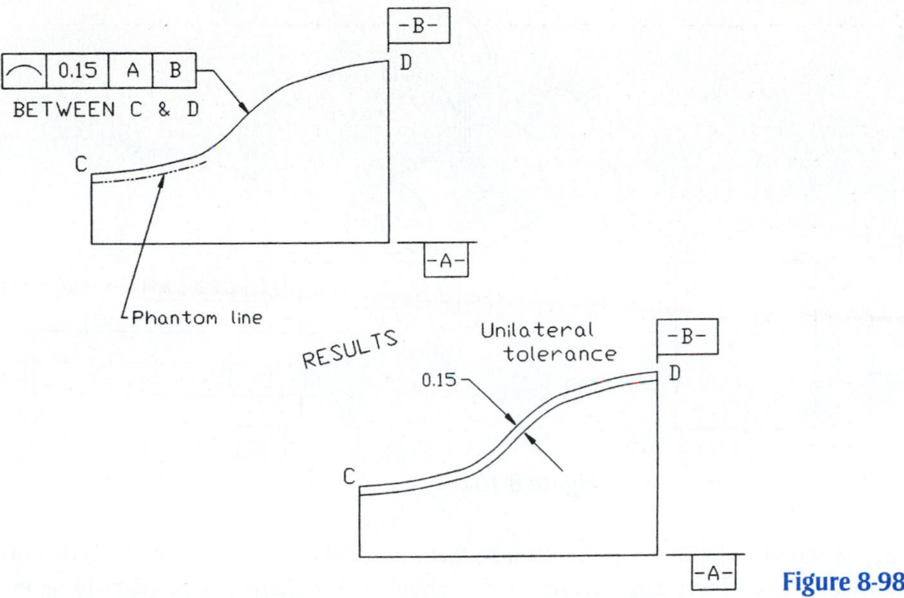

Figure 8-98

RUNOUTS

A ***runout tolerance*** is used to limit the variations between features of an object and a datum. More specifically they are applied to surfaces around a datum axis such as a cylinder or to a surface constructed perpendicular to a datum axis. There are two types of runout tolerances: circular and total.

runout tolerance: A tolerance used to limit the variations between features of an object and a datum.

Figure 8-99 shows a cylinder that includes a circular runout tolerance. The runout requirements are checked by rotating the object about its longitudinal axis or datum axis while holding an indicator gauge in a fixed position on the object's surface.

Runout tolerances may be either bilateral or unilateral. A runout tolerance is assumed to be bilateral unless otherwise indicated. If a runout tolerance is to be unilateral, a phantom line is used to indicate the side of the object's true surface to which the tolerance is to be applied. See Figure 8-100.

Runout tolerances may be applied to tapered areas of cylindrical objects, as shown in Figure 8-101. The tolerance is checked by rotating the object about a datum axis while holding an indicator gauge in place.

A total runout tolerance limits the variation across an entire surface. See Figure 8-102. An indicator gauge is not held in place while the object is rotated, as it is for circular runout tolerances, but is moved about the rotating surface.

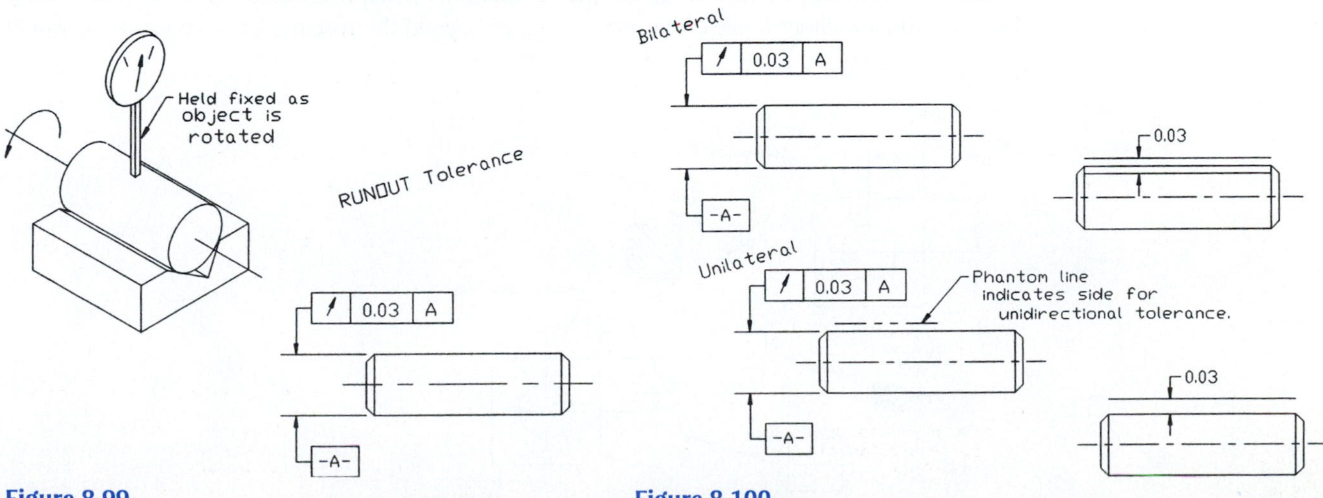

Figure 8-99

Figure 8-100

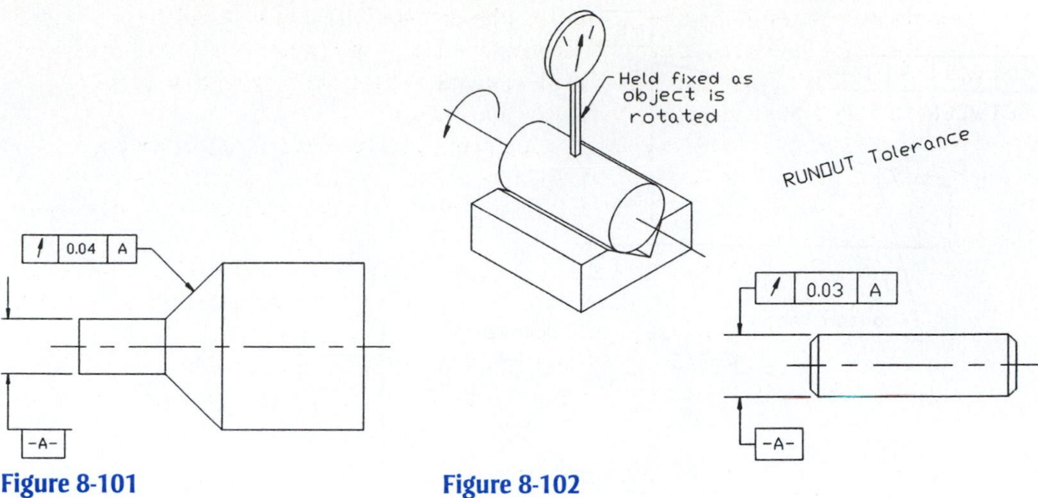

Figure 8-101

Figure 8-102

Figure 8-103 shows a circular runout tolerance that references two datums. The two datums serve as one datum. The object can then be rotated about both datums simultaneously as the runout tolerances are checked.

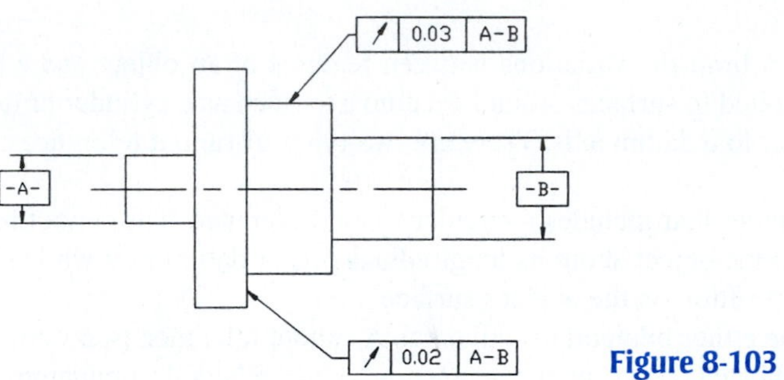

Figure 8-103

Positional Tolerances

As defined earlier, *positional tolerances* are used to locate and tolerance holes. Positional tolerances create a circular tolerance zone for hole center point locations, in contrast with the rectangular tolerance zone created by linear coordinate dimensions. See Figure 8-104. The circular tolerance zone allows for an increase in acceptable tolerance variation without compromising the design integrity of the object. Note how some of the possible hole center points fall in an area outside the rectangular tolerance zone but are still within the circular tolerance zone. If the hole had been located using linear coordinate dimensions, center points located beyond the rectangular tolerance zone would

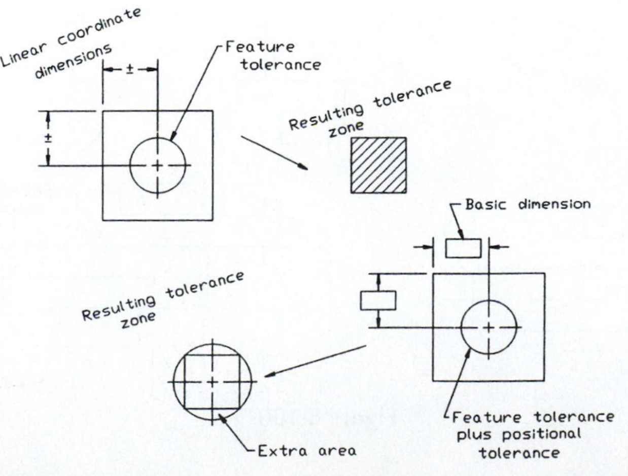

Figure 8-104

have been rejected as beyond tolerance, and yet holes produced using these locations would function correctly from a design standpoint. The center point locations would be acceptable if positional tolerances had been specified. The finished hole is round, so a round tolerance zone is appropriate. The rectangular tolerance zone rejects some holes unnecessarily.

Holes are dimensioned and toleranced using geometric tolerances by a combination of locating dimensions, feature dimensions and tolerances, and positional tolerances. See Figure 8-105. The locating dimensions are enclosed in rectangular boxes and are called *basic dimensions.* Basic dimensions are assumed to be exact.

The feature tolerances for the hole are as presented earlier in the chapter. They can be presented using plus or minus or limit-type tolerances. In the example shown in Figure 8-105 the diameter of the hole is toleranced using a plus and minus 0.05 tolerance.

The basic locating dimensions of 45 and 50 are assumed to be exact. The tolerances that would normally accompany linear locational dimensions are replaced by the positional tolerance.

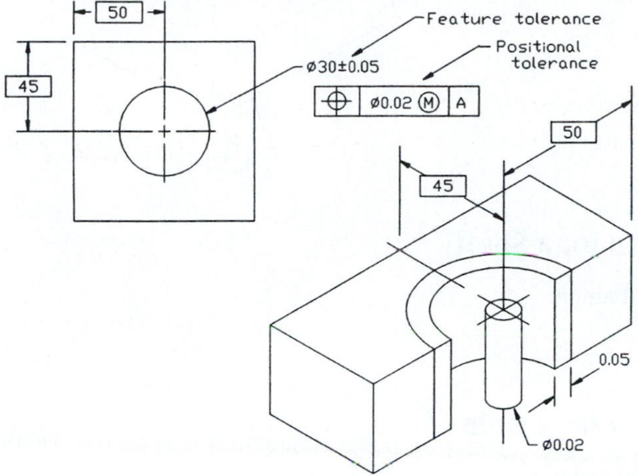

Figure 8-105

The positional tolerance also specifies that the tolerance be applied at the centerline at maximum material condition. The resulting tolerance zones are as shown in Figure 8-105.

Figure 8-106 shows an object containing two holes that are dimensioned and toleranced using positional tolerances. There are two consecutive horizontal basic dimensions. Because basic dimensions are exact, they do not have tolerances that accumulate; that is, there is no tolerance buildup.

Geometric positional tolerances must include basic dimensions.

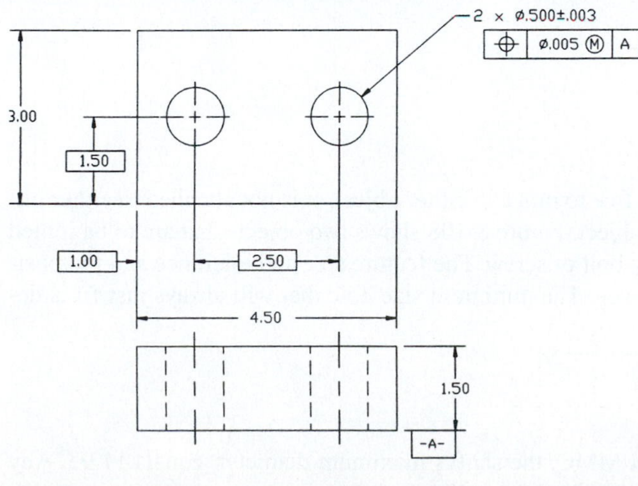

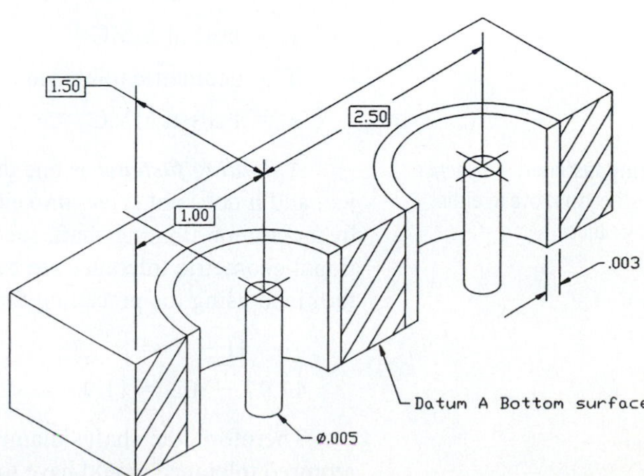

Figure 8-106

Virtual Condition

virtual condition: The combination of a feature's maximum material condition and its geometric tolerance.

Virtual condition is a combination of a feature's MMC and its geometric tolerance. For external features (shafts) it is the MMC plus the geometric tolerance; for internal features (holes) it is the MMC minus the geometric tolerance.

The following calculations are based on the dimensions shown in Figure 8-107.

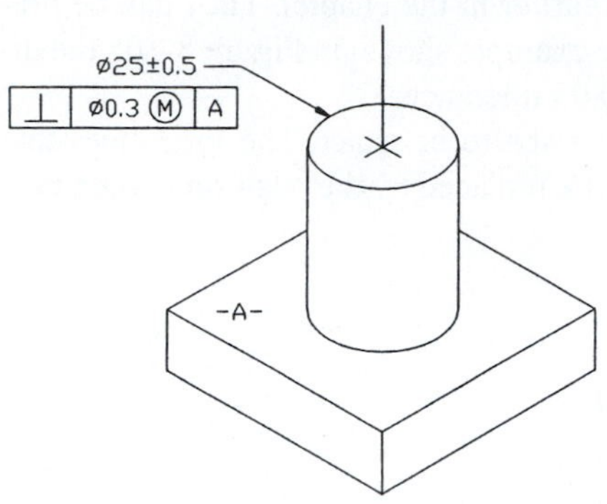

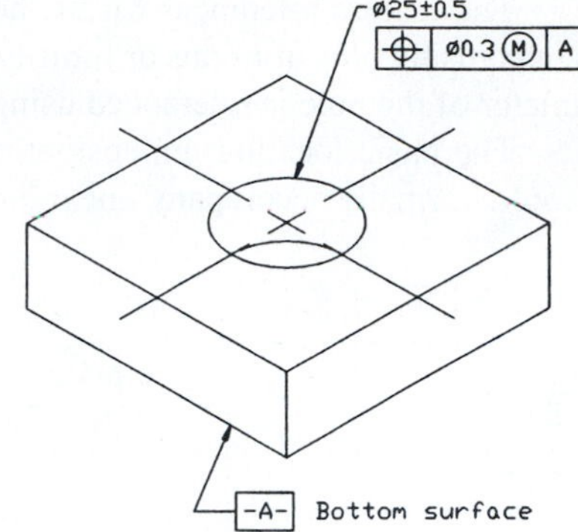

Figure 8-107

Calculating the Virtual Condition for a Shaft

25.5 MMC for shaft — maximum diameter
+0.3 Geometric tolerance
25.8 Virtual condition

Calculating the Virtual Condition for a Hole

24.5 MMC for hole — minimum diameter
−0.3 Geometric tolerance
24.2 Virtual condition

Floating Fasteners

Positional tolerances are particularly helpful when dimensioning matching parts. Because basic locating dimensions are considered exact, the sizing of mating parts is dependent only on the hole and shaft's MMC and the geometric tolerance between them.

The relationship for floating fasteners and holes in objects may be expressed as a formula:

$$H - T = F$$

where:

H = hole at MMC

T = geometric tolerance

F = shaft at MMC

floating fastener: A fastener that is free to move in either mating object.

A ***floating fastener*** is one that is free to move in either object. It is not attached to either object and it does not screw into either object. Figure 8-108 shows two objects that are to be joined by a common floating shaft, such as a bolt or screw. The feature size and tolerance and the positional geometric tolerance are both given. The minimum size hole that will always just fit is determined using the preceding formula.

$$H - T = F$$

$$11.97 - .02 = 11.95$$

Therefore, the shaft's diameter at MMC, the shaft's maximum diameter, equals 11.95. Any required tolerance would have to be subtracted from this shaft size.

The .02 geometric tolerance is applied at the hole's MMC, so as the hole's size expands within its feature tolerance, the tolerance zone for the acceptable matching parts also expands.

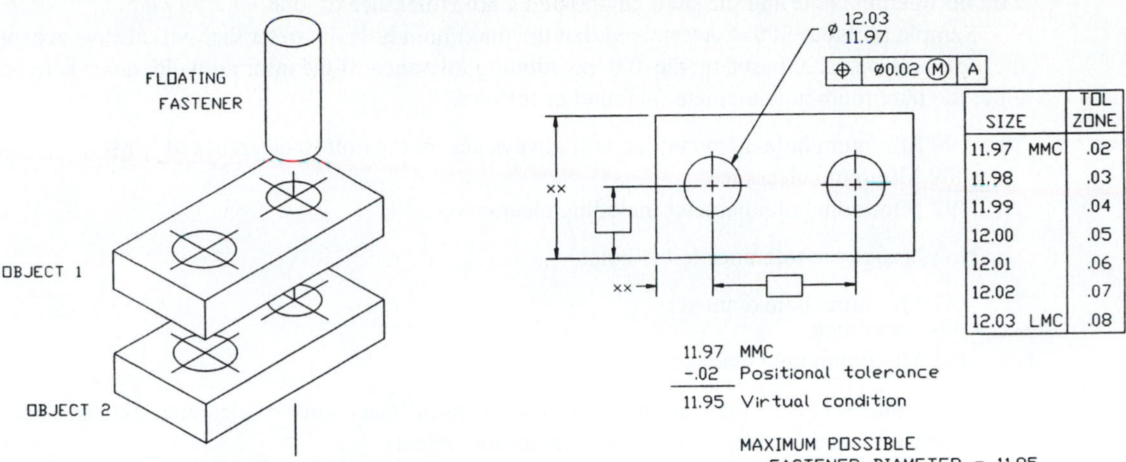

Figure 8-108

Sample Problem SP8-4

The situation presented in Figure 8-108 can be worked in reverse; that is, hole sizes can be derived from given shaft sizes.

The two objects shown in Figure 8-109 are to be joined by a .250-in. bolt. The parts are floating; that is, they are both free to move, and the fastener is not joined to either object. What is the MMC of the holes if the positional tolerance is to be .030?

A manufacturer's catalog specifies that the tolerance for .250 bolts is .2500 to .2600.

Rewriting the formula

$$H - T = F$$

to isolate the H yields

$$\begin{aligned} H &= F + T \\ &= .260 + .030 \\ &= .290 \end{aligned}$$

The .290 value represents the minimum hole diameter, MMC, for all four holes that will always accept the .250 bolt. Figure 8-110 shows the resulting drawing callout.

Any clearance requirements or tolerances for the hole would have to be added to the .290 value.

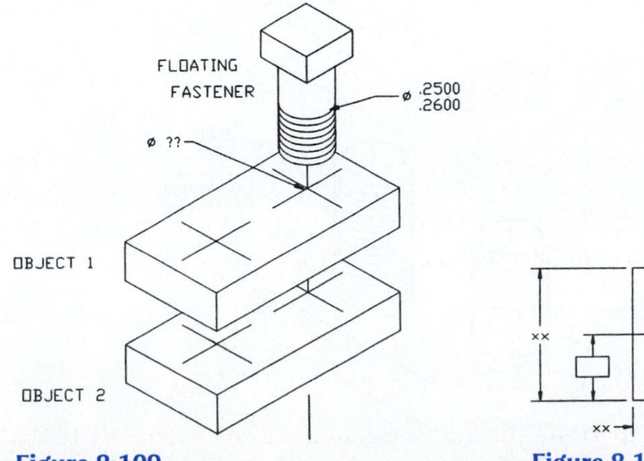

Figure 8-109

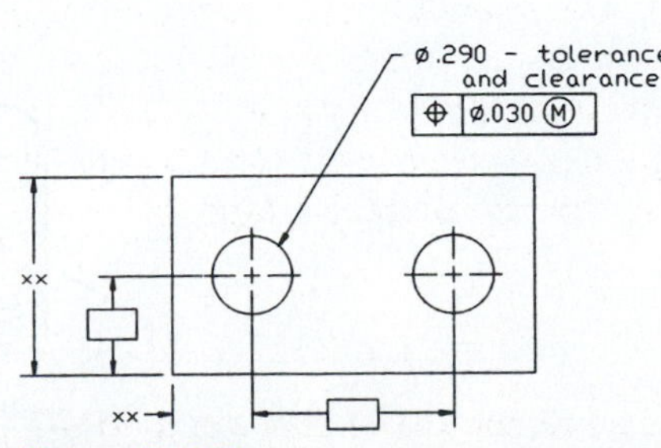

Figure 8-110

SAMPLE PROBLEM SP8-5

Repeat the problem presented in SP8-4 but be sure that there is always a minimum clearance of .002 between the hole and the shaft, and assign a hole tolerance of .008.

Sample problem SP8-4 determined that the maximum hole diameter that will always accept the .250 bolt was .290 based on the .030 positioning tolerance. If the minimum clearance is to be .002, the maximum hole diameter is found as follows:

.290 Minimum hole diameter that will always accept the bolt (0 clearance at MMC)
+.002 Minimum clearance
.292 Minimum hole diameter including clearance

Now, assign the tolerance to the hole:

.292 Minimum hole diameter
+.001 Tolerance
.293 Maximum hole diameter

See Figure 8-111 for the appropriate drawing callout. The choice of clearance size and hole tolerance varies with the design requirements for the objects.

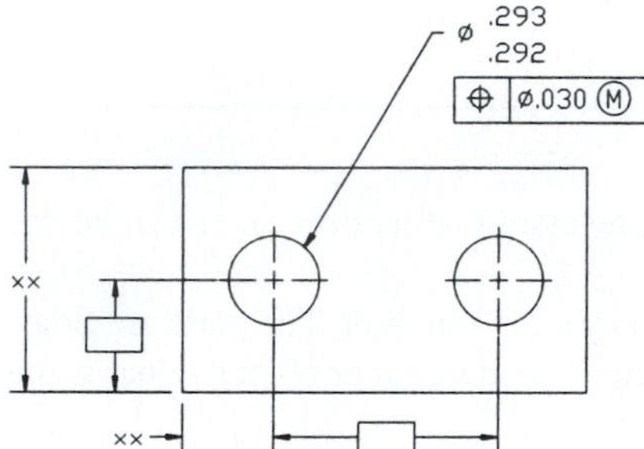

Figure 8-111

FIXED FASTENERS

fixed fastener: A fastener that is attached to one of the mating objects.

A ***fixed fastener*** is one that is attached to one of the mating objects. See Figure 8-112. Because the fastener is fixed to one of the objects, the geometric tolerance zone must be smaller than that used for floating fasteners. The fixed fastener cannot move without moving the object it is attached to. The relationship between fixed fasteners and holes in mating objects is defined by the formula

$$H - 2T = F$$

The tolerance zone is cut in half. This can be demonstrated by the objects shown in Figure 8-113. The same feature sizes that were used in Figure 8-113 are assigned, but in this

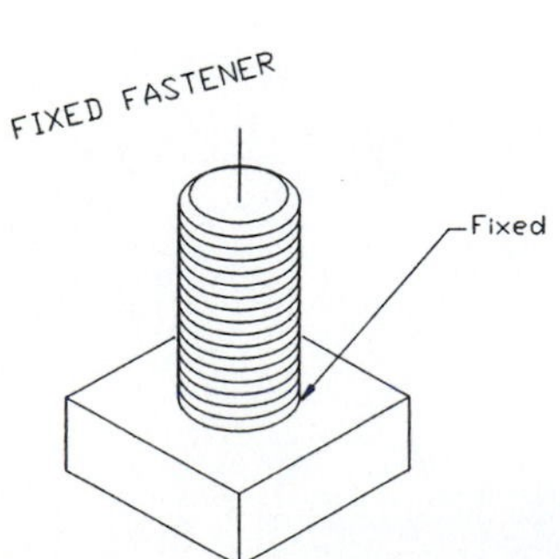

Figure 8-112

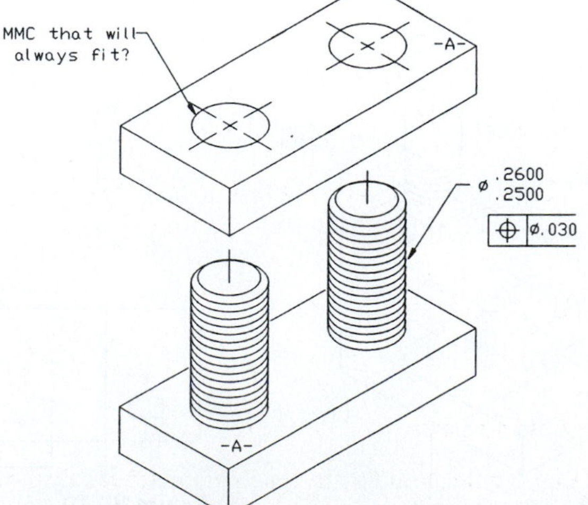

Figure 8-113

example the fasteners are fixed. Solving for the geometric tolerance yields a value as follows:

$$H - F = 2T$$
$$11.97 - 11.95 = 2T$$
$$.02 = 2T$$
$$.01 = T$$

The resulting positional tolerance is half that obtained for floating fasteners.

Sample Problem SP8-6

This problem is similar to sample problem SP8-4, but the given conditions are applied to fixed fasteners rather than floating fasteners. Compare the resulting shaft diameters for the two problems. See Figure 8-114.

A. What is the minimum diameter hole that will always accept the fixed fasteners?
B. If the minimum clearance is .005 and the hole is to have a tolerance of .002, what are the maximum and minimum diameters of the hole?

$$H - 2T = F$$
$$H = F + 2T$$
$$= .260 + 2(.030)$$
$$= .260 + .060$$
$= .320$ Minimum diameter that will always accept the fastener

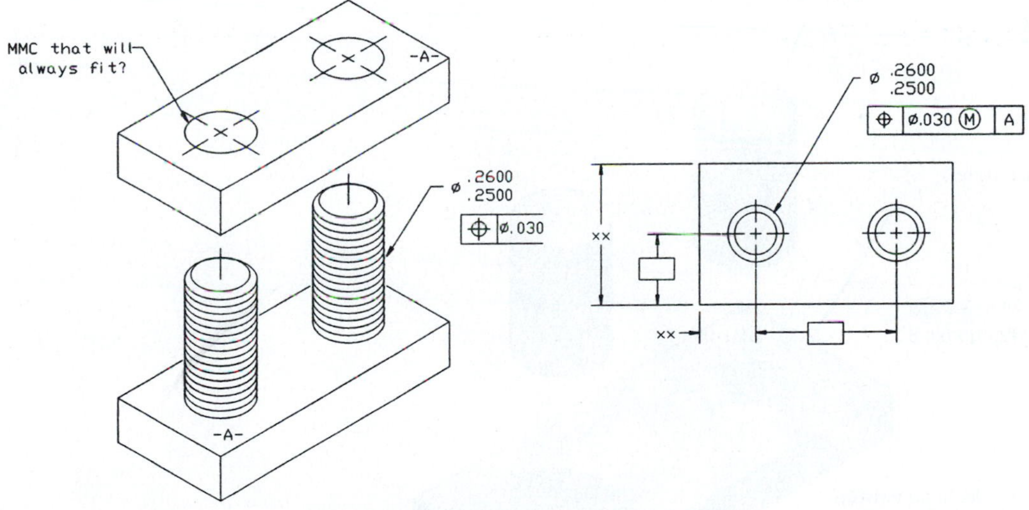

Figure 8-114

If the minimum clearance is .005 and the hole tolerance is .002,

.320 Virtual condition
+.005 Clearance
.325 Minimum hole diameter

.325 Minimum hole diameter
+.002 Tolerance
.327 Maximum hole diameter

The maximum and minimum values for the hole's diameter can then be added to the drawing of the object that fits over the fixed fasteners. See Figure 8-115.

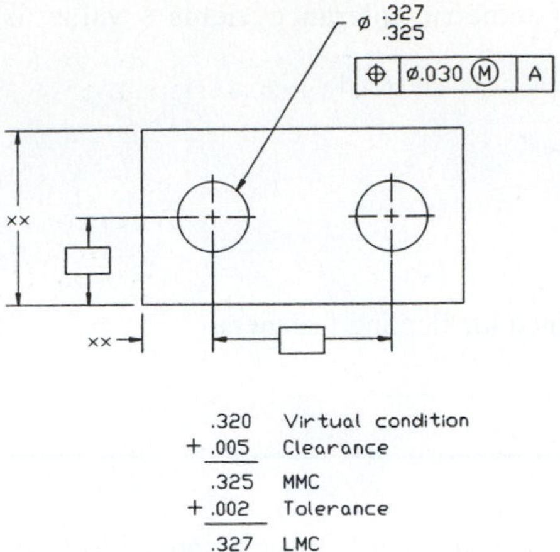

Figure 8-115

Design Problems

This problem was originally done on p. 375 using rectangular tolerances. It is done in this section using positional geometric tolerances so that the two systems can be compared. It is suggested that the previous problem be reviewed before reading this section.

Figure 8-116 shows top and bottom parts that are to be joined in the floating condition. A nominal distance of 50 between hole centers and 20 for the holes has been assigned. In the

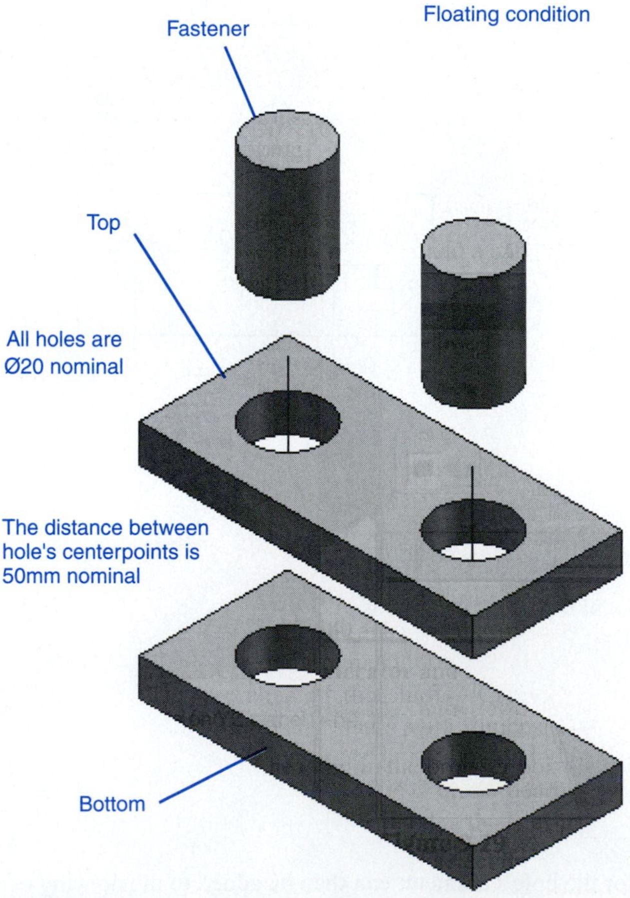

Figure 8-116

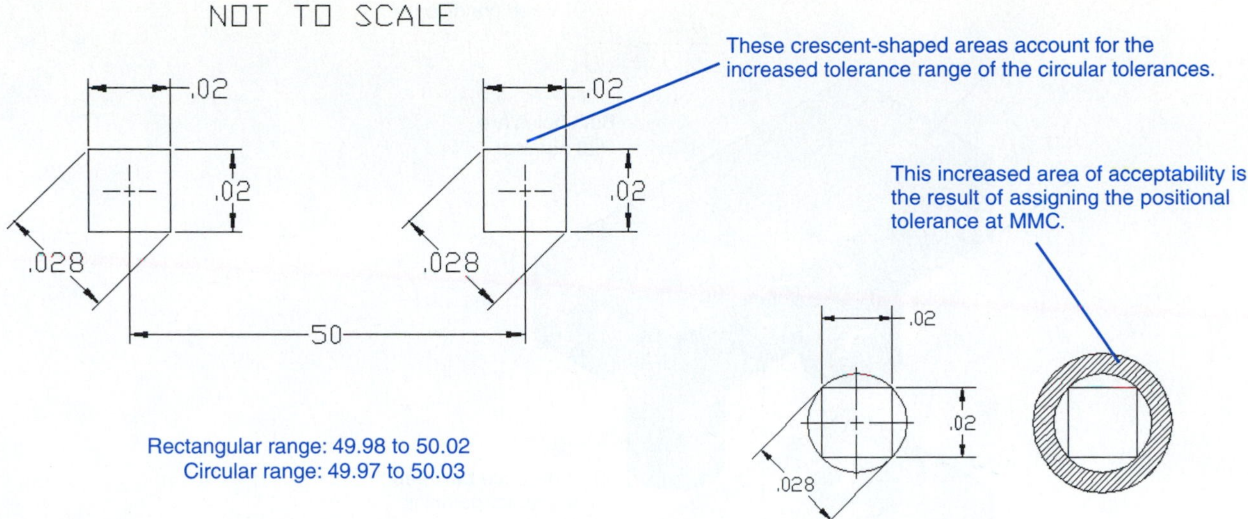

Figure 8-117

previous solution a rectangular tolerance of ±.01 was selected, and there was a minimum hole diameter of 20.00. Figure 8-117 shows the resulting tolerance zones.

The diagonal distance across the rectangular tolerance zone is .028 and was rounded off to .03 to yield a maximum possible fastener diameter of 19.97. If the same .03 value is used to calculate the fastener diameter using positional tolerance, the results are as follows:

$$H - T = F$$
$$20.00 - .03 = 19.97$$

The results seem to be the same, but because of the circular shape of the positional tolerance zone, the manufactured results are not the same. The minimum distance between the inside edges of the rectangular zones is 49.98, or .01 from the center point of each hole. The minimum distance from the innermost points of the circular tolerance zones is 49.97, or .015 (half the rounded-off .03 value) from the center point of each hole. The same value difference also occurs for the maximum distance between center points, where 50.02 is the maximum distance for the rectangular tolerances, and 50.03 is the maximum distance for the circular tolerances. The size of the circular tolerance zone is larger because the hole tolerances are assigned at MMC. Figure 8-117 shows a comparison between the tolerance zones, and Figure 8-118 shows how the positional tolerances would be presented on a drawing of either the top or bottom part.

Figure 8-119 shows the same top and bottom parts joined together in the fixed condition. The initial nominal values are the same. If the same .03 diagonal value is assigned as a positional tolerance, the results are as follows:

$$H - T = F$$
$$20.00 - .06 = 19.94$$

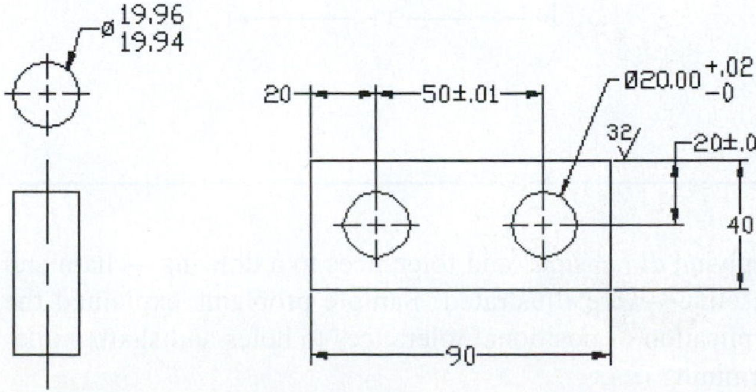

Figure 8-118

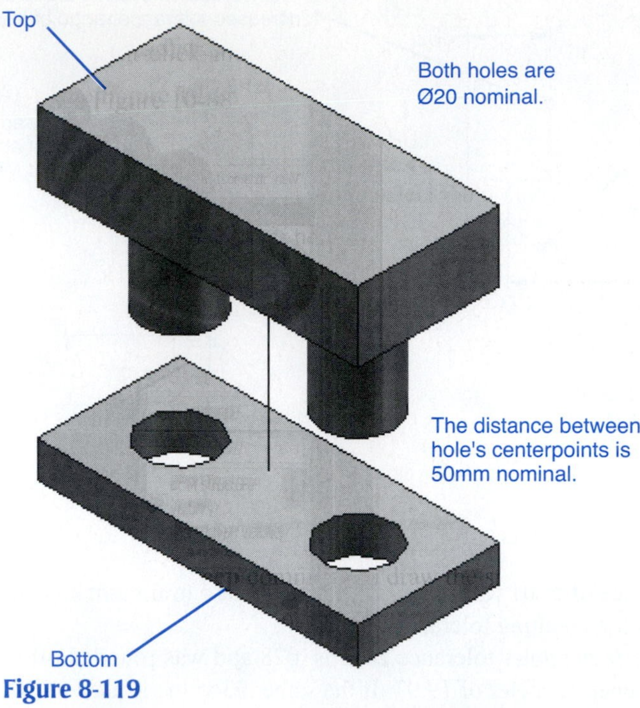

Figure 8-119

These results appear to be the same as those generated by the rectangular tolerance zone, but the circular tolerance zone allows a greater variance in acceptable manufactured parts. Figure 8-120 shows how the positional tolerance would be presented on a drawing.

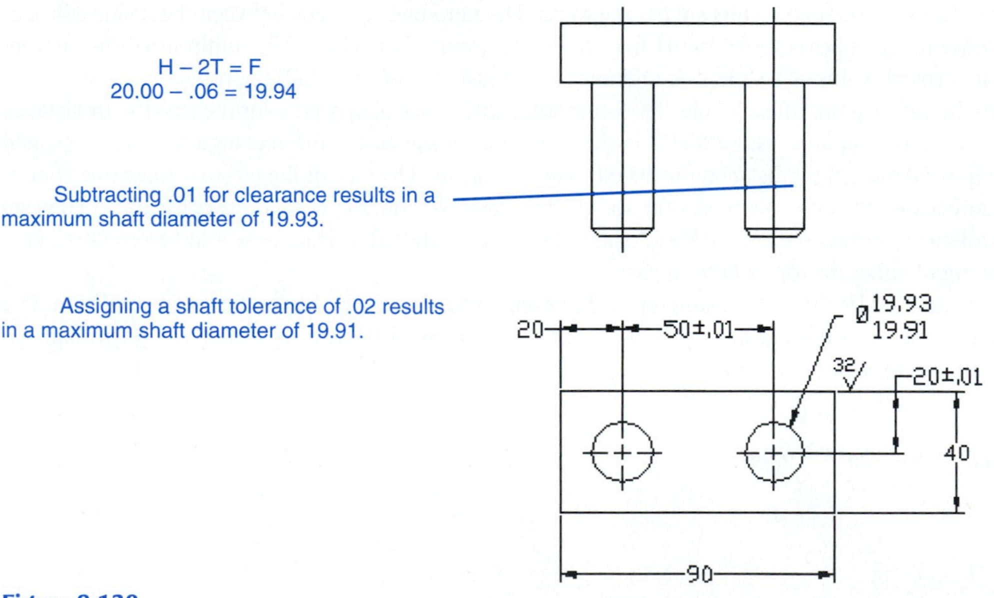

Figure 8-120

SUMMARY

Tolerances define the manufacturing limits for dimensions. This chapter defined the various tolerance conventions and illustrated how to create plus and minus tolerances and limit tolerances, including angular tolerances. The two systems for applying dimensions and tolerances to a drawing—chain and baseline—were illustrated. Sample problems explained the application of positional tolerances to holes and shafts in determining fits.

Surface finishes and the use of surface control symbols were also discussed and illustrated.

Examples were given of the various forms of geometric tolerancing—defining surfaces in terms of their flatness or roundness or in terms of how perpendicular or parallel they are to other surfaces. The four classifications of tolerances of form—flatness, straightness, roundness, and cylindricity—were also illustrated. Sample problems involving both fixed and floating fasteners demonstrated the use of tolerances.

CHAPTER PROJECTS

Project 8-1:

Draw a model of the objects shown in Figures P8-1A through P8-1D using the given dimensions and tolerances. Create a drawing layout with a view of the model as shown. Add the specified dimensions and tolerances.

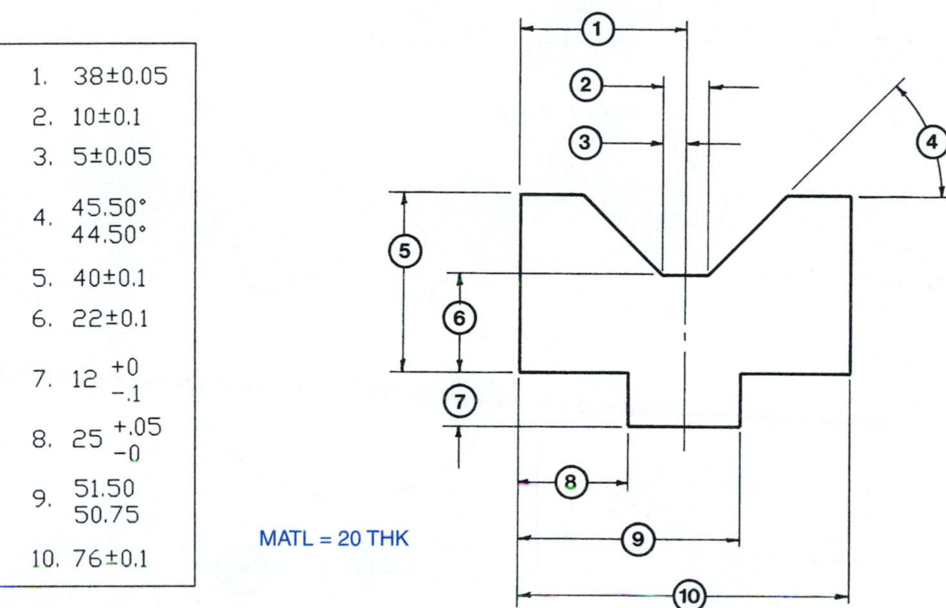

Figure P8-1A MILLIMETERS

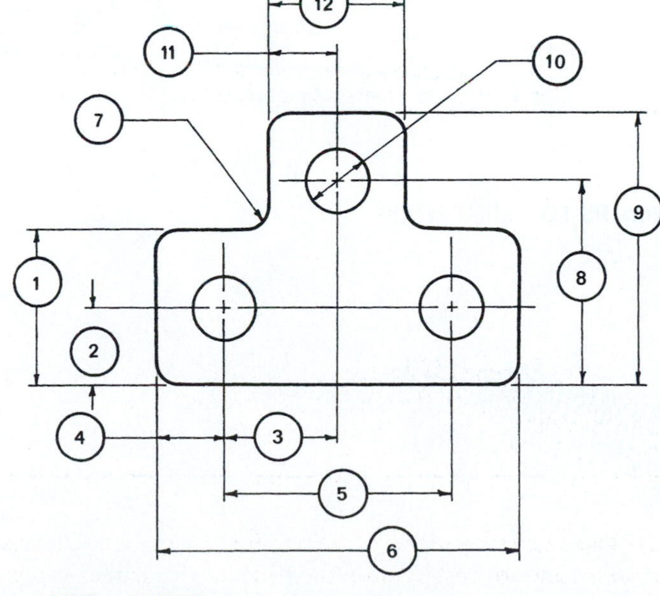

Figure P8-1B MILLIMETERS

1. 3.00±.01
2. 1.56±.01
3. 46.50° / 45.50°
4. .750±.005
5. 2.75 / 2.70
6. 3.625±.010
7. 45°±.5°
8. 2.250±.005

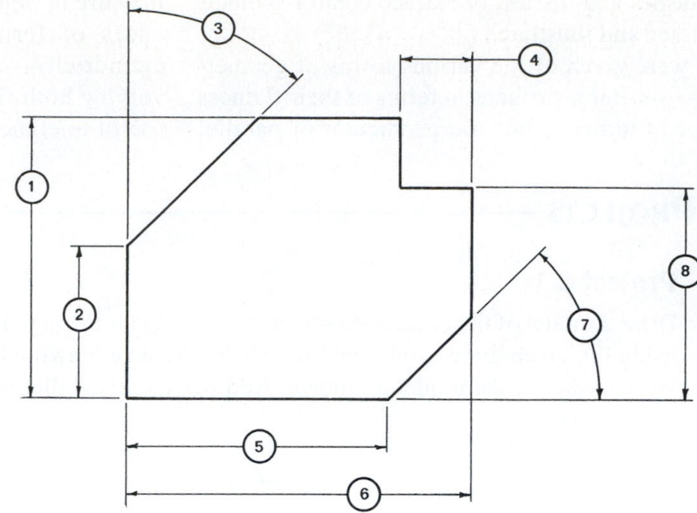

MATL = .75 THK

Figure P8-1C INCHES

1. 50 $^{+.2}_{0}$
2. R45±.1 - 2 PLACES
3. 63.5 $^{0}_{-.2}$
4. 76±.1
5. 38±.1
6. Ø12.00 $^{+.05}_{0}$ - 3 HOLES
7. 30±.03
8. 30±.03
9. 100 $^{+.4}_{0}$

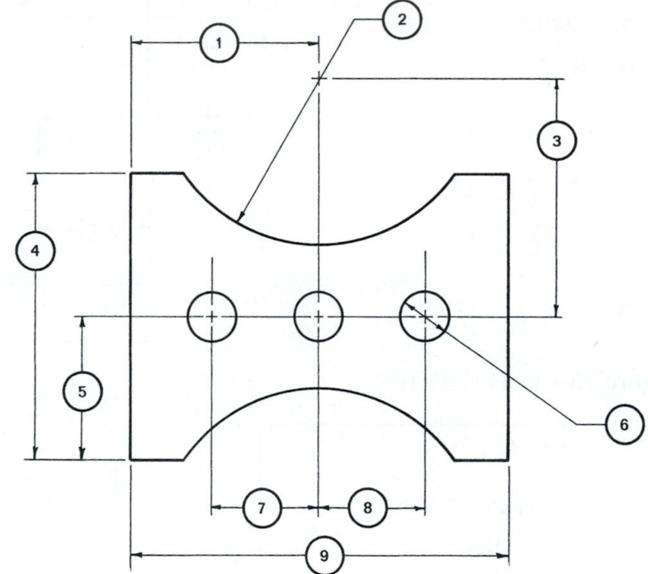

MATL = 10 THK

Figure P8-1D MILLIMETERS

Project 8-2:

Redraw the following object, including the given dimensions and tolerances. Calculate and list the maximum and minimum distances for surface A.

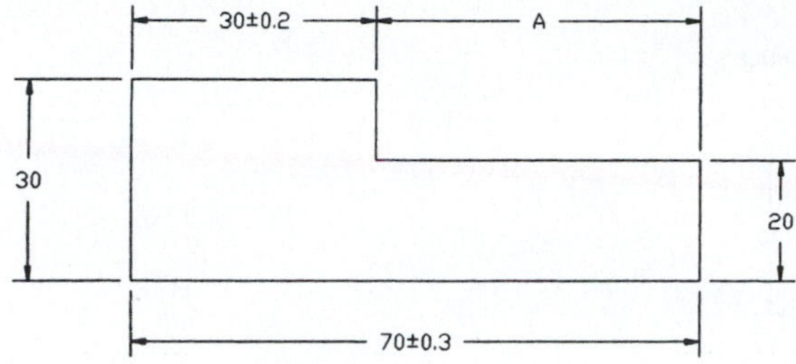

MATL = 25 THK

Figure P8-2 MILLIMETERS

Project 8-3:

A. Redraw the following object, including the dimensions and tolerances. Calculate and list the maximum and minimum distances for surface A.
B. Redraw the given object and dimension it using baseline dimensions. Calculate and list the maximum and minimum distances for surface A.

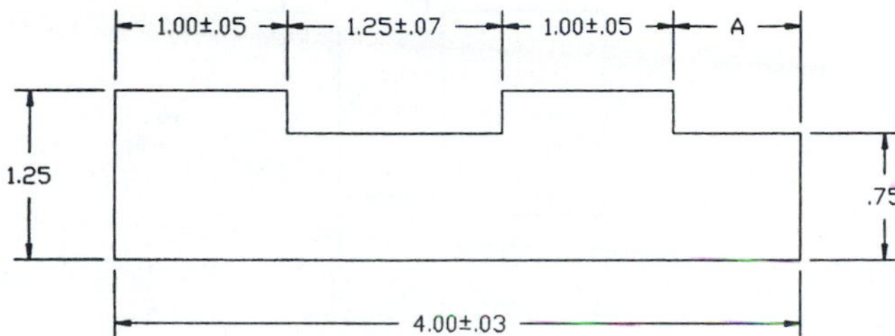

MATL = 1.25 THK

Figure P8-3 INCHES

Project 8-4:

Redraw the following object, including the dimensions and tolerances. Calculate and list the maximum and minimum distances for surfaces D and E.

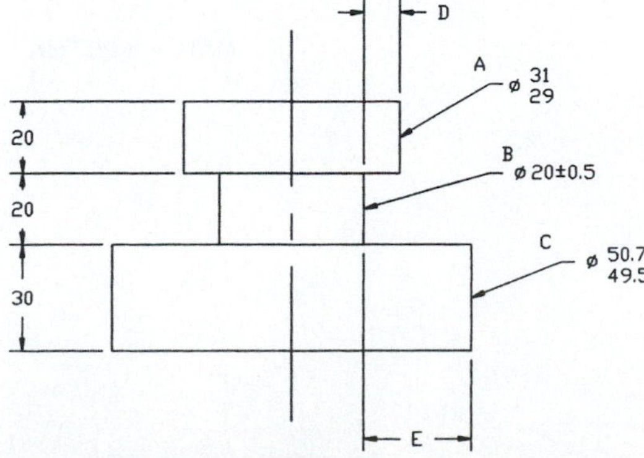

Figure P8-4 MILLIMETERS

Project 8-5:

Dimension the following object twice, once using chain dimensions and once using baseline dimensions. Calculate and list the maximum and minimum distances for surface D for both chain and baseline dimensions. Compare the results.

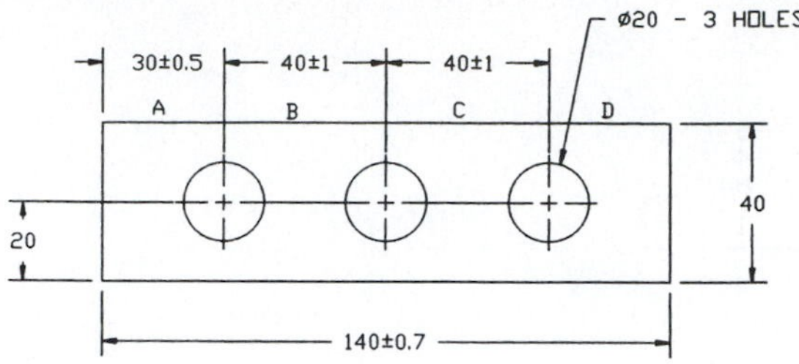

MATL = 20 THK

Figure P8-5 MILLIMETERS

Project 8-6:

Redraw the following shapes, including the dimensions and tolerances. Also list the required minimum and maximum values for the specified distances.

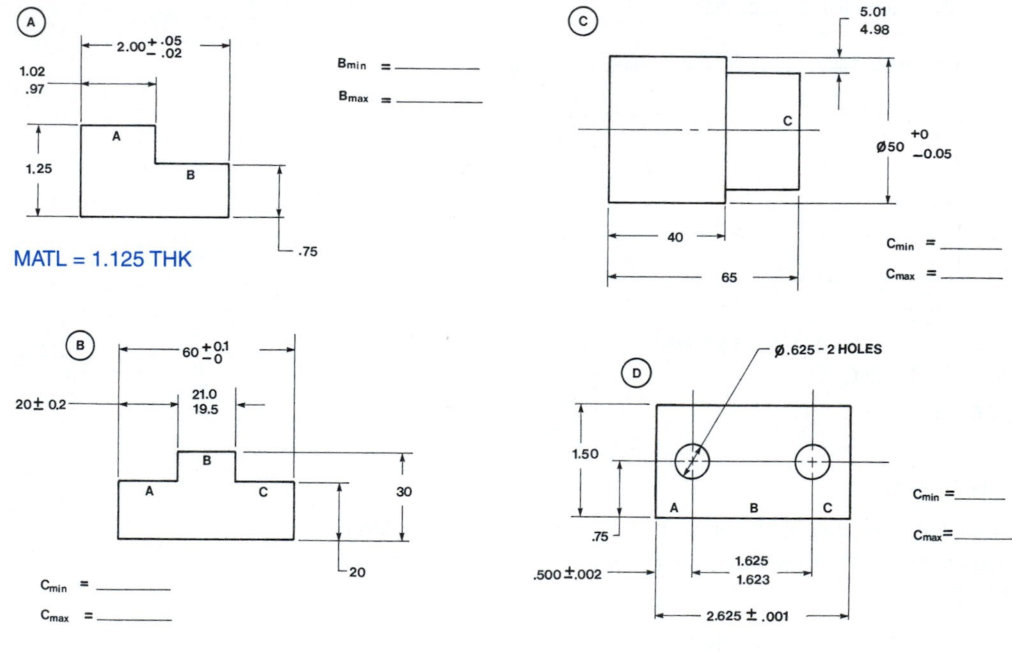

MATL = 45 THK

MATL = .625 THK

Figure P8-6 INCHES

Project 8-7:

Redraw and complete the following inspection report. Under the Results column classify each "AS MEASURED" value as OK if the value is within the stated tolerances, REWORK if the value indicates that the measured value is beyond the stated tolerance but can be reworked to bring it into the acceptable range, or SCRAP if the value is not within the tolerance range and cannot be reworked to make it acceptable.

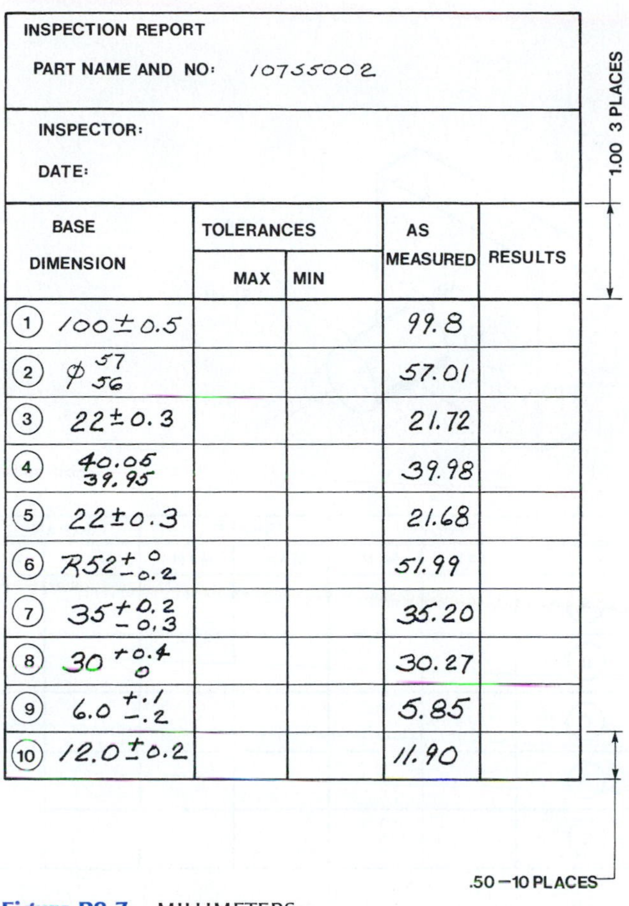

INSPECTION REPORT

PART NAME AND NO: 10755002

INSPECTOR:

DATE:

BASE DIMENSION	TOLERANCES		AS MEASURED	RESULTS
	MAX	MIN		
(1) 100 ± 0.5			99.8	
(2) $\phi\,^{57}_{56}$			57.01	
(3) 22 ± 0.3			21.72	
(4) $^{40.05}_{39.95}$			39.98	
(5) 22 ± 0.3			21.68	
(6) $R52^{+0}_{-0.2}$			51.99	
(7) $35^{+0.2}_{-0.3}$			35.20	
(8) $30^{+0.4}_{0}$			30.27	
(9) $6.0^{+.1}_{-.2}$			5.85	
(10) 12.0 ± 0.2			11.90	

Figure P8-7 MILLIMETERS

Project 8-8:

Redraw the following charts and complete them based on the following information. All values are in millimeters.

A. Nominal = 16, Fit = H8/d8
B. Nominal = 30, Fit = H11/c11
C. Nominal = 22, Fit = H7/g6
D. Nominal = 10, Fit = C11/h11
E. Nominal = 25, Fit = F8/h7
F. Nominal = 12, Fit = H7/k6
G. Nominal = 3, Fit = H7/p6
H. Nominal = 18, Fit = H7/s6
I. Nominal = 27, Fit = H7/u6
J. Nominal = 30, Fit = N7/h6

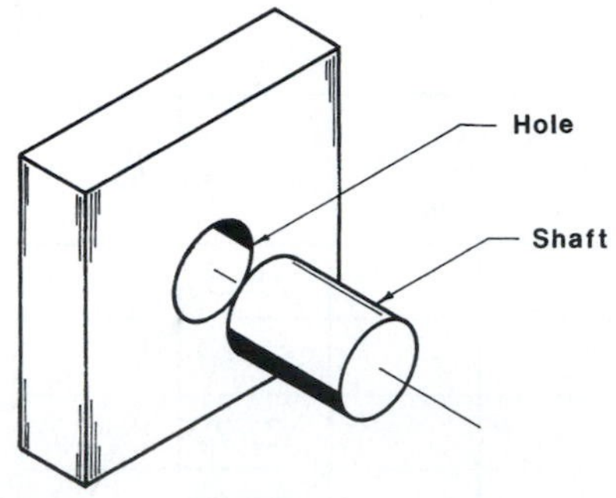

half space

3.75
6 equal
spaces

NOMINAL	HOLE		SHAFT		CLEARANCE	
	MAX	MIN	MAX	MIN	MAX	MIN
A						
B						
C						
D						
E						

1.5 6.0 – 6 equal spaces

NOMINAL	HOLE		SHAFT		INTERFERENCE	
	MAX	MIN	MAX	MIN	MAX	MIN
F						
G						
H						
I						
J						

Use the same dimensions given above

Figure P8-8 MILLIMETERS

Project 8-9:

Redraw the following charts and complete them based on the following information. All values are in inches.

A. Nominal = 0.25, Fit = Class LC5, H7/g6
B. Nominal = 1.00, Fit = Class LC7, H10/e9
C. Nominal = 1.50, Fit = Class LC9, F11/h11
D. Nominal = 0.75, Fit = Class RC3, H7/f6
E. Nominal = 1.75, Fit = Class RC6, H9/e8
F. Nominal = .500, Fit = Class LT2, H8/js7
G. Nominal = 1.25, Fit = Class LT5, H7/n6
H. Nominal = 1.38, Fit = Class LN3, J7/h6
I. Nominal = 1.625, Fit = Class FN, H7/s6
J. Nominal = 2.00, Fit = Class FN4, H7/u6

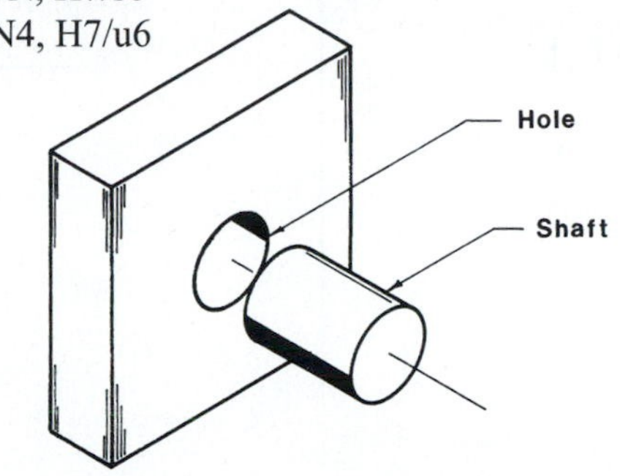

half space

3.75
6 equal
spaces

NOMINAL	HOLE		SHAFT		CLEARANCE	
	MAX	MIN	MAX	MIN	MAX	MIN
A						
B						
C						
D						
E						

1.5 — 6.0 – 6 equal spaces

NOMINAL	HOLE		SHAFT		INTERFERENCE	
	MAX	MIN	MAX	MIN	MAX	MIN
F						
G						
H						
I						
J						

Use the same dimensions given above

Figure P8-9 INCHES

Project 8-10:

Draw the chart shown and add the appropriate values based on the dimensions and tolerances given in Figures P8-10A through P8-10D.

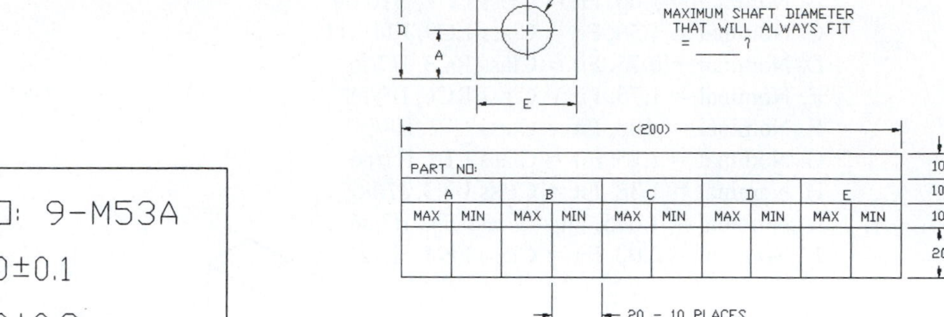

PART NO: 9-M53A

A. 20±0.1

B. 30±0.2

C. Ø20±0.05

D. 40

E. 60

Figure P8-10A MILLIMETERS

PART NO: 9-M53B

A. 32.02 / 31.97

B. 47.52 / 47.50

C. Ø18 $^{+0.05}_{\ \ 0}$

D. 64±0.05

E. 100±0.05

Figure P8-10B MILLIMETERS

PART NO: 9-E47A

A. 2.00±.02

B. 1.75±.03

C. Ø.750±.005

D. 4.00±.05

E. 3.50±.05

Figure P8-10C MILLIMETERS

PART NO: 9-E47B

A. 18 $^{+0}_{-0.02}$

B. 26 $^{+0}_{-0.04}$

C. Ø 24.03 / 23.99

D. 52±0.04

E. 36±0.02

Figure P8-10D MILLIMETERS

Project 8-11:

Prepare front and top views of parts 4A and 4B based on the given dimensions. Add tolerances to produce the stated clearances.

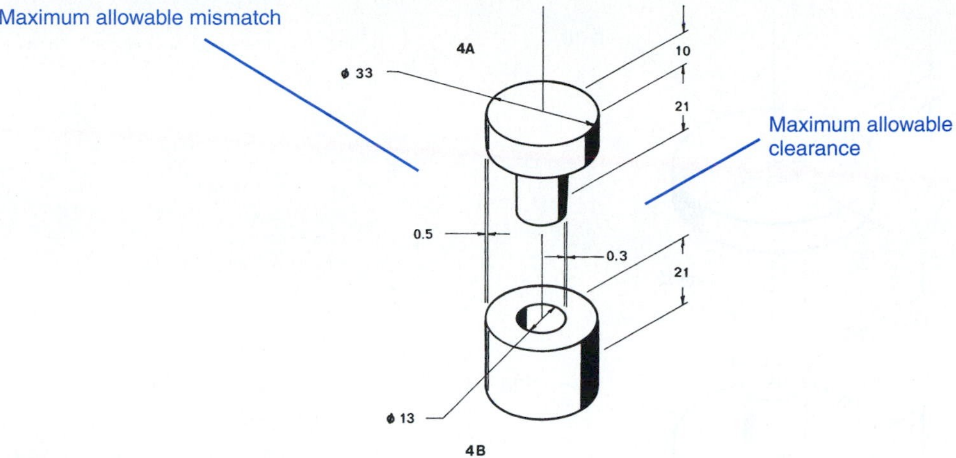

Figure P8-11 MILLIMETERS

Project 8-12:

Redraw parts A and B and dimensions and tolerances to meet the "UPON ASSEMBLY" requirements.

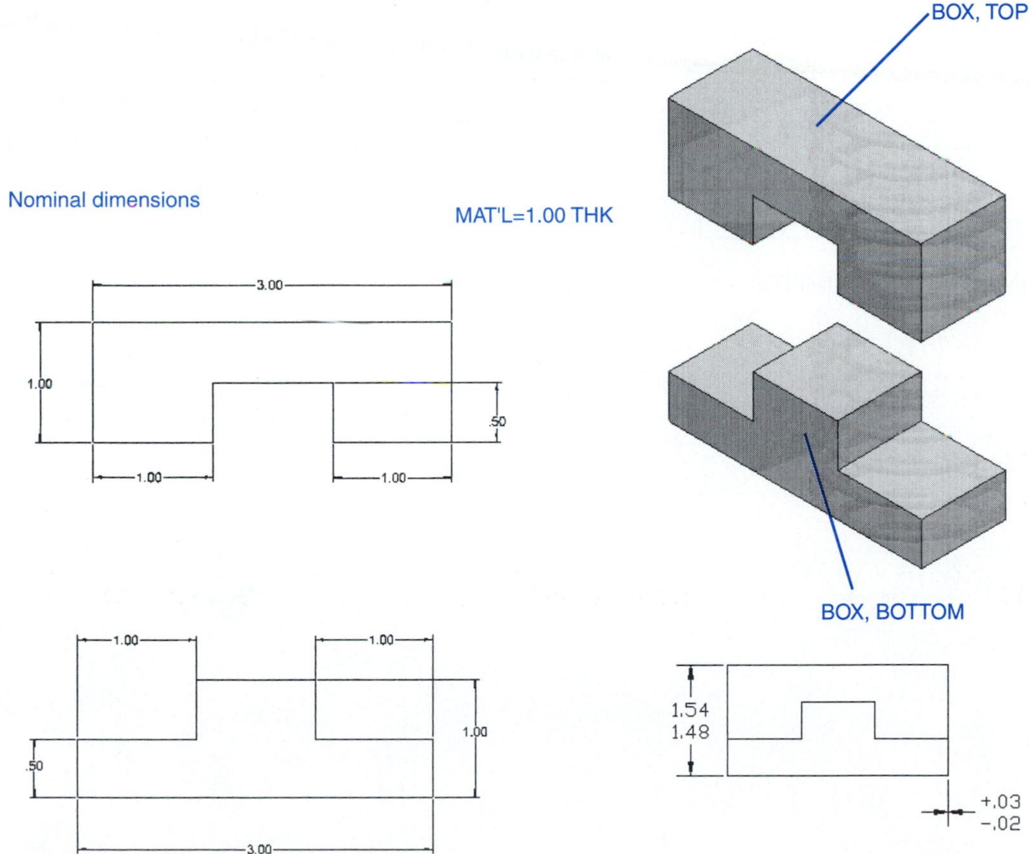

Figure P8-12 INCHES

Project 8-13:

Draw a front and top view of both given objects. Add dimensions and tolerances to meet the "FINAL CONDITION" requirements.

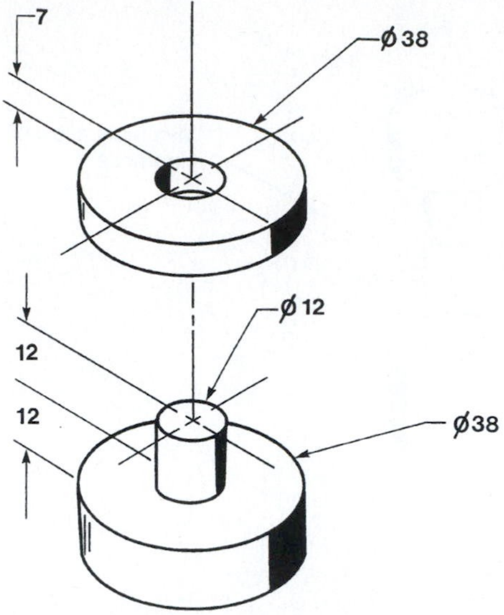

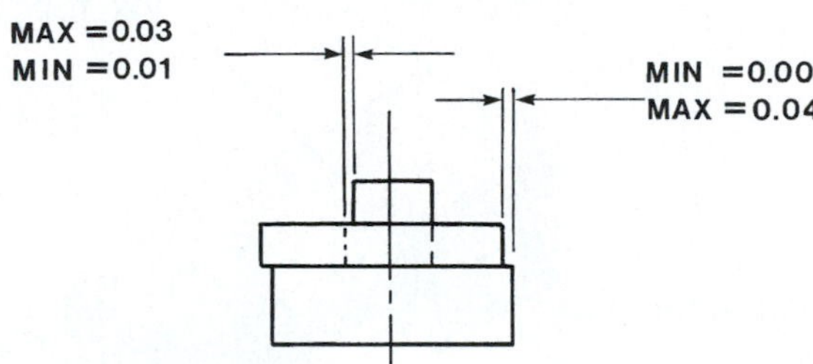

Figure P8-13 MILLIMETERS

Project 8-14:

Given the following nominal sizes, dimension and tolerance parts AM311 and AM312 so that they always fit together regardless of orientation. Further, dimension the overall lengths of each part so that in the assembled condition they will always pass through a clearance gauge with an opening of 80.00±0.02.

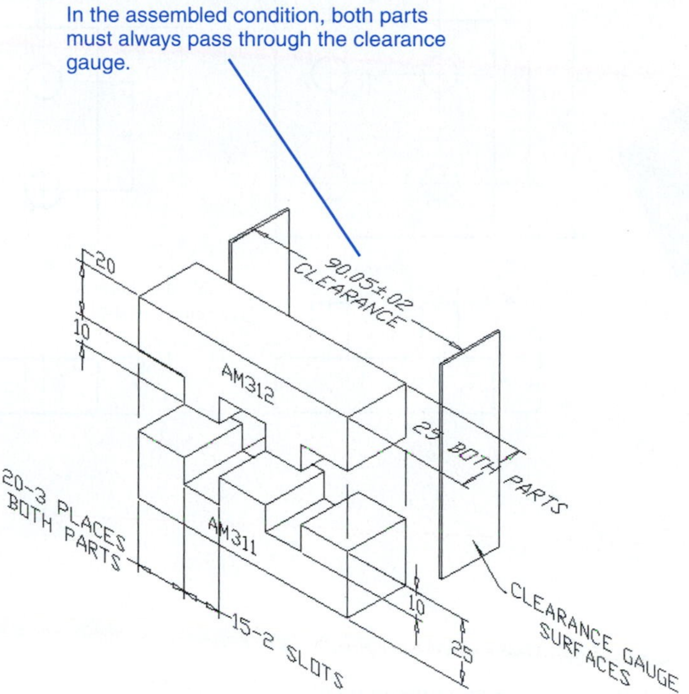

Figure P8-14 MILLIMETERS

Project 8-15:

Given the following rail assembly, add dimensions and tolerances so that the parts always fit together as shown in the assembled position.

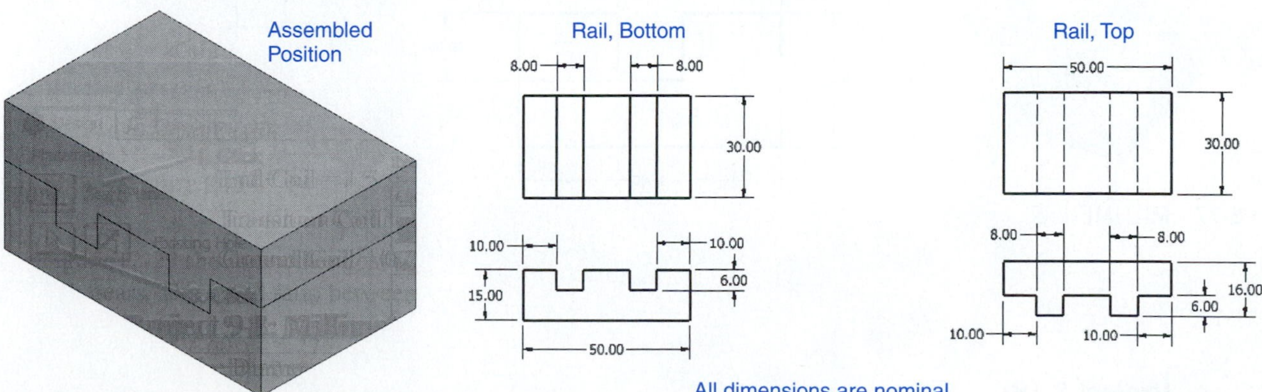

Figure P8-15 MILLIMETERS

Project 8-16:

Given the following peg assembly, add dimensions and tolerances so that the parts always fit together as shown in the assembled position.

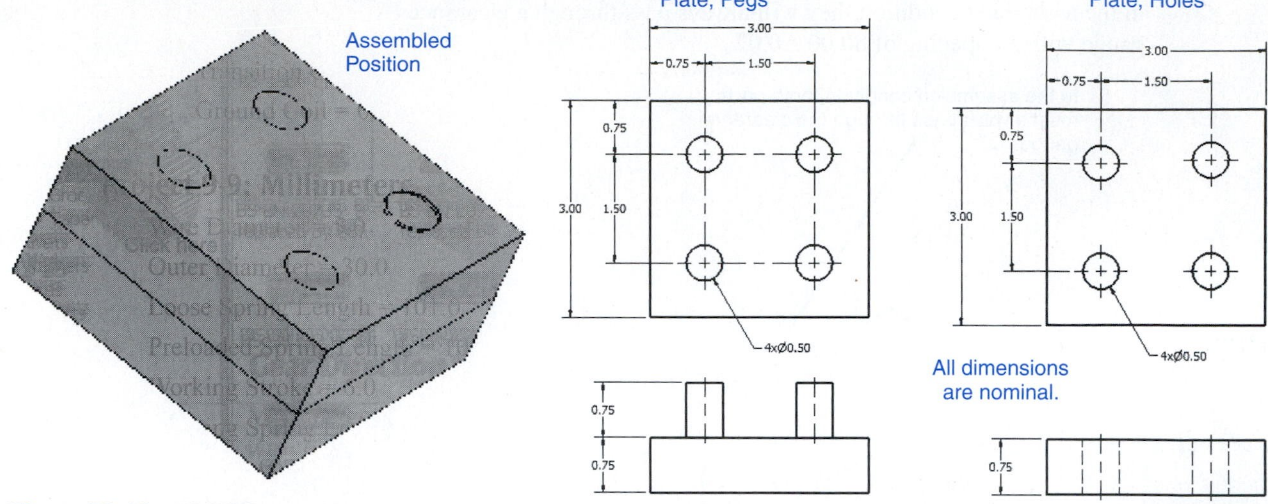

Figure P8-16 INCHES

Project 8-17:

Given the following collar assembly, add dimensions and tolerances so that the parts always fit together as shown in the assembled position.

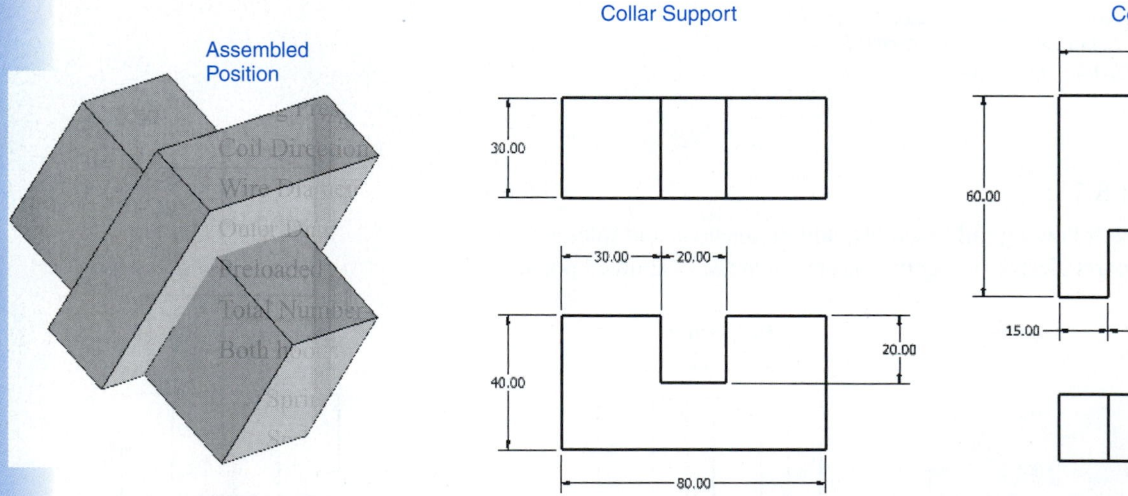

Figure P8-17 MILLIMETERS

Project 8-18:

Given the following vee-block assembly, add dimensions and tolerances so that the parts always fit together as shown in the assembled position. The total height of the assembled blocks must be between 4.45 and 4.55 in.

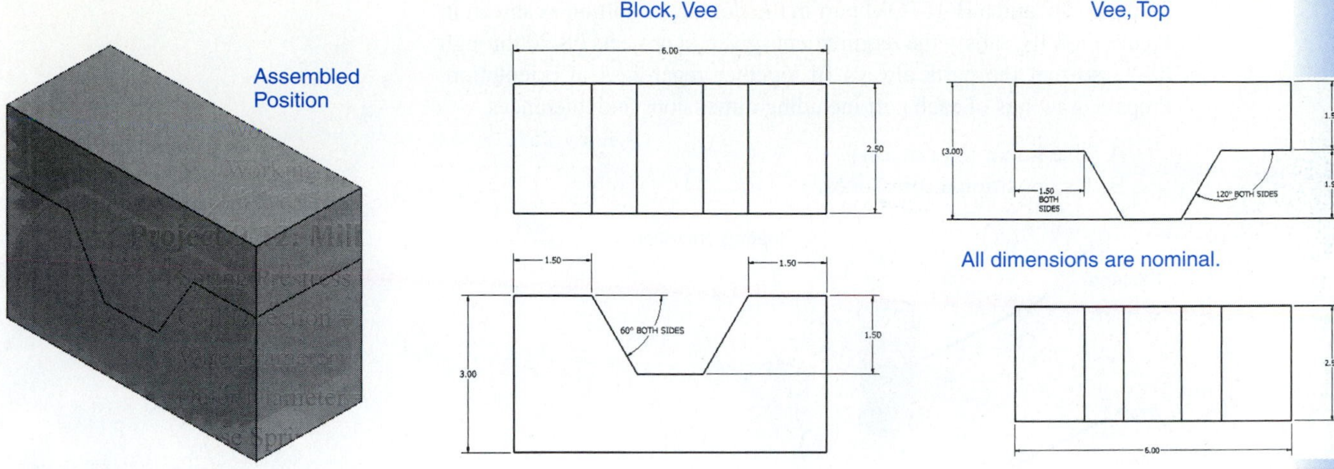

Figure P8-18 INCHES

Project 8-19:

Design a bracket that will support the three Ø100 wheels shown. The wheels will utilize three Ø5.00 ± 0.01 shafts attached to the bracket. The bottom of the bracket must have a minimum of 10 mm from the ground. The wall thickness of the bracket must always be at least 5 mm, and the minimum bracket opening must be at least 15 mm.

1. Prepare a front and a side view of the bracket.
2. Draw the wheels in their relative positions using phantom lines.
3. Add all appropriate dimensions and tolerances.

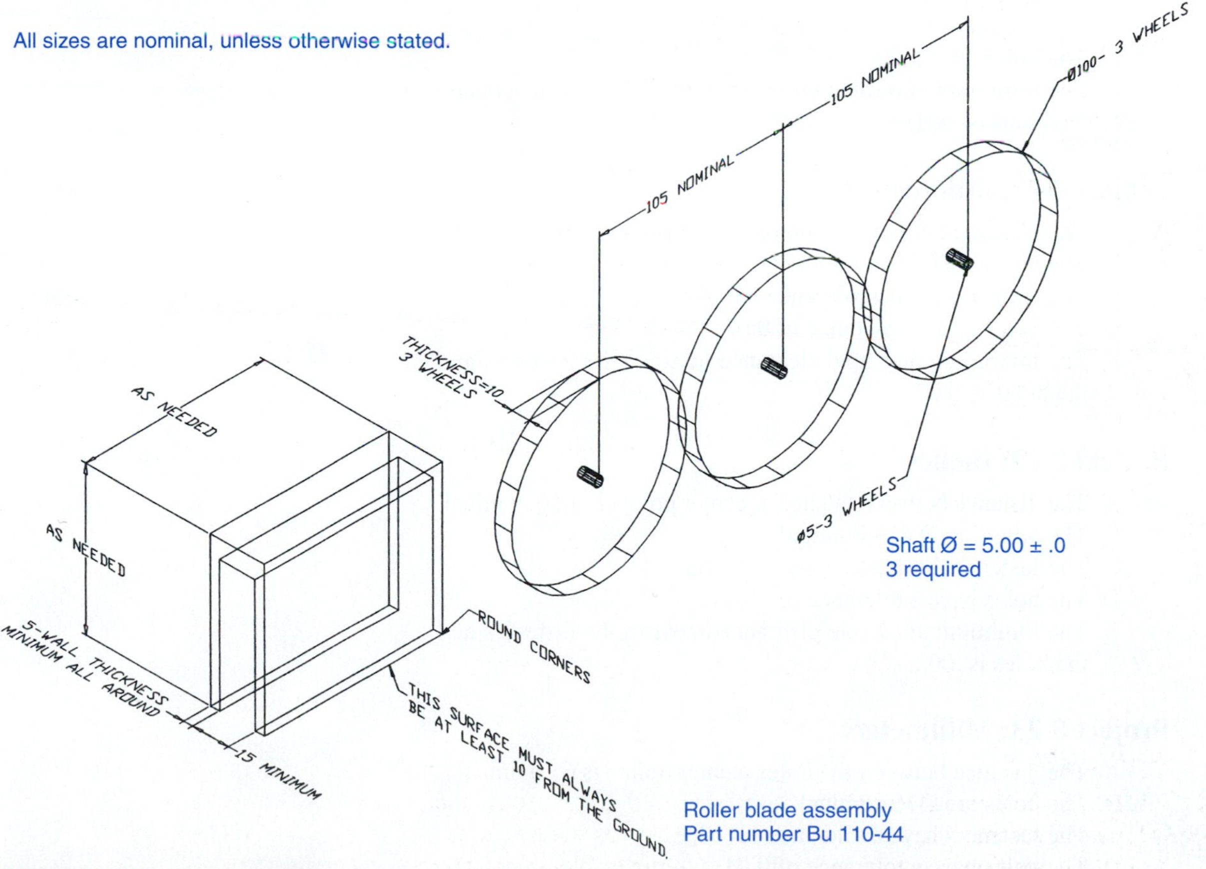

Figure P8-19A MILLIMETERS

Given a TOP and a BOTTOM part in the floating condition as shown in Figure P8-19B, satisfy the requirements given in projects P8-20 through P8-23 so that the parts always fit together regardless of orientation. Prepare drawings of each part including dimensions and tolerances.

A. Use linear tolerances.
B. Use positional tolerances.

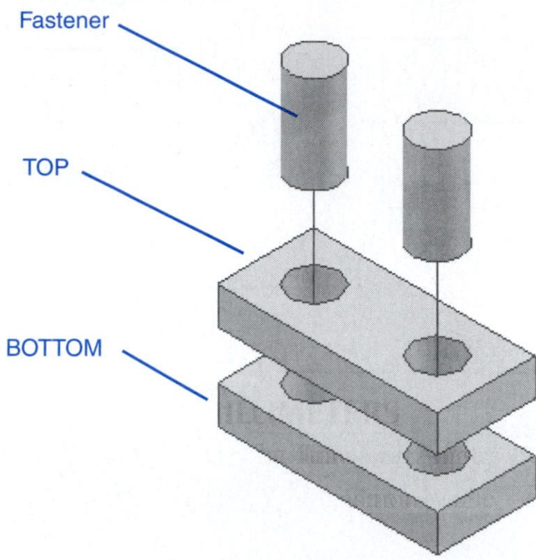

Figure P8-19B

Project 8-20: Inches

A. The distance between the holes' center points is 2.00 nominal.
B. The holes are Ø.375 nominal.
C. The fasteners have a tolerance of .001.
D. The holes have a tolerance of .002.
E. The minimum allowable clearance between the fasteners and the holes is .003.

Project 8-21: Millimeters

A. The distance between the holes' center points is 80 nominal.
B. The holes are Ø12 nominal.
C. The fasteners have a tolerance of 0.05.
D. The holes have a tolerance of 0.03.
E. The minimum allowable clearance between the fasteners and the holes is 0.02.

Project 8-22: Inches

A. The distance between the holes' center points is 3.50 nominal.
B. The holes are Ø.625 nominal.
C. The fasteners have a tolerance of .005.
D. The holes have a tolerance of .003.
E. The minimum allowable clearance between the fasteners and the holes is .002.

Project 8-23: Millimeters

A. The distance between the holes' center points is 65 nominal.
B. The holes are Ø16 nominal.
C. The fasteners have a tolerance of 0.03.
D. The holes have a tolerance of 0.04.
E. The minimum allowable clearance between the fasteners and the holes is 0.03.

Given a top and a bottom part in the fixed condition as shown in Figure P8-23, satisfy the requirements given in projects P8-24 through P8-23 so that the parts fit together regardless of orientation. Prepare drawings of each part including dimensions and tolerances.

A. Use linear tolerances
B. Use positional tolerances.

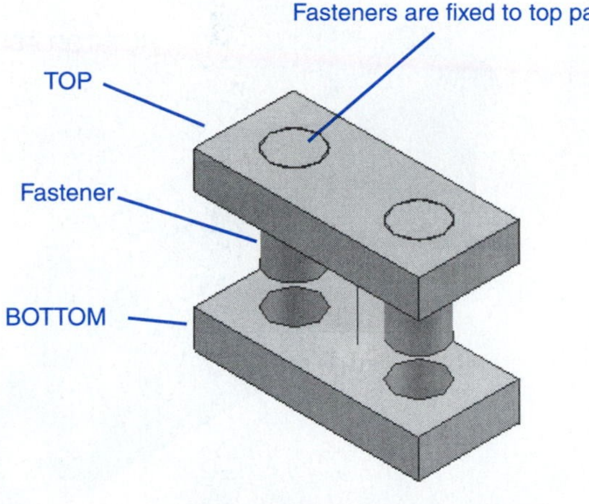

Figure P8-23

Project 8-24: Millimeters

A. The distance between the holes' center points is 60 nominal.
B. The holes are Ø10 nominal.
C. The fasteners have a tolerance of 0.04.
D. The holes have a tolerance of 0.02.
E. The minimum allowable clearance between the fasteners and the holes is 0.02.

Project 8-25: Inches

A. The distance between the holes' center points is 3.50 nominal.
B. The holes are Ø.563 nominal.
C. The fasteners have a tolerance of .005.
D. The holes have a tolerance of .003.
E. The minimum allowable clearance between the fasteners and the holes is .002.

Project 8-26: Millimeters

A. The distance between the holes' center points is 100 nominal.
B. The holes are Ø18 nominal.
C. The fasteners have a tolerance of 0.02.
D. The holes have a tolerance of 0.01.
E. The minimum allowable clearance between the fasteners and the holes is 0.03

Project 8-27: Inches

A. The distance between the holes' center points is 1.75 nominal.
B. The holes are Ø.250 nominal.
C. The fasteners have a tolerance of .002.
D. The holes have a tolerance of .003.
E. The minimum allowable clearance between the fasteners and the holes is .001.

Project 8-28: Millimeters

Dimension and tolerance the rotator assembly shown in Figure P8-28. Use the given dimensions as nominal and add sleeve bearings between the LINKs and both the CROSS-LINK and the PLATE. Create drawings of each part. Modify the dimensions as needed and add the appropriate tolerances. Specify the selected sleeve bearing.

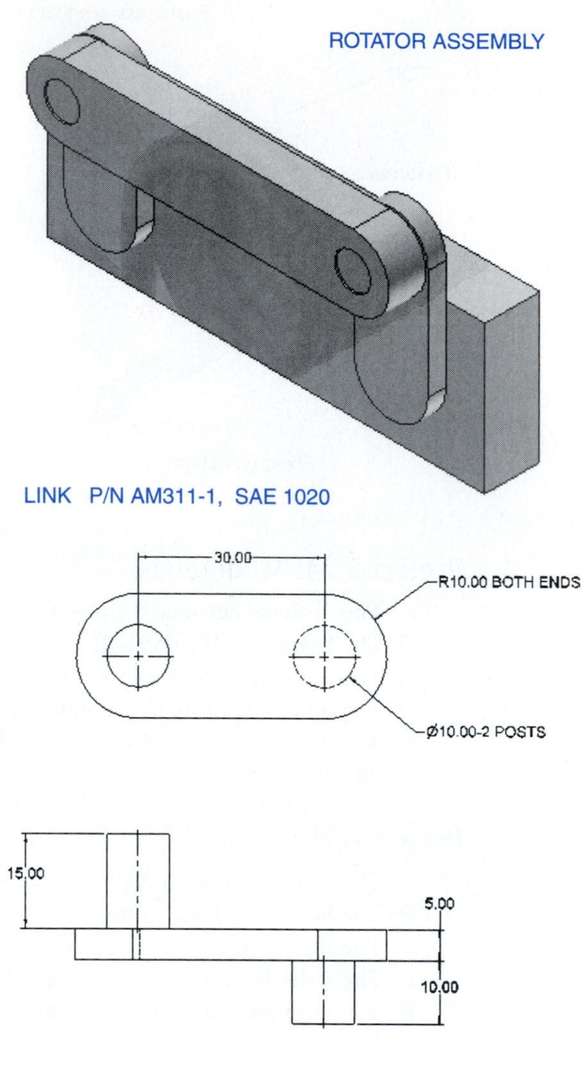

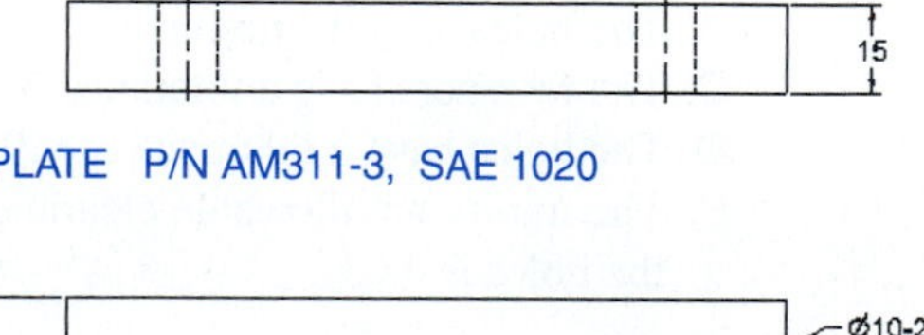

CROSS-LINK P/N AM311-2, SAE 1020

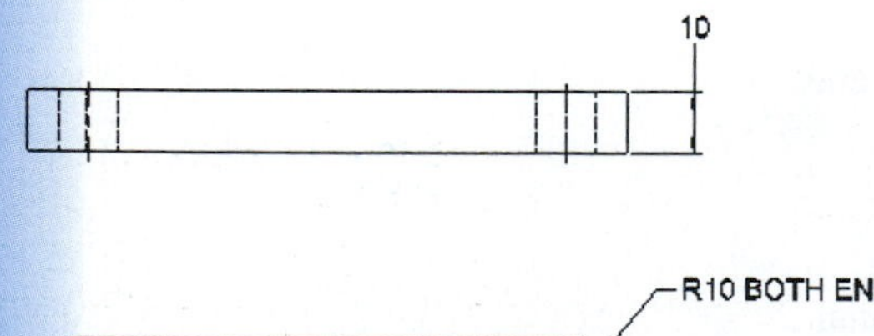

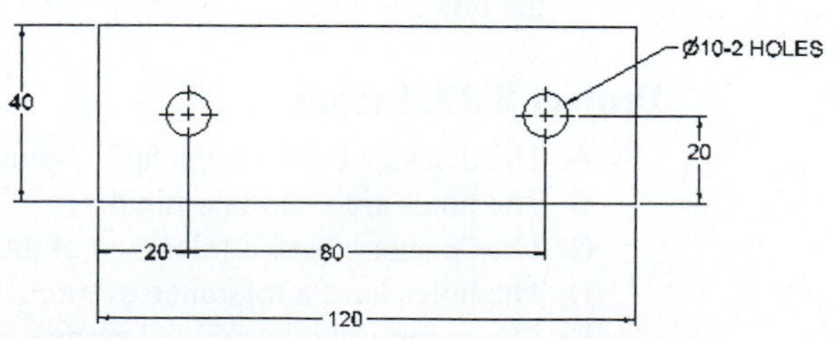

Figure P8-28

Project 8-29:

Dimension and tolerance the rocker assembly shown in Figure P8-29. Use the given dimensions as nominal, and add sleeve bearings between all moving parts. Create drawings of each part. Modify the dimensions as needed and add the appropriate tolerances. Specify the selected sleeve bearing.

ROCKER ASSEMBLY
DRIVE LINK
Ø 10 x 15 PEG
CENTER LINK
Ø 10 x 10 PEG
PLATE, WEB
ROCKER LINK
Ø 10 x 15 PEG

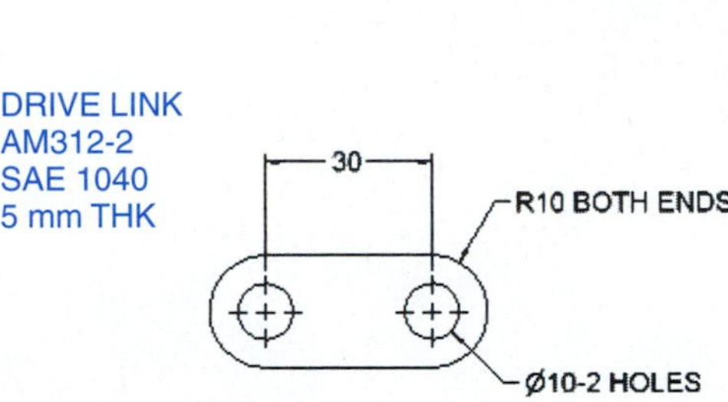

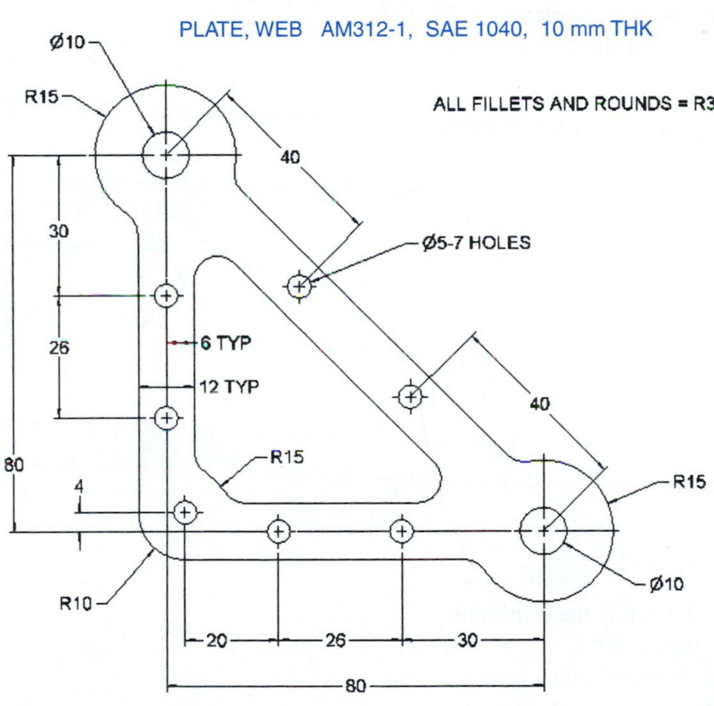

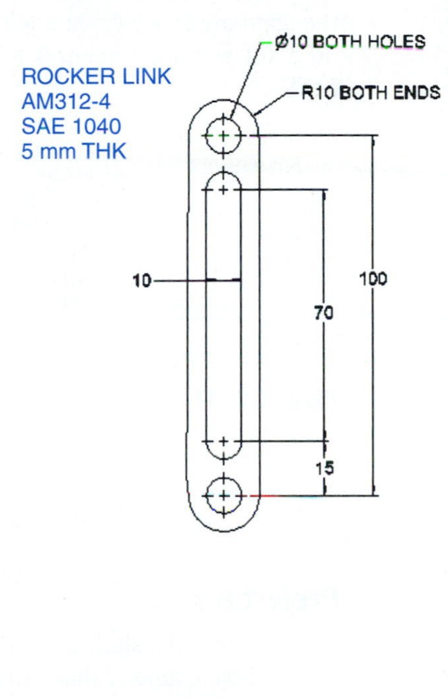

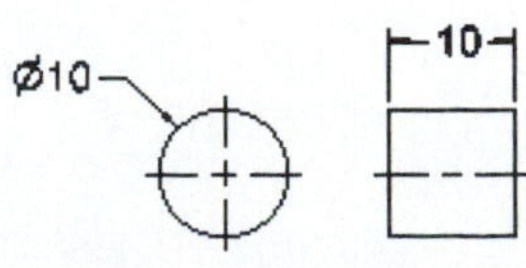

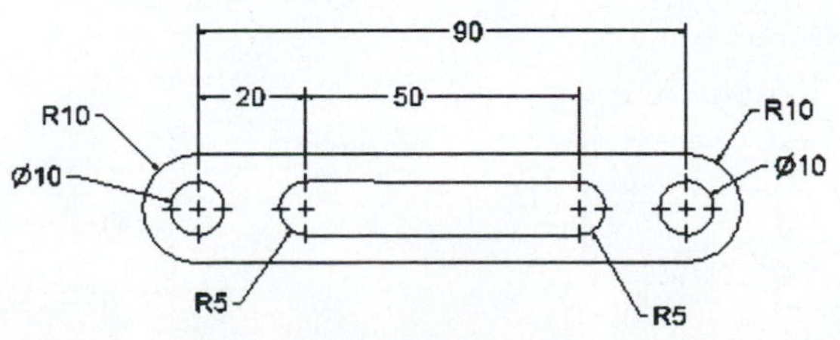

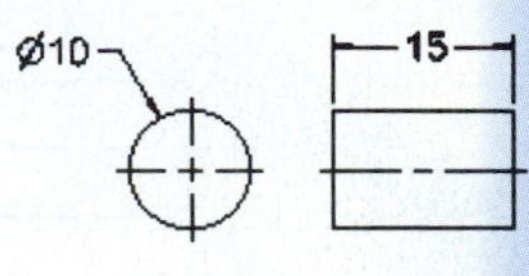

Figure P8-29

Project 8-30:

Draw the model shown in Figure P8-30, create a drawing layout with the appropriate views, and add the specified dimensions and tolerances.

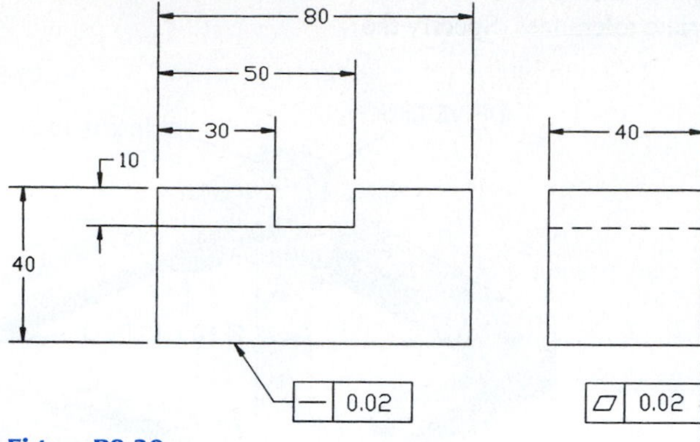

Figure P8-30

Project 8-31:

Redraw the shaft shown in Figure P8-31, create a drawing layout with the appropriate views, and add a feature dimension and tolerance of 36 ± 0.1 and a straightness tolerance of 0.07 about the centerline at MMC.

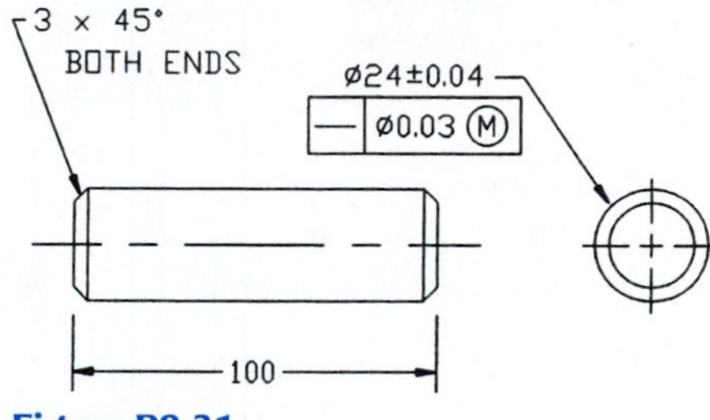

Figure P8-31

Project 8-32:

A. Given the shaft shown in Figure P8-32, what is the minimum hole diameter that will always accept the shaft?

B. If the minimum clearance between the shaft and a hole is equal to 0.02, and the tolerance on the hole is to be 0.6, what are the maximum and minimum diameters for the hole?

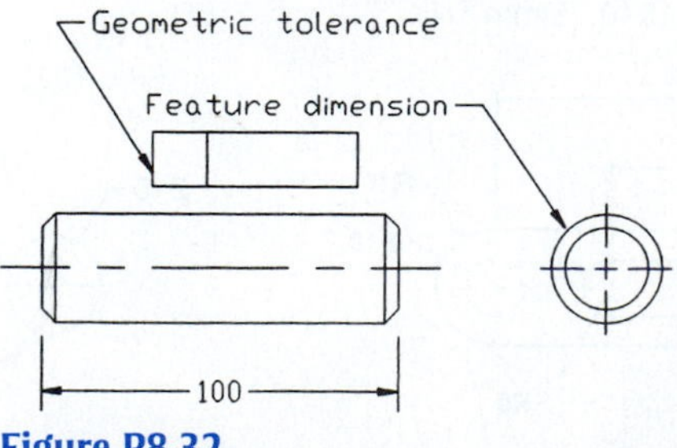

Figure P8-32

Project 8-33:

A. Given the shaft shown in Figure P8-33, what is the minimum hole diameter that will always accept the shaft?

B. If the minimum clearance between the shaft and a hole is equal to .005, and the tolerance on the hole is to be .007, what are the maximum and minimum diameters for the hole?

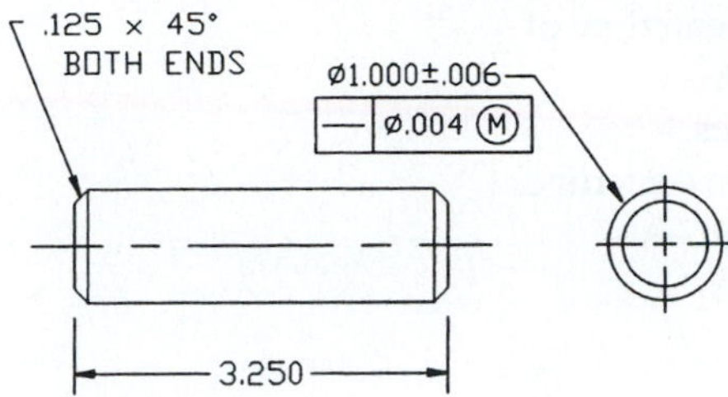

Figure P8-33

Project 8-34:

Draw a front and a right-side view of the object shown in Figure P8-34 and add the appropriate dimensions and tolerances based on the following information. Numbers located next to an edge line indicate the length of the edge.

A. Define surfaces A, B, and C as primary, secondary, and tertiary datums, respectively.

B. Assign a tolerance of ±0.5 to all linear dimensions.

C. Assign a feature tolerance of 12.07 − 12.00 to the protruding shaft.

D. Assign a flatness tolerance of 0.01 to surface –A–.

E. Assign a straightness tolerance of 0.03 to the protruding shaft.

F. Assign a perpendicularity tolerance to the centerline of the protruding shaft of 0.02 at MMC relative to datum –A–.

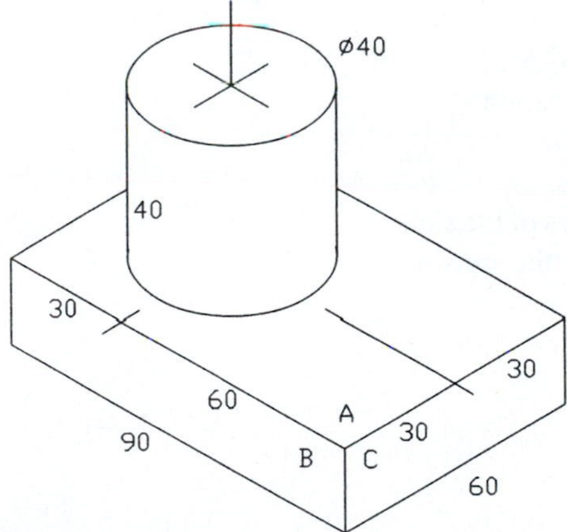

Figure P8-34

Project 8-35:

Draw a front and a right-side view of the object shown in Figure P8-35 and add the following dimensions and tolerances.

A. Define the bottom surface as datum –A–.
B. Assign a perpendicularity tolerance of 0.4 to both sides of the slot relative to datum –A–.
C. Assign a perpendicularity tolerance of 0.2 to the centerline of the 30 diameter hole at MMC relative to datum –A–.
D. Assign a feature tolerance of ±0.8 to all three holes.
E. Assign a parallelism tolerance of 0.2 to the common centerline between the two 20 diameter holes relative to datum –A–.
F. Assign a tolerance of ±0.5 to all linear dimensions.

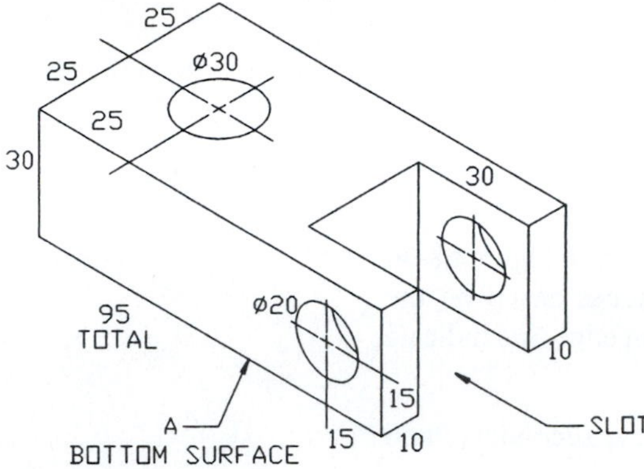

Figure P8-35

Project 8-36:

Draw a circular front and the appropriate right-side view of the object shown in Figure P8-36 and add the following dimensions and tolerances.

A. Assign datum –A– as indicated.
B. Assign the object's longitudinal axis as datum –B–.
C. Assign the object's centerline through the slot as datum –C–.
D. Assign a tolerance of ±0.5 to all linear tolerances.
E. Assign a tolerance of ±0.5 to all circular features.
F. Assign a parallelism tolerance of 0.01 to both edges of the slot.
G. Assign a perpendicularity tolerance of 0.01 to the outside edge of the protruding shaft.

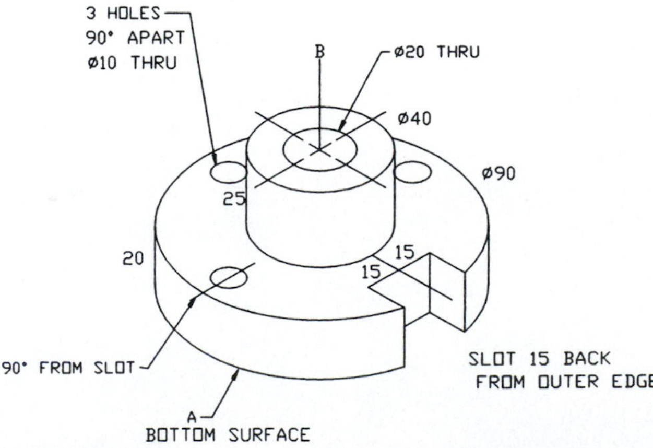

Figure P8-36

Project 8-37:

Given the two objects shown in Figure P8-37, draw a front and a side view of each. Assign a tolerance of ±0.5 to all linear dimensions. Assign a feature tolerance of ±0.4 to the shaft, and also assign a straightness tolerance of 0.2 to the shaft's centerline at MMC.

Tolerance the hole so that it will always accept the shaft with a minimum clearance of 0.1 and a feature tolerance of 0.2. Assign a perpendicularity tolerance of 0.05 to the centerline of the hole at MMC.

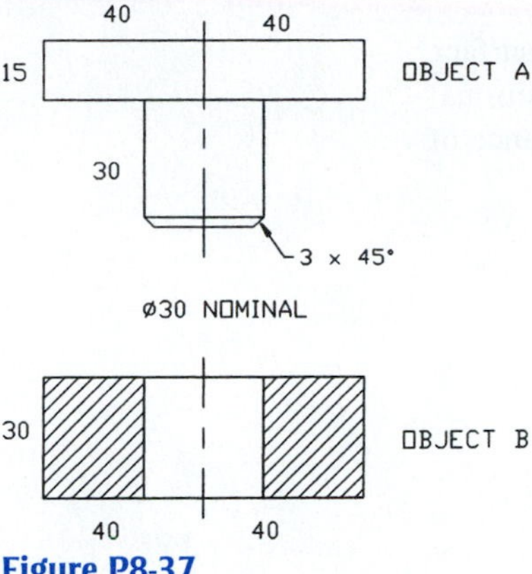

Figure P8-37

Project 8-38:

Given the two objects shown in Figure P8-38, draw a front and a side view of each. Assign a tolerance of ±0.005 to all linear dimensions. Assign a feature tolerance of ±0.004 to the shaft, and also assign a straightness tolerance of 0.002 to the shaft's centerline at MMC.

Tolerance the hole so that it will always accept the shaft with a minimum clearance of 0.001 and a feature tolerance of 0.002.

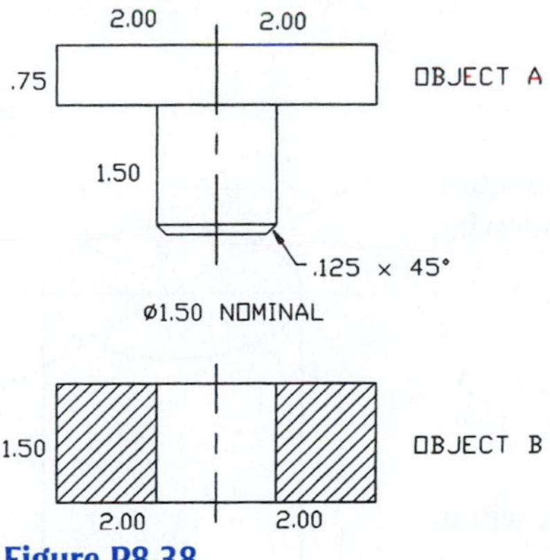

Figure P8-38

Project 8-39:

Draw a model of the object shown in Figure P8-39, then create a drawing layout including the specified dimensions. Add the following tolerances and specifications to the drawing.

A. Surface 1 is datum –A–.
B. Surface 2 is datum –B– and is perpendicular to datum –A– within 0.1 mm.
C. Surface 3 is datum –C– and is parallel to datum A within 0.3 mm.
D. Locate a 16-mm diameter hole in the center of the front surface that goes completely through the object. Use positional tolerances to locate the hole. Assign a positional tolerance of 0.02 at MMC perpendicular to datum –A–.

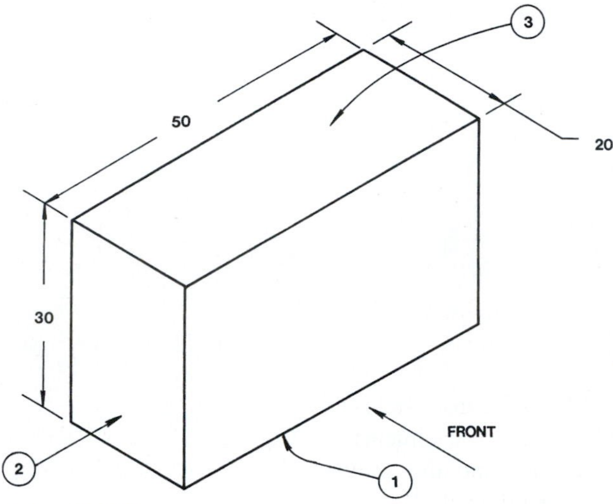

Figure P8-39

Project 8-40:

Draw a model of the object shown in Figure P8-40, then create a drawing layout including the specified dimensions. Add the following tolerances and specifications to the drawing.

A. Surface 1 is datum –A–.
B. Surface 2 is datum –B– and is perpendicular to datum –A– within .003 in.
C. Surface 3 is parallel to datum –A– within .005 in.
D. The cylinder's longitudinal centerline is to be straight within .001 in. at MMC.
E. Surface 2 is to have circular accuracy within .002 in.

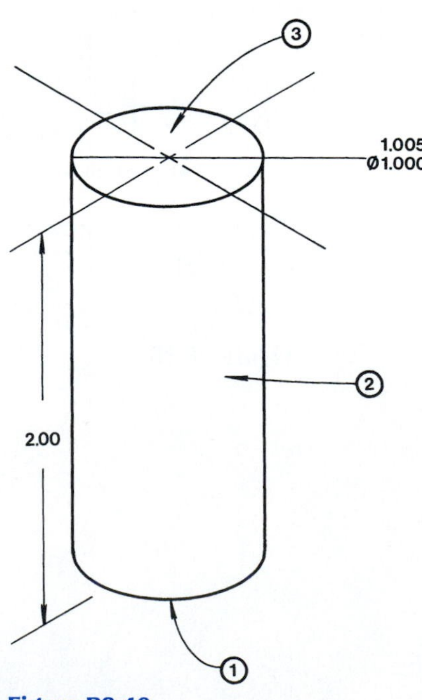

Figure P8-40

Project 8-41:

Draw a model of the object shown in Figure P8-41, then create a drawing layout including the specified dimensions. Add the following tolerances and specifications to the drawing.

A. Surface 1 is datum –A–.
B. Surface 4 is datum –B– and is perpendicular to datum A within 0.08 mm.
C. Surface 3 is flat within 0.03 mm.
D. Surface 5 is parallel to datum A within 0.01 mm.
E. Surface 2 has a runout tolerance of 0.2 mm relative to surface 4.
F. Surface 1 is flat within 0.02 mm.
G. The longitudinal centerline is to be straight within 0.02 at MMC and perpendicular to datum –A–.

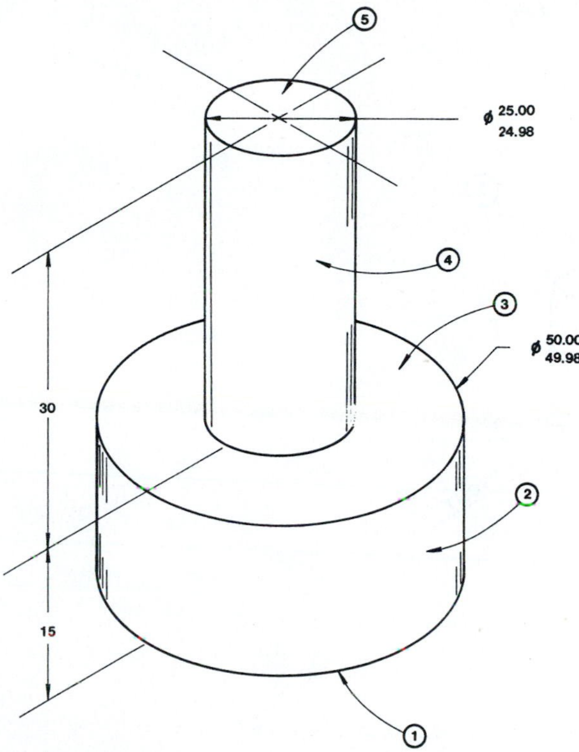

Figure P8-41

Project 8-42:

Draw a model of the object shown in Figure P8-42, then create a drawing layout including the specified dimensions. Add the following tolerances and specifications to the drawing.

A. Surface 2 is datum –A–.
B. Surface 6 is perpendicular to datum –A– with .000 allowable variance at MMC but with a .002 in. MAX variance limit beyond MMC.
C. Surface 1 is parallel to datum –A– within .005.
D. Surface 4 is perpendicular to datum –A– within .004 in.

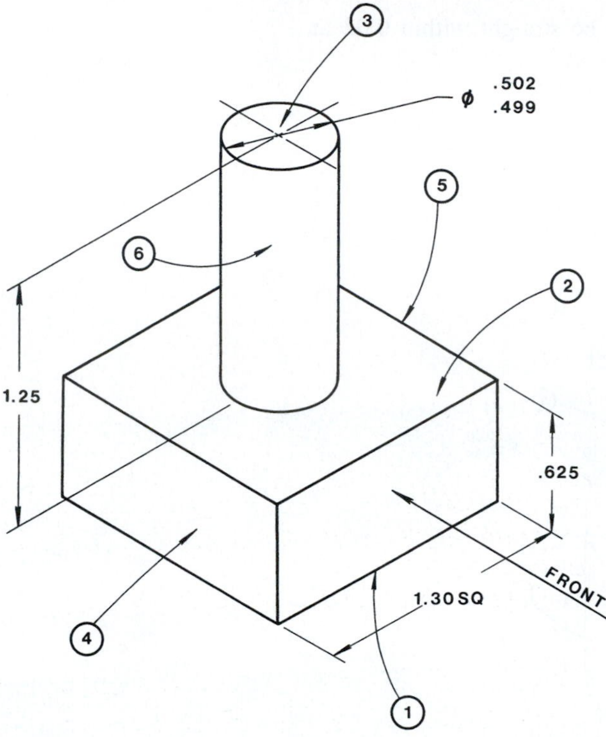

Figure P8-42

Project 8-43:

Draw a model of the object shown in Figure P8-43, then create a drawing layout including the specified dimensions. Add the following tolerances and specifications to the drawing.

A. Surface 1 is datum –A–.
B. Surface 2 is datum –B–.
C. The hole is located using a true position tolerance value of 0.13 mm at MMC. The true position tolerance is referenced to datums –A– and –B–.
D. Surface 1 is to be straight within 0.02 mm.
E. The bottom surface is to be parallel to datum –A– within 0.03 mm.

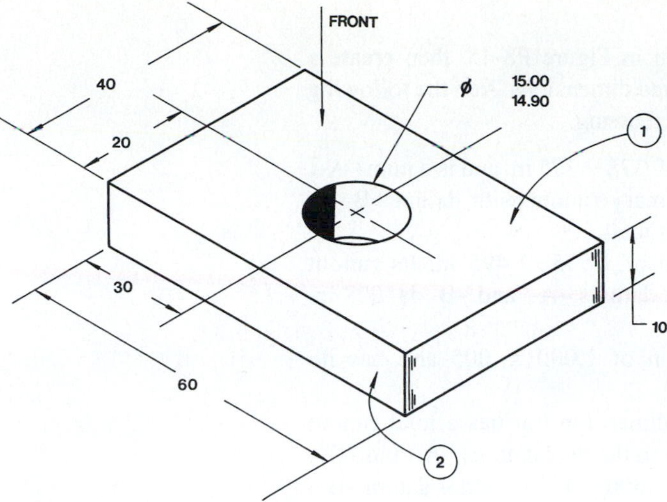

Figure P8-43

Project 8-44:

Draw a model of the object shown in Figure P8-44, then create a drawing layout including the specified dimensions. Add the following tolerances and specifications to the drawing.

A. Surface 1 is datum –A–.
B. Surface 2 is datum –B–.
C. Surface 3 is perpendicular to surface 2 within 0.02 mm.
D. The four holes are to be located using a positional tolerance of 0.07 mm at MMC referenced to datums –A– and –B–.
E. The centerlines of the holes are to be straight within 0.01 mm at MMC.

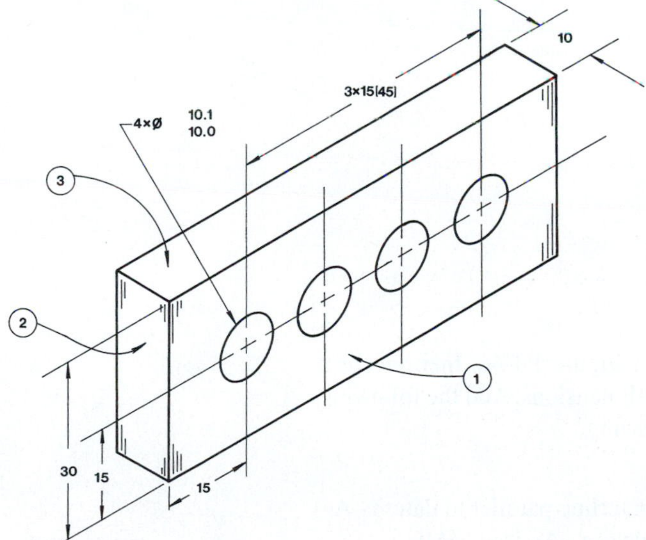

Figure P8-44

Project 8-45:

Draw a model of the object shown in Figure P8-45, then create a drawing layout including the specified dimensions. Add the following tolerances and specifications to the drawing.

A. Surface 1 has a dimension of .378−.375 in. and is datum –A–. The surface has a dual primary runout with datum –B– to within .005 in. The runout is total.
B. Surface 2 has a dimension of 1.505−1.495 in. Its runout relative to the dual primary datums –A– and –B– is .008 in. The runout is total.
C. Surface 3 has a dimension of 1.000 ± .005 and has no geometric tolerance.
D. Surface 4 has no circular dimension but has a total runout tolerance of .006 in. relative to the dual datums –A– and –B–.
E. Surface 5 has a dimension of .500−.495 in. and is datum –B–. It has a dual primary runout with datum –A– within .005 in. The runout is total.

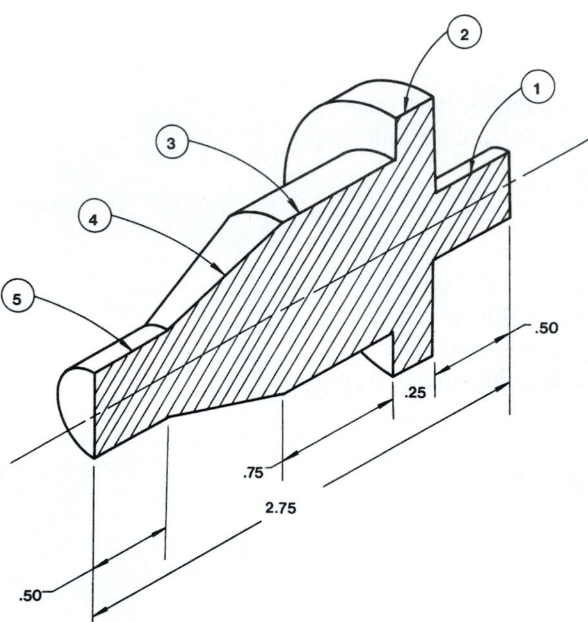

Figure P8-45

Project 8-46:

Draw a model of the object shown in Figure P8-46, then create a drawing layout including the specified dimensions. Add the following tolerances and specifications to the drawing.

A. Hole 1 is datum –A–.
B. Hole 2 is to have its circular centerline parallel to datum –A– within 0.2 mm at MMC when datum –A– is at MMC.
C. Assign a positional tolerance of 0.01 to each hole's centerline at MMC.

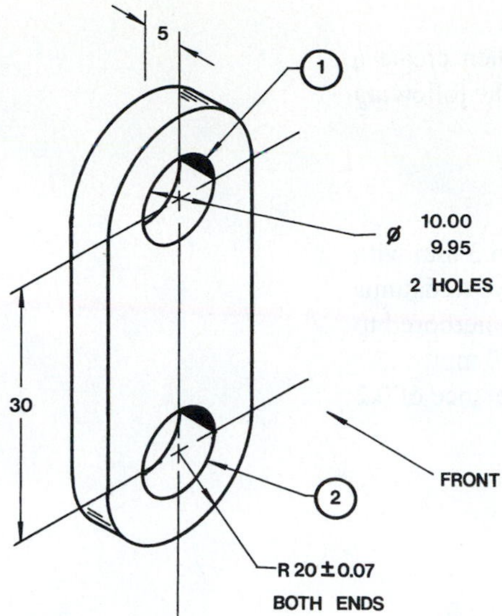

Figure P8-46

Project 8-47:

Draw a model of the object shown in Figure P8-47, then create a drawing layout including the specified dimensions. Add the following tolerances and specifications to the drawing.

A. Surface 1 is datum –A–.
B. Surface 2 is datum –B–.
C. The six holes have a diameter range of .502–.499 in. and are to be located using positional tolerances so that their centerlines are within .005 in. at MMC relative to datums –A– and –B–.
D. The back surface is to be parallel to datum –A– within .002 in.

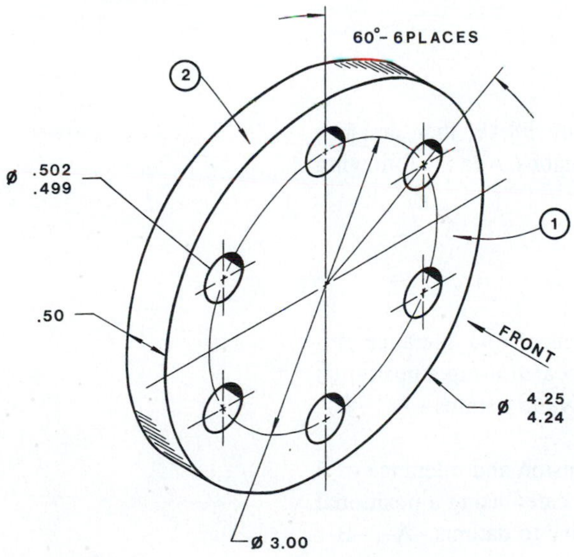

Figure P8-47

Project 8-48:

Draw a model of the object shown in Figure P8-48, then create a drawing layout including the specified dimensions. Add the following tolerances and specifications to the drawing.

A. Surface 1 is datum –A–.
B. Hole 2 is datum –B–.
C. The eight holes labeled 3 have diameters of 8.4−8.3 mm with a positional tolerance of 0.15 mm at MMC relative to datums –A– and –B–. Also, the eight holes are to be counterbored to a diameter of 14.6−14.4 mm and to a depth of 5.0 mm.
D. The large center hole is to have a straightness tolerance of 0.2 at MMC about its centerline.

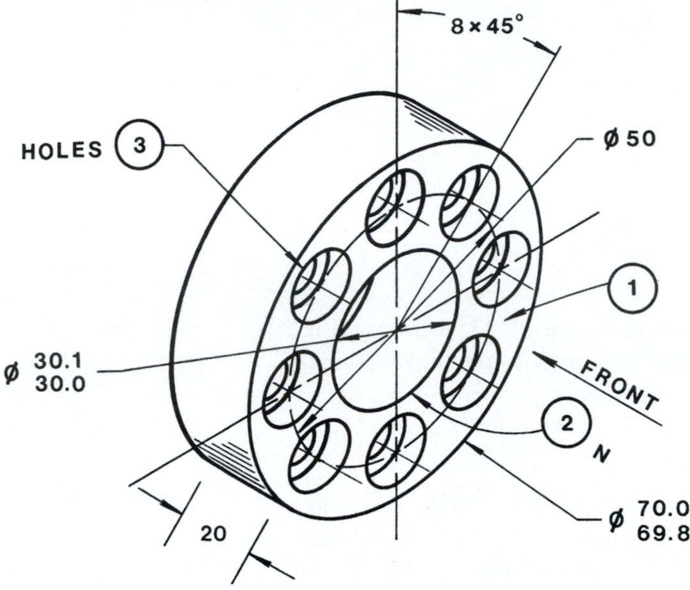

Figure P8-48

Project 8-49:

Draw a model of the object shown in Figure P8-49, then create a drawing layout including the specified dimensions. Add the following tolerances and specifications to the drawing.

A. Surface 1 is datum –A–.
B. Surface 2 is datum –B–.
C. Surface 3 is datum –C–.
D. The four holes labeled 4 have a dimension and tolerance of 8 + 0.3, −0 mm. The holes are to be located using a positional tolerance of 0.05 mm at MMC relative to datums –A–, –B–, and –C–.
E. The six holes labeled 5 have a dimension and tolerance of 6 +0.2, −0 mm. The holes are to be located using a positional tolerance of 0.01 mm at MMC relative to datums –A–, –B–, and –C–.

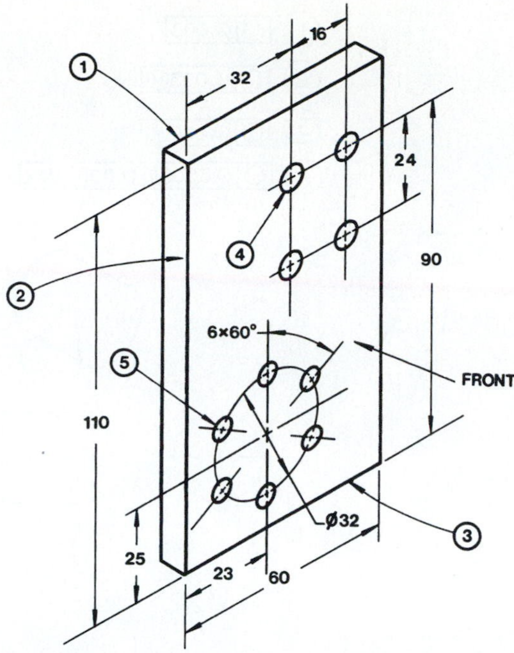

Figure P8-49

Project 8-50:

The objects in Figure P8-51A and P8-51B labeled A and B are to be toleranced using four different tolerances as shown. Redraw the charts shown in Figure P8-50 and list the appropriate allowable tolerance for "as measured" increments of 0.1 mm or .001 in. Also include the appropriate geometric tolerance drawing called out above each chart.

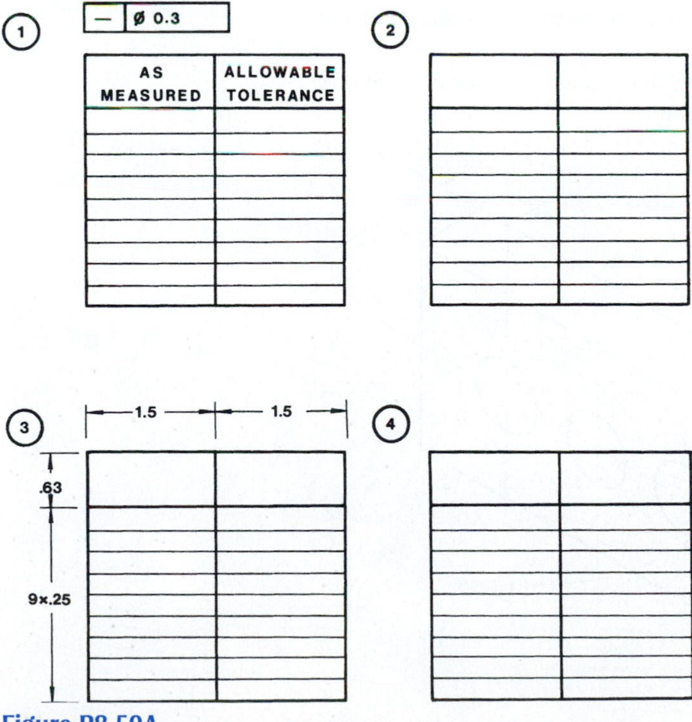

Figure P8-50A

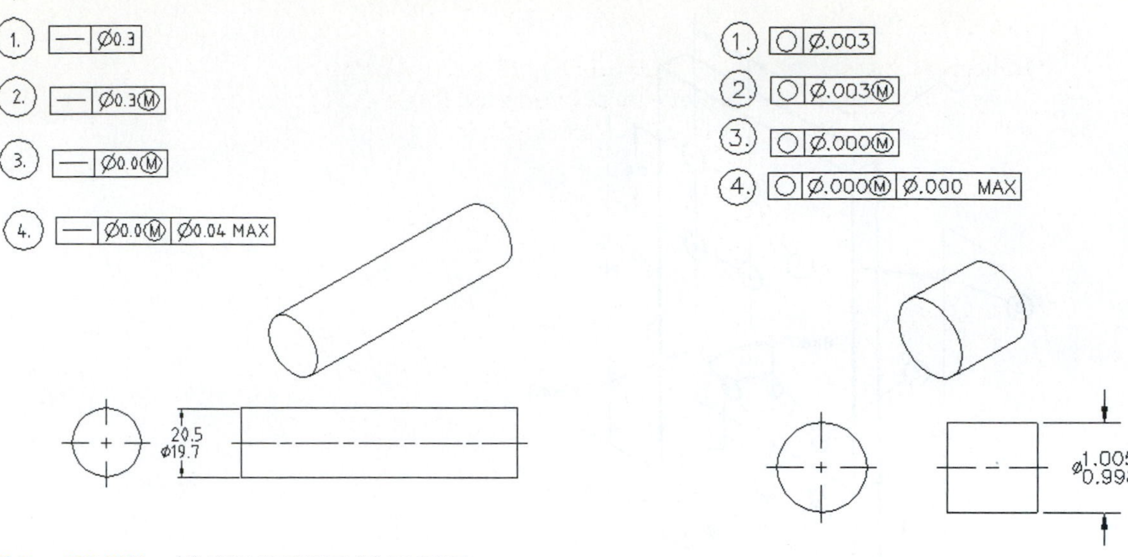

Figure P8-50B (A) MILLIMETERS (B) INCHES

Project 8-51:

Assume that there are two copies of the part in Figure P8-52 and that these parts are to be joined together using four fasteners in the floating condition. Draw front and top views of the object, including dimensions and tolerances. Add the following tolerances and specifications to the drawing, then draw front and top views of a shaft that can be used to join the two objects. The shaft should be able to fit into any of the four holes.

A. Surface 1 is datum –A–.
B. Surface 2 is datum –B–.
C. Surface 3 is perpendicular to surface 2 within 0.02 mm.
D. Specify the positional tolerance for the four holes applied at MMC.
E. The centerlines of the holes are to be straight within 0.01 mm at MMC.
F. The clearance between the shafts and the holes is to be 0.05 minimum and 0.10 maximum.

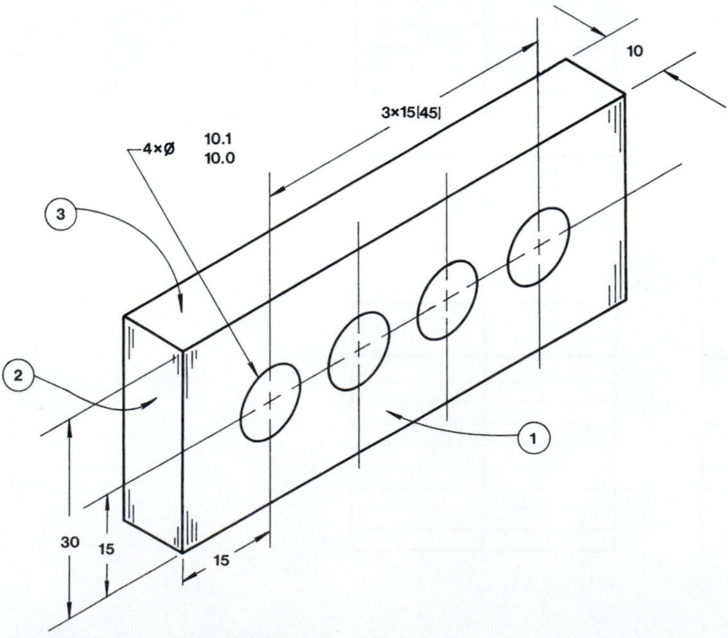

Figure P8-51 MILLIMETERS

Project 8-52:

Dimension and tolerance parts 1 and 2 of Figure P8-53 so that part 1 always fits into part 2 with a minimum clearance of .005 in. The tolerance for part 1's outer matching surface is .006 in.

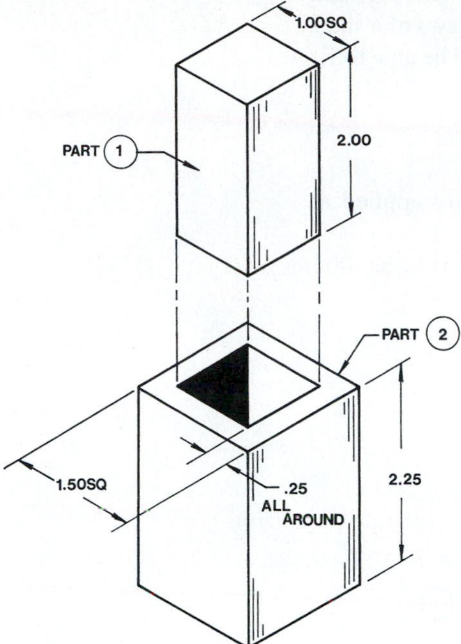

Figure P8-52 INCHES

Project 8-53:

Dimension and tolerance parts 1 and 2 of Figure P8-54 so that part 1 always fits into part 2 with a minimum clearance of 0.03 mm. The tolerance for part 1's diameter is 0.05 mm. Take into account the fact that the interface is long relative to the diameters.

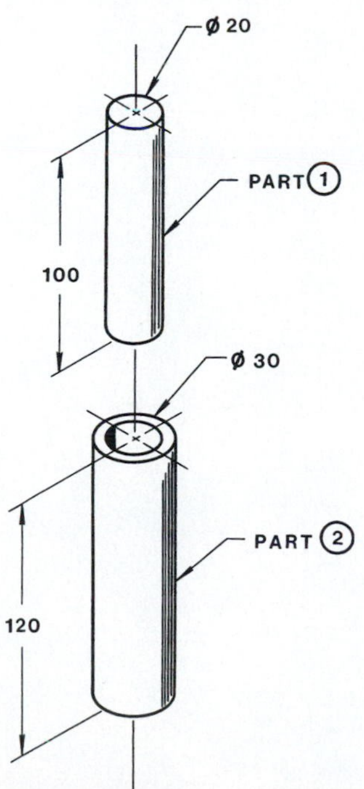

Figure P8-53 MILLIMETERS

Project 8-54:

Assume that there are two copies of the part in Figure P8-55 and that these parts are to be joined together using six fasteners in the floating condition. Draw front and top views of the object, including dimensions and tolerances. Add the following tolerances and specifications to the drawing, then draw front and top views of a shaft that can be used to join the two objects. The shaft should be able to fit into any of the six holes.

A. Surface 1 is datum –A–.
B. Surface 2 is round within .003.
C. Specify the positional tolerance for the six holes applied at MMC.
D. The clearance between the shafts and the holes is to be .001 minimum and .003 maximum.

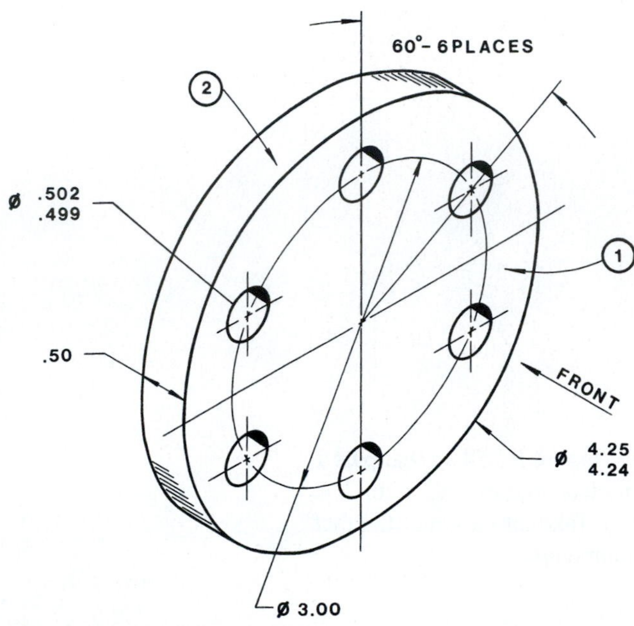

Figure P8-54 INCHES

Project 8-55:

The assembly shown in Figure P8-56 is made from parts defined in Chapter 5.

1. Draw an exploded assembly drawing.
2. Draw a BOM.
3. Use the drawing layout mode and draw orthographic views of each part. Include dimensions and geometric tolerances. The pegs should have a minimum clearance of 0.02. Select appropriate tolerances and define them for each hole using positional tolerance.

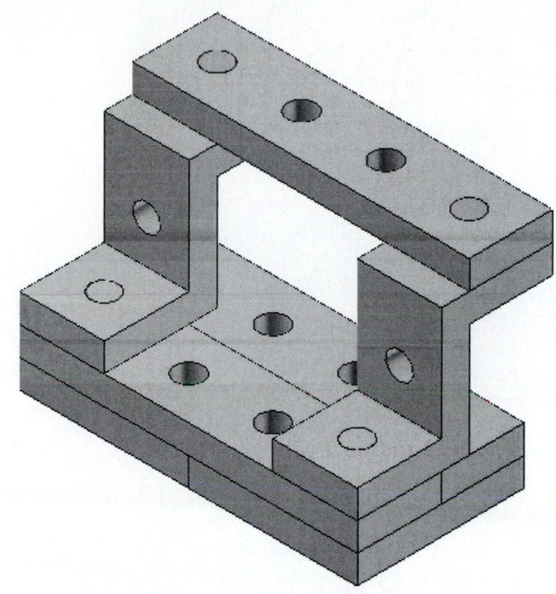

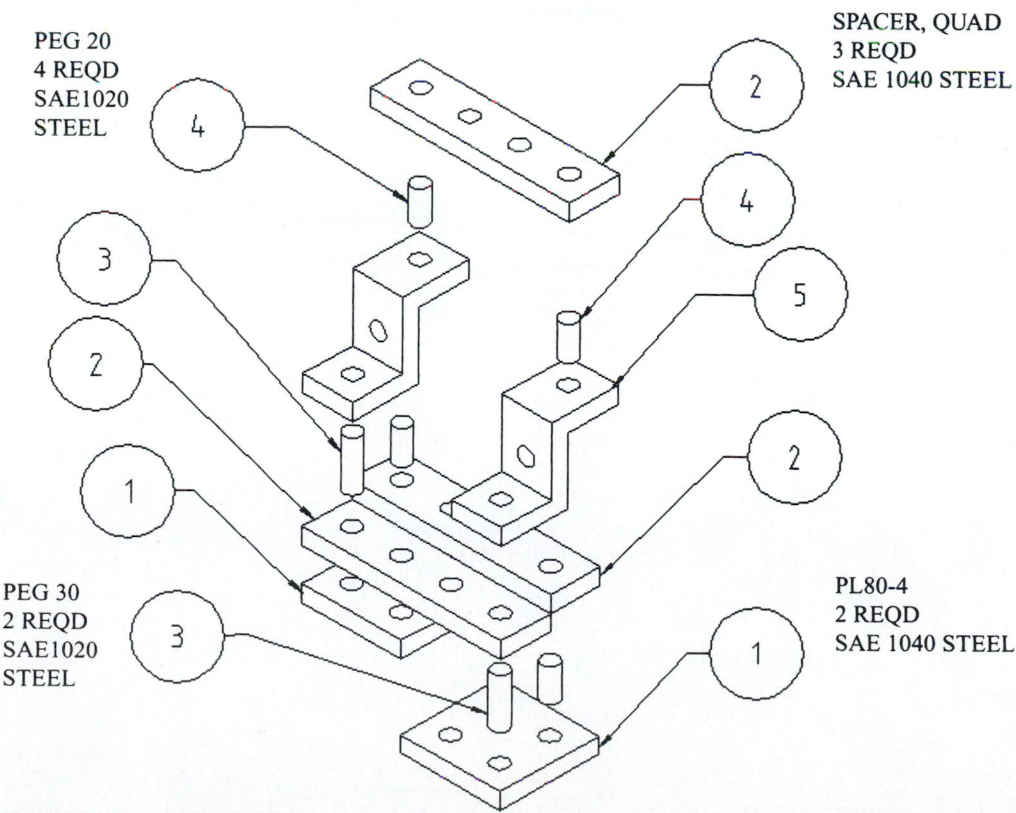

Figure P8-55 MILLIMETERS

Springs 9

Chapter Objectives

- Show how to draw springs using the **Coil** tool and the **Design Accelerator.**
- Show how to draw compression springs.
- Show how to draw extension springs.
- Show how to draw torsion springs.
- Show how to draw Belleville springs.

Introduction

This chapter shows how to draw springs. Both the **Coil** tool and the **Design Accelerator** are used to draw springs. Compression, extension, torsion, and Belleville springs are introduced.

Compression Springs

Exercise 9-1: Drawing a Compression Spring Using the **Coil** Command

1. Start a new **Metric** drawing using the **Standard (mm).ipt** format.
2. Create an isometric view and draw a line and a circle as shown in Figure 9-1.

The circle diameter is the wire's diameter (5), and the distance between the line and the center point of the circle equals the spring's mean diameter (Ø20, R=10).

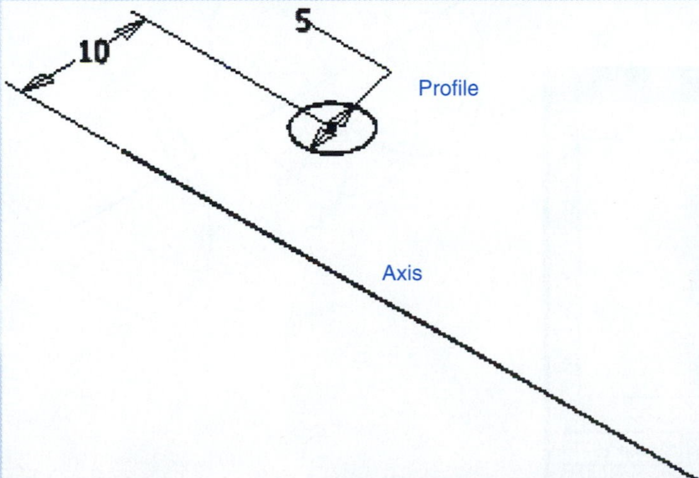

Figure 9-1

3. Right-click the mouse and select the **Finish Sketch** option.
4. Select the **Coil** command on the **Part Features** panel bar.

The **Coil** dialog box will appear. See Figure 9-2.

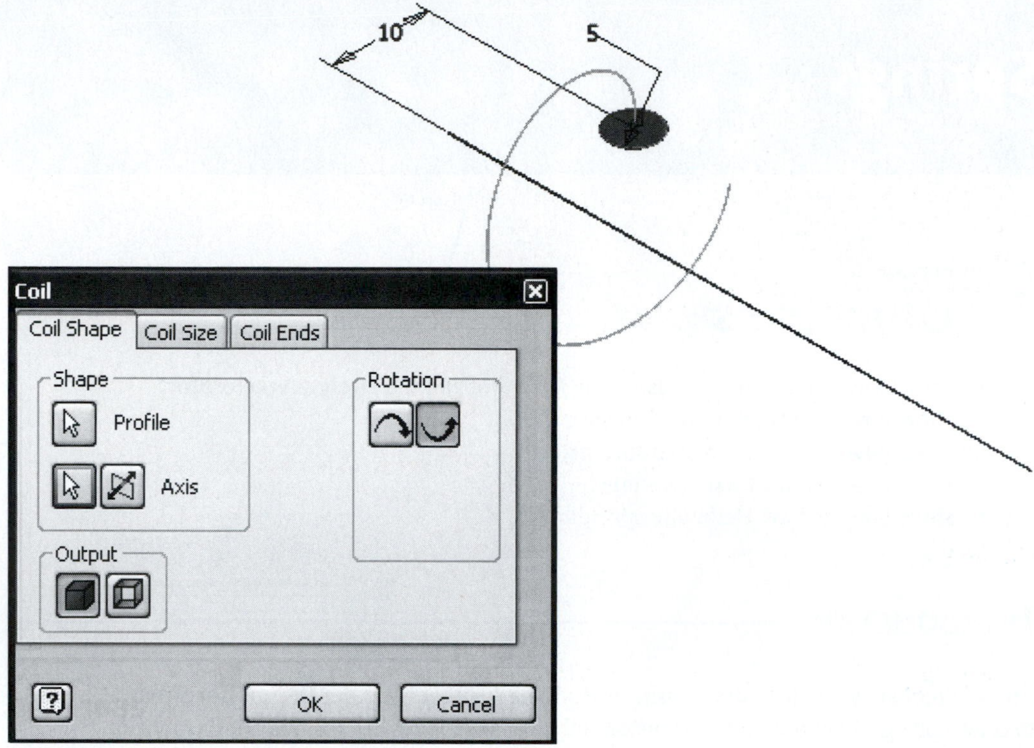

Figure 9-2

5. Select the circle as the **Profile** and the line as the **Axis.**
6. Click on the **Coil Size** tab on the **Coil** dialog box.

See Figure 9-3.

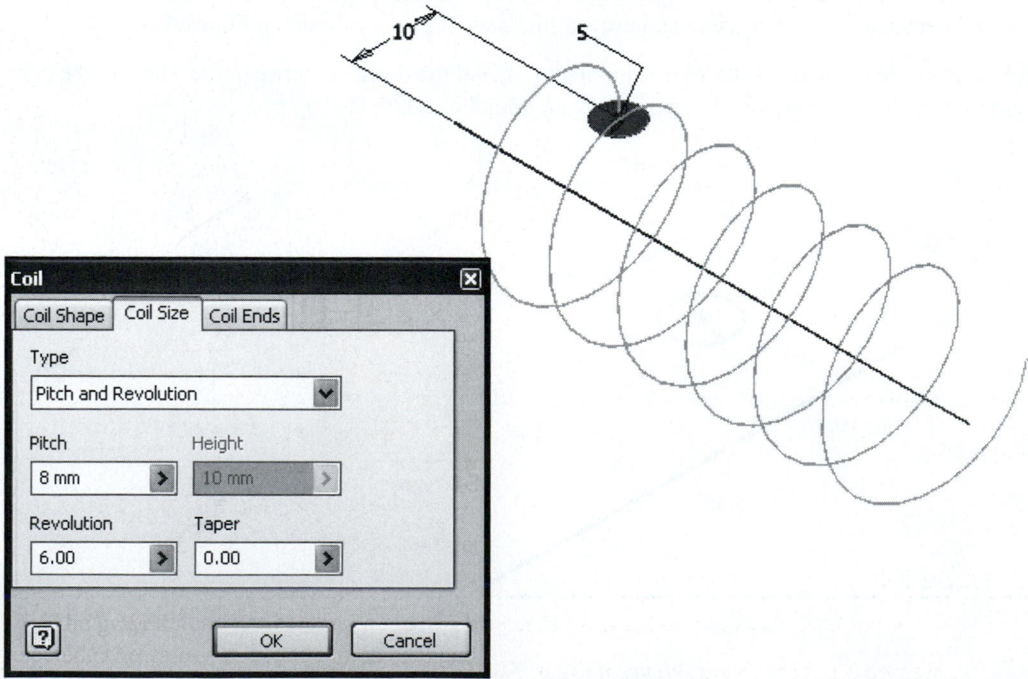

Figure 9-3

7. Set the **Pitch** value for **8** and the **Revolution** for **6.**
8. Click **OK.**

See Figure 9-4.

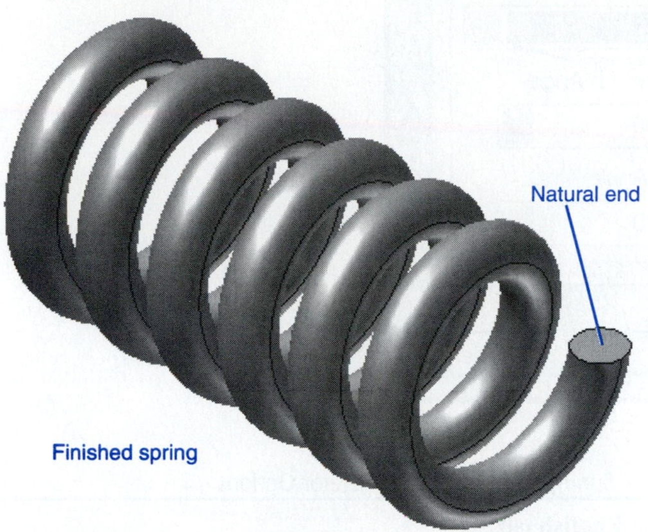

Figure 9-4

Coil Ends

The ends of a spring created using the **Coil** command can be drawn in one of two ways: natural or flat. Figure 9-5 shows three different possible coil end configurations.

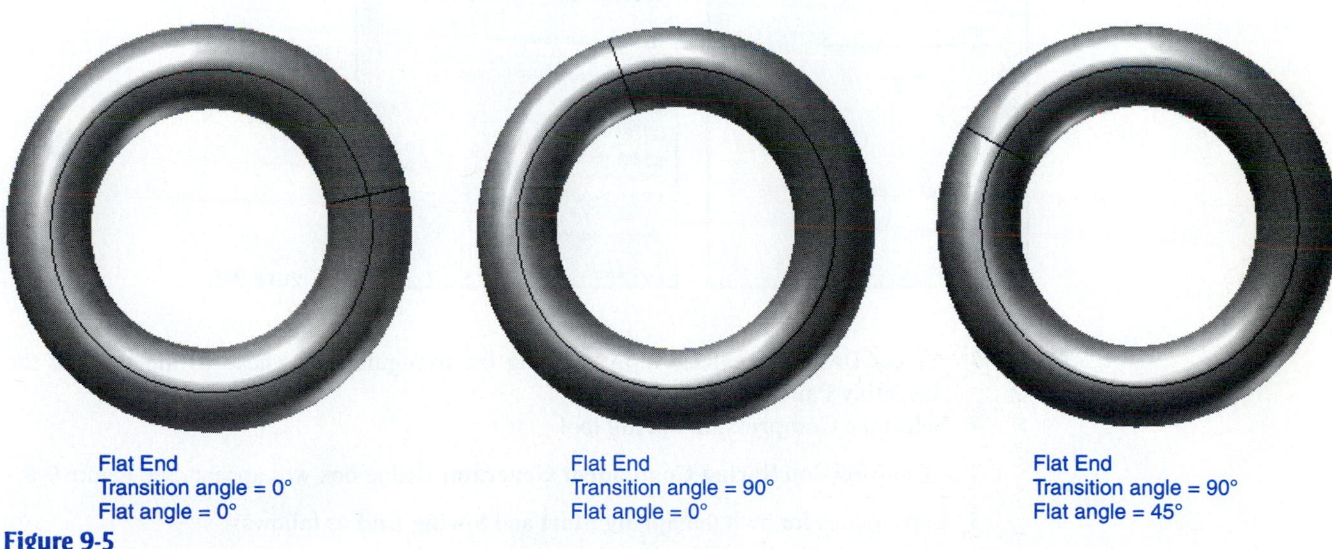

Figure 9-5

Exercise 9-2: Changing the End of a Spring

1. Click the **Coil Ends** tab on the **Coil** dialog box.

See Figure 9-6. The spring shown in Figure 9-4 has natural ends. Springs with flat ends must have their transition and flat angles defined, as was done in Figure 9-5.

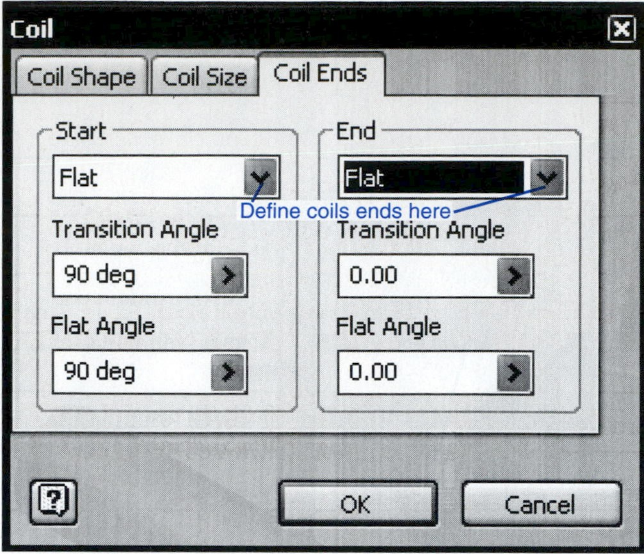

Figure 9-6

Exercise 9-3: Drawing a Compression Spring Using Design Accelerator Options

1. Start a new drawing using the **English** tab.
2. Select the **Standard (in).iam** format.

The **Assembly Panel** will appear. See Figure 9-7.

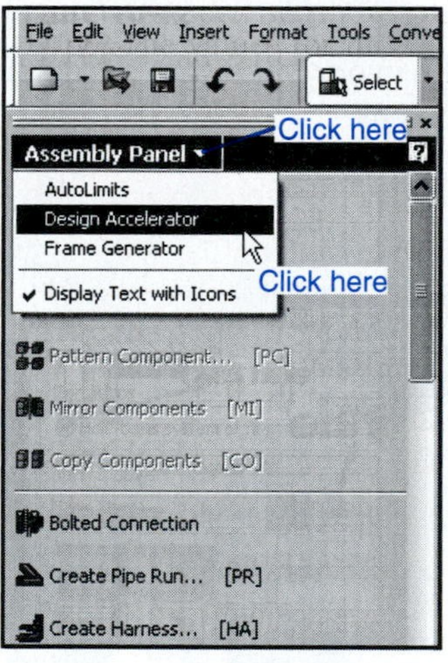

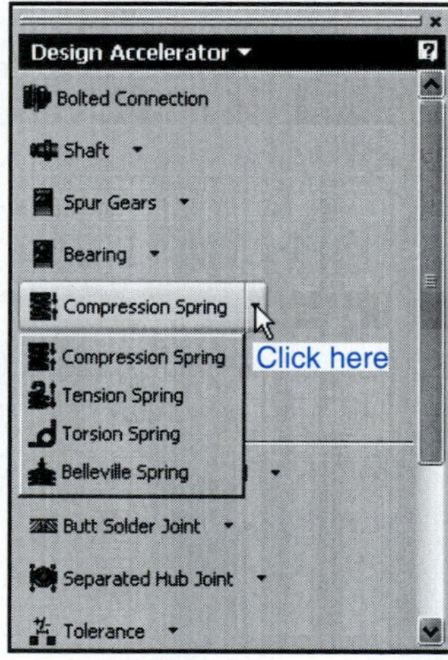

Figure 9-7

3. Access **Design Accelerator** by clicking the triangular arrowhead to the right of the **Assembly Panel** heading.
4. Select the **Compression Spring** tool.

The **Compression Spring Component Generator** dialog box will appear. See Figure 9-8.

5. Enter values for both the **Spring Start** and **Spring End** as follows:

Closed End Coils = 2.000

Transition Coils = 1.000

Ground Coils = 0.750

6. Click the **Calculate** tab.

The dialog box will change. See Figure 9-9.

7. Set the **Spring Strength Calculation** for **Work Forces Calculation.**

8. Set the **Dimension** values as follows:

Wire Diameter = 0.125
Outside Diameter = 0.875
Loose Spring Length = 2.00

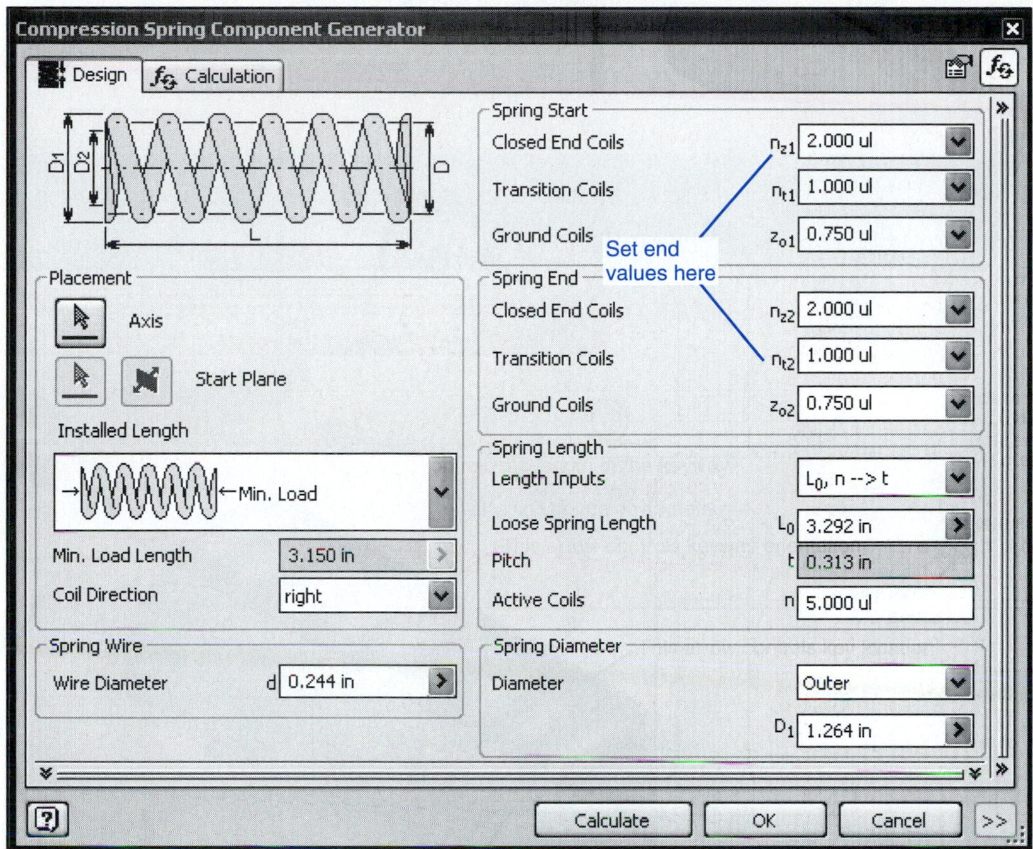

Figure 9-8

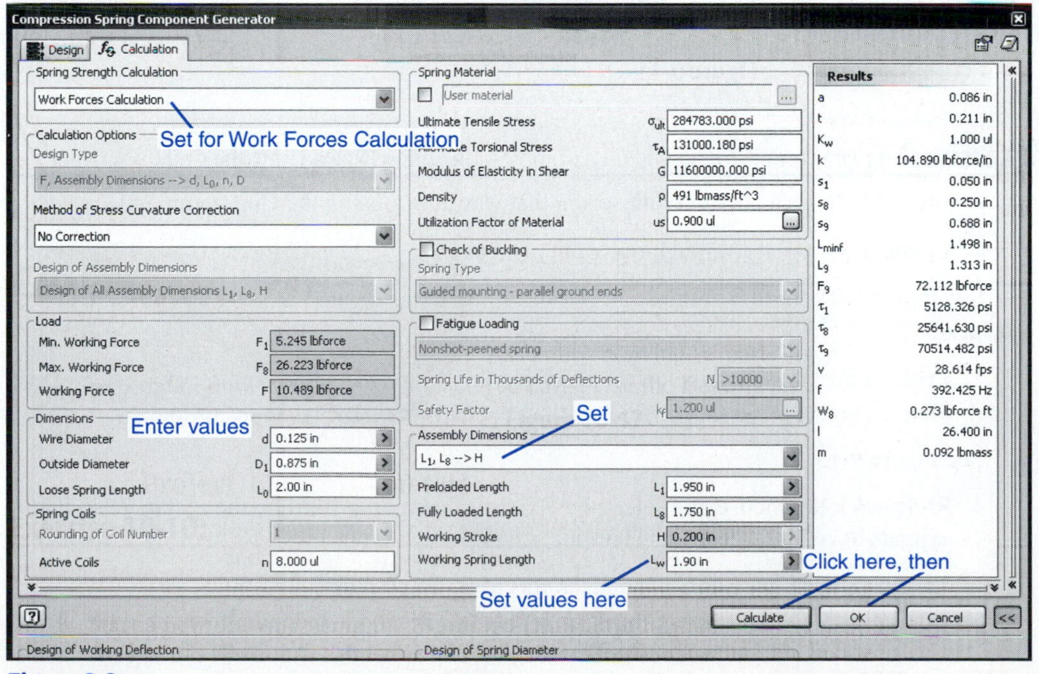

Figure 9-9

9. Set the **Assembly** values for the following.

L_1, L_8--> H

Preloaded Length = 1.950

Fully Loaded Length = 1.750

Working Spring Length = 1.900

10. Click **Calculate,** then click **OK.**

Figure 9-10 shows the finished compression spring.

TIP The spring's ends are square ground. Square-ground ends sit better on flat surfaces and tend not to buckle when put under a load.

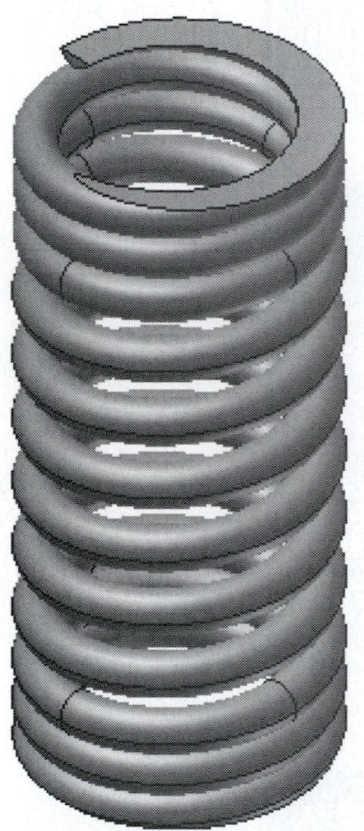

Figure 9-10

Figure 9-11

Exercise 9-4: Drawing a Ground End on a Compression Spring Drawn Using the Coil Command

Figure 9-11 shows a compression spring that was drawn using the **Coil** command.

1. Create a work plane through one of the end coils.

See Figure 9-12. In this example an XZ work plane was created offset 4.00 from the end of the spring.

2. Create a new sketch plane on the work plane. Draw a rectangle on the new sketch plane. Size the rectangle so that it is larger than the outside diameter of the spring.

See Figure 9-13.

3. Right-click the mouse and select the **Extrude** command. Extrude the rectangle so that it extends beyond the end of the spring.

In this example an extrusion value of **10** was used. See Figure 9-14.

4. Use the **Cut** option of the **Extrude** command and cut out the extruded rectangle.

Figure 9-15 shows the finished spring.

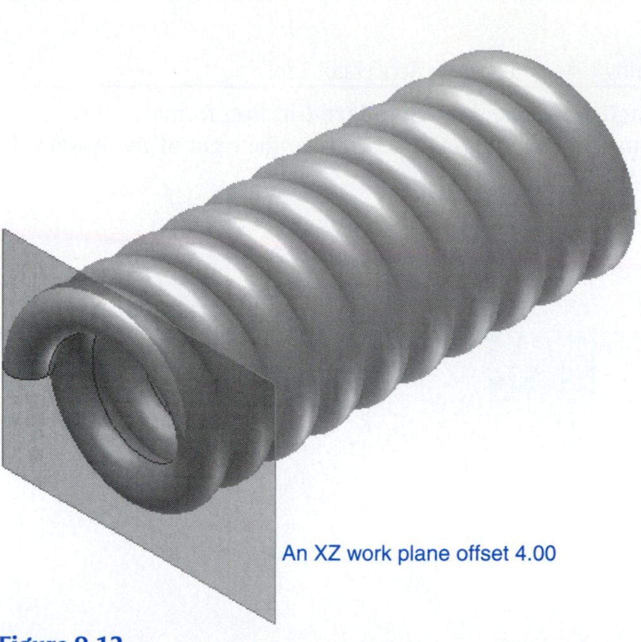

Figure 9-12

A 2D rectangle on the work plane

Figure 9-13

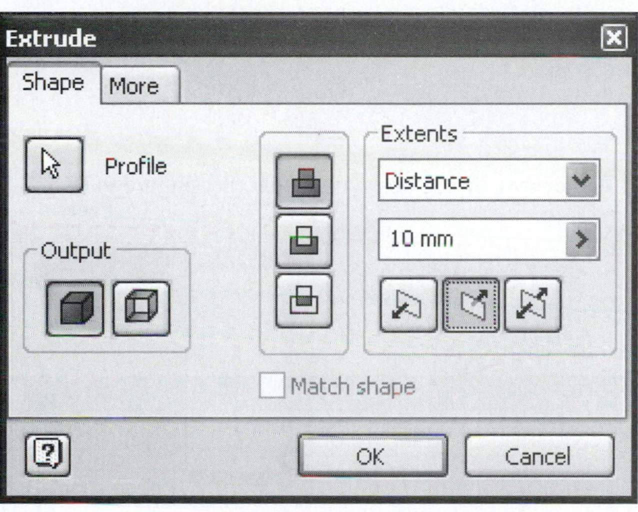

Figure 9-14

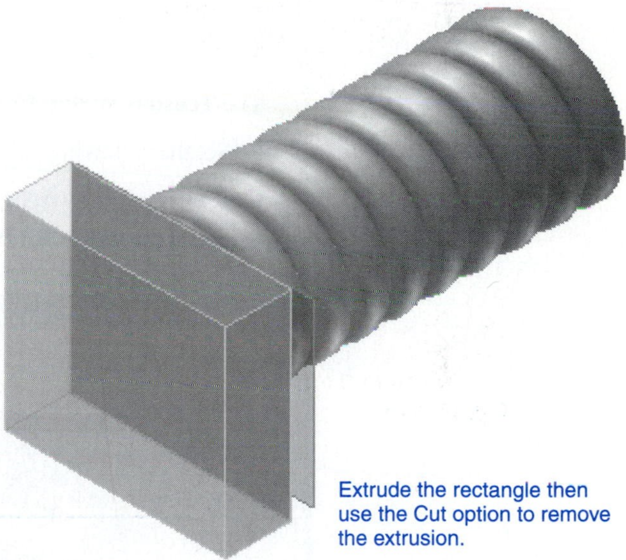

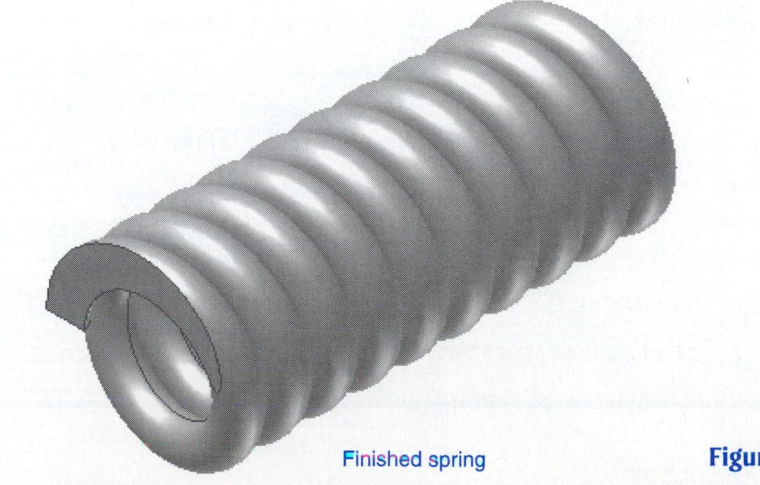

Figure 9-15

Tension Springs

Exercise 9-5: Drawing a Tension Spring Using the Design Accelerator Tool

1. Start a new drawing using **English** units and the **Standard (in).iam** format.
2. Access the **Design Accelerator** tool by clicking the arrow to the right of the **Assembly Panel** heading.

See Figure 9-16.

Figure 9-16

3. Click the **Tension Spring** tool.

The **Tension Spring Component Generator** dialog box will appear. See Figure 9-17.

4. Set the following values and inputs:

Coil Direction = right
Wire Diameter = 0.1205

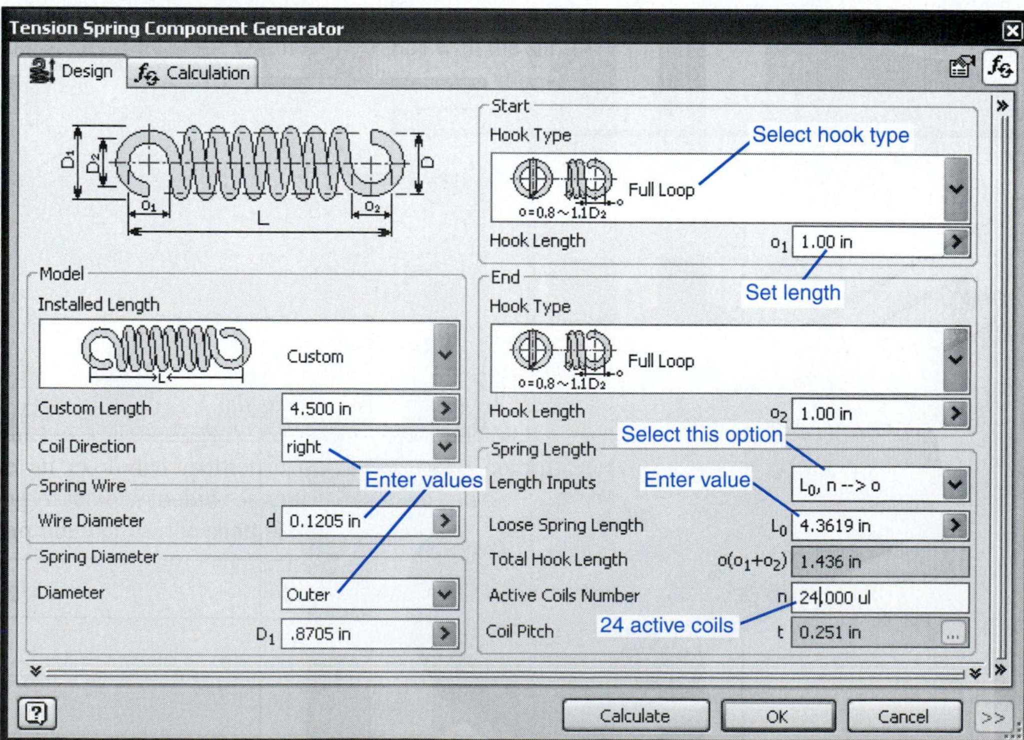

Figure 9-17

Diameter, Outer = .8705
Start Hook Type = Full Loop
Start Hook Length = 1.00
End Hook Type = Full Loop
End Hook Length = 1.00
Length Inputs = L_0**,n --> o**
Loose Spring Length = 4.3619
Active Coils Number = 24

5. Click **OK.**

The finished tension spring will appear. See Figure 9-18.

If an input error is made, when **OK** is clicked, a red line will appear across the bottom of the screen and the incorrect value will be highlighted.

Figure 9-18

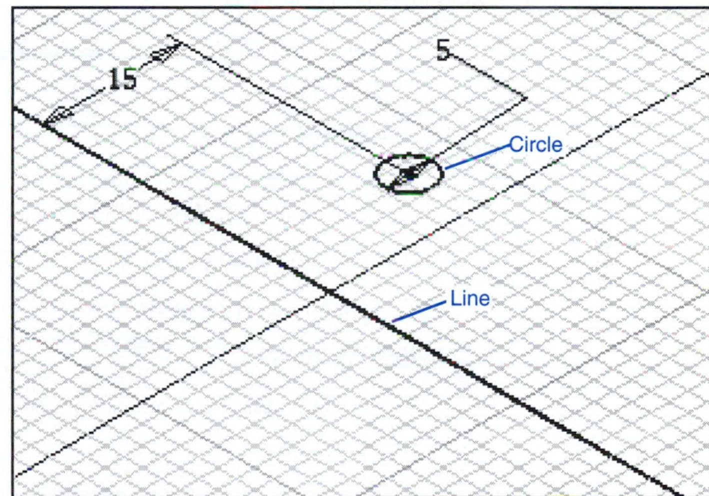

Figure 9-19

Exercise 9-6: Drawing a Tension Spring Using Coil

1. Start a new drawing using the **Standard (mm).ipt** format.
2. Draw a **Ø5**-mm circle **15** mm from a line.

See Figure 9-19.

3. Right-click the mouse and select the **Finish Sketch** option. Select the **Coil** command.

The **Coil** dialog box will appear. See Figure 9-20.

4. Select the circle as the **Profile** and the line as the **Axis.** Click the **Coil Size** tab.

See Figure 9-21.

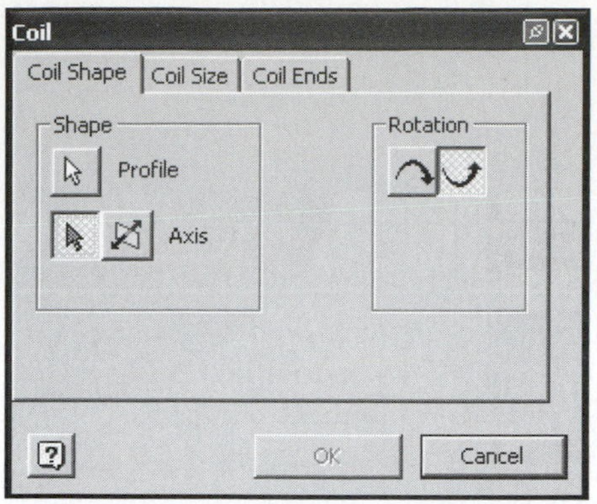

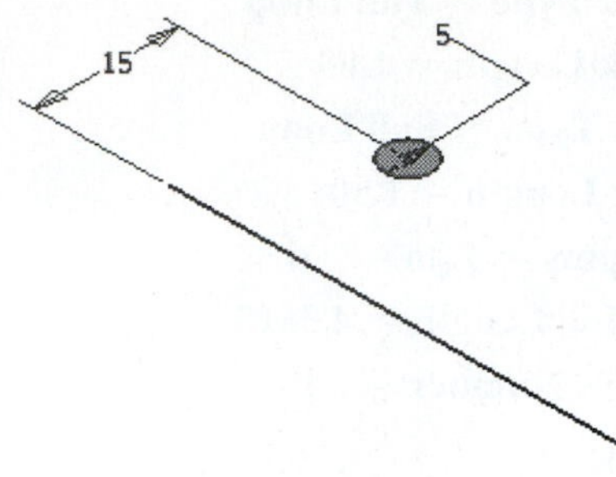

Figure 9-20

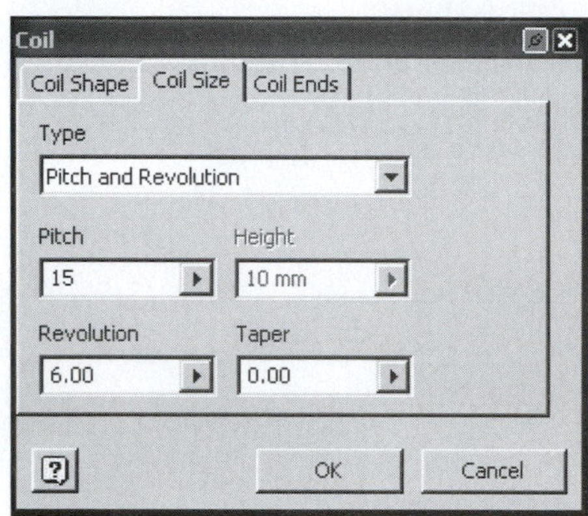

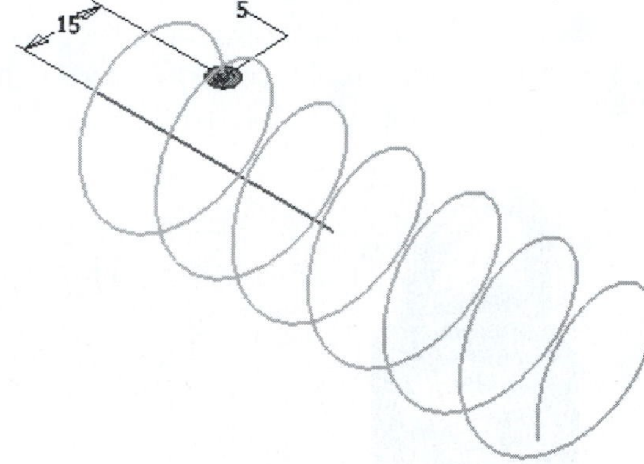

Figure 9-21

5. Set the **Pitch** for **15,** the **Revolution** for **6.00,** and the **Taper** for **0.00.** Click the **Coil Ends** tab.

See Figure 9-22.

6. Set the **Start** for **Flat** and both the **Transition Angle** and **Flat Angle** for **90°.** Set the **End** for **Flat** and the **Transition Angle** and **Flat Angle** for **0.00°.** Click **OK.**

The finished spring will appear. See Figure 9-23.

7. Rotate the spring so that the start end is completely visible.

See Figure 9-24.

8. Create a new sketch plane aligned with the start end of the spring.

See Figure 9-25.

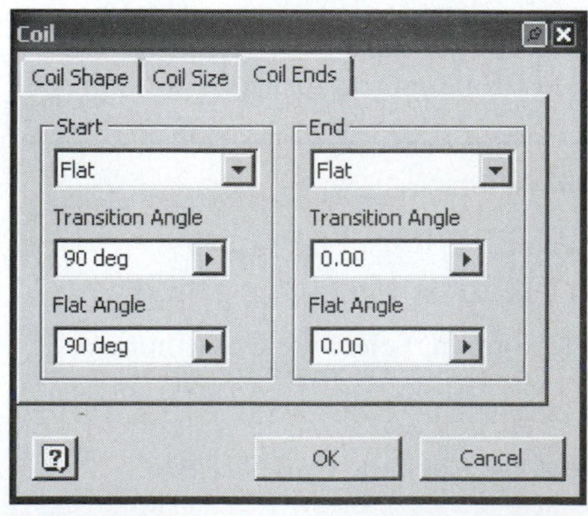

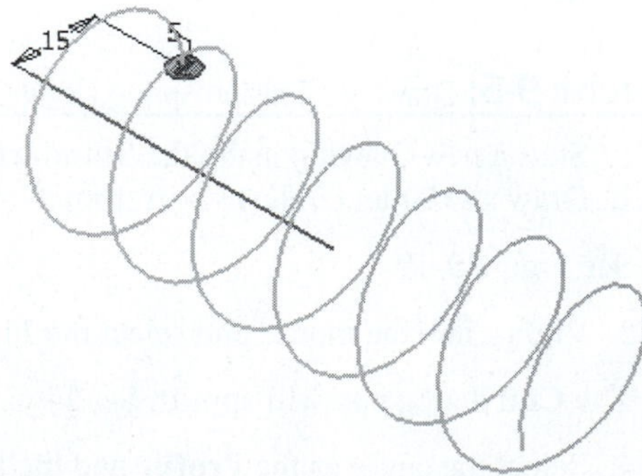

Figure 9-22

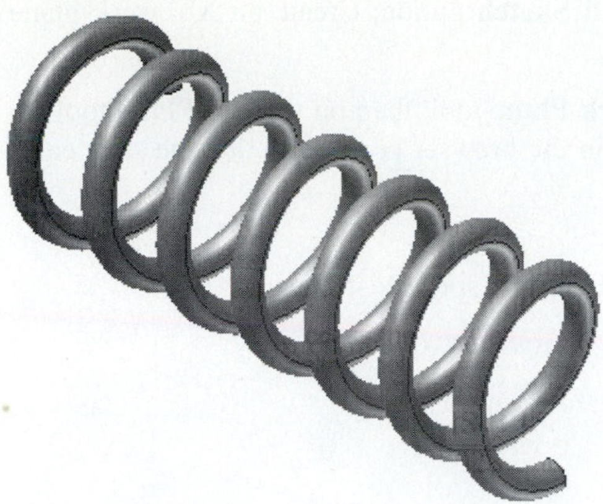

Figure 9-23

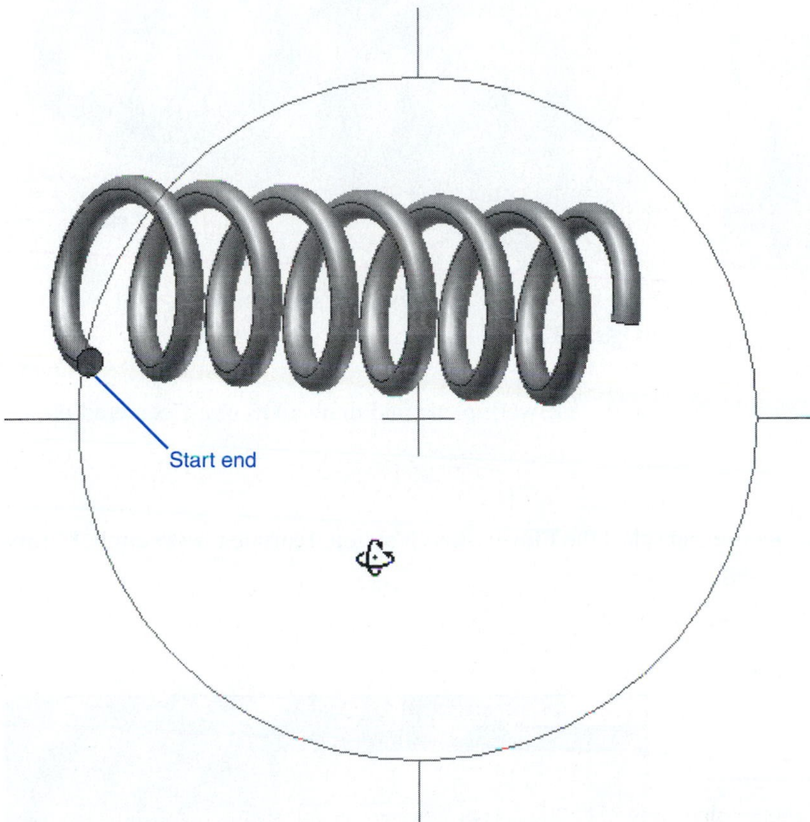

Figure 9-24

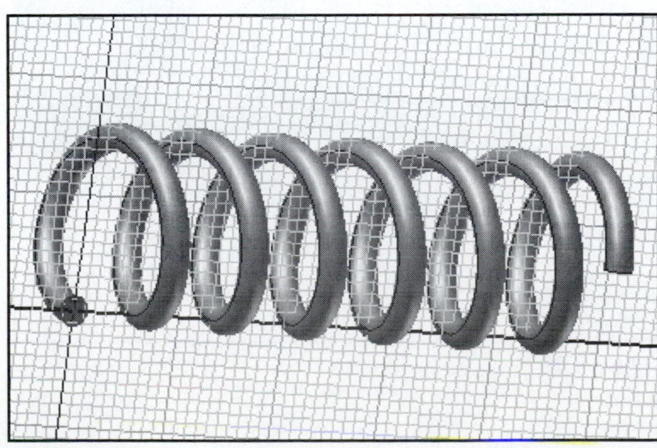

Figure 9-25

9. Right-click the mouse and select the **Finish Sketch** option. Create an XY work plane aligned with the start end of the spring.

To create an XY work plane, click on the **Work Plane** tool, then on the **XY Plane** tool located under the **Origin** heading under **Part name** in the browser box, then click the start end's center point.

See Figure 9-26.

10. Add a YZ work plane through the center point of the spring's start end.

See Figure 9-27.

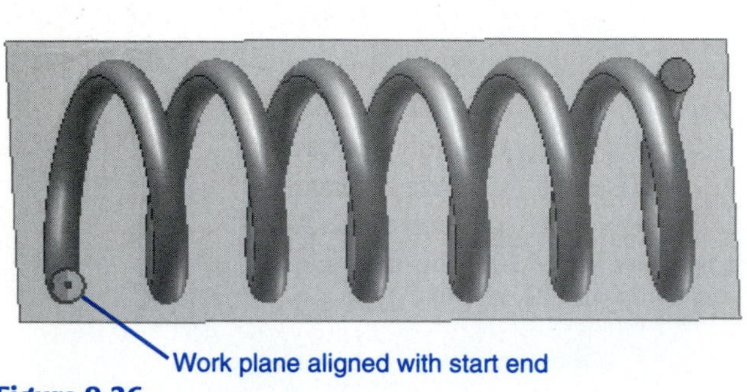

Figure 9-26

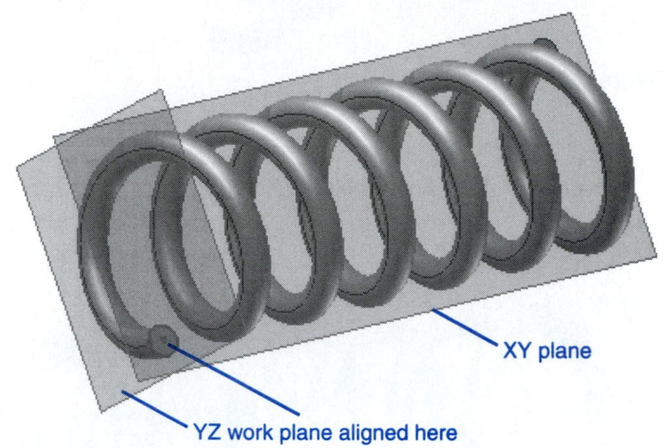

Figure 9-27

11. Create a new sketch plane on the YZ work plane and draw a **Ø5** circle centered on the spring's start end.

See Figure 9-28.

12. Right-click the mouse and select the **Finish Sketch** option. Extrude the Ø5 circle **12 mm** away from the spring.

See Figure 9-29.

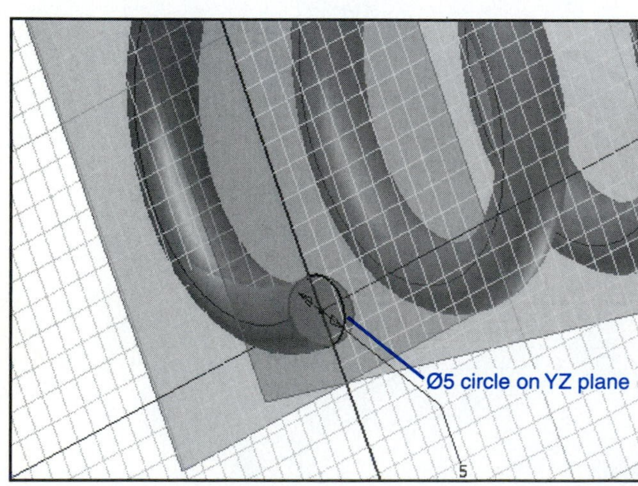

Figure 9-28

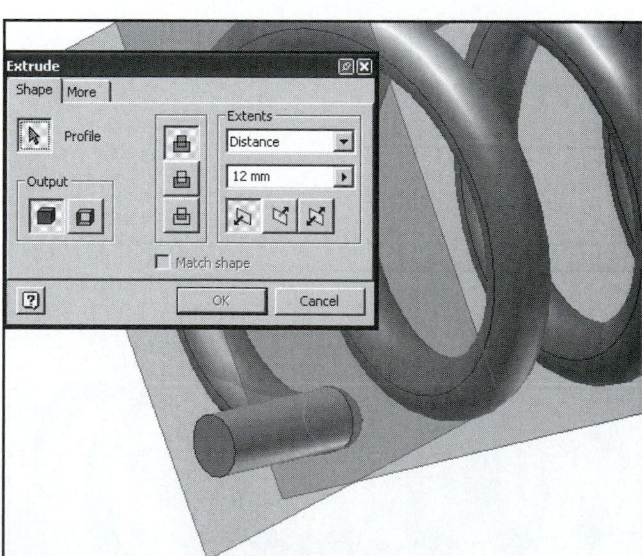

Figure 9-29

13. Rotate the spring and use the **Loft** command to fill in the area between the extruded cylinder and the spring's start end.

See Figure 9-30.

14. Create a new sketch plane on the end of the 12-mm extrusion created in step 12. Draw a **Ø5** circle centered about the center point of the extrusion's end.

See Figure 9-31.

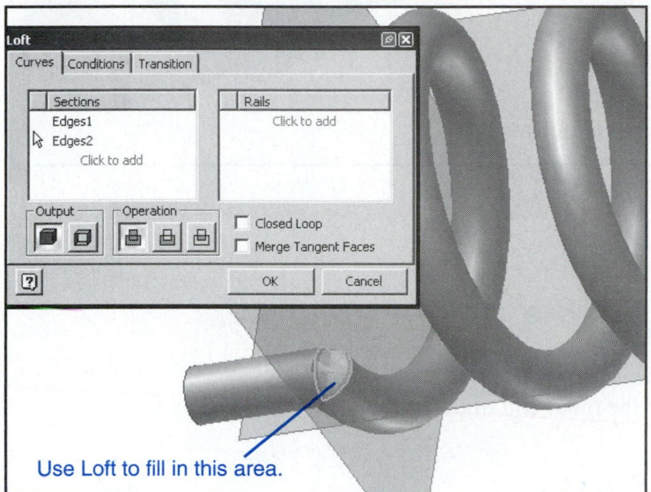

Figure 9-30

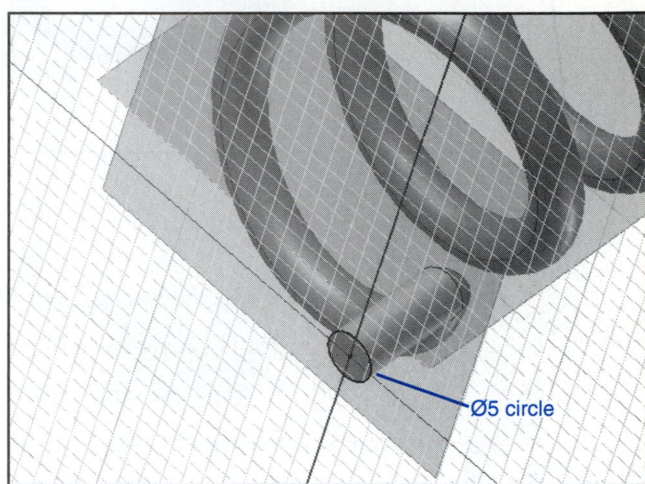

Figure 9-31

15. Create a fixed work point on the extrusion's center point as shown. Create a new sketch plane on the XY work plane and draw a **15-mm, 180°** arc using the **Center point Arc** command.

See Figure 9-32. Use the background grid and visually align the arc's center point with the extrusion's center point.

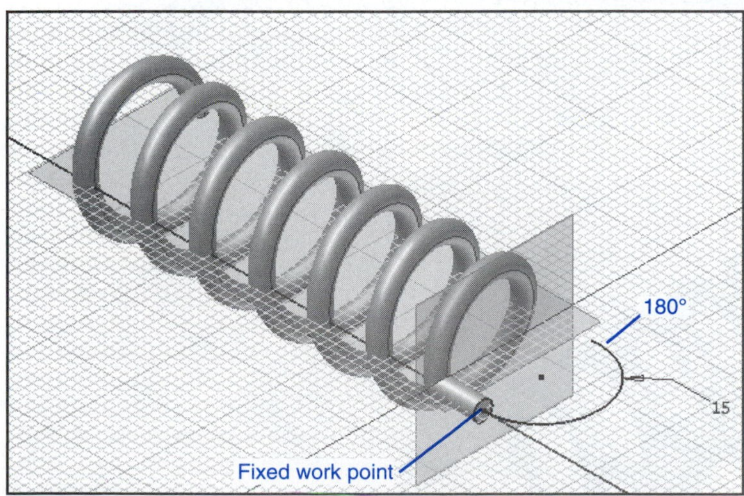

Figure 9-32

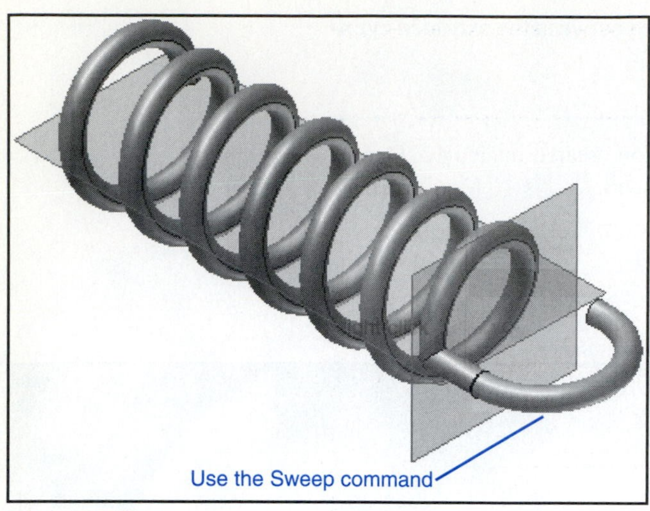

Figure 9-33

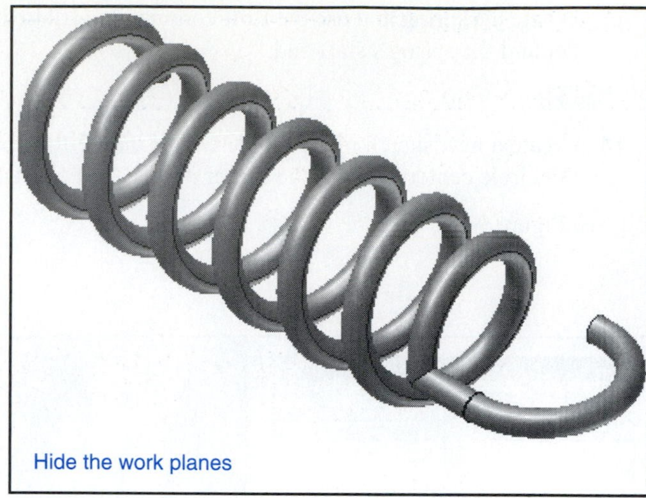

Figure 9-34

16. Use the **Sweep** command to draw the spring's hooked end.

    See Figures 9-33 and 9-34.

17. Apply the same procedure to the other end of the spring.

    See Figures 9-35 and 9-36.

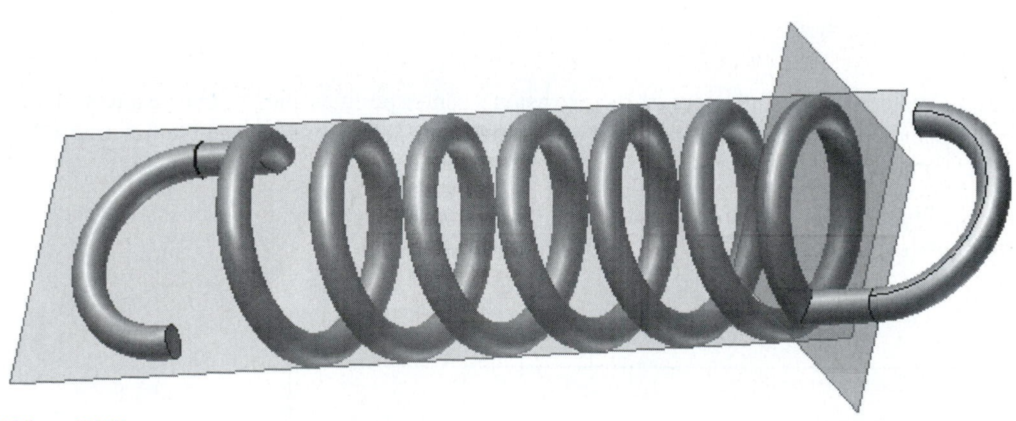

Figure 9-35

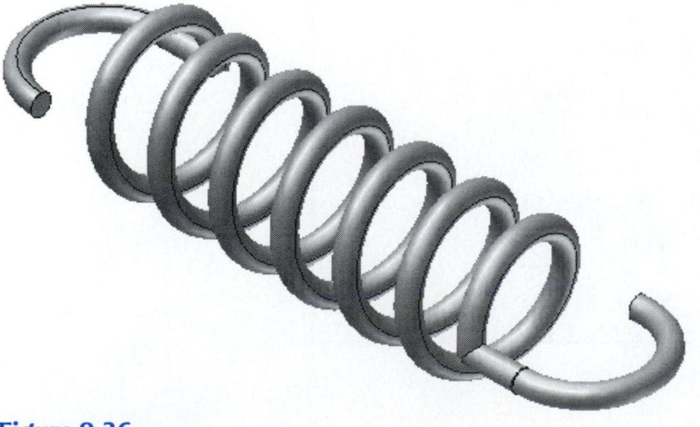

Figure 9-36

TORSION SPRINGS

Exercise 9-7: Drawing a Torsion Spring Using Design Accelerator

1. Start a new drawing using the **Standard (in).iam** format.
2. Access **Design Accelerator** and select the **Torsion Spring** option.

The **Torsion Spring Component Generator** dialog box will appear. See Figure 9-37.

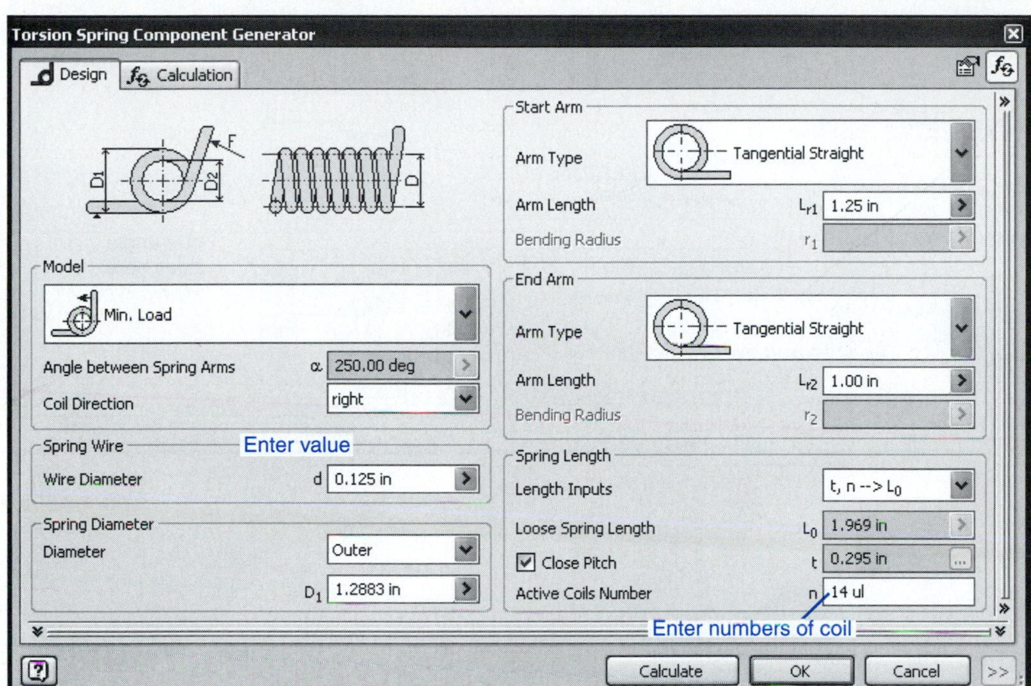

Figure 9-37

3. Enter the following values:

Coil Direction = right

Wire Diameter = 0.125

Diameter, Outer = 1.2883

Start Arm Length = 1.25

End Arm Length = 1.00

Active Coils Number = 14

4. Click the **Calculate** tab and enter the following values and options. See Figure 9-38.

Spring Strength Calculation = Work Force Calculation

Direction of Spring Load = A Load Coils the Spring

Min. Angular Deflection of Working Arm = 12.00

Angle of Working Stroke = 28

Angular Deflection of Working Arm = 12.00

5. Click **OK.**

Figure 9-39 shows the finished torsion spring.

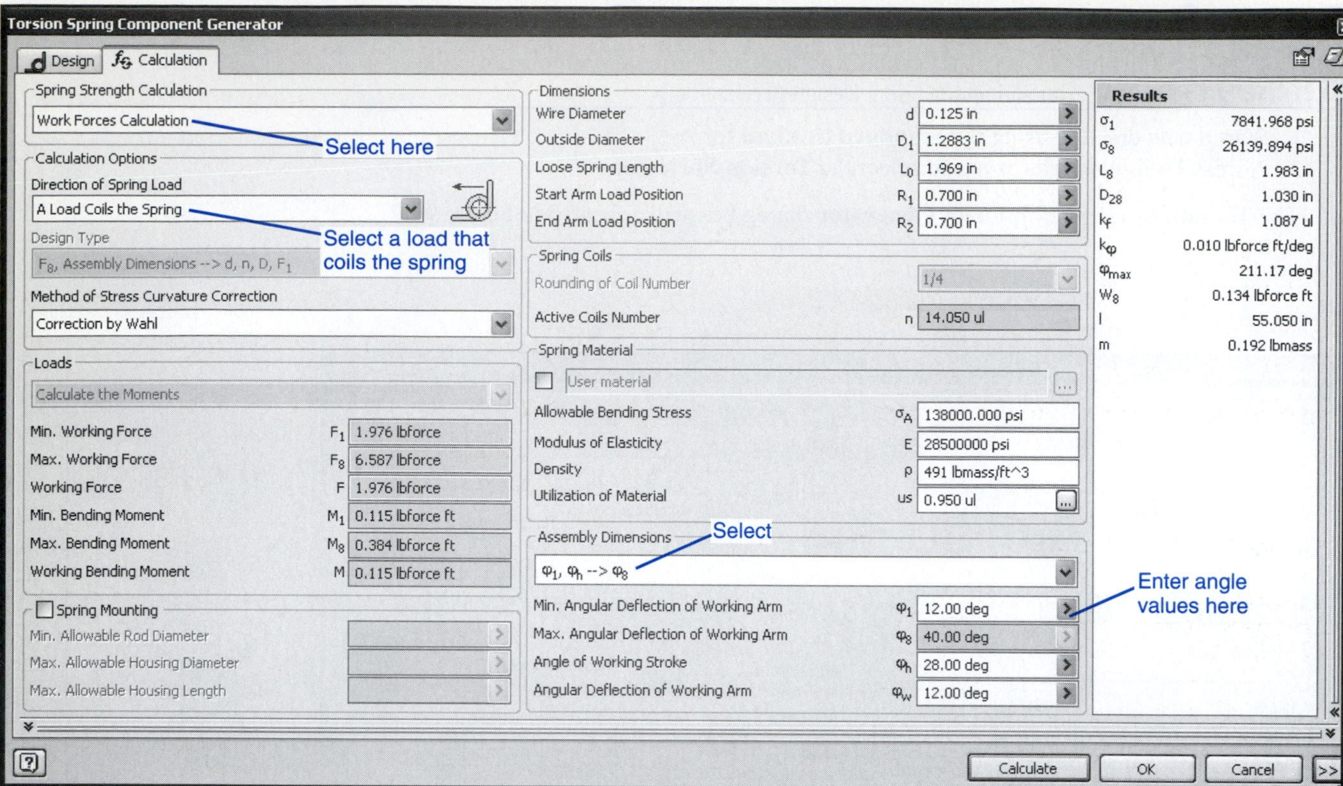

Figure 9-38

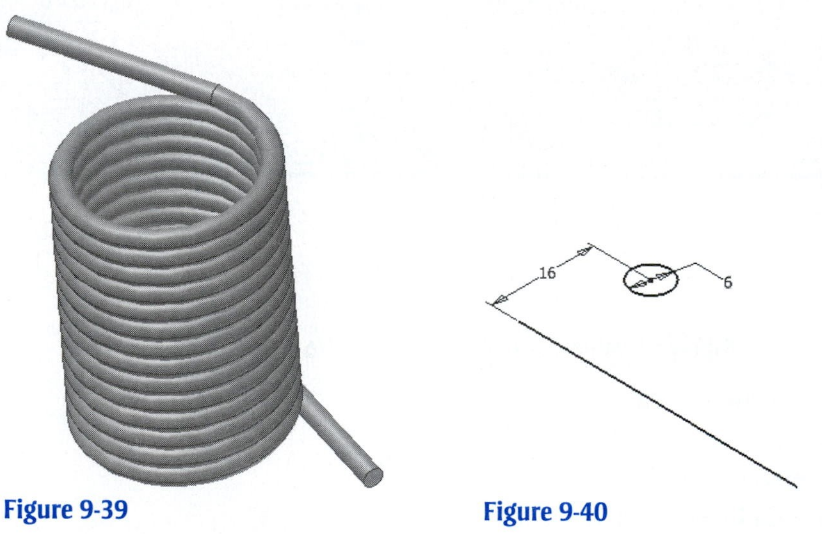

Figure 9-39 **Figure 9-40**

Exercise 9-8: Drawing a Torsion Spring Using the Coil Command

1. Start a new drawing using the **Standard (mm).ipt** format.
2. Define a wire diameter of **6 mm** and a mean diameter of **32 mm.**

See Figure 9-40.

3. Access the **Coil** command and set the **Pitch** for **10** and the **Revolution** for **8.**

See Figure 9-41.

4. Set the **Start** for **Flat** and a **Transition Angle** and **Flat Angle** of **90°.** Set the **End** for **Flat** and the **Transition Angle** and **Flat Angle** for **45°.**

See Figure 9-42.

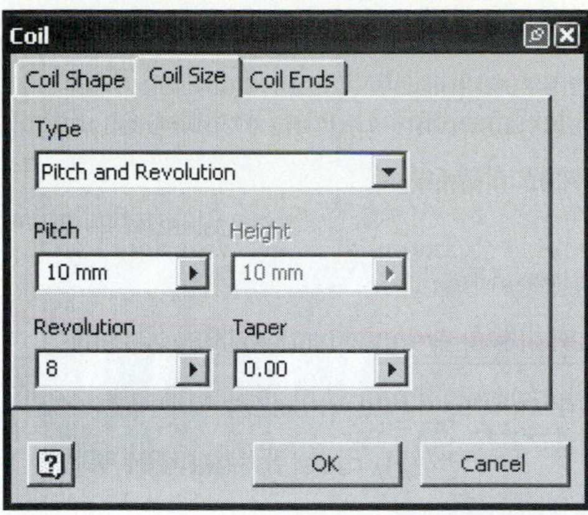

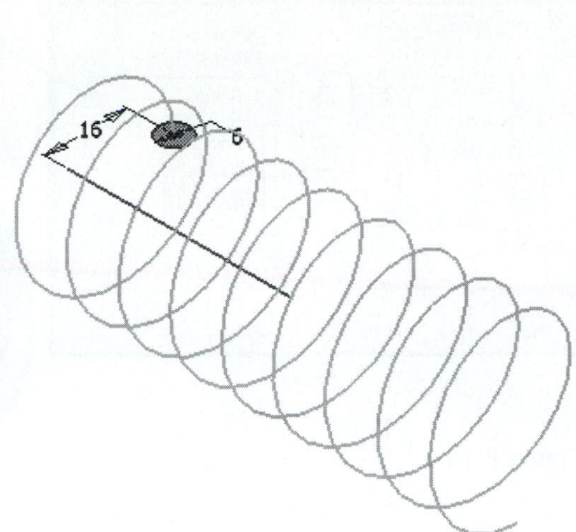

Define the pitch and revolution

Figure 9-41

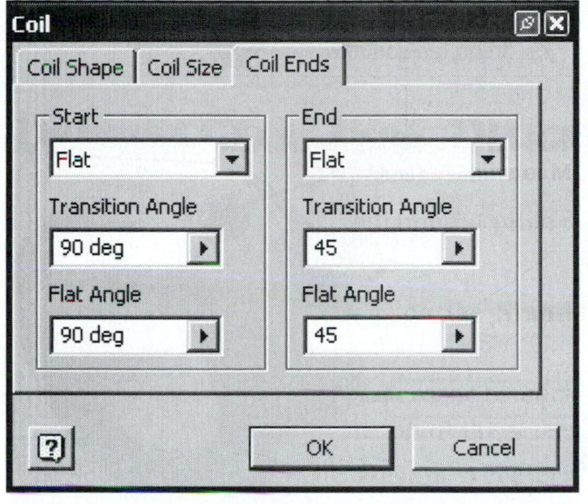

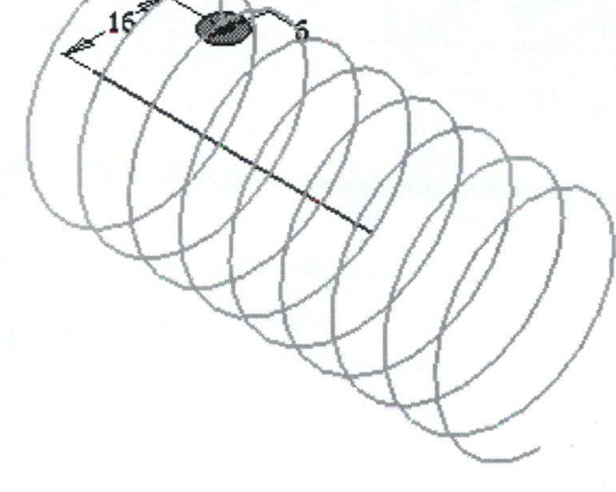

Define the coil ends

Figure 9-42

5. Rotate the spring to access the start end. Create a new sketch plane on the end. Draw a **Ø6**-mm circle on the end centered on the end's center point.

See Figure 9-43.

6. Right-click the mouse and select the **Finish Sketch** option. Extrude the start end **50** mm.

See Figure 9-44.

7. Extrude the other spring end **50** mm.

Figure 9-45 shows the finished spring.

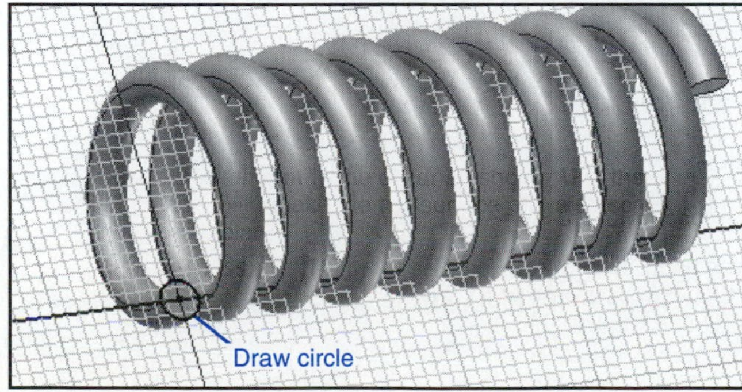

Figure 9-43

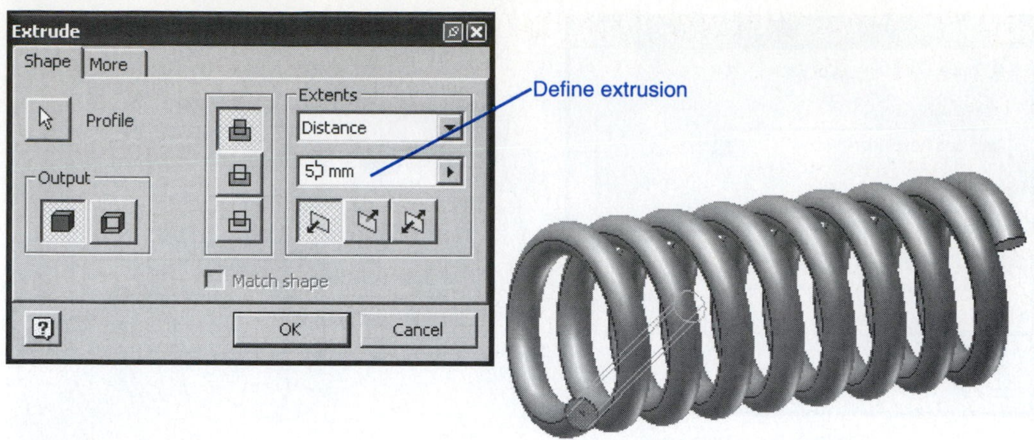

Figure 9-44

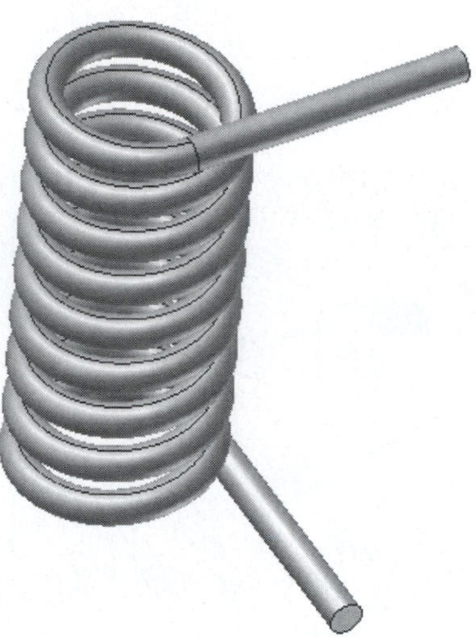

Figure 9-45

Belleville Springs

Belleville spring: A disklike device that resists bending; springs can be stacked bending in the same direction or in opposition.

A ***Belleville spring*** is a disklike device that resists bending. Belleville springs can be stacked together, all bending in the same direction, or stacked in opposition. The **Design Accelerator** can be used to generate either condition.

Exercise 9-9: Drawing a Belleville Spring Using Design Accelerator

1. Start a new drawing using the **Standard (in).ipt** format.
2. Access **Design Accelerator** and click the **Belleville Spring** option.

The **Belleville Spring Generator** dialog box will appear. See Figure 9-46.

3. Click on the arrow to the right of the **Single-disk Spring Dimensions** heading.

A list of options will cascade down.

4. Select the **3.000 in x 1.500 in x 0.100 in x 0.175 in** option.

See Figure 9-47.

5. Enter the **Height** value.

In this example a value of **0.16** was used. See Figure 9-48

The finished spring will appear. See Figure 9-49. Figure 9-50 shows orthographic views of the spring.

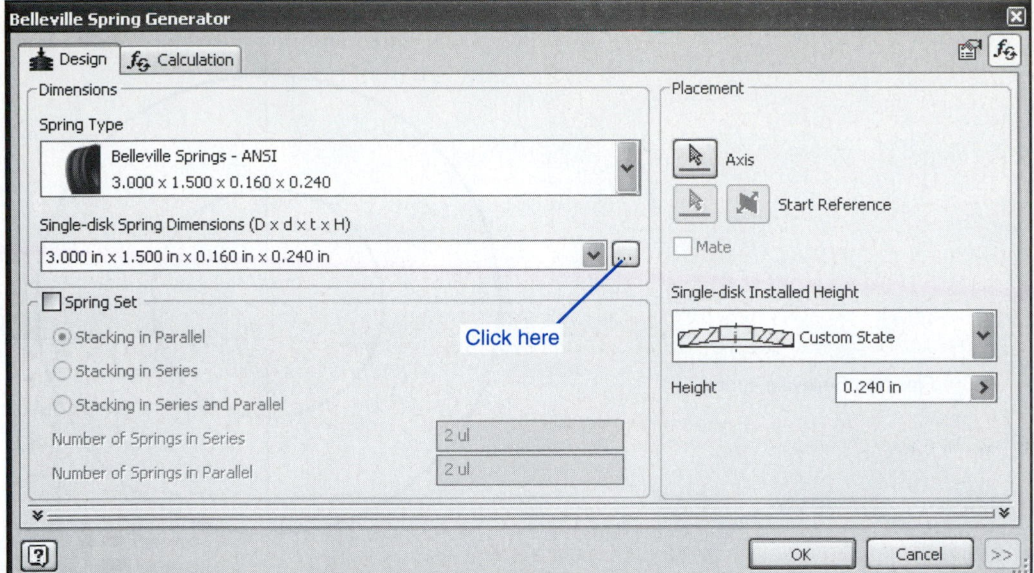

Figure 9-46

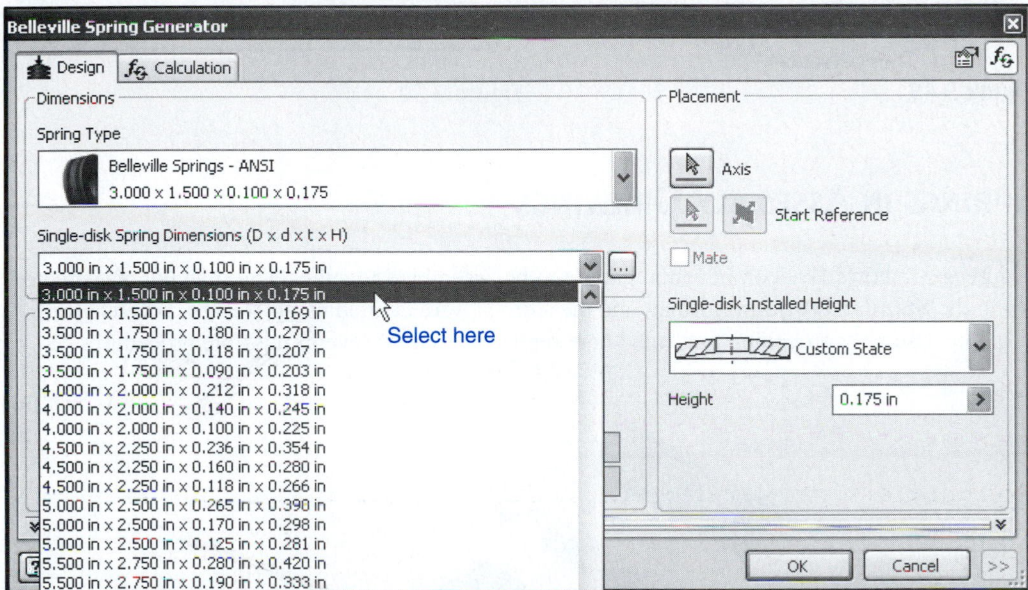

Figure 9-47

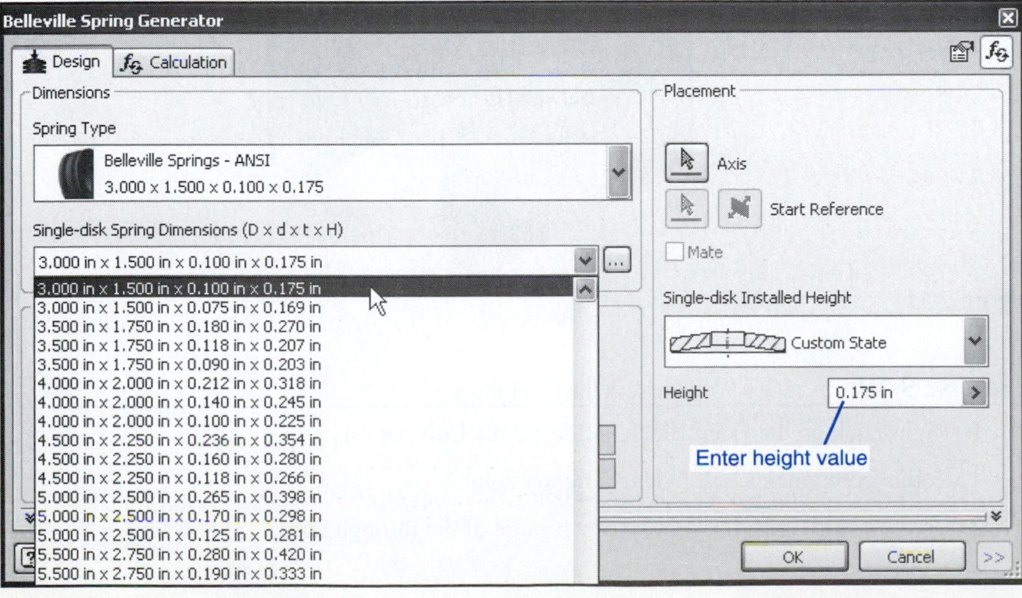

Figure 9-48

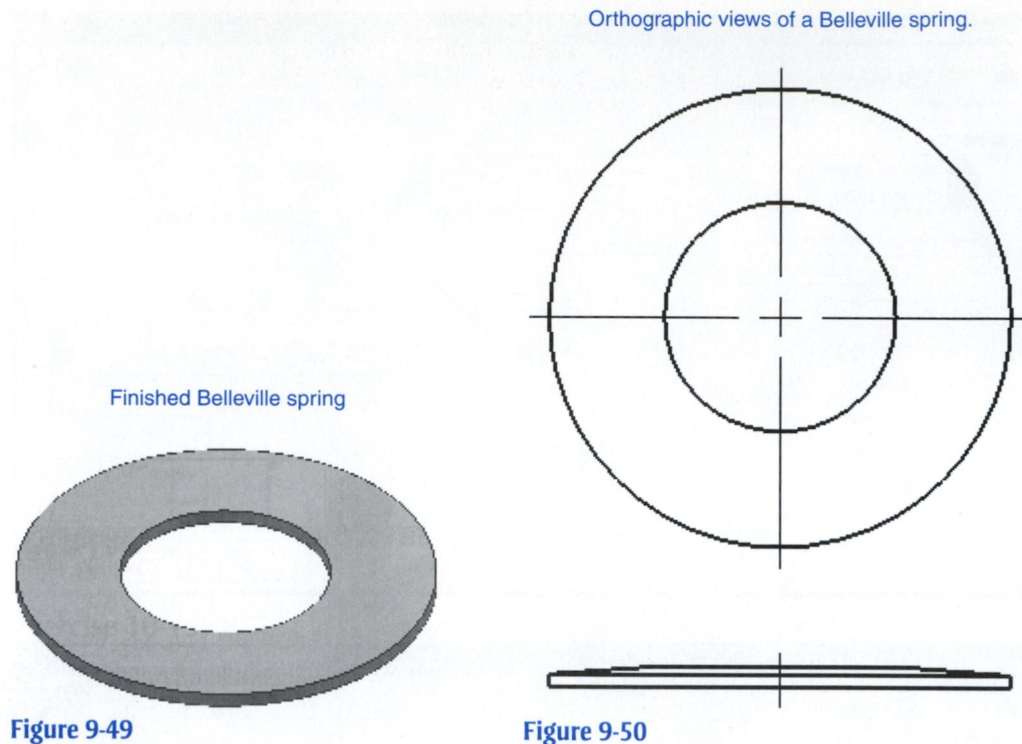

Figure 9-49 **Figure 9-50**

Springs in Assembly Drawings

Figure 9-51 shows five components that are to be assembled together. The drawing was created using the **Standard (in).iam** format, and the springs were created using **Design Accelerator.**

The assembly procedure presented here represents one of several possible procedures.

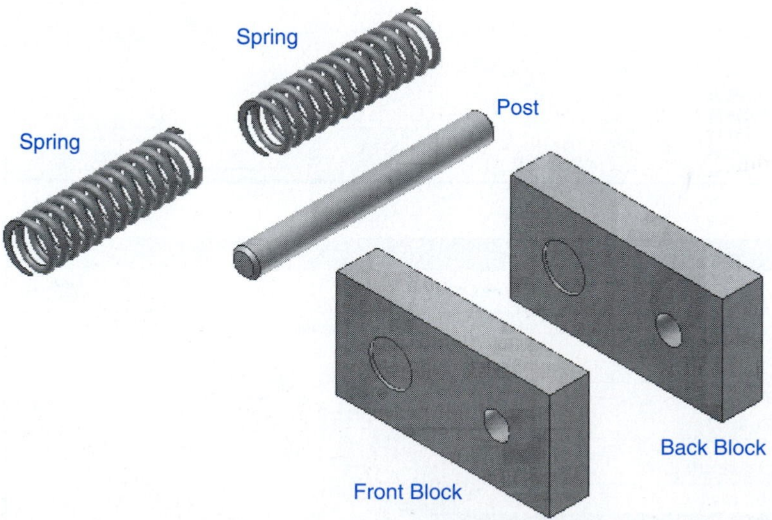

Figure 9-51

Exercise 9-10: Creating a Work Axis on the Front Block

1. Right-click on the Front Block and select the **Edit** option.

The other components will fade to a lighter color.

2. Create a work axis through the center point of the through hole.

3. Create a new sketch plane aligned with the bottom surface of the shallow hole. Click the right mouse button and select the **Finish Sketch** (*not* ***Finish Edit***) option. Create a work point on the center point of the shallow hole.

See Figure 9-52.

4. Create an XY work plane aligned with the work point.
5. Create a work axis perpendicular to the work plane through the work point.

See Figure 9-53.

6. Right-click the mouse and select **Finish Edit.**

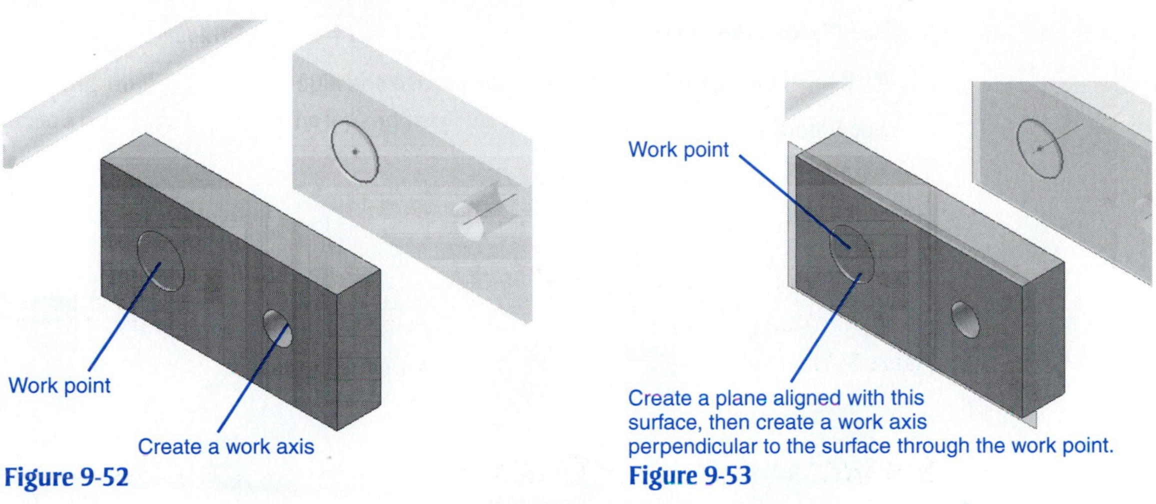

Figure 9-52

Figure 9-53

Exercise 9-11: Adding a Work Axis to the Springs

1. In the browser box, click the **Compression Spring** heading, then click the **Work Axis** heading.

See Figure 9-54.

2. Right-click on **WorkAxis1** and select the **Visibility** option.

The work axis will appear on the spring. See Figure 9-55.

3. Repeat the procedure for the second spring.

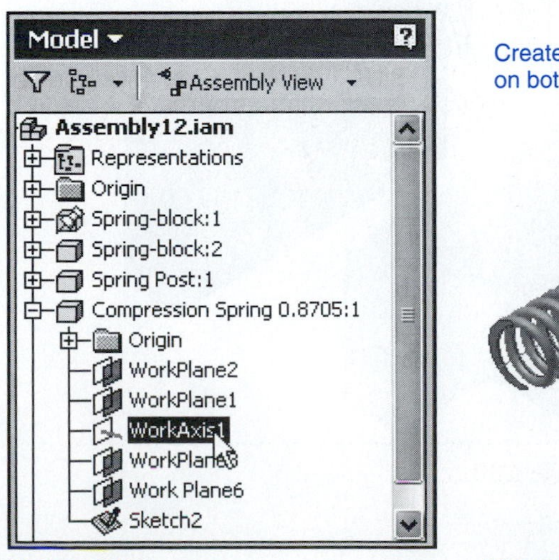

Figure 9-54

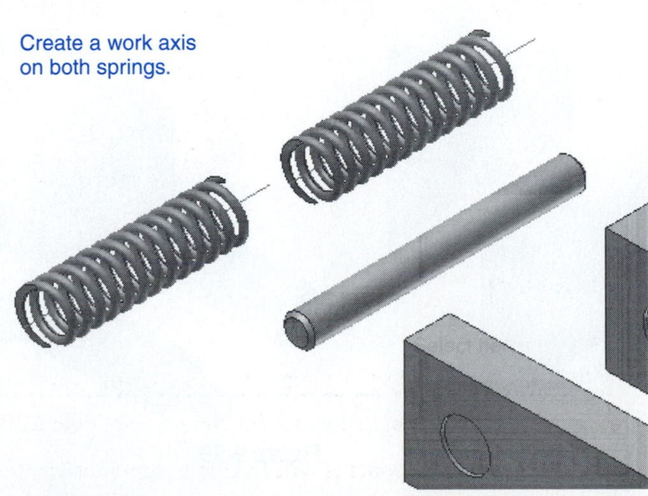

Figure 9-55

Exercise 9-12: Assembling the Components

1. Insert the Post into the Front Block.

See Figure 9-56. If the front surface of the Front Block is used, offset the post **−0.75** so that it aligns with the back surface of the Front Block.

2. Rotate one of the springs so that the ground end is accessible. Use the **Mate Constraint** and align the spring's ground surface with the front surface of the Front Block.

See Figure 9-57.

3. Use the **Mate Constraint** to align the spring's work axis with the work axis through the through hole.

See Figure 9-58.

4. Repeat the procedure to position the second spring.

See Figure 9-59.

5. Add the Rear Block to the assembly.

See Figure 9-60.

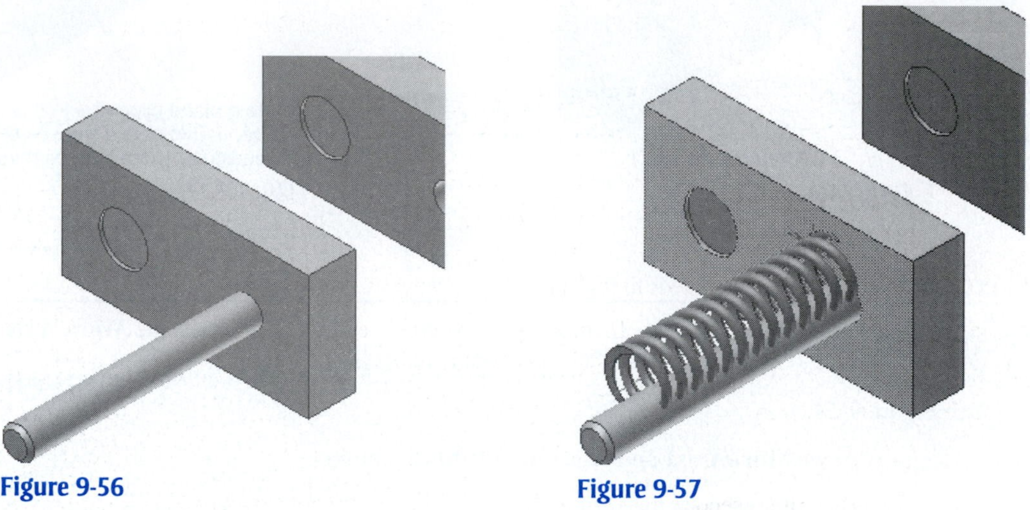

Figure 9-56

Figure 9-57

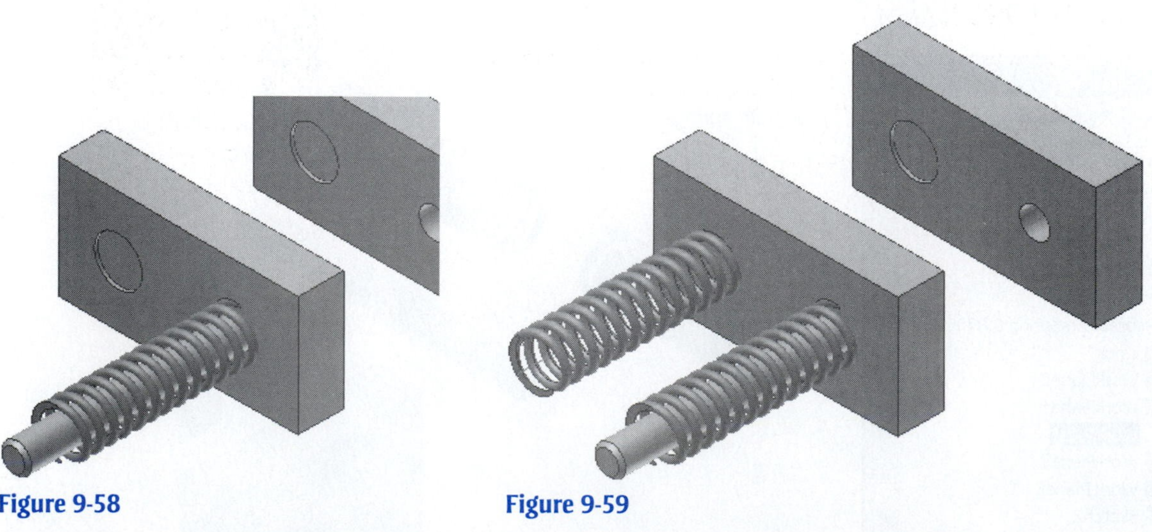

Figure 9-58

Figure 9-59

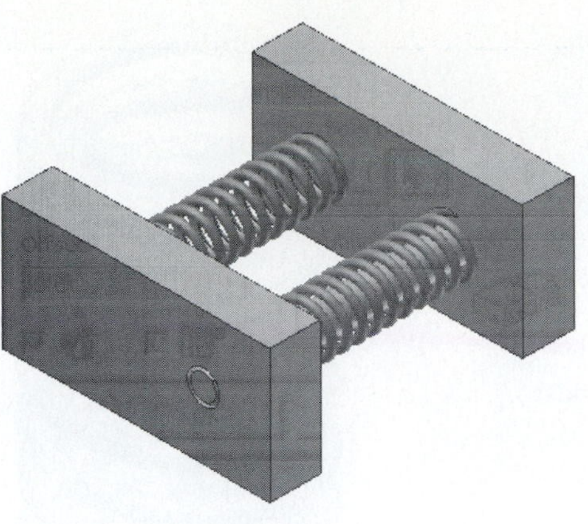

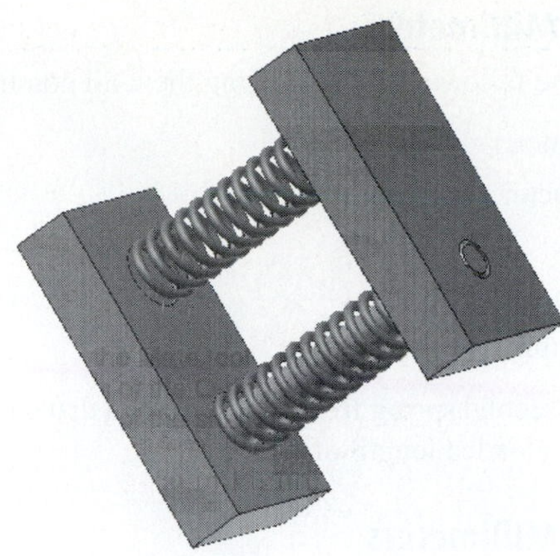

Figure 9-60

Summary

This chapter showed how to draw four types of springs—compression, extension, torsion, and Belleville—using both the **Coil** and the **Design Accelerator** tools. Coil features such as natural ends and flat ends with different transition and flat angles were illustrated.

Chapter Projects

Project 9-1: Inches

A. Draw the following springs using the **Coil** command.

Mean Diameter = 2.00

Wire Diameter = 0.125

Pitch = 0.375

Revolution = 16

Ends = Natural

B. Draw a second spring from the same data that has ground ends and a preloaded length of 5.00.

Figure P9-1

Project 9-2: Inches

A. Draw the following springs using the **Coil** command.

Mean Diameter = 0.50

Wire Diameter = 0.06

Pitch = 0.12

Revolution = 10

Ends = Flat, Transition angle = 90°, Flat angle = 0°

B. Draw a second spring from the same data that has ground ends and a preloaded length of 1.00.

Project 9-3: Millimeters

A. Draw the following springs using the **Coil** command.

Mean Diameter = 8

Wire Diameter = 3

Pitch = 4

Revolution = 20

Ends = Natural

B. Draw a second spring from the same data that has ground ends and a preloaded length of 65.00.

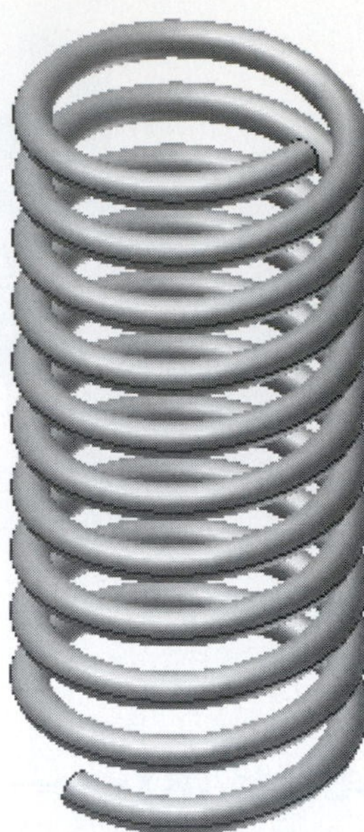

Figure P9-2

Project 9-4: Millimeters

A. Draw the following springs using the **Coil** command.

Mean Diameter = 24

Wire Diameter = 12

Pitch = 14

Revolution = 6

Ends = Flat, Transition Angle = 90°, Flat Angle = 45°

B. Draw a second spring from the same data that has ground ends and a preloaded length of 75.00.

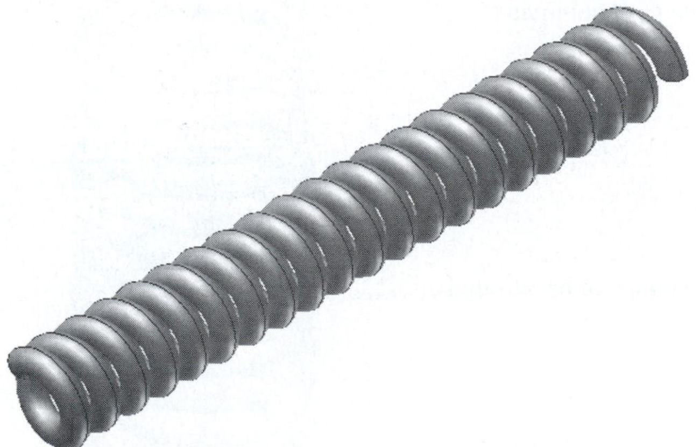

Figure P9-3

Figure P9-4

Project 9-5: Millimeters

A. Draw the following springs using the **Coil** command.

Mean Diameter = 25

Wire = 5 × 5 square

Pitch = 6

Revolution = 8

Ends = Natural

Draw the following compression springs using the **Design Accelerator** tool.

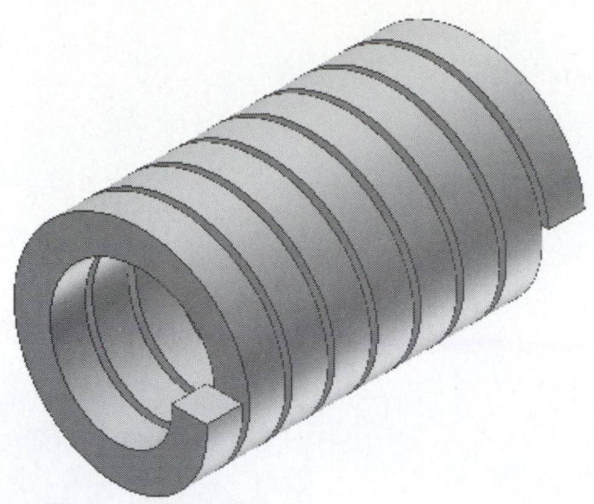

Figure P9-5

Project 9-6: Inches

Wire Diameter = 0.072

Outside Diameter = 0.625

Loose Spring Length = 3.15

Preloaded Spring Length = 3.00

Working Stroke = 0.15

Working Spring Length = 2.95

Coil Direction = Right

Active Coils Number = 16

Both coils

End Coil = 1.5

Transition Coil = 1.0

Ground Coil = 0.75

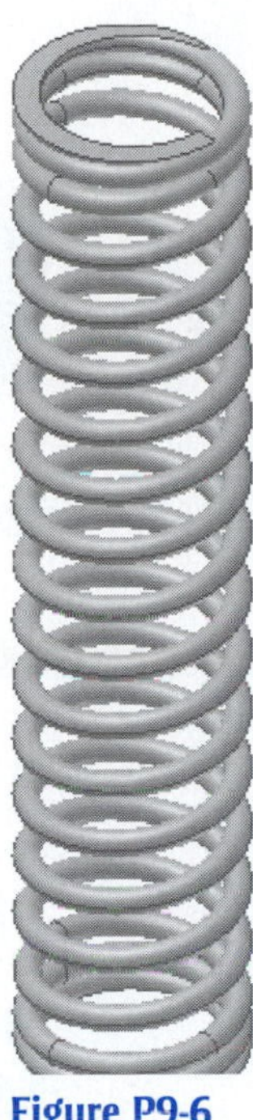

Figure P9-6

Figure P9-7

Project 9-7: Inches

Wire Diameter = 0.2437

Outside Diameter = 1.250

Loose Spring Length = 4.1

Preloaded Spring Length = 3.825

Working Stroke = 0.20

Working Spring Length = 3.82

Coil Direction = Left

Active Coils Number = 9

Both coils

End Coil = 1.5

Transition Coil = 1.0

Ground Coil = 0.75

Project 9-8: Millimeters

Wire Diameter = 3.00

Outside Diameter = 20.0

Loose Spring Length = 81.0

Preloaded Spring Length = 80.0

Working Stroke = 4.0

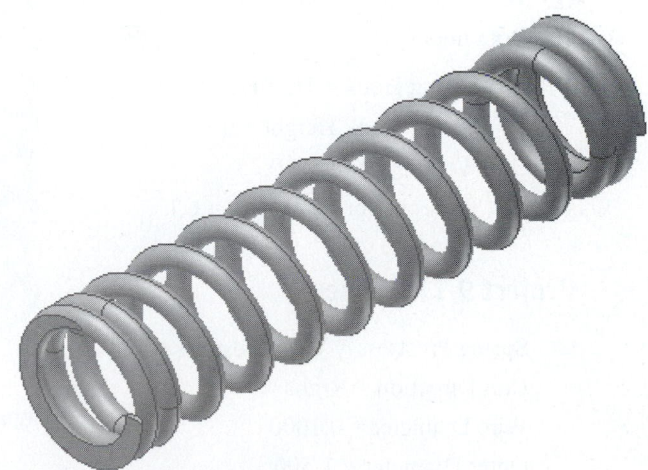

Figure P9-8

Working Spring Length = 80.0

Coil Direction = Right

Active Coils Number = 10

Both coils

End Coil = 1.5

Transition Coil = 1.0

Ground Coil = 0.75

Project 9-9: Millimeters

Wire Diameter = 5.0

Outer Diameter = 30.0

Loose Spring Length = 101.0

Preloaded Spring Length = 100.0

Working Stroke = 6.0

Working Spring Length = 94.0

Coil Direction = Right

Active Coils Number = 9

Both coils

End Coil = 1.5

Transition Coil = 1.0

Ground Coil = 0.75

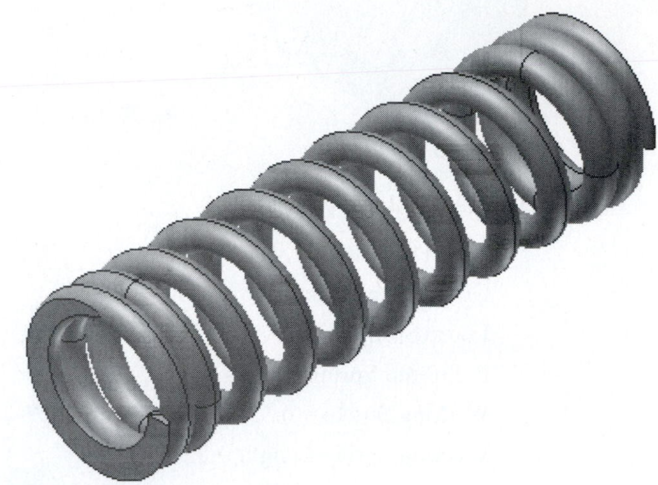

Figure P9-9

Draw the following tension springs using the **Design Accelerator** tool. *Note:* $L_0 < L_c < L_8$. The **Custom Length** of the spring must be greater than the **Loose Spring Length** and less then the **Fully Loaded Spring Length.**

Project 9-10: Inches

Spring Prestress = With Prestress

Coil Direction = Right

Wire Diameter = 0.0938

Outer Diameter = 1.000

Preloaded Spring Length = 6.00

Total Number of Coils = 12

Both hooks

Spring Hook = Half Hook

Spring Hook Height = 0.50

Working Stroke = 0.25

Working Spring Length = 6.1

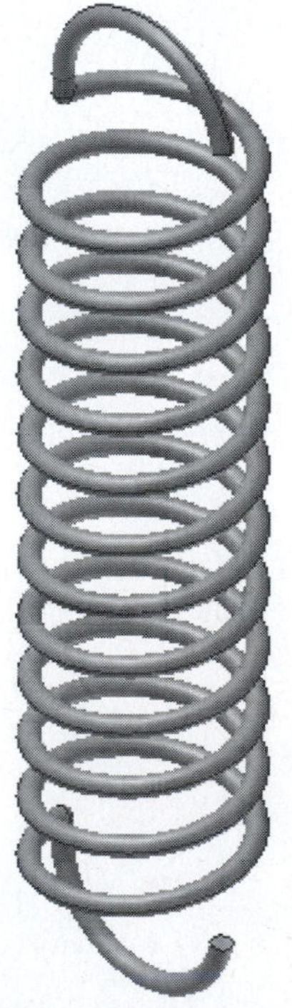

Figure P9-10

Project 9-11: Inches

Spring Prestress = With Prestress

Coil Direction = Right

Wire Diameter = 0.1000

Outer Diameter = 1.500

Loose Spring Length = 6.00

Total Number of Coils = 16

Both hooks

Spring Hook = Non-specified Hook Type

Fully Loaded Spring Length = 6.25

Working Stroke = 0.300

Working Spring Length = 6.00

Project 9-12: Millimeters

Spring Prestress = Without Prestress

Coil Direction = Right

Wire Diameter = 4

Outer Diameter = 22

Loose Spring Length = 130

Total Number of Coils = 8

Both hooks

Spring Hook = Raised Hook

Hook Length = 16.936

Fully Loaded Spring Length = 138

Working Stroke = 4

Working Spring Length = 136

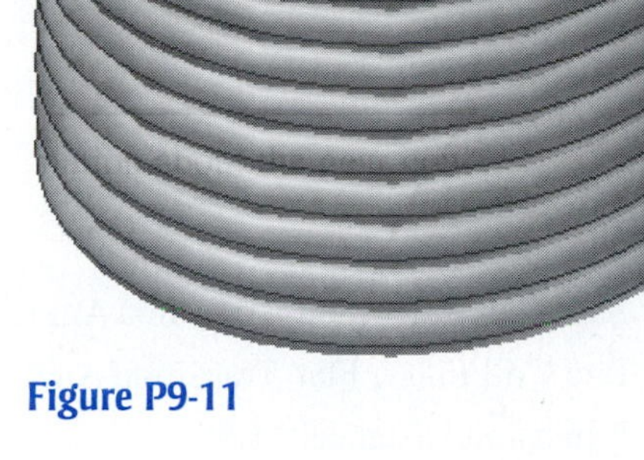

Figure P9-11

Project 9-13: Millimeters

Spring Prestress = With Prestress

Coil Direction = Right

Wire Diameter = 3.75

Outer Diameter = 40

Loose Spring Length = 64.923

Total Number of Coils = 4.18

Both hooks

Spring Hook = Half Hook

Hook Length = 22.514

Fully Loaded Spring Length = 80

Working Stroke = 6

Working Spring Length = 74

Figure P9-12

Draw the following tension springs using the **Coil** command.

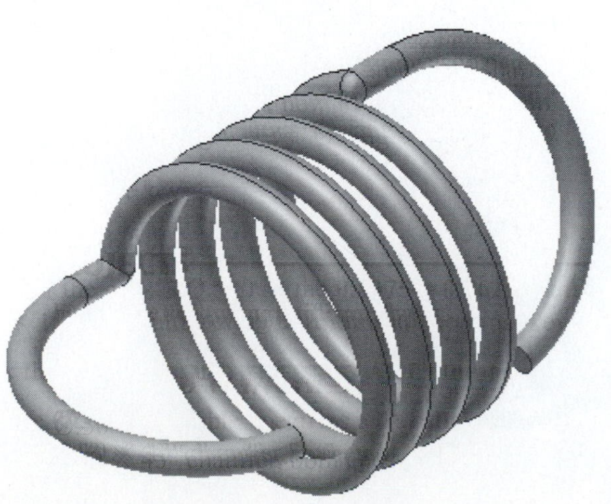

Figure P9-13

Project 9-14: Inches

Wire Diameter = 0.19

Mean Diameter = 1.00

Pitch = .375

Revolution = 12

Taper = 0

Start Coil End = Flat, Transition Angle = 90.0°, Flat Angle = 90.0°

End Coil End = Flat, Transition Angle = 0.0°, Flat Angle = 0.0°

Extension Distance = 0.50

Arc Radius = 1.00

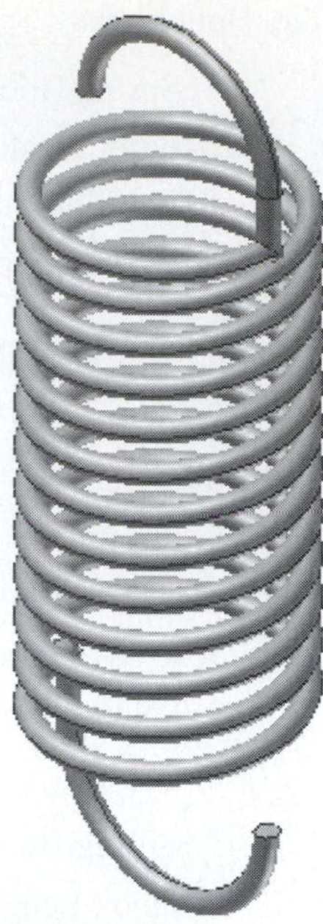

Figure P9-14

Project 9-15: Inches

Wire Diameter = 0.375

Mean Diameter = 1.50

Pitch = .625

Revolution = 24

Taper = 0

Start Coil End = Flat, Transition Angle = 90.0°, Flat Angle = 90.0°

End Coil End = Flat, Transition Angle = 0.0°, Flat Angle = 0.0°

Extension Distance = 1.50

Arc Radius = 1.50

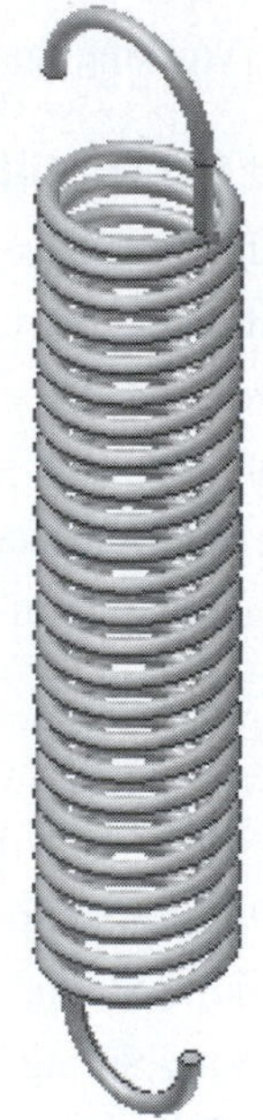

Figure P9-15

Project 9-16: Millimeters

Wire Diameter = 16

Mean Diameter = 30

Pitch = 30

Revolution = 10

Taper = 0

Start Coil End = Flat, Transition Angle = 90.0°, Flat Angle = 90.0°

End Coil End = Flat, Transition Angle = 0.0°, Flat Angle = 0.0°

Extension Distance = 18

Arc Radius = 30

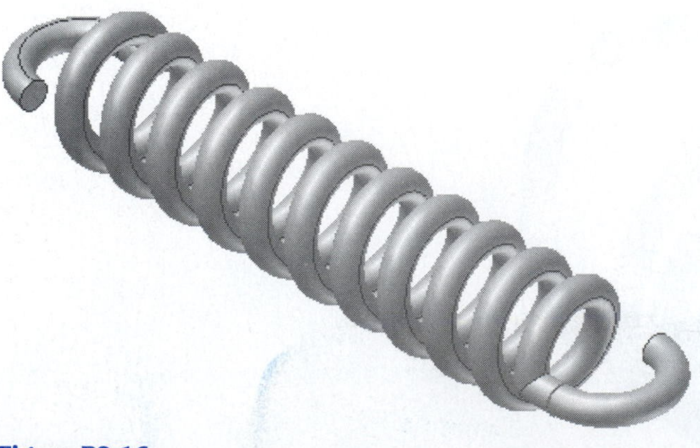

Figure P9-16

Project 9-17: Millimeters

Wire Diameter = 8

Mean Diameter = 24

Pitch = 14

Revolution = 16

Taper = 0

Start Coil End = Flat, Transition Angle = 90.0°, Flat Angle = 90.0°

End Coil End = Flat, Transition Angle = 0.0°, Flat Angle = 0.0°

Extension Distance = 26

Arc Radius = 24

Draw the following torsion springs using the **Design Accelerator** tool. Use the default values for any values not specified.

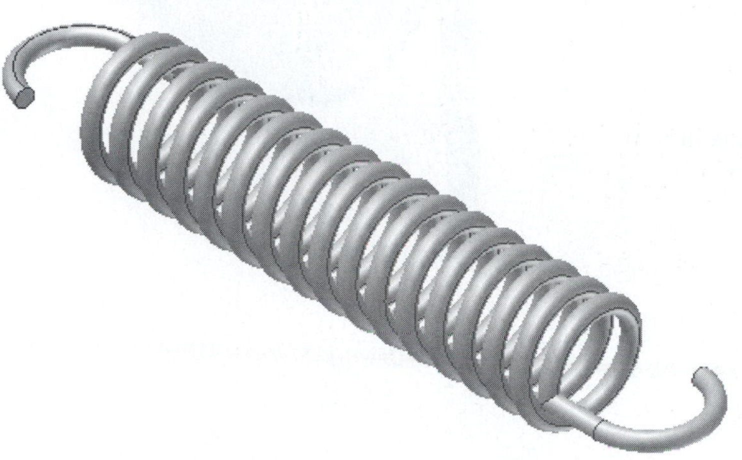

Figure P9-17

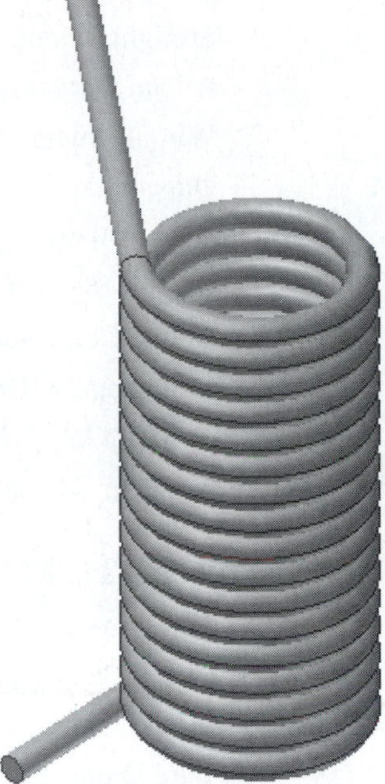

Figure P9-18

Project 9-18: Inches

Coil Direction = Right

Straight torsion arms

A load that coils the spring

Wire Diameter = 0.1055

Outside Diameter = 0.875

Start Arm Length = 1.50

End Arm Length = 0.75

Active Coils Number = 12

Min. Angular Deflection of Working Arm = 12

Angle of Working Stroke = 36

Angle of Deflection of Working Arm = 15

Project 9-19: Inches

Coil Direction = Right

Straight torsion arms

A load that uncoils the spring

Wire Diameter = 0.072

Outside Diameter = 0.500

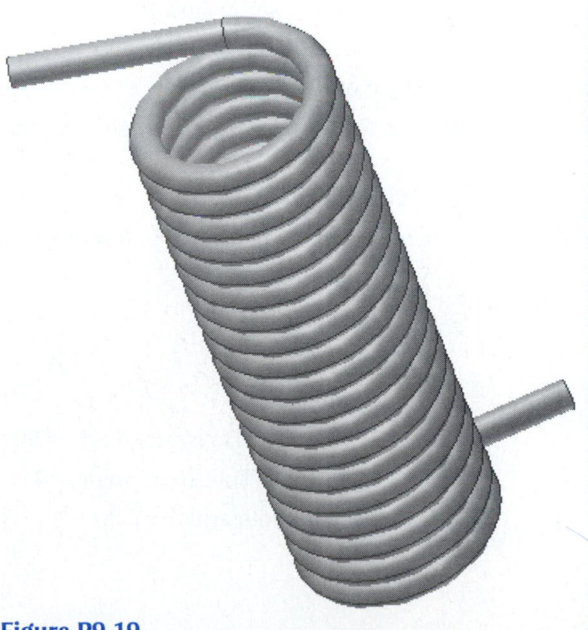

Figure P9-19

Start Arm Length = 0.50
End Arm Length = 0.50
Active Coils Number = 18
Min. Angular Deflection of Working Arm = 20
Angle of Working Stroke = 10
Angle of Deflection of Working Arm = 30

Project 9-20: Millimeters

Coil Direction = Left
Straight torsion arms
A load that coils the spring
Wire Diameter = 2.00
Outside Diameter = 16
Arm of Working Force = 20.0
Arm of Support = 20.0
Active Coils Number = 10
Min. Angular Deflection of Working Arm = 10
Angle of Working Stroke = 40
Angle of Deflection = 10

Figure P9-20

Project 9-21: Millimeters

Coil Direction = Right
Straight torsion arms
A load that uncoils the spring
Wire Diameter = 3.0
Mean Diameter = 40
Arm of Working Force = 35
Arm of Support = 30
Active Coils Number = 20
Min. Angular Deflection of Working Arm = 15
Angle of Working Stroke = 50
Angle of Deflection = 15

Figure P9-21

Draw the following torsion springs using the **Coil** command.

Project 9-22: Inches

Create a drawing using the **Standard (in).ipt** format.
Wire Diameter = 0.19
Mean Diameter = 1.25
Pitch = 0.375
Revolution = 12
Start Coil = Flat, Transition Angle = 90°, Flat Angle = 90°
End Coil = Flat, Transition Angle = 45°, Flat Angle = 45°
Arm Length (both arms) = 2.00

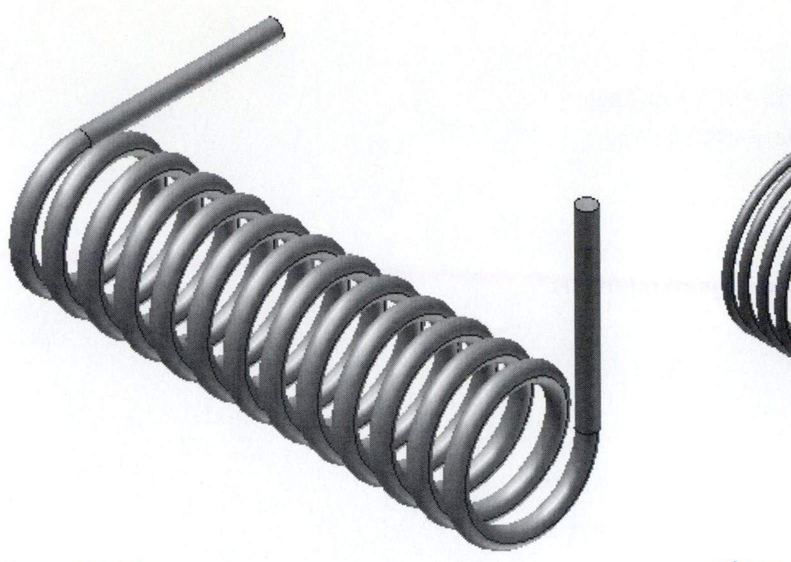

Figure P9-22

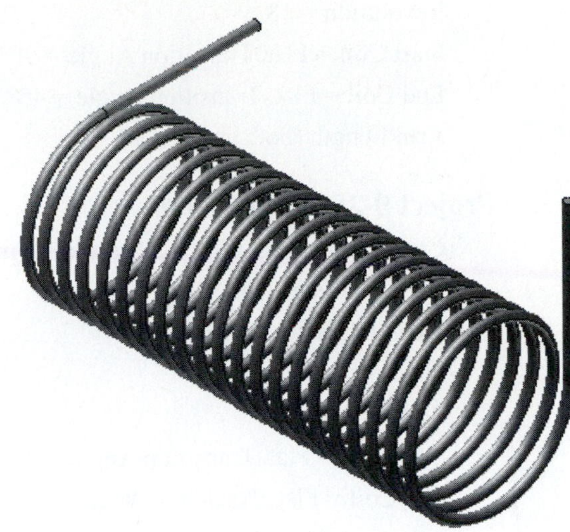

Figure P9-23

Project 9-23: Inches

Create a drawing using the **Standard (in).ipt** format.

Wire Diameter = 0.06

Mean Diameter = 0.50

Pitch = 0.125

Revolution = 20

Start Coil = Flat, Transition Angle = 90°, Flat Angle = 90°

End Coil = Flat, Transition Angle = 45°, Flat Angle = 45°

Arm Length (both arms) = 1.00

Project 9-24: Millimeters

Create a drawing using the **Standard (mm).ipt** format.

Wire Diameter = 4.0

Mean Diameter = 12.0

Figure P9-24

Pitch = 6

Revolution = 18

Start Coil = Flat, Transition Angle = 90°, Flat Angle = 90°

End Coil = Flat, Transition Angle = 45°, Flat Angle = 45°

Arm Length (both arms) = 15

Project 9-25: Millimeters

Create a drawing using the **Standard (mm).ipt** format.

Wire Diameter = 5

Mean Diameter = 40.0

Pitch = 10

Revolution = 18

Start Coil = Flat, Transition Angle = 90°, Flat Angle = 90°

End Coil = Flat, Transition Angle = 45°, Flat Angle = 45°

Arm Length (both arms) = 60

Figure P9-25

Project 9-26: Inches

Draw the Damper Assembly shown in Figure P9-26. This exercise is loosely based on a damper system placed under the seat of helicopter pilots to help minimize the amount of vibration they are subjected to.

Spring data:

Spring Forces Calculation = Work Force Calculation

Wire Diameter = 0.125

Outside Diameter = 1.000

Preloaded Spring Length = 2.10

Working Stroke = 0.25

Working Spring Length = 2.00

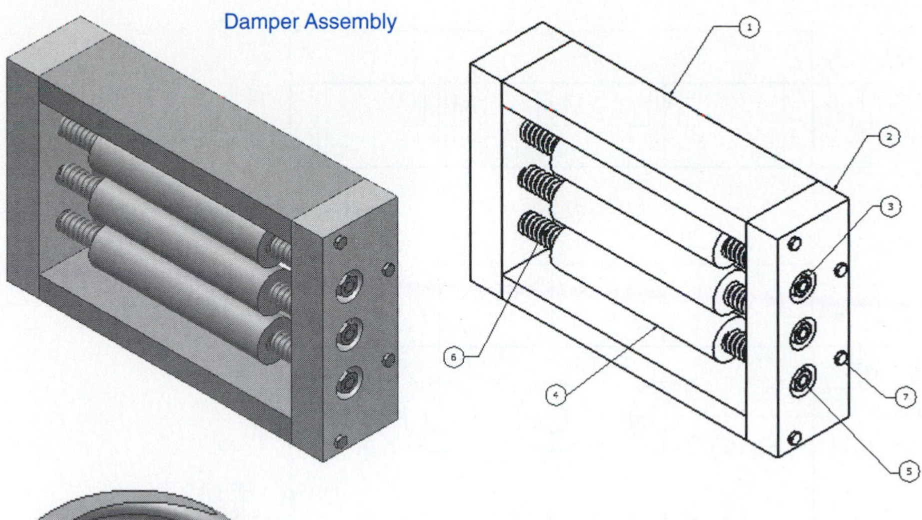

Parts List				
ITEM	PART NUMBER	DESCRIPTION	MATERIAL	QTY
1	AM311-1	BASE	Steel, Mild	1
2	AM311-2	PLATE,END	Steel, Mild	2
3	AM311-3	POST, GUIDE	Steel, Mild	3
4	EK-152	WEIGHT	Steel, Mild	3
5	AS 2465 - 1/2 UNC	HEX NUT	Steel, Mild	6
6		COMPRESSION SPRING	Steel, Mild	6
7	AS 2465 - 1/4 x 2 1/2 UNC	HEX BOLT	Steel, Mild	8

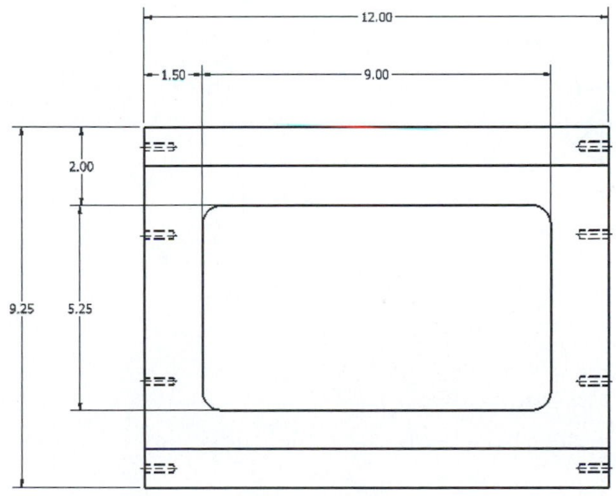

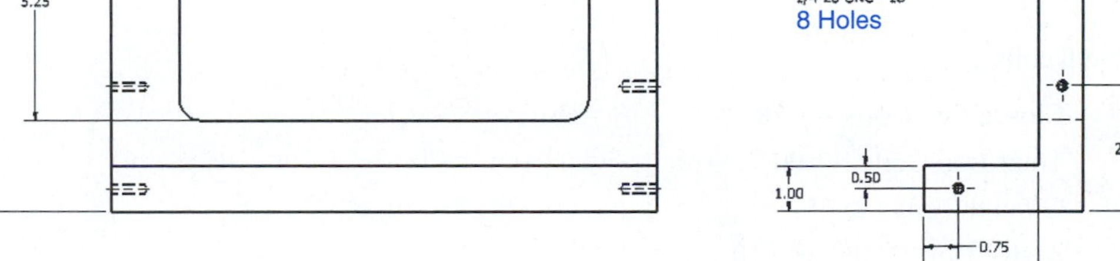

Figure P9-26 *(Continued)*

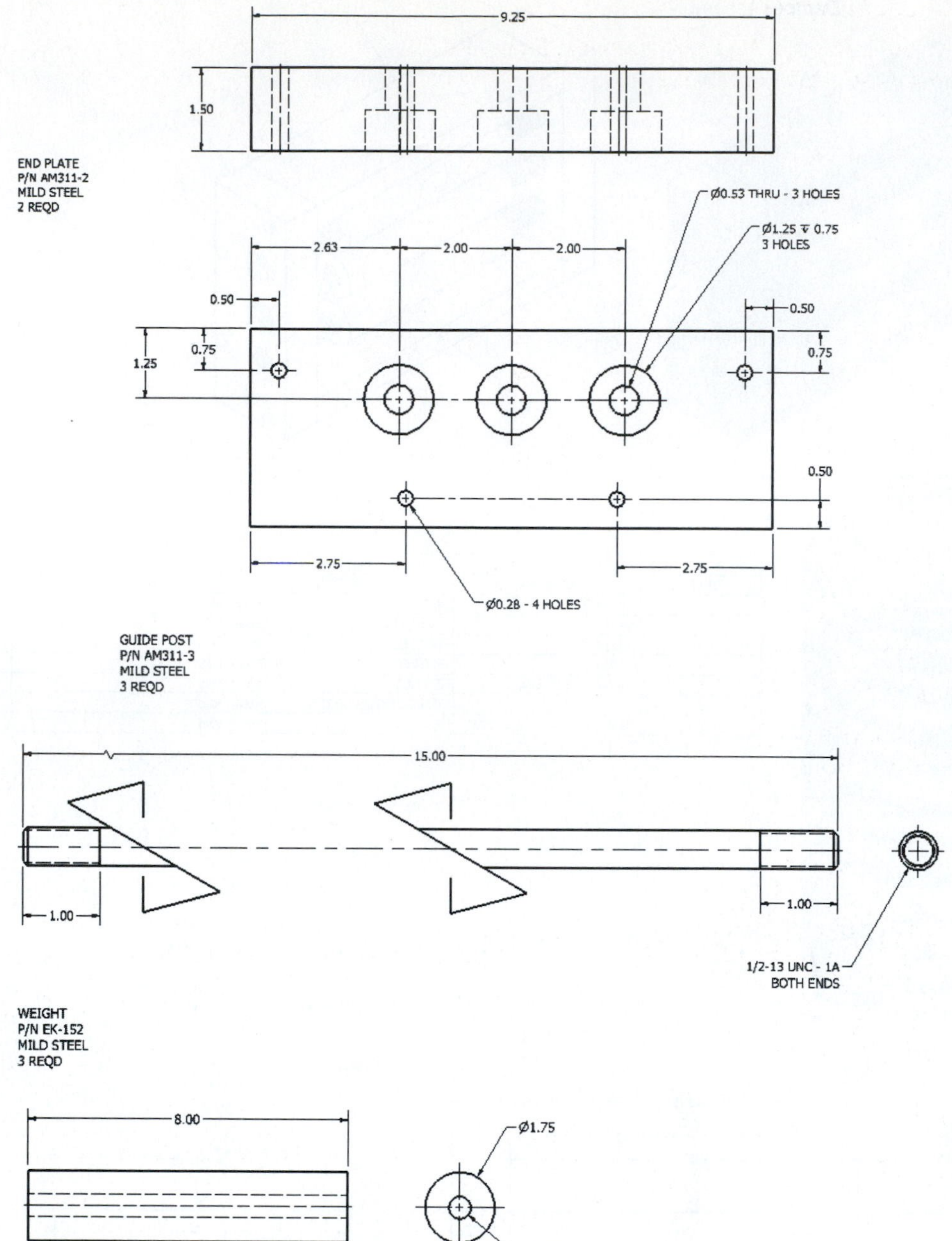

Figure P9-26

Both coils

Closed End Coils = 1.50

Transition Coils = 1.00

Ground Coils = 0.75

Active Coils Number = 10

Draw the following:

A. An assembly drawing
B. A presentation drawing
C. An exploded isometric drawing
D. A parts list

Project 9-27: Millimeters

Draw the Circular Damper Assembly shown in Figure P9-27.
The Threaded Post is M18 × 530 mm.
Spring data:

> Wire Diameter = 3.0
>
> Outer Diameter = 28
>
> Loose Spring Length = 81

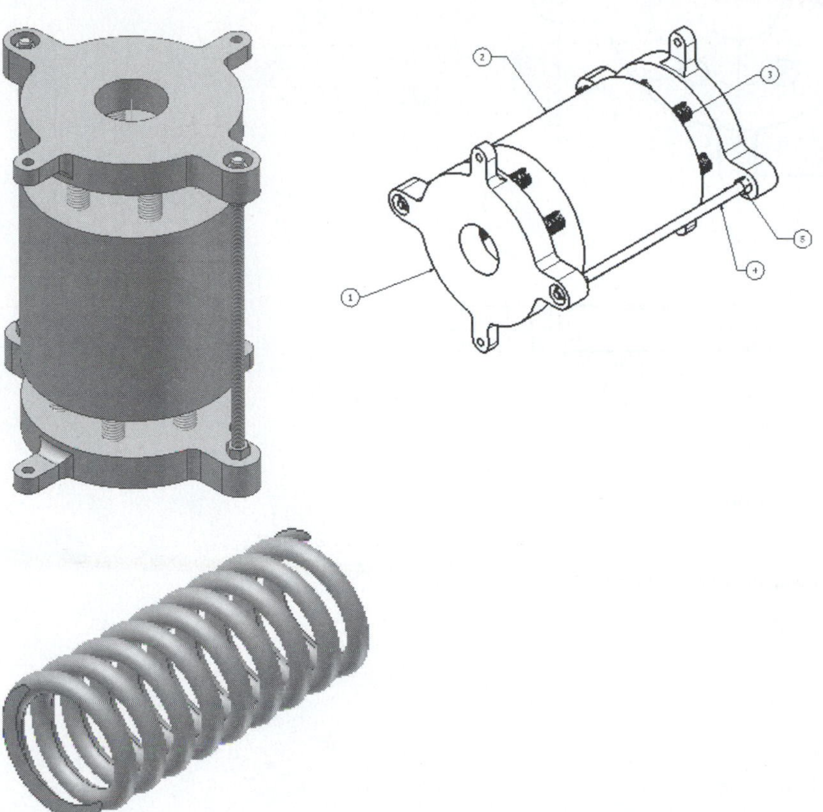

Parts List				
ITEM	PART NUMBER	DESCRIPTION	MATERIAL	QTY
1	BU2008-1	BASE, HOLDER	Steel	2
3	BU2008-2	SPRING, COMPRESSION	Steel, Mild	12
2	BU2008-3	COUNTERWEIGHT	Steel	1
4	AM312-12	POST, THREADED	Steel	2
5	AS 1112 - M18 Type	HEX NUT	Steel, Mild	8

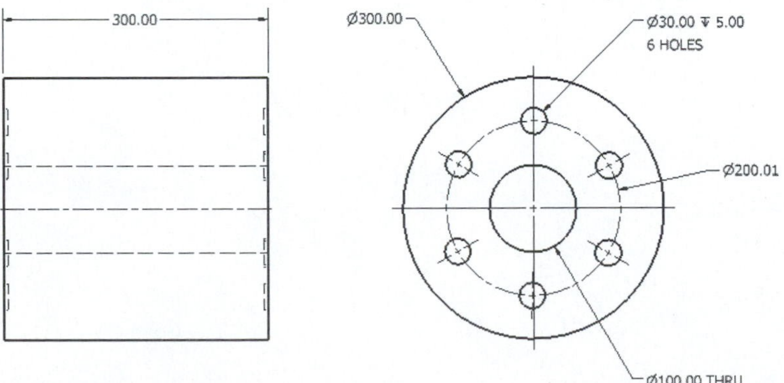

Figure P9-27 *(Continued)*

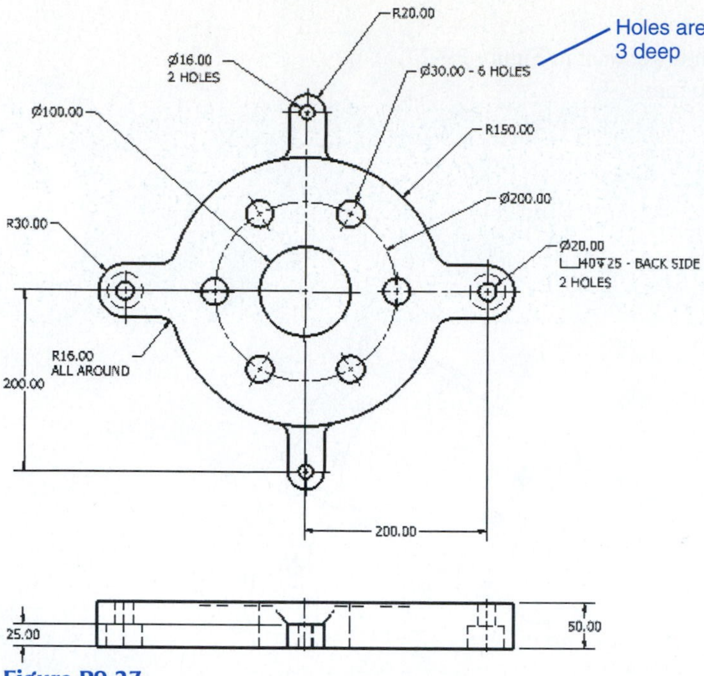

Figure P9-27

Preloaded Spring Length = 80

Working Stroke = 5

Working Spring Length = 78

Coil Direction = Right

End Coil = 1.5

Transition Coil = 1.00

Ground Coil = 0.75

Active Coils Number = 16

Draw the following:

A. An assembly drawing
B. A presentation drawing
C. An exploded isometric drawing
D. A parts list

Shafts 10

Chapter Objectives

- Show how to use **Design Accelerator** to draw shafts.
- Show how to add retaining ring grooves to shafts.
- Show how to add keyways to shafts.
- Show how to add O-ring grooves to a shaft
- Show how to add pin holes to a shaft.
- Show how to use the **Content Center** to add retaining rings, keys, O-rings, and pins to shafts.

INTRODUCTION

The **Design Accelerator** tool can be used to draw many different styles of shafts. Figure 10-1 shows a uniform shaft with chamfered ends that was created using **Design Accelerator.** Shafts may also be created by extruding a circle, but the **Design Accelerator** allows features such as keyways and retaining ring grooves to easily be added to the shaft.

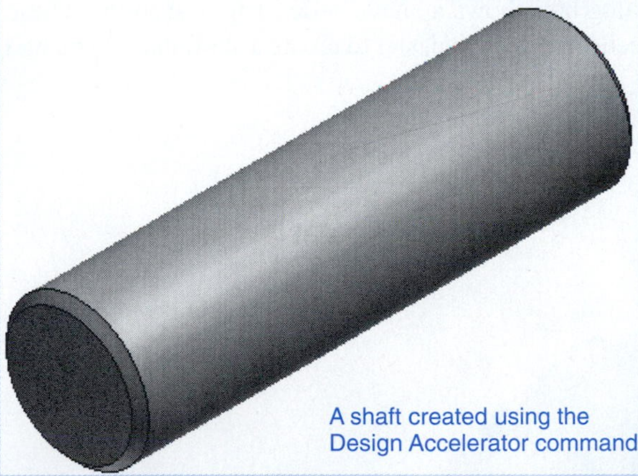

Figure 10-1

Uniform Shafts and Chamfers

Exercise 10-1: Drawing a Uniform Shaft with Chamfered Ends

The shaft is to be Ø1.00 × 4.00 in.

1. Create a new drawing using the **Standard (in).iam** format.
2. Click the arrow to the right of the **Assembly Panel** heading and access the **Design Accelerator** tool.

See Figure 10-2.

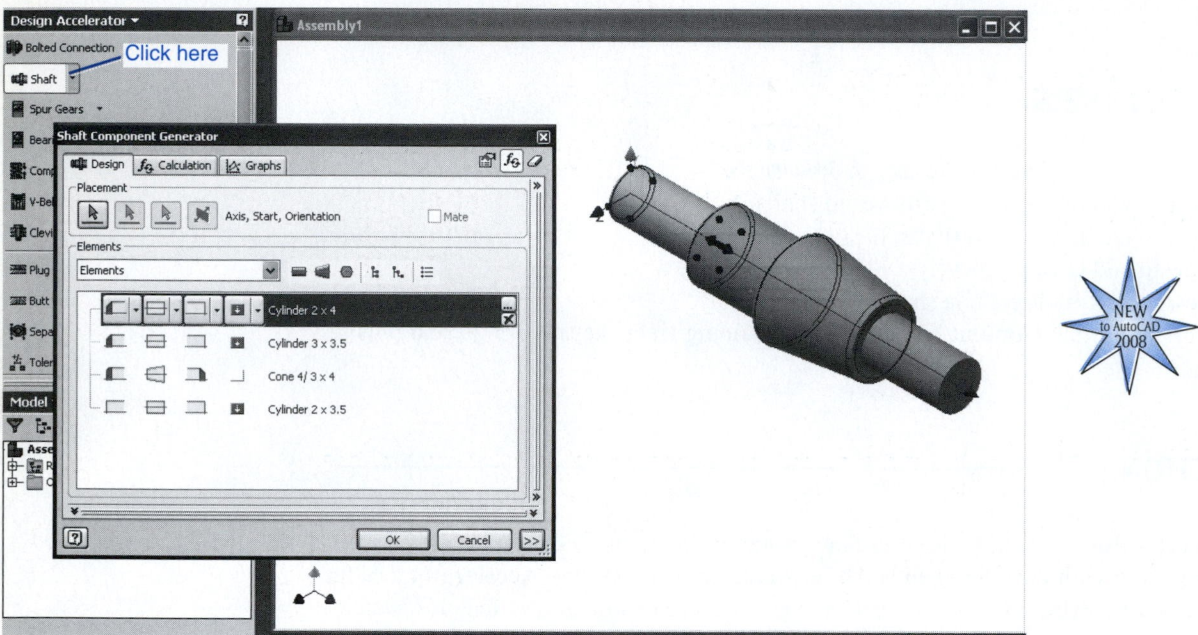

Figure 10-2

The **Shaft Component Generator** dialog box will appear. See Figure 10-3.

The **Shaft Component Generator** dialog box shows a shaft made of four sections. These sections can be manipulated or deleted as needed, making it easier to create a shaft than if you had to start from scratch.

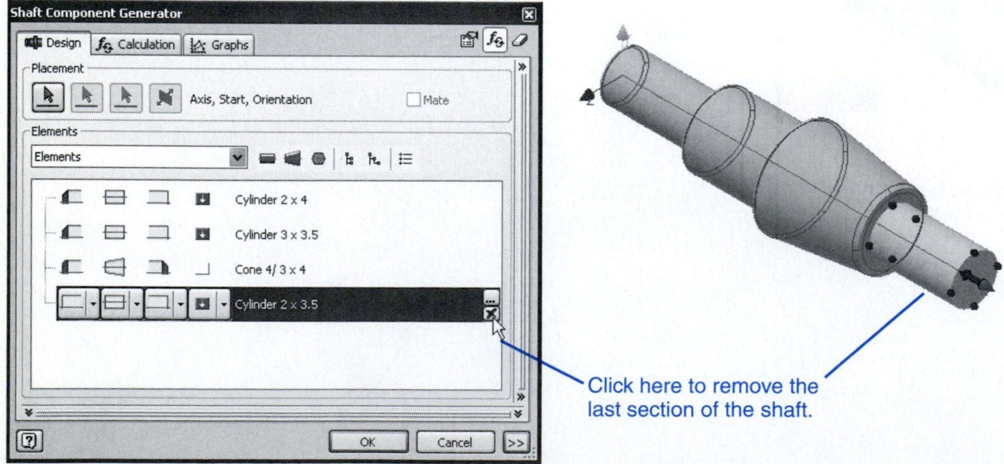

Figure 10-3

Exercise 10-2: Deleting a Shaft Section

1. Move the cursor onto the shaft section to be deleted and click the mouse.

The section's heading will highlight.

2. Click the X box in the lower right corner of the section's heading.

The section will be deleted. See Figure 10-3. In this example the **Cylinder 2 × 3.5** section will be deleted. Figure 10-4 shows all but one section deleted.

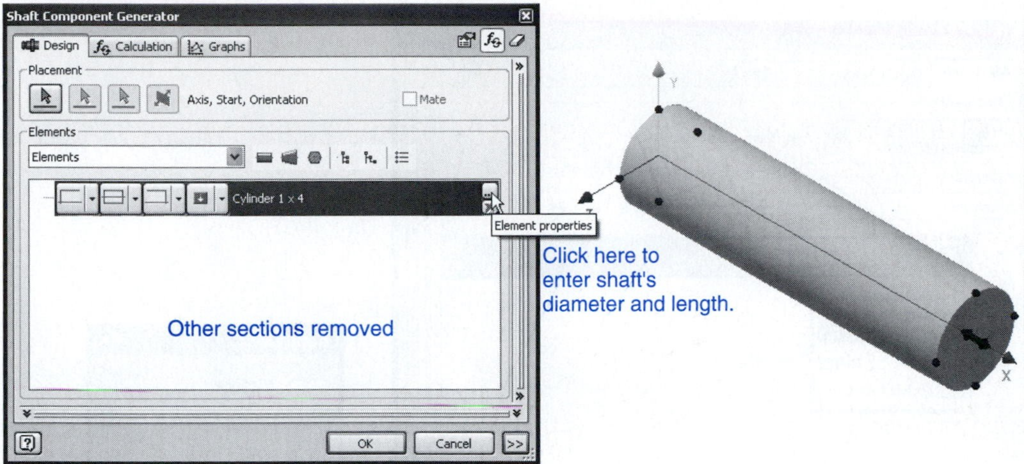

Figure 10-4

Exercise 10-3: Defining the Cylinder's Diameter and Length

1. Click the **Element properties** box.

The **Cylinder Dimensions** dialog box will appear. See Figure 10-4.

2. Click to the right of the default diameter value.

An arrowhead will appear.

3. Click the arrowhead.

A listing of standard shaft diameter values will cascade down.

See Figure 10-5.

4. Select the appropriate diameter.
5. Repeat the procedure for the shaft's length.

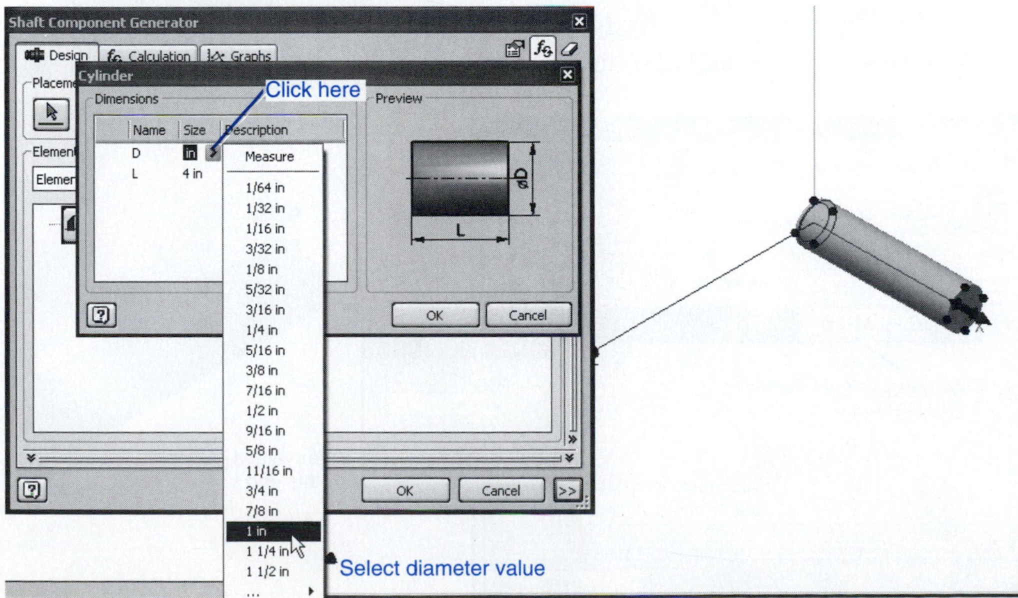

Figure 10-5

Exercise 10-4: Creating Chamfers

1. Click the **Left Edge Feature** box.

See Figure 10-6.

2. Select the **Chamfer** option.

The **Chamfer** dialog box will appear. See Figure 10-7.

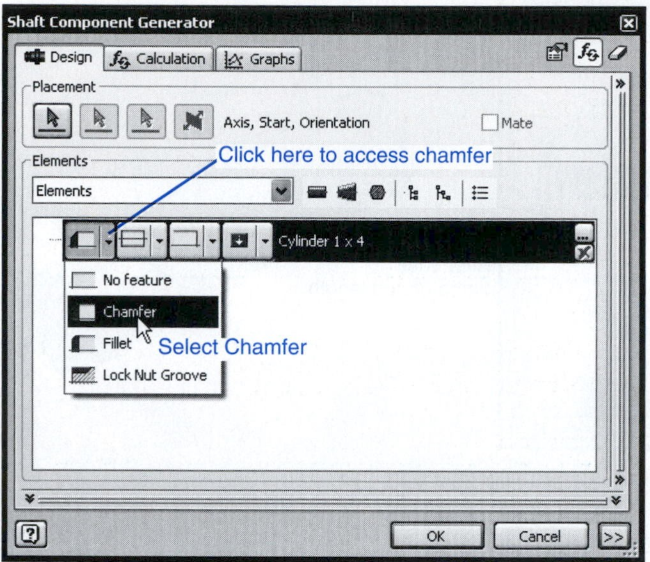

Figure 10-6

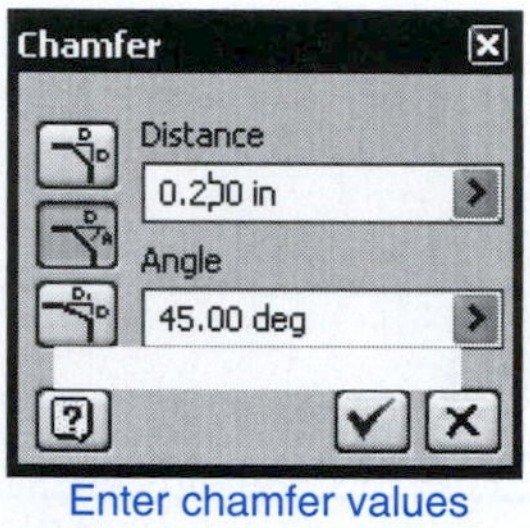

Chamfer
Distance
0.200 in
Angle
45.00 deg
Enter chamfer values

Figure 10-7

3. Enter the chamfer values.
4. Click the check mark box.

This will create a chamfer on the left end of the shaft.

5. Click the **Right Edge Feature** box and create a chamfer on the right edge of the shaft.

See Figure 10-8.

6. Click the check mark box.
7. Click **OK.**

Figure 10-9 shows the finished shaft.

Note:

The preview of the shaft that appears on the screen will show the size and location of the chamfers.

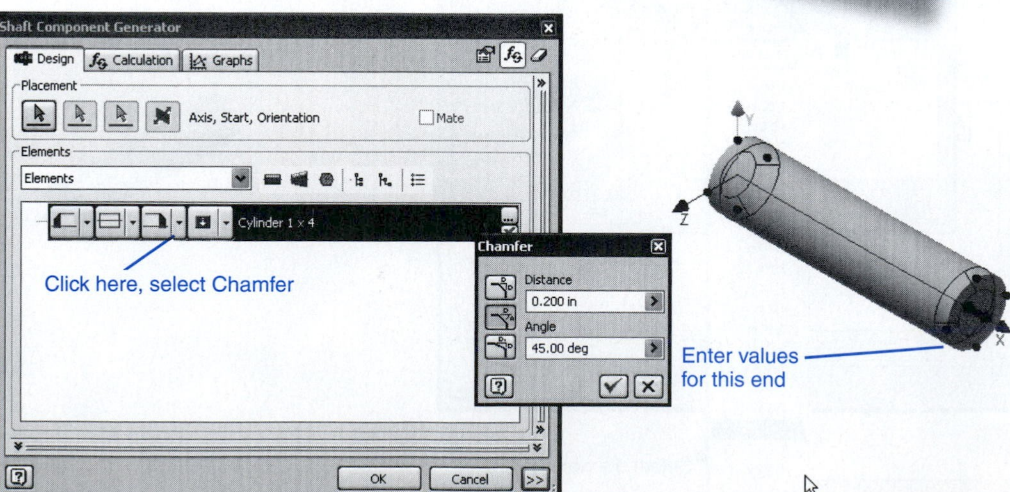

Figure 10-8

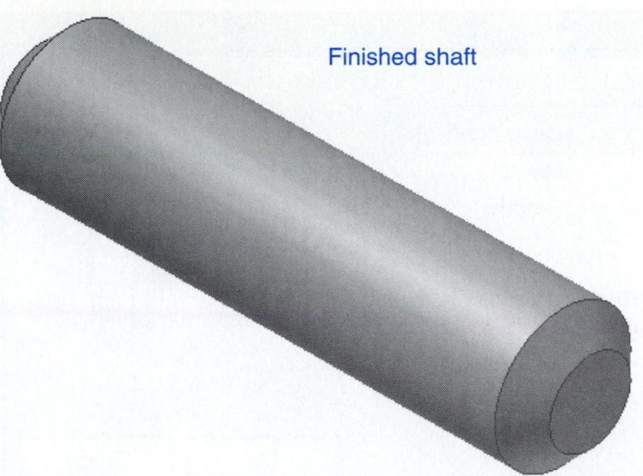

Figure 10-9

Shafts and Retaining Rings

Retaining rings are used to prevent longitudinal movement of shafts. Grooves are cut into the shafts and the rings fitted into the groves. Both internal and external rings are available.

retaining ring: Used to prevent longitudinal movement of a shaft. The ring fits into a groove cut into the shaft.

Exercise 10-5: Drawing a Shaft with Retaining Rings

This example will use a Ø20-mm × 80-mm shaft with two retaining rings, one located 3 mm from each end. The shaft will be made from two Ø20 × 40 shafts. (One Ø20 × 80 shaft could also be used.)

1. Create a new drawing using the **Standard (mm).iam** format.
2. Access the **Design Accelerator** by clicking the arrow to the right of the **Assembly Panel** heading.
3. Click the **Shaft** option.
4. Remove the middle two elements of the default shaft setup.

See Figure 10-10.

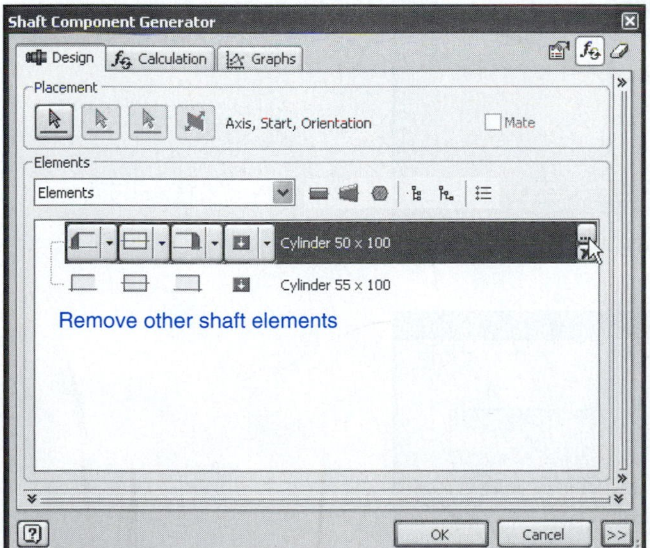

Figure 10-10

5. Enter a diameter value of **20** and a length of **40** for each section of the shaft.

See Figure 10-11.

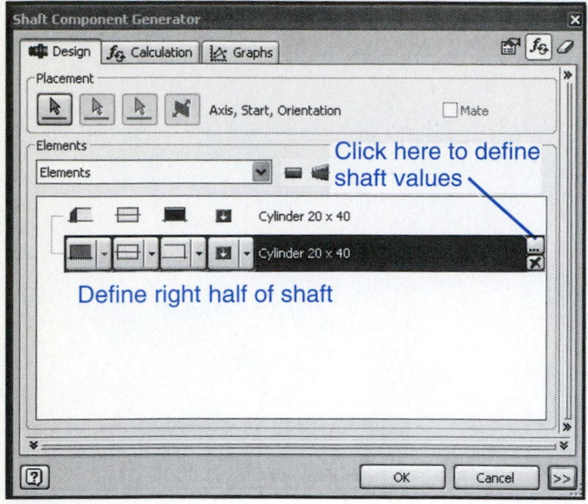

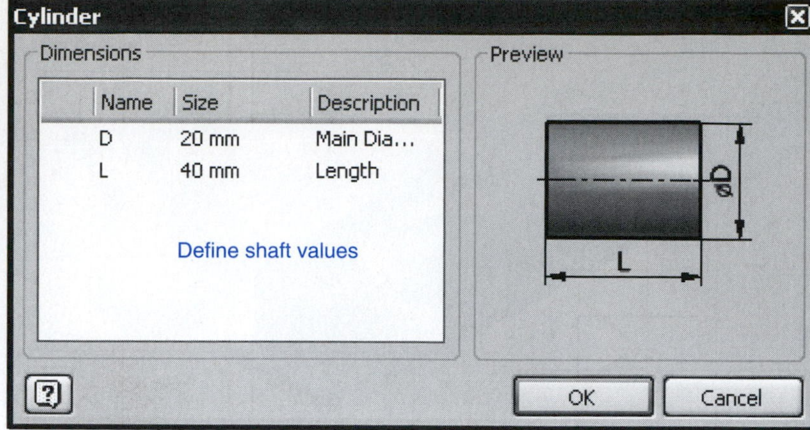

Figure 10-11

6. Click the **Element Features** box and select the **Add Retaining Ring** option.

See Figure 10-12. The **Retaining Ring Groove** dialog box will appear. See Figure 10-13.

7. Enter the distance value (in this example the value is **3**) and click on the arrow on the right side of the **R-Ring** box.

The **Place from Content Center** dialog box will appear. See Figure 10-14.

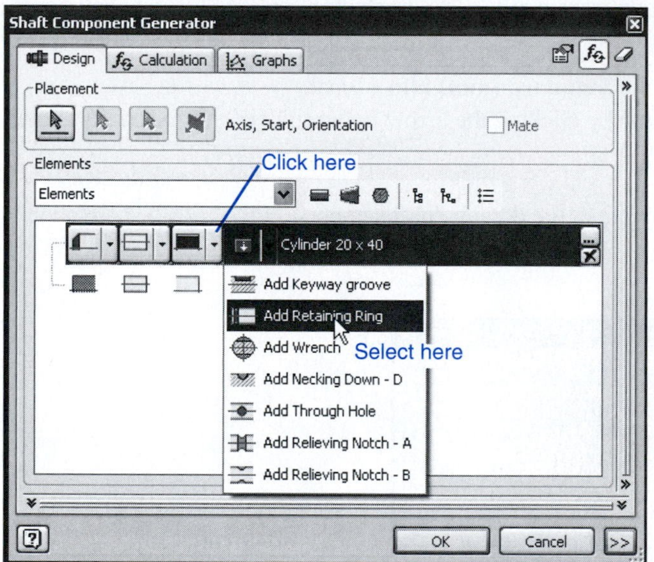

Figure 10-12

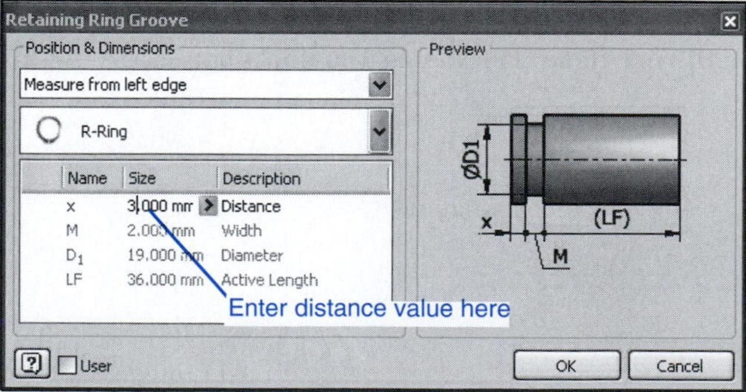

Figure 10-13

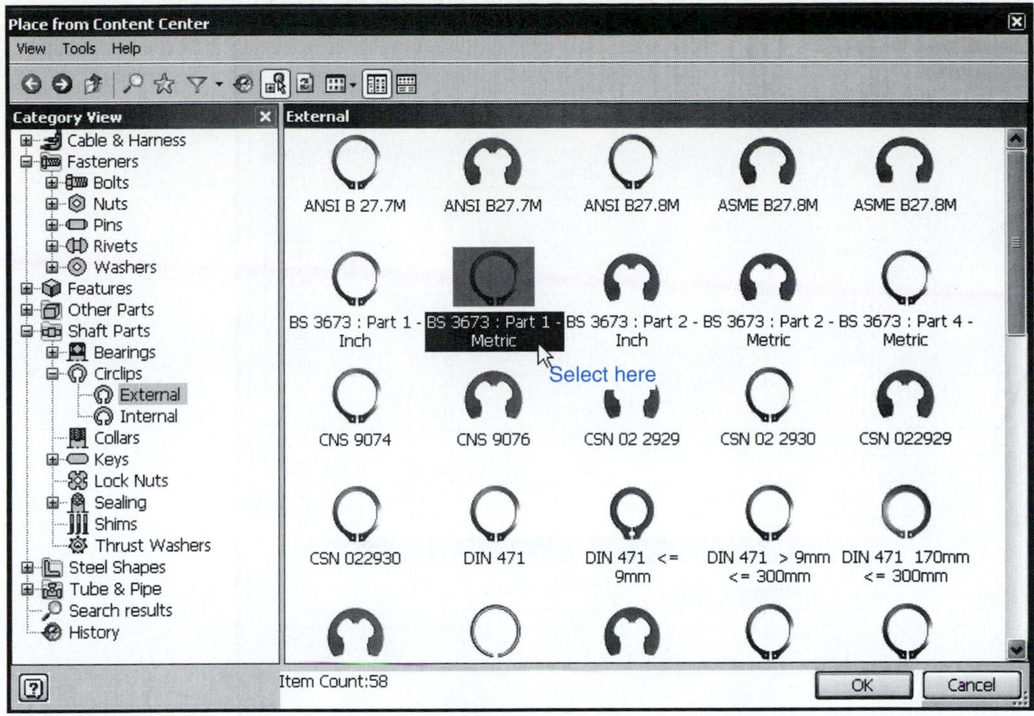

Figure 10-14

8. Select a **BS 3673 : Part 1 - Metric** retaining ring.
9. Select a shaft diameter value of **20.**

See Figure 10-15. Figure 10-16 shows the **Table View** for the ∅20 shaft retaining ring. This table lists the **Width across Open** (2.184) and **Groove Diameter** (19.05) that match the selected retaining ring. These values will be added to the shaft.

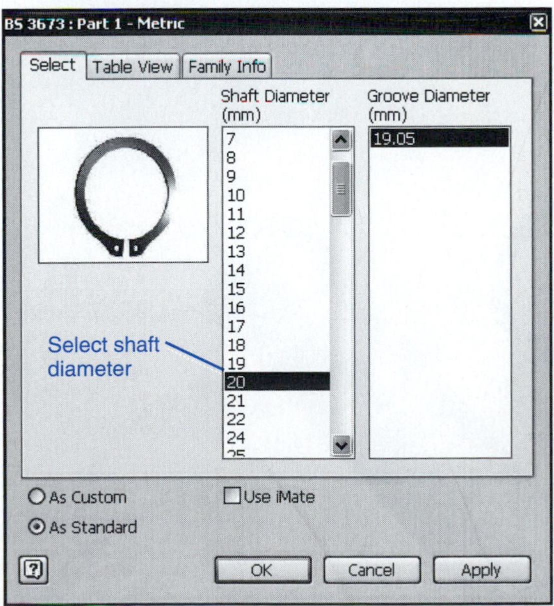

Figure 10-15

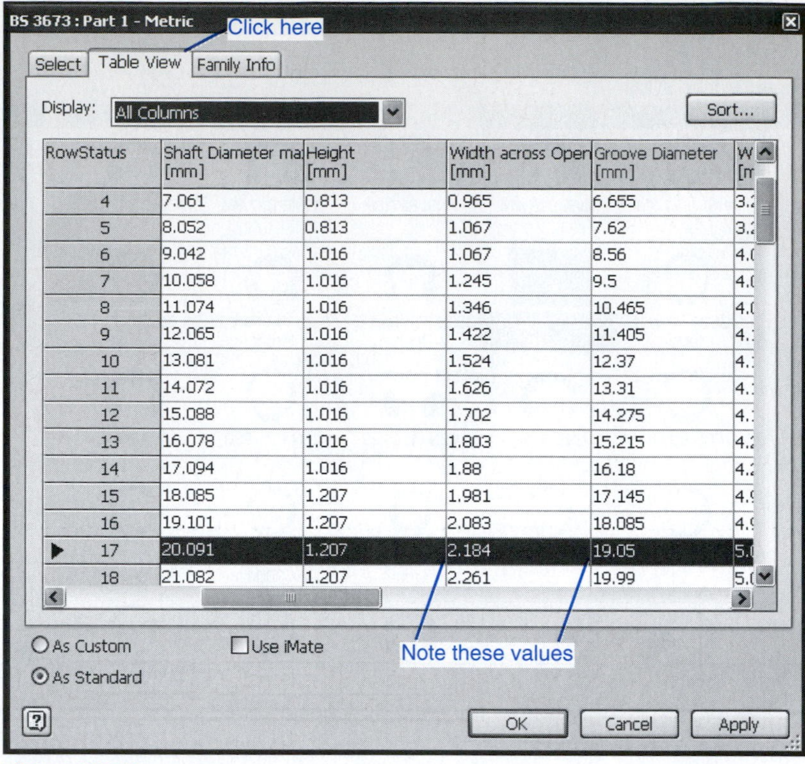

RowStatus	Shaft Diameter ma [mm]	Height [mm]	Width across Open [mm]	Groove Diameter [mm]
4	7.061	0.813	0.965	6.655
5	8.052	0.813	1.067	7.62
6	9.042	1.016	1.067	8.56
7	10.058	1.016	1.245	9.5
8	11.074	1.016	1.346	10.465
9	12.065	1.016	1.422	11.405
10	13.081	1.016	1.524	12.37
11	14.072	1.016	1.626	13.31
12	15.088	1.016	1.702	14.275
13	16.078	1.016	1.803	15.215
14	17.094	1.016	1.88	16.18
15	18.085	1.207	1.981	17.145
16	19.101	1.207	2.083	18.085
▶ 17	20.091	1.207	2.184	19.05
18	21.082	1.207	2.261	19.99

Figure 10-16

10. Repeat the process for the other end of the shaft.

 See Figure 10-17.

11. Add two **BS 3673 : Part 1 - Metric** retaining rings to the drawing.

 See Figure 10-18.

12. Use the **Constraint** tool to position the retaining rings in the grooves on the shaft.

 See Figure 10-19.

Note:
Select the **Measure from right edge** option.

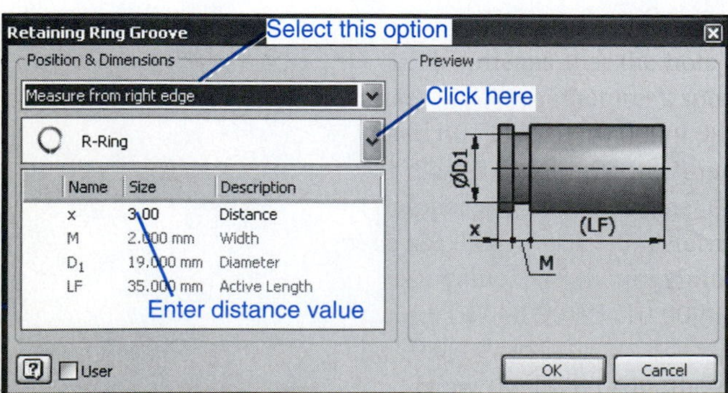

Figure 10-17

Figure 10-18

Figure 10-19

SHAFTS AND KEYS

Keys are used with shafts to transfer rotary motion and torque. Figure 10-20 shows a hub that has been inserted onto a shaft with a square key between the hub and shaft. As the shaft turns, the motion and torque of the shaft will be transferred through the key into the hub.

There are five general types of keys: Pratt and Whitney, square, rectangular, Woodruff, and Gib. See Figure 10-21.

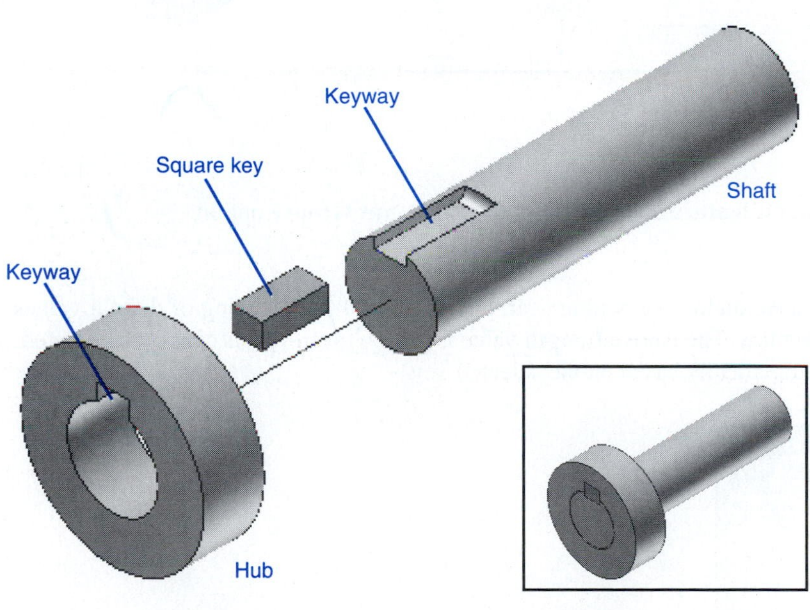

Figure 10-20

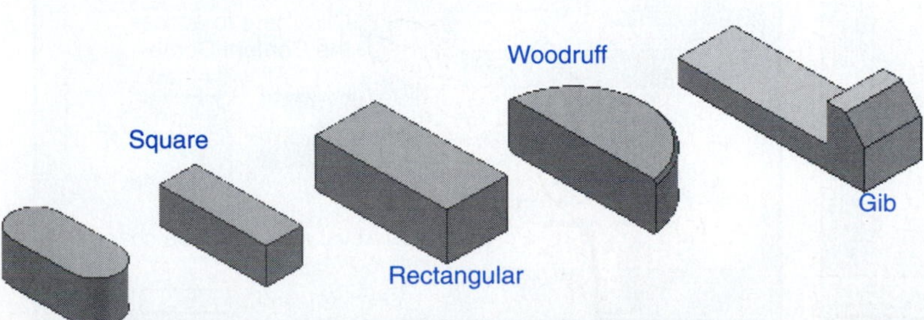

Figure 10-21

Square Keys

Exercise 10-6: Drawing a Keyway on a Shaft

1. Start a new drawing using the **Standard (mm).iam** format.
2. Access the **Design Accelerator** and access the **Shaft** option.
3. Remove all but one shaft element and draw a **∅30 × 60** shaft.

See Figure 10-22.

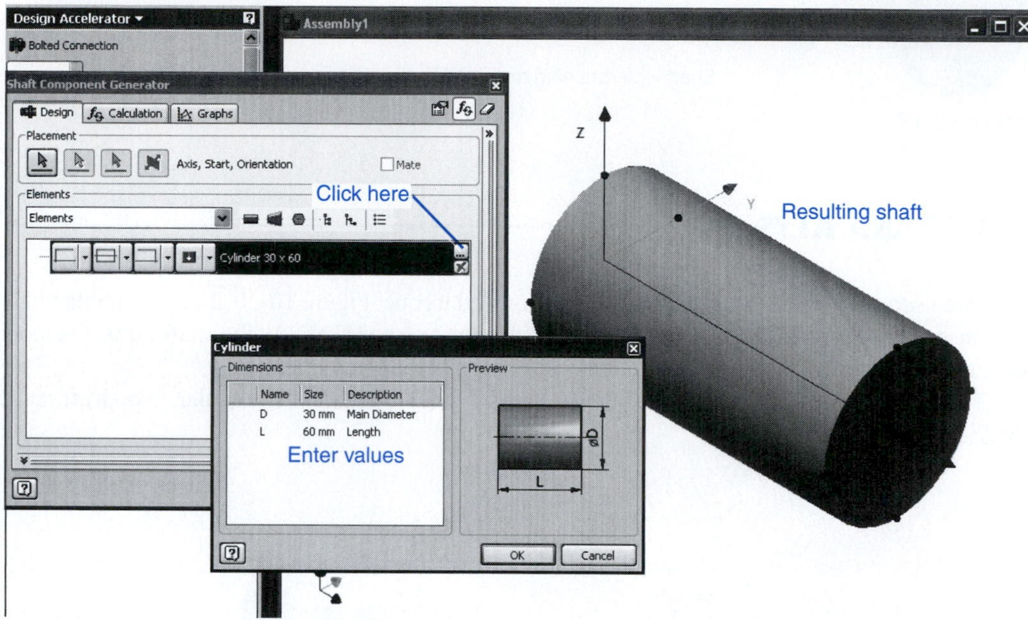

Figure 10-22

4. Click the **Left end feature** box and select the **Lock Nut Groove** option.

See Figure 10-23.

The **Locknut Groove** dialog box will appear. See Figure 10-24. A listing of default values will appear in the dialog box. The **Active Length** value is gray. Grayed values cannot be changed; they are calculated automatically based on the selected key.

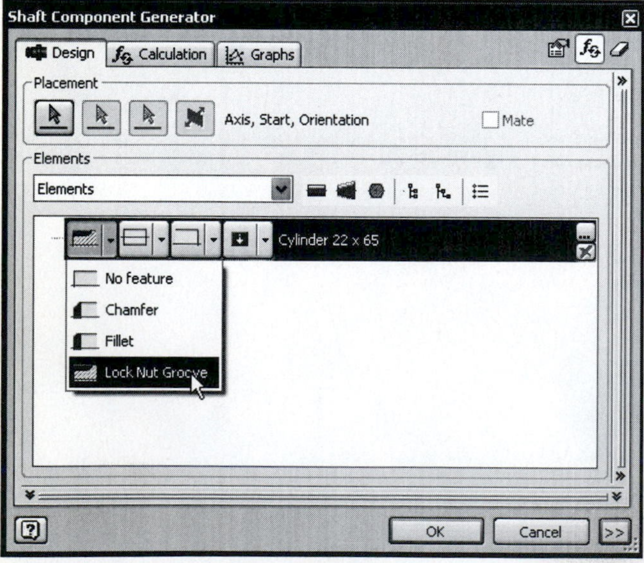

Figure 10-23

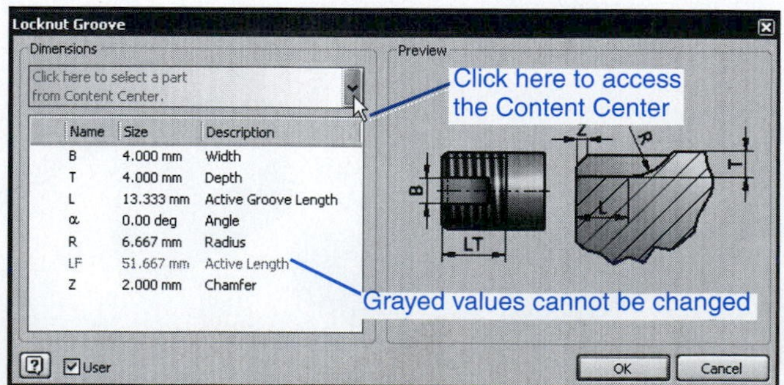

Figure 10-24

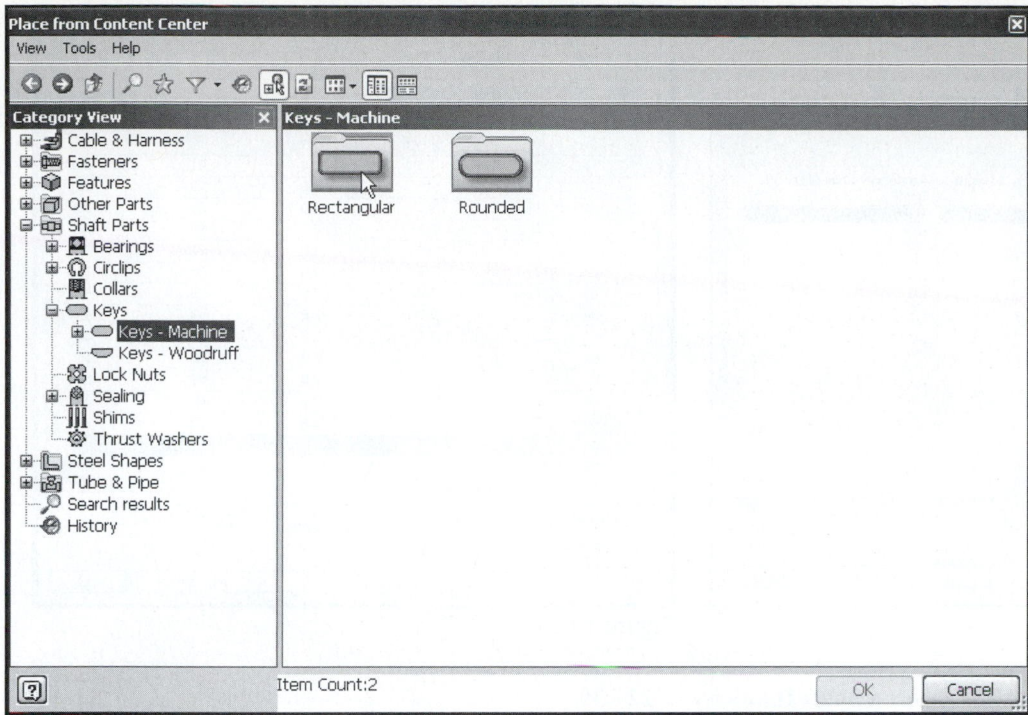

Figure 10-25

5. Click the arrow as shown in Figure 10-24 to access the **Content Center.**

The **Place from Content Center** dialog box will appear. See Figure 10-25.

6. Select the **Rectangular Key** option.
7. Select the **IS 2048 B** key.

See Figure 10-26.

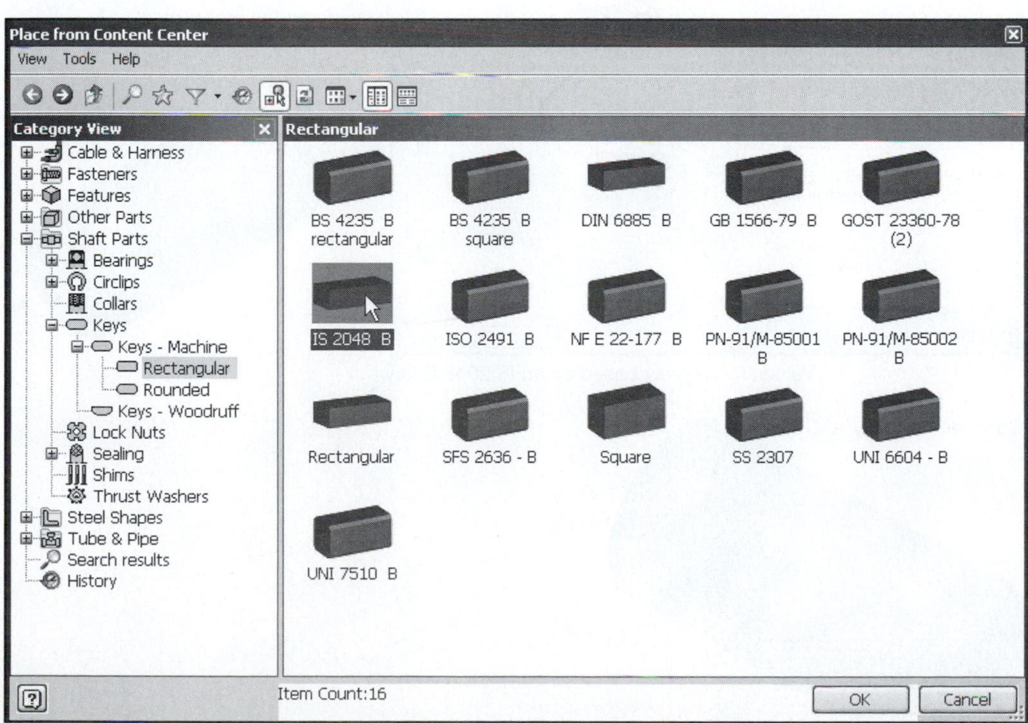

Figure 10-26

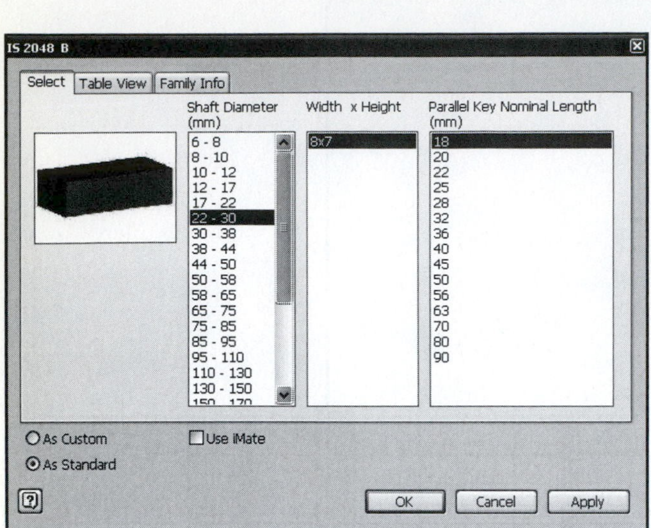

Figure 10-27

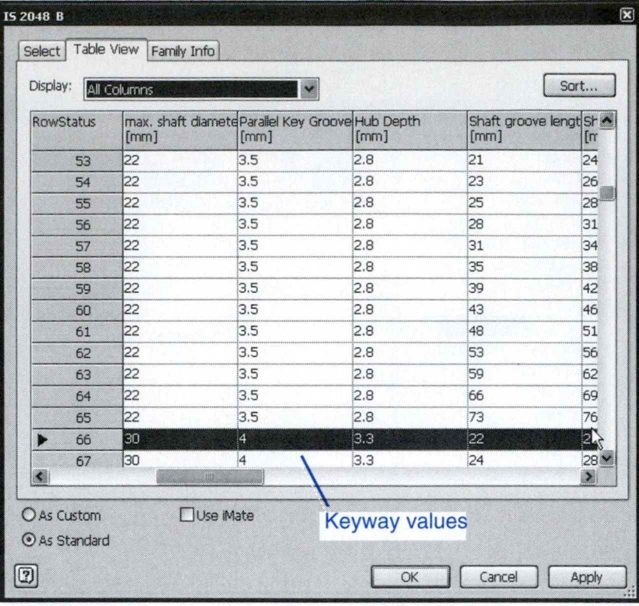

RowStatus	max. shaft diamete [mm]	Parallel Key Groove [mm]	Hub Depth [mm]	Shaft groove lengt [mm]	Sh [m
53	22	3.5	2.8	21	24
54	22	3.5	2.8	23	26
55	22	3.5	2.8	25	28
56	22	3.5	2.8	28	31
57	22	3.5	2.8	31	34
58	22	3.5	2.8	35	38
59	22	3.5	2.8	39	42
60	22	3.5	2.8	43	46
61	22	3.5	2.8	48	51
62	22	3.5	2.8	53	56
63	22	3.5	2.8	59	62
64	22	3.5	2.8	66	69
65	22	3.5	2.8	73	76
66	30	4	3.3	22	2
67	30	4	3.3	24	28

Figure 10-28

8. Select a **Shaft Diameter** of **22 - 30.**

See Figure 10-27. The selected key's width and height are **8 × 7,** and the nominal length is **18.** Figure 10-28 shows the **Table View** for the selected key. This table gives the dimensions for the keyway.

9. Return to the **Design Accelerator** and check the dimension values in the **Locknut Groove** dialog box. The values should match those given in the **Table View.**

See Figure 10-29.

10. Click **OK.**

Figure 10-30 shows the finished keyway.

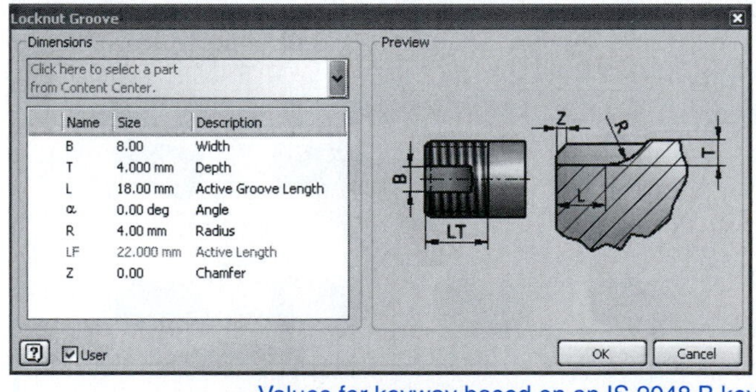

Name	Size	Description
B	8.00	Width
T	4.000 mm	Depth
L	18.00 mm	Active Groove Length
α	0.00 deg	Angle
R	4.00 mm	Radius
LF	22.000 mm	Active Length
Z	0.00	Chamfer

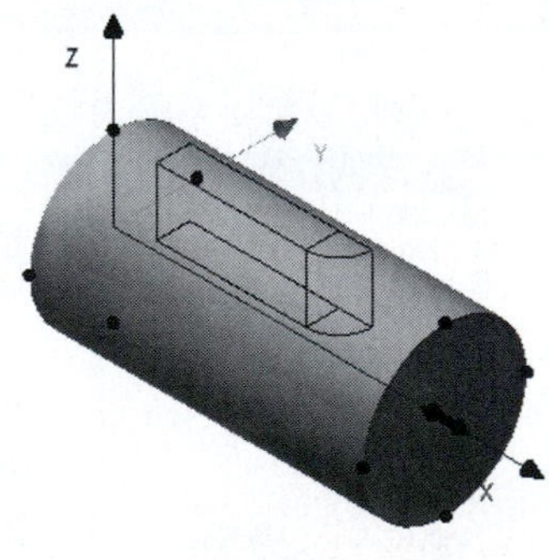

Figure 10-29

Values for keyway based on an IS 2048 B key

Shaft with keyway

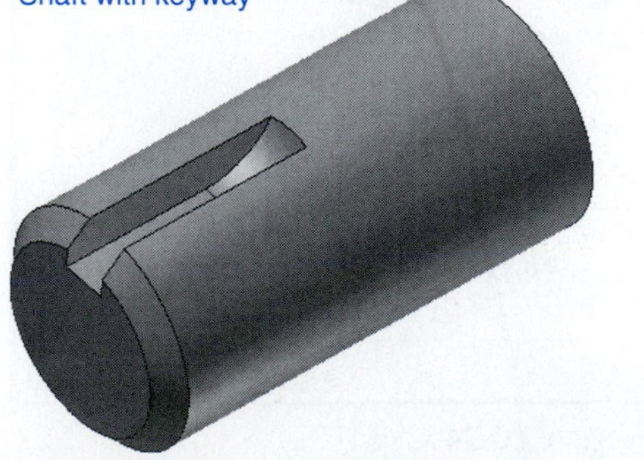

Figure 10-30

Exercise 10-7: Drawing a Keyway

Figure 10-31 shows a Ø30 × 60 shaft. Draw a keyway that is 8 wide, 4 deep, and 18 long with end radius equal to 4.

1. Create a work plane tangent to the edge of the shaft.
2. Create a new sketch plane on the work plane and draw an **8 × 22** rectangle. The 22 value includes the 4 needed to create the radius at the end of the keyway.

See Figure 10-32.

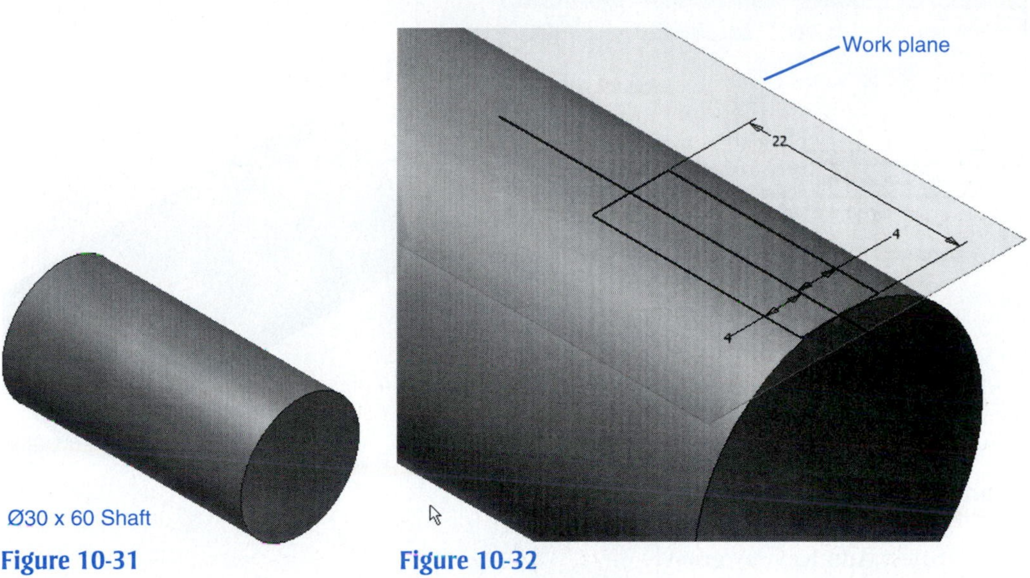

Figure 10-31

Figure 10-32

3. Use the **Cut** option on the **Extrude** tool to cut out the keyway.
4. Draw an **R4** radius at the end of the keyway.

See Figure 10-33. Figure 10-34 shows the finished keyway.

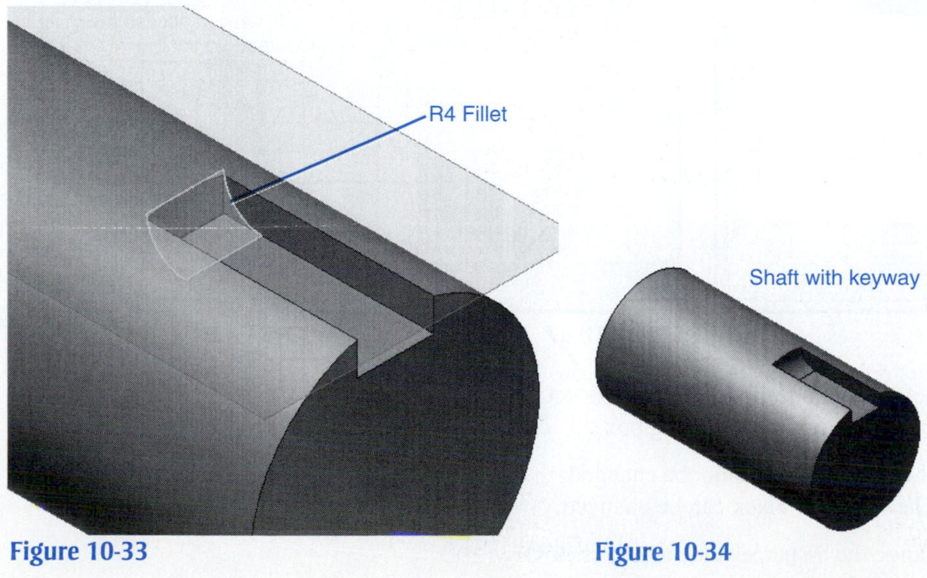

Figure 10-33

Figure 10-34

Pratt and Whitney Keys

Pratt and Whitney key: A key similar to a square key but with rounded ends.

Pratt and Whitney keys are similar to square keys but have rounded ends.

Exercise 10-8: Drawing a Shaft with a Pratt and Whitney Keyway

1. Start a new drawing using the **Standard (mm).iam** format.
2. Access the **Design Accelerator** and select the **Shaft** option.
3. Draw a **∅30 × 65** shaft with **3 × 45°** chamfers at each end.

See Figure 10-35.

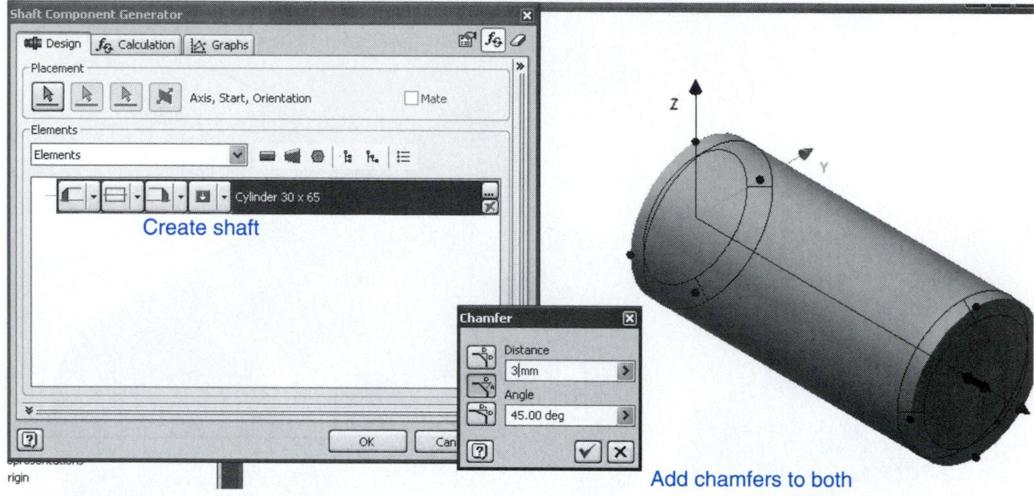

Figure 10-35

4. Select **Add Keyway groove.**

See Figure 10-36. The **Keyway** dialog box will appear. See Figure 10-37.

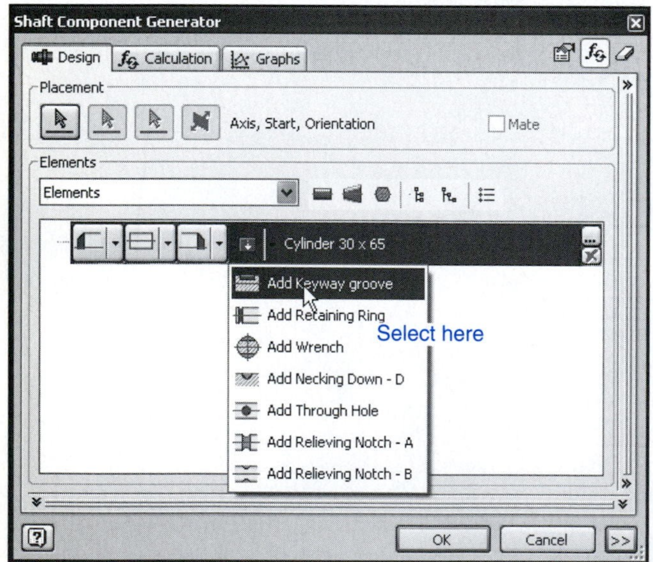

Figure 10-36

Figure 10-37

5. Access the **Content Center** and select an **ISO 2491 A** key.
6. Return to the **Keyway** dialog box.

The grayed numbers cannot be changed; they were generated when the **ISO 2491 A** key was selected. The values in black can be changed.

7. Change the distance value to **16;** click **OK.**

Figure 10-38 shows the finished keyway.

Figure 10-38

8. Access the **Place from Content Center** dialog box and add the key.

See Figure 10-39.

9. Use the **Constraint** tool and place the key into the keyway.

See Figure 10-40.

Figure 10-41 shows the shaft and key mounted into a hub.

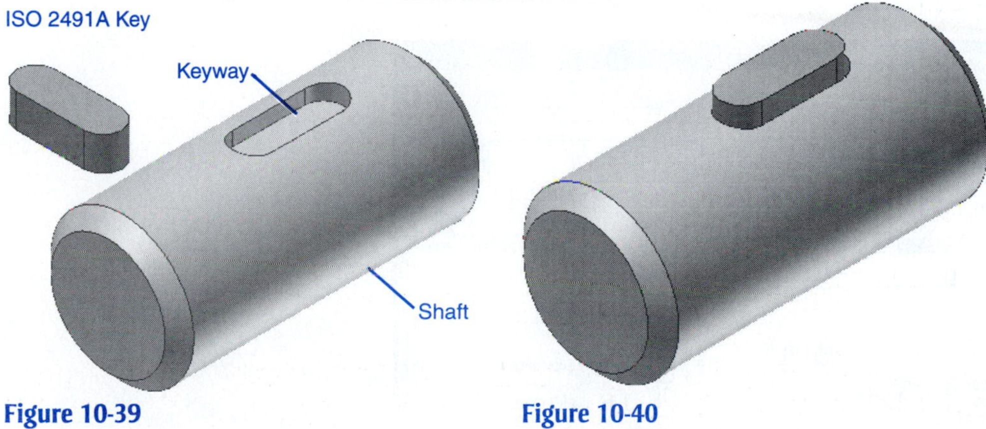

Figure 10-39

Figure 10-40

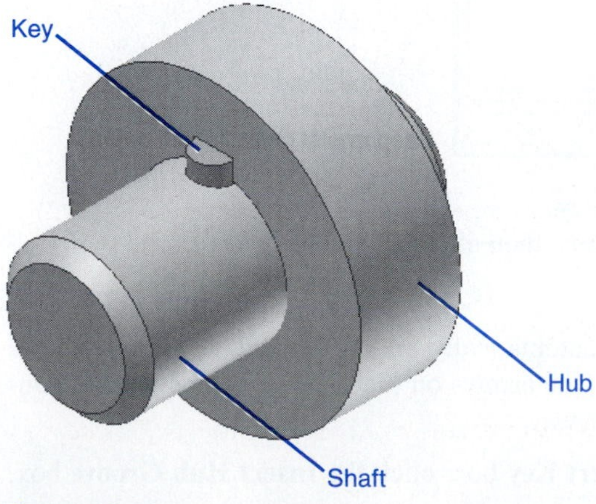

Figure 10-41

Exercise 10-9: Drawing a Pratt and Whitney Keyway Using the Key Connection Option on Design Accelerator

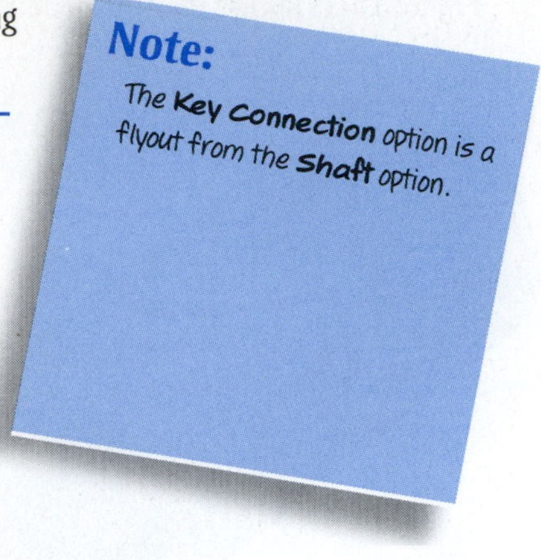

1. Start a new drawing using the **Standard (mm).iam** format.
2. Access the **Design Accelerator** and select the **Shaft** option.
3. Draw a **∅30 × 65** shaft.

See Figure 10-42.

4. Click the **Key Connection** option on the **Design Accelerator.**

See Figure 10-43. The **Parallel Key Connection Generator** box will appear. See Figure 10-44.

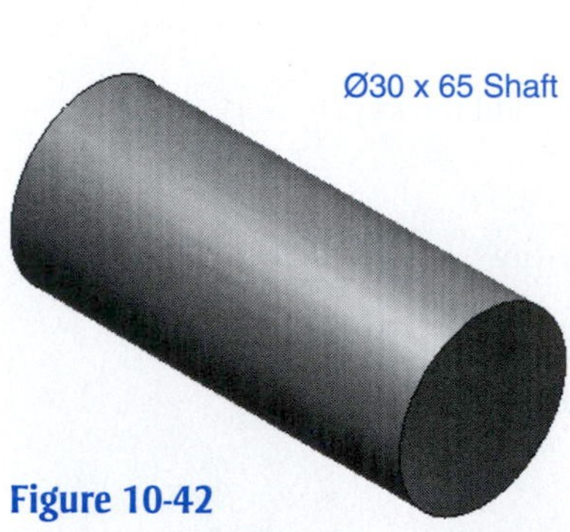

Figure 10-42

Figure 10-43

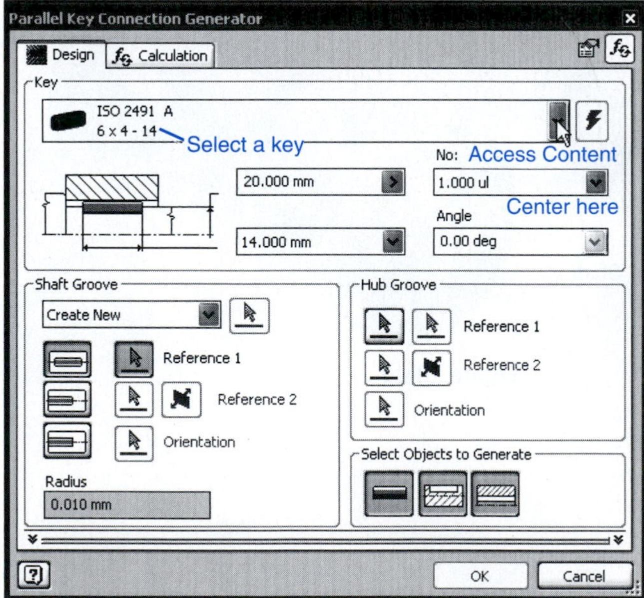

Figure 10-44

5. Access the **Content Center** and select a key.
6. Click the **Groove with Rounded Edges** box, then click the rounded side of the shaft.

See Figure 10-45.

The **Planar Face of Work Plane** box will automatically be highlighted. A preview of the shaft will appear on the shaft. See Figure 10-46. The arrows on the preview can be used to control the position, length, and orientation of the keyway.

7. Edit the keyway as needed, click the **Insert Key** box, click the **Insert Hub Groove** box, then click **OK.**

Figure 10-47 shows the finished keyway.

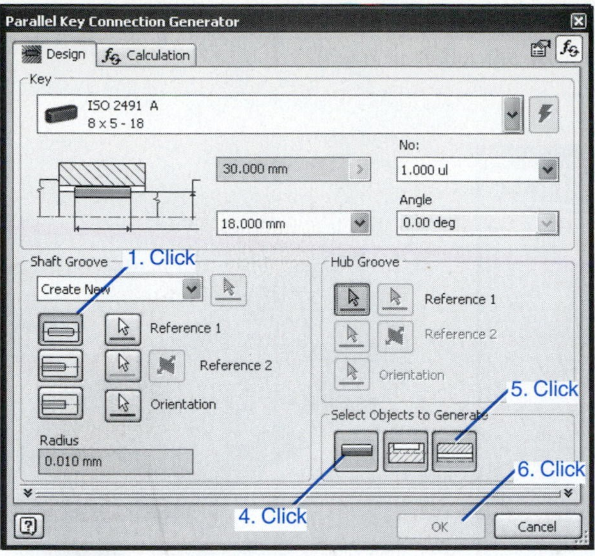

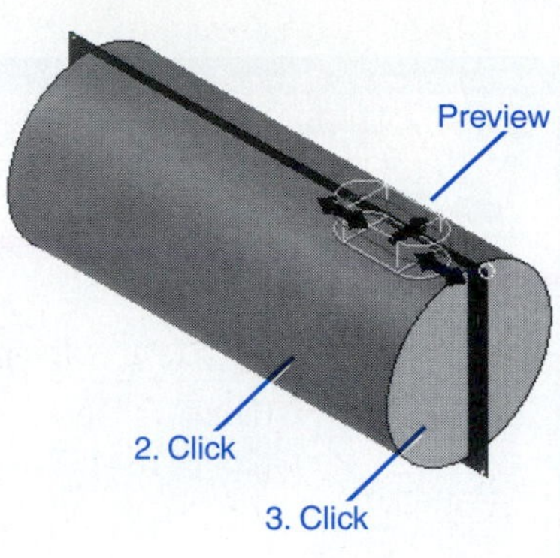

Figure 10-45

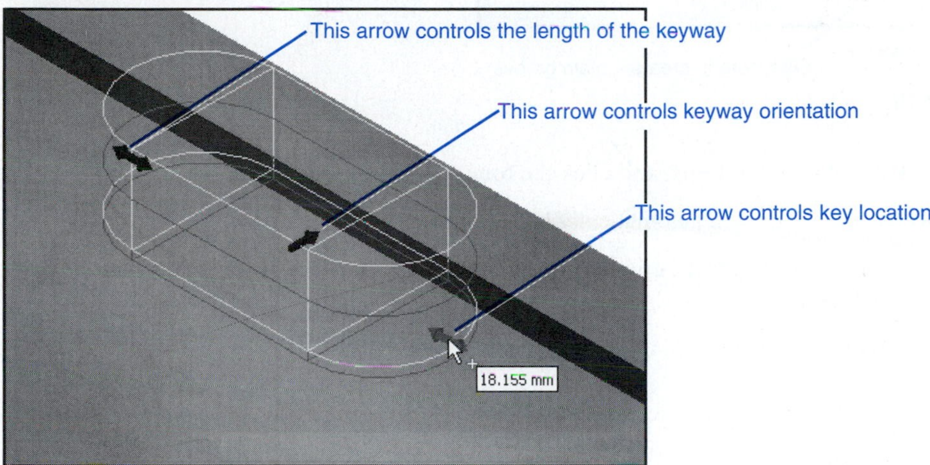

Figure 10-46

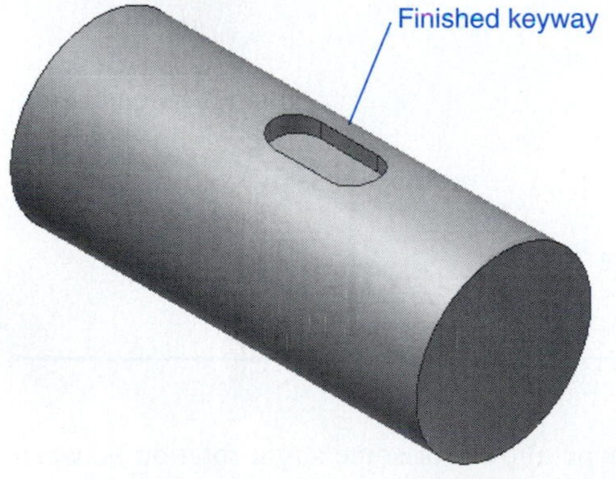

Figure 10-47

Exercise 10-10: Creating a Plain Groove Keyway at the End of a Shaft

Plain grooves are used with square keys.

1. Start a new drawing using the **Standard (mm).iam** format.
2. Access the **Design Accelerator** and select the **Shaft** option.
3. Draw a **∅30 × 65** shaft with **3 × 45°** chamfers at each end.
4. Access the **Key Connection** option on the **Design Accelerator.**

5. Select a key.
6. Click the **Plain Groove** box.

See Figure 10-48.

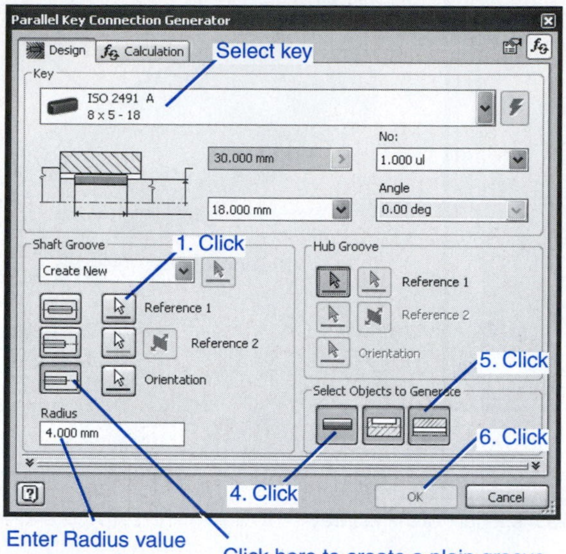

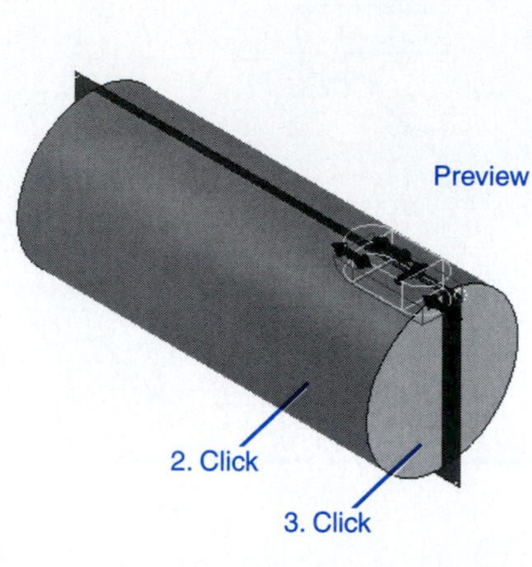

Figure 10-48

7. Click the **Reference 1** box, and click the rounded side of the shaft.

The **Reference 2** box will automatically be highlighted.

8. Click the end of the shaft, click the **Insert Key** box, click the **Insert Hub Groove** box, then click **OK.**

Figure 10-49 shows the resulting plain groove.

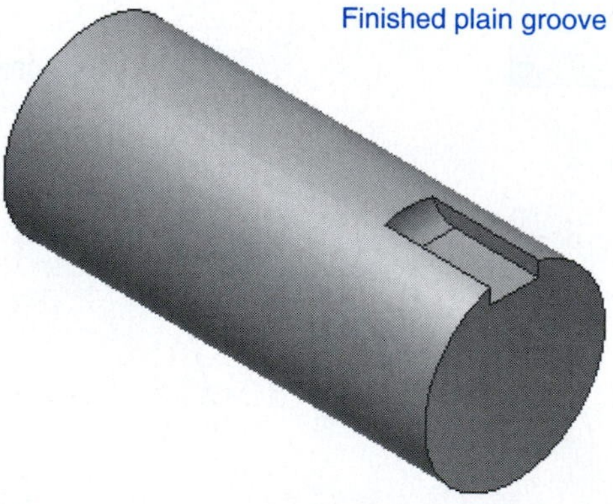

Figure 10-49

WOODRUFF KEYS

Woodruff key: A crescent-shaped key that allows for some slight rotation between the shaft and hub.

Woodruff keys are crescent shaped. The crescent shape allows for some slight rotation between the shaft and hub.

Exercise 10-11: Drawing a Shaft with a Woodruff Key

1. Create a new drawing using the **Standard (mm).iam** format.
2. Access the **Design Center** tool, click **Shaft Parts,** click **Keys,** then click **Keys - Woodruff.**

See Figure 10-50.

3. Select an **IS 2294 (I)** key, then a shaft diameter of **32 - 38.**

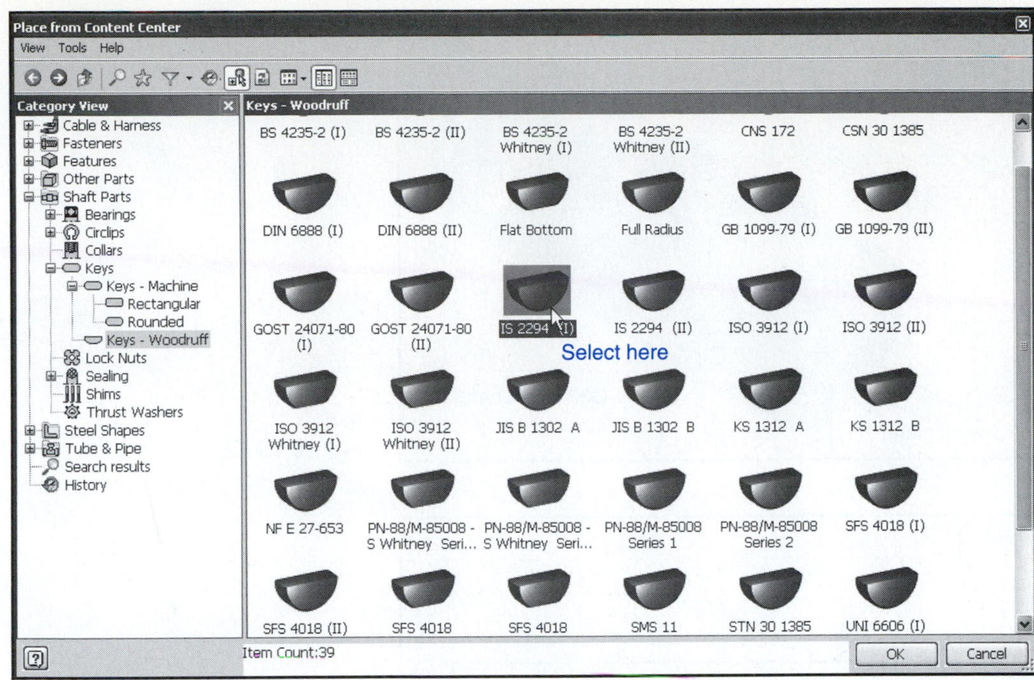

Figure 10-50

See Figure 10-51. The **Table view** listed for the IS 2294 (I) key will be used to create the shaft keyway.

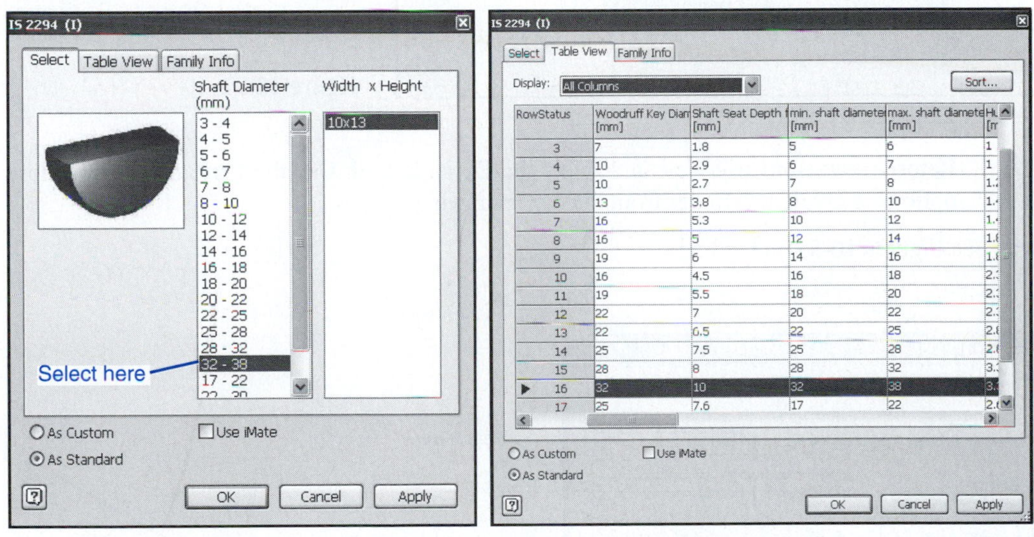

RowStatus	Woodruff Key Diam [mm]	Shaft Seat Depth [mm]	min. shaft diameter [mm]	max. shaft diameter [mm]	H [m
3	7	1.8	5	6	1
4	10	2.9	6	7	1
5	10	2.7	7	8	1.
6	13	3.8	8	10	1.
7	16	5.3	10	12	1.
8	16	5	12	14	1.
9	19	6	14	16	1.
10	16	4.5	16	18	2.
11	19	5.5	18	20	2.
12	22	7	20	22	2.
13	22	6.5	22	25	2.
14	25	7.5	25	28	2.
15	28	8	28	32	3.
▶ 16	32	10	32	38	3.
17	25	7.6	17	22	2.

Figure 10-51

4. Click **Apply.**

The key will appear on the screen. We need to determine the size of the key to create an appropriate keyway in the shaft. Refer to the table values in Figure 10-51. The Woodruff key diameter is **32,** and the shaft seat depth is **10.** The key itself has a height of **13.**

5. Access the **Design Accelerator** and draw a **∅32×65** shaft. Create a YZ work plane through the center of the shaft and create a new sketch plane on the work plane.

See Figure 10-52.

6. Use the **Look At** command to rotate the shaft into a 2D view and add lines and a circle as shown in Figure 10-53.

The 28 distance was selected at random; all other dimensions came from the **Table View** values for the key.

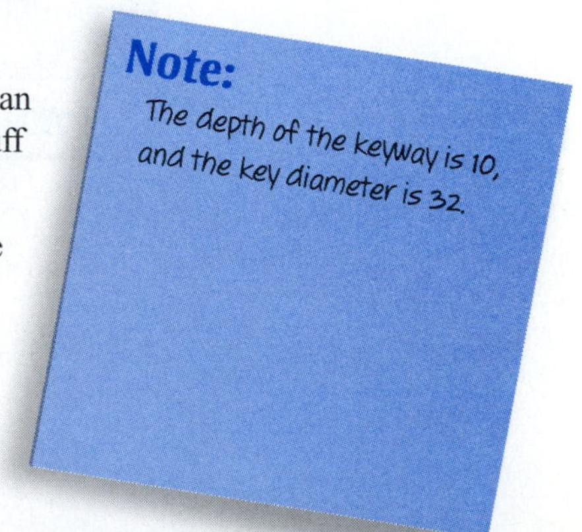

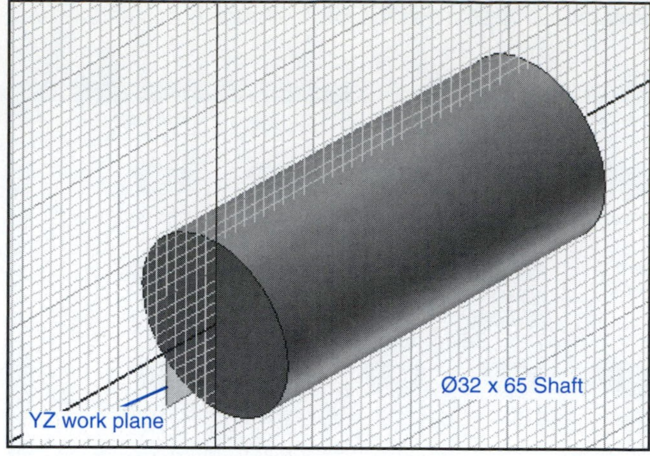

Figure 10-52

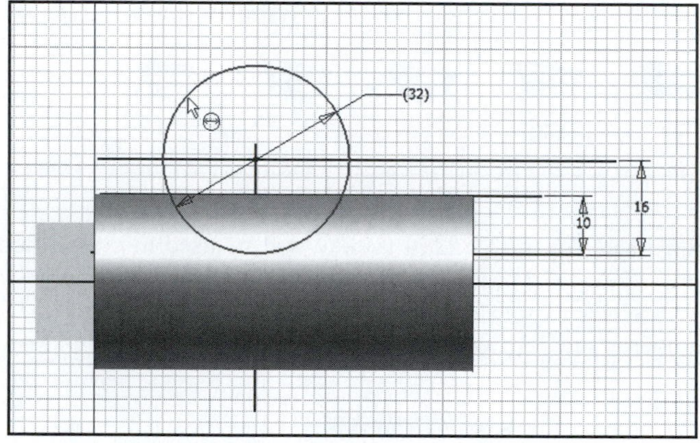

Figure 10-53

7. Return to an isometric view and select the **Extrude** tool. Use the **Cut** and **Intersection** options to create a cylinder from the circle drawn in step 6.

See Figures 10-54 and 10-55.

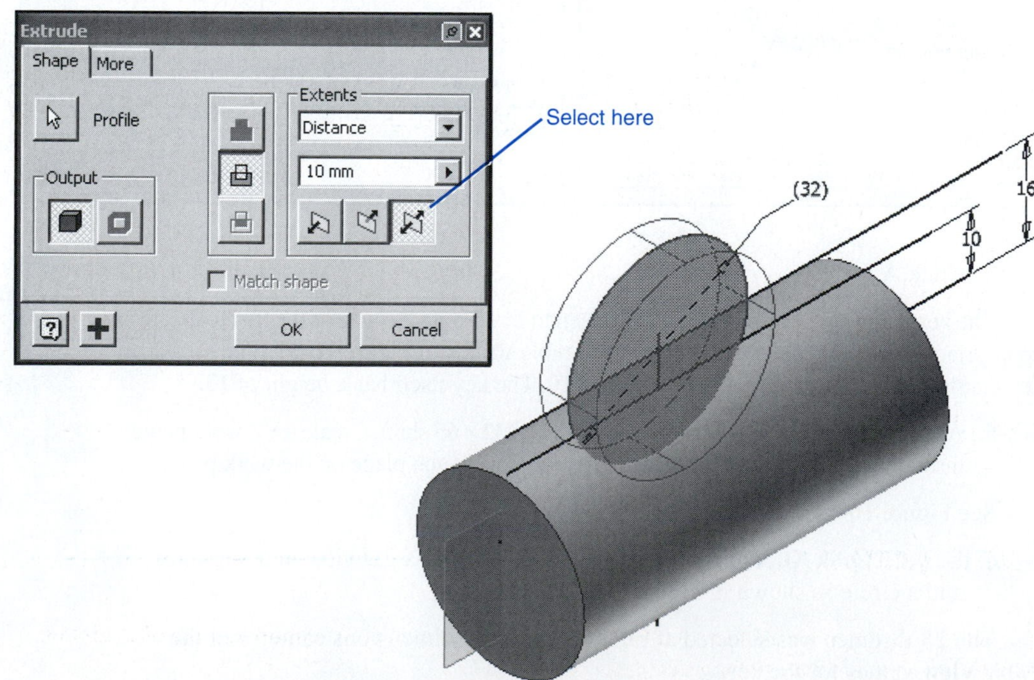

Figure 10-54

8. Use the **Constraint** command to locate the key into the shaft's keyway.

See Figure 10-56.

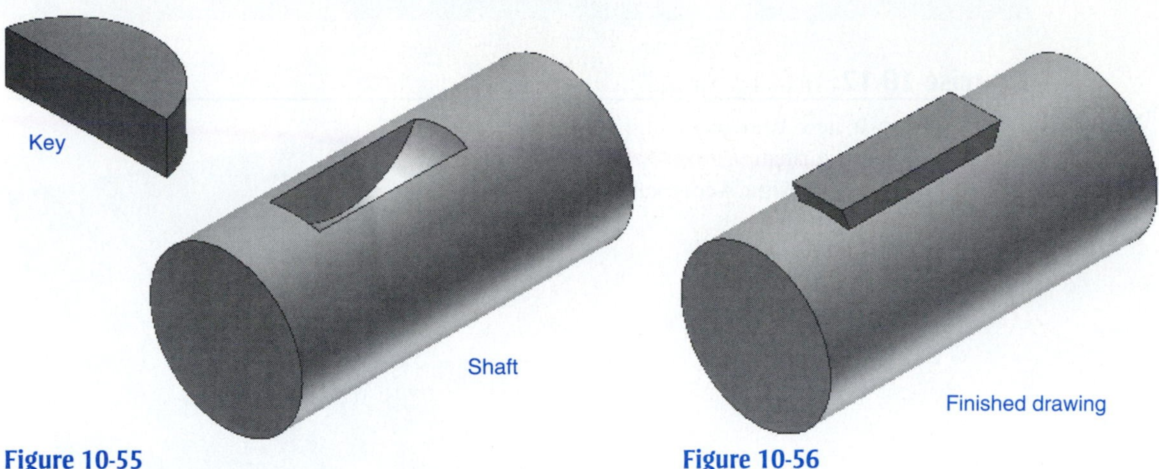

Figure 10-55 **Figure 10-56**

TIP

You may have to use work planes to locate the key into the shaft.

Figure 10-57 shows a hub design to fit over the shaft and Woodruff key shown in Figure 10-56. The 19.3 dimension value is derived from adding the bore hole radius of 16.0 to a hub depth value of 3.3 listed in the **Table View** for the key.

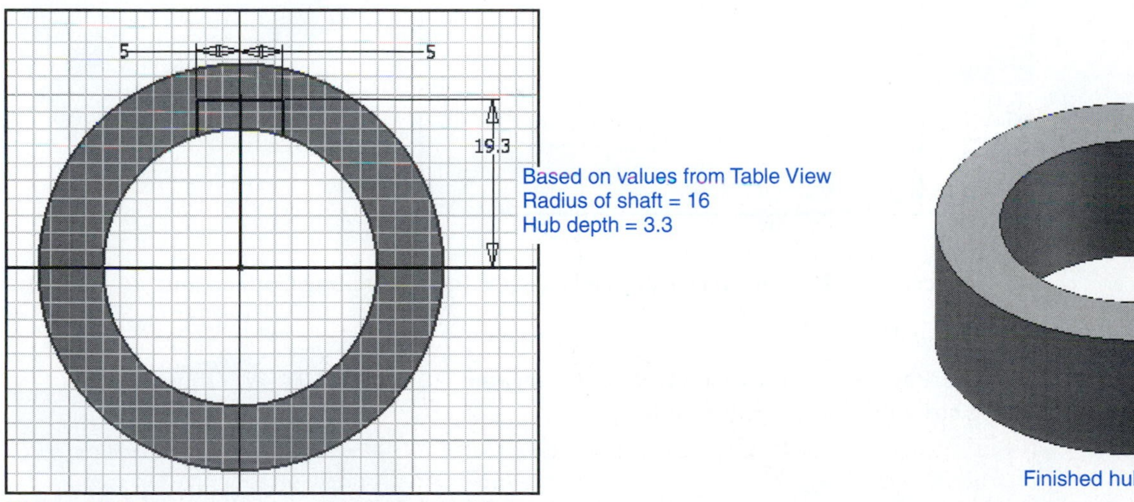

Figure 10-57

Shafts with Splines

Splines are a series of cutouts in a shaft that are sized to match a corresponding set of cutouts in a hub. Splines are used to transmit torque.

splines: A series of cutouts in a shaft that are sized to match a corresponding set of cutouts in a hub; they are used to transmit torque.

Splines are generally used on larger shafts. Smaller shafts use setscrews or pins.

Exercise 10-12: Drawing a Spline

1. Create a new drawing using the **Standard (mm).iam** format.
2. Access the **Design Accelerator** and create a ∅**32** × **65** shaft.

See Figure 10-58.

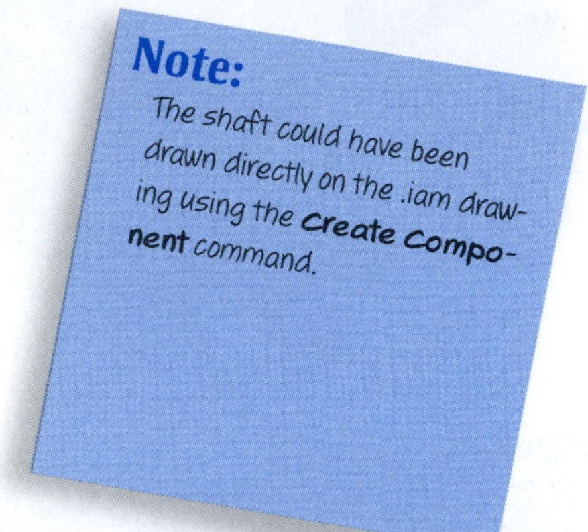

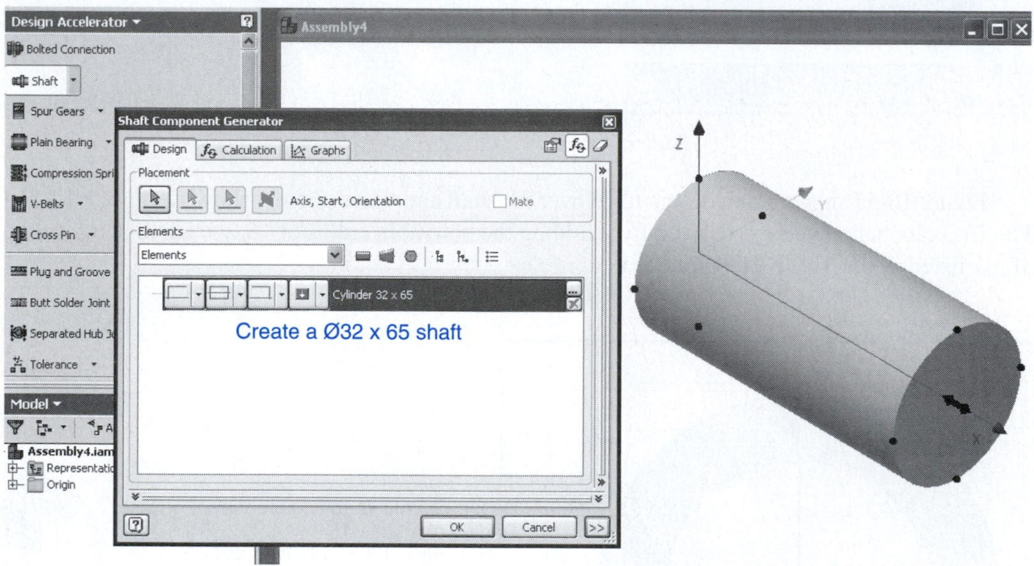

Figure 10-58

3. Access the **Design Accelerator** and click the **Parallel Spline Connection.**

See Figure 10-59.

The **Parallel Splines Connection Generator** dialog box will appear. See Figure 10-60.

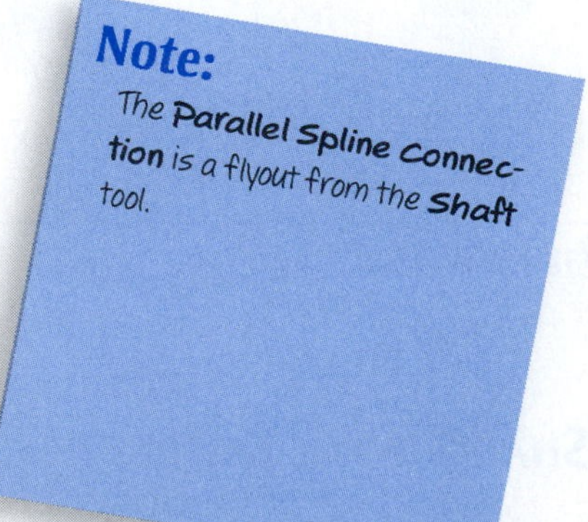

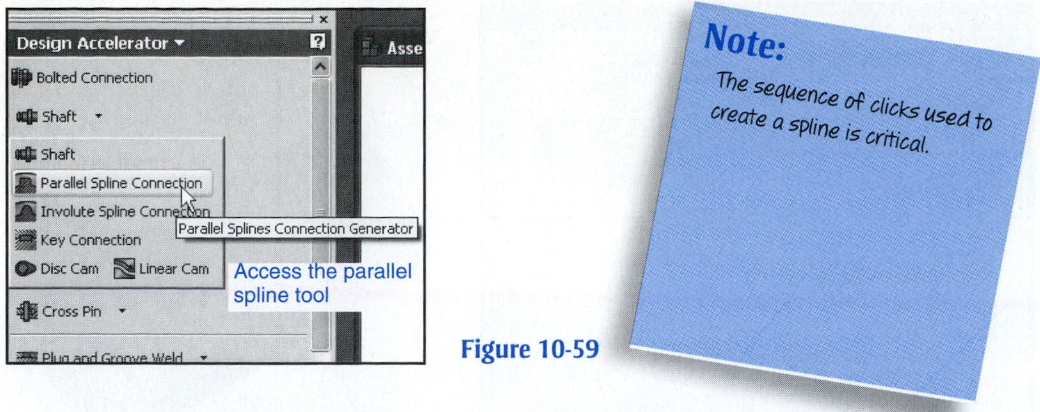

Figure 10-59

4. Set the **Splines Type** for light, the **Spline** dimension for **6×28×32,** the spline **Length** for 12, and **Radius** for 4.

See Figure 10-61.

The outside diameter of the spline must equal the outside diameter of the shaft

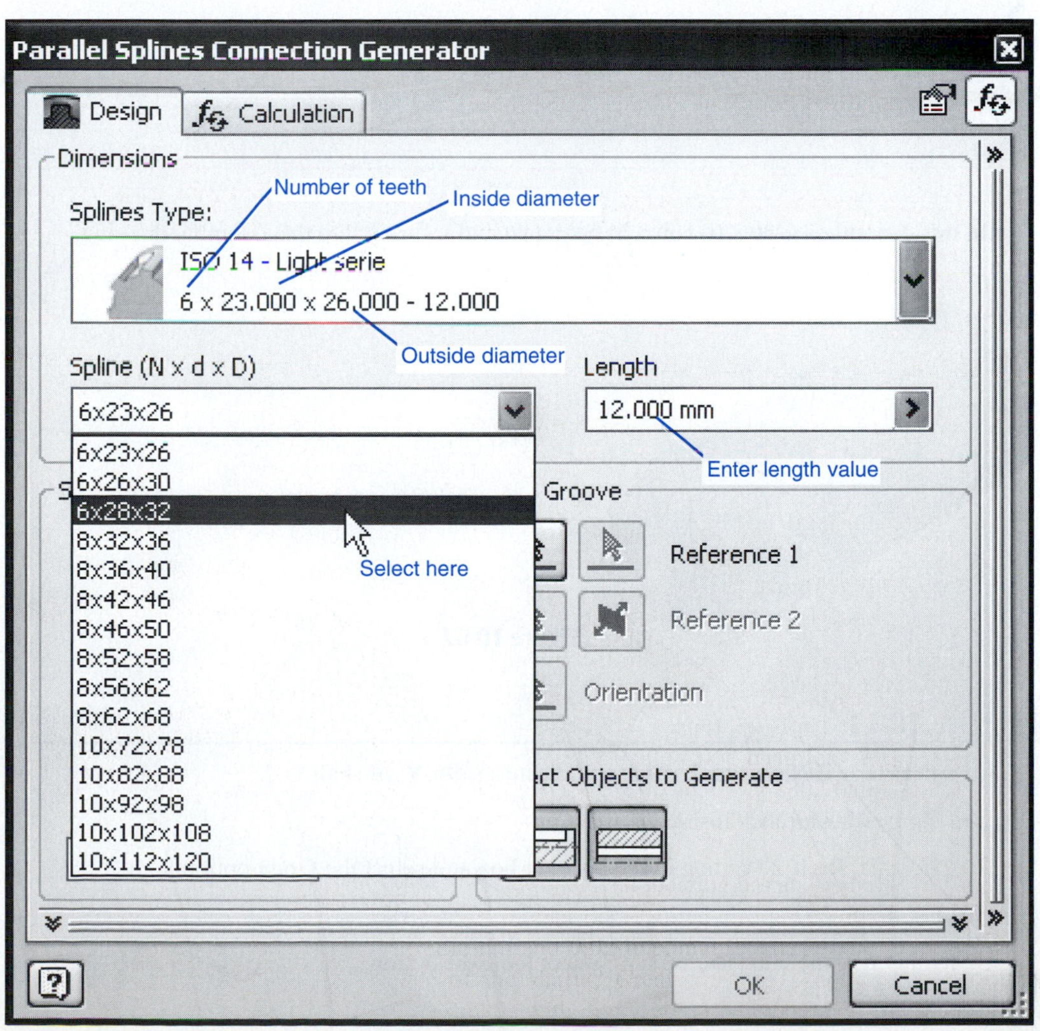

Figure 10-60

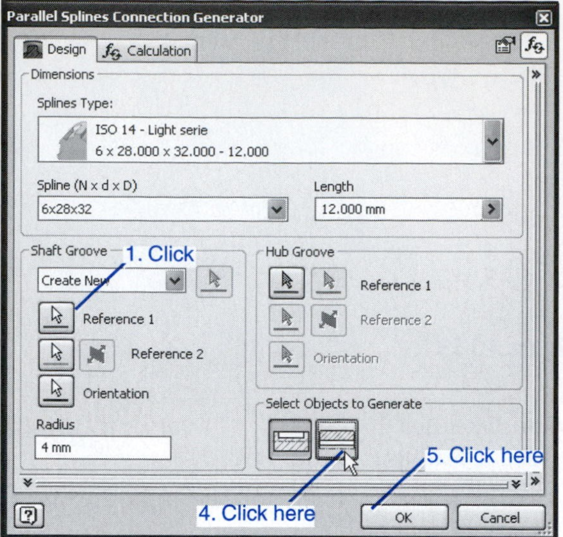

Figure 10-61

5. Click the **Reference 1** box, click the rounded sides of the shaft (the **Reference 2** box will automatically become active), click the flat end of the shaft, click the right box in the **Select Objects to Generate box,** then click **OK.**

Figure 10-62 shows the finished spline.

Splines are defined by N×d×D, where N is the number of teeth on the spline, d is the inside diameter, and D is the outside diameter.

In this example a ∅90×10 hub will be drawn and inserted into the assembly drawing.

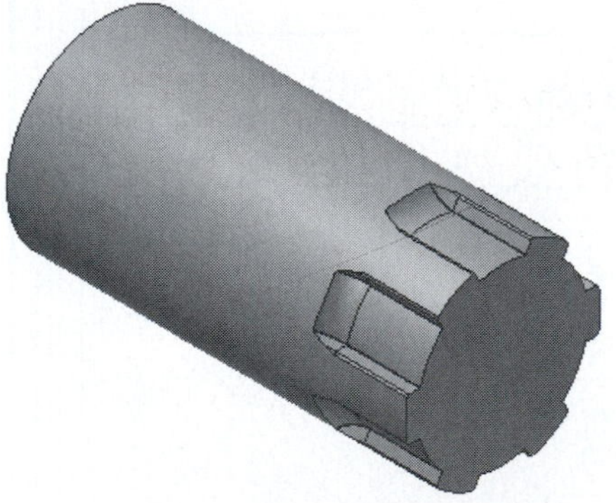

Figure 10-62

Exercise 10-13: Drawing a Hub

1. Access the **Design Accelerator** and create a **∅90 × 10** shaft.

See Figures 10-63 and 10-64.

2. Click the **Shaft 2** heading in the browser box and select the **Open** option.

See Figure 10-65.

3. Right click the hub (Shaft 2) and select the **Edit** option.

See Figure 10-66.

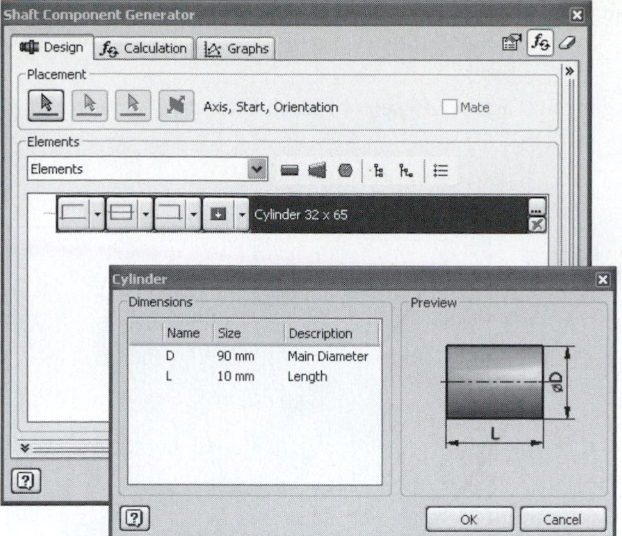

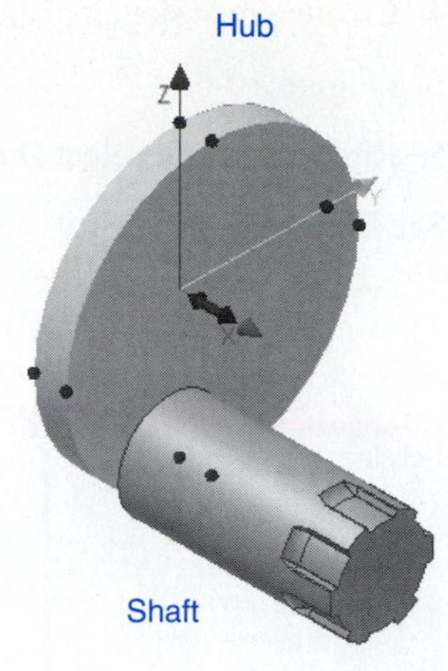

Figure 10-63

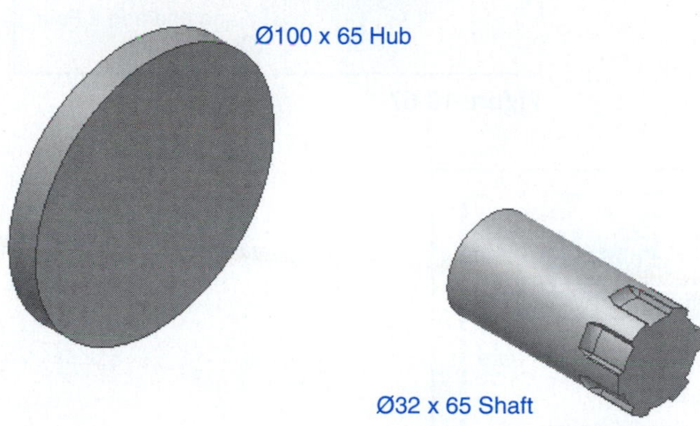

Figure 10-64

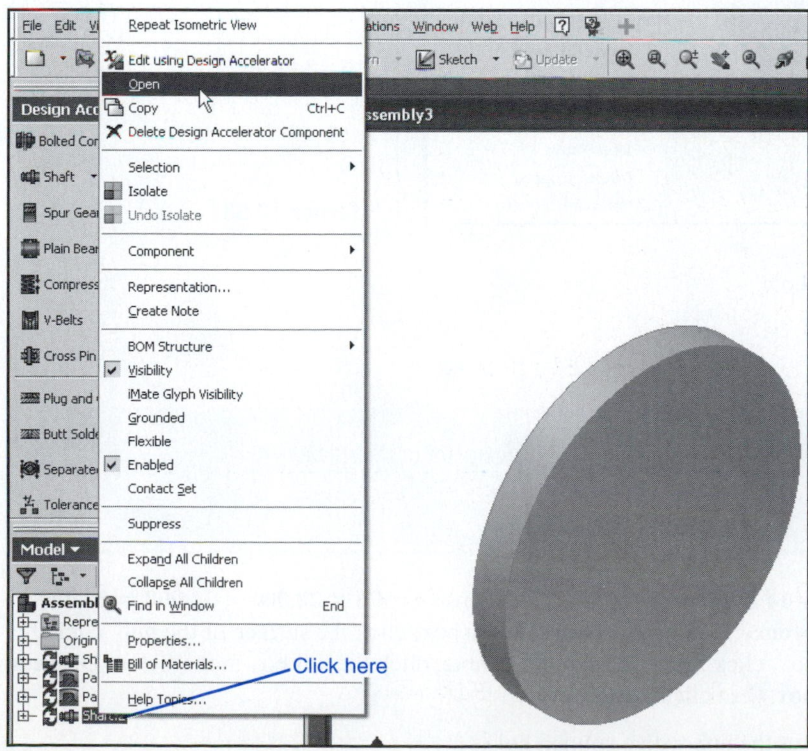

Figure 10-65

4. Create a new sketch plane on the front surface of the hub and add a **Point, Center Point.**

See Figure 10-67.

5. Right-click and select **Done.** Right-click again and select **Finish Sketch.**

See Figure 10-68

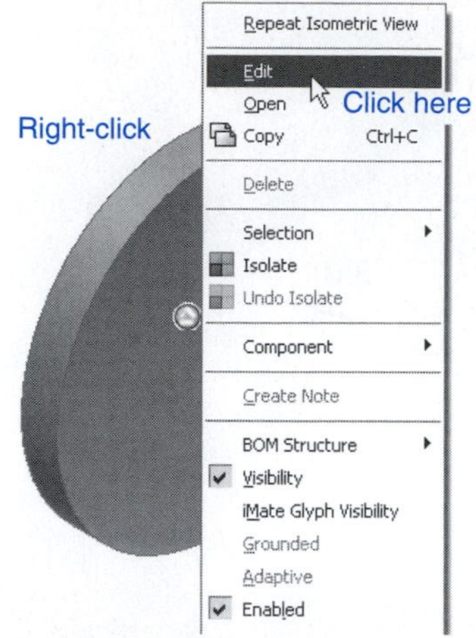

Figure 10-66

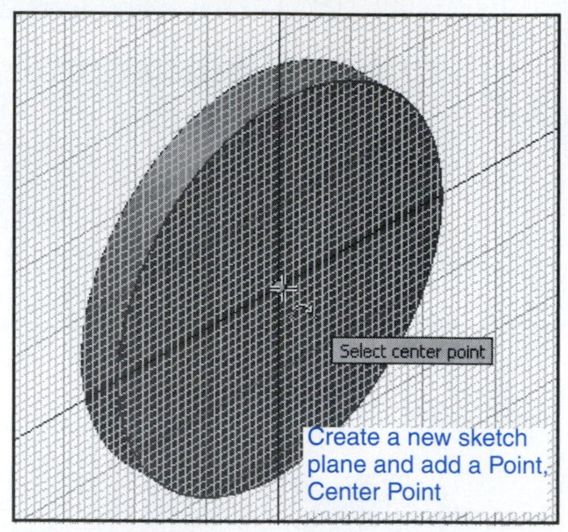

Figure 10-67

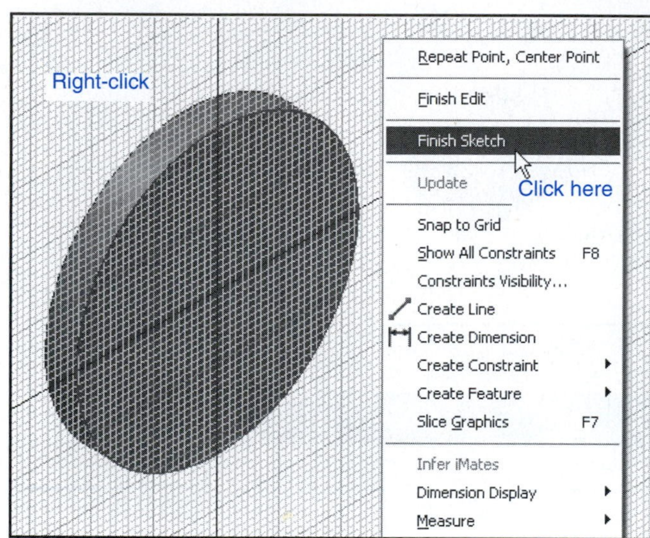

Figure 10-68

6. Create a ∅**28** hole.

See Figure 10–69

The hole diameter must equal the inside diameter of the spline.

7. Select an **ISO - 14 Light** serie spline with dimensions of **6 × 28.000 × 32.000 = 14.000.**
8. Click the **Reference 1** box in the **Hub Groove** box, click the surface of the hub, click the **Reference 2** box, click the edge of the ∅28 hole, click the left box in the **Select Objects to Generate box**, than click **OK.**

Figure 10-70 shows the hub with a splined hole.

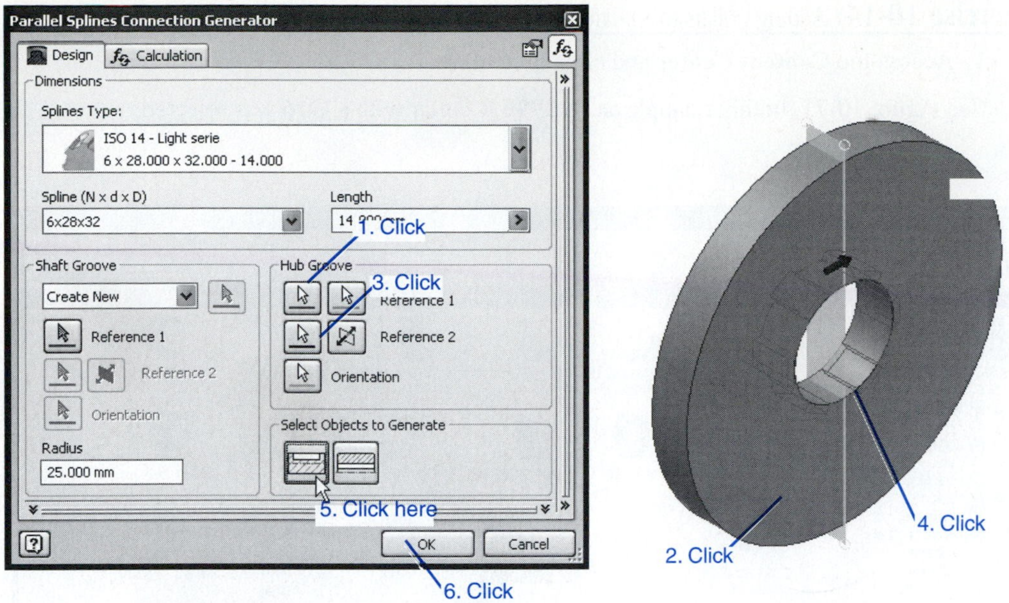

Figure 10-69

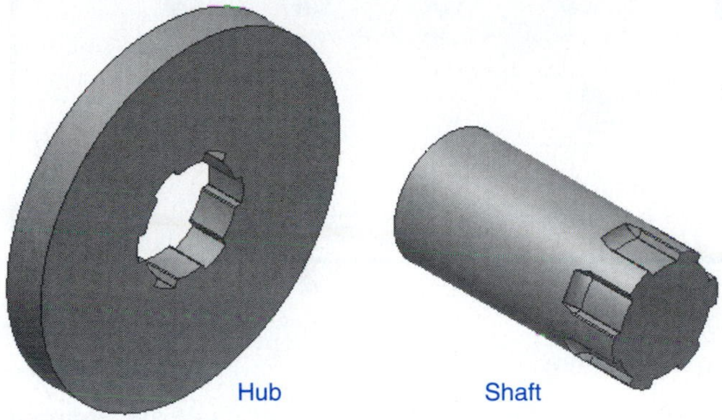

Figure 10-70

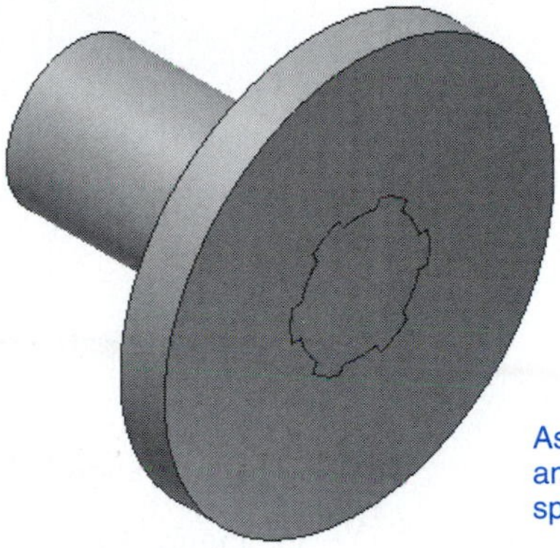

Figure 10-71

9. Use the **Constraint Insert** option and mount the hub onto the shaft.

See Figure 10-71.

Collars

Collars are used to hold shafts in position as they rotate. Collars may be mounted inside or outside the support structure. See Figure 10-72.

collar: Used to hold a shaft in place as it rotates; may be mounted on the inside or outside of the support structure.

Collars are used to hold shafts in position. Figure 10-72

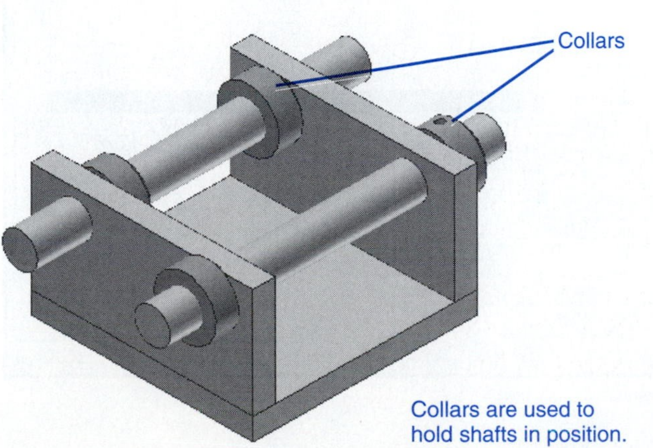

Exercise 10-14: Adding Collars to a Shaft Assembly—Setscrews

1. Access the **Content Center** and click on **Collars.**

See Figure 10-73. In this example an **IS 2995 B** collar with a **∅16** was selected.

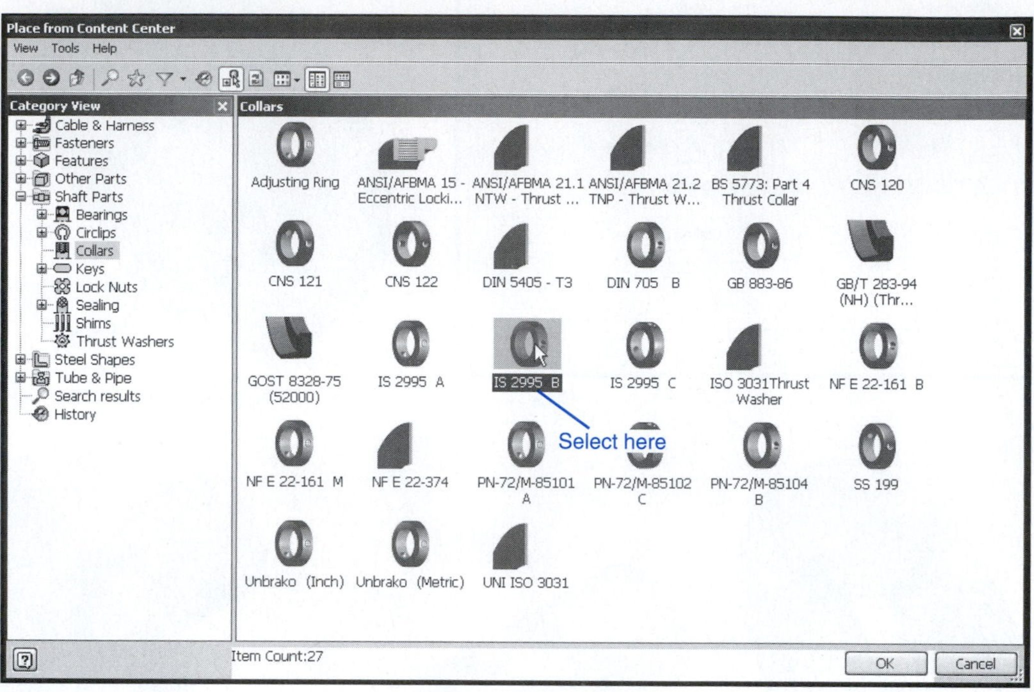

Figure 10-73

2. Click the **Table View** tab, then **All Columns.**

See Figure 10-74. Note the size specifications for the collar. In this example the **Nominal Diameter** is the diameter of the small hole in the collar. The diameter of the hole is **4.**

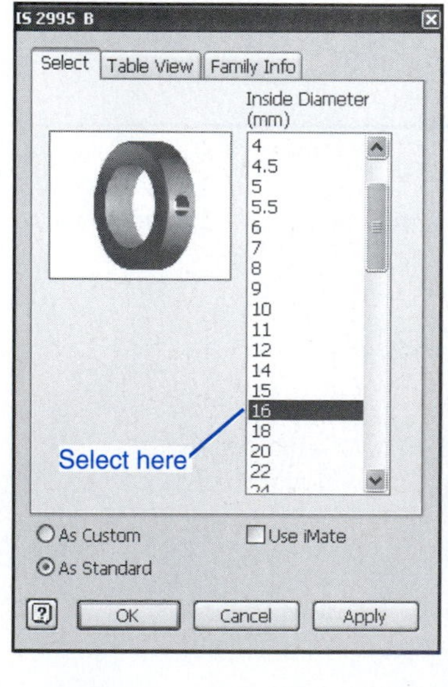

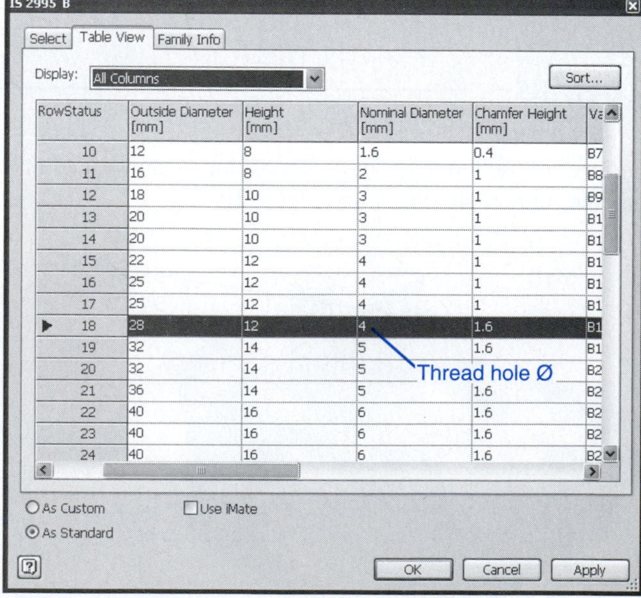

RowStatus	Outside Diameter [mm]	Height [mm]	Nominal Diameter [mm]	Chamfer Height [mm]	Va
10	12	8	1.6	0.4	B7
11	16	8	2	1	B8
12	18	10	3	1	B9
13	20	10	3	1	B1
14	20	10	3	1	B1
15	22	12	4	1	B1
16	25	12	4	1	B1
17	25	12	4	1	B1
18	28	12	4	1.6	B1
19	32	14	5	1.6	B1
20	32	14	5		B2
21	36	14	5	1.6	B2
22	40	16	6	1.6	B2
23	40	16	6	1.6	B2
24	40	16	6	1.6	B2

Figure 10-74

This hole can be threaded and a setscrew added to hold the collar on the shaft.

5. If the hole is threaded (for this example assume the hole is threaded) and has a size of M4, access the **Content Center** and select a setscrew.

See Figure 10-75. In this example a **CSN 02 1185 Slotted Set Screw** with an **M4** thread and a conical point was selected.

6. Use the **Constraint** tool and insert the setscrew into the hole.

The assembly sequence presented here is one of several possibilities.

7. Insert the setscrew into the drawing and position it above the hole.

See Figure 10-75.

7. Draw a work axis through the hole and use the **Constraint Mate** tool to align the hole's work axis with the setscrew's axis.

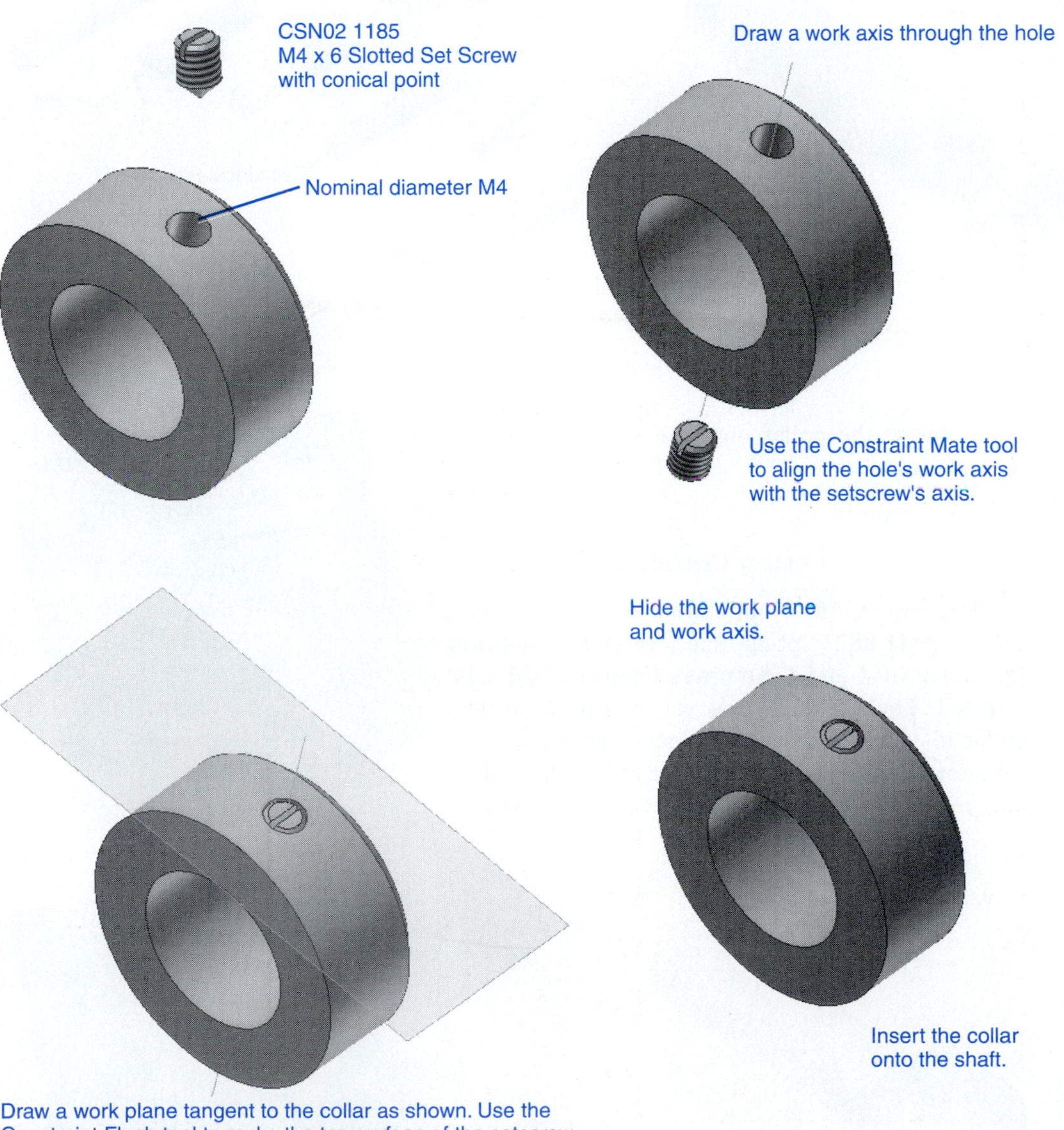

Figure 10-75

8. Draw a work plane tangent to the collar and hole's edge. Use the **Constraint Flush** tool to make the setscrew flush with the work plane.
9. Hide the work plane and work axis.

Adding Collars to a Shaft Assembly—Pins

A collar may be held in place using pins. Figure 10-76 shows a shaft and an IS 2995 B collar, both with ∅4.00 holes. A pin will be inserted through the hole in the collar into the hole in the shaft.

Figure 10-77 shows some of the different types of pins available in the Inventor **Content Center.** For this example a spring-type cylindrical pin will be used. These type pins are squeezed to a smaller diameter, inserted into the hole, and then released back to their original diameter.

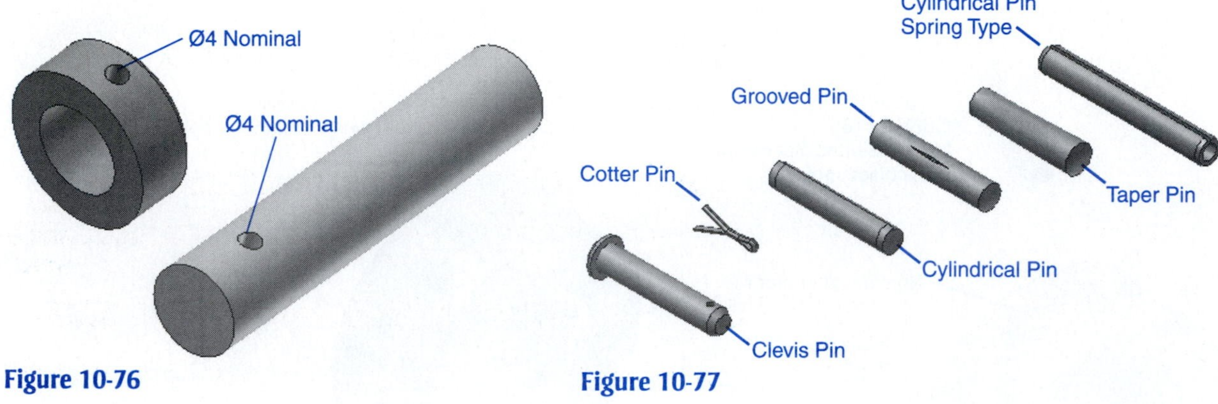

Figure 10-76

Figure 10-77

1. Insert the collar onto the shaft with the holes aligned.

See Figure 10-78.

2. Access the **Content Center** and select a pin.

See Figures 10-79 and 10-80. In this example a **BS EN ISO 8572** spring-type pin with a **Nominal Diameter** of **4** and a **Nominal Length** of **20** was selected. The length 20 was selected because the collar thickness is 6 and the shaft diameter is 16, for a total of 22. There is only one hole in the collar, so the pin length cannot exceed 22 or it will

Note:
Consider defining a work axis through the holes to help with the alignment.

Note:
Consider using work planes and a work axis when constraining the pin and collar.

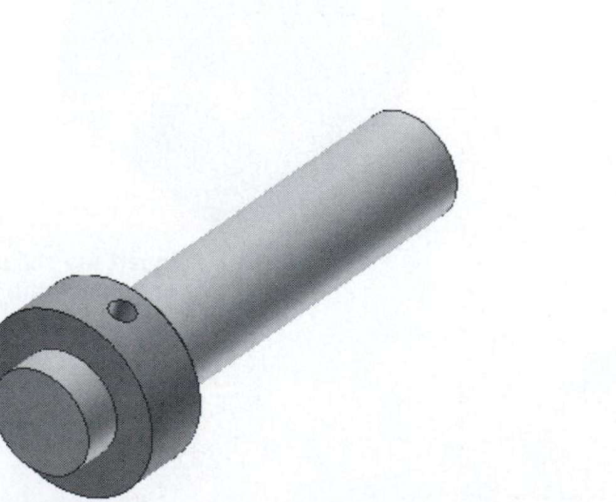

Figure 10-78

extend beyond the collar. A length value of 20 allowed for a clearance of 2. Figure 10-81 shows the selected pin with the collar and shaft.

3. Use the **Constraint** tool to assemble the pin into the collar and shaft.

See Figure 10-82.

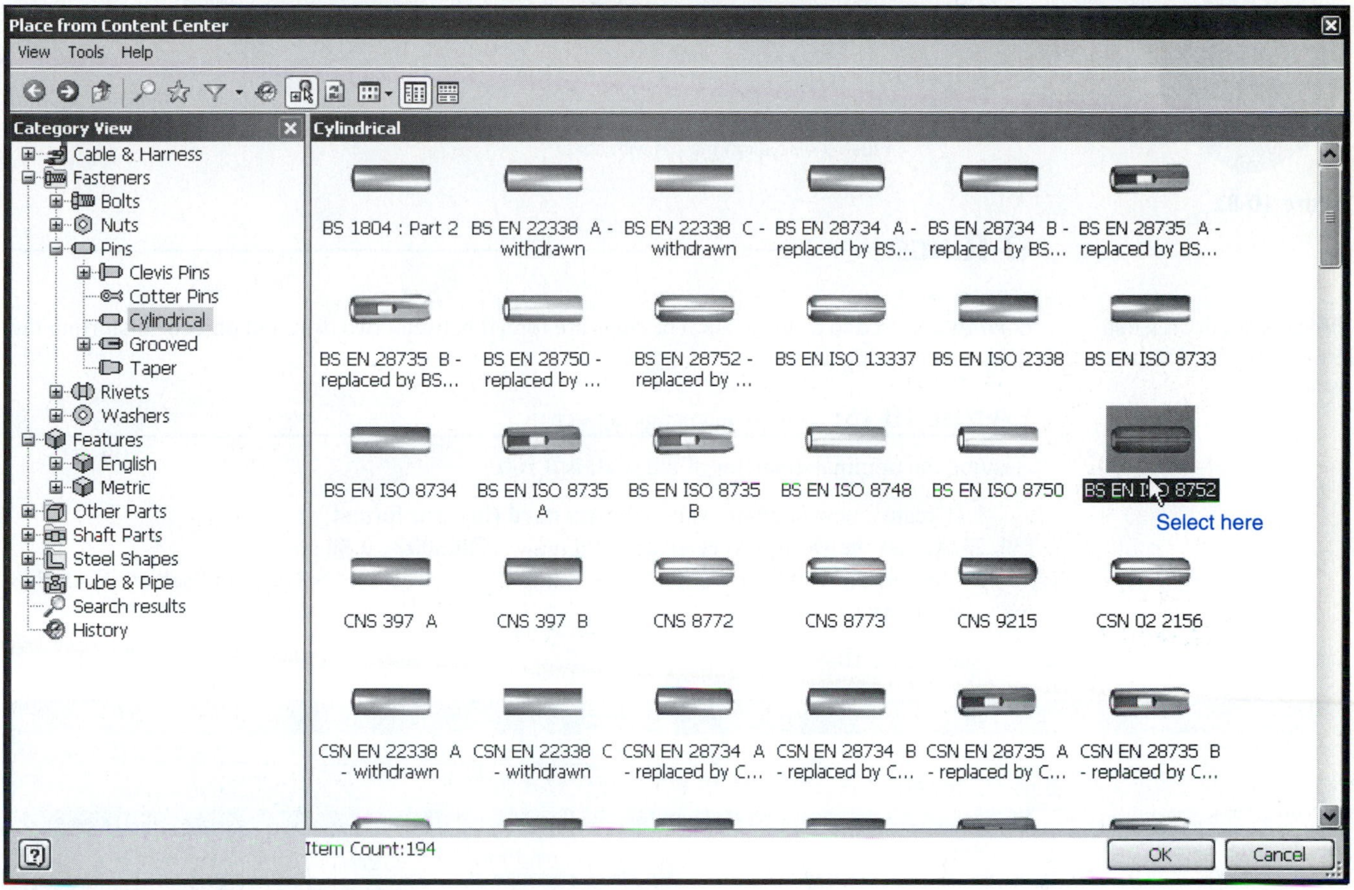

Figure 10-79

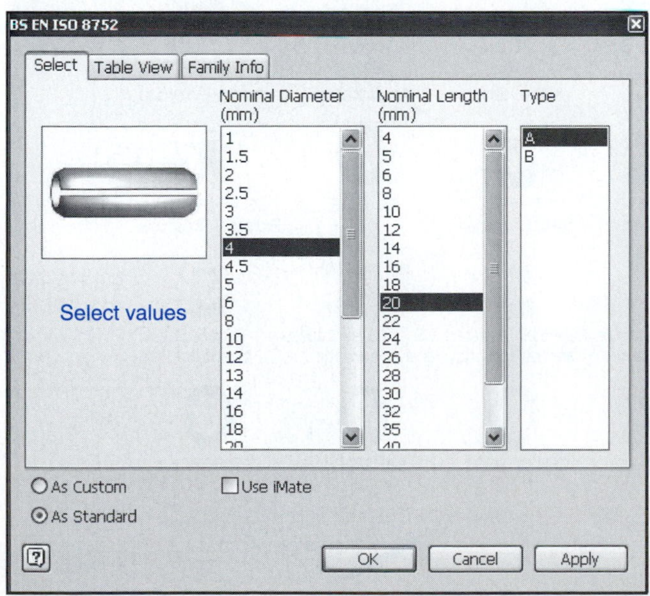

Figure 10-80

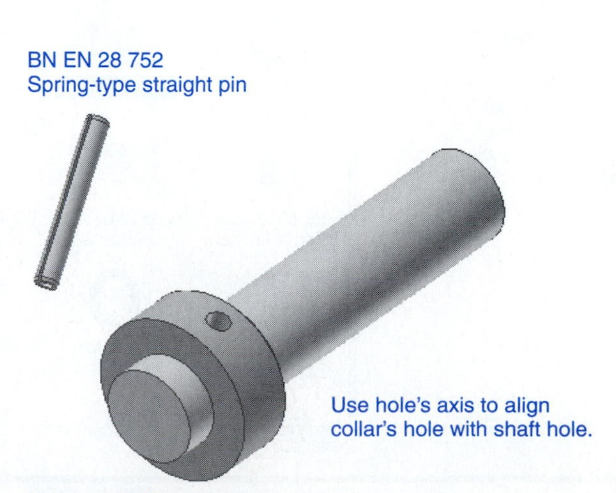

Figure 10-81

Insert pin into collar and shaft
Use a work plane and the Constraint Flush tool to align the pin with collar.
Finished drawing

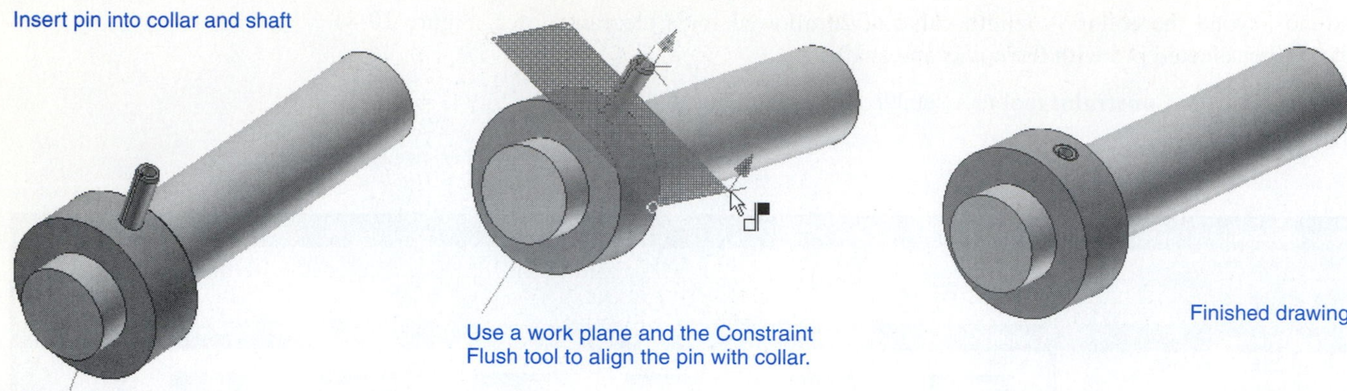

Figure 10-82

O-RINGS

O-ring: A ring that is forced between two objects to create a seal.

O-rings are used to create seals. The rings are forced between two different objects, distorting the rings and creating a seal.

Exercise 10-15: Drawing an O-Ring and a Shaft

Assume the nominal diameter of the shaft is 0.500.

1. Create a new drawing using the **Standard (in).iam** format.
2. Access the **Design Accelerator** and draw a **∅0.500 × 2.50** shaft.
3. Access the **Content Center,** click **Shaft Parts, Sealing, O-Rings** and select an **AS 568 O-Ring**.

See Figure 10-83.

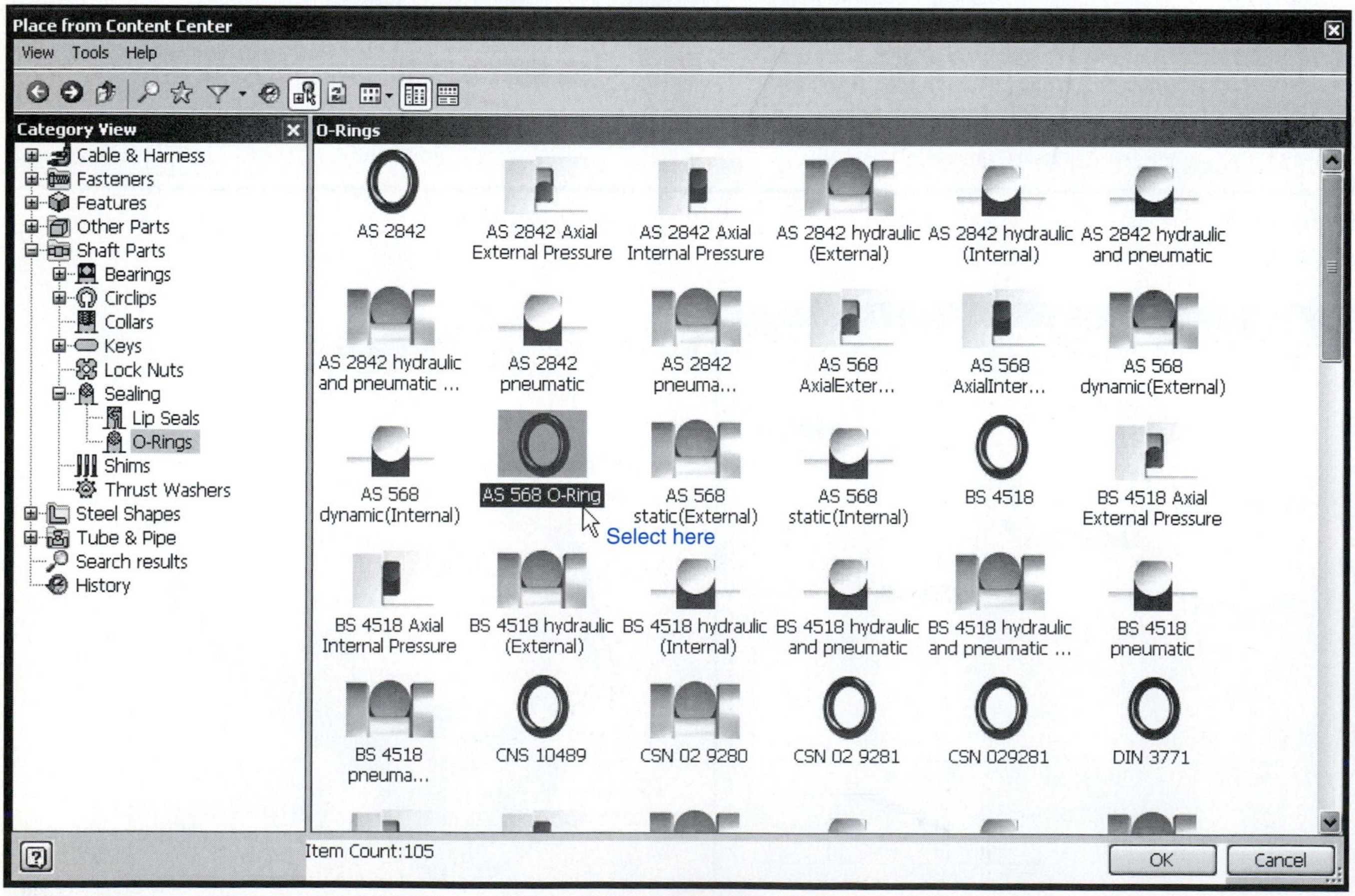

Place from Content Center
View Tools Help
Category View
Cable & Harness
Fasteners
Features
Other Parts
Shaft Parts
Bearings
Circlips
Collars
Keys
Lock Nuts
Sealing
Lip Seals
O-Rings
Shims
Thrust Washers
Steel Shapes
Tube & Pipe
Search results
History
O-Rings
AS 2842
AS 2842 Axial External Pressure
AS 2842 Axial Internal Pressure
AS 2842 hydraulic (External)
AS 2842 hydraulic (Internal)
AS 2842 hydraulic and pneumatic
AS 2842 hydraulic and pneumatic ...
AS 2842 pneumatic
AS 2842 pneuma...
AS 568 AxialExter...
AS 568 AxialInter...
AS 568 dynamic(External)
AS 568 dynamic(Internal)
AS 568 O-Ring
AS 568 static(External)
AS 568 static(Internal)
BS 4518
BS 4518 Axial External Pressure
Select here
BS 4518 Axial Internal Pressure
BS 4518 hydraulic (External)
BS 4518 hydraulic (Internal)
BS 4518 hydraulic and pneumatic
BS 4518 hydraulic and pneumatic ...
BS 4518 pneumatic
BS 4518 pneuma...
CNS 10489
CSN 02 9280
CSN 02 9281
CSN 029281
DIN 3771
Item Count:105
OK
Cancel

Figure 10-83

4. Click the **Table View** tab, then **All Columns.** Select a **DASH # 014 O-Ring** because it has a 0.489 inside diameter.

See Figure 10-84.

5. Insert the ring into the drawing.
6. Right-click the **Shaft** heading in the browser box and select the **Edit using Design Accelerator** option.

See Figure 10-85. The **Shaft Component Generator** dialog box will appear. See Figure 10-86.

7. Select the **Add Necking Down – D** option.

The **Necking-down** option will appear on the screen.

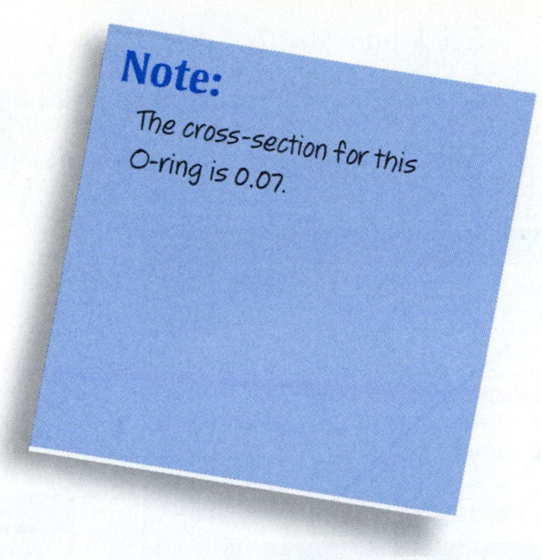

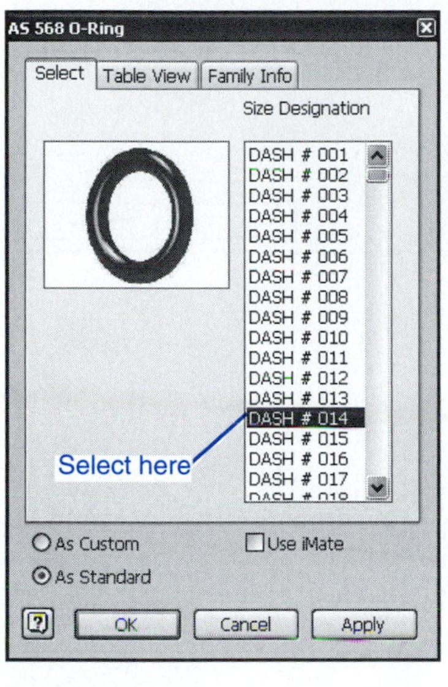

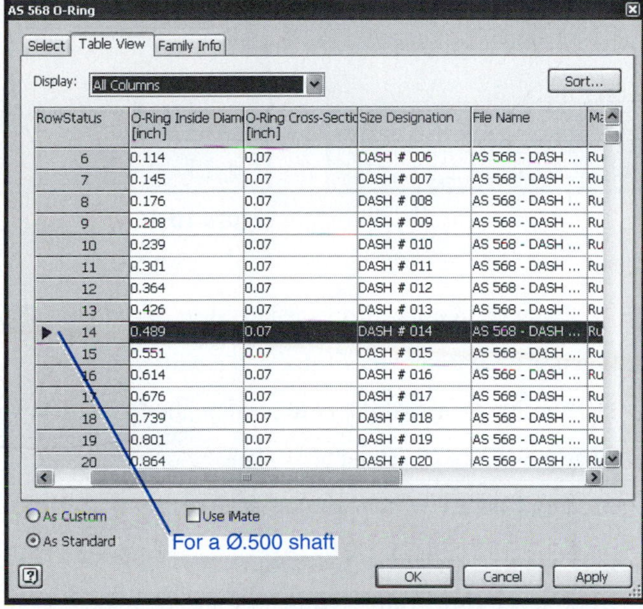

Figure 10-84

Figure 10-85

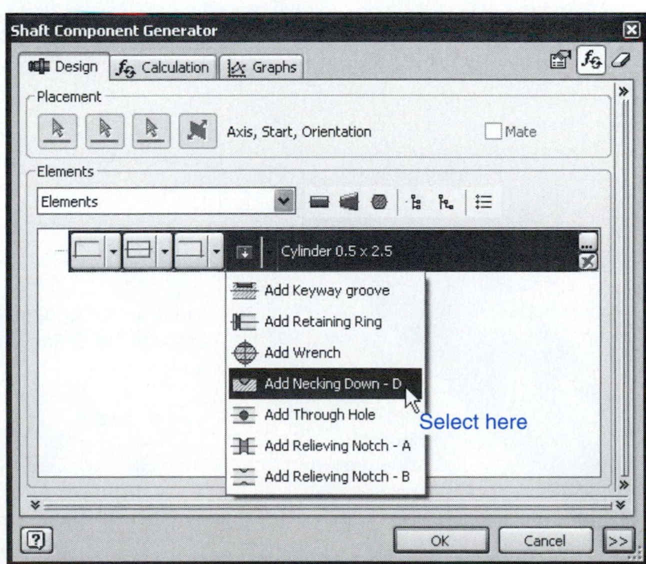

Figure 10-86

See Figure 10-87.

8. Click the **Elements properties** box.

The **Necking Down – D Type** dialog box will appear. See Figure 10-88.

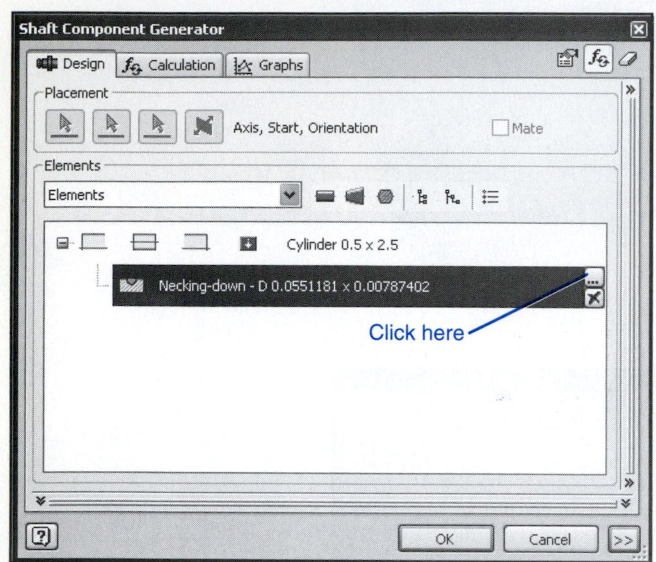

Figure 10-87

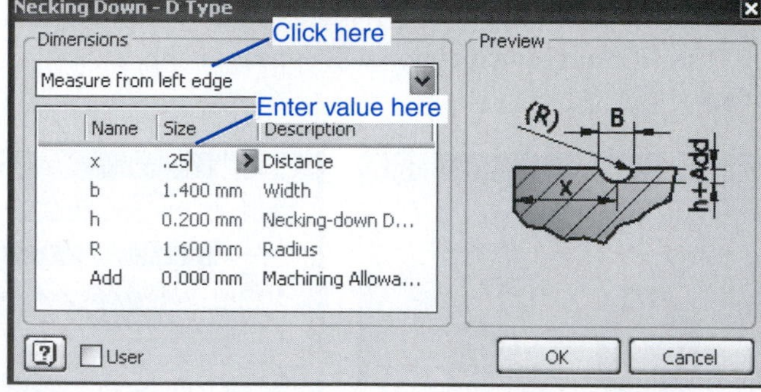

Figure 10-88

9. Enter the distance from the left edge. Accept the other values. Click **OK.**

Figure 10-89 shows the O-ring and modified shaft.

10. Use the **Constraint** tool to position the O-ring onto the shaft's groove.

The procedure presented here is one possibility.

11. Draw an XY work plane through the O-ring.

See Figure 10-90.

12. Use the **Constraint's Mate** tool and align the X axis of the O-ring with the X axis of the shaft.
13. Use the **Flush** command to align the front surface of the shaft with the work plane on the O-ring, then define a **−0.25** offset.

Figure 10-90 shows the finished O-ring and shaft.

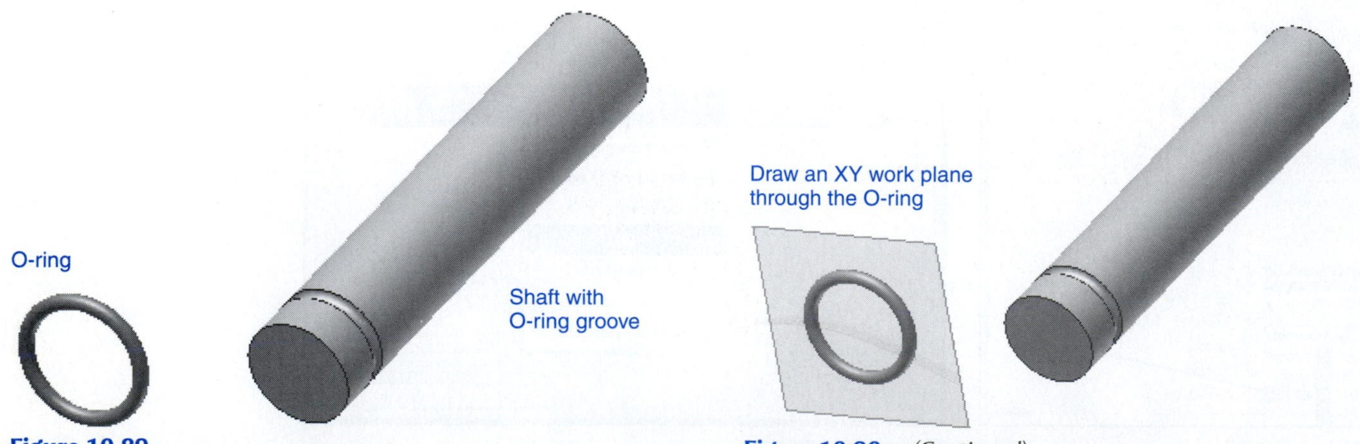

Figure 10-89

Figure 10-90 *(Continued)*

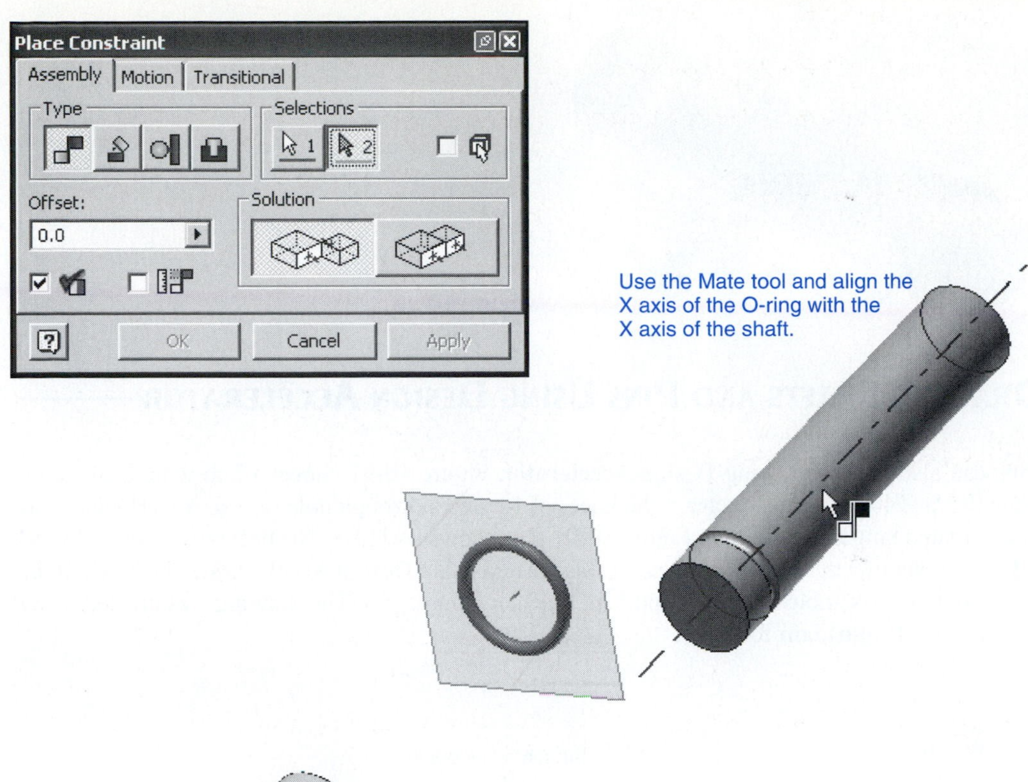

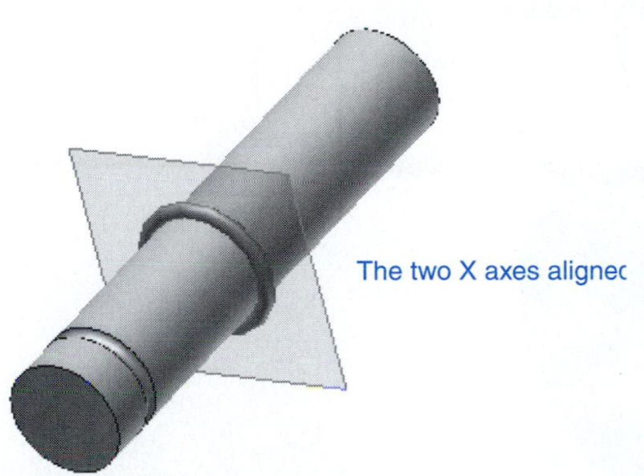

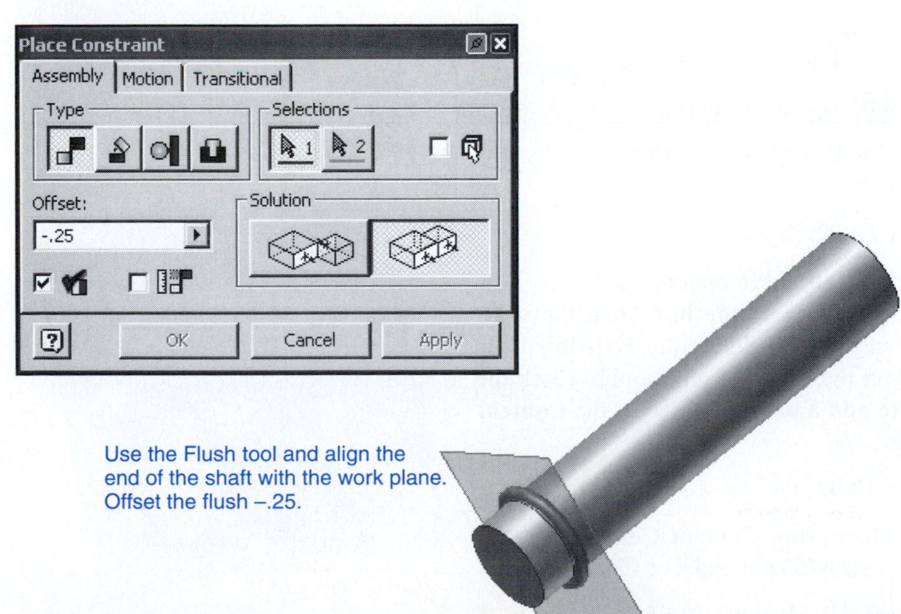

Figure 10-90 *(Continued)*

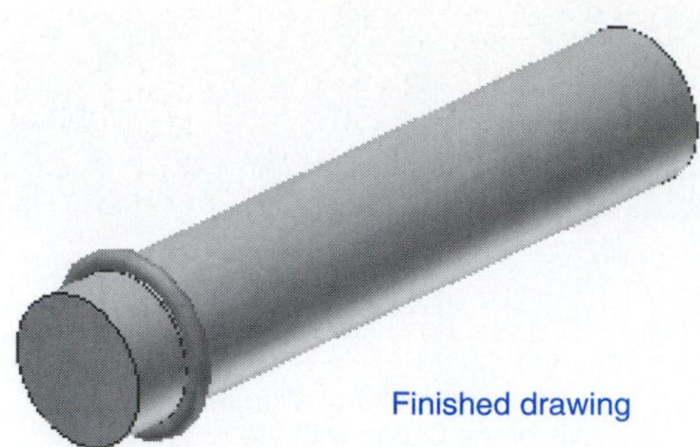

Figure 10-90

DRAWING SHAFTS AND PINS USING DESIGN ACCELERATOR

Pins can also be drawn using **Design Accelerator.** Figure 10-91 shows a hub with ∅20 inside diameter, a ∅40 outside diameter, a thickness of 16, and a through hole of ∅6. A work plane has been created tangent to the hub. Figure 10-91 also shows a ∅20 × 30 shaft with a ∅6 hole and a 1 × 45° chamfer at each end. The shaft was created using **Design Accelerator.** The hub and the shaft are to be assembled together and held together using a pin. The drawing was created using the **Standard (mm).iam** format.

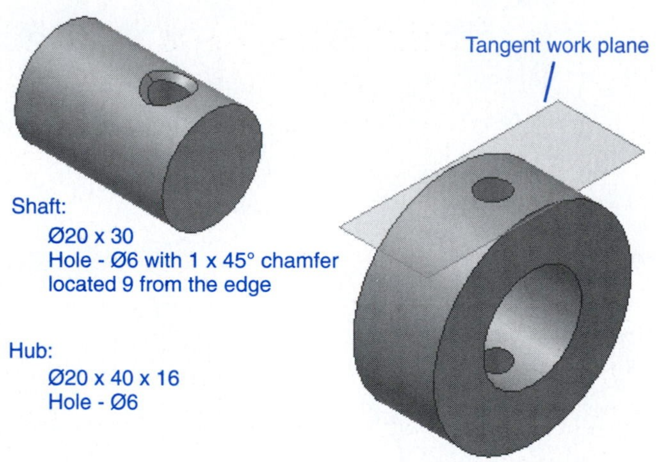

Figure 10-91

Exercise 10-16: Drawing a Pin Using Design Accelerator

1. Access the **Design Accelerator** and select the **Pin** option.

See Figure 10-92.

2. Select the **Radial Pin** option.
3. Select the **Start Plane** box, then the work plane on the hub. Select the **Existing Hole** box, then the **∅6** hole on the hub. Click the **Click to add a pin** box to access the **Content Center.**

See Figure 10-93.

4. In the **Place from Content Center** select a **BS EN ISO 8752** pin. Select a ∅6 × 40 pin.

See Figures 10-94 and 10-95. Figure 10-96 shows the pin in the assembly drawing.

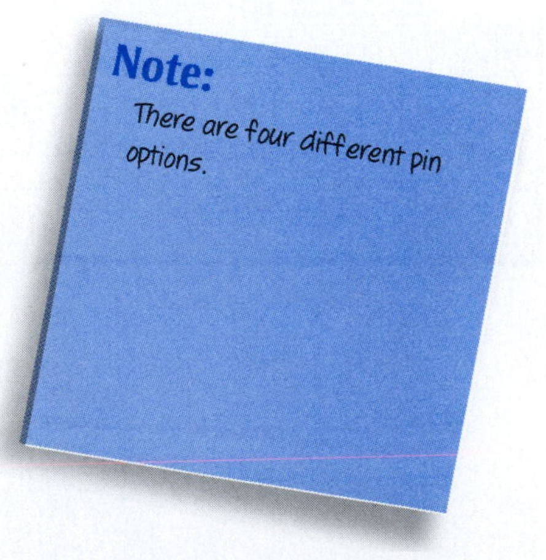

Create work axes for each of the three parts to help align the assembly.

5. Use the **Constraint** tool and assemble the parts.

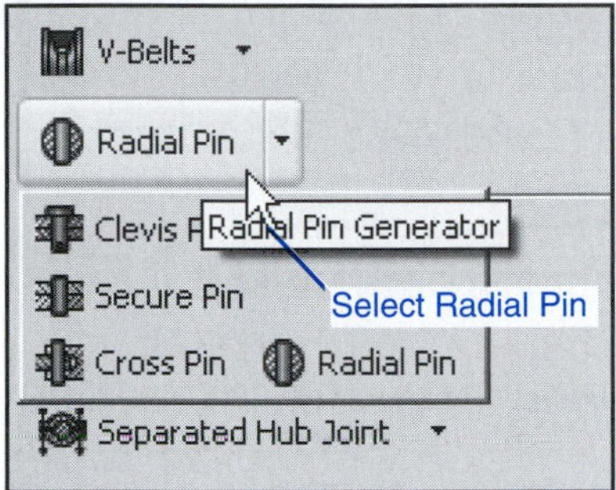

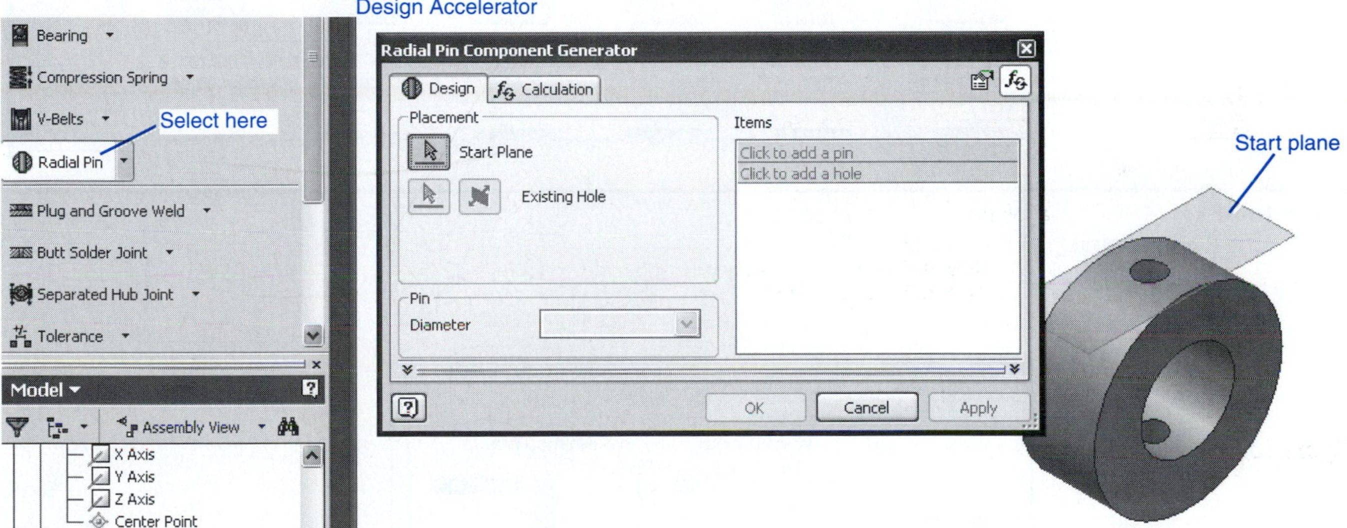

Figure 10-92

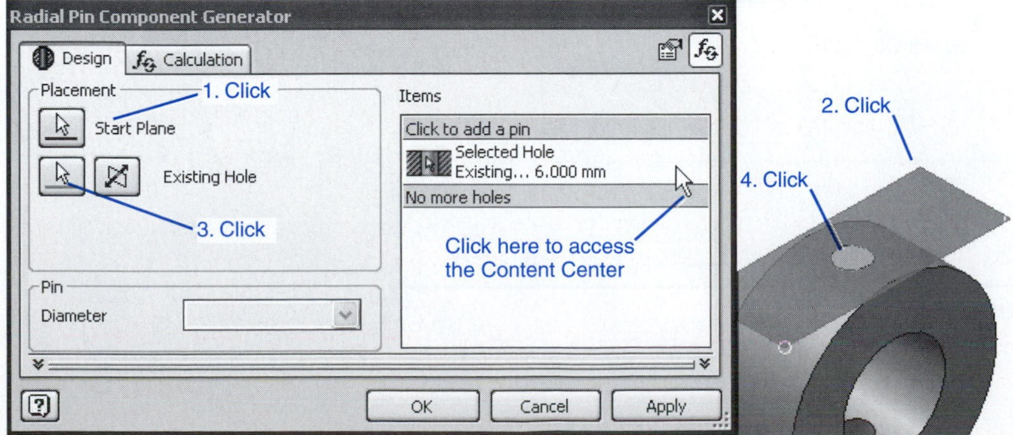

Figure 10-93

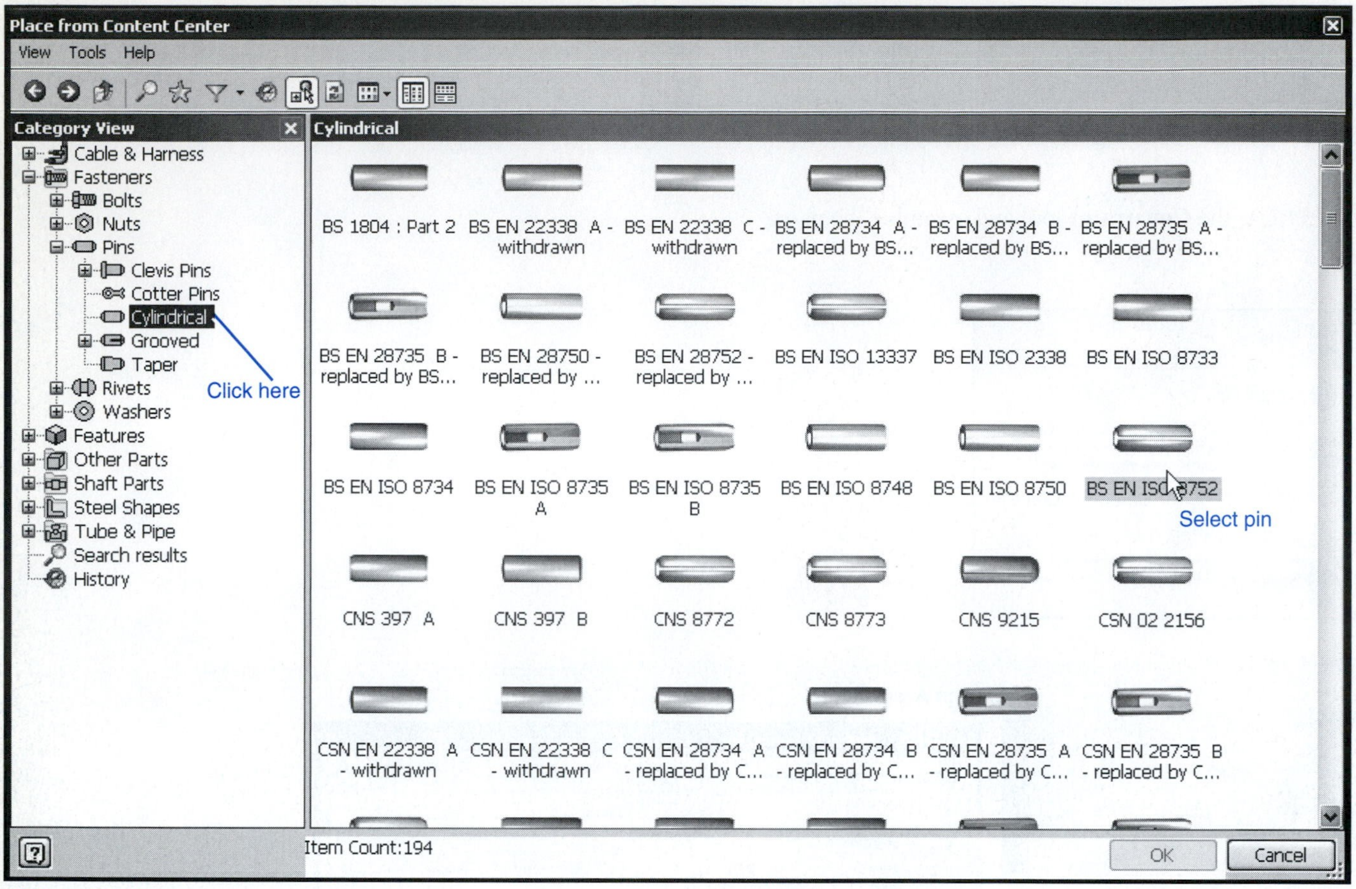

Figure 10-94

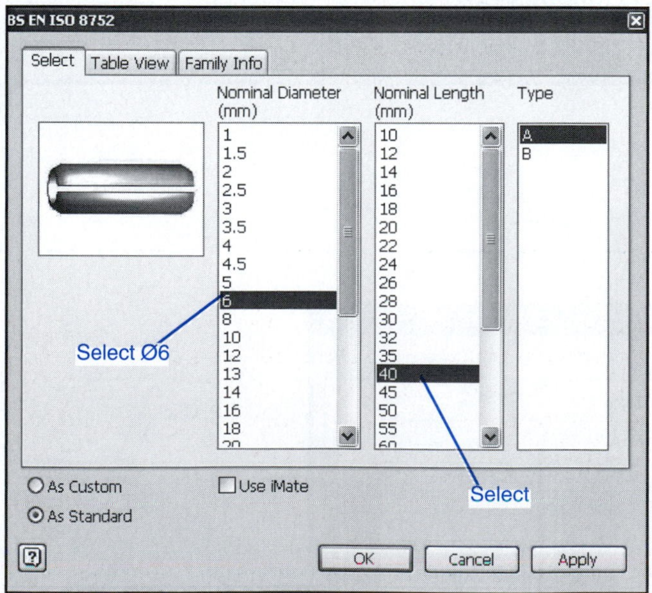

Figure 10-95

Figure 10-97 shows the finished assembly.

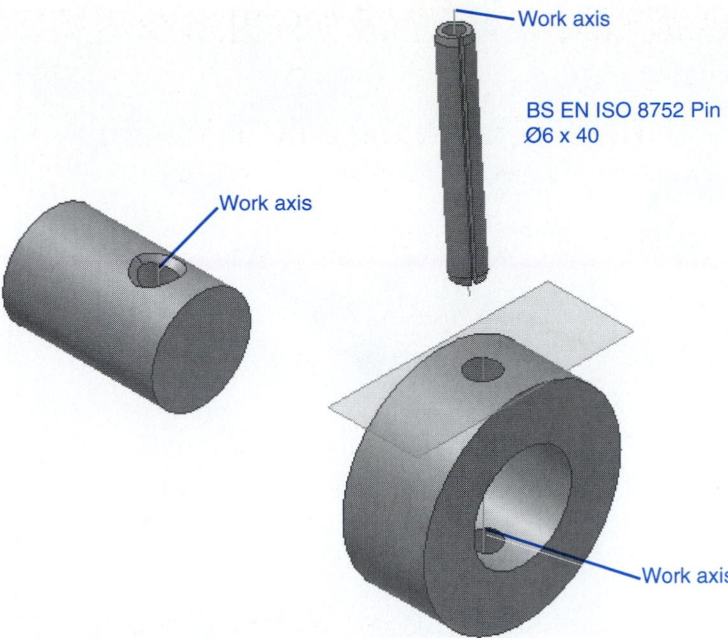

Figure 10-96

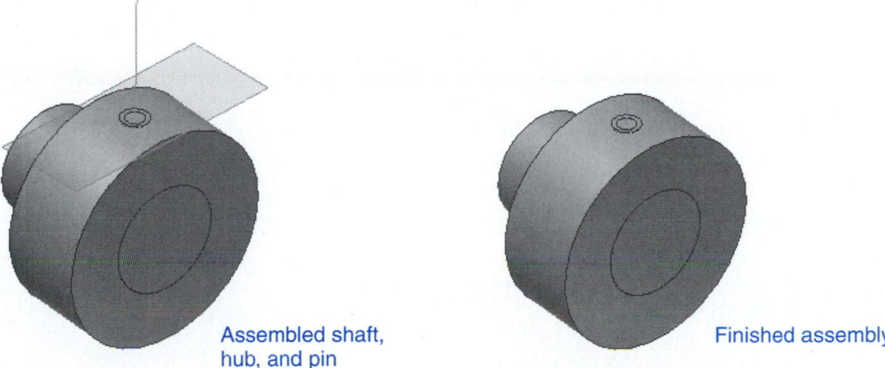

Figure 10-97

SUMMARY

This chapter illustrated how to use **Design Accelerator** to draw shafts, how to add retaining ring grooves to shafts, how to add keyways to shafts, how to add O-ring grooves to shafts, how to add pin holes to shafts, and how to use the **Content Center** to add retaining rings, keys, O-rings, and pins to shafts. The five general types of keys were described and illustrated, namely, Pratt and Whitney, square, rectangular, Woodruff, and Gib.

Splines on a shaft, for transmitting torque, were also introduced and used in a drawing.

CHAPTER PROJECTS

Draw the following uniform shafts.

Project 10-1: INCHES

Ø.50 × 3.50
0.10 × 45° chamfer, both ends

Project 10-2: INCHES

Ø2.00 × 10.00
0.125 × 45° chamfer, both ends

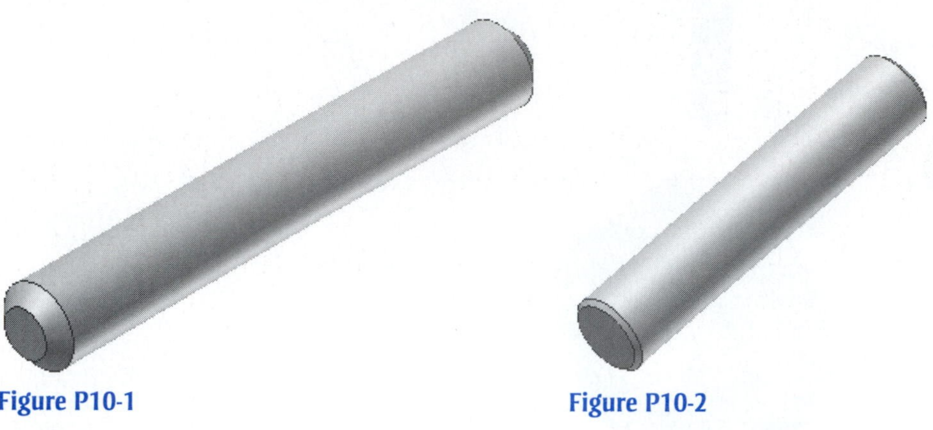

Figure P10-1

Figure P10-2

Project 10-3: MILLIMETERS

Ø20 × 110
3 × 45° chamfer, both ends

Project 10-4: MILLIMETERS

Ø60 × 100
5 × 45° chamfer, both ends

Draw the following shafts and retaining rings. Create the appropriate grooves on the shaft, and position the retaining rings onto the shaft.

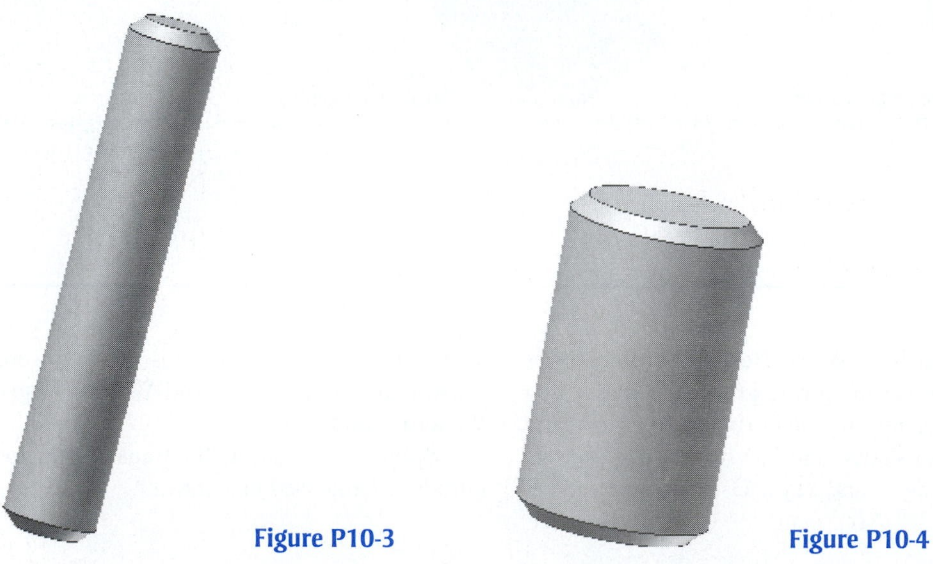

Figure P10-3

Figure P10-4

Project 10 -5: INCHES

Draw a Ø1.00 × 5.00 shaft.
Mount two BS 3673: Part 1 Inch retaining rings located 0.25 from each end.
Draw a 0.06 × 45° chamfer on each end.

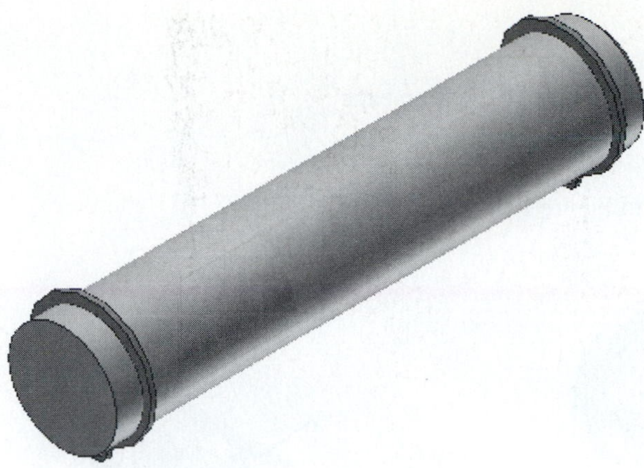

Figure P10-5

Project 10-6: INCHES

Draw a Ø.25 × 6.00 shaft.
Mount two BS 3673: Part 1 Inch retaining rings located 0.375 from each end.
Draw a 0.03 × 45° chamfer on each end.

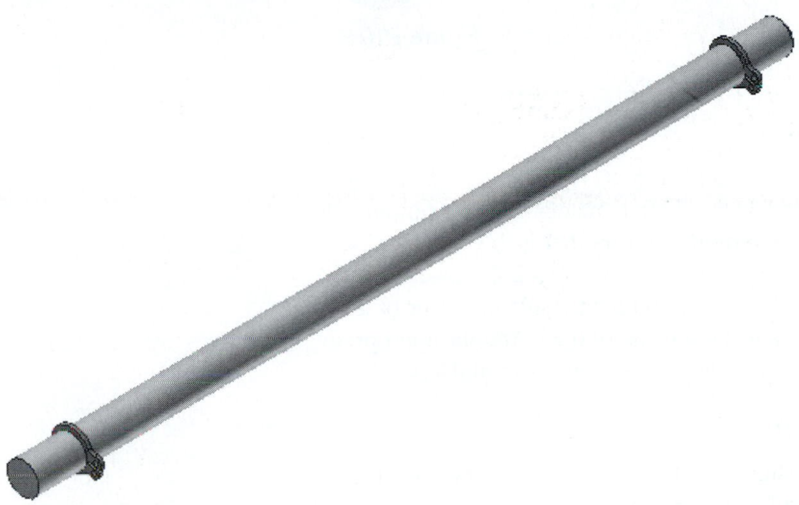

Figure P10-6

Project 10-7: MILLIMETERS

Draw a Ø20 × 50 shaft.
Mount two CSN 02 2930 Metric retaining rings located 2 from each end.
Draw a 3 × 45° chamfer on each end.

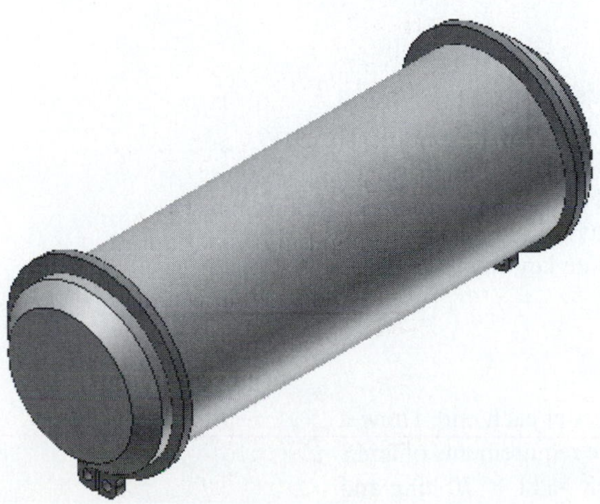

Figure P10-7

Project 10-8: MILLIMETERS

Draw a Ø16 × 64 shaft.
Mount two CSN 02 2930 Metric retaining rings located 10 from each end.
Draw a 4 × 45° chamfer on each end.

Draw the following shafts and keys. Refer to Figure P10-9.

Figure P10-8

Figure P10-9

Project 10-9: INCHES

Draw a Ø2.00 × 5.25 shaft with 0.125 × 45° chamfers at each end. Draw a keyway on one end on the shaft to match the size requirements of a square key (see the **Content Center**) that is 1/2 × 1/2 × 1.25 long and has a nominal diameter range of 1.75 - 2.25. Insert the key into the shaft's keyway. Draw a hub with an inside diameter of 2.00, outside diameter of 3.50, and a thickness of 0.75. Add the appropriate keyway to the hub, then insert the hub over the shaft and key.

Project 10-10: INCHES

Draw a Ø3.50 × 10.00 shaft with 0.25 × 45° chamfers at each end. Draw a keyway on one end on the shaft to match the size requirements of a square key (see the **Content Center**) that is 7/8 × 7/8 × 2.50 long and has a nominal diameter range of 3.25 - 3.75. Insert the key into the shaft's keyway. Draw a hub with an inside diameter of 3.50, outside diameter of 5.00, and a thickness of 1.00. Add the appropriate keyway to the hub, then insert the hub over the shaft and key.

Project 10-11: MILLIMETERS

Draw a Ø26 × 72 shaft with 3 × 45° chamfers at each end. Draw a keyway on one end on the shaft to match the size requirements of an IS 2048 B key (see the **Content Center**) that is 2 × 2 × 22 long and has a nominal diameter range of 22 - 30. Insert the key into the shaft's keyway. Draw a hub with an inside diameter of 26, outside diameter of 40, and a thickness of 14. Add the appropriate keyway to the hub, then insert the hub over the shaft and key.

Project 10-12: MILLIMETERS

Draw a Ø60 × 240 shaft with 5 × 45° chamfers at each end. Draw a keyway on one end on the shaft to match the size requirements of an IS 2048 B key (see the **Content Center**) that is 18 × 11 × 70 long and has a nominal diameter range of 58 - 65. Insert the key into the shaft's keyway. Draw a hub with an inside diameter of 60, outside diameter of 80, and a thickness of 20. Add the appropriate keyway to the hub, then insert the hub over the shaft and key.

Draw the following shafts and keys. Refer to Figure P10-10.

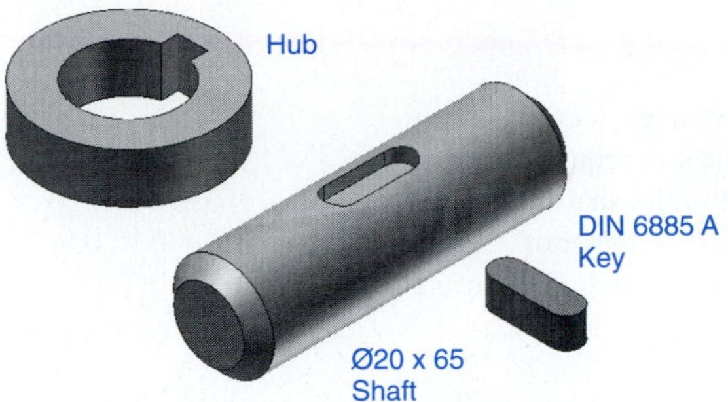

Figure P10-10

Project 10-13: MILLIMETERS

Draw a Ø20 × 65 shaft with 3 × 45° chamfers at each end. Draw a keyway on the shaft to match the size requirements of a DIN 6885 A key (see the **Content Center**). Locate the shaft's keyway 24 from the end of the shaft. Insert the key into the shaft's keyway. Draw a hub with an inside diameter of 20, outside diameter of 35, and a thickness of 10. Add the appropriate keyway to the hub, then insert the hub over the shaft and key.

Project 10-14: MILLIMETERS

Draw a Ø52 × 100 shaft with 6 × 45° chamfers at each end. Draw a keyway on the shaft to match the size requirements of a DIN 6885 A key (see the **Content Center**). Locate the shaft's keyway 34 from the edge of the shaft. Insert the key into the shaft's keyway. Draw a hub with an inside diameter of 52, outside diameter of 70, and a thickness of 16. Add the appropriate keyway to the hub, then insert the hub over the shaft and key.

Project 10-15: INCHES

Draw a Ø0.750 × 4.25 shaft with 0.125 × 45° chamfers at each end. Draw a keyway 0.500 from the end of the shaft to match the size requirements of a 0.1875 × 0.1875 × 0.74 rectangular or square parallel key that has a nominal diameter range of 0.57 - 0.88 (see the **Content Center**). Insert the key into the shaft's keyway. Draw a hub with an inside diameter of 0.75, outside diameter of 2.50, and a thickness of 0.625. Add the appropriate keyway to the hub, then insert the hub over the shaft and key.

Project 10-16: INCHES

Draw a Ø1.50 × 7.50 shaft with 0.19 × 45° chamfers at each end. Draw a keyway 2.50 from the end of the shaft to match the size requirements of a 0.375 × 0.375 × 0.875 rectangular or square parallel key that has a nominal diameter range of 1.38 - 1.75. Insert the key into the shaft's keyway. Draw a hub with an inside diameter of 1.50, outside diameter of 2.75, and a thickness of 0.750. Add the appropriate keyway to the hub, then insert the hub over the shaft and key.

Project 10-17: MILLIMETERS

Draw a Ø163 × 40 shaft with 2 × 45° chamfers at each end. Draw a keyway 10 from the end of the shaft to match the size requirements of a CSN 30 1385 Woodruff key. Insert the key into the shaft's keyway. Draw a hub with an inside diameter of 16, outside diameter of 40, and a thickness of 8. Add the appropriate keyway to the hub, then insert the hub over the shaft and key.

Project 10-18: MILLIMETERS

Draw a Ø40 × 105 shaft with 4 × 45° chamfers at each end. Draw a keyway 27 from the end of the shaft to match the size requirements of a CSN 30 1385 Woodruff key. Insert the key into the shaft's keyway. Draw a hub with an inside diameter of 40, outside diameter of 75, and a thickness of 12. Add the appropriate keyway to the hub, then insert the hub over the shaft and key.

Project 10-19: INCHES

Draw a Ø0.750 × 4.25 shaft with 0.125 × 45° chamfers at each end. Draw a keyway 0.500 from the end of the shaft to match the size requirements of Full Radius No. 606 3/16 × 0.75 Woodruff key (see the **Content Center**). Insert the key into the shaft's keyway. Draw a hub with an inside diameter of 0.75, outside diameter of 2.50, and a thickness of 0.625. Add the appropriate keyway to the hub, then insert the hub over the shaft and key.

Project 10-20: INCHES

Draw a Ø1.50 × 7.50 shaft with 0.19 × 45° chamfers at each end. Draw a keyway 2.50 from the end of the shaft to match the size requirements of a Full Radius No. 1422 I 7/16 × 2 Woodruff key (see the **Content Center**). Insert the key into the shaft's keyway. Draw a hub with an inside diameter of 1.50, outside diameter of 2.75, and a thickness of 0.750. Add the appropriate keyway to the hub, then insert the hub over the shaft and key.

Draw the following shafts and hubs. Refer to Figures P10-11 through P10-14.

Project 10-21: MILLIMETERS

Draw a shaft and hub based on the following data and join them using a spline.

Shaft: Ø120 × 30
Spline Type = Light
Spline Dimensions = 8 × 42 × 46

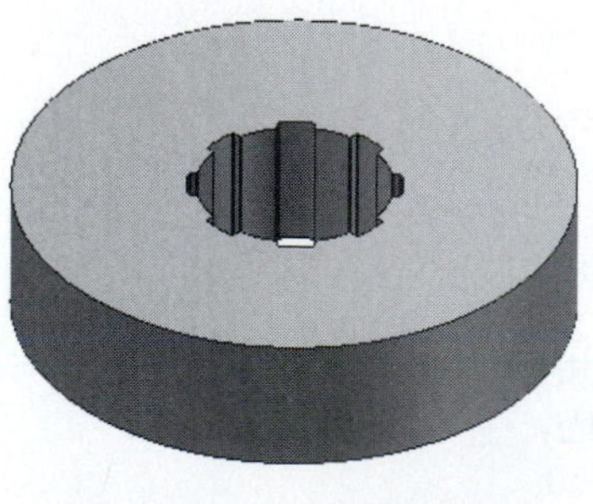

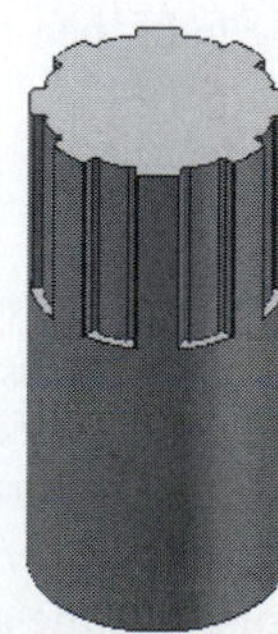

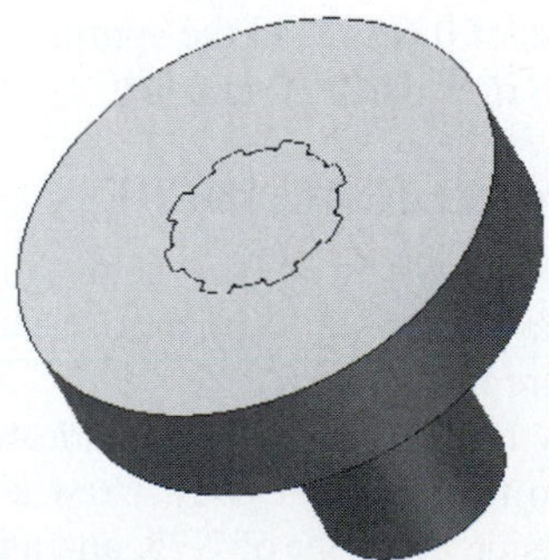

Figure P10-11

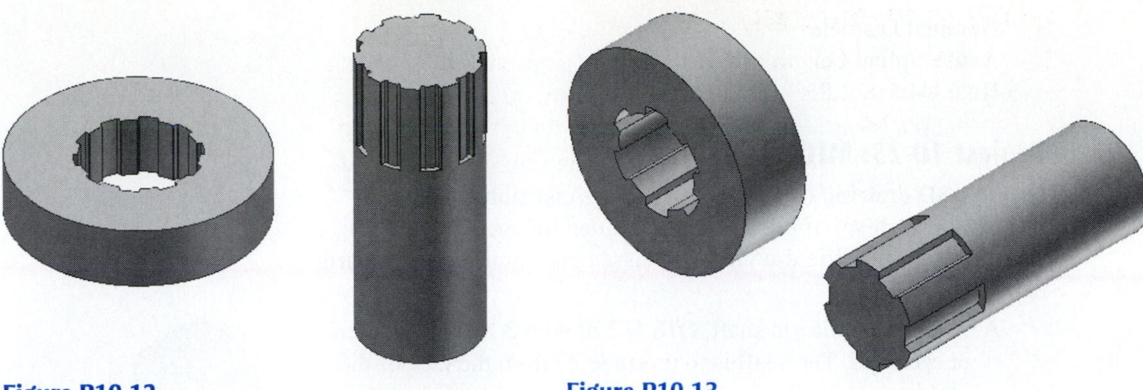

Figure P10-12

Figure P10-13

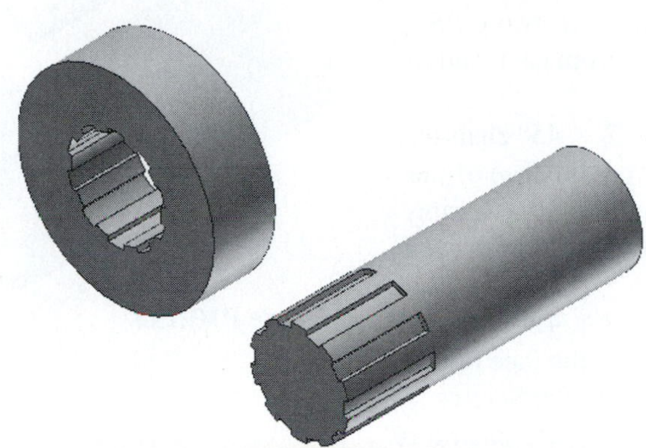

Figure P10-14

Active Spline Length = 38
Hub: Ø46 × 100

Project 10-22: MILLIMETERS

Draw a shaft and hub based on the following data and join them using a spline.

Shaft: Ø78 × 200
Spline Type = Light
Spline Dimensions = 10 × 72 × 78
Active Spline Length = 60
Hub: Ø160 × 40

Project 10-23: INCHES

Draw a shaft and hub based on the following data and join them using a spline.

Shaft: Ø1.50 × 4
Fit Type = 6B To Side - No Load
Nominal Diameter = 1.50
Active Spline Length = 1.25
Hub: Ø3 × 1

Project 10-24: INCHES

Draw a shaft and hub based on the following data and join them using a spline.

Shaft: Ø2 × 6
Fit Type = 10 A Permanent Fit

Nominal Diameter = 2
Active Spline Length = 1.75
Hub: Ø4.5 × 1.25

Project 10-25: MILLIMETERS

Create a 3D drawing of the Shaft Support Assembly in Figure P10-15 for each set of parameters presented below. Also include an exploded isometric drawing with assembly numbers and a parts list.

A. A straight uniform shaft, Ø16 × 320 with 2 × 45° chamfers at each end. The shaft is to protrude 20 from the face of the Holders.
B. A straight uniform shaft, Ø16 × 320 with 2 × 45° chamfers at each end. The shaft is to protrude 20 from the face of the Holders. Add the appropriate grooves and insert two CNS 9074 External Retaining Rings located 17 from each end of the shaft.
C. A straight uniform shaft, Ø16 × 320 with 2 × 45° chamfers at each end. The shaft is to protrude 20 from the face of the Holders. Add the appropriate grooves and insert two E-ring - Type - 3CM External Retaining Rings located 18 from each end of the shaft.
D. A straight uniform shaft, Ø16 × 320 with 2 × 45° chamfers at each end. The shaft is to protrude 20 from the face of the Holders. Add two CNS - 122 Collars with Ø4 Set Screws.
E. A straight uniform shaft, Ø16 × 320 with 2 × 45° chamfers at each end. The shaft is to protrude 20 from the face of the support. Add two CNS - 122 collars with Ø4 Spring-type Straight Pins.
F. A straight uniform shaft, Ø16 × 280 with 2 × 45° chamfers at each end. On one end insert a square key between the shaft and the Holders.

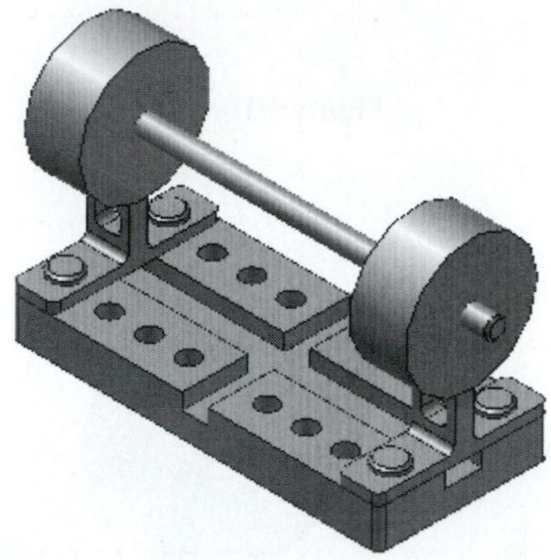

Figure P10-15a

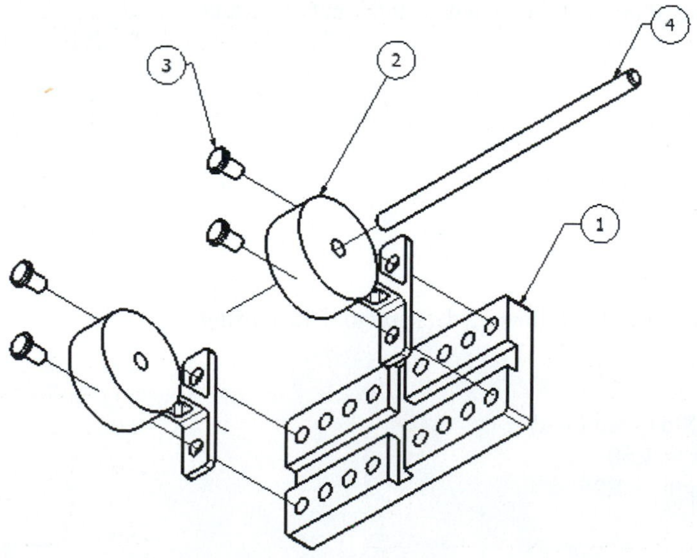

Parts List				
ITEM	PART NUMBER	DESCRIPTION	MATERIAL	QTY
1	ENG-311-1	BASE,CAST	CAST IRON	1
2	ENG-312	HOLDERS	CAST IRON	2
3	ENG-132A	PINS	2040 STEEL	4
4	ENG-131	Shaft	STEEL	1

Figure P10-15b

BASE, CAST
MAT'L = CAST IRON
P/N = ENG-311-1

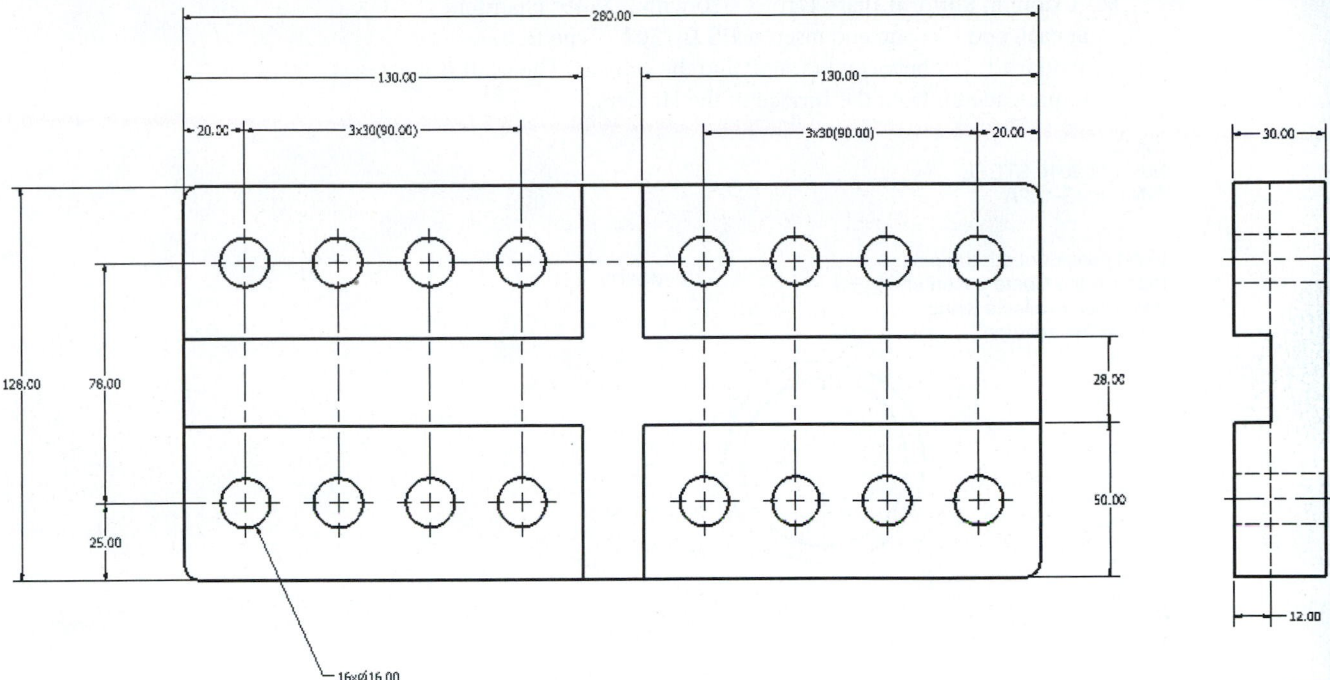

Figure P10-15c

HOLDERS
MAT'L = CAST IRON
P/N = ENG-312

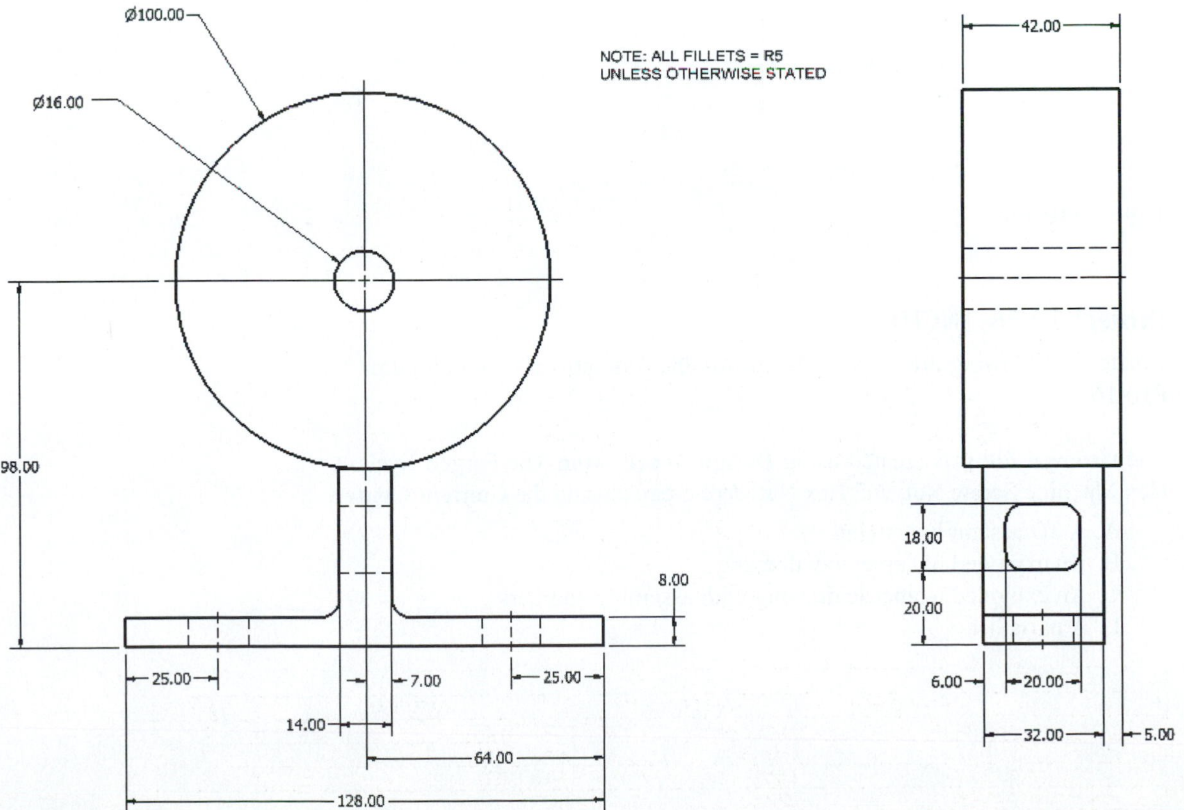

Figure P10-15d

G. A straight uniform shaft, Ø16 × 320 with 2 × 45° chamfers at each end. On one end insert a UNI 7510A Key between the shaft and the support. The shaft is to protrude 20 from the surface of the Holders.

H. A straight uniform shaft, Ø16 × 320 with 2 × 45° chamfers at each end. On one end insert a JIS B 1302 - Type B Woodruff Key between the shaft and the support. The shaft is to protrude 20 from the surface of the Holders.

PINS
MAT'L = 2040 STEEL
P/N = ENG-132A

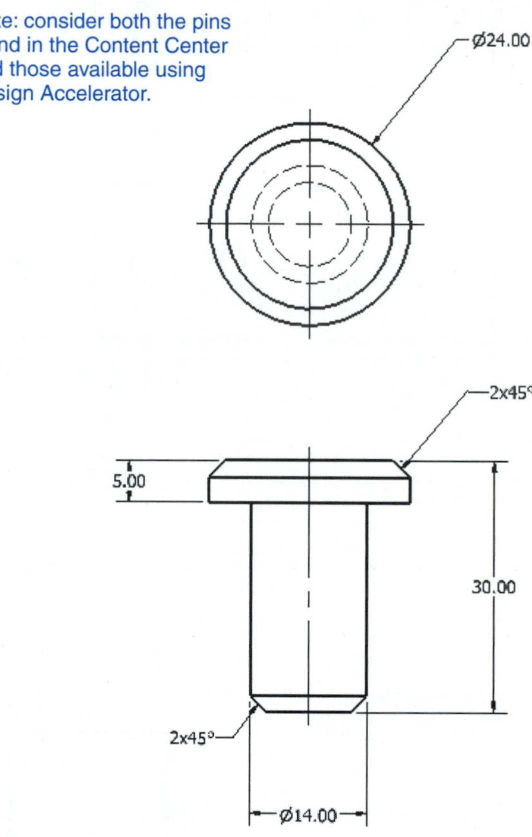

Figure P10-15e

Project 10-26: INCHES

Create the following drawings of the Adjustable Assembly shown in Figure P10-16.

The Grooved Pin was created using **Design Accelerator.** The Forged Eyebolt, Hex Machine Screw Nut, and Hex Nut were created using the **Content Center.**

A. A 3D assembly drawing
B. An exploded 3D assembly drawing
C. An exploded isometric drawing with assembly numbers
D. A parts list

Adjustable Assembly

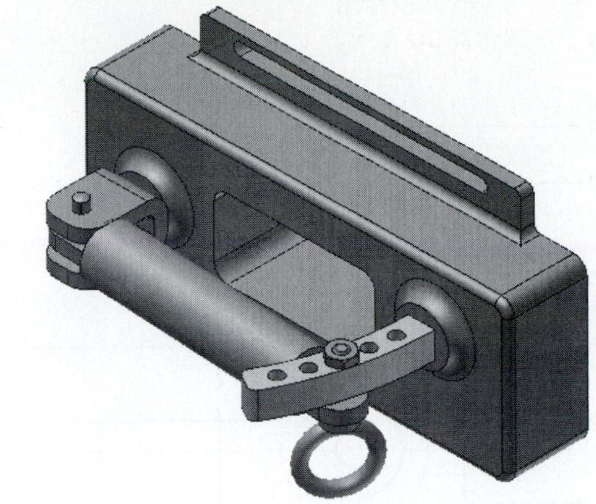

Figure P10-16a

POST, ADJUSTABLE
P/N = ENG 404
MAT'L = MILD STEEL

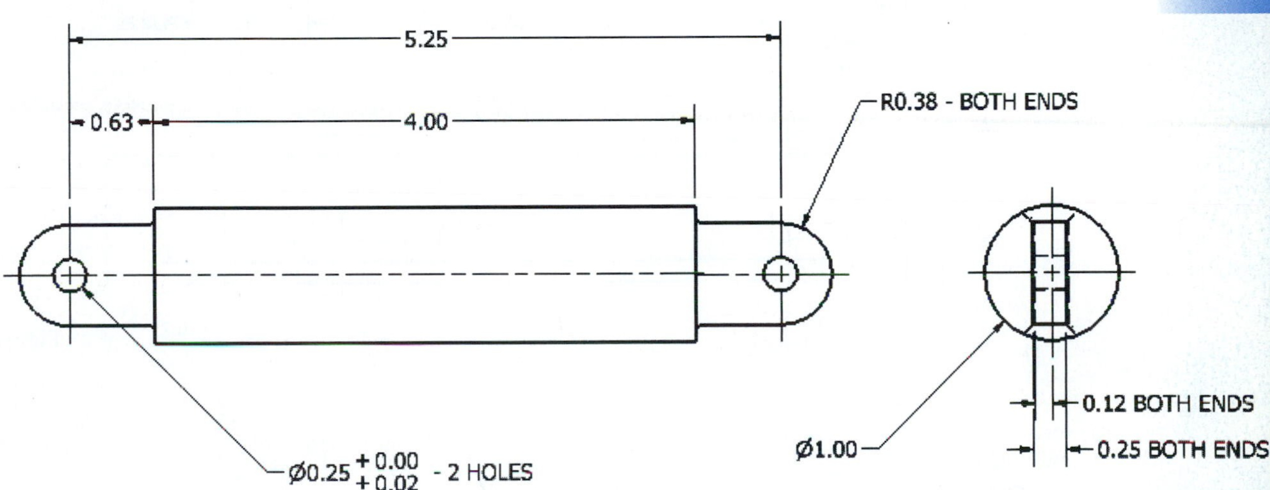

Figure P10-16b

BASE #4, CAST
P/N = ENG 311
MAT'L = CAST IRON

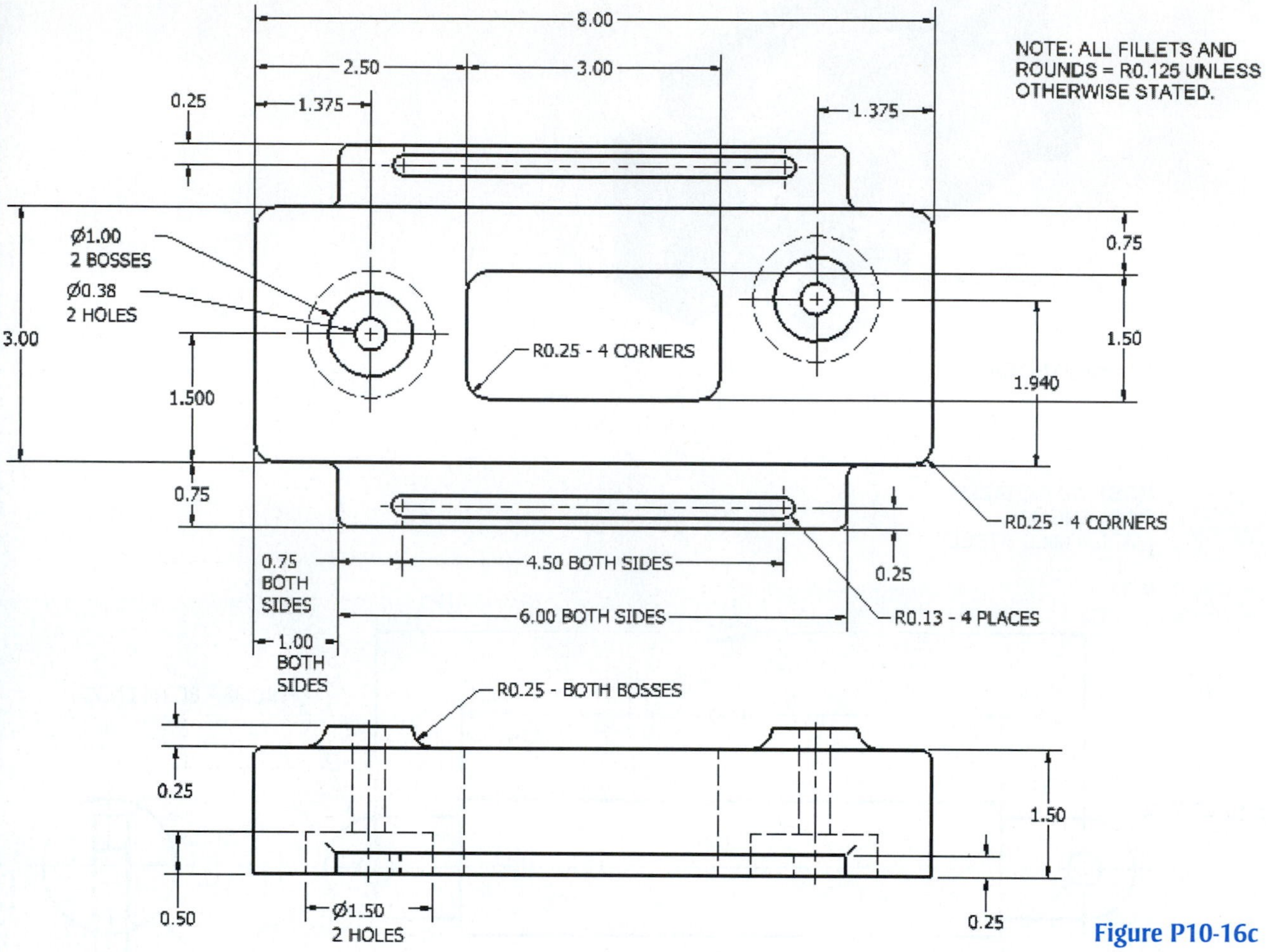

8.00
2.50
3.00
0.25
1.375
1.375
NOTE: ALL FILLETS AND ROUNDS = R0.125 UNLESS OTHERWISE STATED.
Ø1.00 2 BOSSES
Ø0.38 2 HOLES
3.00
1.500
0.75
R0.25 - 4 CORNERS
0.75
1.50
1.940
R0.25 - 4 CORNERS
0.75 BOTH SIDES
4.50 BOTH SIDES
0.25
6.00 BOTH SIDES
R0.13 - 4 PLACES
1.00 BOTH SIDES
R0.25 - BOTH BOSSES
0.25
1.50
0.50
Ø1.50 2 HOLES
0.25

Figure P10-16c

YOKE
P/N = BU 1964
MAT'L = CAST IRON

1.00
Ø0.25 2 HOLES
0.50
R0.38 BOTH SIDES
1.50
1.63
0.53
3/8-16 UNC - 1A
1.00
0.25 ALL AROUND
R0.38
R0.13

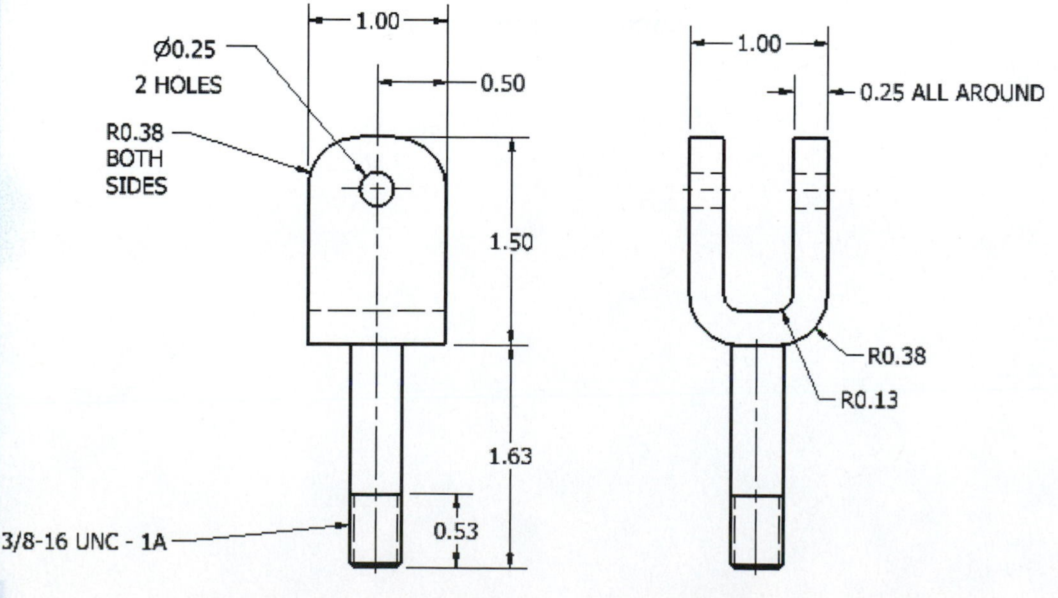

Figure P10-16d

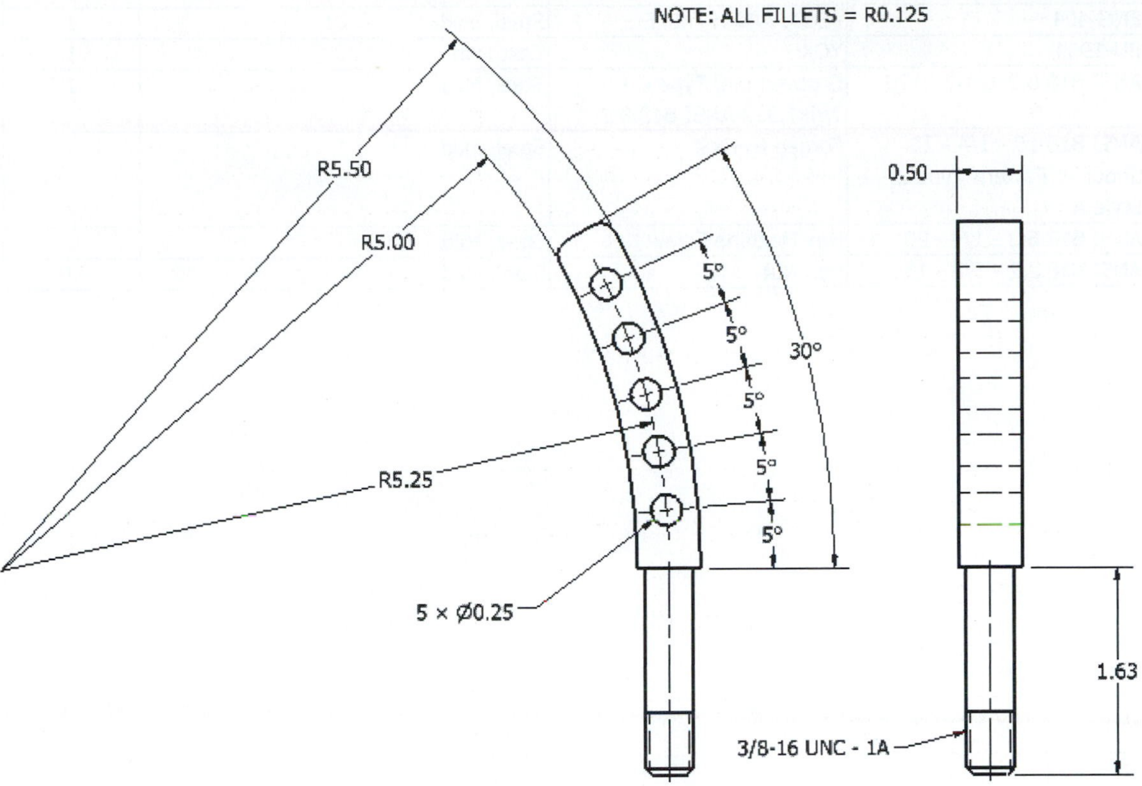

Figure P10-16e

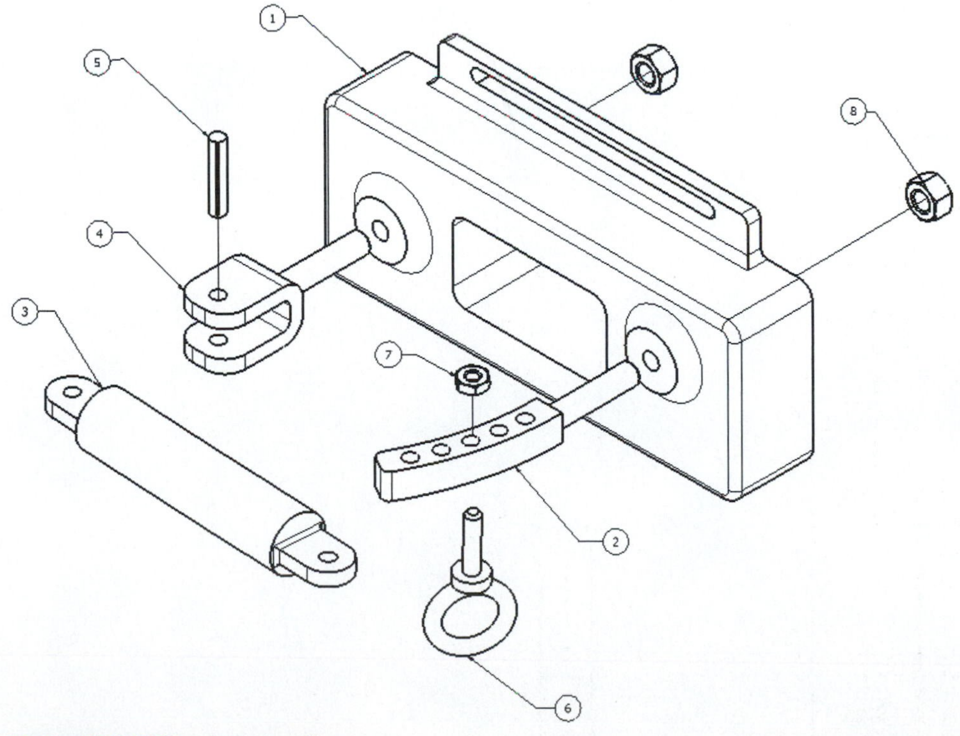

Figure P10-16f *(Continued)*

Parts List				
ITEM	PART NUMBER	DESCRIPTION	MATERIAL	QTY
1	ENG-311	BASE#4, CAST	Cast Iron	1
2	ENG-312	SUPPORT, ROUNDED	SAE 1040 STEEL	1
3	ENG-404	POST, ADJUSTABLE	Steel, Mild	1
4	BU-1964	YOKE	Cast Iron	1
5	ANSI B18.8.2 1/4x1.3120	Grooved pin, Type C - 1/4x1.312 ANSI B18.8.2	Steel, Mild	1
6	ANSI B18.15 - 1/4 - 20. Shoulder Pattern Type 2 - Style A	Forged Eyebolt	Steel, Mild	1
7	ANSI B18.6.3 - 1/4 - 20	Hex Machine Screw Nut	Steel, Mild	1
8	ANSI B18.2.2 - 3/8 - 16	Hex Nut	Steel, Mild	2

Figure P10-16f

Bearings

11

Chapter Objectives

- Understand the different types of bearings: plain, ball, and thrust.
- Show how to select bearing from the **Content Center.**
- Understand how tolerances are used with bearings.
- Show how to use bearings in assemblies.

INTRODUCTION

This chapter discusses three types of bearings: plain, ball, and thrust. See Figure 11-1. **Plain bearings** are hollow cylinders that may have flanges at one end. Plain bearings are made from nylon or Teflon for dry (no lubrication) applications, and impregnated bronze or other materials when lubrications are used.

plain bearing: A hollow cylinder that may have a flange at one end and may or may not be lubricated; also called a *sleeve bearing* or *bushing.*

Figure 11-1

Plain bearings require less space than other types of bearings and are cheaper but have higher friction properties.

Ball bearings are made from two cylindrical rings separated by a row of balls. Shapes other than spheres may be used (rollers), and more than one row of balls may be included. Ball bearings usually take more space than plain bearings and are more expensive. They have better friction properties than plain bearings.

Thrust bearings are used to absorb loads along the axial direction (the length of the shaft). They are similar in size and cost to ball bearings.

ball bearing: A bearing made from two cylindrical rings separated by a row of balls or rollers.

thrust bearing: A bearing used to absorb a load along the axial direction of a shaft.

PLAIN BEARINGS

Figure 11-2 shows a U-bracket that will be used to support a rotating shaft. Plain bearings will be inserted between the shaft and the U-bracket. The bearings will be obtained from the **Content Center.**

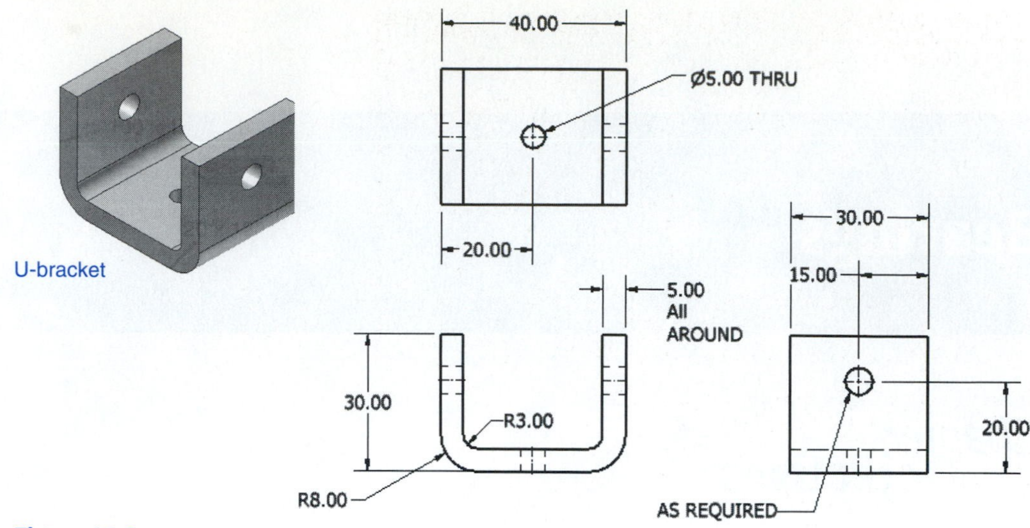

Figure 11-2

The shaft has a nominal diameter of 6.00 mm and the shaft will be inserted into the bearing and the bearing into the U-bracket.

Nomenclature

Plain bearings are also called *sleeve bearings* or *bushings.* The terms are interchangeable

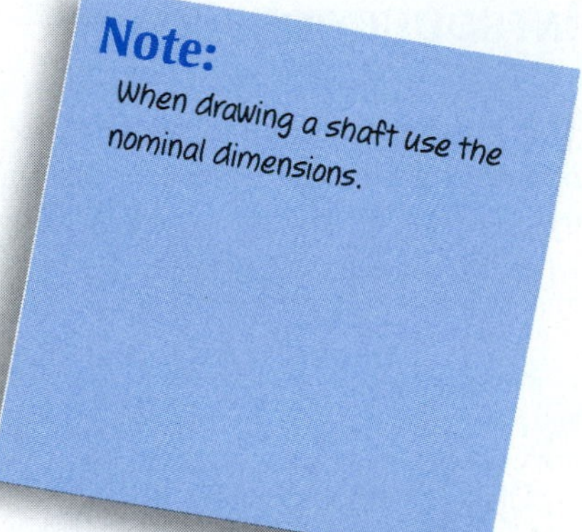

Shaft Tolerances

For this example it is assumed that the shaft will be purchased from a vendor. Tolerances for purchased shafts can be found in the manufacturer's catalogs or from listings posted on the Web. The shaft selected for this example has a tolerance of +0.00/−0.02.

Figure 11-3 shows a Ø6.0000 shaft 50.00 long. Each end has a 0.50 × 45° chamfer. The orthographic views of the shaft include the shaft's diametric tolerance.

TIP Purchased parts do not require individual drawings. They are listed in the parts list by manufacturer and manufacturer's part number or catalog number.

Exercise 11-1: Selecting a Plain Bearing

1. Access the **Content Center,** click **Shaft Part, Bearings-Plain** and select a **CNS 9348 (Cylindrical)** plain bearing.

See Figure 11-4.

2. Select the **6** × **10** × **4** nominal-sized bearing.

See Figure 11-5.

3. Click **OK** and add two bearings to the drawing.

See Figure 11-6.

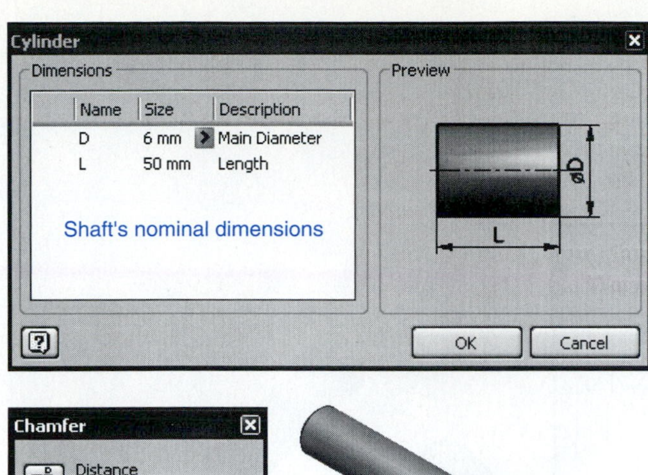

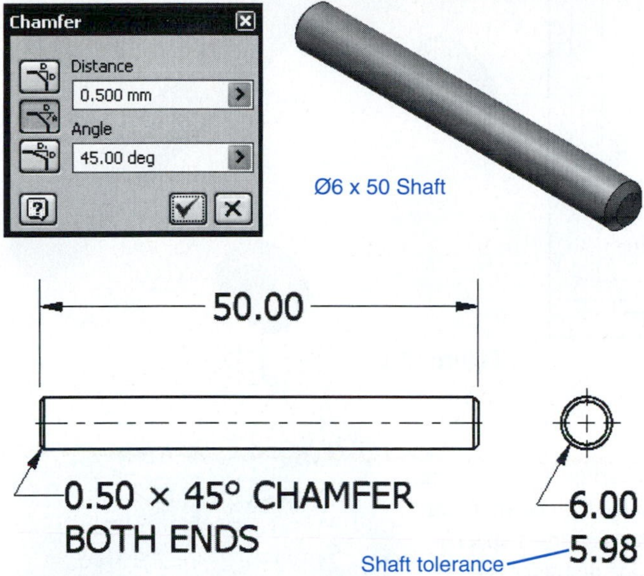

Figure 11-3

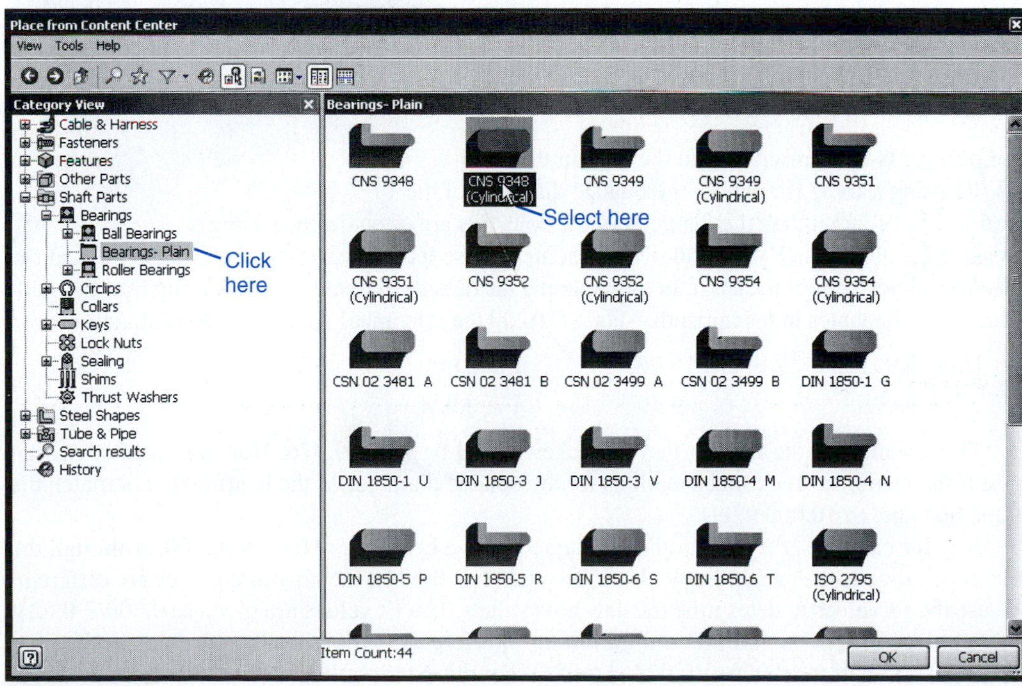

Figure 11-4

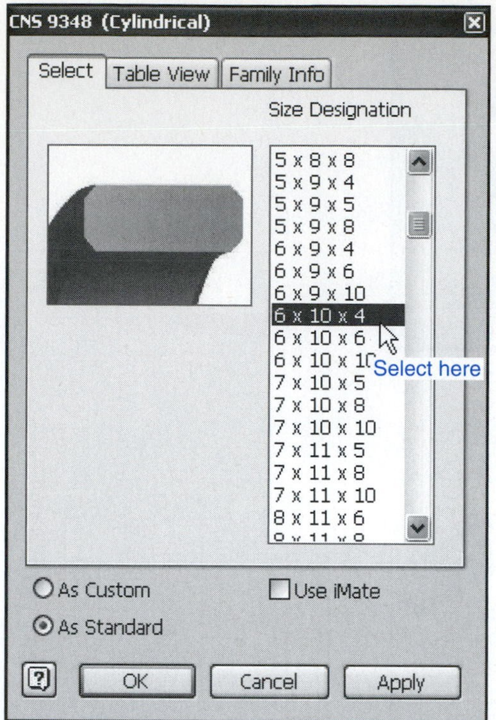

Figure 11-5

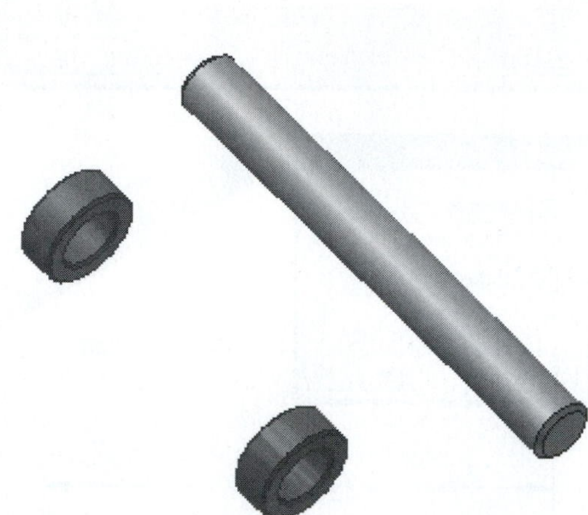

Figure 11-6

Shaft/Bearing Interface

The shaft and inside diameter of the bearing (the *bore*) are generally fitted together using a clearance fit. The shaft has a tolerance of +0.00/−0.02. From manufacturer's specifications we know that the bearing's inside diameter is 6.00 nominal with a tolerance of +0.02/0.00. Therefore, the maximum clearance is 0.04, and the minimum is 0.00.

The range of clearance tolerance depends on the load and speed of the application.

Note:
If the clearance is too large, excessive vibrations could result at high speeds.

The Hole in the U-Bracket

The bearing is to be inserted into the hole in the U-bracket using a force fit; that is, the outside diameter of the bearing will be larger than the diameter of the hole. Fits are often defined using a standard notation such as F7/h6, where the uppercase letter defines the hole tolerance and the lowercase letter defines the shaft. In this example the outside diameter of the bearing uses the shaft values. See the tables in the appendix. For a Ø10.00 the tolerances for an F7/h6 combination are

Hole: 9.991	Shaft: 10.000	Fit: 0.000
9.976	9.991	−0.024

This means that the hole in the U-bracket should be 9.991/9.976. However, a problem can arise if the manufacturer's specifications for the outside diameter of the bearing do not match the stated h6 value of 10.000/9.991.

Say, for example, the given outside diameter for a bearing is 10.009/10.000. Although the tolerance range is the same, 0.009 (10.000 − 9.991 = 0.009), the absolute values are different. We use the fit values to determine the new hole values. The fit values are given as 0.000/−0.024. These values are the maximum and minimum interference. Applying these values to the hole we get 10.000/9.985 (10.009 − 0.024 − 9.985). For this bearing the hole in the U-bracket would be 10.000/9.985.

Draw the hole using the nominal value of 10.00, and dimension the hole using the derived values. See Figure 11-7.

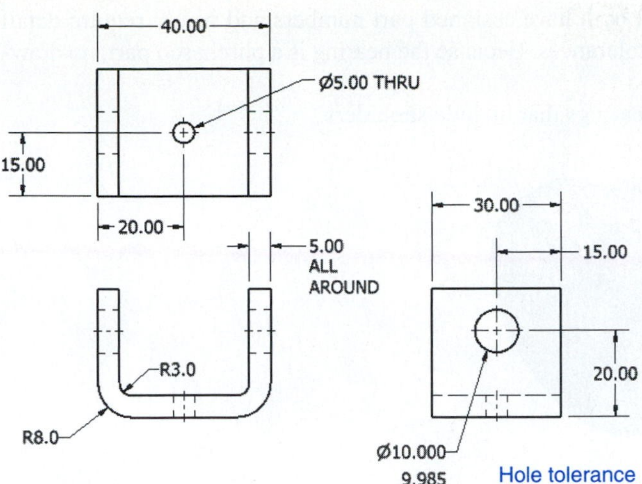

Figure 11-7

Figure 11-8 shows the finished assembly. The components were assembled using the **Constraint** tool. The shaft was offset 5 from the edge of the bearing.

Figure 11-9 shows the bearing assembly along with its parts list. Note that the bearing does not have an assigned part number but uses the manufacturer's number assigned by the **Content**

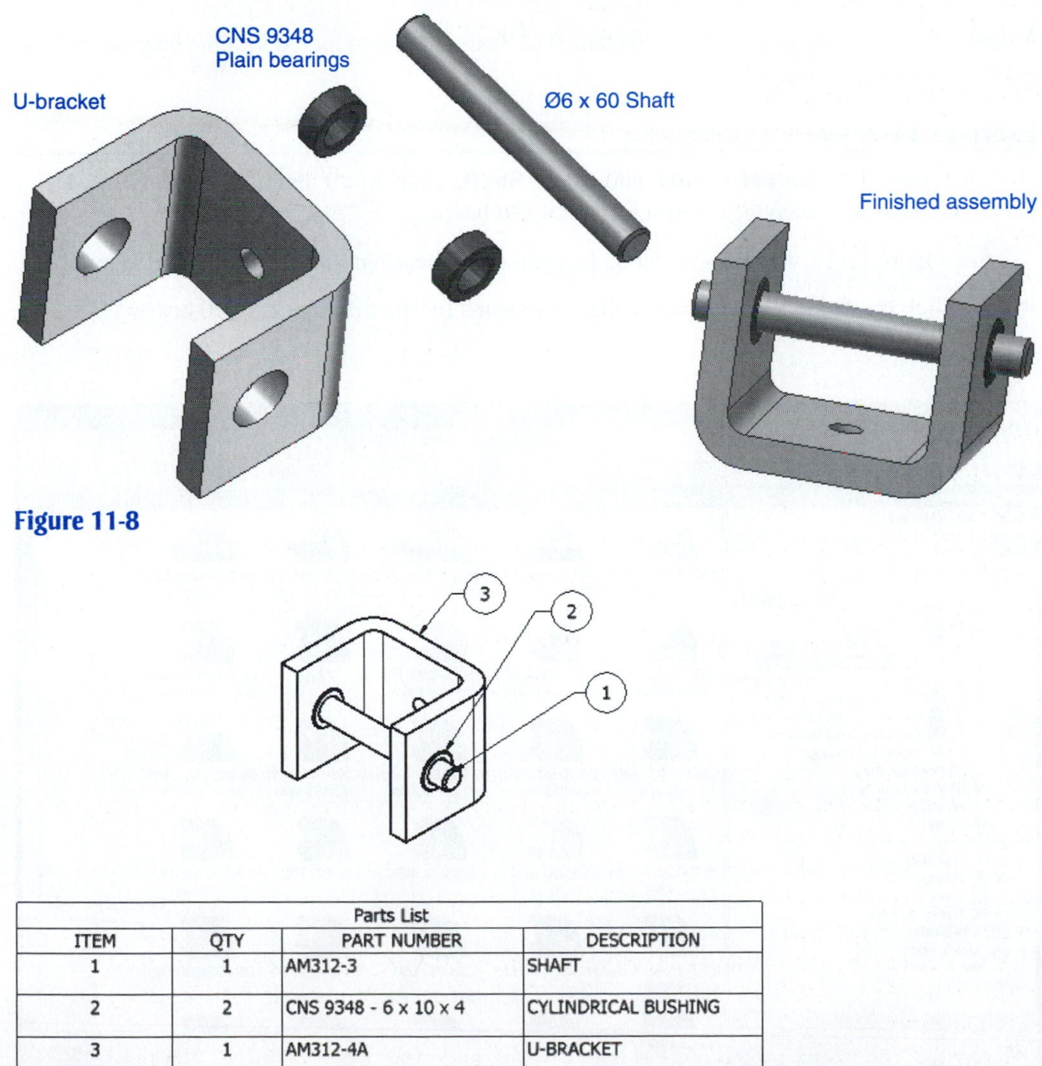

Figure 11-8

Parts List			
ITEM	QTY	PART NUMBER	DESCRIPTION
1	1	AM312-3	SHAFT
2	2	CNS 9348 - 6 x 10 x 4	CYLINDRICAL BUSHING
3	1	AM312-4A	U-BRACKET

Figure 11-9

Center. The U-bracket and the shaft both have assigned part numbers and would require detail drawings including dimensions and tolerances. Because the bearing is a purchased part, no drawing is required.

Figure 11-10 shows two plain bearings that include shoulders.

Figure 11-10

BALL BEARINGS

A bearing is to be selected for a Ø6.00 shaft. The shaft is to be mounted into the bearing and the bearing inserted into a U-bracket.

Exercise 11-2: Selecting a Ball Bearing

1. Access the **Content Center** and select **Shaft**, click **Shaft Part, Ball Bearings, Deep Groove Ball Bearings,** and select a **BS 290** bearing.

See Figure 11-11. The **BS 290** dialog box shows the bearings are defined by coded numbers.

2. Click the **Table View** tab to see the dimensions for the different BS 290 bearings.

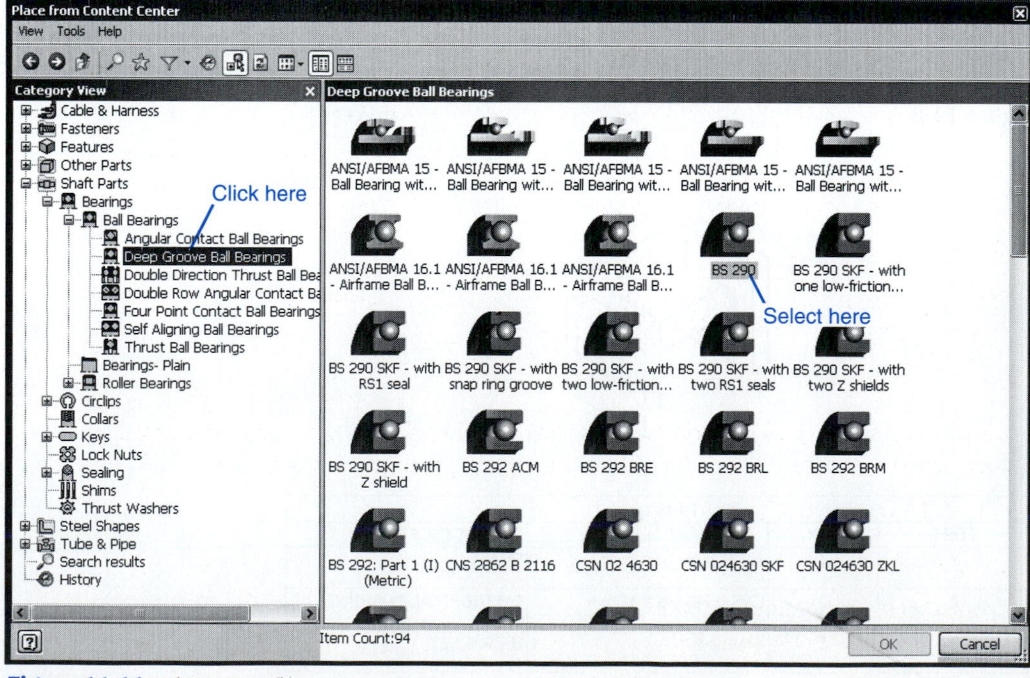

Figure 11-11 *(Continued)*

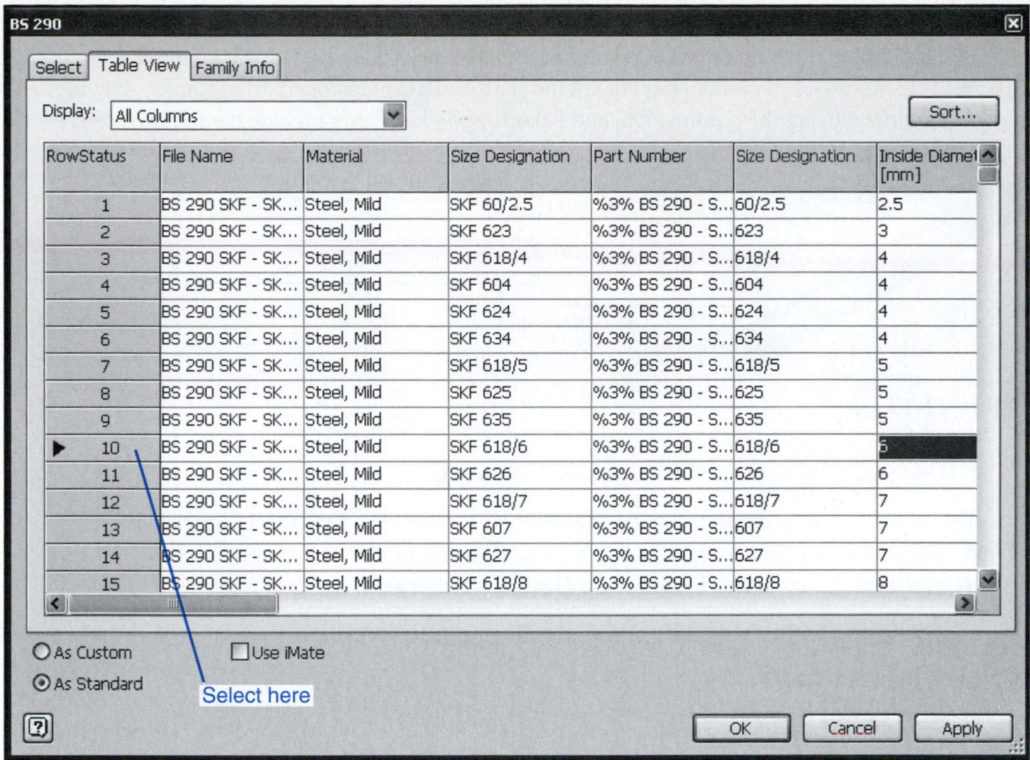

RowStatus	File Name	Material	Size Designation	Part Number	Size Designation	Inside Diamet [mm]
1	BS 290 SKF - SK...	Steel, Mild	SKF 60/2.5	%3% BS 290 - S...	60/2.5	2.5
2	BS 290 SKF - SK...	Steel, Mild	SKF 623	%3% BS 290 - S...	623	3
3	BS 290 SKF - SK...	Steel, Mild	SKF 618/4	%3% BS 290 - S...	618/4	4
4	BS 290 SKF - SK...	Steel, Mild	SKF 604	%3% BS 290 - S...	604	4
5	BS 290 SKF - SK...	Steel, Mild	SKF 624	%3% BS 290 - S...	624	4
6	BS 290 SKF - SK...	Steel, Mild	SKF 634	%3% BS 290 - S...	634	4
7	BS 290 SKF - SK...	Steel, Mild	SKF 618/5	%3% BS 290 - S...	618/5	5
8	BS 290 SKF - SK...	Steel, Mild	SKF 625	%3% BS 290 - S...	625	5
9	BS 290 SKF - SK...	Steel, Mild	SKF 635	%3% BS 290 - S...	635	5
10	BS 290 SKF - SK...	Steel, Mild	SKF 618/6	%3% BS 290 - S...	618/6	6
11	BS 290 SKF - SK...	Steel, Mild	SKF 626	%3% BS 290 - S...	626	6
12	BS 290 SKF - SK...	Steel, Mild	SKF 618/7	%3% BS 290 - S...	618/7	7
13	BS 290 SKF - SK...	Steel, Mild	SKF 607	%3% BS 290 - S...	607	7
14	BS 290 SKF - SK...	Steel, Mild	SKF 627	%3% BS 290 - S...	627	7
15	BS 290 SKF - SK...	Steel, Mild	SKF 618/8	%3% BS 290 - S...	618/8	8

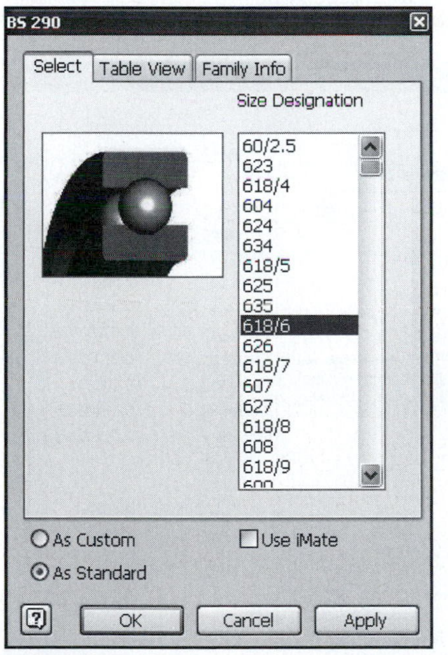

Figure 11-11

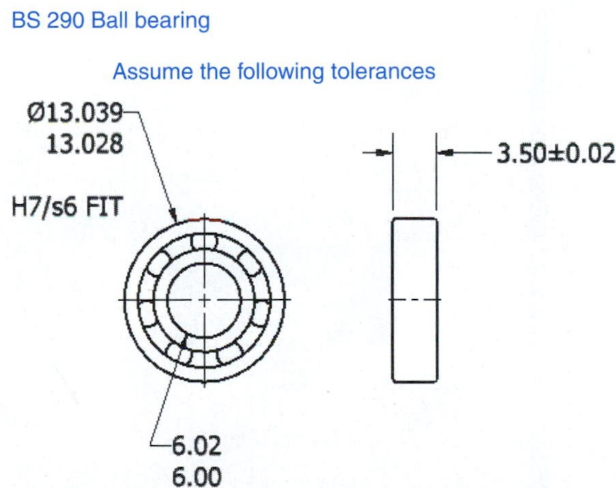

Figure 11-12

In this example, a **618/6** bearing was selected. This bearing has a nominal bore of Ø6.00 and an outside nominal diameter of Ø13.00. Assume the bearing has the tolerances shown in Figure 11-12.

3. Calculate the shaft's diameter.

Say the clearance requirements for the shaft are

Clearance = minimum = 0.00
Clearance = maximum = 0.03

This means that the shaft diameter tolerances are 6.00/5.99.

4. Access the **Design Accelerator** and draw the shaft.

Use the nominal diameter of **Ø6.00** and a length of **50.** Add a **0.50 × 45°** chamfer to each end. Figure 11-13 shows a solid model drawing of the shaft and a dimensioned orthographic drawing. The model was drawn using the nominal Ø6, and the orthographic views include the required tolerances.

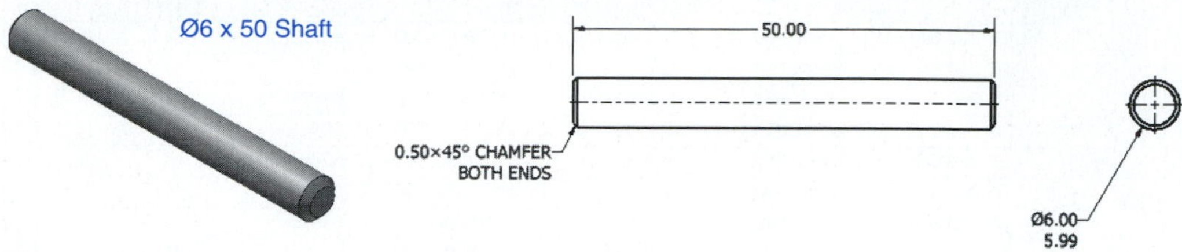

Figure 11-13

5. Determine the size for the hole in the U-bracket.

The interface between the outside diameter of the bearing and the U-bracket is defined as H7/s6. The values for this tolerance can be found in the tables in the appendix or by using the **Limits and Fits** dialog boxes found in the **Design Accelerator.**

6. Access the **Limits and Fits** tool in the **Design Accelerator.**

See Figure 11-14.

NEW to AutoCAD 2008

7. Select the **Hole-basis** condition, enter a **Basic Size** of **13.00,** and select an **Interference** fit.

See Figure 11-15. The tolerance values will appear under the **Results** heading. The hole's values can also be obtained by moving the cursor to the **H** tolerance zone.

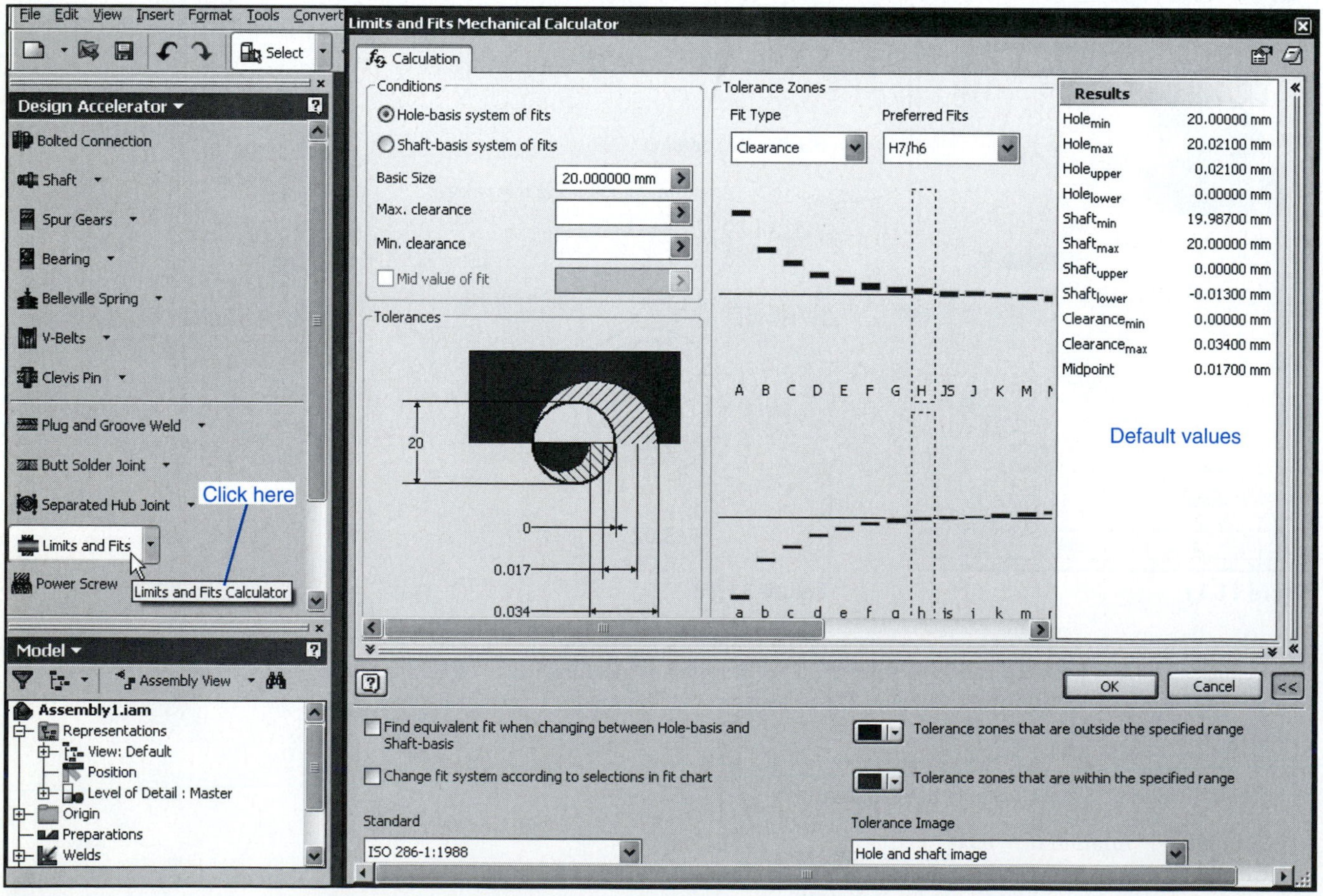

Figure 11-14

The terms **Hole-basis** and **Shaft-basis** refer to how the tolerance was calculated. In this example, note in the **Results** box that the minimum hole diameter is 13.000. This is the starting point for the tolerance calculations, so the calculation is Hole-basis. If the minimum shaft diameter had been defined as 13.000, the tolerance calculations would have started from this value. The results would have been Shaft-basis.

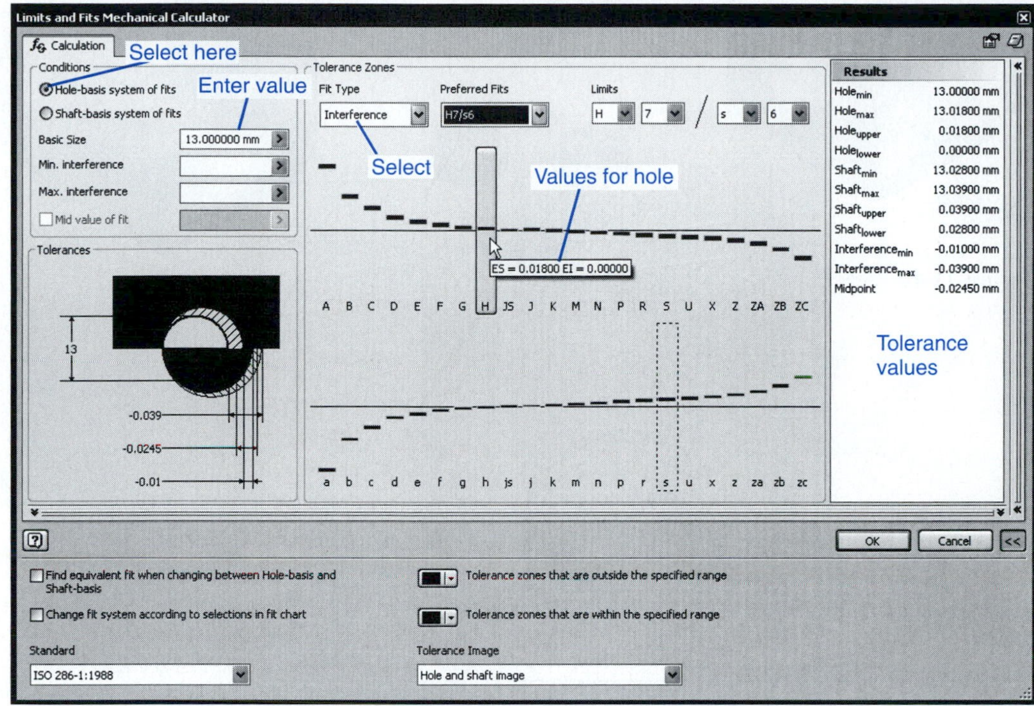

Figure 11-15

From Figure 11-15 we see that the hole has a tolerance of 13.018/13.000

8. Apply the hole tolerance to the hole in the U-bracket.

See Figure 11-16.

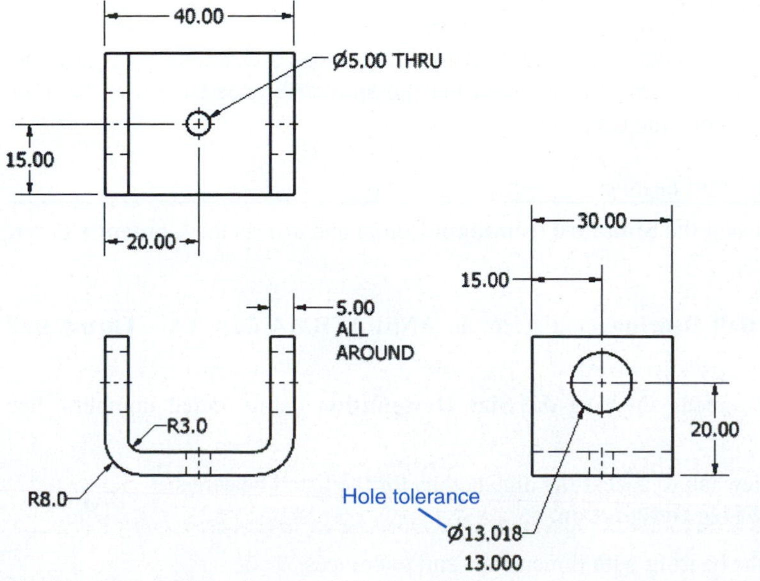

Figure 11-16

Figure 11-17 shows the finished assembly drawing with the bearings. Figure 11-18 shows the parts list for the assembly. Note that no part number was assigned to the bearings, as they are purchased parts. The manufacturer's part numbers were used.

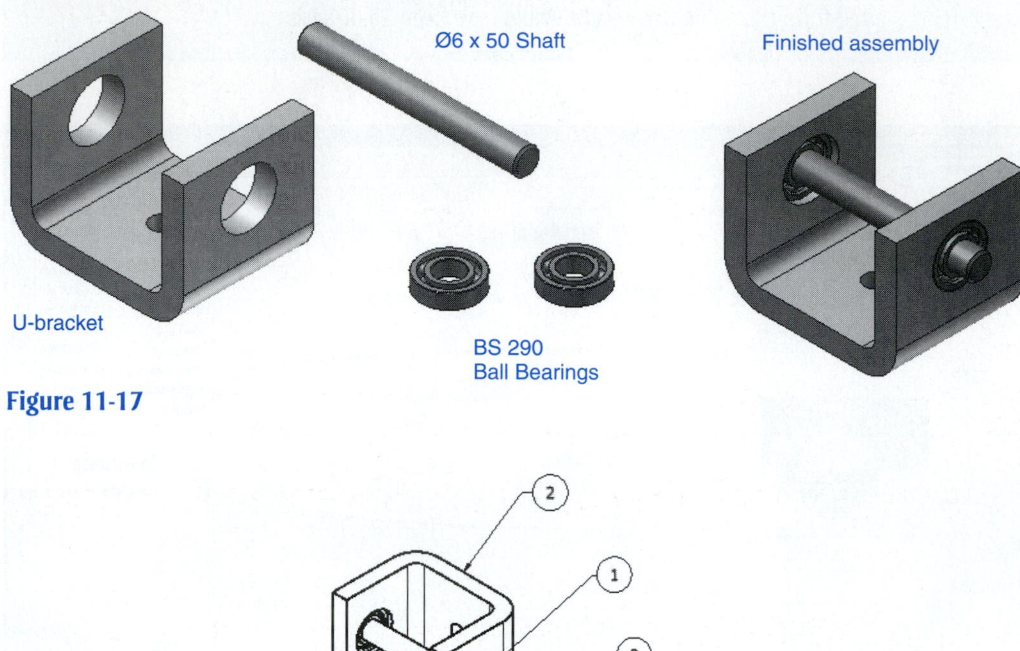

Figure 11-17

Parts List				
ITEM	PART NUMBER	DESCRIPTION	MATERIAL	QTY
1	BS 290 - SKF 618-6	Single row ball bearings	Steel, Mild	2
2	AM-301-1	U-BRACKET-13	Default	1
3	AM-301 -2	SHAFT	Welded Aluminum-6 061	1

Figure 11-18

Thrust Bearings

Tolerances for thrust bearings are similar to ball bearings. In general, a clearance fit is used between the shaft and the bearing's bore (inside diameter) and an interference fit is used between the housing and the bearing's outside diameter.

Exercise 11-3: Selecting Thrust Bearings

1. Create a drawing using the **Standard (mm).iam** format and access the **Content Center.**

See Figure 11-19.

2. Click on **Thrust Ball Bearings** and select an **ANSI/AFBMA 24.1 TA - Thrust Ball Bearing.**

The dialog box will appear showing the **Size Designation** using coded numbers. See Figure 11-20.

3. Click the **Table View** tab to access the dimensions for the listed bearings.
4. Select the **10TA12 Size Designation.**

Figure 11-21 shows the bearing with dimensions and tolerances.

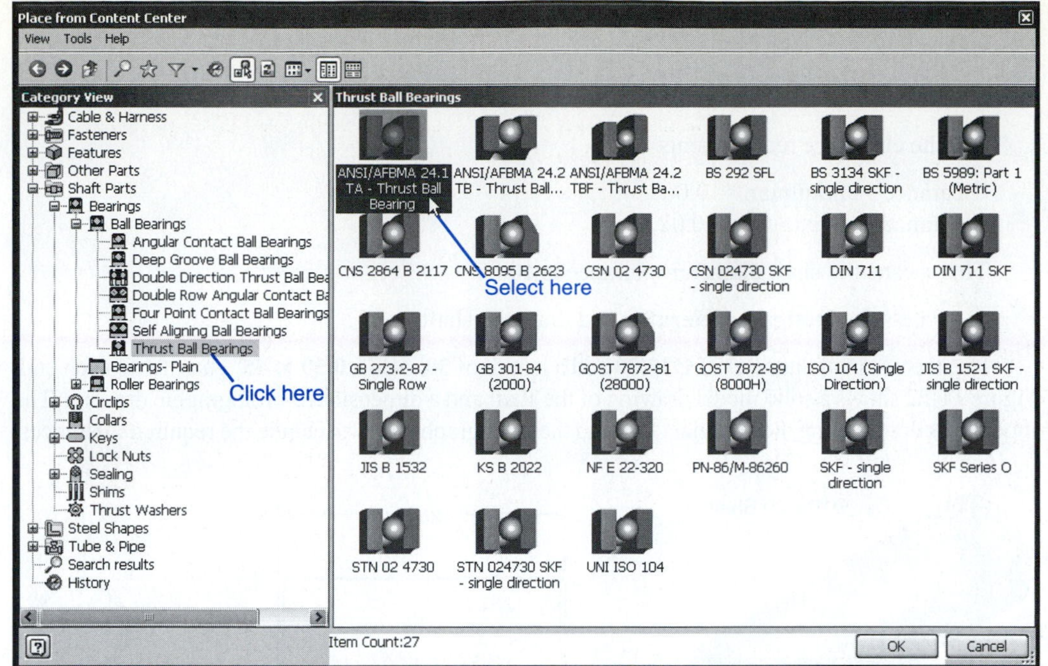

Figure 11-19

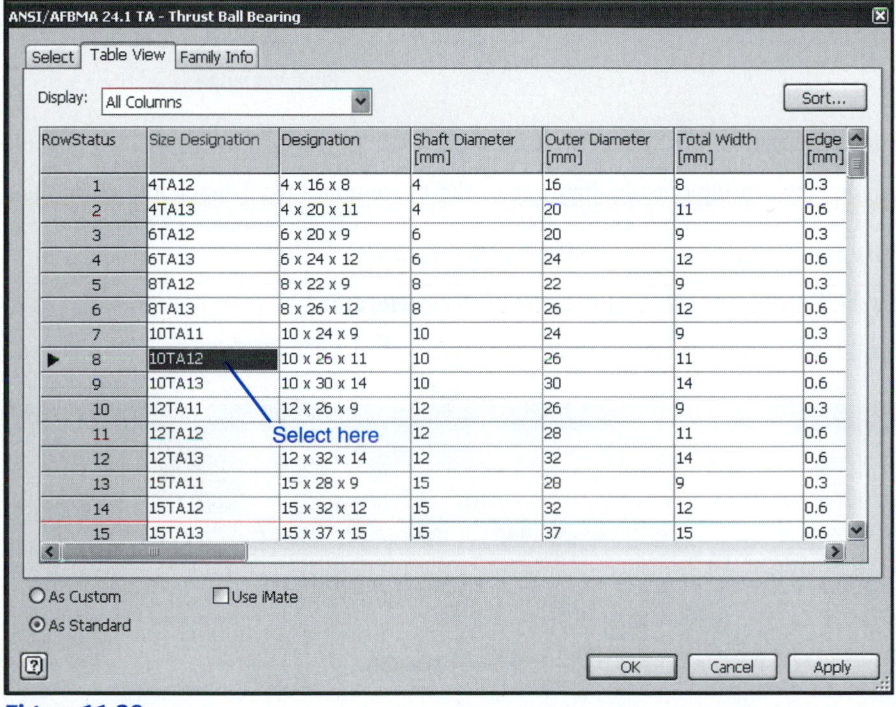

RowStatus	Size Designation	Designation	Shaft Diameter [mm]	Outer Diameter [mm]	Total Width [mm]	Edge [mm]
1	4TA12	4 x 16 x 8	4	16	8	0.3
2	4TA13	4 x 20 x 11	4	20	11	0.6
3	6TA12	6 x 20 x 9	6	20	9	0.3
4	6TA13	6 x 24 x 12	6	24	12	0.6
5	8TA12	8 x 22 x 9	8	22	9	0.3
6	8TA13	8 x 26 x 12	8	26	12	0.6
7	10TA11	10 x 24 x 9	10	24	9	0.3
▶ 8	10TA12	10 x 26 x 11	10	26	11	0.6
9	10TA13	10 x 30 x 14	10	30	14	0.6
10	12TA11	12 x 26 x 9	12	26	9	0.3
11	12TA12		12	28	11	0.6
12	12TA13	12 x 32 x 14	12	32	14	0.6
13	15TA11	15 x 28 x 9	15	28	9	0.3
14	15TA12	15 x 32 x 12	15	32	12	0.6
15	15TA13	15 x 37 x 15	15	37	15	0.6

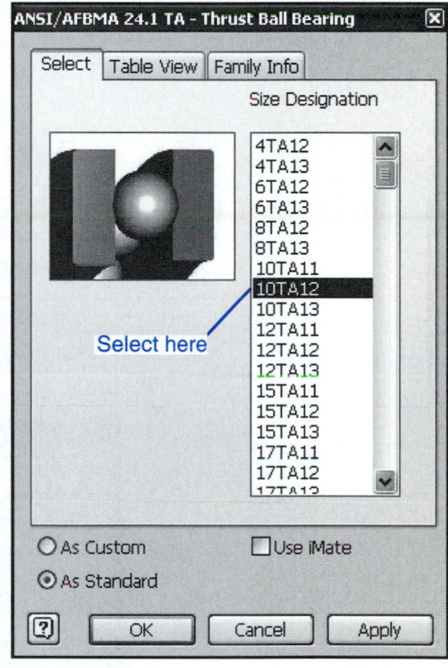

Figure 11-20

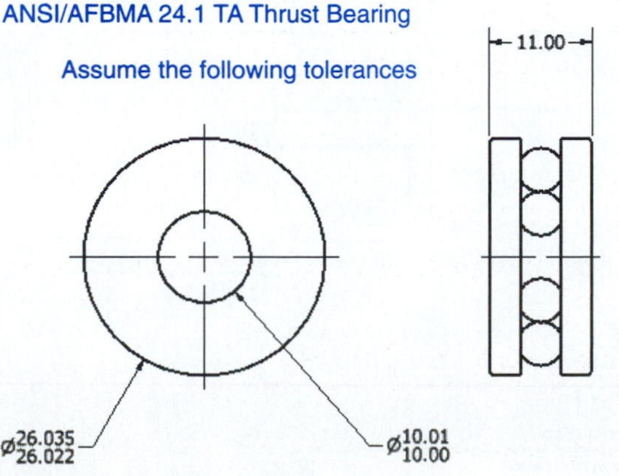

Figure 11-21

5. Calculate the shaft's diameter.

Say the clearance requirements are

Clearance = minimum = 0.00
Clearance = maximum = 0.02

This means the shaft diameter tolerance is 10.00/9.99.

6. Access the **Design Accelerator** and draw the shaft.

Use the nominal diameter of **Ø10.00** and a length of **30.** Add a **0.50 × 45°** chamfer to each end. Figure 11-22 shows a solid model drawing of the shaft and a dimensioned orthographic drawing. The model was drawn using the nominal Ø10, and the orthographic views include the required tolerances.

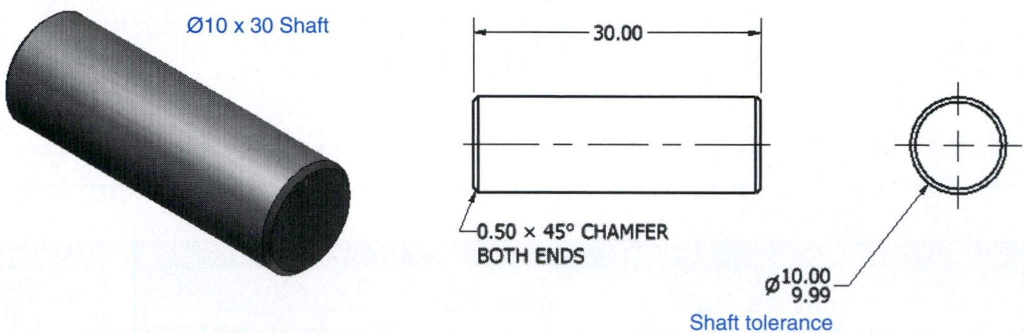

Figure 11-22

7. Determine the size for the hole in the T-bracket.

The interface between the outside diameter of the bearing and the T-bracket is defined as H7/p6. The values for this tolerance can be found in the tables in the appendix or by using the **Limits and Fits** dialog boxes found in the **Design Accelerator.**

8. Access the **Limits and Fits** tool in the **Design Accelerator.**

See Figure 11-23.

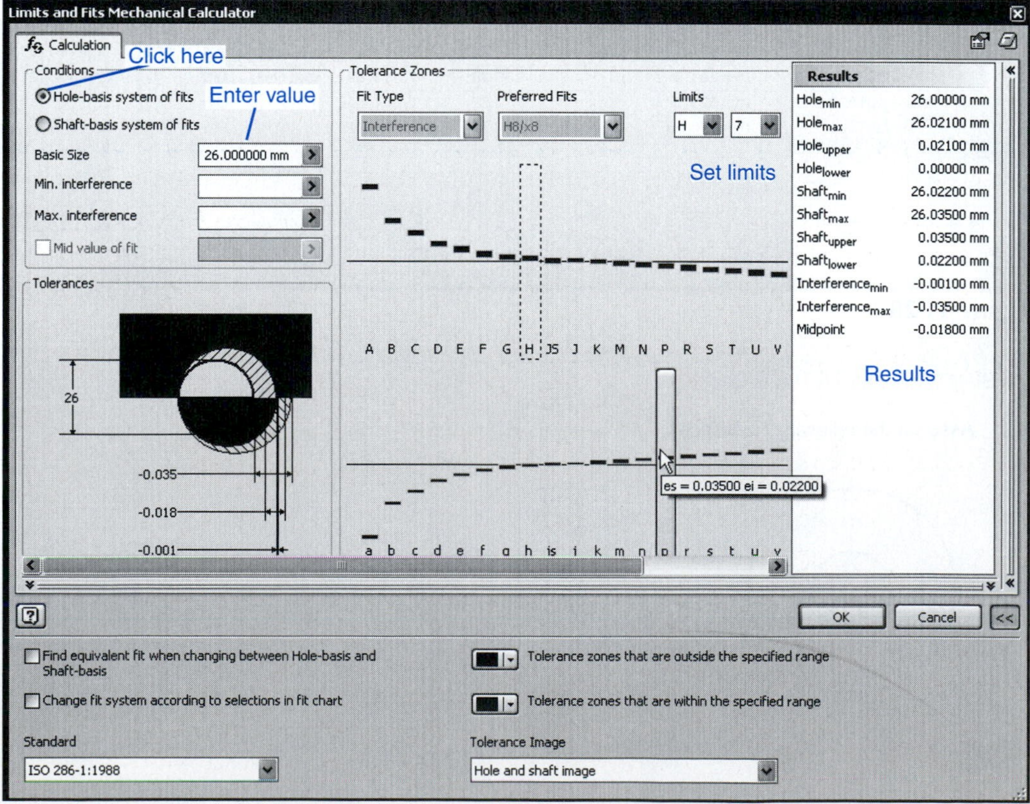

Figure 11-23

9. Select the **Hole-basis** condition, enter a **Basic Size** of **26.00,** and select an **Interference** fit.

The tolerance values will appear under the **Results** heading. The hole's values can also be obtained by moving the cursor to the **H** tolerance zone. The hole's tolerance is Ø26.021/26.000.

In this example the hole in the T-bracket will use a counterbored hole. The bearing has a thickness of 11.00, so the counter bored hole will have a depth of 11 and a diameter of 26.021/26.000.

10. Apply the hole tolerances to the T-bracket.

See Figure 11-24.

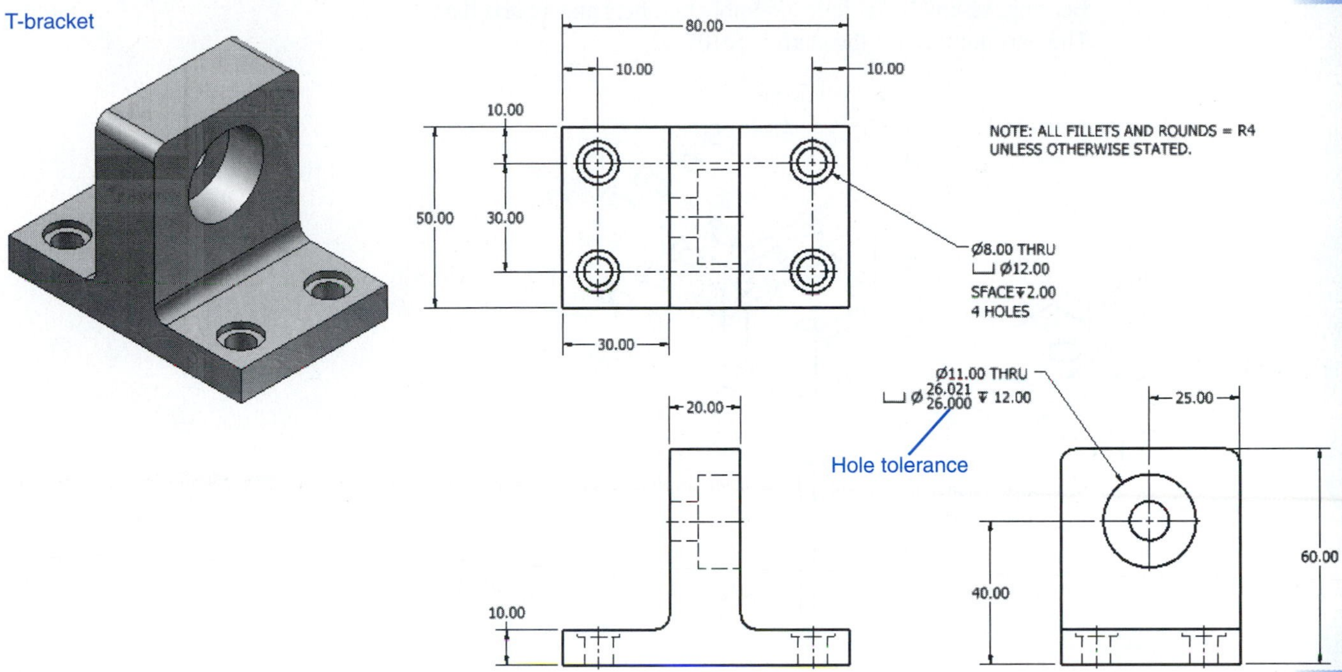

Figure 11-24

Summary

This chapter explained the differences among three types of bearings: plain, ball, and thrust. Various bearings were selected from the **Content Center** and used in drawings. The difference between shaft-basis and hole-basis tolerances was discussed and illustrated. Shafts and bearings were used in assemblies utilizing both clearance and force fits.

Chapter Projects

Figure P11-1 shows a Bearing Assembly drawing, a parts list, and detail drawings of all manufactured parts. Use the Bearing Assembly with Projects 11-1 through 11-8.

Project 11-1: Millimeters

A. Insert two Ø12 × 90 shafts with 0.50 × 45° chamfers at each end. Insert each shaft into a CNS 02 3481 A plain bearing. Insert the two bearing shaft subassemblies into the Bearing Assembly shown in Figure P11-1. See Figure P11-2.

B. Assume the CNS 02 3481 A plain bearing has an inside diameter of 12.01/12.00 and that the required clearance with the shaft is 0.00 minimum and 0.02 maximum. What is the diametric tolerance of the shafts? Create an orthographic drawing of one of the shafts with dimensions and tolerances.

C. Assume the CNS 02 3481 A plain bearing has an outside diameter of 16.039/16.028 and that the bearing will fit into the Side Part of the Bearing Assembly using an H7/s6 interference fit. What is the dimension and tolerance for the holes in the Side Part? Create an orthographic drawing of the Side Part and include all dimensions and tolerances.

D. Create a presentation drawing of the completed Bearing Assembly.

E. Create an exploded isometric drawing of the completed Bearing Assembly. Include assembly numbers and a parts list. The part number for the shaft is SH07-12.

Bearing Assembly

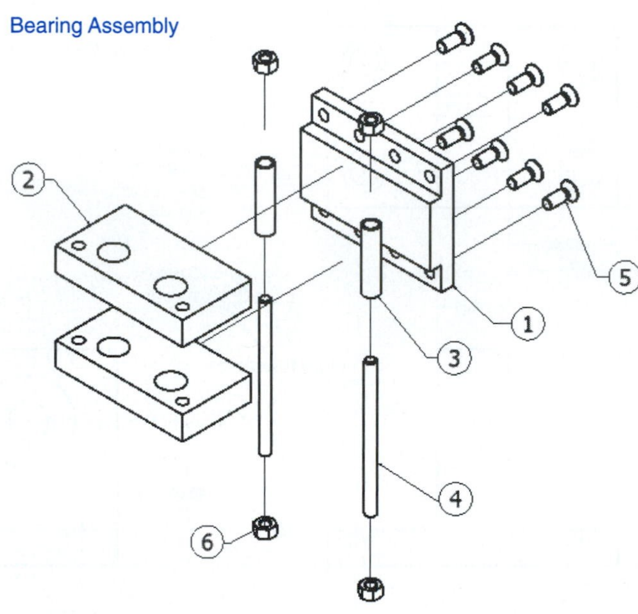

Bearing Assembly
Parts list

Parts List				
ITEM	PART NUMBER	DESCRIPTION	MATERIAL	QTY
1	BU2007-1	BASE	SAE 1020	1
2	BU2007-2	PART, SIDE	SAE 1020	2
3	BUENG-A	POST, GUIDE	SAE 1040	2
4	BUENG-B	POST, THREADED	SAE 1040	2
5	AS 1427 - M8 x 20	Pozidriv ISO metric machine screws	Steel, Mild	8
6	ANSI B18.2.4.2M - M8x1.25	Metric Hex Nuts Styles 2	Steel, Mild	4

Figure P-1 *(Continued)*

Base
P/N BU2007-1
SAE 1020

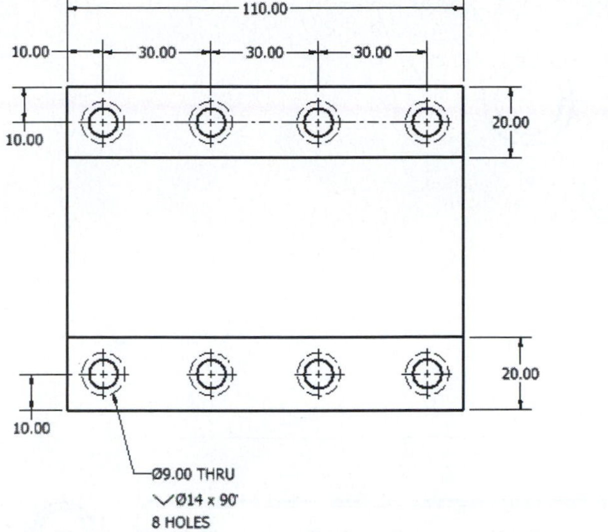

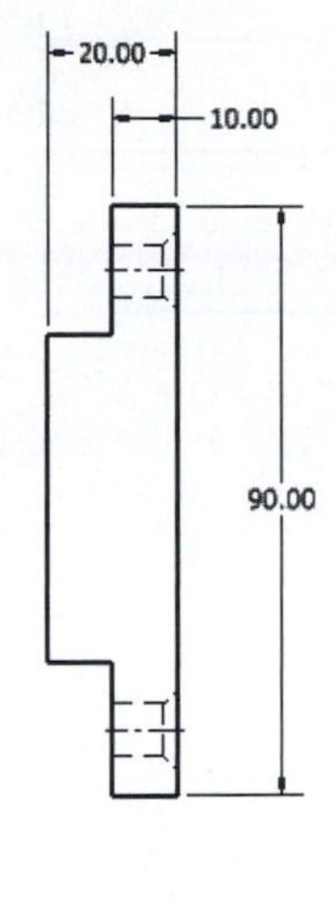

Part, Side
P/N BU2007-2
SAE 1020

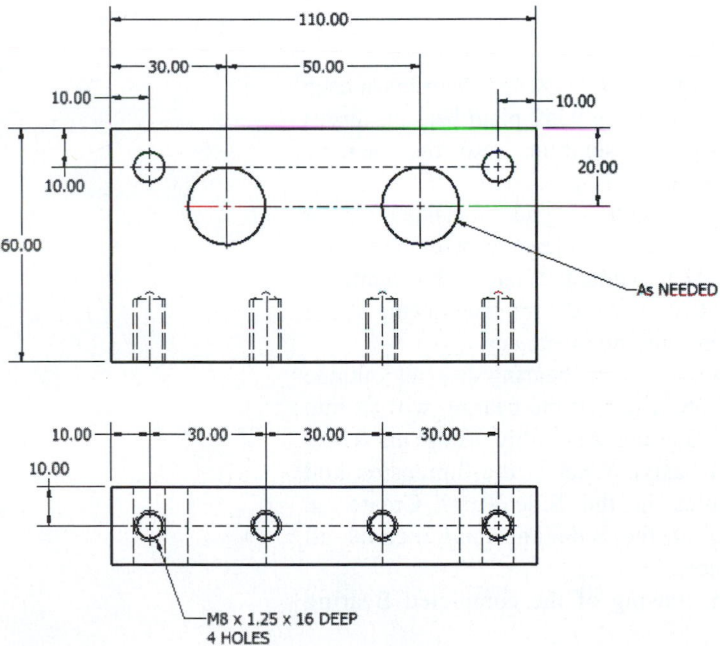

Figure P11-1 *(Continued)*

Post, Guide
BUENG-A
SAE 1040

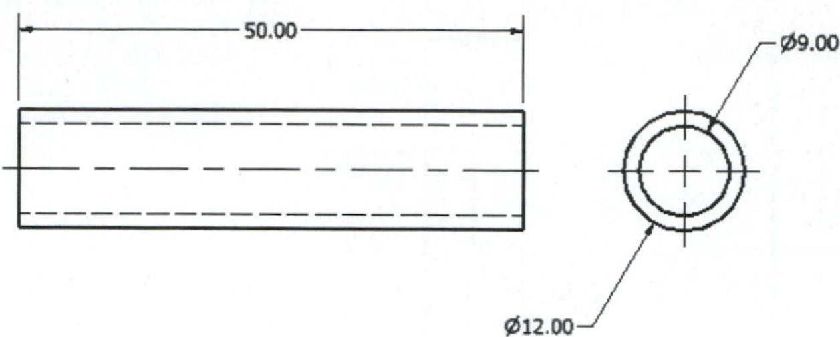

Post, Threaded
P/N BUENG-B
SAE 1040

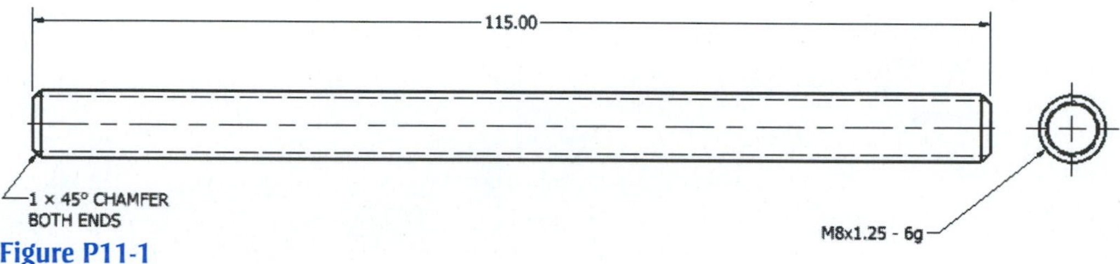

Figure P11-1

Project 11-2: Millimeters

A. Insert two Ø16 × 90 shafts with 0.50 × 45° chamfers at each end. Insert each shaft into a CSN 9352 plain bearing. Insert the two bearing shaft subassemblies into the Bearing Assembly shown in Figure P11-1.

B. Assume the CSN 9352 A plain bearing has an inside diameter of 16.02/12.01 and that the required clearance with the shaft is 0.01 minimum and 0.04 maximum. What is the diametric tolerance of the shafts? Create an orthographic drawing of one of the shafts with dimensions and tolerances.

C. Assume the CSN 9352 A plain bearing has an outside diameter of 22.000/21.987 and that the bearing will fit into the Side Part of the Bearing Assembly using an S7/h6 interference fit (shaft basis). What is the dimension and tolerance for the holes in the Side Part? Create an orthographic drawing of the Side Part and include all dimensions and tolerances.

D. Create a presentation drawing of the completed Bearing Assembly.

E. Create an exploded isometric drawing of the completed Bearing Assembly. Include assembly numbers and a parts list. The part number for the shaft is SH07-16.

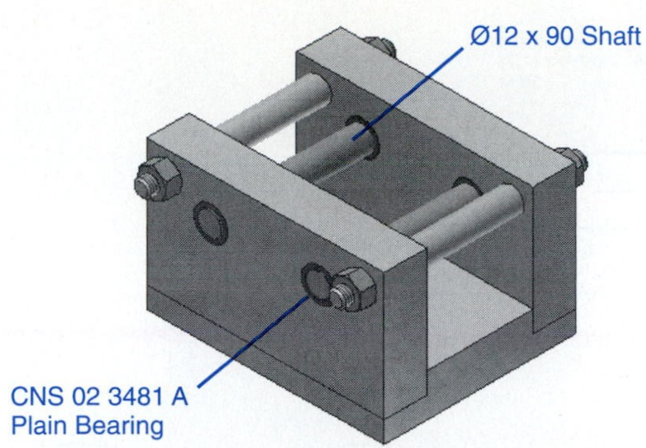

Figure P11-2

Project 11-3: Millimeters

A. Insert two Ø10 × 130 shafts with 0.50 × 45° chamfers at each end. Insert each shaft into a DIN 1850-5 P plain bearing. Insert the two bearing shaft subassemblies into the Bearing Assembly shown in Figure P11-1.

B. Assume the DIN 1850-5 P plain bearing has an inside diameter of 10.02/12.00 and that the required clearance with the shaft is 0.00 minimum and 0.04 maximum. What is the

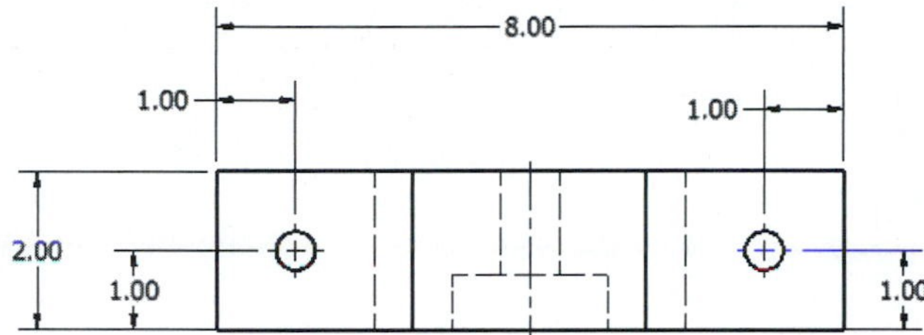

NOTE: ALL FILLETS AND ROUNDS R=0.250
UNLESS OTHERWISE STATED.

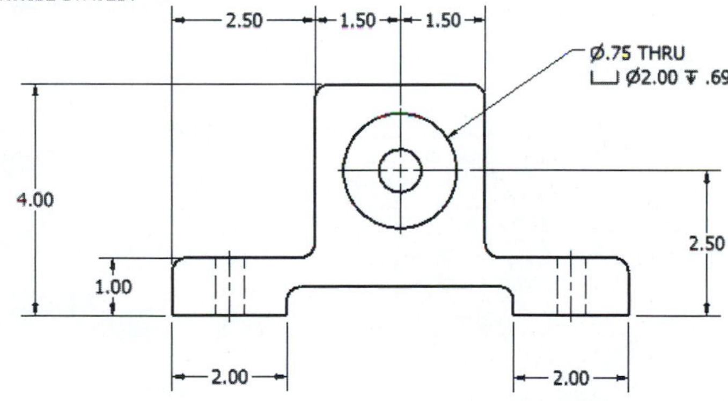

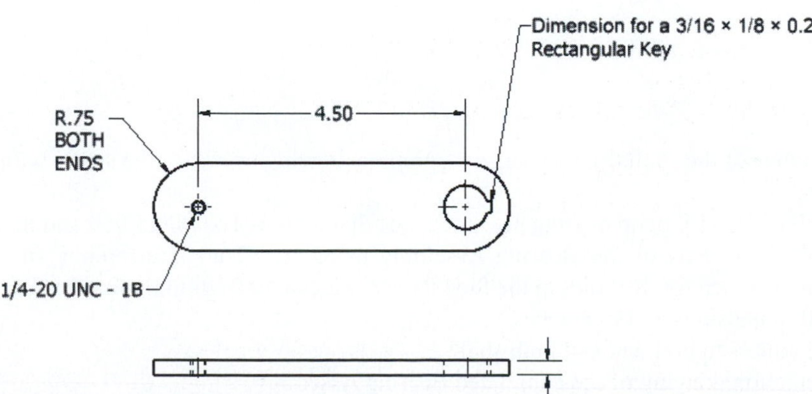

Figure P11-3 *(Continued)*

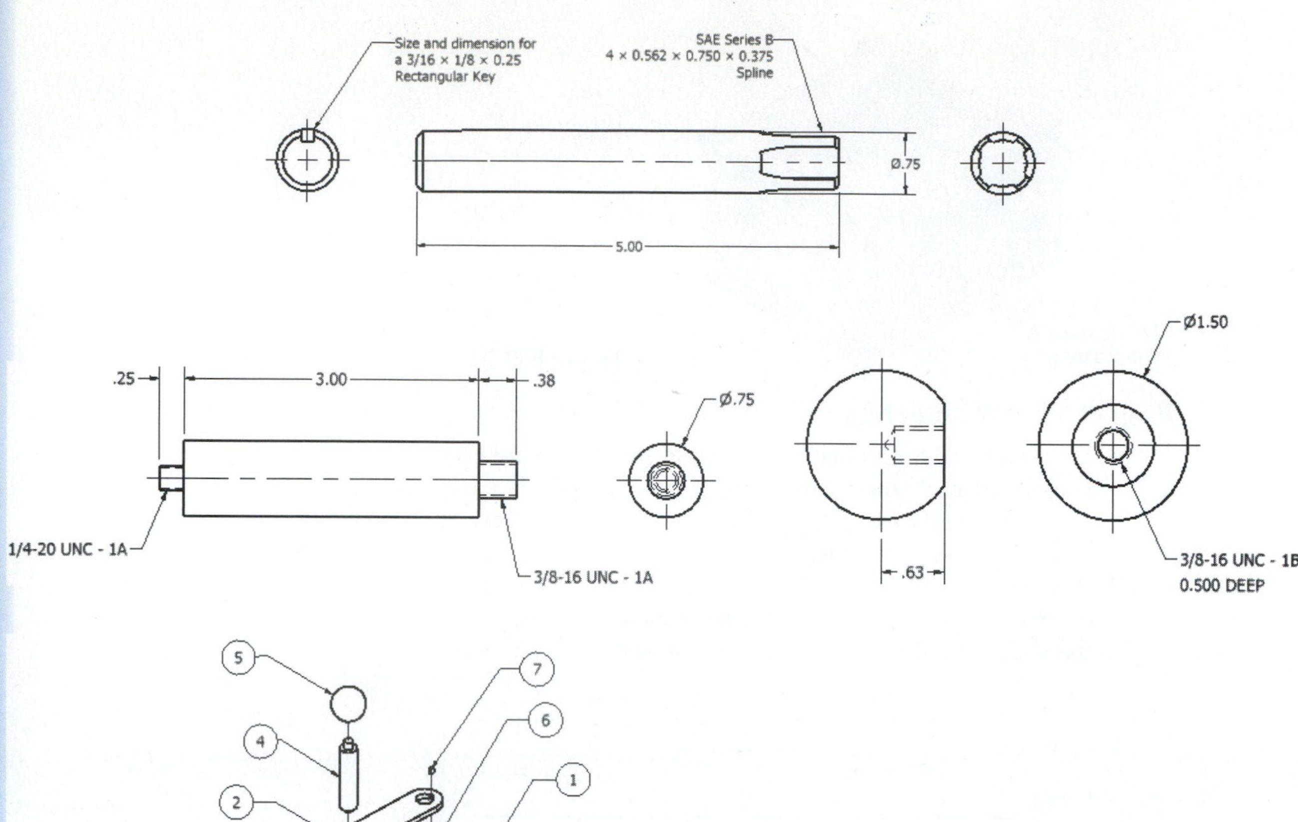

Handle Assembly

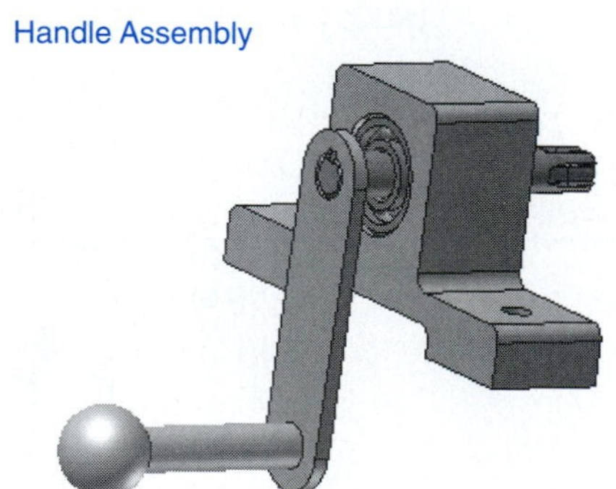

Parts List				
ITEM	PART NUMBER	DESCRIPTION	MATERIAL	QTY
1	EK131-1	SUPPORT	STEEL	1
2	EK131-2	LINK	STEEL	1
3	EK131-3	SHAFT,DRIVE	STEEL	1
4	EK131-4	POST, THREADED	STEEL	1
5	EK131-5	BALL	STEEL	1
6	BS 292 - BRM 3/4	Deep Groove Ball Bearings	STEEL,MILD	1
7	3/16x1/8x1/4	RECTANGULAR KEY	STEEL	1

Figure P11-3

diametric tolerance of the shafts? Create an orthographic drawing of one of the shafts with dimensions and tolerances.

C. Assume the DIN 1850-5 P plain bearing has an outside diameter of 16.000/15.989 and that the bearing will fit into the Side Part of the Bearing Assembly using an S7/h6 interference fit. What is the dimension and tolerance for the holes in the Side Part? Create an orthographic drawing of the Side Part and include all dimensions and tolerances.
D. Add CNS 122 collars to both ends of both shafts.
E. Create a presentation drawing of the completed Bearing Assembly
F. Create an exploded isometric drawing of the completed Bearing Assembly. Include assembly numbers and a parts list. The part number for the shaft is SH08-16.

Project 11-4: Millimeters

A. Insert two Ø20 × 110 shafts with 1.00 × 45° chamfers at each end. Insert each shaft into a JIS B 1582 plain bearing. Insert the two bearing shaft subassemblies into the Bearing Assembly shown in Figure P11-1.

B. Assume the JIS B 1582 plain bearing has an inside diameter of 20.02/20.00 and that the required clearance with the shaft is 0.01 minimum and 0.05 maximum. What is the diametric tolerance of the shaft? Create an orthographic drawing of the shaft with dimensions and tolerances.

C. Assume the JIS B 1582 plain bearing has an outside diameter of 28.039/28.028 and that the bearing will fit into the Side Part of the Bearing Assembly using an H7/s6 interference fit. What is the dimension and tolerance for the holes in the Side Part? Create an orthographic drawing of the Side Part and include all dimensions and tolerances.

D. Add CNS 9074 retaining rings 5 mm from each end of both shafts.

E. Create a presentation drawing of the completed Bearing Assembly.

F. Create an exploded isometric drawing of the completed Bearing Assembly. Include assembly numbers and a parts list. The part number for the shaft is SH07-12.

Project 11-5: Millimeters

A. Insert two Ø10 × 90 shafts with 0.50 × 45° chamfers at each end. Insert each shaft into a BS 290 – 61800 ball bearing. Insert the two bearing shaft subassemblies into the Bearing Assembly shown in Figure P11-1.

B. Assume the BS 290 – 61800 ball bearing has an inside diameter of 10.02/10.00 and that the required clearance with the shaft is 0.00 minimum and 0.04 maximum. What is the diametric tolerance of the shafts? Create an orthographic drawing of one of the shafts with dimensions and tolerances.

C. Assume the BS 290 – 61800 ball bearing has an outside diameter of 16.000/15.989 and that the bearing will fit into the Side Part of the Bearing Assembly using an U7/h6 interference fit. What is the dimension and tolerance for the holes in the Side Part? Create an orthographic drawing of the Side Part and include all dimensions and tolerances.

D. Create a presentation drawing of the completed Bearing Assembly.

E. Create an exploded isometric drawing of the completed Bearing Assembly. Include assembly numbers and a parts list. The part number for the shaft is SH07-12.

Project 11-6: Millimeters

A. Insert two Ø15 × 130 Shafts with 0.50 × 45° chamfers at each end. Insert each shaft into a CSN 02 4630 – 61802 ball bearing. Insert the two bearing shaft subassemblies into the Bearing Assembly shown in Figure P11-1.

B. Assume the CSN 02 4630 – 61802 ball bearing has an inside diameter of 15.01/12.00 and that the required clearance with the shaft is 0.00 minimum and 0.02 maximum. What is the diametric tolerance of the shafts? Create an orthographic drawing of one of the shafts with dimensions and tolerances.

C. Assume the CSN 02 4630 − 61802 ball bearing has an outside diameter of 24.000/24.987 and that the bearing will fit into the Side Part of the Bearing Assembly using an F7/h6 interference fit. What is the dimension and tolerance for the holes in the Side Part? Create an orthographic drawing of the Side Part and include all dimensions and tolerances.
D. Add a DIN 705 B collar to each end of both shafts.
E. Create a presentation drawing of the completed Bearing Assembly.
F. Create an exploded isometric drawing of the completed Bearing Assembly. Include assembly numbers and a parts list. The part number for the shaft is SH07-12.

Project 11-7: Millimeters

A. Insert two Ø12 × 110 shafts with 1.00 × 45° chamfers at each end. Insert each shaft into a CSN 02 4630 − 16103 ball bearing. Insert the two bearing shaft subassemblies into the Bearing Assembly shown in Figure P11-1. For this exercise replace the through hole with a counterbored hole with a nominal size Ø13.00 THRU, Ø30 CBORE 8.00 DEEP.
B. Assume the CSN 02 4630 − 16103 ball bearing has an inside diameter of 12.01/12.00 and that the required clearance with the shaft is 0.00 minimum and 0.02 maximum. What is the diametric tolerance of the shafts? Create an orthographic drawing of one of the shafts with dimensions and tolerances.
C. Assume the CSN 02 4630 − 16103 ball bearing has an outside diameter of 30.000/29.987 and that the bearing will fit into the Side Part of the Bearing Assembly using an S7/h6 interference fit. What is the dimension and tolerance for the holes in the Side Part? Create an orthographic drawing of the Side Part and include all dimensions and tolerances.
D. Add a DIN 471 5.00 mm from each end of both shafts.
E. Create a presentation drawing of the completed Bearing Assembly.
F. Create an exploded isometric drawing of the completed Bearing Assembly. Include assembly numbers and a parts list. The part number for the shaft is SH07-12.

Project 11-8: Millimeters

A. Insert two Ø16 × 100 shafts with 0.75 × 45° chamfers at each end. Insert each shaft into a GB 273.2−87 − 7/70 16×26×5 thrust bearing. Insert the two bearing shaft subassemblies into the Bearing Assembly shown in Figure P11-1. For this exercise replace the through hole with a counterbored hole with a nominal size Ø17.00 THRU, Ø26 CBORE 5.00 DEEP.
B. Assume the GB 273.2−87 − 7/70 16×26×5 thrust bearing has an inside diameter of 16.02/16.00 and that the required clearance with the shaft is 0.00 minimum and 0.04 maximum. What is the diametric tolerance of the shafts? Create an orthographic drawing of one of the shafts with dimensions and tolerances.
C. Assume the GB 273.2−87 − 7/70 16×26×5 thrust bearing has an outside diameter of 26.039/26.028 and that the bearing will fit into the Side Part of the Bearing Assembly using an H7/s6 interference fit. What is the dimension and tolerance

for the holes in the Side Part? Create an orthographic drawing of the Side Part and include all dimensions and tolerances.

D. Create a presentation drawing of the completed Bearing Assembly.
E. Create an exploded isometric drawing of the completed Bearing Assembly. Include assembly numbers and a parts list. The part number for the shaft is SH07-16.

Project 11-9: Inches

Figure P11-3 shows a Handle Assembly.

A. Draw an assembly drawing. The Link is offset .75 from the support.
B. Create a presentation drawing
C. Create an exploded isometric drawing with a parts list.
D. Animate the drawing so that the Ball, Threaded Post, Link, Rectangular Key, and Shaft rotate within the bearing.

Gears

12

Chapter Objectives

- Understand the different types of gears: spur, bevel, and worm.
- Understand gear terminology.
- Show how to select gears from the **Content Center.**
- Show how to draw hubs on gears and to add setscrews.
- Show how to draw keyways on gears and to add keys.
- Show how to use gears in assemblies.

INTRODUCTION

This chapter explains how to draw gears using **Design Accelerator.** Three types of gears are covered: spur, bevel, and worm. See Figure 12-1. The chapter shows how to add hubs and splines to the gears and how to combine gears to create gear trains.

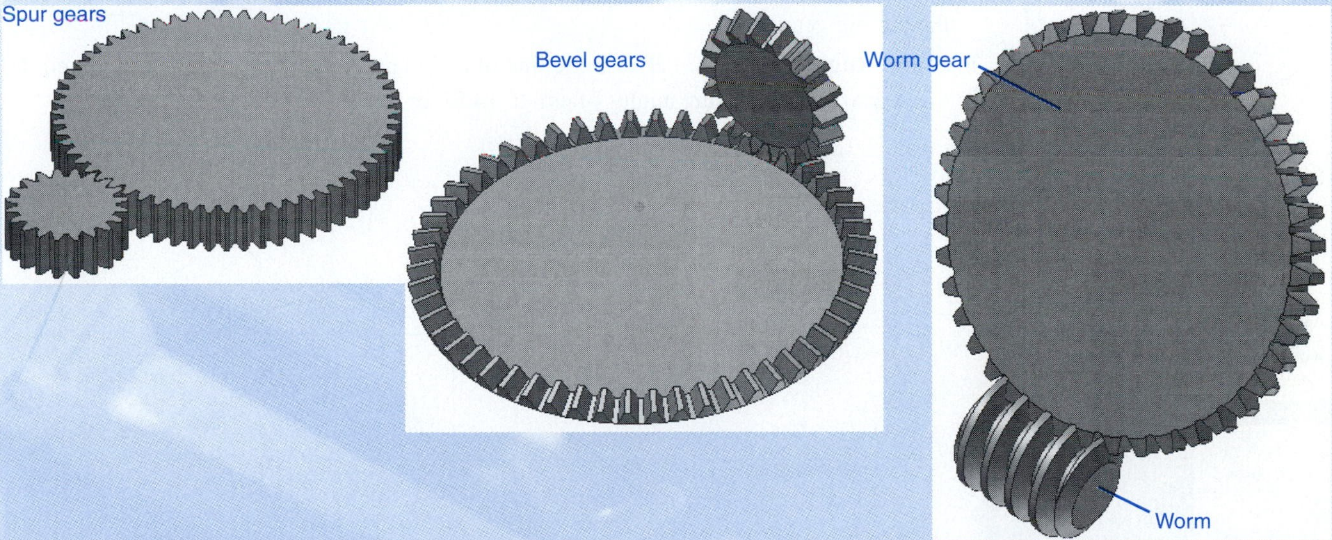

Figure 12-1

Gear Terminology

Pitch Diameter (*D*): The diameter used to define the spacing of gears. Ideally, gears are exactly tangent to each other along their pitch diameters.

Diametral Pitch (*P*): The number of teeth per inch. Meshing gears must have the same diametral pitch. Manufacturers' gear charts list gears with the same diametral pitch.

Module (*M*): The pitch diameter divided by the number of teeth. The metric equivalent of diametral pitch.

Number of Teeth (*N*): The number of teeth of a gear.

Circular Pitch (*CP*): The circular distance from a fixed point on one tooth to the same position on the next tooth as measured along the pitch circle. The circumference of the pitch circle divided by the number of teeth.

Preferred Pitches: The standard sizes available from gear manufacturers. Whenever possible, use preferred gear sizes.

Center Distance (*CD*): The distance between the center points of two meshing gears.

Backlash: The difference between a tooth width and the engaging space on a meshing gear.

Addendum (*a*): The height of a tooth above the pitch diameter.

Dedendum (*d*): The depth of a tooth below the pitch diameter.

Whole Depth: The total depth of a tooth. The addendum plus the dedendum.

Working Depth: The depth of engagement of one gear into another. Equal to the sum of the two gears' adendeums.

Circular Thickness: The distance across a tooth as measured along the pitch circle.

Face Width (*F*): The distance from front to back along a tooth as measured perpendicular to the pitch circle.

Outside Diameter: The largest diameter of the gear. Equal to the pitch diameter plus the addendum.

Root Diameter: The diameter of the base of the teeth. The pitch diameter minus the dedendum.

Clearance: The distance between the addendum of the meshing gear and the dedendum of the mating gear.

Pressure Angle: The angle between the line of action and a line tangent to the pitch circle. Most gears have pressure angles of either 14.5° or 20°.

See Figure 12-2.

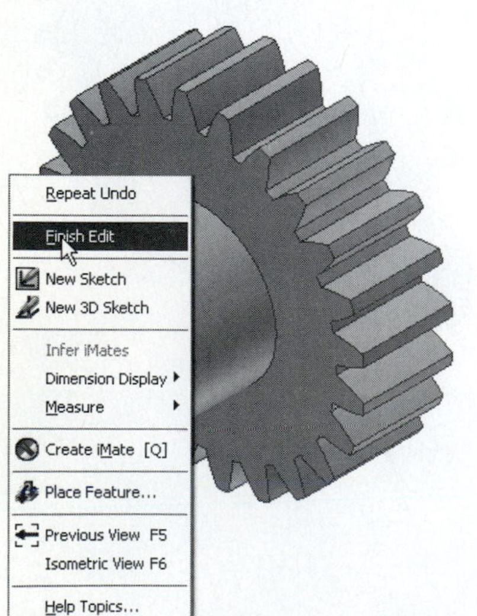

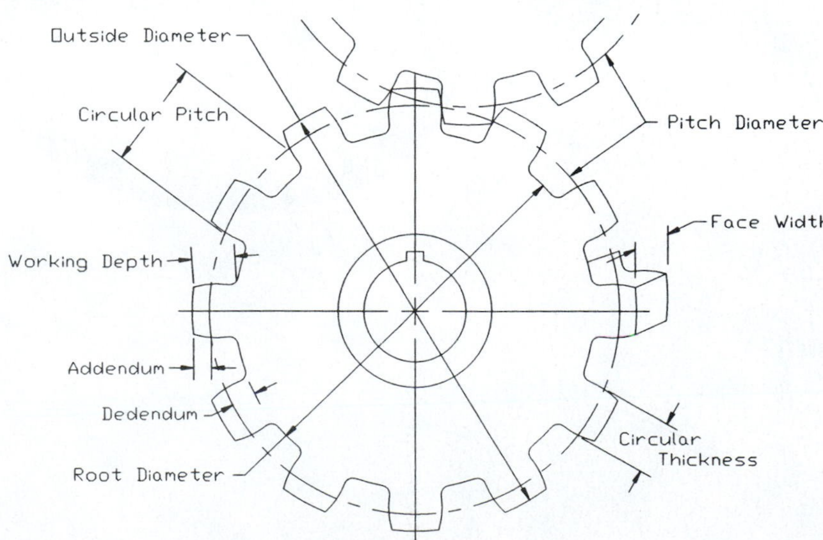

Figure 12-2

Gear Formulas

Figure 12-3 shows a chart of formulas commonly associated with gears. The formulas are for spur gears.

Diametral pitch (P)	$P = \frac{N}{D}$
Pitch diameter (D)	$D = \frac{N}{P}$
Number of teeth (N)	$N = DP$
Addendum (a)	$a = \frac{1}{P}$

Metric

Module (M)	$M = \frac{D}{N}$

Figure 12-3

Drawing Gears Using Design Accelerator

Draw a gear with 60 teeth, a diametral pitch of 24, a pressure angle of 14.5°, and a face width of 0.75 in.

1. Start a new drawing using the **Standard (in).iam** format.
2. Access the **Design Accelerator** by clicking the **Tool** heading at the top of the screen.

See Figure 12-4. The **Design Accelerator** dialog box will appear.

3. Click the **Spur Gears** tool.

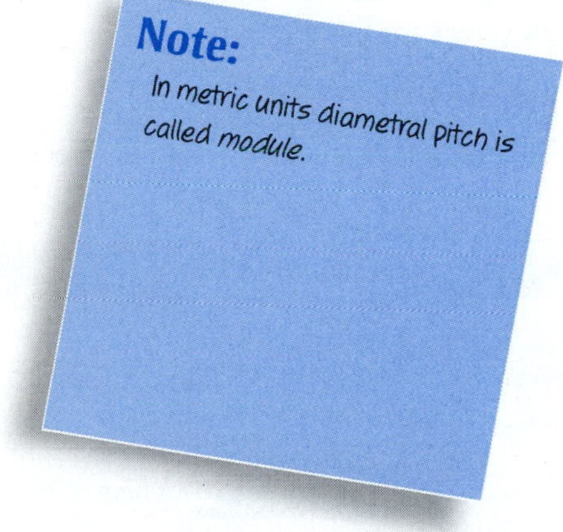

Figure 12-4

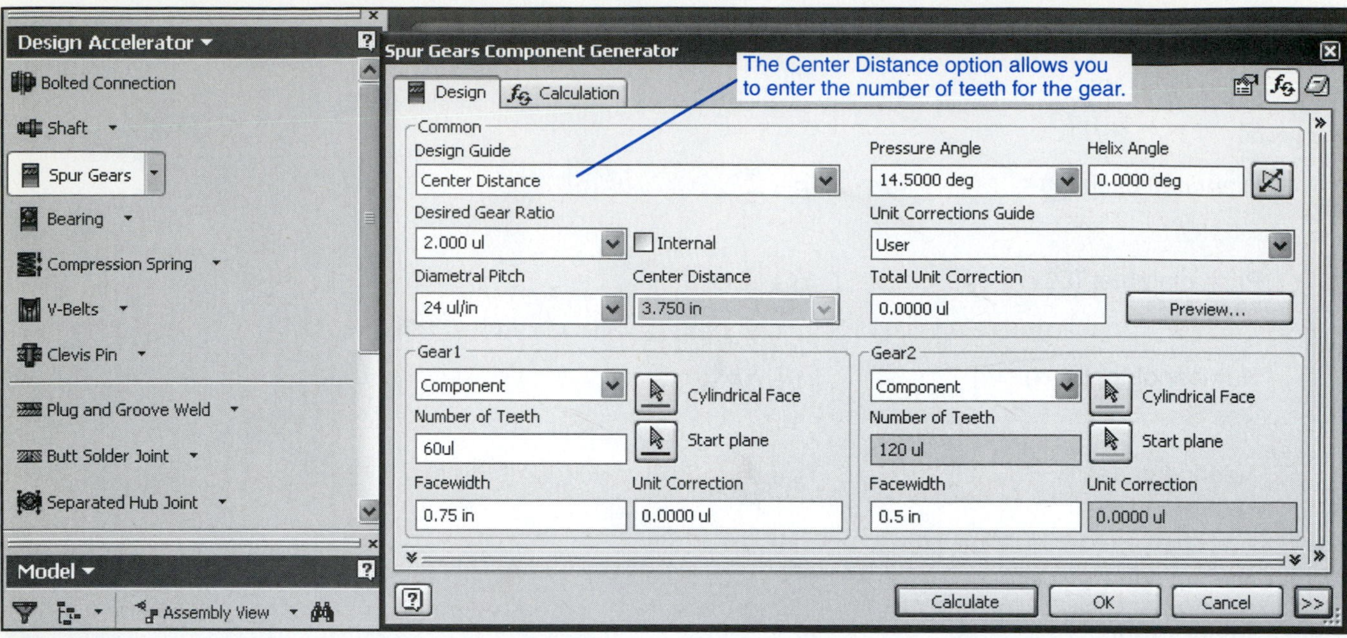

Figure 12-5

The **Spur Gears Component Generator** dialog box will appear. See Figure 12-5. There are several options available under the **Design Guide** option. The **Center Distance** option allows you to enter values for the number of teeth for the gear.

4. Enter the appropriate values.

Design Accelerator will draw two matching gears. Set the gear ratio for **2.000** and accept the default values for **Gear2.**

5. Click **OK.**

The two gears will appear on the screen. See Figure 12-6.

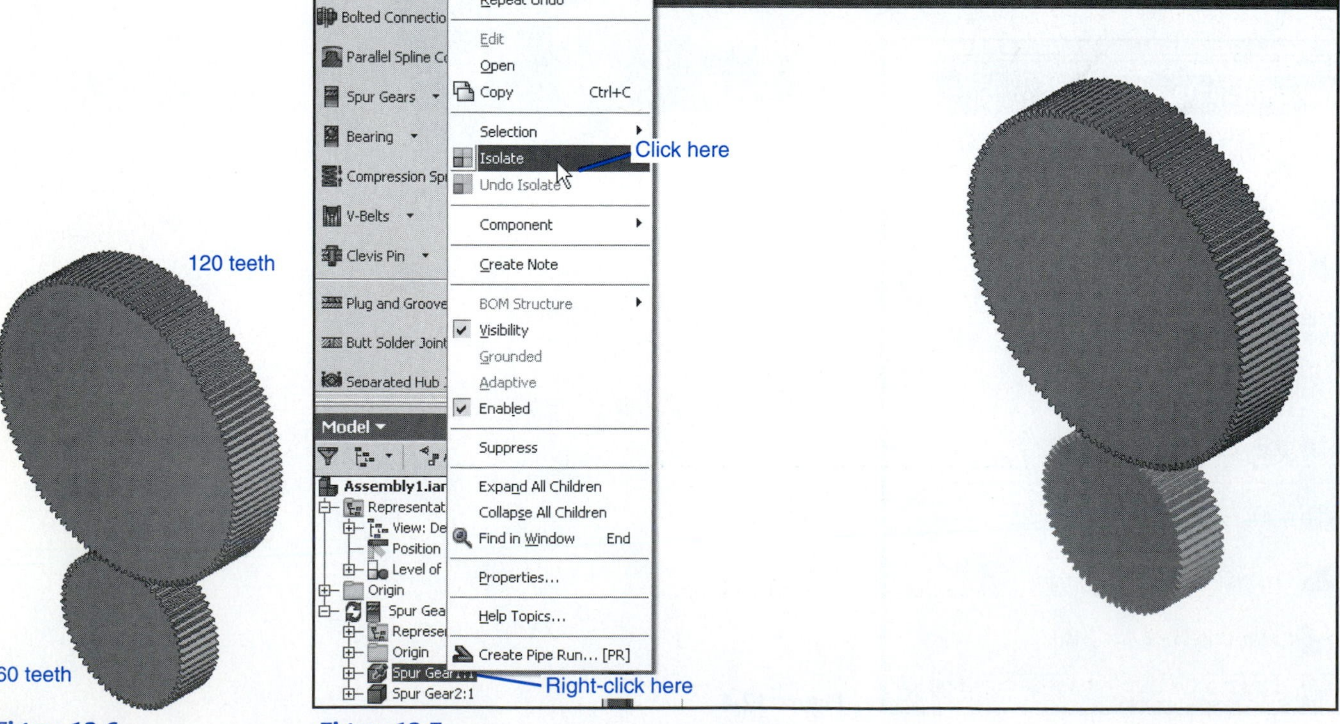

Figure 12-6 **Figure 12-7**

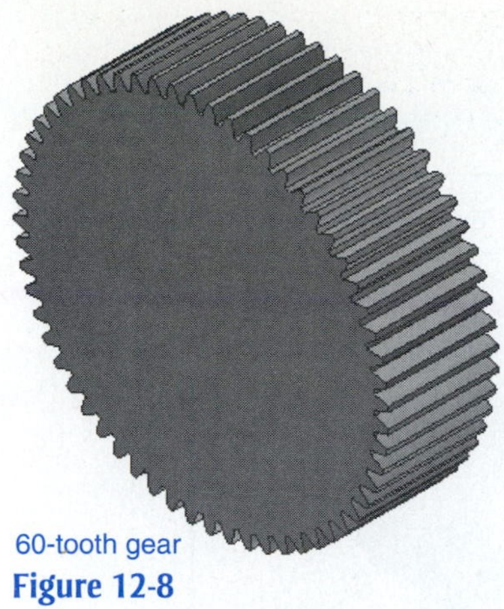

Figure 12-8

6. Right-click **Spur Gear1:1** in the browser box and click the **Isolate** option.

See Figure 12-7. Figure 12-8 shows the finished gear.

GEAR HUBS

Figure 12-9 shows a 24-tooth gear with a 0.375 face width. It was created using the method explained in the previous section. This section will show how to add a hub to the gear. The hub will include a threaded hole for a setscrew.

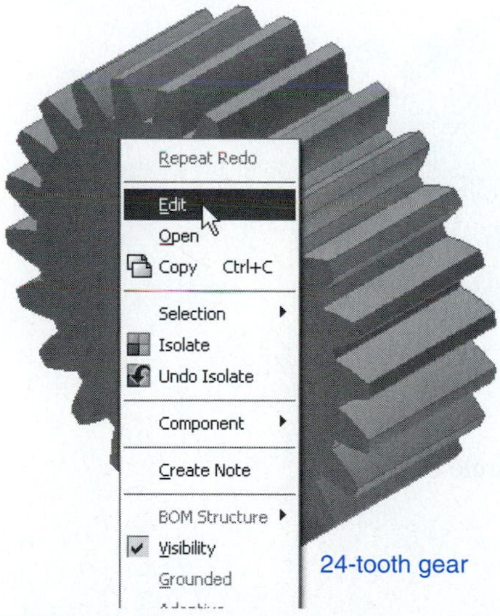

Figure 12-9

Exercise 12-1: Adding a Hub to a Gear

1. Right-click the gear and select the **Edit** option.

   See Figure 12-9.

2. Right-click the gear again and select the **New Sketch** option.

   See Figure 12-10.

3. Draw a **Ø1.250** circle on the gear.

   See Figure 12-11.

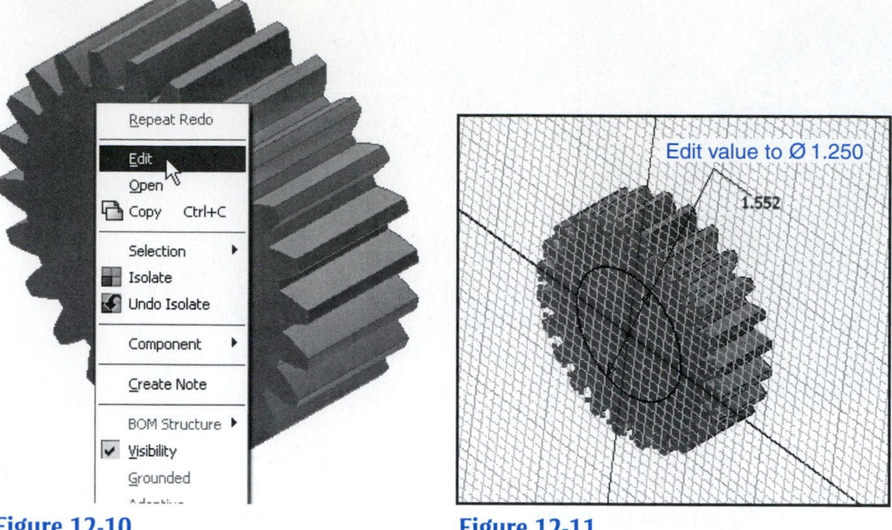

Figure 12-10

Figure 12-11

4. Right-click the mouse and select the **Done** option. Right-click the mouse again and select the **Finish Sketch** option.

See Figures 12-12 and 12-13.

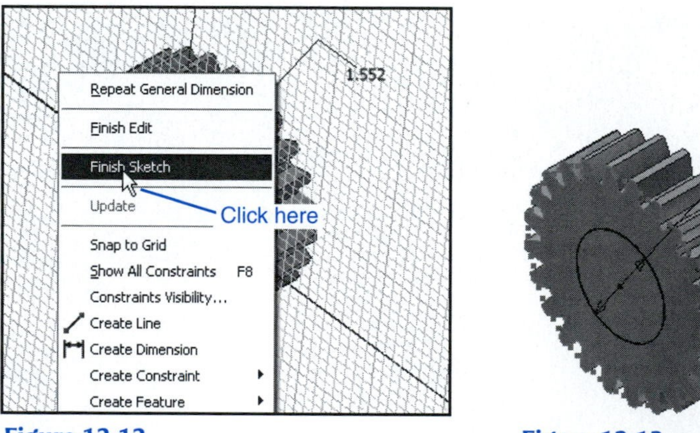

Figure 12-12

Figure 12-13

5. Click the **Extrude** tool and extrude the Ø1.250 circle **0.75.**

See Figure 12-14.

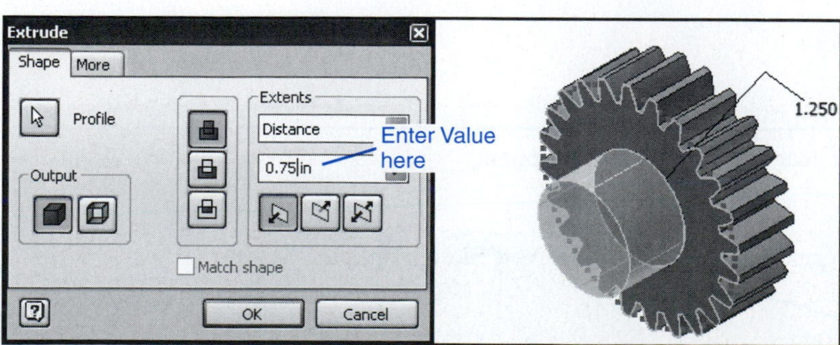

Figure 12-14

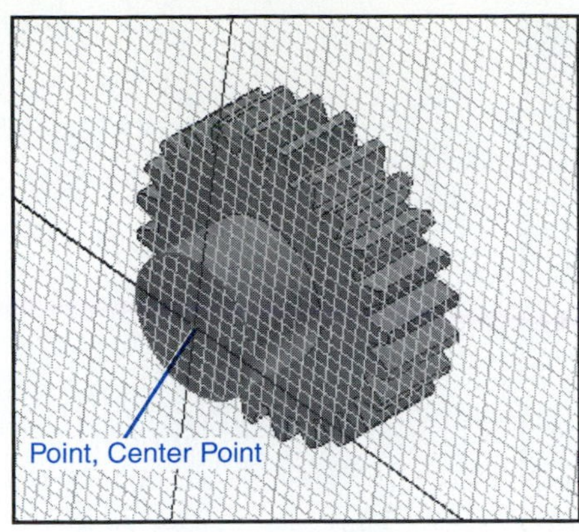

Figure 12-15

Figure 12-16

6. Create a new sketch plane on the top surface of the extrusion and locate a **Point, Center Point.**

See Figure 12-15.

7. Right-click the mouse and select the **Done** option. Right-click the mouse again and select the **Finish Sketch** option.
8. Click the **Hole** tool and add a **Ø.500** hole to the gear.

See Figure 12-16.

9. Create a work plane tangent to the hub.

See Figure 12-17.

The tangent work plane is created by clicking the **Work Plane** tool then the **XZ plane** under the **Origin** heading in the browser box. Move the cursor to the outside surface of the hub. A work plane will appear tangent to the hub.

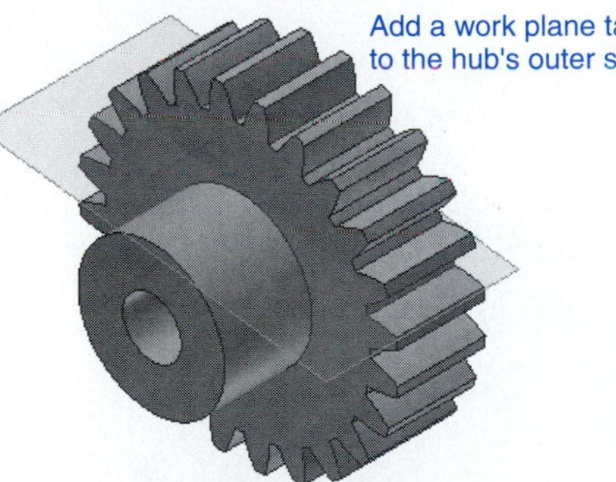

Figure 12-17

10. Create a new sketch on the work plane and locate a **Point, Center Point .375** from the top edge of the hub.

See Figure 12-18.

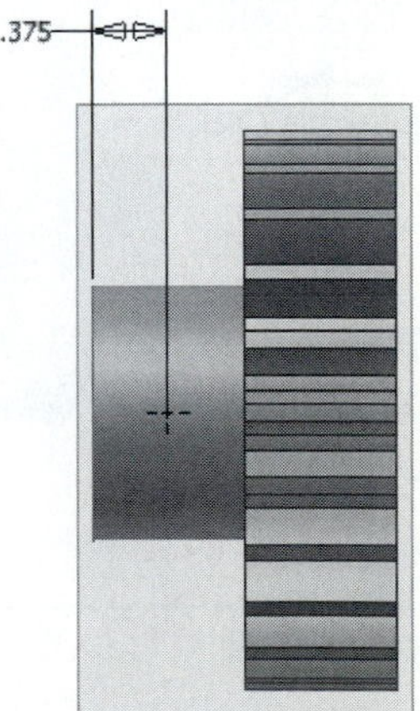

Figure 12-18

11. Right-click the mouse and select the **Done** option. Right-click the mouse again and select the **Finish Sketch** option.
12. Click the **Hole** tool and create a **10-24 UNC** threaded hole in the hub. Hide the work plane.

Do not use the **Through All** distance, as this will create two holes.

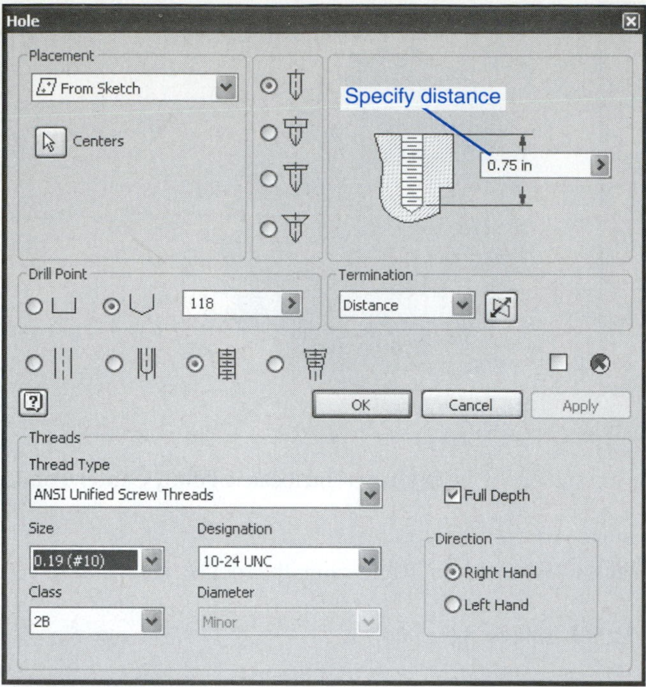

Figure 12-19

See Figure 12-19.

13. Right-click the mouse and select the **Finish Edit** option. Return to the **Assembly Panel** by clicking the arrow to the right of the **Design Accelerator** heading.

See Figure 12-20. Figure 12-21 shows the finished gear.

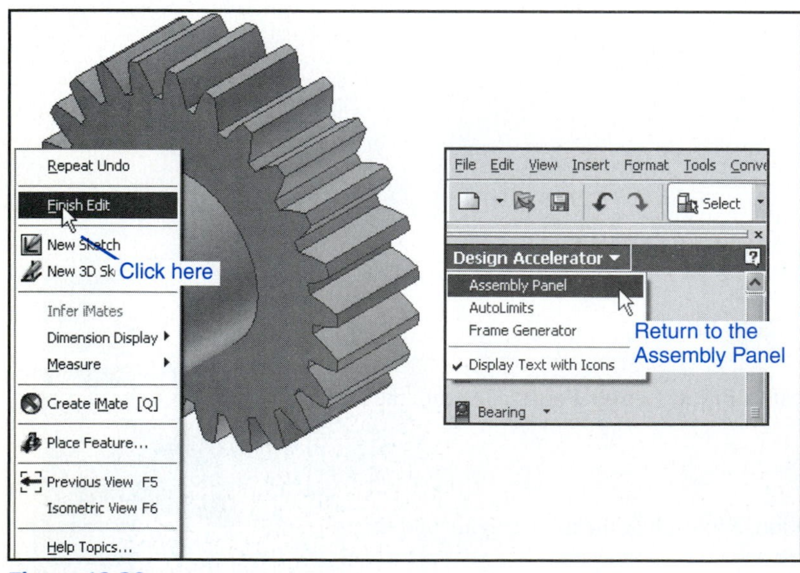

Figure 12-20

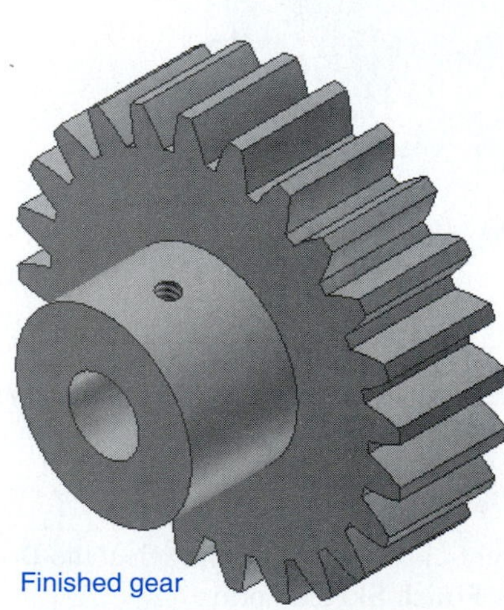

Figure 12-21

14. Click on the **Place from Content Center** tool.
15. Access the **Set Screws** and select a **Hexagon Socket Set Screw – Half Dog Point – Inch** setscrew. Click **OK.**

See Figure 12-22.

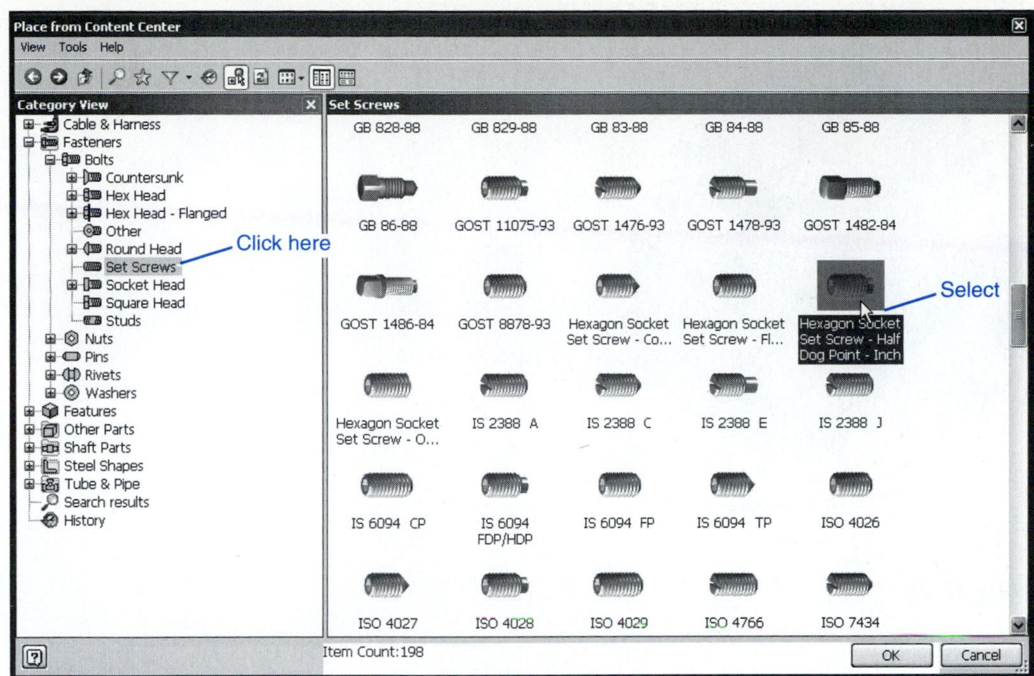

Figure 12-22

16. Select a **#10 UNC** thread that is **0.38** long. Click **OK.**

See Figure 12-23.

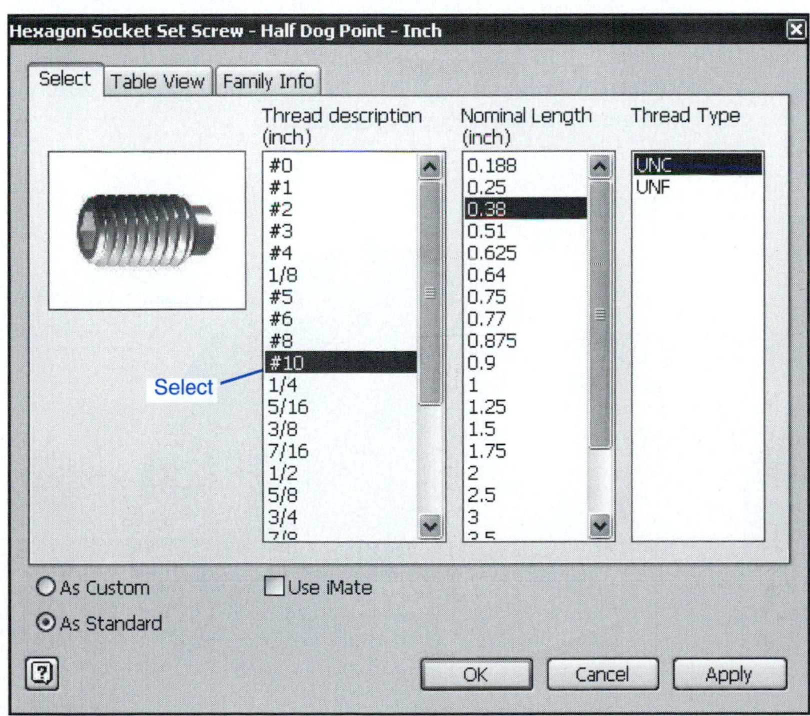

Figure 12-23

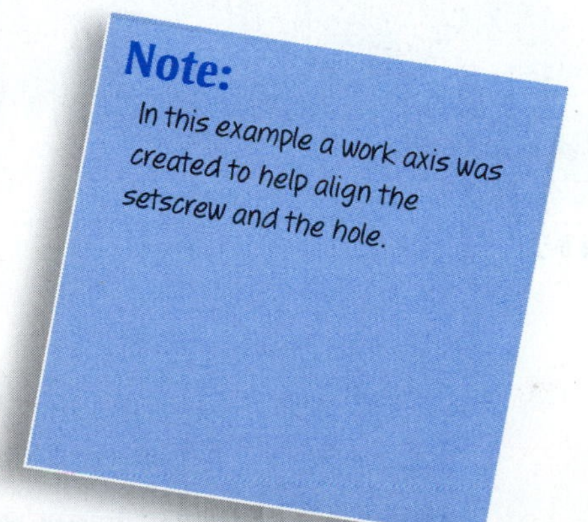

17. Use the **Constraint** tool and position the setscrew.

See Figure 12-24.

Figure 12-24

Gear Ratios

gear ratio: The ratio of number of teeth on the larger of two meshing gears to the number on the smaller gear.

The speed ratio between two gears is determined by the number of teeth on each gear. For example, if two spur gears have 60 teeth and 20 teeth, respectively, the ***gear ratio*** is 60/20 = 3/1. See Figure 12-25. Thus, if the larger gear with 60 teeth is rotating at 30 RPM, the smaller gear with 20 teeth will rotate at 90 RPM. It is a general rule of thumb not to use spur gear ratios greater than 5:1.

Bevel gear ratios are also determined by the number of teeth on each gear. Again, the general rule of thumb is not to use gear ratios greater than 5:1. See Figure 12-26.

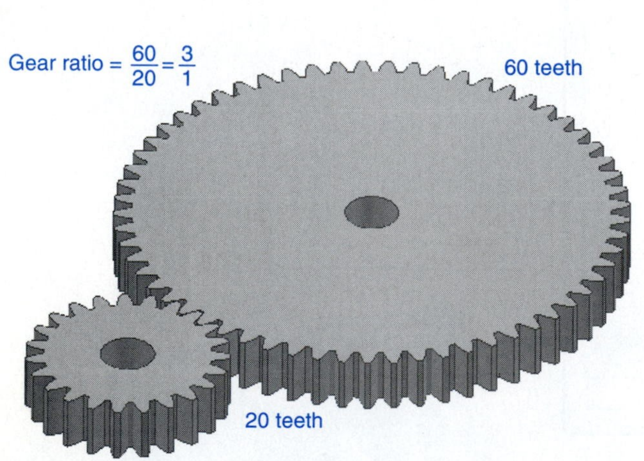

Figure 12-25

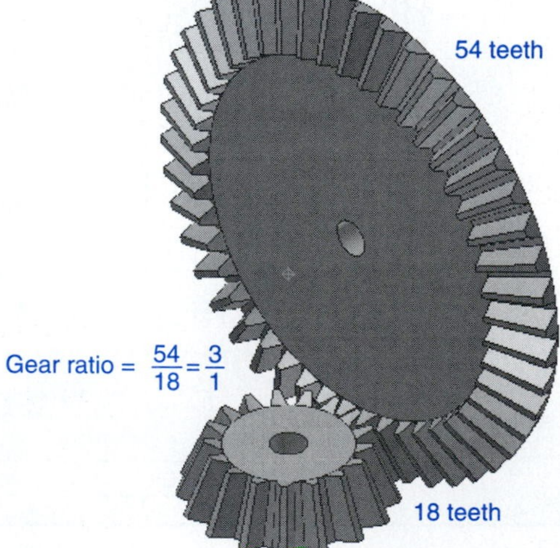

Figure 12-26

The gear ratio for a worm and worm gear combination is determined by the number of teeth on the worm gear. The cylinder-shaped gear is called a *worm,* and the round gear is called a *worm gear.* The worm gear is assumed to be 1. If a worm gear is meshed with a worm, and the worm gear has 42 teeth, the gear ratio will be 42:1. See Figure 12-27.

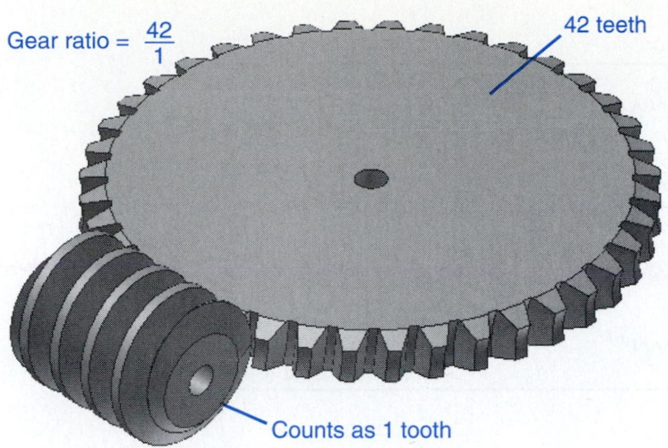

Figure 12-27

Gear Trains

When more than two gears are used in a design the combination is called a ***gear train.*** Figure 12-28 shows a gear train that contains four gears: two 20-tooth gears, one 40-tooth, and one 60-tooth. The speed ratio between the input revolutions per minute (RPM) and output RPM is determined by multiplying the individual gear ratios together. Observe that the 40-tooth gear and one of the 20-tooth gears are mounted on the same shaft. There is no speed ratio between these two gears, as they have the same angular velocity. The speed ratio is

gear train: The combination of more than two meshing gears.

$$\left(\frac{40}{20}\right)\left(\frac{60}{20}\right) = \frac{6}{1}$$

For an input speed of 1750 RPM, the output speed would be

$$1750\left(\frac{1}{6}\right) = 292 \text{ RPM}$$

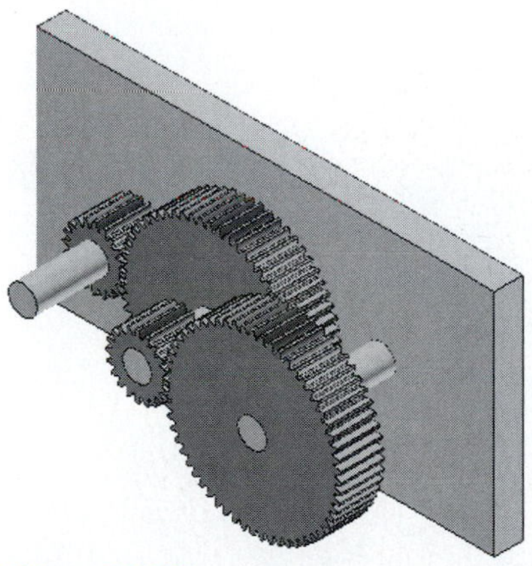

Figure 12-28

Figure 12-29 shows another gear train that includes six gears: three 20-tooth gears and three 60-tooth gears. The speed ratio between input and output speeds is

$$\left(\frac{60}{20}\right)\left(\frac{60}{20}\right)\left(\frac{60}{20}\right) = \frac{27}{1}$$

For an input speed of 1750 RPM, the output speed would be

$$\frac{1750}{27} = 64.8 \text{ RPM}$$

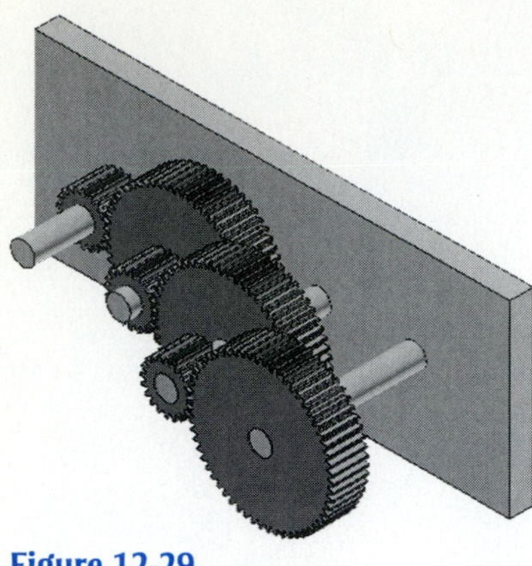

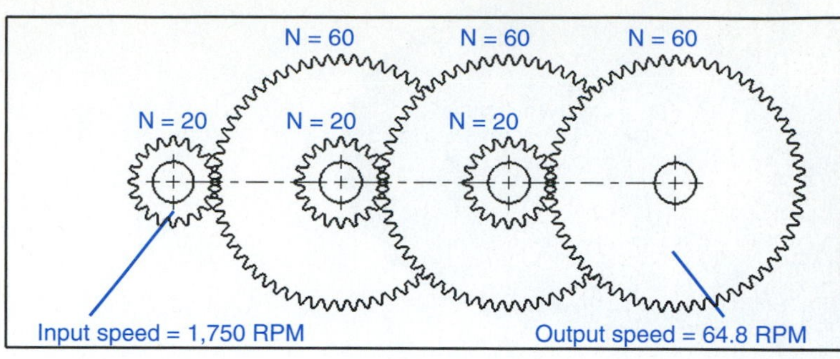

Figure 12-29

Gear Direction

idler gear: A gear added to a gear train for the sole purpose of changing the direction of the final rotation.

Meshing gears always rotate in opposite directions. If gear 1 in Figure 12-29 were to rotate clockwise (CW), then gear 2 would rotate counterclockwise (CCW). Gear 3 would also rotate CCW, driving gear 4 in a CW direction. Gear 5 would rotate in the CW direction and drive gear 6 in the CCW direction. A gear called an ***idler*** may be added to a gear train for the sole purpose of changing the direction of the final rotation. Idler gears are usually identical with one of the gears they are meshing with so as not to affect the final speed ratio. See Figure 12-30.

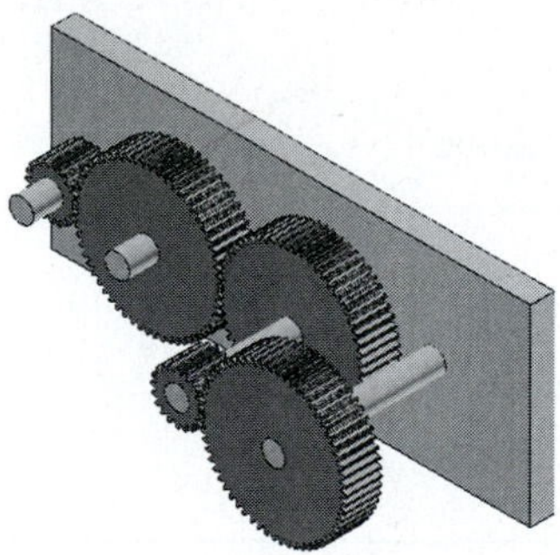

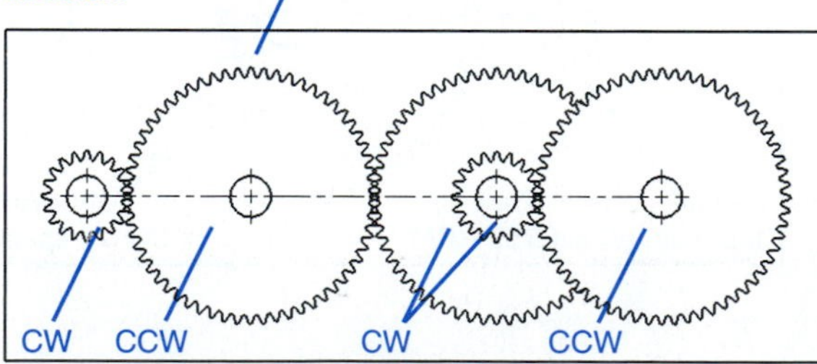

Figure 12-30

Gears with Keyways

Figure 12-31 shows a hubless spur gear. It is to be joined to a shaft using a square key. A keyway must be added to the gear. The gear has a bore of Ø.5000 in. and a face width of .500 in. Its pressure angle is 14.5°, it has 48 teeth, and a diametral pitch of 24. It was drawn using the **Design Accelerator** on a **Standard (in).iam** format drawing.

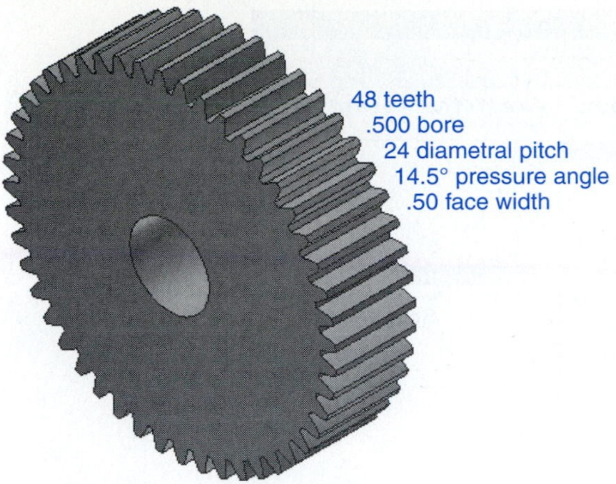

Figure 12-31

Exercise 12-2: Adding a Keyway to a Gear

First, determine the key that will be inserted into the gear. The information about the gear contained in the **Content Center** will specify the size of the keyway.

1. Click the **Place from Content Center** tool.

See Figure 12-32.

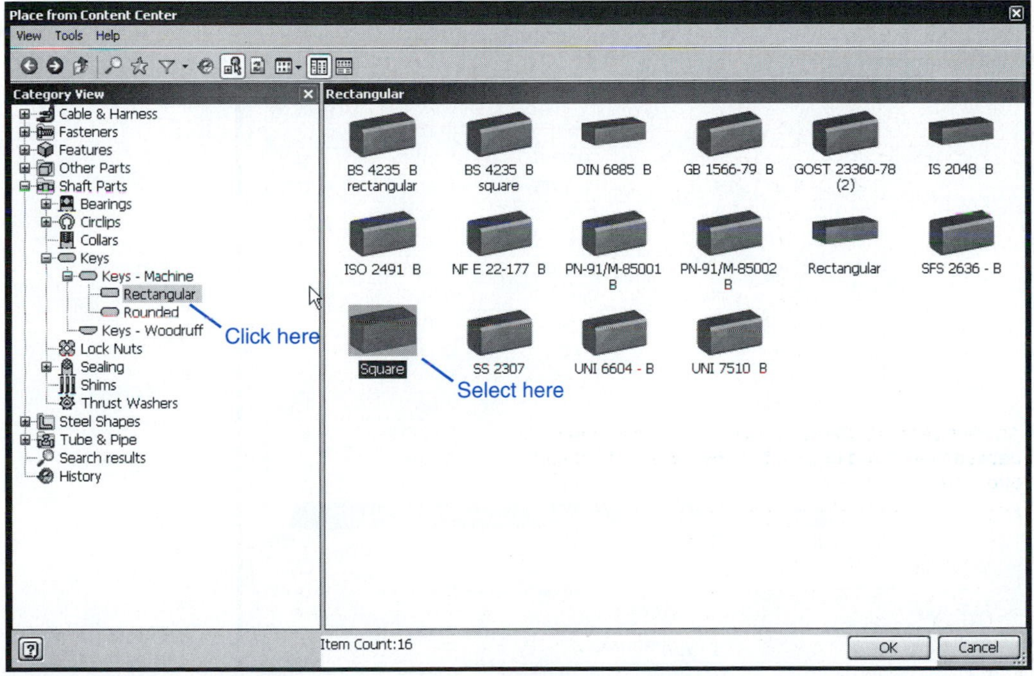

Figure 12-32

2. Select a **Square** key.

See Figure 12-33.

The bore of the gear is Ø.500 so the **0.4375 = 0.5625** shaft diameter range is selected. This yields a 1/8 × 1/8 width and height for the key. A nominal key length of **0.625** was selected.

3. Click the **Table View** tab and access **All Columns.**

See Figure 12-34. The table specifies a **Parallel Key Width** of **0.125** and a **Parallel Key Groove** of **0.0625.**

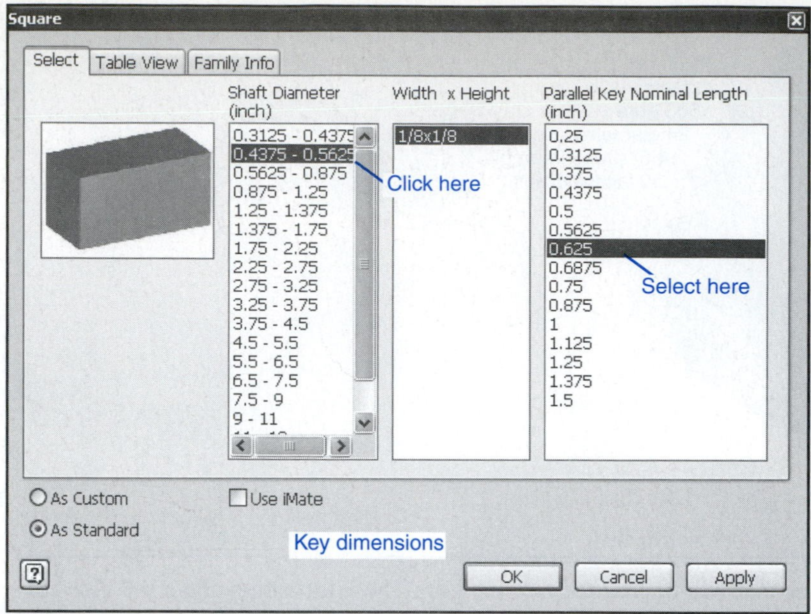

Figure 12-33

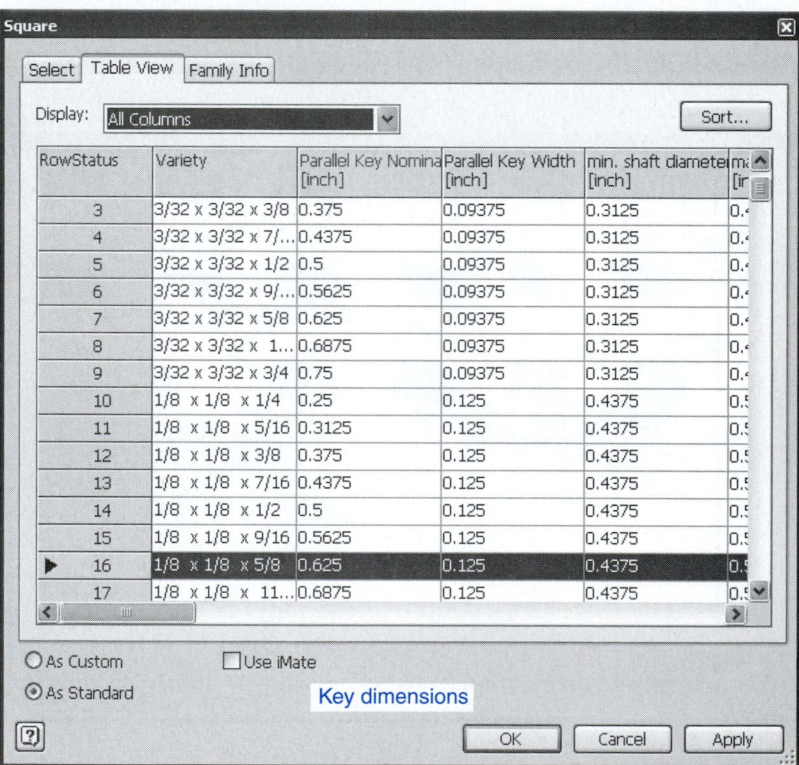

Figure 12-34

4. Create a new sketch plane on the front surface of the gear and draw a two-point rectangle as shown.

See Figure 12-35. The **.3130** value was derived from the bore's radius of .250 and the **Parallel Key Groove** requirement of **0.063.** The width of the keyway is .125 (1/8).

5. Right-click the mouse and select the **Finish Sketch** option.

See Figure 12-36.

Figure 12-35

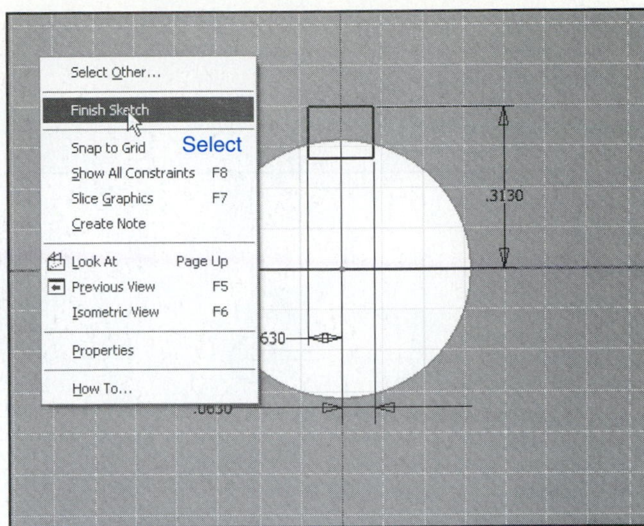

Figure 12-36

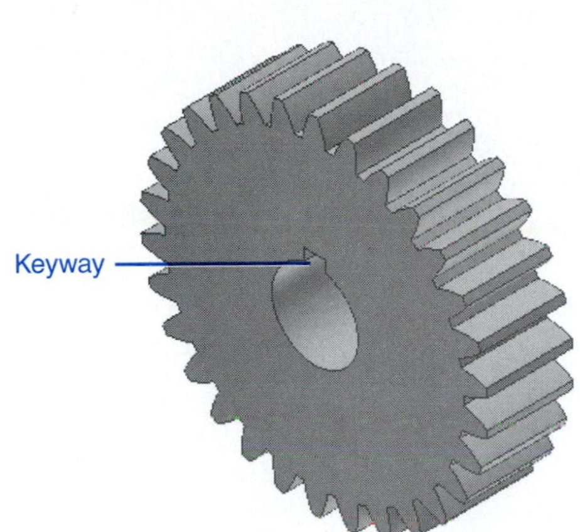

Figure 12-37

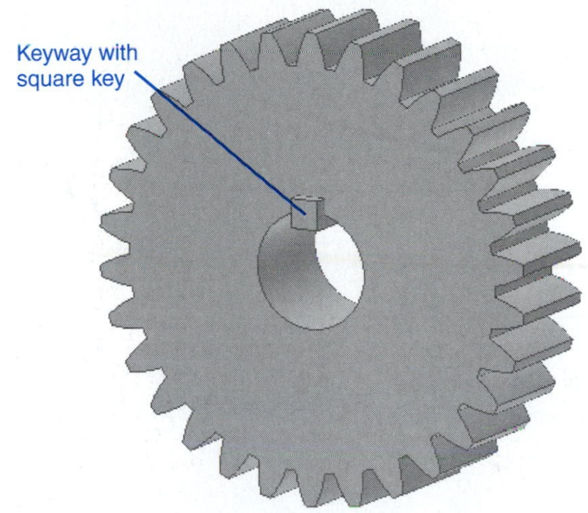

Figure 12-38

6. Use the **Extrude** tool and **Cut** out the rectangle, producing the keyway.

See Figure 12-37. Figure 12-38 shows the square key inserted into the gear.

Gear Assemblies

This section shows how to draw gear assemblies. Two meshing gears will be mounted onto a support plate using two shafts. The support plate and gear shafts were drawn using the dimensions shown in Figure 12-39.

Exercise 12-3: Drawing a Gear Assembly

1. Create a new drawing using the **Standard (mm).iam** format.
2. Access the **Design Accelerator** and click the **Spur Gear** option.
3. Set the values for the dimensions of the gear in the **Spur Gears Component Generator** dialog box as shown in Figure 12-40.
4. Finish the gear drawing.

Figure 12-41 shows the finished gears.

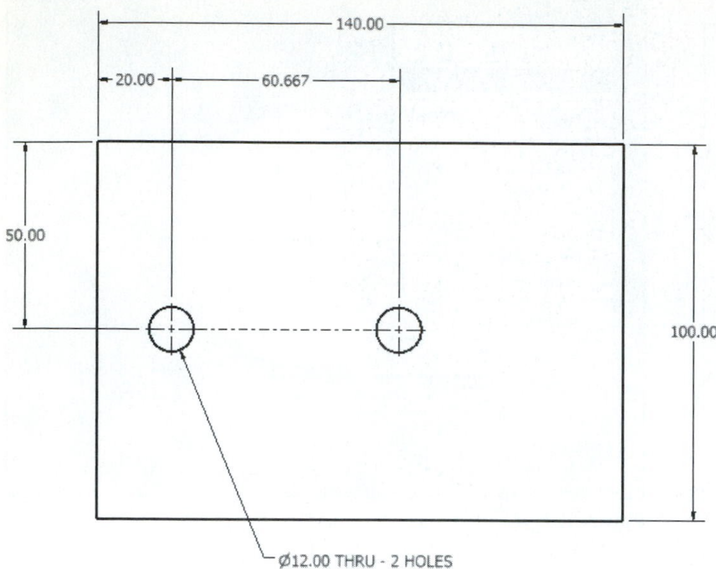

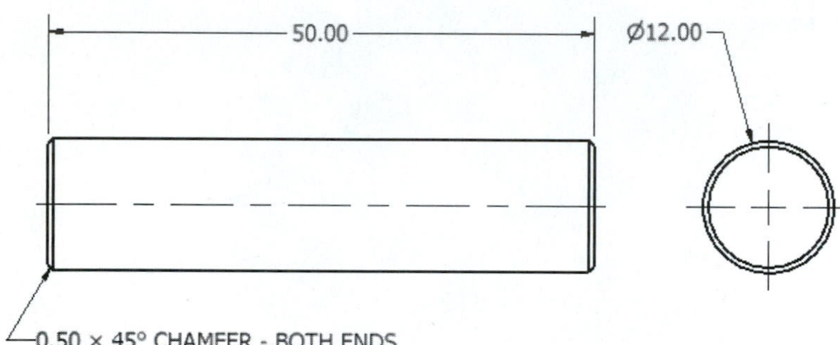

Figure 12-39

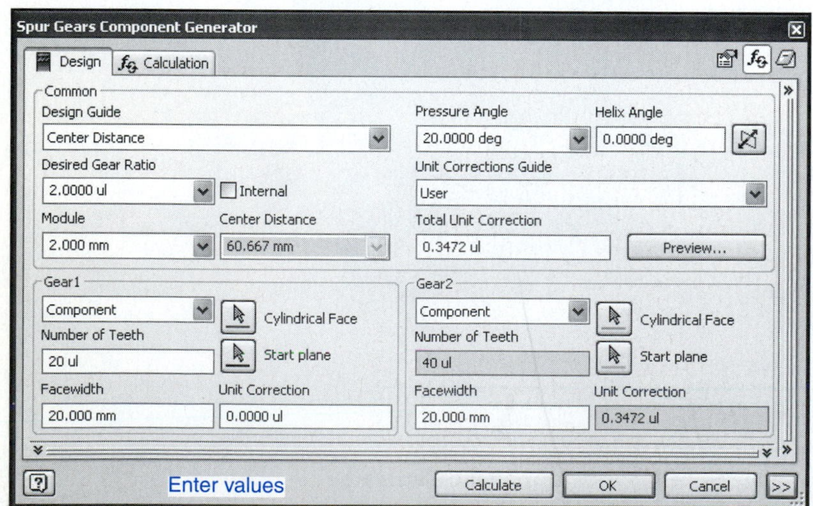

Figure 12-40

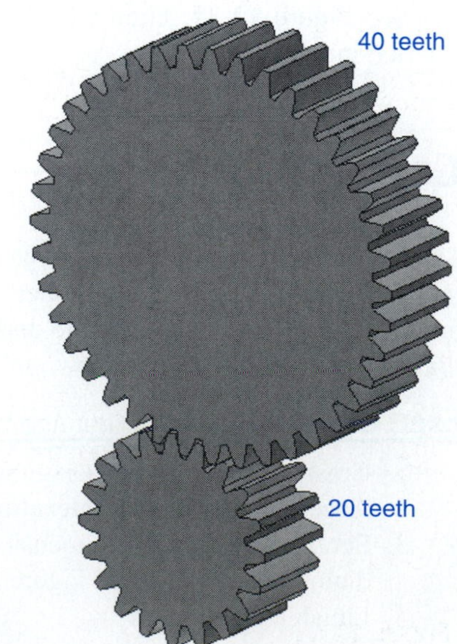

Figure 12-41

5. Right-click the **Spur Gear1:1** in the browser box and select the **Isolate** option.
6. Right-click the 20-tooth gear and select the **Edit** option.
7. Create a new sketch plane and draw a **Ø20.00** circle. Right-click the mouse and select the **Finish Sketch** option.

See Figure 12-42.

Note:

See page 553 for a more detailed explanation of how to draw gear hubs.

Do not select the **Finish Edit** option. Remember that you are working only on the 20-tooth gear, so you want to stay in that sketch mode.

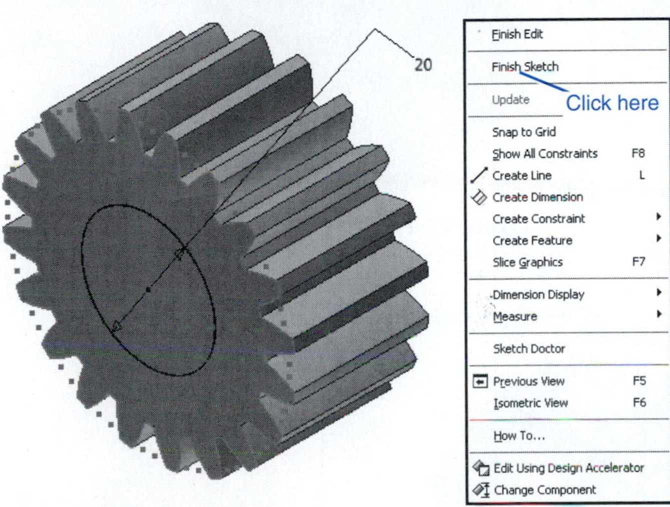

Figure 12-42

8. Extrude the **Ø20** circle through a distance of **12.**

See Figure 12-43.

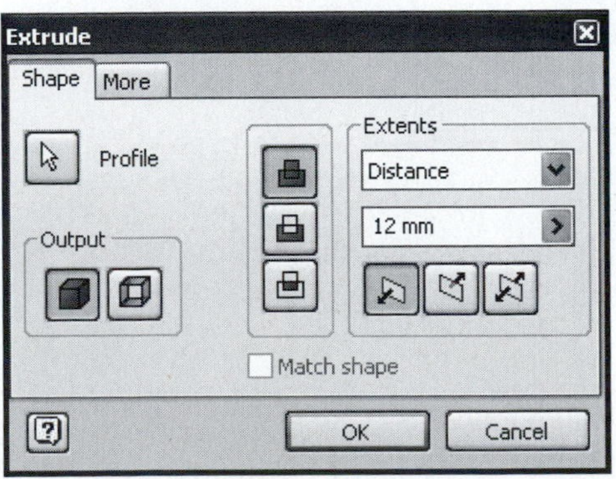

Figure 12-43

9. Create a new sketch plane on the new top surface.

   See Figure 12-44.

10. Draw a **Ø12.00** hole through the gear.

    See Figure 12-45.

11. Create a new work plane tangent to the gear's hub. Locate a center point for a hole.

    See Figure 12-46.

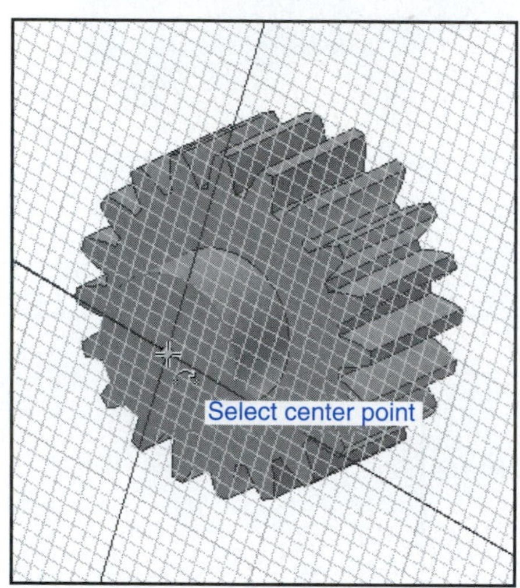

Figure 12-44

Figure 12-45

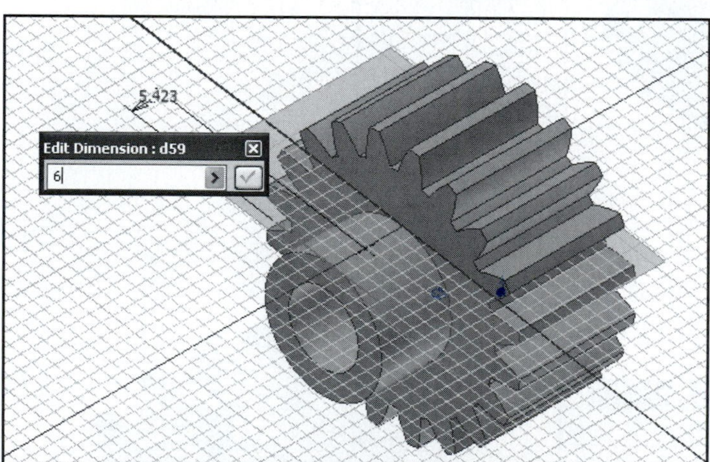

Figure 12-46

12. Draw a threaded hole. Use an **M4 × 0.7** metric thread.

    See Figure 12-47.

13. Right-click the mouse and select the **Finish Edit** option.

    The 40-tooth gear will reappear.

    See Figure 12-48.

14. Create a hub, bore, and **M4 × 0.7** threaded hole on the 40-tooth gear.

    See Figure 12-49.

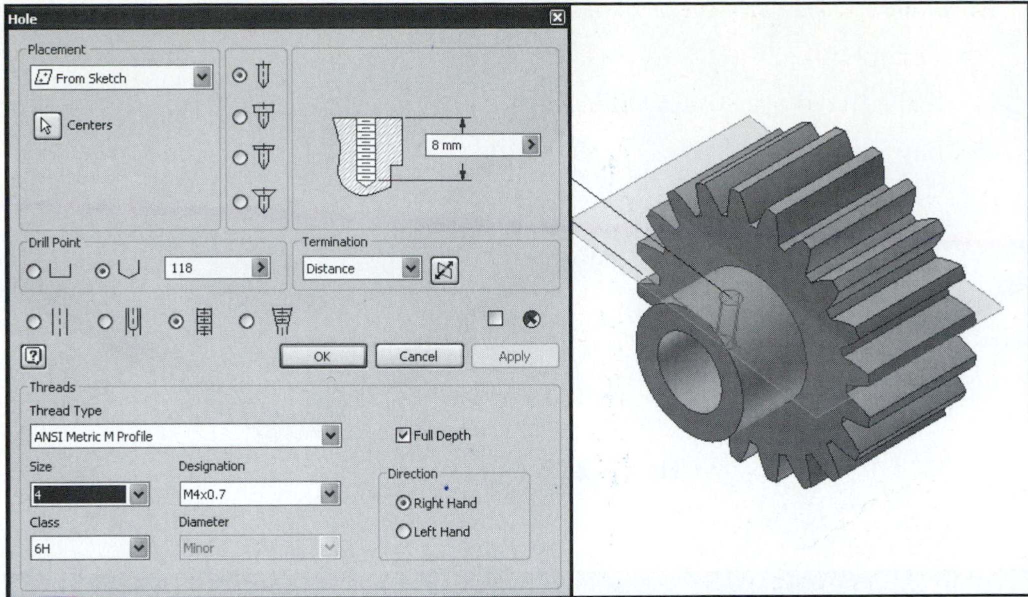

Figure 12-47

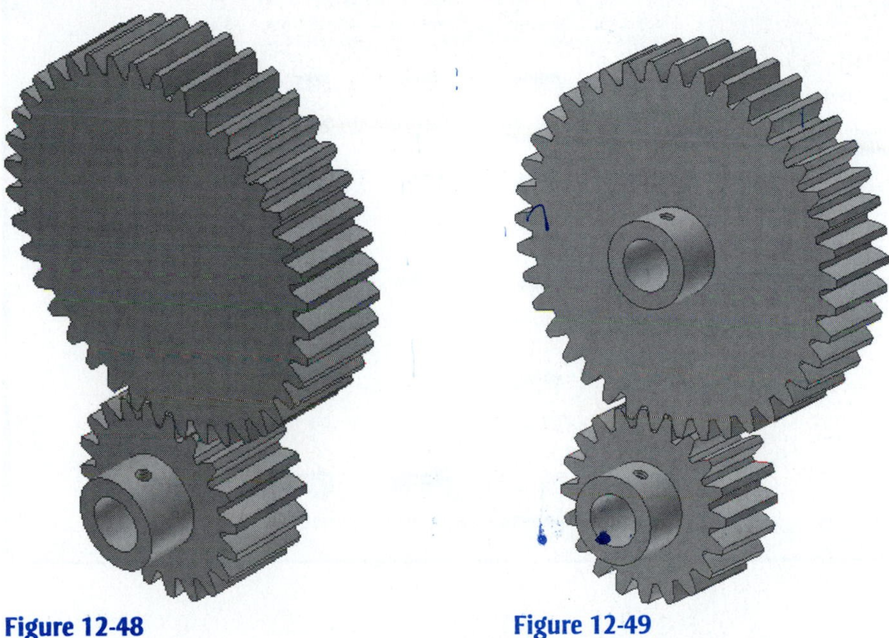

Figure 12-48

Figure 12-49

15. Save the two gears as **Gear Subassembly.**
16. Start a new drawing using the **Standard (mm).iam** format.

You could have continued to work in the gear drawing and either inserted the existing components or created new components using the **Create Component** tool.

17. Use the **Place Component** command and insert a gear support plate, two shafts, and the gear subassembly.

See Figure 12-50.

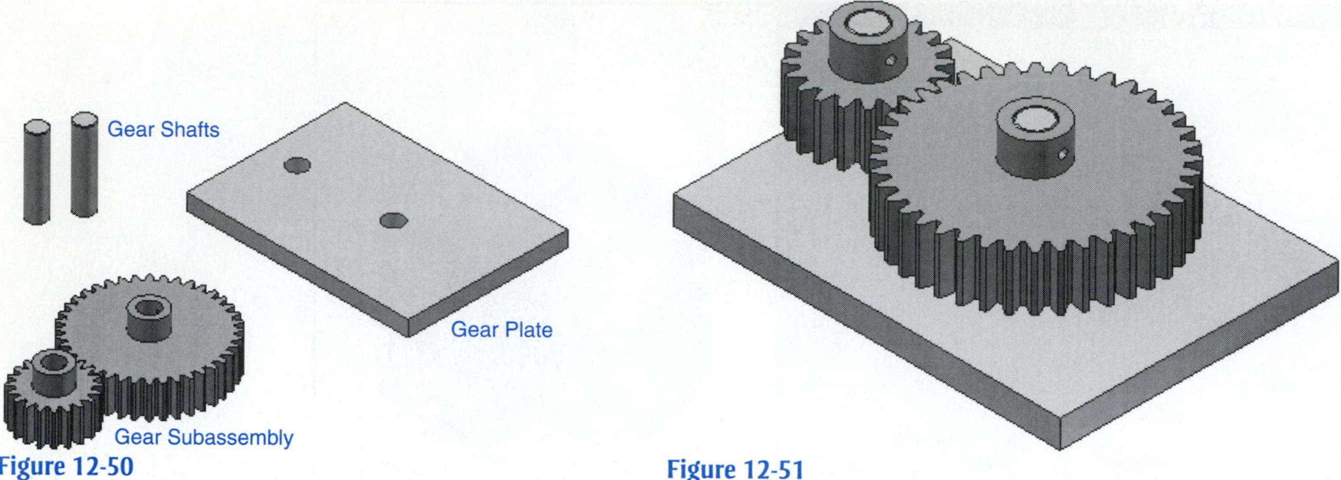

Figure 12-50

Figure 12-51

Figure 12-52

18. Use the **Constraint** tool and assemble the gears, shafts, and support plate.

    See Figure 12-51.

19. Right-click the mouse and access the **Content Center.** Select a **CNS EN 24766 Set Screw** with a point that has an **M4** thread and is **4** long.

    See Figures 12-52 and 12-53.

20. Click **OK.**

    Add two setscrews to the assembly drawing. See Figure 12-54.

21. Insert the setscrews into the gear hubs.

    Figure 12-55 shows the finished assembly.

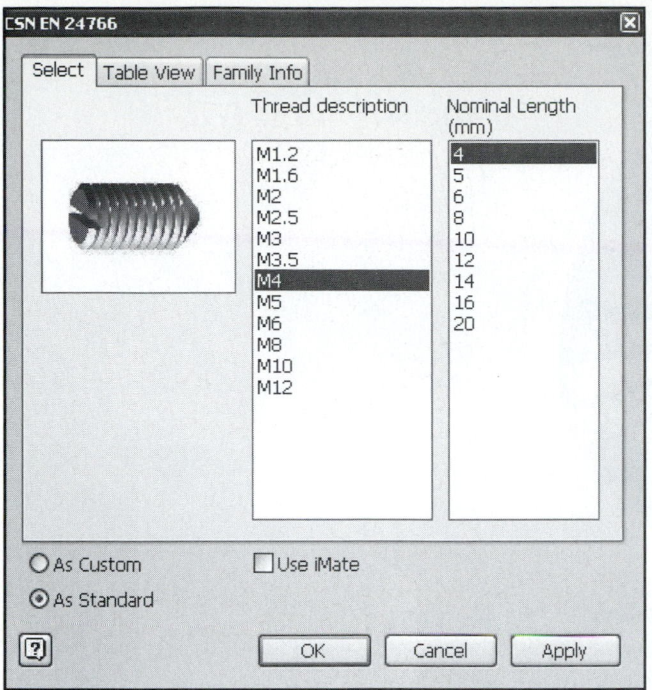

Figure 12-53

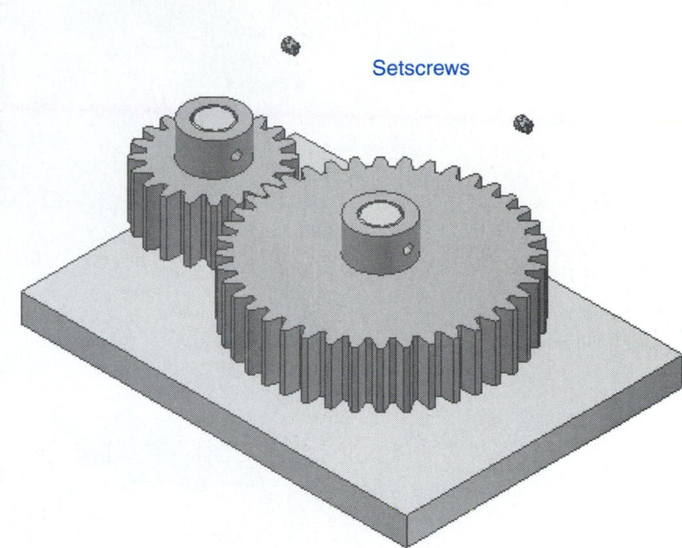

Figure 12-54

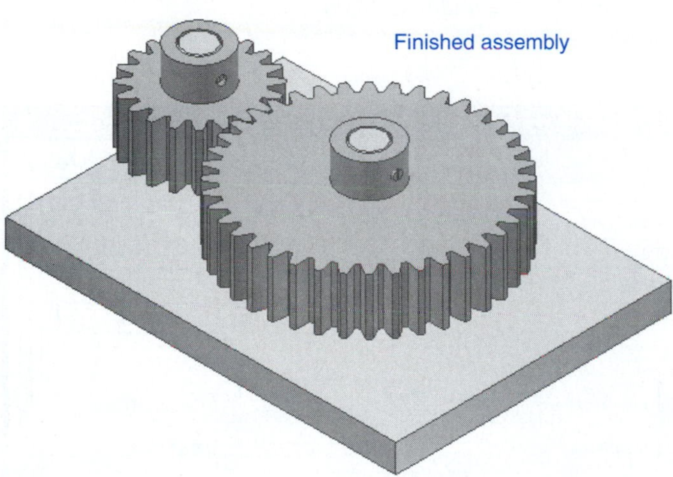

Figure 12-55

Use work planes and work axes if necessary to insert the setscrews into the hubs. Remember to assemble in the assembly drawing and not on individual components.

Bevel Gears

Bevel gears are conical-shaped gears that have intersecting axes. Spur gears have parallel axes. Figure 12-56 shows a set of meshing bevel gears.

bevel gears: Conical-shaped gears that have intersecting axes.

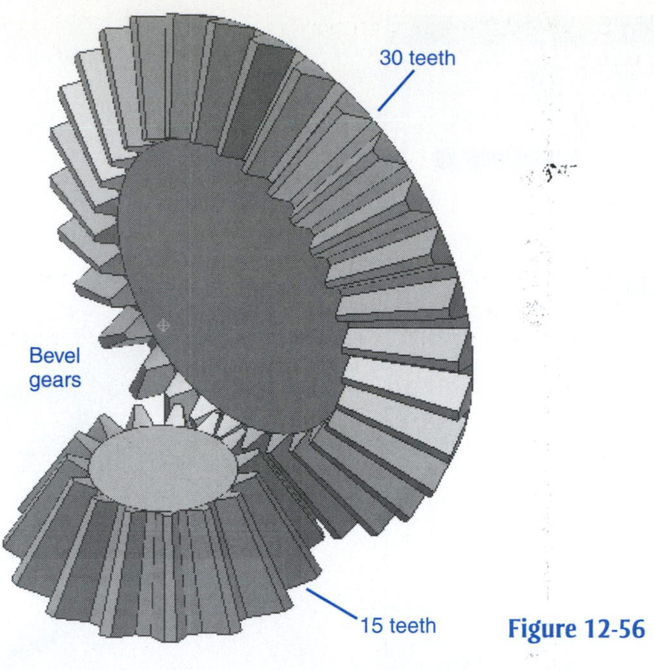

Figure 12-56

Exercise 12-4: Drawing Bevel Gears

Draw a set of bevel gears with 15 and 30 teeth, respectively. Define the module as 3.000, the shaft angle as 90°, the pressure angle as 20, and the face width as 20.00.

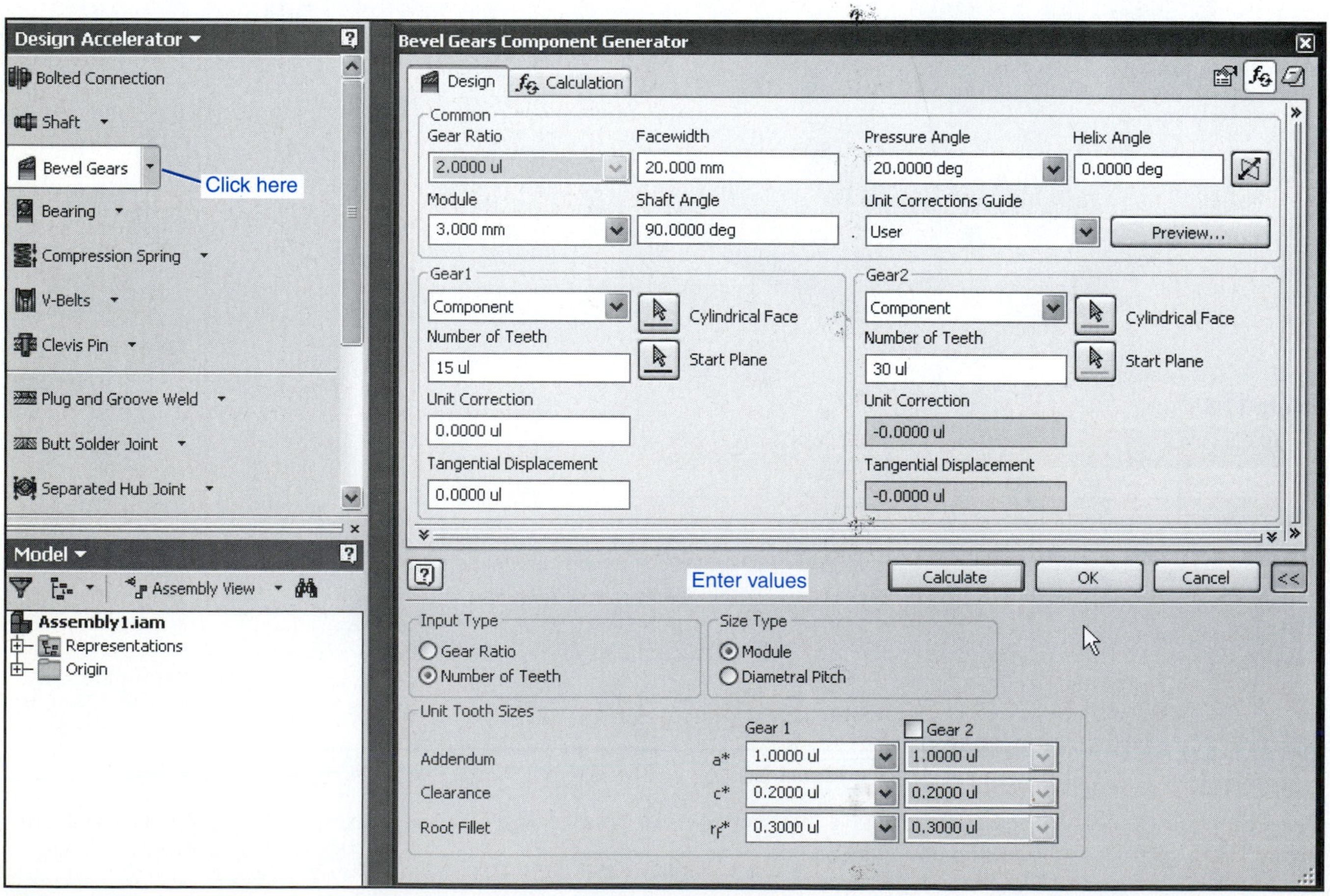

Design Accelerator
Bolted Connection
Shaft
Bevel Gears
Click here
Bearing
Compression Spring
V-Belts
Clevis Pin
Plug and Groove Weld
Butt Solder Joint
Separated Hub Joint
Model
Assembly View
Assembly1.iam
Representations
Origin

Bevel Gears Component Generator
Design | Calculation
Common
Gear Ratio 2.0000 ul | Facewidth 20.000 mm | Pressure Angle 20.0000 deg | Helix Angle 0.0000 deg
Module 3.000 mm | Shaft Angle 90.0000 deg | Unit Corrections Guide User | Preview...
Gear1: Component | Cylindrical Face | Number of Teeth 15 ul | Start Plane | Unit Correction 0.0000 ul | Tangential Displacement 0.0000 ul
Gear2: Component | Cylindrical Face | Number of Teeth 30 ul | Start Plane | Unit Correction -0.0000 ul | Tangential Displacement -0.0000 ul
Enter values
Calculate | OK | Cancel | <<
Input Type: Gear Ratio / Number of Teeth
Size Type: Module / Diametral Pitch
Unit Tooth Sizes: Gear 1 | Gear 2
Addendum a* 1.0000 ul | 1.0000 ul
Clearance c* 0.2000 ul | 0.2000 ul
Root Fillet r_f* 0.3000 ul | 0.3000 ul

Figure 12-57

1. Start a new drawing using the **Standard (mm).iam** format.
2. Access the **Design Accelerator.**

See Figure 12-57.

3. Click **OK.**

The gears will appear on the screen.

Exercise 12-5: Adding Hubs to Bevel Gears

1. Rotate the gears so that the bottom surface of the small gear is in view.
2. Right-click on the **Bevel Gear1:1** heading in the browser box and select the **Edit** option.

See Figure 12-58.

3. Create a new sketch plane on the bottom surface of the smaller gear.

See Figure 12-59.

8left

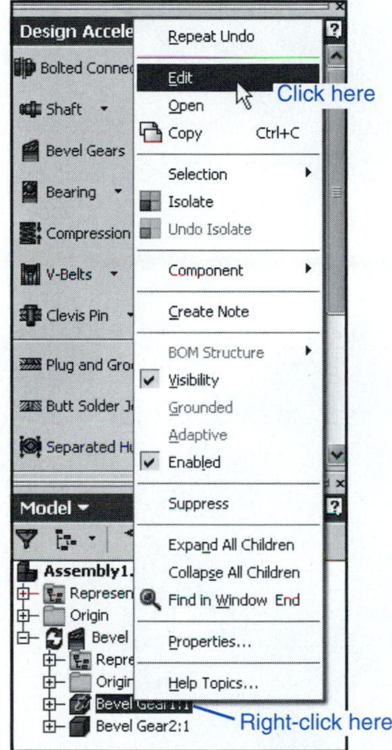

Figure 12-58

Figure 12-59

4. Draw a **Ø24** circle on the new sketch plane.
5. Right-click the mouse and select the **Finish Sketch** option.

See Figure 12-60. Do not click the **Finish Edit** option. You are still working on the smaller gear, so continue working on the sketch.

6. Extrude the circle **12 mm.**
7. Right-click the mouse and create a new sketch plane on the top surface of the hub.
8. Create a **Ø12.00** hole through all.
9. Right-click the mouse and select the **Finish Edit** option. Click the **Isometric** command.

See Figure 12-61.

10. Repeat the procedure for the larger gear. Make the hub **15 mm** high with a **Ø15** hole.

See Figure 12-62. Figure 12-63 shows the finished bevel gears.

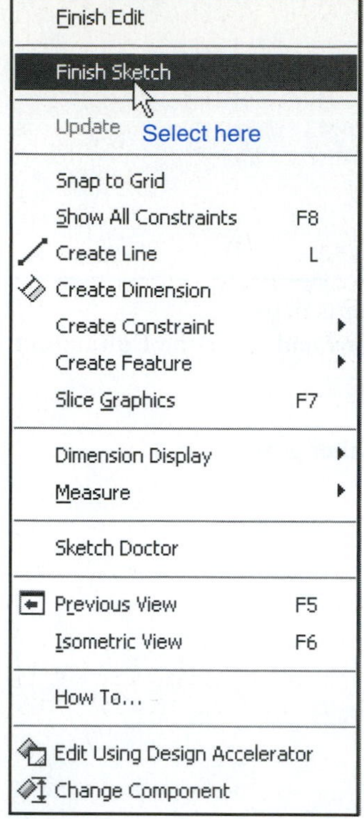

Figure 12-60

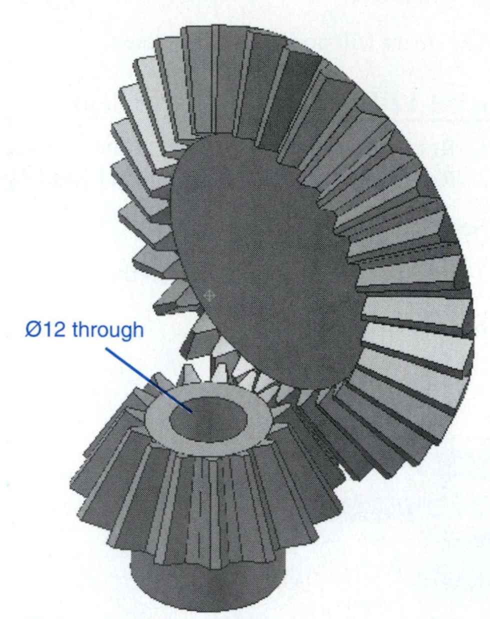

Figure 12-61

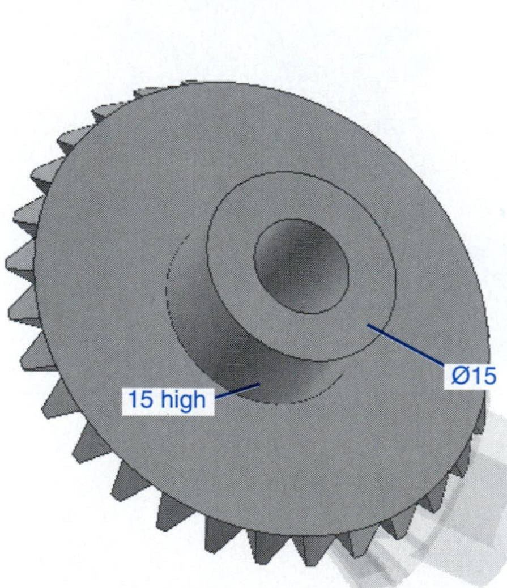

Figure 12-62

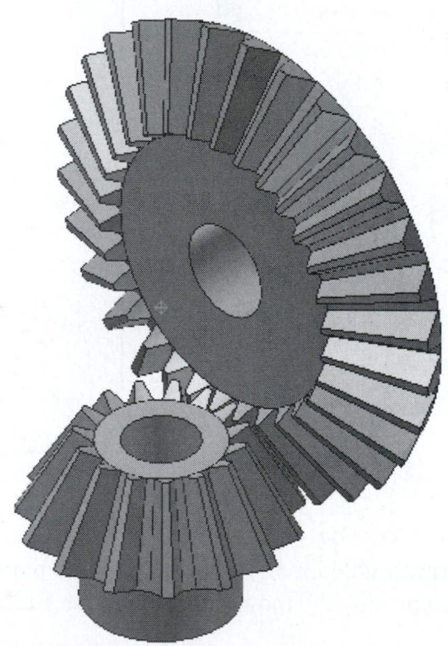

Figure 12-63

Exercise 12-6: Adding Keyways to the Bevel Gears

Before we can draw keyways, we must determine the size of the key. In the previous section the gears were given bore diameters of Ø12 and Ø15, so keys must be selected to match those diameters.

1. Return to the **Assembly Panel** and click the **Place from Content Center** tool.
2. Select the **Shaft Parts** option, **Keys** and **Keys - Machine.**

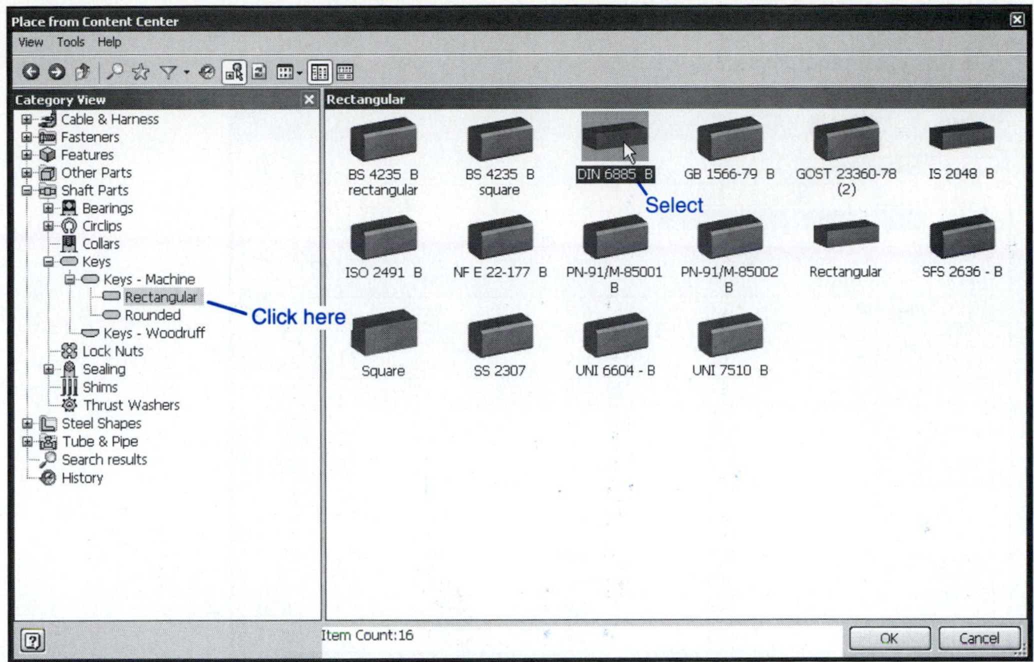

Figure 12-64

3. Select a **DIN 6885 B** square key.

See Figure 12-64.

4. Select the **10 - 12** shaft diameter input.

This will generate a key size of **4×4.**

5. Select a **20** nominal length.

See Figure 12-65.

6. Click the **Table Value** tab.

See Figure 12-66.

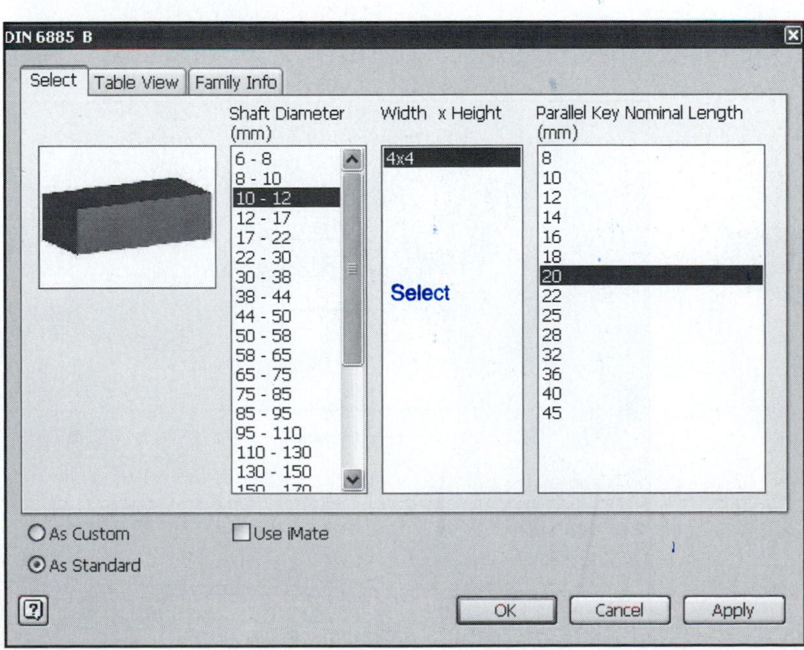

Figure 12-65

7. Scroll the table to determine the recommended **Hub Depth.**

In this example the value is **1.8.**

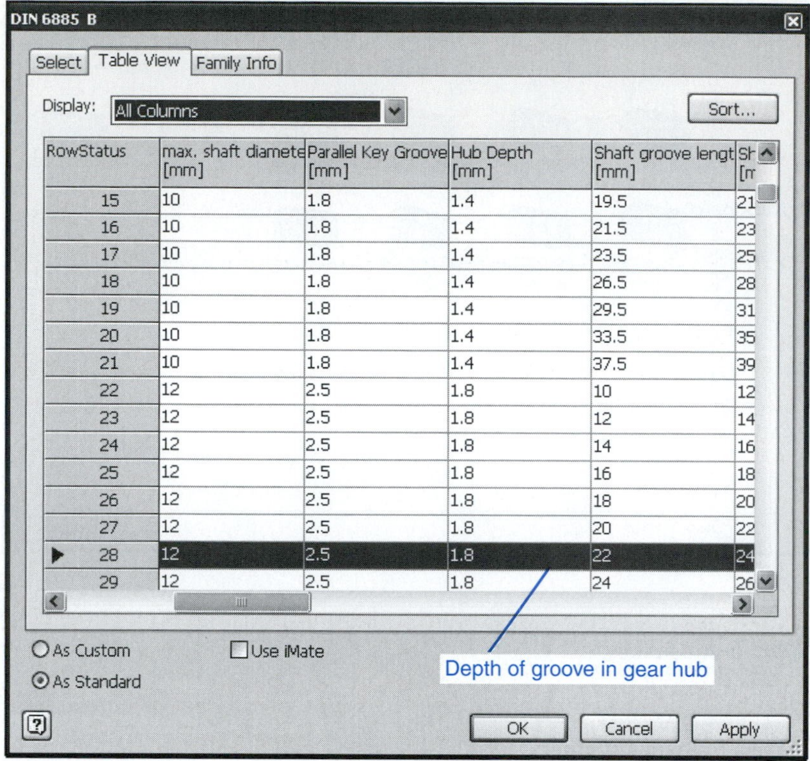

RowStatus	max. shaft diamete [mm]	Parallel Key Groove [mm]	Hub Depth [mm]	Shaft groove lengt [mm]	Sh [m
15	10	1.8	1.4	19.5	21
16	10	1.8	1.4	21.5	23
17	10	1.8	1.4	23.5	25
18	10	1.8	1.4	26.5	28
19	10	1.8	1.4	29.5	31
20	10	1.8	1.4	33.5	35
21	10	1.8	1.4	37.5	39
22	12	2.5	1.8	10	12
23	12	2.5	1.8	12	14
24	12	2.5	1.8	14	16
25	12	2.5	1.8	16	18
26	12	2.5	1.8	18	20
27	12	2.5	1.8	20	22
▶ 28	12	2.5	1.8	22	24
29	12	2.5	1.8	24	26

Figure 12-66

8. Close the **Content Center** and return to the bevel gear drawing.
9. Select the smaller gear, right-click the mouse, and select the **Edit** option. Create a new sketch plane on the top surface of the gear.

See Figure 12-67. In this example the **Look At** command was used to generate a 2D view of the hub's top surface.

10. Draw a rectangle as shown.

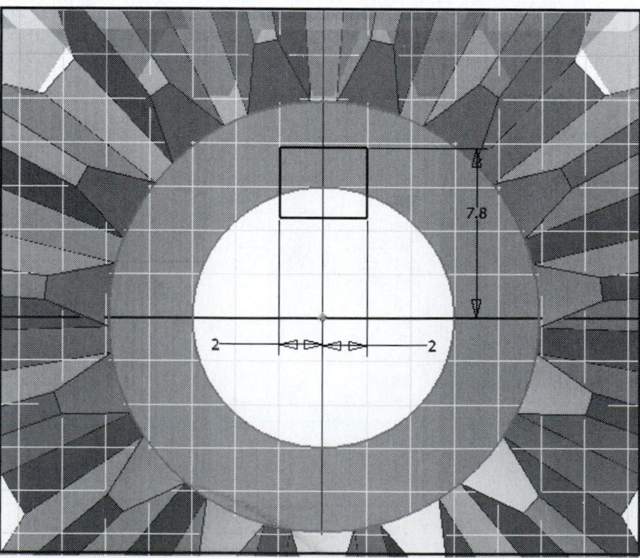

Figure 12-67

The width value of 4 matches the key width of 4 (no tolerances were factored in). The 7.8 value was derived from adding the bore's radius (6.0) to the **Hub Depth** value given in the table (1.8). The keyway depth in the shaft is 2.5, for a total key size of 4.3, or slightly larger than the key height of 4. These values will vary when tolerances are considered.

11. Right-click the mouse, click the **Extrude** tool, and **Cut** the rectangle through the gear, producing a keyway.
12. Repeat the procedure for the larger gear.

Figure 12-68 shows the finished keyways in the bevel gears.

Supports for Bevel Gears

Figure 12-69 shows a dimensioned drawing of the bevel gears created in the last section. The drawing was created using the **ANSI (mm).idw** format. The two distances, 50.66 and 36.80, are needed to position the holes in the support structure.

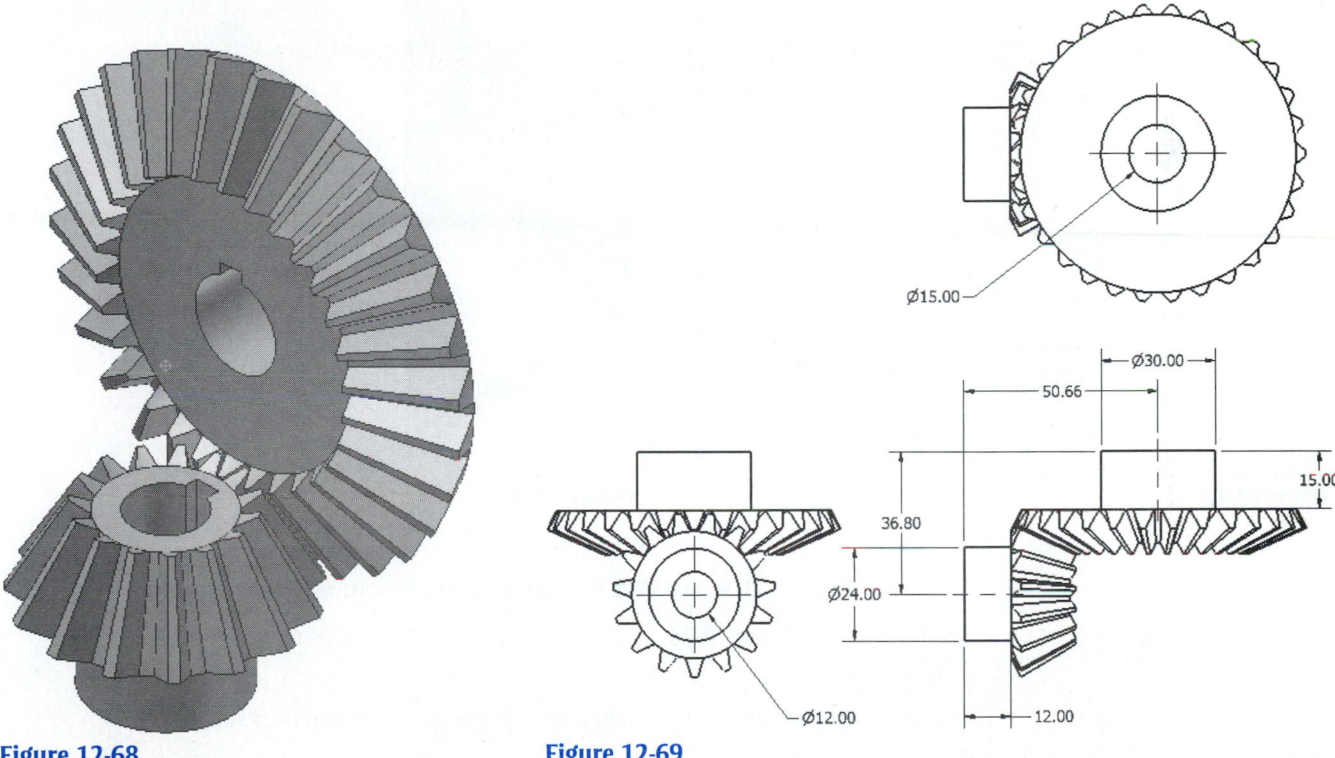

Figure 12-68

Figure 12-69

Exercise 12-7: Designing a Support Structure

1. Draw an L-shaped bracket.

In this example the horizontal portion of the bracket is 120 × 120 × 10, and the vertical portion is 100 × 120 × 10.

2. Draw a **Ø40** circle located as shown.

See Figure 12-70. The 55.66 value is derived from the 50.66 center distance and an additional 5 for a boss.

3. Draw a boss **5** high with a **Ø15** hole.

See Figure 12-71.

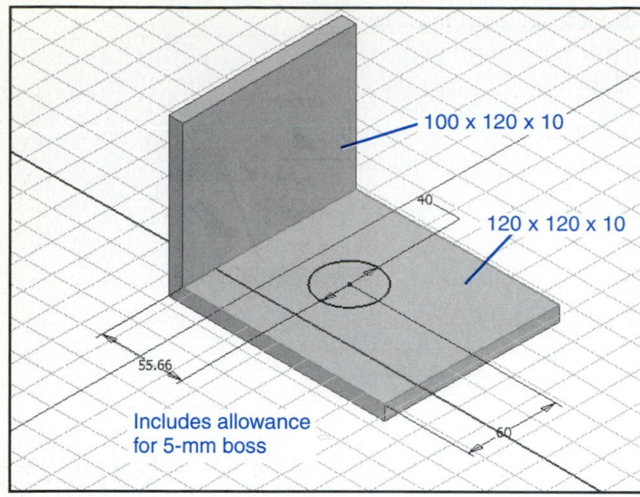

Figure 12-70

Figure 12-71

Figure 12-72

Figure 12-73

4. Draw a boss **5** high with a **Ø12** hole on the vertical portion of the bracket.

See Figures 12-72 and 12-73.

5. Add any fillets required.
6. Use the **Constraint** tool and add the bevel gears to the support bracket.

See Figure 12-74.

Figure 12-74

Worm Gears

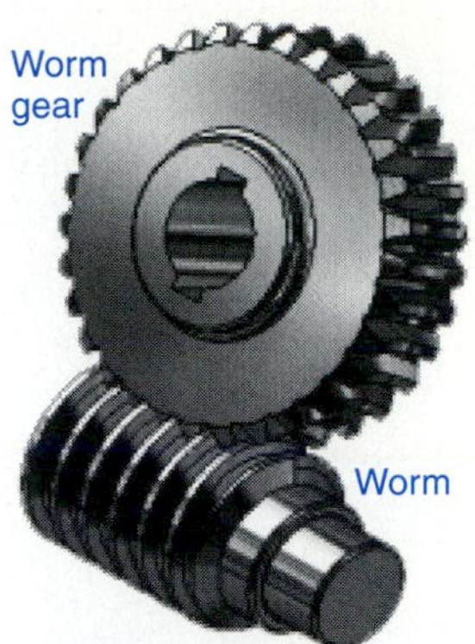

Figure 12-75

Figure 12-75 shows a worm and a worm gear. Worm gear ratios are based on the fact that the worm has a value of 1. If the worm gear has 40 teeth, the gear ratio is 40 to 1.

Exercise 12-8: Drawing a Set of Worm Gears

1. Create a new drawing using the **Standard (mm).iam** format.
2. Access the **Design Accelerator,** then click the **Worm Gears** tool.

See Figure 12-76.

3. Enter the appropriate values and click **OK.**

Figure 12-77 shows the finished worm gears.

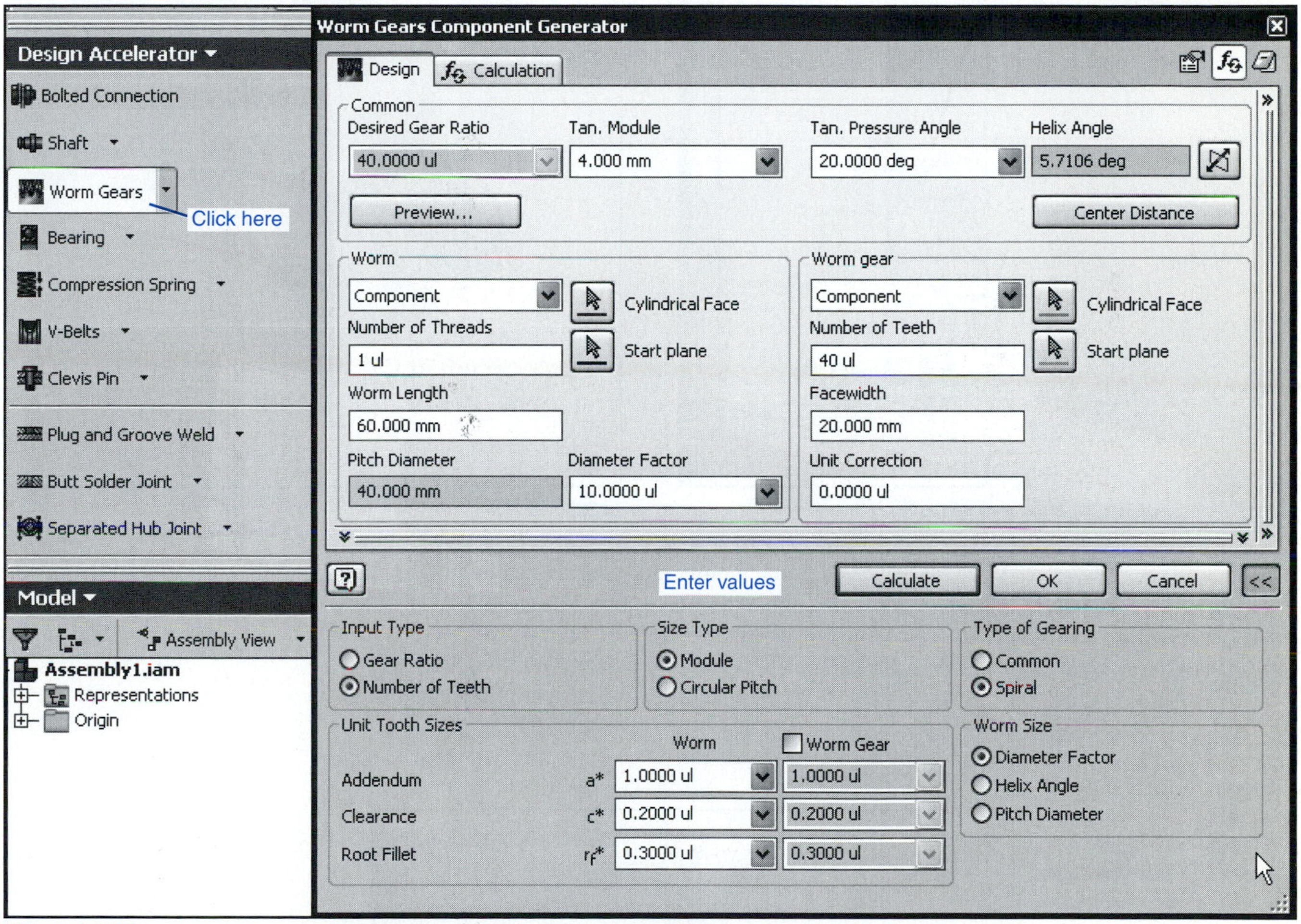

Figure 12-76

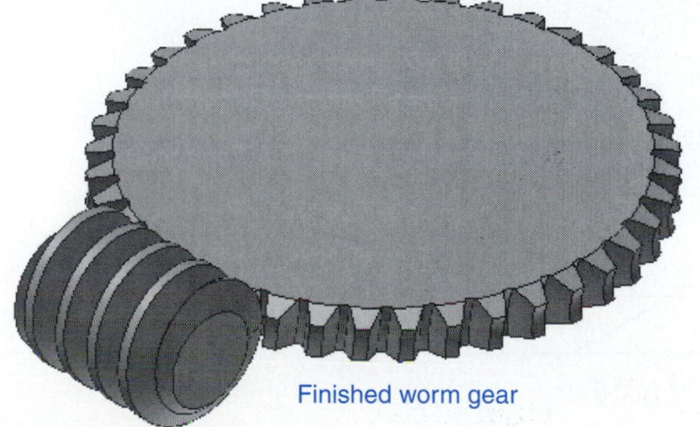

Figure 12-77

4. Right-click the **Worm1:1** heading in the browser box.

See Figure 12-78.

5. Right-click the worm and create a new sketch plane on the end of the worm and draw a **Ø20** circle.

See Figure 12-79.

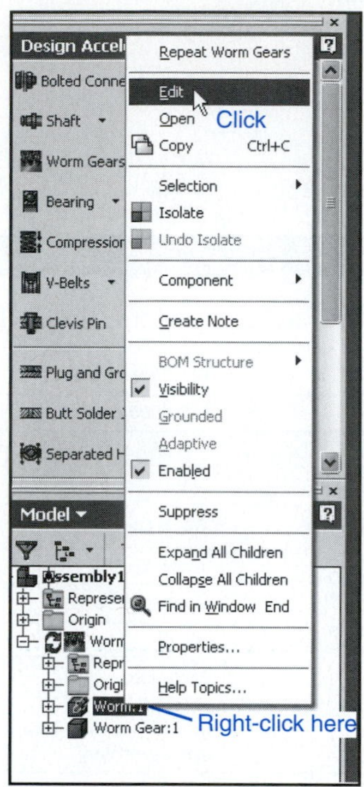

Figure 12-78

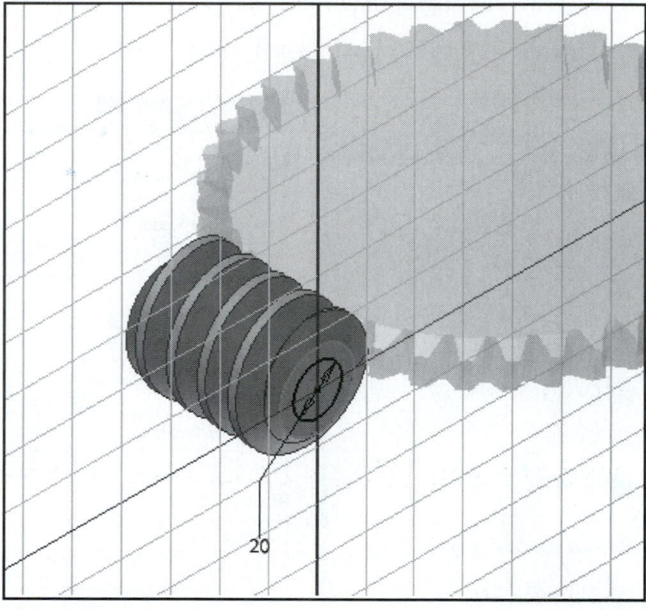

Figure 12-79

6. Right-click the mouse and select the **Done** option, then right-click the mouse again and select the **Finish Sketch** option.
7. Extrude the Ø20 circle a distance of **24** to create a hub. Create a new sketch plane on the top surface of the hub and create a **Ø12.0** hole.
8. Create a work plane tangent to the extruded hub.

See Figure 12-80.

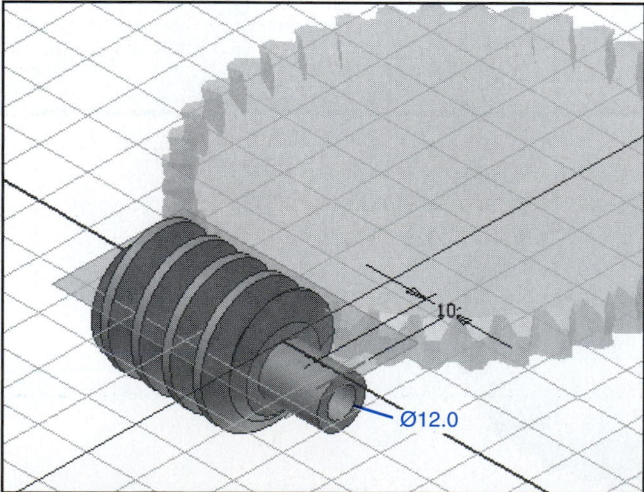

Figure 12-80

9. Add a threaded hole to the hub **10** from the top edge of the hub.

In this example an **M4** thread was added.

10. Right-click the worm and select the **Done** option, then right-click the mouse again and select the **Finish Edit** option.
11. Add a **Ø40** hub with a height of **20** and a **Ø20.0** hole to the worm gear using the same procedure as was used for the worm. Create an **M4** hole **10** from the top edge of the hub.

See Figure 12-81.

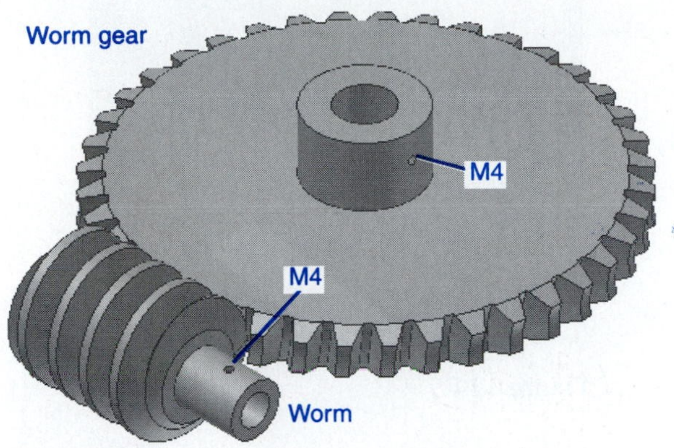

Figure 12-81

Supports for Worm Gears

Figure 12-82 shows orthographic views of the worm gears drawn in the previous section. The dimension defines the distance between the gear's centers as 100.00. This value is used to create the support structure for the gears.

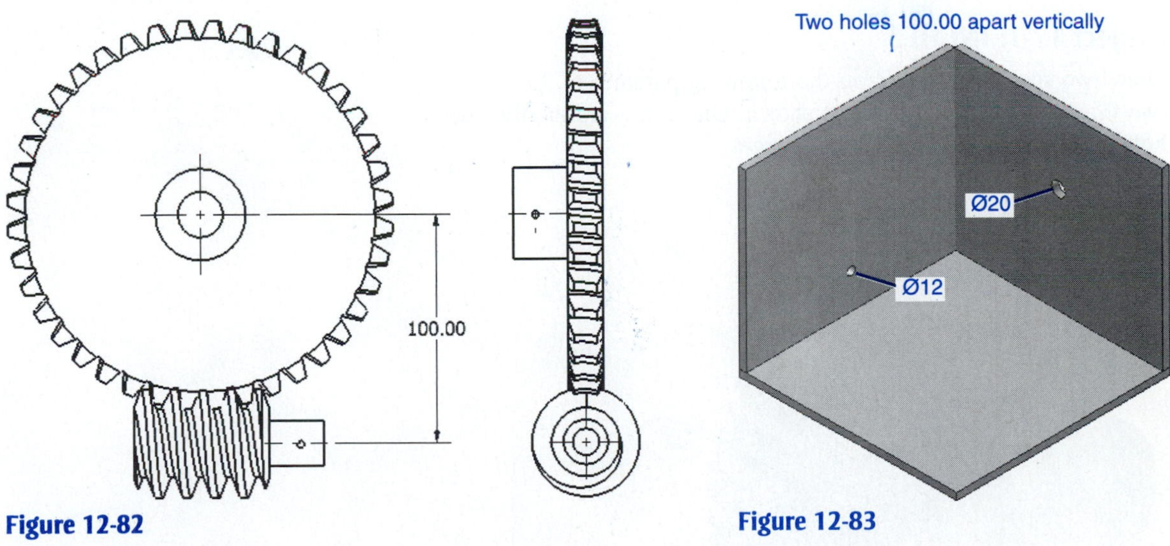

Figure 12-82

Figure 12-83

Exercise 12-9: Drawing Worm Gear Supports.

1. Draw a corner bracket with two holes **100.00** apart.

Figure 12-83 shows two plates with two holes 100.00 apart vertically. The sides are **10** thick.

2. Add the appropriate shafts.

The shafts shown should be supported at both ends and not cantilevered as shown in Figure 12-84. The front two plates were omitted for clarity.

3. Use the **Constraint** command and insert the worm gears.

Figure 12-85 shows the finished worm gear support.

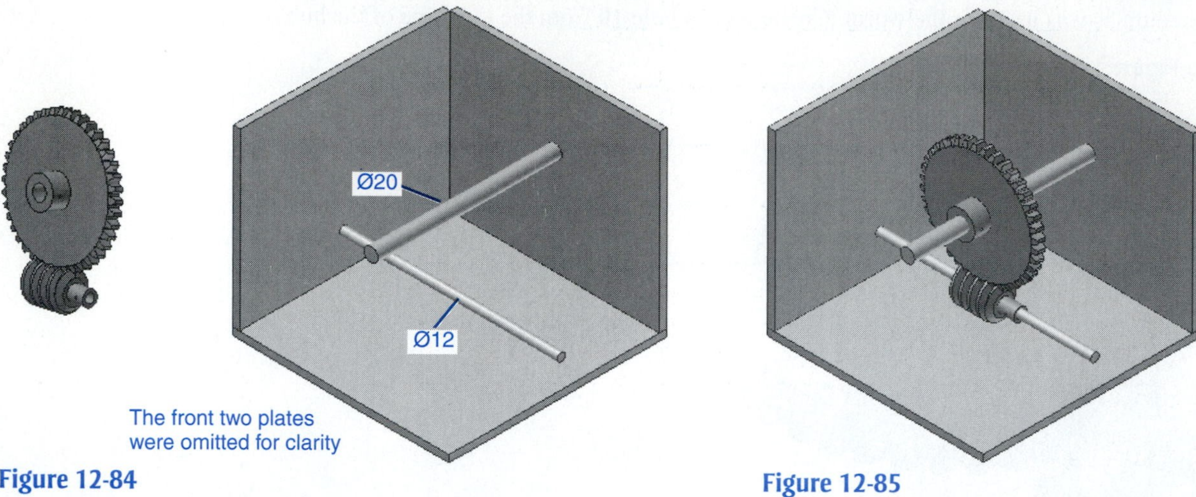

Figure 12-84

Figure 12-85

SUMMARY

This chapter explained the differences among three types of gears: spur, bevel, and worm. Gear terminology, formulas, and ratios were explained.

The addition of hubs, splines, and keyways to gears was illustrated, and gears were assembled into gear trains.

CHAPTER PROJECTS

Project 12-1: INCHES

Draw two spur gears based on the following parameters. Locate the two threaded holes 90° apart as shown. On both gears the threaded hole is located halfway up the hub height.

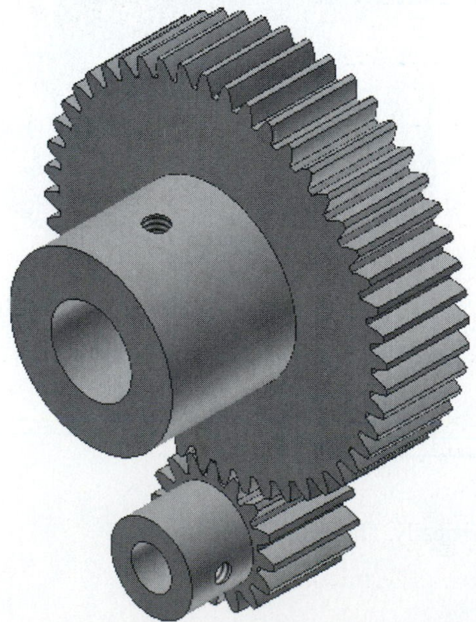

Figure P12-1

	Gear 1	Gear 2
Number of teeth	16	48
Face width	.50	.50
Diametral pitch	24	24
Pressure angle	20	20
Hub Ø	.50	.50
Hub height	.375	.750
Bore	.250	.500
Threaded hole	0.138(#6)UNC	0.164(#8)UNC

Project 12-2: INCHES

A. Draw two spur gears based on the following parameters. On both gears the threaded hole is located halfway up the hub height.

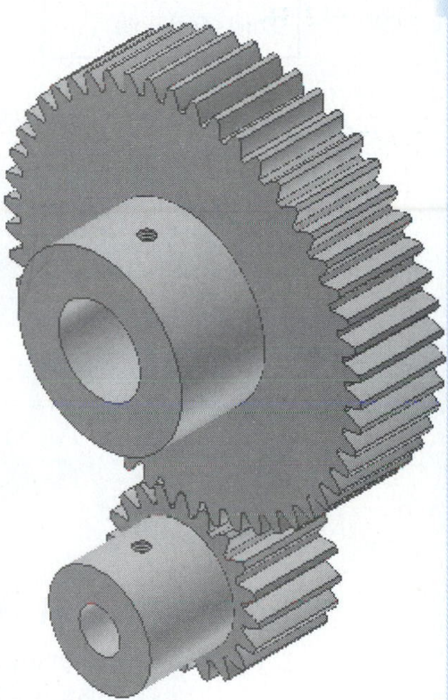

Figure P12-2

	Gear 1	Gear 2
Number of teeth	20	50
Face width	1.00	1.00
Diametral pitch	12	12
Pressure angle	20	20
Hub Ø	1.00	1.00
Hub height	.375	.750
Bore	.500	1.000
Threaded hole	0.216(#12)UNC	0.216(#12)UNC

B. Design Exercise

Create a support plate and shafts for the two gears drawn in part A of this exercise. See the section on gear assemblies. Create two shafts, Ø.500 and Ø1.000 with .05 chamfers at each end. The shafts should be long enough to allow for at least .25 clearance between the gears and the support plate. The shafts should extend from the bottom surface of the plate to the top surface of the gears' hub.

Create a plate .375 thick. Size the plate so that it extends at least .25 beyond the edges on either gear. Mount the shafts in SKF Series RLS ball bearings, and create holes in the plate that will accommodate the outside diameters of the bearings.

Project 12-3: MILLIMETERS

Draw two spur gears based on the following parameters. On both gears the threaded hole is located halfway up the hub height.

	Gear 1	Gear 2
Number of teeth	30	90
Face width	20	20
Diametral pitch	1.5	1.5
Pressure angle	20	20
Hub Ø	30	50
Hub height	20	20
Bore	12	16
Threaded hole	M3	M5

Figure P12-3

Project 12-4: MILLIMETERS

A. Draw two spur gears based on the following parameters. On both gears the threaded hole is located halfway up the hub height.

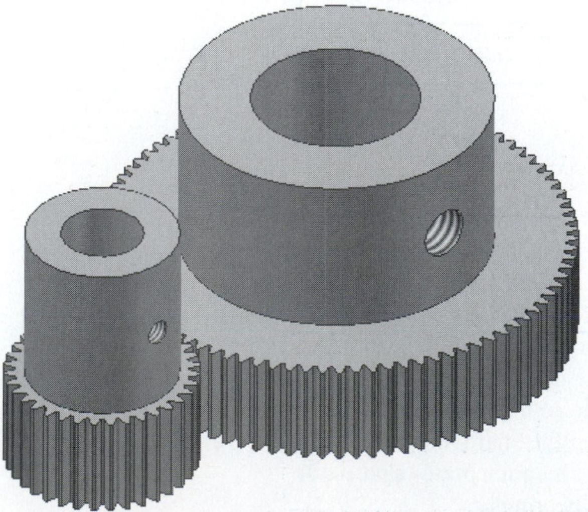

Figure P12-4

	Gear 1	Gear 2
Number of teeth	36	90
Face width	10	10
Diametral pitch	0.5	0.5
Pressure angle	20	20
Hub Ø	15	30
Hub height	16	16
Bore	8	16
Threaded hole	M3	M5

B. Design Exercise

Create a support plate and shafts for the two gears drawn in part A of this project. See the section on gear assemblies. Create two shafts, Ø8.0 and Ø16.0 with 0.50 chamfers at each end. The shafts should be long enough to allow for at least 5.0 clearance between the gears and the support plate. The shafts should extend from the bottom surface of the plate to the top surface of the gears' hub.

Create a plate 10 thick. Size the plate so that it extends at least 5.0 beyond the edges on either gear. Mount the shafts in DIN 1854-4 M plain bearings and create holes in the plate that will accommodate the outside diameters of the bearings.

Project 12-5: INCHES

A. Draw two bevel gears based on the following parameters. On both gears the threaded hole is located halfway up the hub height.

	Gear 1	Gear 2
Shaft angle	90	90
Pressure angle	20	20
Helix angle	20	20
Number of teeth	24	60
Face width	.50	.50
Diametral pitch	16	16
Hub Ø	1.00	1.00
Hub height	.75	.75
Bore	.50	.50
Threaded hole	#8-UNC	#8-UNC

B. Design Exercise

Create an L-bracket support plate and shafts for the two gears drawn in part A of this project. See the section on supports for bevel gear. Create two shafts, Ø.50 with .05 chamfers at each end. The shafts should be long enough to allow for at least .25 clearance between the gears and the support plate. The shafts should extend from the bottom surface of the plate to the top surface of the gears' hub.

Create an L-bracket .375 thick. Size the bracket so that it extends at least .25 beyond the edges on either gear. Make the outside diameters of the bosses at least .025 greater that the outside diameter of the bearings. Make the bosses .25 high.

Mount the shafts in SKF Series RLS ball bearings, and create holes in the plate that will accommodate the outside diameters of the bearings.

Project 12-6: MILLIMETERS

A. Draw two bevel gears based on the following parameters. On both gears the threaded hole is located halfway up the hub height.

	Gear 1	Gear 2
Shaft angle	90	90
Pressure angle	20	20
Helix angle	20	20
Number of teeth	20	40
Face width	12	12
Tangential module	2	2
Hub Ø	20	40
Hub height	16	20
Bore	12	20
Threaded hole	M5	M6

B. Design Exercise

Create an L-bracket support plate and shafts for the two gears drawn in part A of this project. See the section on supports for bevel gear. Create two shafts, Ø12.0 and Ø20.0 with 0.50 chamfers at each end. The shafts should be long enough to allow for at least 5.0 clearance between the gears and the support plate. The shafts should extend from the bottom surface of the plate to the top surface of the gears' hub.

Create an L-bracket 10 thick. Size the plate so that it extends at least 5.0 beyond the edges on either gear. Make the outside diameters of the bosses at least 5 greater that the outside diameter of the bearings. Make the bosses 5.00 high.

Mount the shafts in DIN 1854-4 M plain bearings, and create holes in the plate that will accommodate the outside diameters of the bearings.

Project 12-7: INCHES

A. Draw a worm and a worm gear based on the following parameters. On both gears the threaded hole is located halfway up the hub height.

Worm gear: Number of teeth = 48		
Worm: Number of threads = 1		
	Worm	**Worm Gear**
Face width		.75
Diametral pitch	12	12
Pressure angle	20	20
Worm length	2.50	
Hub Ø	.50	.50
Hub height	.500	.750
Bore	.375	.500
Threaded hole	0.138(#6)UNC	0.164(#8)UNC

B. Design Exercise

Create a corner-bracket support plate and shafts for the two gears drawn in part A of this project. See the section on worm gear supports. Create two shafts, Ø.375 and Ø.500 with .05 chamfers at each end. The shafts should be long enough to allow for at least .25 clearance between the gears and the support plate. The shafts should extend from the back surface of the plate to the top surface of the gears' hub.

Create a corner bracket .375 thick. Size the bracket so that it extends at least .25 beyond the edges on either gear. Mount the shafts in SKF Series RLS ball bearings and create holes in the sides that will accommodate the outside diameters of the bearings.

Project 12-8: INCHES

A. Draw a worm and a worm gear based on the following parameters. On both gears the threaded hole is located halfway up the hub height.

Worm gear: Number of teeth = 60		
Worm: Number of threads = 1		
	Worm	**Worm Gear**
Face width		24
Module	4	4
Pressure angle	14.5	14.5
Worm length	65	
Hub Ø	16	24
Hub height	12	16
Bore	8.0	10.0
Threaded hole	M4	M4

B. Design Exercise

Create a corner-bracket support plate and shafts for the two gears drawn in part A of this project. See the section on worm gear supports. Create two shafts, Ø8.0 and Ø10.0 with .50 chamfers at each end. The shafts should be long enough to allow for at least 5.0 clearance between the gears and the support plate. The shafts should extend from the back surface of the plate to the top surface of the gears' hub.

Create a corner bracket 8.0 thick. Size the bracket so that it extends at least 6.0 beyond the edges on either gear. Mount the shafts in DIN 1854-4 M plain bearings, and create holes in the sides that will accommodate the outside diameters of the bearings.

Project 12-9: INCHES

A. Prepare an assembly drawing of the 2-Gear Assembly shown in Figure P12-5. The gears have the following parameters:

Center distance = 3.00		
	Gear 1	**Gear 2**
Number of teeth	48	96
Face width	.50	.50
Diametral pitch	24	24
Pressure angle	20	20
Hub Ø	.750	1.000
Hub height	.500	.500
Bore	.5.00	.625

B. Prepare a presentation drawing.
C. Animate the presentation drawing.
D. Prepare an exploded isometric drawing with assembly numbers and a parts list.
E. Prepare a detailed dimensioned drawing of each part.

2-Gear Assembly

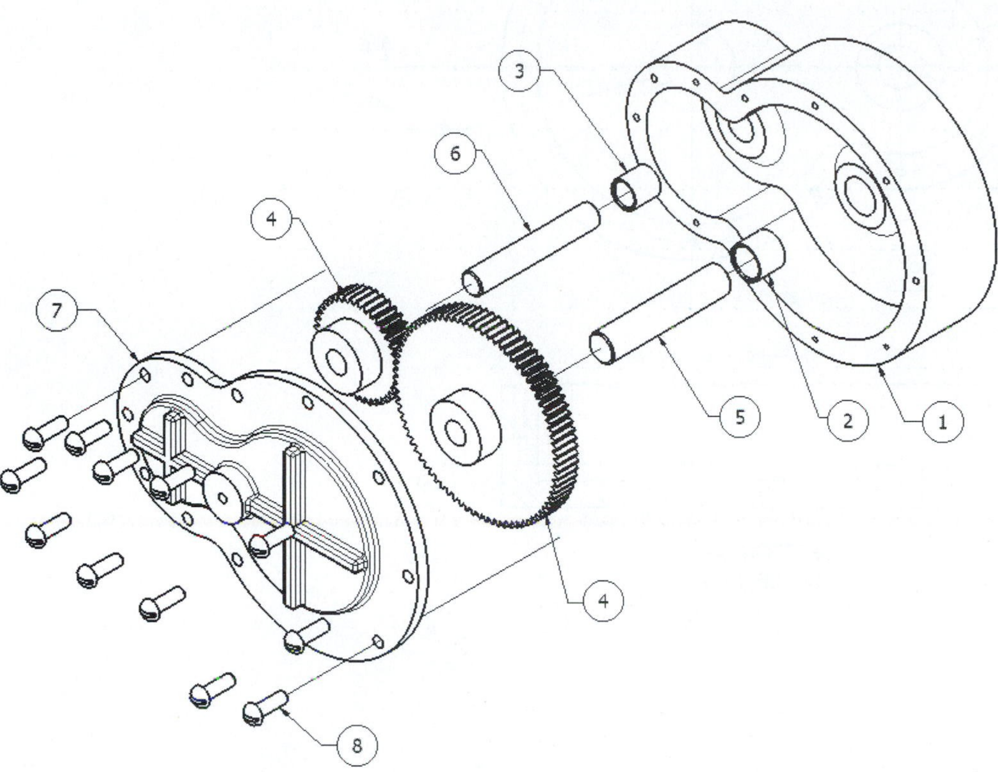

Figure P12-5a

Parts List				
ITEM	PART NUMBER	DESCRIPTION	MATERIAL	QTY
1	ENG-453-A	GEAR, HOUSING	CAST IRON	1
2	BU-1123	BUSHING Ø0.75	Delrin, Black	1
3	BU-1126	BUSHING Ø0.625	Delrin, Black	1
4	ASSEMBLY-6	GEAR ASSEEMBLY	STEEL	1
5	AM-314	SHAFT, GEAR Ø.625	STEEL	1
6	AM-315	SHAFT, GEAR Ø.0.500	STEEL	1
7	ENG -566-B	COVER, GEAR	CAST IRON	1
8	ANSI B18.6.2 - 1/4-20 UNC - 0.75	Slotted Round Head Cap Screw	Steel, Mild	12

Figure P12-5b

Gear, Housing
P/N ENG-453-A
Cast Iron

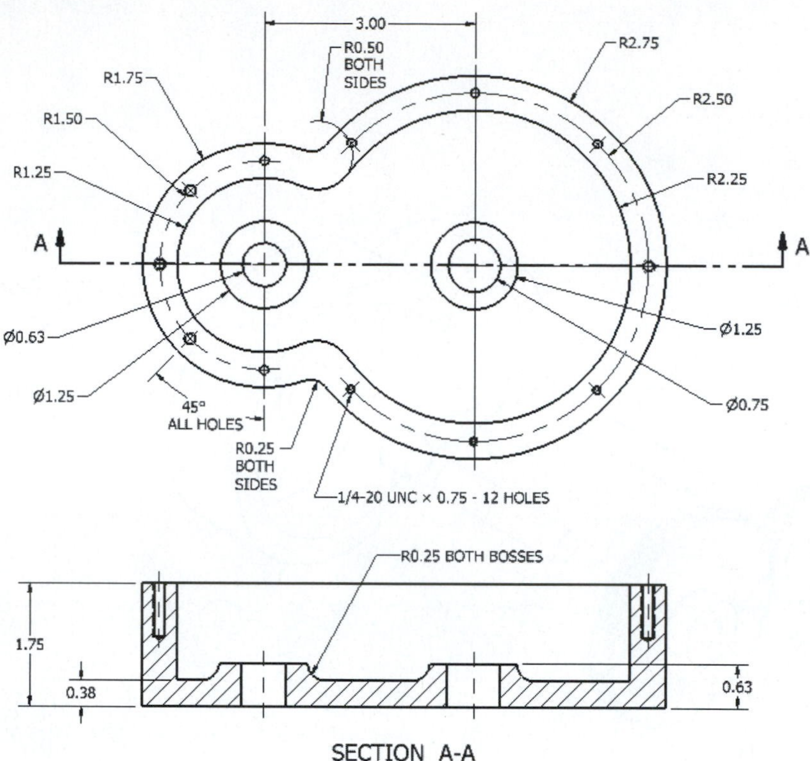

Figure P12-5c

Shaft, Gear Ø.625
P/N AM-314
Steel

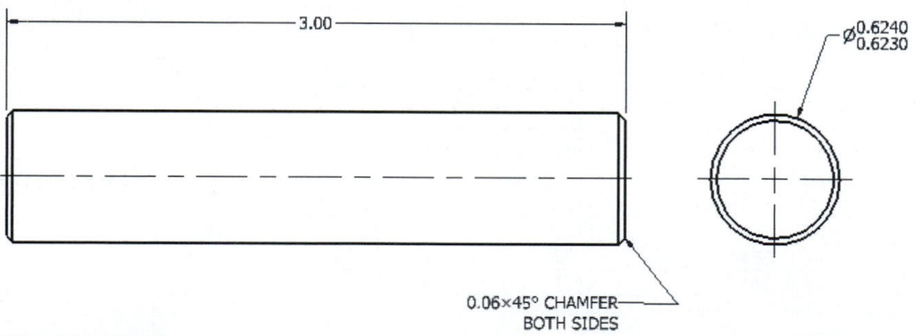

Figure P12-5d

Shaft, Gear Ø.500
P/N AM-315
Steel

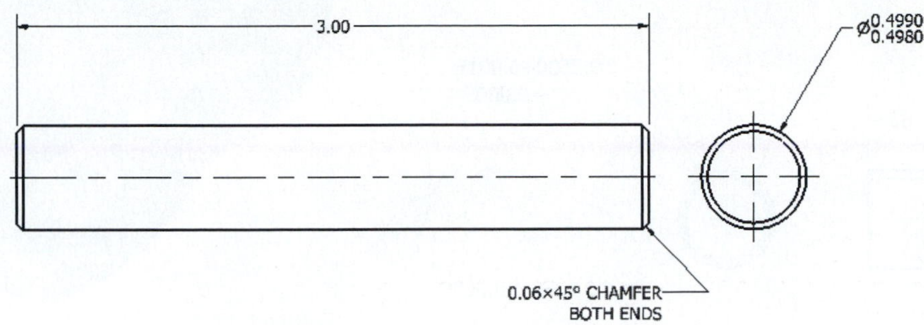

Figure P12-5e

Cover, Gear
P/N ENG-566-B
Cast Iron

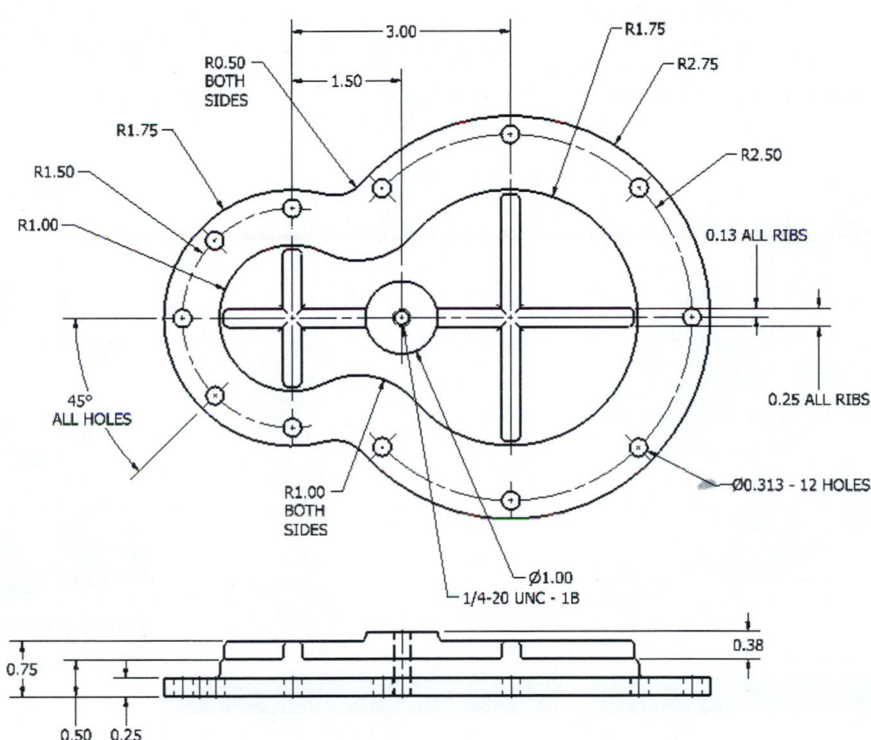

Figure P12-5f

Bushing Ø.625
P/N BU-1123
Delrin, Black

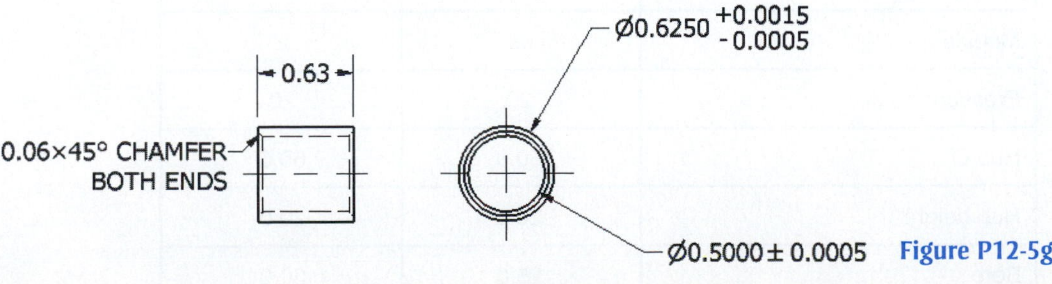

Figure P12-5g

Bushing Ø0.750
P/N BU-1126
Delrin, Black

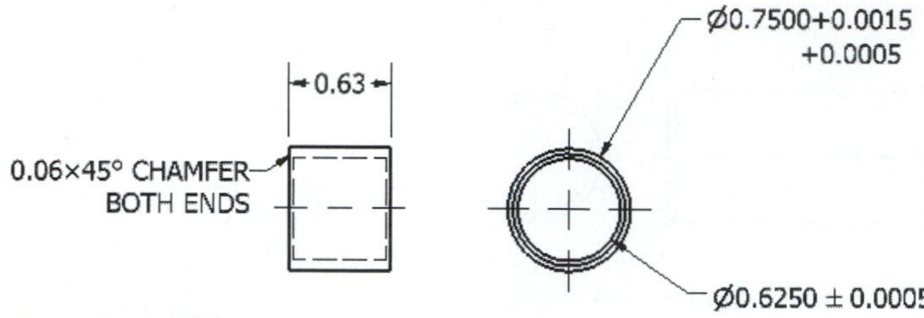

Figure P12-5h

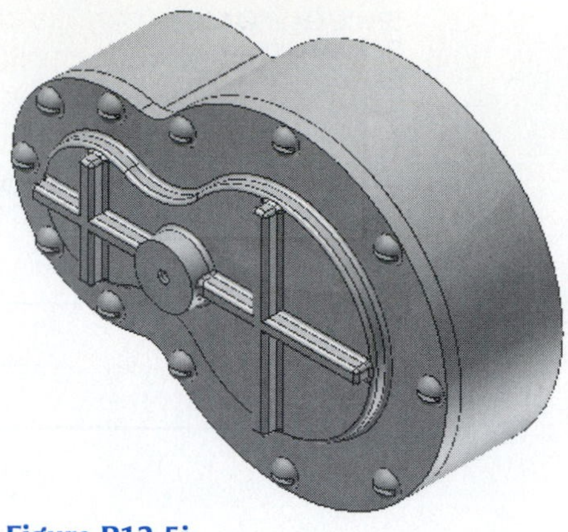

Figure P12-5i

Project 12-10: MILLIMETERS

A. Prepare an assembly drawing of the 4-Gear Assembly shown in Figure P12-6. The gears have the following parameters;

Presentation Drawing

4-Gear Assembly

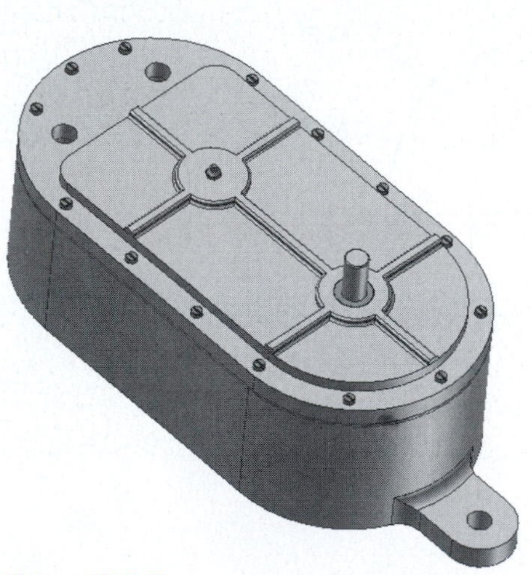

Figure P12-6a

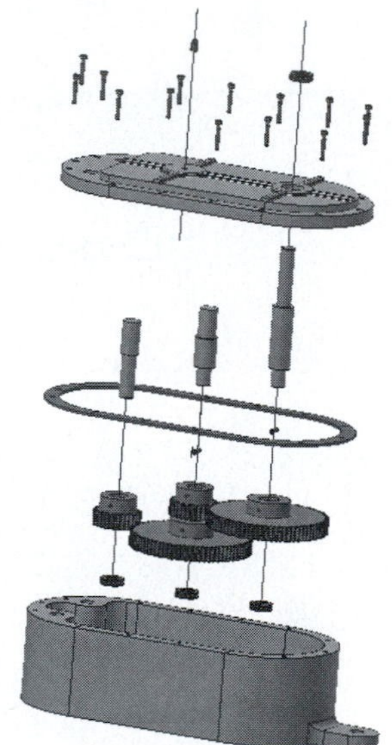

Figure P12-6b

Center distance = 3.00		
	Gears 1, 3	**Gears 2, 4**
Number of teeth	30	96
Face width	20.0	20.0
Module	2	2
Pressure angle	20	20
Hub Ø	50.0	60.0
Hub height	20.0	20.0
Bore	25.0	30.0

B. Prepare a presentation drawing.
C. Animate the presentation drawing.
D. Prepare an exploded isometric drawing with assembly numbers and a parts list.
E. Prepare a detailed dimensioned drawing of each part.

Parts List

Parts List				
ITEM	PART NUMBER	DESCRIPTION	MATERIAL	QTY
1	ENG-311-1	4-GEAR HOUSING	CAST IRON	1
2	BS 5989: Part 1 - 0 10 - 20x32x8	Thrust Thrust Ball Bearing	Steel, Mild	4
3	SH-4002	SHAFT, NEUTRAL	STEEL	1
4	SH-4003	SHAFT, OUTPUT	STEEL	1
5	SH-4004A	SHAFT,INPUT	STEEL	1
6	4-GEAR-ASSEMBLY		STEEL	2
7	CSN 02 1181 - M6 x 16	Slotted Headless Set Screw - Flat Point	Steel, Mild	2
8	ENG-312-1	GASKET	Brass, Soft Yellow	1
9	COVER			1
10	CNS 4355 - M 6 x 35	Slotted Cheese Head Screw	Steel, Mild	14
11	CSN 02 7421 - M10 x 1coned short	Lubricating Nipple, coned Type A	Steel, Mild	1

Figure P12-6c

Housing, Gear
P/N ENG-311-1
Cast Iron

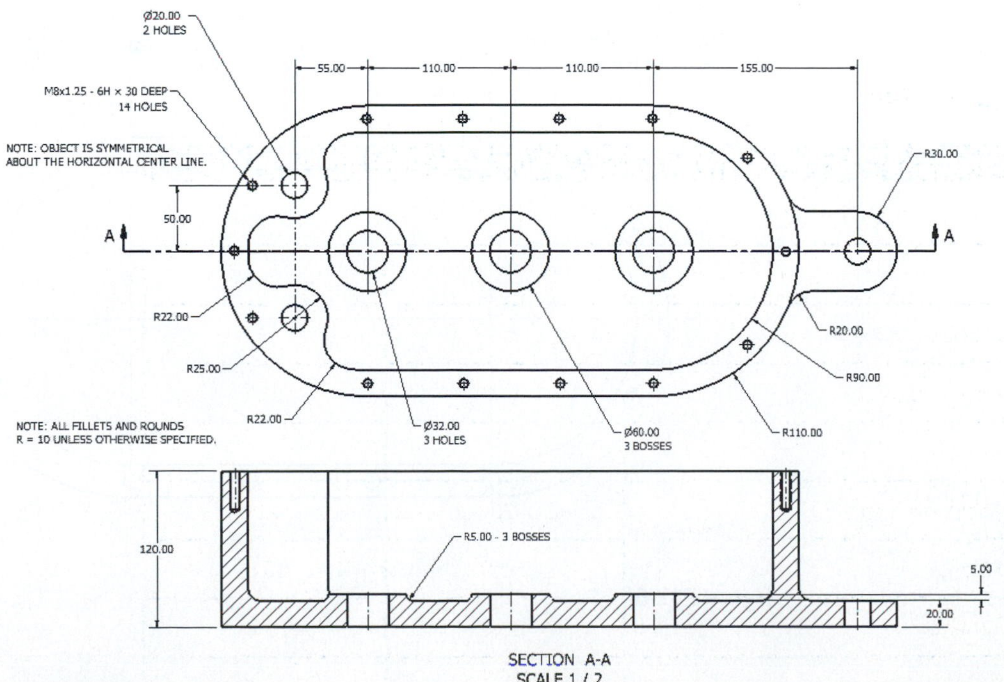

Figure P12-6d

Cover
P/N AM-311-2

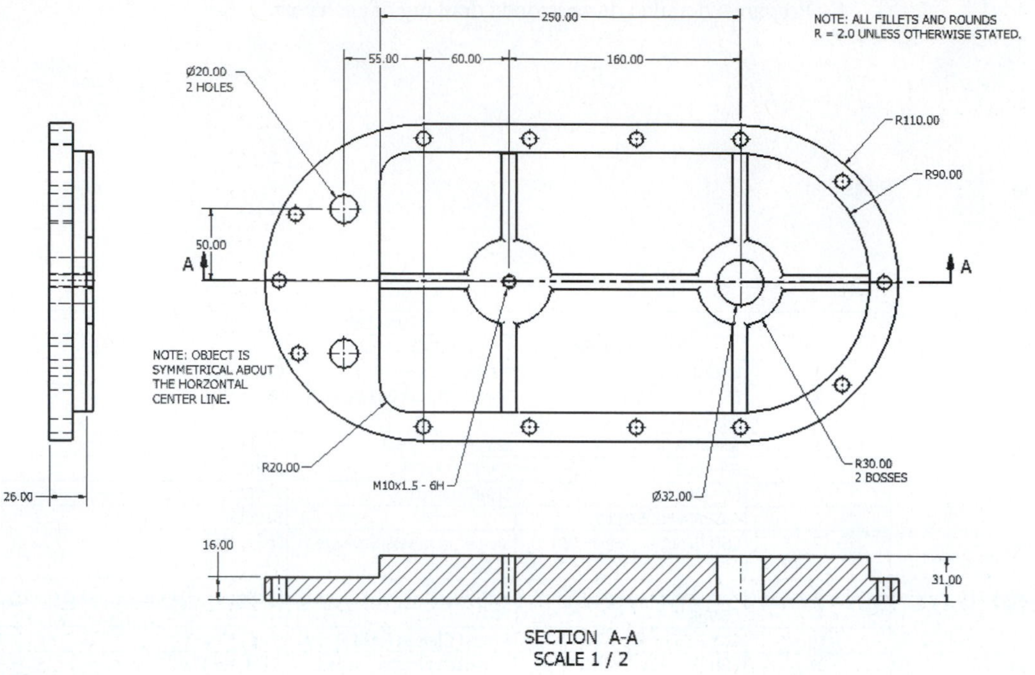

Figure P12-6e

Gasket
P/N ENG-312-1
Brass, soft yellow

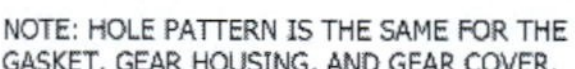

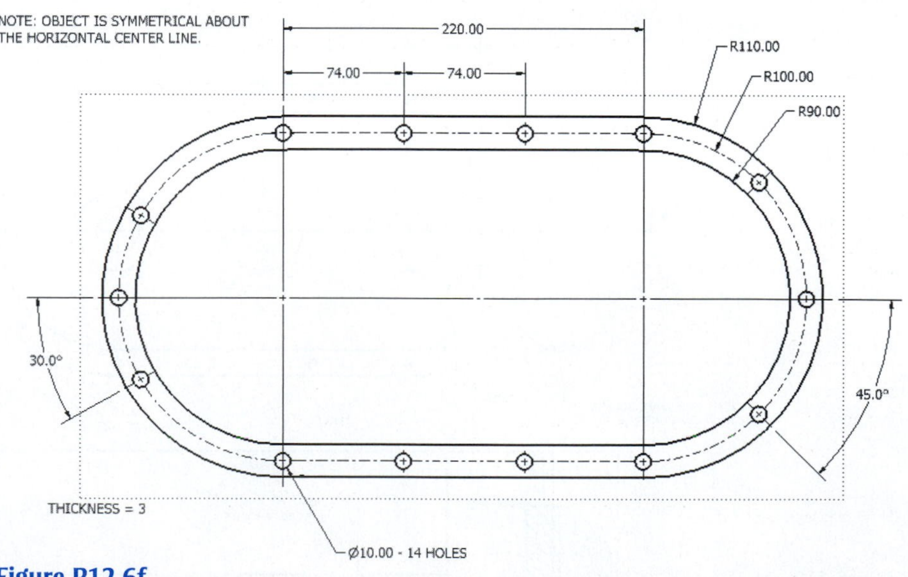

Figure P12-6f

Gear Subassembly

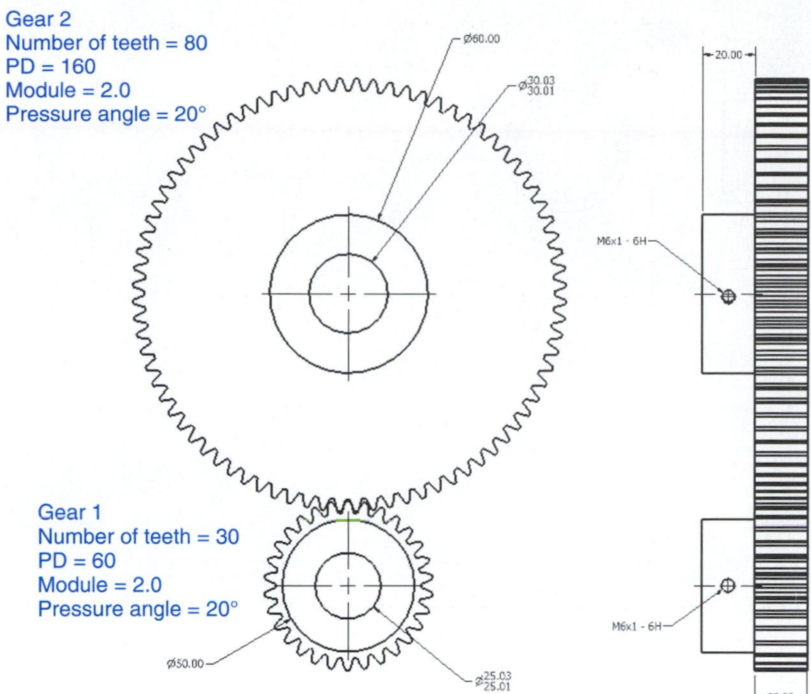

Figure P12-6g

Shaft, Output
P/N SH-4003
Steel

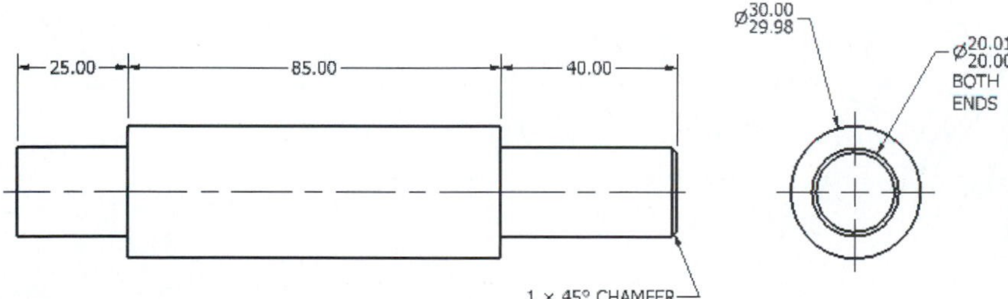

Figure P12-6h

Shaft, Input
P/N SH-4004-A
Steel

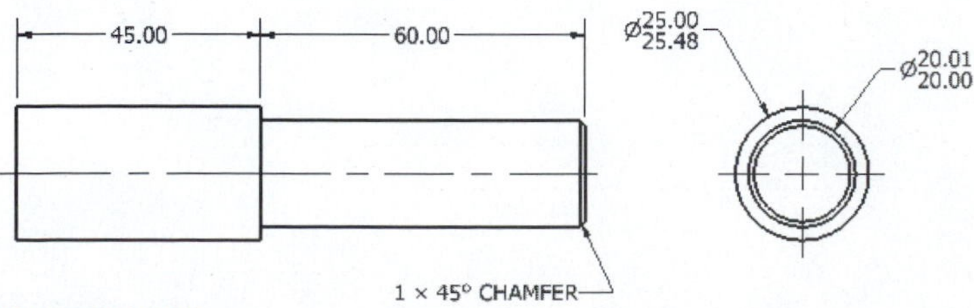

Figure P12-6i

Shaft, Neutral
P/N SH-4002
Steel

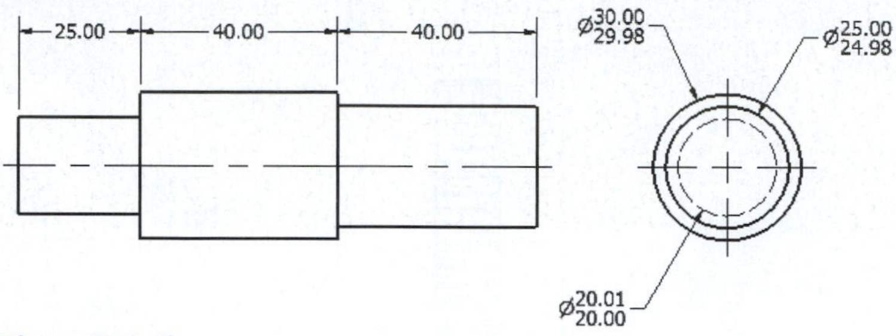

Figure P12-6j

Exploded Isometric Drawing

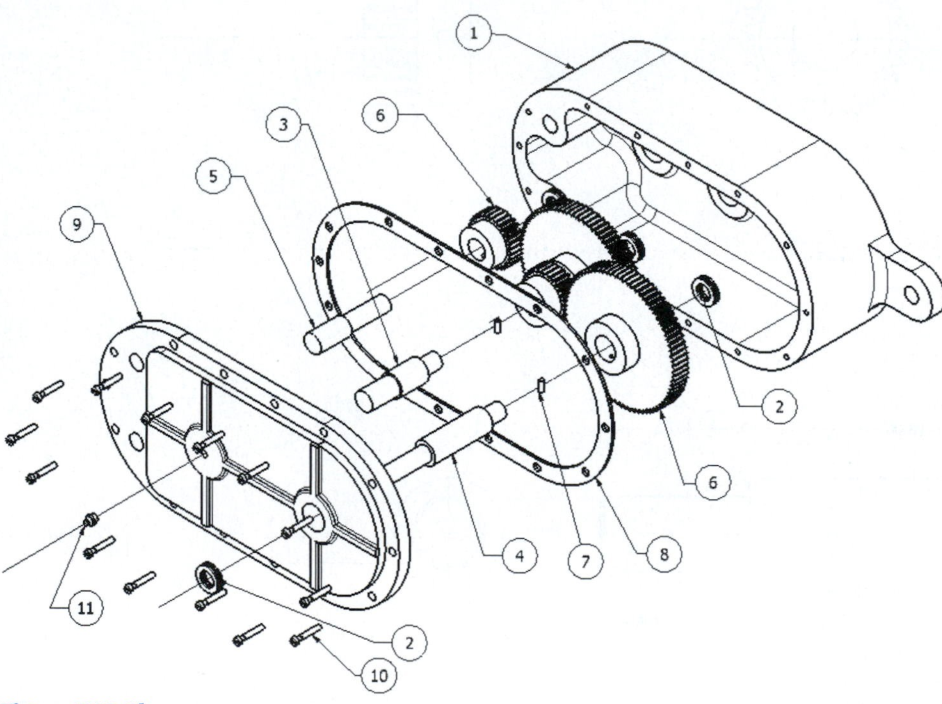

Figure P12-6k

Sheet Metal Drawings

13

Chapter Objectives

- Show how to create sheet metal drawings.
- Explain sheet metal gauges.
- Understand sheet metal terminology.

Introduction

This chapter explains how to create sheet metal drawings. Gauges for sheet metal are presented along with bend radii, flanges, tabs, reliefs, and flat patterns.

Sheet Metal Drawings

Figure 13-1 shows a 3D solid model of a sheet metal part and a dimensioned orthographic drawing of that part. The orthographic drawing was created from the 3D model. The following sections explain how to create the 3D sheet metal drawing.

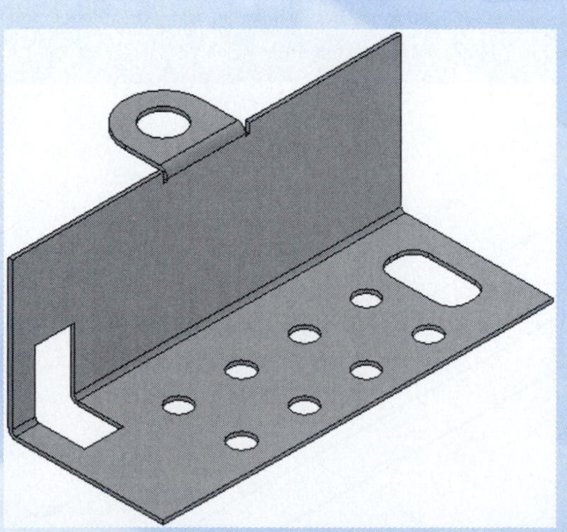

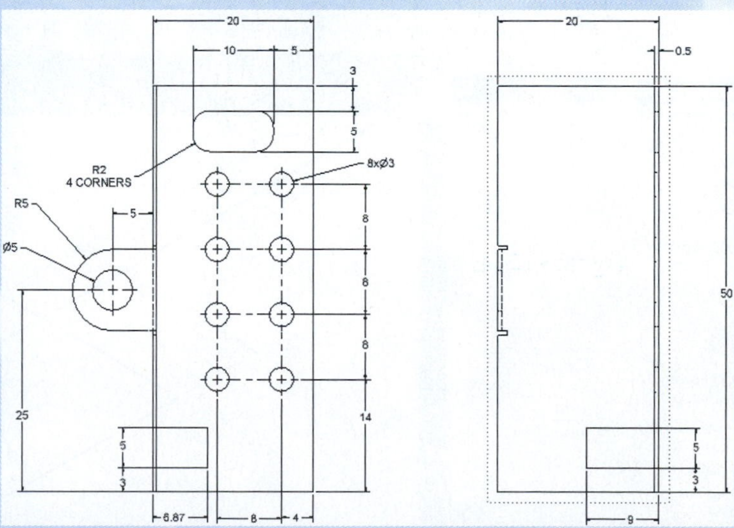

Figure 13-1

Exercise 13-1: Creating a 3D Sheet Metal Drawing

1. Create a new drawing using the **Sheet Metal (mm).ipt** format.

See Figure 13-2. The 2D sketch panel will appear. See Figure 13-3. Sheet metal drawings are initiated as 2D sketches, then developed using a combination of **2D Sketch** and **Sheet Metal Features** panel commands.

2. Use the **Two-point rectangle** command and draw a **20 × 50** rectangle.
3. Right-click the mouse and select the **Isometric** option.

See Figure 13-3.

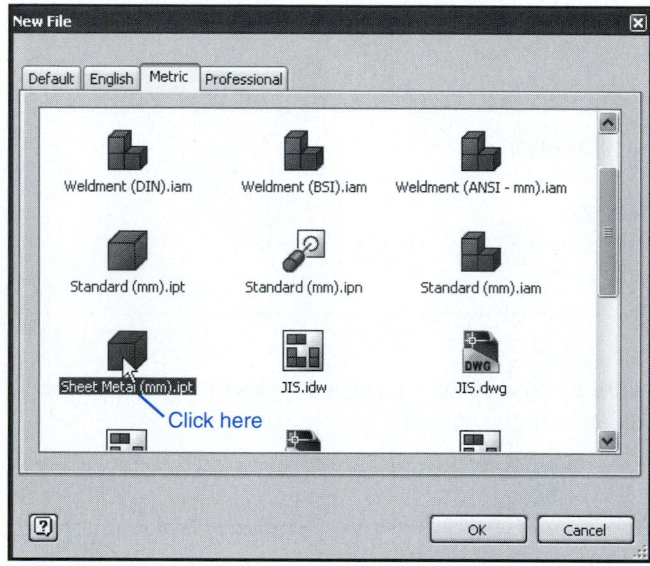

Figure 13-2

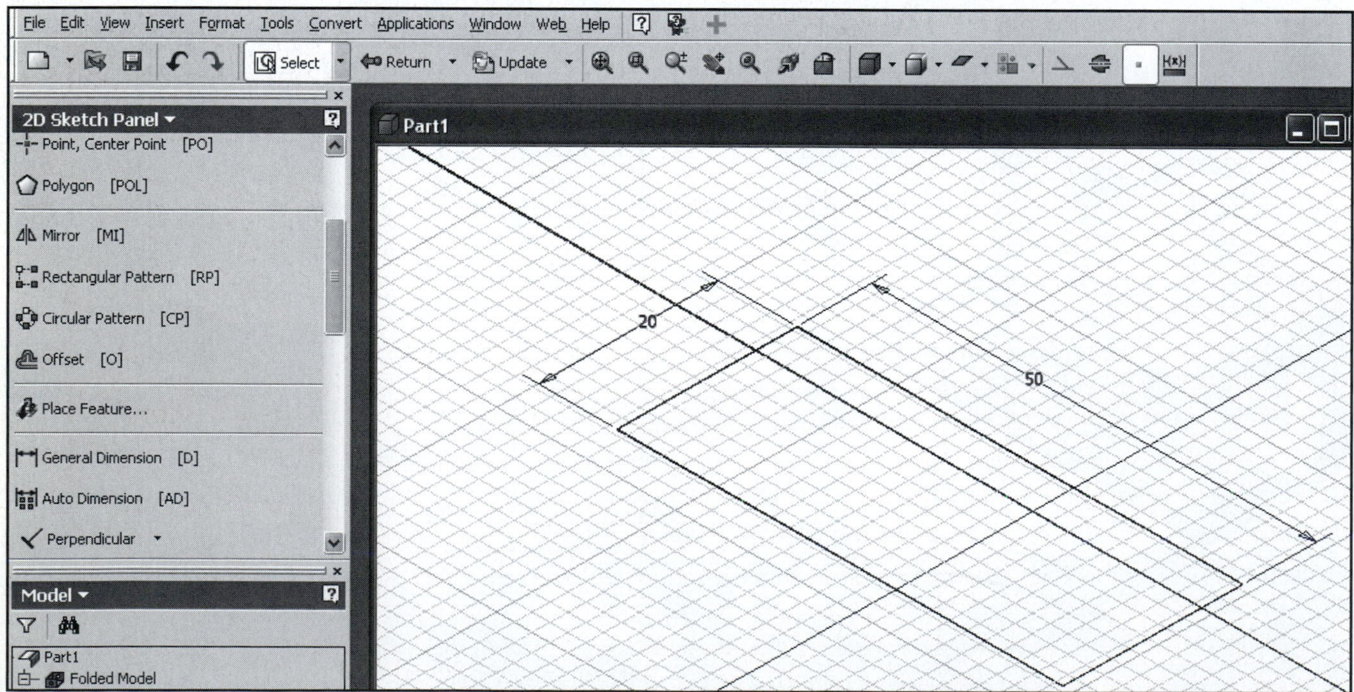

Figure 13-3

Exercise 13-2: Adding Thickness

1. Right-click the mouse and select the **Done** option.
2. Right-click the mouse again and select the **Finish Sketch** option.

The **Sheet Metal Features** panel will appear. See Figure 13-4. Not all commands will be active at this time, but they will become active as the drawing progresses.

Note: The **Sheet Metal** Features Panel may also be accessed by clicking the arrow to the right of the **2D Sketch Panel** heading. See Figure 13-4.

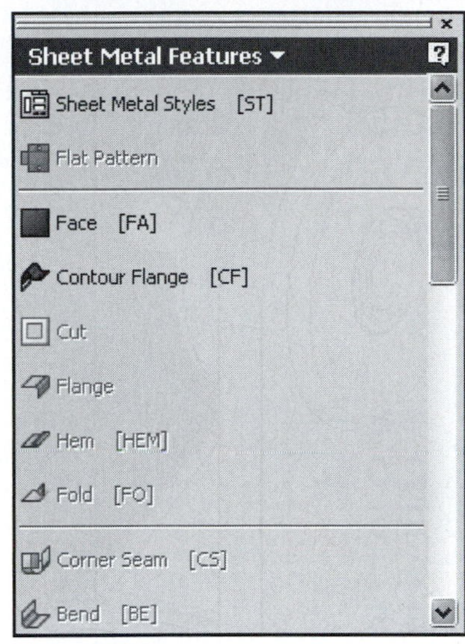

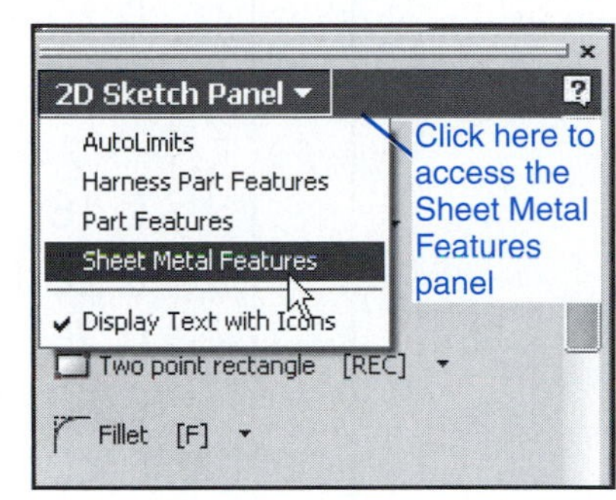

Figure 13-4

3. Select the **Sheet Metal Styles** option from the **Sheet Metal Features** panel.

The **Sheet Metal Styles** dialog box will appear. See Figure 13-5. The **Sheet Metal Styles** box is used to define the thickness, material, and bend characteristics of the part.

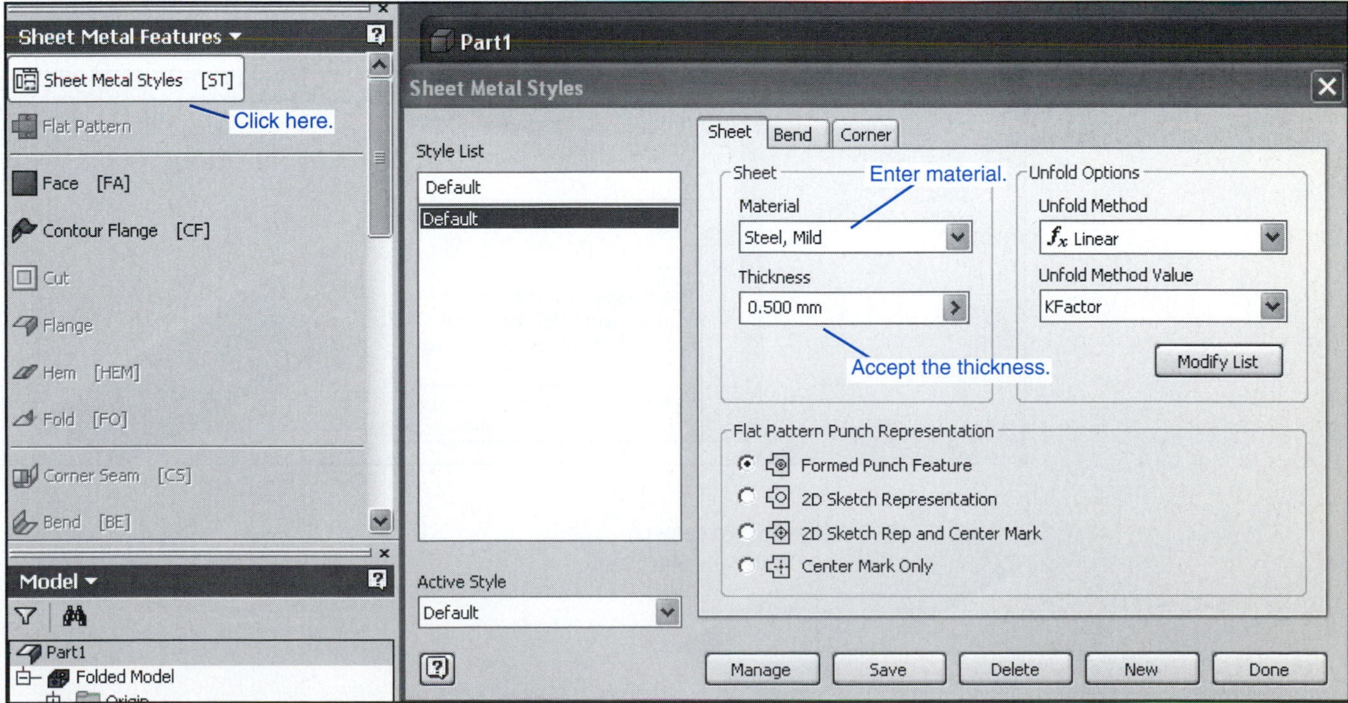

Figure 13-5

Inventor has many default values already in place. Figure 13-5 shows that the default thickness is 0.500 mm. Sheet metal is manufactured in standard thicknesses. Figure 13-6 is a partial listing of available standard sheet metal thicknesses in inches, and Figure 13-7 is a partial listing of sheet metal thicknesses in millimeters.

Figure 13-7 lists 0.500 mm as a standard thickness, so this default value will be used for this example.

Wire and Sheet Metal Gauges

Gauge	Thickness (in.)	Gauge	Thickness (in.)
000 000	0.5800	18	0.0403
00 000	0.5165	19	0.0359
0 000	0.4600	20	0.0320
000	0.4096	21	0.0285
00	0.3648	22	0.0253
0	0.3249	23	0.0226
1	0.2893	24	0.0201
2	0.2576	25	0.0179
3	0.2294	26	0.0159
4	0.2043	27	0.0142
5	0.1819	28	0.0126
6	0.1620	29	0.0113
7	0.1443	30	0.0100
8	0.1285	31	0.0089
9	0.1144	32	0.0080
10	0.1019	33	0.0071
11	0.0907	34	0.0063
12	0.0808	35	0.0056
13	0.0720	36	0.0050
14	0.0641	37	0.0045
15	0.0571	38	0.0040
16	0.0508	39	0.0035
17	0.0453	40	0.0031

Figure 13-6

0.050	0.50	4.0
0.060	0.60	5.0
0.080	0.80	6.0
0.10	1.0	8.0
0.12	1.2	10.0
0.16	1.6	
0.20	2.0	
0.25	2.5	
0.30	3.0	
0.40	3.5	

Figure 13-7

4. Accept the **0.500 mm Thickness** value and enter a material of **Steel Mild.** Click **Done.**
5. Select the **Face** tool from the **Sheet Metal Features** panel.
6. Select the sketch, then click **OK.**

See Figure 13-8.

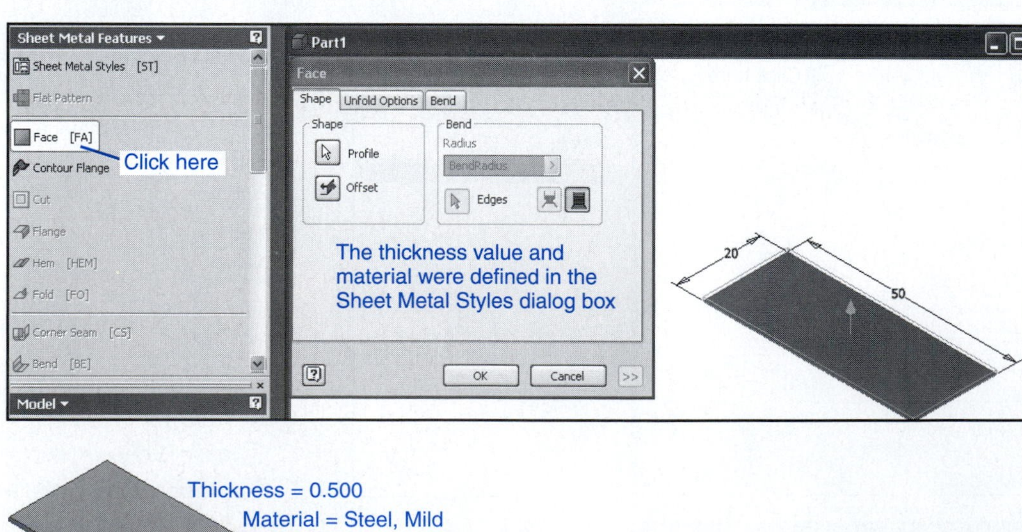

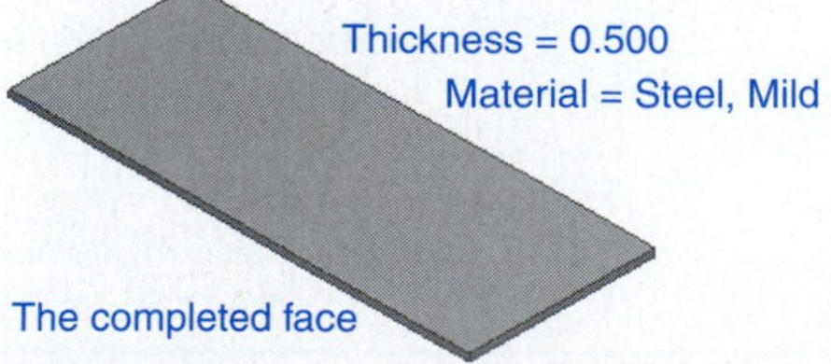

Figure 13-8

Bend Radii

As sheet metal is bent the inside surface is subjected to compression, and the outside surface to tension. These forces cause the material to stretch slightly.

The default value for the inside bend radius is equal to the sheet metal thickness. Figure 13-9 shows the **Sheet Metal Styles** dialog box with the **Bend** tab selected. Note that the **Radius** box is set to **Thickness.** The default outside bend radius is equal to twice the sheet metal thickness.

The default values for **Radius** and **Relief Shape** will be accepted for this example.

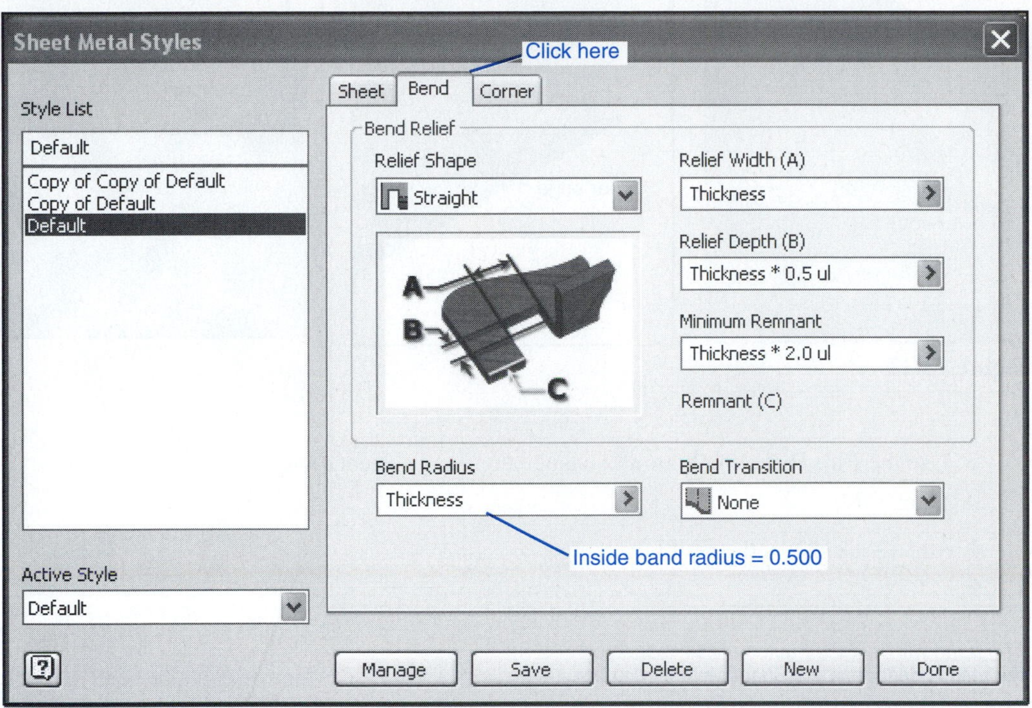

Figure 13-9

Flanges

1. Select the ***Flange*** tool from the **Sheet Metal Features** panel.

The **Flange** dialog box will appear. See Figure 13-10.

flange: A rim formed on the edge of sheet metal for strength.

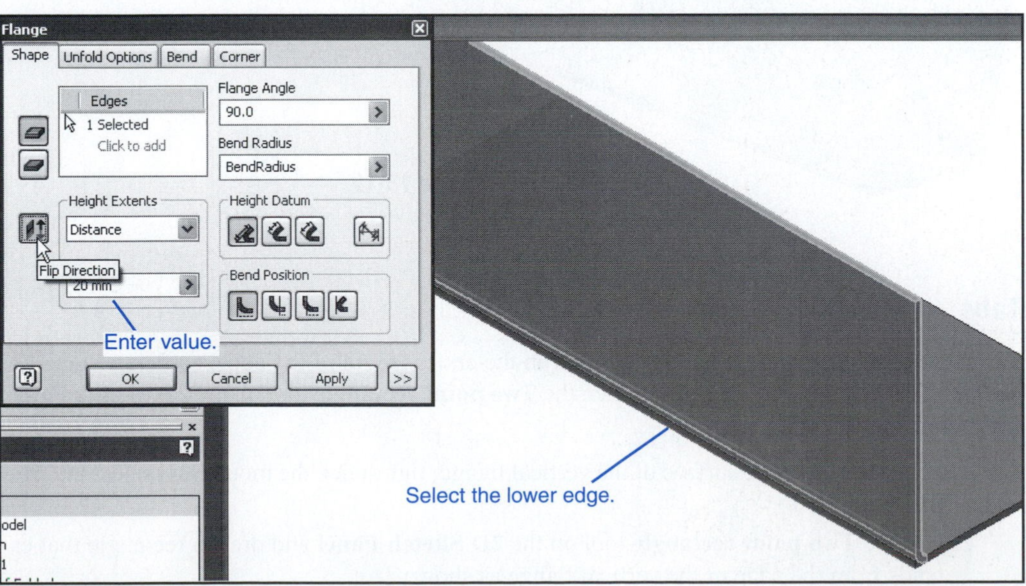

Figure 13-10

2. Set the distance for **20** and select the lower rear edge of the sketch.

The lower edge was chosen because the flange total height is to be 20 mm. If the upper edge were chosen, the total height would be 20.5, the flange height plus the material thickness. Figure 13-11 shows the flange orientation resulting from edge selection.

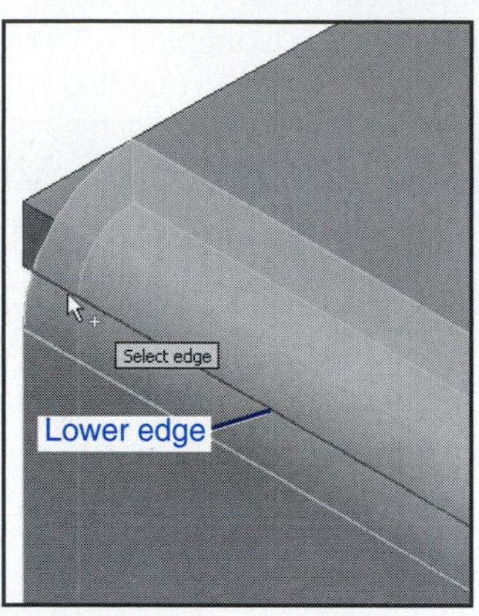

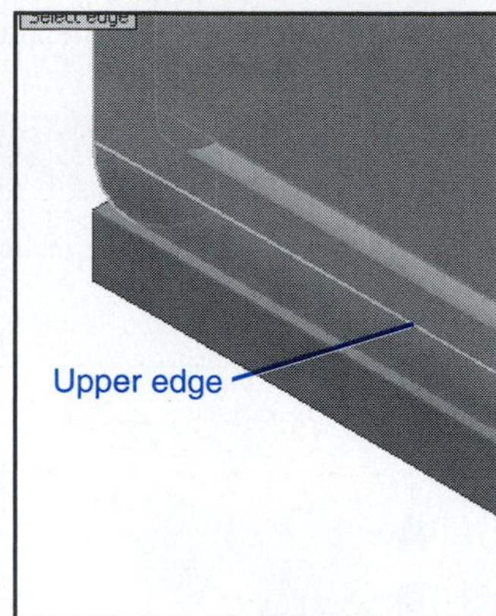

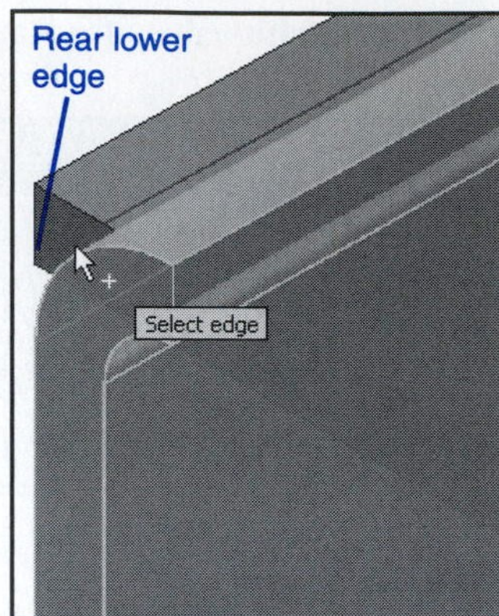

Figure 13-11

3. Use the **Flip Direction** button to change the flange orientation.
4. Click **OK.**

Figure 13-12 shows the resulting flange.

The bend will be added automatically based on the values defined in the Sheet Metal Styles box.

Figure 13-12

Tabs

tab: A feature similar to a flange but that does not run the entire length of the edge.

Tabs are similar to flanges, but tabs do not run the entire length of the edge, as flanges do. Tabs are created using a new sketch plane, then the **Two point rectangle** tool on the **2D Sketch Panel.** See Figure 13-13.

1. Select the top-edge surface of the vertical flange, right-click the mouse and select the **New Sketch** option.
2. Use the **Two point rectangle** tool on the **2D Sketch Panel** and draw a rectangle that extends from the edge of the vertical flange as shown.

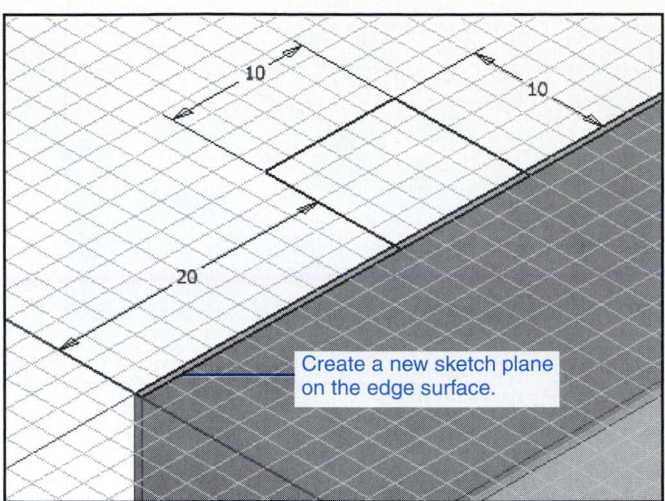

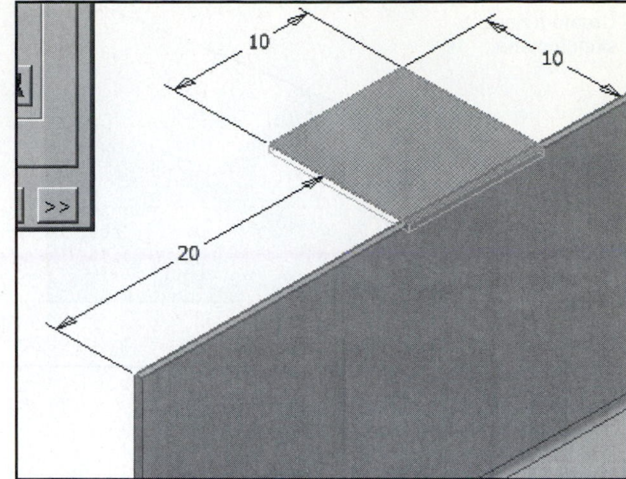

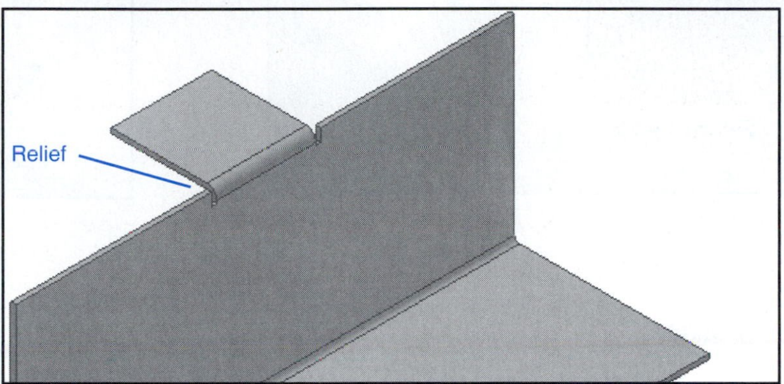

Figure 13-13

3. Use the **General Dimension** tool to size and locate the tab in accordance with the dimensions given in Figure 13-1.
4. Right-click the mouse and select the **Done** option, then right-click the mouse again and select the **Finish Sketch** option.
5. Select the **Face** tool, and define the tab as the **Profile.**
6. Select **OK.**

Figure 13-13 shows the resulting tab.

Reliefs

Reliefs are cut out in material to allow it to be bent. If the material were not relieved, it would tear uncontrollably as the bend was formed.

relief: An area cut out of material to allow it to be bent.

Inventor's default relief value is equal to the thickness of the sheet metal material. Figure 13-13 shows the relief that was automatically created as the tab was formed.

Holes

Holes are added to sheet metal parts in the same manner as they are added to 3D models. See Figure 13-14.

1. Create a new sketch plane on the top surface of the tab.
2. Use the **Point, Center Point** tool to define the hole's center point.
3. Use the **General Dimension** tool to dimension the hole's center point location.
4. Right-click the mouse, click the **Done** option, then click the **Finish Sketch** option.
5. Use the **Hole** tool on the **Part Features** panel to create the hole.

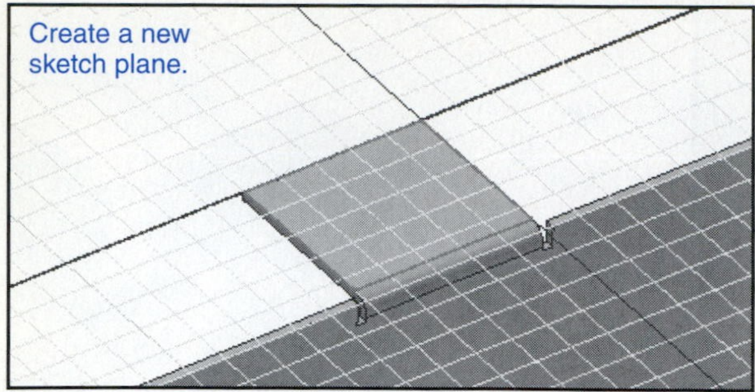

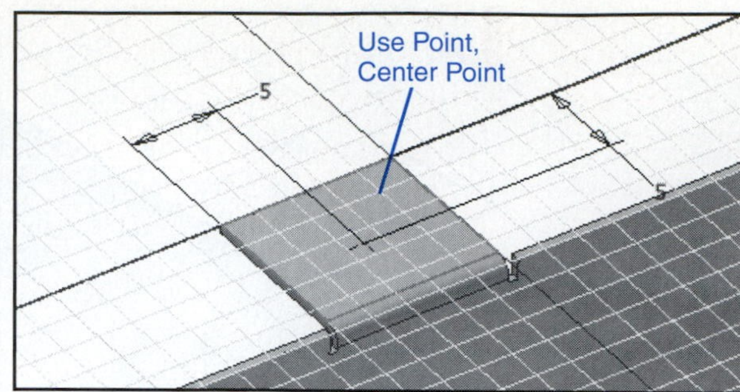

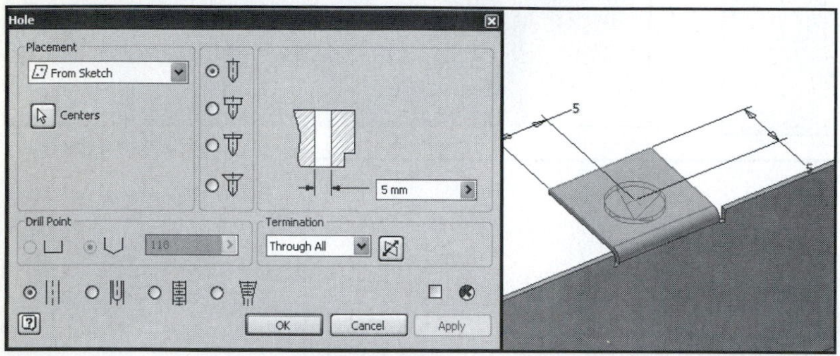

Figure 13-14

Corners

Both internal and external corners are created using the **Corner Round** tool found on the **Sheet Metal Features** panel.

1. Click the **Corner Round** tool on the **Sheet Metal Features** panel.
2. The **Corner Round** dialog box will appear. See Figure 13-15.
3. Set the **Radius** value for **5.**
4. Select the two outside corners of the tab.
5. Click **OK.**

Figure 13-15 shows the resulting rounded corners.

Cuts

Cuts may be any shape, other than a hole, that passes through the sheet metal. In this example a rectangular shape is used. See Figure 13-16.

1. Create a new sketch plane and sketch a rectangle as shown. Use the **General Dimension** tool to size and locate the rectangle.
2. Right-click the mouse and select the **Done** option, then select the **Finish Sketch** option.

The **Sheet Metal Features** panel will appear.

3. Select the **Cut** tool.

The **Cut** dialog box will appear.

4. Select the rectangle as the **Profile.**
5. Ensure that the direction of the cut is correct, then click the **OK** box.

The rectangular area will be removed. The depth of the cut will automatically be set for the thickness evalue.

6. Select the **Corner Round** tool and set the **Radius** value for **2 mm.**

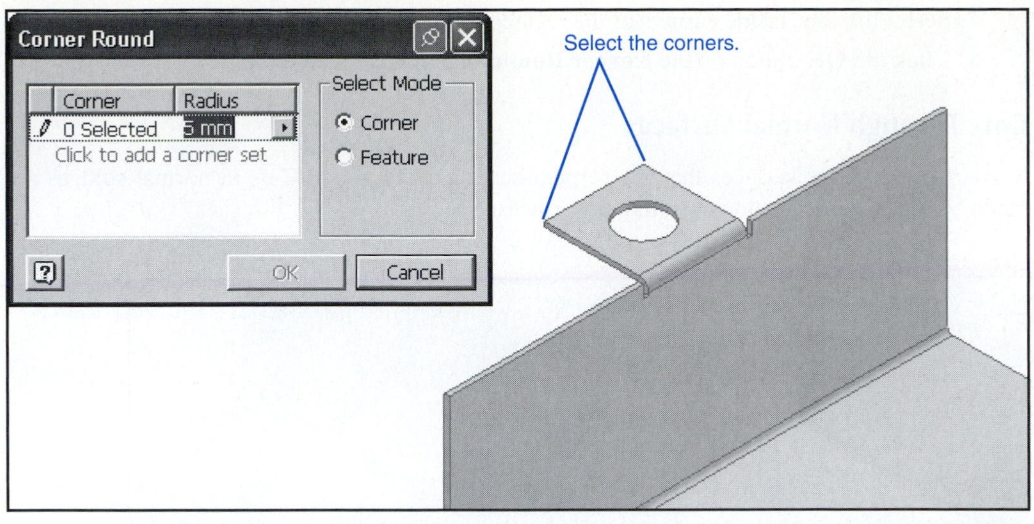

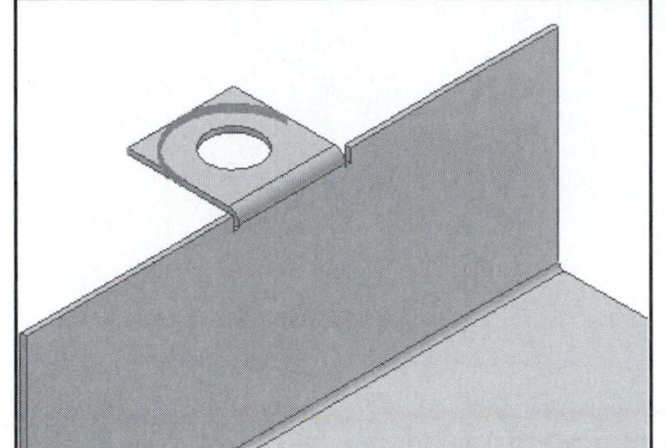

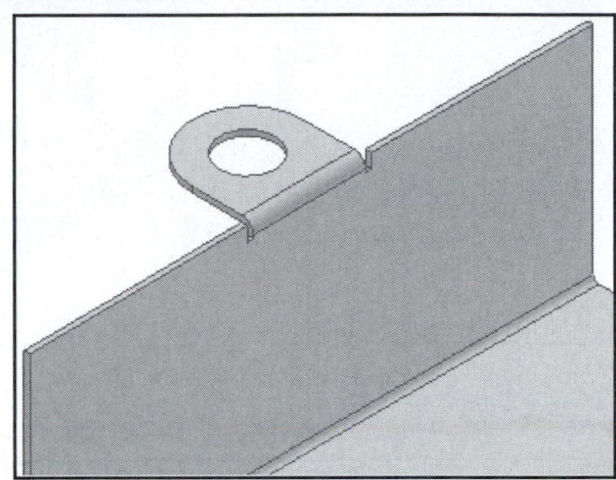

Figure 13-15

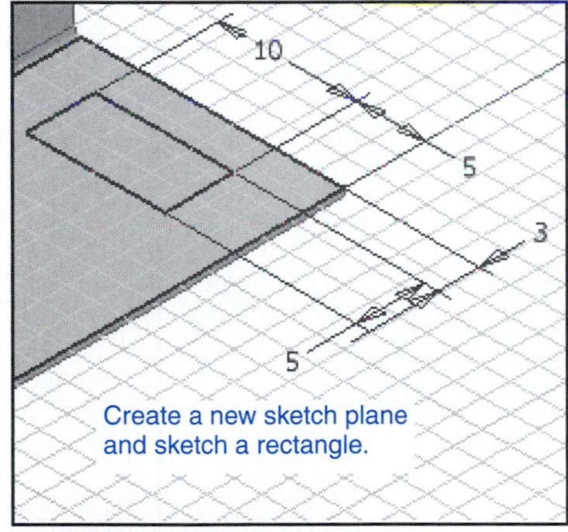

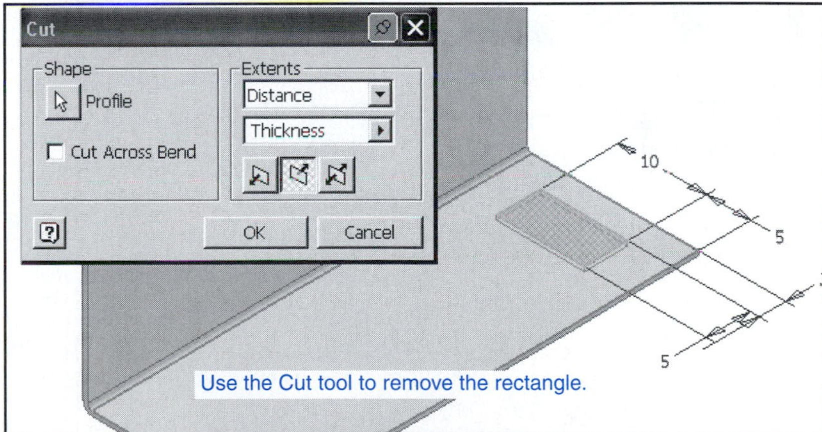

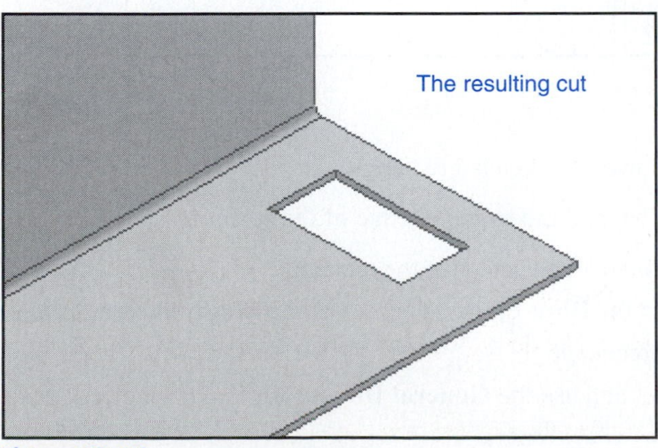

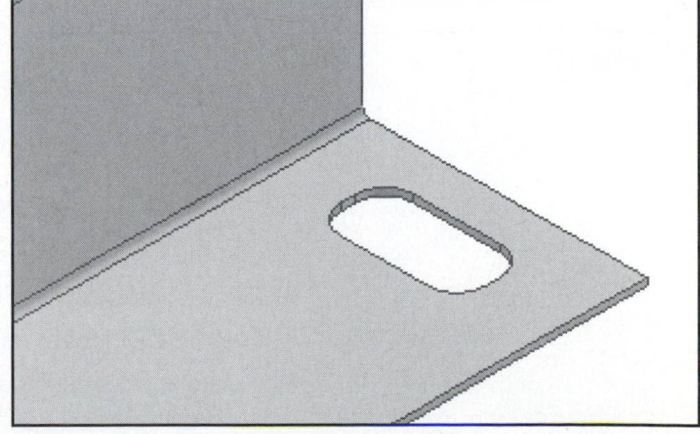

Figure 13-16

7. Select the four inside corners of the rectangular cut.
8. Click the **OK** button on the **Corner Round** dialog box.

Cuts Through Normal Surfaces

Normal surfaces are surfaces that are perpendicular to each other. Cuts in normal sufaces are made by making intersecting cuts in both surfaces. See Figure 13-17.

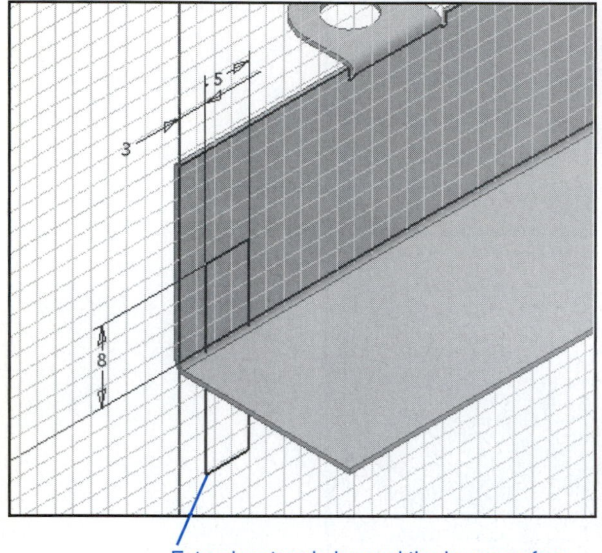

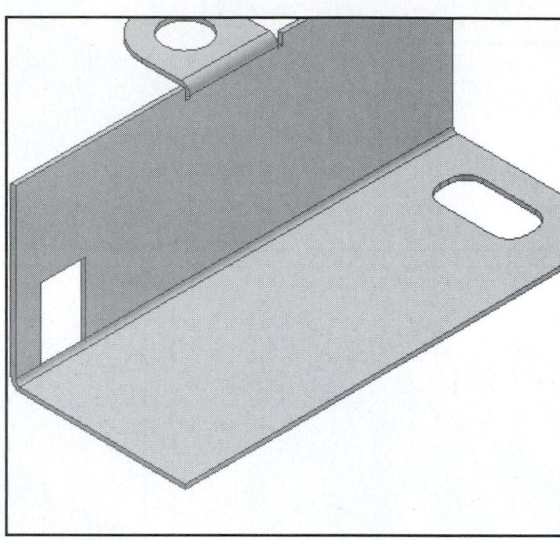

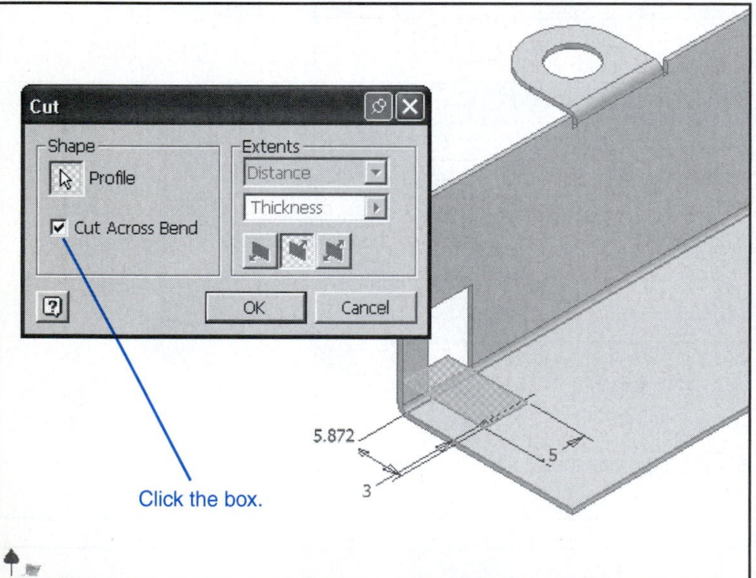

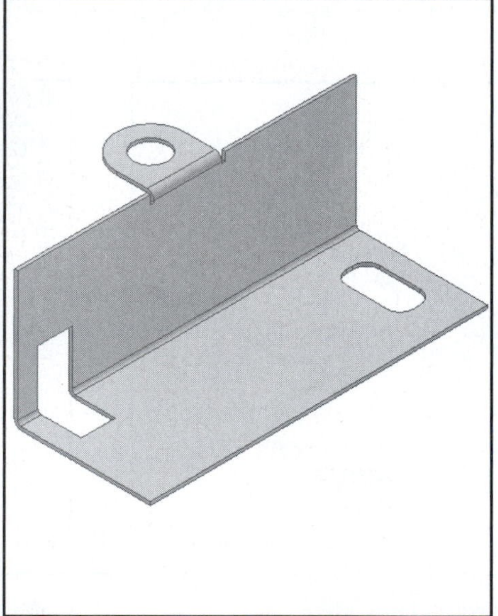

Figure 13-17

1. Create a new sketch plane as shown and sketch a rectangle.

Ensure that the rectangle extends beyond the rounded edge of the surface.

2. Use the **General Dimension** tool to locate and size the rectangle.
3. Right-click the mouse and select the **Done** option, then select the **Finish Sketch** option.
4. Use the **Cut** tool to remove the rectangle.
5. Create another new sketch plane, and use the **General Dimension** tool to size and locate the rectangle.

6. Right-click the mouse and select the **Done** option, then the **Finish Sketch** option.
7. Click the **Cut** tool.

The **Cut** dialog box will appear.

8. Click the **Cut Across Bend** box, then **OK.**

Hole Patterns

A hole pattern is created from an existing hole. See Figure 13-18.

1. Create a new sketch plane, define a hole center, then add the hole.

The dimensions for the hole come from the dimensions given in Figure 13-1.

2. Click the **Rectangular Pattern** tool.

The **Rectangular Pattern** dialog box will appear.

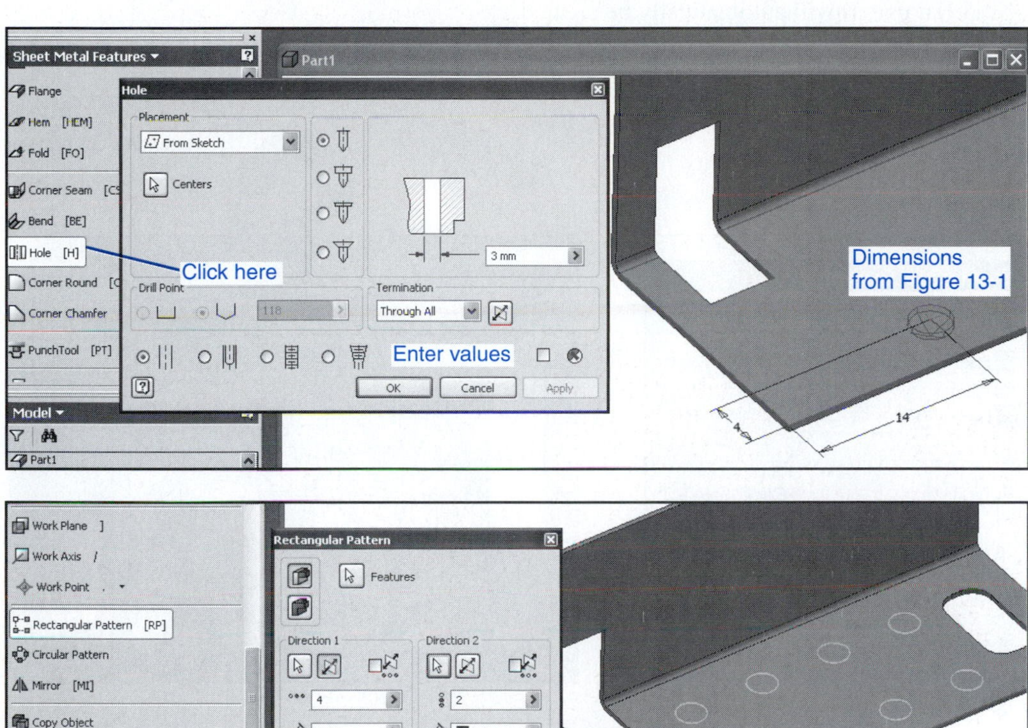

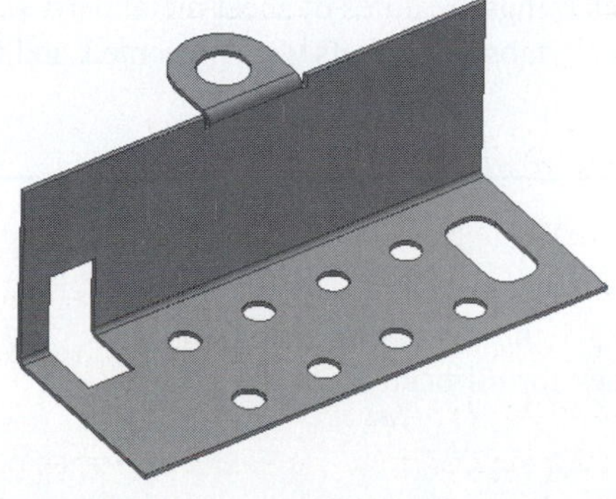

Figure 13-18

3. Define the hole as the **Feature.**
4. Click the arrow under the **Direction 1** heading, then the top front edge of the part. Use the **Flip** button to change directions if necessary.
5. Set the number of holes under **Direction 1** for **4** and the spacing for **8 mm.**
6. Click the arrow under the **Direction 2** heading, then click the left front edge of the part to define the direction.
7. Set the number of holes for **2** and the distance for **8.**
8. Click **OK.**

FLAT PATTERNS

Flat patterns of 3D sheet metal parts can be created using the **Flat Pattern** tool. See Figure 13-19.

1. Click the **Flat Pattern** tool.

A flat patternwill automatically be created.

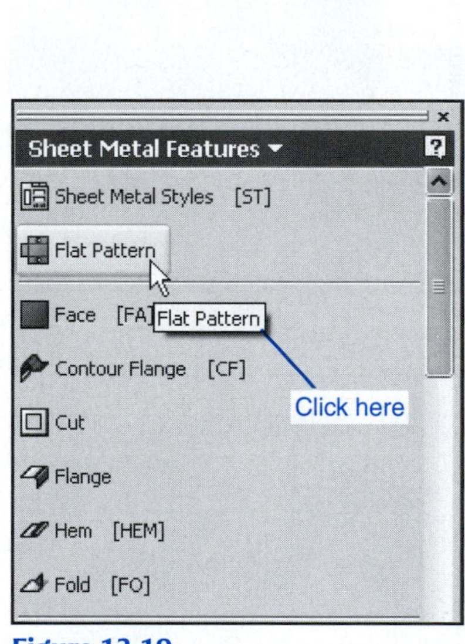

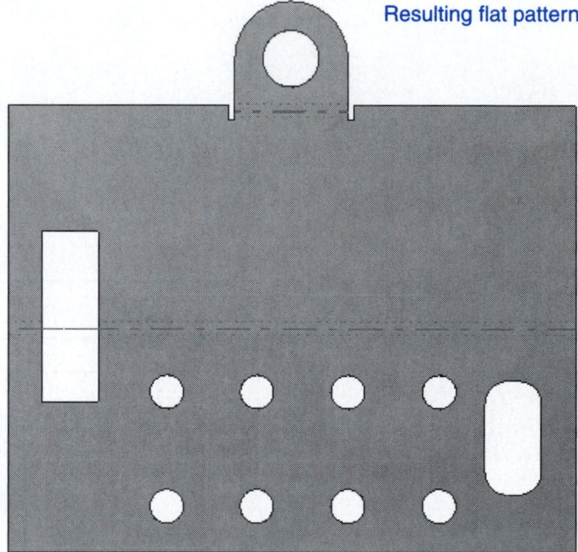

Figure 13-19

SUMMARY

This chapter defined and illustrated how to create sheet metal drawings from 3D models and orthographic drawings. Features of sheet metal parts such as bend radii, flanges, tabs, and reliefs were presented, and flat patterns were created.

CHAPTER PROJECTS

Project 13-1:

Redraw the sheet metal parts in Figures P13-1 through P13-6 using the given dimensions. Use the default values for all bend radii and reliefs.

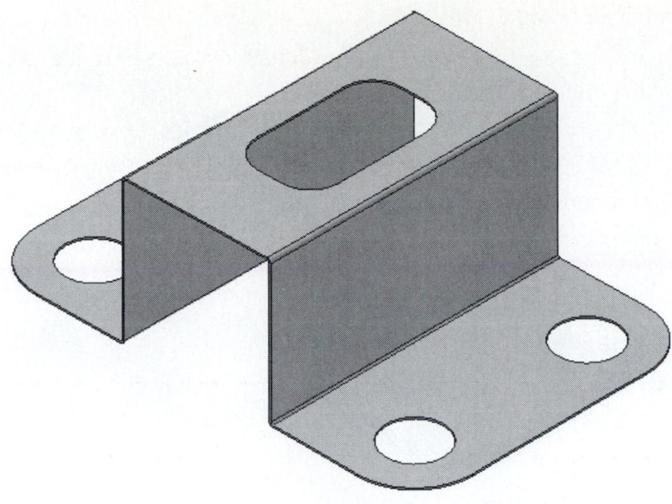

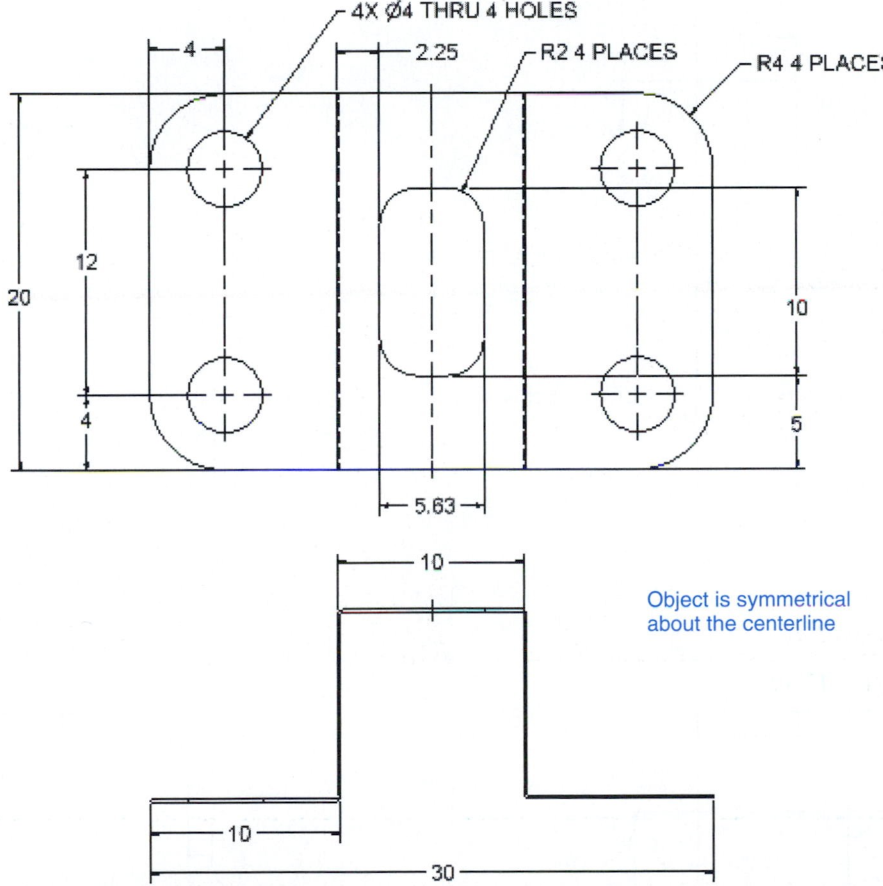

Figure P13-1 MILLIMETERS

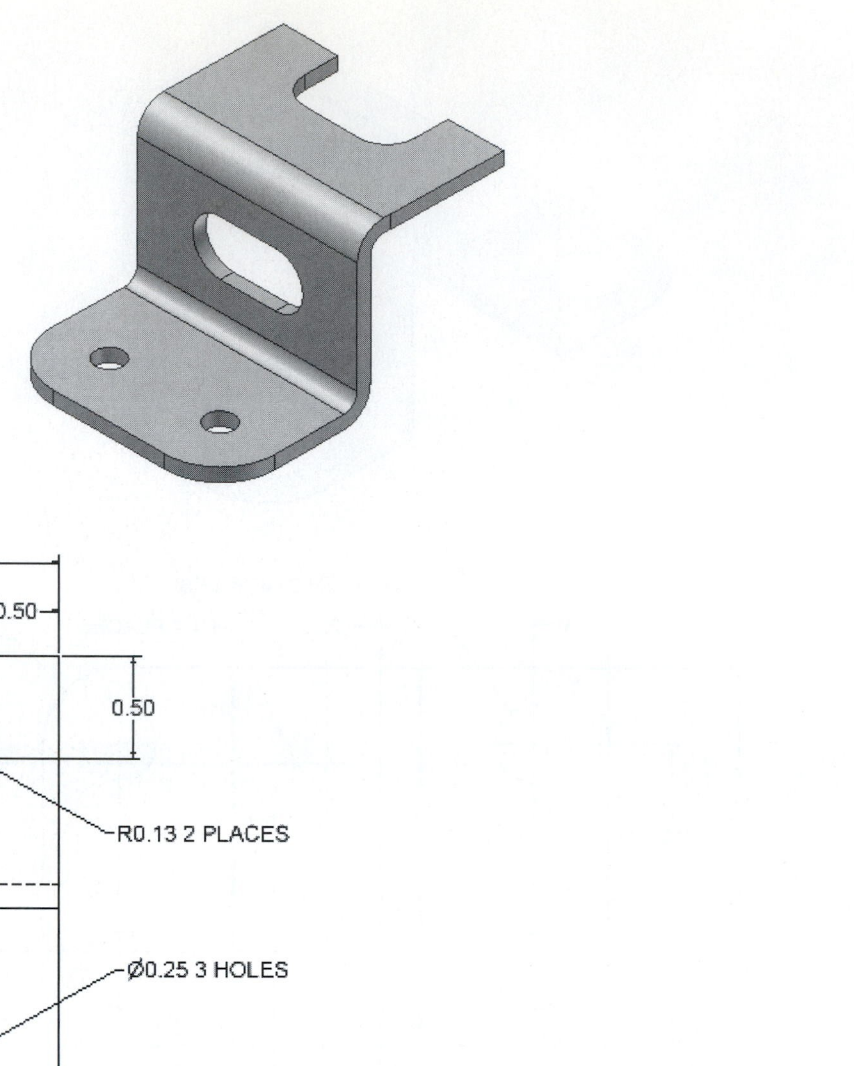

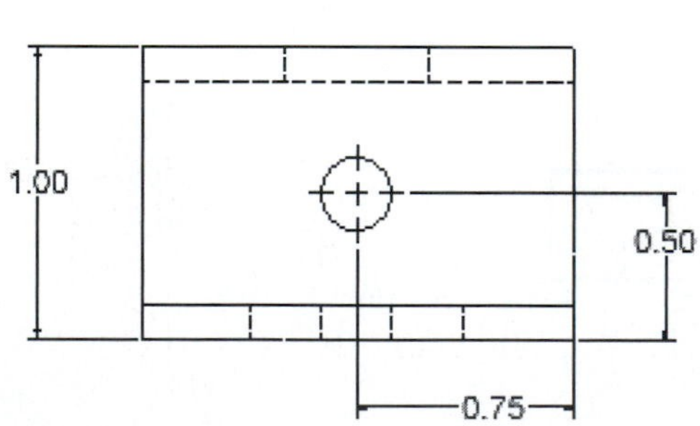

Figure P13-2 INCHES

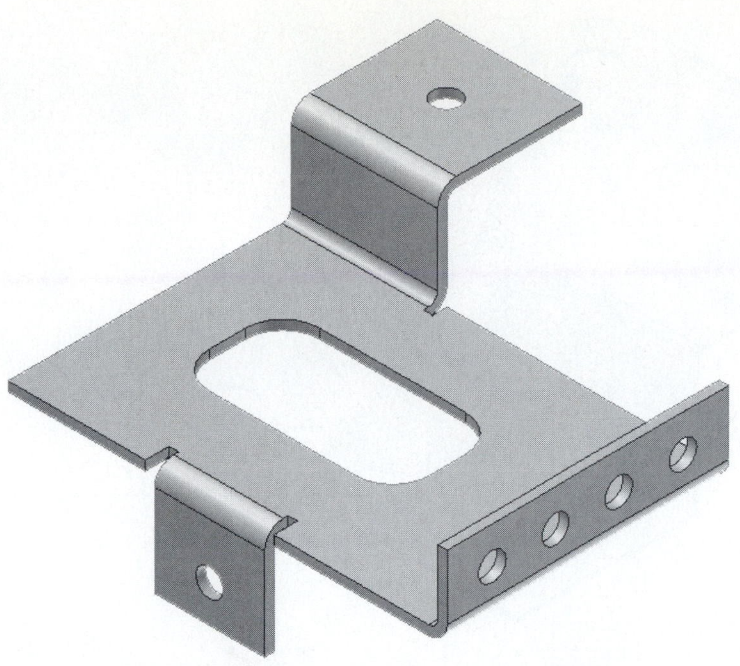

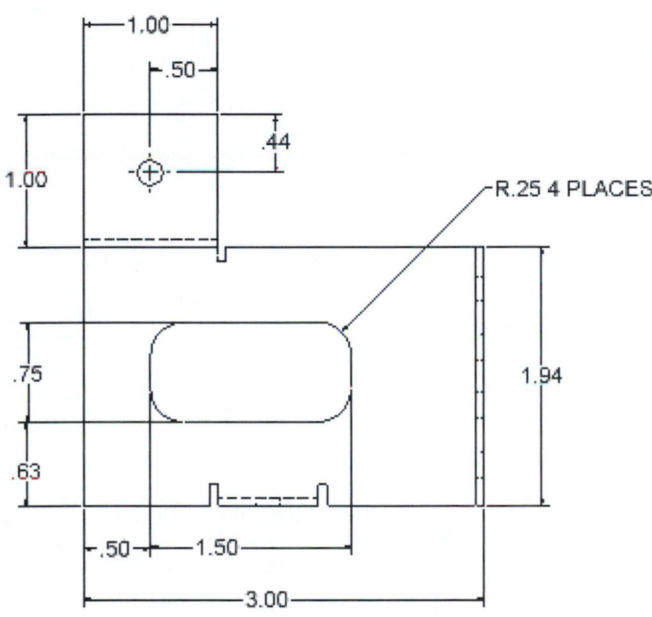

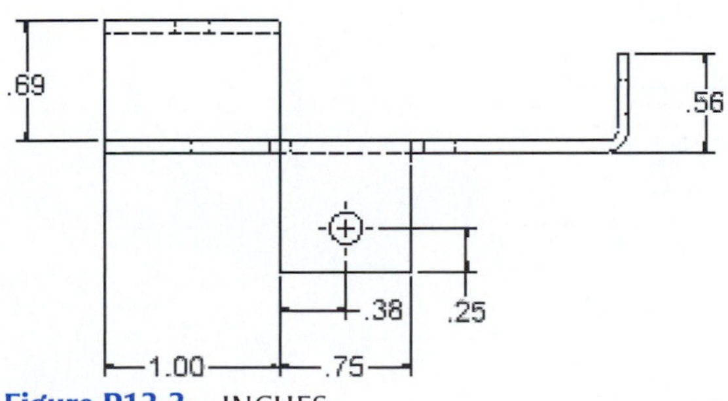

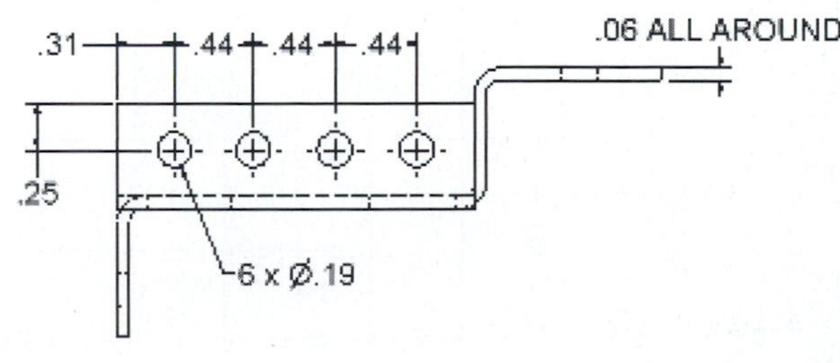

Figure P13-3 INCHES

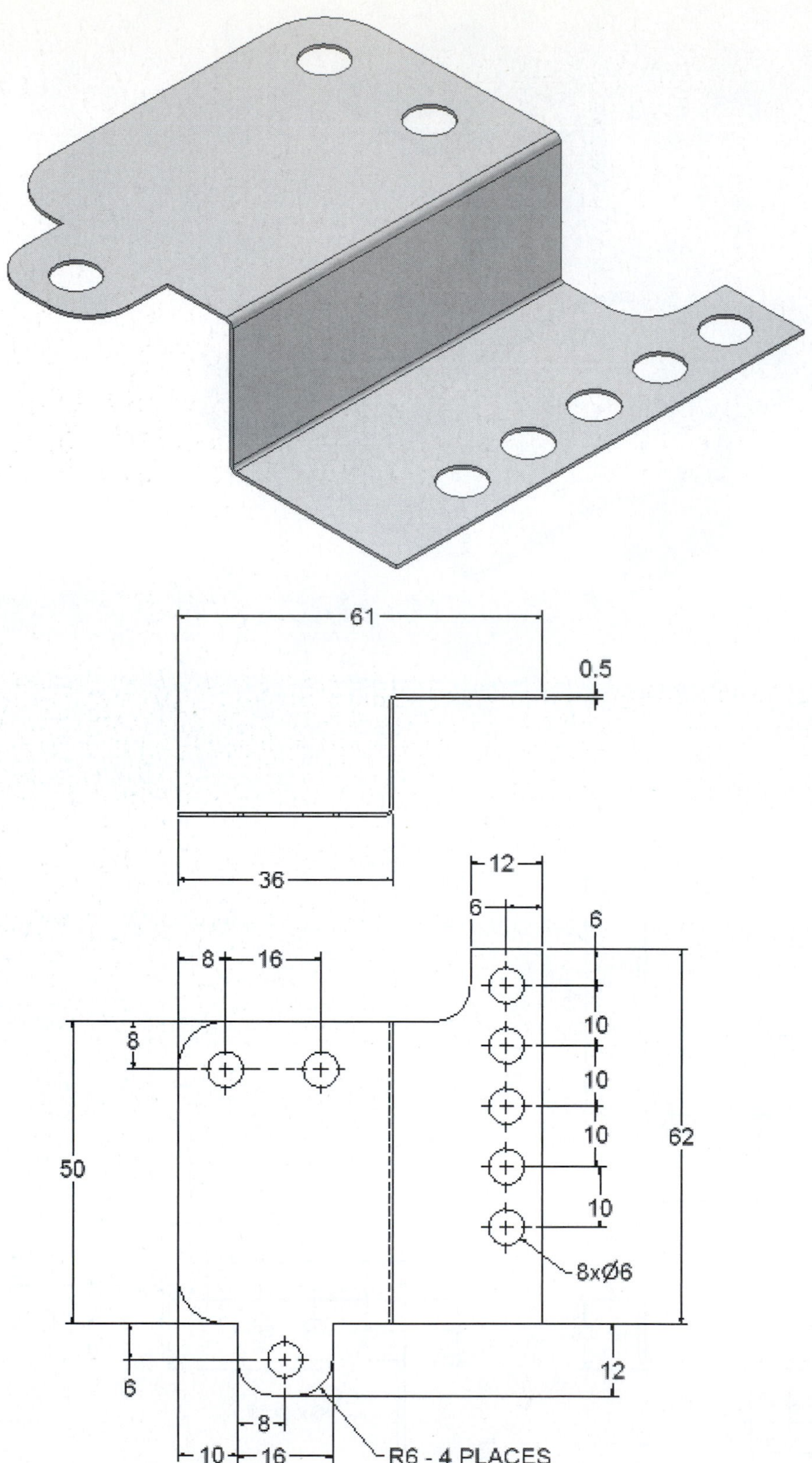

Figure P13-4 MILLIMETERS

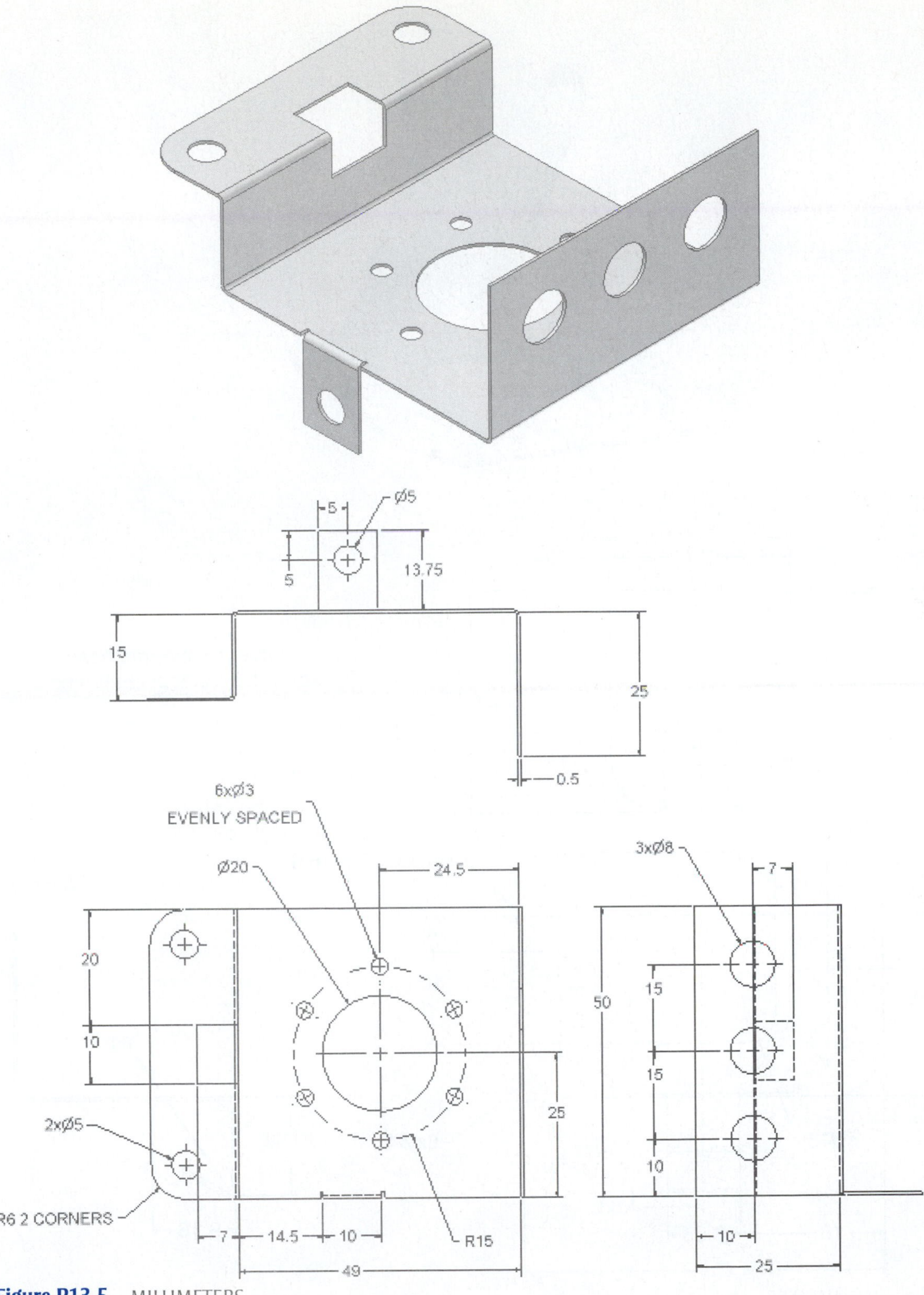

Figure P13-5 MILLIMETERS

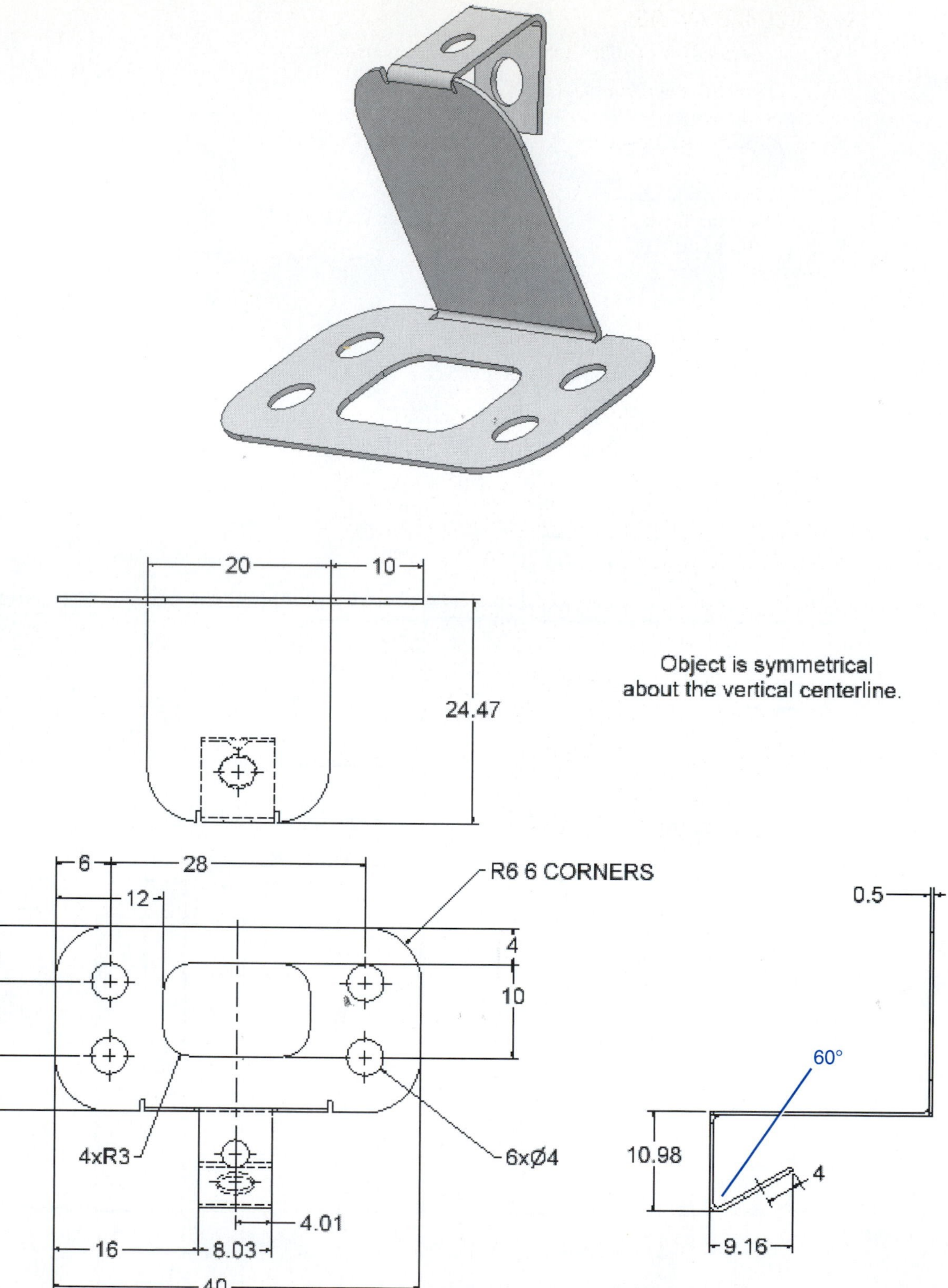

Figure P13-6 MILLIMETERS

Project 13-2: INCHES

Design and draw a box similar to that shown that has a capacity of

A. 100 cubic centimeters and is a cube.
B. 4 fluid ounces.
C. 100 cubic centimeters and is rectangular with the length of one side 2 times the length of the other.
D. 125 cubic inches and is a cube.
E. 125 cubic inches and is rectangular with the length of one side 1.5 times the length of the other.
F. 8 fluid ounces.

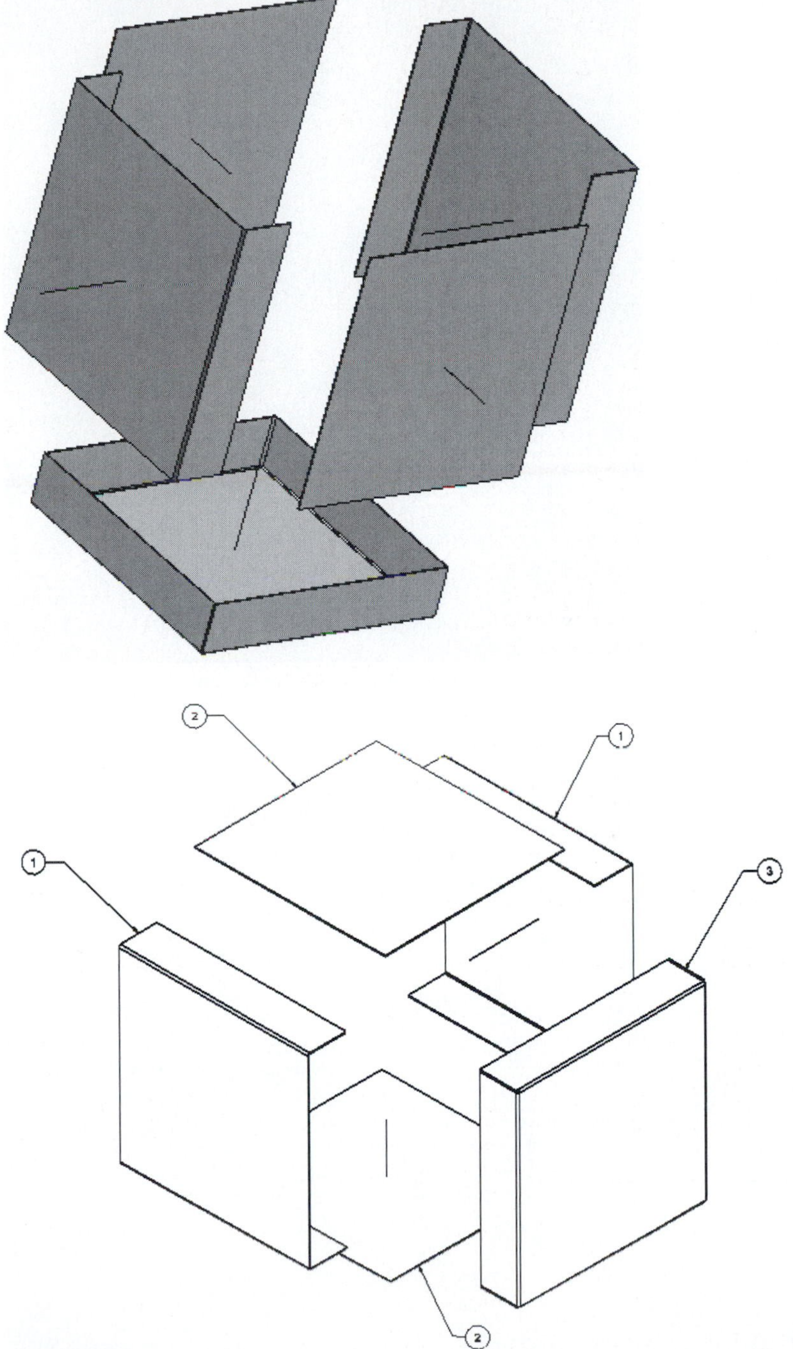

Figure P13-7

Weldment Drawings 14

Chapter Objectives

- Understand how to design and draw weldments.
- Explain fillet and groove welds.
- Show how to create weld symbols.

INTRODUCTION

Weldments are assemblies made from several smaller parts that have been welded together. Weldments are often cheaper to manufacture because they save extensive machining time or replace expensive castings.

weldment: An assembly made from several smaller parts that have been welded together.

FILLET WELDS

Figure 14-1 shows a simple weldment. It was created from two 0.375-in. thick plates and joined by a ***fillet weld.*** The base plate is 2.00 × 4.00 in., and the vertical plate is 1.25 × 4.00 in. Both parts are made from low-carbon steel.

fillet weld: A weld usually created at 45° to join pieces that are perpendicular to each other; may be continuous or intermittent.

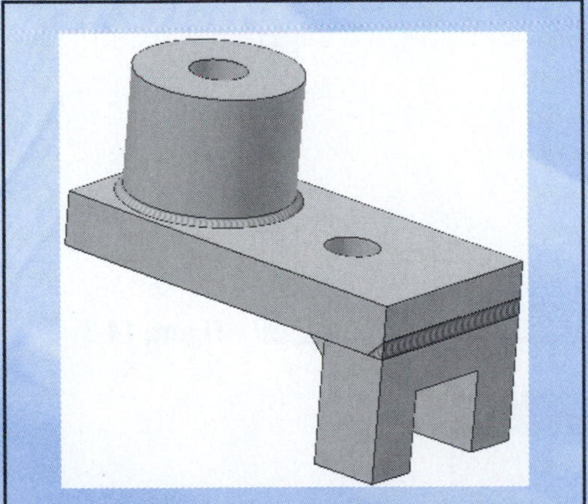

Figure 14-1

Exercise 14-1: Creating the Components

1. Click the **New** tool, then the **English** tab, then **Weldment (ANSI).iam.**
2. Click the **Create Component** tool.

The **Create In-Place Component** dialog box will appear. See Figure 14-2.

3. Define a new component named **Base, Weld.**
4. Click the **Browse Template** box.

The **Open Template** dialog box will appear. See Figure 14-3.

5. Select the **Standard (in).ipt** format. Click **OK.**

The **Create In-Place Component** dialog box will reappear. Note that **English\Standard (in).ipt** is the new template. See Figure 14-4.

6. Click **OK.**

The 2D **Sketch Panel** will appear.

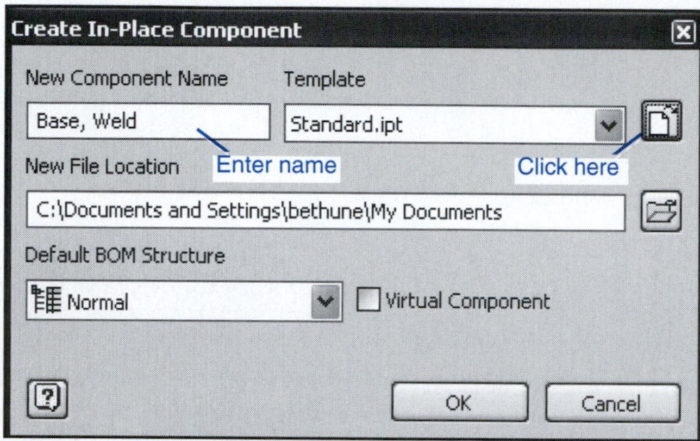

Figure 14-2

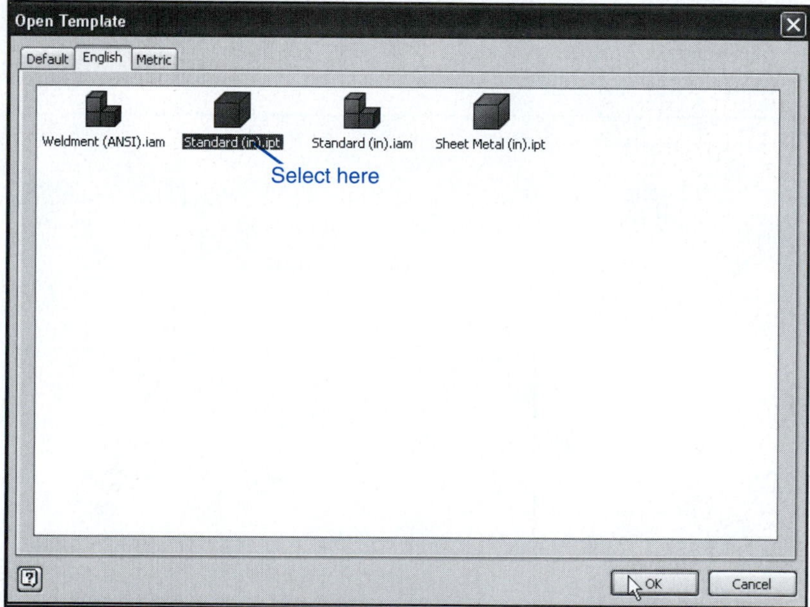

Figure 14-3

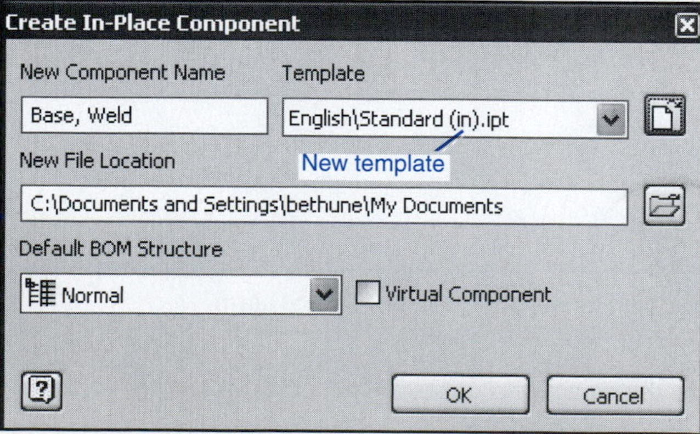

Figure 14-4

7. Sketch a **2.00 × 4.00** rectangle.
8. Right-click the mouse and click **Done.** Right-click the mouse again and click **Finish Sketch.**

The **Part Features** panel will appear.

9. Click the **Extrude** tool and define a thickness of **0.375** for the rectangle.

See Figure 14-5.

Exercise 14-2: Creating a Second Plate

1. Right-click the mouse and select the **Finish Edit** option.

See Figure 14-6.

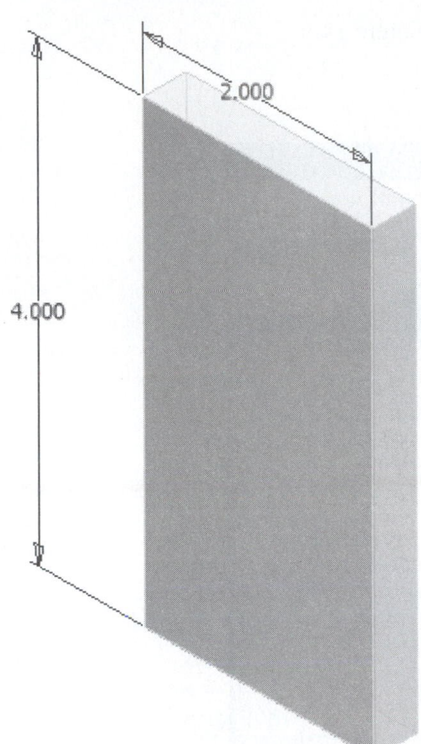

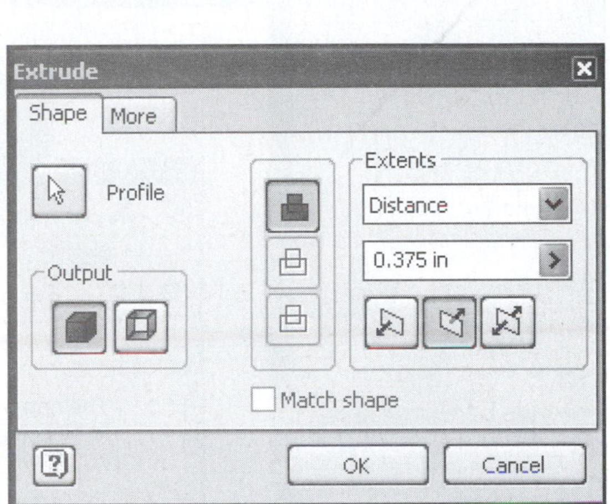

Figure 14-5

Figure 14-6

2. Select the **Copy Components** tool.

See Figure 14-7. The **Copy Components: Status** dialog box will appear.

3. Click the **Base, Weld:1** component.

The component file name will appear in the dialog box.

4. Click **OK.**

The **Copy Components: File Names** dialog box will appear. See Figure 14-8.

The copied component will be assigned a new file name. In this example the new name is **Base, Weld_CPY.ipt.** Another name could be entered.

5. Click the **Increment** box and remove the check mark. Click the **Apply** box.

The component copy will appear on the screen .See Figure 14-9.

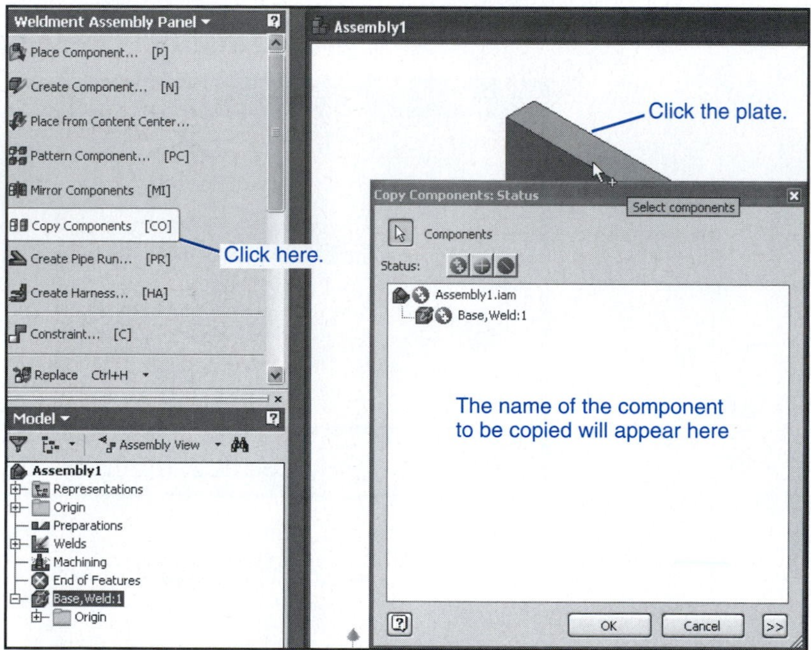

Figure 14-7

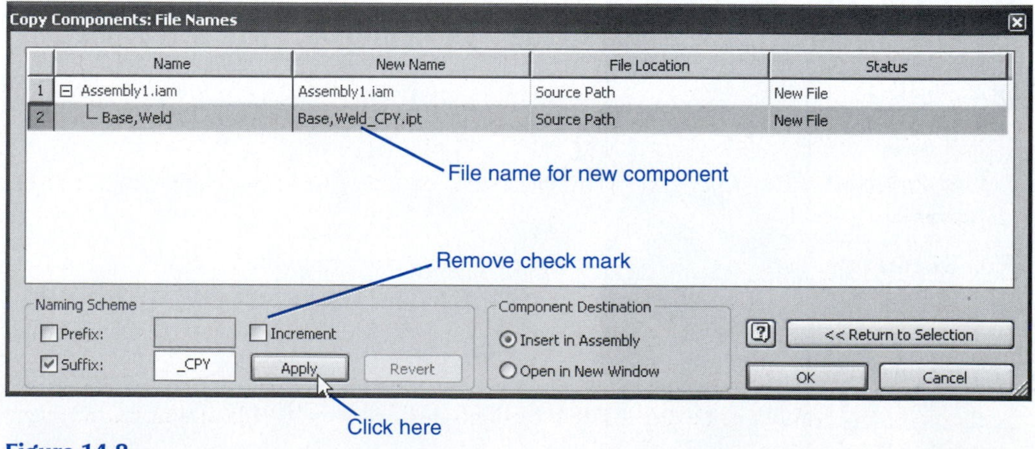

Figure 14-8

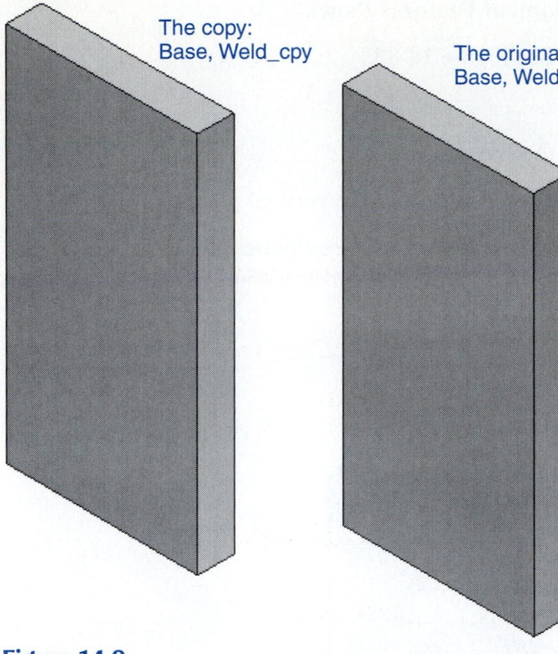

Figure 14-9

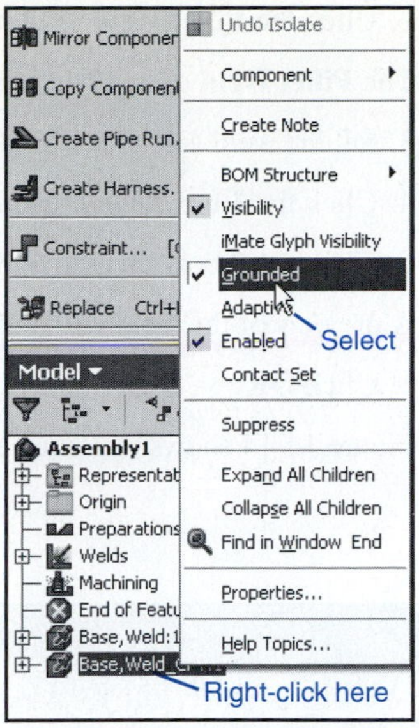

Figure 14-10

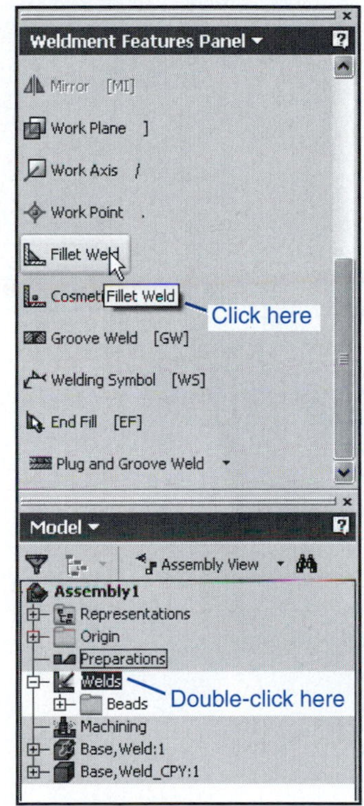

Figure 14-12

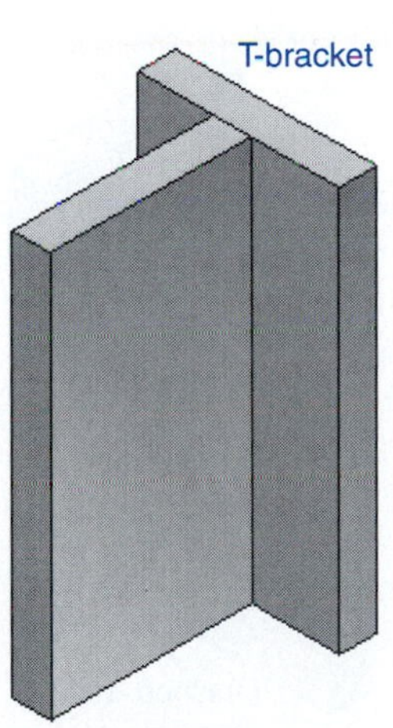

Figure 14-11

Exercise 14-3: Creating a T-Bracket

1. Right-click **Base, Weld_CPY** in the browser box and click the **Grounded** option.

See Figure 14-10. This will remove the grounded constraint. The pushpin icon will disappear from the browser box.

2. Use the **Constraint** tool and assemble the two plates to form a T-bracket.

See Figure 14-11.

Exercise 14-4: Creating the Welds

1. Access the **Weldment Features Panel** by double-clicking the **Welds** tool in the browser.

See Figure 14-12.

2. Click the **Fillet Weld** tool on the **Weldment Features Panel.**

The **Fillet Weld** dialog box will appear. See Figure 14-13.

3. Set the weld size to **0.125,** as shown.
4. Click the **1** box, then click the top of the **Base, Weld.**
5. Click the **2** box, then click the front vertical surface of the vertical plate.

A preview of the weld will appear as small right triangles. See Figure 14-13.

6. Click **OK.**

Figure 14-14 shows the finished weld.

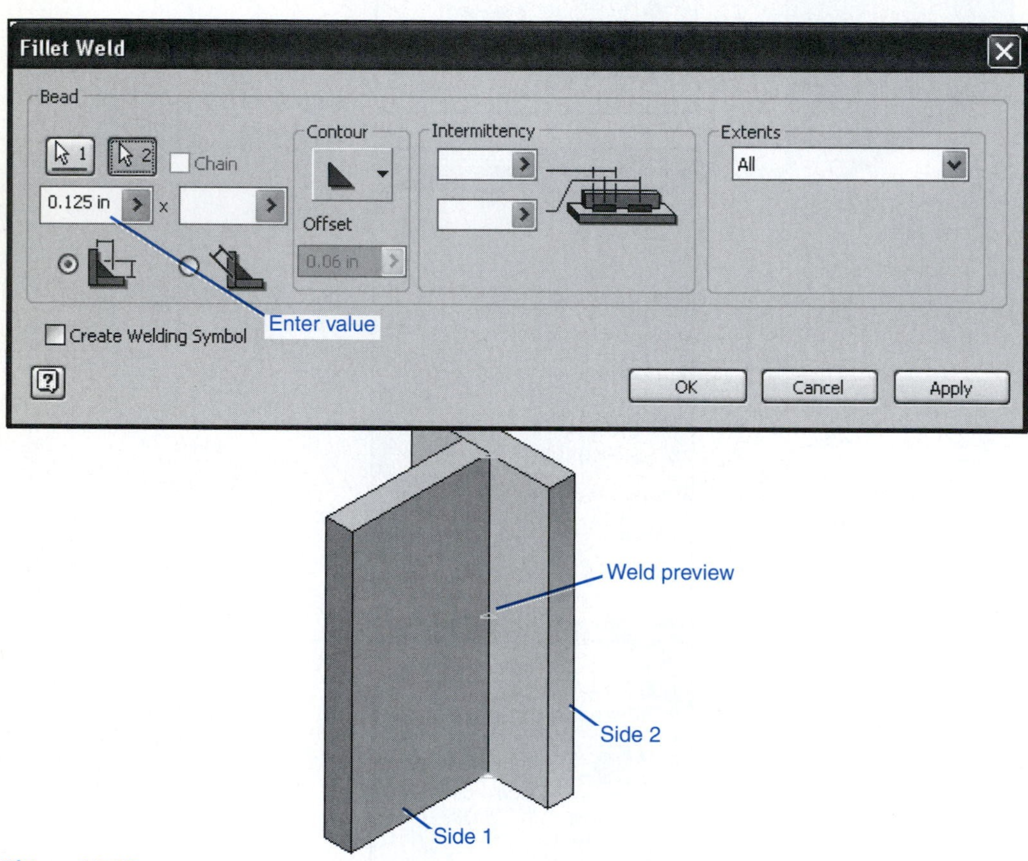

Figure 14-13

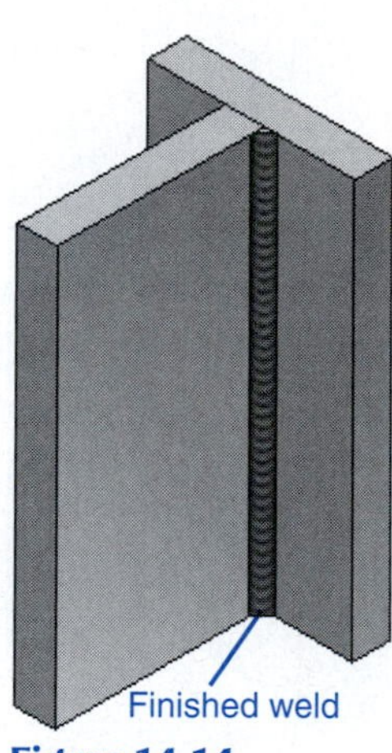

Figure 14-14

Intermittent Fillet Welds

Fillet welds may be located intermittently along a weld line.

1. Rotate the T-bracket created in the previous section.
2. Click the **Fillet Weld** tool.

The **Fillet Weld** dialog box will appear. See Figure 14-15.

3. Enter the appropriate **Intermittency** values.

Note that values **1.00** and **1.50** have been entered in the **Intermittency** box. The 1.00 value is the length of each weld, and the 1.50 value is the distance from the center of one weld to the center of the next.

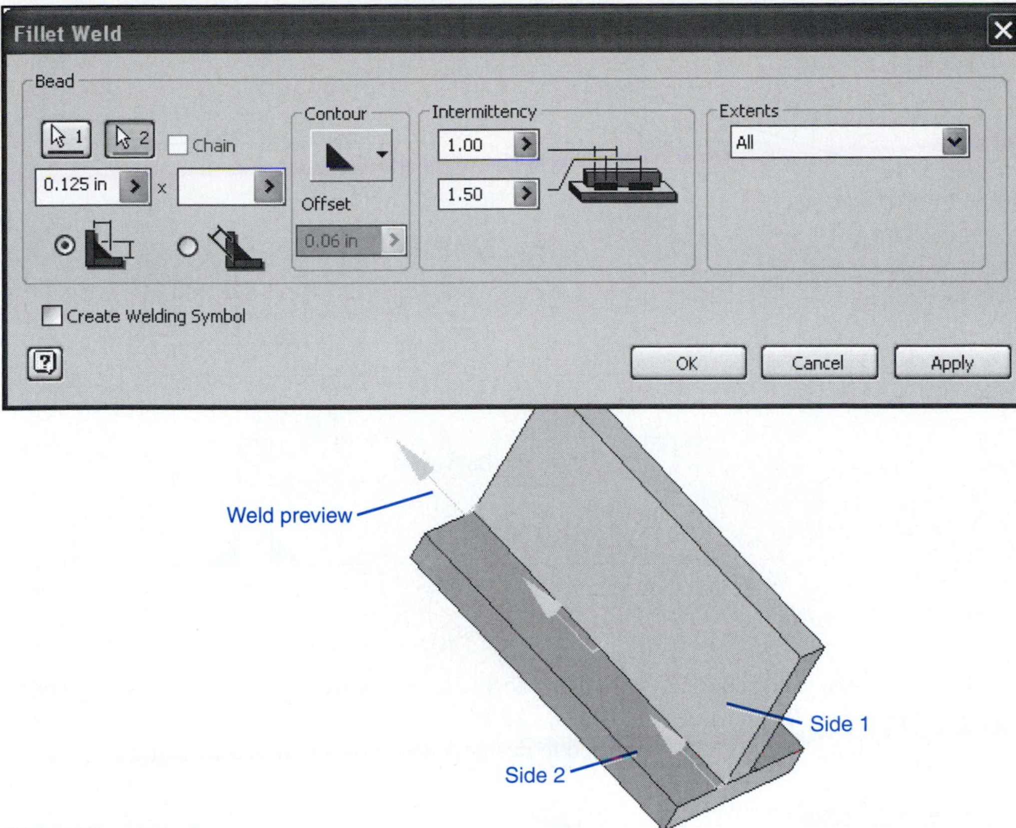

Figure 14-15

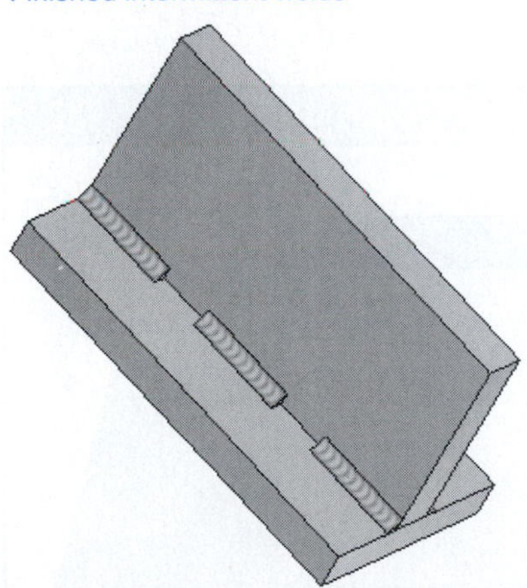

Figure 14-16

4. Click **Apply.**

Figure 14-16 shows the finished intermittent welds.

Weld Symbols

Welds are defined on drawings using symbols. The symbol for a fillet weld is shown in Figure 14-17. Note that the location of the flaglike portion of the symbol defines the location of the weld. It is not always possible to point directly at a weld location, so the **Other side** symbol is very useful.

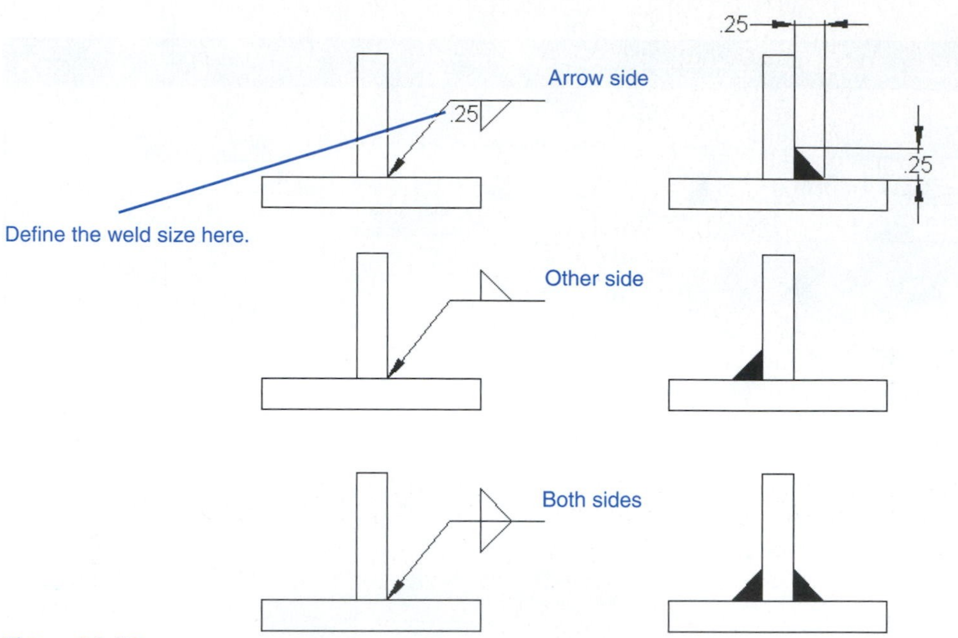

Figure 14-17

The size of the weld is defined as shown. Most fillet welds are created at 45°, although other angles are possible. A fillet weld defined by .25 indicates that the 45° weld is defined by two sides, both .25 long. Metric values are used to define a weld size in the same manner.

Exercise 14-5: Adding a Weld Symbol to a Drawing

1. Access the **Fillet Weld** dialog box and click the **Create Welding Symbol** box or click the **Welding Symbol** tool on the **Weldment Features Panel.**

See Figure 14-18

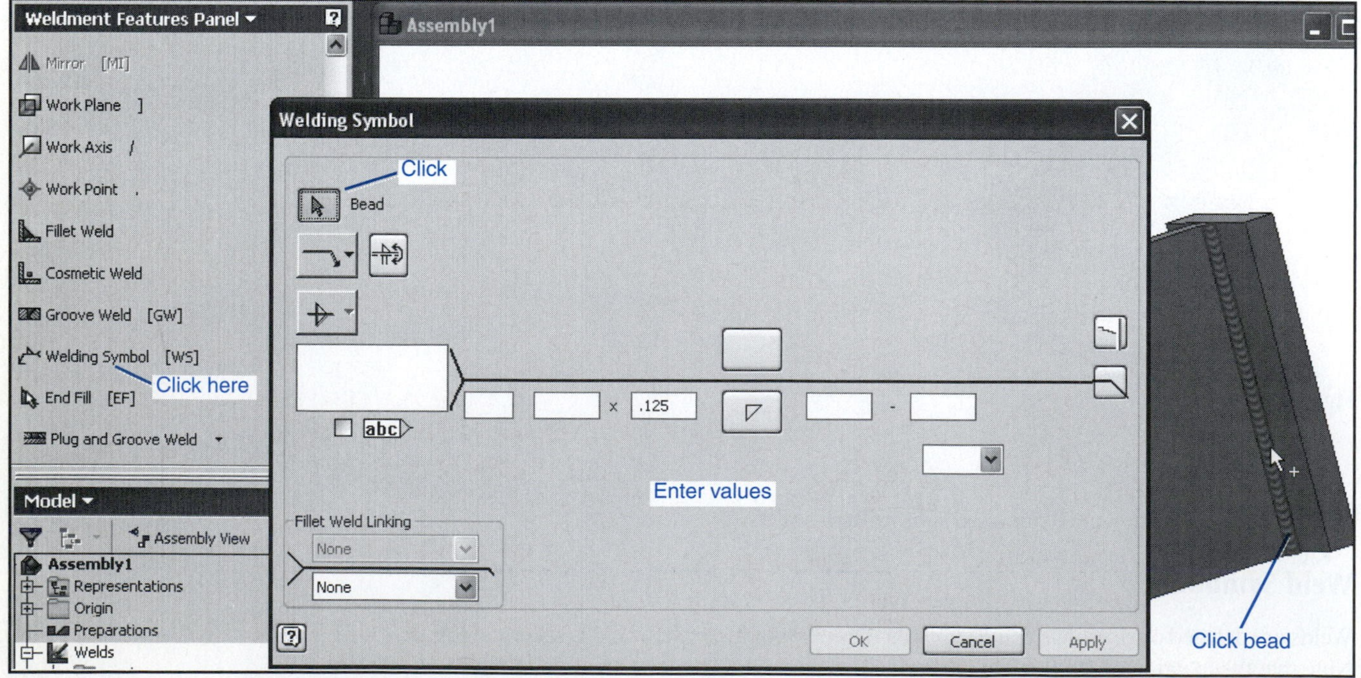

Figure 14-18

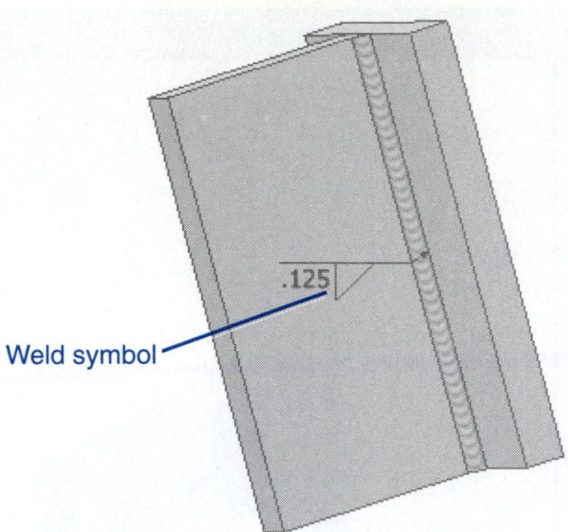

Figure 14-19

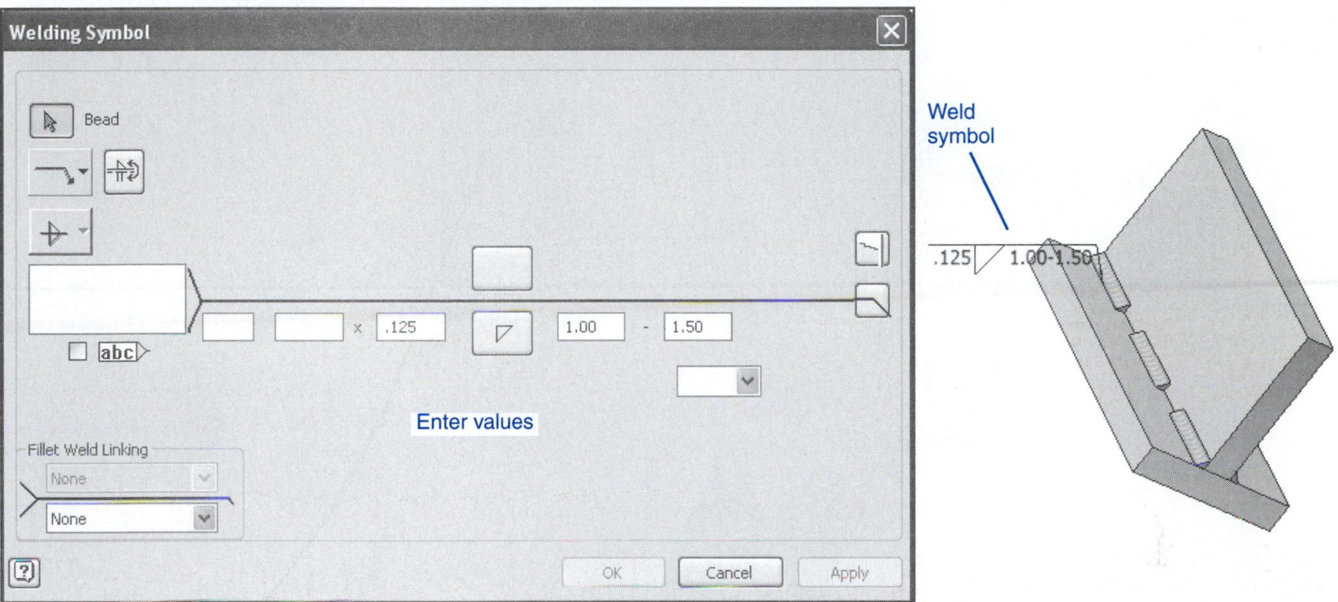

Figure 14-20

2. Define the weld's size, then add the weld-size value at the left of the fillet weld symbol.
3. Click **OK.**

Figure 14-19 shows a weld drawing with a weld symbol.
Figure 14-20 shows the fillet weld symbol for an intermittent fillet weld.

All Around

The addition of a circle to the fillet weld symbol indicates that the weld is to placed *all around* the object. Figure 14-21 shows a cylinder welded to a plate. In this example the fillet weld was defined using millimeters. A 5-mm × 5-mm weld is to be created all the way around the cylinder.

Exercise 14-6: Creating an All-Around Fillet Weld

1. Create a new drawing using the **Weldment ISO.iam** format.
2. Create a **5 × 40 × 40-mm** plate and a **Ø20 × 30** cylinder.
3. Assemble the parts so that the cylinder is centered on the plate.
4. Click **Weld** in the browser box, then click the **Fillet Weld** tool.

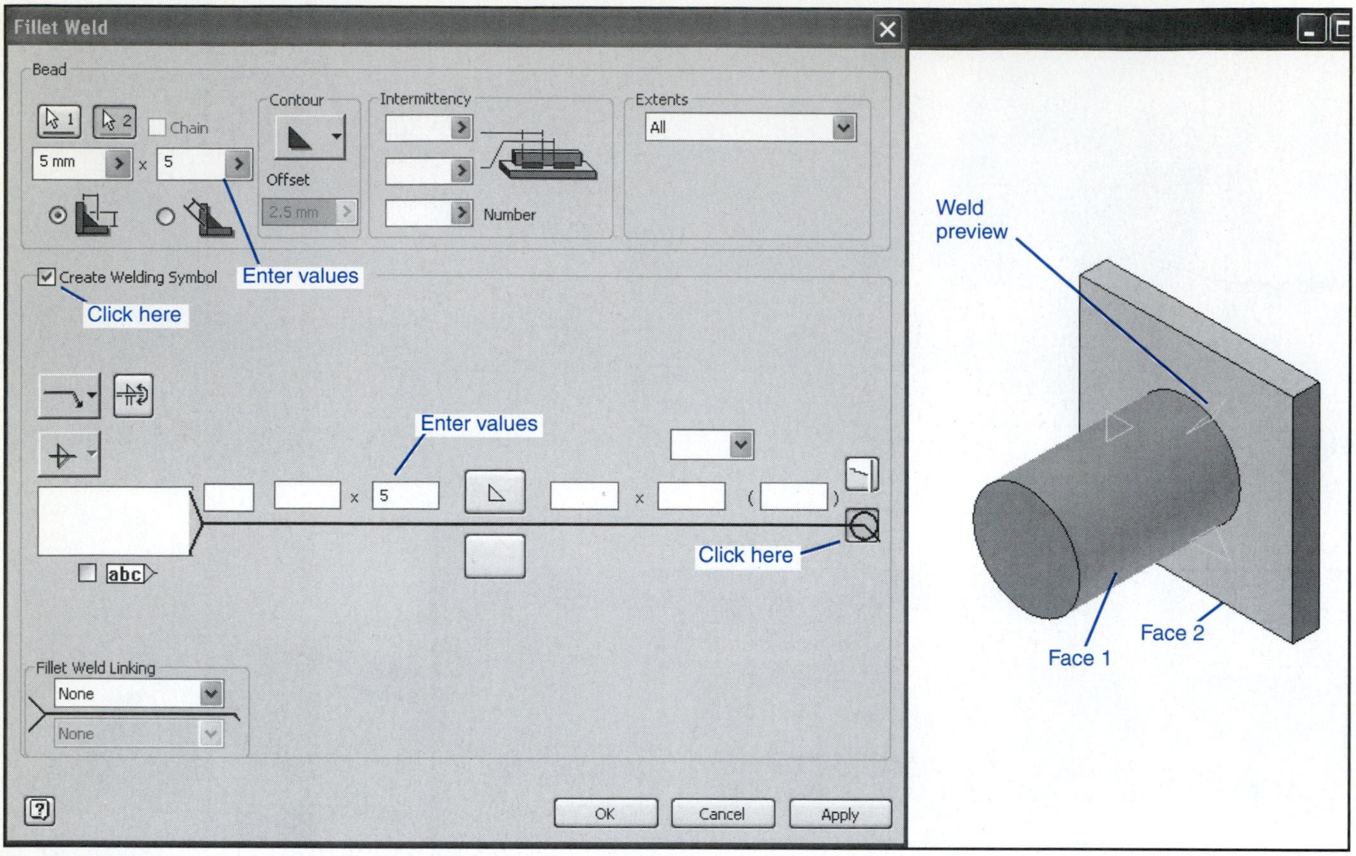

Figure 14-21

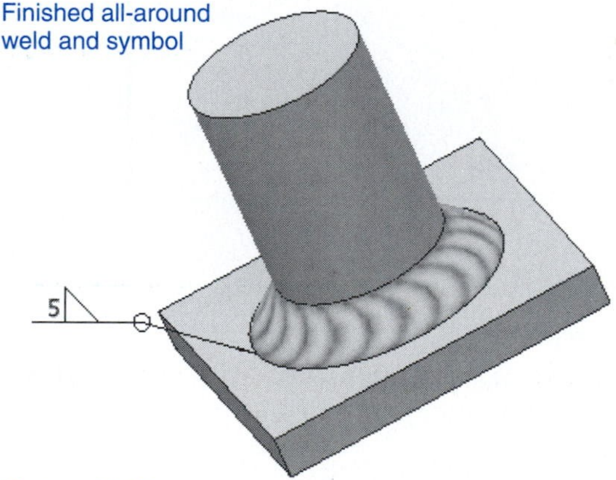

Figure 14-22

The **Fillet Weld** dialog box will appear. See Figure 14-21.

5. Set the weld size for **5 × 5 mm,** then click the **Create Welding Symbol** box.
6. Enter a fillet weld value of **5.**
7. Click the right end of the arrow symbol to create a circle around the bend in the arrow.
8. Click **OK.**

Figure 14-22 shows the finished weld and symbol.

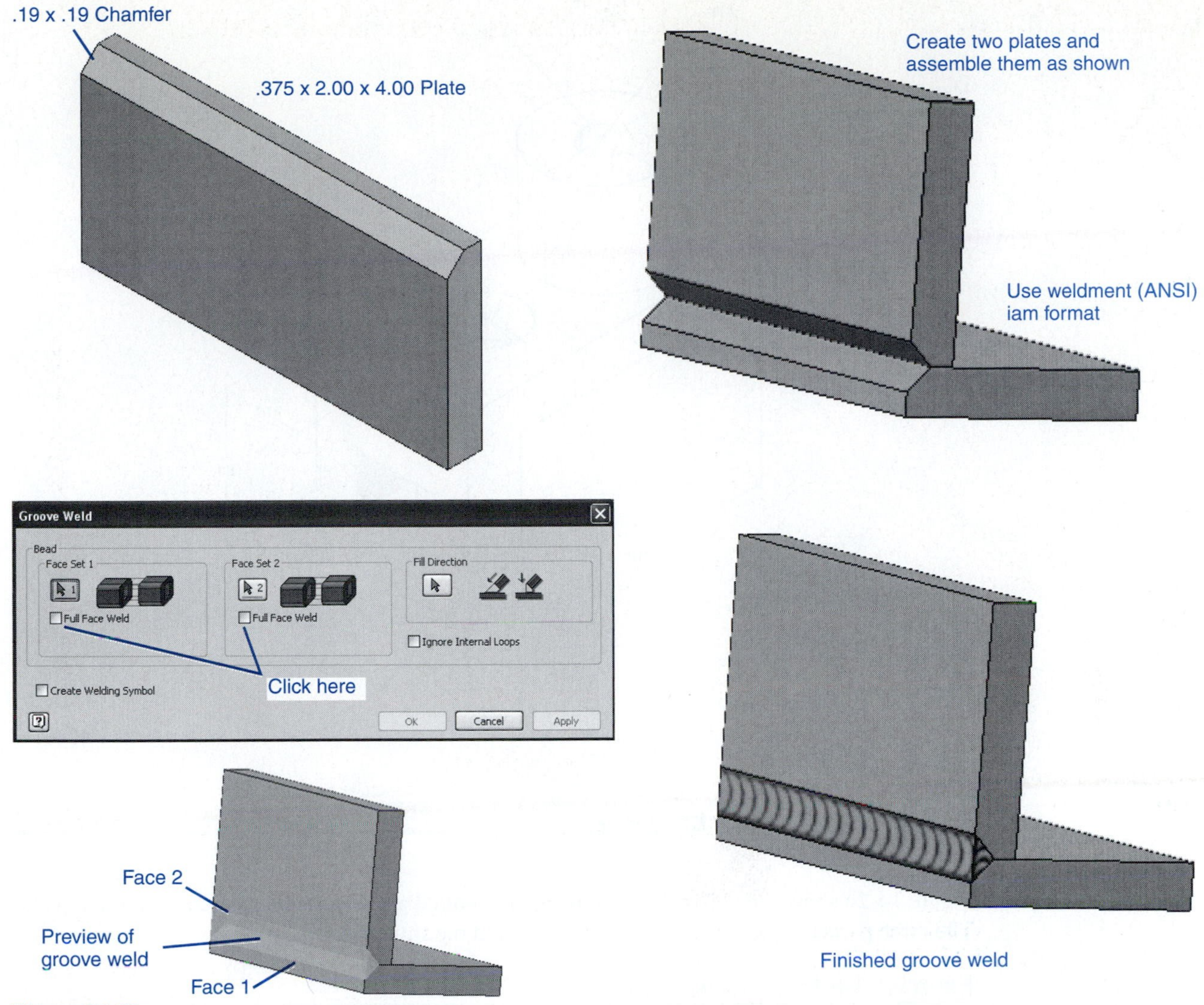

Figure 14-23

WELDMENTS—GROOVE WELDS

Groove welds are used when two parts abut. A chamfer is cut into each part and the weld is placed in the resulting groove. Figure 14-23 shows an L-bracket created as a weldment.

groove weld: A weld used when two parts abut, placed in the groove formed when a chamfer is cut into each part.

It was created as follows.

1. Draw a **.375 × 2.00 × 4.00-in.** plate and cut a **.19 × .19** chamfer as shown.
2. Create a weldment drawing using the **Weldment (ANSI).iam** format.
3. Add two plates to the drawing and assemble them as shown.
4. Double-click the word **Weld** in the browser box, then click the **Groove Weld** tool in the **Weldment Features Panel.**

The **Groove Weld** dialog box will appear.

5. Click the **Full Face Weld** boxes for both **Face Set 1** and **Face Set 2.**
6. Define **Face 1** and **Face 2.**
7. Click **OK.**

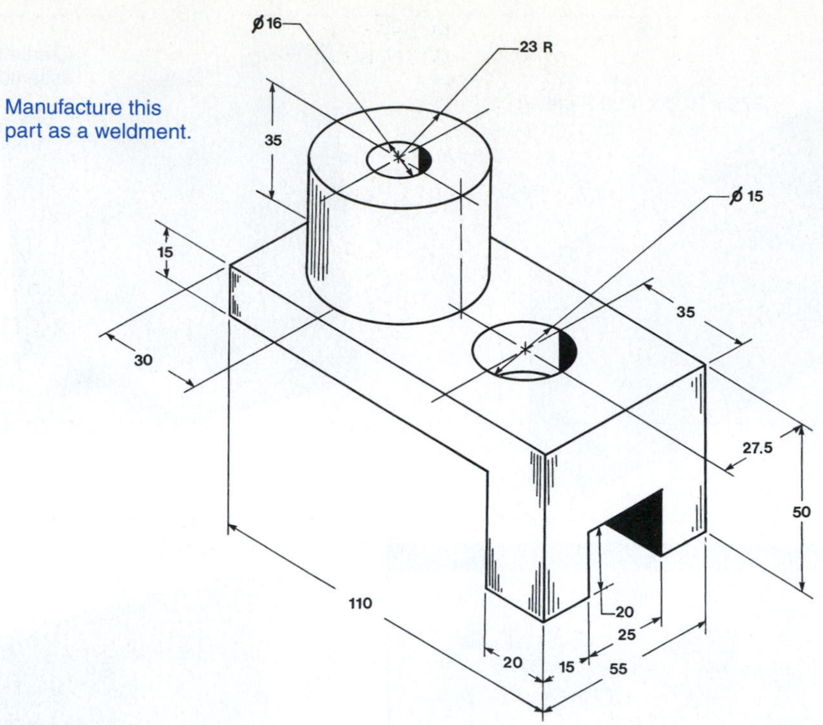

Figure 14-24

Sample Problem SP14-1

Figure 14-24 shows an object that is to be manufactured as a weldment created from three parts. The three parts are the barrel, the center plate, and the front plate.

Exercise 14-7: Creating the Weldment

Use the **Weldment (ISO).iam** format. The dimensions for each part were derived from those given in Figure 14-24.

1. Create a **Metric, Weldment (ISO). iam** drawing.
2. Use the **Create Component** tool (use the **Standard (mm).ipt** format) and create a **15 × 55 × 110** plate with a **5 × 5** chamfer and a **Ø15** as shown.
3. Right-click and select the **Finish Edit** option.
4. Use the **Create Component** tool And create a **Ø46 × 35** barrel with a **Ø16** through-all hole. Use the **Constraint** tool to position the barrel.
5. Reorient the model and use the **Create Component** tool to create a **20 × 55 × 35** flange with a **25 × 20** cutout as shown.
6. Add a grove weld as shown.
7. Add a **5 × 5** fillet weld as shown.
8. Add a **2 × 2** fillet weld around the barrel as shown.

Figure 14-25 shows the weldment.

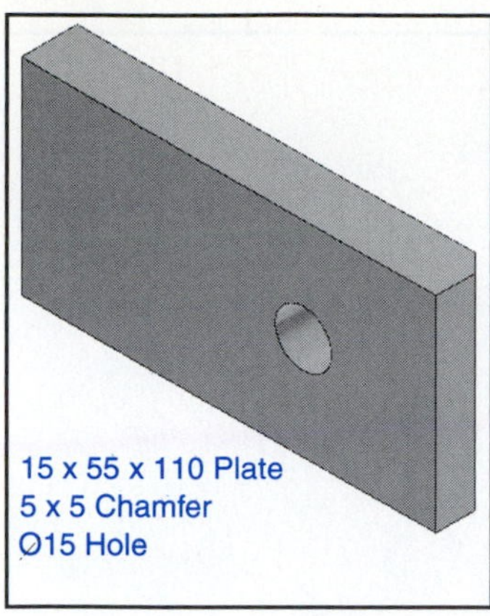

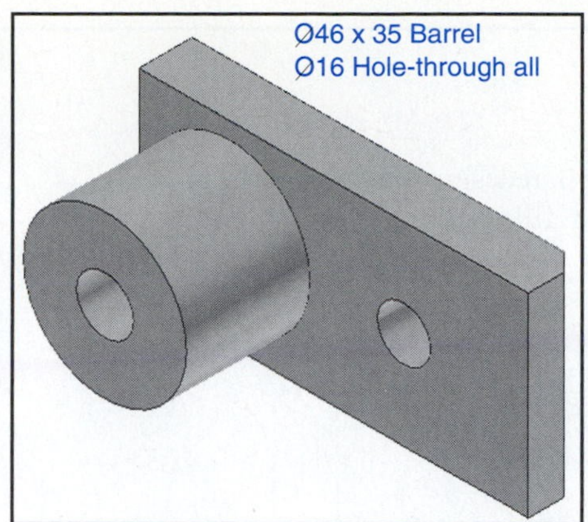

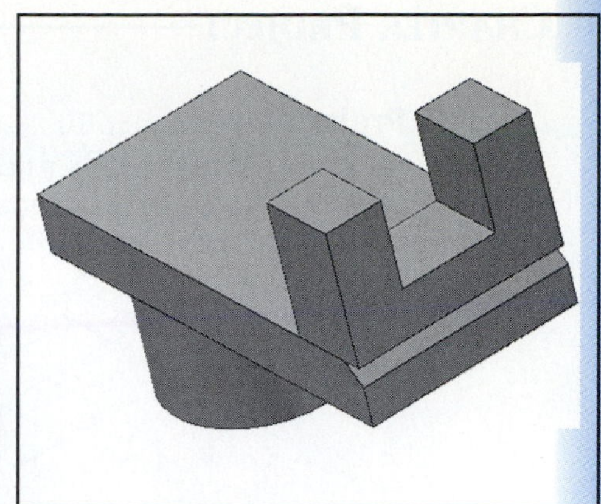

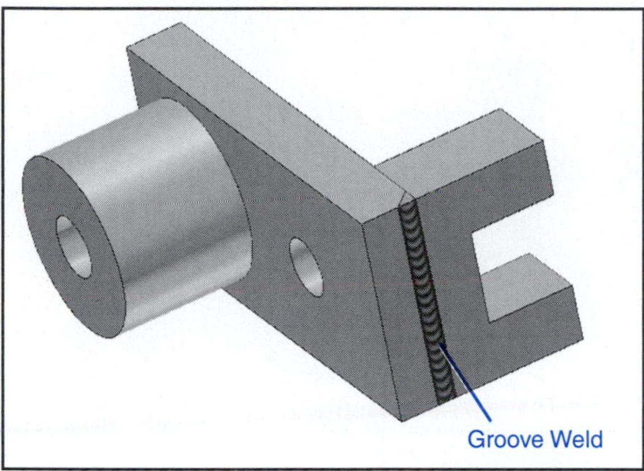

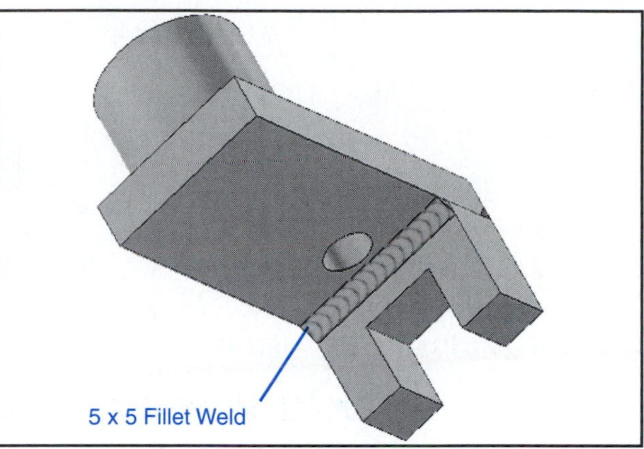

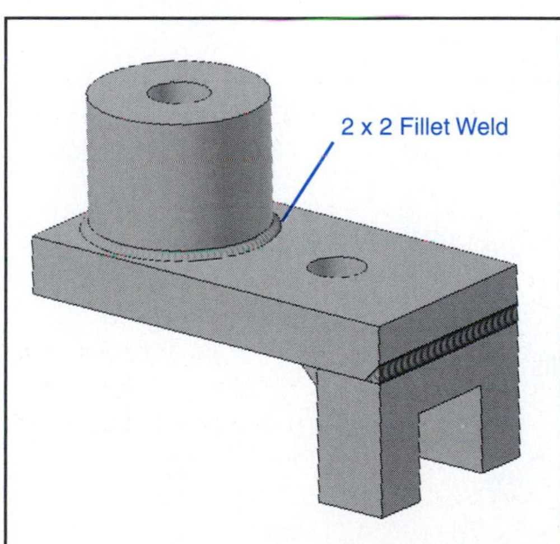

Figure 14-25

SUMMARY

This chapter illustrated how to create and draw weldments, which are assemblies of several parts welded together. Fillet welds, both continuous and intermittent as well as all around, were introduced; welding symbols were added to drawings; and groove welds were illustrated.

Chapter Project

Project 14-1:

For Figures P14-1 through P14-10, redesign the given parts as weldments. Use either 5-mm or .20-in. fillet welds.

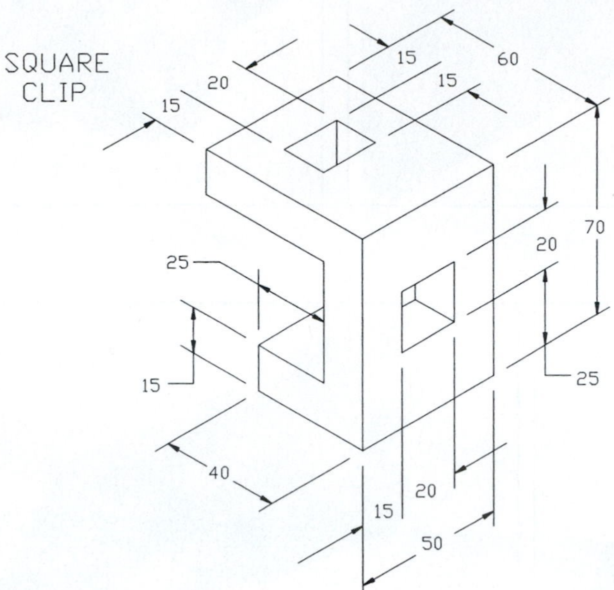

Figure P14-1 MILLIMETERS

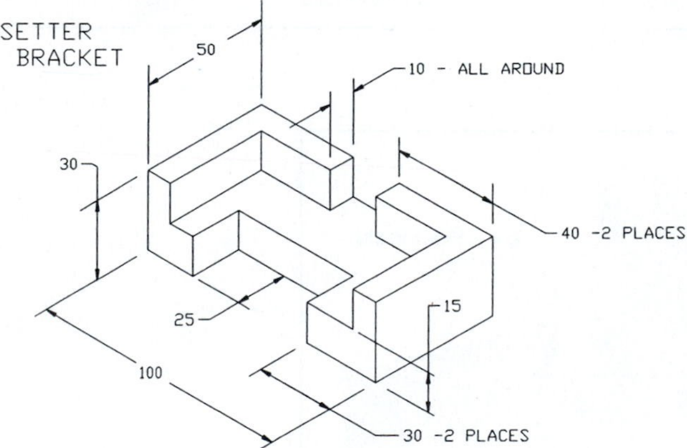

Figure P14-2 MILLIMETERS

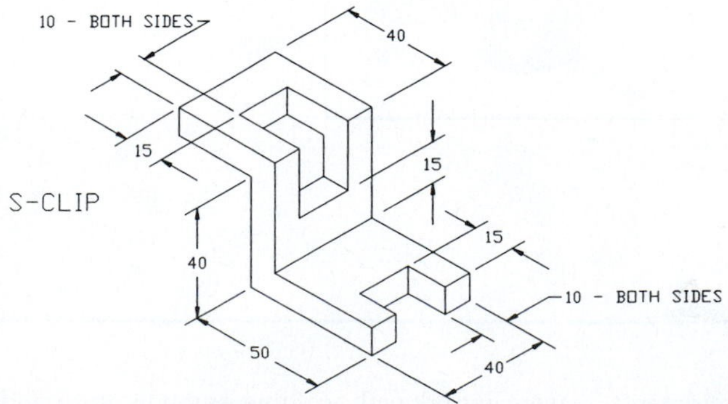

Figure P14-3 MILLIMETERS

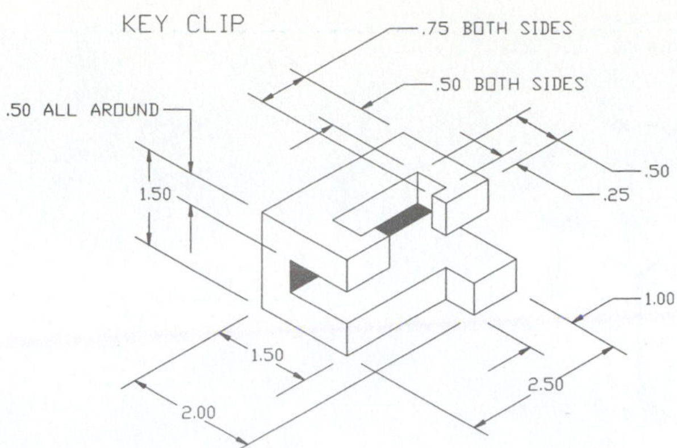

Figure P14-4 INCHES

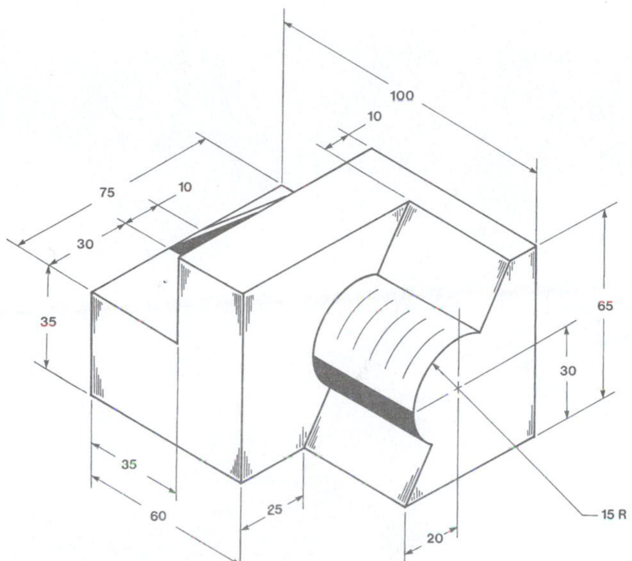

Figure P14-5 MILLIMETERS

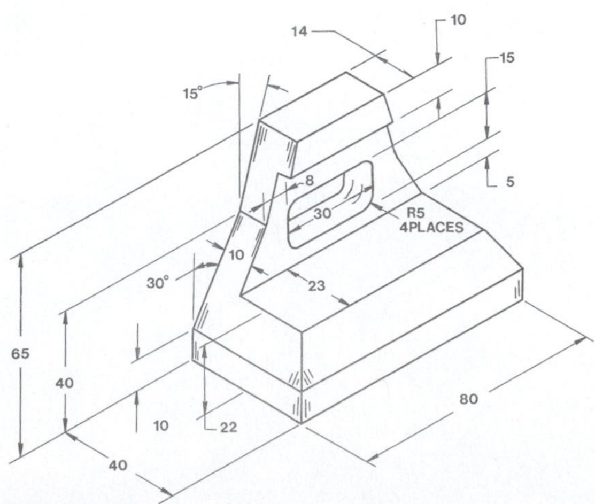

Figure P14-6 MILLIMETERS

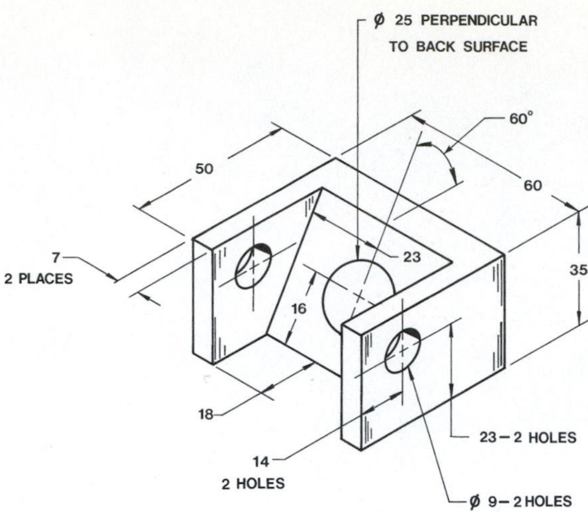

Figure P14-7 MILLIMETERS

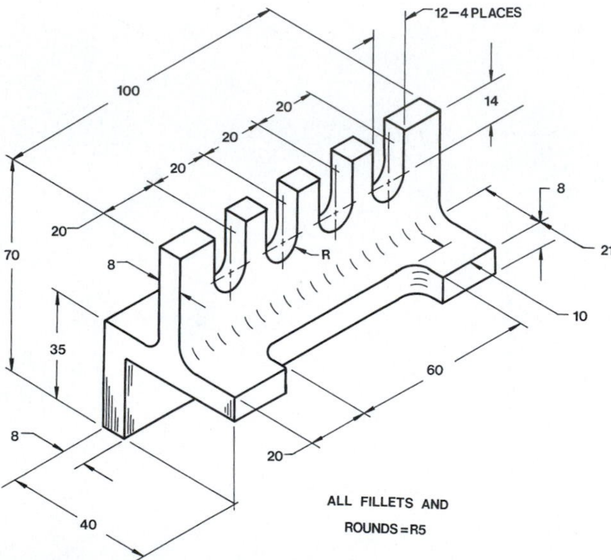

Figure P14-8 MILLIMETERS

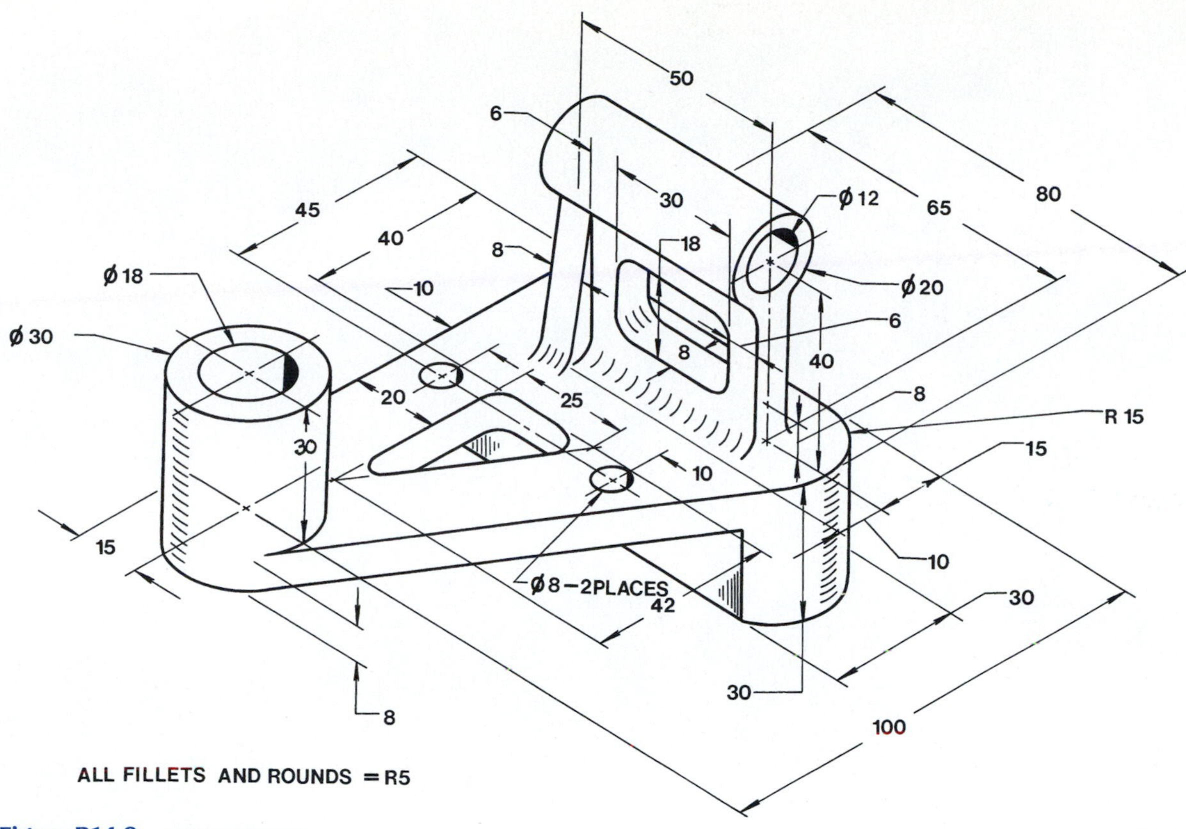

Figure P14-9 MILLIMETERS

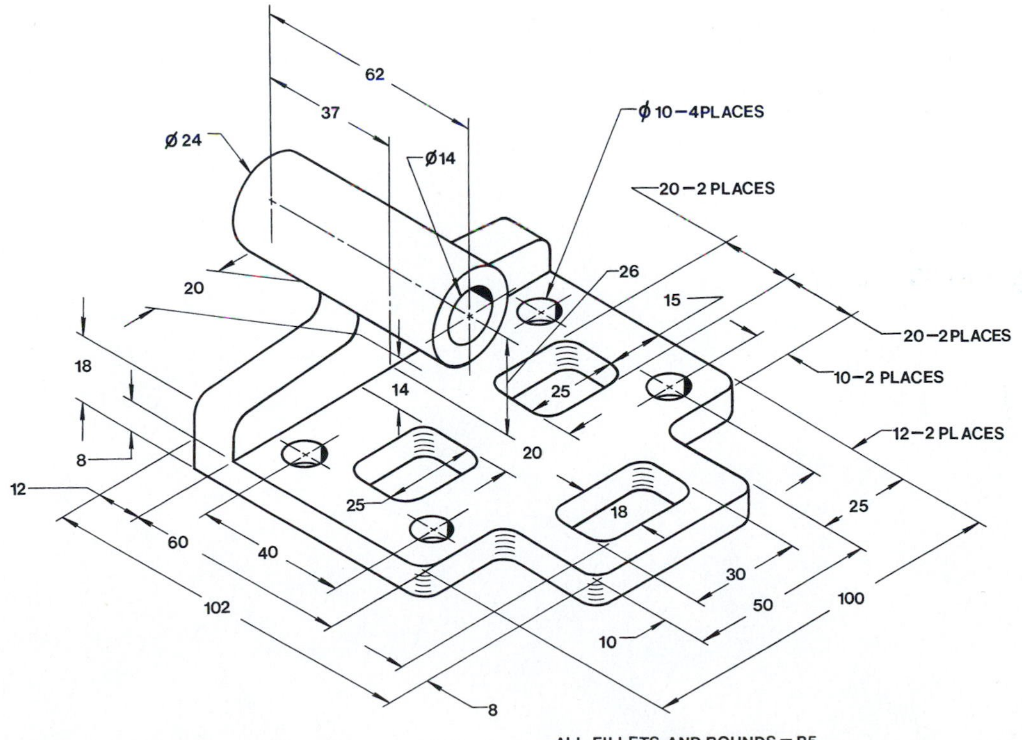

Figure P14-10 MILLIMETERS

Cams 15

Chapter Objectives

- Show how to use **Design Accelerator** to create cams.
- Explain how to create and use displacement diagrams.
- Show how to insert cams into assembly drawings.

INTRODUCTION

This chapter explains how to draw and design cams. ***Cams*** are eccentric objects that convert rotary motion into linear motion. Cams are fitted onto rotating shafts and lift and lower followers as they rotate.

cam: An eccentric object that converts rotary motion into linear motion.

Cams can be designed and drawn using the **Design Accelerator.** Displacement diagrams are defined and cams are generated from the displacement diagrams. The shape of the cam's profile causes the follower to rise and fall as the cam rotates. Changes in the cam's displacement cause the follower to accelerate. Excessive acceleration can generate excessive forces.

Figure 15-1 shows a cam drawn using Inventor. The cam bore includes a keyway.

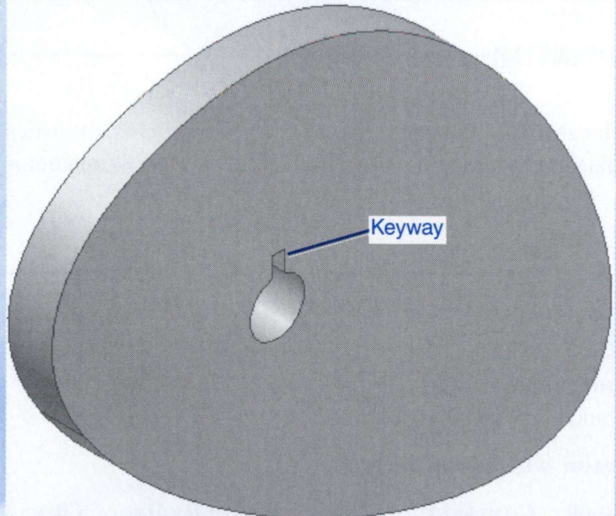

Figure 15-1

DISPLACEMENT DIAGRAMS

Displacement diagrams are used to define the motion of a cam using a linear diagram. The distances are then transferred to a base circle to create the required cam shape. Inventor will automatically create a cam from given displacement information.

displacement diagram: A linear diagram used to define the motion of a cam.

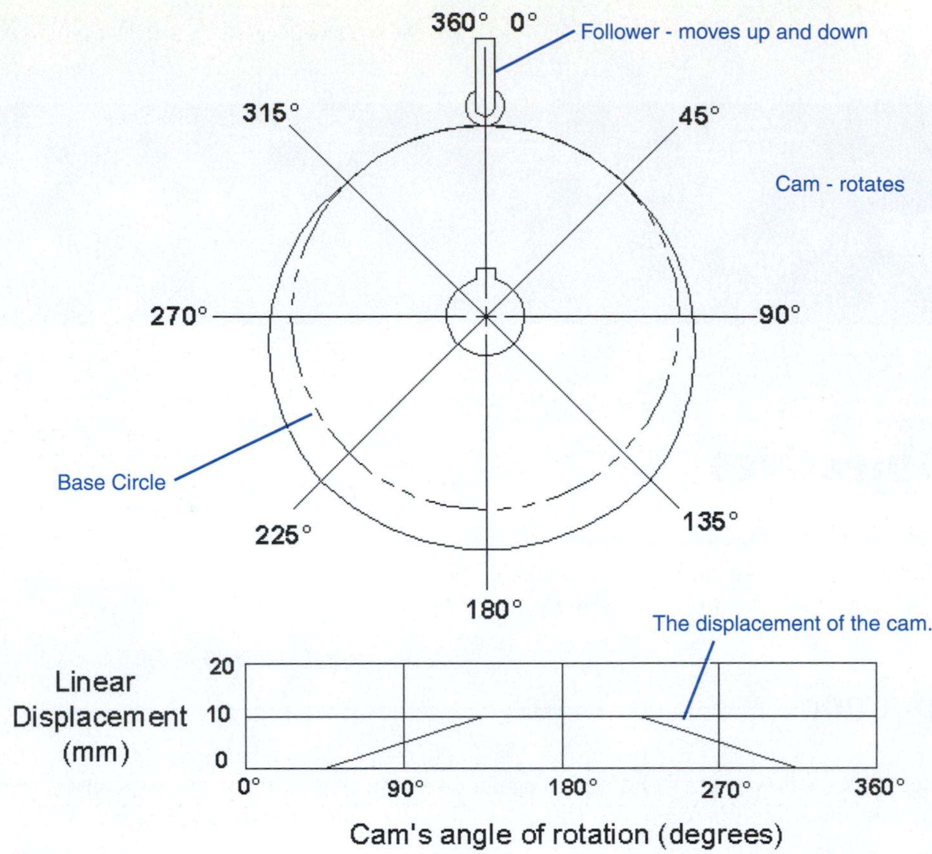

Figure 15-2

Figure 15-2 shows a displacement diagram and a cam shape generated from the information on the diagram. The displacement diagram shown is drawn using only straight lines, which is called *uniform motion.* This diagram results in points of discontinuity that can result in erratic follower motion. Several different shapes can be used to smooth out these areas and create smoother follower motion.

DRAWING A CAM USING INVENTOR

A cam is drawn using the **Design Accelerator** tool by first defining the physical characteristics of the cam then defining the cam's displacement diagram. A drawing of the cam will then automatically be generated.

Exercise 15-1: Drawing a Cam

In this example a Ø30 base circle is used. The cam's thickness is 10 mm, the roller is Ø5.0, and the material is EN C45 heat-treated steel.

1. Create a drawing using the **Standard.iam** format.
2. Access the **Design Accelerator** and click the **Disc Cam** tool.

The **Disc Cam Component Generator** will appear. See Figure 15-3.

3. Enter a **Basic Radius** value of **30.00,** a **Cam Width** of **10.00,** a **Roller Radius** of **5.0,** and a **Roller Width** of **5.0.**
4. Click the first segment and set h_{max} for **10.00.** Click the **Calculate** box.
5. Set the **Segment** angular distances as follows:

Segment 1—0° to 90°

Segment 2—90° to 270°

Segment 3—270° to 360°

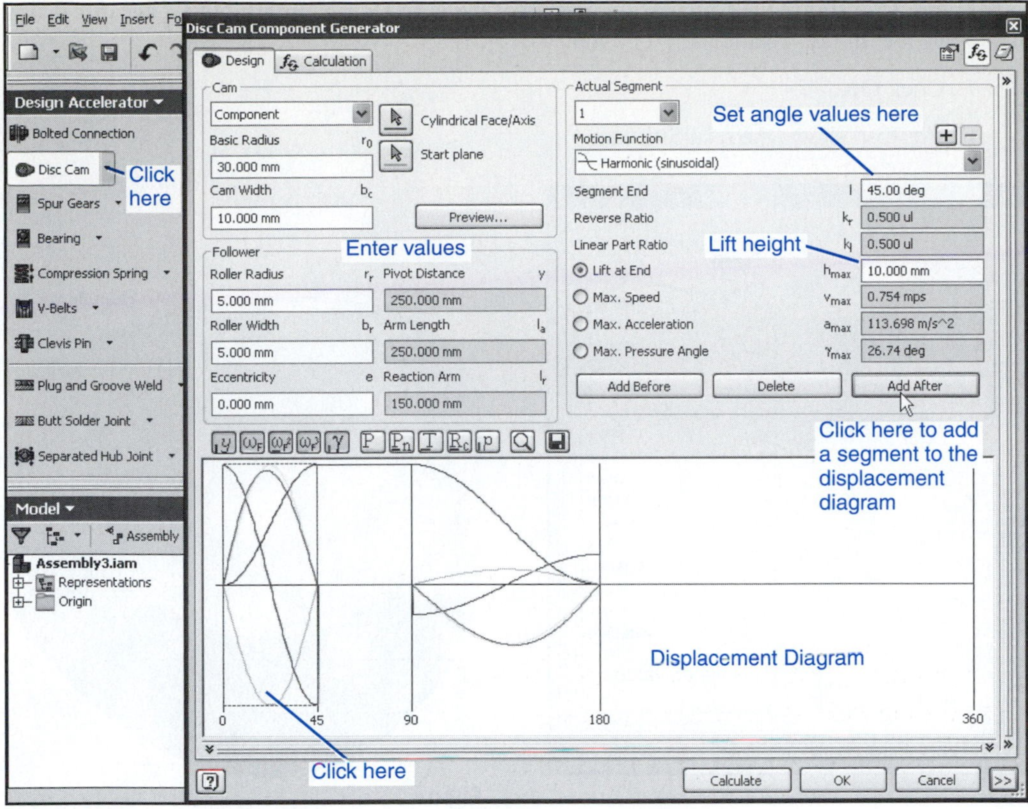

Figure 15-3

See Figure 15-4. Define the segment's end values by entering values into the **Segment End** box and clicking the **Calculate** box.

6. Enter a **Motion Function** of **Cycloidal (extended sinusoidal).**

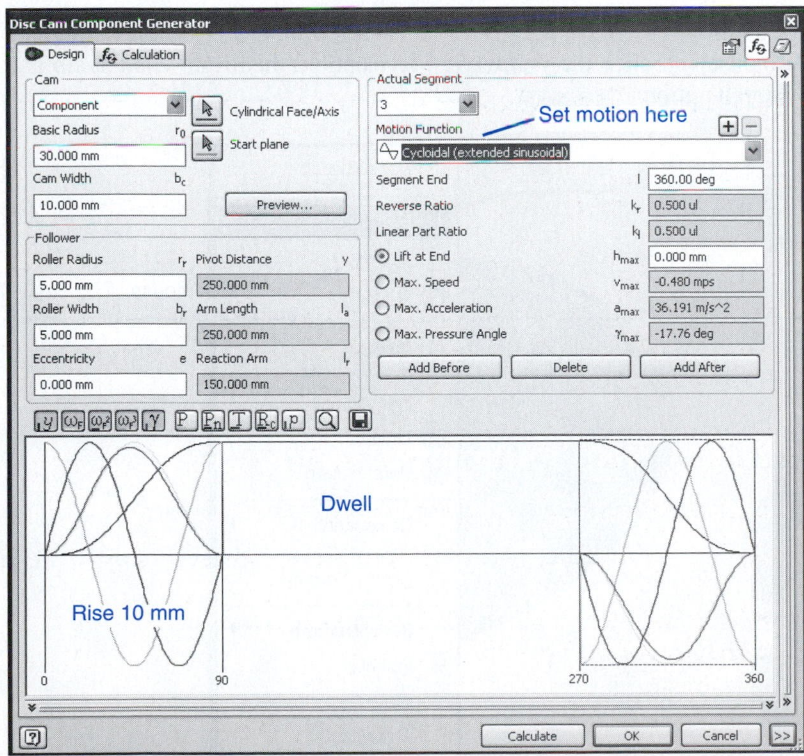

Figure 15-4

Figure 15-5 shows the **Calculation** portion of the **Disc Cam Component Generator** dialog box. Note that the h_{max} value is 10.000 mm.

7. Click **OK.**

Figure 15-6 shows the resulting cam.

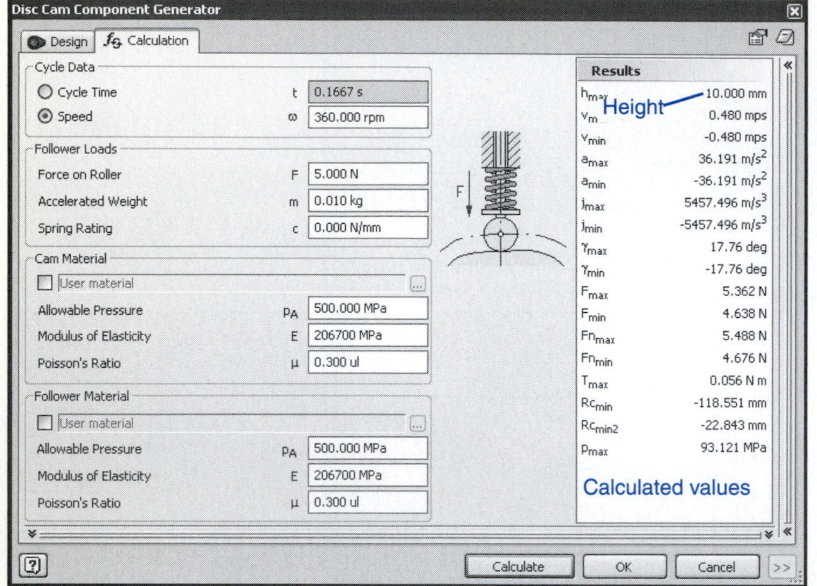

Figure 15-5

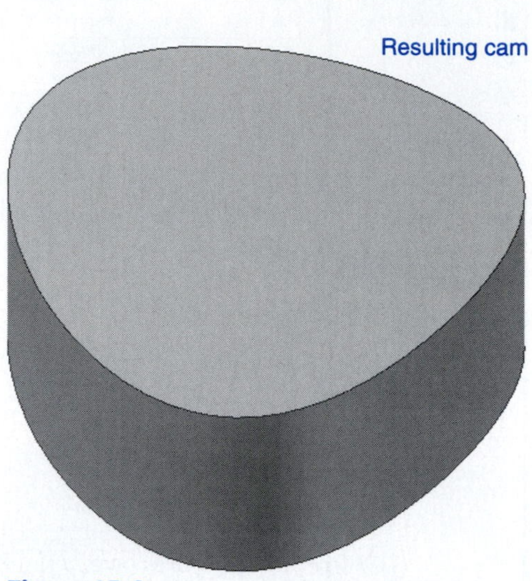

Figure 15-6

Exercise 15-2: Adding a Hole to a Cam

1. Right-click the mouse and click the **Open** option.

See Figure 15-7.

2. Right-click the mouse and click the **Edit** option.

See Figure 15-8.

3. Right-click the mouse and click the **Isometric View** option, then right-click again and click the **New Sketch** option.

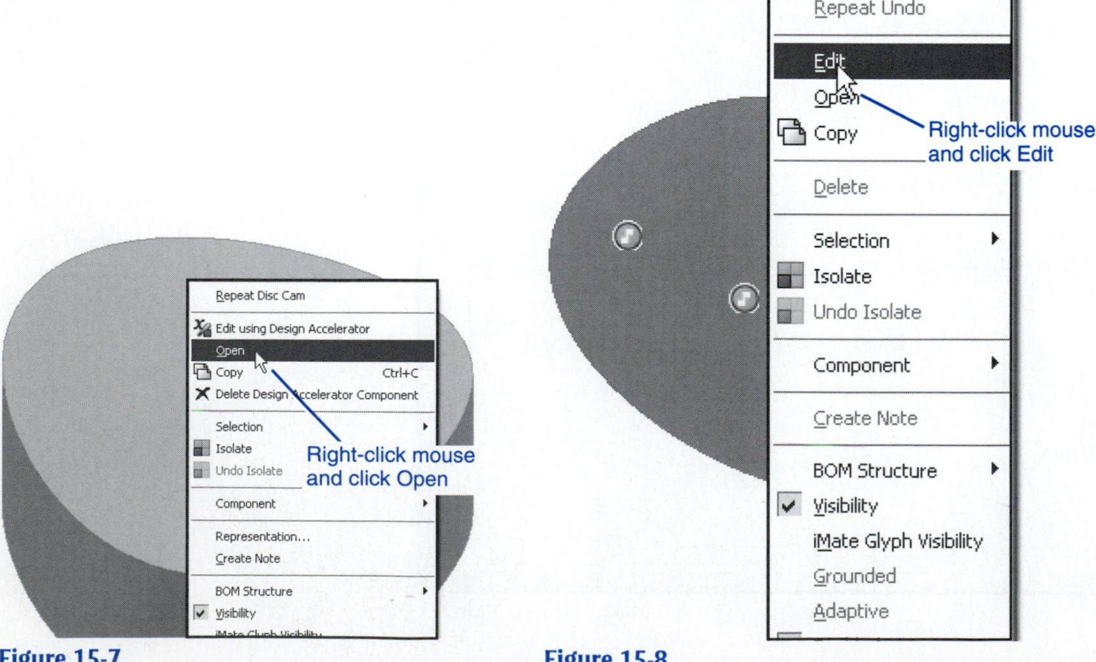

Figure 15-7

Figure 15-8

See Figure 15-9.

4. Create a **Point, Center Point.**

See Figure 15-10.

5. Right-click the mouse and click the **Finish Sketch** option.
6. Click the **Hole** tool and enter the values.

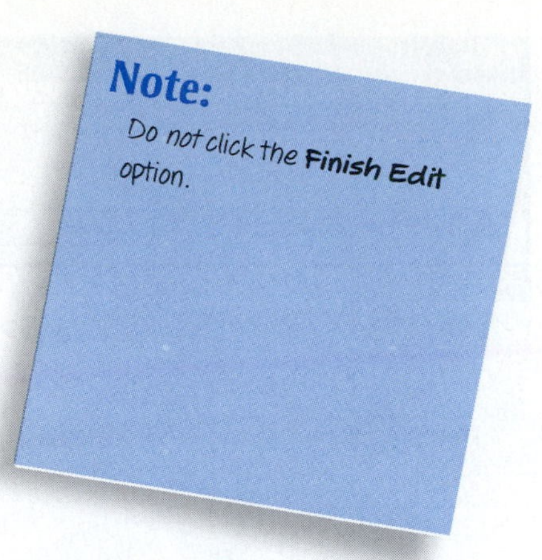

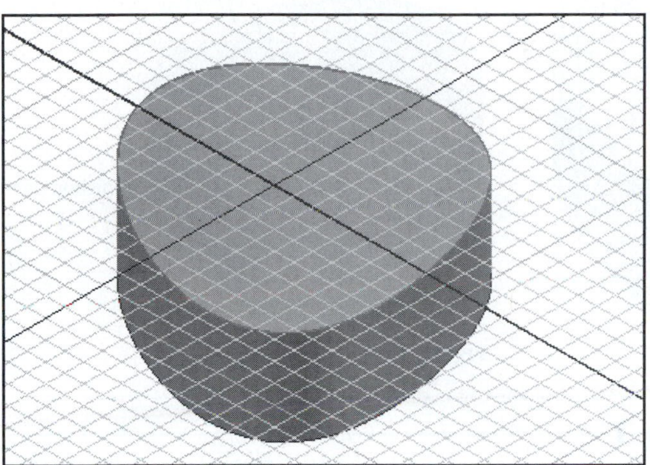

Figure 15-9

Figure 15-10

See Figure 15-11.

7. Click the **X** box in the upper right corner of the screen.

A dialog box will appear. See Figure 15-12.

8. Click the **No** box.

The cam will appear. See Figure 15-13.

9. Click the cam.

Figure 15-14 shows the finished cam.

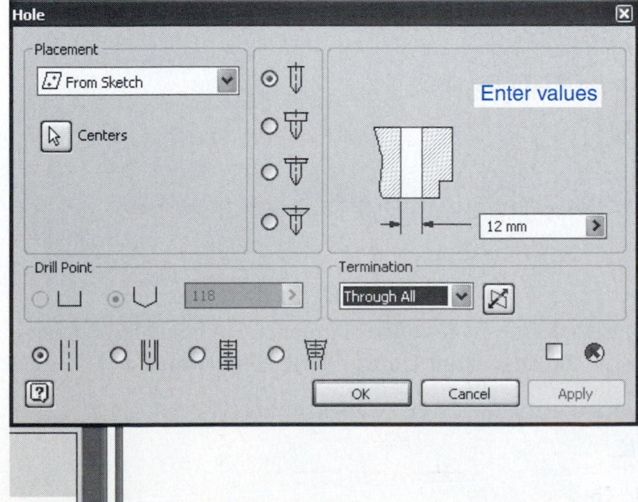

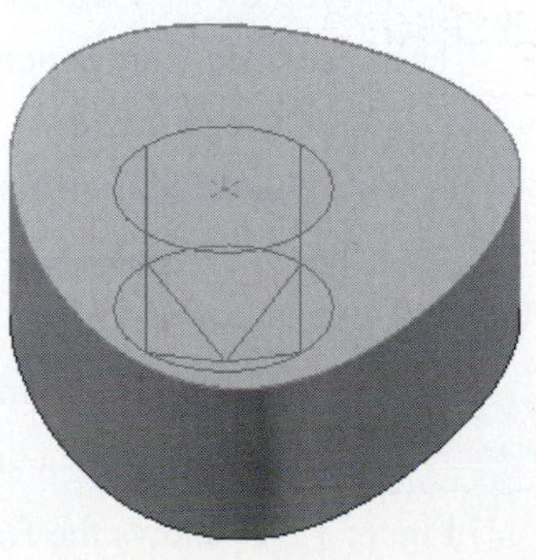

Figure 15-11

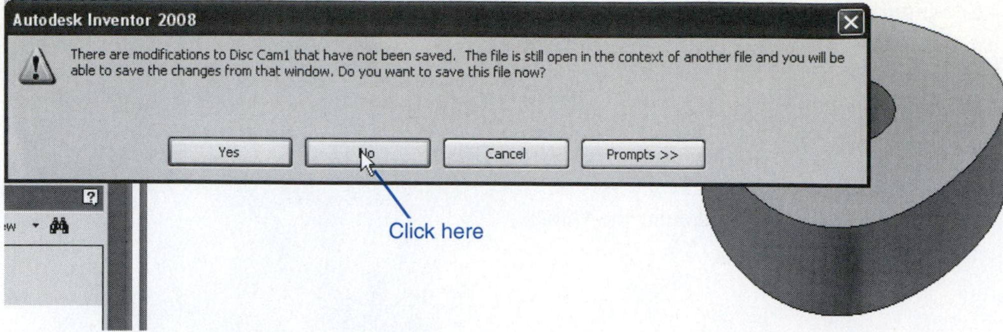

Figure 15-12

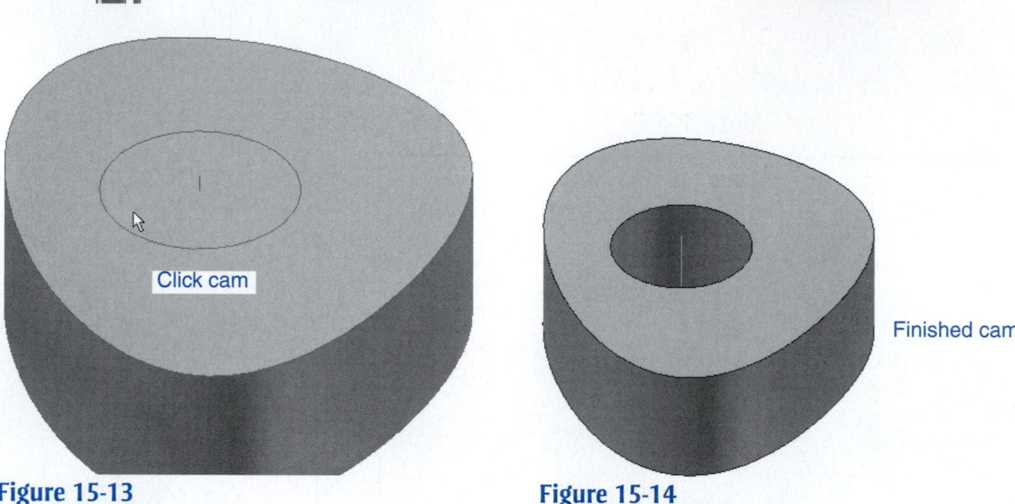

Figure 15-13

Figure 15-14

SAMPLE PROBLEM SP15-1

Design and draw a cam that meets the following specifications.

Base circle = Ø80 mm
Dwell = 45°
Rise 5 mm using harmonic motion over 90°
Dwell 45°
Rise 5 mm using harmonic motion over 90°
Drop 10 mm using parabolic with linear part motion in 90°
Bore = Ø20
Cam width = 10
Roller radius = 8
Roller width = 10

1. Create a drawing using the **Standard.iam** format.
2. Access the **Design Accelerator.**
3. Select the **Disc Cam** tool.

The **Disc Cam Component Generator** dialog box will appear. See Figure 15-15.

4. Enter the cam values.
5. Click the **Add After** box.

The cam will require five segments to define its motion. Segments can be added by using the **Add Before** or **Add After** options on the **Disc Cam Component Generator.**

6. Complete the displacement diagram using the given specifications.
7. Click **Calculate,** then click **OK.**

Figure 15-16 shows the finished cam.

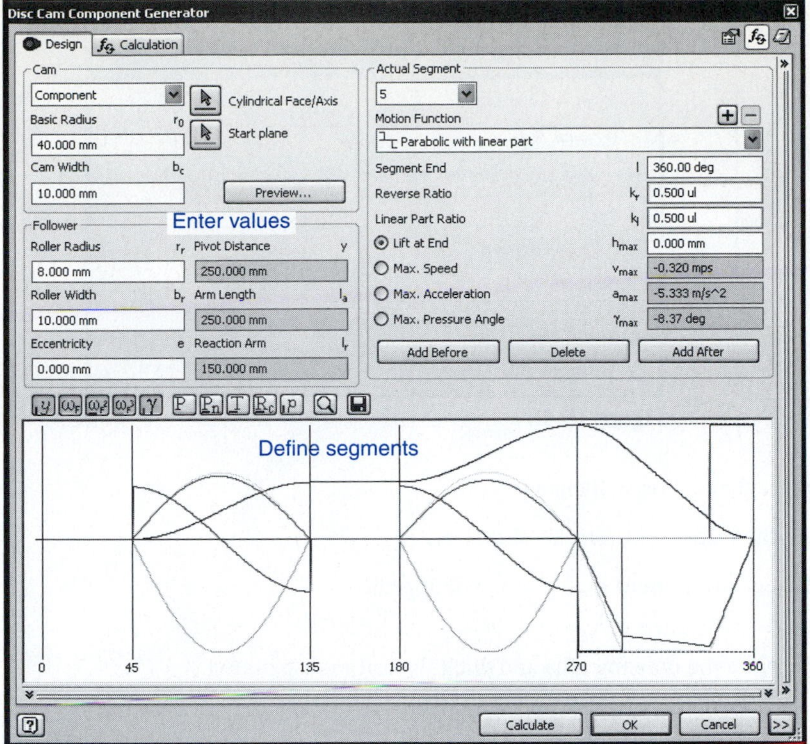

Figure 15-15

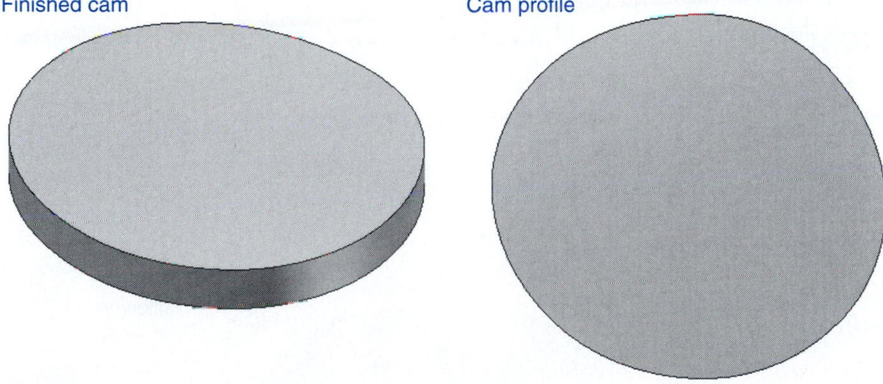

Figure 15-16

Animating Cams and Followers

Cams may be animated, that is, made to rotate on the screen while their followers move linearly. Figure 15-17 shows a cam that was previously created. A new **Standard (mm).iam** drawing was created, and the **Place Component** tool was used to add the cam drawing to the assembly drawing. The assembly drawing was saved as **Cam Assembly.**

Exercise 15-3: Creating a Follower

1. Click the **Create Component** tool and create a new component named **Follower.**

   The **Create In-Place Component** dialog box will appear. See Figure 15-18.

2. Click the **Browser Templates** box.

   The **Open Template** dialog box will appear.

3. Click the **Metric** tab.

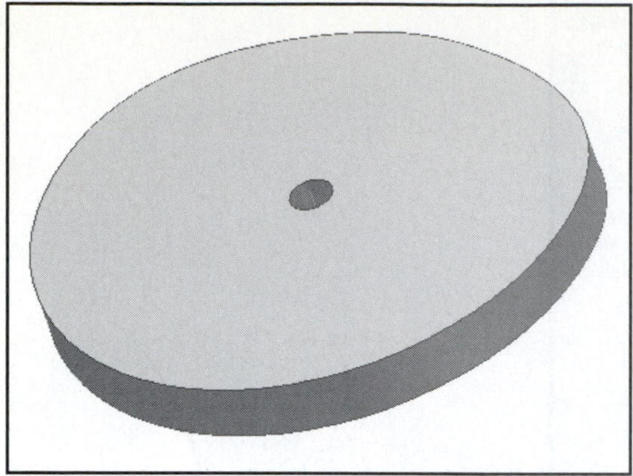

Figure 15-17

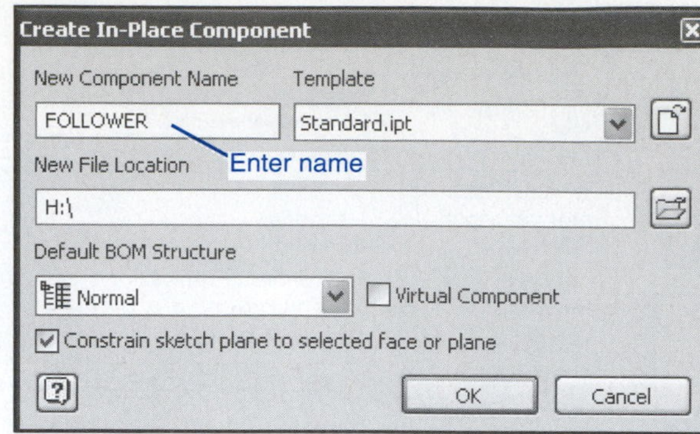

Figure 15-18

The **Open Template** dialog box will change.

4. Select the **Standard (mm).ipt** format; click **OK.**

The **Create In-Place Component** dialog box will appear.

5. Click **OK.**
6. Move the cursor into the drawing area and click the left mouse button.

A grid will appear on the screen. See Figure 15-19.

7. Create a rectangular follower **(10 × 10 × 100)** as shown.
8. Click the right mouse button and select **Done,** then click the right mouse button again and select **Finish Sketch.**
9. Extrude the follower sketch **10 mm.**
10. Right-click the mouse and select **Finish Edit.**

See Figure 15-20.

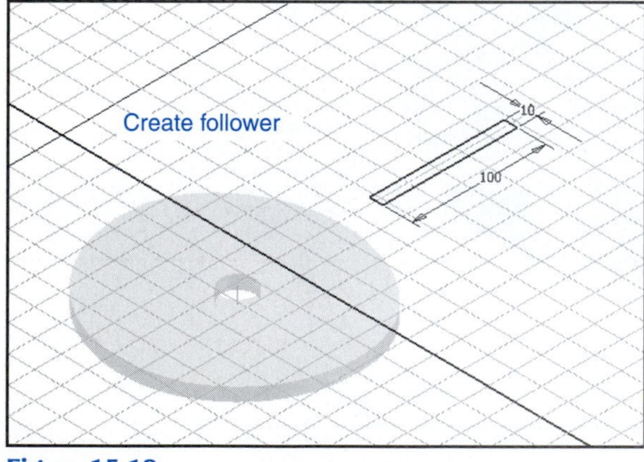

Figure 15-19

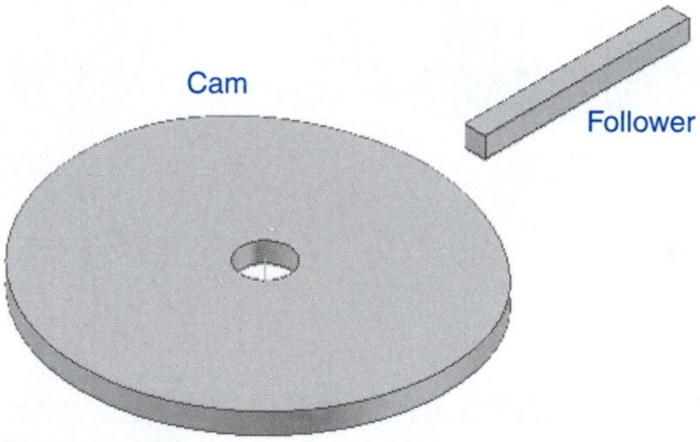

Cam

Follower

Figure 15-20

11. Use the **Constraint** tool and constrain the follower tangent to the cam.

See Figure 15-21.

The **Move and Rotate Component** tool may have to be used to position the follower so it can be constrained.

Exercise 15-4: Creating a Follower Guide

1. Click the **Create Component** tool and create a new component named **Guide.** Use the **Standard (mm).** ipt format.

See Figure 15-22.

2. Click the top surface of the follower to create a new sketch plane.

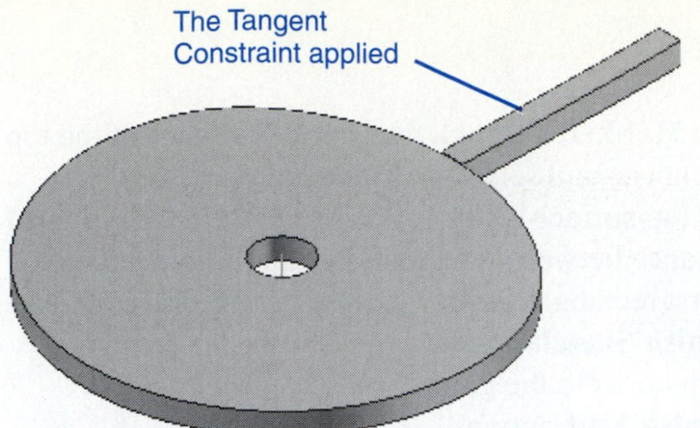

Figure 15-21

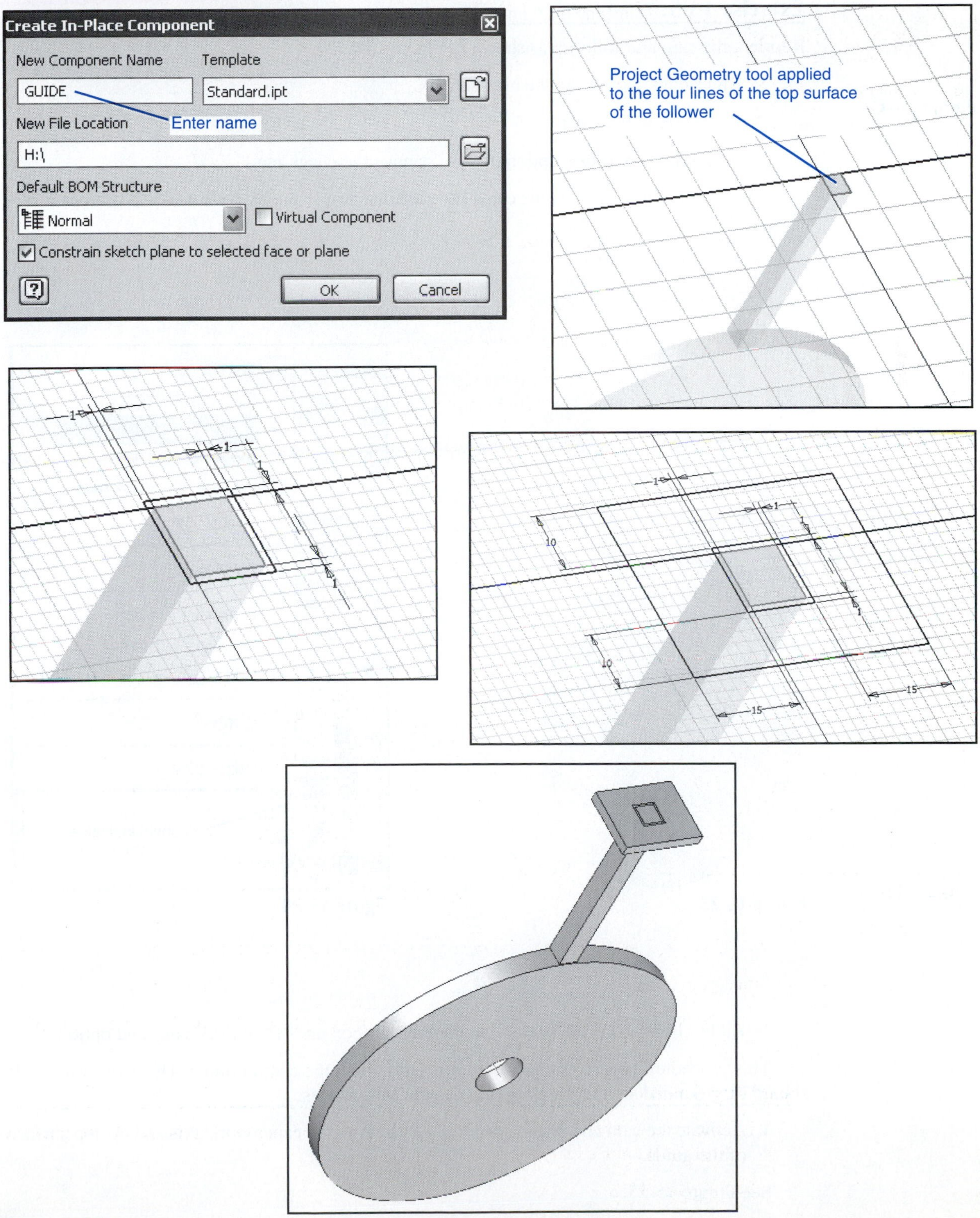

Create In-Place Component

New Component Name: GUIDE — Enter name

Template: Standard.ipt

New File Location: H:\

Default BOM Structure: Normal

Virtual Component

Constrain sketch plane to selected face or plane

OK Cancel

Project Geometry tool applied to the four lines of the top surface of the follower

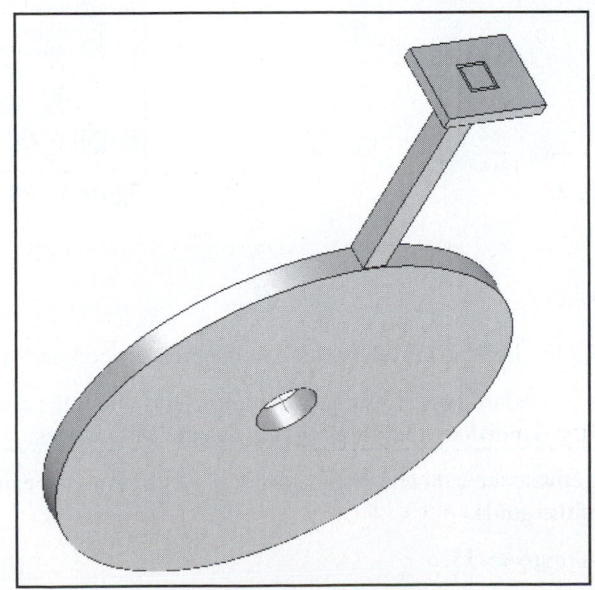

Figure 15-22

3. Click the **Project Geometry** tool on the **2D Sketch Panel,** then click four lines of the top surface of the follower. Right-click the mouse and select the **Done** option.
4. Sketch a rectangle around the projected top surface of the follower, then use the **General Dimension** tool to define a **1-mm** clearance between the guide and the follower.
5. Sketch a second rectangle around the first rectangle as shown, then right-click the mouse and select the **Done** option, then the **Finish Sketch** option.
6. Use the **Extrude** tool to add a **5-mm** thickness to the guide.
7. Right-click the mouse and select the **Finish Edit** option.

Exercise 15-5: Animating the Cam

Reorient the cam and follower as shown in Figure 15-23.

1. Right-click the cam in the browser box.

A dialog box will appear.

2. Select the **Grounded** option; that is, remove the check mark.

This action will unground the cam. The pushpin icon in the browser box will disappear. See Figure 15-24.

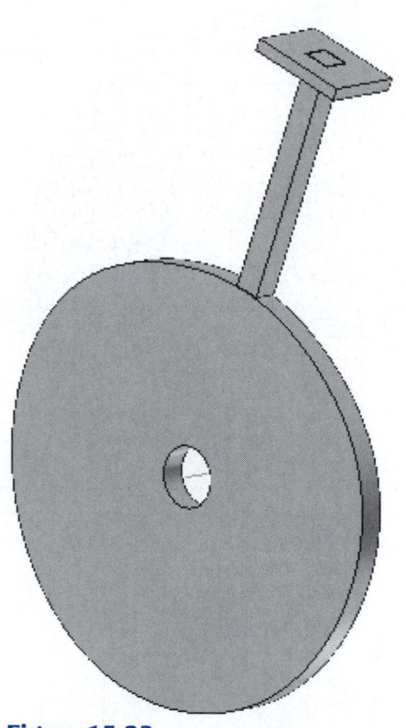

Figure 15-23

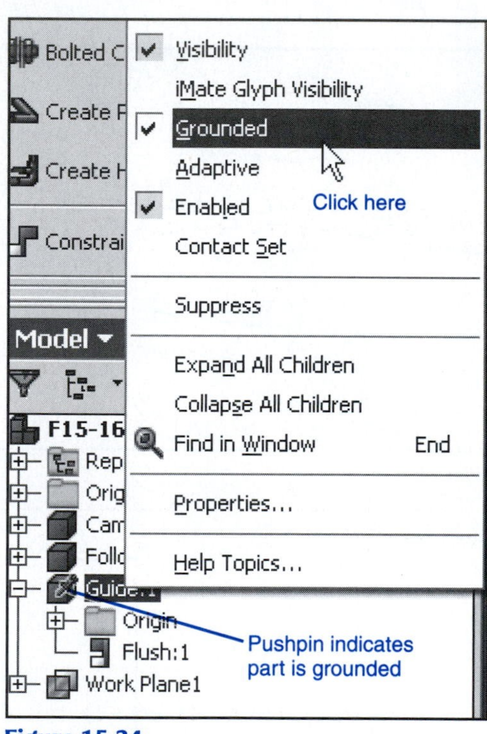

Figure 15-24

3. Right-click the **Guide** heading in the browser box and select the **Grounded** option.

This procedure will fix the guide in place and allow the cam to rotate. The cam's work axis should be grounded so it will stay in place as the cam rotates.

4. Activate the cam and create a work plane through the cam's work axis and the top surface of the guide.

See Figure 15-25.

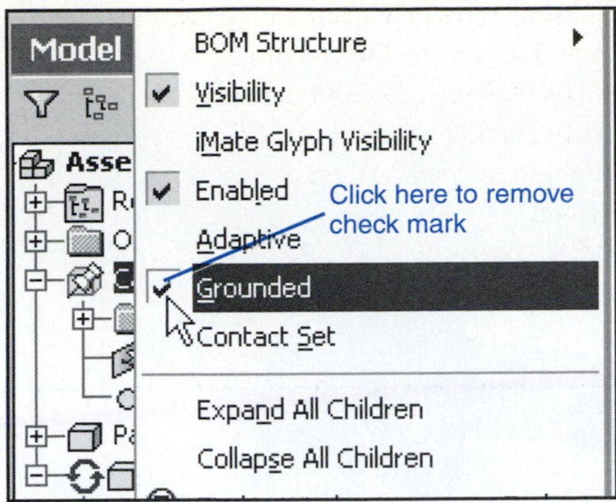

Figure 15-25

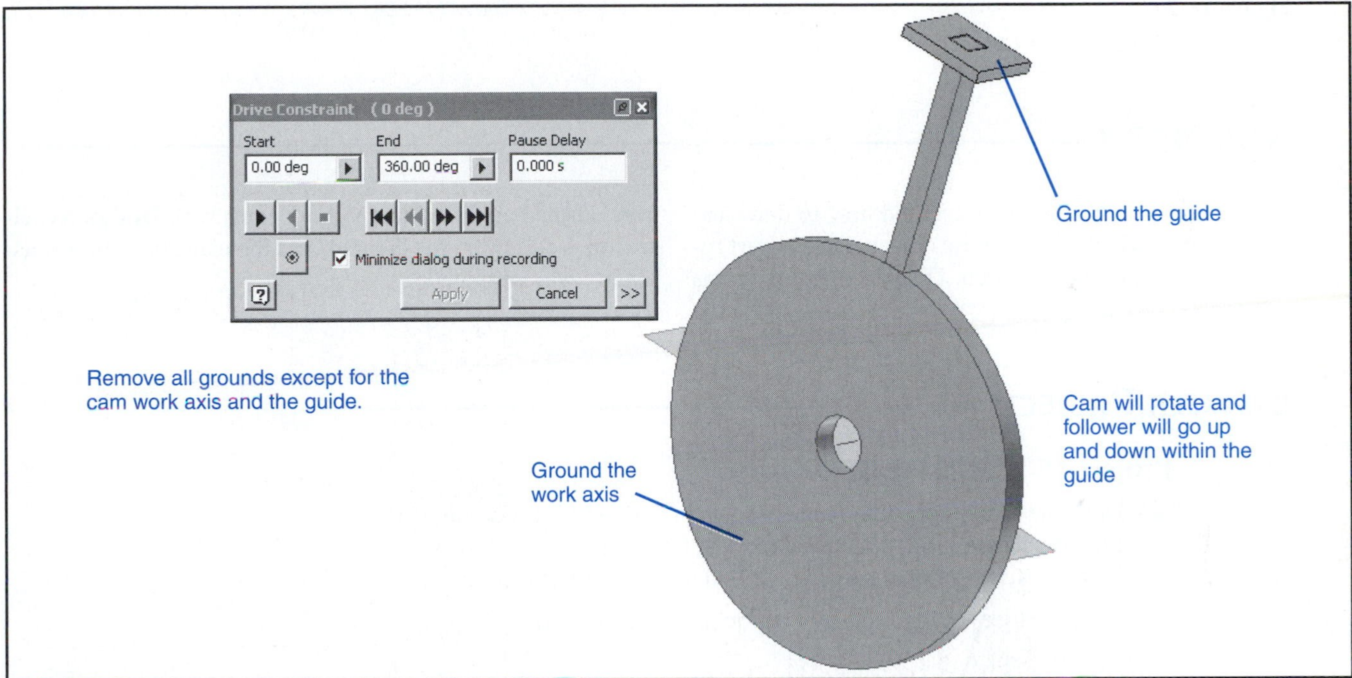

Figure 15-26

5. Click the **Constraint** tool, **Angle** option, and define a **0.00** degree angle between the work plane on the cam and the guide's top surface.
6. Right-click the **Angle** (0.00 deg) listing in the browser box and select the **Drive Constraint** option.

The **Drive Constraint** dialog box will appear. See Figure 15-26.

7. Set the **Start** angle for **0.00 deg** and the **End** angle for **360.00 deg,** then click the forward button on the **Drive Constraint** dialog box.

The cam should rotate and the follower go up and down within the guide block. If the cam does not rotate properly, check for unneeded constraints.

There are many different types of cam followers. Figure 15-27 shows four different types. As a cam turns, the follower is pushed up and down. A spring is often used to force the follower to stay in contact with the cam surface.

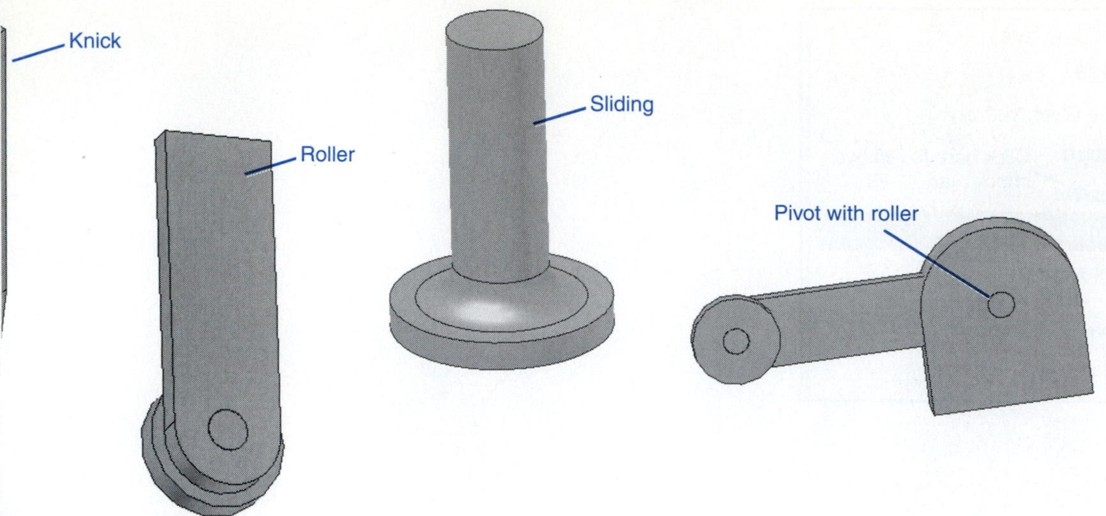

Figure 15-27

Summary

This chapter explained and illustrated how to draw and design cams, which are eccentric objects that convert rotary motion into linear motion. Displacement diagrams were defined and used to generate cams with **Design Accelerator.** A follower was created, and the cam and follower assembly was animated.

Chapter Projects

Project 15-1: MILLIMETERS

Draw the Cam Support Assembly shown in Figure P15-1. For this example the nominal dimensions for the bearings are as follows. More detailed dimensions can be found in the **Content Center.**

DIN625—SKF 6203 (ID ×OD × THK) 17 × 40 × 10

DIN625—SKF 634 4 × 13 × 4

GB 2273.2-87—7/70 8 × 18 × 5

The nominal dimensions for the rectangular key are 5 × 5 × 16. The values for the compression spring are as follows:

Wire diameter = 1.5

Inside diameter = 9.0

Loose spring length = 24

Preload spring length = 23

Fully loaded = 20

Working spring length = 21

Right coil direction

Active coils = 10.125

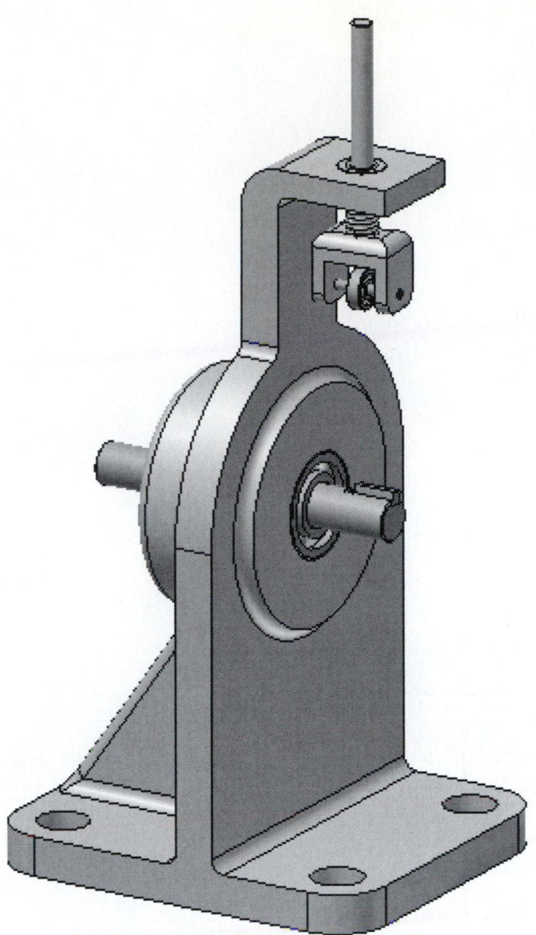

Figure P15-1a

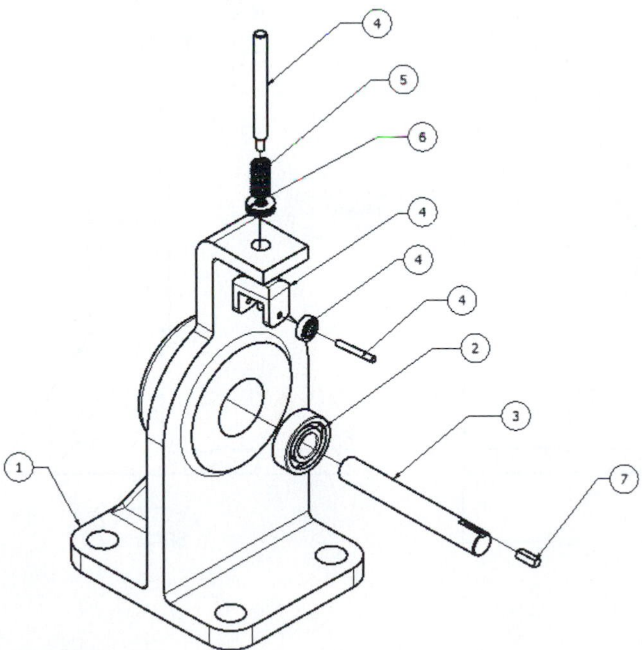

Figure P15-1b

Parts List			
ITEM	QTY	PART NUMBER	DESCRIPTION
1	1	ENG-2008-A	BASE, CAST
2	1	DIN625 - SKF 6203	Single row ball bearings
3	1	SHF-4004-16	SHAFT: Ø16×120,WITH 2.3×5×16 KEYWAY
4	1		SUB-ASSEMBLY, FOLLOWER
5	1	SPR-C22	SPRING,COMPRESSION
6	1	GB 273.2-87 - 7/70 - 8 x 18 x 5	Rolling bearings - Thrust bearings - Plan of boundary dimensions
7	1	IS 2048 - 1983 - Specification for Parallel Keys and Keyways B 5 x 5 x 16	Specification for Parallel Keys and Keyways

Figure P15-1c

Cast Base
P/N ENG-2008-A
MATL = Steel

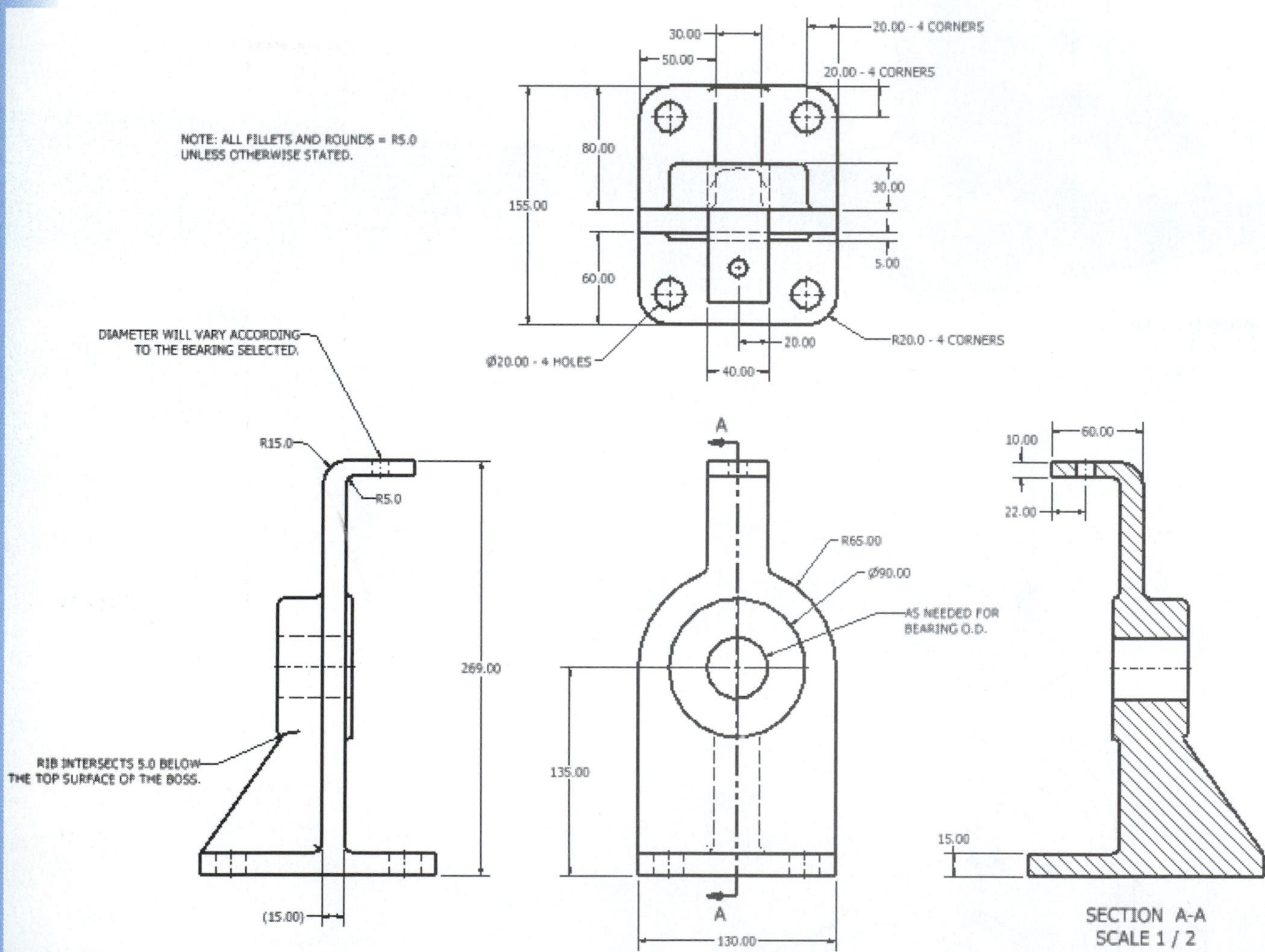

Figure P15-1d

Follower Subassembly

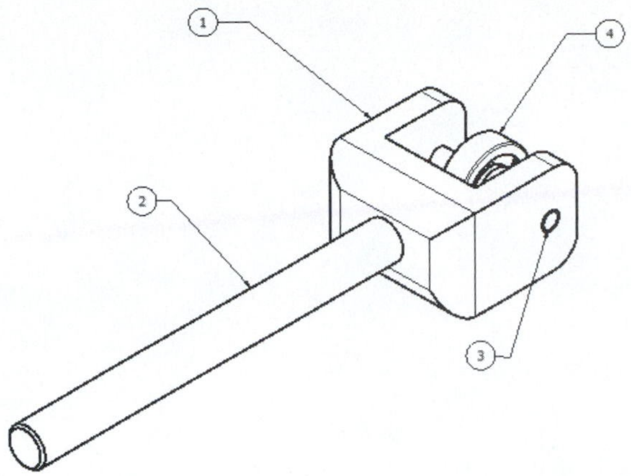

Parts List				
ITEM	PART NUMBER	DESCRIPTION	MATERIAL	QTY
1	AM-232	HOLDER	STEEL	1
2	AM-256	POST, FOLLOWER	STEEL	1
3	BS 1804-2 - 4 x 30	Parallel steel dowel pins - metric series	Steel, Mild	1
4	DIN625- SKF 634	Single row ball bearings	Steel, Mild	1

Figure P15-1e

Follower Post
P/N AM-256
MATL = Steel

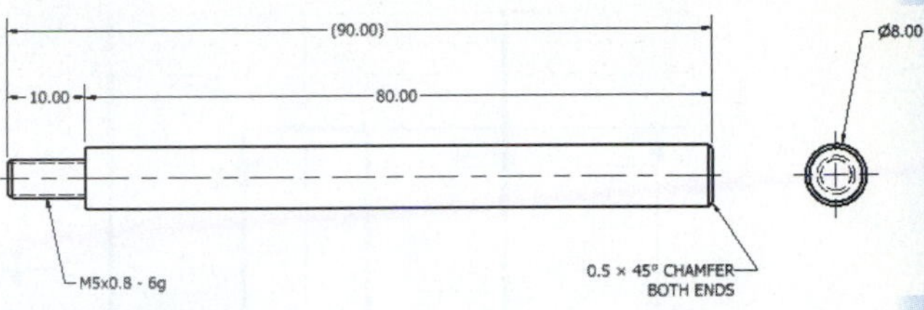

Figure P15-1g

Shaft
P/N SHF-4004-16
MATL=Steel

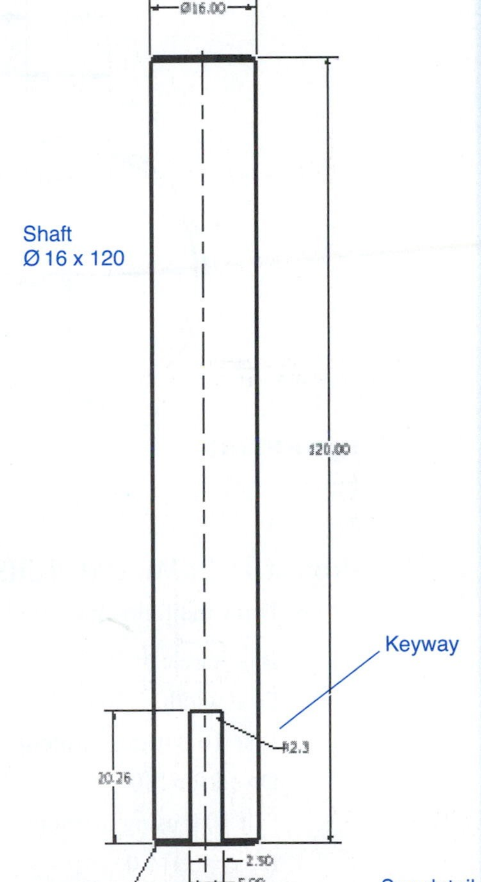

Holder
P/N AM-232
MATL = Steel

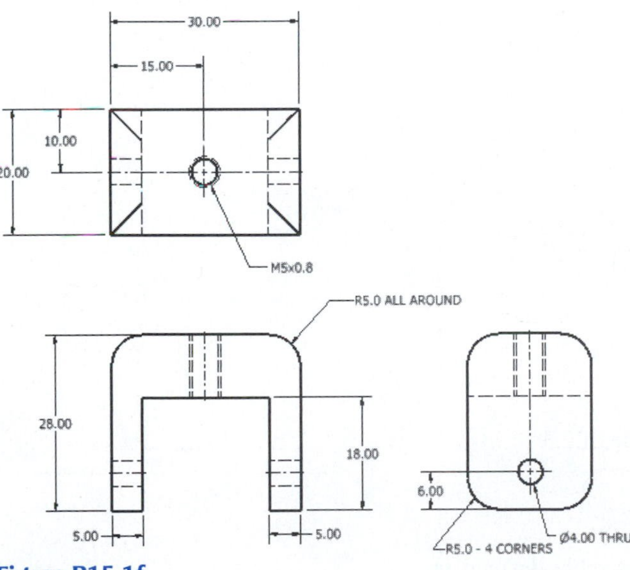

Figure P15-1f

Figure P15-1h

Detail drawing

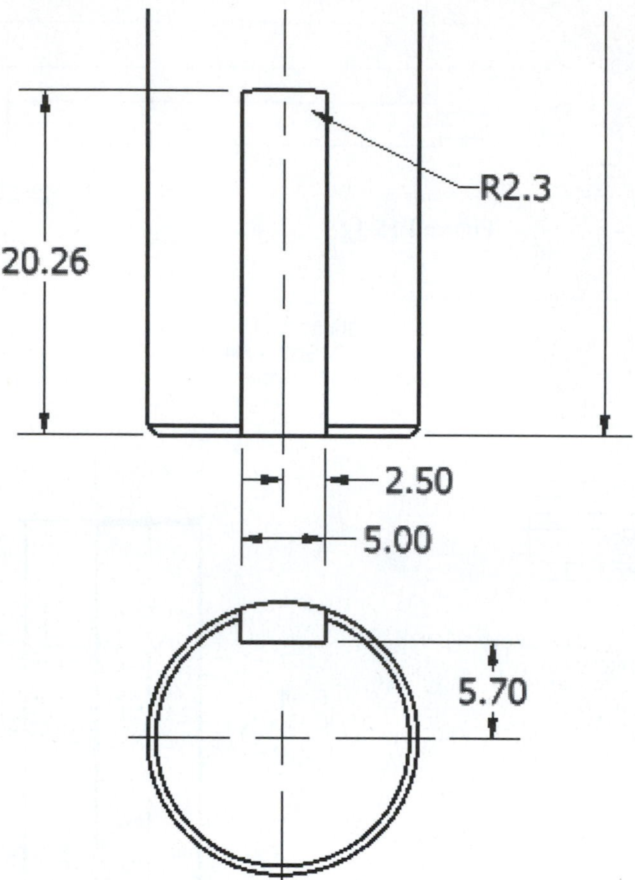

Figure P15-1i

Project 15-2: MILLIMETERS

A. Draw the following cam:

   Base circle = R73.0

   Face width = 16.0

   Rise 10.0 using harmonic motion in 90°

   Dwell for 180°

   Fall 10.0 using harmonic motion in 90°

   Bore = Ø16.0

   Keyway = 2.3 × 5 × 16 with a radius value of 2.3

   Follower diameter = 16

   Follower width = 4

B. Mount the cam into the Cam Support Assembly defined in Project 15-1.

   See Figure P15-2.

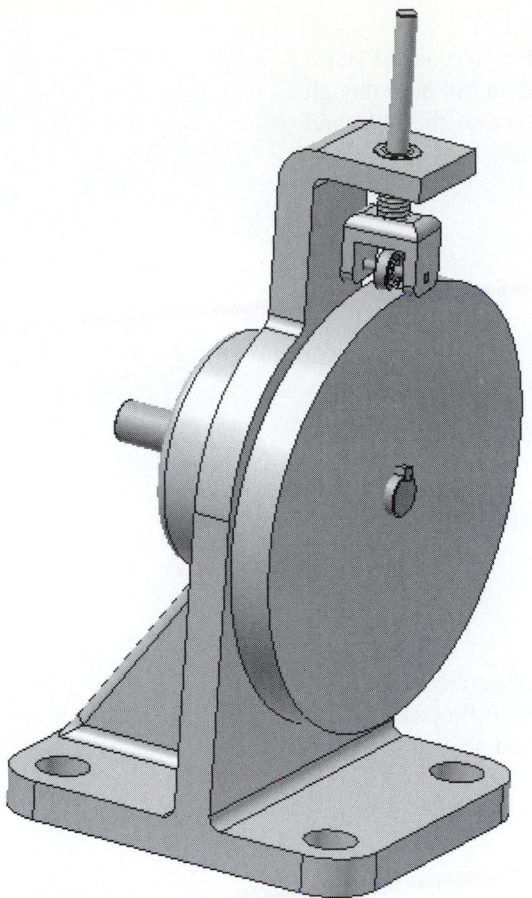

Figure P15-2

Project 15-3: MILLIMETERS

A. Draw the following cam:

Base circle = Ø73.0

Face width = 16.0

Rise 8.0 using double harmonic motion—Part 1 motion in 90°

Dwell for 45°

Rise 6.0 using double harmonic motion—Part 1 motion in 90°

Dwell for 45°

Fall 14.0 using double harmonic motion—Part 1 motion in 90°

Bore = Ø16.0

Keyway = 2.3 × 5 × 16 with a radius value of 2.3

Follower diameter = 16

Follower width = 4

B. Mount the cam into the Cam Support Assembly defined in Project 15-1.

C. Animate the cam and follower.

Project 15-4: MILLIMETERS

A. Draw the following cam and add a Ø24.0 × 16.0 hub. Place a Ø16 hole through the hub and cam. Add an M4 hole though the hub located 10.0 from the top surface of the hub, and insert an M4 × 4.0 CSN 02 1181 Set Screw.

Base circle = Ø73.0

Face width = 16.0

Rise 6.0 using cycloidal motion in 45°

Dwell for 45°

Rise 6.0 using cycloidal motion in 90°

Dwell for 45°

Fall 12.0 using cycloidal motion in 90°

Bore = Ø16.0

Keyway = 2.3 × 5 × 16 with a radius value of 2.3

Follower diameter = 16

Follower width = 4

B. Design
Mount the cam into the Cam Support Assembly defined in Project 15-1. Modify the shaft presented in Project 15-1 by removing the keyway. Assign the mofified shaft a new part number, SHT-466E.

C. Animate the cam and follower.

Project 15-5: MILLIMETERS

A. Draw the following cam:

Base circle = Ø73.0

Face width = 16.0

Rise 12.0 using harmonic motion in 90°

Dwell for 180°

Fall 12.0 using harmonic motion in 90°

Follower diameter = 16

Follower width = 4

B. Design

Make the following modifications, then mount the cam into the Cam Support Assembly defined in Project 15-1.

Make the cam's bore 20.0.

Modify the shaft to have a Ø20.0.

Select a new key based on the Ø20.0 shaft, and add the appropriate keyway to the shaft and cam.

Select a new bearing to accept the Ø20.0 shaft.

Modify the hole in the Cast Base to accept the outside diameter of the selected bearing.

Consider using a counterbored hole to mount the bearing.

C. Animate the cam and follower.

Appendix

Wire and Sheet Metal Gauges

Gauge	Thickness	Gauge	Thickness
000 000	0.5800	18	0.0403
00 000	0.5165	19	0.0359
0 000	0.4600	20	0.0320
000	0.4096	21	0.0285
00	0.3648	22	0.0253
0	0.3249	23	0.0226
1	0.2893	24	0.0201
2	0.2576	25	0.0179
3	0.2294	26	0.0159
4	0.2043	27	0.0142
5	0.1819	28	0.0126
6	0.1620	29	0.0113
7	0.1443	30	0.0100
8	0.1285	31	0.0089
9	0.1144	32	0.0080
10	0.1019	33	0.0071
11	0.0907	34	0.0063
12	0.0808	35	0.0056
13	0.0720	36	0.0050
14	0.0641	37	0.0045
15	0.0571	38	0.0040
16	0.0508	39	0.0035
17	0.0453	40	0.0031

Figure A-1

American Standard Clearance Locational Fits

Nominal Size Range Inches		Class LC1			Class LC2			Class LC3			Class LC4		
		Limits of Clearance	Standard Limits		Limits of Clearance	Standard Limits		Limits of Clearance	Standard Limits		Limits of Clearance	Standard Limits	
Over	To		Hole H6	Shaft h5		Hole H7	Shaft h6		Hole H8	Shaft h7		Hole H10	Shaft h9
0	0.12	0	+0.25	0	0	+0.4	0	0	+0.6	0	0	+1.6	0
		0.45	0	-0.2	0.65	0	-0.25	1	0	-0.4	2.6	0	-1.0
0.12	0.24	0	+0.3	0	0	+0.5	0	0	+0.7	0	0	+1.8	0
		0.5	0	-0.2	0.8	0	-0.3	1.2	0	-0.5	3.0	0	-1.2
0.24	0.40	0	+0.4	0	0	+0.6	0	0	+0.9	0	0	+2.2	0
		0.65	0	-0.25	1.0	0	-0.4	1.5	0	-0.6	3.6	0	-1.4
0.40	0.71	0	+0.4	0	0	+0.7	0	0	+1.0	0	0	+2.8	0
		0.7	0	-0.3	1.1	0	-0.4	1.7	0	-0.7	4.4	0	-1.6
0.71	1.19	0	+0.5	0	0	+0.8	0	0	+1.2	0	0	+3.5	0
		0.9	0	-0.4	1.3	0	-0.5	2	0	-0.8	5.5	0	-2.0
1.19	1.97	0	+0.6	0	0	+1.0	0	0	+1.6	0	0	+4.0	0
		1.0	0	-0.4	1.6	0	-0.6	2.6	0	-1.0	6.5	0	-2.5

Figure A-2A

Nominal Size Range Inches		Class LC5			Class LC6			Class LC7			Class LC8		
		Limits of Clearance	Standard Limits		Limits of Clearance	Standard Limits		Limits of Clearance	Standard Limits		Limits of Clearance	Standard Limits	
Over	To		Hole H7	Shaft g6		Hole H9	Shaft f8		Hole H10	Shaft e9		Hole H10	Shaft d9
0	0.12	0.1	+0..4	-0.1	0.3	+1.0	-0.3	0.6	+1.6	-0.6	1.0	+1.6	-1.0
		0.75	0	-0.35	1.9	0	-0.9	3.2	0	-1.6	3.6	0	-2.0
0.12	0.24	0.15	+0.5	-0.15	0.4	+1.2	-0.4	0.8	+1.8	-0.8	1.2	+1.8	-1.2
		0.95	0	-0.45	2.3	0	-1.1	3.8	0	-2.0	4.2	0	-2.4
0.24	0.40	0.2	+0.6	-0.2	0.5	+1.4	-0.5	1.0	+2.2	-1.0	1.6	+2.2	-1.6
		1.2	0	-0.6	2.8	0	-1.4	4.6	0	-2.4	5.2	0	-3.0
0.40	0.71	0.25	+0.7	-0.25	0.6	+1.6	-0.6	1.2	+2.8	-1.2	2.0	+2.8	-2.0
		1.35	0	-0.65	3.2	0	-1.6	5.6	0	-2.8	6.4	0	-3.6
0.71	1.19	0.3	+0.8	-0.3	0.8	+2.0	-0.8	1.6	+3.5	-1.6	2.5	+3.5	-2.5
		1.6	0	-0.8	4.0	0	-2.0	7.1	0	-3.6	8.0	0	-4.5
1.19	1.97	0.4	+1.0	-0.4	1.0	+2.5	-1.0	2.0	+4.0	-2.0	3.0	+4.0	-3.0
		2.0	0	-1.0	5.1	0	-2.6	8.5	0	-4.5	9.5	0	-5.5

Figure A-2B

American Standard Running and Sliding Fits
(Hole Basis)

Nominal Size Range Inches		Class RC1			Class RC2			Class RC3			Class RC4		
		Limits of Clearance	Standard Limits		Limits of Clearance	Standard Limits		Limits of Clearance	Standard Limits		Limits of Clearance	Standard Limits	
Over	To		Hole H5	Shaft g4		Hole H6	Shaft g5		Hole H7	Shaft f6		Hole H8	Shaft f7
0	0.12	0.1 0.45	+0.2 0	-0.1 -0.25	0.1 0.55	+0.25 0	-0.1 -0.3	0.3 0.95	+0.4 0	-0.3 -0.55	0.3 1.3	+0.6 0	-0.3 -0.7
0.12	0.24	0.15 0.5	+0.2 0	-0.15 -0.3	0.15 0.65	+0.3 0	-0.15 -0.35	0.4 1.12	+0.5 0	-0.4 -0.7	0.4 1.5	+0.7 0	-0.4 -0.0
0.24	0.40	0.2 0.6	+0.25 0	-0.2 -0.35	0.2 0.85	+0.4 0	-0.2 -0.45	0.5 1.5	+0.6 0	-0.5 -0.9	0.5 2.0	+0.9 0	-0.5 -1.1
0.40	0.71	0.25 0.75	+0.3 0	-0.25 -0.45	0.25 0.95	+0.4 0	-0.25 -0.55	0.6 1.7	+0.7 0	-0.6 -1.0	0.6 2.3	+1.0 0	-0.6 -1.3
0.71	1.19	0.3 0.95	+0.4 0	-0.3 -0.55	0.3 1.2	+0.5 0	-0.3 -0.7	0.8 2.1	+0.8 0	-0.8 -1.3	0.8 2.8	+1.2 0	-0.8 -1.6
1.19	1.97	0.4 1.1	+0.4 0	-0.4 -0.7	0.4 1.4	+0.6 0	-0.4 -0.8	1.0 2.6	+1.0 0	-1.0 -1.6	1.0 3.6	+1.6 0	-1.0 -2.0

Figure A-3A

Nominal Size Range Inches		Class RC5			Class RC6			Class RC7			Class RC8		
		Limits of Clearance	Standard Limits		Limits of Clearance	Standard Limits		Limits of Clearance	Standard Limits		Limits of Clearance	Standard Limits	
Over	To		Hole H8	Shaft e7		Hole H9	Shaft e8		Hole H9	Shaft d8		Hole H10	Shaft c9
0	0.12	0.6 1.6	+0.6 0	-0.6 -1.0	0.6 2.2	+1.0 0	-0.6 -1.2	1.0 2.6	+1.0 0	-1.0 -1.6	2.5 5.1	+1.6 0	-2.5 -3.5
0.12	0.24	0.8 2.0	+0.7 0	-0.8 -1.3	0.8 2.7	+1.2 0	-0.8 -1.5	1.2 3.1	+1.2 0	-1.2 -1.9	2.8 5.8	+1.8 0	-2.8 -4.0
0.24	0.40	1.0 2.5	+0.9 0	-1.0 -1.6	1.0 3.3	+1.4 0	-1.0 -1.9	1.6 3.9	+1.4 0	-1.6 -2.5	3.0 6.6	+2.2 0	-3.0 -4.4
0.40	0.71	1.2 2.9	+1.0 0	-1.2 -1.9	1.2 3.8	+1.6 0	-1.2 -2.2	2.0 4.6	+1.6 0	-2.0 -3.0	3.5 7.9	+2.8 0	-3.5 -5.1
0.71	1.19	1.6 3.6	+1.2 0	-1.6 -2.4	1.6 4.8	+2.0 0	-1.6 -2.8	2.5 5.7	+2.0 0	-2.5 -3.7	4.5 10.0	+3.5 0	-4.5 -6.5
1.19	1.97	2.0 4.6	+1.6 0	-2.0 -3.0	2.0 6.1	+2.5 0	-2.0 -3.6	3.0 7.1	+2.5 0	-3.0 -4.6	5.0 11.5	+4.0 0	-5.0 -7.5

Figure A-3B

American Standard Transition Locational Fits

Nominal Size Range Inches		Class LT1			Class LT2			Class LT3		
		Fit	Standard Limits		Fit	Standard Limits		Fit	Standard Limits	
Over	To		Hole H7	Shaft js6		Hole H8	Shaft js7		Hole H7	Shaft k6
0	0.12	−0.10 +0.50	+0.4 0	+0.10 −0.10	−0.2 +0.8	+0.6 0	+0.2 −0.2			
0.12	0.24	−0.15 −0.65	+0.5 0	+0.15 −0.15	−0.25 +0.95	+0.7 0	+0.25 −0.25			
0.24	0.40	−0.2 +0.5	+0.6 0	+0.2 −0.2	−0.3 +1.2	+0.9 0	+0.3 −0.3	−0.5 +0.5	+0.6 0	+0.5 +0.1
0.40	0.71	−0.2 +0.9	+0.7 0	+0.2 −0.2	−0.35 +1.35	+1.0 0	+0.35 −0.35	−0.5 +0.6	+0.7 0	+0.5 +0.1
0.71	1.19	−0.25 +1.05	+0.8 0	+0.25 −0.25	−0.4 +1.6	+1.2 0	+0.4 −0.4	−0.6 +0.7	+0.8 0	+0.6 +0.1
1.19	1.97	−0.3 +1.3	+1.0 0	+0.3 −0.3	−0.5 +2.1	+1.6 0	+0.5 −0.5	+0.7 +0.1	+1.0 0	+0.7 +0.1

Figure A-4A

Nominal Size Range Inches		Class LT4			Class LT5			Class LT6		
		Fit	Standard Limits		Fit	Standard Limits		Fit	Standard Limits	
Over	To		Hole H8	Shaft k7		Hole H7	Shaft n6		Hole H7	Shaft n7
0	0.12				−0.5 +0.15	+0.4 0	+0.5 +0.25	−0.65 +0.15	+0.4 0	+0.65 +0.25
0.12	0.24				−0.6 +0.2	+0.5 0	+0.6 +0.3	−0.8 +0.2	+0.5 0	+0.8 +0.3
0.24	0.40	−0.7 +0.8	+0.9 0	+0.7 +0.1	−0.8 +0.2	+0.6 0	+0.8 +0.4	−1.0 +0.2	+0.6 0	+1.0 +0.4
0.40	0.71	−0.8 +0.9	+1.0 0	+0.8 +0.1	−0.9 +0.2	+0.7 0	+0.9 +0.5	−1.2 +0.2	+0.7 0	+1.2 +0.5
0.71	1.19	−0.9 +1.1	+1.2 0	+0.9 +0.1	−1.1 +0.2	+0.8 0	+1.1 +0.6	−1.4 +0.2	+0.8 0	+1.4 +0.6
1.19	1.97	−1.1 +1.5	+1.6 0	+1.1 +0.1	−1.3 +0.3	+1.0 0	+1.3 +0.7	−1.7 +0.3	+1.0 0	+1.7 +0.7

Figure A-4B

American Standard Interference Locational Fits

Nominal Size Range Inches Over – To	Limits of Interference	Class LN1 Standard Limits Hole H6	Class LN1 Standard Limits Shaft n5	Limits of Interference	Class LN2 Standard Limits Hole H7	Class LN2 Standard Limits Shaft p6	Limits of Interference	Class LN3 Standard Limits Hole H7	Class LN3 Standard Limits Shaft r6
0 - 0.12	0 0.45	+0.25 0	+0.45 +0.25	0 0.65	+0.4 0	+0.63 +0.4	0.1 0.75	+0.4 0	+0.75 +0.5
0.12 - 0.24	0 0.5	+0.3 0	+0.5 +0.3	0 0.8	+0.5 0	+0.8 +0.5	0.1 0.9	+0.5 0	+0.9 +0.6
0.24 - 0.40	0 0.65	+0.4 0	+0.65 +0.4	0 1.0	+0.6 0	+1.0 +0.6	0.2 1.2	+0.6 0	+1.2 +0.8
0.40 - 0.71	0 0.8	+0.4 0	+0.8 +0.4	0 1.1	+0.7 0	+1.1 +0.7	0.3 1.4	+0.7 0	+1.4 +1.0
0.71 - 1.19	0 1.0	+0.5 0	+1.0 +0.5	0 1.3	+0.8 0	+1.3 +0.8	0.4 1.7	+0.8 0	+1.7 +1.2
1.19 - 1.97	0 1.1	+0.6 0	+1.1 +0.6	0 1.6	+1.0 0	+1.6 +1.0	0.4 2.0	+1.0 0	+2.0 +1.4

Figure A-5

American Standard Force and Shrink Fits

Nominal Size Range Inches Over – To	Limits of Interference	Class FN 1 Standard Limits Hole	Class FN 1 Standard Limits Shaft	Limits of Interference	Class FN 2 Standard Limits Hole	Class FN 2 Standard Limits Shaft	Limits of Interference	Class FN 3 Standard Limits Hole	Class FN 3 Standard Limits Shaft	Limits of Interference	Class FN 4 Standard Limits Hole	Class FN 4 Standard Limits Shaft
0 - 0.12	0.05 0.5	+0.25 0	+0.5 +0.3	0.2 0.85	+0.4 0	+0.85 +0.6				0.3 0.95	+0.4 0	+0.95 +0.7
0.12 - 0.24	0.1 0.6	+0.3 0	+0.6 +0.4	0.2 1.0	+0.5 0	+1.0 +0.7				0.4 1.2	+0.5 0	+1.2 +0.9
0.24 - 0.40	0.1 0.75	+0.4 0	+0.75 +0.5	0.4 1.4	+0.6 0	+1.4 +1.0				0.6 1.6	+0.6 0	+1.6 +1.2
0.40 - 0.56	0.1 0.8	+0.4 0	+0.8 +0.5	0.5 1.6	+0.7 0	+1.6 +1.2				0.7 1.8	+0.7 0	+1.8 +1.4
0.56 - 0.71	0.2 0.9	+0.4 0	+0.9 +0.6	0.5 1.6	+0.7 0	+1.6 +1.2				0.7 1.8	+0.7 0	+1.8 +1.4
0.71 - 0.95	0.2 1.1	+0.5 0	+1.1 +0.7	0.6 1.9	+0.8 0	+1.9 +1.4				0.8 2.1	+0.8 0	+2.1 +1.6
0.95 - 1.19	0.3 1.2	+0.5 0	+1.2 +0.8	0.6 1.9	+0.8 0	+1.9 +1.4	0.8 2.1	+0.8 0	+2.1 +1.6	1.0 2.3	+0.8 0	+2.1 +1.8
1.19 - 1.58	0.3 1.3	+0.6 0	+1.3 +0.9	0.8 2.4	+1.0 0	+2.4 +1.8	1.0 2.6	+1.0 0	+2.6 +2.0	1.5 3.1	+1.0 0	+3.1 +2.5
1.58 - 1.97	0.4 1.4	+0.6 0	+1.4 +1.0	0.8 2.4	+1.0 0	+2.4 +1.8	1.2 2.8	+1.0 0	+2.8 +2.2	1.8 3.4	+1.0 0	+3.4 +2.8

Figure A-6

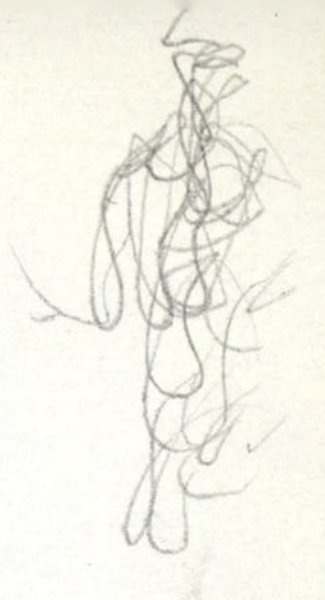

Preferred Clearance Fits — Cylindrical Fits
(Hole Basis; ANSI B4.2)

Basic Size		Loose Running			Free Running			Close Running			Sliding			Locational Clear.		
		Hole H11	Shaft c11	Fit	Hole H9	Shaft d9	Fit	Hole H8	Shaft f7	Fit	Hole H7	Shaft g6	Fit	Hole H7	Shaft h6	Fit
4	Max	4.075	3.930	0.220	4.030	3.970	0.090	4.018	3.990	0.040	4.012	3.996	0.024	4.012	4.000	0.020
	Min	4.000	3.855	0.070	4.000	3.940	0.030	4.000	3.978	0.010	4.000	3.988	0.004	4.000	3.992	0.000
5	Max	5.075	4.930	0.220	5.030	4.970	0.090	5.018	4.990	0.040	5.012	4.996	0.024	5.012	5.000	0.020
	Min	5.000	4.855	0.070	5.000	4.940	0.030	5.000	4.978	0.010	5.000	4.988	0.004	5.000	4.992	0.000
6	Max	6.075	5.930	0.220	6.030	5.970	0.090	6.018	5.990	0.040	6.012	5.996	0.024	6.012	6.000	0.020
	Min	6.000	5.885	0.070	6.000	5.940	0.030	6.000	5.978	0.010	6.000	5.988	0.004	6.000	5.992	0.000
8	Max	8.090	7.920	0.260	8.036	7.960	0.112	8.022	7.987	0.050	8.015	7.995	0.029	8.015	8.000	0.024
	Min	8.000	7.830	0.080	8.000	7.924	0.040	8.000	7.972	0.013	8.000	7.986	0.005	8.000	7.991	0.000
10	Max	10.090	9.920	0.026	10.036	9.960	0.112	10.022	9.987	0.050	10.015	9.995	0.029	10.015	10.000	0.024
	Min	10.000	9.830	0.080	10.000	9.924	0.040	10.000	9.972	0.013	10.000	9.986	0.005	10.000	9.991	0.000
12	Max	12.112	11.905	0.315	12.043	11.950	0.136	12.027	11.984	0.061	12.018	11.994	0.035	12.018	12.000	0.029
	Min	12.000	11.795	0.095	12.000	11.907	0.050	12.000	11.966	0.016	12.000	11.983	0.006	12.000	11.989	0.000
16	Max	16.110	15.905	0.315	16.043	15.950	0.136	16.027	15.984	0.061	16.018	15.994	0.035	16.018	16.000	0.029
	Min	16.000	15.795	0.095	16.000	15.907	0.050	16.000	15.966	0.016	16.000	19.983	0.006	16.000	15.989	0.000
20	Max	20.130	19.890	0.370	20.052	19.935	0.169	20.033	19.980	0.074	20.021	19.993	0.041	20.021	20.000	0.034
	Min	20.000	19.760	0.110	20.000	19.883	0.065	20.000	19.959	0.020	20.000	19.980	0.007	20.000	19.987	0.000
25	Max	25.130	24.890	0.370	25.052	24.935	0.169	25.033	24.980	0.074	25.021	24.993	0.041	25.021	25.000	0.034
	Min	25.000	24.760	0.110	25.000	24.883	0.065	25.000	24.959	0.020	25.000	24.980	0.007	25.000	24.987	0.000
30	Max	30.130	29.890	0.370	30.052	29.935	0.169	30.033	29.980	0.074	30.021	29.993	0.041	30.021	30.000	0.034
	Min	30.000	29.760	0.110	30.000	29.883	0.065	30.000	29.959	0.020	30.000	29.980	0.007	30.000	29.987	0.000

Figure A-7

Preferred Transition and Interference Fits — Cylindrical Fits
(Hole Basis; ANSI B4.2)

Basic Size		Locational Trans.			Locational Trans.			Locational Inter.			Medium Drive			Force		
		Hole H7	Shaft k6	Fit	Hole H7	Shaft n6	Fit	Hole H7	Shaft p6	Fit	Hole H7	Shaft s6	Fit	Hole H7	Shaft u6	Fit
4	Max	4.012	4.009	0.011	4.012	4.016	0.004	4.012	4.020	0.000	4.012	4.027	−0.007	4.012	4.031	-0.011
	Min	4.000	4.001	−0.009	4.000	4.008	−0.016	4.000	4.012	−0.020	4.000	4.019	−0.027	4.000	4.023	-0.031
5	Max	5.012	5.009	0.011	5.012	5.016	0.004	5.012	5.020	0.000	5.012	5.027	−0.007	5.012	5.031	-0.011
	Min	5.000	5.001	−0.009	5.000	5.008	−0.016	5.000	5.012	−0.020	5.000	5.019	−0.027	5.000	5.023	-0.031
6	Max	6.012	6.009	0.011	6.012	6.016	0.004	6.012	6.020	0.000	6.012	6.027	−0.007	6.012	6.031	-0.011
	Min	6.000	6.001	−0.009	6.000	6.008	−0.016	6.000	6.012	−0.020	6.000	6.019	−0.027	6.000	6.023	-0.031
8	Max	8.015	8.010	0.014	8.015	8.019	0.005	8.015	8.024	0.000	8.015	8.032	−0.008	8.015	8.037	-0.013
	Min	8.000	8.001	−0.010	8.000	8.010	−0.019	8.000	8.015	−0.024	8.000	8.023	−0.032	8.000	8.028	-0.037
10	Max	10.015	10.010	0.014	10.015	10.019	0.005	10.015	10.024	0.000	10.015	10.032	−0.008	10.015	10.037	-0.013
	Min	10.000	10.001	−0.010	10.000	10.010	−0.019	10.000	10.015	−0.024	10.000	10.023	−0.032	10.000	10.028	-0.037
12	Max	12.018	12.012	0.017	12.018	12.023	0.006	12.018	12.029	0.000	12.018	12.039	−0.010	12.018	12.044	-0.015
	Min	12.000	12.001	−0.012	12.000	12.012	−0.023	12.000	12.018	−0.029	12.000	12.028	−0.039	12.000	12.033	-0.044
16	Max	16.018	16.012	0.017	16.018	16.023	0.006	16.018	16.029	0.000	16.018	16.039	−0.010	16.018	16.044	-0.015
	Min	16.000	16.001	−0.012	16.000	16.012	−0.023	16.000	16.018	−0.029	16.000	16.028	−0.039	16.000	16.033	-0.044
20	Max	20.021	20.015	0.019	20.021	20.028	0.006	20.021	20.035	−0.001	20.021	20.048	−0.014	20.021	20.054	-0.020
	Min	20.000	20.002	−0.015	20.000	20.015	−0.028	20.000	20.022	−0.035	20.000	20.035	−0.048	20.000	20.041	-0.054
25	Max	25.021	25.015	0.019	25.021	25.028	0.006	25.021	25.035	−0.001	25.021	25.048	−0.014	25.021	25.061	-0.027
	Min	25.000	25.002	−0.015	25.000	25.015	−0.028	25.000	25.022	−0.035	25.000	25.035	−0.048	25.000	25.048	-0.061
30	Max	30.021	30.015	0.019	30.021	30.028	0.006	30.021	30.035	−0.001	30.021	30.048	−0.014	30.021	30.061	-0.027
	Min	30.000	30.002	−0.015	30.000	30.015	−0.028	30.000	30.022	−0.035	30.000	30.035	−0.048	30.000	30.048	-0.061

Figure A-8

Preferred Clearance Fits — Cylindrical Fits
(Shaft Basis; ANSI B4.2)

Basic Size		Loose Running			Free Running			Close Running			Sliding			Locational Clear.		
		Hole C11	Shaft h11	Fit	Hole D9	Shaft h9	Fit	Hole F8	Shaft h7	Fit	Hole G7	Shaft h6	Fit	Hole H7	Shaft h6	Fit
4	Max	4.145	4.000	0.220	4.060	4.000	0.090	4.028	4.000	0.040	4.016	4.000	0.024	4.012	4.000	0.020
	Min	4.070	3.925	0.070	4.030	3.970	0.030	4.010	3.988	0.010	4.004	3.992	0.004	4.000	3.992	0.000
5	Max	5.145	5.000	0.220	5.060	5.000	0.090	5.028	5.000	0.040	5.016	5.000	0.024	5.012	5.000	0.020
	Min	5.070	4.925	0.070	5.030	4.970	0.030	5.010	4.988	0.010	5.004	4.992	0.004	5.000	4.992	0.000
6	Max	6.145	6.000	0.220	6.060	6.000	0.090	6.028	6.000	0.040	6.016	6.000	0.024	6.012	6.000	0.020
	Min	6.070	5.925	0.070	6.030	5.970	0.030	6.010	5.988	0.010	6.004	5.992	0.004	6.000	5.992	0.000
8	Max	8.170	8.000	0.260	8.076	8.000	0.112	8.035	8.000	0.050	8.020	8.000	0.029	8.015	8.000	0.024
	Min	8.080	7.910	0.080	8.040	7.964	0.040	8.013	7.985	0.013	8.005	7.991	0.005	8.000	7.991	0.000
10	Max	10.170	10.000	0.260	10.076	10.000	0.112	10.035	10.000	0.050	10.020	10.000	0.029	10.015	10.000	0.024
	Min	10.080	9.910	0.080	10.040	9.964	0.040	10.013	9.985	0.013	10.005	9.991	0.005	10.000	9.991	0.000
12	Max	12.205	12.000	0.315	12.093	12.000	0.136	12.043	12.000	0.061	12.024	12.000	0.035	12.018	12.000	0.029
	Min	12.095	11.890	0.095	12.050	11.957	0.050	12.016	11.982	0.016	12.006	11.989	0.006	12.000	11.989	0.000
16	Max	16.205	16.000	0.315	16.093	16.000	0.136	16.043	16.000	0.061	16.024	16.000	0.035	16.018	16.000	0.029
	Min	16.095	15.890	0.095	16.050	15.957	0.050	16.016	15.982	0.016	06.006	15.989	0.006	16.000	15.989	0.000
20	Max	20.240	20.000	0.370	20.117	20.000	0.169	20.053	20.000	0.074	20.028	20.000	0.041	20.021	20.000	0.034
	Min	20.110	19.870	0.110	20.065	19.948	0.065	20.020	19.979	0.020	20.007	19.987	0.007	20.000	19.987	0.000
25	Max	25.240	25.000	0.370	25.117	25.000	0.169	25.053	25.000	0.074	25.028	25.000	0.041	25.021	25.000	0.034
	Min	25.110	24.870	0.110	25.065	24.948	0.065	25.020	24.979	0.020	25.007	24.987	0.007	25.000	24.987	0.000
30	Max	30.240	30.000	0.370	30.117	30.000	0.169	30.053	30.000	0.074	30.028	30.000	0.041	30.021	30.000	0.034
	Min	30.110	29.870	0.110	30.065	29.948	0.065	29.979	29.979	0.020	30.007	29.987	0.007	30.000	29.987	0.000

Figure A-9

Preferred Transition and Interference Fits — Cylindrical Fits
(Shaft Basis; ANSI B4.2)

Basic Size		Locational Trans.			Locational Trans.			Locational Inter.			Medium Drive			Force		
		Hole K7	Shaft h6	Fit	Hole N7	Shaft h6	Fit	Hole F7	Shaft h6	Fit	Hole S7	Shaft h6	Fit	Hole U7	Shaft h6	Fit
4	Max	4.003	4.000	0.011	3.996	4.000	0.004	3.992	4.000	0.000	3.985	4.000	-0.007	3.981	4.000	-0.011
	Min	3.991	3.992	-0.009	3.984	3.992	-0.016	3.980	3.992	-0.020	3.973	3.992	-0.027	3.969	3.992	-0.031
5	Max	5.003	5.000	0.011	4.996	5.000	0.004	4.992	5.000	0.000	4.985	5.000	-0.007	4.981	5.000	-0.011
	Min	4.991	4.992	-0.009	4.984	4.992	-0.016	4.980	4.992	-0.020	4.973	4.992	-0.027	4.969	4.992	-0.031
6	Max	6.003	6.000	0.011	5.996	6.000	0.004	5.992	6.000	0.000	5.985	6.000	-0.007	5.981	6.000	-0.011
	Min	5.991	5.991	-0.009	5.984	5.992	-0.016	5.980	5.992	-0.020	5.973	5.992	-0.027	5.969	5.992	-0.031
8	Max	8.005	8.000	0.014	7.996	8.000	0.005	7.991	8.000	0.000	7.983	8.000	-0.008	7.978	8.000	-0.013
	Min	7.990	7.991	-0.010	7.981	7.991	-0.019	7.976	7.991	-0.024	7.968	7.991	-0.032	7.963	7.991	-0.037
10	Max	10.005	10.000	0.014	9.996	10.000	0.005	9.991	10.000	0.000	9.983	10.000	-0.008	9.978	10.000	-0.013
	Min	9.990	9.991	-0.010	9.981	9.991	-0.019	9.976	9.991	-0.024	9.968	9.991	-0.032	9.963	9.991	-0.037
12	Max	12.006	12.000	0.017	11.995	12.000	0.006	11.989	12.000	0.000	11.979	12.000	-0.010	11.974	12.000	-0.015
	Min	11.988	11.989	-0.012	11.977	11.989	-0.023	11.971	11.989	-0.029	11.961	11.989	-0.039	11.956	11.989	-0.044
16	Max	16.006	16.000	0.017	15.995	16.000	0.006	15.989	16.000	0.000	15.979	16.000	-0.010	15.974	16.000	-0.015
	Min	15.988	15.989	-0.012	15.977	15.989	-0.023	15.971	15.989	-0.029	15.961	15.989	-0.039	15.956	15.989	-0.044
20	Max	20.006	20.000	0.019	19.993	20.000	0.006	19.986	20.000	-0.001	19.973	20.000	-0.014	19.967	20.000	-0.020
	Min	19.985	19.987	-0.015	19.972	19.987	-0.028	19.965	19.987	-0.035	19.952	19.987	-0.048	19.946	19.987	-0.054
25	Max	25.006	25.000	0.019	24.993	25.000	0.006	24.986	25.000	-0.001	24.973	25.000	-0.014	24.960	25.000	-0.027
	Min	24.985	24.987	-0.015	24.972	24.987	-0.028	24.965	24.987	-0.035	24.952	24.987	-0.048	24.939	24.987	-0.061
30	Max	30.006	30.000	0.019	29.993	30.000	0.006	29.986	30.000	-0.001	29.973	30.000	-0.014	29.960	30.000	-0.027
	Min	29.985	29.987	-0.015	29.972	29.987	-0.028	29.987	29.987	-0.035	29.952	29.987	-0.048	29.939	29.987	-0.061

Figure A-10

Index

A

B

C

D

E

F

G

H

I

K

L

M

N

O

P

R

S

T

U